Management Science, Business, and Economics

applied Calculus

Seventh Edition

Geoffrey C. Berresford

Long Island University

Andrew M. Rockett

Long Island University

CENGAGE
Learning®

Australia • Brazil • Mexico • Singapore • United Kingdom • United States

Applied Calculus, Seventh Edition
Geoffrey C. Berresford,
Andrew M. Rockett

Vice-President, General Manager: Balraj Kalsi

Product Directors: Richard Stratton and Terrence Boyle

Product Manager: Rita Lombard

Senior Content Developer: Erin Brown

Product Assistant: Katharine Werring

Senior Product Assistant: Jennifer Cordoba

Content Developer: Andrew Coppola

Marketing Manager: Julie Schuster

Content Project Manager: Jill Quinn

Senior Art Director: Linda May

Manufacturing Planner: Doug Bertke

IP Analyst: Christina Ciaramella

IP Project Manager: John Sarantakis

Production Service: MPS Limited

Compositor: MPS Limited

Text Designer: Rokusek Design

Cover Image: © Showmethemoney | Dreamstime.com

About the Cover
We were fortunate to have Lee Berresford's advice on the cover design. We thank her for her contributions.

For product information and technology assistance, contact us at
Cengage Learning Customer & Sales Support, 1-800-354-9706
For permission to use material from this text or product,
submit all requests online at **www.cengage.com/permissions**.
Further permissions questions can be emailed to
permissionrequest@cengage.com.

Library of Congress Control Number: 2014946320
Student Edition:

ISBN: 978-1-305-08531-2

Cengage Learning
20 Channel Center Street
Boston, MA 02210
USA

Cengage Learning is a leading provider of customized learning solutions with office locations around the globe, including Singapore, the United Kingdom, Australia, Mexico, Brazil, and Japan. Locate your local office at **www.cengage.com/global**.

Cengage Learning products are represented in Canada by Nelson Education, Ltd.

To learn more about Cengage Learning Solutions, visit **www.cengage.com**.

Purchase any of our products at your local college store or at our preferred online store **www.cengagebrain.com**.

Printed in the United States of America
Print Number: 01   Print Year: 2014

Contents

Overview

A scientific study of yawning found that more yawns occurred in calculus class than anywhere else.* This book hopes to remedy that situation. Rather than being another dry recitation of standard results, our presentation exhibits many of the fascinating and useful applications of mathematics in business, the sciences, and everyday life. Even beyond its utility, however, there is a beauty to calculus, and we hope to convey some of its elegance and simplicity.

This book is an introduction to calculus and its applications to the management, social, behavioral, and biomedical sciences, and other fields. The seven-chapter *Brief Applied Calculus* contains more than enough material for a one-semester course, and the eleven-chapter *Applied Calculus* contains additional chapters on trignometry, differential equations, sequences and series, and probability for a two-semester course. The only prerequisites are some knowledge of algebra, functions, and graphing, which are reviewed in Chapter 1 and in greater detail in the Algebra Review appendix.

ACCURATE AND ACCESSIBLE

Our foremost goal in writing these books has been to make the content as accessible to as many students as possible. Over time, we have introduced various features to address the changing needs of students as they learn the essential techniques and fundamental concepts of calculus. In order maintain students' interest and provide them with the most accurate and engaging textbook, we have been guided by the following principles.

- *Informal Proofs*   Because this book is applied rather than theoretical, we have preferred intuitive and geometric justifications to formal proofs. We provide a justification or proof for every important mathematical idea. When proofs are given, they are correct and mathematically honest.

- *Integration of Mathematics and Applications*   Every section has applications to motivate the mathematics being developed (see, for example, pages 27–28 and 119–120). There are no "pure math" sections.

- *Rapid Start*   When learning something, it is best to begin doing it as soon as possible. Therefore, we keep the preliminary material brief so that students begin calculus without delay (in Section 2.2). An early start allows more time for interesting applications throughout the course.

- *Just-in-Time Review*   Review material is placed just before it is used, where it is more likely to be remembered, rather than in lengthy early chapters that "review" material that was never mastered in the first place. Exponential and logarithmic functions are reviewed just before they are differentiated in Section 4.3, and the sine and cosine functions are reviewed just before they are differentiated in Section 8.3, as are the other trigonometric functions in Section 8.5.

- *Continual Algebra Reinforcement*   Since many of today's students have weak algebra skills, which impede their understanding of calculus, examples have blue annotations in the right margin giving brief explanations of the steps (see, for example, page 88). For extra support, we also offer a Diagnostic Test (appearing before Chapter 1) to help students identify skills that may

*Ronald Baenninger, "Some Comparative Aspects of Yawning in Betta splendens, Homo sapiens, Panthera leo, and Papoi spinx," *Journal of Comparative Psychology* **101** (4).

need review along with a supplementary Algebra Review appendix for additional reference.

CHANGES IN THE SEVENTH EDITION

New Content

- Section 3.7 *Differentials, Approximations, and Marginal Analysis* is new in the seventh edition. This section is optional and can be omitted without loss of continuity.

- An Algebra Review appendix is keyed to parts of the text (see, for example, page 49).

- A Diagnostic Test has been added to help students identify skills that may need review. This test appears before Chapter 1. Complete solutions are given in the Algebra Review appendix.

- New material on parallel and perpendicular lines has been added to Section 1.1, *Real Numbers, Inequalities, and Lines*.

- New exercises have been added and over 100 updated (including all of the Wall Street financial exercises) with current real-world data and sources. New *Explorations and Excursions* exercises give further details or theoretical underpinnings of the topics in the main text.

- A new "What You'll Explore" paragraph on the opening page of each chapter previews the ideas and applications to come.

Enhanced Learning Support

- Throughout the text there are now (→) LOOKING AHEAD and (←) LOOKING BACK marginal notes that show connections between current material and past or future developments to unify students' understanding of calculus.

- New [Take Note] marginal prompts provide observations that simplify or clarify ideas.

- Newly added ✎ FOR MORE HELP and ✎ FOR HELP GETTING STARTED prompts point students to Examples or parts of the Algebra Review appendix for additional help.

Graphing Calculator

- The graphing calculator screens throughout the book are now in color, based on the TI-84 Plus *C* Silver Edition, although students can still use the TI-83 or TI-84 (regular or Plus) calculators and follow instructions provided to get corresponding black-and-white graphs.

- References to the Internet are now given for graphing calculator programs from sites such as ticalc.org. The programs may be used for Riemann sums (page 332), trapezoidal approximation (page 418), Simpson's rule (page 421), slope fields (pages 594, 596, and 614), and Euler's method (pages 634–635). For Newton's method the authors explain how the calculator may be used to perform the calculations directly with a few keystrokes (page 685). The graphing calculator programs from earlier editions are now available on the Student and the Instructor Companion Sites.

To get the most out of this book, familiarize yourself with the following features—all designed to increase your understanding and mastery of the material. These learning aids, together with any help available through your college, should make your encounter with calculus both successful and enjoyable.

APPLICATIONS

From archaeological finds to physics, from social issues to politics, the applications show that calculus is more than just manipulation of abstract symbols. Rather, it is a powerful tool that can be used to help understand and manage both the natural world and our activities in it.

Application Preview

Following each chapter opener, an Application Preview offers a "mathematics in your world" application. A page with further information on the topic and a related exercise number are often given.

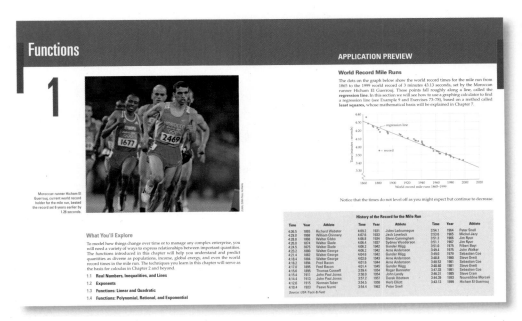

Diverse Applications

Along with an emphasis on business and biomedical sciences, a variety of other fields are represented throughout the text. Applications based on contemporary real-world data are denoted with an icon

GUIDED LEARNING SUPPORT

Annotations

To aid students' understanding of the solution steps within examples or to provide interpretations, blue annotations appear to the right of most mathematical formulas. Calculations presented within annotations provide explanations and justifications for the steps.

EXAMPLE 3   **DEPRECIATING AN ASSET**

A car worth $30,000 depreciates in value by 40% each year. How much is it worth after 3 years?

Solution

The car loses 40% of its value each year, which is equivalent to an interest rate of *negative* 40%. The compound interest formula gives

$$30{,}000(1 - 0.40)^3 = 30{,}000(0.60)^3 = \$6480$$

Using a calculator

$P(1 + r/m)^{mt}$ with $P = 30{,}000$, $r = -0.40$, $m = 1$, and $t = 3$

The exponential function $f(x) = 30{,}000(0.60)^x$, giving the value of the car after x years of depreciation, is graphed on the left. Notice that a yearly loss of 40% means that 60% of the value is retained each year.

Be Careful

The "Be Careful" icon marks places where the authors help students avoid common errors.

4.3   Differentiation of Logarithmic and Exponential Functions   **279**

 Be Careful Do *not* take the derivative of e^x by the Power Rule,

$$\frac{d}{dx}x^n = nx^{n-1}$$

The Power Rule applies to x^n, *a variable to a constant power*, while e^x is *a constant to a variable power*. The two types of functions are quite different, as their graphs show.

Looking Ahead
Looking Back

New in the 7e! These notes appear in the margins and show connections between current material and previous or future developments to solidify and unify understanding of calculus topics.

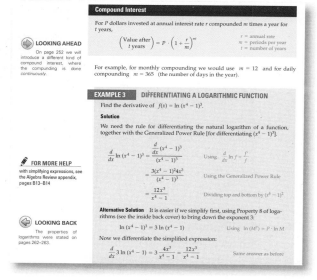

Compound Interest

For P dollars invested at annual interest rate r compounded m times a year for t years,

$$\left(\begin{matrix}\text{Value after} \\ t \text{ years}\end{matrix}\right) = P \cdot \left(1 + \frac{r}{m}\right)^{mt}$$

r = annual rate
m = periods per year
t = number of years

For example, for monthly compounding we would use $m = 12$ and for daily compounding $m = 365$ (the number of days in the year).

LOOKING AHEAD
On page 252 we will introduce a different kind of compound interest, where the compunding is done continuously.

EXAMPLE 3   **DIFFERENTIATING A LOGARITHMIC FUNCTION**

Find the derivative of $f(x) = \ln (x^4 - 1)^3$.

Solution

We need the rule for differentiating the natural logarithm of a function, together with the Generalized Power Rule [for differentiating $(x^4 - 1)^3$].

$$\frac{d}{dx}\ln (x^4 - 1)^3 = \frac{\frac{d}{dx}(x^4 - 1)^3}{(x^4 - 1)^3}$$

Using $\frac{d}{dx}\ln f = \frac{f'}{f}$

$$= \frac{3(x^4 - 1)^2 4x^3}{(x^4 - 1)^3}$$

Using the Generalized Power Rule

$$= \frac{12x^3}{x^4 - 1}$$

Dividing top and bottom by $(x^4 - 1)^2$

FOR MORE HELP
with simplifying expressions, see the Algebra Review appendix, pages B13–B14

Alternative Solution   It is easier if we simplify first, using Property 8 of logarithms (see the inside back cover) to bring down the exponent 3:

$$\ln (x^4 - 1)^3 = 3 \ln (x^4 - 1)$$

Using $\ln (M^P) = P \cdot \ln M$

Now we differentiate the simplified expression:

$$\frac{d}{dx}3 \ln (x^4 - 1) = 3\frac{4x^3}{x^4 - 1} = \frac{12x^3}{x^4 - 1}$$

Same answer as before

LOOKING BACK
The properties of logarithms were stated on pages 262–263.

GUIDED LEARNING SUPPORT

Take Note
New in the 7/e! Appearing in the margins, these prompts include observations to help simplify or clarify ideas in the text.

For More Help
For Help Getting Started
New in the 7/e! These prompts appear within the margins of the text and end-of-section exercises. They direct students to Examples from within the text or parts of the Algebra Review appendix, as a refresher.

PRACTICE AND PREPARE

Practice Problems
Students can check their understanding of a topic as they read the text or do homework by working out a Practice Problem. Complete solutions are found at the end of each section, just before the Section Summary.

Exercises

The exercises that appear at the end of each section are graded from routine drills to significant applications. The *Applied Exercises* are labeled with general and specific titles so instructors can assign problems appropriate for the class. *Conceptual Exercises* develop intuitive insights to solve problems quickly and simply. *Explorations and Excursions* push students further. Just-in-time *Review Exercises* are found in selected sections. They recall skills previously learned that are relevant to content in an upcoming section (see, for example, page 355).

Although this may be used to *estimate* future costs (*about* $15 for each additional unit), it does not mean that one additional unit will increase costs by exactly $15, two more by exactly $30, and so on, since the marginal rate usually changes as production increases. A marginal cost is only an *approximate* predictor of future costs.

2.5 Exercises

1–6. For each function, find:

a. $f'(x)$ **b.** $f''(x)$ **c.** $f'''(x)$ **d.** $f^{(4)}(x)$

1. $f(x) = x^4 - 2x^3 - 3x^2 + 5x - 7$
2. $f(x) = x^4 - 3x^3 + 2x^2 - 8x + 4$
3. $f(x) = 1 + x + \frac{1}{2}x^2 + \frac{1}{6}x^3 + \frac{1}{24}x^4 + \frac{1}{120}x^5$

4. $f(x) = 1 + x + \frac{1}{2}x^2 + \frac{1}{6}x^3 + \frac{1}{24}x^4$
5. $f(x) = \sqrt{x^5}$ **6.** $f(x) = \sqrt{x^3}$

7–12. For each function, find: **a.** $f'(x)$ and **b.** $f''(3)$.

7. $f(x) = \dfrac{x-1}{x}$ **8.** $f(x) = \dfrac{x+2}{x}$

Applied Exercises

33. GENERAL: Velocity After *t* hours a freight train is $s(t) = 18t^2 - 2t^3$ miles due north of its starting point (for $0 \le t \le 9$).
  a. Find its velocity at time $t = 3$ hours.
  b. Find its velocity at time $t = 7$ hours.
  c. Find its acceleration at time $t = 1$ hour.

34. GENERAL: Velocity After *t* hours a passenger train is $s(t) = 24t^2 - 2t^3$ miles due west of its starting point (for $0 \le t \le 12$).
  a. Find its velocity at time $t = 4$ hours.
  b. Find its velocity at time $t = 10$ hours.
  c. Find its acceleration at time $t = 1$ hour.

35. GENERAL: Velocity A rocket can rise to a height of $h(t) = t^3 + 0.5t^2$ feet in *t* seconds. Find its velocity and acceleration 10 seconds after it is launched.

  c. Find the acceleration at any time *t*. (This number is called the *acceleration due to gravity*.)

Conceptual Exercises

47–50. Suppose that the quantity described is represented by a function $f(t)$ where *t* stands for time. Based on the description:
  a. Is the first derivative positive or negative?
  b. Is the second derivative positive or negative?

47. The temperature is dropping increasingly rapidly.

48. The economy is growing, but more slowly.
49. The stock market is declining, but less rapidly.
50. The population is growing increasingly fast.
51. True or False: If $f(x)$ is a polynomial of degree *n*, then $f^{(n+1)}(x) = 0$.

Explorations and Excursions   The following problems extend and augment the material presented in the text.

More About Higher-Order Derivatives

55. Find $\dfrac{d^{100}}{dx^{100}}(x^{100} - 4x^{99} + 3x^{50} + 6)$.

  [*Hint:* You may use the "factorial" notation: $n! = n(n-1) \cdots 1$. For example, $3! = 3 \cdot 2 \cdot 1 = 6$.]

56. Find a general formula for $\dfrac{d^n}{dx^n}x^{-1}$.

  [*Hint:* Calculate the first few derivatives and look for a pattern. You may use the "factorial" notation: $n! = n(n-1) \cdots 1$. For example, $3! = 3 \cdot 2 \cdot 1 = 6$.]

57. Verify the following formula for the *second* derivative of a product, where *f* and *g* are differentiable functions of *x*:

$$\frac{d^2}{dx^2}(f \cdot g) = f'' \cdot g + 2f' \cdot g' + f \cdot g''$$

  [*Hint:* Use the Product Rule repeatedly.]

58. Verify the following formula for the *third* derivative of a product, where *f* and *g* are differentiable functions of *x*:

$$\frac{d^3}{dx^3}(f \cdot g) = f''' \cdot g + 3f'' \cdot g' + 3f' \cdot g'' + f \cdot g'''$$

  [*Hint:* Differentiate the formula in Exercise 57 by the Product Rule.]

PRACTICE AND PREPARE

Section Summary
Found at the end of every section, summaries briefly state the main ideas of the section and provide study tools or reminders for students

---

3.7   **Section Summary**

For an *independent* variable x, the differential dx is any nonzero number. For the *dependent* variable $y = f(x)$, the differential is $dy = f'(x)\,dx$. Values for both x and dx must be known before dy can be evaluated.

The best linear approximation of a differentiable function $y = f(x)$ near x is the *tangent line approximation* given by

$$f(x + \Delta x) \approx f(x) + f'(x)\,dx \qquad (\Delta x = dx)$$

since $\Delta y \approx dy$. This approximation becomes more accurate for values of $\Delta x = dx$ closer to zero.

For a *dependent* variable y, the *error* Δy resulting from a measurement error Δx is sometimes called the *absolute* error, and may be approximated by the differential, $\Delta y \approx dy$. The *relative* error $\Delta y/y \approx dy/y$ compares the absolute error to the actual value, and is usually written as a percentage. Errors are sometimes called "changes" depending on the situation.

Marginals can be used to find approximations of revenue, cost, and profit (see page 234), and indicate how these quantities vary near a particular level of production.

Chapter Summary
Found at the end of every chapter, the Chapter Summary with Hints and Suggestions review the important developments of the chapter and give insights to unify the material to help students prepare for tests and exams.

---

2   **Chapter Summary with Hints and Suggestions**

Reading the text and doing the exercises in this chapter have helped you to master the following concepts and skills, which are listed by section (in case you need to review them) and are keyed to particular Review Exercises. Answers for all Review Exercises are given at the back of the book, and full solutions can be found in the Student Solutions Manual.

2.1 Limits and Continuity
- Find the limit of a function from tables. *(Review Exercises 1–2.)*
- Find left and right limits. *(Review Exercises 3–4.)*
- Find the limit of a function. *(Review Exercises 5–14.)*
- Determine whether a function is continuous or discontinuous. *(Review Exercises 15–22.)*

2.2 Rates of Change, Slopes, and Derivatives
- Find the derivative of a function from the *definition* of the derivative. *(Review Exercises 23–26.)*

$$f'(x) = \lim_{h \to 0} \frac{f(x+h) - f(x)}{h}$$

$MC(x) = C'(x) \qquad MR(x) = R'(x) \qquad MP(x) = P'(x)$

- Find and interpret the derivative of a learning curve. *(Review Exercise 35.)*
- Find and interpret the derivative of an area or volume formula. *(Review Exercises 36–37.)*

2.4 The Product and Quotient Rules
- Find the derivative of a function using the Product Rule or Quotient Rule. *(Review Exercises 38–48.)*

$$\frac{d}{dx}(f \cdot g) = f' \cdot g + f \cdot g'$$

$$\frac{d}{dx}\left(\frac{f}{g}\right) = \frac{g \cdot f' - g' \cdot f}{g^2}$$

- Find the tangent line to a curve at a given point. *(Review Exercise 49.)*
- Use differentiation to solve an applied problem and interpret the answer. *(Review Exercises 50–52.)*

$$MAC(x) = \frac{C(x)}{x}$$

$$MAR(x) = \frac{R(x)}{x}$$

$$MAP(x) = \frac{P(x)}{x}$$

Review Exercises and Chapter Test
Following the Chapter Summary are the Review Exercises and a Chapter Test. Selected questions from the Review Exercises are specially color-coded to indicate that they may be used as a practice Chapter Test. Both even and odd answers are supplied in the back of the book for students to check their proficiency.

---

2   **Review Exercises and Chapter Test**   ◯ indicates a Chapter Test exercise.

2.1 Limits and Continuity

1–2. Complete the tables and use them to find each limit (or state that it does not exist). Round calculations to three decimal places.

1 a. $\lim\limits_{x \to 2^-} (4x + 2)$

x	$4x + 2$
1.9	
1.99	
1.999	

b. $\lim\limits_{x \to 2^+} (4x + 2)$

x	$4x + 2$
2.1	
2.01	
2.001	

c. $\lim\limits_{x \to 2} (4x + 2)$

2 a. $\lim\limits_{x \to 0^-} \dfrac{\sqrt{x+1} - 1}{x}$

x	$\frac{\sqrt{x+1}-1}{x}$	x	$\frac{\sqrt{x+1}-1}{x}$
−0.1		0.1	
−0.01		0.01	
−0.001		0.001	

b. $\lim\limits_{x \to 0^+} \dfrac{\sqrt{x+1} - 1}{x}$

c. $\lim\limits_{x \to 0} \dfrac{\sqrt{x+1} - 1}{x}$

Cumulative Review
Cumulative Review questions appear after every three to four chapters, with all answers supplied in the back of the book.

1–3   **Cumulative Review for Chapters 1–3**

The following exercises review some of the basic techniques that you learned in Chapters 1–3. Answers to all of these cumulative review exercises are given in the answer section at the back of the book.

1. Find an equation for the line through the points $(-4, 3)$ and $(6, -2)$. Write your answer in the form $y = mx + b$.

2. Simplify $\left(\frac{4}{25}\right)^{-1/2}$.

3. Find, correct to three decimal places: $\lim\limits_{x \to 0} (1 + 3x)^{1/x}$.

4. For the function $f(x) = \begin{cases} 4x - 8 & \text{if } x < 3 \\ 7 - 2x & \text{if } x \ge 3 \end{cases}$

   a. Draw its graph.
   b. Find $\lim\limits_{x \to 3^-} f(x)$.
   c. Find $\lim\limits_{x \to 3^+} f(x)$.
   d. Find $\lim\limits_{x \to 3} f(x)$.
   e. Is $f(x)$ continuous or discontinuous, and if it is discontinuous, where?

5. Use the definition of the derivative, $f'(x) = \lim\limits_{h \to 0} \dfrac{f(x+h) - f(x)}{h}$, to find the derivative of $f(x) = 2x^2 - 5x + 7$.

6. Find the derivative of $f(x) = 8\sqrt{x^3} - \dfrac{3}{x^2} + 5$.

14. Find the equation for the tangent line to the curve $y = \dfrac{4(x+3)}{\sqrt{x^2 + 3}}$ at $x = -1$.

15. Make sign diagrams for the first and second derivatives and draw the graph of the function $f(x) = x^3 - 12x^2 - 60x + 400$. Show on your graph all relative extreme points and inflection points.

16. Make sign diagrams for the first and second derivatives and draw the graph of the function $f(x) = \sqrt[3]{x^2} - 1$. Show on your graph all relative extreme points and inflection points.

17. A homeowner wishes to use 600 feet of fence to enclose two identical adjacent pens, as in the diagram below. Find the largest total area that can be enclosed.

18. A store can sell 12 telephone answering machines per day at a price of $200 each. The manager estimates that for each $10 price reduction she can sell 2 more per day. The answering machines cost the store $80

TECHNOLOGY

OPTIONAL! Using this book does not require a graphing calculator, but having one will enable you to do many problems more easily and as the same time deepen your understanding by allowing you to concentrate on concepts. The displays shown in the text are from the Texas Instruments TI-84 Plus *C* Silver Edition, except for a few from the TI-89, but any graphing calculator or computer may be used instead. For those who do not have a graphing calculator, the Explorations have been designed to be read for enrichment.

Similarly, if you have access to a computer, you may wish to do some of the Spreadsheet Explorations.

Graphing Calculator Exploration

The graph of the function from Example 5, $y_1 = 9x - 20x^{3/2}$ (written in x instead of t for ease of entry), is shown on the left on the standard window [−10, 10] by [−10, 10]. This might lead you to believe, erroneously, that the function is maximized at the endpoint (0, 0).

a. Why does this graph not look like the graph at the end of the previous example? [*Hint:* Look at the scale.]

b. Can you find a window on which your graphing calculator will show a graph like the one at the end of the preceding solution?

This example illustrates one of the pitfalls of graphing calculators—the part of the curve where the "action" takes place may be entirely hidden in one pixel. Calculus, on the other hand, will *always* find the critical value, no matter where it is, and then a graphing calculator can be used to confirm your answer by showing the graph on an appropriate window.

on [−10, 10] by [−10, 10]

Graphing Calculator Explorations

To allow for optional use of the graphing calculator, these Explorations are boxed. Most can also be read simply for enrichment. Exercises and examples that are designed to be done with a graphing calculator are marked with an icon.

90. BUSINESS: Movie Prices National average theater admissions prices for recent decades are shows in the following graph.

Movie Price ($)

7.85
5.39
4.22
2.69
1.55
0.78

1960 1970 1980 1990 2000 2010
Year

a. Number the bars with x-values 1–6 (so that x stands for *decades since 1950*) and use quadratic regression to fit a parabola to the data. State the regression function. [*Hint:* See Example 10.]

b. Use the regression function to predict movie prices in the years 2020 and 2030.

Source: Entertainment Weekly

Modeling

Selected application exercises feature regression capabilities of graphing calculators to fit curves to actual data.

Spreadsheet Explorations

Boxed for optional use, these explorations will enhance students' understanding of the material using Excel for those who prefer spreadsheet technology. See "Integrating Excel" on the next page for a list of exercises that can be done with Excel.

Spreadsheet Exploration

Another function that is not differentiable is $f(x) = x^{2/3}$. The following spreadsheet* calculates values of the difference quotient $\dfrac{f(x+h) - f(x)}{h}$ at $x = 0$ for this function. Since $f(0) = 0$, the difference quotient at $x = 0$ simplifies to:

$$\frac{f(x+h) - f(x)}{h} = \frac{f(0+h) - f(0)}{h} = \frac{f(h)}{h} = \frac{h^{2/3}}{h} = h^{-1/3}$$

For example, cell **B5** evaluates $h^{-1/3}$ at $h = \frac{1}{1000}$ obtaining $\left(\frac{1}{1000}\right)^{-1/3} = 1000^{1/3} = \sqrt[3]{1000} = 10$. Column **B** evaluates this different quotient for the *positive* values of h in column **A**, while column **E** evaluates it for the corresponding negative values of h in column **D**.

	B5	▼		=	=A5^(-1/3)	
	A	B	C	D	E	
1	h	(f(0+h)-f(0))/h		h	(f(0+h)-f(0))/h	
2	1.0000000	1.0000000		-1.0000000	-1.0000000	
3	0.1000000	2.1544347		-0.1000000	-2.1544347	
4	0.0100000	4.6415888		-0.0100000	-4.6415888	
5	0.0010000	10.0000000		-0.0010000	-10.0000000	
6	0.0001000	21.5443469		-0.0001000	-21.5443469	
7	0.0000100	46.4158883		-0.0000100	-46.4158883	
8	0.0000010	100.0000000		-0.0000010	-100.0000000	
9	0.0000001	215.4434690		-0.0000001	-215.4434690	

becoming large   becoming small

Notice that the values in column **B** are becoming arbitrarily large, while the values in column **E** are becoming arbitrarily small, so the difference quotient does not approach a limit as $h \to 0$. This shows that the derivative of $f(x) = x^{2/3}$ at 0 *does not exist*, so the function $f(x) = x^{2/3}$ is *not differentiable* at $x = 0$.

INTEGRATING EXCEL

If you would like to use Excel or another spreadsheet software when working the exercises in this text, refer to the chart below. It lists exercises from many sections that you might find instructive to do with spreadsheet technology. If you would like help using Excel, please consider the *Excel Guide* available via CengageBrain.com.

Section	Suggested Exercises	Section	Suggested Exercises
1.1	59–78	6.1	60–64
1.2	103–110	6.2	65, 66, 68
1.3	69–82, 84–90	6.3	41–42
1.4	79–92	6.4	9–18, 27–37
2.1	77–78, 81–82	7.1	29–30, 38–42
2.2	9–16	7.2	47–48, 53–56
2.3	47–50	7.3	29–32
2.4	61–64	7.4	13–18, 27–32
2.5	45–46	7.5	29–36
2.6	65, 69	7.6	31–32, 35–36
2.7	11–12	7.7	41–42
3.1	68–71, 85	8.1	9–20
3.2	61–64	8.2	36–41
3.3	23–40, 52–54	8.3	73–80
3.4	23–24	8.4	49–54
3.5	20	8.5	13–16, 20–26
3.6	69–70	9.1	71
3.7	23–26	9.2	54
4.1	11–12, 47–51	9.3	36–38, 41–48
4.2	31–50	9.4	11–24, 27–30
4.3	97–99	10.1	49–59
4.4	38–39	10.2	9–12, 21–22
5.1	41–42	10.3	35–36
5.2	45–46, 55–58	10.4	11–24, 33–38
5.3	13–18, 83–88	11.1	17–18, 29–36
5.4	32, 35–36, 61, 69	11.2	37–41
5.5	31–32	11.3	23
5.6	77–78	11.4	21–26

SUPPLEMENTS

For the Student	For the Instructor
Student Solutions Manual ISBN: 978-1-305-10795-3 This manual contains fully worked-out solutions to all of the odd-numbered exercises in the text, giving students a way to check their answers and ensure that they took the correct steps to arrive at an answer.	**Complete Solutions Manual** This manual contains solutions to all exercises from the text including Chapter Review Exercises and Cumulative Reviews. It also contains two chapter-level tests for each chapter, one short-answer and one multiple choice, along with answers to each. This manual can be found on the Instructor Companion Site.
CengageBrain.com To access additional course materials, please visit **www.cengagebrain.com.** At the **CengageBrain.com** home page, search for the ISBN (from the back cover of your book) of your title using the search box at the top of the page. This will take you to the product page where these resources can be found.	**Instructor Companion Site** Everything you need for your course in one place! This collection of book-specific lecture and class tools is available online via **www.cengage.com/login.** Access and download PowerPoint® presentations, images, solutions manual, and more.
Enhanced WebAssign® Instant Access Code: 978-1-285-85761-9 Printed Access Card: 978-1-285-85758-9 Enhanced WebAssign combines exceptional mathematics content with the most powerful online homework solution, WebAssign. It now includes QuickPrep content to review key precalculus content, available as a CoursePack of prebuilt assignments to assign at the beginning of the course or where needed most. Enhanced WebAssign engages students with immediate feedback, rich tutorial content, and an interactive, fully customizable eBook, the Cengage YouBook, helping students to develop a deeper conceptual understanding of their subject matter.	**Enhanced WebAssign®** Instant Access Code: 978-1-285-85761-9 Printed Access Card: 978-1-285-85758-9 Enhanced WebAssign combines exceptional mathematics content with the most powerful online homework solution, WebAssign. It now includes QuickPrep content to review key precalculus content, available as a CoursePack of prebuilt assignments to assign at the beginning of the course or where needed most. Enhanced WebAssign engages students with immediate feedback, rich tutorial content, and an interactive, fully customizable eBook, the Cengage YouBook, helping students to develop a deeper conceptual understanding of their subject matter. Visit **www.cengage.com/ewa** to learn more.
	Cengage Learning Testing Powered by Cognero® Instant Access Code: 978-1-305-11229-2 Cognero is a flexible, online system that allows you to author, edit, and manage test bank content, create multiple test versions in an instant, and deliver tests from your LMS, your classroom or wherever you want. This is available online via **www.cengage.com/login**.

ACKNOWLEDGMENTS

We are indebted to many people for their useful suggestions, conversations, and correspondence during the writing and revising of this book. We thank Chris and Lee Berresford, Anne Burns, Richard Cavaliere, Ruch Enoch, Theordore Faticoni, Jeff Goodman, Susan Halter, Brita and Ed Immergut, Ethel Matin, Gary Patric, Shelly Rothman, Charlene Russert, Stuart Saal, Bob Sickles, Michael Simon, John Stevenson, and all of our "Math 6" students at C.W. Post for serving as proofreaders and critics over many years.

We had the good fortune to have had the support of expert editorial, production, and marketing colleagues at Cengage Learning: Richard Stratton; Rita Lombard; Erin Brown; Jennifer Cordoba; Jessica Rasile; Jill Quinn; Linda May; and Julie Schuster. We also express our gratitude to the many others at Cengage Learning who made important contributions, too numerous to mention. We are especially grateful to Magdalena Luca who worked to ensure the accuracy of the seventh edition.

We also wish to acknowledge Christi Verity and Aldena Calden (UMass, Amherst) for their contributions on the solutions manuals. Very special thanks also go to Lee Berresford for her contributions on the cover design.

The following reviewers and readers have contributed greatly to the development of the seventh edition of this text:

Haya Adner	*Queensborough Community College*
Kimberly Benien	*Wharton County Junior College*
Mark Billiris	*St. Petersburg College*
Lynn Cade	*Pensacola State College*
Seo-eun Choi	*Arkansas State University*
Cindy Dickson	*College of Southern Idaho*
Susan Howell	*University of Southern Mississippi*
Magdalena Luca	*MCPHS University*
Kevin Lynch	*Northeast State Community College*
Cornelius Nelan	*Quinnipiac University*
Victor Swaim	*Southeastern Louisiana University*
William Veczko	*St. Johns River State College*

We also thank the reviewers of recent editions:

Frederick Adkins, *Indiana University of Pennsylvania*; David Allen, *Iona College*; Joel M. Berman, *Valencia Community College*; John A. Blake, *Oakwood College*; Dave Bregenzer, *Utah State University*; Kelly Brooks, *Pierce College*; Donald O. Clayton, *Madisonville Community College*; Charles C. Clever, *South Dakota State University*; Julane Crabtree, *Johnson Community College*; Dale L. Craft, *South Florida Community College*; Kent Craghead, *Colby Community College*; Biswa Datta, *Northern Illinois University*; Lloyd David, *Montreat College*; Allan Donsig, *University of Nebraska—Lincoln*; Sally Edwards, *Johnson Community College*; Frank Farris, *Santa Clara University*; Brad Feldser, *Kennesaw State University*; Daria Filippova, *Bowling Green State University*; Abhay Gaur, *Duquesne University*; Jerome Goldstein, *University of Memphis*; John Haverhals, *Bradley University*; John B. Hawkins, *Georgia Southern University*; Randall Helmstutler, *University of Virginia*; Susan Howell, *University of Southern Mississippi*; Heather Hulett, *University of Wisconsin—La Crosse*; David Hutchison, *Indiana State University*; Dan Jelsovsky, *Florida Southern College*; Alan S. Jian, *Solano Community College*; Dr. Hilbert Johs, *Wayne State College*; Hideaki Kaneko, *Old Dominion University*; John Karloff, *University of North Carolina*; Susan Kellicut,

Seminole Community College; Todd King, *Michigan Technical University;* JoAnn Kump, *West Chester University;* Richard Leedy, *Polk Community College;* Michael Longfritz, *Rensselear Polytechnic Institute;* Dr. Hank Martel, *Broward Community College;* Kimberly McGinley Vincent, *Washington State University;* Donna Mills, *Frederick Community College;* Pat Moreland, *Cowley College;* Sanjay Mundkur, *Kennesaw State University;* Sue Neal, *Wichita State University;* Cornelius Nelan, *Quinnipiac University;* David Parker, *Salisbury University;* Shahla Peterman, *University of Missouri—Rolla;* Susan Pfiefer, *Butler Community College;* Daniel Plante, *Stetson University;* Brooke Quinlan, *Hillsborough Community College;* Catherine A. Roberts, *University of Rhode Island;* George W. Schultz, *St. Petersburg College;* Larry Small, *Pierce College;* Paul H. Stanford, *University of Texas—Dallas;* Xingping Sun, *Missouri State University;* Jill Van Valkenburg, *Bowling Green State University;* Jaak Vilms, *Colorado State University;* Erica Voges, *New Mexico State University;* Jane West, *Trident Technical College;* Elizabeth White, *Trident Technical College;* Kenneth J. Word, *Central Texas College;* Wen-Qing Xu, *California State University—Long Beach.*

Finally, and most importantly, we thank our wives, Barbara and Kathryn, for their encouragement and support.

COMMENTS WELCOMED

With the knowledge that any book can always be improved, we welcome corrections, constructive criticisms, and suggestions from every reader.

geoffrey.berresford@liu.edu
andrew.rockett@liu.edu

DIAGNOSTIC TEST

Are you ready to study calculus?

Algebra is the language in which we express the ideas of calculus. Therefore, to understand calculus and express its ideas with precision, you need to know some algebra.

If you are comfortable with the algebra covered in the following problems, you are ready to begin your study of calculus. If not, turn to the *Algebra Review* appendix beginning on page B1 and review the *Complete Solutions* to these problems, and continue reading the other parts of the Appendix that cover anything that you do not know.

Problems

Answers

1. True or False? $\frac{1}{2} < -3$

False

2. Express $\{x \mid -4 < x \le 5\}$ in interval notation.

$(-4, 5]$

3. What is the slope of the line through the points $(6, -7)$ and $(9, 8)$?

5

4. On the line $y = 3x + 4,$ what value of Δy corresponds to $\Delta x = 2$?

6

5. Which sketch shows the graph of the line $y = 2x - 1$?

a

6. True of False? $\left(\dfrac{\sqrt{x}}{y}\right)^{-2} = \dfrac{y^2}{x}$

True

7. Find the zeros of the function $f(x) = 9x^2 - 6x - 1.$

$x = \dfrac{1 \mp \sqrt{2}}{3}$

8. Expand and simplify $x(8 - x) - (3x + 7).$

$-x^2 + 5x - 7$

9. What is the domain of $f(x) = \dfrac{x^2 - 3x + 2}{x^3 + x^2 - 6x}?$

$\{x \mid x \ne -3, x \ne 0, x \ne 2\}$

10. Find the difference quotient $\dfrac{f(x + h) - f(x)}{h}$ for $f(x) = x^2 - 5x.$

$2x - 5 + h$

Functions

1

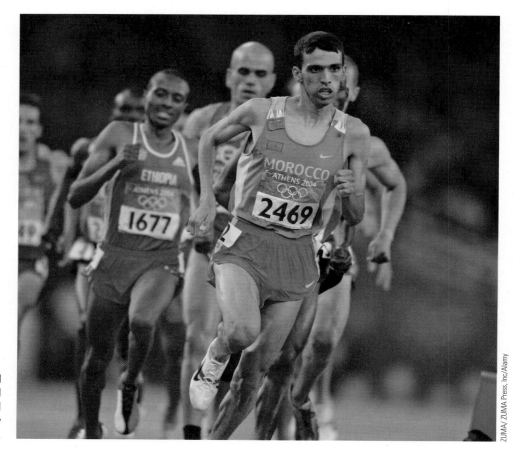

Moroccan runner Hicham El Guerrouj, current world record holder for the mile run, bested the record set 6 years earlier by 1.26 seconds.

ZUMA/ ZUMA Press, Inc/Alamy

What You'll Explore

To model how things change over time or to manage any complex enterprise, you will need a variety of ways to express relationships between important quantities. The functions introduced in this chapter will help you understand and predict quantities as diverse as populations, income, global energy, and even the world record times in the mile run. The techniques you learn in this chapter will serve as the basis for calculus in Chapter 2 and beyond.

1.1   Real Numbers, Inequalities, and Lines

1.2   Exponents

1.3   Functions: Linear and Quadratic

1.4   Functions: Polynomial, Rational, and Exponential

World Record Mile Runs

The dots on the graph below show the world record times for the mile run from 1865 to the 1999 world record of 3 minutes 43.13 seconds, set by the Moroccan runner Hicham El Guerrouj. These points fall roughly along a line, called the **regression line.** In this section we will see how to use a graphing calculator to find a regression line (see Example 9 and Exercises 73–78), based on a method called **least squares,** whose mathematical basis will be explained in Chapter 7.

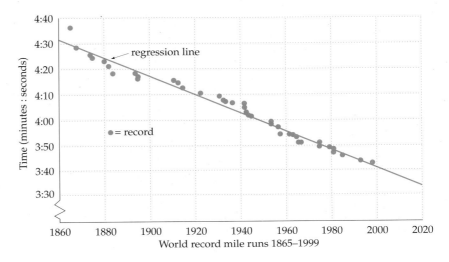

World record mile runs 1865–1999

Notice that the times do not level off as you might expect but continue to decrease.

History of the Record for the Mile Run

Time	Year	Athlete	Time	Year	Athlete	Time	Year	Athlete
4:36.5	1865	Richard Webster	4:09.2	1931	Jules Ladoumegue	3:54.1	1964	Peter Snell
4:29.0	1868	William Chinnery	4:07.6	1933	Jack Lovelock	3:53.6	1965	Michel Jazy
4:28.8	1868	Walter Gibbs	4:06.8	1934	Glenn Cunningham	3:51.3	1966	Jim Ryun
4:26.0	1874	Walter Slade	4:06.4	1937	Sydney Wooderson	3:51.1	1967	Jim Ryun
4:24.5	1875	Walter Slade	4:06.2	1942	Gunder Hägg	3:51.0	1975	Filbert Bayi
4:23.2	1880	Walter George	4:06.2	1942	Arne Andersson	3:49.4	1975	John Walker
4:21.4	1882	Walter George	4:04.6	1942	Gunder Hägg	3:49.0	1979	Sebastian Coe
4:18.4	1884	Walter George	4:02.6	1943	Arne Andersson	3:48.8	1980	Steve Ovett
4:18.2	1894	Fred Bacon	4:01.6	1944	Arne Andersson	3:48.53	1981	Sebastian Coe
4:17.0	1895	Fred Bacon	4:01.4	1945	Gunder Hägg	3:48.40	1981	Steve Ovett
4:15.6	1895	Thomas Conneff	3:59.4	1954	Roger Bannister	3:47.33	1981	Sebastian Coe
4:15.4	1911	John Paul Jones	3:58.0	1954	John Landy	3:46.31	1985	Steve Cram
4:14.4	1913	John Paul Jones	3:57.2	1957	Derek Ibbotson	3:44.39	1993	Noureddine Morceli
4:12.6	1915	Norman Taber	3:54.5	1958	Herb Elliott	3:43.13	1999	Hicham El Guerrouj
4:10.4	1923	Paavo Nurmi	3:54.4	1962	Peter Snell			

Source: USA Track & Field

The equation of the regression line is $y = -0.356x + 257.44$, where x represents years after 1900 and y is the time in seconds. The regression line can be used to predict the world mile record in future years. Notice that the most recent world record would have been predicted quite accurately by this line, since the rightmost dot falls almost exactly on the line.

Linear trends, however, must not be extended too far. The downward slope of this line means that it will eventually "predict" mile runs in a fraction of a second, or even in *negative* time (see Exercises 59 and 60 on pages 17–18). *Moral:* In the real world, linear trends do not continue indefinitely. This and other topics in "linear" mathematics will be developed in Section 1.1.

1.1   Real Numbers, Inequalities, and Lines

Introduction

Quite simply, *calculus is the study of rates of change.* We will use calculus to analyze rates of inflation, rates of learning, rates of population growth, and rates of natural resource consumption.

In this first section we will study **linear** relationships between two variable quantities—that is, relationships that can be represented by **lines.** In later sections we will study **nonlinear** relationships, which can be represented by **curves.**

Real Numbers and Inequalities

In this book the word "number" means **real number,** a number that can be represented by a point on the number line (also called the **real line**).

The *order* of the real numbers is expressed by **inequalities.** For example, $a < b$ means "a is to the *left* of b" or, equivalently, "b is to the *right* of a."

Inequalities		
Inequality	**In Words**	**Brief Examples**
$a < b$	a is less than (smaller than) b	$3 < 5$
$a \leq b$	a is less than or equal to b	$-5 \leq -3$
$a > b$	a is greater than (larger than) b	$\pi > 3$
$a \geq b$	a is greater than or equal to b	$2 \geq 2$

The inequalities $a < b$ and $a > b$ are called **strict inequalities,** and $a \leq b$ and $a \geq b$ are called **nonstrict inequalities.**

IMPORTANT NOTE Throughout this book are many **Practice Problems**— short questions designed to check your understanding of a topic before moving on to new material. Full solutions are given at the end of the section. Solve the following Practice Problem and then check your answer.

PRACTICE PROBLEM 1

Which number is smaller: $\dfrac{1}{100}$ or $-1,000,000$?   Solution on page 15 >

Multiplying or dividing both sides of an inequality by a negative number reverses the direction of the inequality:

$$-3 < 2 \quad \text{but} \quad 3 > -2 \qquad \text{Multiplying by } -1$$

A **double inequality,** such as $a < x < b$, means that *both* the inequalities $a < x$ and $x < b$ hold. The inequality $a < x < b$ can be interpreted graphically as "x is between a and b."

$$a < x < b$$

LOOKING AHEAD

Sets and intervals will be important on page 33 when we define *domains* of functions.

Sets and Intervals

Braces $\{\ \}$ are read "the set of all" and a **vertical bar** $\mid$ is read "such that."

EXAMPLE 1   INTERPRETING SETS

The set of all

a. $\{x \mid x > 3\}$ means "the set of all x such that x is greater than 3."

Such that

b. $\{x \mid -2 < x < 5\}$ means "the set of all x such that x is between -2 and 5."

PRACTICE PROBLEM 2

a. Write in set notation "the set of all x such that x is greater than or equal to -7."
b. Express in words: $\{x \mid x < -1\}$.   Solution on page 15 >

The set $\{x \mid 2 \le x \le 5\}$ can be expressed in **interval notation** by enclosing the endpoints 2 and 5 in **square brackets**, [2, 5], to indicate that the endpoints are *included.* The set $\{x \mid 2 < x < 5\}$ can be written with **parentheses,** (2, 5), to indicate that the endpoints 2 and 5 are *excluded.* An interval is **closed** if it includes both endpoints, and **open** if it includes neither endpoint. The four types of intervals are shown below: a **solid dot** ● on the graph indicates that the point is *included* in the interval; a **hollow dot** ○ indicates that the point is *excluded.*

Finite Intervals

Interval Notation	Set Notation	Graph	Type	Brief Examples
$[a, b]$	$\{x \mid a \le x \le b\}$	●———● a b	Closed (includes endpoints)	$[-2, 5]$ ●———● -2 5
(a, b)	$\{x \mid a < x < b\}$	○———○ a b	Open (excludes endpoints)	$(-2, 5)$ ○———○ -2 5
$[a, b)$	$\{x \mid a \le x < b\}$	●———○ a b	Half-open or	$[-2, 5)$ ●———○ -2 5
$(a, b]$	$\{x \mid a < x \le b\}$	○———● a b	half-closed	$(-2, 5]$ ○———● -2 5

An interval may extend infinitely far to the *right* (indicated by the symbol ∞ for **infinity**) or infinitely far to the *left* (indicated by $-\infty$ for **negative infinity**). Note that ∞ and $-\infty$ are not numbers but are merely symbols to indicate that the interval extends endlessly in that direction. The infinite intervals in the following box are said to be **closed** or **open** depending on whether they *include* or *exclude* their single endpoint.

Infinite Intervals

Interval Notation	Set Notation	Graph	Type	Brief Examples
$[a, \infty)$	$\{\, x \mid x \geq a \,\}$	*(closed at a)*	Closed	$[3, \infty)$ *(closed at 3)*
(a, ∞)	$\{\, x \mid x > a \,\}$	*(open at a)*	Open	$(3, \infty)$ *(open at 3)*
$(-\infty, a]$	$\{\, x \mid x \leq a \,\}$	*(closed at a)*	Closed	$(-\infty, 5]$ *(closed at 5)*
$(-\infty, a)$	$\{\, x \mid x < a \,\}$	*(open at a)*	Open	$(-\infty, 5)$ *(open at 5)*

We use *parentheses* rather than square brackets with ∞ and $-\infty$ since they are not actual numbers.

The interval $(-\infty, \infty)$ extends infinitely far in *both* directions (meaning the entire real line) and is also denoted by $\mathbb{R}$ (the set of all real numbers).

$$\mathbb{R} = (-\infty, \infty)$$

Cartesian Plane

Two real lines or **axes,** one horizontal and one vertical, intersecting at their zero points, define the **Cartesian plane.*** The point where they meet is called the **origin.** The axes divide the plane into four **quadrants,** I through IV, as shown below.

Any point in the Cartesian plane can be specified uniquely by an ordered pair of numbers (x, y); x, called the **abscissa** or **x-coordinate,** is the number on the horizontal axis corresponding to the point; y, called the **ordinate** or **y-coordinate,** is the number on the vertical axis corresponding to the point.

The Cartesian plane

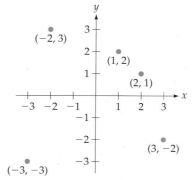

The Cartesian plane with several points.
Order matters: (1, 2) is not the same as (2, 1)

⊳ **LOOKING AHEAD**

We will use the Δ notation again on page 95.

Lines and Slopes

The symbol Δ (read "delta," the Greek letter D) means "the change in." For any two points (x_1, y_1) and (x_2, y_2) we define

*So named because it was originated by the French philosopher and mathematician René Descartes (1596–1650). Following the custom of the day, Descartes signed his scholarly papers with his Latin name Cartesius, hence "Cartesian" plane.

$$\Delta x = x_2 - x_1 \qquad \text{The change in } x \text{ is the difference in the } x\text{-coordinates}$$

$$\Delta y = y_2 - y_1 \qquad \text{The change in } y \text{ is the difference in the } y\text{-coordinates}$$

Any two distinct points determine a line. A nonvertical line has a **slope** that measures the *steepness* of the line, and is defined as *the change in y divided by the change in x for any two points on the line.*

Take Note

One of the main purposes of calculus is to extend the concept of slope from lines to *curves*.

Slope of Line Through (x_1, y_1) and (x_2, y_2)

$$m = \frac{\Delta y}{\Delta x} = \frac{y_2 - y_1}{x_2 - x_1} \qquad \begin{array}{l} \text{Slope is the change in } y \text{ over} \\ \text{the change in } x \ (x_2 \neq x_1) \end{array}$$

⚠ Be Careful In slope, the *x*-values go in the *denominator*.

The changes Δy and Δx are often called, respectively, the "rise" and the "run," with the understanding that a negative "rise" means a "fall." Slope is then "rise over run."

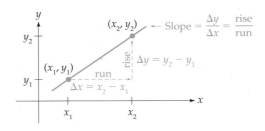

EXAMPLE 2   FINDING SLOPES AND GRAPHING LINES

Find the slope of the line through each pair of points, and graph the line.

a. $(2, 1), (3, 4)$ **b.** $(2, 4), (3, 1)$

c. $(-1, 3), (2, 3)$ **d.** $(2, -1), (2, 3)$

Solution

We use the slope formula $m = \dfrac{y_2 - y_1}{x_2 - x_1}$ for each pair $(x_1, y_1), (x_2, y_2)$.

a. For $(2, 1)$ and $(3, 4)$ the slope is   **b.** For $(2, 4)$ and $(3, 1)$ the slope is

$$\frac{4 - 1}{3 - 2} = \frac{3}{1} = 3. \qquad\qquad\qquad \frac{1 - 4}{3 - 2} = \frac{-3}{1} = -3.$$

c. For $(-1, 3)$ and $(2, 3)$ the slope is $\dfrac{3 - 3}{2 - (-1)} = \dfrac{0}{3} = 0$.

d. For $(2, -1)$ and $(2, 3)$ the slope is *undefined*:

$$\dfrac{3 - (-1)}{2 - 2} = \dfrac{4}{0}.$$

Notice in the preceding graphs that when the x-coordinates are the same [as in part (d)], the line is *vertical*, and when the y-coordinates are the same [as in part (c)], the line is *horizontal*.

If $\Delta x = 1$, as in Examples 2a and 2b, then the slope is just the "rise," giving an alternative definition for slope:

$$\text{Slope} = \left(\begin{array}{c} \text{Amount that the line rises} \\ \text{when } x \text{ increases by 1} \end{array} \right)$$

PRACTICE PROBLEM 3

A company president is considering four different business strategies, called S_1, S_2, S_3, and S_4, each with different projected future profits. The graph on the right shows the annual projected profit for the first few years for each of the strategies.

Which strategy yields:

a. the highest projected profit in year 1?
b. the highest projected profit in the long run?

Solutions on page 15 >

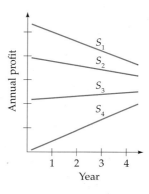

Equations of Lines

The point where a nonvertical line crosses the y-axis is called the **y-intercept** of the line. The y-intercept can be given either as the y-coordinate b or as the point $(0, b)$. Such a line can be expressed very simply in terms of its slope and y-intercept, representing the points by variable coordinates (or "variables") x and y.

Slope-Intercept From of a Line

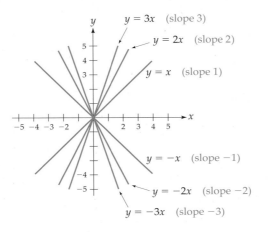

$$y = mx + b$$

slope y-intercept

Brief Example

For the line with slope -2 and y-intercept 4:

$$y = -2x + 4$$

y-intercept $(0, 4)$

Second point 1 unit "over" and 2 "down"

For lines through the origin, the equation takes the particularly simple form, $y = mx$ (since $b = 0$), as illustrated on the left.

The most useful equation for a line is the *point-slope form*.

Point-Slope Form of a Line

$$y - y_1 = m(x - x_1)$$

(x_1, y_1) = point on the line
m = slope

This form comes directly from the slope formula $m = \dfrac{y_2 - y_1}{x_2 - x_1}$ by replacing x_2 and y_2 by x and y, and then multiplying each side by $(x - x_1)$. It is most useful when you know the slope of the line and a point on it.

EXAMPLE 3	USING THE POINT-SLOPE FORM

Find an equation of the line through $(6, -2)$ with slope $-\frac{1}{2}$.

Solution

$$y - (-2) = -\frac{1}{2}(x - 6)$$

$y - y_1 = m(x - x_1)$ with
$y_1 = -2, \ m = -\frac{1}{2}, \ \text{and} \ x_1 = 6$

$$y + 2 = -\frac{1}{2}x + 3$$

Eliminating parentheses

$$y = -\frac{1}{2}x + 1$$

Subtracting 2 from each side

Alternatively, we could have found this equation using $y = mx + b$, replacing m by the given slope $-\frac{1}{2}$, and then substituting the given $x = 6$ and $y = -2$ to evaluate b.

EXAMPLE 4	FINDING AN EQUATION FOR A LINE THROUGH TWO POINTS

Find an equation for the line through the points $(4, 1)$ and $(7, -2)$.

Solution

The slope is not given, so we calculate it from the two points.

$$m = \frac{-2 - 1}{7 - 4} = \frac{-3}{3} = -1 \qquad m = \frac{y_2 - y_1}{x_2 - x_1} \ \text{with} \ (4, 1) \ \text{and} \ (7, -2)$$

Then we use the point-slope formula with this slope and either of the two points.

$$y - 1 = -1(x - 4)$$

$y - y_1 = m(x - x_1)$ with
slope -1 and point $(4, 1)$

$$y - 1 = -x + 4$$

Eliminating parentheses

$$y = -x + 5$$

Adding 1 to each side

PRACTICE PROBLEM 4

Find the slope-intercept form of the line through the points $(2, 1)$ and $(4, 7)$.

Solution on page 15 >

Vertical and horizontal lines have particularly simple equations: a variable equaling a constant.

EXAMPLE 5   GRAPHING VERTICAL AND HORIZONTAL LINES

Graph the lines $x = 2$ and $y = 6$.

Solution

EXAMPLE 6   FINDING EQUATIONS OF VERTICAL AND HORIZONTAL LINES

a. Find an equation for the *vertical* line through $(3, 2)$.

b. Find an equation for the *horizontal* line through $(3, 2)$.

Solution

a. Vertical line $x = 3$ $x = a$, with a being the
                                               x-coordinate from $(3, 2)$

b. Horizontal line $y = 2$ $y = b$, with b being the
                                               y-coordinate from $(3, 2)$

PRACTICE PROBLEM 5

Find an equation for the vertical line through $(-2, 10)$.

Solution on page 15 >

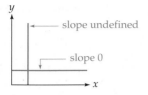

In a vertical line, the x-coordinate does not change, so $\Delta x = 0$, making the slope $m = \Delta y / \Delta x$ *undefined*. Therefore, distinguish carefully between slopes of vertical and horizontal lines:

> Vertical line: Slope is *undefined*.
>
> Horizontal line: Slope *is* defined, and is *zero*.

There is one form that covers *all* lines, vertical and nonvertical.

General Linear Equation

$$ax + by = c$$

For constants a, b, c, with a and b not both zero

Any equation that can be written in this form is called a **linear equation,** and the variables are said to **depend linearly** on each other.

EXAMPLE 7   **FINDING THE SLOPE AND THE Y-INTERCEPT FROM A LINEAR EQUATION**

Find the slope and y-intercept of the line $2x + 3y = 12$.

Solution

We write the line in slope-intercept form. Solving for y:

$$3y = -2x + 12$$

Subtracting $2x$ from both sides of $2x + 3y = 12$

$$y = -\frac{2}{3}x + 4$$

Dividing each side by 3 gives the slope-intercept form $y = mx + b$

Therefore, the slope is $-\frac{2}{3}$ and the y-intercept is $(0, 4)$.

PRACTICE PROBLEM 6

Find the slope and y-intercept of the line $x - \dfrac{y}{3} = 2$.

Solution on page 15 >

Parallel and Perpendicular Lines

Slope can be used to determine whether two lines are parallel or perpendicular.

Slopes of Parallel and Perpendicular Lines

	Brief Examples
Two distinct lines are *parallel* if they have the *same* slope:	The following pairs of numbers are *negative reciprocals* (reciprocals with opposite signs):
$$m_1 = m_2$$	$$2 \quad \text{and} \quad -\frac{1}{2}$$
Two lines are *perpendicular* if their slopes are *negative reciprocals*:	$$-\frac{3}{4} \quad \text{and} \quad \frac{4}{3}$$
$$m_1 = -\frac{1}{m_2}$$	$$\frac{1}{5} \quad \text{and} \quad -5$$

EXAMPLE 8   FINDING PARALLEL AND PERPENDICULAR LINES

Find an equation for the line through the point $(6, -3)$ that is (a) parallel to the line $2x + 3y = 12$ and (b) perpendicular to the line $2x + 3y = 12$.

Solution

a. Ordinarily, we would now find the slope of the line $2x + 3y = 12$. However, in Example 7 we found that the slope of this line is $-\frac{2}{3}$. Therefore, we want the line with slope $m = -\frac{2}{3}$ that passes through $(6, -3)$. We use the slope-intercept form with the above slope and point:

$$y - (-3) = -\frac{2}{3}(x-6) \qquad \text{$y - y_1 = m(x - x_1)$ with $m = -\frac{2}{3}$,} \\ \text{$x_1 = 6$, and $y_1 = -3$}$$

$$y + 3 = -\frac{2}{3}x + 4 \qquad \text{Multiplying out and simplifying}$$

$$y = -\frac{2}{3}x + 1 \qquad \text{Answer (after subtracting 3)}$$

b. The *perpendicular* line will have slope that is the *negative reciprocal* of $-\frac{2}{3}$, which is $m = \frac{3}{2}$. With this slope and the same point, the slope-intercept form gives

$$y - (-3) = \frac{3}{2}(x - 6) \qquad \text{$y - y_1 = m (x - x_1)$ with $m = \frac{3}{2}$,} \\ \text{$x_1 = 6$, and $y_1 = -3$}$$

$$y + 3 = \frac{3}{2}x - 9 \qquad \text{Multiplying out and simplifying}$$

$$y = \frac{3}{2}x - 12 \qquad \text{Answer (after subtracting 3)}$$

The graphs of the three lines are shown on the left.

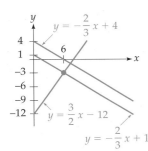

PRACTICE PROBLEM 7

Find an equation for the line through the point (9, 2) that is perpendicular to the line $x - \dfrac{y}{3} = 2$. [*Hint:* Use your answer to Practice Problem 6.]

Solution on page 16 >

Linear Regression

Take Note

You don't need to know about regression to read most of this book.

Given two points, we can find a line through them, as in Example 4. However, some real-world situations involve *many* data points, which may lie approximately but not exactly on a line. How can we find the line that, in some sense, *lies closest* to the points or *best approximates* the points? The most widely used technique is called **linear regression** or **least squares**, and its mathematical basis will be explained in Section 7.4. Even before studying its mathematical basis, however, we can easily find the regression line using a graphing calculator (or spreadsheet or other computer software).

| EXAMPLE 9 | LINEAR REGRESSION USING A GRAPHING CALCULATOR |

The following graph shows the average number of "tweets" per day sent on Twitter in recent years.

Source: Twitter

a. Use linear regression to fit a line to the data.

b. Interpret the slope of the line.

c. Use the regression line to predict the number of tweets per day in the year 2022.

Solution

a. We number the years with *x*-values 0–3, so *x* stands for *years since 2010* (we could choose other *x*-values instead). We enter the data into lists, as shown in the first screen below (as explained in the appendix *Graphing Calculator Basics—Entering Data* on page A3), and use *ZoomStat* to graph the data points.

Then (using STAT, CALC, and LinReg) graph the regression along with the data points.

The regression line, which clearly fits the points quite well, is

$$y = 129x + 39$$

b. Since x is in years, the slope 129 means that the number of tweets per day increases by about 129 million each year.

c. To predict the number of tweets per day in the year 2022, we evaluate Y1 at 12 (since $x = 12$ corresponds to 2022). From the screen on the right, if the current trend continues, about 1.6 billion tweets per day will be sent in 2022.

Y₁(12)

1587

Solutions TO PRACTICE PROBLEMS

1. $-1,000,000$ [the negative sign makes it less than (to the left of) the positive number $\frac{1}{100}$]

2. a. $\{ x \mid x \geq -7 \}$

 b. The set of all x such that x is less than -1

3. a. S_1 **b.** S_4

4. $m = \dfrac{7 - 1}{4 - 2} = \dfrac{6}{2} = 3$ From points $(2, 1)$ and $(4, 7)$

 $y - 1 = 3(x - 2)$ Using the point-slope form with $(x_1, y_1) = (2, 1)$

 $y - 1 = 3x - 6$

 $y = 3x - 5$

5. $x = -2$

6. $x - \dfrac{y}{3} = 2$

 $-\dfrac{y}{3} = -x + 2$ Subtracting x from each side

 $y = 3x - 6$ Multiplying each side by -3

Slope is $m = 3$ and y-intercept is $(0, -6)$.

7. $m = -\dfrac{1}{3}$ The negative reciprocal of the slope found in Practice Problem 6

$$y - 2 = -\dfrac{1}{3}(x - 6)$$ $y - y_1 = m(x - x_1)$ with $m = -\frac{1}{3}$, $x_1 = 6$, and $y_1 = 2$

$$y = -\dfrac{1}{3}x + 4$$ After simplifying

1.1   Section Summary

An **interval** is a set of real numbers corresponding to a section of the real line. The interval is **closed** if it contains all of its endpoints, and **open** if it contains none of its endpoints.

The nonvertical line through two points (x_1, y_1) and (x_2, y_2) has **slope**

$$m = \frac{\Delta y}{\Delta x} = \frac{y_2 - y_1}{x_2 - x_1} \qquad x_1 \neq x_2$$

The slope of a *vertical* line is *undefined* or, equivalently, *does not exist*.

There are five **equations** or **forms** for lines:

$$y = mx + b \qquad \text{Slope-intercept form}$$
m = slope, b = y-intercept

$$y - y_1 = m(x - x_1) \qquad \text{Point-slope form}$$
(x_1, y_1) = point, m = slope

$$x = a \qquad \text{Vertical line (slope undefined)}$$
a = x-intercept

$$y = b \qquad \text{Horizontal line (slope zero)}$$
b = y-intercept

$$ax + by = c \qquad \text{General linear equation}$$

A graphing calculator can find the regression line for a set of points, which can then be used to predict future trends.

1.1   Exercises

1–4. Write each interval in set notation and graph it on the real line.

1. $[0, 6)$ **2.** $(-3, 5]$ **3.** $(-\infty, 2]$ **4.** $[7, \infty)$

5. Given the equation $y = 5x - 12$, how will y change if x:
   a. Increases by 3 units?
   b. Decreases by 2 units?

6. Given the equation $y = -2x + 7$, how will y change if x:
   a. Increases by 5 units?
   b. Decreases by 4 units?

7–14. Find the slope (if it is defined) of the line determined by each pair of points.

7. $(2, 3)$ and $(4, -1)$ **8.** $(3, -1)$ and $(5, 7)$

9. $(-4, 0)$ and $(2, 2)$ **10.** $(-1, 4)$ and $(5, 1)$

11. $(0, -1)$ and $(4, -1)$ **12.** $(-2, \frac{1}{2})$ and $(5, \frac{1}{2})$

13. $(2, -1)$ and $(2, 5)$ **14.** $(6, -4)$ and $(6, -3)$

15–32. For each equation, find the slope m and y-intercept $(0, b)$ (when they exist) and draw the graph.

15. $y = 3x - 4$ **16.** $y = 2x$

17. $y = -\frac{1}{2}x$ **18.** $y = -\frac{1}{3}x + 2$

19. $y = 4$ **20.** $y = -3$

21. $x = 4$ **22.** $x = -3$

23. $2x - 3y = 12$ **24.** $3x + 2y = 18$

25. $x + y = 0$ **26.** $x = 2y + 4$

27. $x - y = 0$ **28.** $y = \frac{2}{3}(x - 3)$

29. $y = \dfrac{x + 2}{3}$ **30.** $\dfrac{x}{2} + \dfrac{y}{3} = 1$

31. $\dfrac{2x}{3} - y = 1$ **32.** $\dfrac{x+1}{2} + \dfrac{y+1}{2} = 1$

33–46. Write an equation of the line satisfying the following conditions. If possible, write your answer in the form $y = mx + b$.

33. Slope -2.25 and y-intercept 3

34. Slope $\frac{2}{3}$ and y-intercept -8

35. Slope 5 and passing through the point $(-1, -2)$

36. Slope -1 and passing through the point $(4, 3)$

37. Horizontal and passing through the point $(1.5, -4)$

38. Horizontal and passing through the point $(\frac{1}{2}, \frac{3}{4})$

39. Vertical and passing through the point $(1.5, -4)$

40. Vertical and passing through the point $(\frac{1}{2}, \frac{3}{4})$

41. Passing through the points $(5, 3)$ and $(7, -1)$

42. Passing through the points $(3, -1)$ and $(6, 0)$

43. Passing through the points $(1, -1)$ and $(5, -1)$

44. Passing through the points $(2, 0)$ and $(2, -4)$

45. Passing through the point $(12, 2)$ that is (a) parallel to the line $4y - 3x = 5$ and (b) perpendicular to the line $4y - 3x = 5$

46. Passing through the point $(-6, 5)$ that is (a) parallel to the line $x + 3y = 7$ (b) perpendicular to the line $x + 3y = 7$

47–50. Write an equation of the form $y = mx + b$ for each line in the following graphs. [*Hint:* Either find the slope and y-intercept or use any two points on the line.]

47.

48.

49.

50.

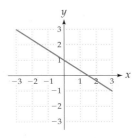

51–52. Write equations for the lines determining the four sides of each figure.

51.

52.

53. Show that $y - y_1 = m(x - x_1)$ simplifies to $y = mx + b$ if the point (x_1, y_1) is the y-intercept $(0, b)$.

54. Show that the linear equation $\dfrac{x}{a} + \dfrac{y}{b} = 1$ has x-intercept $(a, 0)$ and y-intercept $(0, b)$. (The x-intercept is the point where the line crosses the x-axis.)

55. a. Graph the lines $y_1 = -x$, $y_2 = -2x$, and $y_3 = -3x$ on the window $[-5, 5]$ by $[-5, 5]$. Observe how the coefficient of x changes the slope of the line.
   b. Predict how the line $y = -9x$ would look, and then check your prediction by graphing it.

56. a. Graph the lines $y_1 = x + 2$, $y_2 = x + 1$, $y_3 = x$, $y_4 = x - 1$, and $y_5 = x - 2$ on the window $[-5, 5]$ by $[-5, 5]$. Observe how the constant changes the position of the line.
   b. Predict how the lines $y = x + 4$ and $y = x - 4$ would look, and then check your prediction by graphing them.

Applied Exercises

57. BUSINESS: Energy Usage A utility considers demand for electricity "low" if it is below 8 mkW (million kilowatts), "average" if it is at least 8 mkW but below 20 mkW, "high" if it is at least 20 mkW but below 40 mkW, and "critical" if it is 40 mkW or more. Express these demand levels in interval notation. [*Hint:* The interval for "low" is $[0, 8)$.]

58. GENERAL: Grades If a grade of 90 through 100 is an A, at least 80 but less than 90 is a B, at least 70 but less than 80 a C, at least 60 but less than 70 a D, and below 60 an F, write these grade levels in interval form (ignoring rounding). [*Hint:* F would be $[0, 60)$.]

59. ATHLETICS: Mile Run Read the Application Preview on pages 3–4.
   a. Use the regression line $y = -0.356x + 257.44$ to predict the world record in the year 2020. [*Hint:* If x represents years after 1900, what value of x corresponds to the year 2020? The resulting y will be in seconds, and should be converted to minutes and seconds.]

b. According to this formula, when will the record be 3 minutes 30 seconds? [*Hint:* Set the formula equal to 210 seconds and solve. What year corresponds to this x-value?]

60. ATHLETICS: Mile Run Read the Application Preview on pages 3–4. Evaluate the regression line $y = -0.356x + 257.44$ at $x = 720$ and at $x = 722$ (corresponding to the years 2620 and 2622). Does the formula give reasonable times for the mile record in these years? [*Moral:* Linear trends may not continue indefinitely.]

61. BUSINESS: U.S. Computer Sales Recently, tablet computer sales in the United States have been growing approximately linearly. In 2011 sales were 70 million units, and in 2013 sales were 146 million units.
a. Use the two (year, sales) data points $(1, 70)$ and $(3, 146)$ to find the linear relationship $y = mx + b$ between $x =$ years since 2010 and $y =$ sales (in millions).
b. Interpret the slope of the line.
c. Use the linear relationship to predict sales in the year 2020.
Note: Tablet computers include iPads, Kindles, and Nooks.
Source: Standard & Poor's

62. ECONOMICS: Per Capita Personal Income In the short run, per capita personal income (PCPI) in the United States grows approximately linearly. In 2009 PCPI was 38.6, and in 2012 it had grown to 42.8 (both in thousands of dollars).
a. Use the two (year, PCPI) data points $(1, 38.6)$ and $(4, 42.8)$ to find the linear relationship $y = mx + b$ between $x =$ years since 2008 and $y =$ PCPI.
b. Interpret the slope of the line.
c. Use your linear relationship to predict PCPI in 2020.
Source: Bureau of Economic Analysis

63. GENERAL: Temperature On the Fahrenheit temperature scale, water freezes at 32° and boils at 212°. On the Celsius (centigrade) scale, water freezes at 0° and boils at 100°.
a. Use the two (Celsius, Fahrenheit) data points $(0, 32)$ and $(100, 212)$ to find the linear relationship $y = mx + b$ between $x =$ Celsius temperature and $y =$ Fahrenheit temperature.
b. Find the Fahrenheit temperature that corresponds to 20° Celsius.

64. BUSINESS: Financial Engineering Salaries Starting salaries in the United States for associates with master's degrees in financial engineering (FE) have been rising approximately linearly, from $74,800 in 2009 to $89,800 in 2013.
a. Use the two (year, salary) data points $(0, 74.8)$ and $(4, 89.8)$ to find the linear relationship $y = mx + b$ between $x =$ years since 2009 and $y =$ salary in thousands of dollars.

b. Use your formula to predict a new FE associate's salary in 2021. [*Hint:* Since x is years after 2009, what x-value corresponds to 2021?]
Source: salaryquest.com

65–66. BUSINESS: Straight-Line Depreciation
Straight-line depreciation is a method for estimating the value of an asset (such as a piece of machinery) as it loses value ("depreciates") through use. Given the original *price* of an asset, its *useful lifetime,* and its *scrap value* (its value at the end of its useful lifetime), the value of the asset after t years is given by the formula:

$$\text{Value} = (\text{Price}) - \left(\frac{(\text{Price}) - (\text{Scrap value})}{(\text{Useful lifetime})} \right) \cdot t$$

$$\text{for} \quad 0 \le t \le (\text{Useful lifetime})$$

65. a. A farmer buys a harvester for $50,000 and estimates its useful life to be 20 years, after which its scrap value will be $6000. Use the formula above to find a formula for the value V of the harvester after t years, for $0 \le t \le 20$.
b. Use your formula to find the value of the harvester after 5 years.
 c. Graph the function found in part (a) on a graphing calculator on the window [0, 20] by [0, 50,000]. [*Hint:* Use x instead of t.]

66. a. A newspaper buys a printing press for $800,000 and estimates its useful life to be 20 years, after which its scrap value will be $60,000. Use the formula above Exercise 65 to find a formula for the value V of the press after t years, for $0 \le t \le 20$.
b. Use your formula to find the value of the press after 10 years.
 c. Graph the function found in part (a) on a graphing calculator on the window [0, 20] by [0, 800,000]. [*Hint:* Use x instead of t.]

67–68. BUSINESS: Isocost Lines An *isocost line* (*iso* means "same") shows the different combinations of labor and capital (the value of factory buildings, machinery, and so on) a company may buy for the same total cost. An isocost line has equation

$$wL + rK = C \quad \text{for} \quad L \ge 0, \ K \ge 0$$

where L is the units of labor costing w dollars per unit, K is the units of capital purchased at r dollars per unit, and C is the total cost. Since both L and K must be non-negative, an isocost line is a line segment in just the first quadrant.

67. a. Write the equation of the isocost line with $w = 10$, $r = 5$, $C = 1000$, and graph it in the first quadrant.
b. Verify that the following (L, K) pairs all have the same total cost.

$$(100, 0), \quad (75, 50), \quad (20, 160), \quad (0, 200)$$

68. a. Write the equation of the isocost line with $w = 8$, $r = 6$, $C = 15,000$, and graph it in the first quadrant.

b. Verify that the following (L, K) pairs all have the same total cost.

(1875, 0), (1200, 900), (600, 1700), (0, 2500)

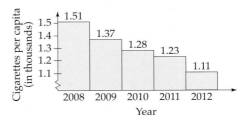

69. SOCIAL SCIENCE: Age at First Marriage Americans are marrying later and later. Based on data for the years 2000 to 2011, the median age at first marriage for men is $y_1 = 0.18x + 26.7$, and for women it is $y_2 = 0.14x + 25$, where x is the number of years since 2000.

a. Graph these lines on the window [0, 30] by [0, 35].
b. Use these lines to predict the median marriage ages for men and women in the year 2020. [*Hint:* Which x-value corresponds to 2020?]
c. Predict the median marriage ages for men and women in the year 2030.

Source: U.S. Census Bureau

70. SOCIAL SCIENCE: Equal Pay for Equal Work Women's pay has often lagged behind men's, although Title VII of the Civil Rights Act requires equal pay for equal work. Based on data from 2000–2011, women's annual earnings as a percent of men's can be approximated by the formula $y = 0.36x + 77$, where x is the number of years since 2000. (For example, $x = 10$ gives $y = 80.6$, so in 2010 women's wages were about 80.6% of men's wages.)

a. Graph this line on the window [0, 30] by [0, 100].
b. Use this line to predict the percentage in the year 2020. [*Hint:* Which x-value corresponds to 2020?]
c. Predict the percentage in the year 2025.

Source: U.S. Department of Labor—Women's Bureau

71. SOCIAL SCIENCES: Smoking and Income Based on a recent study, the probability that someone is a smoker decreases with the person's income. If someone's family income is x thousand dollars, then the probability (expressed as a percentage) that the person smokes is approximately $y = -0.31x + 40$ (for $10 \leq x \leq 100$).

a. Graph this line on the window [0, 100] by [0, 50].
b. What is the probability that a person with a family income of $40,000 is a smoker? [*Hint:* Since x is in thousands of dollars, what x-value corresponds to $40,000?]
c. What is the probability that a person with a family income of $70,000 is a smoker?

Round your answers to the nearest percent.

Source: Journal of Risk and Uncertainty 21(2/3)

72. ECONOMICS: Does Money Buy Happiness? Several surveys in the United States and Europe have asked people to rate their happiness on a scale of 3 = "very happy," 2 = "fairly happy," and 1 = "not too happy," and then tried to correlate the answer with the person's income. For those in one income group (making $25,000 to $55,000) it was found that their "happiness" was approximately given by $y = 0.065x - 0.613$. Find the reported "happiness" of a person

with the following incomes (rounding your answers to one decimal place).

a. $25,000 **b.** $35,000 **c.** $45,000

Source: Review of Economics and Statistics 85(4)

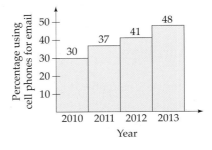

73. BUSINESS: Cigarettes The following graph gives the number of cigarettes per capita sold to adults in the United States in recent years.

a. Number the years (bars) with x-values 1–5 so that x stands for *years since 2007*, and use linear regression to fit a line to the data. State the regression formula. [*Hint:* See Example 9.]
b. Interpret the slope of the line.
c. Use the regression line to predict the number of cigarettes per capita sold in the year 2020.

Source: Centers for Disease Control

74. SOCIAL SCIENCES Email on Cell Phones Although 91% of American adults own a cell phone, not all use it to check their email, many preferring to use computers instead. The percentages of cell phone owners who use their phones for email in recent years are shown in the following graph.

a. Number the years (bars) with x-values 1–4 so that x stands for *years since 2009*, and use linear regression to fit a line to the data. State the regression formula. [*Hint:* See Example 9.]
b. Interpret the slope of the line.
c. Use the regression formula to predict the percentage of Americans who will use their cell phones for email in 2020.

Source: Pew Internet

75–76. BIOMEDICAL SCIENCES: Life Expectancy The following tables give the life expectancy for a newborn child born in the indicated year. (Exercise 75 is for males, Exercise 76 for females.)

75.

Birth Year	1970	1980	1990	2000	2010
Life Expectancy (male)	67.1	70.0	71.8	74.1	75.7

76.

Birth Year	1970	1980	1990	2000	2010
Life Expectancy (female)	74.7	77.4	78.8	79.3	80.8

a. Number the data columns with *x*-values 1–5 and use linear regression to fit a line to the data. State the regression formula. [*Hint:* See Example 9.]

b. Interpret the slope of the line. From your answer, what is the *yearly* change in life expectancy?

c. Use the regression line to predict life expectancy for a child born in 2025.

Source: U.S. Census Bureau

 77. BIOMEDICAL SCIENCES: Future Life Expectancy
Clearly, as people age they should expect to live for fewer additional years. The following graph gives the future life expectancy for Americans of different ages.

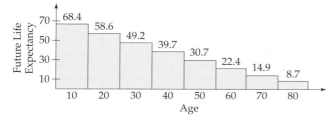

a. Letting *x* = Age, use linear regression to fit a line to the data. State the regression formula. [*Hint:* See Example 9.]

b. Interpret the slope of the line.

c. Use the regression line to estimate future longevity at age 25.

d. Would it make sense to use the regression line to estimate longevity at age 90? What future longevity would the line predict?

Source: National Center for Health Statistics

78. GENERAL: Seat Belt Use Because of driver education programs and stricter laws, seat belt use has increased steadily over recent decades. The following table gives the percentage of automobile occupants using seat belts in selected years.

Year	1995	2000	2005	2010
Seat Belt Use (%)	60	71	81	85

a. Number the data columns with *x*-values 1–4 and use linear regression to fit a line to the data. State the regression formula. [*Hint:* See Example 9.]

b. Interpret the slope of the line. From your answer, what is the *yearly* increase?

c. Use the regression line to predict seat belt use in 2017. [*Hint:* What (decimal) *x*-value corresponds to 2017?]

d. Would it make sense to use the regression line to predict seat belt use in 2025? What percentage would you get?

Source: National Highway Traffic Safety Administration

Conceptual Exercises

79. True or False: ∞ is the largest number.

80. True or False: All negative numbers are smaller than all positive numbers.

81. Give two definitions of *slope*.

82. Fill in the missing words: If a line slants downward as you go to the right, then its _____ is _____.

83. True or False: A vertical line has slope 0.

84. True or False: Every line has a slope.

85. True or False: Every line can be expressed in the form $ax + by = c$.

86. True or False: $x = 3$ is a vertical line.

87. True or False: The slope of a line is $\dfrac{x_2 - x_1}{y_2 - y_1}$.

88. True or False: A vertical line can be expressed in slope-intercept form.

89. A 5-foot-long board is leaning against a wall so that it meets the wall at a point 4 feet above the floor. What is the slope of the board? [*Hint:* Draw a picture.]

90. A 5-foot-long ramp is to have a slope of 0.75. How high should the upper end be elevated above the lower end? [*Hint:* Draw a picture.]

Explorations and Excursions   The following problems extend and augment the material presented in the text.

More About Linear Equations

91. Find the *x*-intercept $(a, 0)$ where the line $y = mx + b$ crosses the *x*-axis. Under what condition on *m* will a single *x*-intercept exist?

92. i. Show that the general linear equation $ax + by = c$ with $b \neq 0$ can be written as $y = -\dfrac{a}{b}x + \dfrac{c}{b}$ which is the equation of a line in slope-intercept form.

ii. Show that the general linear equation $ax + by = c$ with $b = 0$ but $a \neq 0$ can be written as $x = \dfrac{c}{a}$, which is the equation of a vertical line.

[*Note:* Since these steps are reversible, parts (i) and (ii) together show that the general linear equation $ax + by = c$ (for a and b not both zero) includes vertical and nonvertical lines.]

Beverton-Holt Recruitment Curve

93–94. Some organisms exhibit a density-dependent mortality from one generation to the next. Let $R > 1$ be the net reproductive rate (that is, the number of surviving offspring per parent), let $x > 0$ be the density of parents and y be the density of surviving offspring. The *Beverton-Holt recruitment curve* is

$$y = \frac{Rx}{1 + \left(\dfrac{R-1}{K}\right)x}$$

where $K > 0$ is the *carrying capacity* of the environment. Notice that if $x = K$, then $y = K$.

93. Show that if $x < K$, then $x < y < K$. Explain what this means about the population size over successive generations if the initial population is smaller than the carrying capacity of the environment.

94. Show that if $x > K$, then $K < y < x$. Explain what this means about the population size over successive generations if the initial population is larger than the carrying capacity of the environment.

<table>
<tr><td>1.2</td><td></td></tr>
</table>

1.2   Exponents

Introduction

Not all variables are related linearly. In this section we will discuss exponents, which will enable us to express many *nonlinear* relationships.

Positive Integer Exponents

Numbers may be expressed with exponents, as in $2^3 = 2 \cdot 2 \cdot 2 = 8$. More generally, for any positive integer n, x^n means the product of n x's.

$$x^n = \overbrace{x \cdot x \cdots \cdots x}^{n}$$

The number being raised to the power is called the **base** and the power is the **exponent**:

$$x^n \quad \text{Exponent or power} \atop \text{Base}$$

There are several *properties of exponents* for simplifying expressions. The first three are known, respectively, as the addition, subtraction, and multiplication properties of exponents.

Properties of Exponents

		Brief Examples
$x^m \cdot x^n = x^{m+n}$	To *multiply* powers of the same base, *add* the exponents	$x^2 \cdot x^3 = x^5$
$\dfrac{x^m}{x^n} = x^{m-n}$	To *divide* powers of the same base, *subtract* the exponents (top exponent minus bottom exponent)	$\dfrac{x^5}{x^3} = x^2$
$(x^m)^n = x^{m \cdot n}$	To raise a power to a power, *multiply* the powers	$(x^2)^3 = x^6$

$(xy)^n = x^n \cdot y^n$	To raise a product to a power, raise *each factor* to the power	$(2x)^3 = 2^3 \cdot x^3 = 8x^3$
$\left(\dfrac{x}{y}\right)^n = \dfrac{x^n}{y^n}$	To raise a fraction to a power, raise the numerator *and* denominator to the power	$\left(\dfrac{x}{5}\right)^3 = \dfrac{x^3}{5^3} = \dfrac{x^3}{125}$

PRACTICE PROBLEM 1

Simplify: **a.** $\dfrac{x^5 \cdot x}{x^2}$ **b.** $[(x^3)^2]^2$ **c.** $[2x^2 y^4]^3$ Solutions on page 28 >

Remember: For exponents in the form $x^2 \cdot x^3 = x^5$,   *add* exponents.
                 For exponents in the form $(x^2)^3 = x^6$,   *multiply* exponents.

Zero and Negative Exponents

Any number except zero can be raised to a negative or zero power.

Zero and Negative Integer Exponents

For $x \neq 0$		**Brief Examples**
$x^0 = 1$	x to the power 0 is one	$5^0 = 1$
$x^{-1} = \dfrac{1}{x}$	x to the power -1 is one over x	$7^{-1} = \dfrac{1}{7}$
$x^{-2} = \dfrac{1}{x^2}$	x to the power -2 is one over x squared	$3^{-2} = \dfrac{1}{3^2} = \dfrac{1}{9}$
$x^{-n} = \dfrac{1}{x^n}$	x to a negative power is one over x to the positive power	$(-2)^{-3} = \dfrac{1}{(-2)^3} = \dfrac{1}{-8} = -\dfrac{1}{8}$

Note that 0^0 and 0^{-3} are *undefined*.

     The definitions of x^0 and x^{-n} are motivated by the following calculations.

$$1 = \frac{x^2}{x^2} = x^{2-2} = x^0$$

The subtraction property of exponents leads to $x^0 = 1$

$$\frac{1}{x^n} = \frac{x^0}{x^n} = x^{0-n} = x^{-n}$$

$x^0 = 1$ and the subtraction property of exponents lead to $x^{-n} = \dfrac{1}{x^n}$

PRACTICE PROBLEM 2

Evaluate: **a.** 2^0 **b.** 2^{-4} Solutions on page 28 >

A fraction to a negative power means *division* by the fraction, so we "invert and multiply."

$$\left(\frac{x}{y}\right)^{-1} = \frac{1}{\dfrac{x}{y}} = 1 \cdot \frac{y}{x} = \frac{y}{x}$$

                                 Reciprocal of the original fraction

Therefore, for $x \neq 0$ and $y \neq 0$,

$$\left(\frac{x}{y}\right)^{-1} = \frac{y}{x}$$

A fraction to the power -1 is the reciprocal of the fraction

$$\left(\frac{x}{y}\right)^{-n} = \left(\frac{y}{x}\right)^{n}$$

A fraction to the negative power is the reciprocal of the fraction to the positive power

EXAMPLE 1   **SIMPLIFYING FRACTIONS TO NEGATIVE EXPONENTS**

a. $\left(\dfrac{3}{2}\right)^{-1} = \dfrac{2}{3}$ **b.** $\left(\dfrac{1}{2}\right)^{-3} = \left(\dfrac{2}{1}\right)^{3} = \dfrac{2^{3}}{1^{3}} = 8$

↑
Reciprocal of the original

PRACTICE PROBLEM 3

Simplify: $\left(\dfrac{2}{3}\right)^{-2}$

Solution on page 28 >

Roots and Fractional Exponents

We may take the square root of any *nonnegative* number, and the cube root of *any* number.

EXAMPLE 2   **EVALUATING ROOTS**

a. $\sqrt{9} = 3$ **b.** $\sqrt{-9}$ is undefined

Square root of negative numbers are not defined

c. $\sqrt[3]{8} = 2$ **d.** $\sqrt[3]{-8} = -2$

Cube roots of negative numbers *are* defined

e. $\sqrt[3]{\dfrac{27}{8}} = \dfrac{\sqrt[3]{27}}{\sqrt[3]{8}} = \dfrac{3}{2}$

There are *two* square roots of 9, namely 3 and -3, but the radical sign $\sqrt{}$ means just the *positive* one (the "principal" square root).

$\sqrt[n]{a}$ means the principal nth root of a.

Principal means the positive root if there are two

In general, we may take *odd* roots of *any* number, but *even* roots only if the number is positive or zero.

EXAMPLE 3	EVALUATING ROOTS OF POSITIVE AND NEGATIVE NUMBERS

Odd roots of negative numbers *are* defined

a. $\sqrt[4]{81} = 3$ **b.** $\sqrt[5]{-32} = -2$ Since $(-2)^5 = -32$

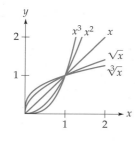

The diagram on the left shows the graphs of some powers and roots of x for $x > 0$. Which of these would *not* be defined for $x < 0$?
[*Hint:* See Example 2.]

Fractional Exponents

Fractional exponents are defined as follows:

Powers of the Form $\frac{1}{n}$

		Brief Examples
$x^{\frac{1}{2}} = \sqrt{x}$	Power $\frac{1}{2}$ means the principal square root	$9^{\frac{1}{2}} = \sqrt{9} = 3$
$x^{\frac{1}{3}} = \sqrt[3]{x}$	Power $\frac{1}{3}$ means the cube root	$125^{\frac{1}{3}} = \sqrt[3]{125} = 5$
$x^{\frac{1}{n}} = \sqrt[n]{x}$	Power $\frac{1}{n}$ means the principal nth root (for a positive integer n)	$(-32)^{\frac{1}{5}} = \sqrt[5]{-32} = -2$

The definition of $x^{\frac{1}{2}}$ is motivated by the multiplication property of exponents:

$$\left(x^{\frac{1}{2}}\right)^2 = x^{\frac{1}{2} \cdot 2} = x^1 = x$$

Taking square roots of each side of $\left(x^{\frac{1}{2}}\right)^2 = x$ gives

$$x^{\frac{1}{2}} = \sqrt{x}$$

x to the half power means the square root of x

EXAMPLE 4	EVALUATING FRACTIONAL EXPONENTS

a. $\left(\dfrac{4}{25}\right)^{\frac{1}{2}} = \sqrt{\dfrac{4}{25}} = \dfrac{\sqrt{4}}{\sqrt{25}} = \dfrac{2}{5}$

b. $\left(-\dfrac{27}{8}\right)^{\frac{1}{3}} = \sqrt[3]{-\dfrac{27}{8}} = -\dfrac{\sqrt[3]{27}}{\sqrt[3]{8}} = -\dfrac{3}{2}$

PRACTICE PROBLEM 4

Evaluate: **a.** $(-27)^{\frac{1}{3}}$ **b.** $\left(\dfrac{16}{81}\right)^{\frac{1}{4}}$ Solutions on page 28 >

To define $x^{\frac{m}{n}}$ for positive integers m and n, the exponent $\frac{m}{n}$ must be fully reduced (for example, $\frac{4}{6}$ must be reduced to $\frac{2}{3}$). Then

$$x^{\frac{m}{n}} = \left(x^{\frac{1}{n}}\right)^m = \left(x^m\right)^{\frac{1}{n}}$$

Since in both cases the exponents multiply to $\frac{m}{n}$

Therefore, we define:

Fractional Exponents

$$x^{\frac{m}{n}} = \left(\sqrt[n]{x}\right)^m = \sqrt[n]{x^m}$$

$x^{m/n}$ means the mth power of the nth root, or equivalently, the nth root of the mth power

LOOKING AHEAD

Fractional exponents will be important when we use the *power rule* on page 100 and beyond.

Both expressions, $\left(\sqrt[n]{x}\right)^m$ and $\sqrt[n]{x^m}$, will give the same answer. In either case *the numerator determines the power* and *the denominator determines the root.*

Power over root

EXAMPLE 5   **EVALUATING FRACTIONAL EXPONENTS**

a. $8^{2/3} = \sqrt[3]{8^2} = \sqrt[3]{64} = 4$ same   First the power, then the root

b. $8^{2/3} = \left(\sqrt[3]{8}\right)^2 = (2)^2 = 4$ First the root, then the power

c. $25^{3/2} = \left(\sqrt{25}\right)^3 = (5)^3 = 125$

d. $\left(\dfrac{-27}{8}\right)^{2/3} = \left(\sqrt[3]{\dfrac{-27}{8}}\right)^2 = \left(\dfrac{-3}{2}\right)^2 = \dfrac{9}{4}$

PRACTICE PROBLEM 5

Evaluate: **a.** $16^{3/2}$ **b.** $(-8)^{2/3}$ Solutions on page 28 >

EXAMPLE 6   **EVALUATING NEGATIVE FRACTIONAL EXPONENTS**

a. $8^{-2/3} = \dfrac{1}{8^{2/3}} = \dfrac{1}{\left(\sqrt[3]{8}\right)^2} = \dfrac{1}{2^2} = \dfrac{1}{4}$

A negative exponent means the reciprocal of the number to the positive exponent, which is then evaluated as before

b. $\left(\dfrac{9}{4}\right)^{-3/2} = \left(\dfrac{4}{9}\right)^{3/2} = \left(\sqrt{\dfrac{4}{9}}\right)^3 = \left(\dfrac{2}{3}\right)^3 = \dfrac{8}{27}$

Interpreting the power 3/2
Reciprocal to the positive exponent
Negative exponent

PRACTICE PROBLEM 6

Evaluate: **a.** $25^{-3/2}$ **b.** $\left(\dfrac{1}{4}\right)^{-1/2}$ **c.** $5^{1.3}$ [*Hint:* Use a calculator.]

Solutions on page 28 >

⚠ **Be Careful** While the square root of a product *is* equal to the product of the square roots,

$$\sqrt{a \cdot b} = \sqrt{a} \cdot \sqrt{b}$$

the corresponding statement for *sums* is *not* true:

$$\sqrt{a + b} \quad \text{is } not \text{ equal to} \quad \sqrt{a} + \sqrt{b}$$

For example,

$$\underbrace{\sqrt{9 + 16}}_{\sqrt{25}} \neq \underbrace{\sqrt{9}}_{3} + \underbrace{\sqrt{16}}_{4}$$

The two sides are not equal: one is 5 and the other is 7

Therefore, do not "simplify" $\sqrt{x^2 + 9}$ into $x + 3$. The expression $\sqrt{x^2 + 9}$ *cannot be simplified*. Similarly,

$$(x + y)^2 \quad \text{is } not \text{ equal to} \quad x^2 + y^2$$

The expression $(x + y)^2$ means $(x + y)$ times itself:

$$(x + y)^2 = (x + y)(x + y) = x^2 + xy + yx + y^2 = x^2 + 2xy + y^2$$

This result is worth remembering, since we will use it frequently in Chapter 2.

$$(x + y)^2 = x^2 + 2xy + y^2$$

$(x + y)^2$ is the first number squared plus twice the product of the numbers plus the second number squared

Learning Curves in Airplane Production

Time
Repetitions

It is a truism that the more you practice a task, the faster you can do it. Successive repetitions generally take less time, following a "learning curve" like that on the left. Learning curves are used in industrial production. For example, it took 150,000 work-hours to build the first Boeing 707 airliner, while later planes $(n = 2, 3, \ldots, 300)$ took less time.*

$$\binom{\text{Time to build}}{\text{plane number } n} = 150\, n^{-0.322} \text{ thousand work-hours}$$

The time for the 10th Boeing 707 is found by substituting $n = 10$:

$$\binom{\text{Time to build}}{\text{plane } 10} = 150(10)^{-0.322}$$

$150n^{-0.322}$
with $n = 10$

$$\approx 71.46 \text{ thousand work-hours} \quad \text{Using a calculator}$$

This shows that building the 10th Boeing 707 took about 71,460 work-hours, which is less than half of the 150,000 work-hours needed for the first. For the 100th 707:

$$\binom{\text{Time to build}}{\text{plane } 100} = 150(100)^{-0.322}$$

$150n^{-0.322}$
with $n = 100$

$$\approx 34.05 \text{ thousand work-hours}$$

*A work-hour is the amount of work that a person can do in 1 hour. For further information on learning curves in industrial production, see J. M. Dutton et al., "The History of Progress Functions as a Managerial Technology," *Business History Review* **58**.

or about 34,050 work-hours, which is less than the half time needed to build the 10th. Such learning curves are used for determining the cost of a contract to build several planes.

Notice that the learning curve graphed on the previous page decreases less steeply as the number of repetitions increases. This means that while construction time continues to decrease, it does so more slowly for later planes. This behavior, called **diminishing returns,** is typical of learning curves.

Power Regression (Optional)

Just as we used *linear* regression to fit a *line* to data points, we can use **power regression** to fit a *power curve* like those shown on page 24 to data points. The procedure is easily accomplished using a graphing calculator (or spreadsheet or other computer software), as in the following Example.

When do you use power regression instead of linear regression (or some other type)? You should look at the data and see if they lie more along a *curve* like those shown on page 24 rather than along a line. Furthermore, sometimes there are *theoretical* reasons to prefer a curve. For example, sales of a product may increase linearly for a short time, but then usually grow more slowly because of market saturation or competition, and so are best modeled by a curve.

| EXAMPLE 7 | POWER REGRESSION USING A GRAPHING CALCULATOR |

The following graph shows Google's net income in billions of dollars for recent years.

Source: Standard & Poor's

a. Use power regression to fit a power curve to the data and state the regression formula.

b. Use the regression formula to predict Google's net income in 2020.

Solution

a. We number the years with x-values $1-4$, so x stands for *years since 2008* (we could choose other x-values instead, but with power regression the x-values must all be *positive*). We enter the data into lists, as shown in the first screen below (as explained in the appendix *Graphing Calculator Basics—Entering Data* on page A3) and use *ZoomStat* to graph the data points.

Then (using STAT, CALC, and PwrReg) we graph the regression curve along with the data points.

The regression curve, which fits the points reasonably well, is

$$y = 6.54x^{0.359} \qquad \text{Rounded}$$

b. To predict income in 2020, we evaluate Y1 at 12 (since $x = 12$ corresponds to 2020). From the screen below, if the current trend continues, net income in 2020 will be approximately $16 billion.

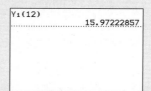

Solutions TO PRACTICE PROBLEMS

1. a. $\dfrac{x^5 \cdot x}{x^2} = \dfrac{x^6}{x^2} = x^4$

   b. $[(x^3)^2]^2 = x^{3 \cdot 2 \cdot 2} = x^{12}$

   c. $[2x^2 y^4]^3 = 2^3 (x^2)^3 (y^4)^3 = 8x^6 y^{12}$

2. a. $2^0 = 1$

   b. $2^{-4} = \dfrac{1}{2^4} = \dfrac{1}{16}$

3. $\left(\dfrac{2}{3}\right)^{-2} = \left(\dfrac{3}{2}\right)^2 = \dfrac{9}{4}$

4. a. $(-27)^{1/3} = \sqrt[3]{-27} = -3$

   b. $\left(\dfrac{16}{81}\right)^{1/4} = \sqrt[4]{\dfrac{16}{81}} = \dfrac{\sqrt[4]{16}}{\sqrt[4]{81}} = \dfrac{2}{3}$

5. a. $16^{3/2} = \left(\sqrt{16}\right)^3 = 4^3 = 64$

   b. $(-8)^{2/3} = \left(\sqrt[3]{-8}\right)^2 = (-2)^2 = 4$

6. a. $25^{-3/2} = \dfrac{1}{25^{3/2}} = \dfrac{1}{\left(\sqrt{25}\right)^3} = \dfrac{1}{5^3} = \dfrac{1}{125}$

   b. $\left(\dfrac{1}{4}\right)^{-1/2} = \left(\dfrac{4}{1}\right)^{1/2} = \sqrt{4} = 2$ **c.** $5^{1.3} \approx 8.103$

| **1.2** | **Section Summary** |

We defined zero, negative, and fractional exponents as follows:

$$x^0 = 1 \qquad \text{for } x \neq 0$$

$$x^{-n} = \frac{1}{x^n} \qquad \text{for } x \neq 0$$

$$x^{\frac{m}{n}} = \left(\sqrt[n]{x}\right)^m = \sqrt[n]{x^m} \qquad m > 0, \quad n > 0, \quad \frac{m}{n} \text{ fully reduced}$$

With these definitions, the following properties of exponents hold for *all* exponents, whether integral or fractional, positive or negative.

$$x^m \cdot x^n = x^{m+n} \qquad (x^m)^n = x^{m \cdot n} \qquad \left(\frac{x}{y}\right)^n = \frac{x^n}{y^n}$$

$$\frac{x^m}{x^n} = x^{m-n} \qquad (xy)^n = x^n \cdot y^n$$

1.2 Exercises

1–48. Evaluate each expression *without* using a calculator.

1. $(2^2 \cdot 2)^2$ **2.** $(5^2 \cdot 4)^2$ **3.** 2^{-4}

4. 3^{-3} **5.** $\left(\frac{1}{2}\right)^{-3}$ **6.** $\left(\frac{1}{3}\right)^{-2}$

7. $\left(\frac{5}{8}\right)^{-1}$ **8.** $\left(\frac{3}{4}\right)^{-1}$ **9.** $4^{-2} \cdot 2^{-1}$

10. $3^{-2} \cdot 9^{-1}$ **11.** $\left(\frac{3}{2}\right)^{-3}$ **12.** $\left(\frac{2}{3}\right)^{-3}$

13. $\left(\frac{1}{3}\right)^{-2} - \left(\frac{1}{2}\right)^{-3}$ **14.** $\left(\frac{1}{3}\right)^{-2} - \left(\frac{1}{2}\right)^{-2}$

15. $\left[\left(\frac{2}{3}\right)^{-2}\right]^{-1}$ **16.** $\left[\left(\frac{2}{5}\right)^{-2}\right]^{-1}$

17. $25^{1/2}$ **18.** $36^{1/2}$ **19.** $25^{3/2}$

20. $16^{3/2}$ **21.** $16^{3/4}$ **22.** $27^{2/3}$

23. $(-8)^{2/3}$ **24.** $(-27)^{2/3}$ **25.** $(-8)^{5/3}$

26. $(-27)^{5/3}$ **27.** $\left(\frac{25}{36}\right)^{3/2}$ **28.** $\left(\frac{16}{25}\right)^{3/2}$

29. $\left(\frac{27}{125}\right)^{2/3}$ **30.** $\left(\frac{125}{8}\right)^{2/3}$ **31.** $\left(\frac{1}{32}\right)^{2/5}$

32. $\left(\frac{1}{32}\right)^{3/5}$ **33.** $4^{-1/2}$ **34.** $9^{-1/2}$

35. $4^{-3/2}$ **36.** $9^{-3/2}$ **37.** $8^{-2/3}$

38. $16^{-3/4}$ **39.** $(-8)^{-1/3}$ **40.** $(-27)^{-1/3}$

41. $(-8)^{-2/3}$ **42.** $(-27)^{-2/3}$ **43.** $\left(\frac{25}{16}\right)^{-1/2}$

44. $\left(\frac{16}{9}\right)^{-1/2}$ **45.** $\left(\frac{25}{16}\right)^{-3/2}$ **46.** $\left(\frac{16}{9}\right)^{-3/2}$

47. $\left(-\frac{1}{27}\right)^{-5/3}$ **48.** $\left(-\frac{1}{8}\right)^{-5/3}$

49–52. Use a calculator to evaluate each expression. Round answers to two decimal places.

49. $7^{0.39}$ **50.** $5^{0.47}$ **51.** $8^{2.7}$ **52.** $5^{3.9}$

53–56. Use a graphing calculator to evaluate each expression.

53. $[(0.1)^{0.1}]^{0.1}$ **54.** $\left(1 + \frac{1}{1000}\right)^{1000}$

55. $\left(1 - \frac{1}{1000}\right)^{-1000}$ **56.** $(1 + 10^{-6})^{10^6}$

57–70. Write each expression in power form ax^b for numbers a and b.

57. $\dfrac{4}{x^5}$ **58.** $\dfrac{6}{2x^3}$

59. $\dfrac{4}{\sqrt[3]{8x^4}}$ **60.** $\dfrac{6}{\sqrt{4x^3}}$

61. $\dfrac{24}{\left(2\sqrt{x}\right)^3}$ **62.** $\dfrac{18}{\left(3\sqrt[3]{x}\right)^2}$

63. $\sqrt{\dfrac{9}{x^4}}$

64. $\sqrt[3]{\dfrac{8}{x^6}}$

65. $\dfrac{5x^2}{\sqrt{x}}$

66. $\dfrac{3\sqrt{x}}{x}$

67. $\dfrac{12\sqrt[3]{x^2}}{3x^2}$

68. $\dfrac{10\sqrt{x}}{2\sqrt[3]{x}}$

69. $\dfrac{\sqrt{36x}}{2x}$

70. $\dfrac{\sqrt[3]{8x^2}}{4x}$

74. $[z(z^3 \cdot z)^2 z^2]^2$

75. $[(x^2)^2]^2$

76. $[(x^3)^3]^3$

77. $(3x^2y^5z)^3$

78. $(2x^4yz^6)^4$

79. $\dfrac{(ww^2)^3}{w^3w}$

80. $\dfrac{(ww^3)^2}{w^3w^2}$

81. $\dfrac{(5xy^4)^2}{25x^3y^3}$

82. $\dfrac{(4x^3y)^2}{8x^2y^3}$

83. $\dfrac{(9xy^3z)^2}{3(xyz)^2}$

84. $\dfrac{(5x^2y^3z)^2}{5(xyz)^2}$

85. $\dfrac{(2u^2vw^3)^2}{4(uw^2)^2}$

86. $\dfrac{(u^3vw^2)^2}{9(u^2w)^2}$

71–86. Simplify.

71. $(x^3 \cdot x^2)^2$ **72.** $(x^4 \cdot x^3)^2$ **73.** $[z^2(z \cdot z^2)^2z]^3$

Applied Exercises

87–88. ALLOMETRY: Dinosaurs The study of size and shape is called "allometry," and many allometric relationships involve exponents that are fractions or decimals. For example, the body measurements of most four-legged animals, from mice to elephants, obey (approximately) the following power law:

$$\left(\begin{array}{c}\text{Average body} \\ \text{thickness}\end{array}\right) = 0.4 \,(\text{hip-to-shoulder length})^{3/2}$$

where body thickness is measured vertically and all measurements are in feet. Assuming that this same relationship held for dinosaurs, find the average body thickness of the following dinosaurs, whose hip-to-shoulder length can be measured from their skeletons:

87. Diplodocus, whose hip-to-shoulder length was 16 feet.

88. Triceratops, whose hip-to-shoulder length was 14 feet.

89–90. BUSINESS: The Rule of .6 Many chemical and refining companies use "the rule of point six" to estimate the cost of new equipment. According to this rule, if a piece of equipment (such as a storage tank) originally cost C dollars, then the cost of similar equipment that is x times as large will be approximately $x^{0.6}C$ dollars. For example, if the original equipment cost C dollars, then new equipment with twice the capacity of the old equipment $(x = 2)$ would cost $2^{0.6}C = 1.516C$ dollars—that is, about 1.5 times as much. Therefore, to increase capacity by 100% costs only about 50% more.*

89. Use the rule of .6 to find how costs change if a company wants to quadruple $(x = 4)$ its capacity.

*Although the rule of .6 is only a rough "rule of thumb," it can be somewhat justified on the basis that the equipment of such industries consists mainly of containers, and the cost of a container depends on its surface area (square units), which increases more slowly than its capacity (cubic units).

90. Use the rule of .6 to find how costs change if a company wants to triple $(x = 3)$ its capacity.

91–92. BUSINESS: Phillips Curves Unemployment and inflation are inversely related, with one rising as the other falls, and an equation giving the relation is called a *Phillips curve* after the economist A. W. Phillips (1914–1975).

91. Phillips used data from 1861 to 1957 to establish that in the United Kingdom the unemployment rate x and the wage inflation rate y were related by

$$y = 9.638x^{-1.394} - 0.900$$

where x and y are both percents. Use this relation to estimate the inflation rate when the unemployment rate was

a. 2 percent   **b.** 5 percent

Source: Economica **25**

92. Between 2000 and 2010, the Phillips curve for the U.S. unemployment rate x and Consumer Price Index (CPI) inflation rate y was

$$y = 45.4x^{-1.54} - 1$$

where x and y are both percents. Use this relation to estimate the inflation rate when the unemployment rate is

a. 3 percent   **b.** 8 percent

Source: Bureau of Labor Statistics

93–94. ALLOMETRY: Heart Rate It is well known that the hearts of smaller animals beat faster than the hearts of larger animals. The actual relationship is approximately

$$(\text{Heart rate}) = 250(\text{Weight})^{-1/4}$$

where the heart rate is in beats per minute and the weight is in pounds. Use this relationship to estimate the heart rate of:

93. A 16-pound dog.

94. A 625-pound grizzly bear.

Source: Biology Review **41**

95–96. BUSINESS: Learning Curves in Airplane Production Recall (pages 26–27) that the learning curve for the production of Boeing 707 airplanes is $150n^{-0.322}$ (thousand work-hours). Find how many work-hours it took to build:

95. The 50th Boeing 707.

96. The 250th Boeing 707.

97–98. GENERAL: Earthquakes The sizes of major earthquakes are measured on the *Moment Magnitude Scale*, or MMS, although the media often still refer to the outdated *Richter* scale. The MMS measures the total *energy released* by an earthquake, in units denoted M_W (W for the *work* accomplished). An increase of 1 M_W means the energy increased by a factor of 32, so an increase from A to B means the energy increased by a factor of 32^{B-A}. Use this formula to find the increase in energy between the following earthquakes:

97. The 1994 Northridge, California, earthquake that measured 6.7 M_W and the 1906 San Francisco earthquake that measured 7.8 M_W. (The San Francisco earthquake resulted in 3000 deaths and a 3-day fire that destroyed 4 square miles of San Francisco.)

98. The 2001 earthquake in India that measured 7.7 M_W and the 2011 earthquake in Japan that measured 9.0 M_W. (The earthquake in Japan generated a 28-foot tsunami wave that traveled six miles inland, killing 24,000 and causing an estimated $300 billion in damage, making it the most expensive natural disaster ever recorded.)

Source for M_W: U.S. Geological Survey

99–100. BUSINESS: Isoquant Curves An *isoquant curve* (*iso* means "same" and *quant* is short for "quantity") shows the various combinations of labor and capital (the invested value of factory buildings, machinery, and raw materials) a company could use to achieve the same total production level. For a given production level, an isoquant curve can be written in the form $K = aL^b$ where K is the amount of capital, L is the amount of labor, and a and b are constants. For each isoquant curve, find the value of K corresponding to the given value of L.

99. $K = 3000L^{-1/2}$ and $L = 225$

100. $K = 4000L^{-2/3}$ and $L = 125$

101–102. GENERAL: Waterfalls Water falling from a waterfall that is x feet high will hit the ground with speed $\frac{60}{11}x^{0.5}$ miles per hour (neglecting air resistance).

101. Find the speed of the water at the bottom of the highest waterfall in the world, Angel Falls in Venezuela (3281 feet high).

102. Find the speed of the water at the bottom of the highest waterfall in the United States, Ribbon Falls in Yosemite, California (1650 feet high).

103–104. ENVIRONMENTAL SCIENCE: Biodiversity It is well known that larger land areas can support larger numbers of species. According to one study, multiplying the land area by a factor of x multiplies the number of species by a factor

of $x^{0.239}$. Use a graphing calculator to graph $y = x^{0.239}$. Use the window [0, 100] by [0, 4].

Source: Robert H. MacArthur and Edward O. Wilson, *The Theory of Island Biogeography*

103. Find the multiple x for the land area that leads to *double* the number of species. That is, find the value of x such that $x^{0.239} = 2$. [*Hint:* Either use TRACE or find where $y_1 = x^{0.239}$ INTERSECTs $y_2 = 2$.]

104. Find the multiple x for the land area that leads to triple the number of species. That is, find the value of x such that $x^{0.239} = 3$. [*Hint:* Either use TRACE or find where $y_1 = x^{0.239}$ INTERSECTs $y_2 = 3$.]

105–106. GENERAL: Speed and Skidmarks Police or insurance investigators often want to estimate the speed of a car from the skidmarks it left while stopping. A study found that for standard tires on dry asphalt, the speed (in mph) is given approximately by $y = 9.4x^{0.37}$, where x is the length of the skidmarks in feet. (This formula takes into account the deceleration that occurs even *before* the car begins to skid.) Estimate the speed of a car if it left skidmarks of:

105. 150 feet.   **106.** 350 feet.

Source: Accident Analysis and Prevention **36**

107. BUSINESS: Semiconductor Sales The following table shows worldwide sales for semiconductors used in cell phones and laptop computers for recent years.

Year	2011	2012	2013
Sales (billions $)	80.2	87.1	93.6

a. Number the data columns with x-values 1–3 (so that x stands for *years since 2010*), use power regression to fit a power curve to the data, and state the regression formula. [*Hint:* See Example 7.]

b. Use the regression formula to predict sales in 2020. [*Hint:* What x-value corresponds to 2020?]

Source: Standard and Poor's

108. SOCIAL SCIENCE: Alcohol and Tobacco Expenditures The following table gives the per capita expenditures for alcohol and tobacco for Americans in recent years.

Year	2009	2010	2011	2012
Expenditures (in $)	607	646	667	685

a. Number the data columns with x-values 1–4 (so that x stands for *years since 2008*), use power regression to fit a power curve to the data, and state the regression formula. [*Hint:* See Example 7.]

b. Use the regression formula to predict these expenditures in the year 2020. [*Hint:* What x-value corresponds to 2020?]

Source: Consumer Americas 2013

109. BUSINESS: Nevada Gambling Winnings The following table shows the winnings in Nevada casinos for recent years.

Year	2010	2011	2012
Winnings (million $)	41.6	42.8	43.4

a. Number the data columns with x-values 1–3 (so that x stands for *years since 2009*), use power regression to fit a power curve to the data, and state the regression formula. [*Hint:* See Example 7.]

b. Use the regression formula to predict Nevada casino winnings in the year 2020.

Source: Standard & Poor's Industry Surveys

110. BUSINESS: American Express Operating Revenues The following table shows the operating revenues (in billions of dollars) for American Express.

Year	2009	2010	2011	2012
Operating revenues (billion $)	26.5	30.2	32.3	33.8

a. Number the data columns with x-values 1–4 (so that x stands for *years since 2008*), use power regression to fit a power curve to the data, and state the regression formula. [*Hint:* See Example 7.]

b. Use the regression formula to predict American Express operating revenues in the year 2020.

Source: Standard & Poor's

Conceptual Exercises

111. Should $\sqrt{9}$ be evaluated as 3 or ± 3?

112–114. For each statement, either state that it is True (and find a property in the text that shows this) or state that it is False (and give an example to show this).

112. $x^m \cdot x^n = x^{m \cdot n}$ **113.** $\dfrac{x^m}{x^n} = x^{m/n}$ **114.** $(x^m)^n = x^{m^n}$

115–117. For each statement, state *in words* the values of x for which each exponential expression is defined.

115. $x^{1/2}$ **116.** $x^{1/3}$ **117.** x^{-1}

118. When defining $x^{m/n}$, why did we require that the exponent $\frac{m}{n}$ be fully reduced?

[*Hint:* $(-1)^{2/3} = \left(\sqrt[3]{-1}\right)^2 = 1$, but with an equal but *unreduced* exponent you get $(-1)^{4/6} = \left(\sqrt[6]{-1}\right)^4$. Is this defined?]

1.3   Functions: Linear and Quadratic

Introduction

In the previous section we saw that the time required to build a Boeing 707 airliner varies, depending on the number that have already been built. Mathematical relationships such as this, in which one number depends on another, are called *functions* and are central to the study of calculus. In this section we define and give some applications of functions.

Functions

A **function** is a rule or procedure for finding, from a given number, a new number.* If the function is denoted by f and the given number by x, then the resulting number is written $f(x)$ (read "f of x") and is called *the value of the function f at x*. We emphasize that $f(x)$ must be a *single* number.

---

*In this chapter the word "function" will mean *function of one variable*. In Chapter 7 we will discuss functions of more than one variable.

The set of numbers x for which a function f is defined is called the **domain** of f, and the set of all resulting function values $f(x)$ is called the **range** of f.

Function

A *function* f is a rule that assigns to each number x in a set exactly one number $f(x)$.

The set of all allowable values of x is called the *domain*.

The set of all values $f(x)$ for x in the domain is called the *range*.

For example, recording the temperature at a given location throughout a particular day would define a *temperature* function:

$$f(x) = \begin{pmatrix} \text{Temperature at} \\ \text{time } x \text{ hours} \end{pmatrix} \qquad \text{Domain would be } [0, 24)$$

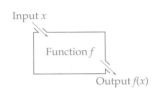

A function f may be thought of as a numerical procedure or "machine" that takes an "input" number x and produces an "output" number $f(x)$, as shown on the left. The permissible input numbers form the domain, and the resulting output numbers form the range.

We will be mostly concerned with functions that are defined by *formulas* for calculating $f(x)$ from x. If the domain of such a function is not stated, then it is always taken to be the *largest* set of numbers for which the function is defined, called the **natural domain** of the function. To **graph** a function f, we plot all points (x, y) such that x is in the domain and $y = f(x)$. We call x the **independent variable** and y the **dependent variable,** since y *depends on* (is calculated from) x. The domain and range can be illustrated graphically.

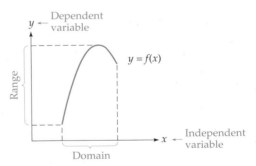

The domain of a function $y = f(x)$ is the set of all allowable x-values, and the range is the set of all corresponding y-values.

PRACTICE PROBLEM 1

Find the domain and range of the function graphed below.

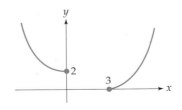

Solution on page 43 >

EXAMPLE 1   FINDING THE DOMAIN AND RANGE

For the function $f(x) = \dfrac{1}{x-1}$, find:

a. $f(5)$ **b.** the domain   **c.** the range

Solution

a. $f(5) = \dfrac{1}{5-1} = \dfrac{1}{4}$

$f(x) = \dfrac{1}{x-1}$ with $x = 5$

$f(x) = \dfrac{1}{x-1}$ is defined for

b. Domain $= \{x \mid x \neq 1\}$

all x except $x = 1$ (since we can't divide by zero)

c. The graph of the function (from a graphing calculator) is shown on the right. From it, and realizing that the curve continues upward and downward (as may be verified by zooming out), it is clear that *every* y-value is taken except for $y = 0$ (since the curve does not touch the x-axis). Therefore:

$f(x) = \dfrac{1}{x-1}$ on $[-5, 5]$ by $[-5, 5]$

Range $= \{y \mid y \neq 0\}$

May also be written $\{z \mid z \neq 0\}$ or with any other letter

The range could also be found by solving $y = \dfrac{1}{x-1}$ for x, giving

$x = \dfrac{1}{y} + 1$, which again shows that y can take any value except 0.

EXAMPLE 2   FINDING THE DOMAIN AND RANGE

For $f(x) = x^2 + 1$, determine:

a. $f(-2)$ **b.** the domain   **c.** the range

Solution

a. $f(-2) = (-2)^2 + 1 = 4 + 1 = 5$

$f(x) = x^2 + 1$ with x replaced by -2

b. Domain $= \mathbb{R}$

$x^2 + 1$ is defined for all real numbers

$y = x^2 + 1$

c. The function $f(x) = x^2 + 1$ is a square (which is positive or zero) plus 1 and so will take every y-value that is 1 or higher. Therefore:

Range $= \{y \mid y \geq 1\}$

The domain and range can be seen from the graph on the left.

Any letters may be used for defining a function or describing the domain and the range. For example, since the range $\{y \mid y \geq 1\}$ is a set of numbers, it could be written $\{w \mid w \geq 1\}$ or, in interval notation without *any* variables, as $[1, \infty)$.

PRACTICE PROBLEM 2

For $g(z) = \sqrt{z - 2}$, determine:

a. $g(27)$ **b.** the domain   **c.** the range

Solutions on page 43 >

For each x in the domain of a function there must be a *single* number $y = f(x)$, so the graph of a function cannot have two points (x, y) with the same x-value but different y-values. This leads to the following *graphical* test for functions.

Vertical Line Test for Functions

A curve in the Cartesian plane is the graph of a *function* if and only if no vertical line intersects the curve at more than one point.

EXAMPLE 3   USING THE VERTICAL LINE TEST

This is *not* the graph of a function of x because there is a vertical line (shown dashed) that intersects the curve twice.

This *is* the graph of a function of x because no vertical line intersects the curve more than once.

A graph that has two or more points (x, y) with the same x-value but different y-values, such as the one on the left above, defines a *relation* rather than a function. We will be concerned exclusively with *functions,* and so we will use the terms "function," "graph," and "curve" interchangeably.

Functions can be classified into several types.

Linear Function

A *linear function* is a function that can be expressed in the form

$$f(x) = mx + b$$

with constants m and b. Its graph is a line with slope m and y-intercept b.

| EXAMPLE 4 | FINDING A COMPANY'S COST FUNCTION |

An electronics company manufactures pocket calculators at a cost of $9 each, and the company's fixed costs (such as rent) amount to $400 per day. Find a function $C(x)$ that gives the total cost of producing x pocket calculators in a day.

Solution

Each calculator costs $9 to produce, so x calculators will cost $9x$ dollars, to which we must add the fixed costs of $400.

$$C(x) \;=\; 9x \;+\; 400$$

Total   Unit Number Fixed
cost   cost of units cost

The graph of $C(x) = 9x + 400$ is a line with slope 9 and y-intercept 400, as shown on the left. Notice that the **slope** is the same as the **rate of change** of the cost (increasing at the rate of $9 per additional calculator), which is also the company's **marginal cost** (the cost of producing one more calculator is $9). The *slope,* the *rate of change,* and the *marginal* cost are always the same, as we will see in Chapter 2.

The most important property of a linear function is that it always changes by the *same amount* (the slope) whenever the independent variable increases by 1. In the preceding Example, the slope has units of *dollars per calculator.*

PRACTICE PROBLEM 3

A trucking company will deliver furniture for a charge of $25 plus 5% of the purchase price of the furniture. Find a function $D(x)$ that gives the delivery charge for a piece of furniture that costs x dollars.

Solution on page 43 >

A mathematical description of a real-world situation is called a **mathematical model.** For example, the cost function $C(x) = 9x + 400$ from the previous Example is a mathematical model for the cost of manufacturing calculators. In this model, x, the number of calculators, should take only whole-number values (0, 1, 2, 3, . . .), and the graph should consist of discrete dots rather than a continuous curve. Instead, we will find it easier to let x take *continuous* values, and round up or down as necessary at the end.

Quadratic Function

A *quadratic function* is a function that can be expressed in the form

$$f(x) = ax^2 + bx + c$$

with constants ("coefficients") $a \neq 0$, b, and c. Its graph is called a *parabola.*

The condition $a \neq 0$ keeps the function from becoming $f(x) = bx + c$, which would be linear.

Many familiar curves are parabolas.

The center of gravity of a
diver describes a parabola.

A stream of water from a hose
takes the shape of a parabola.

The parabola $f(x) = ax^2 + bx + c$ opens *upward* if the constant a is *positive* and opens *downward* if the constant a is *negative*. The **vertex** of a parabola is its "central" point, the *lowest* point on the parabola if it opens *up* and the *highest* point if it opens *down*.

LOOKING AHEAD

Opens up and *opens down* are related to *concave up* and *concave down* on page 178.

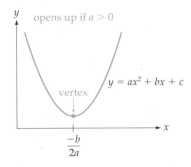

IMPORTANT NOTE The Graphing Calculator Explorations are optional (unless your instructor says otherwise). At a minimum, looking carefully at the graphs and reading the explanations provide useful information about the mathematics being discussed.

Graphing Calculator Exploration

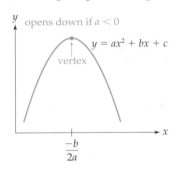

a. Graph the parabolas $y_1 = x^2$, $y_2 = 2x^2$, and $y_3 = 4x^2$ on the window $[-5, 5]$ by $[-10, 10]$. How does the shape of the parabola change when the coefficient of x^2 increases?

b. Graph $y_4 = -x^2$. What did the negative sign do to the parabola?

c. Predict the shape of the parabolas $y_5 = -2x^2$ and $y_6 = \frac{1}{3}x^2$. Then check your predictions by graphing the functions.

The x-coordinate of the vertex of a parabola may be found by the following formula, which is shown in the diagrams above and will be derived in Exercise 63 on page 175.

Vertex Formula for a Parabola

The vertex of the parabola $f(x) = ax^2 + bx + c$ has x-coordinate

$$x = \frac{-b}{2a}$$

For example, the vertex of the parabola $y = 2x^2 + 4x - 5$ has x-coordinate

$$x = \frac{-4}{2 \cdot 2} = \frac{-4}{4} = -1 \qquad\qquad x = \frac{-b}{2a} \text{ with } a = 2 \text{ and } b = 4$$

as we can see from the graph on the left. The y-coordinate of the vertex, -7, comes from evaluating $y = 2x^2 + 4x - 5$ at this x-coordinate.

EXAMPLE 5   GRAPHING A QUADRATIC FUNCTION

Graph the quadratic function $f(x) = 2x^2 - 40x + 104$.

Solution

Graphing using a graphing calculator is largely a matter of finding an appropriate viewing window, as the following three unsatisfactory windows show.

on $[-10, 10]$ by $[-10, 10]$ on $[0, 20]$ by $[-10, 10]$ on $[0, 20]$ by $[-105, -95]$

To find an appropriate viewing window, we use the vertex formula:

$$x = \frac{-(-40)}{2 \cdot 2} = \frac{40}{4} = 10 \qquad\qquad x = \frac{-b}{2a} \text{ with } a = 2 \text{ and } b = -40$$

We move a few units, say 5, to either side of $x = 10$, making the x-window $[5, 15]$. Using the calculator to CALCULATE the given function at $x = 10$ (or evaluating by hand) gives $y(10) = -96$. Since the parabola opens upward (the coefficient of x^2 is positive), the curve rises up from its vertex, so we select a y-interval to include -96 upward, say $[-100, -70]$. Graphing the function on the window $[5, 15]$ by $[-100, -70]$ gives the following result. (Some other graphing windows are just as good.)

$f(x) = 2x^2 - 40x + 104$
on $[5, 15]$ by $[-100, -70]$

LOOKING AHEAD

Graphing a function by finding its high and low points will be important on page 164 and beyond.

Note that even without a graphing calculator we could sketch the graph of $f(x) = 2x^2 - 40x + 104$ by finding the vertex $(10, -96)$ just as we did, calculating two more points (say at $x = 5$ and $x = 15$), and drawing an upward-opening parabola from the vertex through these points.

Solving Quadratic Equations

A value of x that solves an equation $f(x) = 0$ is called a **root of the equation,** or a **zero of the function,** or an *x*-intercept of the graph of $y = f(x)$. The roots of a quadratic equation can often be found by factoring.

EXAMPLE 6 **SOLVING A QUADRATIC EQUATION BY FACTORING**

Solve $2x^2 - 4x = 6$.

Solution

FOR MORE HELP

with factoring, see the Algebra Review appendix, pages B11–B12

$$2x^2 - 4x - 6 = 0 \qquad \text{Subtracting 6 from each side to get zero on the right}$$

$$2(x^2 - 2x - 3) = 0 \qquad \text{Factoring out a 2}$$

$$2(x - 3) \cdot (x + 1) = 0 \qquad \text{Factoring } x^2 - 2x - 3$$

$$\underbrace{}_{\substack{\text{Equals 0} \\ \text{at } x = 3}} \underbrace{}_{\substack{\text{Equals 0} \\ \text{at } x = -1}} \qquad \text{Finding } x\text{-values that make each factor zero}$$

$$x = 3, \quad x = -1 \qquad \text{Solutions}$$

Graphing Calculator Exploration

Find both solutions to the equation in Example 6 by graphing the function $f(x) = 2x^2 - 4x - 6$ and using ZERO or TRACE to find where the curve crosses the *x*-axis. Your answers should agree with those found in Example 6.

PRACTICE PROBLEM 4

Solve by factoring or graphing: $9x - 3x^2 = -30$ Solution on page 43 >

Quadratic equations can often be solved by the *Quadratic Formula,* which is derived on page 44.

The quadratic formula gives these *x*-values

Quadratic Formula

The solutions to $ax^2 + bx + c = 0$ are

$$x = \frac{-b \pm \sqrt{b^2 - 4ac}}{2a}$$

The "plus or minus" ($\pm$) sign means calculate *both* ways, first using the $+$ sign and then using the $-$ sign

In a business, it is often important to find a company's **break-even points,** the numbers of units of production where a company's costs are equal to its revenue.

EXAMPLE 7   FINDING BREAK-EVEN POINTS

A company that installs automobile compact disc (CD) players finds that if it installs x CD players per day, then its costs will be $C(x) = 120x + 4800$ and its revenue will be $R(x) = -2x^2 + 400x$ (both in dollars). Find the company's break-even points. [*Note:* In Section 3.4 we will see how such cost and revenue functions are found.]

Solution

$$120x + 4800 = -2x^2 + 400x \qquad \text{Setting } C(x) = R(x)$$

$$2x^2 - 280x + 4800 = 0 \qquad \text{Combining all terms on one side}$$

$$x = \frac{280 \pm \sqrt{(-280)^2 - 4 \cdot 2 \cdot 4800}}{2 \cdot 2} \qquad \begin{array}{l}\text{Quadratic Formula with}\\ a = 2, \ b = -280, \ \text{and}\\ c = 4800\end{array}$$

$$= \frac{280 \pm \sqrt{40{,}000}}{4} = \frac{280 \pm 200}{4} \qquad \begin{array}{l}\text{Working out the}\\ \text{formula on a}\\ \text{calculator}\end{array}$$

$$= \frac{480}{4} \text{ or } \frac{80}{4} = 120 \text{ or } 20$$

The company will break even when it installs either 20 or 120 CD players. These break-even points may be seen in the graph on the left.

✎ FOR MORE HELP

with evaluating expressions, see the Algebra Review appendix, page B7

Cost and revenue meet at the break-even points

Although it is important for a company to know where its break-even points are, most companies want to do better than break even—they want to maximize their profit. **Profit** is defined as *revenue minus cost* (since profit is what is left over after subtracting expenses from income).

Profit

$$\text{Profit} = \text{Revenue} - \text{Cost}$$

EXAMPLE 8   MAXIMIZING PROFIT

For the CD installer whose daily revenue and cost functions were given in Example 7, find the number of units that maximizes profit, and the maximum profit.

Solution

The profit function is the revenue function minus the cost function.

$$P(x) = \underbrace{-2x^2 + 400x}_{R(x)} - \underbrace{(120x + 4800)}_{C(x)} \qquad \begin{array}{l}P(x) = R(x) - C(x) \text{ with}\\ R(x) = -2x^2 + 400x \text{ and}\\ C(x) = 120x + 4800\end{array}$$

$$= -2x^2 + 280x - 4800 \qquad \text{Simplifying}$$

Since this function represents a parabola opening downward (because of the -2), it is maximized at its vertex, which is found using the vertex formula.

$$x = \frac{-280}{2(-2)} = \frac{-280}{-4} = 70$$

$x = \dfrac{-b}{2a}$ with

$a = -2$ and $b = 280$

Thus, profit is maximized when 70 CD players are installed. For the maximum profit, we substitute $x = 70$ into the profit function:

$$P(70) = -2(70)^2 + 280 \cdot 70 - 4800$$

$P(x) = -2x^2 + 280x - 4800$
with $x = 70$

$$= 5000$$

Multiplying and combining

Therefore, the company will maximize its profit when it installs 70 CD players per day. Its maximum profit will be $5000 per day. This maximum may be seen in the graph on the left.

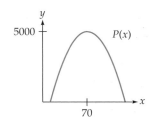

Why doesn't a company make more profit the more it sells? Because to increase its sales it must lower its prices, which eventually leads to lower profits. The relationship among the cost, revenue, and profit functions can be seen graphically as follows.

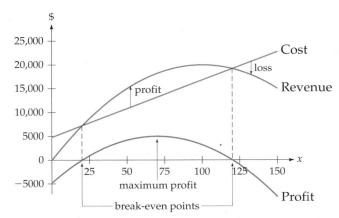

Notice that the break-even points correspond to a profit of zero, and that the maximum profit occurs halfway between the two break-even points.

Not all quadratic equations have (real) solutions.

EXAMPLE 9   USING THE QUADRATIC FORMULA

Solve $\frac{1}{2}x^2 - 3x + 5 = 0$.

Solution

The Quadratic Formula with $a = \frac{1}{2}$, $b = -3$, and $c = 5$ gives

$$x = \frac{3 \pm \sqrt{9 - 4\left(\frac{1}{2}\right)(5)}}{2\left(\frac{1}{2}\right)} = \frac{3 \pm \sqrt{9 - 10}}{1} = 3 \pm \sqrt{-1} \qquad \text{Undefined}$$

Therefore, the equation $\frac{1}{2}x^2 - 3x + 5 = 0$ has *no real solutions* (because of the undefined $\sqrt{-1}$). The geometrical reason that there are no solutions can be seen in the graph on the right: The curve never reaches the x-axis, so the function never equals zero.

$$f(x) = \frac{1}{2}x^2 - 3x + 5$$

The quantity $b^2 - 4ac$, whose square root appears in the Quadratic Formula, is called the **discriminant.** If the discriminant is *positive* (as in Example 7), the equation $ax^2 + bx + c = 0$ has *two* solutions (since the square root is added or subtracted). If the discriminant is *zero,* there is only *one* root (since adding or subtracting zero gives the same answer). If the discriminant is *negative* (as in Example 9), then the equation has *no* real roots. Therefore, the discriminant being positive, zero, or negative corresponds to the parabola meeting the x-axis at 2, 1, or 0 points, as shown below.

Quadratic Regression (Optional)

For data points that seem to lie along a *parabola,* we use **quadratic regression** to fit a parabola to the points, with the procedure carried out on a graphing calculator.

| EXAMPLE 10 | QUADRATIC REGRESSION USING A GRAPHING CALCULATOR |

Worldwide sales of personal computers (PCs) fell in 2011 and 2012 because of weak economies and earthquakes and floods in Asia, but have since rebounded and are expected to continue to grow. Sales (millions of units) for recent years are shown in the following graph.

Source: Standard & Poor's

a. Use quadratic regression to fit a parabola to the data and state the regression function.

b. Use the regression function to predict PC sales in 2020.

Solution

a. We number the years with x-values 0–3, so x stands for *years after 2010*. We enter the data into lists, as shown in the first screen below (as explained in the appendix *Graphing Calculator Basics—Entering Data* on page A3), and use *ZoomStat* to graph the data points.

Then (using STAT, CALC, and QuadReg) we graph the regression along with the data points.

The regression curve, which fits the points reasonably well, is

$$y = 0.85x^2 - 2.09x + 73.06$$

b. To predict sales in 2020, we evaluate Y1 at 10 (since $x = 10$ corresponds to 2020). From the screen on the right, if the current trend continues, worldwide PC sales in 2020 will be approximately 137 million units (rounded).

Solutions TO PRACTICE PROBLEMS

1. Domain: $\{x \mid x \le 0 \text{ or } x \ge 3\}$; range: $\{y \mid y \ge 0\}$

2. a. $g(27) = \sqrt{27 - 2} = \sqrt{25} = 5$

b. Domain: $\{z \mid z \ge 2\}$

c. Range: $\{y \mid y \ge 0\}$

on $[-1, 30]$ by $[-1, 7]$

3. $D(x) = 25 + 0.05x$

4.
$$9x - 3x^2 = -30$$
$$-3x^2 + 9x + 30 = 0$$
$$-3(x^2 - 3x - 10) = 0$$
$$-3(x - 5)(x + 2) = 0$$
$$x = 5, x = -2 \qquad \text{or from}$$

on $[-5, 8]$ by $[-20, 40]$

1.3   Section Summary

In this section we defined and gave examples of *functions,* and saw how to find their domains and ranges. The essential characteristic of a function f is that for any given "input" number x in the domain, there is exactly one "output" number $f(x)$. This requirement is stated geometrically in the *vertical line test,* that no vertical line can intersect the graph of a function at more than one point. We then defined *linear functions* (whose graphs are lines) and *quadratic functions* (whose graphs are parabolas). We solved quadratic equations by factoring, graphing, and using the Quadratic Formula. We maximized and minimized quadratic functions using the vertex formula.

Derivation of the Quadratic Formula

$ax^2 + bx + c = 0$	The quadratic set equal to zero
$ax^2 + bx = -c$	Subtracting c
$4a^2x^2 + 4abx = -4ac$	Multiplying by $4a$
$4a^2x^2 + 4abx + b^2 = b^2 - 4ac$	Adding b^2
$(2ax + b)^2 = b^2 - 4ac$	Since $4a^2x^2 + 4abx + b^2 = (2ax + b)^2$
$2ax + b = \pm\sqrt{b^2 - 4ac}$	Taking square roots
$2ax = -b \pm \sqrt{b^2 - 4ac}$	Subtracting b
$x = \dfrac{-b \pm \sqrt{b^2 - 4ac}}{2a}$	Dividing by $2a$ gives the Quadratic Formula

1.3   Exercises

1–8. Determine whether each graph defines a function of x.

1.

2.

7.

8.

3.

4.

9–10. Find the domain and range of each function graphed below.

9.

10.

5.

6.
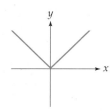

11–22. For each function:

a. Evaluate the given expression.
b. Find the domain of the function.
c. Find the range. [*Hint:* Use a graphing calculator.]

(See instructions on previous page.)

11. $f(x) = \sqrt{x-1}$; find $f(10)$

12. $f(x) = \sqrt{x-4}$; find $f(40)$

13. $h(z) = \dfrac{1}{z+4}$; find $h(-5)$

14. $h(z) = \dfrac{1}{z+7}$; find $h(-8)$

15. $h(x) = x^{1/4}$; find $h(81)$

16. $h(x) = x^{1/6}$; find $h(64)$

17. $f(x) = x^{2/3}$; find $f(-8)$

18. $f(x) = x^{4/5}$; find $f(-32)$

19. $f(x) = \sqrt{4-x^2}$; find $f(0)$

20. $f(x) = \dfrac{1}{\sqrt{x}}$; find $f(4)$

21. $f(x) = \sqrt{-x}$; find $f(-25)$

22. $f(x) = -\sqrt{-x}$; find $f(-100)$

23–30. Graph each function "by hand." [*Note:* Even if you have a graphing calculator, it is important to be able to sketch simple curves by finding a few important points.]

23. $f(x) = 3x - 2$

24. $f(x) = 2x - 3$

25. $f(x) = -x + 1$

26. $f(x) = -3x + 5$

27. $f(x) = 2x^2 + 4x - 16$

28. $f(x) = 3x^2 - 6x - 9$

29. $f(x) = -3x^2 + 6x + 9$

30. $f(x) = -2x^2 + 4x + 16$

31–34. For each quadratic function:

a. Find the vertex using the vertex formula.

 b. Graph the function on an appropriate window. (Answers may differ.)

31. $f(x) = x^2 - 40x + 500$

32. $f(x) = x^2 + 40x + 500$

33. $f(x) = -x^2 - 80x - 1800$

34. $f(x) = -x^2 + 80x - 1800$

35–52. Solve each equation by factoring or the Quadratic Formula, as appropriate.

35. $x^2 - 6x - 7 = 0$

36. $x^2 - x - 20 = 0$

37. $x^2 + 2x = 15$

38. $x^2 - 3x = 54$

39. $2x^2 + 40 = 18x$

40. $3x^2 + 18 = 15x$

41. $5x^2 - 50x = 0$

42. $3x^2 - 36x = 0$

43. $2x^2 - 50 = 0$

44. $3x^2 - 27 = 0$

 FOR HELP GETTING STARTED

with Exercises 11–22, see Examples 1 and 2 on page 34.

45. $4x^2 + 24x + 40 = 4$

46. $3x^2 - 6x + 9 = 6$

47. $-4x^2 + 12x = 8$

48. $-3x^2 + 6x = -24$

49. $2x^2 - 12x + 20 = 0$

50. $2x^2 - 8x + 10 = 0$

51. $3x^2 + 12 = 0$

52. $5x^2 + 20 = 0$

53–62. Solve each equation using a graphing calculator. [*Hint:* Begin with the window $[-10, 10]$ by $[-10, 10]$ or another of your choice (see *Useful Hint* in the *Graphing Calculator Basics* appendix, page A2) and use ZERO or TRACE and ZOOM IN.] (In Exercises 61 and 62, round answers to two decimal places.)

53. $x^2 - x - 20 = 0$

54. $x^2 + 2x - 15 = 0$

55. $2x^2 + 40 = 18x$

56. $3x^2 + 18 = 15x$

57. $4x^2 + 24x + 45 = 9$

58. $3x^2 - 6x + 5 = 2$

59. $3x^2 + 7x + 12 = 0$

60. $5x^2 + 14x + 20 = 0$

61. $2x^2 + 3x - 6 = 0$

62. $3x^2 + 5x - 7 = 0$

 63. Use your graphing calculator to graph the following four equations simultaneously on the window $[-10, 10]$ by $[-10, 10]$:

$$y_1 = 2x + 6$$
$$y_2 = 2x + 2$$
$$y_3 = 2x - 2$$
$$y_4 = 2x - 6$$

a. What do the lines have in common and how do they differ?

b. Write the equation of another line with the same slope that lies 2 units below the lowest line. Then check your answer by graphing it with the others.

64. Use your graphing calculator to graph the following four equations simultaneously on the window $[-10, 10]$ by $[-10, 10]$:

$$y_1 = 3x + 4$$
$$y_2 = x + 4$$
$$y_3 = -x + 4 \quad (\text{Use } \boxed{(-)} \text{ to get } -x.)$$
$$y_4 = -3x + 4$$

a. What do the lines have in common and how do they differ?

b. Write the equation of a line through this y-intercept with slope $\frac{1}{2}$. Then check your answer by graphing it with the others.

Applied Exercises

65. BUSINESS: Cost Functions A lumberyard will deliver wood for $4 per board foot plus a delivery charge of $20. Find a function $C(x)$ for the cost of having x board feet of lumber delivered.

66. BUSINESS: Cost Functions A company manufactures bicycles at a cost of $55 each. If the company's fixed costs are $900, express the company's costs as a linear function of x, the number of bicycles produced.

67. BUSINESS: Salary An employee's weekly salary is $500 plus $15 per hour of overtime. Find a function $P(x)$ giving his pay for a week in which he worked x hours of overtime.

68. BUSINESS: Salary A sales clerk's weekly salary is $300 plus 2% of her total week's sales. Find a function $P(x)$ for her pay for a week in which she sold x dollars of merchandise.

69. GENERAL: Water Pressure At a depth of d feet underwater, the water pressure is $p(d) = 0.45d + 15$ pounds per square inch. Find the pressure at:
 a. The bottom of a 6-foot-deep swimming pool.
 b. The maximum ocean depth of 35,000 feet.

70. GENERAL: Boiling Point At higher altitudes, water boils at lower temperatures. This is why at high altitudes foods must be boiled for longer times—the lower boiling point imparts less heat to the food. At an altitude of h thousand feet above sea level, water boils at a temperature of $B(h) = -1.8h + 212$ degrees Fahrenheit. Find the altitude at which water boils at 98.6 degrees Fahrenheit. (Your answer will show that at a high enough altitude, water boils at normal body temperature. This is why airplane cabins must be pressurized—at high enough altitudes one's blood would boil.)

71–72. GENERAL: Stopping Distance A car traveling at speed v miles per hour on a dry road should be able to come to a full stop in a distance of

$$D(v) = 0.055v^2 + 1.1v \text{ feet}$$

Find the stopping distance required for a car traveling at:

71. 40 mph.

72. 60 mph.

Source: National Transportation Safety Board

73. BIOMEDICAL: Cell Growth The number of cells in a culture after t days is given by $N(t) = 200 + 50t^2$. Find the size of the culture after:
 a. 2 days.
 b. 10 days.

74. ATHLETICS: Juggling If you toss a ball h feet straight up, it will return to your hand after $T(h) = 0.5\sqrt{h}$ seconds. This leads to the *juggler's dilemma:* Juggling more balls means tossing them higher. However, the square root in the above formula means that tossing them twice as high does not gain twice as much time, but only $\sqrt{2} \approx 1.4$ times as much time. Because of this, there is a limit to the number of balls that a person can juggle, which seems to be about ten. Use this formula to find:
 a. How long will a ball spend in the air if it is tossed to a height of 4 feet? 8 feet?
 b. How high must it be tossed to spend 2 seconds in the air? 3 seconds in the air?

75. GENERAL: Impact Velocity If a marble is dropped from a height of x feet, it will hit the ground with velocity $v(x) = \frac{60}{11}\sqrt{x}$ miles per hour (neglecting air resistance). Use this formula to find the velocity with which a marble will strike the ground if it is dropped from the height of the tallest building in the United States, the 1776-foot One World Trade Center in New York City.

76. GENERAL: Tsunamis The speed of a tsunami (popularly known as a tidal wave, although it has nothing whatever to do with tides) depends on the depth of the water through which it is traveling. At a depth of d feet, the speed of a tsunami will be $s(d) = 3.86\sqrt{d}$ miles per hour. Find the speed of a tsunami in the Pacific basin where the average depth is 15,000 feet.

Source: William Bascom, Waves and Beaches

77–78. GENERAL: Impact Time of a Projectile If an object is thrown upward so that its height (in feet) above the ground t seconds after it is thrown is given by the function $h(t)$ below, find when the object hits the ground. That is, find the positive value of t such that $h(t) = 0$. Give the answer correct to two decimal places. [*Hint:* Enter the function in terms of x rather than t. Use the ZERO operation, or TRACE and ZOOM IN, or similar operations.]

77. $h(t) = -16t^2 + 45t + 5$

78. $h(t) = -16t^2 + 40t + 4$

79. BUSINESS: Break-Even Points and Maximum Profit A company that produces tracking devices for computer disk drives finds that if it produces x devices per week, its costs will be $C(x) = 180x + 16{,}000$ and its revenue will be $R(x) = -2x^2 + 660x$ (both in dollars).
 a. Find the company's break-even points.
 b. Find the number of devices that will maximize profit, and the maximum profit.

80. BUSINESS: Break-Even Points and Maximum Profit City and Country Cycles finds that if it sells x racing bicycles per month, its costs will be $C(x) = 420x + 72{,}000$ and its revenue will be $R(x) = -3x^2 + 1800x$ (both in dollars).
 a. Find the store's break-even points.
 b. Find the number of bicycles that will maximize profit, and the maximum profit.

81. **BUSINESS: Break-Even Points and Maximum Profit**   A sporting goods store finds that if it sells x exercise machines per day, its costs will be $C(x) = 100x + 3200$ and its revenue will be $R(x) = -2x^2 + 300x$ (both in dollars).
    a. Find the store's break-even points.
    b. Find the number of sales that will maximize profit, and the maximum profit.

82. **SOCIAL SCIENCE: Health Club Attendance**   A study analyzed how the number of visits a person makes to a health club varies with the monthly membership price. It found that the number of visits per year is given approximately by $v(x) = -0.004x^2 + 0.56x + 42$, where x is the monthly membership price. What monthly price maximizes the number of visits?
    Source: American Economic Review **96(3)**

83. **ATHLETICS: Muscle Contraction**   The fundamental equation of muscle contraction is of the form $(w + a)(v + b) = c$, where w is the weight placed on the muscle, v is the velocity of contraction of the muscle, and a, b, and c are constants that depend upon the muscle and the units of measurement. Solve this equation for v as a function of w, a, b, and c.
    Source: E. Batschelet, Introduction to Mathematics for Life Scientists

84. **GENERAL: Longevity**   When a person reaches age 65, the probability of living for another x decades is approximated by the function $f(x) = -0.077x^2 - 0.057x + 1$ (for $0 \leq x \leq 3$). Find the probability that such a person will live for another:
    a. One decade.
    b. Two decades.
    c. Three decades.
    Source: Statistical Abstracts of the United States, 2007

85. **BEHAVIORAL SCIENCES: Smoking and Education**   According to a study, the probability that a smoker will quit smoking increases with the smoker's educational level. The probability (expressed as a percent) that a smoker with x years of education will quit is approximately $y = 0.831x^2 - 18.1x + 137.3$ (for $10 \leq x \leq 16$).
    a. Graph this curve on the window [10, 16] by [0, 100].
    b. Find the probability that a high school graduate smoker ($x = 12$) will quit.
    c. Find the probability that a college graduate smoker ($x = 16$) will quit.
    Source: Review of Economics and Statistics **LXXVII(1)**

86. **ENVIRONMENTAL SCIENCE: Wind Energy**   The use of wind power is growing rapidly after a slow start, especially in Europe, where it is seen as an efficient and renewable source of energy. Global wind power generating capacity for the years 1996 to 2008 is given approximately by $y = 0.9x^2 - 3.9x + 12.4$ thousand megawatts (MW), where x is the number of years

after 1995. (One megawatt would supply the electrical needs of approximately 100 homes).
    a. Graph this curve on the window [0, 20] by [0, 300].
    b. Use this curve to predict the global wind power generating capacity in the year 2015. [*Hint:* Which x-value corresponds to 2015? Then use TRACE, VALUE, or TABLE.]
    c. Predict the global wind power generating capacity in the year 2020.
    Source: Worldwatch Institute

87. **GENERAL: Newsletters**   A newsletter has a maximum audience of 100 subscribers. The publisher estimates that she will lose 1 reader for each dollar she charges. Therefore, if she charges x dollars, her readership will be $(100 - x)$.
    a. Multiply this readership by x (the price) to find her total revenue. Multiply out the resulting quadratic function.
    b. What price should she charge to maximize her revenue? [*Hint:* Find the value of x that maximizes this quadratic function.]

88. **ATHLETICS: Cardiovascular Zone**   Your maximum heart rate (in beats per minute) may be estimated as 220 minus your age. For maximum cardiovascular effect, many trainers recommend raising your heart rate to between 50% and 70% of this maximum rate (called the *cardio zone*).
    a. Write a linear function to represent this upper limit as a function of x, your age. Then write a similar linear function to represent the lower limit. Use decimals instead of percents.
    b. Use your functions to find the upper and lower cardio limits for a 20-year-old person. Find the cardio limits for a 60-year-old person.
    Source: Mayo Clinic

89. **SOCIAL SCIENCE: Immigration**   The percentage of immigrants in the United States has changed since World War I as shown in the following graph.

    a. Number the bars with x-values 0–8 (so that x stands for *decades since 1930*) and use quadratic regression to fit a parabola to the data. State the regression function. [*Hint:* See Example 10.]
    b. Use your curve to estimate the percentage in 2016.
    Sources: Center for Immigration Studies and U.S. Census Bureau

90. BUSINESS: Movie Prices National average theater admissions prices for recent decades are shows in the following graph.

a. Number the bars with x-values 1–6 (so that x stands for *decades since 1950*) and use quadratic regression to fit a parabola to the data. State the regression function. [*Hint:* See Example 10.]

b. Use the regression function to predict movie prices in the years 2020 and 2030.

Source: Entertainment Weekly

Conceptual Exercises

91. Can the graph of a function have more than one x-intercept? Can it have more than one y-intercept?

92. If a linear function is such that $f(2) = 5$ and $f(3) = 7$, then $f(4) = ?$ [*Hint:* No work necessary.]

93. If a linear function is such that $f(4) = 7$ and $f(6) = 11$, then $f(5) = ?$ [*Hint:* No work necessary.]

94. The Apocryphal Manufacturing Company makes widgets out of blivets. If a linear function $f(x) = mx + b$ gives the number of widgets that can be made from x blivets, what are the units of the slope m (*widgets per blivet* or *blivets per widget*)?

95. In a linear function $f(x) = mx + b$, the slope m has units *blargs per prendle*. What are the units of x? What are the units of y? [*Hint:* One is in blargs and the other is in prendles, but which is which?]

96. For the quadratic function $f(x) = ax^2 + bx + c$, what condition on one of the coefficients will guarantee that the function has a highest value? A lowest value?

97. We have discussed quadratic functions that open *up* or open *down*. Can a quadratic function open *sideways*? Explain.

98. Explain why, if a quadratic function has two x-intercepts, the x-coordinate of the vertex will be halfway between them.

1.4   # Functions: Polynomial, Rational, and Exponential

Introduction

In this section we will define other useful types of functions, including polynomial, rational, exponential, and logarithmic functions, although the latter two types will be discussed more extensively in Sections 4.1 and 4.2. We will also define an important operation, the *composition* of functions.

Polynomial Functions

A **polynomial function** (or simply a **polynomial**) is a function that can be written in the form

$$f(x) = a_n x^n + a_{n-1} x^{n-1} + \cdots + a_2 x^2 + a_1 x + a_0$$

where n is a nonnegative integer and $a_0, a_1, \ldots, a_n$ are (real) numbers, called **coefficients**. The *domain* of a polynomial is $\mathbb{R}$, the set of all (real) numbers. The **degree** of a polynomial is the highest power of the variable. The following are polynomials.

$$f(x) = 2x^8 - 3x^7 + 4x^5 - 5$$

A polynomial of degree 8 (since the highest power of x is 8)

$$f(x) = -4x^2 - \tfrac{1}{3}x + 19$$

A polynomial of degree 2 (a quadratic function)

$$f(x) = x - 1$$

A polynomial of degree 1
(a linear function)

$$f(x) = 6$$

A polynomial of degree 0
(a constant function)

Cost

Units produced

A cost function may
increase at different rates at
different production levels.

A polynomial of degree 3 is called a *cubic function*, or just a *cubic* and a polynomial of degree 4 is called a *quartic function* or just a *quartic*.

Polynomials are used to model many situations in which change occurs at different rates. For example, the polynomial graphed on the left might represent the total cost of manufacturing x units of a product. At first, costs rise steeply because of high start-up expenses, then more slowly as the economies of mass production come into play, and finally more steeply as new production facilities need to be built.

Polynomial equations can often be solved by factoring (just as with quadratic equations).

EXAMPLE 1   **SOLVING A POLYNOMIAL EQUATION BY FACTORING**

Solve $3x^4 - 6x^3 = 24x^2$

Solution

$$3x^4 - 6x^3 - 24x^2 = 0$$

Rewritten with all the
terms on the left side

$$3x^2(x^2 - 2x - 8) = 0$$

Factoring out $3x^2$

$$3x^2 \cdot (x - 4) \cdot (x + 2) = 0$$

Factoring further

Equals   Equals   Equals
zero at   zero at   zero at
$x = 0$ $x = 4$ $x = -2$

Finding the zeros
of each factor

$$x = 0, \quad x = 4, \quad x = -2$$

Solutions

 FOR MORE HELP

with factoring, see the Algebra
Review appendix, pages B11–B12

As in this Example, if a positive power of x can be factored out of a polynomial, then $x = 0$ is one of the roots.

PRACTICE PROBLEM 1

Solve $2x^3 - 4x^2 = 48x$

Solution on page 60 >

Rational Functions

The word "ratio" means fraction or quotient, and the quotient of two polynomials is called a **rational function**. The following are rational functions.

$$f(x) = \frac{3x + 2}{x - 2} \qquad g(x) = \frac{1}{x^2 + 1}$$

A rational function
is a polynomial
over a polynomial

The *domain* of a rational function is the set of numbers for which the denominator is not zero. For example, the domain of the function $f(x)$ on the left above is $\{x \mid x \neq 2\}$ (since $x = 2$ makes the denominator zero), and the domain of $g(x)$ on the right is the set of all real numbers $\mathbb{R}$ (since $x^2 + 1$ is never zero). The graphs of these functions are shown on the next page. Notice that these graphs have **asymptotes**, lines that the graphs *approach* but never actually reach.

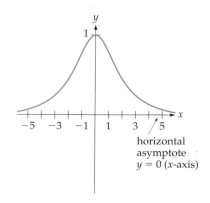

Graph of $f(x) = \dfrac{3x + 2}{x - 2}$

Graph of $g(x) = \dfrac{1}{x^2 + 1}$

PRACTICE PROBLEM 2

What is the domain of $f(x) = \dfrac{18}{(x + 2)(x - 4)}$?

Solution on page 60 >

 Be Careful Simplifying a rational function by canceling a common factor from the numerator and the denominator can change the domain of the function, so that the "simplified" and "original" versions may not be equal (since they have different domains). For example, the rational function on the left below is not defined at $x = 1$, while the simplified version on the right *is* defined at $x = 1$, so that the two functions are technically not equal.

$$\underbrace{\frac{x^2 - 1}{x - 1}}_{\substack{\text{Not defined at } x = 1, \\ \text{so the domain is } \{x \mid x \neq 1\}}} = \frac{(x + 1)(x - 1)}{x - 1} \neq \underbrace{x + 1}_{\substack{\text{Is defined at } x = 1, \\ \text{so the domain is } \mathbb{R}}}$$

 LOOKING AHEAD

We will return to this observation on page 76 when we discuss *limits*.

However, the functions *are* equal at every x-value *except* $x = 1$, and the graphs below are the same except that the rational function omits the point at $x = 1$.

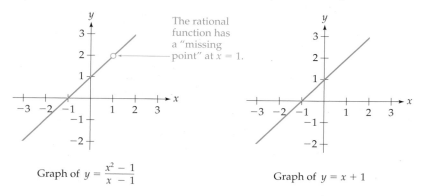

Graph of $y = \dfrac{x^2 - 1}{x - 1}$

Graph of $y = x + 1$

Exponential Functions

A function in which the independent variable appears in the exponent, such as $f(x) = 2^x$, is called an **exponential function.**

EXAMPLE 2   **GRAPHING AN EXPONENTIAL FUNCTION**

Graph the exponential function $f(x) = 2^x$.

Solution

This function is defined for *all* real numbers, so its domain is $\mathbb{R}$. Values of the function are shown in the table on the left below, and plotting these points and drawing a smooth curve through them give the curve on the right below.

x	$y = 2^x$
3	$2^3 = 8$
2	$2^2 = 4$
1	$2^1 = 2$
0	$2^0 = 1$
-1	$2^{-1} = \frac{1}{2}$
-2	$2^{-2} = \frac{1}{4}$
-3	$2^{-3} = \frac{1}{8}$

$f(x) = 2^x$ has domain $\mathbb{R}$ and range $\{y \mid y > 0\}$

$f(x) = 2^x$

Exponential functions are often used to model population growth and decline.

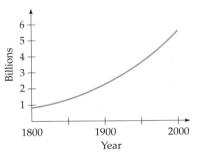

World population since the year 1800 can be approximated by an exponential function.

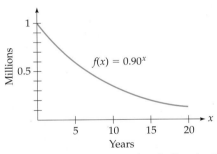

$f(x) = 0.90^x$

A population of 1 million that declines by 10% each year is modeled by an exponential function.

In mathematics the letter e is used to represent a constant whose value is approximately 2.718 (written $e \approx 2.718$). The exponential function $f(x) = e^x$ will be very important beginning in Chapter 4. Another important function is the logarithmic function to the base e, written $f(x) = \ln x$. These functions are graphed below.

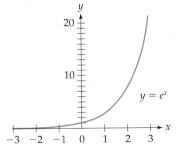

The exponential function e^x has domain $\mathbb{R}$ and range $\{y \mid y > 0\}$.

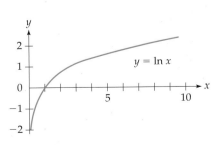

$y = \ln x$

The natural logarithm function has domain $\{x \mid x > 0\}$ and range $\mathbb{R}$.

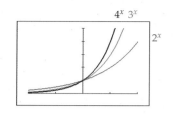

Graph the functions $y_1 = 2^x$, $y_2 = 3^x$, and $y_3 = 4^x$ on the window $[-2, 2]$ by $[0, 5]$.

a. Which function rises most steeply?

b. Between which two curves would e^x lie? Check your prediction by graphing $y_4 = e^x$. [*Hint:* On some calculators, e^x is obtained by pressing [2nd]

[LN] [x] .]

Piecewise Linear Functions

The rule for calculating the values of a function may be given in several parts. If each part is linear, the function is called a **piecewise linear function,** and its graph consists of "pieces" of straight lines.

EXAMPLE 3   GRAPHING A PIECEWISE LINEAR FUNCTION

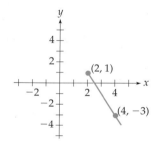

This notation means:

Graph $\quad f(x) = \begin{cases} 5 - 2x & \text{if } x \geq 2 \\ x + 3 & \text{if } x < 2 \end{cases}$ ⟵ Use the top formula for $x \geq 2$
⟵ Use the bottom formula for $x < 2$

Solution

We graph one "piece" at a time.

Step 1: To graph the first part, $f(x) = 5 - 2x$ if $x \geq 2$, we use the "endpoint" $x = 2$ and also $x = 4$ (or any other x-value satisfying $x \geq 2$). The points are $(2, 1)$ and $(4, -3)$, with the y-coordinates calculated from $f(x) = 5 - 2x$. Draw the line through these two points, but only for $x \geq 2$ (from $x = 2$ to the *right*), as shown on the left.

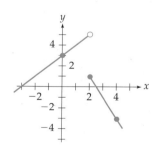

Step 2: For the second part, $f(x) = x + 3$ if $x < 2$, the restriction $x < 2$ means that the line ends just *before* $x = 2$. We mark this "missing point" $(2, 5)$ by an "open circle" o to indicate that it is *not* included in the graph (the y-coordinate comes from $f(x) = x + 3$). For a second point, choose $x = 0$ (or any other $x < 2$), giving $(0, 3)$. Draw the line through these two points, but only for $x < 2$ (to the *left* of $x = 2$), completing the graph of the function.

An important piecewise linear function is the *absolute value* function.

EXAMPLE 4   THE ABSOLUTE VALUE FUNCTION

The absolute value function is $f(x) = |x|$, and is defined as

$$|x| = \begin{cases} x & \text{if } x \geq 0 \\ -x & \text{if } x < 0 \end{cases}$$

The second line, for *negative x*, attaches a *second* negative sign to make the result *positive*

For example, when applied to either 3 or -3, the function gives *positive* 3:

$$|3| = 3$$

Using the top formula (since $3 \geq 0$)

$$|-3| = -(-3) = 3$$

Using the bottom formula (since $-3 < 0$)

To graph the absolute value function, we may proceed as in Example 3, or simply observe that for $x \geq 0$, the function gives $y = x$ (a half-line from the origin with slope 1), and for $x < 0$, it gives $y = -x$ (a half-line on the other side of the origin with slope -1), as shown below.

Absolute Value Function

$$|x| = \begin{cases} x & \text{if } x \geq 0 \\ -x & \text{if } x < 0 \end{cases}$$

The absolute value function $f(x) = |x|$ has a "corner" at the origin.

Examples 3 and 4 show that the "pieces" of a piecewise linear function may or may not be connected.

| EXAMPLE 5 | GRAPHING AN INCOME TAX FUNCTION |

Federal income taxes are "progressive," meaning that they take a higher percentage of higher incomes. For example, the 2013 federal income tax for a single taxpayer whose taxable income was not more than $87,850 was determined by a three-part rule: 10% of income up to $8925, plus 15% of any amount over $8925 up to $36,250, plus 25% of any amount over $36,250 up to $87,850. For an income of x dollars, the tax $f(x)$ may be expressed as follows:

$$f(x) = \begin{cases} 0.10x & \text{if } 0 \leq x \leq 8925 \\ 892.50 + 0.15(x - 8925) & \text{if } 8925 < x \leq 36,250 \\ 4991.25 + 0.25(x - 36,250) & \text{if } 36,250 < x \leq 87,850 \end{cases}$$

Graphing this by the same technique as before leads to the graph shown below. The slopes 0.10, 0.15, and 0.25 are called the *marginal tax rates*.

Source: Internal Revenue Service

**Graphing
Calculator
Exploration**

on $[-1, 8]$ by $[-2, 10]$
$(4, 2)$ is included
$(4, 0)$ is excluded

In the same way, we can define and graph piecewise *nonlinear* functions. To graph

$$f(x) = \begin{cases} 2x + 3 & \text{if } x < 1 \\ 6 - x & \text{if } 1 \leq x \leq 4 \\ (x - 4)^2 & \text{if } x > 4 \end{cases}$$

enter

$$y_1 = (2x + 3)(x < 1) + (6 - x)(x \geq 1 \text{ and } x \leq 4) + (x - 4)^2(x > 4)$$

or, equivalently,

$$y_1 = (2x + 3)(x < 1) + (6 - x)(x \geq 1)(x \leq 4) + (x - 4)^2(x > 4)$$

The inequalities and the word "and" are found in the TEST and LOGIC menus. (This function could also have been graphed *by hand* following the procedure explained in Example 3.)

Composite Functions

Just as we substitute a *number* into a function, we may substitute a *function* into a function. For two functions f and g, evaluating f at $g(x)$ gives $f(g(x))$, called the *composition* of f with g evaluated at x.*

Composite Functions
The *composition* of f with g evaluated at x is $f(g(x))$.

LOOKING AHEAD

Composite functions will be important on page 140 and beyond.

The *domain* of $f(g(x))$ is the set of all numbers x in the domain of g such that $g(x)$ is in the domain of f. If we think of the functions f and g as "numerical machines," then the composition $f(g(x))$ may be thought of as a *combined* machine in which the output of g is connected to the input of f.

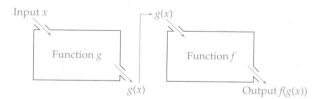

A "machine" for generating the composition of f with g.
A number x is fed into the function g, and the output
$g(x)$ is then fed into the function f, resulting in $f(g(x))$.

*The composition $f(g(x))$ may also be written $(f \circ g)(x)$, although we will not use this notation.

EXAMPLE 6   FINDING COMPOSITE FUNCTIONS

If $f(x) = x^7$ and $g(x) = x^3 - 2x$, find:

a. $f(g(x))$ **b.** $f(f(x))$

Solution

a. $f(g(x))$ $=$ $\underbrace{[g(x)]^7}_{\substack{f(x) = x^7 \text{ with } x \\ \text{replaced by } g(x)}}$ $=$ $\underbrace{(x^3 - 2x)^7}_{\substack{\text{Using} \\ g(x) = x^3 - 2x}}$

b. $f(f(x))$ $=$ $\underbrace{[f(x)]^7}_{\substack{f(x) = x^7 \text{ with } x \\ \text{replaced by } f(x)}}$ $=$ $\underbrace{(x^7)^7}_{\substack{\text{Using} \\ f(x) = x^7}}$ $=$ x^{49}

EXAMPLE 7   FINDING COMPOSITE FUNCTIONS IN DIFFERENT ORDERS

If $f(x) = \dfrac{x+8}{x-1}$ and $g(x) = \sqrt{x}$, find: **a.** $f(g(x))$ **b.** $g(f(x))$

Solution

a. $f(g(x)) = \dfrac{g(x) + 8}{g(x) - 1} = \dfrac{\sqrt{x} + 8}{\sqrt{x} - 1}$

$f(x) = \dfrac{x+8}{x-1}$ with x replaced by $g(x) = \sqrt{x}$

b. $g(f(x)) = \sqrt{f(x)} = \sqrt{\dfrac{x+8}{x-1}}$

$g(x) = \sqrt{x}$ with x replaced by $f(x) = \dfrac{x+8}{x-1}$

Be Careful   The order of composition is important: $f(g(x))$ is not the same as $g(f(x))$. To show this, we evaluate the above $f(g(x))$ and $g(f(x))$ at $x = 4$:

$f(g(4)) = \dfrac{\sqrt{4} + 8}{\sqrt{4} - 1} = \dfrac{2 + 8}{2 - 1} = \dfrac{10}{1} = 10$ $f(g(x)) = \dfrac{\sqrt{x} + 8}{\sqrt{x} - 1}$ at $x = 4$

Different answers

$g(f(4)) = \sqrt{\dfrac{4 + 8}{4 - 1}} = \sqrt{\dfrac{12}{3}} = \sqrt{4} = 2$ $g(f(x)) = \sqrt{\dfrac{x+8}{x-1}}$ at $x = 4$

PRACTICE PROBLEM 3

If $f(x) = x^2 + 1$ and $g(x) = \sqrt[3]{x}$, find: **a.** $f(g(x))$ **b.** $g(f(x))$

Solutions on page 60 >

EXAMPLE 8   **PREDICTING WATER USAGE**

A planning commission estimates that if a city's population is p thousand people, its daily water usage will be $W(p) = 30p^{1.2}$ thousand gallons. The commission further predicts that the population in t years will be $p(t) = 60 + 2t$ thousand people. Express the water usage W as a function of t, the number of years from now, and find the water usage 10 years from now.

Solution

Water usage W as a function of t is the *composition* of $W(p)$ with $p(t)$:

$$W(p(t)) = 30[p(t)]^{1.2} = 30(60 + 2t)^{1.2}$$
$W = 30p^{1.2}$ with p replaced by $p(t) = 60 + 2t$

To find water usage in 10 years, we evaluate $W(p(t))$ at $t = 10$:

$$W(p(10)) = 30(60 + 2 \cdot 10)^{1.2}$$
$30(60 + 2t)^{1.2}$ with $t = 10$

$$= 30(80)^{1.2} \approx 5765$$
Using a calculator

Thousand gallons

Therefore, in 10 years the city will need about 5,765,000 gallons of water per day.

Shifts of Graphs

Sometimes the graph of a composite function is just a horizontal or vertical shift of an original graph. This occurs when one of the functions is simply the addition or subtraction of a constant. The following diagram shows the graph of $y = x^2$ together with various shifts and the functions that generate them.

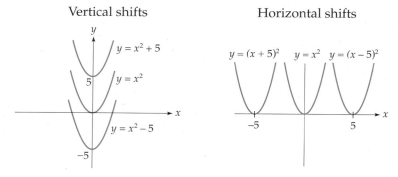

Vertical shifts

Horizontal shifts

In general, adding to or subtracting from the *x-value* means a *horizontal* shift, while adding to or subtracting from the *function* means a *vertical* shift. These same ideas hold for *any* function: Given the graph of $y = f(x)$, adding or subtracting a positive number a to the function $f(x)$ or to the variable x shifts the graph as follows:

Shifts of Graphs

Function	Shift	
$y = f(x) + a$	shifted *up* by a units	⎫ Vertical shifts
$y = f(x) - a$	shifted *down* by a units	⎭
$y = f(x + a)$	shifted *left* by a units	⎫ Horizontal shifts
$y = f(x - a)$	shifted *right* by a units	⎭

Of course, a graph can be shifted both horizontally and vertically, as illustrated by the following shifts of $y = x^2$:

$y = (x + 2)^2 - 7$
(shifted *left* 2 units
and *down* 7 units)

$y = (x - 4)^2 + 1$
(shifted *right* 4 units
and *up* 1 unit)

Such double shifts can be applied to *any* function $y = f(x)$: The graph of $y = f(x + a) + b$ is shifted *left a* units and *up b* units (with the understanding that a *negative a* or *b* means that the direction is reversed).

 Be Careful   Remember that adding a *positive* number to x means a *left shift*.

 Graphing Calculator Exploration

$y = |x - 2| - 6$

The absolute value function $y = |x|$ may be graphed on some graphing calculators as $y_1 = \text{ABS}(x)$.

a. Graph $y_1 = \text{ABS}(x - 2) - 6$ and observe that the absolute value function is shifted *right* 2 units and *down* 6 units. (The graph shown is drawn using ZOOM ZSquare.)

b. Predict the shift of $y_1 = \text{ABS}(x + 4) + 2$ and then verify your prediction by graphing the function on your calculator.

Given a function $f(x)$, to find an algebraic expression for the "shifted" function $f(x + h)$ we simply replace each occurrence of x by $x + h$.

> **EXAMPLE 9**   **FINDING $f(x + h)$ FROM $f(x)$**
>
> If $f(x) = x^2 - 5x$, find $f(x + h)$.
>
> **Solution**
>
> $$f(x + h) = (x + h)^2 - 5(x + h) \qquad \text{$f(x) = x^2 - 5x$ with each x}$$
> $$\text{replaced by $x + h$}$$
>
> $$= x^2 + 2xh + h^2 - 5x - 5h \qquad \text{Expanding}$$

LOOKING BACK

On page 26 we found a formula for squaring a sum, which gives $(x + h)^2 = x^2 + 2xh + h^2$

Difference Quotients

LOOKING AHEAD

The difference quotient will be used on pages 88–92

The quantity $\dfrac{f(x + h) - f(x)}{h}$ is called the **difference quotient,** since it is a quotient whose numerator is a difference. It gives the slope (rise over run) between the points in the curve $y = f(x)$ at x and at $x + h$.

| EXAMPLE 10 | FINDING A DIFFERENCE QUOTIENT |

If $f(x) = x^2 - 4x + 1,$ find and simplify $\dfrac{f(x+h) - f(x)}{h}$ $(h \neq 0)$

Solution

$$\frac{f(x+h) - f(x)}{h} = \frac{\overbrace{(x+h)^2 - 4(x+h) + 1}^{f(x+h)} - \overbrace{(x^2 - 4x + 1)}^{f(x)}}{h}$$

$$= \frac{x^2 + 2xh + h^2 - 4x - 4h + 1 - x^2 + 4x - 1}{h} \qquad \text{Expanding}$$

$$= \frac{x^2 + 2xh + h^2 - 4x - 4h + \cancel{1} - \cancel{x^2} + \cancel{4x} - \cancel{1}}{h} \qquad \text{Canceling}$$

$$= \frac{2xh + h^2 - 4h}{h} = \frac{h(2x + h - 4)}{h} \qquad \begin{array}{l}\text{Factoring an } h \\ \text{from the top}\end{array}$$

$$= \frac{\cancel{h}(2x + h - 4)}{\cancel{h}} = 2x + h - 4 \qquad \begin{array}{l}\text{Canceling } h \text{ from} \\ \text{top and bottom} \\ (\text{since } h \neq 0)\end{array}$$

 FOR MORE HELP

with canceling, see the Algebra
Review appendix, pages B12–B13.

| PRACTICE PROBLEM 4 |

If $f(x) = 3x^2 - 2x + 1,$ find and simplify $\dfrac{f(x+h) - f(x)}{h}.$

Solution on page 60 >

| EXAMPLE 11 | FINDING A DIFFERENCE QUOTIENT |

If $f(x) = \dfrac{1}{x},$ find and simplify $\dfrac{f(x+h) - f(x)}{h}$ $(h \neq 0)$

Solution

$$\frac{f(x+h) - f(x)}{h} = \frac{\overbrace{\dfrac{1}{x+h}}^{f(x+h)} - \overbrace{\dfrac{1}{x}}^{f(x)}}{h}$$

$$= \frac{1}{h}\left(\frac{1}{x+h} - \frac{1}{x}\right) \qquad \begin{array}{l}\text{Multiplying by } 1/h \\ \text{instead of dividing by } h\end{array}$$

$$= \frac{1}{h}\left(\frac{x}{(x+h)x} - \frac{x+h}{(x+h)x}\right) \qquad \begin{array}{l}\text{Using the common} \\ \text{denominator } (x+h)x\end{array}$$

$$= \frac{1}{h} \cdot \frac{\overbrace{x - (x+h)}^{-h}}{(x+h)x} = \frac{1}{h} \cdot \frac{-h}{(x+h)x} \qquad \begin{array}{l}\text{Subtracting fractions,} \\ \text{and simplifying}\end{array}$$

$$= \frac{1}{\cancel{h}} \cdot \frac{\cancel{-h}^{\,-1}}{(x+h)x} = \frac{-1}{(x+h)x} \qquad \text{Canceling } h \ (h \neq 0)$$

FOR MORE HELP

with canceling, see the Algebra
Review appendix, pages B12–B13.

Exponential Regression (Optional)

If data appear to lie along an *exponential* curve, such as those shown on page 51, we may fit an exponential curve to the data using **exponential regression.** The mathematical basis is explained in Section 7.4, with the procedure using a graphing calculator explained below. We will see in Section 4.1 that populations are often modeled by exponential curves, as are quantities that rise and fall with populations, such as health care expenses.

| EXAMPLE 12 | EXPONENTIAL REGRESSION USING A GRAPHING CALCULATOR |

Health care costs represent a growing share of our nation's budget. The following chart gives the per capita national health expenditure amounts in recent years.

Source: Centers for Medicare & Medicaid Services

a. Use exponential regression to fit a curve to the data and state the regression function.

b. Use the regression to predict per capita national health expenditure in 2020.

Solution

a. We number the years with *x*-values 1–3, so *x* stands for the *number of decades since 1980.* We enter the data into lists, as shown in the first screen below (as explained in the appendix *Graphing Calculator Basics—Entering Data,* page A3), and use *ZoomStat* to graph the data points.

Then (using STAT, CALC, and ExpReg) we graph the regression curve along with the data points.

The regression curve, which fits the points quite well, is

$$y = 1661 \cdot 1.716^x$$

b. To predict the per capita national health expenditure in 2020, we evaluate Y1 at 4 (since $x = 4$ means 4 decades after 1980). From the screen on the right, if the current trend continues, the per capita national health expenditure will be about \$14,415 in 2020.

```
Y1(4)
              14414.73627
```

Solutions TO PRACTICE PROBLEMS

1. $2x^3 - 4x^2 - 48x = 0$

$\quad 2x(x^2 - 2x - 24) = 0$

$\quad 2x(x + 4)(x - 6) = 0$

$\quad x = 0, \quad x = -4, \quad x = 6$

2. $\{x \mid x \neq -2, x \neq 4\}$

3. a. $f(g(x)) = [g(x)]^2 + 1 = \left(\sqrt[3]{x}\right)^2 + 1 \quad \text{or} \quad x^{2/3} + 1$

$\quad$ **b.** $g(f(x)) = \sqrt[3]{f(x)} = \sqrt[3]{x^2 + 1} \quad \text{or} \quad (x^2 + 1)^{1/3}$

4. $\dfrac{f(x + h) - f(x)}{h} = \dfrac{3(x + h)^2 - 2(x + h) + 1 - (3x^2 - 2x + 1)}{h}$

$\quad\quad\quad = \dfrac{3x^2 + 6xh + 3h^2 - 2x - 2h + \cancel{1} - 3x^2 + 2x - \cancel{1}}{h}$

$\quad\quad\quad = \dfrac{h(6x + 3h - 2)}{h} = 6x + 3h - 2$

1.4 Section Summary

We have introduced a variety of functions: polynomials (which include linear and quadratic functions), rational functions, exponential functions, and piecewise linear functions. Examples of these are shown below and on the next page. You should be able to identify these basic types of functions from their algebraic forms. We also added constants to perform horizontal and vertical *shifts* of graphs of functions, and combined functions by using the "output" of one as the "input" of the other, resulting in *composite* functions.

A Gallery of Functions

POLYNOMIALS

Linear function
$f(x) = mx + b$

Quadratic functions
$f(x) = ax^2 + bx + c$

$f(x) = ax^4 + bx^3 + cx^2 + dx + e$

RATIONAL FUNCTIONS

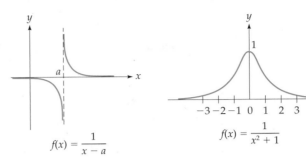

$$f(x) = \frac{1}{x - a}$$

$$f(x) = \frac{1}{x^2 + 1}$$

EXPONENTIAL FUNCTIONS

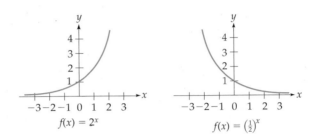

$$f(x) = 2^x$$

$$f(x) = \left(\tfrac{1}{2}\right)^x$$

PIECEWISE LINEAR FUNCTIONS

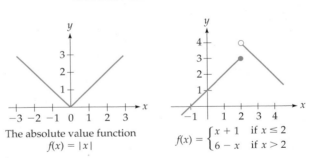

The absolute value function
$$f(x) = |x|$$

$$f(x) = \begin{cases} x + 1 & \text{if } x \le 2 \\ 6 - x & \text{if } x > 2 \end{cases}$$

1.4 Exercises

1–2. Find the domain and range of each function graphed below.

1.

2.

3–10. For each function:

    a. Evaluate the given expression.

    b. Find the domain of the function.

    c. Find the range. [*Hint*: Use a graphing calculator. You may have to ignore some false lines on the graph. Graphing in "dot mode" will also eliminate false lines.]

3. $f(x) = \dfrac{1}{x + 4}$; find $f(-3)$

4. $f(x) = \dfrac{1}{(x - 1)^2}$; find $f(-1)$

5. $f(x) = \dfrac{x^2}{x - 1}$; find $f(-1)$

6. $f(x) = \dfrac{x^2}{x + 2}$; find $f(2)$

7. $f(x) = \dfrac{12}{x(x + 4)}$; find $f(2)$

8. $f(x) = \dfrac{16}{x(x - 4)}$; find $f(-4)$

9. $g(x) = 4^x$; find $g\left(-\tfrac{1}{2}\right)$

10. $g(x) = 8^x$; find $g\left(-\tfrac{1}{3}\right)$

11–22. Solve each equation by factoring. [*Hint for Exercises 19–22*: First factor out a fractional power.]

11. $x^5 + 2x^4 - 3x^3 = 0$

12. $x^6 - x^5 - 6x^4 = 0$

13. $5x^3 - 20x = 0$

14. $2x^5 - 50x^3 = 0$

🖊 **FOR MORE HELP**

with factoring in Exercises 11–22, see the Algebra Review appendix, pages B11–B12.

15. $2x^3 + 18x = 12x^2$

16. $3x^4 + 12x^2 = 12x^3$

17. $6x^5 = 30x^4$

18. $5x^4 = 20x^3$

19. $3x^{5/2} - 6x^{3/2} = 9x^{1/2}$

20. $2x^{7/2} + 8x^{5/2} = 24x^{3/2}$

21. $2x^{5/2} + 4x^{3/2} = 6x^{1/2}$

22. $3x^{7/2} - 12x^{5/2} = 36x^{3/2}$

 23–24. Solve each equation using a graphing calculator. Round answers to two decimal places.

23. $x^5 - x^4 - 5x^3 = 0$

24. $x^6 + 2x^5 - 5x^4 = 0$

25–30. Graph each function.

25. $f(x) = 3^x$ **26.** $f(x) = \left(\frac{1}{3}\right)^x$

27. $f(x) = \begin{cases} 2x - 7 & \text{if } x \ge 4 \\ 2 - x & \text{if } x < 4 \end{cases}$

28. $f(x) = \begin{cases} 8 - 2x & \text{if } x \ge 2 \\ x + 2 & \text{if } x < 2 \end{cases}$

29. $f(x) = |x - 3| - 3$

30. $f(x) = |x + 2| - 2$

31–32. Use a graphing calculator to graph each piecewise *nonlinear* function on the window $[-2, 10]$ by $[-5, 5]$. Where parts of the graph do not touch, state which point is included and which is excluded.

31. $f(x) = \begin{cases} x^2 & \text{if } x \le 2 \\ 6 - x & \text{if } 2 < x < 6 \\ 2x - 17 & \text{if } x \ge 6 \end{cases}$

32. $f(x) = \begin{cases} 4 - x^2 & \text{if } x < 3 \\ 2x - 11 & \text{if } 3 \le x < 7 \\ 8 - x & \text{if } x \ge 7 \end{cases}$

33–46. Identify each function as a polynomial, a rational function, an exponential function, a piecewise linear function, or none of these. (Do not graph them; just identify their types.)

33. $f(x) = x^5$ **34.** $f(x) = 4^x$

35. $f(x) = 5^x$ **36.** $f(x) = x^4$

37. $f(x) = x + 2$

38. $f(x) = \begin{cases} 3x - 1 & \text{if } x \ge 2 \\ 1 - x & \text{if } x < 2 \end{cases}$

39. $f(x) = \dfrac{1}{x + 2}$ **40.** $f(x) = x^2 + 9$

41. $f(x) = \begin{cases} x - 2 & \text{if } x < 3 \\ 7 - 4x & \text{if } x \ge 3 \end{cases}$

42. $f(x) = \dfrac{x}{x^2 + 9}$ **43.** $f(x) = 3x^2 - 2x$

44. $f(x) = x^3 - x^{2/3}$ **45.** $f(x) = x^2 + x^{1/2}$

46. $f(x) = 5$

47. For the functions $y_1 = \left(\frac{1}{3}\right)^x$, $y_2 = \left(\frac{1}{2}\right)^x$, $y_3 = 2^x$, and $y_4 = 3^x$:

 a. Predict which curve will be the highest for large values of x.

 b. Predict which curve will be the lowest for large values of x.

 c. Check your predictions by graphing the functions on the window $[-3, 3]$ by $[0, 5]$.

 d. From your graph, what is the common y-intercept? Why do all such exponential functions meet at this point?

48. Graph the parabola $y_1 = 1 - x^2$ and the semicircle $y_2 = \sqrt{1 - x^2}$ on the window $[-1, 1]$ by $[0, 1]$. (You may want to adjust the window to make the semicircle look more like a semicircle.) Use TRACE to determine which is the "inside" curve (the parabola or the semicircle) and which is the "outside" curve. These graphs show that when you graph a parabola, you should draw the curve near the vertex to be slightly more "pointed" than a circular curve.

49–56. For each pair of functions $f(x)$ and $g(x)$, find
a. $f(g(x))$ **b.** $g(f(x))$ and **c.** $f(f(x))$

49. $f(x) = x^5$; $g(x) = 7x - 1$

50. $f(x) = x^8$; $g(x) = 2x + 5$

51. $f(x) = \dfrac{1}{x}$; $g(x) = x^2 + 1$

52. $f(x) = \sqrt{x}$; $g(x) = x^3 - 1$

53. $f(x) = \sqrt{x} - 1$; $g(x) = x^3 - x^2$

54. $f(x) = x^2 + 1$; $g(x) = x - \sqrt{x}$

55. $f(x) = x^2 - x$; $g(x) = \dfrac{x^3 - 1}{x^3 + 1}$

56. $f(x) = x^3 + x$; $g(x) = \dfrac{x^4 + 1}{x^4 - 1}$

57–58. For each pair of functions $f(x)$ and $g(x)$, find and *fully simplify* **a.** $f(g(x))$ and **b.** $g(f(x))$

57. $f(x) = 2x - 6$; $g(x) = \dfrac{x}{2} + 3$

58. $f(x) = x^3 + 1$; $g(x) = \sqrt[3]{x - 1}$

59–62. For each function, find and simplify $f(x + h)$.

59. $f(x) = 5x^2$

60. $f(x) = 3x^2$

61. $f(x) = 2x^2 - 5x + 1$

62. $f(x) = 3x^2 - 5x + 2$

63–74. For each function, find and simplify
$\dfrac{f(x + h) - f(x)}{h}$. (Assume $h \ne 0$.)

(See instructions on previous page.)

63. $f(x) = 5x^2$

64. $f(x) = 3x^2$

65. $f(x) = 2x^2 - 5x + 1$

66. $f(x) = 3x^2 - 5x + 2$

67. $f(x) = 7x^2 - 3x + 2$

68. $f(x) = 4x^2 - 5x + 3$

69. $f(x) = x^3$
[*Hint:* Use $(x + h)^3 = x^3 + 3x^2h + 3xh^2 + h^3$.]

70. $f(x) = x^4$
[*Hint:* Use $(x + h)^4 = x^4 + 4x^3h + 6x^2h^2 + 4xh^3 + h^4$.]

71. $f(x) = \dfrac{2}{x}$

72. $f(x) = \dfrac{3}{x}$

73. $f(x) = \dfrac{1}{x^2}$

74. $f(x) = \sqrt{x}$

[*Hint for Exercise 74:* Multiply top and bottom of the fraction by $\sqrt{x + h} + \sqrt{x}$.]

75. Find, rounding to five decimal places:

a. $\left(1 + \dfrac{1}{100}\right)^{100}$

FOR HELP GETTING STARTED
on Exercises 63–74, see pages 57–58.

b. $\left(1 + \dfrac{1}{10{,}000}\right)^{10{,}000}$

c. $\left(1 + \dfrac{1}{1{,}000{,}000}\right)^{1{,}000{,}000}$

d. Do the resulting numbers seem to be approaching a limiting value? Estimate the limiting value to five decimal places. The number that you have approximated is denoted e, and will be used extensively in Chapter 4.

76. Use the TABLE feature of your graphing calculator to evaluate $\left(1 + \dfrac{1}{x}\right)^x$ for values of x such as 100, 10,000, 1,000,000, and higher values. Do the resulting numbers seem to be approaching a limiting value? Estimate the limiting value to five decimal places. The number that you have approximated is denoted e, and will be used extensively in Chapter 4.

77. How will the graph of $y = (x + 3)^3 + 6$ differ from the graph of $y = x^3$? Check by graphing both functions together.

78. How will the graph of $y = -(x - 4)^2 + 8$ differ from the graph of $y = -x^2$? Check by graphing both functions together.

Applied Exercises

79–80. SOCIAL SCIENCES: World Population The world population (in millions) since the year 1700 is approximated by the exponential function $P(x) = 522(1.0053)^x$, where x is the number of years since 1700 (for $0 \le x \le 200$). Using a calculator, estimate the world population in the year:

79. 1750

80. 1800

Source: World Almanac

81. ECONOMICS: Income Tax The following function expresses an income tax that is 10% for incomes below $5000, and otherwise is $500 plus 30% of income in excess of $5000.

$$f(x) = \begin{cases} 0.10x & \text{if } 0 \le x < 5000 \\ 500 + 0.30(x - 5000) & \text{if } x \ge 5000 \end{cases}$$

a. Calculate the tax on an income of $3000.

b. Calculate the tax on an income of $5000.

c. Calculate the tax on an income of $10,000.

d. Graph the function.

82. ECONOMICS: Income Tax The following function expresses an income tax that is 15% for incomes below $6000, and otherwise is $900 plus 40% of income in excess of $6000.

$$f(x) = \begin{cases} 0.15x & \text{if } 0 \le x < 6000 \\ 900 + 0.40(x - 6000) & \text{if } x \ge 6000 \end{cases}$$

a. Calculate the tax on an income of $3000.

b. Calculate the tax on an income of $6000.

c. Calculate the tax on an income of $10,000.

d. Graph the function.

83–84. GENERAL: Dog-Years The usual estimate that each human-year corresponds to 7 dog-years is not very accurate for young dogs, since they quickly reach adulthood. Exercises 83 and 84 give more accurate formulas for converting human-years x into dog-years. For each conversion formula:

a. Find the number of dog-years corresponding to the following amounts of human time: 8 months, 1 year and 4 months, 4 years, 10 years.

b. Graph the function.

Source: Bull. Acad. Vet. France **26**

83. The following function expresses dog-years as $10\frac{1}{2}$ dog-years per human-year for the first 2 years and then 4 dog-years per human-year for each year thereafter.

$$f(x) = \begin{cases} 10.5x & \text{if } 0 \le x \le 2 \\ 21 + 4(x - 2) & \text{if } x > 2 \end{cases}$$

84. The following function expresses dog-years as 15 dog-years per human-year for the first year, 9 dog-years per human-year for the second year, and then 4 dog-years per human-year for each year thereafter.

$$f(x) = \begin{cases} 15x & \text{if } 0 \le x \le 1 \\ 15 + 9(x - 1) & \text{if } 1 < x \le 2 \\ 24 + 4(x - 2) & \text{if } x > 2 \end{cases}$$

85–86. BUSINESS: Isocosts and Isoquants The intersection of an isocost line $wL + rK = C$ and an isoquant curve $K = aL^b$ (see pages 18 and 31) gives the amounts of labor L and capital K for fixed production and cost. Find the intersection point (L, K) of each isocost and isoquant. [*Hint:* After substituting the second expression into the first, multiply through by L and factor.]

85. $3L + 8K = 48$ and $K = 24 \cdot L^{-1}$

86. $5L + 4K = 120$ and $K = 180 \cdot L^{-1}$

87. BUSINESS: Insurance Reserves An insurance company keeps reserves (money to pay claims) of $R(v) = 2v^{0.3}$, where v is the value of all of its policies, and the value of its policies is predicted to be $v(t) = 60 + 3t$, where t is the number of years from now. (Both R and v are in millions of dollars.) Express the reserves R as a function of t, and evaluate the function at $t = 10$.

88. BUSINESS: Research Expenditures An electronics company's research budget is $R(p) = 3p^{0.25}$, where p is the company's profit, and the profit is predicted to be $p(t) = 55 + 4t$, where t is the number of years from now. (Both R and p are in millions of dollars.) Express the research expenditure R as a function of t, and evaluate the function at $t = 5$.

89. BIOMEDICAL: Cell Growth One leukemic cell in an otherwise healthy mouse will divide into two cells every 12 hours, so that after x days the number of leukemic cells will be $f(x) = 4^x$.

  a. Find the approximate number of leukemic cells after 10 days.
  b. If the mouse will die when its body has a billion leukemic cells, will it survive beyond day 15?

  Source: Annals of NY Academy of Sciences **54**

90. BUSINESS: E-Commerce Electronic commerce or e-commerce, buying and selling over the Internet, has been growing rapidly. The total value of U.S. e-commerce in recent years in billions of dollars is given by the exponential function $f(x) = 226(1.11)^x$, where x is the number of years since 2012. Predict total e-commerce in the year 2020.

  Source: Wikipedia

91. BUSINESS: Computer Affordability As prices of computers fall and wages rise, it takes less time to earn enough money to buy a computer. The following graph shows the number of weeks that an American with an average income had to work to buy a notebook computer in recent years.

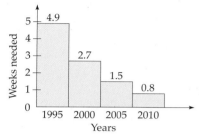

  a. Number the years (bars) with x-values 1–4 so that 1 stands for 1995, 2 stands for 2000, and so on. Use exponential regression to fit a curve to the data. State the regression formula. [*Hint:* See Example 12.]
  b. Use the regression formula to predict the number of weeks of income needed to buy a computer in 2020. [*Hint:* What number corresponds to 2020?]

  Source: Standard & Poor's, Intel

92. SOCIAL SCIENCES: Population of Texas The following table gives the population of Texas in recent years.

Year	2000	2005	2010
Population (in millions)	20.9	22.8	25.1

  a. Number the years with x-values 1–3 so that 1 stands for 2000, 2 stands for 2005, and so on. Use exponential regression to fit a curve to the data. State the regression formula. [*Hint:* See Example 12.]
  b. Use the regression formula to predict the population of Texas in 2020.

  Source: U.S. Census Bureau

Conceptual Exercises

93. How do two graphs differ if their functions are the same except that the domain of one excludes some x-values from the domain of the other?

94. Which of the following is *not* a polynomial, and why?
$$x^2 + \sqrt{2} \qquad x^{\sqrt{2}} + 1 \qquad \sqrt{2}x^2 + 1$$

95. The income tax function graphed on page 53 has segments with various slopes. What would it mean about an income tax if a segment had slope 1?

96. If $f(x) = ax$, then $f(f(x)) = ?$

97. If $f(x) = x + a$, then $f(f(x)) = ?$

98. How do the graphs of $f(x)$ and $f(x) + 10$ differ?

99. How do the graphs of $f(x)$ and $f(x + 10)$ differ?

100. How do the graphs of $f(x)$ and $f(x + 10) + 10$ differ?

101. True or False: If $f(x) = x^2$, then $f(x + h) = x^2 + h^2$.

102. True or False: If $f(x) = mx + b$, then $f(x + h) = f(x) + mh$.

Explorations and Excursions   The following problems extend and augment the material presented in the text.

Greatest Integer Function

103. For any x, the function INT(x) is defined as the greatest integer less than or equal to x. For example, INT(3.7) = 3 and INT(−4.2) = −5.
  a. Use a graphing calculator to graph the function $y_1 = \text{INT}(x)$. (You may need to graph it in DOT mode to eliminate false connecting lines.)
  b. From your graph, what are the domain and range of this function?

104. a. Use a graphing calculator to graph the function $y_1 = 2\,\text{INT}(x)$. [See the previous exercise for a definition of INT(x).]
  b. From your graph, what are the domain and range of this function?

More About Compositions

105. a. Find the composition $f(g(x))$ of the two linear functions $f(x) = ax + b$ and $g(x) = cx + d$ (for constants $a, b, c,$ and d).
  b. Is the composition of two linear functions always a linear function?

106. a. Is the composition of two quadratic functions always a quadratic function? [*Hint:* Find the composition of $f(x) = x^2$ and $g(x) = x^2$.]
  b. Is the composition of two polynomials always a polynomial?

---

1   Chapter Summary with Hints and Suggestions

Reading the text and doing the exercises in this chapter have helped you to master the following concepts and skills, which are listed by section (in case you need to review them) and are keyed to particular Review Exercises. Answers for all Review Exercises are given at the back of the book, and full solutions can be found in the Student Solutions Manual.

1.1   Real Numbers, Inequalities, and Lines

- Translate an interval into set notation and graph it on the real line. (*Review Exercises 1–4.*)

$$[a, b] \quad (a, b) \quad [a, b) \quad (a, b]$$

$$(-\infty, b] \quad (-\infty, b) \quad [a, \infty) \quad (a, \infty) \quad (-\infty, \infty)$$

- Express given information in interval form. (*Review Exercises 5–6.*)

- Find an equation for a line that satisfies certain conditions. (*Review Exercises 7–12.*)

$$m = \frac{y_2 - y_1}{x_2 - x_1} \qquad y = mx + b$$

$$y - y_1 = m(x - x_1) \qquad x = a \qquad y = b$$

$$ax + by = c$$

- Find an equation of a line from its graph. (*Review Exercises 13–14.*)

- Use straight-line depreciation to find the value of an asset. (*Review Exercises 15–16.*)

- Use real-world data to find a regression line and make a prediction. (*Review Exercise 17.*)

1.2   Exponents

- Evaluate negative and fractional exponents without a calculator. (*Review Exercises 18–25.*)

$$x^0 = 1 \qquad x^{-n} = \frac{1}{x^n} \qquad x^{m/n} = \sqrt[n]{x^m} = \left(\sqrt[n]{x}\right)^m$$

- Evaluate an exponential expression using a calculator. (*Review Exercises 26–29.*)

- Use real-world data to find a power regression curve and make a prediction. (*Review Exercise 30.*)

1.3   Functions: Linear and Quadratic

- Evaluate and find the domain and range of a function. (*Review Exercises 31–34.*)

  A function f is a rule that assigns to each number x in a set (the domain) a (single) number $f(x)$. The range is the set of all resulting values $f(x)$.

- Use the vertical line test to see if a graph defines a function. (*Review Exercises 35–36.*)

- Graph a linear function: $f(x) = mx + b$ (*Review Exercises 37–38.*)

- Graph a quadratic function: $f(x) = ax^2 + bx + c$ (*Review Exercises 39–40.*)

- Solve a quadratic equation by factoring and by the Quadratic Formula. (*Review Exercises 41–44.*)

Vertex

$$x = \frac{-b}{2a}$$

x-intercepts

$$x = \frac{-b \pm \sqrt{b^2 - 4ac}}{2a}$$

- Use a graphing calculator to graph a quadratic function. *(Review Exercises 45–46.)*

- Construct a linear function from a word problem or from real-life data, and then use the function in an application. *(Review Exercises 47–50.)*

- For given cost and revenue functions, find the break-even points and maximum profit. *(Review Exercises 51–52.)*

- Use real-world data to find a quadratic regression curve and make a prediction. *(Review Exercise 53.)*

1.4 Functions: Polynomial, Rational, and Exponential

- Evaluate and find the domain and range of a more complicated function. *(Review Exercises 54–57.)*

- Solve a polynomial equation by factoring. *(Review Exercises 58–61.)*

- Graph an exponential function. *(Review Exercise 62.)*

- Graph a "shifted" function. *(Review Exercise 63.)*

- Graph a piecewise linear function. *(Review Exercises 64–65.)*

- Given two functions, find their composition. *(Review Exercises 66–69.)*

$$f(g(x)) \qquad g(f(x))$$

- For a given function $f(x)$, find and simplify the difference quotient $\dfrac{f(x + h) - f(x)}{h}$. *(Review Exercises 70–71.)*

- Solve an applied problem involving the composition of functions. *(Review Exercise 72.)*

- Solve a polynomial equation. *(Review Exercises 73–74.)*

- Use real-world data to find an exponential regression curve and make a prediction. *(Review Exercise 75.)*

Hints and Suggestions

- *(Overview)* In reviewing this chapter, notice the difference between *geometric* objects (points, curves, etc.) and *analytic* objects (numbers, functions, etc.), and the connections between them. You should be able to express geometric objects analytically, and vice versa. For example, given a *graph* of a line, you should be able to find an *equation* for it, and given a quadratic *function*, you should be able to *graph* it.

- A graphing calculator or a computer with appropriate software can help you to *explore* a concept (for example, seeing how a curve changes as a coefficient or exponent changes), and also to *solve* a problem (for example, eliminating the point-plotting aspect of graphing, or finding a regression line or curve).

- The Practice Problems help you to check your mastery of the skills presented. Complete solutions are given at the end of each section.

- The Student Solutions Manual, available separately from your bookstore, provides fully worked-out solutions to selected exercises.

| **1** | **Review Exercises and Chapter Test** | ◯ indicates a Chapter Test exercise. |

1.1 Real Numbers, Inequalities, and Lines

1–4. Write each interval in set notation and graph it on the real line.

◯**1** $(2, 5]$ **2.** $[-2, 0)$ **3.** $[100, \infty)$ **4.** $(-\infty, 6]$

5. **GENERAL: Wind Speed** The United States Coast Guard defines a "hurricane" as winds of at least 74 mph, a "storm" as winds of at least 55 mph but less than 74 mph, a "gale" as winds of at least 38 mph but less than 55 mph, and a "small craft warning" as winds of at least 21 mph but less than 38 mph. Express each of these wind conditions in interval form. [*Hint:* A small craft warning is [21, 38).]

6. State in interval form:

   a. The set of all positive numbers.
   b. The set of all negative numbers.
   c. The set of all nonnegative numbers.
   d. The set of all nonpositive numbers.

7–12. Write an equation of the line satisfying each of the following conditions. If possible, write your answer in the form $y = mx + b$.

7. Slope 2 and passing through the point $(1, -3)$

8. Slope -3 and passing through the point $(-1, 6)$

◯**9** Vertical and passing through the point $(2, 3)$

10. Horizontal and passing through the point $(2, 3)$

◯**11** Passing through the point $(-1, 3)$ and $(2, -3)$

12. Passing through the point $(6, -1)$ and perpendicular to the line $x + 2y = 8$

13–14. Write an equation of the form $y = mx + b$ for each line graphed below.

(13)

14.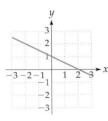

(15) **BUSINESS: Straight-Line Depreciation** A contractor buys a backhoe for $25,000 and estimates its useful life to be 8 years, after which its scrap value will be $1000.
  a. Use straight-line depreciation to find a formula for the value V of the backhoe after t years, for $0 \le t \le 8$.
  b. Use your formula to find the value of the backhoe after 4 years.

16. BUSINESS: Straight-Line Depreciation A trucking company buys a satellite communication system for $78,000 and estimates its useful life to be 15 years, after which its scrap value will be $3000.
  a. Use straight-line depreciation to find a formula for the value V of the system after t years, for $0 \le t \le 15$.
  b. Use your formula to find the value of the system after 8 years.

 17. BUSINESS: 3D Movies The number of 3D movie screens in the United States in recent years is shown in the following table.

Year	2009	2010	2011	2012
Number (in thousands)	2.6	7.7	12.5	17.8

  a. Number the data columns with x-values 1–4 (so that x stands for the *number of years since 2008*) and use linear regression to fit a line to the data and state the regression formula. [*Hint:* See Example 9 on pages 14–15.]
  b. Interpret the slope of the line.
  c. Use the regression line to predict the number of 3D movie screens in the year 2020.

Sources: Motion Picture Association of America, Standard & Poor's.

1.2 Exponents

18–25. Evaluate each expression without using a calculator.

(18) $\left(\frac{1}{6}\right)^{-2}$ 　　**19.** $\left(\frac{4}{3}\right)^{-1}$ 　　**20.** $64^{1/2}$

21. $1000^{1/3}$ 　　**22.** $81^{-3/4}$ 　　**23.** $100^{-3/2}$

24. $\left(-\frac{8}{27}\right)^{-2/3}$ 　　**25.** $\left(\frac{9}{16}\right)^{-3/2}$

26–27. Use a calculator to evaluate each expression. Round answers to two decimal places.

26. $3^{2.4}$ 　　　　　　**27.** $12^{1.9}$

28–29. ENVIRONMENTAL SCIENCE: Animal Size It is well known that larger islands or continents can support larger animals. One study found the following relationships between $x =$ land size (in square miles) and $y =$ weight (in pounds) of the "top animal" ever to live on that land mass. Use the formula to estimate the size of the top animal for:
  a. Hawaii (4000 square miles).
  b. North America (9,400,000 square miles).

28. $y = 0.86x^{0.47}$ (for cold-blooded meat-eating animals, such as lizards and crocodiles)

29. $y = 1.7x^{0.52}$ (for warm-blooded plant-eating animals, which includes many mammals)

Source: Proceedings of the National Academy of Sciences **98**

 30. BUSINESS: Gambling Revenues The operating revenues for Las Vegas Sands (the largest gaming enterprise in Nevada) for recent years are given in the following table.

Year	2009	2010	2011	2012
Operating Revenues (billion $)	4.6	6.9	9.4	11.1

  a. Number the data columns with x-values 1–4 (so that x stands for *number of years since 2008*), use power regression to fit a curve to the data, and state the regression formula. [*Hint:* See Example 7 on pages 27–28.]
  b. Use the regression function to predict the operating revenue in the year 2020.

Source: Standard & Poor's.

1.3 Functions: Linear and Quadratic

31–34. For each function:
a. Evaluate the given expression.
b. Find the domain.
 c. Find the range.

31. $f(x) = \sqrt{x - 7}$; find $f(11)$

32. $g(t) = \dfrac{1}{t + 3}$; find $g(-1)$

(33) $h(w) = w^{-3/4}$; find $h(16)$

34. $w(z) = z^{-4/3}$; find $w(8)$

35–36. Determine whether each graph defines a function of x.

(35)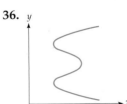

36.

37–40. Graph each function.

37 $f(x) = 4x - 8$ **38.** $f(x) = 6 - 2x$

39 $f(x) = -2x^2 - 4x + 6$ **40.** $f(x) = 3x^2 - 6x$

41–44. Solve each equation by **a.** factoring and **b.** the Quadratic Formula.

41. $3x^2 + 9x = 0$ **42.** $2x^2 - 8x - 10 = 0$

43. $3x^2 + 3x + 5 = 11$ **44.** $4x^2 - 2 = 2$

45–46. For each quadratic function:

a. Find the vertex using the vertex formula.
b. Graph the function on an appropriate window. (Answers may vary.)

45 $f(x) = x^2 - 10x - 25$ **46.** $f(x) = x^2 + 14x - 15$

47. BUSINESS: Car Rentals   A rental company rents cars for $45 per day and $0.25 per mile. Find a function $C(x)$ for the cost of a rented car driven for x miles in a day.

48. BUSINESS: Simple Interest   If money is borrowed for less than a year, the interest is often calculated as *simple* interest, according to the formula Interest $= P \cdot r \cdot t$, where P is the principal, r is the rate (expressed as a decimal), and t is the time (in years). Find a function $I(t)$ for the interest charged on a loan of $10,000 at an interest rate of 8% for t years. Simplify your answer.

49 ENVIRONMENTAL SCIENCES: Air Temperature The air temperature decreases by about 1 degree Fahrenheit for each 300 feet of altitude. Find a function $T(x)$ for the temperature at an altitude of x feet if the sea level temperature is 70°.

Source: Federal Aviation Administration

50. ENVIRONMENTAL SCIENCES: Carbon Dioxide Pollution   The burning of fossil fuels (such as oil and coal) added 20.3 billion tons of carbon dioxide to the atmosphere during 2010, and this annual amount is growing by 0.45 billion tons per year. Find a function $C(t)$ for the amount of carbon dioxide added during the year t years after 2010, and use the formula to find how soon this annual amount will reach 25 billion tons. [*Note:* Carbon dioxide traps solar heat, increasing the earth's temperature, and may lead to flooding of lowland areas by melting the polar ice.]

Source: U.S. Environmental Protection Agency

51 BUSINESS: Break-Even Points and Maximum Profit   A store that installs satellite TV receivers finds that if it installs x receivers per week, its costs will be $C(x) = 80x + 1950$ and its revenue will be $R(x) = -2x^2 + 240x$ (both in dollars).

a. Find the store's break-even points.
b. Find the number of receivers the store should install to maximize profit, and the maximum profit.

52. BUSINESS: Break-Even Points and Maximum Profit   An air conditioner outlet finds that if it sells x air conditioners per month, its costs will be $C(x) = 220x + 202,500$

and its revenue will be $R(x) = -3x^2 + 2020x$ (both in dollars).

a. Find the outlet's break-even points.
b. Find the number of air conditioners the outlet should sell to maximize profit, and the maximum profit.

53. BUSINESS: Google's Operating Revenues   Google's operating revenues in recent years are given in the following table.

Year	2009	2010	2011	2012
Revenues (billion $)	23.7	29.3	37.9	50.2

a. Number the data columns with x-values 1–4 (so that x stands for *years since 2008*), use quadratic regression to fit a curve to the data, and state the regression formula. [*Hint:* See Example 10 on pages 42–43.]
b. Use the regression function to predict Google's revenues in the year 2020.

Source: Standard & Poor's

1.4   Functions: Polynomial, Rational, and Exponential

54–57. For each function:

a. Evaluate the given expression.
b. Find the domain.
c. Find the range.

54. $f(x) = \dfrac{3}{x(x-2)}$; find $f(-1)$

55. $f(x) = \dfrac{16}{x(x+4)}$; find $f(-8)$

56. $g(x) = 9^x$; find $g\left(\dfrac{3}{2}\right)$

57. $g(x) = 8^x$; find $g\left(\dfrac{5}{3}\right)$

58–61. Solve each equation by factoring.

58 $5x^4 + 10x^3 = 15x^2$ **59.** $4x^5 + 8x^4 = 32x^3$

60. $2x^{5/2} - 8x^{3/2} = 10x^{1/2}$ **61.** $3x^{5/2} + 3x^{3/2} = 18x^{1/2}$

62–65. Graph each function.

62. $f(x) = 4^x$ **63.** $f(x) = (x-2)^2 - 4$

64 $f(x) = \begin{cases} 3x - 7 & \text{if } x \geq 2 \\ -x - 1 & \text{if } x < 2 \end{cases}$

(If you use a graphing calculator for Exercises 64 and 65, be sure to indicate any missing points.)

65. $f(x) = \begin{cases} 6 - 2x & \text{if } x > 2 \\ 2x - 1 & \text{if } x \leq 2 \end{cases}$

66–69. For each pair of functions $f(x)$ and $g(x)$, find: **a.** $f(g(x))$, **b.** $g(f(x))$.

66. $f(x) = x^2 + 1$; $g(x) = \dfrac{1}{x}$

67. $f(x) = \sqrt{x}$; $g(x) = 5x - 4$

68 $f(x) = \dfrac{x + 1}{x - 1}$; $g(x) = x^3$

69. $f(x) = 2^x$; $g(x) = x^2$

70–71. For each function, find and simplify the difference quotient $\dfrac{f(x + h) - f(x)}{h}$.

70 $f(x) = 2x^2 - 3x + 1$ **71.** $f(x) = \dfrac{5}{x}$

72. BUSINESS: Advertising Budget A company's advertising budget is $A(p) = 2p^{0.15}$, where p is the company's profit, and the profit is predicted to be $p(t) = 18 + 2t$, where t is the number of years from now. (Both A and p are in millions of dollars.) Express the advertising budget A as a function of t, and evaluate the function at $t = 4$.

73. a. Solve the equation $x^4 - 2x^3 - 3x^2 = 0$ by factoring.

  b. Use a graphing calculator to graph $y = x^4 - 2x^3 - 3x^2$ and find the x-intercepts of the graph. Be sure that you understand why your answers to parts (a) and (b) agree.

74 a. Solve the equation $x^3 + 2x^2 - 3x = 0$ by factoring.

  b. Use a graphing calculator to graph $y = x^3 + 2x^2 - 3x$ and find the x-intercepts of the graph. Be sure that you understand why your answers to parts (a) and (b) agree.

 75. SOCIAL SCIENCES: Crime Rates For several decades, crime in the United States has been declining. The numbers of property crimes per 100,000 people for recent decades are shown in the following table.

Decade	1990	2000	2010
Crimes per 100,000	5.07	3.62	2.94

a. Number the data columns with x-values 1–3 (so that x stands for the *number of decades since 1980*) and use exponential regression to fit a curve to the data. State the regression formula. [*Hint:* See Example 12 on pages 59–60.]

b. Use the regression function to predict the crime rate in the year 2020.

Source: U.S. Census Bureau

Derivatives and Their Uses

2

Yoav Levy/PhotoTake

What You'll Explore

In a world where so many things are changing, being able to calculate the *rate* of the change can be very useful. In this chapter you will see how to do this using *derivatives*, one of the two central ideas of calculus. You will learn how to calculate the rate of growth of a company's profit, the rate of decline of a city's population, the velocity and acceleration of a car or train, and the marginal cost and revenue of production for a company, all using derivatives.

Temperature, Superconductivity, and Limits

It has long been known that there is a coldest possible temperature, called **absolute zero,** the temperature of an object if all heat could be removed from it. On the Fahrenheit temperature scale, absolute zero is about 460 degrees below zero. On the "absolute" or "Kelvin" temperature scale (named after the nineteenth-century scientist Lord Kelvin), absolute zero temperature is assigned the value 0. Absolute zero is a temperature that can be *approached* but never actually *reached.* At temperatures approaching absolute zero, some metals become increasingly able to conduct electricity, with efficiencies approaching 100%, a state called *superconductivity.* The following graph gives the electrical conductivity of aluminum, showing the remarkable fact that as temperature decreases to absolute zero, aluminum becomes *superconducting*—its conductivity approaches a limit of 100% (see Exercise 83 on page 86).

Conductivity of aluminum as a function of temperature

Source: Michael Tinkham, *Introduction to Superconductivity*

If $f(t)$ stands for the percent conductivity of aluminum at temperature t degrees Kelvin, then the fact that conductivity approaches 100 as temperature decreases to zero may be written

$$\lim_{t \to 0^+} f(t) = 100 \qquad \text{Limit as } t \to 0^+ \text{ of } f(t) \text{ is } 100$$

This is an example of *limits*, which are discussed in Section 2.1.

A magnet will "float" above a superconductor, leading to many commercial applications such as high-speed "maglev" trains that float silently above the tracks.

Introduction

This chapter introduces the **derivative,** one of the most important concepts in all of calculus. We begin by discussing two preliminary topics, **limits** and **continuity,** both of which will be treated intuitively rather than formally and will be useful when we define the derivative in the next section.

Limits

The word "limit" is used in everyday conversation to describe the ultimate behavior of something, as in the "limit of one's endurance" or the "limit of one's patience." In mathematics, the word "limit" has a similar but more precise meaning. The notation $x \to 3$ (read: "x approaches 3") means that x takes values *arbitrarily close to 3 without ever equaling 3*. Given a function $f(x)$, if x approaching 3 causes the function to take values approaching (or equaling) some particular number, such as 10, then we will call 10 *the limit of the function* and write

$$\lim_{x \to 3} f(x) = 10 \qquad \text{Limit of } f(x) \text{ as } x \text{ approaches 3 is 10}$$

 Be Careful $x \to 3$ means that x takes values closer and closer to 3 *but never equals 3*.

In practice, the two simplest ways we can approach 3 are *from the left* or *from the right*. For example, the numbers 2.9, 2.99, 2.999, … approach 3 *from the left*, which we denote by $x \to 3^-$, and the numbers 3.1, 3.01, 3.001, … approach 3 *from the right*, denoted by $x \to 3^+$. Such limits are called *one-sided limits*.

$$
\begin{array}{cc}
x \to 3^- & x \to 3^+ \\
\text{(approaching 3} & \text{(approaching 3} \\
\text{from the left)} & \text{from the right)}
\end{array}
$$

| | 2.9 | | 2.99 3 3.01 | | 3.1 |

The following Example shows how to find limits from tables of values of the function.

EXAMPLE 1   FINDING A LIMIT BY TABLES

Use tables to find $\lim_{x \to 3} (2x + 4)$.   Limit of $2x + 4$ as x approaches 3

Solution

We make two tables, as shown below, one with x approaching 3 *from the left*, and the other with x approaching 3 *from the right*.

x	$2x + 4$
2.9	9.8
2.99	9.98
2.999	9.998

(Approaching 3 from the left → Limit is 10)

x	$2x + 4$
3.1	10.2
3.01	10.02
3.001	10.002

(Approaching 3 from the right → Limit is 10)

This table shows $\lim_{x \to 3^-} (2x + 4) = 10$ This table shows $\lim_{x \to 3^+} (2x + 4) = 10$

Choosing x-values even closer to 3 (such as 2.9999 or 3.0001) would result in values of $2x + 4$ *even closer to 10*, so that both one-sided limits equal 10:

$$\lim_{x \to 3^-} (2x + 4) = 10 \qquad \text{and} \qquad \lim_{x \to 3^+} (2x + 4) = 10$$

Since approaching 3 from *either* side causes $2x + 4$ to approach the same number, 10, we may state that *the limit* is 10:

$$\lim_{x \to 3} (2x + 4) = 10 \qquad \text{Limit of } 2x + 4 \text{ as } x \text{ approaches 3 is 10}$$

Take Note

Saying that the limit *equals* 10 means that 10 is the only number that values of $2x + 4$ get arbitrary close to as x approaches 3.

This limit says that as x approaches 3 from *either side*, or even *alternating sides*, as in 2.9, 3.01, 2.999, 3.0001, ... , the values of $2x + 4$ will approach 10. This may be seen in the succession of graphs below: as $x \to 3$ (on the x-axis), $f(x) \to 10$ (on the y-axis).

Limits can be defined with greater rigor,* but we will use the following intuitive definition to express the idea that $\lim_{x \to c} f(x) = L$ means that $f(x)$ *approaches the number L as x approaches the number c.*

LOOKING AHEAD

Limits will be used on pages 88–92 to define the *derivative*.

Limits

The statement

$$\lim_{x \to c} f(x) = L \qquad \text{Limit of } f(x) \text{ as } x \text{ approaches } c \text{ is } L$$

means that the values of $f(x)$ can be made arbitrarily close to L by taking values of x sufficiently close (but not equal) to c.

The one-sided limits

$$\lim_{x \to c^-} f(x) = L \qquad \text{Limit of } f(x) \text{ as } x \text{ approaches } c \text{ from the } \textit{left} \text{ is } L$$

$$\lim_{x \to c^+} f(x) = L \qquad \text{Limit of } f(x) \text{ as } x \text{ approaches } c \text{ from the } \textit{right} \text{ is } L$$

have similar meanings but with the x-values restricted to, respectively, $x < c$ and $x > c$.

The limit $\lim_{x \to c} f(x)$ is sometimes called a *two-sided limit* to distinguish it from one-sided limits. A limit may fail to exist, as we will see in Example 5, but if the limit *does* exist, it must be a *single number*. As we saw in the preceding Example if both one-sided limits exist and have the same value, then the (two-sided) limit will exist and *have this same value*. In the Application Preview on page 71, we used the notation $t \to 0^+$ to indicate the one-sided limit as the temperature approached absolute zero through positive values.

The correct limit in Example 1 could have been found simply by *evaluating* the function at $x = 3$:

$$f(3) = 2 \cdot 3 + 4 = 10 \qquad \begin{array}{l} f(x) = 2x + 4 \text{ evaluated at } x = 3 \\ \text{gives the correct limit, } 10 \end{array}$$

However, finding limits by this technique of *direct substitution* is not always possible, as the next Example shows.

---

*More precisely, the definition of $\lim_{x \to c} f(x) = L$ is that for every number $\varepsilon > 0$ there is a number $\delta > 0$ such that $|f(x) - L| < \varepsilon$ whenever $0 < |x - c| < \delta$. Notice that the last condition excludes $x = c$, so the behavior of f at c has no bearing on the limit as x approaches c.

EXAMPLE 2	FINDING A LIMIT BY TABLES

Find $\lim\limits_{x \to 0} (1 + x)^{1/x}$ correct to three decimal places.

Solution

As before, we make tables with x approaching 0 (the given x-value) from the left and from the right, with the values of $(1 + x)^{1/x}$ found from a calculator.

x	$(1 + x)^{1/x}$	x	$(1 + x)^{1/x}$
0.1	2.594	−0.1	2.868
0.01	2.705	−0.01	2.732
0.001	2.717	−0.001	2.720
0.0001	2.718	−0.0001	2.718

To use a graphing calculator to find these numbers, enter the function as $(1 + x)^{1/x}$, and use the TABLE feature. See Appendix A, page A2.

Since the last row of the tables show that the two one-sided limits have the same value, this common value of the right and left limits is *the limit*:

$$\lim_{x \to 0} (1 + x)^{1/x} \approx 2.718$$

From the agreement in the last columns of the tables

This limit may be seen from the graph of $f(x) = (1 + x)^{1/x}$ shown on the left.

Notice that the limiting point on the y-axis is *missing* since the function is not defined at $x = 0$ (because the exponent would be 1/0).

$y = (1 + x)^{1/x}$

$(0, e)$

PRACTICE PROBLEM 1

Evaluate $(1 + x)^{1/x}$ at $x = 0.000001$ and $x = -0.000001$. Based on the results, give a better approximation for $\lim\limits_{x \to 0} (1 + x)^{1/x}$. Solution on page 82 >

The actual value of this particular limit is the number that was discussed on page 51, $e \approx 2.71828$, which will be very important in our later work. Since $(1 + x)^{1/x}$ cannot be evaluated *at* $x = 0$, this limit *requires* the limit process.

Which limits can be evaluated by **direct substitution** (as in Example 1) and which cannot (as in Example 2)? The answer comes from the following "Rules of Limits."

Rules of Limits

For any constants a and c, and any positive integer n:

1. $\lim\limits_{x \to c} a = a$ The limit of a constant is just the constant

2. $\lim\limits_{x \to c} x^n = c^n$ The limit of a power is the power of the limit

3. $\lim\limits_{x \to c} \sqrt[n]{x} = \sqrt[n]{c}$ $(c > 0$ if n is even$)$ The limit of a root is the root of the limit

4. If $\lim\limits_{x \to c} f(x)$ and $\lim\limits_{x \to c} g(x)$ both exist, then

    a. $\lim\limits_{x \to c} [f(x) + g(x)] = \lim\limits_{x \to c} f(x) + \lim\limits_{x \to c} g(x)$ The limit of a sum is the sum of the limits

b. $\lim\limits_{x \to c} [f(x) - g(x)] = \lim\limits_{x \to c} f(x) - \lim\limits_{x \to c} g(x)$ The limit of a difference is the difference of the limits

c. $\lim\limits_{x \to c} [f(x) \cdot g(x)] = [\lim\limits_{x \to c} f(x)] \cdot [\lim\limits_{x \to c} g(x)]$ The limit of a product is the product of the limits

d. $\lim\limits_{x \to c} \dfrac{f(x)}{g(x)} = \dfrac{\lim\limits_{x \to c} f(x)}{\lim\limits_{x \to c} g(x)}$ $(\text{if } \lim\limits_{x \to c} g(x) \neq 0)$ The limit of a quotient is the quotient of the limits

These rules, which may be proved from the definition of limit, can be summarized as follows.

Summary of Rules of Limits

For functions composed of additions, subtractions, multiplications, divisions, powers, and roots, limits may be evaluated by direct substitution, provided that the resulting expression is defined.

$$\lim\limits_{x \to c} f(x) = f(c)$$ Limit evaluated by direct substitution

The *Rules of Limits* and also the above *summary* hold for one-sided limits as well as for regular (two-sided) limits.

EXAMPLE 3   FINDING LIMITS BY DIRECT SUBSTITUTION

a. $\lim\limits_{x \to 4} \sqrt{x} = \sqrt{4} = 2$ Direct substitution of $x = 4$ using Rule 3 or the Summary

b. $\lim\limits_{x \to 6} \dfrac{x^2}{x + 3} = \dfrac{6^2}{6 + 3} = \dfrac{36}{9} = 4$ Direct substitution of $x = 6$ (Rules 4, 2, and 1 or the Summary)

c. $\lim\limits_{h \to 0} \dfrac{-1}{(2 + h) \cdot 2} = \dfrac{-1}{(2 + 0) \cdot 2} = \dfrac{-1}{2 \cdot 2} = -\dfrac{1}{4}$ Direct substitution of $h = 0$

LOOKING AHEAD

This limit will be used on page 91.

PRACTICE PROBLEM 2

Find $\lim\limits_{x \to 3} (2x^2 - 4x + 1)$.   Solution on page 82 >

Direct substitution into a quotient may give the *indeterminate form* $\frac{0}{0}$, which is undefined. However, the limit may still exist and may often be found by factoring, simplifying, and *then* using direct substitution.

EXAMPLE 4   FINDING A LIMIT BY SIMPLIFYING

Find $\lim\limits_{x \to 1} \dfrac{x^2 - 1}{x - 1}$.

Solution

Direct substitution of $x = 1$ into $\dfrac{x^2 - 1}{x - 1}$ gives the indeterminate form $\frac{0}{0}$, which is undefined. But factoring and simplifying give

LOOKING BACK

On page 50 we saw that $\dfrac{x^2 - 1}{x - 1}$ and $x + 1$ are not the same. But their *limits* are the same, since $x \to 1$ uses values close but not equal to 1.

$$\lim_{x \to 1} \frac{x^2 - 1}{x - 1} = \lim_{x \to 1} \frac{(x+1)(x-1)}{x - 1} = \lim_{x \to 1} \frac{(x+1)(x-1)}{x-1} = \lim_{x \to 1} (x + 1) = 2$$

| Factoring the numerator | Canceling the $(x - 1)$'s (since $x \neq 1$) | Now use direct substitution |

Therefore, the limit *does* exist and equals 2.

Graphing Calculator Exploration

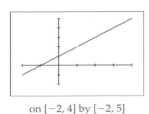

on $[-2, 4]$ by $[-2, 5]$

The graph of $f(x) = \dfrac{x^2 - 1}{x - 1}$ (from Example 4) is shown on the left.

a. Can you explain why the graph appears to be a straight line?

b. From your knowledge of rational functions, should the graph really be a (complete) line? [*Hint:* See page 50.] Can you see that a point is indeed missing from the graph?*

c. Use the graph on the left (where each tick mark is 1 unit) or enter the function and use TRACE on your calculator to verify that the limit as x approaches 1 is indeed 2.

*To have your calculator show the point as missing, choose a window such that the x-value in question is midway between XMIN and XMAX, or see the *Useful Hint* in *Graphing Calculator Basics* in Appendix A on page A2.

PRACTICE PROBLEM 3

Find $\lim\limits_{x \to 5} \dfrac{2x^2 - 10x}{x - 5}$

Solution on page 82 >

A limit (or the fact that a limit does not exist) may be found from a graph.

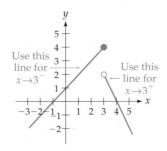

Use this line for $x \to 3^-$

Use this line for $x \to 3^+$

EXAMPLE 5   FINDING THAT A LIMIT DOES NOT EXIST

For the piecewise linear function $f(x) = \begin{cases} x + 1 & \text{if } x \leq 3 \\ 8 - 2x & \text{if } x > 3 \end{cases}$ graphed on the left, find the following limits or state that they do not exist.

a. $\lim\limits_{x \to 3^-} f(x)$ **b.** $\lim\limits_{x \to 3^+} f(x)$ **c.** $\lim\limits_{x \to 3} f(x)$

We give two solutions, one using the graph and one using the expression for the function. Both methods are important.

Solution (Using the graph)

a. $\lim\limits_{x \to 3^-} f(x) = 4$

Approaching 3 *from the left* means using the line on the *left* of $x = 3$, which approaches height 4

b. $\lim\limits_{x \to 3^+} f(x) = 2$

Approaching 3 *from the right* means using the line on the *right* of $x = 3$, which approaches height 2

c. $\lim\limits_{x \to 3} f(x)$ does not exist

The two one-sided limits both exist, but they have different values (4 and 2), so the limit does not exist

Solution $\left(\text{Using}\quad f(x) = \begin{cases} x+1 & \text{if } x \le 3 \\ 8-2x & \text{if } x > 3 \end{cases} \right)$

a. For $x \to 3^-$ we have $x < 3$, so $f(x)$ is given by the *upper* line of the function:

$$\lim_{x \to 3^-} f(x) = \lim_{x \to 3^-} (x+1) = 3+1 = 4 \qquad f(x) = \begin{cases} x+1 & \text{if } x \le 3 \\ 8-2x & \text{if } x > 3 \end{cases}$$

b. For $x \to 3^+$ we have $x > 3$, so $f(x)$ is given by the *lower* line of the function:

$$\lim_{x \to 3^+} f(x) = \lim_{x \to 3^+} (8-2x) = 8 - 2 \cdot 3 = 2 \qquad f(x) = \begin{cases} x+1 & \text{if } x \le 3 \\ 8-2x & \text{if } x > 3 \end{cases}$$

c. $\lim\limits_{x \to 3} f(x)$ does not exist since although the two one-sided limits exist, they are not equal.

Notice that the two methods found the same answers.

PRACTICE PROBLEM 4

Explain the difference among $x \to 3^-$, $x \to -3$, and $x \to -3^-$.

Solution on page 83 >

Limits Involving Infinity

A limit statement such as $\lim\limits_{x \to c} f(x) = \infty$ (the symbol ∞ is read "infinity") does *not* mean that the function takes values near "infinity," since there is no number "infinity." It means, instead, that the values of the function *become arbitrarily large* (the graph rises arbitrarily high) near the number c. Similarly, the limit statement $\lim\limits_{x \to c} f(x) = -\infty$ means that the values of the function *become arbitrarily small* (the graph falls arbitrarily low) near the number c. Such statements, which may also be written with one-sided limits, mean that the graph has a *vertical asymptote* at $x = c$, as in the following Example.

EXAMPLE 6   FINDING LIMITS INVOLVING $\pm\infty$

Describe the asymptotic behavior of $f(x) = \dfrac{3x+2}{x-2}$ using limits involving $\pm\infty$.

Solution

The graph of $f(x) = \dfrac{3x+2}{x-2}$ is shown on the next page. It is undefined at $x = 2$ (the denominator would be zero), as indicated by the vertical dashed line. Notice that as x approaches 2 *from the right*, the curve rises arbitrarily *high*, and as x approaches 2 *from the left*, the curve falls arbitrarily *low*, as expressed in the following limit statements. The tables on the right show these results numerically.

Graph of $f(x) = \dfrac{3x + 2}{x - 2}$

x	$\dfrac{3x + 2}{x - 2}$
2.1	83
2.01	803
2.001	8003

This table shows $\lim\limits_{x \to 2^{+}} f(x) = \infty$

x	$\dfrac{3x + 2}{x - 2}$
1.9	−77
1.99	−797
1.999	−7997

This table shows $\lim\limits_{x \to 2^{-}} f(x) = -\infty$

Since the limit from one side was ∞ and from the other was $-\infty$, the (two-sided) limit $\lim\limits_{x \to 2} f(x)$ *does not exist.* (If *both* one-sided limits had yielded ∞, we could have stated $\lim\limits_{x \to 2} f(x) = \infty$, and if both had yielded $-\infty$ we could have stated $\lim\limits_{x \to 2} f(x) = -\infty$.)

For x-values *arbitrarily far out to the right* (denoted $x \to \infty$), the curve levels off approaching height 3, which we express as $\lim\limits_{x \to \infty} f(x) = 3$, and for x-values *arbitrarily far out to the left* (denoted $x \to -\infty$), the curve again levels off at height 3, which we express as $\lim\limits_{x \to -\infty} f(x) = 3$. These results are shown below both graphically and numerically: the graph has a *horizontal asymptote* at $y = 3$ and the tables have (rounded) function values approaching 3.

Graph of $f(x) = \dfrac{3x + 2}{x - 2}$

x	$\dfrac{3x + 2}{x - 2}$
100	3.082
1000	3.008
10000	3.001

This table shows $\lim\limits_{x \to \infty} f(x) = 3$

x	$\dfrac{3x + 2}{x - 2}$
−100	2.922
−1000	2.992
−10000	2.999

This table shows $\lim\limits_{x \to -\infty} f(x) = 3$

Summarizing:

Limits Involving Infinity

$\lim\limits_{x \to c^{+}} f(x) = \infty$	means that the values of $f(x)$ grow arbitrarily large as x approaches c from the right
$\lim\limits_{x \to c^{-}} f(x) = \infty$	means that the values of $f(x)$ grow arbitrarily large as x approaches c from the left
$\lim\limits_{x \to c} f(x) = \infty$	means that *both* of the above statements are true

Similar statements hold if ∞ is replaced by $-\infty$ and the words "arbitrarily large" by "arbitrarily small."

$$\lim_{x \to \infty} f(x) = L \qquad \text{means that the values of } f(x) \text{ become arbitrarily close to the number } L \text{ as } x \text{ becomes arbitrarily large}$$

$$\lim_{x \to -\infty} f(x) = L \qquad \text{means that the values of } f(x) \text{ become arbitrarily close to the number } L \text{ as } x \text{ becomes arbitrarily small}$$

Be Careful To say that a limit *exists* means that the limit is a *number,* and since ∞ and $-\infty$ are not numbers, a statement such as $\lim\limits_{x \to c} f(x) = \infty$ means that *the limit does not exist.* The limit statement $\lim\limits_{x \to c} f(x) = \infty$ goes further to explain *why* the limit does not exist: the function values become arbitrarily large and so do not approach any limit.

PRACTICE PROBLEM 5

Use limits involving $\pm\infty$ to describe the asymptotic behavior of the function $f(x) = \dfrac{1}{(x + 2)^2}$ from its graph.

Solution on page 83 >

Limits of Functions of Two Variables

In the next section we will be finding limits of functions of *two* variables, with only one variable approaching a limit.

EXAMPLE 7   **FINDING A LIMIT OF A FUNCTION OF TWO VARIABLES**

Find $\lim\limits_{h \to 0} (4x + 2h - 9)$.

Solution

Only h is approaching zero, so x remains unchanged. According to the Rules of Limits on pages 74–75, we may evaluate the limit by direct substitution of $h = 0$:

$$\lim_{h \to 0} (4x + 2h - 9) = 4x + 2 \cdot \underbrace{0}_{0} - 9 = 4x - 9$$

LOOKING AHEAD

This limit will be used on page 92.

PRACTICE PROBLEM 6

Find $\lim\limits_{h \to 0} (3x^2 + 3xh + h^2)$.

Solution on page 83 >

Continuity

Intuitively, a function is said to be **continuous** at c if its graph passes through the point at $x = c$ *without a "hole" or a "jump."* For example, the first function on the next page is *continuous* at c (it has no hole or jump at $x = c$), while the second and third are **discontinuous** at c.

In other words, a function is *continuous at c* if the curve *approaches the point at* $x = c$, which may be stated in terms of limits:

$$\lim_{x \to c} f(x) = f(c)$$

Height of the *curve* approaches the height of the *point*

This equation means that the quantities on both sides must exist and be *equal*, which we make explicit as follows:

Continuity

A function *f* is continuous at *c* if the following three conditions hold:

    1. $f(c)$ is defined   Function is *defined* at *c*

    2. $\lim_{x \to c} f(x)$ exists   Left and right limits exist and agree

    3. $\lim_{x \to c} f(x) = f(c)$ Limit and value *at c* agree

f is *discontinuous* at *c* if one or more of these conditions *fails* to hold.

Condition 3, which is just the statement that the expressions in Conditions 1 and 2 are equal to each other, may by itself be taken as the definition of continuity.

EXAMPLE 8   FINDING DISCONTINUITIES FROM A GRAPH

Each function below is *discontinuous at c* for the indicated reason.

$f(c)$ is not defined
(the point at $x = c$ is missing)

$\lim_{x \to c} f(x)$ does not exist

(the left and right limits do not agree)

$\lim_{x \to c} f(x) \neq f(c)$

(the limit exists and $f(c)$ is defined, but they are *not* equal)

PRACTICE PROBLEM 7

For each graph below, determine whether the function is continuous at *c*. If it is *not* continuous, indicate the *first* of the three conditions in the definition of continuity (given above) that is violated.

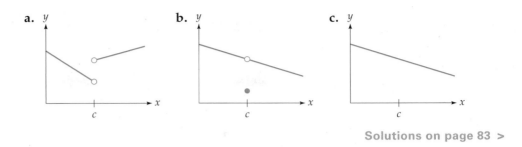

Solutions on page 83 >

Continuity on Intervals

A function is continuous on an *open interval* (a, b) if it is continuous at each point of the interval. A function is continuous on a *closed interval* $[a, b]$ if it is continuous on the open interval (a, b) and has "one-sided continuity" at the endpoints: $\lim_{x \to a^+} f(x) = f(a)$ and $\lim_{x \to b^-} f(x) = f(b)$. A function that is continuous on the entire real line $(-\infty, \infty)$ is said to be *continuous everywhere*, or simply *continuous*.

Which Functions Are Continuous?

Which functions are continuous? *Linear* and *quadratic* functions are continuous, since their graphs are, respectively, straight lines and parabolas, with no holes or jumps. Similarly, *exponential* functions are continuous and *logarithmic* functions are continuous on their domains, as may be seen from the graphs on page 51. These and other continuous functions can be combined as follows to give other continuous functions.

Continuous Functions

If functions f and g are continuous at c, then the following are also continuous at c:

1. $f \pm g$ — Sums and differences of continuous functions are continuous

2. $a \cdot f$ [for any constant a] — Constant multiples of continuous functions are continuous

3. $f \cdot g$ — Products of continuous functions are continuous

4. f/g [if $g(c) \neq 0$] — Quotients of continuous functions are continuous

5. $f(g(x))$ [for f continuous at $g(c)$] — Compositions of continuous functions are continuous

These statements, which can be proved from the Rules of Limits, show that the following types of functions are continuous:

Every polynomial function is continuous.

Every rational function is continuous except where the denominator is zero.

EXAMPLE 9   DETERMINING CONTINUITY

Determine whether each function is continuous or discontinuous. If discontinuous, state *where* it is discontinuous.

a. $f(x) = x^3 - 3x^2 - x + 3$ **b.** $f(x) = \dfrac{1}{(x+1)^2}$ **c.** $f(x) = e^{x^2-1}$

Solution

a. The first function is continuous since it is a polynomial.

b. The second function is discontinuous at $x = -1$ (where the denominator is zero). It is continuous at all other x-values.

c. The third function is continuous since it is the composition of the exponential function e^x and the polynomial $x^2 - 1$.

Calculator-drawn graphs of the functions in Example 9 are shown below. The polynomial (on the left) and the exponential function (on the right) are continuous, although you can't really tell from such graphs. The rational function (in the middle) exhibits the discontinuity at $x = -1$.

$f(x) = x^3 - 3x^2 - x + 3$
on $[-5, 5]$ by $[-5, 10]$

$f(x) = \dfrac{1}{(x+1)^2}$
on $[-5, 5]$ by $[-5, 10]$

$f(x) = e^{x^2-1}$
on $[-2, 2]$ by $[-1, 4]$

Relationship Between Limits and Continuity

We first defined continuity intuitively, saying that a function is continuous at c if its graph passes through the point at $x = c$ without a "hole" or a "jump." We then used limits to define continuity more precisely, saying that to be continuous at c, f must satisfy

$$\lim_{x \to c} f(x) = f(c)$$

However, this is just the equation that we encountered on page 75 in finding which limits may be evaluated by *direct substitution*. We may now use continuity to restate this result more succinctly: The limit $\lim_{x \to c} f(x)$ may be evaluated by direct substitution if and only if f is continuous at c.

Solutions TO PRACTICE PROBLEMS

1. $\lim_{x \to 0} (1 + x)^{1/x} \approx 2.71828$

2. $\lim_{x \to 3} (2x^2 - 4x + 1) = 2 \cdot 3^2 - 4 \cdot 3 + 1 = 18 - 12 + 1 = 7$

3. $\lim_{x \to 5} \dfrac{2x^2 - 10x}{x - 5} = \lim_{x \to 5} \dfrac{2x(x - 5)}{x - 5} = \lim_{x \to 5} \dfrac{2x(x - 5)}{x - 5} = \lim_{x \to 5} 2x = 10$

4. $x \to 3^-$ means: x approaches (positive) 3 *from the left*.

$x \to -3$ means: x approaches -3 (the ordinary two-sided limit).

$x \to -3^-$ means: x approaches -3 *from the left*.

5. $\lim\limits_{x \to -\infty} f(x) = 0$; $\lim\limits_{x \to 2^-} f(x) = \infty$ and $\lim\limits_{x \to 2^+} f(x) = \infty$, so that $\lim\limits_{x \to 2} f(x) = \infty$; and

$\lim\limits_{x \to \infty} f(x) = 0$.

6. $\lim\limits_{h \to 0} (3x^2 + 3xh + h^2) = 3x^2 + 3x \cdot 0 + 0^2$ (using direct substitution)

$= 3x^2$

LOOKING AHEAD

This limit will be used on page 94.

7. a. Discontinuous, $f(c)$ is not defined

b. Discontinuous, $\lim\limits_{x \to c} f(x) \neq f(c)$ **c.** Continuous

2.1   Section Summary

Intuitively, the limit $\lim\limits_{x \to c} f(x)$ is the number (if it exists) that $f(x)$ approaches as x approaches the number c. More precisely, $\lim\limits_{x \to c} f(x) = L$ if the values of f can be made as close as desired to L by taking values of x sufficiently close (but not equal to) c. We may find limits by letting x approach c from just one side and then from the other: The limit exists only if these two *one-sided limits* agree (that is, have the same value). Other ways to find limits are by direct substitution (if the function is continuous) or by algebraic simplification. We then used limits with the symbols ∞ and $-\infty$ to indicate that a value becomes arbitrarily large or arbitrarily small.

We defined continuity geometrically (no holes or jumps) and analytically ($\lim\limits_{x \to c} f(x) = f(c)$). We discussed continuity for two of the most useful types of functions: Polynomials are continuous everywhere, and rational functions are continuous everywhere except where their denominators are zero.

2.1   Exercises

1–4. Complete the tables and use them to find the given limits. Round calculations to three decimal places.

FOR HELP GETTING STARTED

See Example 2 on page 74

1. a. $\lim\limits_{x \to 2^-} (5x - 7)$

b. $\lim\limits_{x \to 2^+} (5x - 7)$

c. $\lim\limits_{x \to 2} (5x - 7)$

X	5x − 7		X	5x − 7
1.9			2.1	
1.99			2.01	
1.999			2.001	

2. a. $\lim\limits_{x \to 4^-} (2x + 1)$

b. $\lim\limits_{x \to 4^+} (2x + 1)$

c. $\lim\limits_{x \to 4} (2x + 1)$

X	2x + 1		X	2x + 1
3.9			4.1	
3.99			4.01	
3.999			4.001	

3. a. $\lim\limits_{x \to 1^-} \left(\dfrac{x^3 - 1}{x - 1} \right)$

b. $\lim\limits_{x \to 1^+} \left(\dfrac{x^3 - 1}{x - 1} \right)$

c. $\lim\limits_{x \to 1} \left(\dfrac{x^3 - 1}{x - 1} \right)$

X	$\dfrac{X^3 - 1}{X - 1}$		X	$\dfrac{X^3 - 1}{X - 1}$
0.9			1.1	
0.99			1.01	
0.999			1.001	

4. a. $\lim\limits_{x \to 1^-} \left(\dfrac{x^4 - 1}{x - 1} \right)$

b. $\lim\limits_{x \to 1^+} \left(\dfrac{x^4 - 1}{x - 1} \right)$

c. $\lim\limits_{x \to 1} \left(\dfrac{x^4 - 1}{x - 1} \right)$

X	$\dfrac{X^4 - 1}{X - 1}$		X	$\dfrac{X^4 - 1}{X - 1}$
0.9			1.1	
0.99			1.01	
0.999			1.001	

5–8. Find each limit by constructing tables similar to those in Exercises 1–4.

5. $\lim\limits_{x \to 0} (1 + 2x)^{1/x}$

6. $\lim\limits_{x \to 0} (1 - x)^{1/x}$

7. $\lim\limits_{x \to 2} \dfrac{\frac{1}{x} - \frac{1}{2}}{x - 2}$

8. $\lim\limits_{x \to 1} \dfrac{\sqrt{x} - 1}{x - 1}$

9–12. Find each limit by graphing the function and using TRACE or TABLE to examine the graph near the indicated x-value.

9. $\lim\limits_{x \to 1} \dfrac{\frac{1}{x} - 1}{1 - x}$

Use window $[0, 2]$ by $[0, 5]$.

10. $\lim\limits_{x \to 1.5} \dfrac{2x^2 - 4.5}{x - 1.5}$

Use window $[0, 3]$ by $[0, 10]$.

11. $\lim\limits_{x \to 4} \dfrac{x^{1.5} - 4x^{0.5}}{x^{1.5} - 2x}$ 12. $\lim\limits_{x \to 1} \dfrac{x - 1}{x - \sqrt{x}}$

[*Hint:* Choose a window whose *x*-values are centered at the limiting *x*-value.]

13–32. Find the following limits *without* using a graphing calculator or making tables.

13. $\lim\limits_{x \to 3} (4x^2 - 10x + 2)$ 14. $\lim\limits_{x \to 7} \dfrac{x^2 - x}{2x - 7}$

15. $\lim\limits_{x \to 5} \dfrac{3x^2 - 5x}{7x - 10}$ 16. $\lim\limits_{t \to 3} \sqrt[3]{t^2 + t - 4}$

17. $\lim\limits_{x \to 3} \sqrt{2}$ 18. $\lim\limits_{q \to 9} \dfrac{8 + 2\sqrt{q}}{8 - 2\sqrt{q}}$

19. $\lim\limits_{t \to 25} [(t + 5)t^{-1/2}]$ 20. $\lim\limits_{s \to 4} (s^{3/2} - 3s^{1/2})$

21. $\lim\limits_{h \to 0} (5x^3 + 2x^2 h - xh^2)$ 22. $\lim\limits_{h \to 0} (2x^2 + 4xh + h^2)$

23. $\lim\limits_{x \to 2} \dfrac{x^2 - 4}{x - 2}$ 24. $\lim\limits_{x \to 1} \dfrac{x - 1}{x^2 + x - 2}$

25. $\lim\limits_{x \to -3} \dfrac{x + 3}{x^2 + 8x + 15}$ 26. $\lim\limits_{x \to -4} \dfrac{x^2 + 9x + 20}{x + 4}$

27. $\lim\limits_{x \to -1} \dfrac{3x^3 - 3x^2 - 6x}{x^2 + x}$ 28. $\lim\limits_{x \to 0} \dfrac{x^2 - x}{x^2 + x}$

29. $\lim\limits_{h \to 0} \dfrac{2xh - 3h^2}{h}$ 30. $\lim\limits_{h \to 0} \dfrac{5x^4 h - 9xh^2}{h}$

31. $\lim\limits_{h \to 0} \dfrac{4x^2 h + xh^2 - h^3}{h}$ 32. $\lim\limits_{h \to 0} \dfrac{x^2 h - xh^2 + h^3}{h}$

33–36. For each piecewise linear function $f(x)$ graphed below, find:

a. $\lim\limits_{x \to 2^-} f(x)$ b. $\lim\limits_{x \to 2^+} f(x)$ c. $\lim\limits_{x \to 2} f(x)$

33.

34.

35.

36.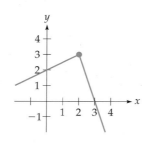

37–40. For each piecewise linear function, find:

a. $\lim\limits_{x \to 4^-} f(x)$ b. $\lim\limits_{x \to 4^+} f(x)$ c. $\lim\limits_{x \to 4} f(x)$

37. $f(x) = \begin{cases} 3 - x & \text{if } x \le 4 \\ 10 - 2x & \text{if } x > 4 \end{cases}$

38. $f(x) = \begin{cases} 5 - x & \text{if } x < 4 \\ 2x - 5 & \text{if } x \ge 4 \end{cases}$

39. $f(x) = \begin{cases} 2 - x & \text{if } x \le 4 \\ x - 6 & \text{if } x > 4 \end{cases}$

40. $f(x) = \begin{cases} 2 - x & \text{if } x < 4 \\ 2x - 10 & \text{if } x \ge 4 \end{cases}$

41–44. For each function, find:

a. $\lim\limits_{x \to 0^-} f(x)$ b. $\lim\limits_{x \to 0^+} f(x)$ c. $\lim\limits_{x \to 0} f(x)$

41. $f(x) = |x|$ 42. $f(x) = -|x|$

43. $f(x) = \dfrac{|x|}{x}$ 44. $f(x) = -\dfrac{|x|}{x}$

45–52. Use limits involving $\pm\infty$ to describe the asymptotic behavior of each function from its graph.

45. $f(x) = \dfrac{1}{3 - x}$ 46. $f(x) = \dfrac{1}{(x + 3)^2}$

47. $f(x) = \dfrac{1}{x^2}$ 48. $f(x) = \dfrac{1}{x}$

49. $f(x) = \dfrac{2x + 1}{x - 1}$ 50. $f(x) = \dfrac{x - 3}{x + 3}$

51. $f(x) = \dfrac{x^2}{(x+2)^2}$ **52.** $f(x) = \dfrac{2x^2}{(x-1)^2}$

53–60. Determine whether each function is continuous or discontinuous at *c.* If it is discontinuous, indicate the *first* of the three conditions in the definition of continuity (page 80) that is violated.

53.   **54.**

55.   **56.**

57.   **58.**

59.   **60.**

61–64. For each piecewise linear function:

a. Draw its graph (by hand or using a graphing calculator).

b. Find the limits as *x* approaches 3 from the left and from the right.

c. Is it continuous at $x = 3$? If not, indicate the first of the three conditions in the definition of continuity (page 80) that is violated.

61. $f(x) = \begin{cases} x & \text{if } x \le 3 \\ 6 - x & \text{if } x > 3 \end{cases}$

62. $f(x) = \begin{cases} 5 - x & \text{if } x \le 3 \\ x - 2 & \text{if } x > 3 \end{cases}$

63. $f(x) = \begin{cases} x & \text{if } x \le 3 \\ 7 - x & \text{if } x > 3 \end{cases}$

64. $f(x) = \begin{cases} 5 - x & \text{if } x \le 3 \\ x - 1 & \text{if } x > 3 \end{cases}$

65–78. Determine whether each function is continuous or discontinuous. If discontinuous, state where it is discontinuous.

65. $f(x) = 7x - 5$

66. $f(x) = 5x^3 - 6x^2 + 2x - 4$

67. $f(x) = \dfrac{x+1}{x-1}$ **68.** $f(x) = \dfrac{x^3}{(x+7)(x-2)}$

69. $f(x) = \dfrac{12}{5x^3 - 5x}$ **70.** $f(x) = \dfrac{x+2}{x^4 - 3x^3 - 4x^2}$

71. $f(x) = \begin{cases} 3 - x & \text{if } x \le 4 \\ 10 - 2x & \text{if } x > 4 \end{cases}$
 [*Hint:* See Exercise 37.]

72. $f(x) = \begin{cases} 5 - x & \text{if } x < 4 \\ 2x - 5 & \text{if } x \ge 4 \end{cases}$
 [*Hint:* See Exercise 38.]

73. $f(x) = \begin{cases} 2 - x & \text{if } x \le 4 \\ x - 6 & \text{if } x > 4 \end{cases}$
 [*Hint:* See Exercise 39.]

74. $f(x) = \begin{cases} 2 - x & \text{if } x < 4 \\ 2x - 10 & \text{if } x \ge 4 \end{cases}$
 [*Hint:* See Exercise 40.]

75. $f(x) = |x|$
 [*Hint:* See Exercise 41.]

76. $f(x) = \dfrac{|x|}{x}$
 [*Hint:* See Exercise 43.]

 77. $f(x) = \begin{cases} x^2 & \text{if } x \le 2 \\ 6 - x & \text{if } 2 < x < 6 \\ 2x - 17 & \text{if } x \ge 6 \end{cases}$

[*Hint for Exercises 77 and 78:* You may have graphed these piecewise nonlinear functions in Exercises 31 and 32 on page 62.]

 78. $f(x) = \begin{cases} 4 - x^2 & \text{if } x < 3 \\ 2x - 11 & \text{if } 3 \le x < 7 \\ 8 - x & \text{if } x \ge 7 \end{cases}$

79. By canceling the common factor,
$\dfrac{(x-1)(x+2)}{x-1}$ simplifies to $x + 2$. At $x = 1$,

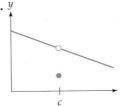

however, the function $\dfrac{(x-1)(x+2)}{x-1}$ is *dis*continuous (since it is undefined where the denominator is zero), while $x+2$ is *continuous*. Are these two functions, one obtained from the other by simplification, equal to each other? Explain.

Applied Exercises

80. GENERAL: Relativity According to Einstein's special theory of relativity, under certain conditions a 1-foot-long object moving with velocity v will appear to an observer to have length $\sqrt{1-(v/c)^2}$, in which c is a constant equal to the speed of light. Find the limiting value of the apparent length as the velocity of the object approaches the speed of light by finding

$$\lim_{v \to c^-} \sqrt{1-\left(\frac{v}{c}\right)^2}$$

81. BUSINESS: Interest Compounded Continuously If you deposit $1 into a bank account paying 10% interest compounded continuously (see Section 4.1), a year later its value will be

$$\lim_{x \to 0}\left(1+\frac{x}{10}\right)^{1/x}$$

Find the limit by making a TABLE of values correct to two decimal places, thereby finding the value of the deposit in dollars and cents.

82. BUSINESS: Interest Compounded Continuously If you deposit $1 into a bank account paying 5% interest compounded continuously (see Section 4.1), a year later its value will be

$$\lim_{x \to 0}\left(1+\frac{x}{20}\right)^{1/x}$$

Find the limit by making a TABLE of values correct to two decimal places, thereby finding the value of the deposit in dollars and cents.

83. GENERAL: Superconductivity The conductivity of aluminum at temperatures near absolute zero is approximated by the function

$$f(x) = \frac{100}{1+0.001x^2}$$

which expresses the conductivity as a percent. Find the limit of this conductivity percent as the temperature x approaches 0 (absolute zero) from the right. (See the Application Preview on page 71.)

84. GENERAL: First-Class Mail In 2014, the U.S. Postal Service would deliver a first-class letter weighing 3.5 ounces or less for the following prices: not more than 1 ounce, 49¢; more than 1 ounce but not more than 2 ounces, 70¢; more than 2 ounces but not more than 3 ounces, 91¢; and more than 3 ounces but not more than 3.5 ounces, $1.12. This information determines the price (in cents) as a function of the weight (in ounces). At which values in open interval $(0, 3.5)$ is this function discontinuous?

Source: U.S. Postal Service

Conceptual Exercises

85. A student once said: "The limit is where a function is *going*, even if it never *gets* there." Explain what the student meant.

86. A student once said: "A continuous function *gets* where it's going; a discontinuous function *doesn't*." Explain what the student meant.

87. True or False: If $f(2)$ is not defined, then $\lim_{x \to 2} f(x)$ does not exist.

88. True or False: If $f(2) = 5$, then $\lim_{x \to 2} f(x) = 5$.

89. True or False: If $\lim_{x \to 2^+} f(x) = 7$, then $\lim_{x \to 2^-} f(x) = 7$.

90. True or False: If $\lim_{x \to 2} f(x) = 7$, then $\lim_{x \to 2^+} f(x) = 7$.

91. True or False: $\lim_{x \to 1}\dfrac{x^2-1}{x-1} = \dfrac{\lim_{x \to 1}(x^2-1)}{\lim_{x \to 1}(x-1)}$

92. True or False: If a function is not defined at $x = 5$, then the function is not continuous at $x = 5$.

93. If $\lim_{x \to 2} f(x) = 7$ and $f(x)$ is continuous at $x = 2$, then $f(2) = 7$.

94. True or False: If a function is defined and continuous at every x-value, then its graph has no jumps or breaks.

95. For the function graphed below, find:

   a. $\lim_{x \to 0^+} f(x)$ **b.** $\lim_{x \to 0^-} f(x)$ **c.** $\lim_{x \to 0} f(x)$

2.2   Rates of Change, Slopes, and Derivatives

Introduction

In this section we will define the **derivative,** one of the two most important concepts in all of calculus, which measures the **rate of change** of a function or, equivalently, the **slope** of a curve.* We begin by discussing rates of change.

Average and Instantaneous Rate of Change

We often speak in terms of *rates of change* to express how one quantity changes with another. For example, in the morning the temperature might be "rising at the rate of 3 degrees per hour" and in the evening it might be "falling at the rate of 2 degrees per hour." For simplicity, suppose that in some location the temperature at time x hours is $f(x) = x^2$ degrees. We shall calculate the *average rate of change of temperature* over various time intervals—the change in temperature divided by the change in time. For example, at time 1 the temperature is $1^2 = 1$ degrees, and at time 3 the temperature is $3^2 = 9$ degrees, so the temperature went up by $9 - 1 = 8$ degrees in 2 hours, for an average rate of:

$$\begin{pmatrix} \text{Average rate} \\ \text{of change} \\ \text{from 1 to 3} \end{pmatrix} = \frac{3^2 - 1^2}{3 - 1} = \frac{9 - 1}{2} = \frac{8}{2} = 4 \qquad \begin{array}{l} \text{Average rate} \\ \text{of change} \\ \text{over 2 hours} \end{array}$$

Similarly,   degrees per hour

$$\begin{pmatrix} \text{Average rate} \\ \text{of change} \\ \text{from 1 to 2} \end{pmatrix} = \frac{2^2 - 1^2}{2 - 1} = \frac{4 - 1}{1} = 3 \qquad \begin{array}{l} \text{Average rate} \\ \text{of change} \\ \text{over 1 hour} \end{array}$$

$$\begin{pmatrix} \text{Average rate} \\ \text{of change} \\ \text{from 1 to 1.5} \end{pmatrix} = \frac{1.5^2 - 1^2}{1.5 - 1} = \frac{2.25 - 1}{0.5} = \frac{1.25}{0.5} = 2.5 \qquad \begin{array}{l} \text{Average rate} \\ \text{of change over} \\ \text{0.5 hour} \end{array}$$

degrees per hour

$$\begin{pmatrix} \text{Average rate} \\ \text{of change} \\ \text{from 1 to 1.1} \end{pmatrix} = \frac{1.1^2 - 1^2}{1.1 - 1} = \frac{1.21 - 1}{0.1} = \frac{0.21}{0.1} = 2.1 \qquad \begin{array}{l} \text{Average rate} \\ \text{of change over} \\ \text{0.1 hour} \end{array}$$

We see that rate of change of temperature over shorter and shorter time intervals is decreasing from 4 to 3 to 2.5 to 2.1, numbers that seem to be approaching 2 degrees per hour. To verify that this is indeed true, we generalize the process. We have been finding the average over a time interval from 1 to a slightly later time that we now call $1 + h,$ where h is a small positive number. Notice that in each step we calculated the following expression for successively smaller values of h.

$$\begin{pmatrix} \text{Average rate} \\ \text{of change} \\ \text{from 1 to 1 + h} \end{pmatrix} = \frac{(1 + h)^2 - 1^2}{h} \qquad \begin{array}{l} \text{For } h \text{ we used} \\ 2, 1, 0.5, \text{ and } 0.1 \end{array}$$

$$\text{From } 1 + h - 1 = h$$

---

*The second most important concept in calculus is the definite integral, which will be defined in Section 5.3.

If we now use our limit notation to let h approach zero, the amount of time will shrink to an instant, giving what is called the *instantaneous rate of change*:

$$\left(\begin{array}{c}\text{Instantaneous}\\\text{rate of change}\\\text{at time 1}\end{array}\right) = \lim_{h \to 0} \frac{(1+h)^2 - 1^2}{h}$$ Taking the limit as h approaches zero

We have been using the function $f(x) = x^2$ and the time $x = 1$, but the same procedure applies to *any* function f and number x, leading to the following general definition:

Average and Instantaneous Rate of Change

LOOKING BACK

We defined the *difference quotient* on page 57.

The *average* rate of change of a function f between x and $x + h$ is

$$\frac{f(x+h) - f(x)}{h}$$ Difference quotient gives the *average* rate of change

The *instantaneous* rate of change of a function f at the number x is

$$\lim_{h \to 0} \frac{f(x+h) - f(x)}{h}$$ Taking the limit makes it *instantaneous*

In the difference quotient, the numerator is the change in the *function* between two x-values, $f(x + h) - f(x)$, and the denominator is the change between the two x-values, $(x + h) - (x) = h$.

We will now use the second formula to check our guess that the average rate of change of temperature is indeed approaching 2 degrees per hour.

EXAMPLE 1   FINDING AN INSTANTANEOUS RATE OF CHANGE

Find the instantaneous rate of change of the temperature function $f(x) = x^2$ at time $x = 1$.

Solution

$$\lim_{h \to 0} \frac{f(x+h) - f(x)}{h}$$ Formula for the instantaneous rate of change of f at x

$$\lim_{h \to 0} \frac{f(1+h) - f(1)}{h}$$ Substituting $x = 1$

$$= \lim_{h \to 0} \frac{(1+h)^2 - 1^2}{h}$$ $f(x) = x^2$ gives $f(1 + h) = (1 + h)^2$ and $f(1) = 1^2$

$$= \lim_{h \to 0} \frac{1 + 2h + h^2 - 1}{h}$$ Expanding $(1 + h)^2 = 1 + 2h + h^2$

$$= \lim_{h \to 0} \frac{\cancel{1} + 2h + h^2 - \cancel{1}}{h}$$ Simplifying

$$= \lim_{h \to 0} \frac{h(2 + h)}{h} = \lim_{h \to 0} \frac{\cancel{h}(2 + h)}{\cancel{h}}$$ Factoring out h and canceling (since $h \neq 0$)

$$= \lim_{h \to 0} (2 + h) = 2$$ Evaluating the limit by direct substitution

FOR MORE HELP

with expanding, factoring, and canceling, see the Algebra Review appendix, pages B8 and B12–B13.

Since $f(x)$ gives the temperature at time x hours, this means that *the instantaneous rate of change of temperature at time $x = 1$ hour is 2 degrees per hour* (just as we had guessed earlier).

Be Careful Make sure you understand how to find the *units* of the rate of change. In this Example, f gave the temperature (degrees) at time x (hours), so the units of the rate of change were *degrees per hour*. In general, for a function $f(x)$, the units of the instantaneous rate of change are *function units per x unit*.

PRACTICE PROBLEM 1

If $f(x)$ gives the population of a city in year x, what are the units of the instantaneous rate of change? Solution on page 96 >

Secant and Tangent Lines

How can we "see" the average and instantaneous rates of change on the graph of a function? First, some terminology. A **secant line** to a curve is a line that passes through two points of the curve. A **tangent line** is a line that passes through a point of the curve and matches exactly the *steepness* of the curve at that point.* These are shown in the following diagram.

Tangent and secant lines to a curve

If the curve is the graph of a function f and the points P and Q have x-coordinates x and $x + h$, respectively (and so y-coordinates $f(x)$ and $f(x + h)$, respectively), then the above graph takes the following form:

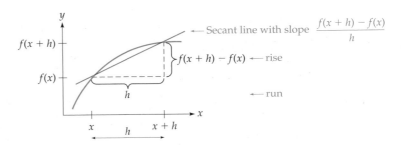

Observe that the *slope* (rise over run) of the secant line is the difference quotient $\frac{f(x + h) - f(x)}{h}$, exactly the same as our earlier definition of the average rate of change of the function between x and $x + h$.

Furthermore, the two points where the secant line meets the curve are separated by a distance h along the x-axis. Letting $h \to 0$ forces the second point to approach the first, causing the *secant* line to approach the *tangent* line, the slope of which is then $\lim\limits_{h \to 0} \frac{f(x + h) - f(x)}{h}$. But this is exactly the same as our earlier definition of the *instantaneous rate of change of the function at x*, as shown in the following diagram.

---

*The word "secant" comes from the Latin *secare*, "to cut," suggesting that the secant line "cuts" the curve at two points. The word "tangent" comes from the Latin *tangere*, "to touch," suggesting that the tangent line just "touches" the curve at one point.

The last two graphs have shown that the slope of the *secant* line is the *average* rate of change of the function, and, more importantly, that the slope of the *tangent* line is the *instantaneous* rate of change of the function.

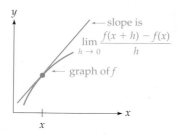

Slope of the Tangent Line

The slope of the tangent line to the graph of f at x is

$$\lim_{h \to 0} \frac{f(x+h) - f(x)}{h}$$

The fact that rates of change are related to slopes (and in fact are given by the same formula) should come as no surprise: If a function has a large rate of change, its graph will rise rapidly, giving it a large slope; if a function has only a small rate of change, its graph will rise slowly, giving it a small slope. In fact, *rate of change* and *slope* are just the numerical and graphical versions of exactly the same idea.

In Example 1 we found the instantaneous rate of change of a function. We now use the same formula to find the slope of the tangent line to the graph of a function.

EXAMPLE 2   FINDING THE SLOPE OF A TANGENT LINE

Find the slope of the tangent line to $f(x) = \dfrac{1}{x}$ at $x = 2$.

Solution

$$\lim_{h \to 0} \frac{f(x+h) - f(x)}{h}$$
Formula for the slope of the tangent line at x

$$\lim_{h \to 0} \frac{f(2+h) - f(2)}{h}$$
Substituting $x = 2$

$$= \lim_{h \to 0} \frac{\dfrac{1}{2+h} - \dfrac{1}{2}}{h}$$
Using the function $f(x) = \dfrac{1}{x}$

$$= \lim_{h \to 0} \frac{1}{h}\left(\frac{1}{2+h} - \frac{1}{2}\right)$$
Since multiplying by $1/h$ is equivalent to dividing by h

$$= \lim_{h \to 0} \frac{1}{h} \cdot \frac{2 - (2+h)}{(2+h) \cdot 2}$$
Combining fractions using the common denominator $(2+h) \cdot 2$

$$= \lim_{h \to 0} \frac{1}{h} \cdot \frac{-h}{(2+h) \cdot 2}$$
Simplifying the numerator

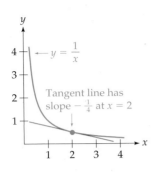

$$= \lim_{h \to 0} \frac{1}{\cancel{h}} \cdot \frac{\cancel{-h}^{-1}}{(2+h) \cdot 2} = \lim_{h \to 0} \frac{-1}{(2+h) \cdot 2} \qquad \text{Canceling}$$

$$= \frac{-1}{(2+0) \cdot 2} = \frac{-1}{4} = -\frac{1}{4} \qquad \begin{array}{l}\text{Evaluating the limit by}\\ \text{direct substitution of } h = 0\\ \text{(as we did on page 75)}\end{array}$$

Therefore, the curve $y = \frac{1}{x}$ has slope $-\frac{1}{4}$ at $x = 2$, as shown in the graph on the left.

| **EXAMPLE 3** | **FINDING A TANGENT LINE** |

Use the result of the preceding Example to find the equation of the tangent line to $f(x) = \frac{1}{x}$ at $x = 2$.

Solution

From Example 2 we know that the *slope* of the tangent line is $m = -\frac{1}{4}$. The point on the curve at $x = 2$ is $(2, \frac{1}{2})$, the *y*-coordinate coming from $y = f(2) = \frac{1}{2}$. The point-slope formula then gives:

$$y - \tfrac{1}{2} = -\tfrac{1}{4}(x - 2) \qquad \begin{array}{l} y - y_1 = m(x - x_1) \text{ with } m = -\frac{1}{4},\\ x_1 = 2, \text{ and } y_1 = \frac{1}{2}\end{array}$$

$$y - \tfrac{1}{2} = -\tfrac{1}{4}x + \tfrac{1}{2} \qquad \text{Multiplying out}$$

$$y = 1 - \tfrac{1}{4}x \qquad \text{Simplifying}$$

Equation of the tangent line

The first of the following graphs shows the curve $y = \frac{1}{x}$ along with the tangent line $y = 1 - \frac{1}{4}x$ that we found in the preceding Example. The next two graphs show the results of successively "zooming in" near the point of tangency, showing that the curve seems to straighten out and almost *become its own tangent line*. For this reason, the tangent line is called *the best linear approximation to the curve* near the point of tangency. A verification of this is given in Exercise 75 on page 113.

$y = 1/x$ and its tangent line at $x = 2$ on [0, 4] by [0, 4]

After one "zoom in" near (2, 1/2)

After a second "zoom in" centered at (2, 1/2)

Graphing Calculator Exploration

Use a graphing calculator to graph $y = x^3 - 2x^2 - 3x + 4$ (or any function of your choice). Then "zoom in" a few times around a point to see the curve straighten out and almost *become* its own tangent line.

The Derivative

In Examples 1 and 2 we found an instantaneous rate of change and the slope of a tangent line *at a particular number.* It is much more efficient to carry out the same calculation but keeping x as a *variable,* obtaining a new function that gives the instantaneous rate of change or the slope of the tangent line at *any* value of x. This new function is denoted with a *prime,* $f'(x)$ (read: "f prime of x"), and is called *the derivative of f at x.*

Derivative

For a function f, the *derivative of f at x* is defined as

$$f'(x) = \lim_{h \to 0} \frac{f(x + h) - f(x)}{h} \qquad \text{Limit of the difference quotient}$$

(provided that the limit exists). The derivative $f'(x)$ gives the *instantaneous rate of change of f at x* and also the *slope of the graph of f at x.*

In general, the units of the derivative are *function units per x unit.*

EXAMPLE 4   **FINDING A DERIVATIVE FROM THE DEFINITION**

Find the derivative of $f(x) = 2x^2 - 9x + 39$.

Solution

$$f'(x) = \lim_{h \to 0} \frac{f(x + h) - f(x)}{h} \qquad \text{Definition of the derivative}$$

$$= \lim_{h \to 0} \frac{\overbrace{2(x + h)^2 - 9(x + h) + 39}^{f(x+h)} - \overbrace{(2x^2 - 9x + 39)}^{f(x)}}{h} \qquad \text{Using } f(x) = 2x^2 - 9x + 39$$

$$= \lim_{h \to 0} \frac{2x^2 + 4xh + 2h^2 - 9x - 9h + 39 - 2x^2 + 9x - 39}{h} \qquad \text{Expanding and simplifying}$$

$$= \lim_{h \to 0} \frac{4xh + 2h^2 - 9h}{h} = \lim_{h \to 0} \frac{h(4x + 2h - 9)}{h} \qquad \text{Simplifying (since } h \neq 0)$$

$$= \lim_{h \to 0} (4x + 2h - 9) = 4x - 9 \qquad \text{Evaluating the limit by direct substitution}$$

Therefore, the derivative of $f(x) = 2x^2 - 9x + 39$ is $f'(x) = 4x - 9$.

⬅ LOOKING BACK

We evaluated this limit on page 79.

Calculating derivatives should be thought of as an *operation on functions,* taking one function [such as $f(x) = 2x^2 - 9x + 39$] and giving another function [$f'(x) = 4x - 9$]. The resulting function is called "the derivative" because it is *derived* from the first, and the process of obtaining it is called "differentiation." If the derivative is defined at x, then the original function is said to be **differentiable at x.**

EXAMPLE 5   **USING A DERIVATIVE IN AN APPLICATION**

The following graph shows the annual sales of Apple Macintosh computers between 2000 and 2010 (in hundred thousands of units).

Source: Apple

These annual sales are approximated by the function $f(x) = 2x^2 - 9x + 39$, where x stands for *years since 2000*. Find the instantaneous rate of change of this function at $x = 10$ and at $x = 1$ and interpret the results.

Solution

The instantaneous rate of change means the *derivative*, so ordinarily we would now take the derivative of the sales function $f(x) = 2x^2 - 9x + 39$. However, this is just what we did in the previous Example, so we will use the result that the derivative is $f'(x) = 4x - 9$. Therefore, we need only evaluate the derivative at the given x-values and interpret the results.

$$f'(10) = 4 \cdot 10 - 9 = 31 \qquad \text{Evaluating } f'(x) = 4x - 9 \text{ at } x = 10$$

Interpretation: $x = 10$ represents the year 2010 (10 years after 2000). Therefore, in 2010 Macintosh sales were growing at the rate of 31 hundred thousand, or 3.1 million units per year.

$$f'(1) = 4 \cdot 1 - 9 = -5 \qquad \text{Evaluating } f'(x) = 4x - 9 \text{ at } x = 1$$

Interpretation: In the year 2001 (corresponding to $x = 1$), Macintosh sales were falling at the rate of 5 hundred thousand, or 500,000 units per year.

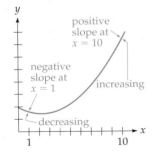

A derivative that is *positive* means that the original quantity is *increasing*, and a derivative that is *negative* means that the original quantity is *decreasing*.

The graph of $f(x) = 2x^2 - 9x + 39$ on the left shows that the slope of the curve is indeed negative at $x = 1$ and positive at $x = 10$.

EXAMPLE 6   **FINDING A DERIVATIVE FROM THE DEFINITION**

Find the derivative of $f(x) = x^3$.

Solution

In our solution we will use the expansion $(x + h)^3 = x^3 + 3x^2h + 3xh^2 + h^3$ (found by multiplying together three copies of $(x + h)$).

$$f'(x) = \lim_{h \to 0} \frac{f(x+h) - f(x)}{h} \qquad \text{Definition of } f'(x)$$

$$= \lim_{h \to 0} \frac{(x+h)^3 - x^3}{h} \qquad \text{Using } f(x) = x^3$$

$$= \lim_{h \to 0} \frac{x^3 + 3x^2h + 3xh^2 + h^3 - x^3}{h} \qquad \text{Using the expansion of } (x+h)^3$$

$$= \lim_{h \to 0} \frac{\cancel{x^3} + 3x^2h + 3xh^2 + h^3 - \cancel{x^3}}{h} \qquad \text{Canceling}$$

$$= \lim_{h \to 0} \frac{h(3x^2 + 3xh + h^2)}{h} \qquad \text{Factoring out an } h$$

$$= \lim_{h \to 0} \frac{\cancel{h}(3x^2 + 3xh + h^2)}{\cancel{h}} = \lim_{h \to 0} 3x^2 + 3xh + h^2 \qquad \text{Canceling again}$$

$$= 3x^2 \qquad \text{Evaluating the limit by direct substitution}$$

Therefore, the derivative of $f(x) = x^3$ is $f'(x) = 3x^2$.

 LOOKING BACK

We evaluated this limit on page 83.

Leibniz's Notation for the Derivative

Calculus was developed by Isaac Newton (1642–1727) and Gottfried Wilhelm Leibniz (1646–1716) in two different countries, so there naturally developed two different notations for the derivative. Newton denoted derivatives by a dot over the function, $\dot{f}$, a notation that has been largely replaced by our "prime" notation.

Leibniz wrote the derivative of $f(x)$ by writing $\dfrac{d}{dx}$ in front of the function: $\dfrac{d}{dx} f(x)$. In Leibniz's notation, the fact that the derivative of x^3 is $3x^2$ is written

$$\frac{d}{dx} x^3 = 3x^2 \qquad \text{The derivative of } x^3 \text{ is } 3x^2$$

The following table shows equivalent expressions in the two notations.

Prime Notation		Leibniz's Notation	
$f'(x)$	$=$	$\dfrac{d}{dx} f(x)$	Prime and $\dfrac{d}{dx}$ both mean the derivative
y'	$=$	$\dfrac{dy}{dx}$	For y a function of x

Each notation has its own advantages, and we will use both.* Leibniz's notation comes from writing the definition of the derivative as:

---

*Other notations for the derivative are $Df(x)$ and $D_x f(x)$, but we will not use them.

LOOKING BACK

On page 7 we used the Δ notation to define slope of a *line*, and now to define the slope of a *curve*.

$$\frac{dy}{dx} = \lim_{\Delta x \to 0} \frac{f(x + \Delta x) - f(x)}{\Delta x}$$

Definition of the derivative (page 92) with the change in x written as Δx

or

$$\frac{dy}{dx} = \lim_{\Delta x \to 0} \frac{\Delta y}{\Delta x}$$

$f(x + \Delta x) - f(x)$ is the change in y, and so can be written Δy

It is as if the limit turns Δ (read "Delta," the Greek letter D) into d, changing the $\frac{\Delta y}{\Delta x}$ into $\frac{dy}{dx}$. That is, Leibniz's notation reminds us that the derivative $\frac{dy}{dx}$ is the limit of the slope $\frac{\Delta y}{\Delta x}$.

 Be Careful   Some functions are *not differentiable* (the derivative does not exist) at certain x-values. For example, the following diagram shows a function that has a "corner point" at $x = 1$. At this point the slope (and therefore the derivative) cannot be defined, so the function is not differentiable at $x = 1$. Other nondifferentiable functions will be discussed in Section 2.7.

 LOOKING AHEAD

On pages 151–152 we will analyze the "corner point" of a similar function in greater detail.

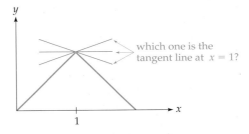

which one is the tangent line at $x = 1$?

Since the tangent line cannot be uniquely defined at $x = 1$, the slope, and therefore the derivative, is undefined at $x = 1$.

The following diagram shows the geometric relationship between a function (upper graph) and its derivative (lower graph). Observe carefully how the *slope of f* is shown by the *value of f'*.

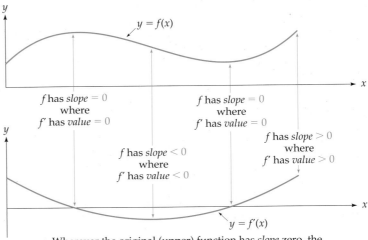

$y = f(x)$

f has *slope* = 0 where f' has *value* = 0

f has *slope* = 0 where f' has *value* = 0

f has *slope* < 0 where f' has *value* < 0

f has *slope* > 0 where f' has *value* > 0

$y = f'(x)$

Wherever the original (upper) function has *slope* zero, the derivative (lower) has *value* zero (it crosses the x-axis).

PRACTICE PROBLEM 2

The graph shows a function and its derivative. Which is the original function (#1 or #2) and which is its derivative? [*Hint:* Which curve has *slope* zero where the other has *value* zero?]

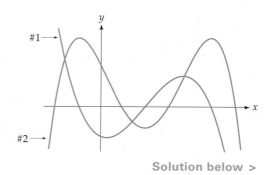

Solution below >

Solutions TO PRACTICE PROBLEMS

1. The units are *people per year*, measuring the rate of growth of the population.

2. #2 is the original function and #1 is its derivative.

2.2   Section Summary

The derivative of the function f at x is

$$f'(x) = \lim_{h \to 0} \frac{f(x + h) - f(x)}{h}$$ Provided that the limit exists

The steps of the calculation can be renumbered by **DESL** (pronounced "diesel"): Write the **D**efinition, **E**xpress the numerator in terms of the function, **S**implify, and take the **L**imit.

   The derivative $f'(x)$ gives both the *slope of the graph* of the function at x and the *instantaneous rate of change of the function* at x. In other words, "derivative," "slope," and "instantaneous rate of change" are merely the mathematical, the geometric, and the analytic versions of the same idea.

   The derivative gives the rate of change *at a particular instant*, not an actual change over a period of time. Instantaneous rates of changes are like the speeds on an automobile speedometer—a reading of 50 mph at one moment does not mean that you will travel exactly 50 miles in the next hour, since the actual distance depends upon your speed during the entire hour. The derivative, however, may be interpreted as the *approximate* change resulting from a 1-unit increase in the independent variable. For example, if your speedometer reads 50 mph, then you may say that you will travel *about* 50 miles during the next hour, meaning that this will be true provided that your speed remains steady throughout the hour. (In Chapter 5 we will see how to calculate actual changes from rates of change that do not stay constant.)

2.2 Exercises

1–4. By imagining tangent lines at points P_1, P_2, and P_3, state whether the slopes are positive, zero, or negative at these points.

1.

2.

3.

4.

5–6. Use the tangent lines shown at points P_1 and P_2 to find the slopes of the curve at these points.

5.

6.

7–8. Use the graph of each function $f(x)$ to make a rough sketch of the derivative $f'(x)$ showing where $f'(x)$ is positive, negative, and zero. (Omit scale on y-axis.)

7.

8.

9–10. Find the average rate of change of the given function between the following pairs of x-values.

a. $x = 1$ and $x = 3$
b. $x = 1$ and $x = 2$
c. $x = 1$ and $x = 1.5$
d. $x = 1$ and $x = 1.1$
e. $x = 1$ and $x = 1.01$
f. What number do your answers seem to be approaching?

FOR HELP GETTING STARTED
see page 87.

9. $f(x) = x^2 + x$ **10.** $f(x) = 2x^2 + 5$

11–12. Find the average rate of change of the given function between the following pairs of x-values.

a. $x = 2$ and $x = 4$
b. $x = 2$ and $x = 3$
c. $x = 2$ and $x = 2.5$
d. $x = 2$ and $x = 2.1$
e. $x = 2$ and $x = 2.01$
f. What number do your answers seem to be approaching?

FOR HELP GETTING STARTED
see page 87.

11. $f(x) = 2x^2 + x - 2$ **12.** $f(x) = x^2 + 2x - 1$

13–14. Find the average rate of change of the given function between the following pairs of x-values.

a. $x = 3$ and $x = 5$
b. $x = 3$ and $x = 4$
c. $x = 3$ and $x = 3.5$
d. $x = 3$ and $x = 3.1$
e. $x = 3$ and $x = 3.01$
f. What number do your answers seem to be approaching?

FOR HELP GETTING STARTED
see page 87.

13. $f(x) = 5x + 1$ **14.** $f(x) = 7x - 2$

15–16. Find the average rate of change of the given function between the following pairs of x-values.

a. $x = 4$ and $x = 6$
b. $x = 4$ and $x = 5$
c. $x = 4$ and $x = 4.5$
d. $x = 4$ and $x = 4.1$
e. $x = 4$ and $x = 4.01$
f. What number do your answers seem to be approaching?

15. $f(x) = \sqrt{x}$ **16.** $f(x) = \dfrac{4}{x}$

17–20. Use the formula on page 88 to find the instantaneous rate of change of the function at the given x-value. If you did the related problem in Exercises 9–16, compare your answers. [*Hint:* See Example 1.]

17. $f(x) = x^2 + x$ at $x = 1$

18. $f(x) = x^2 + 2x - 1$ at $x = 2$

19. $f(x) = 5x + 1$ at $x = 3$

20. $f(x) = \dfrac{4}{x}$ at $x = 4$

21–24. Use the formula on page 90 to find the slope of the tangent line to the curve at the given x-value. If you did the related problem in Exercises 9–16, compare your answers. [*Hint:* See Example 2.]

21. $f(x) = 2x^2 + x - 2$ at $x = 2$

22. $f(x) = 2x^2 + 5$ at $x = 1$

23. $f(x) = \sqrt{x}$ at $x = 4$

24. $f(x) = 7x - 2$ at $x = 3$

25–44. Find $f'(x)$ by using the definition of the derivative. [*Hint:* See Example 4.]

25. $f(x) = x^2 - 3x + 5$ **26.** $f(x) = 2x^2 - 5x + 1$

27. $f(x) = 1 - x^2$ **28.** $f(x) = \frac{1}{2}x^2 + 1$

29. $f(x) = 9x - 2$ **30.** $f(x) = -3x + 5$

31. $f(x) = \dfrac{x}{2}$ **32.** $f(x) = 0.01x + 0.05$

33. $f(x) = 4$ **34.** $f(x) = \pi$

35. $f(x) = ax^2 + bx + c$
(a, b, and c are constants)

36. $f(x) = (x + a)^2$
(a is a constant.) [*Hint:* First expand $(x + a)^2$.]

37. $f(x) = x^5$
[*Hint:* Use $(x + h)^5 =$
$x^5 + 5x^4h + 10x^3h^2 + 10x^2h^3 + 5xh^4 + h^5$]

38. $f(x) = x^4$
[*Hint:* Use $(x + h)^4 =$
$x^4 + 4x^3h + 6x^2h^2 + 4xh^3 + h^4$]

39. $f(x) = \dfrac{2}{x}$ **40.** $f(x) = \dfrac{1}{x^2}$

41. $f(x) = \sqrt{x}$ **42.** $f(x) = \dfrac{1}{\sqrt{x}}$

[*Hint:* Multiply the numerator and denominator of the difference quotient by $(\sqrt{x + h} + \sqrt{x})$ and then simplify.]

[*Hint:* Multiply the numerator and denominator of the difference quotient by $(\sqrt{x} + \sqrt{x + h})$ and then simplify.]

43. $f(x) = x^3 + x^2$ **44.** $f(x) = \dfrac{1}{2x}$

45. a. Find the equation for the tangent line to the curve $f(x) = x^2 - 3x + 5$ at $x = 2$, writing the equation in slope-intercept form.
[*Hint:* Use your answer to Exercise 25.]

b. Use a graphing calculator to graph the curve together with the tangent line to verify your answer.

46. a. Find the equation for the tangent line to the curve $f(x) = 2x^2 - 5x + 1$ at $x = 2$, writing the equation in slope-intercept form.
[*Hint:* Use your answer to Exercise 26.]

b. Use a graphing calculator to graph the curve together with the tangent line to verify your answer.

47. a. Graph the function $f(x) = x^2 - 3x + 5$ on the window $[-10, 10]$ by $[-10, 10]$. Then use the DRAW menu to graph the TANGENT line at $x = 2$. Your screen should also show the *equation* of the tangent line. (If you did Exercise 45, this equation for the tangent line should agree with the one you found there.)

b. Add to your graph the tangent line at $x = 1$, and the tangent lines at any other x-values that you choose.

48. a. Graph the function $f(x) = 2x^2 - 5x + 1$ on the window $[-10, 10]$ by $[-10, 10]$. Then use the DRAW menu to graph the TANGENT line at $x = 2$. Your screen should also show the *equation* of the tangent line. (If you did Exercise 46, this equation for the tangent line should agree with the one you found there.)

b. Add to your graph the tangent line at $x = 0$, and the tangent lines at any other x-values that you choose.

49–54. For each function:

a. Find $f'(x)$ using the definition of the derivative.
b. Explain, by considering the original function, why the derivative is a constant.

49. $f(x) = 3x - 4$ **50.** $f(x) = 2x - 9$

51. $f(x) = 5$ **52.** $f(x) = 12$

53. $f(x) = mx + b$ (m and b are constants)

54. $f(x) = b$ (b is a constant)

Applied Exercises

55. BUSINESS: Temperature The temperature in an industrial pasteurization tank is $f(x) = x^2 - 8x + 110$ degrees centigrade after x minutes (for $0 \le x \le 12$).

a. Find $f'(x)$ by using the definition of the derivative.
b. Use your answer to part (a) to find the instantaneous rate of change of the temperature after 2 minutes. Be sure to interpret the sign of your answer.
c. Use your answer to part (a) to find the instantaneous rate of change after 5 minutes.

56. GENERAL: Population The population of a town is $f(x) = 3x^2 - 12x + 200$ people after x weeks (for $0 \le x \le 20$).

a. Find $f'(x)$ by using the definition of the derivative.
b. Use your answer to part (a) to find the instantaneous rate of change of the population after 1 week. Be sure to interpret the sign of your answer.
c. Use your answer to part (a) to find the instantaneous rate of change of the population after 5 weeks.

57. BEHAVIORAL SCIENCE: Learning Theory In a psychology experiment, a person could memorize x words in $f(x) = 2x^2 - x$ seconds (for $0 \le x \le 10$).

a. Find $f'(x)$ by using the definition of the derivative.
b. Find $f'(5)$ and interpret it as an instantaneous rate of change in the proper units.

58. BUSINESS: Advertising An automobile dealership finds that the number of cars that it sells on day x of an advertising campaign is $S(x) = -x^2 + 10x$ (for $0 \le x \le 7$).

a. Find $S'(x)$ by using the definition of the derivative.
b. Use your answer to part (a) to find the instantaneous rate of change on day $x = 3$.
c. Use your answer to part (a) to find the instantaneous rate of change on day $x = 6$.

Be sure to interpret the signs of your answers.

59. SOCIAL SCIENCE: Immigration The percentage of people in the United States who are immigrants (that is, were born elsewhere) for different decades is shown below.

Sources: Center for Immigration Studies and U.S. Census Bureau

These percentages are approximated by the function $f(x) = \frac{1}{2}x^2 - 3.7x + 12$, where x stands for *the number of decades since 1930* (so that, for example, $x = 5$ would stand for 1980).

a. Find $f'(x)$ using the definition of the derivative.
b. Evaluate the derivative at $x = 1$ and interpret the result.
c. Find the rate of change of the immigrant percentage in the year 2010.

60. BUSINESS: Gambling Income Net income for Bally Technologies (a manufacturer of gambling machinery) for recent years is shown at the top of the next column, where income is in millions of dollars.

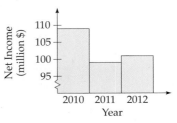

Source: Standard & Poor's Industry Surveys

The income is approximated by the function $f(x) = 6x^2 - 16x + 109$, where x stands for *years since 2010*.

a. Find $f'(x)$ using the definition of the derivative.
b. Evaluate the derivative at $x = 0$ and interpret the result.
c. Find the rate of change of the index in 2012.

Conceptual Exercises

61. Describe the difference between the *average* rate of change and the *instantaneous* rate of change of a function. What formula would you use to find the instantaneous rate of change?

62. Describe the difference between a *secant* line and a *tangent* line for the graph of a function. What formula would you use to find the slope of the secant? What formula for the tangent?

63. When we calculate the derivative using the formula $f'(x) = \lim_{h \to 0} \dfrac{f(x + h) - f(x)}{h}$, we eventually evaluate the limit by direct substitution of $h = 0$. Why don't we just substitute $h = 0$ into the formula to begin with?

64. For a function $f(x)$, if f is in widgets and x is in blivets, what are the units of the derivative $f'(x)$, *widgets per blivet* or *blivets per widget*?

65. The derivative $f'(x)$ of a function is in *prendles per blarg*. What are the units of x? What are the units of f? [*Hint:* One is *prendles* and the other is *blargs*.]

66. The population of a city in year x is given by a function whose derivative is negative. What does this mean about the city?

67. A patient's temperature at time x hours is given by a function whose derivative is positive for $0 < x < 24$ and negative for $24 < x < 48$. Assuming that the patient begins and ends with a normal temperature, is the patient's health improving or deteriorating during the first day? During the second day?

68. Suppose that the temperature outside your house x hours after midnight is given by a function whose derivative is negative for $0 < x < 6$ and positive for $6 < x < 12$. What can you say about the temperature at time 6 a.m. compared to the temperature throughout the first half of the day?

2.3   Some Differentiation Formulas

Introduction

In Section 2.2 we defined the *derivative* of a function and used it to calculate instantaneous rates of change and slopes. Even for a function as simple as $f(x) = x^2$, however, calculating the derivative from the definition was rather involved. Calculus would be of limited usefulness if all derivatives had to be found in this way.

In this section we will learn several **rules of differentiation** that will simplify finding the derivatives of many useful functions. The rules are derived from the definition of the derivative, which is why we studied the definition first. We will also learn another important use for differentiation: calculating **marginals** (marginal revenue, marginal cost, and marginal profit), which are used extensively in business and economics.

Derivative of a Constant

The first rule of differentiation shows how to differentiate a constant function.

Constant Rule	
	Brief Example
For any constant c, $$\frac{d}{dx}c = 0$$	$$\frac{d}{dx}7 = 0$$

$f(x) = c$
derivative (slope) is zero

In words: *the derivative of a constant is zero.* This rule is obvious geometrically, as shown in the diagram: The graph of a constant function $f(x) = c$ is the horizontal line $y = c$. Since the slope of a horizontal line is zero, the derivative of $f(x) = c$ is zero. It is also follows from the *rate of change* interpretation of derivatives: The rate of change is zero since a constant never changes.

This rule is easily proved from the definition of the derivative. The constant function $f(x) = c$ has the same value c for *any* value of x, so, in particular, $f(x + h) = c$ and $f(x) = c$. Substituting these into the definition of the derivative gives

$$f'(x) = \lim_{h \to 0} \frac{\overbrace{f(x + h)}^{c} - \overbrace{f(x)}^{c}}{h} = \lim_{h \to 0} \frac{c - c}{h} = \lim_{h \to 0} \frac{\overbrace{0}^{0}}{h} = \lim_{h \to 0} 0 = 0$$

Therefore, the derivative of a constant function $f(x) = c$ is $f'(x) = 0$.

Power Rule

One of the most useful differentiation formulas in all of calculus is the **Power Rule.** It tells how to differentiate powers such as x^7 or x^{100}.

Power Rule	
For any constant exponent n, $$\frac{d}{dx}x^n = n \cdot x^{n-1}$$	To differentiate x^n, bring down the exponent as a multiplier and then decrease the exponent by 1

A derivation of the Power Rule for positive integer exponents is given at the end of this section. We will use the Power Rule for *all* real numbers n, since more general proofs will be given later (see pages 124, 226, and 283–284).

EXAMPLE 1	**USING THE POWER RULE**

a. $$\frac{d}{dx}x^7 = 7x^{7-1} = 7x^6$$

Bring down the exponent   Decrease the exponent by 1

b. $$\frac{d}{dx}x^{100} = 100x^{100-1} = 100x^{99}$$

c. $\dfrac{d}{dx}\dfrac{1}{x^2} = \dfrac{d}{dx}x^{-2} = -2x^{-2-1} = -2x^{-3} = -\dfrac{2}{x^3}$ The Power Rule holds for negative exponents

FOR MORE HELP

with fractional and negative exponents, see pages 22–25.

d. $\dfrac{d}{dx}\sqrt{x} = \dfrac{d}{dx}x^{\frac{1}{2}} = \dfrac{1}{2}x^{\frac{1}{2}-1} = \dfrac{1}{2}x^{-\frac{1}{2}} = \dfrac{1}{2\sqrt{x}}$ and for fractional exponents

e. $\dfrac{d}{dx}x = \dfrac{d}{dx}x^1 = 1x^{1-1} = x^0 = 1$

This last result is used so frequently that it should be remembered separately.

$$\frac{d}{dx}x = 1 \qquad\qquad \text{The derivative of } x \text{ is } 1$$

derivative (slope) is 1

This result is obvious geometrically, as shown in the diagram.

From now on we will skip the middle step in these examples, differentiating powers in one step:

$$\frac{d}{dx}x^{50} = 50x^{49}$$

$$\frac{d}{dx}x^{2/3} = \frac{2}{3}x^{-1/3} \qquad \tfrac{2}{3} - 1$$

PRACTICE PROBLEM 1

Find

a. $\dfrac{d}{dx}x^2$ **b.** $\dfrac{d}{dx}\dfrac{1}{x^5}$ **c.** $\dfrac{d}{dx}\sqrt[4]{x}$ Solutions on page 107 >

Constant Multiple Rule

The Power Rule shows how to differentiate a power such as x^3. The *Constant Multiple Rule* extends this result to functions such as $5x^3$, a constant *times* a function. Briefly, to differentiate a constant times a function, we simply "carry along" the constant and differentiate the function.

Constant Multiple Rule

For any constant c,

$$\frac{d}{dx}[c \cdot f(x)] = c \cdot f'(x)$$

The derivative of a constant times a function is the constant times the derivative of the function

(provided, of course, that the derivative $f'(x)$ exists). A derivation of this rule is given at the end of this section.

EXAMPLE 2   **USING THE CONSTANT MULTIPLE RULE**

a. $\dfrac{d}{dx}5x^3 = 5 \cdot 3x^2 = 15x^2$ **b.** $\dfrac{d}{dx}3x^{-4} = 3(-4)x^{-5} = -12x^{-5}$

Carry along the constant ⌎ ⌐ Derivative of x^3

Again we will skip the middle step, bringing down the exponent and immediately multiplying it by the number in front of the x.

EXAMPLE 3   **CALCULATING DERIVATIVES MORE QUICKLY**

a. $\dfrac{d}{dx}8x^{-1/2} = -4x^{-3/2}$ **b.** $\dfrac{d}{dx}7x = 7 \cdot 1 = 7$

⌎ $8(-\tfrac{1}{2})$ ⌎ Derivative of x

This last Example, showing that the derivative of $7x$ is just 7, leads to a very useful general rule.

For any constant c,

$$\frac{d}{dx}(cx) = c$$ The derivative of a constant times x is just the constant

---

EXAMPLE 4   **FINDING DERIVATIVES INVOLVING CONSTANTS**

a. $\dfrac{d}{dx}(7x) = 7$ Using $\dfrac{d}{dx}(cx) = c$

b. $\dfrac{d}{dx}7 = 0$ For a constant alone, the derivative is zero

c. $\dfrac{d}{dx}(7x^2) = 7 \cdot 2x = 14x$ But for a constant *times a function,* the derivative is the constant times the derivative of the function

---

  Be Careful   Be sure to understand the difference between the last two examples: for a constant *alone* [as in part (b) above], the derivative is *zero;* but for a constant *times a function* [as in part (c) above], you *carry along the constant and differentiate the function.*

Sum Rule

The *Sum Rule* extends differentiation to sums of functions. Briefly, to differentiate a *sum* of two functions, just differentiate the functions separately and add the results.

Sum Rule

	Brief Example
$$\frac{d}{dx}[f(x) + g(x)] = f'(x) + g'(x)$$	$$\frac{d}{dx}(x^3 + x^5) = 3x^2 + 5x^4$$

(provided, of course, that both the derivatives $f'(x)$ and $g'(x)$ exist). In words: *the derivative of a sum is the sum of the derivatives.* A derivation of the Sum Rule is given at the end of this section.

A similar rule holds for the *difference* of two functions,

$$\frac{d}{dx}[f(x) - g(x)] = f'(x) - g'(x)$$

The derivative of a difference is the difference of the derivatives

(provided that $f'(x)$ and $g'(x)$ exist). These two rules may be combined:

Sum-Difference Rule

$$\frac{d}{dx}[f(x) \pm g(x)] = f'(x) \pm g'(x)$$

Use both upper signs or both lower signs

Similar rules hold for sums and differences of any finite number of terms. Using these rules, we may differentiate any polynomial or, more generally, functions with variables raised to *any* constant powers.

EXAMPLE 5   USING THE SUM-DIFFERENCE RULE

a. $\dfrac{d}{dx}(x^3 - x^5) = 3x^2 - 5x^4$

Derivatives taken separately

b. $\dfrac{d}{dx}(5x^{-2} - 6x^{1/3} + 4) = -10x^{-3} - 2x^{-2/3}$

The constant 4 has derivative 0

Leibniz's Notation and Evaluation of Derivatives

Leibniz's derivative notation $\dfrac{d}{dx}$ is often read *the derivative with respect to x* to emphasize that the independent variable is x. To differentiate a function of some *other* variable, the x in $\dfrac{d}{dx}$ is replaced by the other variable. For example:

Function	Derivative	
$f(t)$	$\dfrac{d}{dt} f(t)$	Use $\dfrac{d}{dt}$ for the derivative with respect to t
w^3	$\dfrac{d}{dw} w^3$	Use $\dfrac{d}{dw}$ for the derivative with respect to w

LOOKING AHEAD

The vertical *evaluation bar* is a standard mathematical notation. It is used again on page 333.

The following two notations both mean the derivative *evaluated* at $x = 2$.

Derivative

$$f'(2)$$

Evaluated at $x = 2$

Derivative

$$\left.\frac{df}{dx}\right|_{x=2}$$

Evaluated at $x = 2$

Bar | means "evaluated at"

 Be Careful Both notations mean *first* differentiate and *then* evaluate.

EXAMPLE 6   EVALUATING A DERIVATIVE

If $f(x) = 2x^3 - 5x^2 + 7$, find $f'(2)$.

Solution

$$f'(x) = 6x^2 - 10x \qquad\qquad \text{First differentiate}$$

$$f'(2) = 6 \cdot 2^2 - 10 \cdot 2 = 24 - 20 = 4 \qquad \text{Then evaluate}$$

PRACTICE PROBLEM 2

If $f(x) = 5x^4 + 1$, find $\left.\dfrac{df}{dx}\right|_{x=-1}$

Solution on page 107 >

EXAMPLE 7   FINDING A TANGENT LINE

Find the equation of the tangent line to $f(x) = 2x^3 - 5x^2 + 7$ at $x = 2$.

Solution

As we know from page 92, the slope of the tangent line at $x = 2$ is the slope of the curve at $x = 2$, which is the value of the derivative f' at $x = 2$. Ordinarily, we would first differentiate and then evaluate to find $f'(2)$. However, this is exactly what we did in the preceding Example, finding $f'(2) = 4$, so the slope of the tangent line is $m = 4$. The point on the curve at $x = 2$ is $(2, 3)$, the y-coordinate coming from the original function: $y = f(2) = 3$. The point-slope formula then gives

$$y - 3 = 4(x - 2) \qquad y - y_1 = m(x - x_1) \text{ with } m = 4, y_1 = 3, \text{ and } x_1 = 2$$

$$y - 3 = 4x - 8 \qquad \text{Multiplying out}$$

$$y = 4x - 5 \qquad \text{Adding 3 to each side}$$

Equation of the tangent line

The graph on the left shows that $y = 4x - 5$ *is* the tangent line to the original function at $x = 2$.

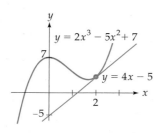

Marginal Analysis: Derivatives in Business and Economics

There is another interpretation for the derivative, one that is particularly important in business and economics. Suppose that a company has calculated its revenue, cost, and profit functions, as defined below.

$$R(x) = \left(\begin{array}{c}\text{Total revenue (income)}\\ \text{from selling } x \text{ units}\end{array}\right) \qquad \text{Revenue function}$$

$$C(x) = \left(\begin{array}{c}\text{Total cost of}\\ \text{producing } x \text{ units}\end{array}\right) \qquad \text{Cost function}$$

$$P(x) = \left(\begin{array}{c}\text{Total profit from producing}\\ \text{and selling } x \text{ units}\end{array}\right) \qquad \text{Profit function}$$

The term **marginal cost** means the additional cost of producing one more unit, $C(x + 1) - C(x)$, which may be written $\frac{C(x+1) - C(x)}{1}$, which is just the difference quotient $\frac{C(x+h) - C(x)}{h}$ with $h = 1$. If many units are being produced, then $h = 1$ is a relatively small number compared with x, so this difference quotient may be approximated by its limit as $h \to 0$, that is, by the *derivative* of the cost function. In view of this approximation, in calculus the marginal cost is *defined* to be the derivative of the cost function:

> **Take Note**
>
> The marginal cost is the rate of change of the cost function *at a particular production level*, and may not give the exact cost of the next unit, just as driving at 60 mph now does not guarentee that you will travel exactly 60 miles in the next hour.

$$MC(x) = C'(x) \qquad \text{Marginal cost is the derivative of cost}$$

The **marginal revenue** function $MR(x)$ and the **marginal profit** function $MP(x)$ are similarly defined as the derivatives of the revenue and cost functions.

$$MR(x) = R'(x) \qquad \text{Marginal revenue is the derivative of revenue}$$

$$MP(x) = P'(x) \qquad \text{Marginal profit is the derivative of profit}$$

All three formulas can be summarized very briefly: "marginal" means "derivative of." The use of derivatives to find the changes in revenue, cost, or profit resulting from one additional unit is called *marginal analysis*.

We now have three interpretations for the derivative: *slopes, instantaneous rates of change,* and *marginals.*

LOOKING AHEAD

These three interpretations of the derivative will be used throughout this book.

EXAMPLE 8   **FINDING AND INTERPRETING MARGINAL COST**

A company produces a miniature key chain flashlight based on LED (light-emitting diode) technology. The cost function (the total cost of producing x flashlights) is

$$C(x) = 8\sqrt[4]{x^3} + 300 \qquad \text{Cost function}$$

dollars, where x is the number of flashlights produced.

a. Find the marginal cost function $MC(x)$.

b. Find the marginal cost when 81 flashlights have been produced and interpret your answer.

Solution

a. The marginal cost function is the derivative of the cost function $C(x) = 8x^{3/4} + 300$, so

$$MC(x) = 6x^{-1/4} = \frac{6}{\sqrt[4]{x}} \qquad \text{Derivative of } C(x)$$

b. To find the marginal cost when 81 flashlights have been produced, we evaluate the marginal cost function $MC(x)$ at $x = 81$:

$$MC(81) = \frac{6}{\sqrt[4]{81}} = \frac{6}{3} = 2 \qquad MC(x) = \frac{6}{\sqrt[4]{x}} \text{ evaluated at } x = 81$$

Interpretation: When 81 flashlights have been produced, the marginal cost is $2, meaning that to produce one more flashlight costs about $2.

Source: BereLite

y

slope 2 at 81

300

$C(x) = 8\sqrt[4]{x^3} + 300$

0   81   x

Cost function showing
the marginal cost
(slope) at $x = 81$

EXAMPLE 9   **FINDING A LEARNING RATE**

A psychology researcher finds that the number of names that a person can memorize in x minutes is approximately $f(x) = 6\sqrt[3]{x^2}$. Find the instantaneous rate of change of this function after 8 minutes and interpret your answer.

Solution

$$f(x) = 6x^{2/3} \qquad\qquad 6\sqrt[3]{x^2} \text{ in exponential form}$$

$$f'(x) = 6 \cdot \frac{2}{3}x^{-1/3} = 4x^{-1/3} \qquad \begin{array}{l}\text{The instantaneous rate of}\\ \text{change is } f'(x)\end{array}$$

$$f'(8) = 4(8)^{-1/3} = 4\left(\frac{1}{\sqrt[3]{8}}\right) = 4\left(\frac{1}{2}\right) = 2 \qquad \text{Evaluating at } x = 8$$

Interpretation: After 8 minutes the person can memorize about two additional names per minute.

Functions as Single Objects

You may have noticed that calculus requires a more abstract point of view than precalculus mathematics. In earlier courses you looked at functions and graphs as collections of individual points, to be plotted one at a time. Now, however, we are operating on *whole functions* all at once (for example, differentiating the function x^3 to obtain the function $3x^2$). In calculus, the basic objects of interest are *functions*, and a function should be thought of as a *single* object.

    This is in keeping with a trend toward increasing abstraction as you learn mathematics. You first studied single numbers, then points (pairs of numbers), then functions (collections of points), and now collections of functions (polynomials, differentiable functions, and so on). Each stage has been a generalization of the previous stage as you have reached higher levels of sophistication. This process of generalization or "chunking" of knowledge enables you to express ideas of wider applicability and power.

Derivatives on a Graphing Calculator

Graphing calculators have an operation called NDERIV (or something similar), standing for **numerical derivative,** which gives an *approximation* of the derivative of a function. Most do so by evaluating the **symmetric difference quotient,** $\frac{f(x + h) - f(x - h)}{2h}$ for a small value of h, such as $h = 0.001$. The numerator represents the change in the function when x changes by $2h$ (from $x - h$ to $x + h$), and the denominator divides by this change in x. Geometrically, the symmetric difference quotient gives the slope of the secant line through two points on the curve h units on either side of the point at x. While NDERIV usually approximates the derivative quite closely, it sometimes gives erroneous results, as we will see in later sections. For this reason, using a graphing calculator effectively requires an understanding of both the calculus that underlies it and the technology that limits it.

Graphing Calculator Exploration

a. Use a graphing calculator to graph $y_1 = x^3 - x^2 - 6x + 3$ on $[-5, 5]$ by $[-10, 10]$, as shown on the left.

b. Define y_2 as the derivative of y_1 (using NDERIV) and graph both functions.

c. Observe that where y_1 is horizontal, the *value* of y_2 is zero; where y_1 slopes *upward*, y_2 is *positive;* and where y_1 slopes *downward*, y_2 is *negative.* Would you be able to use these observations to identify which curve is the original function and which is the derivative?

d. Now check the answer to Example 9 as follows: Redefine y_1 as $y_1 = 6x^{2/3}$, reset the window to $[0, 10]$ by $[-10, 30]$, GRAPH y_1 and y_2, and find the VALUE of y_2 at $x = 8$. Your answer should agree with that of Example 9.

Solutions TO PRACTICE PROBLEMS

1. a. $\dfrac{d}{dx} x^2 = 2x^{2-1} = 2x$

b. $\dfrac{d}{dx} \dfrac{1}{x^5} = \dfrac{d}{dx} x^{-5} = -5x^{-5-1} = -5x^{-6} = -\dfrac{5}{x^6}$

c. $\dfrac{d}{dx} \sqrt[4]{x} = \dfrac{d}{dx} x^{1/4} = \dfrac{1}{4} x^{(1/4) - 1} = \dfrac{1}{4} x^{-3/4} = \dfrac{1}{4\sqrt[4]{x^3}}$

2. $\dfrac{df}{dx} = 20x^3$

$\left. \dfrac{df}{dx} \right|_{x=-1} = 20(-1)^3 = -20$

2.3   **Section Summary**

Our development of calculus has followed two quite different lines—one technical (the *rules* of derivatives) and the other conceptual (the *meaning* of derivatives).

On the conceptual side, derivatives have three meanings:

- Instantaneous rates of change
- Slopes
- Marginals

The fact that the derivative represents all three of these ideas simultaneously is one of the reasons that calculus is so useful.

On the technical side, although we have learned several differentiation rules, we really know how to differentiate only one kind of function, *x to a constant power:*

$$\frac{d}{dx} x^n = nx^{n-1}$$

The other rules,

$$\frac{d}{dx} [c \cdot f(x)] = c \cdot f'(x)$$

and

$$\frac{d}{dx} [f(x) \pm g(x)] = f'(x) \pm g'(x)$$

simply extend the Power Rule to sums, differences, and constant multiples of such powers. Therefore, any function to be differentiated must first be expressed in terms of powers. This is why we reviewed exponential notation so carefully in Chapter 1.

Verification of the Rules of Differentiation

Verification of the Power Rule for Positive Integer Exponents

Multiplying $(x + h)$ times itself repeatedly gives

$$(x + h)^2 = x^2 + 2xh + h^2$$

$$(x + h)^3 = x^3 + 3x^2h + 3xh^2 + h^3$$

and in general, for any positive integer n,

$$(x + h)^n = x^n + nx^{n-1}h + \tfrac{1}{2}n(n - 1)x^{n-2}h^2 + \cdots + nxh^{n-1} + h^n$$

$$= x^n + nx^{n-1}h + h^2[\tfrac{1}{2}n(n - 1)x^{n-2} + \cdots + nxh^{n-3} + h^{n-2}] \quad \text{Factoring out } h^2$$

$$= x^n + nx^{n-1}h + h^2 \cdot P$$

The resulting formula

$$(x + h)^n = x^n + nx^{n-1}h + h^2 \cdot P$$

P stands for the polynomial in the square bracket above

will be useful in the following verification. To prove the Power Rule for any positive integer n, we use the definition of the derivative to differentiate $f(x) = x^n$.

$$f'(x) = \lim_{h \to 0} \frac{f(x + h) - f(x)}{h} \qquad \text{Definition of the derivative}$$

$$= \lim_{h \to 0} \frac{(x + h)^n - x^n}{h} \qquad \text{Since } f(x + h) = (x + h)^n \text{ and } f(x) = x^n$$

$$= \lim_{h \to 0} \frac{x^n + nx^{n-1}h + h^2 \cdot P - x^n}{h} \qquad \text{Expanding, using the formula derived earlier}$$

$$= \lim_{h \to 0} \frac{nx^{n-1}h + h^2 \cdot P}{h} \qquad \text{Canceling the } x^n \text{ and the } -x^n$$

$$= \lim_{h \to 0} \frac{h(nx^{n-1} + h \cdot P)}{h} \qquad \text{Factoring out an } h$$

$$= \lim_{h \to 0} (nx^{n-1} + h \cdot P) \qquad \text{Canceling the } h \text{ (since } h \neq 0)$$

$$= nx^{n-1} \qquad \text{Evaluating the limit by direct substitution}$$

This shows that for any positive integer n, the derivative of x^n is nx^{n-1}.

Verification of the Constant Multiple Rule

For a constant c and a function f, let $g(x) = c \cdot f(x)$. If $f'(x)$ exists, we may calculate the derivative $g'(x)$ as follows:

$$g'(x) = \lim_{h \to 0} \frac{g(x+h) - g(x)}{h} \qquad \text{Definition of the derivative}$$

$$= \lim_{h \to 0} \frac{c \cdot f(x+h) - c \cdot f(x)}{h} \qquad \begin{array}{l}\text{Since } g(x+h) = c \cdot f(x+h) \\ \text{and } g(x) = c \cdot f(x)\end{array}$$

$$= \lim_{h \to 0} \frac{c \cdot [f(x+h) - f(x)]}{h} \qquad \text{Factoring out the } c$$

$$= c \cdot \lim_{h \to 0} \frac{f(x+h) - f(x)}{h} \qquad \begin{array}{l}\text{Taking } c \text{ outside the limit} \\ \text{leaves just the definition} \\ \text{of the derivative } f'(x)\end{array}$$

$$= c \cdot f'(x) \qquad \text{Constant Multiple Rule}$$

This shows that the derivative of a constant times a function, $c \cdot f(x)$, is the constant times the derivative of the function, $c \cdot f'(x)$.

Verification of the Sum Rule

For two functions f and g, let their sum be $s = f + g$. If $f'(x)$ and $g'(x)$ exist, we may calculate $s'(x)$ as follows:

$$s'(x) = \lim_{h \to 0} \frac{s(x+h) - s(x)}{h} \qquad \text{Definition of the derivative}$$

$$= \lim_{h \to 0} \frac{[f(x+h) + g(x+h)] - [f(x) + g(x)]}{h} \qquad \begin{array}{l}\text{Since } s(x+h) = \\ f(x+h) + g(x+h) \\ \text{and } s(x) = f(x) + g(x)\end{array}$$

$$= \lim_{h \to 0} \frac{f(x+h) + g(x+h) - f(x) - g(x)}{h} \qquad \begin{array}{l}\text{Eliminating} \\ \text{the brackets}\end{array}$$

$$= \lim_{h \to 0} \frac{f(x+h) - f(x) + g(x+h) - g(x)}{h} \qquad \begin{array}{l}\text{Rearranging} \\ \text{the numerator}\end{array}$$

$$= \lim_{h \to 0} \left[\frac{f(x+h) - f(x)}{h} + \frac{g(x+h) - g(x)}{h} \right] \qquad \begin{array}{l}\text{Separating the} \\ \text{fraction into two} \\ \text{parts}\end{array}$$

$$= \underbrace{\lim_{h \to 0} \frac{f(x+h) - f(x)}{h}}_{f'(x)} + \underbrace{\lim_{h \to 0} \frac{g(x+h) - g(x)}{h}}_{g'(x)} \qquad \begin{array}{l}\text{Using Limit Rule} \\ \text{4a on page 74} \\[4pt] \text{Recognizing the definition} \\ \text{of the derivatives of } f \text{ and } g\end{array}$$

$$= f'(x) + g'(x) \qquad \text{Sum Rule}$$

This shows that the derivative of a sum $f(x) + g(x)$ is the sum of the derivatives $f'(x) + g'(x)$.

2.3 Exercises

1–30. Find the derivative of each function.

1. $f(x) = x^4$

2. $f(x) = x^5$

3. $f(x) = x^{500}$

4. $f(x) = x^{1000}$

5. $f(x) = x^{1/2}$

6. $f(x) = x^{1/3}$

7. $g(x) = \frac{1}{2}x^4$

8. $f(x) = \frac{1}{3}x^9$

9. $g(w) = 6\sqrt[3]{w}$

10. $g(w) = 12\sqrt{w}$

> ✏️ **FOR HELP GETTING STARTED**
>
> with Exercises 9–12, see Example 1 on pages 100–101.

11. $h(x) = \dfrac{3}{x^2}$

12. $h(x) = \dfrac{4}{x^3}$

13. $f(x) = 4x^2 - 3x + 2$

14. $f(x) = 3x^2 - 5x + 4$

15. $f(x) = \dfrac{1}{x^{1/2}}$

16. $f(x) = \dfrac{1}{x^{2/3}}$

17. $f(x) = \dfrac{6}{\sqrt[3]{x}}$

18. $f(x) = \dfrac{4}{\sqrt{x}}$

19. $f(r) = \pi r^2$

20. $f(r) = \dfrac{4}{3}\pi r^3$

21. $f(x) = \dfrac{1}{6}x^3 + \dfrac{1}{2}x^2 + x + 1$

22. $f(x) = \dfrac{1}{24}x^4 + \dfrac{1}{6}x^3 + \dfrac{1}{2}x^2 + x + 1$

23. $g(x) = \sqrt{x} - \dfrac{1}{x}$

24. $g(x) = \sqrt[3]{x} - \dfrac{1}{x}$

25. $h(x) = 6\sqrt[3]{x^2} - \dfrac{12}{\sqrt[3]{x}}$

26. $h(x) = 8\sqrt{x^3} - \dfrac{8}{\sqrt[4]{x}}$

27. $f(x) = \dfrac{10}{\sqrt{x}} - \dfrac{9\sqrt[3]{x^5}}{5} + 17$

28. $f(x) = \dfrac{9}{2\sqrt[3]{x^2}} - 16\sqrt{x^5} - 14$

29. $f(x) = \dfrac{x^2 + x^3}{x}$

30. $f(x) = x^2(x + 1)$

31–38. Find the indicated derivatives.

31. If $f(x) = x^5$, find $f'(-2)$.

32. If $f(x) = x^4$, find $f'(-3)$.

33. If $f(x) = 6\sqrt[3]{x^2} - \dfrac{48}{\sqrt[3]{x}}$, find $f'(8)$.

34. If $f(x) = 12\sqrt[3]{x^2} + \dfrac{48}{\sqrt[3]{x}}$, find $f'(8)$.

35. If $f(x) = x^3$, find $\dfrac{df}{dx}\bigg|_{x=-3}$

36. If $f(x) = x^4$, find $\dfrac{df}{dx}\bigg|_{x=-2}$

37. If $f(x) = \dfrac{16}{\sqrt{x}} + 8\sqrt{x}$, find $\dfrac{df}{dx}\bigg|_{x=4}$

38. If $f(x) = \dfrac{54}{\sqrt{x}} + 12\sqrt{x}$, find $\dfrac{df}{dx}\bigg|_{x=9}$

39. a. Find the equation of the tangent line to $f(x) = x^2 - 2x + 2$ at $x = 3$.

b. Graph the function and the tangent line on the window $[-1, 6]$ by $[-10, 20]$.

40. a. Find the equation of the tangent line to $f(x) = x^2 - 4x + 6$ at $x = 1$.

b. Graph the function and the tangent line on the window $[-1, 5]$ by $[-2, 10]$.

41. a. Find the equation of the tangent line to $f(x) = x^3 - 3x^2 + 2x - 2$ at $x = 2$.

b. Graph the function and the tangent line on the window $[-1, 4]$ by $[-7, 5]$.

42. a. Find the equation of the tangent line to $f(x) = 3x^2 - x^3$ at $x = 1$.

b. Graph the function and the tangent line on the window $[-1, 3]$ by $[-2, 5]$.

43. Use a graphing calculator to verify that the derivative of a constant is zero, as follows. Define y_1 to be a constant (such as $y_1 = 5$) and then use NDERIV to define y_2 to be the derivative of y_1. Then graph the two functions together on an appropriate window and use TRACE to observe that the derivative y_2 is zero (graphed as a line along the x-axis), showing that the derivative of a constant is zero.

44. Use a graphing calculator to verify that the derivative of a linear function is a constant, as follows. Define y_1 to be a linear function (such as $y_1 = 3x - 4$) and then use NDERIV to define y_2 to be the derivative of y_1. Then graph the two functions together on an appropriate window and observe that the derivative y_2 is a constant (graphed as a horizontal line, such as $y_2 = 3$), verifying that the derivative of $y_1 = mx + b$ is $y_2 = m$.

Applied Exercises

45–46. BUSINESS: Phillips Curves Unemployment and inflation are inversely related, with one rising as the other falls, and an equation giving the relation is called a *Phillips curve* after the economist A. W. Phillips (1914–1975).

45. Phillips used data from 1861 to 1957 to establish that in the United Kingdom the unemployment rate x and the wage inflation rate y were related by

$$y = 9.638x^{-1.394} - 0.900$$

where x and y are both in percents. Find the derivative of this function at each x-value and interpret your results.

a. 2 percent **b.** 5 percent

Source: Economica **25**

46. Between 2000 and 2010, the Phillips curve for the U.S. unemployment rate x and Consumer Price Index (CPI) inflation rate y was

$$y = 45.4x^{-1.54} - 1$$

where x and y are both in percents. Find the derivative of this function at each x-value and interpret your results.

a. 3 percent
b. 8 percent

Source: Bureau of Labor Statistics

47. BUSINESS: Software Costs Businesses can buy multiple licenses for PowerZip data-compression software at a total cost of approximately $C(x) = 24x^{2/3}$ dollars for x licenses. Find the derivative of this cost function at:

a. $x = 8$ and interpret your answer.
b. $x = 64$ and interpret your answer.

Source: Trident Software

48. BUSINESS: Software Costs Media companies can buy multiple licenses for AudioTime audio-recording software at a total cost of approximately $C(x) = 168x^{5/6}$ dollars for x licenses. Find the derivative of this cost function at:

a. $x = 1$ and interpret your answer.
b. $x = 64$ and interpret your answer.

Source: NCH Swift Sound

49. BUSINESS: Marginal Cost (*47 continued*) Use a calculator to find the actual cost of the 64th license by evaluating $C(64) - C(63)$ for the cost function in Exercise 47. Is your answer close to the $4 that you found for part (b) of that Exercise?

50. BUSINESS: Marginal Cost (*48 continued*) Use a calculator to find the actual cost of the 64th license by evaluating $C(64) - C(63)$ for the cost function in Exercise 48. Is your answer close to the $70 that you found for part (b) of that Exercise?

51. BUSINESS: Marketing to Young Adults Companies selling products to young adults often try to predict the size of that population in future years. According to predictions by the Census Bureau, the 18- to 24-year-old population in the United States will follow the curve shown below.

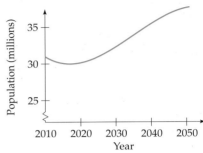

This population is given by

$$P(x) = -\tfrac{1}{3}x^3 + 25x^2 - 300x + 31{,}000$$

(in thousands), where x is the number of years after 2010. Find the rate of change of this population:

a. In the year 2030 and interpret your answer.
b. In the year 2010 and interpret your answer.

Source: Statistical Abstract of the United States

52. BIOMEDICAL: Flu Epidemic The number of people newly infected on day t of a flu epidemic is $f(t) = 13t^2 - t^3$ (for $0 \le t \le 13$). Find the instantaneous rate of change of this number on:

a. Day 5 and interpret your answer.
b. Day 10 and interpret your answer.

53. BUSINESS: Advertising It has been estimated that the number of people who will see a newspaper advertisement that has run for x consecutive days is of the form $N(x) = T - \tfrac{1}{2}T/x$ for $x \ge 1$, where T is the total readership of the newspaper. If a newspaper has a circulation of 400,000, an ad that runs for x days will be seen by

$$N(x) = 400{,}000 - \frac{200{,}000}{x}$$

people. Find how fast this number of potential customers is growing when this ad has run for 5 days.

54. BIOMEDICAL: Blood Flow Nitroglycerin is often prescribed to enlarge blood vessels that have become too constricted. If the cross-sectional area of a blood vessel t hours after nitroglycerin is administered is $A(t) = 0.01t^2$ square centimeters (for $1 \le t \le 5$), find the instantaneous rate of change of the cross-sectional area 4 hours after the administration of nitroglycerin.

55. BUSINESS: 3D Movies The number of 3D movies in the United States in recent years is released in the United States in recent years is approximated by $f(x) = \tfrac{1}{2}x^2 + 11x + 8$, where x is the number of years after 2010. Find the rate of change of this number in the year 2020 and interpret your answer.

Source: Motion Picture Association of America

56. GENERAL: Traffic Safety Traffic fatalities have decreased over recent decades, with the number of fatalities per hundred million vehicle miles traveled given approximately by $f(x) = \dfrac{1}{\sqrt{x}} + 1$, where x stands for the number of five-year intervals since 1985 (so, for example, $x = 2$ would mean 1995). Find the rate of change of this function at:

a. $x = 4$ and interpret your answer.
b. $x = 9$ and interpret your answer.

Source: National Highway Traffic Safety Administration

57. PSYCHOLOGY: Learning Rates A language school has found that its students can memorize $p(t) = 24\sqrt{t}$ phrases in t hours of class (for $1 \le t \le 10$). Find the instantaneous rate of change of this quantity after 4 hours of class.

58. ENVIRONMENTAL SCIENCE: Water Quality Downstream from a waste treatment plant the amount of dissolved oxygen in the water usually decreases for some distance (due to bacteria consuming the oxygen) and then increases (due to natural purification). A graph of the dissolved oxygen at various distances downstream looks like the curve below (known as the "oxygen sag"). The amount of dissolved oxygen is usually taken as a measure of the health of the river.

Suppose that the amount of dissolved oxygen x miles downstream is $D(x) = 0.2x^2 - 2x + 10$ mpl (milligrams per liter) for $0 \le x \le 20$. Use this formula to find the instantaneous rate of change of the dissolved oxygen:

a. 1 mile downstream.
b. 10 miles downstream.

Interpret the signs of your answers.

59–60. ECONOMICS: Marginal Utility Generally, the more you have of something, the less valuable each additional unit becomes. For example, a dollar is less valuable to a millionaire than to a beggar. Economists define a person's "utility function" $U(x)$ for a product as the "perceived value" of having x units of that product. The *derivative* of $U(x)$ is called the *marginal utility function*, $MU(x) = U'(x)$. Suppose that a person's utility function for money is given by the function below. That is, $U(x)$ is the utility (perceived value) of x dollars.

a. Find the marginal utility function $MU(x)$.
b. Find $MU(1)$, the marginal utility of the first dollar.

c. Find $MU(1,000,000)$, the marginal utility of the millionth dollar.

59. $U(x) = 100\sqrt{x}$ **60.** $U(x) = 12\sqrt[3]{x}$

61. GENERAL: Smoking and Education According to a study, the probability that a smoker will quit smoking increases with the smoker's educational level. The probability (expressed as a percent) that a smoker with x years of education will quit is approximated by the equation $f(x) = 0.831x^2 - 18.1x + 137.3$ (for $10 \le x \le 16$).

a. Find $f(12)$ and $f'(12)$ and interpret these numbers. [*Hint:* $x = 12$ corresponds to a high school graduate.]
b. Find $f(16)$ and $f'(16)$ and interpret these numbers. [*Hint:* $x = 16$ corresponds to a college graduate.]

Source: Review of Economics and Statistics **LXXVII**

 62. BIOMEDICAL: Lung Cancer Asbestos has been found to be a potent cause of lung cancer. According to one study of asbestos workers, the number of lung cancer cases in the group depended on the number t of years of exposure to asbestos according to the function $N(t) = 0.00437t^{3.2}$.

a. Graph this function on the window [0, 15] by [−10, 30].
b. Find $N(10)$ and $N'(10)$ and interpret these numbers.

Source: British Journal of Cancer **45**

63–64. GENERAL: College Tuition The following graph shows the average annual college tuition costs (tuition and fees) for a year at a private nonprofit or public four-year college. The data are given for five-year intervals.

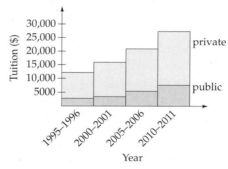

Source: The College Board

63. The tuition for a *private* college is approximated by the function $f(x) = 650x^2 + 3000x + 12,000$, where x is the number of five-year intervals since the academic year 1995–96 (so the years in the graph are numbered $x = 0$ through $x = 3$).

a. Use this function to predict tuition in the academic year 2020–21. [*Hint:* What x-value corresponds to that year?]
b. Find the derivative of this function for the x-value that you used in part (a) and interpret it as a rate of change in the proper units.
c. From your answer to part (b), estimate how rapidly tuition will be increasing *per year* in 2020–21.

64. The tuition for a *public* college is approximated by the function $f(x) = 400x^2 + 500x + 2700$, where x is the number of five-year intervals since the academic year 1995–96 (so the years in the graph are numbered $x = 0$ through $x = 3$).

   a. Use this function to predict tuition in the academic year 2020–21. [*Hint:* What x-value corresponds to that year?]

b. Find the derivative of this function for the x-value that you used in part (a) and interpret it as a rate of change in the proper units.

c. From your answer to part (b), estimate how rapidly tuition will be increasing *per year* in 2020–21.

Conceptual Exercises

65. Explain, in *two different ways*, without using the rules of differentiation, why the derivative of the constant function $f(x) = 2$ must be $f'(x) = 0$. [*Hint:* Think of the slope of the graph of a constant function, and also of the instantaneous rate of change of a function that stays constant.]

66. Explain, in *two different ways*, without using the rules of differentiation, why the derivative of the linear function $f(x) = 3x - 5$ must be $f'(x) = 3$. [*Hint:* Think of the slope of the line $y = mx + b$ that represents this function, and also of the instantaneous rate of change of a function that increases linearly.]

67. Give an intuitive explanation of the Constant Multiple Rule (page 101) by thinking that if $f(x)$ has a certain rate of change, then what will be the rate of change of the function $2 \cdot f(x)$ or of the function $c \cdot f(x)$?

68. Give an intuitive explanation of the Sum Rule (page 103) by thinking that if $f(x)$ and $g(x)$ have certain rates of change, then what will be the rate of change of $f(x) + g(x)$?

69. How will the slopes of f and of $-f$ differ? Explain intuitively and in terms of the rules of differentiation.

70. How will the slopes of f and $f + 10$ differ? Explain intuitively and in terms of the rules of differentiation.

71. We have said that the expression $f'(2)$ means *first* differentiate and *then* evaluate. What if we were to first evaluate the function at 2 and *then* differentiate? What would we get? Would the answer depend on the particular function or on the particular number?

72. Give an example to show that if a function is positive (at a particular x-value) its derivative (at that same x-value) need not be positive.

73. A recent article studying the relationship between life expectancy and education found that if a function $l(s)$ gives the life expectancy of a person who has had s years of schooling, then $\dfrac{dl}{ds} = 1.7$. Interpret this result.

Source: Review of Economic Studies, 72

74. A recent article studying the relationship between the probability of an accident and driving speed found that if a function $p(s)$ gives the probability (as a percent) of an accident for a driver who is exceeding the speed limit by s%, then $\dfrac{dp}{ds} = 13$. Interpret this result.

Source: Accident Analysis and Prevention, 38

Explorations and Excursions The following problem extends and augments the material presented in the text.

Tangent lines are the best linear approximations

75. To find the line $y = L(x)$ that best approximates a curve $y = f(x)$ near $x = c$, we calculate the difference between them

$$D(x) = f(x) - L(x)$$

and show that minimizing it near and at the point $(c, f(c))$ leads to the tangent line.

  a. Show that if the difference $D(x)$ is zero at $x = c$, then $f(c) = L(c)$.

  b. Use the result of part (a) to show that the line passes through the point $(c, f(c))$, and use the point-slope form for the equation of a line (see page 9) to show that the line has the form

$$L(x) = m(x - c) + f(c)$$

  c. Substitute this expression for $L(x)$ into the formula for $D(x)$ to show that if $D(c) = 0$, then

$$D(x) = f(x) - m(x - c) - f(c)$$

  d. Since the best linear approximation will be the line for which $D(x)$ changes from zero as little as possible near $x = c$, the derivative of $D(x)$ should be zero at c. Find $D'(c)$ and then find the m that makes $D'(c) = 0$.

  e. Find $L(x)$ for this best slope $m = f'(c)$ and show that the line $y = L(x)$ is the tangent line to the curve $y = f(x)$ at $x = c$.

  f. Explain why this shows that the tangent line to a curve at a point is the best linear approximation to the curve at that point.

2.4   The Product and Quotient Rules

Introduction

In the previous section we learned how to differentiate the sum and difference of two functions—we simply take the sum or difference of the derivatives. In this section we learn how to differentiate the *product* and *quotient* of two functions. Unfortunately, we do not simply take the product or quotient of the derivatives. Matters are a little more complicated.

Product Rule

To differentiate the product of two functions, $f(x) \cdot g(x)$, we use the *Product Rule*.

Product Rule

$$\frac{d}{dx}[f(x) \cdot g(x)] = f'(x) \cdot g(x) + f(x) \cdot g'(x)$$

The derivative of a product is the derivative of the first times the second plus the first times the derivative of the second.

[provided, of course, that the derivatives $f'(x)$ and $g'(x)$ both exist]. A derivation of the Product Rule is given at the end of this section.

The formula is clearer if we write the functions simply as f and g.

$$\frac{d}{dx}(f \cdot g) = f' \cdot g + f \cdot g'$$

Derivative   Second   First   Derivative
of the first   of the second

EXAMPLE 1   USING THE PRODUCT RULE

Use the Product Rule to calculate $\dfrac{d}{dx}(x^3 \cdot x^5)$.

Solution

$$\frac{d}{dx}(x^3 \cdot x^5) = 3x^2 \cdot x^5 + x^3 \cdot 5x^4 = 3x^7 + 5x^7 = 8x^7$$

Derivative   Second   First   Derivative
of the first   of the second

We may check this answer by simplifying the original product, $x^3 \cdot x^5 = x^8$, and then differentiating:

$$\frac{d}{dx}(\underbrace{x^3 \cdot x^5}_{x^8}) = \frac{d}{dx}x^8 = 8x^7 \qquad \text{Agrees with above answer}$$

Be Careful The derivative of a product is *not* the product of the derivatives: $(f \cdot g)' \neq f' \cdot g'$. For $x^3 \cdot x^5$ the product of the derivatives would be $3x^2 \cdot 5x^4 = 15x^6$, which is *not* the correct answer $8x^7$ that we found above. The Product Rule shows the correct way to differentiate a product.

EXAMPLE 2   **USING THE PRODUCT RULE**

Use the Product Rule to find $\dfrac{d}{dx}[(x^2 - x + 2)(x^3 + 3)]$.

Solution

$$\frac{d}{dx}[(x^2 - x + 2)(x^3 + 3)]$$

$$= \underbrace{(2x - 1)}(x^3 + 3) + (x^2 - x + 2)\underbrace{(3x^2)}$$

Derivative of   Derivative
$x^2 - x + 2$ of $x^3 + 3$

$$= 2x^4 + 6x - x^3 - 3 + 3x^4 - 3x^3 + 6x^2 \qquad \text{Multiplying out}$$

$$= 5x^4 - 4x^3 + 6x^2 + 6x - 3 \qquad \text{Simplifying}$$

In the preceding Example we could have *first* multiplied out the original function and *then* differentiated. However, our way was slightly easier. Furthermore, we will soon have problems that can *only* be done by the Product Rule, so the practice will be useful.

PRACTICE PROBLEM 1

Use the Product Rule to find $\dfrac{d}{dx}[x^3(x^2 - x)]$. Solution on page 121 >

EXAMPLE 3   **USING THE PRODUCT RULE**

Use the Product Rule to find the derivative of $f(x) = \sqrt{x}(2x - 4)$.

Solution

Writing the function in power form

$$f(x) = x^{1/2}(2x - 4)$$

the Product Rule gives

$$f'(x) = \tfrac{1}{2}x^{-1/2}(2x - 4) + x^{1/2}(2) = x^{1/2} - 2x^{-1/2} + 2x^{1/2} \qquad \text{Multiplying out}$$

Derivative   Derivative
of $x^{1/2}$ of $2x - 4$

$$= 3x^{1/2} - 2x^{-1/2} = 3\sqrt{x} - \frac{2}{\sqrt{x}} \qquad \begin{array}{l}\text{Combining terms}\\\text{and writing in}\\\text{radical form}\end{array}$$

FOR MORE HELP

with multiplying out and combining powers, see the Algebra Review appendix, page B5.

Quotient Rule

The *Quotient Rule* shows how to differentiate a quotient of two functions.

Quotient Rule

$$\frac{d}{dx}\left(\frac{f(x)}{g(x)}\right) = \frac{g(x)\cdot f'(x) - g'(x)\cdot f(x)}{[g(x)]^2}$$

The bottom times the derivative of the top, minus the derivative of the bottom times the top

The bottom squared

[provided that the derivatives $f'(x)$ and $g'(x)$ both exist and that $g(x) \neq 0$]. A derivation of the Quotient Rule is given at the end of this section.

The Quotient Rule looks less formidable if we write the functions simply as f and g,

$$\frac{d}{dx}\left(\frac{f}{g}\right) = \frac{g\cdot f' - g'\cdot f}{g^2}$$

or even as*

$$\frac{d}{dx}\left(\frac{\text{top}}{\text{bottom}}\right) = \frac{(\text{bottom})\cdot\left(\dfrac{d}{dx}\,\text{top}\right) - \left(\dfrac{d}{dx}\,\text{bottom}\right)\cdot(\text{top})}{(\text{bottom})^2}$$

EXAMPLE 4   **USING THE QUOTIENT RULE**

Use the Quotient Rule to find $\dfrac{d}{dx}\left(\dfrac{x^9}{x^3}\right)$.

Solution

Bottom   Derivative of the top   Derivative of the bottom   Top

$$\frac{d}{dx}\left(\frac{x^9}{x^3}\right) = \frac{(x^3)(9x^8) - (3x^2)(x^9)}{(x^3)^2} = \frac{9x^{11} - 3x^{11}}{x^6} = \frac{6x^{11}}{x^6} = 6x^5$$

Bottom squared

We may check this answer by simplifying the original quotient and then differentiating:

$$\frac{d}{dx}\left(\frac{x^9}{x^3}\right) = \frac{d}{dx}x^6 = 6x^5$$ Agrees with above answer

 Be Careful The derivative of a quotient is *not* the quotient of the derivatives:

$$\left(\frac{f}{g}\right)' \text{ is } not \text{ equal to } \frac{f'}{g'}$$

---

*Some students remember this by the mnemonic *Lo D Hi − Hi D Lo over Lo Lo* where *Lo* and *Hi* stand for the denominator and numerator, respectively.

For the quotient $\dfrac{x^9}{x^3}$ taking the quotient of the derivatives would give $\dfrac{9x^8}{3x^2} = 3x^6$, which is *not* the correct answer $6x^5$ that we found above. The Quotient Rule shows the correct way to differentiate a quotient.

EXAMPLE 5   USING THE QUOTIENT RULE

Find $\dfrac{d}{dx}\left(\dfrac{x^2}{x+1}\right)$.

Solution

The Quotient Rule gives

$$\underset{\text{Bottom squared}}{\underbrace{\dfrac{d}{dx}\left(\dfrac{x^2}{x+1}\right) = \dfrac{(x+1)(2x) - (1)(x^2)}{(x+1)^2} = \dfrac{2x^2 + 2x - x^2}{(x+1)^2} = \dfrac{x^2 + 2x}{(x+1)^2}}}$$

Bottom · Derivative of the top · Derivative of the bottom · Top

PRACTICE PROBLEM 2

Find $\dfrac{d}{dx}\left(\dfrac{2x^2}{x^2+1}\right)$.

Solution on page 121 >

EXAMPLE 6   FINDING THE COST OF CLEANER WATER

Practically every city must purify its drinking water and treat its wastewater. The cost of the treatment rises steeply for higher degrees of purity. If the cost of purifying a gallon of water to a purity of x percent is

$$C(x) = \frac{2}{100 - x} \qquad \text{for } 80 < x < 100$$

dollars, find the rate of change of the purification costs when the purity is:

a. 90%   **b.** 98%

Solution

The rate of change of cost is the *derivative* of the cost function:

$$C'(x) = \frac{d}{dx}\left(\frac{2}{100 - x}\right) = \frac{(100 - x)(0) - (-1)(2)}{(100 - x)^2}$$

Derivative of 2 · Derivative of $100 - x$

Differentiating by the Quotient Rule

$$= \frac{0 + 2}{(100 - x)^2} = \frac{2}{(100 - x)^2}$$

Simplifying (the derivative is undefined at $x = 100$)

a. For 90% purity we evaluate at $x = 90$:

$$C'(90) = \frac{2}{(100 - 90)^2} = \frac{2}{10^2} = \frac{2}{100} = 0.02$$

$C'(x) = \dfrac{2}{(100 - x)^2}$

evaluated at $x = 90$

Interpretation: At 90% purity, the rate of change of the cost is 0.02 dollars per percentage point, meaning that the costs increase by about *2 cents for each additional percentage of purity.*

b. For 98% purity we evaluate $C'(x)$ at $x = 98$:

$$C'(98) = \frac{2}{(100 - 98)^2} = \frac{2}{2^2} = \frac{2}{4} = \frac{1}{2} = 0.50$$

$C'(x) = \dfrac{2}{(100 - x)^2}$ evaluated at $x = 98$

Interpretation: At 98% purity, the rate of change of the cost is 0.50 dollars per percentage point, meaning that the costs increase by about *50 cents for each additional percentage of purity.*

Notice that an extra percentage of purity above the 98% level is 25 times as costly as an extra percentage above the 90% purity level.

Graphing Calculator Exploration

└ calculator error!

a. On a graphing calculator, enter the cost function from Example 6 as $y_1 = 2/(100 - x)$. Then use NDERIV to define y_2 to be the derivative of y_1.

b. Graph both y_1 and y_2 on the window [80,100] by [−1, 5]. Your graph should resemble the one on the left (but you may have an additional "false" vertical line on the right).

c. Verify the results of Example 6 by evaluating y_2 at $x = 90$ and at $x = 98$.

d. Evaluate y_2 at $x = 100$, giving (supposedly) the derivative of y_1 at $x = 100$. However, in Example 6 we saw that the derivative of y_1 is *undefined* at $x = 100$. Your calculator is giving you a "false value" for the derivative, resulting from NDERIV's use of a symmetric difference quotient (see page 107) and a small positive value for *h*. Therefore, to use your calculator effectively, you must also understand calculus.

Not every quotient requires the Quotient Rule. Some are simple enough to be differentiated by the Power Rule.

EXAMPLE 7 **DIFFERENTIATING A QUOTIENT BY THE POWER RULE**

Find the derivative of $y = \dfrac{5}{x^2}$.

Solution

$$\frac{d}{dx}\left(\frac{5}{x^2}\right) = \frac{d}{dx}(5x^{-2}) = -10x^{-3} = -\frac{10}{x^3}$$

Differentiated by the Power Rule

In this Example we rewrote the expression before and after the differentiation:

Begin	Rewrite	Differentiate	Rewrite
$y = \dfrac{5}{x^2}$	$y = 5x^{-2}$	$\dfrac{dy}{dx} = -10x^{-3}$	$\dfrac{dy}{dx} = -\dfrac{10}{x^3}$

Take Note

If the numerator or denominator is a constant, rewrite the fraction to avoid having to use the Quotient Rule.

This way is much easier than using the Quotient Rule if the numerator or denominator is a constant.

Marginal Average Cost

It is often useful to calculate not just the *total* cost of producing x units of some product, but also the **average cost per unit,** denoted $AC(x)$, which is found by dividing the total cost $C(x)$ by the number of units x.

$$AC(x) = \frac{C(x)}{x}$$ Average cost per unit is total cost divided by the number of units

The derivative of the average cost function is called the **marginal average cost,** *MAC.**

$$MAC(x) = \frac{d}{dx}\left[\frac{C(x)}{x}\right]$$ Marginal average cost is the derivative of average cost

Marginal average revenue, *MAR,* and **marginal average profit,** *MAP,* are defined similarly as the derivatives of average revenue per unit, $\dfrac{R(x)}{x}$, and average profit per unit, $\dfrac{P(x)}{x}$.

$$MAR(x) = \frac{d}{dx}\left[\frac{R(x)}{x}\right]$$ Marginal average revenue is the derivative of average revenue $\dfrac{R(x)}{x}$

$$MAP(x) = \frac{d}{dx}\left[\frac{P(x)}{x}\right]$$ Marginal average profit is the derivative of average profit $\dfrac{P(x)}{x}$

EXAMPLE 8   **FINDING AND INTERPRETING MARGINAL AVERAGE COST**

POD, or *printing on demand,* is a recent development in publishing that makes it feasible to print small quantities of books (even a single copy), thereby eliminating overstock and storage costs. For example, POD-publishing a typical 200-page book would cost $18 per copy, with fixed costs of $1500. Therefore, the cost function is

$$C(x) = 18x + 1500$$ Total cost of producing x books

a. Find the average cost function.

b. Find the marginal average cost function.

c. Find the marginal average cost at $x = 100$ and interpret your answer.

Source: e-booktime.com

---

*The marginal average cost function is sometimes denoted $\overline{C}'(x)$, with similar notations used for marginal average revenue and marginal average profit, although we will not use this notation.

Solution

a. The average cost function is

$$AC(x) = \underbrace{\frac{18x + 1500}{x}}_{\substack{\text{Total cost divided} \\ \text{by number of units}}} = \underbrace{18 + \frac{1500}{x}}_{\text{Simplifying}} = \underbrace{18 + 1500x^{-1}}_{\text{In power form}}$$

b. The *marginal* average cost is the derivative of average cost. We could use the Quotient Rule on the first expression above, but it is easier to use the Power Rule on the last expression:

$$MAC(x) = \frac{d}{dx}(18 + 1500x^{-1}) = -1500x^{-2} = -\frac{1500}{x^2}$$

c. Evaluating at $x = 100$:

$$MAC(100) = -\frac{1500}{100^2} = -\frac{1500}{10,000} = -0.15 \qquad -\frac{1500}{x^2} \text{ at } x = 100$$

Interpretation: When 100 books have been produced, the average cost per book is decreasing (because of the negative sign) by about *15 cents per additional book produced*. This reflects the fact that while *total* costs rise when you produce more, the *average cost per unit* decreases, because of the economies of mass production.

Graphing Calculator Exploration

Use a graphing calculator to investigate further the effects of mass production in the preceding Example.

Graph the average cost function [any of the expressions for $AC(x)$ from part (a) of the solution] on the window [0,400] by [0,50]. Your graph should resemble that shown on the left, showing that the average cost drops from the 30s down to the 20s as the number of books increases from 100 to 400. Note that although average cost falls, it does so more slowly as the number of units increases (the *law of diminishing returns*). Using dy/dx at $x = 100$ we get the value of $MAC(100)$ that we found in part (c).

EXAMPLE 9   **FINDING TIME SAVED BY SPEEDING**

A certain mathematics professor drives 25 miles to his office every day, mostly on highways. If he drives at constant speed v miles per hour, his travel time (distance divided by speed) is

$$T(v) = \frac{25}{v}$$

hours. Find $T'(55)$ and interpret this number.

Solution

Since $T(v) = \dfrac{25}{v}$ is a quotient, we could differentiate it by the Quotient Rule. However, it is easier to write $\dfrac{25}{v}$ as a *power*,

$$T(v) = 25v^{-1}$$

and differentiate using the Power Rule:

$$T'(v) = -25v^{-2}$$

This gives the rate of change of the travel time with respect to driving speed. $T'(v)$ is negative, showing that as speed increases, travel time *decreases*. Evaluating this at speed $v = 55$ gives

$$T'(55) = -25(55)^{-2} = \frac{-25}{(55)^2} \approx -0.00826 \qquad \text{Using a calculator}$$

This number, the rate of change of travel time with respect to driving speed, means that when driving at 55 miles per hour, you save only 0.00826 hour for each extra mile per hour of speed. Multiplying by 60 gives the saving in *minutes*:

$$(-0.00826)(60) \approx -0.50 = -\frac{1}{2}$$

That is, each extra mile per hour of speed saves only about half a minute, or 30 seconds. For example, speeding by 10 mph would save only about $\frac{1}{2} \cdot 10 = 5$ minutes. One must then decide whether this slight savings in time is worth the risk of an accident or a speeding ticket.

Solutions TO PRACTICE PROBLEMS

1. $\dfrac{d}{dx}[x^3(x^2 - x)] = 3x^2(x^2 - x) + x^3(2x - 1)$

$$= 3x^4 - 3x^3 + 2x^4 - x^3$$

$$= 5x^4 - 4x^3$$

2. $\dfrac{d}{dx}\left(\dfrac{2x^2}{x^2 + 1}\right) = \dfrac{(x^2 + 1)4x - 2x \cdot 2x^2}{(x^2 + 1)^2}$

$$= \frac{4x^3 + 4x - 4x^3}{(x^2 + 1)^2} = \frac{4x}{(x^2 + 1)^2}$$

2.4   Section Summary

The following is a list of the differentiation formulas that we have learned so far. The letters c and n stand for constants, and f and g stand for differentiable functions of x.

$$\frac{d}{dx}c = 0$$

$$\frac{d}{dx}x^n = nx^{n-1} \qquad \text{Special case: } \frac{d}{dx}x = 1$$

$$\frac{d}{dx}(c \cdot f) = c \cdot f' \qquad \text{Special case: } \frac{d}{dx}(cx) = c$$

$$\frac{d}{dx}(f \pm g) = f' \pm g'$$

$$\frac{d}{dx}(f \cdot g) = f' \cdot g + f \cdot g'$$

$$\frac{d}{dx}\left(\frac{f}{g}\right) = \frac{g \cdot f' - g' \cdot f}{g^2} \qquad g \neq 0$$

The preceding formulas are used extensively throughout calculus, and you should not proceed to the next section until you have mastered them.

Verification of the Differentiation Formulas

We conclude this section with derivations of the Product and Quotient Rules, and the Power Rule in the case of *negative* integer exponents. First, however, we need to establish a preliminary result about an arbitrary function g:

$$\text{If } g'(x) \text{ exists, then } \lim_{h \to 0} g(x + h) = g(x).$$

We begin with $\lim_{h \to 0} g(x + h)$ and show that it is equal to $g(x)$:

$$\lim_{h \to 0} g(x + h) = \lim_{h \to 0} [g(x + h) - g(x) + g(x)] \qquad \text{Subtracting and adding } g(x)$$

$$= \lim_{h \to 0} \left[\frac{g(x + h) - g(x)}{h} \cdot h + g(x) \right] \qquad \text{Dividing and multiplying by } h$$

$$= \lim_{h \to 0} \left[\frac{g(x + h) - g(x)}{h} \cdot h \right] + \lim_{h \to 0} g(x) \qquad \text{The limit of a sum is the sum of the limits}$$

$$= \underbrace{\lim_{h \to 0} \frac{g(x + h) - g(x)}{h}}_{g'(x)} \cdot \underbrace{\lim_{h \to 0} h}_{0} + g(x) \qquad \text{The limit of a product is the product of the limits}$$

$$= g'(x) \cdot 0 + g(x) \qquad \text{Since the first limit above is the definition of } g'(x) \text{ and the second limit is zero}$$

$$= g(x) \qquad \text{Simplifying}$$

This proves the result that if $g'(x)$ exists, then $\lim_{h \to 0} g(x + h) = g(x)$. Replacing $x + h$ by a new variable y, this equation becomes

$$\lim_{y \to x} g(y) = g(x) \qquad y = x + h, \text{ so } h \to 0 \text{ implies } y \to x$$

According to the definition of continuity on page 80 (but with different letters), this equation means that the function g is *continuous* at x. Therefore, the result that we have shown can be stated simply:

If a function is *differentiable* at x, then it is *continuous* at x.

Or, even more briefly:

> Differentiability implies continuity.

Verification of the Product Rule

For two functions f and g, let their product be $p(x) = f(x) \cdot g(x)$. If $f'(x)$ and $g'(x)$ exist, we may calculate $p'(x)$ as follows.

$$p'(x) = \lim_{h \to 0} \frac{p(x + h) - p(x)}{h}$$ Definition of the derivative

$$= \lim_{h \to 0} \frac{f(x + h)g(x + h) - f(x)g(x)}{h}$$ $p(x + h) = f(x + h) \cdot g(x + h)$ and $p(x) = f(x) \cdot g(x)$

$$= \lim_{h \to 0} \frac{f(x + h)g(x + h) - f(x)g(x + h) + f(x)g(x + h) - f(x)g(x)}{h}$$ Subtracting and adding $f(x)g(x + h)$

$$= \lim_{h \to 0} \left[\frac{f(x + h)g(x + h) - f(x)g(x + h)}{h} + \frac{f(x)g(x + h) - f(x)g(x)}{h} \right]$$ Separating the fraction into two parts

$$= \lim_{h \to 0} \frac{[f(x + h) - f(x)]g(x + h)}{h} + \lim_{h \to 0} \frac{f(x)[g(x + h) - g(x)]}{h}$$ Using Limit Rule 4a on page 74 and factoring

$$= \underbrace{\lim_{h \to 0} \frac{f(x + h) - f(x)}{h}}_{f'(x)} \underbrace{\lim_{h \to 0} g(x + h)}_{g(x)} + \underbrace{f(x)}_{f(x)} \underbrace{\lim_{h \to 0} \frac{g(x + h) - g(x)}{h}}_{g'(x)}$$ Using Limit Rule 4c on page 75

Recognizing the definitions of $f'(x)$ and $g'(x)$

$$= f'(x) \cdot g(x) + f(x) \cdot g'(x)$$ Product Rule

Verification of the Quotient Rule

For two functions f and g with $g(x) \neq 0$, let the quotient be $q(x) = \dfrac{f(x)}{g(x)}$. If $f'(x)$ and $g'(x)$ exist, we may calculate $q'(x)$ as follows.

$$q'(x) = \lim_{h \to 0} \frac{q(x + h) - q(x)}{h}$$ Definition of the derivative

$$= \lim_{h \to 0} \frac{\dfrac{f(x + h)}{g(x + h)} - \dfrac{f(x)}{g(x)}}{h}$$ $q(x + h) = \dfrac{f(x + h)}{g(x + h)}$

and $q(x) = \dfrac{f(x)}{g(x)}$

$$= \lim_{h \to 0} \frac{1}{h} \left[\frac{f(x + h)}{g(x + h)} - \frac{f(x)}{g(x)} \right]$$ Since dividing by h is equivalent to multiplying by $1/h$

$$= \lim_{h \to 0} \left[\frac{1}{h} \cdot \frac{g(x)f(x + h) - g(x + h)f(x)}{g(x + h)g(x)} \right]$$ Subtracting the fractions, using the common denominator $g(x + h)g(x)$

$$= \lim_{h \to 0} \left[\frac{1}{h} \cdot \frac{g(x)f(x + h) - g(x)f(x) - [g(x + h)f(x) - g(x)f(x)]}{g(x + h)g(x)} \right]$$ Subtracting and adding $g(x)f(x)$

$$= \lim_{h \to 0} \left[\frac{1}{g(x + h)g(x)} \cdot \frac{g(x)[f(x + h) - f(x)] - [g(x + h) - g(x)]f(x)}{h} \right]$$ Factoring in the numerator; switching the denominators

$$= \lim_{h \to 0} \left[\underbrace{\frac{1}{g(x + h)g(x)}}_{\substack{\text{Approaches} \\ g(x)}} \left(g(x) \underbrace{\lim_{h \to 0} \frac{f(x + h) - f(x)}{h}}_{f'(x)} - \underbrace{\lim_{h \to 0} \frac{g(x + h) - g(x)}{h}}_{g'(x)} f(x) \right) \right]$$ Using Limit Rules 4b and 4c on page 75

$$= \frac{1}{[g(x)]^2} [g(x)f'(x) - g'(x)f(x)]$$ Using Limit Rules 1 and 4d on pages 74–75

$$= \frac{g(x)f'(x) - g'(x)f(x)}{[g(x)]^2}$$ Quotient Rule

Verification of the Power Rule for Negative Integer Exponents

On page 108 we proved the Power Rule for *positive* integer exponents. Using the Quotient Rule, we may now prove the Power Rule for *negative* integer exponents. Any negative integer n may be written as $n = -p$, where p is a *positive* integer. Then

$$\frac{d}{dx}x^n = \frac{d}{dx}\left(\frac{1}{x^p}\right) \qquad \text{Since } x^n = x^{-p} = \frac{1}{x^p}$$

$$= \frac{x^p \cdot 0 - px^{p-1} \cdot 1}{x^{2p}} \qquad \text{Using the Quotient Rule, with } \frac{d}{dx}1 = 0 \text{ and } \frac{d}{dx}x^p = px^{p-1}$$

$$= \frac{-px^{p-1}}{x^{2p}} \qquad \text{Simplifying}$$

$$= -px^{p-1-2p} = \underbrace{-p}_{n}\underbrace{x^{-p-1}}_{n-1} \qquad \text{Subtracting exponents and simplifying}$$

$$\underbrace{}_{-p-1} \qquad \qquad \text{Since } -p = n$$

$$= nx^{n-1} \qquad \text{Power Rule}$$

This proves the Power Rule, $\dfrac{d}{dx}x^n = nx^{n-1}$, for negative integer exponents n.

2.4 Exercises

1–4. Find the derivative of each function in two ways:

a. Using the *Product* Rule.
b. Multiplying out the function and using the *Power* Rule. Your answers to parts (a) and (b) should agree.

1. $x^4 \cdot x^6$ **2.** $x^7 \cdot x^2$

3. $x^4(x^5 + 1)$ **4.** $x^5(x^4 + 1)$

5–26. Find the derivative of each function by using the Product Rule. Simplify your answers.

5. $f(x) = x^2(x^3 + 1)$ **6.** $f(x) = x^3(x^2 + 1)$

7. $f(x) = x(5x^2 - 1)$ **8.** $f(x) = 2x(x^4 + 1)$

9. $f(x) = \sqrt{x}(6x + 2)$ **10.** $f(x) = 6\sqrt[3]{x}\,(2x + 1)$

11. $f(x) = (x^2 + 1)(x^2 - 1)$

12. $f(x) = (x^3 - 1)(x^3 + 1)$

13. $f(x) = (x^2 + x)(3x + 1)$

14. $f(x) = (x^2 + 2x)(2x + 1)$

15. $f(x) = x^2(x^2 + 3x - 1)$

16. $f(x) = x^3(x^2 - 4x + 3)$

17. $f(x) = (2x^2 + 1)(1 - x)$

18. $f(x) = (2x - 1)(1 - x^2)$

19. $f(x) = (\sqrt{x} - 1)(\sqrt{x} + 1)$

20. $f(x) = (\sqrt{x} + 2)(\sqrt{x} - 2)$

21. $f(t) = 6t^{4/3}(3t^{2/3} + 1)$

22. $f(t) = 4t^{3/2}(2t^{1/2} - 1)$

23. $f(z) = (z^4 + z^2 + 1)(z^3 - z)$

24. $f(z) = (\sqrt[4]{z} + \sqrt{z})(\sqrt[4]{z} - \sqrt{z})$

25. $f(z) = (2z + 4\sqrt{z} - 1)(2\sqrt{z} + 1)$

26. $f(z) = (z + 6\sqrt{z})(z - 2\sqrt{z} + 1)$

27–30. Find the derivative of each function in two ways:

a. Using the *Quotient* rule.
b. Simplifying the original function and using the *Power* Rule.

Your answers to parts (a) and (b) should agree.

27. $\dfrac{x^8}{x^2}$ **28.** $\dfrac{x^9}{x^3}$

29. $\dfrac{1}{x^3}$ **30.** $\dfrac{1}{x^4}$

31–46. Find the derivative of each function by using the Quotient Rule. Simplify your answers.

31. $f(x) = \dfrac{x^4 + 1}{x^3}$ **32.** $f(x) = \dfrac{x^5 - 1}{x^2}$

33. $f(x) = \dfrac{x + 1}{x - 1}$ **34.** $f(x) = \dfrac{x - 1}{x + 1}$

35. $f(x) = \dfrac{3x + 1}{2 + x}$ **36.** $f(x) = \dfrac{x + 1}{2x^2 + 1}$

37. $f(t) = \dfrac{t^2 - 1}{t^2 + 1}$ **38.** $f(t) = \dfrac{t^2 + 1}{t^2 - 1}$

39. $f(s) = \dfrac{s^3 - 1}{s + 1}$ **40.** $f(s) = \dfrac{s^3 + 1}{s - 1}$

41. $f(x) = \dfrac{x^2 - 2x + 3}{x + 1}$ **42.** $f(x) = \dfrac{x^2 + 3x - 1}{x - 1}$

43. $f(x) = \dfrac{x^4 + x^2 + 1}{x^2 + 1}$ **44.** $f(x) = \dfrac{x^5 + x^3 + x}{x^3 + x}$

45. $f(t) = \dfrac{t^2 + 2t - 1}{t^2 + t - 3}$ **46.** $f(t) = \dfrac{2t^2 + t - 5}{t^2 - t + 2}$

47–52. Differentiate each function by rewriting before and after differentiating, as on page 118.

Begin	**Rewrite**	**Differentiate**	**Rewrite**

47. $y = \dfrac{3}{x}$

48. $y = \dfrac{x^2}{4}$

Begin	**Rewrite**	**Differentiate**	**Rewrite**

49. $y = \dfrac{3x^4}{8}$

50. $y = \dfrac{3}{2x^2}$

51. $y = \dfrac{x^2 - 5x}{3}$

52. $y = \dfrac{4}{\sqrt{x}}$

53–58. Find the derivative of each function.

53. $(x^3 + 2)\dfrac{x^2 + 1}{x + 1}$ **54.** $(x^5 + 1)\dfrac{x^3 + 2}{x + 1}$

55. $\dfrac{(x^2 + 3)(x^3 + 1)}{x^2 + 2}$ **56.** $\dfrac{(x^3 + 2)(x^2 + 2)}{x^3 + 1}$

57. $\dfrac{\sqrt{x} - 1}{\sqrt{x} + 1}$ **58.** $\dfrac{\sqrt{x} + 1}{\sqrt{x} - 1}$

Applied Exercises

59. BUSINESS: Marginal Average Revenue Use the Quotient Rule to find a general expression for the marginal average revenue. That is, calculate

$$\dfrac{d}{dx}\left[\dfrac{R(x)}{x}\right] \text{ and simplify your answer.}$$

60. BUSINESS: Marginal Average Profit Use the Quotient Rule to find a general expression for the marginal average profit. That is, calculate

$$\dfrac{d}{dx}\left[\dfrac{P(x)}{x}\right] \text{ and simplify your answer.}$$

61. ENVIRONMENTAL SCIENCE: Water Purification If the cost of purifying a gallon of water to a purity of x percent is

$$C(x) = \dfrac{100}{100 - x} \text{ cents} \quad \text{for } 50 \le x < 100$$

  a. Find the instantaneous rate of change of the cost with respect to purity.
  b. Evaluate this rate of change for a purity of 95% and interpret your answer.
  c. Evaluate this rate of change for a purity of 98% and interpret your answer.

62. BUSINESS: Marginal Average Cost A company can produce computer flash memory devices at a cost of $6 each, while fixed costs are $50 per day. Therefore, the company's cost function is $C(x) = 6x + 50$.

  a. Find the average cost function

$$AC(x) = \dfrac{C(x)}{x}.$$

  b. Find the marginal average cost function $MAC(x)$.

(continues)

  c. Evaluate $MAC(x)$ at $x = 25$ and interpret your answer.

  Source: Adco Marketing

63. ENVIRONMENTAL SCIENCE: Water Purification *(61 continued)*
  a. Use a graphing calculator to graph the cost function $C(x)$ from Exercise 61 on the window [50, 100] by [0, 20]. TRACE along the curve to see how rapidly costs increase for purity (x-coordinate) increasing from 50 to near 100.
  b. To check your answers to Exercise 61, use the "dy/dx" or SLOPE feature of your calculator to find the slope of the cost curve at $x = 95$ and at $x = 98$. The resulting rates of change of the cost should agree with your answers to Exercise 61(b) and (c). Note that further purification becomes increasingly expensive at higher purity levels.

64. BUSINESS: Marginal Average Cost *(62 continued)*
  a. Graph the average cost function $AC(x)$ that you found in Exercise 62(a) on the window [0, 50] by [0, 50]. TRACE along the average cost curve to see how the average cost falls as the number of devices increases. Note that although average cost falls, it does so more slowly as the number of units increases.
  b. To check your answer to Exercise 62, use the "dy/dx" or SLOPE feature of your calculator to find the slope of the average cost curve at $x = 25$. This slope gives the rate of change of the cost, which should agree with your answer to Exercise 62(c). Find the slope (rate of change) for other x-values to see that the rate of change of average cost tends toward zero (the law of diminishing returns).

65. **BUSINESS:** Marginal Average Cost A company can produce LCD digital alarm clocks at a cost of $8 each while fixed costs are $45. Therefore, the company's cost function is $C(x) = 8x + 45$.
    a. Find the average cost function $AC(x) = \dfrac{C(x)}{x}$.
    b. Find the marginal average cost function $MAC(x)$.
    c. Evaluate $MAC(x)$ at $x = 30$ and interpret your answer.

    Source: Casad Company

66. **BUSINESS:** Sales The number of bottles of whiskey that a store will sell in a month at a price of p dollars per bottle is

    $$N(p) = \frac{2250}{p + 7} \qquad (p \geq 5)$$

    Find the rate of change of this quantity when the price is $8 and interpret your answer.

67. **GENERAL:** Body Temperature If a person's temperature after x hours of strenuous exercise is $T(x) = x^3(4 - x^2) + 98.6$ degrees Fahrenheit (for $0 \leq x \leq 2$), find the rate of change of the temperature after 1 hour.

68. **BUSINESS:** CD Sales Suppose that after x months, monthly sales of a compact disc are predicted to be $S(x) = x^2(8 - x^3)$ thousand (for $0 \leq x \leq 2$). Find the rate of change of the sales after 1 month.

69. **GENERAL:** Body Temperature *(67 continued)*
    a. Graph the temperature function $T(x)$ given in Exercise 67 on the window [0, 2] by [90, 110]. TRACE along the temperature curve to see how the temperature rises and then falls as time increases.
    b. To check your answer to Exercise 67, use the *"dy/dx"* or SLOPE feature of your calculator to find the slope (rate of change) of the curve at $x = 1$. Your answer should agree with your answer to Exercise 67.
    c. TRACE along the temperature curve to estimate the maximum temperature.

70. **BUSINESS:** CD Sales *(68 continued)*
    a. Graph the sales function $S(x)$ given in Exercise 68 on the window [0, 2] by [0, 12]. TRACE along the sales curve to see how the sales rise and then fall as x, the number of months, increases.
    b. To check your answer to Exercise 68, use the *"dy/dx"* or SLOPE feature of your calculator to find the slope (rate of change) of the curve at $x = 1$. Your answer should agree with your answer to Exercise 68.
    c. TRACE along the curve to estimate the maximum sales.

71. **ECONOMICS:** National Debt The national debt (the amount of money that the federal government has borrowed from and therefore owes to the public) is approximately $D(x) = -73.1x^2 + 1270x + 16{,}280$ billion dollars, where x is the number of years since 2012. The population of the United States is approximately $P(x) = 2.3x + 314$ million.
    a. Enter these functions into your calculator as y_1 and y_2, respectively, and define y_3 to be $y_1 \div y_2$, the national debt divided by the population, so

that y_3 is the *per capita national debt*, in thousands of dollars (since it is billions divided by millions). Evaluate y_3 at 8 and at 13 to find the *per capita national debt in* the years 2020 and 2025. This is the amount that the government would owe each of its citizens if the debt were divided equally among them.
    b. Use the numerical derivative operation NDERIV to find the derivative of y_3 at 8 and at 13 and interpret your answers.

    Sources: U.S. Treasury Department, U.S. Census Bureau

72. **GENERAL:** Fuel Economy The gas mileage (in miles per gallon) of a subcompact car is approximately

    $$g(x) = \frac{-15x^2 + 1125x}{x^2 - 110x + 3500}$$

    where x is the speed in miles per hour (for $35 \leq x \leq 65$).
    a. Find $g'(x)$.
    b. Find $g'(40)$, $g'(50)$, and $g'(60)$ and interpret your answers.
    c. What does the sign of $g'(40)$ tell you about whether gas mileage increases or decreases with speed when driving at 40 mph? Do the same for $g'(60)$ and 60 mph. Then do the same for $g'(50)$ and 50 mph. From your answers, what do you think is the most economical speed for a subcompact car?

    Source: U.S. Environmental Protection Agency

73. **BUSINESS:** Car Loan Rates From 2008 to 2011, used car loan interest rates at auto finance companies were approximately

    $$I(x) = 0.45(x - 1.7)\,(x^2 - 12.5x + 43)$$

    percent where x is the number of years after 2005. Differentiating using the Product Rule, find $I'(5)$ and $I'(6)$ and interpret your answers.

    Source: Federal Reserve Board

74. **BUSINESS:** Car Loan Rates From 2008 to 2011, new car loan interest rates at auto finance companies were approximately

    $$I(x) = 0.106(x^2 - 6.85x + 14)\,(x^2 - 14.5x + 56)$$

    percent where x is the number of years after 2005. Differentiating using the Product Rule, find $I'(3)$ and $I'(5)$ and interpret your answers.

    Source: Federal Reserve Board

75. **BUSINESS:** Median Weekly Earnings From 2005 to 2013, the median usual weekly earnings of full-time wage and salary workers were approximately

    $$E(x) = \frac{261x}{x + 8.84} + 647$$

    dollars where x is the number of years after 2005. Differentiating using the Quotient Rule, find $E'(4)$ and $E'(8)$ and interpret your answers.

    Source: Bureau of Labor Statistics

76. **BUSINESS:** Consumer Credit From 2002 to 2013, outstanding consumer credit was approximately

    $$C(x) = \frac{75(44 - 5x)}{x^2 - 18.3x + 85} + 115x + 1605$$

millions dollars where x is the number of years after 2000. Differentiating using the Quotient Rule, find $C'(7)$, $C'(9)$, and $C'(11)$ and interpret your answers.

Source: Federal Reserve Board

 77–78. PITFALLS OF NDERIV ON A GRAPHING CALCULATOR

a. Find the derivative (by hand) of each function below, and observe that the derivative is undefined at $x = 0$.
b. Find the derivative of each function below by using NDERIV on a graphing calculator and evaluate the

derivative at $x = 0$. If your calculator gives you an answer, this is a "false value" for the derivative, since in part (a) you showed that the derivative is undefined at $x = 0$. [For an explanation, see the Graphing Calculator Exploration part (d) on page 118.]

77. $y = \dfrac{1}{x^2}$

78. $y = \dfrac{1}{x}$

Conceptual Exercises

79–84. Let f and g be differentiable functions of x. Assume that denominators are not zero.

79. True or False: $\dfrac{d}{dx}(f \cdot g) = f' \cdot g'$

80. True or False: $\dfrac{d}{dx}\left(\dfrac{f}{g}\right) = \dfrac{f'}{g'}$

81. True or False: $\dfrac{d}{dx}(x \cdot f) = f + x \cdot f'$

82. True or False: $\dfrac{d}{dx}\left(\dfrac{f}{x}\right) = \dfrac{x \cdot f' - f}{x^2}$

83. Show that the Product Rule may be written in the following form:
$$\frac{d}{dx}(f \cdot g) = (f \cdot g)\left(\frac{f'}{f} + \frac{g'}{g}\right)$$

[*Hint:* Multiply out the right-hand side.]

84. Show that the Quotient Rule may be written in the following form:
$$\frac{d}{dx}\left(\frac{f}{g}\right) = \left(\frac{f}{g}\right)\left(\frac{f'}{f} - \frac{g'}{g}\right)$$

[*Hint:* Multiply out the right-hand side and combine it into a single fraction.]

85–86. Imagine a country in which everyone is equally wealthy—call the identical amount of money that each person has the *personal wealth.* The *national wealth* of the entire country is then this personal wealth times the population. However, these three quantities, personal wealth, population, and national wealth, may change with time.

85. True or False: To find the rate of change of national wealth, you would multiply the rate of change of the population times the rate of change of the personal wealth.

86. True or False: To find the rate of change of personal wealth, you would divide the rate of change of the national wealth by the rate of change of the population.

Explorations and Excursions   The following problems extend and augment the material presented in the text.

More About Differentiation Formulas

87. **Product Rule For Three Functions** Show that if f, g, and h are differentiable functions of x, then
$$\frac{d}{dx}(f \cdot g \cdot h) = f' \cdot g \cdot h + f \cdot g' \cdot h + f \cdot g \cdot h'$$

[*Hint:* Write the function as $f \cdot (g \cdot h)$ and apply the Product Rule twice.]

88. Derive the Quotient Rule from the Product Rule as follows.

a. Define the quotient to be a single function,
$$Q(x) = \frac{f(x)}{g(x)}$$

b. Multiply both sides by $g(x)$ to obtain the equation $Q(x) \cdot g(x) = f(x)$.

c. Differentiate each side, using the Product Rule on the left side.
d. Solve the resulting formula for the derivative $Q'(x)$.
e. Replace $Q(x)$ by $\dfrac{f(x)}{g(x)}$ and show that the resulting formula for $Q'(x)$ is the same as the Quotient Rule.

Note that in this derivation when we differentiated $Q(x)$ we *assumed* that the derivative of the quotient exists, while in the derivation on page 123 we *proved* that the derivative exists.

89. Find a formula for $\dfrac{d}{dx}[f(x)]^2$ by writing it as $\dfrac{d}{dx}[f(x)f(x)]$ and using the Product Rule. Be sure to simplify your answer.

90. Find a formula for $\frac{d}{dx}[f(x)]^{-1}$ by writing it as $\frac{d}{dx}\left[\frac{1}{f(x)}\right]$ and using the Quotient Rule. Be sure to simplify your answer.

Two Biomedical Applications

91. Beverton-Holt Recruitment Curve Some organisms exhibit a density-dependent mortality from one generation to the next. Let $R > 1$ be the net reproductive rate (that is, the number of surviving offspring per parent), let $x > 0$ be the density of parents, and y be the density of surviving offspring. The *Beverton-Holt recruitment curve* is

$$y = \frac{Rx}{1 + \left(\dfrac{R-1}{K}\right)x}$$

where $K > 0$ is the *carrying capacity* of the organism's environment. Show that $\frac{dy}{dx} > 0$, and interpret this as a statement about the parents and the offspring.

92. Murrell's Rest Allowance Work-rest cycles for workers performing tasks that expend more than 5 kilocalories per minute (kcal/min) are often based on Murrell's formula

$$R(w) = \frac{w-5}{w-1.5} \qquad \text{for} \quad w \geq 5$$

for the number of minutes $R(w)$ of rest for each minute of work expending w kcal/min. Show that $R'(w) > 0$ for $w \geq 5$ and interpret this fact as a statement about the additional amount of rest required for more strenuous tasks.

2.5   Higher-Order Derivatives

Introduction

We have seen that from one function we can calculate a new function, the *derivative* of the original function. This new function, however, can itself be differentiated, giving what is called the **second derivative** of the original function. Differentiating again gives the **third derivative** of the original function, and so on. In this section we will calculate and interpret such **higher-order derivatives**.

Calculating Higher-Order Derivatives

EXAMPLE 1   FINDING HIGHER DERIVATIVES OF A POLYNOMIAL

From $f(x) = x^3 - 6x^2 + 2x - 7$ we may calculate

$$f'(x) = 3x^2 - 12x + 2 \qquad \text{"First" derivative of } f$$

Differentiating again gives

$$f''(x) = 6x - 12 \qquad \text{Second derivative of } f,\text{ read "} f \text{ double prime"}$$

and a third time:

$$f'''(x) = 6 \qquad \text{Third derivative of } f,\text{ read "} f \text{ triple prime"}$$

and a fourth time:

$$f''''(x) = 0 \qquad \text{Fourth derivative of } f,\text{ read "} f \text{ quadruple prime"}$$

All further derivatives of this function will, of course, be zero.

We also denote derivatives by replacing the primes by the number of differentiations in *parentheses*. For example, the fourth derivative may be denoted $f^{(4)}(x)$.

PRACTICE PROBLEM 1

If $f(x) = x^3 - x^2 + x - 1$, find:

a. $f'(x)$ **b.** $f''(x)$ **c.** $f'''(x)$ **d.** $f^{(4)}(x)$

Solutions on page 135 >

Be Careful A 4 in *parentheses*, $f^{(4)}(x)$, means the fourth *derivative* of the function, while a 4 *without* parentheses, $f^4(x)$, means the fourth *power* of the function.

Example 1 showed that a polynomial can be differentiated "down to zero," but the same is not true for all functions.

EXAMPLE 2	FINDING HIGHER DERIVATIVES OF A RATIONAL FUNCTION

Find the first five derivatives of $f(x) = \dfrac{1}{x}$.

Solution

$f(x) = x^{-1}$	$f(x)$ in power form
$f'(x) = -x^{-2}$	First derivative
$f''(x) = 2x^{-3}$	Second derivative
$f'''(x) = -6x^{-4}$	Third derivative
$f^{(4)}(x) = 24x^{-5}$	Fourth derivative
$f^{(5)}(x) = -120x^{-6}$	Fifth derivative

Clearly, we will never get to zero no matter how many times we differentiate.

PRACTICE PROBLEM 2

If $f(x) = 16x^{-1/2}$, find: **a.** $f'(x)$ **b.** $f''(x)$ Solutions on page 135 >

In Leibniz's notation, the second derivative $\dfrac{d}{dx}\dfrac{df}{dx}$ is written $\dfrac{d^2f}{dx^2}$. The superscript goes after the d in the numerator and after the dx in the denominator. The following table shows equivalent statements in the two notations.

Prime Notation		Leibniz's Notation	
$f''(x)$	=	$\dfrac{d^2}{dx^2}f(x)$	Second derivative
y''	=	$\dfrac{d^2y}{dx^2}$	
$f'''(x)$	=	$\dfrac{d^3}{dx^3}f(x)$	Third derivative
y'''	=	$\dfrac{d^3y}{dx^3}$	
$f^{(n)}(x)$	=	$\dfrac{d^n}{dx^n}f(x)$	nth derivative
$y^{(n)}$	=	$\dfrac{d^ny}{dx^n}$	

Calculating higher derivatives merely requires repeated use of the same differentiation rules that we have been using.

EXAMPLE 3   FINDING A SECOND DERIVATIVE USING THE QUOTIENT RULE

Find $\dfrac{d^2}{dx^2}\left(\dfrac{x^2+1}{x}\right)$.

Solution

LOOKING BACK

The Quotient Rule was introduced on page 116.

$$\frac{d}{dx}\left(\frac{x^2+1}{x}\right) = \frac{x(2x)-(x^2+1)}{x^2}$$ First derivative, using the Quotient Rule

$$= \frac{2x^2-x^2-1}{x^2} = \frac{x^2-1}{x^2}$$ Simplifying

Differentiating this answer gives the *second* derivative of the original:

$$\frac{d}{dx}\left(\frac{x^2-1}{x^2}\right) = \frac{x^2(2x)-2x(x^2-1)}{x^4}$$ Second derivative (derivative of the derivative)

$$= \frac{2x^3-2x^3+2x}{x^4} = \frac{2x}{x^4} = \frac{2}{x^3}$$ Simplifying

Answer: $\dfrac{d^2}{dx^2}\left(\dfrac{x^2+1}{x}\right) = \dfrac{2}{x^3}$

The function in this Example was a quotient, so it was perhaps natural to use the Quotient Rule. It is easier, however, to simplify the original function first,

$$\frac{x^2+1}{x} = \frac{x^2}{x} + \frac{1}{x} = x + x^{-1}$$

Take Note

Always try to simplify *before* differentiating.

and then differentiate by the Power Rule. So, the first derivative of $x + x^{-1}$ is $1 - x^{-2}$, and differentiating again gives $2x^{-3}$, agreeing with the answer found by the Quotient Rule.

PRACTICE PROBLEM 3

Find $f''(x)$ if $f(x) = \dfrac{x+1}{x}$.

Solution on page 135 >

EXAMPLE 4   EVALUATING A SECOND DERIVATIVE

If $f(x) = \dfrac{1}{\sqrt{x}}$, find $f''\left(\dfrac{1}{4}\right)$.   First differentiate, then evaluate

Solution

$$f(x) = x^{-1/2}$$ $f(x)$ in power form

$$f'(x) = -\frac{1}{2}x^{-3/2}$$ Differentiating once

$$f''(x) = \frac{3}{4}x^{-5/2} \qquad \text{Differentiating again}$$

$$f''\left(\frac{1}{4}\right) = \frac{3}{4}\left(\frac{1}{4}\right)^{-5/2} = \frac{3}{4}(4)^{5/2} \qquad \text{Evaluating } f''(x) \text{ at } \frac{1}{4}$$

$$= \frac{3}{4}(\sqrt{4})^5 = \frac{3}{4}(2)^5 = \frac{3}{4}(32) = 24$$

FOR MORE HELP

with negative fractional
exponents, see pages 22–25.

LOOKING BACK

The vertical bar
evaluation notation was
introduced on page 104.

PRACTICE PROBLEM 4

Find $\left.\dfrac{d^2}{dx^2}(x^4 + x^3 + 1)\right|_{x=-1}$

Solution on page 135 >

Velocity and Acceleration

There is another important interpretation for the derivative, one that also gives
a meaning to the *second* derivative. Imagine that you are driving along a straight
road, and let $s(t)$ stand for your distance (in miles) from your starting point after
t hours of driving. Then the derivative $s'(t)$ gives the instantaneous rate of change
of distance with respect to time (miles per hour). However, "miles per hour" means
speed or velocity, so the derivative of the **distance** function $s(t)$ is just the **velocity**
function $v(t)$, giving your velocity at any time t.

In general, for an object moving along a straight line, with distance measured
from some fixed point, measured positively in one direction and negatively in the
other (sometimes called "directed distance"),

$$\text{if} \qquad s(t) = \begin{pmatrix} \text{Distance} \\ \text{at time } t \end{pmatrix}$$

$$\text{then} \qquad s'(t) = \begin{pmatrix} \text{Velocity} \\ \text{at time } t \end{pmatrix}$$

Letting $v(t)$ stand for the velocity at time t, we may state this simply as:

$$v(t) = s'(t) \qquad \text{The velocity function is the}$$
$$\text{derivative of the distance function}$$

The units of velocity come directly from the distance and time units of $s(t)$. For
example, if distance is measured in feet and time in seconds, then the velocity is
in *feet per second*, while if distance is in miles and time is in hours, then velocity is
in *miles per hour*.

In everyday speech, the word "accelerating" means "speeding up." That is,
acceleration means the rate of increase of speed, and since rates of increase are
just derivatives, **acceleration** *is the derivative of velocity*. Since velocity is itself the
derivative of distance, acceleration is the *second* derivative of distance. Letting $a(t)$
stand for the acceleration at time t, we have:

Distance, Velocity, and Acceleration

$$s(t) = \begin{pmatrix} \text{Distance} \\ \text{at time } t \end{pmatrix}$$

$$v(t) = s'(t)$$ Velocity is the derivative of distance

$$a(t) = v'(t) = s''(t)$$ Acceleration is the derivative of velocity, and the second derivative of distance

Therefore, we now have an interpretation for the *second* derivative: If $s(t)$ represents distance, then the *first* derivative represents velocity, and the *second* derivative represents *acceleration*. (In physics, there is even an interpretation for the third derivative, which gives the rate of change of acceleration: It is called the "jerk," since it is related to motion being "jerky."*)

If the units of velocity are *miles per hour*, then the units of acceleration (the rate of change of velocity) are miles per hour *per hour* or *miles per hour squared*, written mi/hr^2. In general the units of acceleration are *distance units per time unit squared*.

EXAMPLE 5 **FINDING AND INTERPRETING VELOCITY AND ACCELERATION**

A delivery truck is driving along a straight road, and after t hours its distance (in miles) east of its starting point is

$$s(t) = 24t^2 - 4t^3 \quad \text{for } 0 \le t \le 6$$

a. Find the velocity of the truck after 2 hours.

b. Find the velocity of the truck after 5 hours.

c. Find the acceleration of the truck after 1 hour.

Solution

a. To find velocity, we differentiate distance:

$$v(t) = 48t - 12t^2$$ Differentiating $s(t) = 24t^2 - 4t^3$

$$v(2) = 48 \cdot 2 - 12 \cdot (2)^2$$ Evaluating $v(t)$ at $t = 2$

$$= 96 - 48 = 48 \text{ miles per hour}$$ Velocity after 2 hours

b. At $t = 5$ hours:

$$v(5) = 48 \cdot 5 - 12 \cdot (5)^2$$ Evaluating $v(t)$ at $t = 5$

$$= 240 - 300 = -60 \text{ miles per hour}$$ Velocity after 5 hours

*See T. R. Sandlin, "The Jerk," *Physics Teacher* **28**: 36–40.

What does the negative sign mean? Since distances are measured *eastward* (according to the original problem), the "positive" direction is east, so a negative velocity means a *westward* velocity. Therefore, at time $t = 5$ the truck is driving *westward at 60 miles per hour* (that is, back toward its starting point).

c. The acceleration is

$$a(t) = 48 - 24t$$
Differentiating $v(t) = 48t - 12t^2$

$$a(1) = 48 - 24 = 24$$
Acceleration after 1 hour

Therefore, after 1 hour the acceleration of the truck is 24 mi/hr² (it is "speeding up").

Graphing Calculator Exploration

Distance, velocity, and acceleration on [0, 6] by [−150, 150]. Which is which?

Use a graphing calculator to graph the distance function $y_1 = 24x^2 - 4x^3$ (use x instead of t), the velocity function $y_2 = 48x - 12x^2$, and the acceleration $y_3 = 48 - 24x$. (Alternatively, you could define y_2 and y_3 using NDERIV.) Your display should look like the one on the left. By looking at the graph, can you determine which curve represents distance, which represents velocity, and which represents acceleration? [*Hint:* Which curve gives the slope of which other curve?] Check your answer by using TRACE to identify functions y_1, y_2, and y_3.

PRACTICE PROBLEM 5

A helicopter rises vertically, and after t seconds its height above the ground is $s(t) = 6t^2 - t^3$ feet (for $0 \le t \le 6$).

a. Find its velocity after 2 seconds.

b. Find its velocity after 5 seconds.

c. Find its acceleration after 1 second.

[*Hint:* Distances are measured *upward*, so a negative velocity means *downward*.]

Solutions on page 135 >

Other Interpretations of Second Derivatives

The second derivative has other meanings besides acceleration. In general, second derivatives measure how the rate of change is itself changing. That is, if the first derivative measures the rate of growth, then the second derivative tells whether the growth is "speeding up" or "slowing down."

EXAMPLE 6 **PREDICTING POPULATION GROWTH**

United Nations demographers predict that t years from the year 2010 the population of the world will be:

$$P(t) = 6900 + 198\, t^{2/3}$$

million people. Find $P'(8)$ and $P''(8)$ and interpret these numbers.

Solution The derivative is

$$P'(t) = 132t^{-1/3} \qquad\qquad \text{Derivative of } P(t)$$

so

$$P'(8) = 132(8)^{-1/3} = 132 \cdot \frac{1}{2} = 66 \qquad\qquad \text{Evaluating at } t = 8$$

Interpretation: In 2018 ($t = 8$ years from 2010) the world population will be growing at the rate of 66 million people per year.
 The second derivative is

$$P''(t) = -44t^{-4/3} \qquad\qquad \text{Derivative of } P'(t) = 132t^{-1/3}$$

$$P''(16) = -44(8)^{-4/3} \qquad\qquad \text{Evaluating at } t = 8$$

$$= -44 \cdot \frac{1}{16} = -\frac{44}{16} = -\frac{11}{4} = -2.75$$

The fact that the first derivative is positive and the second derivative (the rate of change of the derivative) is negative means that the growth is continuing but more slowly.

Interpretation: After 8 years, the growth rate is *decreasing* by about 2.75 million people per year each year. In other words, in the following year the population will continue to grow, but at the slower rate of (rounding 2.75 to 3) about $66 - 3 = 63$ million people per year.

Source: U.N. Department of Economic and Social Affairs

Graphing Calculator Exploration

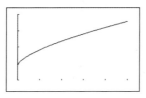
$y_1 = 6900 + 198x^{2/3}$
on [0, 50] by [6000, 10,000]

Use a graphing calculator to graph the population function $y_1 = 6900 + 198x^{2/3}$ (using x instead of t). Your graph should resemble the one on the left. Can you see from the graph that the first derivative is positive (sloping upward), and that the second derivative is negative (slope decreasing)? You may check these facts numerically using NDERIV.

SLOWER GROWTH OF HEALTH COSTS NARROWS DEFICIT

WASHINGTON—A sharp and surprisingly persistent slowdown in the growth of health care costs is helping to narrow the federal deficit, leaving budget experts

Statements about first and second derivatives occur frequently in everyday life, and may even make headlines. For example, the front page newspaper headline on the left (*New York Times*, February 12, 2013) says that health care costs grew (the first derivative is positive) but more slowly (the second derivative is negative), following a curve like that shown below.

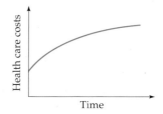

If y represents health care costs, then

$$\frac{dy}{dx} > 0 \quad \text{and} \quad \frac{d^2y}{dx^2} < 0$$

PRACTICE PROBLEM 6

The following headlines appeared recently in the *New York Times*. For each headline, sketch a curve representing the type of growth described and indicate the correct signs of the first and second derivatives.

a.   Rate of Growth of Jobs Slowed for U.S. in April (May 5, 2012)
b.   Economy Grew at Quicker Pace in 4th Quarter (January 28, 2012)

Solutions below >

PRACTICE PROBLEM 7

A mathematics journal* included the following statement: "In the fall of 1972 President Nixon announced that the rate of increase of inflation was decreasing. This was the first time a sitting president used the third derivative to advance his case for reelection." Explain why Nixon's announcement involved a third derivative.

Solution below >

Solutions TO PRACTICE PROBLEMS

1. a. $f'(x) = 3x^2 - 2x + 1$ **b.** $f''(x) = 6x - 2$
   c. $f'''(x) = 6$ **d.** $f^{(4)}(x) = 0$

2. a. $f'(x) = -8x^{-3/2}$ **b.** $f''(x) = 12x^{-5/2}$

3. $f(x) = \dfrac{x}{x} + \dfrac{1}{x} = 1 + x^{-1}$ Simplifying first
   $f'(x) = -x^{-2}$
   $f''(x) = 2x^{-3}$

4. $\dfrac{d}{dx}(x^4 + x^3 + 1) = 4x^3 + 3x^2$

   $\dfrac{d^2}{dx^2}(x^4 + x^3 + 1) = \dfrac{d}{dx}(4x^3 + 3x^2) = 12x^2 + 6x$

   $(12x^2 + 6x)|_{x=-1} = 12 - 6 = 6$

5. a. $v(t) = 12t - 3t^2$
       $v(2) = 24 - 12 = 12 = 12$ ft/sec
   b. $v(5) = 60 - 75 = -15$ ft/sec or 15 ft/sec *downward*
   c. $a(t) = 12 - 6t$ so $a(1) = 12 - 6 = 6$ ft/sec^2

6. a.

$f' > 0, f'' < 0$

b.

$f' > 0, f'' > 0$

7. Inflation is itself a derivative since it is the rate of change of the consumer price index. Therefore, its growth rate is a second derivative, and the slowing of this growth would be a third derivative.

Notices of the American Mathematical Society, **43**(10): 1108.

By simply repeating the process of differentiation we can calculate second, third, and higher derivatives. We also have another interpretation for the derivative, one that gives an interpretation for the second derivative as well. For distance measured along a straight line from some fixed point:

If $s(t) = $ *distance* at time t

then $s'(t) = $ *velocity* at time t

and $s''(t) = $ *acceleration* at time t.

Therefore, whenever you are driving along a straight road, your *speedometer gives the derivative of your odometer reading.*

Velocity is the derivative of distance

We now have *four* interpretations for the derivative: **instantaneous rate of change, slope, marginals,** and **velocity.** It has been said that science is at its best when it unifies, and the derivative, unifying these four different concepts, is one of the most important ideas in all of science. We also saw that the second derivative, which measures the rate of change of the rate of change, can show whether growth is speeding up or slowing down.

 Be Careful   Remember that derivatives measure just what an automobile speedometer measures: the velocity at a particular *instant*. Although this statement may be obvious for velocities, it is easy to forget when dealing with marginals. For example, suppose that the marginal cost for a product is $15 when 100 units have been produced [which may be written $C'(100) = 15$]. Therefore, costs are increasing at the rate of $15 per additional unit, but only at the instant when $x = 100$. Although this may be used to *estimate* future costs (*about* $15 for each additional unit), it does not mean that one additional unit will increase costs by exactly $15, two more by exactly $30, and so on, since the marginal rate usually changes as production increases. A marginal cost is only an *approximate* predictor of future costs.

2.5 Exercises

1–6. For each function, find:

a. $f'(x)$ **b.** $f''(x)$ **c.** $f'''(x)$ **d.** $f^{(4)}(x)$

1. $f(x) = x^4 - 2x^3 - 3x^2 + 5x - 7$
2. $f(x) = x^4 - 3x^3 + 2x^2 - 8x + 4$
3. $f(x) = 1 + x + \frac{1}{2}x^2 + \frac{1}{6}x^3 + \frac{1}{24}x^4 + \frac{1}{120}x^5$

4. $f(x) = 1 + x + \frac{1}{2}x^2 + \frac{1}{6}x^3 + \frac{1}{24}x^4$
5. $f(x) = \sqrt{x^5}$ **6.** $f(x) = \sqrt{x^3}$

7–12. For each function, find: **a.** $f''(x)$ and **b.** $f''(3)$.

7. $f(x) = \dfrac{x-1}{x}$ **8.** $f(x) = \dfrac{x+2}{x}$

9. $f(x) = \dfrac{x+1}{2x}$ **10.** $f(x) = \dfrac{x-2}{4x}$

11. $f(x) = \dfrac{1}{6x^2}$ **12.** $f(x) = \dfrac{1}{12x^3}$

13–18. Find the *second* derivative of each function.

13. $f(x) = (x^2 - 2)(x^2 + 3)$

14. $f(x) = (x^2 - 1)(x^2 + 2)$

15. $f(x) = \dfrac{27}{\sqrt[3]{x}}$ **16.** $f(x) = \dfrac{32}{\sqrt[4]{x}}$

17. $f(x) = \dfrac{x}{x-1}$ **18.** $f(x) = \dfrac{x}{x-2}$

19–26. Evaluate each expression.

19. $\dfrac{d^2}{dr^2}(\pi r^2)$ **20.** $\dfrac{d^3}{dr^3}\left(\dfrac{4}{3}\pi r^3\right)$

21. $\dfrac{d^2}{dx^2}x^{10}\Big|_{x=-1}$ **22.** $\dfrac{d^2}{dx^2}x^{11}\Big|_{x=-1}$

23. $\dfrac{d^3}{dx^3}x^{10}\Big|_{x=-1}$ **24.** $\dfrac{d^3}{dx^3}x^{11}\Big|_{x=-1}$

25. $\dfrac{d^2}{dx^2}\sqrt{x^3}\Big|_{x=1/16}$ **26.** $\dfrac{d^2}{dx^2}\sqrt[3]{x^4}\Big|_{x=1/27}$

27–32. Find the second derivative of each function.

27. $(x^2 - x + 1)(x^3 - 1)$

28. $(x^3 + x - 1)(x^3 + 1)$

29. $\dfrac{x}{x^2 + 1}$ **30.** $\dfrac{x}{x^2 - 1}$

31. $\dfrac{2x-1}{2x+1}$ **32.** $\dfrac{3x+1}{3x-1}$

Applied Exercises

33. GENERAL: Velocity After t hours a freight train is $s(t) = 18t^2 - 2t^3$ miles due north of its starting point (for $0 \le t \le 9$).

 a. Find its velocity at time $t = 3$ hours.
 b. Find its velocity at time $t = 7$ hours.
 c. Find its acceleration at time $t = 1$ hour.

34. GENERAL: Velocity After t hours a passenger train is $s(t) = 24t^2 - 2t^3$ miles due west of its starting point (for $0 \le t \le 12$).

 a. Find its velocity at time $t = 4$ hours.
 b. Find its velocity at time $t = 10$ hours.
 c. Find its acceleration at time $t = 1$ hour.

35. GENERAL: Velocity A rocket can rise to a height of $h(t) = t^3 + 0.5t^2$ feet in t seconds. Find its velocity and acceleration 10 seconds after it is launched.

36. GENERAL: Velocity After t hours a car is a distance $s(t) = 60t + \dfrac{100}{t+3}$ miles from its starting point. Find the velocity after 2 hours.

37. GENERAL: Impact Velocity If a steel ball is tossed from the top of the Burj Khalifa in Dubai, the tallest building in the world, its height above the ground t seconds later will be $s(t) = 2717 - 16t^2$ feet (neglecting air resistance).

 a. How long will it take to reach the ground? [*Hint*: Find when the height equals zero.]
 b. Use your answer to part (a) to find the velocity with which it will strike the ground. (This is called the *impact velocity.*)

 c. Find the acceleration at any time t. (This number is called the *acceleration due to gravity.*)

© Fingerhut/Shutterstock.com

38. GENERAL: Impact Velocity If a marble is dropped from the top of the Sears Tower in Chicago, its height above the ground t seconds after it is dropped will be $s(t) = 1454 - 16t^2$ feet (neglecting air resistance).

 a. How long will it take to reach the ground?
 b. Use your answer to part (a) to find the velocity with which it will strike the ground.
 c. Find the acceleration at any time t. (This number is called the *acceleration due to gravity.*)

39. GENERAL: Maximum Height If a bullet from a 9-millimeter pistol is fired straight up from the ground, its height t seconds after it is fired will be $s(t) = -16t^2 + 1280t$ feet (neglecting air resistance) for $0 \le t \le 80$.

(continues)

a. Find the velocity function.
b. Find the time t when the bullet will be at its maximum height. [*Hint:* At its maximum height the bullet is moving neither up nor down, and has velocity zero. Therefore, find the time when the velocity $v(t)$ equals zero.]
c. Find the maximum height the bullet will reach. [*Hint:* Use the time found in part (b) together with the height function $s(t)$.]

40. BIOMEDICAL: Fever The temperature of a patient t hours after taking a fever reducing medicine is $T(t) = 98 + 8/\sqrt{t}$ degrees Fahrenheit. Find $T(2)$, $T'(2)$, and $T''(2)$, and interpret these numbers.

41. ECONOMICS: National Debt The national debt of a South American country t years from now is predicted to be $D(t) = 65 + 9t^{4/3}$ billion dollars. Find $D'(8)$ and $D''(8)$ and interpret your answers.

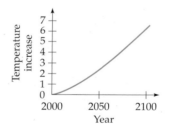

42. ENVIRONMENTAL SCIENCE: Global Temperatures The burning of oil, coal, and other fossil fuels generates "greenhouse gasses" that trap heat and raise global temperatures. Although predictions depend upon assumptions of countermeasures, one study predicts an increase in global temperature (above the 2000 level) of $T(t) = 0.25t^{1.4}$ degrees Fahrenheit, where t is the number of decades since 2000 (so, for example, $t = 2$ means the year 2020). Find $T(10)$, $T'(10)$, and $T''(10)$, and interpret your answers. [*Note:* Rising temperatures could adversely affect weather patterns and crop yields in many areas.]

Source: Intergovernmental Panel on Climate Change

43. ENVIRONMENTAL SCIENCE: Sea Level Increasing global temperatures raise sea levels by thermal expansion and the melting of polar ice. Precise predictions are difficult, but a United Nations study predicts a rise in sea level (above the 2000 level) of $L(x) = 0.02x^3 - 0.07x^2 + 8x$ centimeters, where x is

the number of decades since 2000 (so, for example, $x = 2$ means the year 2020). Find $L(10)$, $L'(10)$, and $L''(10)$, and interpret your answers. [*Note:* Rising sea levels could flood many islands and coastal regions.]

Source: U.N. Environment Programme

44. BUSINESS: Profit The annual profit of the Digitronics company x years from now is predicted to be $P(x) = 5.27x^{0.3} - 0.463x^{1.52}$ million dollars (for $0 \leq x \leq 8$). Evaluate the profit function and its first and second derivatives at $x = 3$ and interpret your answers. [*Hint:* Enter the given function in y_1, define y_2 to be the derivative of y_1 (using NDERIV), and define y_3 to be the derivative of y_2. Then evaluate each at the stated x-value.]

45. GENERAL: Windchill Index The windchill index (revised in 2001) for a temperature of 32 degrees Fahrenheit and wind speed x miles per hour is $W(x) = 55.628 - 22.07x^{0.16}$.
a. Graph the windchill index on a graphing calculator using the window [0, 50] by [0, 40]. Then find the windchill index for wind speeds of $x = 15$ and $x = 30$ mph.
b. Notice from your graph that the windchill index has first derivative negative and second derivative positive. What does this mean about how successive 1-mph increases in wind speed affect the windchill index?
c. Verify your answer to part (b) by defining y_2 to be the derivative of y_1 (using NDERIV), evaluating it at $x = 15$ and $x = 30$, and interpreting your answers.

Source: National Weather Service

46. BIOMEDICAL: AIDS The cumulative number of cases of AIDS (acquired immunodeficiency syndrome) in the United States between 1981 and 2000 is given approximately by the function

$$f(x) = -0.0182x^4 + 0.526x^3 - 1.3x^2 + 1.3x + 5.4$$

in thousands of cases, where x is the number of years since 1980.
a. Graph this function on your graphing calculator on the window [1, 20] by [0, 800]. Notice that at some time in the 1990s the rate of growth began to slow.
b. Find when the rate of growth began to slow. [*Hint:* Find where the second derivative of $f(x)$ is zero, and then convert the x-value to a year.]

Source: Centers for Disease Control

Conceptual Exercises

47–50. Suppose that the quantity described is represented by a function $f(t)$ where t stands for time. Based on the description:
a. Is the first derivative positive or negative?
b. Is the second derivative positive or negative?

47. The temperature is dropping increasingly rapidly.

48. The economy is growing, but more slowly.

49. The stock market is declining, but less rapidly.

50. The population is growing increasingly fast.

51. True or False: If $f(x)$ is a polynomial of degree n, then $f^{(n+1)}(x) = 0$.

52. At time $t = 0$ a helicopter takes off gently and then 60 seconds later it lands gently. Let $f(t)$ be its altitude above the ground at time t seconds.

    a. Will $f'(1)$ be positive or negative? Same question for $f''(1)$.

    b. Will $f'(59)$ be positive or negative? Same question for $f''(59)$.

53. GENERAL: Velocity Each of the following three "stories," labeled **a**, **b**, and **c**, matches one of the velocity graphs, labeled (i), (ii), and (iii). For each story, choose the most appropriate graph.

    a. I left my home and drove to meet a friend, but I got stopped for a speeding ticket. Afterward I drove on more slowly.

    b. I started driving but then stopped to look at the map. Realizing that I was going the wrong way, I drove back the other way.

    c. After driving for a while I got into some stop-and-go driving. Once past the tie-up I could speed up again.

54. BUSINESS: Profit Each of the following three descriptions of a company's profit over time, labeled **a**, **b**, and **c**, matches one of the graphs, labeled (i), (ii), and (iii). For each description, choose the most appropriate graph.

    a. Profits were growing increasingly rapidly.

    b. Profits were declining but the rate of decline was slowing.

    c. Profits were rising, but more and more slowly.

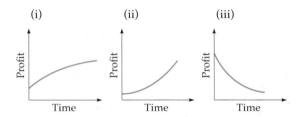

Explorations and Excursions

The following problems extend and augment the material presented in the text.

More About Higher-Order Derivatives

55. Find $\dfrac{d^{100}}{dx^{100}} (x^{100} - 4x^{99} + 3x^{50} + 6)$.

[*Hint:* You may use the "factorial" notation: $n! = n(n - 1) \cdots 1$. For example, $3! = 3 \cdot 2 \cdot 1 = 6$.]

56. Find a general formula for $\dfrac{d^n}{dx^n} x^{-1}$.

[*Hint:* Calculate the first few derivatives and look for a pattern. You may use the "factorial" notation: $n! = n(n - 1) \cdots 1$. For example, $3! = 3 \cdot 2 \cdot 1 = 6$.]

57. Verify the following formula for the *second* derivative of a product, where f and g are differentiable functions of x:

$$\frac{d^2}{dx^2} (f \cdot g) = f'' \cdot g + 2f' \cdot g' + f \cdot g''$$

[*Hint:* Use the Product Rule repeatedly.]

58. Verify the following formula for the *third* derivative of a product, where f and g are differentiable functions of x:

$$\frac{d^3}{dx^3} (f \cdot g) = f''' \cdot g + 3f'' \cdot g' + 3f' \cdot g'' + f \cdot g'''$$

[*Hint:* Differentiate the formula in Exercise 57 by the Product Rule.]

2.6   The Chain Rule and the Generalized Power Rule

Introduction

In this section we will learn the last of the general rules of differentiation, the **Chain Rule** for differentiating **composite functions.** We will then prove a very useful special case of it, the **Generalized Power Rule** for differentiating powers of functions. We begin by reviewing composite functions.

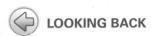 **LOOKING BACK**

We defined *composite functions* on page 54.

Composite Functions

Composite functions are simply functions of functions: The composition of f with g evaluated at x is $f(g(x))$.

EXAMPLE 1   FINDING A COMPOSITE FUNCTION

For $f(x) = x^2$ and $g(x) = 4 - x$, find $f(g(x))$.

Solution

$$f(g(x)) = (4 - x)^2$$

$f(x) = x^2$ with x replaced by $g(x) = 4 - x$

PRACTICE PROBLEM 1

For the same $f(x) = x^2$ and $g(x) = 4 - x$, find $g(f(x))$.

Solution on page 146 >

 Graphing Calculator Exploration

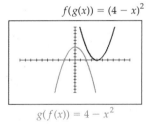

$f(g(x)) = (4 - x)^2$

$g(f(x)) = 4 - x^2$

Use a graphing calculator to verify that the compositions $f(g(x))$ and $g(f(x))$ above are different.

a. Enter $y_1 = x^2$ and $y_2 = 4 - x$.

b. Then define y_3 and y_4 to be the compositions in the two orders by entering: $y_3 = y_1(y_2)$ and $y_4 = y_2(y_1)$.

c. Graph y_3 and y_4 (but turn "off" y_1 and y_2) on the window $[-5, 8]$ by $[-7, 10]$ and notice that the graphs are very different.

Besides building compositions out of simpler functions, we can also *de*compose functions into compositions of simpler functions.

EXAMPLE 2   DECOMPOSING A COMPOSITE FUNCTION

Find functions $f(x)$ and $g(x)$ such that $(x^2 + 1)^5$ is the composition $f(g(x))$.

Solution

Think of $(x^2 + 1)^5$ as an inside function $x^2 + 1$ followed by an outside operation $(\quad)^5$. We match the "inside" and "outside" parts of $(x^2 + 1)^5$ and $f(g(x))$.

Outside function

$$(x^2 + 1)^5 = f(g(x))$$

Inside function

Therefore, $(x^2 + 1)^5$ can be written as $f(g(x))$ with $\begin{cases} f(x) = x^5 \\ g(x) = x^2 + 1 \end{cases}$

(Other answers are possible.)

Note that expressing a function as a composition involves thinking of the function in terms of "blocks," an inside block that starts a calculation and an outside block that completes it.

PRACTICE PROBLEM 2

Find $f(x)$ and $g(x)$ such that $\sqrt{x^5 - 7x + 1}$ is the composition $f(g(x))$.

Solution on page 146 >

The Chain Rule

If we were asked to differentiate the function $(x^2 - 5x + 1)^{10}$, we could first multiply together ten copies of $x^2 - 5x + 1$ (certainly a long, tedious, and error-prone process), and then differentiate the resulting polynomial. There is, however, a much easier way, using the *Chain Rule*, which shows how to differentiate a composite function of the form $f(g(x))$.

Chain Rule

$$\frac{d}{dx} f(g(x)) = f'(g(x)) \cdot g'(x)$$

To differentiate $f(g(x))$, differentiate $f(x)$, then replace each x by $g(x)$, and finally multiply by the derivative of $g(x)$

(provided that the derivatives on the right-hand side of the equation exist). The name comes from thinking of compositions as "chains" of functions. A verification of the Chain Rule is given at the end of this section.

EXAMPLE 3   DIFFERENTIATING USING THE CHAIN RULE

Use the Chain Rule to find $\dfrac{d}{dx}(x^2 - 5x + 1)^{10}$.

Solution

$(x^2 - 5x + 1)^{10}$ is $f(g(x))$ with $\begin{cases} f(x) = x^{10} & \text{Outside function} \\ g(x) = x^2 - 5x + 1 & \text{Inside function} \end{cases}$

Since $f'(x) = 10x^9$, we have

$$f'(g(x)) = 10(g(x))^9 \qquad\qquad \begin{array}{l} f'(x) = 10x^9 \text{ with } x \\ \text{replaced by } g(x) \end{array}$$

$$= 10(x^2 - 5x + 1)^9 \qquad \text{Using } g(x) = x^2 - 5x + 1$$

Substituting this last expression into the Chain Rule gives:

$$\frac{d}{dx} f(g(x)) = f'(g(x))\, g'(x) \qquad\qquad \text{Chain Rule}$$

$$\frac{d}{dx}(x^2 - 5x + 1)^{10} = 10(x^2 - 5x + 1)^9 (2x - 5) \qquad \begin{array}{l} \text{Using } f'(g(x)) = \\ 10(x^2 - 5x + 1)^9 \text{ and} \\ g'(x) = 2x - 5 \end{array}$$

This result says that to differentiate $(x^2 - 5x + 1)^{10}$, we bring down the exponent 10, reduce the exponent to 9 (steps familiar from the Power Rule), and finally multiply by the derivative of the inside function.

$$\frac{d}{dx}(x^2 - 5x + 1)^{10} = 10(x^2 - 5x + 1)^9 (2x - 5)$$

Inside function Bring down the power n Power $n - 1$ Derivative of the inside function

Generalized Power Rule

Example 3 suggests a general rule for differentiating a function to a power.

Generalized Power Rule

LOOKING BACK

The *generalized* power rule is based on the power rule from page 100.

$$\frac{d}{dx}[g(x)]^n = n \cdot [g(x)]^{n-1} \cdot g'(x)$$

To differentiate a function to a power, bring down the power as a multiplier, reduce the exponent by 1, and then multiply by the derivative of the inside function

(provided, of course, that the derivative $g'(x)$ exists). The Generalized Power Rule follows from the Chain Rule by reasoning similar to that of Example 3: the derivative of $f(x) = x^n$ is $f'(x) = nx^{n-1}$, so

$$\frac{d}{dx}f(g(x)) = f'(g(x))\,g'(x) \qquad\qquad \text{Chain Rule}$$

gives

$$\frac{d}{dx}[g(x)]^n = n[g(x)]^{n-1}g'(x) \qquad\qquad \text{Generalized Power Rule}$$

EXAMPLE 4   DIFFERENTIATING USING THE GENERALIZED POWER RULE

Find $\dfrac{d}{dx}\sqrt{x^4 - 3x^3 - 4}$.

Solution

$$\frac{d}{dx}(x^4 - 3x^3 - 4)^{1/2} = \frac{1}{2}(x^4 - 3x^3 - 4)^{-1/2}(4x^3 - 9x^2) = \frac{4x^3 - 9x^2}{2\sqrt{x^4 - 3x^3 - 4}}$$

Inside function Bring down the n Power $n - 1$ Derivative of the inside function

Think of the Generalized Power Rule "from the outside in." That is, first bring down the outer exponent and reduce the exponent by 1, and only then multiply by the derivative of the inside function.

 Be Careful It is the *original function* (not the differentiated function) that is raised to the power $n - 1$. Only at the end do you multiply by the derivative of the inside function.

EXAMPLE 5	SIMPLIFYING AND DIFFERENTIATING

Find $\dfrac{d}{dx}\left(\dfrac{1}{x^2+1}\right)^3$.

Solution

Writing the function as $(x^2+1)^{-3}$ gives

$$\frac{d}{dx}(x^2+1)^{-3} = -3(x^2+1)^{-4}(2x) = -6x(x^2+1)^{-4} = -\frac{6x}{(x^2+1)^4}$$

Inside Bring down Power   Derivative of
function   the n $n-1$ the inside function

PRACTICE PROBLEM 3

Find $\dfrac{d}{dx}(x^3-x)^{-1/2}$.

Solution on page 146 >

EXAMPLE 6	FINDING THE GROWTH RATE OF AN OIL SLICK

An oil tanker hits a reef, and after t days the radius of the oil slick is $r(t) = \sqrt{4t+1}$ miles. How fast is the radius of the oil slick expanding after 2 days?

Solution

To find the rate of change of the radius, we differentiate:

$$\frac{d}{dt}(4t+1)^{1/2} = \frac{1}{2}(4t+1)^{-1/2}(4) = 2(4t+1)^{-1/2}$$

Derivative of $4t+1$

At $t=2$ this is

$$2(4\cdot2+1)^{-1/2} = 2\cdot9^{-1/2} = 2\cdot\frac{1}{3} = \frac{2}{3}$$

Interpretation: After 2 days the radius of the oil slick is growing at the rate of $\frac{2}{3}$ mile per day.

Some problems require the Generalized Power Rule in combination with another differentiation rule, such as the Product or Quotient Rule.

EXAMPLE 7 DIFFERENTIATING USING TWO RULES

Find $\dfrac{d}{dx}[(3x+2)^3(2x-1)^4]$.

Solution

Since this is a product of powers, $(3x+2)^3$ times $(2x-1)^4$, we use the Product Rule together with the Generalized Power Rule.

$$\frac{d}{dx}[(3x+2)^3(2x-1)^4] = \underbrace{3(3x+2)^2(3)}_{\substack{\text{Derivative of}\\(3x+2)^3}}(2x-1)^4 + (3x+2)^3\underbrace{4(2x-1)^3(2)}_{\substack{\text{Derivative of}\\(2x-1)^4}}$$

$$= 9(3x+2)^2(2x-1)^4 + 8(3x+2)^3(2x-1)^3 \quad \text{Simplifying}$$
$$\underset{3\cdot 3}{\LARGE\llcorner} \qquad\qquad\qquad \underset{4\cdot 2}{\LARGE\llcorner}$$

This answer, written in *sum* form, can be factored and simplified by removing the common factor $(3x+2)^2(2x-1)^3$ as follows:

$$(3x+2)^2(2x-1)^3[9(2x-1)+8(3x+2)]$$

$$= (3x+2)^2(2x-1)^3[18x-9+24x+16] \qquad\qquad \text{Multiplying out}$$

$$= (3x+2)^2(2x-1)^3\underbrace{[42x+7]}_{7(6x+1)} \qquad\qquad \text{Combining terms}$$

$$= 7(3x+2)^2(2x-1)^3(6x+1) \qquad\qquad \text{Factored final answer}$$

> ⏚ **Take Note**
>
> Most of the work in this Example was *algebraic*, with calculus occurring only in the first step.

EXAMPLE 8 DIFFERENTIATING USING TWO RULES

Find $\dfrac{d}{dx}\left(\dfrac{x}{x+1}\right)^4$.

Solution

Since the function is a quotient raised to a power, we use the Quotient Rule together with the Generalized Power Rule. Working from the outside in, we obtain

$$\frac{d}{dx}\left(\frac{x}{x+1}\right)^4 = 4\left(\frac{x}{x+1}\right)^3\underbrace{\frac{(x+1)(1)-(1)(x)}{(x+1)^2}}_{\substack{\text{Derivative of the}\\\text{inside function } \frac{x}{x+1}}}$$

$$= 4\left(\frac{x}{x+1}\right)^3\frac{x+1-x}{(x+1)^2} = 4\left(\frac{x}{x+1}\right)^3\frac{1}{(x+1)^2} \qquad \text{Simplifying}$$

$$= 4\frac{x^3}{(x+1)^3}\frac{1}{(x+1)^2} = \frac{4x^3}{(x+1)^5} \qquad \begin{array}{l}\text{Simplifying}\\\text{further}\end{array}$$

| **EXAMPLE 9** | **DIFFERENTIATING USING A RULE TWICE** |

Find $\dfrac{d}{dz}[z^2 + (z^2 - 1)^3]^5$.

Solution

Since this is a function to a power, where the inside function also contains a function to a power, we must use the Generalized Power Rule *twice*.

$$\frac{d}{dz}[z^2 + (z^2 - 1)^3]^5 = 5[z^2 + (z^2 - 1)^3]^4\underbrace{[2z + 3(z^2 - 1)^2(2z)]}_{\substack{\text{Derivative of}\\ z^2 + (z^2 - 1)^3}}$$

$$= 5[z^2 + (z^2 - 1)^3]^4[2z + 6z(z^2 - 1)^2] \qquad \text{Simplifying}$$

$$\underset{3 \cdot 2}{\overset{}{}}$$

Chain Rule in Leibniz's Notation

A composition may be written in two parts:

$$y = f(g(x)) \quad \text{is equivalent to} \quad y = f(u) \quad \text{and} \quad u = g(x)$$

The derivatives of these last two functions are:

$$\frac{dy}{du} = f'(u) \quad \text{and} \quad \frac{du}{dx} = g'(x)$$

The Chain Rule

$$\underbrace{\frac{d}{dx}\underbrace{f(g(x))}_{y} = \underbrace{f'(g(x))}_{\frac{dy}{du}} \cdot \underbrace{g'(x)}_{\frac{du}{dx}}}_{\frac{dy}{dx}} \qquad \text{Chain Rule}$$

with the indicated substitutions then becomes:

Chain Rule in Leibniz's Notation

For $y = f(u)$ with $u = g(x)$,

$$\frac{dy}{dx} = \frac{dy}{du} \cdot \frac{du}{dx}$$

In this form the Chain Rule is easy to remember, since it looks as if the du in the numerator and the denominator cancel:

$$\frac{dy}{dx} = \frac{dy}{d\cancel{u}} \cdot \frac{d\cancel{u}}{dx}$$

However, since derivatives are not really fractions (they are *limits* of fractions), this is only a convenient device for remembering the Chain Rule.

⚠️ **Be Careful** The Product Rule (page 114) showed that the derivative of a product is not the product of the derivatives. We now see where the product of the derivatives *does* appear: it appears in the Chain Rule, when differentiating composite functions. In other words, the product of the derivatives comes not from *products* but from *compositions* of functions.

A Simple Example of the Chain Rule

The derivation of the Chain Rule is rather technical, but we can show the basic idea in a simple example. Suppose that your company produces steel, and you want to calculate your company's total revenue in dollars per year. You would take the revenue from a ton of steel (dollars per ton) and multiply by your company's output (tons per year). In symbols:

$$\frac{\$}{\text{year}} = \frac{\$}{\text{ton}} \cdot \frac{\text{ton}}{\text{year}}$$ Note that "ton" cancels

If we were to express these rates as derivatives, the equation above would become the Chain Rule.

Solutions TO PRACTICE PROBLEMS

1. $g(f(x)) = 4 - f(x) = 4 - x^2$

2. $f(x) = \sqrt{x}, \quad g(x) = x^5 - 7x + 1$

3. $\dfrac{d}{dx}(x^3 - x)^{-1/2} = -\dfrac{1}{2}(x^3 - x)^{-3/2}(3x^2 - 1) = -\dfrac{3x^2 - 1}{2\sqrt{(x^3 - x)^3}}$

2.6   Section Summary

To differentiate a composite function (a function of a function), we have the Chain Rule:

$$\frac{d}{dx}f(g(x)) = f'(g(x)) \cdot g'(x)$$

or, in Leibniz's notation, writing $y = f(g(x))$ as $y = f(u)$ and $u = g(x)$,

$$\frac{dy}{dx} = \frac{dy}{du} \cdot \frac{du}{dx}$$ The derivative of a composite function is the product of the derivatives

To differentiate a function to a power, $[f(x)]^n$, we have the *Generalized Power Rule* (a special case of the Chain Rule when the "outer" function is a power):

$$\frac{d}{dx}[f(x)]^n = n \cdot [f(x)]^{n-1} \cdot f'(x)$$

For now, the Generalized Power Rule is more useful than the Chain Rule, but in Chapter 5 we will make important use of the Chain Rule.

Verification of the Chain Rule

Let $f(x)$ and $g(x)$ be differentiable functions. We define k by

$$k = g(x + h) - g(x)$$

or, equivalently,

$$g(x + h) = g(x) + k$$

Then

$$\lim_{h \to 0} [g(x + h) - g(x)] = \lim_{h \to 0} k = 0$$

$$\underbrace{}_{k}$$

showing that $h \to 0$ implies $k \to 0$ (see page 122). With these relations we may calculate the derivative of the composition $f(g(x))$.

$$\frac{d}{dx} f(g(x)) = \lim_{h \to 0} \frac{f(g(x + h)) - f(g(x))}{h}$$ Definition of the derivative of $f(g(x))$

$$= \lim_{h \to 0} \left[\frac{f(g(x + h)) - f(g(x))}{g(x + h) - g(x)} \cdot \frac{g(x + h) - g(x)}{h} \right]$$ Dividing and multiplying by $g(x + h) - g(x)$

$$= \lim_{h \to 0} \frac{f(g(x + h)) - f(g(x))}{g(x + h) - g(x)} \lim_{h \to 0} \frac{g(x + h) - g(x)}{h}$$ The limit of a product is the product of the limits (Limit Rule 4c on page 75)

$$= \underbrace{\lim_{k \to 0} \frac{f(g(x) + k) - f(g(x))}{k}}_{f'(g(x))} \underbrace{\lim_{h \to 0} \frac{g(x + h) - g(x)}{h}}_{g'(x)}$$ Using the relations $g(x + h) = g(x) + k$ and $k = g(x + h) - g(x)$ and that $h \to 0$ implies $k \to 0$

$$= f'(g(x)) \cdot g'(x)$$ Chain Rule

The last step comes from recognizing the first limit as the definition of the derivative f' at $g(x)$, and the second limit as the definition of the derivative $g'(x)$. This verifies the Chain Rule,

$$\frac{d}{dx} f(g(x)) = f'(g(x)) \cdot g'(x)$$

Strictly speaking, this verification requires an additional assumption, that the denominator $g(x + h) - g(x)$ is never zero. This assumption can be avoided by using Carathéodory's definition of the derivative, as shown in Exercises 85–86.

2.6 Exercises

1–10. Find functions f and g such that the given function is the composition $f(g(x))$.

1. $\sqrt{x^2 - 3x + 1}$ **2.** $(5x^2 - x + 2)^4$

3. $(x^2 - x)^{-3}$ **4.** $\dfrac{1}{x^2 + x}$ **5.** $\dfrac{x^3 + 1}{x^3 - 1}$

6. $\dfrac{\sqrt{x} - 1}{\sqrt{x} + 1}$ **7.** $\left(\dfrac{x + 1}{x - 1}\right)^4$ **8.** $\sqrt{\dfrac{x - 1}{x + 1}}$

9. $\sqrt{x^2 - 9} + 5$ **10.** $\sqrt[3]{x^3 + 8} - 5$

11–46. Use the Generalized Power Rule to find the derivative of each function.

11. $f(x) = (x^2 + 1)^3$ **12.** $f(x) = (x^3 + 1)^4$

13. $g(x) = (2x^2 - 7x + 3)^4$

14. $g(x) = (3x^3 - x^2 + 1)^5$

15. $h(z) = (3z^2 - 5z + 2)^4$

16. $h(z) = (5z^2 + 3z - 1)^3$

17. $f(x) = \sqrt{x^4 - 5x + 1}$ **18.** $f(x) = \sqrt{x^6 + 3x - 1}$

19. $w(z) = \sqrt[3]{9z - 1}$ **20.** $w(z) = \sqrt[5]{10z - 4}$

21. $y = (4 - x^2)^4$ **22.** $y = (1 - x)^{50}$

23. $y = \left(\dfrac{1}{w^3 - 1}\right)^4$ **24.** $y = \left(\dfrac{1}{w^4 + 1}\right)^5$

25. $y = x^4 + (1 - x)^4$

26. $f(x) = (x^2 + 4)^3 - (x^2 + 4)^2$

27. $f(x) = \dfrac{1}{\sqrt{3x^2 - 5x + 1}}$ **28.** $f(x) = \dfrac{1}{\sqrt{2x^2 - 7x + 1}}$

29. $f(x) = \dfrac{1}{\sqrt[3]{(9x + 1)^2}}$ **30.** $f(x) = \dfrac{1}{\sqrt[3]{(3x - 1)^2}}$

31. $f(x) = \dfrac{1}{\sqrt[3]{(2x^2 - 3x + 1)^2}}$

32. $f(x) = \dfrac{1}{\sqrt[3]{(x^2 + x - 9)^2}}$

33. $f(x) = [(x^2 + 1)^3 + x]^3$

34. $f(x) = [(x^3 + 1)^2 - x]^4$

35. $f(x) = 3x^2(2x + 1)^5$ **36.** $f(x) = 2x(x^3 - 1)^4$

37. $f(x) = (2x + 1)^3(2x - 1)^4$

38. $f(x) = (2x - 1)^3(2x + 1)^4$

39. $g(z) = 2z(3z^2 - z + 1)^4$

40. $g(z) = z^2(2z^3 - z + 5)^4$

41. $f(x) = \left(\dfrac{x + 1}{x - 1}\right)^3$ **42.** $f(x) = \left(\dfrac{x - 1}{x + 1}\right)^5$

43. $f(x) = x^2\sqrt{1 + x^2}$ **44.** $f(x) = x^2\sqrt{x^2 - 1}$

45. $f(x) = \sqrt{1 + \sqrt{x}}$ **46.** $f(x) = \sqrt[3]{1 + \sqrt[3]{x}}$

47–50. Find the equation for the tangent line to the curve $y = f(x)$ at the given x-value.

47. $f(x) = (2x + 1)^4$ at $x = -1$

48. $f(x) = \sqrt{x^2 + 3}$ at $x = 1$

49. $f(x) = [(x - 1)^2 - x]^2$ at $x = 2$

50. $f(x) = (2x + 3)^3 + (3x + 2)^2 + 4x - 5$ at $x = -2$

51. Find the derivative of $(x^2 + 1)^2$ in two ways:
   a. By the Generalized Power Rule.
   b. By "squaring out" the original expression and then differentiating.
   Your answers should agree.

52. Find the derivative of $\dfrac{1}{x^2}$ in three ways:
   a. By the Quotient Rule.
   b. By writing $\dfrac{1}{x^2}$ as $(x^2)^{-1}$ and using the Generalized Power Rule.
   c. By writing $\dfrac{1}{x^2}$ as x^{-2} and using the (ordinary) Power Rule.
   Your answers should agree.

53. Find the derivative of $\dfrac{1}{3x + 1}$ in two ways:
   a. By the Quotient Rule.
   b. By writing the function as $(3x + 1)^{-1}$ and using the Generalized Power Rule.
   Your answers should agree. Which way was easier? Remember this for the future.

54. Find an expression for the derivative of the composition of three functions, $\dfrac{d}{dx} f(g(h(x)))$.
   [*Hint:* Use the Chain Rule twice.]

55–56. Find the *second* derivative of each function.

55. $f(x) = (x^2 + 1)^{10}$ **56.** $f(x) = (x^3 - 1)^5$

Applied Exercises

57. BUSINESS: Cost A company's cost function is
$C(x) = \sqrt{4x^2 + 900}$ dollars, where x is the number of units. Find the marginal cost function and evaluate it at $x = 20$.

58. BUSINESS: Cost (*continuation*) Graph the cost function $y_1 = \sqrt{4x^2 + 900}$ on the window [0, 30] by [−10, 70]. Then use NDERIV to define y_2 as the derivative of y_1. Verify the answer to Exercise 57 by evaluating the marginal cost function y_2 at $x = 20$.

59. BUSINESS: Cost (*continuation*) Find the number x of units at which the marginal cost is 1.75.
[*Hint:* TRACE along the marginal cost function y_2 to find where the y-coordinate is 1.75, giving your answer as the x-coordinate rounded to the nearest whole number.]

60. SOCIOLOGY: Educational Status A study estimated how a person's social status (rated on a scale where 100 indicates the status of a college graduate) depended on years of education. Based on this study, with e years of education, a person's status is $S(e) = 0.22(e + 4)^{2.1}$. Find $S'(12)$ and interpret your answer.
Source: Sociometry 34

61. SOCIOLOGY: Income Status A study estimated how a person's social status (rated on a scale where 100 indicates the status of a college graduate) depended upon income. Based on this study, with an income of i thousand dollars, a person's status is $S(i) = 17.5(i - 1)^{0.53}$. Find $S'(25)$ and interpret your answer.
Source: Sociometry 34

62. BUSINESS: Compound Interest If $1000 is deposited in a bank paying $r\%$ interest compounded annually, 5 years later its value will be

$$V(r) = 1000(1 + 0.01r)^5 \quad \text{dollars}$$

Find $V'(6)$ and interpret your answer. [*Hint: r = 6* corresponds to 6% interest.]

63. BIOMEDICAL: Drug Sensitivity The strength of a patient's reaction to a dose of x milligrams of a certain drug is $R(x) = 4x\sqrt{11 + 0.5x}$ for $0 \le x \le 140$. The derivative $R'(x)$ is called the *sensitivity* to the drug. Find $R'(50)$, the sensitivity to a dose of 50 mg.

64. GENERAL: Population The population of a city x years from now is predicted to be $P(x) = \sqrt[4]{x^2 + 1}$ million people for $1 \le x \le 5$. Find when the population will be growing at the rate of a quarter of a million people per year. [*Hint:* On a graphing calculator, enter the given population function in y_1, use NDERIV to define y_2 to be the derivative of y_1, and graph both on the window [1, 5] by [0, 3]. Then TRACE along y_2 to find the x-coordinate (rounded to the nearest tenth of a unit) at which the y-coordinate is 0.25. You may have to ZOOM IN to find the correct x-value.]

65. BIOMEDICAL: Drug Sensitivity *(63 continued)* For the reaction function given in Exercise 63, find the dose at which the sensitivity is 25. [*Hint:* On a graphing calculator, enter the reaction function $y_1 = 4x\sqrt{11 + 0.5x}$, and use NDERIV to define y_2 to be the derivative of y_1. Then graph y_2 on the window [0, 140] by [0, 50] and TRACE along y_2 to find the x-coordinate (rounded to the nearest whole number) at which the y-coordinate is 25. You may have to ZOOM IN to find the correct x-value.]

66. BIOMEDICAL: Blood Flow It follows from *Poiseuille's Law* that blood flowing through certain arteries will encounter a resistance of $R(x) = 0.25(1 + x)^4$, where x is the distance (in meters) from the heart. Find the instantaneous rate of change of the resistance at:
a. 0 meters. **b.** 1 meter.

67. ENVIRONMENTAL SCIENCE: Pollution The carbon monoxide level in a city is predicted to be $0.02x^{3/2} + 1$ ppm (parts per million), where x is the population in thousands. In t years the population of the city is predicted to be $x(t) = 12 + 2t$ thousand people. Therefore, in t years the carbon monoxide level will be

$$P(t) = 0.02(12 + 2t)^{3/2} + 1 \quad \text{ppm}$$

Find $P'(2)$, the rate at which carbon monoxide pollution will be increasing in 2 years.

68. PSYCHOLOGY: Learning After p practice sessions, a subject could perform a task in $T(p) = 36(p + 1)^{-1/3}$ minutes for $0 \le p \le 10$. Find $T'(7)$ and interpret your answer.

69. ENVIRONMENTAL SCIENCE: Greenhouse Gases and Global Warming The following graph shows the concentration (in parts per million, or *ppm*) of carbon dioxide (CO_2), the most common greenhouse gas, in the atmosphere for recent years, showing that, at least in the short run, CO_2 levels rise linearly.

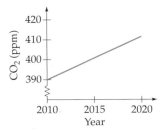

This CO_2 concentration is approximated by the function $y_1 = 2.16x + 390$ where x is the number of years since 2010. The function $y_2 = 0.024x + 51.3$ is an estimate of the average global temperature (in degrees Fahrenheit) if the CO_2 level is x.

a. Enter y_1 and y_2 into your calculator and define y_3 to be the composition $y_3 = y_2(y_1)$, so that y_3 gives the average global temperature for any x (years since 2010).
b. Turn "off" functions y_1 and y_2 and graph y_3 on the window [0, 20] by [60, 62]. Use the operation NDERIV or dy/dx to find the slope, giving the predicted temperature rise per year in the short run.
c. Use this slope to find how long it will take global temperatures to rise by 1.8 degrees. (It has been estimated that a 1.8-degree temperature rise could raise sea levels by a foot, inundating coastal areas and disrupting world food supplies.*)

Sources: NOAA, Worldwatch Institute

70. GENERAL: Happiness and Temperature Based on a recent study, the "happiness" of people who live in a country whose average temperature is t degrees Fahrenheit is given by $h(t) = 8.2 - (0.01t - 2.8)^2$, for $35 \le t \le 72$. ("Happiness" was rated from $1 =$ "not at all happy" to $4 =$ "very happy".) Find $h(40)$ and $h'(40)$. Interpret your answers.

Source: Ecological Economics **52**

Conceptual Exercises

71. True or False: $\dfrac{d}{dx} f(g(x)) = f'(g'(x)) \cdot g'(x)$

72. True or False: $\dfrac{d}{dx} [g(x)]^n = n \cdot [g'(x)]^{n-1} \cdot g'(x)$

73. Explain the difference between the Chain Rule and the Generalized Power Rule.

74. Explain the difference between the Generalized Power Rule and the Power Rule.

---

*The function y_1 giving CO_2 levels is widely accepted as accurate. The function y_2 giving temperature based on CO_2 levels is less well established since factors other than CO_2 affect temperature.

75. True or False: $\dfrac{d}{dx} f(5x) = 5 \cdot f'(5x)$

76. True or False: $\dfrac{d}{dx} f(x/2) = \dfrac{f'(x/2)}{2}$

77. True or False: $\dfrac{d}{dx} \sqrt{g(x)} = \sqrt{g(x)} \cdot g'(x)$

78. True or False: $\dfrac{d}{dx} f(x+5) = f'(x+5)$

79. Imagine a square whose sides are expanding at a given rate (so that the area is also increasing). Since the area is the *square* of the length of a side, is it true that the rate of change of the area is the *square* of the rate of change of the side?

80. Imagine a cube whose sides are expanding at a given rate (so that the volume is also increasing). Since the volume is the *cube* of the length of a side, is it true that the rate of change of the volume is the *cube* of the rate of change of the side?

Explorations and Excursions   The following problems extend and augment the material presented in the text.

Further Uses of the Chain Rule

81. Suppose that $L(x)$ is a function such that $L'(x) = \dfrac{1}{x}$. Use the Chain Rule to show that the derivative of the composite function $L(g(x))$ is

$$\frac{d}{dx} L(g(x)) = \frac{g'(x)}{g(x)}$$

82. Suppose that $E(x)$ is a function such that $E'(x) = E(x)$. Use the Chain Rule to show that the derivative of the composite function $E(g(x))$ is

$$\frac{d}{dx} E(g(x)) = E(g(x)) \cdot g'(x)$$

83. Chain Rule For Three Functions Show that if f, g, and h are three differentiable functions, then

$$\frac{d}{dx} f(g(h(x))) = f'(g(h(x))) \cdot g'(h(x)) \cdot h'(x)$$

[*Hint:* Let $G(x) = g(h(x))$, so that the Chain Rule gives $G'(x) = g'(h(x)) \cdot h'(x)$. Then use the Chain Rule to find $\dfrac{d}{dx} f(G(x))$ and substitute the expressions found in the previous sentence for $G(x)$ and $G'(x)$ to obtain the desired formula.]

84. Using The Chain Rule For Three Functions

a. Use the *Chain Rule for three functions* derived in Exercise 83 to find

$$\frac{d}{dx} ((x^2)^3)^4$$

b. Show that $((x^2)^3)^4$ simplifies to x^{24}, and verify that $\dfrac{d}{dx} x^{24}$ gives the same result as found in part (a).

Carathéodory's Definition of the Derivative and Proof of the Chain Rule

85. The following is an alternate definition of the derivative, due to Constantine Carathéodory.*

> A function f is differentiable at x if there is a function F that is continuous at 0 and such that $f(x+h) - f(x) = F(h) \cdot h$. In this case, $F(0) = f'(x)$.

Show that Carathéodory's definition of the derivative is equivalent to the definition on page 92.

86. The Chain Rule (page 141) states that the derivative of $f(g(x))$ is $f'(g(x)) \cdot g'(x)$. Use Carathéodory's definition of the derivative to prove the Chain Rule by giving reasons for the following steps.

a. Since g is differentiable at x, there is a function G that is continuous at 0 and such that $g(x+h) - g(x) = G(h) \cdot h$, and $G(0) = g'(x)$.

b. Since f is differentiable at $g(x)$, there is a function F that is continuous at 0 and such that $f(g(x) + h) - f(g(x)) = F(h) \cdot h$, and $F(0) = f'(g(x))$.

c. For the function $f(g(x))$ we have

$$f(g(x+h)) - f(g(x))$$
$$= f(g(x) + g(x+h) - g(x)) - f(g(x))$$
$$= f(g(x) + (g(x+h) - g(x))) - f(g(x))$$
$$= F(g(x+h) - g(x)) \cdot (g(x+h) - g(x))$$
$$= F(g(x+h) - g(x)) \cdot G(h) \cdot h$$

d. Therefore, the derivative of $f(g(x))$ is

$$F(g(x+0) - g(x)) \cdot G(0) = F(0) \cdot G(0)$$
$$= f'(g(x)) \cdot g'(x)$$

as was to be proved.

*Greek mathematician, 1873–1950.

Nondifferentiable Functions

Introduction

In spite of all of the rules of differentiation, there are **nondifferentiable functions**—functions that cannot be differentiated at certain values. We begin this section by exhibiting such a function (the absolute value function) and showing that it is not differentiable at $x = 0$. We will then discuss general geometric conditions for a function to be nondifferentiable. Knowing where a function is not differentiable is important for understanding graphs and for interpreting answers from a graphing calculator.

Absolute Value Function

The absolute value function is defined as

⬅ **LOOKING BACK**

We defined the *absolute value function* on page 53.

$$|x| = \begin{cases} x & \text{if } x \geq 0 \\ -x & \text{if } x < 0 \end{cases}$$

Although the absolute value function is *defined* for *all* values of x, we will show that it is *not* differentiable at $x = 0$.

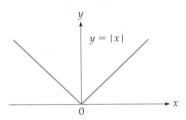

The graph of the absolute value function $f(x) = |x|$ has a "corner" at the origin.

EXAMPLE 1   **SHOWING NONDIFFERENTIABILITY**

Show that $f(x) = |x|$ is not differentiable at $x = 0$.

Solution   We have no "rules" for differentiating the absolute value function, so we must use the definition of the derivative:

⬅ **LOOKING BACK**

The definition of the derivative was given on page 92.

$$f'(x) = \lim_{h \to 0} \frac{f(x + h) - f(x)}{h}$$

provided that this limit exists. It is this provision, which until now we have steadfastly ignored, that will be important in this example. We will show that this limit, and hence the derivative, does not exist at $x = 0$.

For $x = 0$ the definition becomes

$$\lim_{h \to 0} \frac{f(0 + h) - f(0)}{h} = \lim_{h \to 0} \frac{f(h) - f(0)}{h} = \lim_{h \to 0} \underbrace{\frac{|h| - |0|}{h}}_{\substack{\text{Using} \\ f(x) = |x|}} = \lim_{h \to 0} \frac{|h|}{h}$$

For this limit to exist, the two one-sided limits must exist and have the same value. The limit from the right is

$$\lim_{h \to 0^+} \frac{|h|}{h} = \lim_{h \to 0^+} \frac{h}{h} = \lim_{h \to 0^+} 1 = 1$$

Since $|h| = h$
for $h > 0$ $\frac{h}{h} = 1$

The limit from the left is

$$\lim_{h \to 0^-} \frac{|h|}{h} = \lim_{h \to 0^-} \frac{-h}{h} = \lim_{h \to 0^-} (-1) = -1$$

For $h < 0$, $|h| = -h$
(the negative sign makes
the negative h positive) $\frac{-h}{h} = -1$

Since the one-sided limits do not agree (one is $+1$ and the other is -1), the limit $\lim_{h \to 0} \frac{|h|}{h}$ does not exist, so *the derivative does not exist*. This is what we wanted to show—that the absolute value function is not differentiable at $x = 0$.

Geometric Explanation of Nondifferentiability

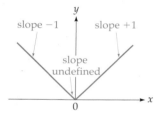

slope -1 slope $+1$
slope undefined

We can give a geometric and intuitive reason why the absolute value function is not differentiable at $x = 0$. Its graph consists of two straight lines with slopes $+1$ and -1 that meet in a corner at the origin. To the right of the origin the slope is $+1$ and to the left of the origin the slope is -1, but *at* the origin the two conflicting slopes make it impossible to define a *single* slope. Therefore, the slope (and hence the derivative) is undefined at $x = 0$.

**Graphing
Calculator
Exploration**

Calculator error!

If your graphing calculator has an operation like ABS or something similar for the absolute value function, graph $y_1 = \text{ABS}(x)$ on the window $[-2, 2]$ by $[-1, 2]$. Use NDERIV to "find" the derivative of y_1 at $x = 0$. Your calculator may give a "false value" such as 0, resulting from its *approximating* the derivative by the symmetric difference quotient (see page 107). The correct answer is that the derivative is *undefined* at $x = 0$, as we just showed. This is why it is important to understand nondifferentiability—your calculator may give a misleading answer.

Other Nondifferentiable Functions

For the same reason, at any **corner point** of a graph, where two different slopes conflict, the function will not be differentiable.

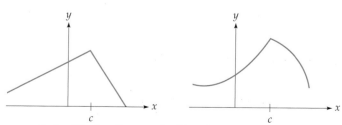

Each of the functions graphed here has a "corner point"
at $x = c$, and so is not differentiable at $x = c$.

There are other reasons, besides a corner point, why a function may not be differentiable. If a curve has a **vertical tangent** line at a point, the slope will not be defined at that *x*-value, since the slope of a vertical line is undefined.

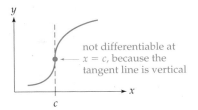

We showed on page 122 that if a function is differentiable, then it is continuous. Therefore, if a function is discontinuous (has a "jump") at some point, then it will not be differentiable at that *x*-value.

If a function *f* satisfies *any* of the following conditions:

1. *f* has a corner point at $x = c$,
2. *f* has a vertical tangent at $x = c$,
3. *f* is discontinuous at $x = c$,

then *f* will not be differentiable at *c*.

Take Note

These three conditions, any one of which will make the derivative fail to exist, will be important when we graph functions in Chapter 3.

PRACTICE PROBLEM

For the function graphed below, find the *x*-values at which the derivative is undefined.

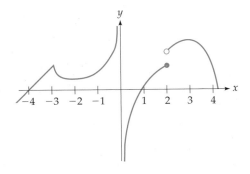

Solution on next page >

Be Careful All differentiable functions are continuous (see page 122), but *not* all continuous functions are differentiable—for example, $f(x) = |x|$. These facts are shown in the following diagram.

 Spreadsheet Exploration

Another function that is not differentiable is $f(x) = x^{2/3}$. The following spreadsheet* calculates values of the difference quotient $\dfrac{f(x + h) - f(x)}{h}$ at $x = 0$ for this function. Since $f(0) = 0$, the difference quotient at $x = 0$ simplifies to:

$$\frac{f(x + h) - f(x)}{h} = \frac{f(0 + h) - f(0)}{h} = \frac{f(h)}{h} = \frac{h^{2/3}}{h} = h^{-1/3}$$

For example, cell **B5** evaluates $h^{-1/3}$ at $h = \frac{1}{1000}$ obtaining $(\frac{1}{1000})^{-1/3} = 1000^{1/3} = \sqrt[3]{1000} = 10$. Column **B** evaluates this different quotient for the *positive* values of h in column **A**, while column **E** evaluates it for the corresponding negative values of h in column **D**.

B5		▼		=	=A5^(-1/3)	
	A	B	C	D	E	
1	h	(f(0+h)-f(0))/h		h	(f(0+h)-f(0))/h	
2	1.0000000	1.0000000		-1.0000000	-1.0000000	
3	0.1000000	2.1544347		-0.1000000	-2.1544347	
4	0.0100000	4.6415888		-0.0100000	-4.6415888	
5	0.0010000	10.0000000		-0.0010000	-10.0000000	
6	0.0001000	21.5443469		-0.0001000	-21.5443469	
7	0.0000100	46.4158883		-0.0000100	-46.4158883	
8	0.0000010	100.0000000		-0.0000010	-100.0000000	
9	0.0000001	215.4434690		-0.0000001	-215.4434690	

becoming large   becoming small

Notice that the values in column **B** are becoming arbitrarily large, while the values in column **E** are becoming arbitrarily small, so the difference quotient does not approach a limit as $h \to 0$. This shows that the derivative of $f(x) = x^{2/3}$ at 0 *does not exist*, so the function $f(x) = x^{2/3}$ is *not differentiable* at $x = 0$.

Solution TO PRACTICE PROBLEM

$x = -3, \quad x = 0, \quad$ and $\quad x = 2$

*To obtain this and other Spreadsheet Explorations, go to www.cengagebrain.com, search for this textbook and then scroll down to access the "free materials" tab.

2.7 Exercises

1–4. For each function graphed below, find the *x*-values at which the derivative does not exist.

1.

2.

3.

4.

5–8. Use the definition of the derivative to show that the following functions are not differentiable at $x = 0$.

5. $f(x) = |2x|$

6. $f(x) = |3x|$

7. $f(x) = x^{2/5}$

8. $f(x) = x^{4/5}$

🖊 **FOR HELP GETTING STARTED**

with Exercises 5 and 6, modify Example 1 on pages 151–152.

 9–10. Use NDERIV to "find" the derivative of each function at $x = 0$. Is the result correct?

9. $f(x) = \dfrac{1}{x}$

10. $f(x) = \dfrac{1}{x^2}$

 11. a. Show that the definition of the derivative applied to the function $f(x) = \sqrt[5]{x}$ at $x = 0$ gives
$$f'(0) = \lim_{h \to 0} \frac{\sqrt[5]{h}}{h}.$$

b. Use a calculator to evaluate the difference quotient $\dfrac{\sqrt[5]{h}}{h}$ for the following values of *h*: 0.1, 0.001, and 0.00001. [*Hint:* Enter the calculation into your calculator with *h* replaced by 0.1, and then change the value of *h* by inserting zeros.]

c. From your answers to part (b), does the limit exist? Does the derivative of $f(x) = \sqrt[5]{x}$ at $x = 0$ exist?

d. Graph $f(x) = \sqrt[5]{x}$ on the window $[-1, 1]$ by $[-1, 1]$. Do you see why the slope at $x = 0$ does not exist?

12. a. Show that the definition of the derivative applied to the function $f(x) = \sqrt[3]{x}$ at $x = 0$ gives
$$f'(0) = \lim_{h \to 0} \frac{\sqrt[3]{h}}{h}.$$

b. Use a calculator to evaluate the difference quotient $\dfrac{\sqrt[3]{h}}{h}$ for the following values of *h*: 0.1, 0.0001, and 0.0000001. [*Hint:* Enter the calculation into your calculator with *h* replaced by 0.1, and then change the value of *h* by inserting zeros.]

c. From your answers to part (b), does the limit exist? Does the derivative of $f(x) = \sqrt[3]{x}$ at $x = 0$ exist?

d. Graph $f(x) = \sqrt[3]{x}$ on the window $[-1, 1]$ by $[-1, 1]$. Do you see why the slope at $x = 0$ does not exist?

Conceptual Exercises

13. Using your own words, explain geometrically why the derivative is undefined where a curve has a *corner point*.

14. Using your own words, explain geometrically why the derivative is undefined where a curve has a *vertical tangent*.

15. Using your own words, explain in terms of instantaneous rates of change why the derivative is undefined where a function has a *discontinuity*.

16. True or False: If a function is *continuous* at a number, then it is *differentiable* at that number.

17. True or False: If a function is *differentiable* at a number, then it is *continuous* at that number.

18. True or False: If a function is not differentiable at a point, then its graph cannot have a tangent line at that point.

19. Using only straight lines, sketch a function that (a) is continuous everywhere and (b) is differentiable everywhere *except* at $x = 1$ and $x = 3$.

20. Sketch a function that (a) is continuous everywhere, (b) has a tangent line at every point, and (c) is differentiable everywhere *except* at $x = 2$ and $x = 4$.

| **2** | **Chapter Summary with Hints and Suggestions** |

Reading the text and doing the exercises in this chapter have helped you to master the following concepts and skills, which are listed by section (in case you need to review them) and are keyed to particular Review Exercises. Answers for all Review Exercises are given at the back of the book, and full solutions can be found in the Student Solutions Manual.

2.1 Limits and Continuity

- Find the limit of a function from tables. (*Review Exercises 1–2.*)

- Find left and right limits. (*Review Exercises 3–4.*)

- Find the limit of a function. (*Review Exercises 5–14.*)

- Determine whether a function is continuous or discontinuous. (*Review Exercises 15–22.*)

2.2 Rates of Change, Slopes, and Derivatives

- Find the derivative of a function from the *definition* of the derivative. (*Review Exercises 23–26.*)

$$f'(x) = \lim_{h \to 0} \frac{f(x+h) - f(x)}{h}$$

2.3 Some Differentiation Formulas

- Find the derivative of a function using the rules of differentiation. (*Review Exercises 27–32.*)

$$\frac{d}{dx} c = 0 \qquad \frac{d}{dx} x^n = nx^{n-1}$$

$$\frac{d}{dx}(c \cdot f) = c \cdot f' \qquad \frac{d}{dx}(f \pm g) = f' \pm g'$$

- Find the tangent line to a curve at a given point. (*Review Exercise 33.*)

- Calculate and interpret a company's marginal cost. (*Review Exercise 34.*)

$$MC(x) = C'(x) \qquad MR(x) = R'(x) \qquad MP(x) = P'(x)$$

- Find and interpret the derivative of a learning curve. (*Review Exercise 35.*)

- Find and interpret the derivative of an area or volume formula. (*Review Exercises 36–37.*)

2.4 The Product and Quotient Rules

- Find the derivative of a function using the Product Rule or Quotient Rule. (*Review Exercises 38–48.*)

$$\frac{d}{dx}(f \cdot g) = f' \cdot g + f \cdot g'$$

$$\frac{d}{dx}\left(\frac{f}{g}\right) = \frac{g \cdot f' - g' \cdot f}{g^2}$$

- Find the tangent line to a curve at a given point. (*Review Exercise 49.*)

- Use differentiation to solve an applied problem and interpret the answer. (*Review Exercises 50–52.*)

$$MAC(x) = \frac{C(x)}{x}$$

$$MAR(x) = \frac{R(x)}{x}$$

$$MAP(x) = \frac{P(x)}{x}$$

2.5 Higher-Order Derivatives

- Calculate the second derivative of a function. (*Review Exercises 53–62.*)

- Find and interpret the first and second derivatives in an applied problem. (*Review Exercise 63.*)

- Find the velocity and acceleration of a rocket. (*Review Exercise 64.*)

$$v(t) = s'(t) \qquad a(t) = v'(t) = s''(t)$$

- Find the maximum height of a projectile. (*Review Exercise 65.*)

2.6 The Chain Rule and the Generalized Power Rule

- Find the derivative of a function using the Generalized Power Rule. *(Review Exercises 66–75.)*

$$\frac{d}{dx} f^n = n \cdot f^{n-1} \cdot f'$$

$$\frac{d}{dx} f(g(x)) = f'(g(x)) \cdot g'(x)$$

$$\frac{dy}{dx} = \frac{dy}{du} \cdot \frac{du}{dx}$$

- Find the derivative of a function using *two* differentiation rules. *(Review Exercises 76–87.)*

- Find the tangent line to a curve at a given point. *(Review Exercise 88.)*

- Find the *second* derivative of a function using the Generalized Power Rule. *(Review Exercises 89–92.)*

- Find the derivative of a function in several different ways. *(Review Exercises 93–94.)*

- Use the Generalized Power Rule to find the derivative in an applied problem and interpret the answer. *(Review Exercises 95–96.)*

- Compare the profit from one unit to the marginal profit found by differentiation. *(Review Exercise 97.)*

- Find where the marginal profit equals a given number. *(Review Exercise 98.)*

- Use the Generalized Power Rule to solve an applied problem and interpret the answer. *(Review Exercises 99–100.)*

2.7 Nondifferentiable Functions

- See from a graph where the derivative is undefined. *(Review Exercises 101–104.)*

$$f' \text{ is undefined at } \begin{cases} \text{corner points} \\ \text{vertical tangents} \\ \text{discontinuities} \end{cases}$$

- Prove that a function is not differentiable at a given value. *(Review Exercises 105–106.)*

Hints and Suggestions

- (*Overview*) This chapter introduced one of the most important concepts in all of calculus, the *derivative*. First we defined it (using limits), then we developed several "rules of differentiation" to simplify its calculation.

- Remember the four interpretations of the derivative— *slopes, instantaneous rates of change, marginals,* and *velocities.*

- The *second* derivative gives the rate of change of the rate of change, and acceleration.

- Graphing calculators help to find limits, graph curves and their tangent lines, and calculate derivatives (using NDERIV) and second derivatives (using NDERIV twice). NDERIV, however, provides only an *approximation* to the derivative, and therefore sometimes gives a misleading result.

- The units of the derivative are important in applied problems. For example, if $f(x)$ gives the temperature in degrees at time x hours, then the derivative $f'(x)$ is in *degrees per hour.* In general, the units of the derivative $f'(x)$ are "f-units" per "x-unit."

| **2** | **Review Exercises and Chapter Test** | ◯ indicates a Chapter Test exercise. |

2.1 Limits and Continuity

1–2. Complete the tables and use them to find each limit (or state that it does not exist). Round calculations to three decimal places.

①a. $\lim_{x \to 2^-} (4x + 2)$

b. $\lim_{x \to 2^+} (4x + 2)$

c. $\lim_{x \to 2} (4x + 2)$

x	$4x + 2$	x	$4x + 2$
1.9		2.1	
1.99		2.01	
1.999		2.001	

②a. $\lim_{x \to 0^-} \frac{\sqrt{x+1} - 1}{x}$

b. $\lim_{x \to 0^+} \frac{\sqrt{x+1} - 1}{x}$

c. $\lim_{x \to 0} \frac{\sqrt{x+1} - 1}{x}$

x	$\frac{\sqrt{x+1}-1}{x}$	x	$\frac{\sqrt{x+1}-1}{x}$
−0.1		0.1	
−0.01		0.01	
−0.001		0.001	

3–4. For each piecewise linear function, find:

a. $\lim\limits_{x \to 5^-} f(x)$ **b.** $\lim\limits_{x \to 5^+} f(x)$ **c.** $\lim\limits_{x \to 5} f(x)$

③ $f(x) = \begin{cases} 2x - 7 & \text{if } x \le 5 \\ 3 - x & \text{if } x > 5 \end{cases}$

④ $f(x) = \begin{cases} 4 - x & \text{if } x < 5 \\ 2x - 11 & \text{if } x \ge 5 \end{cases}$

5–12. Find the following limits (*without* using limit tables).

5. $\lim\limits_{x \to 4} \sqrt{x^2 + x + 5}$ **6.** $\lim\limits_{x \to 0} \pi$

⑦ $\lim\limits_{s \to 16} \left(\dfrac{1}{2} s - s^{1/2} \right)$ **8.** $\lim\limits_{r \to 8} \dfrac{r}{r^2 - 30\sqrt[3]{r}}$

⑨ $\lim\limits_{x \to 1} \dfrac{x^2 - x}{x^2 - 1}$ **10.** $\lim\limits_{x \to -1} \dfrac{3x^3 - 3x}{2x^2 + 2x}$

⑪ $\lim\limits_{h \to 0} \dfrac{2x^2 h - xh^2}{h}$ **12.** $\lim\limits_{h \to 0} \dfrac{6xh^2 - x^2 h}{h}$

13–14. Use limits involving $\pm\infty$ to describe the asymptotic behavior of each function from its graph.

13. $f(x) = \dfrac{1}{(x+1)^2}$ ⑭ $f(x) = \dfrac{3x}{x - 2}$

15–22. For each function, state whether it is continuous or discontinuous. If it is discontinuous, state the values of x at which it is discontinuous.

15. $f(x) = 2x + 5$ **16.** $f(x) = x^2 - 1$

⑰ $f(x) = \dfrac{1}{x + 1}$ **18.** $f(x) = \dfrac{1}{x^2 + 1}$

19. $f(x) = \dfrac{x - 1}{x^2 + x}$ **20.** $f(x) = \dfrac{1}{|x| - 3}$

㉑ $f(x) = \begin{cases} 2x - 7 & \text{if } x \le 5 \\ 3 - x & \text{if } x > 5 \end{cases}$

[*Hint:* See Exercise 3.]

㉒ $f(x) = \begin{cases} 4 - x & \text{if } x < 5 \\ 2x - 11 & \text{if } x \ge 5 \end{cases}$

[*Hint:* See Exercise 4.]

2.2   Rates of Change, Slopes, and Derivatives

23–26. Find the derivative of each function using the definition of the derivative (that is, as you did in Section 2.2).

㉓ $f(x) = 2x^2 + 3x - 1$ **24.** $f(x) = 3x^2 + 2x - 3$

25. $f(x) = \dfrac{3}{x}$ **26.** $f(x) = 4\sqrt{x}$

2.3   Some Differentiation Formulas

27–28. Find the derivative of each function.

㉗ $f(x) = 6\sqrt[3]{x^5} - \dfrac{4}{\sqrt{x}} + 1$

28. $f(x) = 4\sqrt{x^5} - \dfrac{6}{\sqrt[3]{x}} + 1$

29–32. Evaluate each expression.

29. If $f(x) = \dfrac{1}{x^2}$, find $f'\left(\dfrac{1}{2}\right)$.

30. If $f(x) = \dfrac{1}{x}$, find $f'\left(\dfrac{1}{3}\right)$.

31. If $f(x) = 12\sqrt[3]{x}$, find $f'(8)$.

32. If $f(x) = 6\sqrt[3]{x}$, find $f'(-8)$.

㉝ Find the equation for the tangent line to the curve $y = \dfrac{1}{x} + x^2$ at $x = 1$.

㉞ **BUSINESS: Educational Software Costs** Educational institutions can buy multiple licenses for Word Search Maker software (for making word puzzles) at a total cost of $C(x) = 24x^{5/6}$ dollars for x software disks. Find the derivative of this cost function at:
a. $x = 1$ and interpret your answer.
b. $x = 64$ and interpret your answer.
Source: Variety Games, Inc.

35. BUSINESS: Learning Curves in Industry From page 26, the learning curve for building Boeing 707 airplanes is $f(x) = 150x^{-0.322}$, where $f(x)$ is the time (in thousands of hours) that it took to build the xth Boeing 707. Find the instantaneous rate of change of this production time for the tenth plane, and interpret your answer.

36. GENERAL: Geometry The formula for the area of a circle is $A = \pi r^2$, where r is the radius of the circle and π is a constant.

a. Show that the derivative of the area formula is $2\pi r$, the formula for the circumference of a circle.
b. Give an explanation for this in terms of rates of change.

37. GENERAL: Geometry The formula for the volume of a sphere is $V = \frac{4}{3}\pi r^3$, where r is the radius of the sphere and π is a constant.

a. Show that the derivative of the volume formula is $4\pi r^2$, the formula for the surface area of a sphere.
b. Give an explanation for this in terms of rates of change.

2.4 The Product and Quotient Rules

38–47. Find the derivative of each function.

38. $f(x) = 2x(5x^3 + 3)$ **39.** $f(x) = x^2(3x^3 - 1)$

40 $f(x) = (x^2 + 5)(x^2 - 5)$

41. $f(x) = (x^2 + 3)(x^2 - 3)$

42. $y = (x^4 + x^2 + 1)(x^5 - x^3 + x)$

43. $y = (x^5 + x^3 + x)(x^4 - x^2 + 1)$

44. $y = \dfrac{x - 1}{x + 1}$ **45.** $y = \dfrac{x + 1}{x - 1}$

46 $y = \dfrac{x^5 + 1}{x^5 - 1}$ **47.** $y = \dfrac{x^6 - 1}{x^6 + 1}$

48. Find the derivative of $f(x) = \dfrac{2x + 1}{x}$ in three different ways, and check that the answers agree:
a. By the Quotient Rule
b. By writing the function in the form $f(x) = (2x + 1)(x^{-1})$ and using the Product Rule.
c. By thinking of another way, which is the easiest of all.

49. Find the equation for the tangent line to the curve $y = \dfrac{x^3 - 1}{x^4 + 1}$ at $x = -1$.

50. BUSINESS: Sales The manager of an electronics store estimates that the number of flash drives that they will sell at a price of x dollars is

$$S(x) = \frac{2250}{x + 9}$$

Find the rate of change of this quantity when the price is \$6 per flash drive, and interpret your answer.

51 BUSINESS: Marginal Average Profit A company profit function is $P(x) = 6x - 200$ dollars, where x is the number of units.
a. Find the average profit function.
b. Find the marginal average profit function.
c. Evaluate the marginal average profit function at $x = 10$ and interpret your answer.

52. BUSINESS: Marginal Average Cost A company can produce a mini optical computer mouse at a cost of \$7.50 each while fixed costs are \$50. Therefore, the company's cost function is $C(x) = 7.5x + 50$.
a. Find the average cost function $AC(x) = \dfrac{C(x)}{x}$.
b. Find the marginal average cost function $MAC(x)$.
c. Evaluate $MAC(x)$ at $x = 50$ and interpret your answer.

Source: Green Pearle International

2.5 Higher-Order Derivatives

53–56. Find the *second* derivative of each function.

53 $f(x) = 12\sqrt{x^3} - 9\sqrt[3]{x}$ **54.** $f(x) = 18\sqrt[3]{x^2} - 4\sqrt{x^3}$

55. $f(x) = \dfrac{1}{3x^2}$ **56.** $f(x) = \dfrac{1}{2x^3}$

57–62. Evaluate each expression.

57. If $f(x) = \dfrac{2}{x^3}$, find $f''(-1)$.

58. If $f(x) = \dfrac{3}{x^4}$, find $f''(-1)$.

59. $\dfrac{d^2}{dx^2} x^6 \Big|_{x = -2}$ **60.** $\dfrac{d^2}{dx^2} x^{-2} \Big|_{x = -2}$

61 $\dfrac{d^2}{dx^2} \sqrt{x^5} \Big|_{x = 16}$ **62.** $\dfrac{d^2}{dx^2} \sqrt{x^7} \Big|_{x = 4}$

63. **GENERAL:** Population The population of a city t years from now is predicted to be $P(t) = 0.25t^3 - 3t^2 + 5t + 200$ thousand people. Find $P(10)$, $P'(10)$, and $P''(10)$ and interpret your answers.

64. **GENERAL:** Velocity A rocket rises $s(t) = 8t^{5/2}$ feet in t seconds. Find its velocity and acceleration after 25 seconds.

65. **GENERAL:** Velocity The fastest baseball pitch on record (thrown by Lynn Nolan Ryan of the California Angels on August 20, 1974) was clocked at 100.9 miles per hour (148 feet per second).
   a. If this pitch had been thrown straight up, its height after t seconds would have been $s(t) = -16t^2 + 148t + 5$ feet. Find the maximum height the ball would have reached.

   b. Verify your answer to part (a) by graphing the height function $y_1 = -16x^2 + 148x + 5$ on the window [0, 10] by [0, 400]. Then TRACE along the curve to find its highest point (or use the MAXIMUM feature of your calculator).

2.6   The Chain Rule and the Generalized Power Rule

66–87. Find the derivative of each function.

66. $h(z) = (4z^2 - 3z + 1)^3$ **67.** $h(z) = (3z^2 - 5z - 1)^4$

68. $g(x) = (100 - x)^5$ **69.** $g(x) = (1000 - x)^4$

70. $f(x) = \sqrt{x^2 - x + 2}$ **71.** $f(x) = \sqrt{x^2 - 5x - 1}$

72. $w(z) = \sqrt[3]{6z - 1}$ **73.** $w(z) = \sqrt[3]{3z + 1}$

74. $h(x) = \dfrac{1}{\sqrt[5]{(5x + 1)^2}}$ **75.** $h(x) = \dfrac{1}{\sqrt[5]{(10x + 1)^3}}$

76. $g(x) = x^2(2x - 1)^4$ **77.** $g(x) = 5x(x^3 - 2)^4$

78. $y = x^3 \sqrt[3]{x^3 + 1}$ **79.** $y = x^4 \sqrt{x^2 + 1}$

80. $f(x) = [(2x^2 + 1)^4 + x^4]^3$

81. $f(x) = [(3x^2 - 1)^3 + x^3]^2$

82. $f(x) = \sqrt{(x^2 + 1)^4 - x^4}$

83. $f(x) = \sqrt{(x^3 + 1)^2 + x^2}$

84. $f(x) = (3x + 1)^4(4x + 1)^3$

85. $f(x) = (x^2 + 1)^3(x^2 - 1)^4$

86. $f(x) = \left(\dfrac{x + 5}{x}\right)^4$ **87.** $f(x) = \left(\dfrac{x + 4}{x}\right)^5$

88. Find the equation for the tangent line to the curve $y = \sqrt[3]{x^2 + 4}$ at $x = 2$.

89–92. Find the *second* derivative of each function.

89. $h(w) = (2w^2 - 4)^5$ **90.** $h(w) = (3w^2 + 1)^4$

91. $g(z) = z^3(z + 1)^3$ **92.** $g(z) = z^4(z + 1)^4$

93. Find the derivative of $(x^3 - 1)^2$ in two ways:
   a. By the Generalized Power Rule.
   b. By "squaring out" the original expression and then differentiating.
   Your answers should agree.

94. Find the derivative of $g(x) = \dfrac{1}{x^3 + 1}$ in two ways:
   a. By the Quotient Rule.
   b. By the Generalized Power Rule.
   Your answers should agree.

95. **BUSINESS:** Marginal Profit A company's profit from producing x tons of polyurethane is $P(x) = \sqrt{x^3 - 3x + 34}$ thousand dollars (for $0 \le x \le 10$). Find $P'(5)$ and interpret your answer.

96. **BUSINESS:** Compound Interest If $500 is deposited in an account earning interest at r percent annually, after 3 years its value will be $V(r) = 500(1 + 0.01r)^3$ dollars. Find $V'(8)$ and interpret your answer.

97. **BUSINESS:** Marginal Profit (95 continued) Using a graphing calculator, graph the profit function from Exercise 95 on the window [0, 10] by [0, 30].
   a. Evaluate the profit function at 4, 5, and 6 and calculate the following actual costs:

   $P(5) - P(4)$ (actual cost of the fifth ton)

   $P(6) - P(5)$ (actual cost of the sixth ton)

   Compare these results with the "instantaneous" marginal profit at $x = 5$ found in Exercise 95.
   b. Define another function to be the derivative of the previously entered profit function (using NDERIV) and graph both together. Find (to the nearest tenth of a unit) the x-value where the marginal profit reaches 4.

98. **BUSINESS:** Marginal Cost On a graphing calculator, graph the cost function $C(x) = \sqrt[3]{x^2 + 25x + 8}$ as y_1 on the window [0, 20] by [−2, 10]. Use NDERIV to define y_2 to be its derivative, graphing both. TRACE along the derivative to find the x-value (to the nearest unit) where the marginal cost is 0.25.

99. **BIOMEDICAL:** Blood Flow Blood flowing through an artery encounters a resistance of $R(x) = 0.25(0.01x + 1)^4$, where x is the distance (in centimeters) from the heart. Find the instantaneous rate of change of the resistance 100 centimeters from the heart.

100. **GENERAL:** Survival Rate Suppose that for a group of 10,000 people, the number who survive to age x is $N(x) = 1000\sqrt{100 - x}$. Find $N'(96)$ and interpret your answer.

2.7 Nondifferentiable Functions

101–104. For each function graphed below, find the values of x at which the derivative does not exist.

101

102.

103.

104.

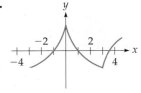

105–106. Use the definition of the derivative to show that the following functions are not differentiable at $x = 0$.

105. $f(x) = |5x|$ **106.** $f(x) = x^{3/5}$

Further Applications of Derivatives

3

©pryzmat/Shutterstock.com

What You'll Explore

A graph is a *picture* of a function, showing how steeply the function is rising or falling along with any high or low points. Such information is very useful for interpreting a company's profit function, a medicine's dose-response function, or any of the stimulus-response functions graphed on the next page.

Finding only the highest and lowest values of the function is called *optimization*, allowing us to maximize revenue, minimize cost, and to find the optimal point for any quantity that can be expressed as a function.

Stevens' Law of Psychophysics

How accurately can you judge how heavy something is? If you give someone two weights, with one *twice* as heavy as the other, most people will judge the heavier weight as being *less* than twice as heavy. This is one of the oldest problems in experimental psychology—how sensation (perceived weight) varies with stimulus (actual weight). Similar experiments can be performed for perceived brightness of a light compared with actual brightness, perceived effort compared with actual work, and so on. The results will vary somewhat from person to person, but the following diagram shows some typical stimulus-response curves (in arbitrary units).

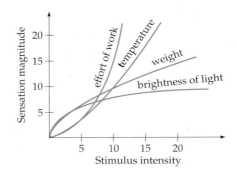

Notice, for example, that perceived effort of work increases more rapidly than actual work, which suggests that a 10% increase in an employee's work should be rewarded with a *greater* than 10% increase in pay.

Such stimulus-response curves were studied by the psychologist S. S. Stevens* at Harvard, who expressed them as power functions:

$$\text{Response} = a(\text{stimulus})^b$$

or

$$f(x) = ax^b \qquad \text{for constants } a \text{ and } b$$

In this chapter we will see that calculus can be very helpful for graphing such functions, including showing whether they "curl upward" like the work and temperature curves or "curl downward" like the weight and brightness curves (see page 188).

3.1   Graphing Using the First Derivative

Introduction

In this chapter we will put derivatives to two major uses: **graphing** and **optimization.** Graphing involves using calculus to find the most important points on a curve, and then sketching the curve either by hand or using a graphing calculator. Optimization means finding the largest or smallest values of a function (for example, maximizing profit or minimizing risk). We begin with graphing, since it will form the basis for optimization later in the chapter.

We saw in Chapter 2 that the *derivative* of a function gives the *slope* of its graph. On an interval:

$$f' > 0 \quad \text{means } f \text{ is } increasing.$$

$$f' < 0 \quad \text{means } f \text{ is } decreasing.$$

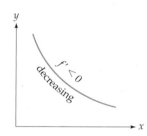

"Increasing" and "decreasing" on a graph mean rising and falling as you move *from left to right,* the same direction as you read these words.

Relative Extreme Points and Critical Numbers

On a graph, a **relative maximum point** is a point that is at least as *high* as the neighboring points of the curve on either side, and a **relative minimum point** is a point that is at least as *low* as the neighboring points on either side.

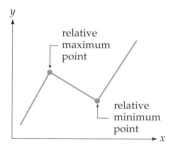

The word "relative" means that although these points may not be the highest and lowest on the *entire* curve, they are the highest and lowest *relative to points nearby.* (Later we will use the terms "absolute maximum" and "absolute minimum" to mean the highest and lowest points on the *entire* curve.) A curve may have any number of relative maximum and minimum points (collectively, **relative extreme points**), even none. For a function f, the relative extreme points may be defined more formally in terms of the values of f.

f has a **relative maximum value** at c if $f(c) \geq f(x)$ for all x near c.

f has a **relative minimum value** at c if $f(c) \leq f(x)$ for all x near c.

By "near c" we mean in some open interval containing c.

In the first of the two graphs below, the relative extreme points occur where the slope is *zero* (where the tangent line is horizontal), and in the second graph they occur where the slope is *undefined* (at corner points). The *x*-coordinates of such points are called **critical numbers.**

Critical Number

A *critical number* of a function f is an x-value in the domain of f at which either

$$f'(x) = 0 \qquad \text{Derivative is zero}$$

or

$$f'(x) \text{ is undefined} \qquad \text{or undefined}$$

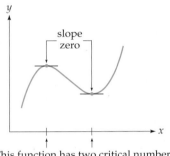

This function has two critical numbers (where the derivative is zero).

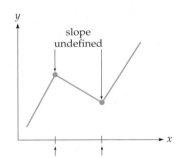

This function also has two critical numbers (but where the derivative is undefined).

Graphing Functions

We graph a function by finding its critical numbers, making a **sign diagram** for the derivative to show the intervals of increase and decrease and the relative extreme points, and then drawing the curve on a graphing calculator or "by hand." Obtaining a reasonable graph even with a graphing calculator requires more than just pushing buttons, as shown in the unsatisfactory graph of the function $f(x) = x^3 - 12x^2 - 60x + 36$ on the left. We will improve on this graph in the following Example.

A "useless" graph of $f(x) = x^3 - 12x^2 - 60x + 36$ on $[-10, 10]$ by $[-10, 10]$

FOR MORE HELP

with factoring and zeros, see the Algebra Review appendix, pages B5 and B12.

| **EXAMPLE 1** | **GRAPHING A FUNCTION** |

Graph the function $f(x) = x^3 - 12x^2 - 60x + 36$.

Solution

Step 1　**Find critical numbers.**

$$f'(x) = 3x^2 - 24x - 60 \qquad \text{Derivative of } f(x) = x^3 - 12x^2 - 60x + 36$$

$$= 3(x^2 - 8x - 20) = 3(x - 10)(x + 2) = 0 \qquad \text{Factoring and setting equal to zero}$$

Zero at　Zero at
$x = 10$　$x = -2$

The derivative is zero at $x = 10$ and at $x = -2$, and there are no numbers at which the derivative is undefined (it is a polynomial), so the critical numbers (CNs) are

$$\text{CN} \begin{cases} x = 10 \\ x = -2 \end{cases} \qquad \text{Both are in the domain of the original function}$$

Step 2 **Make a sign diagram for the derivative.** A sign diagram for f' begins with a copy of the x-axis with the critical numbers written below it and the behavior of f' indicated above it.

Behavior of f'

$$f' = 0 \qquad\qquad f' = 0$$
$$x = -2 \qquad\qquad x = 10$$

Critical numbers

Since f' is continuous, it can change sign only at critical numbers, so f' must keep the same sign between consecutive critical numbers. We determine the sign of f' in each interval by choosing a **test point** in each interval and substituting it into f'. We use the factored form: $f'(x) = 3(x - 10)(x + 2)$. For the first interval, choosing -3 for the test point,

$$f'(-3) = 3(-3 - 10)(-3 + 2) = 3(\text{negative})(\text{negative}) = (\text{positive}).$$

We indicate the sign of f' (the slope of f) by arrows: $\nearrow$ for positive slope, $\rightarrow$ for zero slope, and $\searrow$ for negative slope.

Using test point $-3, f'(-3) = 39$, so f' is *positive*

Using test point $0, f'(0) = -60$, so f' is *negative*

Using test point $11, f'(11) = 39$, so f' is *positive*

$$f' > 0 \qquad f' = 0 \qquad f' < 0 \qquad f' = 0 \qquad f' > 0$$
$$x = -2 \qquad\qquad x = 10$$

The sign diagram shows that to the left of -2 the function increases, then between -2 and 10 it decreases, and then to the right of 10 it increases again. Therefore, the open intervals of increase are $(-\infty, -2)$ and $(10, \infty)$, and the open interval of decrease is $(-2, 10)$.

Arrows $\nearrow \rightarrow \searrow$ indicate a relative *maximum* point, and arrows $\searrow \rightarrow \nearrow$ indicate a relative *minimum* point. We then list these points under the critical numbers.

$$f' > 0 \qquad f' = 0 \qquad f' < 0 \qquad f' = 0 \qquad f' > 0$$
$$x = -2 \qquad\qquad x = 10$$

rel max $(-2, 100)$

rel min $(10, -764)$

The y-coordinates of the points were found by evaluating the *original* function at the x-coordinate: $f(-2) = 100$ and $f(10) = -764$. [*Hint:* Use ▨ with TABLE or CALCULATE.]

Step 3   **Sketch the graph.** Our arrows ↗ → ↘ → ↗ show the general shape of the curve: going up, level, down, level, and up again. The critical numbers show that the graph should include x-values from before $x = -2$ to after $x = 10$, suggesting an interval such as $[-10, 20]$. The y-coordinates show that we want to go from above 100 to below -764, suggesting an interval of y-values such as $[-800, 200]$.

Using a graphing calculator, you then would graph on the window $[-10, 20]$ by $[-800, 200]$ (or some other reasonable window), as shown below.	**By hand,** you would plot the relative maximum point (with a "cap" ⌢) and the relative minimum point (with a "cup" ⌣), and then draw an "up-down-up" curve through them.

Two important observations:

1. Even with a graphing calculator, you still need calculus to find the relative extreme points that determine an appropriate window.

2. Given a calculator-drawn graph such as that on the left above, you should be able to make a hand-drawn graph such as that on the right, including numbers on the axes and coordinates of important points (using TRACE or TABLE if necessary).

Graphing Calculator Exploration

Do you see where the "useless" graph shown on page 165 fits into the "useful" graph above? Check this by graphing the function $y_1 = x^3 - 12x^2 - 60x + 36$ first on the window $[-10, 20]$ by $[-800, 200]$ (obtaining the graph shown) and then on the *standard* window $[-10, 10]$ by $[-10, 10]$.

PRACTICE PROBLEM 1

Find the critical numbers of $f(x) = x^3 - 12x + 8$.   Solution on page 173 >

In Example 1 there were no critical numbers at which the derivative was *undefined* (it was a polynomial). For an example of a function with a critical number where the derivative is *undefined*, think of the absolute value function $f(x) = |x|$. The graph has a "corner point" at the origin (as shown on the left), and so the derivative is undefined at the critical number $x = 0$.

$x = 0$ is a critical number of
$f(x) = |x|$

First-Derivative Test for Relative Extreme Values

The graphical idea from the sign diagram, that ↗ → ↘ (up, level, and down) indicates a relative *maximum* and ↘ → ↗ (down, level, and up) indicates a relative *minimum*, can be stated more formally in terms of the derivative.

max

min

First-Derivative Test

If a function f has a critical number c, then at $x = c$ the function has a
$\quad$ *relative maximum* if $f' > 0$ just before c and $f' < 0$ just after c.
$\quad$ *relative minimum* if $f' < 0$ just before c and $f' > 0$ just after c.

⚠ **Be Careful** Remember the order: Slopes that are *positive then negative* mean a *max*,
and slopes that are *negative then positive* mean a *min*.

If the derivative has the *same* sign on both sides of c, then the function has *neither* a
relative maximum nor a relative minimum at $x = c$.

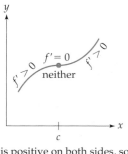

f' is positive on both sides, so
f has *neither* at $x = c$.

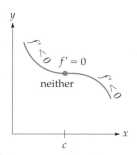

f' is negative on both sides, so
f has *neither* at $x = c$.

The following diagrams show that the first-derivative test applies even at critical
numbers where the derivative is *undefined* (abbreviated: f' und).

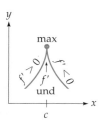

f' is positive then
negative, so f has a
relative *maximum*
at $x = c$.

f' is negative then
positive, so f has a
relative *minimum*
at $x = c$.

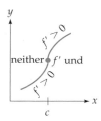

f' is positive on
both sides, so
f has *neither*
at $x = c$.

f' is negative on
both sides, so
f has *neither*
at $x = c$.

EXAMPLE 2   GRAPHING A FUNCTION

Graph $f(x) = -x^4 + 4x^3 - 20$.

Solution

$$f'(x) = -4x^3 + 12x^2 \qquad \text{Differentiating}$$

$$= -4x^2(x - 3) = 0 \qquad \text{Factoring and setting equal to zero}$$

Critical numbers:

$$\text{CN} \begin{cases} x = 0 \\ x = 3 \end{cases} \qquad \begin{array}{l} \text{From } -4x^2 = 0 \\ \text{From } (x - 3) = 0 \end{array}$$

We make a sign diagram for the derivative:

$$f' = 0 \qquad\qquad f' = 0 \qquad\qquad \text{Behavior of } f'$$
$$x = 0 \qquad\qquad x = 3 \qquad\qquad \text{Critical numbers}$$

We determine the sign of $f'(x) = -4x^2(x - 3)$ using test points in each interval (such as -1 in the leftmost interval), and then add arrows.

$$f' > 0 \qquad f' = 0 \qquad f' > 0 \qquad f' = 0 \qquad f' < 0$$
$$x = 0 \qquad\qquad x = 3$$

Finally, we interpret the arrows to describe the behavior of the function, which we state under the horizontal arrows.

$$f' > 0 \qquad f' = 0 \qquad f' > 0 \qquad f' = 0 \qquad f' < 0$$
$$x = 0 \qquad\qquad x = 3$$

rel max
$(3, 7)$

neither
$(0, -20)$

The open intervals of increase are $(-\infty, 0)$ and $(0, 3)$, and the open interval of decrease is $(3, \infty)$.

Using a graphing calculator, we would choose a window such as $[-2, 5]$ by $[-30, 10]$ to include the points $(0, -20)$ and $(3, 7)$, and graph the function.

By hand, we would plot $(0, -20)$ and $(3, 7)$, and join them by a curve that goes, according to the arrows, "up-level-up-level-down."

With a graphing calculator, an incomplete sign diagram may be enough to find an appropriate window.

Graphing Rational Functions

Recall from page 49 that a rational function is a quotient of polynomials $\dfrac{p(x)}{q(x)}$.

On pages 77–78 we graphed a rational function (shown on the left) with its *vertical asymptote* (where the denominator is zero and near which the curve becomes arbitrarily high or low) and *horizontal asymptote* (which the curve approaches as $x \to \infty$ or $x \to -\infty$).

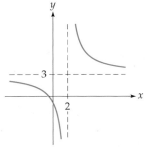

Graph of $f(x) = \dfrac{3x + 2}{x - 2}$

Why do vertical asymptotes occur where the denominator is zero? If the polynomial in the denominator is zero at $x = c$, by continuity it will be *near* zero for x-values *near c*, and reciprocals of numbers near zero are very large or very small (for example, $\frac{1}{0.001} = 1000$ and $\frac{1}{-0.001} = -1000$, as you may verify on a calculator). This leads to the following criterion for vertical asymptotes.

Vertical Asymptotes

A rational function $\dfrac{p(x)}{q(x)}$ has a **vertical asymptote** $x = c$ if $q(c) = 0$ but $p(c) \neq 0$.

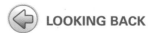

LOOKING BACK

For more on horizontal and vertical asymptotes, see pages 49–50 and 77–79.

A function has a *horizontal* asymptote if it has a limit as $x \to \infty$ or $x \to -\infty$.

Horizontal Asymptotes

A function $f(x)$ has a **horizontal asymptote** $y = c$ if

$$\lim_{x \to \infty} f(x) = c \quad \text{or} \quad \lim_{x \to -\infty} f(x) = c$$

Finding horizontal asymptotes by taking limits as x approaches $\pm\infty$ involves using some particular limits, each of which comes from the fact that *the reciprocal of a large number is a small number* (for example, $\frac{1}{1000} = 0.001$):

$$\lim_{x \to \infty} \frac{1}{x} = 0, \quad \lim_{x \to \infty} \frac{1}{x^2} = 0, \quad \text{or more generally,} \quad \lim_{x \to \infty} \frac{1}{x^n} = 0$$

for any positive integer n. These results also hold if the 1 in the numerator is replaced by any other number, and the same results hold for limits as x approaches *negative* infinity.

EXAMPLE 3   **GRAPHING A RATIONAL FUNCTION**

Graph $f(x) = \dfrac{2x + 3}{x - 1}$.

Solution

Since this is a rational function, we first look for vertical asymptotes. The denominator is zero at $x = 1$, and the numerator is not, so the line $x = 1$ is a vertical asymptote, which we show on the graph on the right.

To find *horizontal* asymptotes, take the limit as x approaches ∞ or $-\infty$. Dividing the numerator and denominator of the function by x and simplifying makes it easy to find the limit:

$$\lim_{x \to \infty} \frac{2x + 3}{x - 1} = \lim_{x \to \infty} \frac{\frac{2x}{x} + \frac{3}{x}}{\frac{x}{x} - \frac{1}{x}} = \lim_{x \to \infty} \frac{2 + \frac{3}{x}}{1 - \frac{1}{x}} = \frac{2 + 0}{1 - 0} = \frac{2}{1} = 2$$

<u>Dividing numerator and denominator by x and then simplifying</u>   <u>Since $\frac{3}{x}$ and $\frac{1}{x}$ both approach 0</u>   <u>Simplifying: the limit is 2</u>

Since the limit is 2, $y = 2$ is a horizontal asymptote, as drawn on the right. Letting $x \to -\infty$ gives the same limit and so the same horizontal asymptote.

Having found the asymptotes, we take the derivative to find the slope of the function. The quotient rule gives:

$$f'(x) = \frac{(x-1)(2) - (1)(2x+3)}{(x-1)^2} = \frac{2x - 2 - 2x - 3}{(x-1)^2} = \frac{-5}{(x-1)^2} \quad \text{From} \quad f(x) = \frac{2x+3}{x-1}$$

This derivative is undefined at $x = 1$, and elsewhere it is always negative since it is a negative number over a square. Therefore, all parts of the curve slope *downward.* The only way that the curve can slope downward everywhere and have the vertical and horizontal asymptotes we found is to have the graph shown on the right. Evaluating at $x = 0$ and plotting the y-intercept $(0, -3)$ also helps.

Graph of $f(x) = \dfrac{2x+3}{x-1}$

In the preceding Example we divided numerator and denominator by x before taking the limit. In other cases we may divide by a *higher* power of x, often the highest power that appears in the function. Doing so makes it easy to find the limit.

EXAMPLE 4   GRAPHING A RATIONAL FUNCTION

Graph $f(x) = \dfrac{3x^2}{x^2 - 4}$.

Solution

Factoring the denominator into $(x - 2)(x + 2)$ shows that there are two vertical asymptotes: $x = 2$ and $x = -2$, as shown on the right. To find *horizontal* asymptotes, we take the limit, first dividing numerator and denominator by x^2, the highest power of x appearing in the function:

$$\lim_{x \to \infty} \frac{3x^2}{x^2 - 4} = \lim_{x \to \infty} \frac{\frac{3x^2}{x^2}}{\frac{x^2}{x^2} - \frac{4}{x^2}} = \lim_{x \to \infty} \frac{3}{1 - \frac{4}{x^2}} = \frac{3}{1 - 0} = 3 \qquad \text{The limit is 3}$$

$\underbrace{\qquad\qquad}$ Dividing by x^2 and simplifying $\underbrace{\quad}$ Since $\frac{4}{x^2}$ approaches 0

Since the limit is 3 (and the limit as $x \to -\infty$ is also 3), $y = 3$ is a horizontal asymptote, as drawn on the right.

After finding the asymptotes, we take the derivative using the quotient rule:

$$f'(x) = \frac{(x^2 - 4)(6x) - (2x)(3x^2)}{(x^2 - 4)^2} = \frac{6x^3 - 24x - 6x^3}{(x^2 - 4)^2} = \frac{-24x}{(x^2 - 4)^2} \quad \text{From } f(x) = \frac{3x^2}{x^2 - 4}$$

Since the derivative changes sign, we make a sign diagram, remembering that *the function is undefined at* $x = 2$ *and at* $x = -2$ (indicated by vertical dashed lines on the graph) and showing that the derivative is zero at $x = 0$:

$$\begin{array}{ccccccc}
f' > 0 & f \text{ und} & f' > 0 & f' = 0 & f' < 0 & f \text{ und} & f' < 0 \\
\hline
 & x = -2 & & x = 0 & & x = 2 & \\
 & & & \longrightarrow & & & \\
\nearrow & & \nearrow & & \searrow & & \searrow \\
 & & & \text{rel max} & & & \\
 & & & (0, 0) & & &
\end{array}$$

Plotting the relative maximum point $(0, 0)$ and drawing the curve to have the asymptotes we found and the slopes shown in the sign diagram gives the curve on the right.

Graph of $f(x) = \dfrac{3x^2}{x^2 - 4}$

The rational functions in the last two examples had both vertical and horizontal asymptotes. However, a rational function may have just one type or no asymptotes at all. For example, the rational function $f(x) = \dfrac{x^2}{1} = x^2$ is in fact a parabola and so has no asymptotes.

PRACTICE PROBLEM 2

Below are four rational functions and below them four graphs numbered 1–4. Match the functions with the correct graphs. [*Hint:* Little or no calculation is necessary. Think of asymptotes and intercepts.]

$$f(x) = \frac{1}{x^2 - 4} \qquad g(x) = \frac{1}{x^2 + 4} \qquad h(x) = \frac{x}{x^2 - 4} \qquad i(x) = \frac{x^2}{x^2 + 4}$$

This function has graph ___   This function has graph ___   This function has graph ___   This function has graph ___

1)   2)   3)   4)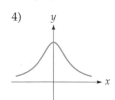

Solutions on next page >

Solutions TO PRACTICE PROBLEMS

1. $f'(x) = 3x^2 - 12 = 3(x^2 - 4) = 3(x + 2)(x - 2)$

$$\text{CN} \begin{cases} x = -2 \\ x = 2 \end{cases}$$

2. $f(x)$ is 3, $g(x)$ is 4, $h(x)$ is 1, and $i(x)$ is 2.

3.1	**Section Summary**

We have graphed two types of functions: polynomials and rational functions (quotients of polynomials). We used the fact that the derivative gives slope: Where the derivative is *positive*, the curve slopes *up*; where the derivative is *negative*, the curve slopes *down*; and where it does one followed by the other, the curve has a relative maximum or minimum point. Such information is most easily displayed on a *sign diagram* for the derivative. We saw that rational functions may have asymptotes (lines that the graph approaches arbitrarily closely), while polynomials do not have asymptotes.

In curve sketching, when do you use f and when do you use f'? Use the derivative f' to find useful x-values, such as critical numbers, but always return to the *original function f* for the y-coordinate of a point to be plotted on the graph.

To sketch the graph of a function f:

- Find the domain (by excluding any x-values at which the function is not defined).

- For a *rational* function, find all asymptotes:

  If the denominator is zero at an x-value c and the numerator is not, then $x = c$ is a *vertical* asymptote.

  If the function approaches a limit c as $x \to \infty$ or as $x \to -\infty$, then $y = c$ is a *horizontal* asymptote.

- Find the critical numbers (where f' is zero or undefined but where f is defined).

- List all of these on a sign diagram for the derivative, indicating the behavior of f' on each interval. Add arrows and relative extreme points.

- Finally, sketch the curve. If you use a graphing calculator, the sign diagram will suggest an appropriate window. If you graph by hand, your sign diagram will show you the shape of the curve.

If there are no relative extreme points, choose a few x-values, including any of special interest. On a graphing calculator, use these x-values to determine an x-interval, and use TABLE or TRACE to find a y-interval; then graph the function on the resulting window. By hand, plot the points corresponding to the chosen x-values and draw an appropriate curve through the points using your sign diagram.

3.1 Exercises

1–2. For the functions graphed on the right:

a. Find the intervals on which the derivative is positive.

b. Find the intervals on which the derivative is negative.

1.

2.

3. Which of the numbers 1, 2, 3, 4, 5, and 6 are critical numbers of the function graphed below?

4. The first four diagrams **a–d** below show the graphs of four functions, followed by the graphs of their derivatives, but not necessarily in the same order. Write below each derivative the correct function from which it came.

a.   **b.**

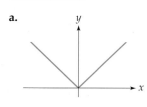

function f_1

function f_2

c.   **d.**

function f_3

function f_4

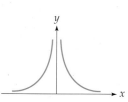

(a) derivative of function: ____

(b) derivative of function: ____

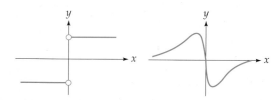

(c) derivative of function: ____

(d) derivative of function: ____

5–16. Find the critical numbers of each function.

5. $f(x) = x^3 - 48x$

6. $f(x) = x^3 - 27x$

7. $f(x) = x^3 - 6x^2 - 15x + 30$

8. $f(x) = x^3 + 6x^2 - 36x - 60$

9. $f(x) = x^4 + 4x^3 - 8x^2 + 1$

10. $f(x) = x^4 + 4x^3 - 20x^2 - 12$

11. $f(x) = (2x - 6)^4$ **12.** $f(x) = (x^2 + 6x - 7)^2$

13. $f(x) = 3x + 5$ **14.** $f(x) = 4x - 12$

15. $f(x) = x^3 + x^2 - x + 4$

16. $f(x) = x^3 - x^2 + 15$

17–32. Sketch the graph of each function "by hand" after making a sign diagram for the derivative and finding all open intervals of increase and decrease.

17. $f(x) = x^3 - 3x^2 - 9x + 10$

18. $f(x) = x^3 + 3x^2 - 9x - 11$

19. $f(x) = x^4 + 4x^3 - 8x^2 + 64$

20. $f(x) = x^4 - 4x^3 - 8x^2 + 64$

21. $f(x) = -x^4 + 4x^3 - 4x^2 + 1$

22. $f(x) = -x^4 - 4x^3 - 4x^2 + 1$

23. $f(x) = 3x^4 - 8x^3 + 6x^2$

24. $f(x) = 3x^4 + 8x^3 + 6x^2$

25. $f(x) = (x - 1)^6$ **26.** $f(x) = (x - 1)^5$

27. $f(x) = (x^2 - 4)^2$ **28.** $f(x) = (x^2 - 2x - 8)^2$

29. $f(x) = x^2(x - 4)^2$ **30.** $f(x) = x(x - 4)^3$

31. $f(x) = x^2(x - 5)^3$ **32.** $f(x) = x^3(x - 5)^2$

33–62. Sketch the graph of each rational function after making a sign diagram for the derivative and finding all relative extreme points and asymptotes.

33. $f(x) = \dfrac{6}{x + 3}$ **34.** $f(x) = \dfrac{4}{x - 2}$

35. $f(x) = \dfrac{3x + 6}{x - 2}$ **36.** $f(x) = \dfrac{2x - 6}{x + 3}$

37. $f(x) = \dfrac{12x - 24}{3x + 6}$ **38.** $f(x) = \dfrac{10x + 30}{2x - 6}$

39. $f(x) = \dfrac{8}{(x - 2)^2}$ **40.** $f(x) = \dfrac{18}{(x + 3)^2}$

41. $f(x) = \dfrac{12}{x^2 - 2x - 3}$ **42.** $f(x) = \dfrac{72}{x^2 - 2x - 8}$

43. $f(x) = \dfrac{8}{x^2 + 4}$ This curve is called the Witch of Agnesi

44. $f(x) = \dfrac{4x}{x^2 + 4}$ This curve is called Newton's serpentine

45. $f(x) = \dfrac{x^2}{x^2 + 1}$ **46.** $f(x) = \dfrac{x^2 - 1}{x^2 + 1}$

47. $f(x) = \dfrac{2x}{x^2 + 1}$ **48.** $f(x) = \dfrac{6x}{x^2 + 9}$

49. $f(x) = \dfrac{3}{x^2 - 1}$ **50.** $f(x) = \dfrac{9}{x^2 - 9}$

51. $f(x) = \dfrac{5x}{x^2 - 9}$ **52.** $f(x) = \dfrac{11x}{x^2 - 25}$

53. $f(x) = \dfrac{2x^2 - 4x + 6}{x^2 - 2x - 3}$ **54.** $f(x) = \dfrac{3x^2 + 6x + 3}{x^2 + 2x - 3}$

55. $f(x) = \dfrac{2x^2}{x^2 - 1}$ **56.** $f(x) = \dfrac{x^2 + 4}{x^2 - 4}$

57. $f(x) = \dfrac{16}{(x + 2)^3}$ **58.** $f(x) = \dfrac{27}{(x - 3)^3}$

59. $f(x) = \dfrac{4}{x^2\,(x - 3)}$ **60.** $f(x) = \dfrac{4}{x\,(x - 3)^2}$

61. $f(x) = \dfrac{2x^2}{x^4 + 1}$ **62.** $f(x) = \dfrac{x^4 - 2x^2 + 1}{x^4 + 2}$

63. Derive the formula $x = \dfrac{-b}{2a}$ for the x-coordinate of the vertex of parabola $y = ax^2 + bx + c$. [*Hint:* The slope is zero at the vertex, so finding the vertex means finding the critical number.]

64. Derive the formula $x = -b$ for the x-coordinate of the vertex of parabola $y = a(x + b)^2 + c$. [*Hint:* The slope is zero at the vertex, so finding the vertex means finding the critical number.]

Applied Exercises

65. BIOMEDICAL: Bacterial Growth A population of bacteria grows to size $p(x) = x^3 - 9x^2 + 24x + 10$ after x hours (for $x \geq 0$). Graph this population curve (based on, if you wish, a calculator graph), showing the coordinates of the relative extreme points.

66. BEHAVIORAL SCIENCE: Learning Curves A *learning curve* is a function $L(x)$ that gives the amount of time that a person requires to learn x pieces of information. Many learning curves take the form $L(x) = (x - a)^n + b$ (for $x \geq 0$), where a, b, and n are constants. Graph the learning curve $L(x) = (x - 2)^3 + 8$ (based on, if you wish, a calculator graph), showing the coordinates of all points corresponding to critical numbers.

67. GENERAL: Airplane Flight Path A plane is to take off and reach a level cruising altitude of 5 miles after a horizontal distance of 100 miles, as shown in the following diagram. Find a polynomial flight path of the form $f(x) = ax^3 + bx^2 + cx + d$ by following Steps i to iv to determine the constants a, b, c, and d.

  i. Use the fact that the plane is on the ground at $x = 0$ [that is, $f(0) = 0$] to determine the value of d.

  ii. Use the fact that the path is horizontal at $x = 0$ [that is, $f'(0) = 0$] to determine the value of c.

  iii. Use the fact that at $x = 100$ the height is 5 and the path is horizontal to determine the values of a and b. State the function $f(x)$ that you have determined.

  iv. Use a graphing calculator to graph your function on the window [0, 100] by [0, 6] to verify its shape.

68. GENERAL: Aspirin Clinical studies have shown that the analgesic (pain-relieving) effect of aspirin is approximately $f(x) = \dfrac{100x^2}{x^2 + 0.02}$ where $f(x)$ is the percentage of pain relief from x grams of aspirin.
  a. Graph this *dose-response* curve for doses up to 1 gram, that is, on the window [0, 1] by [0, 100].

  b. TRACE along the curve to see that the curve is very close to its maximum height of 100% or 1 by the time x reaches 0.65. This means that there is very little added effect in going above 650 milligrams, the amount of aspirin in two regular tablets, notwithstanding the aspirin companies' promotion of "extra strength" tablets. [*Note:* Aspirin's dose-response curve is extremely unusual in that it levels off quite early, and, even more unusual, aspirin's effect in protecting against heart attacks even *decreases* as the dosage x increases above about 80 milligrams.]
  Source: Consumer Reports

69. GENERAL: Drug Interception Suppose that the cost of a border patrol that intercepts x percent of the illegal drugs crossing a state border is $C(x) = \dfrac{600}{100 - x}$ million dollars (for $x < 100$).
  a. Graph this function on [0, 100] by [0, 100].
  b. Observe that the curve is at first rather flat but then rises increasingly steeply as x nears 100. Predict what the graph of the derivative would look like.
  c. Check your prediction by defining y_2 to be the derivative of y_1 (using NDERIV) and graphing both y_1 and y_2.

70. BIOMEDICAL: Heart Medication A cardiac medication is injected into the arm of a patient, and t minutes later the concentration in the heart is $f(t) = \dfrac{4t}{t^2 + 4}$ (milligrams per deciliter of blood). Graph this function on the interval [0, 6], showing the coordinates when the concentration is greatest.

71. BUSINESS: Long-Run Average Cost Suppose that a software company produces CDs (computer disks) at a cost of \$3 each, and fixed costs are \$50. The *cost function*, the total cost of producing x disks, will then be $C(x) = 3x + 50$, and the *average cost per unit* will be the total cost divided by the number of units:
$$AC(x) = \frac{3x + 50}{x}.$$
  a. Show that $\lim\limits_{x \to \infty} AC(x) = 3$, which is the *unit* or *marginal* cost.
  b. Sketch the graph of $AC(x)$, showing the horizontal asymptote. [*Note:* Your graph should be an illustration of the general business principle for linear cost

functions: *In the long run, average cost approaches marginal cost.*]

72. BUSINESS: Long-Run Average Cost Suppose that a company has a linear cost function (the total cost of producing x units) $C(x) = ax + b$ for constants a and b, where a is the *unit* or *marginal* cost and b is the *fixed* cost. Then the *average cost per unit* will be the total cost divided by the number of units:

$AC(x) = \dfrac{ax + b}{x}$. Show that $\lim\limits_{x \to \infty} AC(x) = a$.

[*Note:* Since a is the marginal cost, you have proved

the general business principle for linear cost functions: *In the long run, average cost approaches marginal cost.*]

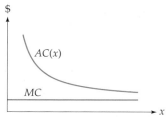

Conceptual Exercises

73. True or False: A *critical number* is where the derivative is zero or undefined.

74. True or False: At a critical number the function must be *defined*.

75. True or False: A relative maximum point is the highest point on the entire curve.

76. True or False: A relative minimum point is the lowest point on the entire curve.

77. From the graph of the *derivative* shown below, find the open intervals of increase and of decrease for the original function.

78. From the graph of the *derivative* shown below, find the open intervals of increase and of decrease for the original function.

79. Explain why, for a (simplified) rational function, the function and its derivative will be undefined at exactly the same x-values.

80. We saw on page 167 that the absolute value function is defined at $x = 0$ but its derivative is not. Can a *rational* function have an x-value where the function is defined but the derivative is not?

81. True or False: If the derivative has the *same sign* immediately on either side of an x-value, the function has neither a maximum nor a minimum at that x-value.

82. True or False: If a function is defined and has a positive derivative for all values of x, then $f(a) < f(b)$ whenever $a < b$.

83. Why does the criterion for $\dfrac{p(x)}{q(x)}$ to have a vertical asymptote at c require that $q(c) = 0$ but $p(c) \neq 0$ (see page 170)? Why not just $q(c) = 0$? Can you think of a rational function that does *not* have a vertical asymptote at a point where its denominator is *zero*? [*Hint:* Make both $p(c) = 0$ and $q(c) = 0$.]

84. Can a rational function have different horizontal asymptotes as $x \to \infty$ and as $x \to -\infty$? [*Hint:* To have a horizontal asymptote other than the x-axis, the highest power of x in the numerator and denominator must be the same, such as in $f(x) = \dfrac{ax^2 + bx + c}{Ax^2 + Bx + C}$. What are the two limits? Can you do the same for higher powers?]

Explorations and Excursions   The following problems extend and augment the material presented in the text.

Business: Average and Marginal Cost

85. A company's cost function is

$$C(x) = x^2 + 2x + 4$$

dollars, where x is the number of units.

a. Enter the cost function in y_1 on a graphing calculator.

b. Define y_2 to be the *marginal* cost function by defining y_2 to be the derivative of y_1 (using NDERIV).

c. Define y_3 to be the company's *average* cost function,

$$AC(x) = \frac{C(x)}{x},$$

by defining $y_3 = \dfrac{y_1}{x}$.

d. Turn off the function y_1 so that it will not be graphed, but graph the marginal cost function y_2 and the average cost function y_3 on the window [0, 10] by [0, 10]. Observe that the marginal cost function pierces the average cost function at its minimum point (use TRACE to see which curve is which function).

e. To see that the final sentence of part (d) is true in general, change the coefficients in the cost function $C(x)$, or change the cost function to a cubic or some other function [so that $C(x)/x$ has a minimum]. Again turn off the cost function and graph the other two to see that the marginal cost function pierces the average cost function at its minimum.

86. Prove that the marginal cost function intersects the average cost function at the point where the average cost is minimized (as shown in the following diagram) by justifying each numbered step ①–⑥ in the following series of equations.

At the x-value where the average cost function $AC(x) = \dfrac{C(x)}{x}$ is minimized, we must have

$$0 \overset{①}{=} \frac{d}{dx} AC(x) \overset{②}{=} \frac{xC'(x) - C(x)}{x^2}$$

$$\overset{③}{=} \frac{1}{x}\left[\frac{xC'(x) - C(x)}{x}\right]$$

$$\overset{④}{=} \frac{1}{x}\left[C'(x) - \frac{C(x)}{x}\right] \overset{⑤}{=} \frac{1}{x}[MC(x) - AC(x)].$$

⑥ Finally, explain why the equality of the first and last expressions in the series of equations proves the result.

The marginal cost function pierces the average cost function at its minimum.

Graphing Using the First and Second Derivatives

Introduction

In the previous section we used the first derivative to find the function's slope and relative extreme points and to draw its graph. In this section we will use the *second* derivative to find the **concavity** or **curl** of the curve and to define the important concept of **inflection point.** The second derivative also gives us a very useful way to distinguish between maximum and minimum points of a curve.

Concavity and Inflection Points

A curve that curls upward is said to be **concave up,** and a curve that curls downward is said to be **concave down.** A point where the concavity *changes* (from up to down or down to up) is called an **inflection point.**

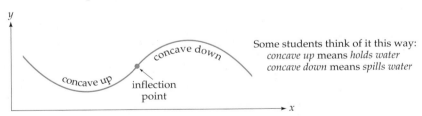

Some students think of it this way:
concave up means *holds water*
concave down means *spills water*

Concavity shows how a curve *curls* or *bends* away from straightness.

A straight line (with any slope) has *no concavity.*

However, bending the two ends *upward* makes it *concave up,*

and bending the two ends *downward* makes it *concave down.*

As these pictures show, a curve that is concave *up* lies *above* its tangent, while a curve that is concave *down* lies *below* its tangent (except, of course, at the point of tangency).

PRACTICE PROBLEM 1

For each of the following curves, label the parts that are concave up and the parts that are concave down. Then find all inflection points.

a.

b.

Solutions on page 185 >

How can we use calculus to determine concavity? The key is the second derivative. The second derivative, being the derivative of the derivative, gives the *rate of change* of the slope, showing whether the slope is *increasing* or *decreasing*. That is, $f'' > 0$ means that the slope is increasing, and so the curve must be *concave up* (as in the following diagram on the left below). Similarly, $f'' < 0$ means that the slope is decreasing, and so the curve must be *concave down* (as on the right below).

 LOOKING BACK

Concave up and *concave down* are generalizations of the idea that parabolas *open up* or *open down*, as illustrated on page 37. See also Exercises 83–84.

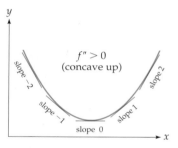

$f'' > 0$ means that the slope is increasing, so *f* is *concave up*.

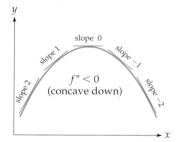

$f'' < 0$ means that the slope is decreasing, so *f* is *concave down*.

Since an inflection point is where the concavity changes, the second derivative must be negative on one side and positive on the other. Therefore, *at* an inflection point, f'' must be either zero or undefined. All of this may be summarized as follows.

Concavity and Inflection Points

On an interval:

  $f'' > 0$ means that *f* is *concave up* (curls upward).

  $f'' < 0$ means that *f* is *concave down* (curls downward).

  An *inflection point* is a point on the curve where the concavity changes (f'' must be zero or undefined).

Graphing Calculator Exploration

f is concave *up*   *f* is concave *down*
   f″ is positive   *f″* is negative

a. Use a graphing calculator to graph $y_1 = \sqrt[3]{x}$ on the window $[-2, 2]$ by $[-2, 2]$. Observe where the curve is concave up and where it is concave down.

b. Use NDERIV to define y_2 to be the derivative of y_1, and y_3 to be the derivative of y_2. Graph y_1 and y_3 (but turn off y_2 so that it will not be graphed).

c. Verify that y_3 (the second derivative) is positive where y_1 is concave up, and negative where y_1 is concave down.

d. Now change y_1 to $y_1 = \dfrac{x^2 + 2}{x^2 + 1}$ and observe that where this curve is concave up or down agrees with where y_3 is positive or negative. According to y_3, how many inflection points does y_1 have? Can you see them on y_1?

To find inflection points, we make a sign diagram for the *second* derivative to show where the concavity changes (where f'' changes sign). An example will make the method clear.

| **EXAMPLE 1** | **GRAPHING AND INTERPRETING A COMPANY'S ANNUAL PROFIT FUNCTION** |

A company's annual profit after x years is $f(x) = x^3 - 9x^2 + 24x$ million dollars (for $x \geq 0$). Graph this function, showing all relative extreme points and inflection points. Interpret the inflection points.

Solution

$$f'(x) = 3x^2 - 18x + 24 \qquad \text{Differentiating}$$
$$= 3(x^2 - 6x + 8) = 3(x - 2)(x - 4) \qquad \text{Factoring}$$

The critical numbers are $x = 2$ and $x = 4$, and the sign diagram for f' (found in the usual way) is

$$f' > 0 \qquad f' = 0 \qquad f' < 0 \qquad f' = 0 \qquad f' > 0$$
$$x = 2 \qquad\qquad x = 4$$

rel max
(2, 20)

rel min
(4, 16)

To find the inflection points, we calculate the second derivative:

$$f''(x) = 6x - 18 = 6(x - 3) \qquad \text{Differentiating } f'(x) = 3x^2 - 18x + 24$$

This is zero at $x = 3$, which we enter on a sign diagram for the *second* derivative.

$$f'' = 0 \qquad\qquad \leftarrow \text{Behavior of } f''$$
$$x = 3 \qquad\qquad \leftarrow \text{Where } f'' \text{ is zero or undefined}$$

We use test points to determine the sign of $f''(x) = 6(x - 3)$ on either side of 3, just as we did for the first derivative.

$$f''(2) = 6(2 - 3) < 0 \qquad f''(4) = 6(4 - 3) > 0$$

$f'' < 0$	$f'' = 0$	$f'' > 0$
	$x = 3$	
con dn		con up

Concave down, concave up
(so concavity *does* change)

IP (3, 18)

IP means *inflection point.* The 18 comes from substituting $x = 3$ into $f(x) = x^3 - 9x^2 + 24x$

Using a graphing calculator, we would choose an *x*-interval such as [0, 6] (to include the *x*-values on the sign diagrams) and a *y*-interval such as [0, 30] (to include the origin and the *y*-values on the sign diagrams), and graph the function.

By hand, we would plot the relative maximum (⌢), minimum (⌣), and inflection point and sketch the curve according to the sign diagrams, being sure to show the concavity changing at the inflection point.

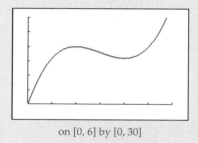

on [0, 6] by [0, 30]

Interpretation of the inflection point: Observe what the graph shows—that the company's profit increased (up to year 2), then decreased (up to year 4), and then increased again. The inflection point at $x = 3$ is where the profit *first began to show signs of improvement.* It marks the end of the period of increasingly steep decline and the first sign of an "upturn," where a clever investor might begin to "buy in."

 Be Careful At an inflection point, the concavity (that is, the sign of f'') must *actually change.* For example, a second derivative sign diagram such as

$f'' < 0$	$f'' = 0$	$f'' < 0$
	$x = 3$	
con dn		con dn

sign of f'' (concavity) does *not* change

would mean that there is *not* an inflection point at $x = 3,$ since the concavity is the same on both sides. For there to be an inflection point, the signs of f'' on the two sides *must be different.*

For each curve, is there an inflection point? [*Hint:* Does the concavity change?]

a.

b.

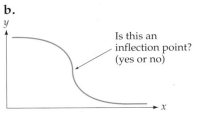

Solutions on page 185 >

Inflection Points in the Real World

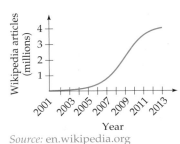

Inflection points occur in many everyday situations.* The graph on the left is typical of a company that at first prospers, then falters, and then recovers. The inflection point is the *first sign of recovery*. It is where a smart investor might purchase the company before the recovery becomes obvious.

The next graph shows the number of articles in *Wikipedia* in English for recent years. The inflection point (in 2008) marks the end of increasingly rapid growth and the beginning of slowing growth. That is, in later years there will still be gains, but they will be smaller and smaller.

In business, a *point of diminishing returns* is a point on a revenue or profit curve where further investment will bring decreasing rather than increasing returns. Geometrically, it is an inflection point where the curve changes from concave up to concave down, as in the second graph. It represents the end of growth that is accelerating and the beginning of growth that is slowing.

Source: en.wikipedia.org

In practice, given a particular function, we find its point of diminishing returns by finding where $f''(x) = 0$ and checking that $f''(x)$ changes from *positive* to *negative*.

EXAMPLE 2   FINDING A POINT OF DIMINISHING RETURNS

A company predicts that if it invests x hundred thousand dollars in advertising, its revenue will be $R(x) = 30x^2 - 2x^3 + 100$ hundred thousand dollars, for $0 \leq x \leq 9$. Find the company's point of diminishing returns.

Solution

The first and second derivatives are

$$R'(x) = 60x - 6x^2 \qquad \text{Differentiating } R(x) = 30x^2 - 2x^3 + 100$$

and

$$R''(x) = 60 - 12x \qquad \text{Differentiating again}$$

$$= 12(5 - x) \qquad \text{Factoring (equals zero at } x = 5)$$

The second derivative is zero at $x = 5$. Using test points $x = 4$ and $x = 6$ on either side of $x = 5$ shows that the second derivative changes from positive (on the left) to negative (on the right), so the revenue curve changes from concave *up* to concave *down*, giving a point of diminishing returns. Therefore, since x is in hundreds of thousands of dollars, the company's point of diminishing returns is $500,000.

---

*For an interesting application of concavity to economics, see Harry M. Markowitz, "The Utility of Wealth," *Journal of Applied Political Economy* **60**(2). In this article, "concave" means "concave up" and "convex" means "concave down." Markowitz won the Nobel Memorial Prize in Economics in 1990.

Distinguish carefully between slope and concavity: *Slope* measures *steepness,* while *concavity* measures *curl.* All combinations of slope and concavity are possible. A graph may be

Increasing and concave *up*
($f' > 0, f'' > 0$), such as

Increasing and concave *down*
($f' > 0, f'' < 0$), such as

Decreasing and concave *up*
($f' < 0, f'' > 0$), such as

Decreasing and concave *down*
($f' < 0, f'' < 0$), such as

The four quarters of a circle illustrate all four possibilities, as shown on the left.

EXAMPLE 3   GRAPHING A FRACTIONAL POWER FUNCTION

Graph $f(x) = 18x^{1/3}$.

Solution

The derivative is

$$f'(x) = 6x^{-2/3} = \frac{6}{\sqrt[3]{x^2}} \qquad \text{\textit{Undefined} at } x = 0 \text{ (zero denominator)}$$

The sign diagram for f' is

$\underline{\quad f' > 0 \qquad f' \text{ und} \qquad f' > 0 \quad}$
$\qquad\qquad\quad x = 0$
$\qquad\qquad\quad$ neither
$\qquad\qquad\quad (0, 0)$

f' is undefined at $x = 0$ and positive on either side (using test points)

The *second* derivative is

$$f''(x) = -4x^{-5/3} = \frac{-4}{\sqrt[3]{x^5}} \qquad \text{Also undefined at } x = 0$$

The sign diagram for f'' is

$\underline{\quad f'' > 0 \qquad f'' \text{ und} \qquad f'' < 0 \quad}$
$\text{con up} \qquad\; x = 0 \qquad \text{con dn}$
$\qquad\qquad\quad \text{IP } (0, 0)$

Concavity is different on either side of $x = 0$ (using test points), so there *is* an inflection point at $x = 0$

Based on this information, we may graph the function with a graphing calculator or by hand:

Using a graphing calculator, we experiment with viewing windows centered at the inflection point $(0, 0)$ until we find one that shows the curve effectively. The graph on $[-2, 2]$ by $[-20, 20]$ is shown below. Many other windows would be just as good.

By hand, we use the sign diagrams to draw the curve to the *left* of $x = 0$ as *increasing* and concave *up*, and to the right of $x = 0$ as increasing and concave *down*, with the two parts meeting at the origin. The scale comes from calculating the points $(1, 18)$ and $(-1, -18)$.

The fact that the derivative is *undefined* at $x = 0$ is shown in the above graph by the *vertical tangent* at the origin (since the slope of a vertical line is undefined). This function is the stimulus-response curve for brightness of light (for $x \geq 0$; see page 163), and the vertical tangent indicates the disproportionately large effect of small increases in dim light.

  EXAMPLE 4   **GRAPHING USING A GRAPHING CALCULATOR**

Use a graphing calculator to graph $f(x) = 36 \sqrt[3]{(x - 1)^2}$.

Solution

Using a standard window $[-10, 10]$ by $[-10, 10]$ gives the graph on the left, which is useless.

Instead, we begin by differentiating $f(x) = 36(x - 1)^{2/3}$ and making its sign diagram.

A useless graph of
$f(x) = 36 \sqrt[3]{(x - 1)^2}$
on $[-10, 10]$ by $[-10, 10]$

$$f'(x) = 24(x - 1)^{-1/3} = \frac{24}{\sqrt[3]{x - 1}}$$

Undefined at $x = 1$

$$\frac{f' < 0 \qquad f' \text{ und} \qquad f' > 0}{x = 1}$$
$$\text{rel min}$$
$$(1, 0)$$

Using test points on either side of $x = 1$

The y-coordinate in $(1, 0)$ comes from evaluating the original function at $x = 1$

From the sign diagram we choose an x-interval centered at the critical number $x = 1$, such as $[-4, 6]$. After some experimenting, we choose the y-interval $[0, 100]$ (beginning at $y = 0$ because the minimum point has y-coordinate 0) so that the curve fills the screen.

$y = 36 \sqrt[3]{(x-1)^2}$ on
$[-4, 6]$ by $[0, 100]$

A sharp point on a graph such as the one at $x = 1$ is called a *cusp*, where the function is not differentiable.

Second-Derivative Test

Determining whether a twice-differentiable function has a relative maximum or minimum at a critical number is merely a question of concavity: concave *up* means a relative *minimum*, and concave *down* means a relative *maximum*.

Concave *up* at a critical number: relative *minimum*.

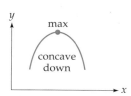

Concave *down* at a critical number: relative *maximum*.

Since the second derivative determines concavity, we have the following *second-derivative test*, which will be very useful in the next two sections.

Second-Derivative Test for Relative Extreme Points

If $x = c$ is a critical number of f at which f'' is defined, then

$f''(c) > 0$ means that f has a relative *minimum* at $x = c$.

$f''(c) < 0$ means that f has a relative *maximum* at $x = c$.

To use the second-derivative test, first find all critical numbers, substitute each into the second derivative (if possible), and determine the sign of the result: A *positive* result means a *minimum* at the critical number, and a *negative* result means a *maximum*. (If the second derivative is zero, then the test is inconclusive, and you should use the First-Derivative Test (page 168) or make a sign diagram for f'.)

EXAMPLE 5   USING THE SECOND-DERIVATIVE TEST

Use the second-derivative test to find all relative extreme points of $f(x) = x^3 - 9x^2 + 24x$.

Solution

$$f'(x) = 3x^2 - 18x + 24 \qquad \text{The derivative}$$

$$= 3(x^2 - 6x + 8) = 3(x - 2)(x - 4) \qquad \text{Factoring}$$

$$\text{CN} \begin{cases} x = 2 \\ x = 4 \end{cases} \qquad \text{Critical numbers}$$

We substitute each critical number into $f''(x) = 6x - 18$.

At $x = 2$: $\quad f''(2) = 6 \cdot 2 - 18 = -6 \qquad$ (negative) $\qquad \begin{array}{l} f''(x) = 6x - 18 \\ \text{at } x = 2 \end{array}$

Therefore, f has a relative *maximum* at $x = 2$.

At $x = 4$: $\quad f''(4) = 6 \cdot 4 - 18 = 6 \qquad$ (positive) $\qquad \begin{array}{l} f''(x) = 6x - 18 \\ \text{at } x = 4 \end{array}$

Therefore, f has a relative *minimum* at $x = 4$.

The relative maximum at $x = 2$ and the relative minimum at $x = 4$ are exactly what we found before when we graphed this function on page 180 or as shown on the right.

Be Careful   The second-derivative test tells us *nothing* if the second derivative is zero at a critical number. This is shown by the following three functions. Each function has a critical number $x = 0$ at which f'' is zero (as you may check), but one has a maximum, one has a minimum, and one has neither.

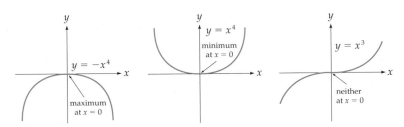

Three functions showing that a critical number where $f'' = 0$
may be a maximum, a minimum, or neither.

Solutions TO PRACTICE PROBLEMS

1. a.   **b.**

2. a. No (concave up on both sides)

   b. Yes (concave down then concave up)

3.2   Section Summary

The main developments of this section were the use of the second derivative to determine the *concavity* or *curl* of a function, and finding and interpreting *inflection points.* On an interval:

$$f'' > 0 \quad \text{means that } f \text{ is concave } up.$$

$$f'' < 0 \quad \text{means that } f \text{ is concave } down.$$

To locate inflection points (points at which the concavity changes), we find where the second derivative is zero or undefined and then make a sign diagram for f'' to see whether the concavity *actually changes.*

To graph a function, we use the *first*-derivative sign diagram to find *slope* and *relative extreme points,* and the *second*-derivative sign diagram to find *concavity* and *inflection points.* Then we graph the function, either by hand or on a graphing calculator (using the relative extreme points and inflection points to determine the window).

The following steps may be useful when graphing a function by hand.

Curve Sketching

1. Find the domain.
2. For a rational function, find and graph any vertical or horizontal asymptotes.
3. Find the first derivative and all x-values at which it is zero or undefined.
4. Make a sign diagram for the first derivative, showing "max," "min," and "neither" points (with y-coordinates found from f).
5. Find the second derivative and all x-values at which it is zero or undefined. (This and the following step may not be necessary for rational functions.)
6. Make a sign diagram for the second derivative, showing inflection points (where the concavity *changes*).
7. Based on the sign diagrams, sketch the graph, labeling all maximum, minimum, and inflection points as well as any asymptotes.

The *second-derivative test* shows whether a function has a relative maximum or minimum at a critical value (provided that f'' is defined and not zero):

$$f'' > 0 \quad \text{means a relative minimum.}$$

$$f'' < 0 \quad \text{means a relative maximum.}$$

The second-derivative test should be thought of as a simple application of concavity.

3.2 Exercises

1–6. For each graph, which of the numbered points are inflection points?

1.

2.

3.

4.

5.

6.

7–30. For each function:
a. Make a sign diagram for the first derivative.
b. Make a sign diagram for the second derivative.
c. Sketch the graph by hand, showing all relative extreme points and inflection points.

7. $f(x) = x^3 + 3x^2 - 9x + 5$

8. $f(x) = x^3 - 3x^2 - 9x + 7$

9. $f(x) = x^3 - 3x^2 + 3x + 4$

10. $f(x) = x^3 + 3x^2 + 3x + 6$

11. $f(x) = x^4 - 8x^3 + 18x^2 + 2$

12. $f(x) = x^4 + 8x^3 + 18x^2 + 8$

13. $f(x) = x^4 + 4x^3 + 15$ 　　14. $f(x) = 2x^4 - 8x^3 + 30$

15. $f(x) = 5x^4 - x^5$ 　　16. $f(x) = (x - 2)^3 + 2$

17. $f(x) = (2x + 4)^5$ 　　18. $f(x) = (3x - 6)^6 + 1$

19. $f(x) = x(x - 3)^2$ 　　20. $f(x) = x^3(x - 4)$

21. $f(x) = x^{3/5}$ 　　22. $f(x) = x^{1/5}$

23. $f(x) = \sqrt[5]{x^4} + 2$ 　　24. $f(x) = \sqrt[5]{x^2} - 1$

25. $f(x) = \sqrt[4]{x^3}$ 　　26. $f(x) = \sqrt{x^5}$

27. $f(x) = \sqrt[3]{(x - 1)^2}$ 　　28. $f(x) = \sqrt[5]{x + 2} + 3$

29. $f(x) = \dfrac{x}{x^2 - 1}$ 　　30. $f(x) = \dfrac{54(x^2 - 1)}{x^2 + 27}$

 31–40. Graph each function using a graphing calculator by first making a sign diagram for just the first derivative. Make a sketch from the screen, showing the coordinates of all relative extreme points and inflection points. Graphs may vary depending on the window chosen.

31. $f(x) = x^{1/2}$ 　　32. $f(x) = x^{3/2}$

33. $f(x) = x^{-1/2}$ 　　34. $f(x) = x^{-3/2}$

35. $f(x) = 9x^{2/3} - 6x$ 　　36. $f(x) = 30x^{1/3} - 10x$

37. $f(x) = 8x - 10x^{4/5}$ 　　38. $f(x) = 6x - 10x^{3/5}$

39. $f(x) = 3x^{2/3} - x^2$ 　　40. $f(x) = 3x^{4/3} - 2x^2$

41–46. For each function, find all critical numbers and then use the second-derivative test to determine whether the function has a relative maximum or minimum at each critical number.

41. $f(x) = x^3 - 6x^2 + 9x - 2$ 　　42. $f(x) = x^3 - 12x + 4$

43. $f(x) = x^4 - 4x^3 + 4x^2 + 1$ 　　44. $f(x) = x^4 - 2x^2 + 1$

45. $f(x) = x + \dfrac{9}{x}$ 　　46. $f(x) = \dfrac{x}{4} + \dfrac{1}{x}$

 47–48. FINDING INFLECTION POINTS Use a graphing calculator to estimate the x-coordinates of the inflection points of each function, rounding your answers to two decimal places. [*Hint:* Graph the second derivative, either calculating it directly or using NDERIV twice, and see where it crosses the x-axis.]

47. $f(x) = x^5 - 2x^3 + 3x + 4$

48. $f(x) = x^5 - 3x^3 + 6x + 2$

49–56. Sketch the graph of a function $f(x)$ that satisfies the stated conditions. Mark any inflection points by writing *IP* on your graph. [*Note:* There is more than one possible answer.]

49. a. f is continuous and differentiable everywhere.
   b. $f(0) = 3$
   c. $f'(x) > 0$ on $(-\infty, -4)$ and $(0, \infty)$
   d. $f'(x) < 0$ on $(-4, 0)$
   e. $f''(x) < 0$ on $(-\infty, -2)$
   f. $f''(x) > 0$ on $(-2, \infty)$

50. a. f is continuous and differentiable everywhere.
   b. $f(0) = 4$
   c. $f'(x) < 0$ on $(-\infty, -1)$ and $(3, \infty)$
   d. $f'(x) > 0$ on $(-1, 3)$
   e. $f''(x) > 0$ on $(-\infty, 1)$
   f. $f''(x) < 0$ on $(1, \infty)$

51. a. f is continuous and differentiable everywhere except at $x = 2$, where it is undefined.
   b. $f(0) = -1$
   c. Horizontal asymptote $y = 1$ and vertical asymptote $x = 2$
   d. $f'(x) < 0$ wherever f' is defined

52. a. f is continuous and differentiable everywhere except at $x = -5$ and $x = 5$, where it is undefined.
   b. $f(0) = -1$
   c. Horizontal asymptote $y = 2$ and vertical asymptotes $x = -5$ and $x = 5$
   d. $f''(x) > 0$ on $(-\infty, -5)$ and $(5, \infty)$
   e. $f''(x) < 0$ on $(-5, 5)$

53. a. f is continuous everywhere and differentiable everywhere except at $x = 0$.
   b. $f(0) = 3$
   c. $f'(x) > 0$ on $(-\infty, -2)$ and $(0, 2)$
   d. $f'(x) < 0$ on $(-2, 0)$ and $(2, \infty)$
   e. $f''(x) < 0$ on $(-\infty, 0)$ and $(0, \infty)$

54. a. f is continuous everywhere and differentiable everywhere except at $x = 0$.
   b. $f(0) = 5$
   c. $f'(x) < 0$ on $(-\infty, -3)$ and $(0, 3)$
   d. $f'(x) > 0$ on $(-3, 0)$ and $(3, \infty)$
   e. $f''(x) > 0$ on $(-\infty, 0)$ and $(0, \infty)$

(See instructions on the previous page.)

55. a. f is continuous and differentiable everywhere.
  b. $f(0) = 6$
  c. $f'(x) < 0$ on $(-\infty, -6)$ and $(0, 6)$
  d. $f'(x) > 0$ on $(-6, 0)$ and $(6, \infty)$
  e. $f''(x) > 0$ on $(-\infty, -3)$ and $(3, \infty)$
  f. $f''(x) < 0$ on $(-3, 3)$

56. a. f is continuous and differentiable everywhere.
  b. $f(0) = 2$
  c. $f'(x) > 0$ on $(-\infty, -8)$ and $(0, 8)$
  d. $f'(x) < 0$ on $(-8, 0)$ and $(8, \infty)$
  e. $f''(x) < 0$ on $(-\infty, -4)$ and $(4, \infty)$
  f. $f''(x) > 0$ on $(-4, 4)$

Applied Exercises

57. BUSINESS: Revenue A company's annual revenue after x years is $f(x) = x^3 - 9x^2 + 15x + 25$ thousand dollars (for $x \geq 0$).
  a. Make sign diagrams for the first and second derivatives.
  b. Sketch the graph of the revenue function, showing all relative extreme points and inflection points.
  c. Give an interpretation of the inflection point.

58. BUSINESS: Sales A company's weekly sales (in thousands) after x weeks are given by $f(x) = -x^4 + 4x^3 + 70$ (for $0 \leq x \leq 3$).
  a. Make sign diagrams for the first and second derivatives.
  b. Sketch the graph of the sales function, showing all relative extreme points and inflection points.
  c. Give an interpretation of the positive inflection point.

59. GENERAL: Temperature The temperature in a refining tower is $f(x) = x^4 - 4x^3 + 112$ degrees Fahrenheit after x hours (for $0 \leq x \leq 5$).
  a. Make sign diagrams for the first and second derivatives.
  b. Sketch the graph of the temperature function, showing all relative extreme points and inflection points.
  c. Give an interpretation of the positive inflection point.

60. BIOMEDICAL: Dosage Curve The dose-response curve for x grams of a drug is $f(x) = 8(x - 1)^3 + 8$ (for $x \geq 0$).
  a. Make sign diagrams for the first and second derivatives.
  b. Sketch the graph of the response function, showing all relative extreme points and inflection points.
  c. Give an interpretation of the inflection point.

61. PSYCHOLOGY: Stimulus and Response Sketch the graph of the brightness response curve $f(x) = x^{2/5}$ for $x \geq 0$, showing all relative extreme points and inflection points.
  Source: S. S. Stevens, On the Psychophysical Law

62. PSYCHOLOGY: Stimulus and Response Sketch the graph of the loudness response curve $f(x) = x^{4/5}$ for $x \geq 0$, showing all relative extreme points and inflection points.
  Source: S. S. Stevens, On the Psychophysical Law

63–64. SOCIOLOGY: Status Sociologists have estimated how a person's "status" in society (as perceived by others) depends on the person's income and education level. One estimate is that status S depends on income i according to the formula $S(i) = 16\sqrt{i}$ (for $i \geq 0$), and that status depends upon education level e according to the formula $S(e) = \frac{1}{4} e^2$ (for $e \geq 0$).
Source: Social Forces **50**

63. a. Sketch the graph of the preceding function $S(i)$.
  b. Is the curve concave up or down? What does this signify about the rate at which status increases at higher income levels?

64. a. Sketch the graph of the preceding function $S(e)$.
  b. Is the curve concave up or down? What does this signify about the rate at which status increases at higher education levels?

65. BUSINESS: Sales Worldwide semiconductor sales are approximated by the function $S(x) = 2x^3 - 18x^2 + 48x + 220$, in billions of dollars, where x stands for the number of *years since 2005* (so that, for example, $x = 5$ would correspond to 2010).

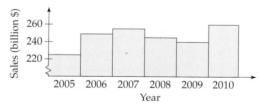

  a. Make sign diagrams for the first and second derivatives.
  b. Sketch the graph of $S(x)$, showing all relative extreme points and inflection points.
  c. Interpret the meaning of the inflection point and determine the year in which it occurred.
  Source: Standard & Poor's

66. BUSINESS: Sales Sales of dynamic random access memory (DRAM) chips are approximated by the function $S(x) = \frac{1}{3}x^3 - 3x^2 + 5x + 32$, in billions of dollars, where x stands for the number of *years since 2004* (so that, for example, $x = 6$ would correspond to 2010).

a. Make sign diagrams for the first and second derivatives.

b. Sketch the graph of $S(x)$, showing all relative extreme points and inflection points.

c. Interpret the meaning of the inflection point and determine the year in which it occurred.

Source: Standard & Poor's

67–68. BUSINESS: Point of Diminishing Returns Find the point of diminishing returns for each revenue function where x is the amount spent on advertising, both in hundred thousand dollars.

67. $R(x) = 10 - 3x + 24x^2 - x^3$ for $0 \le x \le 15$

68. $R(x) = 40x + 72x^2 - 2x^3$ for $0 \le x \le 20$

69–70. BUSINESS: Point of Diminishing Returns Find the point of diminishing returns for each profit function where x is the amount spent on marketing, both in million dollars.

69. $P(x) = 6x + 18x^2 - 1.5x^3$ for $0 \le x \le 6$

70. $P(x) = 20 - 2.4x + 3x^2 - 0.5x^3$ for $0 \le x \le 4$

71. GENERAL: Airplane Flight Path In Exercise 67 on page 175 it was found that the flight path that satisfies the conditions in the following diagram was $y = -0.00001x^3 + 0.0015x^2$. Find the inflection point of this path, and explain why it represents the point of steepest ascent of the airplane.

72. BIOMEDICAL: Dose-Response Curves The relationship between the dosage, x, of a drug and the resulting change in body temperature is given by $f(x) = x^2(3 - x)$ for $0 \le x \le 3$. Make sign diagrams for the first and second derivatives and sketch this dose-response curve, showing all relative extreme points and inflection points.

Conceptual Exercises

73. For which values of the constant a is the function $f(x) = ax^2$ concave *up*? For which value of a is it concave *down*?

74. True or False: If the graph of $f(x)$ is concave up, then the graph of $-f(x)$ will be concave down.

75. True or False: Inflection points are simply points where the second derivative equals zero.

76. True or False: A polynomial of degree n can have at most $n - 2$ inflection points.

77. From the graph of the *second derivative* shown below, is the point at $x = 1$ an inflection point of the original function?

78. From the graph of the *second derivative* shown below, is the point at $x = 1$ an inflection point of the original function?

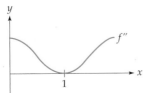

79. A company president is looking at a graph of her company's daily sales during the first quarter of the year. On January 15 sales hit an all-time low and then began to rise; on February 15 there was an inflection point (the only inflection point on the graph); on March 15 the sales hit an all-time high and then began to decline. What was the significance of the inflection point? (Assume that the graph has no straight segments.)

80. An investor is looking at a graph of the value of his portfolio for the first quarter of the year. On January 10 the value was at an all-time high and then began to decline; on February 10 there was an inflection point (the only inflection point on the graph); on March 10 the value was at its lowest point and then began to rise. What was the significance of the inflection point? (Assume that the graph has no straight segments.)

81. True or False: If a graph is concave *up* before an inflection point and concave *down* after it, then the curve has its greatest slope *at* the inflection point.

82. True or False: If a graph is concave *down* before an inflection point and concave *up* after it, then the curve has its smallest slope *at* the inflection point.

Explorations and Excursions
The following problems extend and augment the material presented in the text.

83. CONCAVITY OF A PARABOLA Show that the quadratic function $f(x) = ax^2 + bx + c$ is concave up if $a > 0$ and is concave down if $a < 0$. Therefore, the rule that a parabola opens up if $a > 0$ and down if $a < 0$ is merely an application of concavity. [*Hint:* Find the second derivative.]

84. INFLECTION POINT OF A CUBIC Show that the general "cubic" (third degree) function
$$f(x) = ax^3 + bx^2 + cx + d \quad (\text{with } a \neq 0) \text{ has an inflec-}$$
tion point at $x = \dfrac{-b}{3a}$.

85. INFLECTION POINTS Explain why, at an inflection point, a curve must cross its tangent line (assuming that the tangent line exists).

86. INFLECTION POINTS For a twice-differentiable function, explain why the slope must have a relative maximum or minimum value at an inflection point. [*Hint:* Use the fact that the concavity changes at an inflection point, and then interpret concavity in terms of increasing and decreasing slope.]

3.3   Optimization

Introduction

Many problems consist of **optimizing** a function—that is, finding its maximum or minimum value. For example, you might want to maximize your profit, or to minimize the time required to do a task. If you could express your happiness as a function, you would want to maximize it.* One of the principal uses of calculus is that it provides a very general technique for optimizing functions.

We will concentrate on *applications* of optimization. Accordingly, we will optimize continuous functions that are defined on closed intervals, or functions that have only one critical number in their domain. Most applications fall into these two categories, and the wide range of examples and exercises in this and the following sections will demonstrate the power of these techniques.

Absolute Extreme Values

The **absolute maximum value** of a function is the *largest* value of the function on its domain. Similarly, the **absolute minimum value** of a function is the *smallest* value of the function on its domain. An **absolute extreme value** is a value that is either the absolute maximum or the absolute minimum value of the function. (This use of the word "absolute" has nothing to do with its use in the *absolute value function* defined on page 53.) The maximum and minimum values of the function correspond to the highest and lowest points on its graph.

For a given function, both absolute extreme values may exist, or one or both may fail to exist, as the following graphs show.

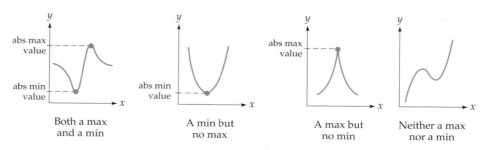

Both a max and a min   A min but no max   A max but no min   Neither a max nor a min

*Expressing happiness in numbers has an honorable past: Plato (*Republic* **IX,** 587) calculated that a king is exactly 729 ($= 3^6$) times as happy as a tyrant.

When will both extreme values exist? For a *continuous* function (one whose graph is a single unbroken curve) defined on a *closed* interval (that is, including both endpoints), the absolute maximum and minimum values are *guaranteed* to exist. We know from graphing functions that maximum and minimum values can occur only at critical numbers, unless they occur at endpoints, where the curve is "cut off" from rising or falling further. To summarize:

Optimizing Continuous Functions on Closed Intervals

A continuous function f on a closed interval $[a, b]$ has both an *absolute maximum value* and an *absolute minimum value.* To find them:

1. Find all critical numbers of f in $[a, b]$.
2. Evaluate f at the critical numbers and at the endpoints a and b.

The largest and smallest values found in Step 2 will be the absolute maximum and minimum values of f on $[a, b]$.

Take Note

Extremes occur only at critical numbers and endpoints.

EXAMPLE 1   **OPTIMIZING A CONTINUOUS FUNCTION ON A CLOSED INTERVAL**

Find the absolute extreme values of $f(x) = x^3 - 9x^2 + 15x$ on $[0, 3]$.

Solution

The function is continuous (it is a polynomial) and the interval is closed, so both extreme values exist. First we find the critical numbers.

$$f'(x) = 3x^2 - 18x + 15 \qquad \text{The derivative}$$

$$= 3(x^2 - 6x + 5) = 3(x - 1)(x - 5) \qquad \text{Factoring}$$

$$\text{CN} \begin{cases} x = 1 \\ x = 5 \end{cases} \quad \leftarrow \text{Not in the given domain} \\ \qquad\qquad\qquad [0, 3], \text{ so we eliminate it}$$

We evaluate the original function $f(x) = x^3 - 9x^2 + 15x$ at the remaining critical number and at the endpoints (EP).

CN: $x = 1$ $f(1) = 1 - 9 + 15$ $= 7$ $\leftarrow$ Largest function value

$\text{EP} \begin{cases} x = 0 \\ x = 3 \end{cases}$ $\begin{matrix} f(0) = 0 - 0 + 0 \\ f(3) = 27 - 9 \cdot 9 + 15 \cdot 3 \end{matrix}$ $\begin{matrix} = 0 \\ = -9 \end{matrix}$ $\leftarrow$ Smallest function value

The largest (7) and the smallest (-9) of the resulting values of the function are the absolute extreme values of f on $[0, 3]$:

    Maximum value of f is 7 (occurring at $x = 1$).
    Minimum value of f is -9 (occurring at $x = 3$).

The absolute maximum and minimum values may be seen in the graph on the left.

For this function, one extreme value occurred at a critical number ($x = 1$), and the other at an endpoint ($x = 3$). In other problems, both extreme values might occur at critical numbers or both might occur at endpoints.

Notice how calculus helped in this example. The absolute extreme values could have occurred at *any* x-values in [0, 3]. Calculus reduced this infinite list of possibilities (*all* numbers between 0 and 3, not just integers) to a mere three numbers (0, 1, and 3). We then had only to "test" these numbers by substituting them into the function to find which numbers made f largest and smallest.

Be Careful   The derivative helped us to find the x-values where the extremes could occur, but the actual extreme values (7 and -9 in the preceding Example) come from evaluating the *original function*.

Second-Derivative Test for Absolute Extreme Values

Recall from page 184 that the second-derivative test is a way to determine whether a function had a *maximum* or *minimum* at a particular critical number and amounts simply to checking the sign of the second derivative at the critical number.

Second-Derivative Test

If $x = c$ is a critical number at which f'' is defined, then

$f''(c) > 0$ means that f has a relative *minimum* at $x = c$.

$f''(c) < 0$ means that f has a relative *maximum* at $x = c$.

Earlier we used the second-derivative test to find *relative* maximum and minimum values. However, if a continuous function has only one critical number in its domain, then the second-derivative test can be used to find *absolute* extreme values (since without a second critical number the function must continue to increase or decrease away from the relative extreme point). Some students refer to this as "the only critical point in town" test.

Applications of Optimization

If a timber forest is allowed to grow for t years, the value of the timber increases in proportion to the square root of t, while maintenance costs are proportional to t. Therefore, the value of the forest after t years is of the form

$$V(t) = a\sqrt{t} - bt \qquad \text{\small{a and b are constants}}$$

| EXAMPLE 2 | OPTIMIZING THE VALUE OF A TIMBER FOREST |

The value of a timber forest after t years is $V(t) = 96\sqrt{t} - 6t$ thousand dollars (for $t > 0$). Find when its value is maximized.

Solution

$$V(t) = 96t^{1/2} - 6t \qquad \text{\small{$V(t)$ in exponential form}}$$

$$V'(t) = 48t^{-1/2} - 6 = 0 \qquad \text{\small{Setting the derivative equal to zero}}$$

$$48t^{-1/2} = 6 \qquad\qquad \text{Adding 6 to each side of } 48t^{-1/2} - 6 = 0$$

$$t^{-1/2} = \frac{6}{48} = \frac{1}{8} \qquad\qquad \text{Dividing by 48}$$

$$\frac{1}{\sqrt{t}} = \frac{1}{8} \qquad\qquad \text{Expressing } t^{-1/2} \text{ in radical form}$$

$$\sqrt{t} = 8 \qquad\qquad \text{Inverting both sides}$$

$$t = 64 \qquad\qquad \text{Squaring both sides}$$

Since there is a *single* critical number ($t = 0$, which makes the derivative undefined, is not in the domain), we may use the second-derivative test. The second derivative is

$$V''(t) = -24t^{-3/2} = -\frac{24}{\sqrt{t^3}} \qquad \text{Differentiating } V'(t) = 48t^{-1/2} - 6$$

At $t = 64$ this is clearly negative, so by the second-derivative test $V(t)$ is *maximized* at $t = 64$. Finally, we state the answer clearly:

The value of the forest is maximized after 64 years. The maximum value is $V(64) = 96\sqrt{64} - 6 \cdot 64 = 384$ thousand dollars, or $384,000.

This maximum may be seen in the graph on the left.

Maximizing Profit

The famous economist John Maynard Keynes (1883–1946) said, "The engine that drives Enterprise is Profit." Many management problems consist of maximizing profit, and require *constructing* the profit function before maximizing it. Such problems have three economic ingredients. The first is that profit is defined as *revenue minus cost:*

$$\text{Profit} = \text{Revenue} - \text{Cost}$$

The second ingredient is that revenue is *price times quantity.* For example, if a company sells 100 toasters for $25 each, the revenue will obviously be $25 \cdot 100 = \$2500$.

$$\text{Revenue} = \left(\begin{array}{c}\text{Unit} \\ \text{price}\end{array}\right) \cdot (\text{Quantity})$$

The third economic ingredient reflects the fact that, in general, price and quantity are inversely related: Increasing the price decreases sales, while decreasing the price increases sales. To put this another way, "flooding the market" with a product drives the price down, while creating a shortage drives the price up. If the relationship between the price p and the quantity x that consumers will buy at that price is expressed as a function $p(x)$, it is called the **price function.***

The price function $p(x)$ shows the inverse relation between price p and quantity x.

Price Function

$p(x)$ gives the price p at which consumers will buy exactly x units of the product.

*We will use *lowercase p* for price and *capital P* for profit.

The price function, relating price and quantity, may be linear or curved (as shown on the previous page), but it will always be a *decreasing* function.

In actual practice, price functions are very difficult to determine, requiring extensive (and expensive) market research. In this section we will be given the price function. In the next section we will see how to do without price functions, at least in simple cases.

EXAMPLE 3   **MAXIMIZING A COMPANY'S PROFIT**

It costs the American Automobile Company $8000 to produce each automobile, and fixed costs (rent and other expenses that do not depend on the amount of production) are $20,000 per week. The company's price function is $p(x) = 22{,}000 - 70x$, where p is the price at which exactly x cars will be sold.

a. How many cars should be produced each week to maximize profit?

b. For what price should they be sold?

c. What is the company's maximum profit?

Solution

Revenue is price times quantity, $R = p \cdot x$:

$$R = p \cdot x = \underbrace{(22{,}000 - 70x)}_{p(x)}x = \underbrace{22{,}000x - 70x^2}_{\substack{\text{Revenue} \\ \text{function } R(x)}}$$

Replacing p by the price function $p = 22{,}000 - 70x$

Cost is the cost per car ($8000) times the number of cars (x) plus the fixed cost ($20,000):

$$C(x) = 8000x + 20{,}000 \qquad \text{(Unit cost)} \cdot \text{(Quantity)} + \text{(Fixed cost)}$$

Profit is revenue minus cost:

$$P(x) = \underbrace{(22{,}000x - 70x^2)}_{R(x)} - \underbrace{(8000x + 20{,}000)}_{C(x)}$$

$$= -70x^2 + 14{,}000x - 20{,}000 \qquad \substack{\text{Profit function} \\ \text{(after simplification)}}$$

a. We maximize the profit by setting its derivative equal to zero:

$$P'(x) = -140x + 14{,}000 = 0 \qquad \substack{\text{Differentiating} \\ P = -70x^2 + 14{,}000x - 20{,}000}$$

$$-140x = -14{,}000 \qquad \text{Solving}$$

$$x = \frac{-14{,}000}{-140} = 100 \qquad \text{Only one critical number}$$

$$P''(x) = -140 \qquad \text{From } P'(x) = -140x + 14{,}000$$

The second derivative is negative, so the profit is maximized at the critical number. (If the second derivative had involved x, we would have substituted the critical number $x = 100$.) Since x is the number of cars, the company should produce 100 cars per week (the time period stated in the problem).

b. The selling price p is found from the price function:

$$p(100) = 22,000 - 70 \cdot 100 = \$15,000$$

$p(x) = 22,000 - 70x$
evaluated at $x = 100$

c. The maximum profit is found from the profit function:

$$P(100) = -70(100)^2 + 14,000(100) - 20,000$$

$$= \$680,000$$

$P(x) = -70x^2 +$
$14,000x - 20,000$
evaluated at $x = 100$

Finally, state the answer clearly in words.

The company should make 100 cars per week and sell them for $15,000 each. The maximum profit will be $680,000.

Actually, automobile dealers seem to prefer prices like $14,999 as if $1 makes a difference.

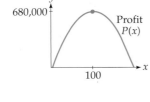

Typical graphs of the revenue and cost functions are shown on the left. At x-values where revenue is above cost, there is a profit, and where the cost is above the revenue, there is a loss.

The height of the profit function at any x is the amount by which the revenue is above the cost in the graph. Since profit equals revenue minus cost, we may differentiate each side of $P(x) = R(x) - C(x)$,   obtaining

$$P'(x) = R'(x) - C'(x)$$

This shows that setting $P'(x) = 0$ (which we do to maximize profit) is equivalent to setting $R'(x) - C'(x) = 0$,   which is equivalent to $R'(x) = C'(x)$.   This last equation may be expressed in marginals, $MR = MC$,   which is a classic economic criterion for maximum profit.

Classic Economic Criterion for Maximum Profit

At maximum profit:

$$\left(\begin{array}{c}\text{Marginal} \\ \text{revenue}\end{array}\right) = \left(\begin{array}{c}\text{Marginal} \\ \text{cost}\end{array}\right)$$

LOOKING AHEAD

Marginal analysis will be explored further in Section 3.7.

We may understand this equation intuitively as follows: If marginal revenue is *greater* than marginal cost, we should *increase* production because the higher marginal revenue means more profit. However, if marginal revenue is *less* than marginal cost, we are losing money on each additional item and so should *decrease* production to reduce the loss. Therefore, we are at maximum profit only where *marginal revenue equals marginal cost*. This reasoning is the beginning of what is called *marginal analysis*.

We could use this criterion to maximize profit, but for simplicity we will maximize profit just as we usually do: by solving $P'(x) = 0$ and then verifying that $P''(x) < 0$ so that the second-derivative test assures that we are *maximizing* profit (thereby avoiding the costly mistake of *minimizing* profit).

EXAMPLE 4   **MAXIMIZING THE AREA OF AN ENCLOSURE**

A farmer has 1000 feet of fence and wants to build a rectangular enclosure along a straight wall. If the side along the wall needs no fence, find the dimensions that make the enclosure as large as possible. Also find the maximum area.

Solution

The largest enclosure means, of course, the largest area.

We let variables stand for the length and width, as shown in the diagram on the right.

$$x = \text{length (parallel to wall)}$$

$$y = \text{width (perpendicular to wall)}$$

The problem becomes

Maximize $A = xy$	Area is length times width
subject to $x + 2y = 1000$	One x side and two y sides from 1000 feet of fence

We must express the area $A = xy$ in terms of one variable. We use

$$x = 1000 - 2y \qquad \text{Solving } x + 2y = 1000 \text{ for } x$$

$$A = xy = \underbrace{(1000 - 2y)}_{x}y = 1000y - 2y^2 \qquad \begin{array}{l}\text{Substituting } x = 1000 - 2y \\ \text{into } A = xy\end{array}$$

$$A' = 1000 - 4y = 0 \qquad \begin{array}{l}\text{Maximizing } A = 1000y - 2y^2 \\ \text{by setting the derivative} \\ \text{equal to zero}\end{array}$$

$$y = 250 \qquad \text{Solving } 1000 - 4y = 0 \text{ for } y$$

Since $A'' = -4$,   the second-derivative test shows that the area is indeed *maximized* when $y = 250$.   The length x is

$$x = 1000 - 2 \cdot 250 = 500 \qquad \text{Evaluating } x = 1000 - 2y \text{ at } y = 250$$

Length (parallel to the wall) is 500 feet, width (perpendicular to the wall) is 250 feet, and area (length times width) is 125,000 square feet.

Spreadsheet Exploration

In the preceding Example you might think that it does not matter how the fence is laid out as long as all 1000 feet are used. To see that the area enclosed really *does* change, and that the maximum occurs at $y = 250$,   the following spreadsheet* calculates the area $A(y) = 1000y - 2y^2$ for *y*-values from 245 to 255. Notice that the area is largest for $y = 250$,   and that for each change of 1 in the *y*-value the change in the area is smallest for widths closest to 250. This verifies *numerically* that the derivative (rate of change) becomes zero as *y* approaches 250.

	B8 ▼	= =1000*A8-2*A8^2
	A	**B**
1	Width (perpendicular to wall)	Area Enclosed
2		
3	245	124950
4	246	124968
5	247	124982
6	248	124992
7	249	124998
8	250	125000
9	251	124998
10	252	124992
11	253	124982
12	254	124968
13	255	124950

← largest area

Based on this spreadsheet, how much area will the farmer lose if he mistakenly makes the width 249 or 251 feet instead of 250 (and therefore the length 502 or 498 feet)? Is this loss significant based on an area of 125,000 square feet? This is characteristic of maximization problems where the slope is zero—being *near* the maximizing value is essentially as good as being *at* it.

To obtain this and other Spreadsheet Explorations, go to www.cengagebrain.com, search for this textbook and then scroll down to access the "free materials" tab.

EXAMPLE 5 MAXIMIZING THE VOLUME OF A BOX

An open-top box is to be made from a square sheet of metal 12 inches on each side by cutting a square from each corner and folding up the sides, as in the diagrams below. Find the volume of the largest box that can be made in this way.

Square sheet   Corners removed   Side flaps folded up to make open-top box.

Solution

Let x = the length of the side of the square cut from each corner.

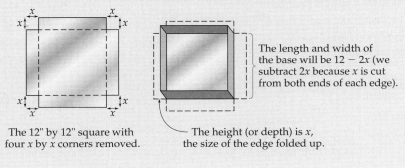

The 12" by 12" square with four x by x corners removed.

The length and width of the base will be $12 - 2x$ (we subtract $2x$ because x is cut from both ends of each edge).

The height (or depth) is x, the size of the edge folded up.

Therefore, the volume is

$$V(x) = (12 - 2x)(12 - 2x)x \qquad \text{(Length)} \cdot \text{(Width)} \cdot \text{(Height)}$$

Since x is a length, $x > 0$, and since x inches are cut from *both* sides of each 12-inch edge, we must have $2x < 12$, so $x < 6$. The problem becomes

Maximize
$$\begin{aligned} V(x) &= (12 - 2x)(12 - 2x)x \qquad &\text{on } 0 < x < 6 \\ &= (144 - 48x + 4x^2)x \\ &= 4x^3 - 48x^2 + 144x \end{aligned} \left.\rule{0pt}{24pt}\right\}\text{Multiplying out}$$

$$\begin{aligned} V'(x) &= 12x^2 - 96x + 144 \qquad &\text{Differentiating} \\ &= 12(x^2 - 8x + 12) \\ &= 12(x - 2)(x - 6) \end{aligned} \left.\rule{0pt}{18pt}\right\}\text{Factoring}$$

$$\text{CN} \begin{cases} x = 2 \\ x = \cancel{6} \end{cases} \qquad \leftarrow \text{Not in the domain, so we eliminate it}$$

The second derivative is $V''(x) = 24x - 96$, which at $x = 2$ is

$$V''(2) = 48 - 96 < 0$$

Therefore, the volume is *maximized* at $x = 2$.

Maximum volume is 128 cubic inches. From $V(x)$ evaluated at $x = 2$

 Be Careful Be sure to use only critical numbers that are *in the domain of the original function*. This was important here and in Example 1 on page 191.

 Graphing Calculator Exploration

Do the previous Example (and then modify it) on a graphing calculator as follows:

a. Enter the volume as $y_1 = (12 - 2x)(12 - 2x)x$ and graph it on [0, 6] by [0, 150].

b. Use MAXIMUM to maximize y_1. Your answer should agree with that found above. (While this may seem faster than doing the problem "by hand," you would still need to have found the volume function and the domain, which was most of the work.)

c. What if the beginning size were not a 12-inch by 12-inch square, but a standard 8.5-inch by 11-inch sheet of paper? Go back to y_1 and replace the two 12's by 8.5 and 11 and find the x and the maximum volume. (*Answer:* Volume about 66 cubic inches)

d. What about a 3 by 5 card?

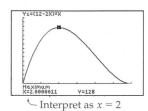
Interpret as $x = 2$

Parts (c) and (d), which would be more difficult to solve by hand, show how useful a graphing calculator is for solving related problems after the first has been analyzed using calculus.

3.3 Section Summary

There is no single, all-purpose procedure for solving word problems. You must think about the problem, draw a picture if possible, and express the quantity to be maximized or minimized in terms of some appropriate variable. You can become good at it only with practice.

We have two procedures for optimizing continuous functions on intervals:

1. If the function has only one critical number in the interval, we find it and use the second-derivative test to show whether the function is maximized or minimized there.

2. If the interval is closed, we evaluate the function at all critical numbers and endpoints in the interval; the largest and smallest resulting values will be the maximum and minimum values of the function.

If neither procedure finds the desired optimal value(s), graph the function. The maximum (or minimum) value will be the y-coordinate of the highest (or lowest) point on the graph.

3.3 Exercises

1–20. Find (*without* using a calculator) the absolute extreme values of each function on the given interval.

FOR HELP GETTING STARTED

with Exercises 1–12, see Example 1 on page 191.

1. $f(x) = x^3 - 6x^2 + 9x + 8$ on $[-1, 2]$

2. $f(x) = x^3 - 6x^2 + 22$ on $[-2, 2]$

3. $f(x) = x^3 - 12x$ on $[-3, 3]$

4. $f(x) = x^3 - 27x$ on $[-2, 2]$

5. $f(x) = x^4 + 4x^3 + 4x^2$ on $[-2, 1]$

6. $f(x) = x^4 - 4x^3 + 4x^2$ on $[0, 3]$

7. $f(x) = x^4 + 4x^3 + 15$ on $[-4, 1]$

8. $f(x) = 2x^4 - 8x^3 + 30$ on $[-1, 4]$

9. $f(x) = 2x^5 - 5x^4$ on $[-1, 3]$

10. $f(x) = 4x^5 - 5x^4$ on $[0, 2]$

11. $f(x) = 3x^2 - x^3$ on $[0, 5]$

12. $f(x) = 6x^2 - x^3$ on $[0, 5]$

13. $f(x) = 5 - x$ on $[0, 5]$

14. $f(x) = x(100 - x)$ on $[0, 100]$

15. $f(x) = x^2(3 - x)$ on $[-1, 3]$

16. $f(x) = x^3(4 - x)$ on $[-1, 4]$

17. $f(x) = (x^2 - 1)^2$ on $[-1, 1]$

18. $f(x) = \sqrt[3]{x^2}$ on $[-1, 8]$

19. $f(x) = \dfrac{x}{x^2 + 1}$ on $[-3, 3]$

20. $f(x) = \dfrac{1}{x^2 + 1}$ on $[-3, 3]$

21. Find the number in the interval $[0, 3]$ such that the number minus its square is:

   a. As large as possible.
   b. As small as possible.

22. Find the number in the interval $[\frac{1}{3}, 2]$ such that the sum of the number and its reciprocal is:

   a. As large as possible.
   b. As small as possible.

23. **ONE FUNCTION, DIFFERENT DOMAINS**

   a. Graph the function $y_1 = x^3 - 15x^2 + 63x$ on the window $[0, 10]$ by $[0, 130]$. By visual inspection of this function on this domain, where do the absolute maximum and minimum values occur: both at critical numbers, both at endpoints, or one at a critical number and one at an endpoint?

   b. Now change the domain to $[0, 8]$ and answer the same question.

   c. Now change the domain to $[2, 8]$ and answer the same question.

   d. Can you find a domain such that the minimum occurs at a critical number and the maximum at an endpoint?

24. EXISTENCE OF EXTREME VALUES

a. Graph the function $y_1 = 5/x$ on the window [1, 10] by [0, 10]. By visual inspection, does the function have an absolute maximum value on this domain? An absolute minimum value?

b. Now change the x window to [0, 10] and answer the same questions. (We cannot now call [0, 10] the domain, since the function is not defined at 0.)

(continues)

c. Based on the screen display, answer the same questions for the domain $(0, \infty)$.

d. Is there a domain on which this function has an absolute maximum but no absolute minimum?

e. It was stated in the box on page 191 that a continuous function on a closed interval will always have absolute maximum and minimum values. Is this claim violated by the function $f(x) = 5/x$ on [0, 10] [part (b)]? Is it violated by $f(x) = 5/x$ on $(0, \infty)$ [part (c)]?

Applied Exercises

25. BIOMEDICAL: Pollen Count The average pollen count in New York City on day x of the pollen season is $P(x) = 8x - 0.2x^2$ (for $0 < x < 40$). On which day is the pollen count highest?

26. GENERAL: Fuel Economy The fuel economy (in miles per gallon) of an average American compact car is $E(x) = -0.015x^2 + 1.14x + 8.3$, where x is the driving speed (in miles per hour, $20 \le x \le 60$). At what speed is fuel economy greatest?

Source: U.S. Environmental Protection Agency

27. GENERAL: Fuel Economy The fuel economy (in miles per gallon) of an average American midsized car is $E(x) = -0.01x^2 + 0.62x + 10.4$, where x is the driving speed (in miles per hour, $20 \le x \le 60$). At what speed is fuel economy greatest?

Source: U.S. Environmental Protection Agency

28. BUSINESS: Copier Repair A copier company finds that copiers that are x years old require, on average, $f(x) = 1.2x^2 - 4.7x + 10.8$ repairs annually for $0 \le x \le 5$. Find the year that requires the least repairs, rounding your answer to the nearest year.

29. GENERAL: Driving and Age Studies have shown that the number of accidents a driver has varies with the age of the driver and is highest for very young and very old drivers. The number of serious accidents for drivers of age x during a recent year was approximately $f(x) = 0.013x^2 - 1.35x + 48$ for $16 \le x \le 85$. Find the age that has the least accidents, rounding your answer to the nearest year.

Source: Insurance Information Institute

30. GENERAL: Water Power The proportion of a river's energy that can be obtained from an undershot waterwheel is $E(x) = 2x^3 - 4x^2 + 2x$, where x is the speed of the waterwheel relative to the speed of the river. Find the maximum value of this function on the interval [0, 1], thereby showing that only about 30% of a river's energy can be captured. Your answer should agree with the old millwright's rule that the speed of the wheel should be about one-third of the speed of the river.

Source: U.S. Government Printing Office, The Energy Book

31. ENVIRONMENTAL SCIENCE: Wind Power Electrical generation from wind turbines (large "windmills") is increasingly popular in Europe and the United States. The

fraction of the wind's energy that can be extracted by a turbine is $f(x) = \frac{1}{2}x(2 - x)^2$, where x is the fraction by which the wind is slowed in passing through the turbine. Find the fraction x that maximizes the energy extracted.

Source: U.S. Government Printing Office, The Energy Book

32. GENERAL: Airplane Accidents A pilot's likelihood of an accident varies with the number of hours flown. For an instrument-rated commercial pilot who has flown x hundred hours, the likelihood of a serious or fatal accident is proportional to $A(x) = \dfrac{x^2}{x^3 + 256}$. Find the value of x for which this accident rate is maximized and interpret your answer.

Source: Accident Analysis and Prevention 60, November 2013.

33. BUSINESS: Timber Value The value of a timber forest after t years is $V(t) = 480\sqrt{t} - 40t$ (for $0 \le t \le 50$). Find when its value is maximized.

34. BIOMEDICAL: Longevity and Exercise A study of the exercise habits of 17,000 Harvard alumni found that the death rate (deaths per 10,000 person-years) was approximately $R(x) = 5x^2 - 35x + 104$, where x is the weekly amount of exercise in thousands of calories ($0 \le x \le 4$). Find the exercise level that minimizes the death rate.

Source: Journal of the American Medical Association 273:15

35. ENVIRONMENTAL SCIENCE: Pollution Two chemical factories are discharging toxic waste into a large lake, and the pollution level at a point x miles from factory A toward factory B is $P(x) = 3x^2 - 72x + 576$ parts per million (for $0 \le x \le 50$). Find where the pollution is the least.

36. BUSINESS: Maximum Profit City Cycles Incorporated finds that it costs $70 to manufacture each bicycle, and fixed costs are $100 per day. The price function is $p(x) = 270 - 10x$, where p is the price (in dollars) at which exactly x bicycles will be sold. Find the quantity City Cycles should produce and the price it should charge to maximize profit. Also find the maximum profit.

37. BUSINESS: Maximum Profit Country Motorbikes Incorporated finds that it costs $200 to produce each motorbike, and that fixed costs are $1500 per day. The price function is $p(x) = 600 - 5x$, where p is the price (in dollars) at which exactly x motorbikes will be sold. Find the quantity Country Motorbikes should produce and the price it should charge to maximize profit. Also find the maximum profit.

38. BUSINESS: Maximum Profit A retired potter can produce china pitchers at a cost of $5 each. She estimates her price function to be $p = 17 - 0.5x$, where p is the price at which exactly x pitchers will be sold per week. Find the number of pitchers that she should produce and the price that she should charge in order to maximize profit. Also find the maximum profit.

39. BUSINESS: Maximum Revenue BoxedNGone truck rentals calculates that its price function is $p(x) = 240 - 3x$, where p is the price (in dollars) at which exactly x trucks will be rented per day. Find the number of trucks that BoxedNGone should rent and the price it should charge to maximize revenue. Also find the maximum revenue.

40. BUSINESS: Maximum Revenue NRG-SUP, a supplier of energy supplements for athletes, determines that its price function is $p(x) = 60 - \frac{1}{2}x$, where p is the price (in dollars) at which exactly x boxes of supplements will be sold per day. Find the number of boxes that NRG-SUP will sell per day and the price it should charge to maximize revenue. Also find the maximum revenue.

41. GENERAL: Parking Lot Design A real estate company wants to build a parking lot along the side of one of its buildings using 800 feet of fence. If the side along the building needs no fence, what are the dimensions of the largest possible parking lot?

parking lot

42. GENERAL: Fences A farmer wants to make two identical rectangular enclosures along a straight river, as in the diagram shown below. If he has 600 yards of fence, and if the sides along the river need no fence, what should be the dimensions of each enclosure if the total area is to be maximized?

river

43. GENERAL: Fences A farmer wants to make three identical rectangular enclosures along a straight river, as in the diagram shown below. If he has 1200 yards of fence, and if the sides along the river need no fence, what should be the dimensions of each enclosure if the total area is to be maximized?

river

44. GENERAL: Area What is the area of the largest rectangle whose perimeter is 100 feet?

45. GENERAL: Package Design An open-top box is to be made from a square piece of cardboard that measures 18 inches by 18 inches by removing a square from each corner and folding up the sides. What are the dimensions and volume of the largest box that can be made in this way?

46. GENERAL: Gutter Design A long gutter is to be made from a 12-inch-wide strip of metal by folding up the two edges. How much of each edge should be folded up in order to maximize the capacity of the gutter? [*Hint:* Maximizing the capacity means maximizing the cross-sectional area, shown below.]

area
x
total: 12"

47. GENERAL: Maximizing a Product Find the two numbers whose sum is 50 and whose product is a maximum.

48. GENERAL: Maximizing Area Show that the largest rectangle with a given perimeter is a square.

49. ATHLETICS: Athletic Fields A running track consists of a rectangle with a semicircle at each end, as shown below. If the perimeter is to be exactly 440 yards, find the dimensions (x and r) that maximize the area of the rectangle. [*Hint:* The perimeter is $2x + 2\pi r$.]

50. GENERAL: Window Design A *Norman window* consists of a rectangle topped by a semicircle, as shown below. If the perimeter is to be 18 feet, find the dimensions (x and r) that maximize the area of the window. [*Hint:* The perimeter is $2x + 2r + \pi r$.]

51. BIOMEDICAL: Bacterial Growth A chemical reagent is introduced into a bacterial population, and t hours later the number of bacteria (in thousands) is $N(t) = 1000 + 15t^2 - t^3$ (for $0 \le t \le 15$).

a. When will the population be the largest, and how large will it be?

b. When will the population be growing at the fastest rate, and how fast? (What word applies to such a point?)

 52. BUSINESS: Value of a Pulpwood Forest The value of a pulpwood forest after growing for x years is predicted to be $V(t) = 400x^{0.4} - 40x$ thousand dollars (for $0 \le x \le 25$). Use a graphing calculator with MAXIMUM to find when the value will be maximized, and what the maximum value will be.

 53–54. GENERAL: Package Design Use a graphing calculator (as explained on page 198) to find the side of the square removed and the volume of the box described in Example 5 (pages 197–198) if the square piece of metal is replaced by a:

53. 5 by 7 card (5 inches by 7 inches)

54. 6 by 8 card (6 inches by 8 inches)

You might try constructing such a box.

Conceptual Exercises

55. What is the difference between an *absolute* maximum value and a *relative* maximum value?

56. True or False: If a function is defined and continuous on a closed interval, then it has both an absolute maximum value and an absolute minimum value.

57. True or False: If a function is continuous on an *open* interval, then it will not have absolute maximum or minimum values.

58. Draw the graph of a function defined on $(-\infty, \infty)$ that has no absolute maximum or minimum value. Draw one that has *both* an absolute maximum *and* an absolute minimum value.

59. Draw the graph of a function defined on the closed interval [0, 1] that has no absolute maximum or minimum value.

60. You are to maximize a function on the interval $[-5, 5]$ and find critical numbers $-1, 0,$ and 6. Which critical numbers should you use?

61. If $f'(5) = 0$ and $f''(5) > 0,$ what conclusion can you draw?

62. If $f'(5) = 0$ and $f''(5) = 0,$ what conclusion can you draw?

63. True or False: If f has an absolute maximum value, then $-f$ will have an absolute minimum value.

64. True or False: A polynomial of degree n can have at most $n - 1$ critical numbers.

Explorations and Excursions   The following problems extend and augment the material presented in the text.

65. BIOMEDICAL: Coughing When you cough, you are using a high-speed stream of air to clear your trachea (windpipe). During a cough your trachea contracts, forcing the air to move faster, but also increasing the friction. If a trachea contracts from a normal (rest) radius of 3 centimeters to a radius of r centimeters, the velocity of the airstream is $V(r) = c(3 - r)r^2,$ where c is a positive constant depending on the length and the elasticity of the trachea. Find the radius r that maximizes this velocity. (X-ray pictures verify that the trachea does indeed contract to this radius.)

Source: Lung **152**

66. BIOMEDICAL: "Efishency" At what speed should a fish swim upstream so as to reach its destination with the least expenditure of energy? The energy depends on the friction of the fish through the water and on the duration of the trip. If the fish swims with velocity v, the

energy has been found experimentally to be proportional to v^k (for constant $k > 2$) times the duration of the trip. A distance of s miles against a current of speed c requires time $\dfrac{s}{v - c}$ (distance divided by speed). The energy required is then proportional to $\dfrac{v^k s}{v - c}.$

For $k = 3,$ minimizing energy is equivalent to minimizing

$$E(v) = \frac{v^3}{v - c}$$

Find the speed v with which the fish should swim in order to minimize its expenditure $E(v)$. (Your answer will depend on c, the speed of the current.)

Source: E. Batschelet, Introduction to Mathematics for Life Scientists

| 3.4 | **Further Applications of Optimization** |

Introduction

In this section we continue to solve optimization problems. In particular, we will see how to maximize a company's profit if we are not given the price function, provided that we are given information describing how price changes will affect sales. We will also see that sometimes x should be chosen as something *other* than the quantity sold.

EXAMPLE 1   **FINDING PRICE AND QUANTITY FUNCTIONS**

A store can sell 20 bicycles per week at a price of $400 each. The manager estimates that for each $10 price reduction she can sell two more bicycles per week. The bicycles cost the store $200 each. If x stands for *the number of $10 price reductions*, express the price p and the quantity q as functions of x.

Solution

Let

$$x = \text{the number of \$10 price reductions}$$

For example, $x = 4$ means that the price is reduced by $40 (four $10 price reductions). Therefore, in general, if there are x $10 price reductions from the original $400 price, then the price $p(x)$ is

$$p(x) = 400 - 10x \qquad\qquad \text{Price}$$

Original price — Less x $10 price reductions

The quantity sold $q(x)$ will be

$$q(x) = 20 + 2x \qquad\qquad \text{Quantity}$$

Original quantity — Plus two for each price reduction

We will return to this example and maximize the store's profit after a practice problem.

PRACTICE PROBLEM

A computer manufacturer can sell 1500 personal computers per month at a price of $3000 each. The manager estimates that for each $200 price reduction he will sell 300 more each month. If x stands for *the number of $200 price reductions*, express the price p and the quantity q as functions of x.

Solution on page 208 >

EXAMPLE 2   **MAXIMIZING PROFIT (*CONTINUATION OF EXAMPLE 1*)**

Using the information in Example 1, find the price of the bicycles and the quantity that maximize profit. Also find the maximum profit.

Solution

In Example 1 we found

$$p(x) = 400 - 10x \qquad \text{Price}$$

$$q(x) = 20 + 2x \qquad \text{Quantity sold at that price}$$

Revenue is price times quantity, $p(x) \cdot q(x)$:

$$R(x) = (400 - 10x)(20 + 2x) \qquad p(x)q(x)$$

$$= 8000 + 600x - 20x^2 \qquad \text{Multiplying out and simplifying}$$

The cost function is unit cost times quantity:

$$C(x) = \underbrace{200}_{\substack{\text{Unit}\\\text{cost}}} \underbrace{(20 + 2x)}_{\substack{\text{Quantity}\\ q(x)}} = 4000 + 400x \qquad \begin{array}{l}\text{If there were a fixed cost,}\\\text{we would add it in}\end{array}$$

Profit is revenue minus cost:

$$P(x) = \underbrace{(8000 + 600x - 20x^2)}_{R(x)} - \underbrace{(4000 + 400x)}_{C(x)}$$

$$= 4000 + 200x - 20x^2 \qquad \text{Simplifying}$$

We maximize profit by setting the derivative equal to zero:

$$200 - 40x = 0 \qquad \text{Differentiating}\ \ P = 4000 + 200x + 20x^2$$

The critical number is $x = 5$. The second derivative, $P''(x) = -40$, shows that the profit is *maximized* at $x = 5$. Since $x = 5$ is the number of $10 price reductions, the original price of $400 should be lowered by $50 ($10 five times), from $400 to $350. The quantity sold is found from the quantity function:

$$q(5) = 20 + 2 \cdot 5 = 30 \qquad q(x) = 20 + 2x\ \ \text{at}\ \ x = 5$$

Finally, we state the answer clearly.

Sell the bicycles for $350 each.

Quantity sold: 30 per week.

Maximum profit: $4500.   From $P(x) = 4000 + 200x - 20x^2$ at $x = 5$

Although we did not need to graph the profit function, the diagram on the left does verify that the maximum occurs at $x = 5$.

Exercise 23 will show how a graphing calculator enables you to modify the problem (such as changing the cost per bicycle) and then immediately recalculate the new answer.

Choosing Variables

Notice that in Examples 1 and 2 we did not choose x to be the quantity sold, but instead to be *the number of $10 price reductions*. We chose this x because from it we could easily calculate both the new price and the new quantity. Other choices for x are also possible, but in situations where a price change will make one quantity rise and another fall, it is often easiest to choose x to be the *number of such changes.*

Be Careful If x is the number of price *increases*, then the negative number $x = -2$ means two price *decreases*. Similarly, a negative number of price *decreases* would mean a price *increase*.

EXAMPLE 3	**MAXIMIZING HARVEST SIZE**

An orange grower finds that if he plants 80 orange trees per acre, each tree will yield 60 bushels of oranges. He estimates that for each additional tree that he plants per acre, the yield of each tree will decrease by 2 bushels. How many trees should he plant per acre to maximize his harvest?

Solution

We take x equal to the number of "changes"—that is, let

$$x = \text{the number of added trees per acre}$$

With x extra trees per acre,

Trees per acre: $80 + x$ Original 80 plus x more

Yield per tree: $60 - 2x$ Original yield less 2 per extra tree

Therefore, the total yield per acre will be

$$Y(x) = \underbrace{(60 - 2x)}_{\substack{\text{Yield} \\ \text{per tree}}}\underbrace{(80 + x)}_{\substack{\text{Trees} \\ \text{per acre}}} = 4800 - 100x - 2x^2$$

We maximize this by setting the derivative equal to zero:

$$-100 - 4x = 0 \qquad \text{Differentiating } Y = 4800 - 100x - 2x^2$$
$$x = -25 \qquad \text{Negative!}$$

The number of *added* trees is negative, meaning that the grower should plant 25 *fewer* trees per acre. The second derivative, $Y''(x) = -4$, shows that the yield is indeed maximized at $x = -25$. Therefore:

Plant 55 trees per acre. $80 - 25 = 55$

$x = -25$
number of "added" trees

Be Careful Don't forget to use the second-derivative test to be sure you have maximized (or minimized). *Minimizing* your yield is seldom a good idea!

Earlier problems involved maximizing areas and volumes using only a fixed amount of material (such as a fixed length of fence). Instead, we could minimize the amount of materials for a fixed area or volume.

EXAMPLE 4   **MINIMIZING PACKAGE MATERIALS**

A moving company wishes to design an open-top box with a square base whose volume is exactly 32 cubic feet. Find the dimensions of the box requiring the least amount of materials.

Solution

The base is square, so we define

x = length of side of base

y = height

The volume (length · width · height) is $x \cdot x \cdot y$ or $x^2 y$, which (according to the problem) must equal 32 cubic feet:

$$x^2 y = 32$$

The box consists of a bottom (area x^2) and four sides (each of area xy). Minimizing the amount of materials means minimizing the surface area of the bottom and four sides:

$$A = x^2 + 4xy$$

$$\left(\begin{array}{c}\text{Area of}\\\text{bottom}\end{array}\right) + \left(\begin{array}{c}\text{Area of}\\\text{four sides}\end{array}\right)$$

As usual, we must express this area in terms of just *one* variable, so we use the volume requirement to express y in terms of x:

$$y = \frac{32}{x^2} \qquad \text{Solving } x^2 y = 32 \text{ for } y$$

The area function becomes

$$A = x^2 + 4x\,\frac{32}{x^2} \qquad A = x^2 + 4xy \text{ with } y \text{ replaced by } \frac{32}{x^2}$$

$$= x^2 + \frac{128}{x} \qquad \text{Simplifying}$$

$$= x^2 + 128x^{-1} \qquad \text{Writing } \frac{1}{x} \text{ as } x^{-1}$$

We minimize this by finding the critical number:

$$A'(x) = 2x - 128x^{-2} \qquad \text{Differentiating } A = x^2 + 128x^{-1}$$

$$2x - \frac{128}{x^2} = 0 \qquad \text{Setting the derivative equal to zero}$$

$$2x^3 - 128 = 0 \qquad \text{Multiplying by } x^2 \text{ (since } x > 0\text{)}$$

$$x^3 = 64 \qquad \text{Adding 128 and then dividing by 2}$$

$$x = 4 \qquad \text{Taking cube roots}$$

The second derivative

$$A''(x) = 2 + 256x^{-3} = 2 + \frac{256}{x^3} \qquad \text{From } A'(x) = 2x - 128\,x^{-2}$$

is positive at $x = 4$, so the area is minimized. Therefore, the dimensions using the least materials are:

Base: 4 feet on each side

Height: 2 feet   Height from $y = 32/x^2$ at $x = 4$

Graphing Calculator Exploration

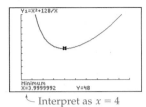

Interpret as $x = 4$

Use a graphing calculator to solve the previous Example as follows:

a. Enter the area function to be minimized as $y_1 = x^2 + 128/x$.

b. Graph y_1 for x-values [0, 10], using TABLE or TRACE to find where y_1 seems to "bottom out" to determine an appropriate y-interval.

c. Graph the function on the window determined in part (b) and then use MINIMUM to find the minimum value. Your answer should agree with that found above.

Notice that either way required first finding the function to be minimized.

Minimizing the Cost of Materials

How would the preceding problem have changed if the material for the bottom of the box had been more costly than the material for the sides? If, for example, the material for the sides cost \$2 per square foot and the material for the base, needing greater strength, cost \$4 per square foot, then instead of simply minimizing the surface area, we would minimize *total cost:*

$$\text{Cost} = \begin{pmatrix}\text{Area of}\\\text{bottom}\end{pmatrix}\begin{pmatrix}\text{Cost of bottom}\\\text{per square foot}\end{pmatrix} + \begin{pmatrix}\text{Area of}\\\text{sides}\end{pmatrix}\begin{pmatrix}\text{Cost of sides}\\\text{per square foot}\end{pmatrix}$$

Since the areas would be just as before, this cost would be

$$\text{Cost} = (x^2)(4) + (4xy)(2) = 4x^2 + 8xy$$

From here on we would proceed just as before, eliminating the y (using the volume relationship $x^2y = 32$) and then setting the derivative equal to zero.

Maximizing Tax Revenue

Governments raise money by collecting taxes. If a sales tax or an import tax is too high, trade will be discouraged and tax revenues will fall. If, on the other hand, the tax rate is too low, trade may flourish but tax revenues will again fall. Economists often want to determine the tax rate that maximizes revenue for the government. To do this, they must first predict the relationship between the tax on an item and the total sales of the item.

Suppose, for example, that the relationship between the tax rate t on an item and its total sales S is

$$S(t) = 9 - 20\sqrt{t}$$

t = tax rate $(0 \le t \le 0.20)$
S = total sales (millions of dollars)

If the tax rate is $t = 0$ (0%), then the total sales will be

$$S(0) = 9 - 20\sqrt{0} = 9 \qquad\qquad \text{\$9 million}$$

If the tax rate is raised to $t = 0.16$ (16%), then sales will be

$$S(0.16) = 9 - 20\sqrt{0.16}$$
$$= 9 - (20)(0.4) = 9 - 8 = 1 \qquad\qquad \text{\$1 million}$$

That is, raising the tax rate from 0% to 16% will discourage \$8 million worth of sales. The graph of $S(t)$ on the left shows how total sales decrease as the tax rate increases. With such information (which may be found from historical data), one can find the tax rate that maximizes revenue.

EXAMPLE 5   **MAXIMIZING TAX REVENUE**

Economists estimate that the relationship between the tax rate t on an item and the total sales S of that item (in millions of dollars) is

$$S(t) = 9 - 20\sqrt{t} \qquad \text{For } 0 \le t \le 0.20$$

Find the tax rate that maximizes revenue to the government.

Solution

The government's revenue R is the tax rate t times the total sales $S(t) = 9 - 20\sqrt{t}$:

$$R(t) = t \cdot \underbrace{(9 - 20t^{1/2})}_{S(t)} = 9t - 20t^{3/2}$$

To maximize this function, we set its derivative equal to zero:

$9 - 30t^{1/2} = 0$	Derivative of $9t - 20t^{3/2}$
$9 = 30t^{1/2}$	Adding $30t^{1/2}$ to each side
$t^{1/2} = \dfrac{9}{30} = 0.3$	Switching sides and dividing by 30
$t = 0.09$	Squaring both sides

This gives a tax rate of $t = 9\%$. The second derivative,

$$R''(t) = -30 \cdot \frac{1}{2} t^{-1/2} = -\frac{15}{\sqrt{t}} \qquad \text{From } R' = 9 - 30t^{1/2}$$

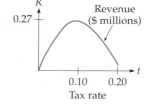

is negative at $t = 0.09$, showing that the revenue is maximized. Therefore,

A tax rate of 9% maximizes revenue for the government.

Graphing Calculator Exploration

on $[-10, 10]$ by $[-10, 10]$

The graph of the function from Example 5, $y_1 = 9x - 20x^{3/2}$ (written in x instead of t for ease of entry), is shown on the left on the standard window $[-10, 10]$ by $[-10, 10]$. This might lead you to believe, erroneously, that the function is maximized at the endpoint (0, 0).

a. Why does this graph not look like the graph at the end of the previous example? [*Hint:* Look at the scale.]

b. Can you find a window on which your graphing calculator will show a graph like the one at the end of the preceding solution?

This example illustrates one of the pitfalls of graphing calculators—the part of the curve where the "action" takes place may be entirely hidden in one pixel. Calculus, on the other hand, will *always* find the critical value, no matter where it is, and then a graphing calculator can be used to confirm your answer by showing the graph on an appropriate window.

Solution   **TO PRACTICE PROBLEM**

Price: $p(x) = 3000 - 200x$

Quantity: $q(x) = 1500 + 300x$

3.4 Exercises

1. **BUSINESS: Maximum Profit** An automobile dealer can sell 12 cars per day at a price of $15,000. He estimates that for each $300 price reduction he can sell two more cars per day. If each car costs him $12,000, and fixed costs are $1000, what price should he charge to maximize his profit? How many cars will he sell at this price? [*Hint:* Let x = the number of $300 price reductions.]

2. **BUSINESS: Maximum Profit** An automobile dealer can sell four cars per day at a price of $12,000. She estimates that for each $200 price reduction she can sell two more cars per day. If each car costs her $10,000, and her fixed costs are $1000, what price should she charge to maximize her profit? How many cars will she sell at this price? [*Hint:* Let x = the number of $200 price reductions.]

3. **BUSINESS: Maximum Revenue** An airline finds that if it prices a cross-country ticket at $200, it will sell 300 tickets per day. It estimates that each $10 price reduction will result in 30 more tickets sold per day. Find the ticket price (and the number of tickets sold) that will maximize the airline's revenue.

4. **ECONOMICS: Oil Prices** An oil-producing country can sell 7 million barrels of oil a day at a price of $90 per barrel. If each $1 price increase will result in a sales decrease of 100,000 barrels per day, what price will maximize the country's revenue? How many barrels will it sell at that price?

5. **BUSINESS: Maximum Revenue** Rent-A-Reck Incorporated finds that it can rent 60 cars if it charges $80 for a weekend. It estimates that for each $5 price increase it will rent three fewer cars. What price should it charge to maximize its revenue? How many cars will it rent at this price?

6. **ENVIRONMENTAL SCIENCES: Maximum Yield** A peach grower finds that if he plants 40 trees per acre, each tree will yield 60 bushels of peaches. He also estimates that for each additional tree that he plants per acre, the yield of each tree will decrease by 2 bushels. How many trees should he plant per acre to maximize his harvest?

7. **ENVIRONMENTAL SCIENCES: Maximum Yield** An apple grower finds that if she plants 20 trees per acre, each tree will yield 90 bushels of apples. She also estimates that for each additional tree that she plants per acre, the yield of each tree will decrease by 3 bushels. How many trees should she plant per acre to maximize her harvest?

8. **GENERAL: Fencing** A farmer has 1200 feet of fence and wishes to build two identical rectangular enclosures, as in the diagram. What should be the dimensions of each enclosure if the total area is to be a maximum?

9. **GENERAL: Minimum Materials** An open-top box with a square base is to have a volume of 4 cubic feet. Find the dimensions of the box that can be made with the smallest amount of material.

10. **GENERAL: Minimum Materials** An open-top box with a square base is to have a volume of 108 cubic inches. Find the dimensions of the box that can be made with the smallest amount of material.

11. **GENERAL: Largest Postal Package** The U.S. Postal Service will accept a package if its length plus its girth (the distance all the way around) does not exceed 84 inches. Find the dimensions and volume of the largest package with a square base that can be mailed.
Source: U.S. Postal Service

girth (distance all the way around)

square base

12. **GENERAL: Fencing** A homeowner wants to build, along his driveway, a garden surrounded by a fence. If the garden is to be 800 square feet, and the fence along the driveway costs $6 per foot while on the other three sides it costs only $2 per foot, find the dimensions that will minimize the cost. Also find the minimum cost.

$2 per foot

$6 per foot

driveway

13. **GENERAL: Fencing** A homeowner wants to build, along her driveway, a garden surrounded by a fence. If the garden is to be 5000 square feet, and the fence along the driveway costs $6 per foot while on the other three sides it costs only $2 per foot, find the dimensions that will minimize the cost. Also find the minimum cost. (See the preceding diagram.)

14–15. ECONOMICS: Tax Revenue Suppose that the relationship between the tax rate t on imported shoes and the total sales S (in millions of dollars) is given by the function below. Find the tax rate t that maximizes revenue for the government.

14. $S(t) = 4 - 6\sqrt[3]{t}$ 15. $S(t) = 8 - 15\sqrt[3]{t}$

16. BIOMEDICAL: Drug Concentration If the amount of a drug in a person's blood after t hours is $f(t) = t/(t^2 + 9)$, when will the drug concentration be the greatest?

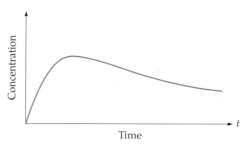

17. GENERAL: Wine Appreciation A case of vintage wine appreciates in value each year, but there is also an annual storage charge. The value of a typical case of investment-grade wine after t years is $V(t) = 2000 + 80\sqrt{t} - 10t$ dollars (for $0 \leq t \leq 25$). Find the storage time that will maximize the value of the wine.

Source: wineeducation.com

18. GENERAL: Bus Shelter Design A bus stop shelter, consisting of two square sides, a back, and a roof, as shown in the following diagram, is to have volume 1024 cubic feet. What are the dimensions that require the least amount of materials?

19. GENERAL: Area Show that the rectangle of fixed area whose perimeter is a minimum is a square.

20. POLITICAL SCIENCE: Campaign Expenses A politician estimates that by campaigning in a county for x days, she will gain $2x$ (thousand) votes, but her campaign expenses will be $5x^2 + 500$ dollars. She wants to campaign for the number of days that maximizes the number of votes per dollar, $f(x) = \dfrac{2x}{5x^2 + 500}$. For how many days should she campaign?

21. GENERAL: Page Design A page of 96 square inches is to have margins of 1 inch on either side and $1\frac{1}{2}$ inches at the top and bottom, as in the diagram. Find the dimensions of the page that maximize the print area.

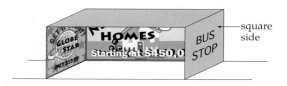

total area: 96 in.^{2}

22. GENERAL: Page Design A page of 54 square inches is to have margins of 1 inch on either side and $1\frac{1}{2}$ inches at the top and bottom, as in the preceding diagram. Find the dimensions of the page that maximize the print area.

23. BUSINESS: Exploring a Profit Maximization Problem Use a graphing calculator to further explore Example 2 (page 204) as follows:

a. Enter the price function $y_1 = 400 - 10x$ and the quantity function $y_2 = 20 + 2x$ into your graphing calculator.

b. Make y_3 the revenue function by defining $y_3 = y_1 y_2$ (price times quantity).

c. Make y_4 the cost function by defining $y_4 = 200 y_2$ (unit cost times quantity).

d. Make y_5 the profit function by defining $y_5 = y_3 - y_4$ (revenue minus cost).

e. Turn off y_1, y_2, y_3, and y_4 and graph the profit function y_5 on the window [0, 10] by [0, 10,000] and then use MAXIMUM to maximize it. Your answer should agree with that found in Example 2.

Now change the problem!

f. What if the store finds that it can buy the bicycles from another wholesaler for $150 instead of $200? In y_4, change the 200 to 150. Then graph the profit y_5 (you may have to turn off y_4 again) and maximize it. Find the new price and quantity by evaluating y_1 and y_2 (using CALCULATE) at the new x-value.

g. What if cycling becomes more popular and the manager estimates that she can sell 30 instead of 20 bicycles per week at the original $400 price? Go back to y_2 and change 20 to 30 (keeping the change made earlier) and graph and find the price and quantity that maximize profit now.

Notice how flexible this setup is for changing any of the numbers.

24. BUSINESS: Maximizing Profit An electronics store can sell 35 cell phones per week at a price of $200. The manager estimates that for each $20 price reduction she can sell 9 more per week. The phones cost the store $100 each, and fixed costs are $700 per week.

a. If x is the number of $20 price reductions, find the price $p(x)$ and enter it in y_1. Then enter the quantity function $q(x)$ in y_2.

b. Make y_3 the revenue function by defining $y_3 = y_1 y_2$ (price times quantity).

c. Make y_4 the cost function by defining y_4 as unit cost times y_2 plus fixed costs.

d. Make y_5 the profit function by defining $y_5 = y_3 - y_4$ (revenue minus cost).

e. Turn off y_1, y_2, y_3, and y_4 and graph the profit function y_5 for x-values $[-10, 10]$, using TABLE or TRACE to find an appropriate y-interval. Then use MAXIMUM to maximize it.

f. Use CALCULATE to find the price and the quantity for this maximum profit.

Conceptual Exercises

25. What is the relationship between *price, quantity,* and *revenue*?

26. What is the relationship between *revenue, cost,* and *profit*?

27. If the current price is $50 and you find that profit is maximized when there are -3 price reductions of $5, what should the price be?

28. If the current price is $75 and you find that profit is maximized when there are -5 price increases of $3, what should the price be?

29. What will be the result of minimizing cost if the cost function is linear?

30. Suppose that, as a variation on Example 4 on page 206, you wanted to find the open-top box with volume 32 cubic feet and a square base that requires the *greatest* amount of materials. Would such a problem have a solution? [*Hint:* No calculation necessary.]

31. If a company wants to increase its profit, why not just produce more of whatever it makes? Explain.

32. If a government wants to increase revenue, why not just increase the tax rate? Explain.

33. In a revenue, cost, and profit problem, is maximizing the revenue the same as maximizing the profit? Explain.

34. In a revenue, cost, and profit problem, is minimizing the cost the same as maximizing the profit? Explain.

Explorations and Excursions   The following problems extend and augment the material presented in the text.

Sources: F. Happensteadt and C. Peskin, *Mathematics in Medicine and the Life Sciences* and R. Anderson and R. May, *Infectious Diseases of Humans: Dynamics and Control*

35. BIOMEDICAL: Contagion If an epidemic spreads through a town at a rate that is proportional to the number of uninfected people and to the square of the number of infected people, then the rate is $R(x) = cx^2(p - x)$, where x is the number of infected people and c and p (the population) are positive constants. Show that the rate $R(x)$ is greatest when two-thirds of the population is infected.

36. BIOMEDICAL: Contagion If an epidemic spreads through a town at a rate that is proportional to the number of infected people and to the number of uninfected people, then the rate is $R(x) = cx(p - x)$, where x is the number of infected people and c and p (the population) are positive constants. Show that the rate $R(x)$ is greatest when half of the population is infected.

3.5   Optimizing Lot Size and Harvest Size

Introduction

In this section we discuss two important applications of optimization, one economic and one ecological. The first concerns the most efficient way for a business to order merchandise (or for a manufacturer to produce merchandise), and the second concerns the preservation of animal populations that are harvested by people. Either of these applications may be read independently of the other.

Minimizing Inventory Costs

A business encounters two kinds of costs in maintaining inventory: **storage costs** (warehouse and insurance costs for merchandise not yet sold) and **reorder costs** (delivery and bookkeeping costs for each order). For example, if a furniture store expects to sell 250 sofas in a year, it could order all 250 at once (incurring high storage costs), or it could order them in many small lots, say 50 orders of five each, spaced throughout the year (incurring high reorder costs). Obviously, the best order size (or **lot size**) is the one that minimizes the total of storage plus reorder costs.

EXAMPLE 1   **MINIMIZING INVENTORY COSTS**

A furniture showroom expects to sell 250 sofas a year. Each sofa costs the store $300, and there is a fixed charge of $500 per order. If it costs $100 to store a sofa for a year, how large should each order be and how often should orders be placed to minimize inventory costs?

Solution

Let

$$x = \text{lot size}$$ Number of sofas in each order

Storage Costs   If the sofas sell steadily throughout the year, and if the store reorders x more whenever the stock runs out, then its inventory during the year looks like the following graph.

Notice that the inventory level varies from the lot size x down to zero, with an average inventory of $x/2$ sofas throughout the year. Because it costs $100 to store a sofa for a year, the total (annual) storage costs are

$$\begin{pmatrix}\text{Storage}\\\text{costs}\end{pmatrix} = \begin{pmatrix}\text{Storage}\\\text{per item}\end{pmatrix} \cdot \begin{pmatrix}\text{Average num-}\\\text{ber of items}\end{pmatrix}$$

$$= \quad 100 \cdot \frac{x}{2} \quad = 50x$$

Reorder Costs   Each sofa costs $300, so an order of lot size x costs $300x$, plus the fixed order charge of $500:

$$\begin{pmatrix}\text{Cost}\\\text{per order}\end{pmatrix} = 300x + 500$$

The yearly supply of 250 sofas, with x sofas in each order, requires $\dfrac{250}{x}$ orders. (For example, 250 sofas at 5 per order require $\dfrac{250}{5} = 50$ orders.) Therefore, the yearly reorder costs are

$$\begin{pmatrix}\text{Reorder}\\\text{costs}\end{pmatrix} = \begin{pmatrix}\text{Cost}\\\text{per order}\end{pmatrix} \cdot \begin{pmatrix}\text{Number}\\\text{of orders}\end{pmatrix} = (300x + 500) \cdot \left(\frac{250}{x}\right)$$

Total Cost $C(x)$ is storage costs plus reorder costs:

$$C(x) = \begin{pmatrix}\text{Storage}\\\text{costs}\end{pmatrix} + \begin{pmatrix}\text{Reorder}\\\text{costs}\end{pmatrix}$$

$$= 100\frac{x}{2} + (300x + 500)\left(\frac{250}{x}\right) \qquad \text{Using the storage and reorder costs found earlier}$$

$$= 50x + 75{,}000 + 125{,}000x^{-1} \qquad \text{Simplifying}$$

To minimize $C(x)$, we differentiate:

$$C'(x) = 50 - 125{,}000x^{-2} = 50 - \frac{125{,}000}{x^2}$$

Differentiating
$C = 50x + 75{,}000 + 125{,}000x^{-1}$

$$50 - \frac{125{,}000}{x^2} = 0$$

Setting the derivative equal to zero

$$50x^2 = 125{,}000$$

Multiplying by x^2 and adding 125,000 to each side

$$x^2 = \frac{125{,}000}{50} = 2500$$

Dividing each side by 50

$$x = 50$$

Taking square roots ($x > 0$) gives lot size 50

$$C''(x) = 250{,}000x^{-3} = 250{,}000\,\frac{1}{x^3}$$

C'' is positive, so C is minimized at $x = 50$

At 50 sofas per order, the yearly 250 will require $\frac{250}{50} = 5$ orders. Therefore, Lot size is 50 sofas, with orders placed five times a year.

 Graphing Calculator Exploration

$Y_1=50X+75000+125000/X$

Minimum
X=50.000015   Y=80000

⌐ Interpret as $x = 50$

a. Verify the answer to the previous Example by graphing the total cost function $y_1 = 50x + 75{,}000 + 125{,}000/x$ on the window [0, 200] by [0, 150,000] and using MINIMUM.

b. Notice that the curve is rather flat to the right of $x = 50$. From this observation, if you cannot order exactly 50 at a time, would it be better to order somewhat more than 50 or somewhat less than 50?

Modifications and Assumptions

If the number of orders per year is not a whole number, say 7.5 orders per year, we just interpret it as 15 orders in 2 years, and handle it accordingly.

We made two major assumptions in Example 1. We assumed that there was a steady demand, and that orders were equally spaced throughout the year. These are reasonable assumptions for many products, while for seasonal products such as bathing suits or winter coats, separate calculations can be done for the "on" and "off" seasons.

Production Runs

Similar analysis applies to manufacturing. For example, if a book publisher can estimate the yearly demand for a book, she may print the yearly total all at once, incurring high storage costs, or she may print them in several smaller runs throughout the year, incurring setup costs for each run. Here the setup costs for each printing run play the role of the reorder costs for a store.

EXAMPLE 2   **MINIMIZING INVENTORY COSTS FOR A PUBLISHER**

A publisher estimates the annual demand for a book to be 4000 copies. Each book costs $8 to print, and setup costs are $1000 for each printing. If storage costs are $2 per book per year, find how many books should be printed per run and how many printings will be needed if costs are to be minimized.

Solution

Let

$$x = \text{the number of books in each run}$$

Storage Costs   As in Example 1, an average of $\dfrac{x}{2}$ books are stored throughout the year, at a cost of $2 each, so annual storage costs are

$$\left(\begin{matrix}\text{Storage} \\ \text{costs}\end{matrix}\right) = \left(\frac{x}{2}\right) \cdot 2 = x$$

Production Costs   The cost per run is

$$\left(\begin{matrix}\text{Costs} \\ \text{per run}\end{matrix}\right) = 8x + 1000 \qquad \begin{array}{l}x \text{ books at \$8 each, plus} \\ \text{\$1000 setup costs}\end{array}$$

The 4000 books at x books per run will require $\dfrac{4000}{x}$ runs. Therefore, production costs are

$$\left(\begin{matrix}\text{Production} \\ \text{costs}\end{matrix}\right) = (8x + 1000)\left(\frac{4000}{x}\right) \qquad \begin{array}{l}\text{Cost per run times} \\ \text{number of runs}\end{array}$$

Total Cost   The total cost is storage costs plus production costs:

$$C(x) = x + (8x + 1000)\left(\frac{4000}{x}\right) \qquad \text{Storage + production}$$

$$= x + 32{,}000 + 4{,}000{,}000x^{-1} \qquad \text{Multiplying out}$$

We differentiate, set the derivative equal to zero, and solve, just as before (omitting the details), obtaining $x = 2000$. The second-derivative test will show that costs are minimized at 2000 books per run. The 4000 books require $\frac{4000}{2000} = 2$ printings. Therefore, the publisher should:

       Print 2000 books per run, with two printings.

Maximum Sustainable Yield

The second application involves industries such as fishing, in which a naturally occurring animal population is "harvested." Harvesting some of the population means more food and other resources for the remaining population so that it will expand and replace the harvested portion. But taking too large a harvest will kill off the animal population (like the bowhead whale, hunted almost to extinction in the nineteenth century). We want to find the **maximum sustainable yield,** the largest amount that may be harvested year after year and still have the population return to its previous level the following year.

     For some animals one can determine a **reproduction function** $f(p)$ which gives the expected population a year from now if the present population is p.

Reproduction Function

A reproduction function $f(p)$ gives the population
a year from now if the current population is p.

For example, the reproduction function $f(p) = -\frac{1}{4}p^2 + 3p$ (where p and $f(p)$ are measured in thousands) means that if the population is now $p = 6$ (thousand), then a year from now the population will be

$$f(6) = -\frac{1}{4}6^2 + 3 \cdot 6 = -9 + 18 = 9 \qquad \text{(thousand)}$$

Therefore, during the year the population will increase from 6000 to 9000.

If, on the other hand, the present population is $p = 10$ (thousand), a year later the population will be

$$f(10) = -\frac{1}{4} \cdot 10^2 + 3 \cdot 10 = -25 + 30 = 5 \qquad \text{(thousand)}$$

That is, during the year the population will decline from 10,000 to 5000 (perhaps because of inadequate food to support such a large population). In actual practice, reproduction functions are very difficult to calculate but can sometimes be estimated by analyzing previous population and harvest data.*

Suppose that we have a reproduction function f and a current population of size p, which will therefore grow to size $f(p)$ next year. The *amount of growth* in the population during that year is

$$\left(\begin{array}{c} \text{Amount} \\ \text{of growth} \end{array} \right) = f(p) - p$$

Current population
Next year's population

Harvesting this amount removes only the *growth*, returning the population to its former size p. The population will then repeat this growth, and taking the same harvest $f(p) - p$ will cause this situation to repeat itself year after year. The quantity $f(p) - p$ is called the sustainable yield.

Sustainable Yield

For reproduction function $f(p)$, the sustainable yield is

$$Y(p) = f(p) - p$$

We want the population size p that maximizes the sustainable yield $Y(p)$. To maximize $Y(p)$, we set its derivative equal to zero:

$$Y'(p) = f'(p) - 1 = 0 \qquad \text{Derivative of } Y = f(p) - p$$

$$f'(p) = 1 \qquad \text{Solving for } f'(p)$$

*For more information, see J. Blower, L. Cook, and J. Bishop, *Estimating the Size of Animal Populations* (London: George Allen and Unwin Ltd.).

For a given reproduction function $f(p)$, we find the maximum sustainable yield by solving this equation (provided that the second-derivative test gives $Y''(p) = f''(p) < 0$).

Maximum Sustainable Yield

For reproduction function $f(p)$, the population p that results in the maximum sustainable yield is the solution to

$$f'(p) = 1$$

(provided that $f''(p) < 0$). The maximum sustainable yield is then

$$Y(p) = f(p) - p$$

Once we calculate the population p that gives the maximum sustainable yield, we wait until the population reaches this size and then harvest, year after year, an amount $Y(p)$.

 Be Careful Note that to find the maximum sustainable yield we set the derivative $f'(p)$ equal to 1, not 0. This is because we are maximizing not the reproductive function $f(p)$ but rather the yield function $Y(p) = f(p) - p$.

EXAMPLE 3   FINDING MAXIMUM SUSTAINABLE YIELD

The reproduction function for the American lobster in an East Coast fishing area is $f(p) = -0.02p^2 + 2p$ (where p and $f(p)$ are in thousands). Find the population p that gives the maximum sustainable yield and find the size of the yield.

Solution

We set the derivative of the reproduction function equal to 1:

$$f'(p) = -0.04p + 2 = 1 \qquad \text{Differentiating } f(p) = -0.02p^2 + 2p$$
$$-0.04p = -1 \qquad \text{Subtracting 2 from each side}$$
$$p = \frac{-1}{-0.04} = 25 \qquad \text{Dividing by } -0.04$$

The second derivative is $f''(p) = -0.04$, which is negative, showing that $p = 25$ (thousand) is the population that gives the maximum sustainable yield. The actual yield is found from the yield function $Y(p) = f(p) - p$:

$$Y(p) = \underbrace{-0.02p^2 + 2p}_{f(p)} - p = -0.02p^2 + p$$

$$Y(25) = -0.02(25)^2 + 25 \qquad \text{Evaluating at } p = 25$$
$$= -12.5 + 25 = 12.5 \qquad \text{(thousand)}$$

The population size for the maximum sustainable yield is 25,000, and the yield is 12,500 lobsters.   25,000 from $p = 25$

Graphing Calculator Exploration

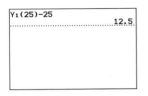

Solve the previous Example on a graphing calculator as follows:

a. Enter the reproduction function as $y_1 = -0.02x^2 + 2x$.

b. Define y_2 to be the derivative of y_1 (using NDERIV).

c. Define $y_3 = 1$.

d. Turn off y_1 and graph y_2 and y_3 on the window [0, 40] by [0, 2].

e. Use INTERSECT to find where y_2 and y_3 meet, thereby solving $f' = 1$. (You should find $x = 25$, as above.)

f. Find the yield by evaluating $y_1 - x$ at the x-value found in part (e).

In this problem, solving "by hand" was probably easier, but this graphing calculator method may be preferable if the reproduction function is more complicated.

3.5 Exercises

BUSINESS: Lot Size

1. A supermarket expects to sell 4000 boxes of sugar in a year. Each box costs $2, and there is a fixed delivery charge of $20 per order. If it costs $1 to store a box for a year, what is the order size and how many times a year should the orders be placed to minimize inventory costs?

2. A supermarket expects to sell 5000 boxes of rice in a year. Each box costs $2, and there is a fixed delivery charge of $50 per order. If it costs $2 to store a box for a year, what is the order size and how many times a year should the orders be placed to minimize inventory costs?

3. A liquor warehouse expects to sell 10,000 bottles of scotch whiskey in a year. Each bottle costs $12, plus a fixed charge of $125 per order. If it costs $10 to store a bottle for a year, how many bottles should be ordered at a time and how many orders should the warehouse place in a year to minimize inventory costs?

4. A wine warehouse expects to sell 30,000 bottles of wine in a year. Each bottle costs $9, plus a fixed charge of $200 per order. If it costs $3 to store a bottle for a year, how many bottles should be ordered at a time and how many orders should the warehouse place in a year to minimize inventory costs?

5. An automobile dealer expects to sell 800 cars a year. The cars cost $9000 each plus a fixed charge of $1000 per delivery. If it costs $1000 to store a car for a year, find the order size and the number of orders that minimize inventory costs.

6. An automobile dealer expects to sell 400 cars a year. The cars cost $11,000 each plus a fixed charge of

$500 per delivery. If it costs $1000 to store a car for a year, find the order size and the number of orders that minimize inventory costs.

BUSINESS: Production Runs

7. A toy manufacturer estimates the demand for a game to be 2000 per year. Each game costs $3 to manufacture, plus setup costs of $500 for each production run. If a game can be stored for a year for a cost of $2, how many should be manufactured at a time and how many production runs should there be to minimize costs?

8. A toy manufacturer estimates the demand for a doll to be 10,000 per year. Each doll costs $5 to manufacture, plus setup costs of $800 for each production run. If it costs $4 to store a doll for a year, how many should be manufactured at a time and how many production runs should there be to minimize costs?

9. A circuit board manufacturer estimates the yearly demand to be 1,000,000. It costs $800 to set up the 3D printer for the circuit board, plus $10 for each one produced. If it costs the company $1 to store a circuit board for a year, how many should be produced at a time and how many production runs will be needed to minimize costs?

10. A manufacturer of Blu-ray Discs (BDs) estimates the yearly demand for a BD to be 10,000. It costs $400 to set up the machinery to burn the BDs, plus $3 for each one produced. If it costs the company $2 to store a BD for a year, how many should be burned at a time and how many production runs will be needed to minimize costs?

ENVIRONMENTAL SCIENCE: Maximum Sustainable Yield

11. Marine ecologists estimate the reproduction curve for swordfish in the Georges Bank fishing grounds to be $f(p) = -0.01p^2 + 5p$, where p and $f(p)$ are in hundreds. Find the population that gives the maximum sustainable yield, and the size of the yield.

12. The reproduction function for the Hudson Bay lynx is estimated to be $f(p) = -0.02p^2 + 5p$, where p and $f(p)$ are in thousands. Find the population that gives the maximum sustainable yield, and the size of the yield.

13. The reproduction function for the Antarctic blue whale is estimated to be $f(p) = -0.0004p^2 + 1.06p$, where p and $f(p)$ are in thousands. Find the population that gives the maximum sustainable yield, and the size of the yield.

14. The reproduction function for the Canadian snowshoe hare is estimated to be $f(p) = -0.025p^2 + 4p$, where p and $f(p)$ are in thousands. Find the population that gives the maximum sustainable yield, and the size of the yield.

15. The reproduction function for king salmon in the Bering Sea is $f(p) = -\frac{1}{8}p^2 + 7.5p$, where p and $f(p)$ are in thousand metric tons. Find the population that gives the maximum sustainable yield, and the size of the yield.

Source: Ecological Modelling **199**

16. The reproduction function for the anchovy in the Peruvian fishery is $f(p) = -0.011p^2 + 1.33p$, where p and $f(p)$ are in million metric tons. Find the population that gives the maximum sustainable yield, and the size of the yield. [*Note:* This catch was exceeded

in the early 1970s, and this, along with other factors, caused a collapse of the Peruvian fishing economy.]

Source: Journal of the Fisheries Research Board of Canada **30**

17. A conservation commission estimates the reproduction function for rainbow trout in a large lake to be $f(p) = 50\sqrt{p}$, where p and $f(p)$ are in thousands and $p \leq 1000$. Find the population that gives the maximum sustainable yield, and the size of the yield.

18. The reproduction function for oysters in a large bay is $f(p) = 30\sqrt[3]{p^2}$, where p and $f(p)$ are in pounds and $p \leq 10,000$. Find the size of the population that gives the maximum sustainable yield, and the size of the yield.

19. The reproduction function for the Pacific sardine (from Baja California to British Columbia) is $f(p) = -0.0005p^2 + 2p$, where p and $f(p)$ are in hundred metric tons. Find the population that gives the maximum sustainable yield, and the size of the yield.

Source: Ecology **48**

20. **BUSINESS: Exploring a Lot Size Problem**
Use a graphing calculator to explore Example 1 (pages 212–213) as follows:
    a. Enter the total cost function, unsimplified, as $y_1 = 100(x/2) + (300x + 500)(250/x)$.
    b. Graph y_1 on the window [0, 200] by [0, 150,000] and use MINIMUM to minimize it. Your answer should agree with the $x = 50$ found in Example 1.
    c. Suppose that business improves, and the showroom expects to sell 350 per year instead of 250. In y_1, change the 250 to 350 and minimize it.
    d. Suppose that a modest recession decreases sales (so change the 350 back to 250), and that inflation has driven the cost of storage up to $125 (so change the 100 to 125), and minimize y_1.

Conceptual Exercises

21. If lot size becomes very large, will storage costs increase or decrease? Will reorder costs increase or decrease?

22. If lot size becomes very small, will storage costs increase or decrease? Will reorder costs increase or decrease?

23. If computer printing technology reduces setup costs (as it already has), would you expect print-run size to increase or decrease?

24. If lot size is x, we say that the average inventory size is $x/2$. What assumptions are we making?

25. For a reproduction function $f(p)$, explain the serious consequences that will result if $f(p) < p$ for all values of p.

26. Explain why a reproduction function $f(p)$ must satisfy $f(0) = 0$.

27. Although reproduction functions can sometimes be estimated by analyzing past population and harvest data, they are, in practice, very difficult to find. Give one reason for this difficulty.

28. Is it realistic to have a reproduction function that is increasing for all values of $p > 0$, such as $f(p) = 5p$ or $f(p) = p^2$?

3.6   **Implicit Differentiation and Related Rates**

Introduction

A function written in the form $y = f(x)$ is said to be defined *explicitly*, meaning that y is defined by a rule or *formula f(x) in x alone*. A function may instead be defined **implicitly,** meaning that y is defined by an *equation in x and y*, such as $x^2 + y^2 = 25$. In this section we will see how to differentiate such **implicit functions** when ordinary "explicit" differentiation is difficult or impossible. We will then use implicit differentiation to find rates of change.

Implicit Differentiation

The equation $x^2 + y^2 = 25$ defines a circle. While a circle is not the graph of a function (it violates the Vertical Line Test, see page 35), the top half by itself defines a function, as does the bottom half by itself. To find these two functions, we solve $x^2 + y^2 = 25$ for y:

$$y^2 = 25 - x^2$$

Subtracting x^2 from each side of $x^2 + y^2 = 25$

$$y = \pm\sqrt{25 - x^2}$$

Plus or minus since when squared either one gives $25 - x^2$

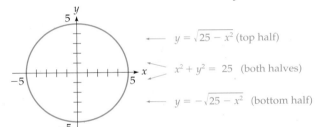

— $y = \sqrt{25 - x^2}$ (top half)

— $x^2 + y^2 = 25$ (both halves)

— $y = -\sqrt{25 - x^2}$ (bottom half)

The *positive* square root defines the top half of the circle (where y is positive), and the *negative* square root defines the bottom half (where y is negative). The equation $x^2 + y^2 = 25$ defines *both* functions at the same time.

To find the slope anywhere on the circle, we could differentiate the "top" and "bottom" functions separately. However, it is easier to find both answers at once by differentiating *implicitly*, that is by differentiating both sides of the equation $x^2 + y^2 = 25$ with respect to x. Remember, however, that y is a *function* of x, so differentiating y^2 means differentiating a *function* squared, which requires the Generalized Power Rule:

$$\frac{d}{dx}\, y^n = n \cdot y^{n-1}\, \frac{dy}{dx}$$

EXAMPLE 1   DIFFERENTIATING IMPLICITLY

Use implicit differentiation to find $\dfrac{dy}{dx}$ when $x^2 + y^2 = 25$.

Solution

We differentiate both sides of the equation with respect to x:

$$\frac{d}{dx}x^2 + \frac{d}{dx}y^2 = \frac{d}{dx}25$$

Differentiating $x^2 + y^2 = 25$

$$2x + 2y\frac{dy}{dx} = 0$$

Using the Generalized Power Rule on y^2

Take Note

y is a *function*, so differentiating it with respect to x gives a *dy/dx*.

Solving for $\dfrac{dy}{dx}$:

$$2y\frac{dy}{dx} = -2x \qquad \text{Subtracting } 2x$$

$$\frac{dy}{dx} = -\frac{x}{y} \qquad \text{Canceling the 2's}$$
$$\text{and dividing by } y$$

Therefore, $\dfrac{dy}{dx} = -\dfrac{x}{y}$ when x and y are related by $x^2 + y^2 = 25$.

Notice that the formula for $\dfrac{dy}{dx}$ involves both x and y. Implicit differentiation enables us to find derivatives that would otherwise be difficult or impossible to calculate, but at a "cost"—the result may depend on both x and y.

 Be Careful Remember that x and y play different roles: x is the *independent* variable, and y is a *function*. Therefore, we must include a $\dfrac{dy}{dx}$ (from the Generalized Power Rule) when differentiating y^n, but not when differentiating x^n since $\dfrac{dx}{dx} = 1$.

EXAMPLE 2   **EVALUATING AN IMPLICIT DERIVATIVE**
(*CONTINUATION OF EXAMPLE 1*)

Find the slope of the circle $x^2 + y^2 = 25$ at the points $(3, 4)$ and $(3, -4)$.

Solution

We simply evaluate the derivative $\dfrac{dy}{dx} = -\dfrac{x}{y}$ (found in Example 1) at the given points.

At $(3, 4)$: $\quad \dfrac{dy}{dx} = -\dfrac{3}{4} \begin{matrix}\leftarrow x \\ \leftarrow y\end{matrix}$

At $(3, -4)$: $\quad \dfrac{dy}{dx} = -\dfrac{3}{-4} = \dfrac{3}{4}$

Note that the negative sign in $\dfrac{dy}{dx} = -\dfrac{x}{y}$ gives the slope the correct sign: negative at $(3, 4)$ and positive at $(3, -4)$.

 Graphing Calculator Exploration

a. Graph the entire circle of the previous Example by graphing $y_1 = \sqrt{25 - x^2}$ and $y_2 = -\sqrt{25 - x^2}$. You may have to adjust the window (or use Zoom ZSquare) to make the circle look "circular."

b. Verify the answer to Example 2 by finding the derivatives of y_1 and y_2 at $x = 3$.

c. Can you find the derivative at $x = 5$? Why not?

Be Careful Derivatives should be evaluated only at points on the curve, so we evaluate $\dfrac{dy}{dx}$ only at x- and y-values *satisfying the original equation.* (It is easy to check that $x = 3$ and $y = \pm 4$ *do* satisfy $x^2 + y^2 = 25$.) Evaluating at a point not on the curve, such as (2, 3), would give a meaningless result.

The following are typical "pieces" that might appear in implicit differentiation problems. Studying them carefully will help you to do longer problems.

EXAMPLE 3   FINDING DERIVATIVES—IMPLICIT AND EXPLICIT

a. $\dfrac{d}{dx}y^3 = 3y^2\dfrac{dy}{dx}$ Differentiating y^3, so *include* $\dfrac{dy}{dx}$

b. $\dfrac{d}{dx}x^3 = 3x^2$ Differentiating x^3, so *no* $\dfrac{dx}{dx}$

c. $\dfrac{d}{dx}(x^3y^5) = \underbrace{3x^2}_{\frac{d}{dx}x^3} \cdot y^5 + x^3 \cdot \underbrace{5y^4\dfrac{dy}{dx}}_{\frac{d}{dx}y^5}$ Using the Product Rule

$\qquad\qquad = 3x^2y^5 + 5x^3y^4\dfrac{dy}{dx}$ Try to do problems such as this in one step, putting the constants in front from the start

---

PRACTICE PROBLEM

Find: **a.** $\dfrac{d}{dx}x^4$ **b.** $\dfrac{d}{dx}y^2$ **c.** $\dfrac{d}{dx}(x^2y^3)$ **Solutions on page 225 >**

Implicit differentiation involves three steps.

Finding $\dfrac{dy}{dx}$ by Implicit Differentiation

1. Differentiate both sides of the equation *with respect to x.* When differentiating a y, include $\dfrac{dy}{dx}$.

2. Collect all terms involving $\dfrac{dy}{dx}$ on one side, and all others on the other side.

3. Factor out the $\dfrac{dy}{dx}$ and solve for it by dividing.

EXAMPLE 4   FINDING AND EVALUATING AN IMPLICIT DERIVATIVE

For $y^4 + x^4 - 2x^2y^2 = 9$ **a.** find $\dfrac{dy}{dx}$ **b.** evaluate it at $x = 2, y = 1$

Solution

$4y^3\dfrac{dy}{dx} + 4x^3 - 4xy^2 - 4x^2y\dfrac{dy}{dx} = 0$ Differentiating with respect to x, putting constants first

$$4y^3 \frac{dy}{dx} - 4x^2 y \frac{dy}{dx} = -4x^3 + 4xy^2$$

Collecting dy/dx terms on the left, others on the right

$$(4y^3 - 4x^2 y) \frac{dy}{dx} = -4x^3 + 4xy^2$$

Factoring out $\frac{dy}{dx}$

$$\frac{dy}{dx} = \frac{-4x^3 + 4xy^2}{4y^3 - 4x^2 y}$$

Dividing by $4y^3 - 4x^2 y$ to solve for dy/dx

$$= \frac{-x^3 + xy^2}{y^3 - x^2 y}$$ ← Answer for part (a)

Dividing by 4

$$\frac{dy}{dx} = \frac{-(2)^3 + (2)(1)^2}{(1)^3 - (2)^2(1)} = \frac{-6}{-3} = 2$$

Evaluating at $x = 2, y = 1$ gives the answer for part (b)

Note that in the preceding Example the given point *is* on the curve, since $x = 2$ and $y = 1$ satisfy the original equation:

$$1^4 + 2^4 - 2 \cdot 2^2 \cdot 1 = 1 + 16 - 8 = 9$$

In economics, a **demand equation** is the relationship between the price p of an item and the quantity x that consumers will demand at that price. (All prices are in dollars unless otherwise stated.)

EXAMPLE 5   **FINDING AND INTERPRETING AN IMPLICIT DERIVATIVE**

For the demand equation $x = \sqrt{1900 - p^3}$, use implicit differentiation to find dp/dx. Then evaluate it at $p = 10$ and interpret your answer.

Solution

$$x^2 = 1900 - p^3$$

Simplifying by squaring both sides of $x = \sqrt{1900 - p^3}$

$$2x = -3p^2 \frac{dp}{dx}$$

Differentiating both sides with respect to x

$$\frac{dp}{dx} = -\frac{2x}{3p^2}$$

Solving for $\frac{dp}{dx}$

To find the value of x we substitute $p = 10$ into the original equation:

$$x = \sqrt{1900 - 10^3} = \sqrt{1900 - 1000} = \sqrt{900} = 30$$ $x = \sqrt{1900 - p^3}$ with $p = 10$

Then

$$\frac{dp}{dx} = -\frac{60}{300} = -0.2$$ $\frac{dp}{dx} = -\frac{2x}{3p^2}$ with $p = 10, x = 30$

Interpretation: $dp/dx = -0.2$ says that the rate of change of price with respect to quantity is -0.2, so that increasing the quantity by 1 means decreasing the price by 0.20 (or 20 cents). Therefore, each 20-cent price decrease brings approximately one more sale (at the given value of p).

Our first step of squaring both sides of the original equation was not necessary, but it made the differentiation easier by avoiding the Generalized Power Rule.

Notice that this particular demand function $x = \sqrt{1900 - p^3}$ can be solved *explicitly* for p:

$$x^2 = 1900 - p^3 \qquad \text{Squaring}$$

$$p^3 = 1900 - x^2 \qquad \text{Adding } p^3 \text{ and subtracting } x^2$$

$$p = (1900 - x^2)^{1/3} \qquad \text{Taking cube roots}$$

We can differentiate this *explicitly* with respect to x:

$$\frac{dp}{dx} = \frac{1}{3}(1900 - x^2)^{-2/3}(-2x) \qquad \text{Using the Generalized Power Rule}$$

$$= -\frac{2}{3}x(1900 - x^2)^{-2/3} \qquad \text{Simplifying}$$

Evaluating at the value of x found in Example 5 gives

$$\frac{dp}{dx} = -\frac{2}{3} \cdot 30(1900 - 30^2)^{-2/3} \qquad \text{Substituting } x = 30$$

$$= -20(1000)^{-2/3} = -\frac{20}{100} = -0.2 \qquad \text{Simplifying}$$

This agrees with the answer by implicit differentiation. Which way was easier?

Related Rates

Sometimes *both* variables in an equation will be functions of a *third* variable, usually t for time. For example, for a seasonal product such as winter coats, the price p and weekly sales x will be related by a demand equation, and both price p and quantity x will depend on the time of year. Differentiating both sides of the demand equation with respect to time t will give an equation relating the derivatives dp/dt and dx/dt. Such "related rates" equations show how fast one quantity is changing relative to another. First, an "everyday" example.

| EXAMPLE 6 | FINDING RELATED RATES |

A pebble thrown into a pond causes circular ripples to radiate outward. If the radius of the outer ripple is growing by 2 feet per second, how fast is the area of its circle growing at the moment when the radius is 10 feet?

Solution

The formula for the area of a circle is $A = \pi r^2$. Both the area A and the radius r of the circle increase with time, so both are functions of t. We are told that the radius is increasing by 2 feet per second $(dr/dt = 2)$, and we want to know how fast the area is changing (dA/dt). To find the relationship between dA/dt and dr/dt, we differentiate both sides of $A = \pi r^2$ with respect to t.

$$\frac{dA}{dt} = 2\pi r \cdot \frac{dr}{dt} \qquad \text{From } A = \pi r^2, \text{ writing the 2 before the } \pi$$

Pebble thrown in

Radius growing by 2 ft/sec

pond

$$\frac{dA}{dt} = 2\pi \cdot \underbrace{10}_{r} \cdot \underbrace{2}_{\frac{dr}{dt}} = \underbrace{40\pi \approx 125.6}_{\substack{\text{Using} \\ \pi \approx 3.14}} \qquad \text{Substituting } r = 10 \text{ and } \frac{dr}{dt} = 2$$

Therefore, at the moment when the radius is 10 feet, the area of the circle is growing at the rate of about 126 square feet per second.

We should be ready to interpret any *rate* as a derivative, just as we interpreted the radius growing by 2 feet per second as $dr/dt = 2$ and the growth rate of the area as dA/dt. The general procedure is as follows.

To Solve a Related Rate Problem

1. Determine the quantities that are changing with time.
2. Find an equation that relates these quantities (a diagram may be helpful).
3. Differentiate both sides of this equation implicitly with respect to t.
4. Substitute into the new equation any given values for the variables and for the derivatives (interpreted as rates of change).
5. Solve for the remaining derivative and interpret the answer as a rate of change.

EXAMPLE 7   USING RELATED RATES TO FIND PROFIT GROWTH

A boat yard's total profit from selling x outboard motors is $P = -x^2 + 1000x - 2000$. If the outboards are selling at the rate of 20 per week, how fast is the profit changing when 400 motors have been sold?

Solution

Profit P and quantity x both change with time, so both are functions of t. We differentiate both sides of $P = -x^2 + 1000x - 2000$ with respect to t and then substitute the given data.

$$\frac{dP}{dt} = -2x \frac{dx}{dt} + 1000 \frac{dx}{dt} \qquad \text{Differentiating with respect to } t$$

$$= \underbrace{-2 \cdot \underbrace{400}_{x} \cdot \underbrace{20}_{dx/dt}} + 1000 \cdot \underbrace{20}_{dx/dt} \qquad \begin{array}{l} \text{Substituting } x = 400 \text{ (number sold)} \\ \text{and } dx/dt = 20 \text{ (sales per week)} \end{array}$$

$$= -16{,}000 + 20{,}000 = 4000$$

Therefore, the company's profits are growing at the rate of \$4000 per week.

EXAMPLE 8   **USING RELATED RATES TO PREDICT POLLUTION**

A study of urban pollution predicts that sulfur oxide emissions in a city will be $S = 2 + 20x + 0.1x^2$ tons, where x is the population (in thousands). The population of the city t years from now is expected to be $x = 800 + 20\sqrt{t}$ thousand people. Find how rapidly the sulfur oxide pollution will be increasing 4 years from now.

Solution

Finding the rate of increase of pollution means finding $\dfrac{dS}{dt}$.

$$\frac{dS}{dt} = 20\frac{dx}{dt} + 0.2x\frac{dx}{dt}$$

$S = 2 + 20x + 0.1x^2$
differentiated with respect to t
(x is also a function of t)

We then find dx/dt from the other given equation:

$$\frac{dx}{dt} = 10t^{-1/2}$$

$x = 800 + 20t^{1/2}$ differentiated with respect to t

$$= 10 \cdot 4^{-1/2} = 10 \cdot \frac{1}{2} = 5$$

Substituting the given $t = 4$ gives $dx/dt = 5$

$\dfrac{dS}{dt}$ then becomes

$$\frac{dS}{dt} = 20 \cdot 5 + 0.2\underbrace{(800 + 20\sqrt{4})}_{x}\underbrace{5}_{dx/dt}$$

$\dfrac{dS}{dt} = 20\dfrac{dx}{dt} + 0.2x\dfrac{dx}{dt}$
with $dx/dt = 5$ and
$x = 800 + 20t^{1/2}$ at $t = 4$

$$= 100 + 0.2(840)5 = 100 + 840 = 940$$

Therefore, in 4 years the sulfur oxide emissions will be increasing at the rate of 940 tons per year.

Graphing Calculator Exploration

Interpret as 940

Verify the answer to the preceding Example on a graphing calculator as follows:

a. Define $y_1 = 2 + 20x + 0.1x^2$ (the sulfur oxide function).

b. Define $y_2 = 800 + 20\sqrt{x}$ (the population function, using x for ease of entry).

c. Define $y_3 = y_1(y_2)$ (the composition of y_1 and y_2, giving pollution in year x).

d. Define y_4 to be the derivative of y_3 (using NDERIV, giving rate of change of pollution).

e. Graph these on the window [0, 10] by [0, 1500]. (Which function does *not* appear on the screen?)

f. Evaluate y_4 at $x = 4$ to verify the answer to Example 8.

Solutions TO PRACTICE PROBLEM

a. $\dfrac{d}{dx}x^4 = 4x^3$ **b.** $\dfrac{d}{dx}y^2 = 2y\dfrac{dy}{dx}$ **c.** $\dfrac{d}{dx}(x^2y^3) = 2xy^3 + 3x^2y^2\dfrac{dy}{dx}$

3.6 Section Summary

An equation in x and y may define one or more functions $y = f(x)$, which we may need to differentiate. Instead of solving the equation for y, which may be difficult or impossible, we can differentiate *implicitly*, differentiating both sides of the original equation with respect to x (writing a dy/dx or y' whenever we differentiate y) and solving for the derivative dy/dx. The derivative at any point of the curve may then be found by substituting the coordinates of that point.

Implicit differentiation is especially useful when several variables in an equation depend on an underlying variable, usually t for time. Differentiating the equation implicitly with respect to this underlying variable gives an equation involving the rates of change of the original variables. Numbers may then be substituted into this "related rate equation" to find a particular rate of change.

Verification of the Power Rule for Rational Powers

On page 100 we stated the Power Rule for differentiation:

$$\frac{d}{dx} x^n = nx^{n-1}$$

Although we have proved it only for *integer* powers, we have been using the Power Rule for *all* constant powers n. Using implicit differentiation, we may now prove the Power Rule for *rational* powers. (Recall that a rational number is of the form p/q, where p and q are integers with $q \neq 0$.) Let $y = x^n$ for a rational exponent $n = p/q$, and let x be a number at which $x^{p/q}$ is differentiable. Then

$$y = x^n = x^{p/q} \qquad \text{Since } n = p/q$$

$$y^q = x^p \qquad \text{Raising each side to the power } q$$

$$qy^{q-1}\frac{dy}{dx} = px^{p-1} \qquad \begin{array}{l}\text{Differentiating each side}\\\text{implicitly with respect to } x\end{array}$$

$$\frac{dy}{dx} = \frac{px^{p-1}}{qy^{q-1}} \qquad \text{Dividing each side by } qy^{q-1}$$

$$= \frac{px^{p-1}}{q\left(x^{\frac{p}{q}}\right)^{q-1}} = \frac{px^{p-1}}{qx^{p-\frac{p}{q}}} \qquad \begin{array}{l}\text{Using } y = x^{p/q} \text{ and}\\\text{multiplying out the exponents}\\\text{in the denominator}\end{array}$$

$$= \frac{p}{q}x^{\underbrace{p-1-\left(p-\frac{p}{q}\right)}_{-1+\frac{p}{q}}} = \frac{p}{q}x^{\frac{p}{q}-1} = nx^{n-1} \qquad \begin{array}{l}\text{Subtracting powers, simplifying,}\\\text{and replacing } p/q \text{ by } n \text{ (twice)}\end{array}$$

This is what we wanted to show, that the derivative of $y = x^n$ is $dy/dx = nx^{n-1}$ for any rational exponent $n = p/q$. This proves the Power Rule for rational exponents.

3.6 Exercises

1–20. For each equation, use implicit differentiation to find dy/dx.

✎ **FOR HELP GETTING STARTED**
with Exercises 1–28, see Example 4 on pages 221–222.

1. $y^3 - x^2 = 4$

2. $y^2 = x^4$

3. $x^3 = y^2 - 2$

4. $x^2 + y^2 = 1$

5. $y^4 - x^3 = 2x$

6. $y^2 = 4x + 1$

7. $(x + 1)^2 + (y + 1)^2 = 18$

8. $xy = 12$

9. $x^2y = 8$

10. $x^2y + xy^2 = 4$

11. $xy - x = 9$

12. $x^3 + 2xy^2 + y^3 = 1$

13. $x(y-1)^2 = 6$ **14.** $(x-1)(y-1) = 25$

15. $y^3 - y^2 + y - 1 = x$ **16.** $x^2 + y^2 = xy + 4$

17. $\dfrac{1}{x} + \dfrac{1}{y} = 2$ **18.** $\sqrt[3]{x} + \sqrt[3]{y} = 2$

19. $x^3 = (y-2)^2 + 1$ **20.** $\sqrt{xy} = x + 1$

21–28. For each equation, find dy/dx evaluated at the given values.

21. $y^2 - x^3 = 1$ at $x = 2, y = 3$

22. $x^2 + y^2 = 25$ at $x = -3, y = 4$

23. $y^2 = 6x - 5$ at $x = 1, y = -1$

24. $xy = 12$ at $x = 6, y = 2$

25. $x^2y + y^2x = 0$ at $x = -2, y = 2$

26. $y^2 + y + 1 = x$ at $x = 1, y = -1$

27. $x^2 + y^2 = xy + 7$ at $x = 3, y = 2$

28. $\sqrt[3]{x} + \sqrt[3]{y} = 3$ at $x = 1, y = 8$

29–36. For each demand equation, use implicit differentiation to find dp/dx.

29. $p^2 + p + 2x = 100$ **30.** $p^3 + p + 6x = 50$

31. $12p^2 + 4p + 1 = x$ **32.** $8p^2 + 2p + 100 = x$

33. $xp^3 = 36$ **34.** $xp^2 = 96$

35. $(p+5)(x+2) = 120$ **36.** $(p-1)(x+5) = 24$

37–40. Find the equation of the tangent line to the curve at the given point using implicit differentiation.

37. *Elliptic curve*
$y^2 = x^3 - 4x + 1$ at $(-2, 1)$
(See diagram in next column.)

38. *Trisectrix of Maclaurin*
$x(x^2 + y^2) = 3x^2 - y^2$ at $(1, 1)$

39. *Cardioid*
$(x^2 + y^2 - y)^2 = x^2 + y^2$ at $(-1, 0)$

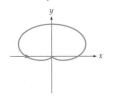

40. *Kepler's trifolium*
$(x^2 + y^2)^2 + 2x^3 = 6xy^2$ at $(1, -1)$

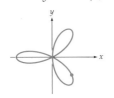

Applied Exercises on Implicit Differentiation

41. BUSINESS: Demand Equation A company's demand equation is $x = \sqrt{2000 - p^2}$, where p is the price in dollars. Find dp/dx when $p = 40$ and interpret your answer.

42. BUSINESS: Demand Equation A company's demand equation is $x = \sqrt{2900 - p^2}$, where p is the price in dollars. Find dp/dx when $p = 50$ and interpret your answer.

43. BUSINESS: Sales The number x of MP3 music players that a store will sell and their price p (in dollars) are related by the equation $2x^2 = 15,000 - p^2$. Find dx/dp at $p = 100$ and interpret your answer. [*Hint:* You will have to find the value of x by substituting the given value of p into the original equation.]

44. BUSINESS: Supply The number x of automobile tires that a factory will supply and their price p (in dollars) are related by the equation $x^2 = 8000 + 5p^2$. Find

dx/dp at $p = 80$ and interpret your answer. [*Hint:* You will have to find the value of x by substituting the given value of p into the original equation.]

45. BUSINESS: Work Hours A management consultant estimates that the number h of hours per day that employees will work and their daily pay of p dollars are related by the equation $60h^5 + 2,000,000 = p^3$. Find dh/dp at $p = 200$ and interpret your answer.

46. BIOMEDICAL: Bacteria The number x of bacteria of type X and the number y of type Y that can coexist in a cubic centimeter of nutrient are related by the equation $2xy^2 = 4000$. Find dy/dx at $x = 5$ and interpret your answer.

47. BUSINESS: Sales If a company spends r million dollars on research, its sales will be s million dollars, where r and s are related by $s^2 = r^3 - 55$.   *(continues)*

a. Find ds/dr by implicit differentiation and evaluate it at $r = 4$, $s = 3$. [*Hint:* Differentiate the equation with respect to r.]

b. Find dr/ds by implicit differentiation and evaluate it at $r = 4$, $s = 3$. [*Hint:* Differentiate the original equation with respect to s.]

c. Interpret your answers to parts (a) and (b) as rates of change.

48. BIOMEDICAL: Muscle Contraction When a muscle lifts a load, it does so according to the "fundamental equation of muscle contraction," also known as *Hill's equation*, $(L + m)(V + n) = k$, where L is the load that the muscle is lifting, V is the velocity of contraction of the muscle, and m, n, and k are constants. Use implicit differentiation to find dV/dL.

Source: E. Batschelet, *Introduction to Mathematics for Life Scientists*

49–50. BUSINESS: Marginal Rate of Substitution Given a Cobb-Douglas production relation

$$P = aL^bK^{1-b} \qquad a > 0 \text{ and } 0 < b < 1$$

giving the total production P from L units of labor and K units of capital, along any isoquant (that is, for a fixed production level), $\dfrac{dK}{dL}$ may be found by implicit differentiation. The absolute value of this derivative is called the *marginal rate of technical substitution* of labor for capital and is denoted *MRTS*.

49. Find *MRTS* for $144 = 12L^{2/3}K^{1/3}$ when $L = 8$ and $K = 27$, and interpret this value.

50. Show that $MRST = \dfrac{b}{1-b}\dfrac{K}{L}$ along any given isoquant.

Exercises on Related Rates

51–58. In each equation, x and y are functions of t. Differentiate with respect to t to find a relation between dx/dt and dy/dt.

51. $x^3 + y^2 = 1$

52. $x^5 - y^3 = 1$

53. $x^2y = 80$

54. $xy^2 = 96$

55. $3x^2 - 7xy = 12$

56. $2x^3 - 5xy = 14$

57. $x^2 + xy = y^2$

58. $x^3 - xy = y^3$

Applied Exercises on Related Rates

59. GENERAL: Snowballs A large snowball is melting so that its radius is decreasing at the rate of 2 inches per hour. How fast is the volume decreasing at the moment when the radius is 3 inches? [*Hint:* The volume of a sphere of radius r is $V = \frac{4}{3}\pi r^3$.]

60. GENERAL: Hailstones A hailstone (a small sphere of ice) is forming in the clouds so that its radius is growing at the rate of 1 millimeter per minute. How fast is its volume growing at the moment when the radius is 2 millimeters? [*Hint:* The volume of a sphere of radius r is $V = \frac{4}{3}\pi r^3$.]

61. BIOMEDICAL: Tumors The radius of a spherical tumor is growing by $\frac{1}{2}$ centimeter per week. Find how rapidly the volume is increasing at the moment when the radius is 4 centimeters. [*Hint:* The volume of a sphere of radius r is $V = \frac{4}{3}\pi r^3$.]

62. BUSINESS: Profit A company's profit from selling x units of an item is $P = 1000x - \frac{1}{2}x^2$ dollars. If sales are growing at the rate of 20 per day, find how rapidly profit is growing (in dollars per day) when 600 units have been sold.

63. BUSINESS: Revenue A company's revenue from selling x units of an item is given as $R = 1000x - x^2$ dollars. If sales are increasing at the rate of 80 per day, find how rapidly revenue is growing (in dollars per day) when 400 units have been sold.

64. SOCIAL SCIENCE: Accidents The number of traffic accidents per year in a city of population p is predicted to be $T = 0.002p^{3/2}$. If the population is growing by 500 people a year, find the rate at which traffic accidents will be rising when the population is $p = 40,000$.

65. SOCIAL SCIENCE: Welfare The number of welfare cases in a city of population p is expected to be $W = 0.003p^{4/3}$. If the population is growing by 1000 people per year, find the rate at which the number of welfare cases will be increasing when the population is $p = 1,000,000$.

66. GENERAL: Rockets A rocket fired straight up is being tracked by a radar station 3 miles from the launching pad. If the rocket is traveling at 2 miles per second, how fast is the distance between the rocket and the tracking station changing at the moment when the rocket is 4 miles up? [*Hint:* The distance D in the illustration satisfies $D^2 = 9 + y^2$. To find the value of D, solve $D^2 = 9 + 4^2$.]

$$D^2 = 9 + y^2$$

D

y

3

67. BIOMEDICAL: Poiseuille's Law Blood flowing through an artery flows faster in the center of the artery and more slowly near the sides (because of friction). The speed of the blood is $V = c(R^2 - r^2)$ millimeters (mm) per second, where R is the radius of the artery, r is the distance of the blood from the center of the artery, and c is a constant.

Artery

R

Suppose that arteriosclerosis is narrowing the artery at the rate of $dR/dt = -0.01$ mm per year. Find the rate at which blood flow is being reduced in an artery whose radius is $R = 0.05$ mm with $c = 500$. [*Hint:* Find dV/dt, considering r to be a constant. The units of dV/dt will be mm per second per year.]

Source: E. Batschelet, *Introduction to Mathematics for Life Scientists*

68. IMPLICIT AND EXPLICIT DIFFERENTIATION The equation $x^2 + 4y^2 = 100$ describes an ellipse.

 a. Use implicit differentiation to find its slope at the points $(8, 3)$ and $(8, -3)$.
 b. Solve the equation for y, obtaining *two* functions, and differentiate both to find the slopes at $x = 8$. [Answers should agree with part (a).]   *(continues)*

 c. Use a graphing calculator to graph the two functions found in part (b) on an appropriate window. Then use NDERIV to find the derivatives at $x = 8$. [Your answers should agree with parts (a) and (b).]

Notice that differentiating implicitly was easier than solving for y and then differentiating.

69. RELATED RATES: Speeding A traffic patrol helicopter is stationary a quarter of a mile directly above a highway, as shown in the diagram below. Its radar detects a car whose line-of-sight distance from the helicopter is half a mile and is increasing at the rate of 57 mph. Is the car exceeding the highway's speed limit of 60 mph?

0.25

z

x

70. RELATED RATES: Speeding *(69 continued)*
In Exercise 69 you found that the car's speed was $\dfrac{(0.5)(57)}{\sqrt{(0.5)^2 - (0.25)^2}}$ mph. Enter this expression into a graphing calculator and then replace both occurrences of the line-of-sight distance 0.5 by 0.4 and calculate the new speed of the car. What if the line-of-sight distance were 0.3 mile?

71. BUSINESS: Sales The number x of printer cartridges that a store will sell per week and their price p (in dollars) are related by the equation $x^2 = 4500 - 5p^2$. If the price is falling at the rate of $1 per week, find how the sales will change if the current price is $20.

72. BUSINESS: Supply The number x of handbags that a manufacturer will supply per week and their price p (in dollars) are related by the equation $5x^3 = 20,000 + 2p^2$. If the price is rising at the rate of $2 per week, find how the supply will change if the current price is $100.

Conceptual Exercises

73. For the function defined explicitly by $y = \sqrt[3]{x^2 - 1}$, define it *implicitly* by an equation without square roots and with zero on the right-hand side.

74. For the function defined implicitly by $2x + 3y = 1$, define y *explicitly* as a function of x.

75. Suppose that you have a formula that relates the amount of gas used (denoted by x) to the distance driven (denoted by y) in your car. State, in everyday language, what $\dfrac{dy}{dx}$ and $\dfrac{dx}{dy}$ would mean.

76. Suppose that you have a formula that relates the value of an investment (denoted by x) to the day of the year (denoted by y). State, in everyday language, what $\dfrac{dy}{dx}$ and $\dfrac{dx}{dy}$ would mean.

77. In Example 1 on pages 219–220 we found that for $x^2 + y^2 = 25$ the slope was given by $\dfrac{dy}{dx} = -\dfrac{x}{y}$. Does this mean that at the point $(10, 10)$ the slope is -1? Explain.

78. In general, implicit differentiation gives an expression for the derivative that involves both x and y. Under what conditions will the expression involve only x?

79. Write in calculus notation: The population of a city is shrinking at the rate of 1500 per year. (Be sure to define your variables.)

80. Write in calculus notation: The value of my car is falling at the rate of $2000 per year. (Be sure to define your variables.)

81. Write in calculus notation: The rate of change of revenue is twice as great as the rate of change in profit. (Be sure to define your variables.)

82. If two quantities, x and y, are related by a linear equation $y = mx + b$, how are the rates of change $\dfrac{dx}{dt}$ and $\dfrac{dy}{dt}$ related?

3.7 Differentials, Approximations, and Marginal Analysis

This section may be omitted without loss of continuity at the discretion of your instructor.

Introduction

We often want to know how one change affects another. For example, a manager might want to estimate the added profit from a small increase in production, or a medical practitioner might want to estimate the additional pain relief from a minor change in medication. Mathematically, this means estimating the change in a function $y = f(x)$ resulting from changing x by a small amount Δx.

In this section, we estimate such changes by using the tangent line to approximate the function near the point of tangency. In Chapter 2 we defined the derivative by approximating the tangent-line slope by secant-line slopes, and now we reverse our viewpoint and estimate secant-line changes by tangent-line changes.

Differentials

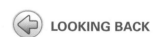

LOOKING BACK

We used Δx and Δy for horizontal and vertical changes on page 95.

The notations Δx and Δy represent changes in the horizontal and vertical directions along the curve $y = f(x)$.

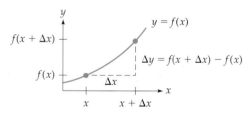

Δx and Δy are changes along the *curve*.

Similarly, we use the notations dx and dy to represent changes in the horizontal and vertical directions along the *tangent line*, as shown in the diagram below. We call them "differentials," and their ratio $\frac{dy}{dx}$ is the slope of the tangent line, $\frac{dy}{dx} = f'(x)$, in agreement with Leibniz's notation for the derivative.

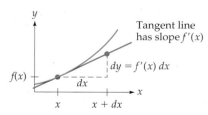

dx and dy are changes along the *tangent line*.

Differentials

Let the function $y = f(x)$ be differentiable at x. The differential dx is any nonzero number, and the differential dy is

$$dy = f'(x)\, dx$$

 Be Careful dx is *not* a product of a "d" and an "x" but is a *single quantity dx*, just as Δx is a single quantity, a change in x, and not a product of "Δ" and "x." Notice also that to calculate dy, values for both x and dx must be given.

EXAMPLE 1

Let $y = x^3 - 2x^2 + 3$. Find dy and evaluate it at $x = 2$ and $dx = 0.5$.

Solution

$$
\begin{aligned}
dy &= (x^3 - 2x^2 + 3)'\, dx && dy = f'(x)\, dx \text{ with } f(x) = x^3 - 2x^2 + 3 \\
&= (3x^2 - 4x)\, dx && \text{Differentiating} \\
&= (3 \cdot 2^2 - 4 \cdot 2) \cdot 0.5 && \text{Using } x = 2 \text{ and } dx = 0.5 \\
&= 4 \cdot 0.5 && \text{Simplifying} \\
&= 2 && \text{Answer}
\end{aligned}
$$

Interpretation: This answer means that the tangent line to the curve rises 2 units after a shift of $dx = 0.5$ units to the right from the point at $x = 2$, as shown in the diagram on the left.

PRACTICE PROBLEM 1

For $y = x^2 + 3$, find dy and evaluate it at $x = 3$ and $dx = -0.05$.

Solution on page 236 >

Note that dx and Δx have similar but different meanings: They both represent changes in the horizontal direction (that is, changes in the independent variable), but dx is the change along the *tangent line* and Δx is the change along the *curve*. Assigning them the same value $dx = \Delta x$ enables us to compare dy and Δy, the changes in the *vertical* direction (the *dependent* variable) along the tangent line and along the curve.

EXAMPLE 2

Find and compare the values of dy and Δy for $y = (x^2 + 2)(3 - x)$ at $x = 1$ and $dx = \Delta x = 0.1$.

Solution

For dy, we proceed as in Example 1:

$$dy = [(x^2 + 2)(3 - x)]'\, dx$$

$dy = f'(x)\, dx$ with
$f(x) = (x^2 + 2)(3 - x)$

$$= [(2x)(3-x) + (x^2+2)(-1)]\, dx \qquad \text{Using the Product Rule}$$

$$= [2(3-1) + (1^2+2)(-1)] \cdot 0.1 \qquad \text{Substituting } x = 1,\, dx = 0.1$$

$$= (4-3) \cdot 0.1 = 0.1 \qquad \text{Evaluating}$$

For Δy, we have:

$$\Delta y = f(x + \Delta x) - f(x) \qquad \text{Definition of } \Delta y$$

$$= [(1.1^2+2)(3-1.1)] - [(1^2+2)(3-1)] \qquad \begin{array}{l} f(x) = (x^2+2)(3-x) \text{ with} \\ x = 1 \text{ and } \Delta x = 0.1 \end{array}$$

$$= 6.099 - 6 = 0.099 \qquad \text{Evaluating}$$

The values of dy and Δy, 0.1 and 0.099, are indeed very close.

PRACTICE PROBLEM 2

Using the same function as in Example 2, find and compare the values of dy and Δy for $x = 1$ and $dx = \Delta x = 0.01$.

Solution on page 236 >

Since the $dy,\, dx$ pairs are changes along the tangent line, their ratio dy/dx is always $f'(x)$. And by the definition of the derivative,

$$\lim_{\Delta x \to 0} \frac{\Delta y}{\Delta x} = f'(x) \qquad \text{Using Leibniz's notation}$$

This means that for small values of Δx, we have the approximation

$$\frac{\Delta y}{\Delta x} \approx f'(x)$$

and so

$$\Delta y \approx f'(x) \cdot \Delta x \qquad \text{Multiplying by } \Delta x$$

Choosing $dx = \Delta x$, this becomes

$$\Delta y \approx f'(x) \cdot dx = dy \qquad \text{Since } dy = f'(x)\, dx$$

Thus for small $\Delta x = dx$, we have the approximation $\Delta y \approx dy$, exactly as we found in Example 2.

Approximation

We saw on page 91 that the tangent line is the best linear approximation to the curve near the point of tangency. Therefore, for small values of $\Delta x = dx$, we may use differentials to approximate changes along a *curve* by changes along the *tangent line*. Rewriting $\Delta y = f(x + \Delta x) - f(x)$ as $f(x + \Delta x) = f(x) + \Delta y$, we obtain the following result.

Linear Approximation

Let the function $y = f(x)$ be differentiable at x. Then

$$\Delta y \approx dy \qquad \text{and} \qquad f(x + \Delta x) \approx f(x) + f'(x)\, dx$$

with these approximations becoming more accurate for $\Delta x = dx$ closer to zero.

EXAMPLE 3

Use differentials to approximate $\sqrt[3]{63}$ using the fact that $\sqrt[3]{64} = 4$.

Solution

To have $\sqrt[3]{63} = f(x + \Delta x)$, we take $f(x) = \sqrt[3]{x}$ with $x = 64$ and $\Delta x = dx = -1$ in the second approximation formula in the preceding box.

$$\sqrt[3]{x + \Delta x} \approx \sqrt[3]{x} + (\sqrt[3]{x})' \, dx \qquad \begin{array}{l} f(x + \Delta x) \approx f(x) + f'(x)\,dx \\ \text{with } f(x) = \sqrt[3]{x} \end{array}$$

$$= \sqrt[3]{x} + \tfrac{1}{3} x^{-2/3} \, dx \qquad \text{Differentiating } \sqrt[3]{x} = x^{1/3}$$

$$= \sqrt[3]{x} + \frac{1}{3(\sqrt[3]{x})^2} \, dx \qquad \text{Rewriting}$$

so

$$\sqrt[3]{64 + (-1)} \approx \sqrt[3]{64} + \frac{1}{3(\sqrt[3]{64})^2} \cdot (-1) \qquad \text{Evaluating at } x = 64 \text{ and } dx = -1$$

Thus

$$\sqrt[3]{63} \approx 4 - \frac{1}{48} \approx 3.97917 \qquad \text{Using } \sqrt[3]{64} = 4 \text{ and a calculator}$$

The actual value of $\sqrt[3]{63}$, correct to five decimal places, is 3.97906, so this approximation is quite good.

PRACTICE PROBLEM 3

Use differentials to approximate $\sqrt[3]{65}$ using the fact that $\sqrt[3]{64} = 4$.

Solution on page 236 >

Error Propagation

A measurement x of a "real-world" quantity is never exact, but comes with an *accuracy* or *error* that we denote by $\Delta x = dx$. Any calculation based on this measurement will also not be exact, and we want to know how the error in the calculation depends on the error dx in the measurement x. Since any calculation can be represented as a function $y = f(x)$, we can estimate the resulting error Δy using the differential $dy = f'(x)\,dx$.

We often judge an error by comparing it to the size of the original measurement, since, for instance, an error of 2 inches in 75 inches is not as significant as an error of 2 inches in 5 inches. Therefore, we define the *relative error* as $\Delta y / y \approx dy / y$, usually expressed as a percentage. Now that we have the term *relative* error, we will sometimes refer to Δy or dy as the *absolute* error. Depending on the application, we may refer to these as "changes" rather than "errors."

EXAMPLE 4

Rated on a scale of 0 ("no relief") to 100 ("complete relief"), the analgesic effect of aspirin is given by

$$A(x) = \frac{100x^2}{x^2 + 0.02}$$

where x denotes the dosage in grams $(0 \leq x \leq 1)$. One "baby aspirin" contains 0.081 g of aspirin. Estimate the effect of a change of ± 0.005 g in this dosage on the pain relief, both as an absolute error and as a relative error.

Source: Consumer Reports

Solution

An error of ± 0.005 means that Δx might be any value from -0.005 to 0.005, but we need only calculate dA at the extreme possibilities $dx = \pm 0.005$.

$$dA = \left(\frac{100x^2}{x^2 + 0.02}\right)' dx \qquad\qquad dA = A'(x)\,dx$$

$$= \frac{(x^2 + 0.02)\,200x - (100x^2)\,2x}{(x^2 + 0.02)^2}\,dx \qquad \text{Using the Quotient Rule}$$

$$= \frac{4x}{(x^2 + 0.02)^2}\,dx \qquad\qquad \text{After simplifying}$$

$$= \frac{4(0.081)}{(0.081^2 + 0.02)^2} \cdot \pm 0.005 \qquad \begin{array}{l}\text{Substituting } x = 0.081 \\ \text{and } dx = \pm 0.005\end{array}$$

$$= \pm 2.3 \qquad\qquad \text{Using a calculator}$$

Interpretation: Changing the dosage by 0.005 g (up or down) from 0.081 will change the pain relief by 2.3 points (up or down, on the scale of 100).

To find the *relative* error, we first calculate the original amount

$$A(0.081) = \frac{100(0.081)^2}{(0.081)^2 + 0.02} \qquad \begin{array}{l}A(x) = \dfrac{100x^2}{x^2 + 0.02} \\ \text{with } x = 0.081\end{array}$$

$$= 24.7 \qquad\qquad \text{Using a calculator}$$

We then divide the last two results to find dA/A:

$$\frac{dA}{A} = \frac{\pm 2.3}{24.7} \qquad \begin{array}{l}\text{Substituting } \Delta A \approx dA = \pm 2.3 \\ \text{and } A = 24.7\end{array}$$

$$= \pm 0.093 \qquad\qquad \text{Using a calculator}$$

Therefore, a ± 0.005 g change in the dosage results in a relative error of $\pm 9.3\%$.

Marginal Analysis

In business and economics, *marginal analysis* refers to the use of differentials to approximate changes in cost, revenue, and profit caused by small changes in production. On page 105, we defined the *marginals* of cost, revenue, and profit functions as the derivatives of these functions. For these functions, we may therefore rewrite their approximation formulas (from page 232) using marginals instead of derivatives.

Marginal Approximation

For cost $C(x)$, revenue $R(x)$, and profit $P(x)$ functions of x units of goods,

$$\Delta C \approx MC \cdot dx \qquad \text{and} \qquad C(x + \Delta x) \approx C(x) + MC(x) \cdot dx$$

$$\Delta R \approx MR \cdot dx \qquad \text{and} \qquad R(x + \Delta x) \approx R(x) + MR(x) \cdot dx$$

$$\Delta P \approx MP \cdot dx \qquad \text{and} \qquad P(x + \Delta x) \approx P(x) + MP(x) \cdot dx$$

with these approximations becoming more accurate for $\Delta x = dx$ closer to zero.

EXAMPLE 5

The cost of manufacturing a WiFi-enabled smart phone is $C(x) = \sqrt{2x+9}$ (in millions of dollars), where x denotes the number produced (in thousands). Use $C(8)$ and $MC(8)$, the cost and marginal cost of manufacturing 8000 phones, to approximate the cost of manufacturing 8300 phones, and give an interpretation of $MC(8)$.

Solution

$$C(8) = \sqrt{2 \cdot 8 + 9} \qquad C = \sqrt{2x+9} \text{ with } x = 8$$
$$= \sqrt{25} = 5 \qquad \text{Evaluating}$$

and

$$MC(x) = \left(\sqrt{2x+9}\right)' \qquad MC = C'$$
$$= \frac{1}{2\sqrt{2x+9}} \cdot 2 \qquad \begin{array}{l}\text{Using the Generalized Power Rule on} \\ C(x) = (2x+9)^{1/2} \text{ and simplifying}\end{array}$$

giving

$$MC(8) = \frac{1}{2\sqrt{2 \cdot 8 + 9}} \cdot 2 \qquad \text{Substituting } x = 8$$
$$= \frac{1}{10} \cdot 2 = \frac{1}{5} = 0.2 \qquad \text{Simplifying and evaluating}$$

Since x is in thousands, $\Delta x = 1$ corresponds to 1000 phones. We are interested in only 300 more, so we take $\Delta x = \frac{300}{1000} = 0.3$ and the marginal approximation formula for cost (on the previous page) becomes

$$C(8.3) \approx C(8) + MC(8) \cdot 0.3 \quad \begin{array}{l}C(x + \Delta x) \approx C(x) + MC(x) \cdot dx \\ \text{with } x = 8 \text{ and } \Delta x = dx = 0.3\end{array}$$
$$= 5 + 0.06 = 5.06 \qquad \text{Using } C(8) = 5, MC(8) = 0.2, \text{ and evaluating}$$

The cost of manufacturing 8300 phones is approximately $5,060,000.

Interpretation of $MC(8) = \frac{1}{5} = 0.2$:

Near this level of production, costs (in millions of dollars) are changing by about *one-fifth* of the change in production (in thousands). This is why a change in production of 0.3 (thousand) caused a change in costs of $0.3 \cdot 0.2 = 0.06$ (million dollars), or $60,000.

The actual value of $C(8.3)$, correct to four decimal places, is 5.0596, so our estimate of 5.06 is quite good. Moreover, knowing that $MC(8) = 0.2$ allows us to quickly estimate the change in cost from changes in production by simply multiplying the production change by 0.2. For instance, 7600 phones would mean a *reduction* of $\Delta x = \frac{-400}{1000} = -0.4$ in production, for a cost change of $(-0.4) \cdot 0.2 = -0.08$ (million dollars), or $-80,000$, reducing costs from $5,000,000 to $4,920,000.

This is a very important use of differentials in business and economics: The marginal cost gives a "feel" for how costs increase or decrease as production is raised or lowered from a given value, with a similar use for revenue, profit, or any other quantity.

PRACTICE PROBLEM 4

Approximate the cost of manufacturing 8250 phones in the situation given in Example 5.

Solution on next page >

Solutions TO PRACTICE PROBLEMS

1. $dy = 2x\,dx$. At $x = 3$ and $dx = 0.01$, $dy = 2 \cdot 3 \cdot (-0.05) = -0.3$.

2. $dy = [2(3-1) + (1^2 + 2)(-1)] \cdot 0.01 = 0.01$

   $\Delta y = (1.01^2 + 2)(3 - 1.01) - (1^2 + 2)(3-1) = 6.009999 - 6 = 0.009999$

   so the values of dy and Δy are even closer for this smaller value of $dx = \Delta x$.

3. The only difference between this problem and Example 3 is that Δx is now $+1$, so $\sqrt[3]{65} \approx 4 + \frac{1}{48} \approx 4.02083$

4. Now $\Delta x = \frac{250}{1000} = 0.25 = \frac{1}{4}$, so the change is $\frac{1}{4} \cdot 0.2 = 0.05$. Therefore, the cost is $C(8.25) \approx 5 + 0.05 = 5.05$, or \$5,500,000. Note that the actual cost $C(8.25)$, correct to five decimal places, is 5.04975, so our estimate is very good.

3.7 Section Summary

For an *independent* variable x, the differential dx is any nonzero number. For the *dependent* variable $y = f(x)$, the differential is $dy = f'(x)\,dx$. Values for both x and dx must be known before dy can be evaluated.

The best linear approximation of a differentiable function $y = f(x)$ near x is the *tangent line approximation* given by

$$f(x + \Delta x) \approx f(x) + f'(x)\,dx \qquad (\Delta x = dx)$$

since $\Delta y \approx dy$. This approximation becomes more accurate for values of $\Delta x = dx$ closer to zero.

For a *dependent* variable y, the *error* Δy resulting from a measurement error Δx is sometimes called the *absolute* error, and may be approximated by the differential, $\Delta y \approx dy$. The *relative* error $\Delta y / y \approx dy / y$ compares the absolute error to the actual value, and is usually written as a percentage. Errors are sometimes called "changes" depending on the situation.

Marginals can be used to find approximations of revenue, cost, and profit (see page 234), and indicate how these quantities vary near a particular level of production.

3.7 Exercises

1–6. Find the differential of each function and evaluate it at the given values of x and dx.

> **FOR HELP GETTING STARTED**
> with Exercises 1–6, see Example 1 on page 231.

1. $y = x^2 - 4x + 5$ at $x = 3$ and $dx = 0.25$.

2. $y = (3x + 1)(x^3 + x + 1)$ at $x = 1$ and $dx = 0.01$.

3. $y = \dfrac{x+1}{x-1}$ at $x = 2$ and $dx = -0.15$.

4. $y = (x + \sqrt{x} - 1)^3$ at $x = 1$ and $dx = 0.2$.

5. $y = \dfrac{x+3}{x^2+1}$ at $x = 2$ and $dx = 0.25$.

6. $y = \dfrac{x}{\sqrt{x+3}}$ at $x = 6$ and $dx = 0.5$.

7–12. Find and compare the values of dy and Δy for each function at the given values of x and $dx = \Delta x$.

> **FOR HELP GETTING STARTED**
> with Exercises 7–12, see Example 2 on pages 231–232.

7. $y = x^3 + x^2 + 3$ at $x = 1$ and $dx = \Delta x = -0.05$.

8. $y = (x + 3)(x^2 - x + 1)$ at $x = -1$ and $dx = \Delta x = 0.1$.

9. $y = \dfrac{x+5}{x+1}$ at $x = 3$ and $dx = \Delta x = 0.4$.

10. $y = \left(\sqrt{x+2} - \dfrac{1}{x}\right)^2$ at $x = 2$ and $dx = \Delta x = 0.1$.

11. $y = \dfrac{3x+5}{x^2+1}$ at $x = 1$ and $dx = \Delta x = 0.2$.

12. $y = \dfrac{x}{\sqrt{3 - x}}$ at $x = 2$ and $dx = \Delta x = 0.05$.

13–16. Find the linear approximation to each function and evaluate it at the given values of x and dx.

> **FOR HELP GETTING STARTED**
>
> with Exercises 13-16, see Example 3 on page 233.

13. x^2 at $x = 10$ and $dx = 2$.

14. $\sqrt{x}$ at $x = 16$ and $dx = -2$.

15. $\dfrac{1}{x}$ at $x = 5$ and $dx = 1$.

16. $\sqrt[3]{x^2}$ at $x = 8$ and $dx = -1$.

17–20. Find the linear approximation for each root by choosing the closest value of x for which the calculation is easy and then using the corresponding value of Δx.

17. $\sqrt{50}$

18. $\sqrt{48}$

19. $\sqrt[3]{9}$

20. $\sqrt[3]{26}$

Applied Exercises

21–22. BUSINESS: Phillips Curves Between 2000 and 2010, the Phillips curve for the U.S. unemployment rate u and the Consumer Price Index inflation rate I was

$$I(u) = 45.4u^{-1.54} - 1$$

where u and I are both percents. Find and interpret dI for each value of u and du.

Source: Bureau of Labor Statistics

21. $u = 3$ and $du = 0.5$

22. $u = 8$ and $du = -0.7$

23–24. BUSINESS: Compound Interest The value $V(r)$ of $1000 deposited in a savings account earning $r\%$ interest compounded annually for 5 years is

$$V(r) = 1000(1 + 0.01r)^5$$

dollars. Find and compare dV and ΔV for each value of r and $dr = \Delta r$.

23. $r = 6$ and $dr = \Delta r = 0.75$

24. $r = 9$ and $dr = \Delta r = -0.35$

25. BIOMEDICAL: Longevity and Exercise A study of the exercise habits of Harvard alumni found that the death rate (deaths per 10,000 person-years) was

$$R(x) = 5x^2 - 35x + 104$$

where x was the weekly amount of exercise (in kilocalories) with $0 \le x \le 4$. Find and compare dR and ΔR for $x = 2$ and $dx = \Delta x = 0.4$.

Source: Journal of the American Medical Association 273

26. BIOMEDICAL: DWI and Crash Risk The crash risk of an intoxicated driver relative to a similar driver with zero blood alcohol is

$$R(x) = 51{,}500x^{4.14} + 1.09$$

where x is the blood alcohol level as a percent $(0.01 \le x \le 0.15)$. Find and compare dR and ΔR for $x = 0.08$ and $dx = \Delta x = 0.005$. [Note: $x \ge 0.08$ defines "driving while intoxicated" and $R(0.08) = 2.6$

means that such a driver is 2.6 times more likely to be involved in an accident than a similar driver who is not impaired.]

Source: Dunlap and Associates, Inc.

27–28. BUSINESS: Wine Appreciation The value $V(t)$ of a case of investment-grade wine after t years is

$$V(t) = 2000 + 80\sqrt{t} - 10t$$

dollars   (for $0 \le t \le 25$).   For each number of years, find dV and use it to estimate the value one year later.

Source: wineeducation.com

27. 9 years   **28.** 16 years

29. GENERAL: Geometry The side of a cube is measured to be 10 inches, with an error of ± 0.01 inch. Find the error and the relative error in the claim that the volume of the cube is 1000 cubic inches.

30. BUSINESS: Manufacturing Tolerances One-inch steel ball bearings are machined to within a tolerance of ± 0.005 inch. Find the error and relative error in the volume.

31–32. GENERAL: Speed and Skid Marks For a car with standard tires stopping on dry asphalt, its speed S (in mph) can be found from the length of its skid marks according to

$$S = 9.4L^{0.37}$$

where L is the length (in feet) of the skid marks. Find the error and relative error in the speed for each skid-mark length if the error in measuring the length is ± 10 feet.

Source: Accident Analysis and Prevention 36

31. 150 feet   **32.** 350 feet

33. BUSINESS: Cost Approximation For the cost function

$$C(x) = \dfrac{125x + 375}{x + 7}$$

where C is in dollars and x is the number produced in hundreds, use $C(13)$ and $MC(13)$ to approximate the cost of producing 1360 items. Give an interpretation of the marginal cost value.

34. BUSINESS: Cost Approximation For the cost function

$$C(x) = x + 10\sqrt{x + 9}$$

where C is in dollars and x is the number produced in hundreds, use $C(7)$ and $MC(7)$ to approximate the cost of producing 620 items. Give an interpretation of the marginal cost value.

35. BUSINESS: Revenue Approximation For the price function

$$p(x) = 10 - \sqrt{3x}$$

where p is in dollars and x is the number sold in hundreds $(0 \le x \le 30)$, use $R(12)$ and $MR(12)$ to approximate the revenue when 1220 items are sold, and interpret the marginal revenue value. How much will the price need to be changed to have this level of sales? [*Hint:* Revenue is price times quantity.]

36. BUSINESS: Revenue Approximation For the price function

$$p(x) = \frac{500}{(x + 3)^2}$$

where p is in dollars and x is the number sold in thousands, use $R(7)$ and $MR(7)$ to approximate the revenue when 7250 items are sold, and interpret the marginal revenue value. How much will the price need to be

changed to have this level of sales? [*Hint:* Revenue is price times quantity.]

37. BUSINESS: Profit Approximation For the price function

$$p(x) = -0.5x + 15$$

where p is in dollars and x is the number sold in hundreds $(0 \le x < 30)$ and cost function

$$C(x) = \frac{80x + 160}{x + 12}$$

where C is in hundreds of dollars and x is the number sold in hundreds, use $P(8)$ and $MP(8)$ to approximate the profit when 830 items are sold, and interpret the marginal profit value.

38. BUSINESS: Profit Approximation For the price function

$$p(x) = 15 - \sqrt{5x}$$

where p is in dollars and x is the number sold in hundreds $(0 \le x \le 45)$ and cost function

$$C(x) = x + 25$$

where C is in hundreds of dollars and x is the number sold in hundreds, use $P(5)$ and $MP(5)$ to approximate the profit when 520 items are sold, and interpret the marginal profit value.

Conceptual Exercises

39. For what kind of a function $y = f(x)$ will $\Delta y = dy$?

40. Does it make sense to use differentials to approximate the change in a function at a point where the tangent line is horizontal?

Explorations and Excursions   The following problems extend and augment the material presented in the text.

41. Show that if $y = ax^b$ (where a and b are constants), then a 1% change in x results in a b% change in y.

42. BUSINESS: Manufacturing Tolerances Steel ball bearings are manufactured with a diameter error of no more than $\pm 0.1\%$. What is the relative error in the volume?

43. BIOMEDICAL: Lung Cancer and Asbestos The number $N(t)$ of lung cancer cases in a group of asbestos workers was given by

$$N(t) = 0.00437t^{3.2}$$

where t denotes the number of years of exposure. By what percent did the number of lung cancer cases change with a 10% longer exposure?

Source: British Journal of Cancer 45

Differentials and the Chain Rule

44–45. Let $y = f(u)$ be differentiable at u and $u = g(x)$ be differentiable at x. The Chain Rule (see page 141) states that $\frac{dy}{dx} = f'(g(x)) \cdot g'(x)$.

44. Identify the "hidden assumption" in the following "proof" of the Chain Rule using the differentials $dy = f'(u)\,du$ and $du = g'(x)\,dx$.

$$dy = f'(u)\,du = f'(g(x)) \cdot g'(x)\,dx$$

and so

$$\frac{dy}{dx} = f'(g(x)) \cdot g'(x)$$

45. On page 145 we stated the Chain Rule in Leibniz's notation as

$$\frac{dy}{dx} = \frac{dy}{du} \cdot \frac{du}{dx}$$

and then remarked that it looked as if the du's simply canceled. Do they?

| **3** | **Chapter Summary with Hints and Suggestions** |

Reading the text and doing the exercises in this chapter have helped you to master the following concepts and skills, which are listed by section (in case you need to review them) and are keyed to particular Review Exercises. Answers for all Review Exercises are given at the back of the book, and full solutions can be found in the Student Solutions Manual.

3.1 Graphing Using the First Derivative

3.2 Graphing Using the First and Second Derivatives

- Graph a polynomial, showing all relative extreme points and inflection points. *(Review Exercises 1−8.)*

$$f' \text{ gives slope, } f'' \text{ gives concavity}$$

- Graph a fractional power function, showing all relative extreme points and inflection points. *(Review Exercises 9−12.)*

- Graph a rational function showing all asymptotes and relative extreme points. *(Review Exercises 13−18.)*

- Graph a rational function showing all asymptotes, relative extreme points, and inflection points. *(Review Exercises 19−20.)*

3.3 Optimization

- Find the absolute extreme values of a given function on a given interval. *(Review Exercises 21−30.)*

- Maximize the efficiency of a tugboat or a flying bird. *(Review Exercises 31−32.)*

3.4 Further Applications of Optimization

- Solve a geometric optimization problem. *(Review Exercises 33−38.)*

- Maximize profit for a company. *(Review Exercise 39.)*

- Maximize revenue from an orchard. *(Review Exercise 40.)*

- Minimize the cost of a power cable. *(Review Exercise 41.)*

- Maximize tax revenue to the government. *(Review Exercise 42.)*

- Minimize the materials used in a container. *(Review Exercises 43−45.)*

3.5 Optimizing Lot Size and Harvest Size

- Find the lot size that minimizes production costs or inventory costs. *(Review Exercises 46−47.)*

- Find the population size that allows the maximum sustainable yield. *(Review Exercises 48−49.)*

3.6 Implicit Differentiation and Related Rates

- Find a derivative by implicit differentiation. *(Review Exercises 50−53.)*

- Find a derivative by implicit differentiation and evaluate it. *(Review Exercises 54−57.)*

- Find the tangent line to a curve at a given point using implicit differentiation. *(Review Exercises 58−59.)*

- Solve a geometric related rates problem. *(Review Exercise 60.)*

- Use related rates to find the growth in profit or revenue. *(Review Exercises 61−62.)*

3.7 Differentials, Approximations, and Marginal Analysis

- Find the differential of $y = f(x)$ and evaluate it at x and dx. *(Review Exercise 64.)*

$$dx = \Delta x \qquad dy = f'(x)\, dx$$

- Compare Δy, dy for given values of x and $dx = \Delta x$. *(Review Exercise 65.)*

$$\Delta y = f(x + \Delta x) - f(x) \qquad \Delta y \approx dy$$

- Find a linear approximation of a required quantity. *(Review Exercise 66.)*

$$f(x + \Delta x) \approx f(x) + f'(x)\, dx$$

- Find the absolute error and relative error resulting from a measurement error. *(Review Exercise 67.)*

$$\Delta y \approx dy \qquad \qquad \frac{\Delta y}{y} \approx \frac{dy}{y}$$

- Use marginals to make linear approximations. *(Review Exercise 68.)*

Hints and Suggestions

- Do not confuse *relative* and *absolute* extremes: a function may have several *relative* maximum points (high points compared to their neighbors), but it can have at most one *absolute* maximum value (the largest value of the function on its entire domain). *Relative* extremes are used in graphing, and *absolute* extremes are used in optimization.

- Graphing calculators can be very helpful for graphing functions. However, you must first find a window that shows the interesting parts of the curve (relative extreme points and inflection points), and that is where calculus is essential.

- We have two procedures for optimizing continuous functions. Both begin by finding all critical numbers of the function in the domain. Then:

    1. If the function has only one critical number, find the sign of the second derivative there: a *positive* sign means an absolute *minimum,* and a *negative* sign means an absolute *maximum.*

    2. If the interval is closed, the maximum and minimum values may be found by evaluating the function at all critical numbers in the interval and endpoints—the largest and smallest resulting values are the maximum and minimum values of the function.

- A good strategy is this: Find all critical numbers in the domain. If there is only one, use procedure 1 above. Otherwise, try to define the function on a closed interval and use procedure 2. If all else fails, make a sketch of the graph.

- Don't forget to use the second-derivative test in applied problems. Your employer will not be happy if you accidentally *minimize* your company's profits.

- In implicit differentiation problems, remember which is the *function* (usually y) and which is the independent variable (usually x).

- In related rate problems, begin by looking for an equation that relates the variables, and then differentiate it with respect to the underlying variable (usually t).

---

3 Review Exercises and Chapter Test   ◯ indicates a Chapter Test exercise.

3.1 and 3.2 Graphing

1–12. Graph each function showing all relative extreme points and inflection points.

1. $f(x) = x^3 - 3x^2 - 9x + 12$
2. $f(x) = x^3 + 3x^2 - 9x - 7$
③ $f(x) = x^4 - 4x^3 + 15$ 4. $f(x) = x^4 + 4x^3 + 17$
5. $f(x) = x(x + 3)^2$ 6. $f(x) = x(x - 6)^2$
7. $f(x) = x(x - 4)^3$ 8. $f(x) = x(x + 4)^3$
9. $f(x) = \sqrt[7]{x^5} + 1$ 10. $f(x) = \sqrt[7]{x^6} + 1$
⑪ $f(x) = \sqrt[7]{x^4} + 1$ 12. $f(x) = \sqrt[7]{x^3} + 1$

13–18. Graph each function showing all asymptotes and relative extreme points.

13. $f(x) = \dfrac{3 - x}{x - 1}$ 14. $f(x) = \dfrac{2x - 4}{x + 1}$
15. $f(x) = \dfrac{1}{(x + 1)^3}$ 16. $f(x) = \dfrac{x(x - 2)}{(x - 1)^2}$
⑰ $f(x) = \dfrac{x^2}{x^2 - 4}$ 18. $f(x) = \dfrac{8}{(x + 1)(x - 2)^2}$

19–20. Graph each function showing all asymptotes, relative extreme points, and inflection points.

⑲ $f(x) = \dfrac{4x^2}{x^2 + 3}$ 20. $f(x) = \dfrac{36x}{(x - 1)^2}$

3.3 and 3.4 Optimization

21–30. Find the absolute extreme values of each function on the given interval.

21. $f(x) = 2x^3 - 6x$ on $[0, 5]$

22. $f(x) = 2x^3 - 24x$ on $[0, 5]$

㉓ $f(x) = x^4 - 4x^3 - 8x^2 + 64$ on $[-1, 5]$

24. $f(x) = x^4 - 4x^3 + 4x^2 + 1$ on $[0, 10]$

25. $h(x) = (x - 1)^{2/3}$ on $[0, 9]$

26. $f(x) = \sqrt{100 - x^2}$ on $[-10, 10]$

㉗ $g(w) = (w^2 - 4)^2$ on $[-3, 3]$

28. $g(x) = x(8 - x)$ on $[0, 8]$

29. $f(x) = \dfrac{x}{x^2 + 1}$ on $[-3, 3]$

30. $f(x) = \dfrac{x}{x^2 + 4}$ on $[-4, 4]$

31. **GENERAL:** Fuel Efficiency At what speed should a tugboat travel upstream so as to use the least amount of fuel to reach its destination? If the tugboat's speed through the water is v and the speed of the current (relative to the land) is c, then the energy used is proportional to $E(v) = \dfrac{v^2}{v - c}$. Find the velocity v that minimizes the energy $E(v)$. Your answer will depend upon c, the speed of the current.
Source: UMAP Modules, COMAP

32. **BIOMEDICAL:** Bird Flight Let v be the flying speed of a bird, and let w be its weight. The power P that the bird must maintain during flight is $P = \dfrac{aw^2}{v} + bv^3$, where a and b are positive constants depending on the shape of the bird and the density of the air. Find the speed v that minimizes the power P.
Source: E. Batschelet, Introduction to Mathematics for Life Scientists

33. **GENERAL:** Fencing A homeowner wants to enclose three adjacent rectangular pens of equal size along a straight wall, as in the following diagram. If the side

along the wall needs no fence, what is the largest total area that can be enclosed using only 240 feet of fence?

(34) GENERAL: Maximum Area A homeowner wants to enclose three adjacent rectangular pens of equal size, as in the following diagram. What is the largest total area that can be enclosed using only 240 feet of fence?

35. GENERAL: Unicorns To celebrate the acquisition of Styria in 1261, Ottokar II sent hunters into the Bohemian woods to capture a unicorn. To display the unicorn at court, the king built a rectangular cage. The material for three sides of the cage cost 3 ducats per running cubit, while the fourth was to be gilded and cost 51 ducats per running cubit. In 1261 it was well known that a happy unicorn requires an area of 2025 square cubits. Find the dimensions that would keep the unicorn happy at the lowest cost.

(36) GENERAL: Box Design An open-top box with a square base is to have a volume of exactly 500 cubic inches. Find the dimensions of the box that can be made with the smallest amount of materials.

37. GENERAL: Packaging Find the dimensions of the cylindrical tin can with volume 16π cubic inches that can be made from the least amount of tin. (*Note:* $16\pi \approx 50$ cubic inches.)

$V = \pi r^2 h$
$A = 2\pi r^2 + 2\pi rh$

38. GENERAL: Packaging Find the dimensions of the open-top cylindrical tin can with volume 8π cubic inches that can be made from the least amount of tin. [*Note:* $8\pi \approx 25$ cubic inches.]

$V = \pi r^2 h$
$A = \pi r^2 + 2\pi rh$

(39) BUSINESS: Maximum Profit A computer dealer can sell 12 personal computers per week at a price of $2000 each. He estimates that each $400 price decrease will result in three more sales per week. If the computers cost him $1200 each, what price should he charge to maximize his profit? How many will he sell at that price?

40. ENVIRONMENTAL SCIENCES: Farming A peach tree will yield 100 pounds of peaches now, which will sell for 40 cents a pound. Each week that the farmer waits will increase the yield by 10 pounds, but the selling price will decrease by 2 cents per pound. How long should the farmer wait to pick the fruit in order to maximize her revenue?

41. GENERAL: Minimum Cost A cable is to connect a power plant to an island that is 1 mile offshore and 3 miles downshore from the power plant. It costs $5000 per mile to lay a cable underwater and $3000 per mile to lay it underground. If the cost of laying the cable is to be minimized, find the distance x along the shore from the island where the cable should meet the land.

42. ECONOMICS: Tax Revenue Economists[*] have found that if cigarettes are taxed at rate t, then cigarette sales will be $S(t) = (64 - 51.26t)/(1 + t)$ (billion dollars annually, before taxes).
a. Use a graphing calculator to find the tax rate t that maximizes revenue to the government.
b. Multiply this tax rate times $4 (a typical pre-tax price of a pack of cigarettes) to find the actual tax per pack.

Source: Chaloupka (ed.), *The Economic Analysis of Substance Use and Abuse*

(43) GENERAL: Packaging A 12-ounce soft drink can has volume 21.66 cubic inches. If the top and bottom are twice as thick as the sides, find the dimensions (radius and height) that minimize the amount of metal used in the can.

44. GENERAL: Box Design A standard 8.5- by 11-inch piece of paper can be made into a box with a lid by cutting x- by x-inch squares from two corners and x- by 5.5-inch rectangles from the other corners and then folding, as shown below. What value of x maximizes the volume of the box, and what is the maximum volume?

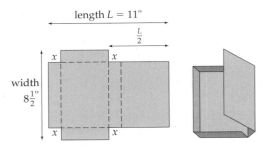

[*]Michael Grossman, Gary Becker (winner of the 1992 Nobel Memorial Prize in Economics), Frank Chaloupka, and Kevin Murphy.

45. GENERAL: Box Design A standard 6- by 8-inch card can be made into a box with a lid by cutting x- by x-inch squares from two corners and x- by 4-inch rectangles from the other corners and then folding. (See the diagram above, but use length 8 and width 6.) What value of x maximizes the volume of the box, and what is the maximum volume?

3.5   Optimizing Lot Size and Harvest Size

46. BUSINESS: Production Runs A wallpaper company estimates the demand for a certain pattern to be 900 rolls per year. It costs $800 to set up the presses to print the pattern, plus $200 to print each roll. If the company can store a roll of wallpaper for a year at a cost of $4, how many rolls should it print at a time and how many printing runs will it need in a year to minimize production costs?

47. BUSINESS: Lot Size A motorcycle shop estimates that it will sell 500 motorbikes in a year. Each bike costs $300, plus a fixed charge of $500 per order. If it costs $200 to store a motorbike for a year, what is the order size and how many orders will be needed in a year to minimize inventory costs?

48. ENVIRONMENTAL SCIENCE: Maximum Sustainable Yield The reproduction function for the North American duck is estimated to be $f(p) = -0.02p^2 + 7p$, where p and $f(p)$ are measured in thousands. Find the size of the population that allows the maximum sustainable yield, and also find the size of the yield.

49. ENVIRONMENTAL SCIENCE: Maximum Sustainable Yield Ecologists estimate the reproduction function for striped bass in an East Coast fishing ground to be $f(p) = 60\sqrt{p}$, where p and $f(p)$ are measured in thousands and $p \le 1000$. Find the size of the population that allows the maximum sustainable yield, and also the size of the yield.

3.6   Implicit Differentiation and Related Rates

50–53. For each equation, use implicit differentiation to find dy/dx.

50. $6x^2 + 8xy + y^2 = 100$

51. $8xy^2 - 8y = 1$

52. $2xy^2 - 3x^2y = 0$

53. $\sqrt{x} - \sqrt{y} = 10$

54–57. For each equation, find dy/dx evaluated at the given values.

54. $x + y = xy$ at $x = 2, y = 2$

55. $y^3 - y^2 - y = x$ at $x = 2, y = 2$

56. $xy^2 = 81$ at $x = 9, y = 3$

57. $x^2y^2 - xy = 2$ at $x = -1, y = 1$

58–59. Find the equation of the tangent line to the curve at the given point using implicit differentiation.

58. *Elliptic curve (see diagram in next column.)*
$y^2 = x^3 - 4x + 4$ at $(2, 2)$

59. *Tilted ellipse*
$2x^2 + 3y^2 = 2xy + 15$ at $(-2, 1)$

60. GENERAL: Melting Ice A cube of ice is melting so that each edge is decreasing at the rate of 2 inches per hour. Find how fast the volume of the ice is decreasing at the moment when each edge is 10 inches long.

61. BUSINESS: Profit A company's profit from selling x units of a product is $P = 2x^2 - 20x$ dollars. If sales are growing at the rate of 30 per day, find the rate of change of profit when 40 units have been sold.

62. BUSINESS: Revenue A company finds that its revenue from selling x units of a product is $R = x^2 + 500x$ dollars. If sales are increasing at the rate of 50 per month, find the rate of change of revenue when 200 units have been sold.

63. BIOMEDICAL: Medication You swallow a spherical pill whose radius is 0.5 centimeter (cm), and it dissolves in your stomach so that its radius decreases at the rate of 0.1 cm per minute. Find the rate at which the volume is decreasing (the rate at which the medication is being made available to your system) when the radius is

a. 0.5 cm

b. 0.2 cm

3.7   Differentials, Approximations, and Marginal Analysis

64. Let $y = x - \sqrt{x}$. Find dy and evaluate it at $x = 4$ and $dx = 0.02$.

65. Find and compare the values of dy and Δy for
$$y = x^2 + \frac{4}{x} \quad \text{at} \quad x = 2 \quad \text{and} \quad dx = \Delta x = 0.01.$$

66. Use differentials to approximate $\sqrt{35}$ from the fact that $\sqrt{36} = 6$.

67. GENERAL: Geometry The side of a cube is measured to be 10 inches, with an error of ± 0.01 inch. Find the error and the relative error in the claim that the surface area of the cube is 600 square inches.

68. BUSINESS: Cost Approximation For the cost function

$$C(x) = 135\sqrt[3]{2x + 3}$$

where C is in dollars and x is the number produced in thousands, use $C(12)$ and $MC(12)$ to approximate the cost of producing 11,600 items. Interpret the marginal cost value.

1–3 Cumulative Review for Chapters 1–3

The following exercises review some of the basic techniques that you learned in Chapters 1–3. Answers to all of these cumulative review exercises are given in the answer section at the back of the book.

1. Find an equation for the line through the points $(-4, 3)$ and $(6, -2)$. Write your answer in the form $y = mx + b$.

2. Simplify $(\frac{4}{25})^{-1/2}$.

3. Find, correct to three decimal places: $\lim\limits_{x \to 0} (1 + 3x)^{1/x}$.

4. For the function $f(x) = \begin{cases} 4x - 8 & \text{if } x < 3 \\ 7 - 2x & \text{if } x \geq 3 \end{cases}$

  a. Draw its graph.

  b. Find $\lim\limits_{x \to 3^-} f(x)$.

  c. Find $\lim\limits_{x \to 3^+} f(x)$.

  d. Find $\lim\limits_{x \to 3} f(x)$.

  e. Is $f(x)$ continuous or discontinuous, and if it is discontinuous, where?

5. Use the definition of the derivative, $f'(x) = \lim\limits_{h \to 0} \dfrac{f(x + h) - f(x)}{h}$, to find the derivative of

$$f(x) = 2x^2 - 5x + 7.$$

6. Find the derivative of $f(x) = 8\sqrt{x^3} - \dfrac{3}{x^2} + 5$.

7. Find the derivative of $f(x) = (x^5 - 2)(x^4 + 2)$.

8. Find the derivative of $f(x) = \dfrac{2x - 5}{3x - 2}$.

9. Find the equation for the tangent line to the curve

$$y = \sqrt{x} - \dfrac{16}{x^2} \quad \text{at} \quad x = 4.$$

10. The population of a city x years from now is predicted to be $P(x) = 3600x^{2/3} + 250{,}000$ people. Find $P'(8)$ and $P''(8)$ and interpret your answers.

11. Find $\dfrac{d}{dx}\sqrt{2x^2 - 5}$ and write your answer in radical form.

12. Find $\dfrac{d}{dx}[(3x + 1)^4(4x + 1)^3]$.

13. Find $\dfrac{d}{dx}\left(\dfrac{x - 2}{x + 2}\right)^3$ and simplify your answer.

14. Find the equation for the tangent line to the curve

$$y = \dfrac{4(x + 3)}{\sqrt{x^2 + 3}} \quad \text{at} \quad x = -1.$$

15. Make sign diagrams for the first and second derivatives and draw the graph of the function $f(x) = x^3 - 12x^2 - 60x + 400$. Show on your graph all relative extreme points and inflection points.

16. Make sign diagrams for the first and second derivatives and draw the graph of the function $f(x) = \sqrt[3]{x^2} - 1$. Show on your graph all relative extreme points and inflection points.

17. A homeowner wishes to use 600 feet of fence to enclose two identical adjacent pens, as in the diagram below. Find the largest total area that can be enclosed.

18. A store can sell 12 telephone answering machines per day at a price of $200 each. The manager estimates that for each $10 price reduction she can sell 2 more per day. The answering machines cost the store $80 each. Find the price and the quantity sold per day to maximize the company's profit.

19. For y defined implicitly by

$$x^3 + 9xy^2 + 3y = 43$$

find $\dfrac{dy}{dx}$ and evaluate it at the point $(1, 2)$.

20. A large spherical balloon is being inflated at the rate of 32 cubic feet per minute. Find how fast the radius is increasing at the moment when the radius is 2 feet.

21. Find and compare the values of dy and Δy for $y = \sqrt{x^3 + 6x - 11}$ at $x = 2$ and $dx = \Delta x = -0.04$.

22. If a stone dropped into an abandoned mine shaft hits the bottom in t seconds, the depth of the shaft is

$$D(t) = 16t^2$$

feet. For a time of 2.5 seconds with an error of ± 0.1 second, find the error and the relative error in the claim that the mine shaft is 100 feet deep.

Exponential and Logarithmic Functions

4

What You'll Explore

Some quantities, such as populations and bank balances, grow exponentially. Others, such as depreciated values of cars and radioactive isotopes used in medicine and carbon-14 dating, *decline* exponentially. The functions necessary to analyze exponential growth and decay are developed in this chapter. They are then put to important uses such as finding doubling times for investments, calculating effective dosages for medications, dating ancient artifacts, and determining prices that maximize consumer expenditure in an economy.

4.1 **Exponential Functions**

4.2 **Logarithmic Functions**

4.3 **Differentiation of Logarithmic and Exponential Functions**

4.4 **Two Applications to Economics: Relative Rates and Elasticity of Demand**

Carbon-14 Dating and the Shroud of Turin

Carbon-14 dating is a method for estimating the age of ancient plants and animals. While a plant is alive, it absorbs carbon dioxide, and with it both "ordinary" carbon and carbon-14, a radioactive form of carbon produced by cosmic rays in the upper atmosphere. When the plant dies, it stops absorbing both types of carbon, and the carbon-14 decays exponentially at a known rate, as shown in the following graph.

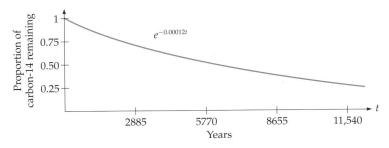

The time since the plant died can then be estimated by comparing the amount of carbon-14 that remains with the amount of ordinary carbon remaining, since they were originally in a known proportion. The comparison technique involves exponential and logarithmic functions, and is explained on pages 267–268. Animal remains can also be dated in this way, since their diets contain plant matter.

In 1988, the Roman Catholic Church used carbon-14 dating to determine the authenticity of the Shroud of Turin, believed by many to be the burial cloth of Jesus. This was done by estimating the age of the flax plants from which the linen shroud was woven. In Exercise 32 on page 272 you will find whether the shroud is old enough to be the burial cloth of Christ.

Bettmann/CORBIS

The usual method to find the amount of carbon-14 remaining in a sample is to burn the sample near a Geiger counter to measure the radioactivity. For the shroud, this would have required a handkerchief-sized sample, and so instead a more advanced method was employed, using a tandem accelerator mass spectrometer and requiring only a postage-stamp-sized sample.

Carbon-14 dating assumes, however, that the ratio of carbon-14 to ordinary carbon in the atmosphere has remained steady over the centuries. This assumption seemed difficult to verify until the discovery in 1955 of a bristlecone pine tree in the White Mountains of California that was over 2000 years old (according to its growth rings). Each annual ring had absorbed carbon and carbon-14 from the air and so provided a year-by-year record of their ratio. The analysis showed that the ratio has remained relatively steady over the centuries, and where variations did occur, it enabled scientists to construct a table for making corrections in the original carbon-14 dates. (Incidentally, bristlecone pine trees that are 4900 years old have been found, making them the oldest living things on earth.) For extremely old remains, such as dinosaur fossils, they use longer lasting radioactive elements, such as potassium-40.

Carbon-14 dating, for which its inventor Willard Libby received a Nobel prize in 1960, has become an invaluable tool in the social and biological sciences.

4.1   Exponential Functions

LOOKING BACK

We first saw exponential functions on page 51.

Introduction

Exponential and **logarithmic functions** are two of the most useful functions in all of mathematics. In this chapter we develop their properties and apply them to a wide variety of problems. We begin with exponential functions, showing how they are used to model the processes of growth and decay.

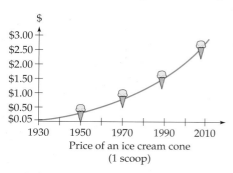

Price of an ice cream cone
(1 scoop)

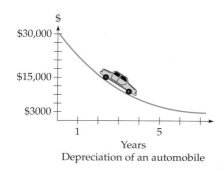

Years
Depreciation of an automobile

We will also define the very important mathematical constant *e*.

Exponential Functions

A function that has a variable in an exponent, such as $f(x) = 2^x$, is called an *exponential function*. The number being raised to the power is called the **base.**

$$f(x) = 2^x$$

Exponent

Base

More formally:

Exponential Functions

For any number $a > 0$, the function

$$f(x) = a^x$$

is called an *exponential function* with base a and exponent (or power) x.

Brief Examples

$f(x) = 2^x$ has base 2

$f(x) = \left(\frac{1}{2}\right)^x$ has base $\frac{1}{2}$

The table below shows some values of the exponential function $f(x) = 2^x$, and its graph (based on these points) is shown on the right.

FOR MORE HELP

with negative exponents, see page 22.

x	$y = 2^x$
-3	$2^{-3} = \frac{1}{8}$
-2	$2^{-2} = \frac{1}{4}$
-1	$2^{-1} = \frac{1}{2}$
0	$2^0 = 1$
1	$2^1 = 2$
2	$2^2 = 4$
3	$2^3 = 8$

$f(x) = 2^x$ has domain $\mathbb{R} = (-\infty, \infty)$ and range $(0, \infty)$

Graph of $y = 2^x$

Clearly, the exponential function 2^x is quite different from the parabola x^2. In particular, it is not symmetric about the y-axis and has the x-axis as a horizontal asymptote: $\lim\limits_{x \to -\infty} 2^x = 0$.

The following table shows some values of the exponential function $f(x) = \left(\frac{1}{2}\right)^x$ and its graph is shown to the right of the table. Notice that it is the mirror image of the curve $y = 2^x$, with both passing through the point $(0, 1)$.

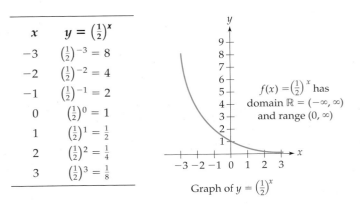

x	$y = \left(\frac{1}{2}\right)^x$
-3	$\left(\frac{1}{2}\right)^{-3} = 8$
-2	$\left(\frac{1}{2}\right)^{-2} = 4$
-1	$\left(\frac{1}{2}\right)^{-1} = 2$
0	$\left(\frac{1}{2}\right)^{0} = 1$
1	$\left(\frac{1}{2}\right)^{1} = \frac{1}{2}$
2	$\left(\frac{1}{2}\right)^{2} = \frac{1}{4}$
3	$\left(\frac{1}{2}\right)^{3} = \frac{1}{8}$

$f(x) = \left(\frac{1}{2}\right)^x$ has domain $\mathbb{R} = (-\infty, \infty)$ and range $(0, \infty)$

Graph of $y = \left(\frac{1}{2}\right)^x$

We can define an exponential function $f(x) = a^x$ for *any* positive base a. We always take the base to be positive, so for the rest of this section *the letter "a" will stand for a positive constant*.

Exponential functions with bases $a > 1$ are used to model *growth*, as in populations or savings accounts, and exponential functions with bases $a < 1$ are used to model *decay*, as in depreciation. (For base $a = 1$, the graph is a horizontal line, since $1^x = 1$ for all x.)

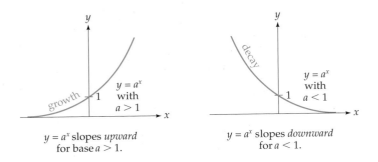

$y = a^x$ slopes *upward* for base $a > 1$.

$y = a^x$ slopes *downward* for $a < 1$.

Compound Interest

Money invested at compound interest grows exponentially. (The word "compound" means that the interest is added to the account, earning more interest.) Banks always state *annual* interest rates, but the compounding may be done more frequently. For example, if a bank offers 8% compounded quarterly, then each quarter you get 2% (one quarter of the annual 8%), so that 2% of your money is added to the account each quarter. If you begin with P dollars (the **principal**), at the end of the first quarter you would have P dollars plus 2% of P dollars:

$$\begin{pmatrix} \text{Value after} \\ \text{1 quarter} \end{pmatrix} = P + 0.02P = P \cdot (1 + 0.02) \qquad \text{Factoring out the } P$$

Notice that increasing a quantity by 2% is the same as multiplying it by $(1 + 0.02)$. Since a year has 4 quarters, t years will have $4t$ quarters. Therefore, to find the value of your account after t years, we simply multiply the principal by $(1 + 0.02)$ a total of $4t$ times, obtaining:

$$\overbrace{\qquad\qquad\qquad\qquad}^{4t \text{ times}}$$

$$\binom{\text{Value after}}{t \text{ years}} = P \cdot (1 + 0.02) \cdot (1 + 0.02) \cdots (1 + 0.02)$$

$$= P \cdot (1 + 0.02)^{4t}$$

The 8%, which gave the $\frac{0.08}{4} = 0.02$ quarterly rate, can be replaced by any interest rate r (written in decimal form), and the 4 can be replaced by any number m of compounding periods per year, leading to the following general formula.

Compound Interest

For P dollars invested at annual interest rate r compounded m times a year for t years,

$$\binom{\text{Value after}}{t \text{ years}} = P \cdot \left(1 + \frac{r}{m}\right)^{mt}$$

r = annual rate
m = periods per year
t = number of years

LOOKING AHEAD

On page 252 we will introduce a different kind of compound interest, where the compunding is done *continuously*.

For example, for monthly compounding we would use $m = 12$ and for daily compounding $m = 365$ (the number of days in the year).

EXAMPLE 1   **FINDING A VALUE UNDER COMPOUND INTEREST**

Find the value of $4000 invested for 2 years at 12% compounded quarterly.

Solution

$$4000 \cdot \left(1 + \underbrace{\frac{0.12}{4}}_{0.03}\right)^{4 \cdot 2} = 4000(1 + 0.03)^8$$

$P \cdot \left(1 + \dfrac{r}{m}\right)^{mt}$
with $P = 4000$,
$r = 0.12$, $m = 4$,
and $t = 2$

$$= 4000 \cdot 1.03^8 \approx 5067.08$$

Using a calculator

The value after 2 years will be $5067.08.

Be Careful Always enter the interest rate into your calculator as a *decimal*.

We may interpret the formula $4000(1 + 0.03)^8$ intuitively as follows: Multiplying the $4000 principal by $(1 + 0.03)$ means that you keep the original amount (the "1") plus some interest (the 0.03), and the exponent 8 means that this is done a total of 8 times.

PRACTICE PROBLEM 1

Find the value of $2000 invested for 3 years at 24% compounded monthly.

Solution on page 255 >

Graphing Calculator Exploration

In the long run, which is more important—principal or interest rate? The graph shows the value of $1000 at 5% interest, together with the value of a mere $200 at the higher rate of 10% (both compounded annually). The fact that the (initially) lower curve eventually surpasses the higher one illustrates a general fact: The higher interest rate will eventually prevail, regardless of the size of the initial investment.

Present Value

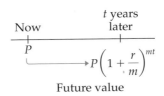

The value to which a sum will grow under compound interest is often called its *future value*. That is, Example 1 showed that the future value of $4000 (at 12% compounded quarterly for 2 years) is $5067.08.

Reversing the order, we can speak of the *present value* of a future payment. Example 1 shows that if a payment of $5067.08 is to be made in 2 years, its *present value* is $4000 (at this interest rate). That is, a promise to pay you $5067.08 in 2 years is worth exactly $4000 now, since $4000 deposited in a bank now would be worth that much in 2 years (at the stated interest rate). To find the *future* value we *multiply* by $(1 + r/m)^{mt}$, and so to find the *present* value, we *divide* by $(1 + r/m)^{mt}$.

Present Value

For a future payment of P dollars at annual interest rate r compounded m times a year to be paid in t years,

$$\begin{pmatrix} \text{Present} \\ \text{value} \end{pmatrix} = \frac{P}{\left(1 + \dfrac{r}{m}\right)^{mt}}$$

r = annual rate
m = periods per year
t = number of years

EXAMPLE 2   FINDING PRESENT VALUE

Find the present value of $5000 to be paid 8 years from now at 10% interest compounded semiannually.

Solution

For semiannual compounding $(m = 2)$, the formula gives

$$\frac{P}{\left(1 + \dfrac{r}{m}\right)^{mt}} = \frac{5000}{\left(1 + \dfrac{0.10}{2}\right)^{2 \cdot 8}} \qquad \begin{array}{l} P = 5000, \ r = 0.10, \\ m = 2, \ \text{and} \ t = 8 \end{array}$$

$$= \frac{5000}{(1 + 0.05)^{16}} \approx 2290.56 \qquad \text{Using a calculator}$$

Therefore, the present value of the $5000 is just $2290.56.

Depreciation by a Fixed Percentage

Depreciation by a fixed percentage means that a piece of equipment loses the same percentage of its current value each year. Losing a percentage of value is like compound interest but with a *negative* interest rate. Therefore, we use the compound interest formula $P(1 + r/m)^{mt}$ with $m = 1$ (since depreciation is annual) and with r being negative.

EXAMPLE 3   **DEPRECIATING AN ASSET**

A car worth $30,000 depreciates in value by 40% each year. How much is it worth after 3 years?

Solution

The car loses 40% of its value each year, which is equivalent to an interest rate of *negative* 40%. The compound interest formula gives

$$30{,}000(1 - 0.40)^3 = 30{,}000(0.60)^3 = \underbrace{\$6480}_{\text{Using a calculator}}$$

$P(1 + r/m)^{mt}$ with $P = 30{,}000$, $r = -0.40$, $m = 1$, and $t = 3$

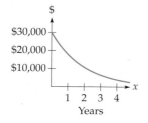

The exponential function $f(x) = 30{,}000(0.60)^x$, giving the value of the car after x years of depreciation, is graphed on the left. Notice that a yearly loss of 40% means that 60% of the value is retained each year.

PRACTICE PROBLEM 2

A printing press, originally worth $50,000, loses 20% of its value each year. What is its value after 4 years?   Solution on page 255 >

The above graph shows that depreciation by a fixed percentage is quite different from "straight-line" depreciation (discussed in Exercises 65–66 on page 18). Under straight-line depreciation the same *dollar* value is lost each year, while under fixed-percentage depreciation the same *percentage* of value is lost each year, resulting in larger dollar losses in the early years and smaller dollar losses in later years. Depreciation by a fixed percentage (also called the **declining balance method**) is one type of **accelerated depreciation**. The method of depreciation that one uses depends on how one chooses to estimate value, and in practice is often determined by the tax laws.

The Number *e*

Imagine that a bank offers 100% interest, and that you deposit $1 for 1 year. Let us see how the value changes under different types of compounding.

For *annual* compounding, your $1 would in a year grow to $2 (the original dollar plus a dollar interest).

For *quarterly* compounding, we use the compound interest formula with $P = 1$, $r = 1$ (for 100%), $m = 4$, and $t = 1$:

$$1\left(1 + \frac{1}{4}\right)^{4 \cdot 1} = 1(1 + 0.25)^4 = (1.25)^4 \approx 2.44 \qquad P\left(1 + \frac{r}{m}\right)^{mt}$$

or $2.44, an improvement of 44 cents over annual compounding.

For *daily* compounding, the value after a year would be

$$\left(1 + \frac{1}{365}\right)^{365} \approx 2.71$$

$m = 365$ periods
$\dfrac{r}{m} = \dfrac{100\%}{365} = \dfrac{1}{365}$

an increase of 27 cents over quarterly compounding. Clearly, if the interest rate, the principal, and the amount of time stay the same, the value increases as the compounding is done more frequently.

In general, if the compounding is done m times a year, the value of the dollar after a year will be

$$\left(\begin{array}{c}\text{Value of \$1 after 1 year at 100\%}\\ \text{interest compounded } m \text{ times a year}\end{array}\right) = \left(1 + \frac{1}{m}\right)^m$$

The following table shows the value of $\left(1 + \dfrac{1}{m}\right)^m$ for various values of m.

Value of $1 at 100% Interest Compounded m Times a Year for 1 Year		
m	$\left(1 + \dfrac{1}{m}\right)^m$	Answer (rounded)	
1	$\left(1 + \dfrac{1}{1}\right)^1 = 2.00000$	Annual compounding
4	$\left(1 + \dfrac{1}{4}\right)^4 \approx 2.44141$	Quarterly compounding
365	$\left(1 + \dfrac{1}{365}\right)^{365} \approx 2.71457$	Daily compounding
10,000	$\left(1 + \dfrac{1}{10,000}\right)^{10,000} \approx 2.71815$	
100,000	$\left(1 + \dfrac{1}{100,000}\right)^{100,000} \approx 2.71827$	
1,000,000	$\left(1 + \dfrac{1}{1,000,000}\right)^{1,000,000} \approx 2.71828$	← Answers agree to five decimal
10,000,000	$\left(1 + \dfrac{1}{10,000,000}\right)^{10,000,000} \approx 2.71828$	← places

Notice that as m increases, the values in the right-hand column seem to be settling down to a definite value, approximately 2.71828. That is, as m approaches infinity, the limit of $\left(1 + \dfrac{1}{m}\right)^m$ is approximately 2.71828. This particular number is very important in mathematics, and is given the name e (just as 3.14159… is given the name π). In the following definition we use the letter n to state the definition in its traditional form.

The Constant e

$$e = \lim_{n \to \infty} \left(1 + \frac{1}{n}\right)^n = 2.71828\ldots$$

The dots mean that the decimal expansion goes on forever

The same e appears in probability and statistics in the formula for the "bell-shaped" or "normal" curve. Its value has been calculated to several million decimal places, and its value to 15 decimal places is $e \approx 2.718281828459045$.

Continuous Compounding of Interest

This kind of compound interest, the limit as the compounding frequency approaches infinity, is called **continuous compounding.** We have shown that $1 at 100% interest compounded continuously for 1 year would be worth precisely e dollars (about $2.72). The formula for continuous compound interest at other rates is as follows (a justification for it is given at the end of this section).

Continuous Compounding

For P dollars invested at annual interest rate r compounded continuously for t years,

$$\begin{pmatrix} \text{Value after} \\ t \text{ years} \end{pmatrix} = Pe^{rt}$$

EXAMPLE 4   FINDING VALUE WITH CONTINUOUS COMPOUNDING

Find the value of $1000 at 8% interest compounded continuously for 20 years.

Solution

We use the formula Pe^{rt} with $P = 1000$, $r = 0.08$, and $t = 20$.

$$Pe^{rt} = \underbrace{1000}_{P} \cdot e^{(0.08)(20)} = 1000 \cdot \underbrace{e^{1.6}}_{4.95303} \approx \$4953.03$$

$e^{1.6}$ is usually found using the [2nd] and [LN] e^x keys

Present Value with Continuous Compounding

As before, the value that an amount will grow to in t years is often called its *future* value, and the current value of a future amount is called its *present* value. Under continuous compounding, to find future value we multiply P by e^{rt}, and so to find *present* value we *divide* by e^{rt}.

Present Value with Continuous Compounding

For a future payment of P dollars at annual interest rate r compounded continuously to be paid in t years,

$$\begin{pmatrix} \text{Present} \\ \text{value} \end{pmatrix} = \frac{P}{e^{rt}} = Pe^{-rt}$$

EXAMPLE 5   FINDING PRESENT VALUE WITH CONTINUOUS COMPOUNDING

The present value of $5000 to be paid in 10 years at 7% interest compounded continuously is

$$\frac{5000}{e^{0.07 \cdot 10}} = \frac{5000}{e^{0.7}} \approx \$2482.93$$

Using a calculator

Intuitive Meaning of Continuous Compounding

Under quarterly compounding, your money is, in a sense, earning interest throughout the quarter, but the interest is not added to your account until the end of the quarter. Under continuous compounding, the interest is added to your account *as it is earned,* with no delay. The extra earnings in continuous compounding come from this "instant crediting" of interest, since then your interest starts earning more interest immediately.

How to Compare Interest Rates

How do you compare different interest rates, such as 12% compounded quarterly and 11.9% compounded continuously? You simply see what each will do for a deposit of $1 for 1 year.

EXAMPLE 6   COMPARING INTEREST RATES

Which gives a better return: 12% compounded quarterly or 11.9% compounded continuously?

Solution

For 12% compounded quarterly (on $1 for 1 year),

$$1 \cdot \left(1 + \frac{0.12}{4}\right)^4 = (1 + 0.03)^4 \approx 1.1255 \qquad \text{Using } P\left(1 + \frac{r}{m}\right)^{mt}$$

For 11.9% compounded continuously,

$$1 \cdot e^{0.119 \cdot 1} = e^{0.119} \approx 1.1264 \quad \swarrow \text{Better} \qquad \text{Rounded, using } Pe^{rt}$$

Therefore, 11.9% compounded continuously is better. (The difference is only a tenth of a cent, but it would be more impressive for principals larger than $1 or periods longer than 1 year.)

The actual percentage increase during 1 year is called the **effective rate of interest,** the **annual percentage rate (APR)**, or the **annual percentage yield (APY).** For example, the "continuous" rate of 11.9% used previously gives a return of $1.126, and after subtracting the dollar invested, leaves a gain of 0.126 on the original $1, which means an APR of 12.6%. The *stated* rate (here, 11.9%) is called the **nominal rate** of interest. That is, a nominal rate of 11.9% compounded continuously is equivalent to an APR of 12.6%. The 1993 Truth in Savings Act requires that banks always state the annual percentage rate.

PRACTICE PROBLEM 3

From the previous Example, what is the annual percentage rate for 12% compounded quarterly? [*Hint:* No calculation is needed.]

Solution on page 255 >

The Function $y = e^x$

The number e gives us a new exponential function $f(x) = e^x$. This function is used extensively in business, economics, and all areas of science. The table below shows the values of e^x for various values of x. These values lead to the graph of $f(x) = e^x$ shown at the right.

x	$y = e^x$
-3	$e^{-3} \approx 0.05$
-2	$e^{-2} \approx 0.14$
-1	$e^{-1} \approx 0.37$
0	$e^0 = 1$
1	$e^1 \approx 2.72$
2	$e^2 \approx 7.39$
3	$e^3 \approx 20.09$

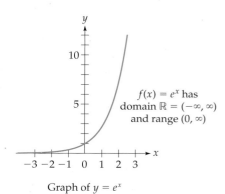

$f(x) = e^x$ has domain $\mathbb{R} = (-\infty, \infty)$ and range $(0, \infty)$

Graph of $y = e^x$

Notice that e^x is never zero and is positive for all values of x, even when x is negative. We restate this important observation as follows:

e to any power is positive.

The following graph shows the function $f(x) = e^{kx}$ for various values of the constant k. For positive values of k the curve rises, and for negative values of k the curve falls (as you move to the right). For higher values of k the curve rises more steeply. Each curve has the x-axis as a horizontal asymptote and crosses the y-axis at 1.

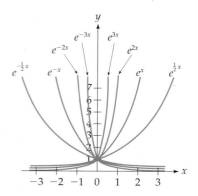

$f(x) = e^{kx}$ for various values of k.

 Graphing Calculator Exploration

In 2013, the two largest countries in the world were China, with 1.36 billion people, and India, with 1.24 billion. According to data from the United Nations, x years after 2013 the population of China will be $1.36e^{0.00557x}$ and the population of India will be $1.24e^{0.0148x}$, both in billions.

a. Graph these two functions on the window [0, 15] by [0, 1.8]. [Use the [2nd] e^x and [LN] keys for entering e to powers.]

b. Use INTERSECT to find the x-value where the curves intersect.

c. From your answer to part (b), predict the year in which India will overtake China as the largest country. [*Hint:* x is years after 2013.]

Source: United Nations

Exponential Growth

All exponential growth, whether continuous or discrete, has one common characteristic: The amount of growth is proportional to the size. For example, the interest that a bank account earns is proportional to the size of the account, and the growth of a population is proportional to the size of the population. This is in contrast, for example, to a person's height, which does not increase exponentially. That is, exponential growth occurs in those situations where a quantity grows *in proportion to its size*. This is why exponential functions are so important.

Solutions TO PRACTICE PROBLEMS

1. $2000(1 + 0.24/12)^{12\cdot3} = 2000(1 + 0.02)^{36} = 2000(1.02)^{36} \approx 2000(2.039887) \approx$ \$4079.77

2. $50{,}000(1 - 0.20)^4 = 50{,}000(0.8)^4 = 50{,}000(0.4096) = \$20{,}480$

3. 12.55% (from 1.1255)

4.1   ## Section Summary

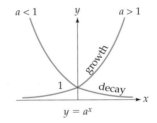

Exponential functions have exponents that involve variables. The exponential functions $f(x) = a^x$ slope *upward* or *downward* depending upon whether the base a (which must be positive) satisfies $a > 1$ or $a < 1$.

The formula $P(1 + r/m)^{mt}$ gives the value after t years for an investment of P dollars that increases at annual rate r compounded m times a year. This same formula with $m = 1$ and a *negative* growth rate r governs depreciation by a fixed percentage.

By considering a 100% increase not given all at once but divided up into smaller and smaller successive increases, we defined a new constant e:

$$e = \lim_{n \to \infty} \left(1 + \frac{1}{n}\right)^n \approx 2.71828$$

Exponential functions with base e are used extensively in modeling many types of growth, such as the growth of populations and interest that is compounded continuously. For interest compounded *continuously*, the formula is Pe^{rt}.

Justification of the Formula for Continuous Compounding

The compound interest formula Pe^{rt} is derived as follows: P dollars invested for t years at interest rate r compounded m times a year yields

$$P\left(1 + \frac{r}{m}\right)^{mt} \qquad \text{See page 248}$$

Define $n = \dfrac{m}{r}$, so that $m = rn$. Replacing m by rn and letting m (and therefore n) approach ∞, this becomes

$$P\left(1 + \frac{r}{rn}\right)^{rnt} = P\left[\left(1 + \frac{1}{n}\right)^n\right]^{rt} \to Pe^{rt} \qquad \text{Letting } n \to \infty$$

$\underbrace{}_{r \text{ cancels}}$ $\underbrace{}_{\substack{\text{Approaches} \\ e \text{ as } n \to \infty}}$

The limit shown on the right is the continuous compounding formula Pe^{rt}.

4.1 Exercises

1–2. Use a calculator to evaluate, rounding to three decimal places.

1. a. e^2 **b.** e^{-2} **c.** $e^{1/2}$

2. a. e^3 **b.** e^{-3} **c.** $e^{1/3}$

3–4. Express as a power of e.

3. a. $e^5 e^{-2}$ **b.** $\dfrac{e^5}{e^3}$ **c.** $\dfrac{e^5 e^{-1}}{e^{-2} e}$

4. a. $e^{-3} e^5$ **b.** $\dfrac{e^4}{e}$ **c.** $\dfrac{e^2 e}{e^{-2} e^{-1}}$

5–8. Graph each function. If you are using a graphing calculator, make a hand-drawn sketch from the screen.

5. $y = 3^x$ **6.** $y = 5^x$ **7.** $y = \left(\frac{1}{3}\right)^x$ **8.** $y = \left(\frac{1}{5}\right)^x$

9–10. Evaluate each expression using a calculator.

9. $e^{1.74}$ **10.** $e^{-0.09}$

11. e^x Versus x^n Which curve is eventually higher, x to a power or e^x?   *(continues)*

a. Graph x^2 and e^x on the window [0, 5] by [0, 20]. Which curve is higher?

b. Graph x^3 and e^x on the window [0, 6] by [0, 200]. Which curve is higher for large values of x?

c. Graph x^4 and e^x on the window [0, 10] by [0, 10,000]. Which curve is higher for large values of x?

d. Graph x^5 and e^x on the window [0, 15] by [0, 1,000,000]. Which curve is higher for large values of x?

e. Do you think that e^x will exceed x^6 for large values of x? Based on these observations, can you make a conjecture about e^x and *any* power of x?

12. Linear Versus Exponential Growth

a. Graph $y_1 = x$ and $y_2 = e^{0.01x}$ in the window [0, 10] by [0, 10]. Which curve is higher for x near 10?

b. Then graph the same curves on the window [0, 1000] by [0, 1000]. Which curve is higher for x near 1000?

A function such as y_1 represents *linear* growth, and y_2 represents *exponential* growth, and the result here is true in general: Exponential growth always beats linear growth (eventually, no matter what the constants).*

Applied Exercises

13. BUSINESS: Interest Find the value of $1000 deposited in a bank at 10% interest for 8 years compounded:

a. annually.   **b.** quarterly.   **c.** continuously.

14. BUSINESS: Interest Find the value of $1000 deposited in a bank at 12% interest for 8 years compounded:

a. annually.   **b.** quarterly.   **c.** continuously.

15. PERSONAL FINANCE: Interest A loan shark lends you $100 at 2% compound interest per week (this is a *weekly*, not an annual rate).

a. How much will you owe after 3 years?

b. In "street" language, the profit on such a loan is known as the "vigorish" or the "vig." Find the shark's vig.

16. GENERAL: Compound Interest In 1626, Peter Minuit purchased Manhattan Island from the native Americans for $24 worth of trinkets and beads. Find what the $24 would be worth in the year 2020 if it had been deposited in a bank paying 5% interest compounded quarterly.

17. PERSONAL FINANCE: Annual Percentage Rate (APR) Find the error in the ad shown below, which appeared in a New York newspaper. [*Hint:* Check that the nominal rate is equivalent to the effective rate. For daily compounding, some banks use 365 days and some use 360 days in the year. Try it both ways.]

> At T&M Bank, flexibility is the key word. You can choose the length of time and the amount of your deposit, which will earn an annual yield of 9.825% based on a rate of 9.25% compounded daily.

18. PERSONAL FINANCE: Annual Percentage Rate (APR) Find the error in the following ad, which appeared in a Washington, D.C., newspaper. Assume that the compounding is done daily. [*Hint:* Check that the nominal rate is equivalent to the effective rate. For daily compounding, some banks use 365 days and some use 360 days in the year. Try it both ways.]

> **Your Money's in 7th Heaven** when you get Hans Johnson's "sky high" return.
> No time restrictions or withdrawal penalties! Funds available when you want them.
>
> | **Current Annual Yield: 7.19%** | **7% Regular Passbooks** |

19. PERSONAL FINANCE: Present Value A rich uncle wants to make you a millionaire. How much money must he deposit in a trust fund paying 8% compounded quarterly at the time of your birth to yield $1,000,000 when you retire at age 60?

20. PERSONAL FINANCE: Present Value The cost of a four-year private college education (after financial aid) has been estimated to be $50,000. How large a trust fund, paying 6% compounded quarterly, must be established at a child's birth to ensure sufficient funds at age 18?

Source: College Board

---

*The realization that populations grow exponentially while food supplies grow only linearly caused the great nineteenth-century essayist Thomas Carlyle (1795–1881) to dub economics the "dismal science." He was commenting not on how interesting economics is, but on the grim conclusions that follow from populations outstripping their food supplies.

21. GENERAL: Consumer Fraud Buying a "vacation time-share" means buying the right to use a vacation property for a fixed period each year. Suppose that you pay $500 for a vacation timeshare and receive a "money-back guarantee" that at any time the company will buy back your timeshare, or if not, give you a $1000 bond. The deception, however, is that the bond is not redeemable for 45 years. Find the real value of the "guarantee"—that is, find the present value of $1000 in 45 years (assume a 6% interest rate compounded annually).

22. PERSONAL FINANCE: Zero-Coupon Bonds Miami-Dade County recently sold $1000 zero-coupon bonds* maturing in 29 years with an annual yield of 5.8%. Find the price. [*Hint:* The price is the present value of $1000 29 years from now at the stated interest rate.]

Source: fmsbonds.com

23. BUSINESS: Zero-Coupon Bonds The City of Chicago recently sold zero-coupon $1000 bonds* maturing in 12 years with an annual yield of 6%. Find the price. [*Hint:* The price is the present value of $1000 6 years from now at the stated interest rate.]

Source: Reuters

24. PERSONAL FINANCE: Baseball Cards A mint-condition 1910 Honus Wagner** baseball card sold at auction in 2007 for $2.35 million, its value having grown by 11% annually. Predict its value in the year 2027.

Source: Los Angeles Times, February 28, 2007

25. BUSINESS: Compound Interest Which is better: 10% interest compounded quarterly or 9.8% compounded continuously?

26. BUSINESS: Compound Interest Which is better: 8% interest compounded quarterly or 7.8% compounded continuously?

27. PERSONAL FINANCE: Depreciation A Toyota Corolla automobile lists for $19,700 and depreciates by 35% per year. Find its value after:

a. 4 years.　　**b.** 6 months.

Source: Toyota

28. BUSINESS: Art Appreciation In 2011, The *Lady Blunt* Stradivarius voilin sold at auction for $16 million, its value having grown by 13% annually. Predict its value in the year 2021.

Source: Bloomberg

29. SOCIAL SCIENCE: Population The population of the world x years after the year 2010 is predicted to be $6.9e^{0.00743x}$ billion people (for $0 \leq x \leq 30$). Use this formula to predict the world population in the year 2030.

Source: United Nationals Population Fund

* A zero-coupon bond makes no payments (coupons) until it matures, at which time it pays its face value of $1000, having sold originally at a lower price.

** Honus Wagner (1874–1955), regarded as the greatest player of his time, played for the Pittsburgh Pirates. Baseball cards were then sold with cigarettes, and Wagner became even more famous when he refused to accept payment for the use of his picture as a protest against the use of tobacco.

30. BUSINESS: Economies In 2012, the United States had the largest economy in the world, $15.7 trillion, with China second, at $8.2 trillion. The U.S. and Chinese economies are predicted to be $15.7e^{0.023x}$ and $8.2e^{0.0634x}$, respectively, x years after 2012. Assuming that these growth rates continue, which economy will be larger in the year 2030?

Source: World Bank, Pricewaterhousecoopers

31. GENERAL: Nuclear Meltdown The probability of a "severe core meltdown accident" at a nuclear reactor in the United States within the next n years is $1 - (0.9997)^{100n}$. Find the probability of a meltdown:

a. within 25 years.　　**b.** within 40 years.

Source: Nuclear Regulatory Commission

(The 1986 core meltdown in the Chernobyl reactor in the Soviet Union spread radiation over much of Eastern Europe, leading to an undetermined number of fatalities.)

32. BIOMEDICAL SCIENCE: Mosquitoes Female mosquitoes (*Culex pipiens*) feed on blood (only the females drink blood) and then lay several hundred eggs. In this way each mosquito can, on the average, breed another 300 mosquitoes in about 9 days. Find the number of great-grandchildren mosquitoes that will be descended from one female mosquito, assuming that all eggs hatch and mature.

33. ENVIRONMENTAL SCIENCE: Light According to the *Bouguer–Lambert Law,* the proportion of light that penetrates ordinary seawater to a depth of x feet is $e^{-0.44x}$. Find the proportion of light that penetrates to a depth of:

a. 3 feet.　　**b.** 10 feet.

34–35. BIOMEDICAL: Drug Dosage If a dosage d of a drug is administered to a patient, the amount of the drug remaining in the tissues t hours later will be $f(t) = de^{-kt}$, where k (the "absorption constant") depends on the drug.

Source: T. R. Harrison, ed., Principles of Internal Medicine

34. For the immunosuppressant cyclosporine, the absorption constant is $k = 0.012$. For a dose of $d = 400$ milligrams, use the previous formula to find the amount of cyclosporine remaining in the tissues after:

a. 24 hours.　　**b.** 48 hours.

35. For the cardioregulator digoxin, the absorption constant is $k = 0.018$. For a dose of $d = 2$ milligrams, use the previous formula to find the amount remaining in the tissues after:

a. 24 hours.　　**b.** 48 hours.

36. BIOMEDICAL: Bacterial Growth A colony of bacteria in a petri dish doubles in size every hour. At noon the petri dish is just covered with bacteria. At what time was the petri dish:

a. 50% covered? [*Hint:* No calculation needed.]

b. 25% covered?

37. BUSINESS: Advertising A company finds that x days after the conclusion of an advertising campaign the daily sales of a new product are $S(x) = 100 + 800e^{-0.2x}$. Find the daily sales 10 days after the end of the advertising campaign.

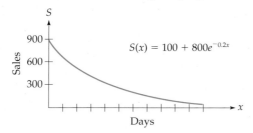

38. BUSINESS: Quality Control A company finds that the proportion of its light bulbs that will burn continuously for longer than t weeks is $e^{-0.01t}$. Find the proportion of bulbs that burn for longer than 10 weeks.

39. GENERAL: Temperature A covered mug of coffee originally at 200 degrees Fahrenheit, if left for t hours in a room whose temperature is 70 degrees, will cool to a temperature of $70 + 130e^{-1.8t}$ degrees. Find the temperature of the coffee after:

  a. 15 minutes.   **b.** half an hour.

40. BEHAVIORAL SCIENCE: Learning In many psychology experiments the percentage of items that are remembered after t time units is

$$p(t) = 100\,\frac{1+e}{1+e^{t+1}}$$

Such curves are called *forgetting curves*. Find the percentage remembered after:

  a. 0 time units.   **b.** 2 time units.

41. BIOMEDICAL: Epidemics The *Reed–Frost* model for the spread of an epidemic predicts that the number I of newly infected people is $I = S(1 - e^{-rx})$, where S is the number of susceptible people, r is the effective contact rate, and x is the number of infectious people. Suppose that a school reports an outbreak of measles with $x = 10$ cases, and that the effective contact rate is $r = 0.01$. If the number of susceptibles is $S = 400$, use the Reed–Frost model to estimate how many students will be newly infected during this stage of the epidemic.

Source: Human Biology **24**

42. SOCIAL SCIENCE: Election Cost The cost of winning a seat in the House of Representatives in recent years has been approximately $1.67e^{0.056x}$ million dollars, where x is the number of years since 2012. Estimate the cost of winning a House seat in the year 2022.

Source: Center for Responsive Politics

43. PERSONAL FINANCE: Rate of Return An investment of $8000 grows to $10,291.73 in 4 years. Find the annual rate of return for annual compounding. [*Hint:* Use $P(1 + r/m)^{mt}$ with $m = 1$ and solve for r (rounded).]

44. PERSONAL FINANCE: Rate of Return An investment of $9000 grows to $10,380.65 in 2 years. Find the annual rate of return for quarterly compounding. [*Hint:* Use $P(1 + r/m)^{mt}$ with $m = 4$ and solve for r (rounded).]

45. ATHLETICS: Olympic Games When the Olympic Games were held near Mexico City in the summer of 1968, many athletes were concerned that the high elevation would affect their performance. If air pressure decreases exponentially by 0.4% for each 100 feet of altitude, by what percentage did the air pressure decrease in moving from Tokyo (the site of the 1964 Summer Olympics, at altitude 30 feet) to Mexico City (altitude 7347 feet)?

Source: CRC Handbook of Chemistry and Physics

46. BIOMEDICAL: Radioactive Contamination The core melt-down and explosions at the nuclear reactor in Chernobyl in 1986 released large amounts of strontium-90, which decays exponentially at the rate of 2.5% per year. Areas downwind of the reactor will be uninhabitable for 100 years. What percent of the original strontium-90 contamination will still be present after:

  a. 50 years?   **b.** 100 years?

Source: Environmental Protection Agency

47. GENERAL: Population The most populous state is California, with Texas second but gaining. According to the Census Bureau, x years after 2012 the population of California will be $38e^{0.01x}$ and the population of Texas will be $26e^{0.019x}$ (both in millions).

  a. Graph these two functions on the window [0, 60] by [0, 80].

  b. In which year is Texas projected to overtake California as the most populous state? [*Hint:* Use INTERSECT.]

Source: U.S. Census Bureau

48. GENERAL: St. Louis Arch The Gateway Arch in St. Louis is built around a mathematical curve called a "cate-nary." The height of this catenary above the ground at a point x feet from the center line is

$$y = 688 - 31.5(e^{0.01033x} + e^{-0.01033x})$$

  a. Graph this curve on the window [−400, 400] by [0, 700].

  b. Find the height of the Gateway Arch at its highest point, using the fact that the top of the arch is 5 feet higher than the top of the central catenary.

Source: National Park Service

49. PERSONAL FINANCE: Payday Loans Unsecured loans intended for short time periods at very high interest rates are called *payday loans*, because they are supposed to be repaid at the next payday. *Wonga*, the largest payday loan maker in Britain, typically charges 1% interest per day. Find Wonga's annual rate of interest. [*Hint*: Find the rate of interest corresponding to 1% per day compounded for 365 days.] [*Note:* Such loans are illegal in the United States.]

Source: New York Times, November 30, 2013

50. PERSONAL FINANCE: Art Appreciation In 2013, Jackson Pollock's drip painting *No. 19, 1948* sold at auction for $58.3 million, its value having increased by 17% annually since its previous sale. Estimate the sale price in 2020.
Source: Christie's

 51. PERSONAL FINANCE: Earnings and Calculus A recent study found that one's earnings are affected by the mathematics courses one has taken. In particular, compared to someone making $40,000 who had taken *no* calculus, a comparable person who had taken *x* years of calculus would be earning $40,000e^{0.195x}$. Find the salary of a person who has taken $x = 2$ years of calculus. [*Note:* Other mathematics courses were included in the study, but calculus courses brought the greatest increase in salary.]
Source: Review of Economics and Statistics **86**

Conceptual Exercises

52. In the box on page 246, why do we require that the base *a* be *positive*? [*Hint:* How would we define $f(x) = a^x$ for $x = 0.5$ if *a* were negative?]

53. Does the graph of a^x for $a > 1$ have a horizontal asymptote? [*Hint:* Look at the graph on page 247.]

54. Does the graph of a^x for $a < 1$ have a horizontal asymptote? [*Hint:* Look at the graph on page 247.]

55. Which will be largest for very large values of *x*: x^2, e^x, or x^{1000}?

56. True or False: If $x < y$, then $e^x < e^y$.

57. What is the most important characteristic of exponential growth?

58. Assuming the same fixed nominal interest rate, put the following types of compounding in order of increasing benefit to the depositor: daily, continuously, quarterly, semiannually.

59. Explain why 5% compounded *continuously* is better than 5% compounded *monthly*.

60. A banker once told one of the authors that continuous compounding means the same thing as daily compounding. Was he right? Explain.

61. A company is considering two ways to depreciate a truck: *straight line* or *by a fixed percentage*. If each way begins with a value of $25,000 and ends 5 years later with a value of $1000, which way will result in the bigger change in value during the *first* year? During the *last* year?

62. Suppose that you have a bank account with interest compounded continuously, but you can't remember the continuously compounded interest rate. If at the end of the year you had 10% more than you began with, was the continuously compounded rate *more* than 10% or *less* than 10%?

4.2   Logarithmic Functions

Introduction

In this section we introduce logarithmic functions, emphasizing the *natural* logarithm function. We then apply natural logarithms to a wide variety of problems, from doubling money under compound interest to calculating proper drug dosage.

Logarithms

The word **logarithm** (abbreviated **log**) means *power* or *exponent*. The number being raised to the power is called the **base** and is written as a subscript. For example, the expression

$$\log_{10} 1000 \qquad \text{Read "log (base 10) of 1000"}$$

Base

means the *exponent* to which we raise 10 to get 1000. Since $10^3 = 1000$, the exponent is 3, so the *logarithm* is 3:

$$\log_{10} 1000 = 3 \qquad \text{Since } 10^3 = 1000$$

We find logarithms by writing them as exponents and then finding the exponent.

| EXAMPLE 1 | FINDING A LOGARITHM |

Evaluate $\log_{10} 100$.

Solution

$\log_{10} 100 = y$ is equivalent to $10^y = 100$ Exponent 2 works, so 2 is the logarithm

The logarithm y is the exponent that solves

Therefore

$$\log_{10} 100 = 2$$ Since $10^2 = 100$

| EXAMPLE 2 | FINDING A LOGARITHM |

Evaluate $\log_{10} \dfrac{1}{10}$.

Solution

$\log_{10} \dfrac{1}{10} = y$ is equivalent to $10^y = \dfrac{1}{10}$ Exponent -1 works, so -1 is the logarithm

The logarithm y is the exponent that solves

Therefore

$$\log_{10} \frac{1}{10} = -1$$ Since $10^{-1} = \dfrac{1}{10}$

PRACTICE PROBLEM 1

Evaluate $\log_{10} 10,000$.

 (not applicable)

Solution on page 269 >

Logarithms to the base 10 are called **common logarithms** and are often written without the base, so that $\log 100$ means $\log_{10} 100$. The [LOG] key on many calculators evaluates common logarithms.

   We may find logarithms to other bases as well. Any positive number other than 1 may be used as a base.

| EXAMPLE 3 | FINDING A LOGARITHM |

Find $\log_9 3$.

Solution

$\log_9 3 = y$ is equivalent to $9^y = 3$ Exponent $\frac{1}{2}$ works, so $\frac{1}{2}$ is the logarithm

Therefore

$$\log_9 3 = \frac{1}{2}$$ Since $9^{1/2} = \sqrt{9} = 3$

In general, for any positive base a other than 1,

Logarithms

$\log_a x = y$ is equivalent to $a^y = x$

$\log_a x$ means the exponent to which we raise a to get x

PRACTICE PROBLEM 2

Find: **a.** $\log_5 125$ **b.** $\log_8 2$ Solutions on page 269 >

Natural Logarithms

The most widely used of all bases is e, the number (approximately 2.718) that we defined on page 251. Logarithms to the base e are called **natural logarithms** or **Napierian logarithms.*** The natural logarithm of x is written $\ln x$ ("n" for "natural") instead of $\log_e x$, and may be found using the 〔 LN 〕 key on a calculator.

Natural Logarithms

$\ln x$ = logarithm of x to the base e

$\ln x$ means $\log_e x$

In words: $\ln x$ is the power to which we raise e to get x.

 Be Careful Since e to any power is positive, $\ln x$ is defined only for *positive values of x.*

PRACTICE PROBLEM 3

Use a calculator to find $\ln 8.34$.   Solution on page 269 >

The following table shows some values of the natural logarithm function $f(x) = \ln x$, and its graph (based on these points) is shown on the right.

x	$y = \ln x$
0.1	-2.30
0.5	-0.69
1	0
2	0.69
5	1.61
10	2.30
15	2.71

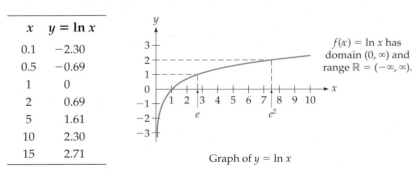

$f(x) = \ln x$ has domain $(0, \infty)$ and range $\mathbb{R} = (-\infty, \infty)$.

Graph of $y = \ln x$

*After John Napier (1550–1617), a Scottish mathematician and, incidentally, the inventor of the decimal point.

Notice that the graph of $\ln x$ is always increasing and has the y-axis as a vertical asymptote: $\lim\limits_{x \to 0^+} \ln x = -\infty$. A definition of natural logarithms without recourse to exponents is given in Exercise 74 on page 342. The natural logarithm function may be used for modeling growth that continually slows.

The following graph shows logarithm functions for several different bases. Notice that each passes through the point (1, 0), since $a^0 = 1$. We will concentrate on the *natural* logarithm function, since it is the one most used in applications.

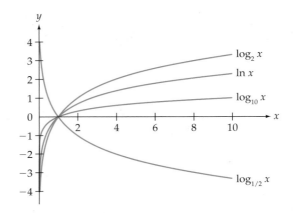

Since logs are exponents, each property of exponents (see pages 21–22) can be restated as a property of logarithms. The first three properties show that some natural logarithms can be found without using a calculator.

Properties of Natural Logarithms

1. $\ln 1 = 0$ The natural log of 1 is 0 (since $e^0 = 1$)

2. $\ln e = 1$ The natural log of e is 1 (since $e^1 = e$)

3. $\ln e^x = x$ The natural log of e to a power is just the power (since $e^x = e^x$)

4. $e^{\ln x} = x$ $x > 0$ e raised to the natural log of a number is just the number

The first two properties are special cases of the third, with $x = 0$ and $x = 1$. Since logarithms are exponents, the third property simply says that the exponent of e that gives e^x is x, which is obvious when you think about it. Since $\ln x$ is the power of e that gives x, raising e to that power must give x, which is the fourth property.

EXAMPLE 4   USING THE PROPERTIES OF NATURAL LOGARITHMS

a. $\ln e^7 = 7$

b. $\ln e^{3x} = 3x$ The natural log of e to a power is just the power

c. $\ln \sqrt[3]{e^2} = \ln e^{2/3} = \frac{2}{3}$

d. $e^{\ln 5} = 5$ e raised to the natural log of 5 is just 5

**Graphing
Calculator
Exploration**

a. Evaluate $\ln e^{17}$ to verify that the answer is 17. Change 17 to other numbers (positive, negative, or zero) in order to verify that $\ln e^x = x$ for any x.

b. Evaluate $e^{\ln 29}$ to verify that the answer is 29. Change the 29 to another positive number to verify that $e^{\ln x} = x$. What about negative numbers, or zero?

The next four properties enable us to simplify logs of products, quotients, and powers. For positive numbers M and N and any number P:

Properties of Natural Logarithms

5. $\ln (M \cdot N) = \ln M + \ln N$

The log of a *product* is the *sum* of the logs

6. $\ln \left(\dfrac{1}{N} \right) = -\ln N$

The log of 1 *over* a number is the *negative* of the log of the number

7. $\ln \left(\dfrac{M}{N} \right) = \ln M - \ln N$

The log of a *quotient* is the *difference* of the logs

8. $\ln (M^P) = P \cdot \ln M$

The log of a number to a power is the power *times* the log

Property 8 can be stated simply: *Logarithms bring down exponents.* A justification for these four properties is given at the end of this section.

EXAMPLE 5   USING THE PROPERTIES OF NATURAL LOGARITHMS

a. $\ln (2 \cdot 3) = \ln 2 + \ln 3$ Property 5

b. $\ln \frac{1}{7} = -\ln 7$ Property 6

c. $\ln \frac{2}{3} = \ln 2 - \ln 3$ Property 7

d. $\ln (2^3) = 3 \ln 2$ $\ln(2^3) = 3 \ln 2$

Properties 1 through 8 are very useful for simplifying functions.

EXAMPLE 6   SIMPLIFYING A FUNCTION

$$f(x) = \ln (2x) - \ln 2$$

$$= \ln 2 + \ln x - \ln 2 \qquad \text{Since } \ln (2x) = \ln 2 + \ln x \text{ by Property 5}$$

$$= \ln x \qquad \text{Canceling}$$

EXAMPLE 7	SIMPLIFYING A FUNCTION

$$f(x) = \ln\left(\frac{x}{e}\right) + 1$$

$$= \ln x - \ln e + 1$$ Since $\ln (x/e) = \ln x - \ln e$ by Property 7

$$= \ln x - 1 + 1$$ Since $\ln e = 1$ by Property 2

$$= \ln x$$ Canceling

EXAMPLE 8	SIMPLIFYING A FUNCTION

$$f(x) = \ln (x^5) - \ln (x^3)$$ Bringing down exponents by Property 8

$$= 5 \ln x - 3 \ln x$$

$$= 2 \ln x$$ Combining

Graphing Calculator Exploration

a. Graph $y = \ln x^2$ on the window $[-5, 5]$ by $[-5, 5]$.

b. Change the function to $y = 2 \ln x$.

Explain why the two graphs are different. (Doesn't Property 8 on the previous page say that $\ln x^2 = 2 \ln x$?)

$y = e^x$ and $y = \ln x$
are inverse functions.

On page 262 we saw the natural logarithm properties $\ln e^x = x$ and $e^{\ln x} = x$. These show that the natural logarithm function $\ln x$ and the exponential function e^x are *inverse functions* in that either one "undoes" or "inverts" the other. (Notationally, e^x raises x up to the exponent, and logarithms bring down exponents, so it is reasonable that they "undo" each other.) This inverse relationship between e^x and $\ln x$ is equivalent to the geometric fact that their graphs are reflections of each other in the diagonal line $y = x$,   as shown on the left.

Doubling Under Compound Interest

How soon will money invested at compound interest double in value? The solution to this question makes important use of the property $\ln (M^P) = P \cdot \ln M$ ("logs bring down exponents").

EXAMPLE 9   FINDING DOUBLING TIME

A sum is invested at 12% interest compounded quarterly. How soon will it double in value?

Solution

We use the formula $P(1 + r/m)^{mt}$ with $r = 0.12$ and $m = 4$. Since double P dollars is $2P$ dollars, we want to solve

$$P\left(1 + \underbrace{\frac{0.12}{4}}_{0.03}\right)^{4t} = 2P$$ t is the number of years

P dollars doubled

$$1.03^{4t} = 2$$ Simplifying $1 + 0.03$ to 1.03 and canceling the P's

The variable is in the *exponent*, so we take logarithms to bring it down.

$$\ln(1.03^{4t}) = \ln 2$$ Taking natural logs of both sides

$$4t \cdot \ln 1.03 = \ln 2$$ Bringing down the exponent (Property 8 of logarithms)

$$t = \frac{\ln 2}{4 \cdot \ln 1.03}$$ Dividing each side by 4 and by $\ln 1.03$ to solve for t

$$t = \frac{0.6931}{4 \cdot 0.0296} \approx 5.9$$ Using a calculator to evaluate the logs and the rest of the expression

A sum at 12% compounded quarterly doubles in about 5.9 years.

Of course, for quarterly compounding you will not get the $2P$ dollars until the end of the last quarter, by which time it will be slightly more.

Notice that the principal P canceled out after the first step, showing that *any* sum will double in the same amount of time: 1 dollar will double into 2 dollars in exactly the same time that a million dollars will double into 2 million dollars.

To find how soon the value *triples*, we would solve $P(1 + r/m)^{mt} = 3P$, and similarly for any other multiple. To find how soon the value would *increase by 50%* (that is, become 1.5 times its original value), we would solve

$$P\left(1 + \frac{r}{m}\right)^{mt} = 1.5P$$ Multiplying by 1.5 increases P by 50%

PRACTICE PROBLEM 4

A sum is invested at 10% interest compounded semiannually (twice a year). How soon will it increase by 60%?

Solution on page 269 >

EXAMPLE 10   **FINDING TRIPLING TIME WITH CONTINUOUS COMPOUNDING**

A sum is invested at 15% compounded continuously. How soon will it triple?

Solution

The idea is the same as before, but for continuous compounding we use the formula Pe^{rt}:

$$Pe^{0.15t} = 3P$$ $3P$ since the value triples

$$e^{0.15t} = 3$$ Canceling the P's

$$\ln e^{0.15t} = \ln 3$$ Taking the natural log of both sides
$$\underbrace{\qquad}_{0.15t}$$

$$0.15t = \ln 3$$ Since the natural log of e to a power is just the power

$$t = \frac{\ln 3}{0.15}$$ Dividing by 0.15 to solve for t

$$\approx \frac{1.0986}{0.15} \approx 7.3$$ Using a calculator

A sum at 15% compounded continuously triples in about 7.3 years.

Drug Dosage

Drug concentration
in the blood

The amount of a drug that remains in a person's bloodstream decreases exponentially with time. If the initial concentration is c (milligrams per milliliter of blood), the concentration t hours later will be

$$C(t) = ce^{-kt}$$

where the "absorption constant" k measures how rapidly the drug is absorbed.

Every medicine has a minimum concentration below which it is not effective. When the concentration falls to this level, another dose should be administered. If doses are administered regularly every T hours over a period of time, the concentration will look like the following graph.

Drug concentration with repeated doses

The problem is to determine the time T at which the dose should be repeated so as to maintain an effective concentration.

EXAMPLE 11   **CALCULATING DRUG DOSAGE**

The absorption constant for penicillin is $k = 0.11$, and the minimum effective concentration is 2 milligrams per milliliter. If the original concentration is $c = 5$, find when another dose should be administered in order to maintain an effective concentration.

Solution

The concentration formula $C(t) = ce^{-kt}$ with $c = 5$ and $k = 0.11$ is

$$C(t) = 5e^{-0.11t}$$

To find the time when this concentration reaches the minimum effective level of 2, we solve

$$5e^{-0.11t} = 2$$

$$e^{-0.11t} = 0.4 \qquad \text{Dividing by 5}$$

$$\underbrace{\ln e^{-0.11t}}_{-0.11t} = \ln 0.4 \qquad \begin{array}{l}\text{Taking natural logs to} \\ \text{bring down the exponent}\end{array}$$

$$-0.11t = \ln 0.4$$

$$t = \frac{\ln 0.4}{-0.11} \approx \frac{-0.9163}{-0.11} \approx 8.3 \qquad \begin{array}{l}\text{Solving for } t \text{ and} \\ \text{using a calculator}\end{array}$$

The concentration will reach the minimum effective level in 8.3 hours, so another dose should be taken approximately every 8 hours.

The last two examples have led to equations of the form $e^{at} = b$. Such equations occur frequently in applications and are solved by taking the natural log of each side to bring down the exponent. Such equations may also be solved on a graphing calculator by graphing the function as y_1 and the constant as y_2 and using INTERSECT to find where they meet.

Carbon-14 Dating

As discussed at the beginning of this chapter, all living things absorb small amounts of radioactive carbon-14 from the atmosphere. When they die, the carbon-14 stops being absorbed and decays exponentially into ordinary carbon. Therefore, the proportion of carbon-14 still present in a fossil or other ancient remain can be used to estimate how old it is. The proportion of the original carbon-14 that will be present after t years is

$$\left(\begin{array}{c}\text{Proportion of carbon-14} \\ \text{remaining after } t \text{ years}\end{array}\right) = e^{-0.00012t}$$

EXAMPLE 12   **DATING BY CARBON-14**

The Dead Sea Scrolls, discovered in a cave near the Dead Sea in what was then Jordan, are among the earliest documents of Western civilization. Estimate the age of the Dead Sea Scrolls if the animal skins on which some were written contain 78% of their original carbon-14.

Solution

The proportion of carbon-14 remaining after t years is $e^{-0.00012t}$. We equate this formula to the actual proportion (expressed as a decimal):

$$e^{-0.00012t} = 0.78 \qquad \text{Equating the proportions}$$
$$\ln e^{-0.00012t} = \ln 0.78 \qquad \text{Taking natural logs}$$
$$-0.00012t = \ln 0.78 \qquad \begin{array}{l}\ln e^{-0.00012t} = -0.00012t \\ \text{by Property 3}\end{array}$$
$$t = \frac{\ln 0.78}{-0.00012} \approx \frac{-0.24846}{-0.00012} \approx 2071 \qquad \begin{array}{l}\text{Solving for } t \text{ and} \\ \text{using a calculator}\end{array}$$

Therefore, the Dead Sea Scrolls are approximately 2070 years old.

Graphing Calculator Exploration

Solve Example 12 by graphing $y_1 = e^{-0.00012x}$ and $y_2 = 0.78$ on the window [0, 4000] by [−0.3, 1] and using INTERSECT to find where they meet. (Solving with a graphing calculator requires choosing an appropriate window, a difficulty that does not occur when solving by logarithms.)

Behavioral Science: Learning Theory

Your ability to do a task generally improves with practice. Frequently, your skill after t units of practice is given by a function of the form

$$S(t) = c(1 - e^{-kt})$$

where c and k are positive constants.

EXAMPLE 13   **ESTIMATING LEARNING TIME**

After t weeks of training, your secretary can type

$$S(t) = 100(1 - e^{-0.25t})$$

words per minute. How many weeks will he take to reach 80 words per minute?

Solution

We solve for t in the following equation:

$$100(1 - e^{-0.25t}) = 80 \qquad \text{Setting } S(t) \text{ equal to 80}$$

$$1 - e^{-0.25t} = 0.80 \qquad \text{Dividing by 100}$$

$$-e^{-0.25t} = -0.20 \qquad \text{Subtracting 1}$$

$$e^{-0.25t} = 0.20 \qquad \text{Multiplying by } -1$$

$$-0.25t = \ln 0.20 \qquad \text{Taking natural logs}$$

$$t = \frac{\ln 0.20}{-0.25} \qquad \text{Solving for } t$$

$$\approx \frac{-1.6094}{-0.25} \approx 6.4 \qquad \text{Using a calculator}$$

He will reach 80 words per minute in about $6\frac{1}{2}$ weeks.

Social Science: Diffusion of Information by Mass Media

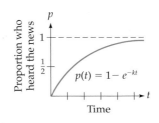

When a news bulletin is repeatedly broadcast over radio and television, the proportion of people who hear the bulletin within t hours is

$$p(t) = 1 - e^{-kt}$$

for some constant k.

EXAMPLE 14   PREDICTING THE SPREAD OF INFORMATION

A storm warning is broadcast, and the proportion of people who hear the bulletin within t hours of its first broadcast is $p(t) = 1 - e^{-0.30t}$. When will 75% of the people have heard the bulletin?

Solution

Equating the proportions gives $1 - e^{-0.30t} = 0.75$. Solving this equation as in the previous examples (omitting the details) gives $t \approx 4.6$. Therefore, it takes about $4\frac{1}{2}$ hours for 75% of the people to hear the news.

Solutions TO PRACTICE PROBLEMS

1. $\log_{10} 10{,}000 = 4$ \qquad Since $10^4 = 10{,}000$

2. a. $\log_5 125 = 3$ \qquad Since $5^3 = 125$

   b. $\log_8 2 = \frac{1}{3}$ \qquad Since $8^{1/3} = \sqrt[3]{8} = 2$

3. $\ln 8.34 \approx 2.121$ (rounded)

4. $P\left(1 + \dfrac{0.10}{2}\right)^{2t} = 1.60 \cdot P$

$$1.05^{2t} = 1.6$$

$$\ln(1.05^{2t}) = \ln 1.6$$

$$2t \cdot \ln 1.05 = \ln 1.6$$

$$t = \frac{\ln 1.6}{2 \cdot \ln 1.05} \approx 4.8$$

In about 4.8 years.

4.2	**Section Summary**

We define logarithms as follows:

$$\log_a x = y \text{ is equivalent to } a^y = x \qquad \text{Logarithm to the base } a$$

$$\ln x = y \text{ is equivalent to } e^y = x \qquad \text{Natural logarithm (base } e)$$

Natural logarithms have the following properties, which are also listed inside the back cover.

1. $\ln 1 = 0$ **2.** $\ln e = 1$

3. $\ln e^x = x$ **4.** $e^{\ln x} = x$

5. $\ln (M \cdot N) = \ln M + \ln N$ **6.** $\ln \left(\dfrac{1}{N}\right) = -\ln N$

7. $\ln \left(\dfrac{M}{N}\right) = \ln M - \ln N$ **8.** $\ln (M^P) = P \cdot \ln M$

Justification of the Properties of Logarithms

Properties 1–4 were justified on page 262.

Property 5 follows from the addition property of exponents, $e^x \cdot e^y = e^{x+y}$ ("the exponent of a product is the sum of the exponents"). Since logs are exponents, this can be restated "the log of a product is the sum of the logs," which is Property 5.

Property 7 follows from the subtraction property of exponents, $e^x/e^y = e^{x-y}$ ("the exponent of a quotient is the difference of the exponents"). This translates into "the log of a quotient is the difference of the logs," which is Property 7.

Property 6 is just a special case of Property 7 (with $M = 1$), along with Property 1.

Property 8 comes from the property of exponents, $(e^x)^y = e^{x \cdot y}$, which says that the exponent y can be "brought down" and multiplied by the x. Since logs are exponents, this says that in $\ln (M^P)$ the exponent P can be brought down and multiplied by the logarithm, which gives $P \cdot \ln M$, and this is just Property 8.

Each property of logs is equivalent to a property of exponents:

Logarithmic Property	*Exponential Property*
$\ln 1 = 0$	$e^0 = 1$
$\ln (M \cdot N) = \ln M + \ln N$	$e^x \cdot e^y = e^{x+y}$
$\ln \left(\dfrac{1}{N}\right) = -\ln N$	$\dfrac{1}{e^y} = e^{-y}$
$\ln \left(\dfrac{M}{N}\right) = \ln M - \ln N$	$\dfrac{e^x}{e^y} = e^{x-y}$
$\ln M^P = P \cdot \ln M$	$(e^x)^y = e^{xy}$

4.2   Exercises

1–4. Find each logarithm *without* using a calculator or tables.

1. a. $\log_5 25$
   b. $\log_3 81$
   c. $\log_3 \frac{1}{3}$
   d. $\log_3 \frac{1}{9}$
   e. $\log_4 2$
   f. $\log_4 \frac{1}{2}$

2. a. $\log_3 27$
   b. $\log_2 16$
   c. $\log_{16} 4$
   d. $\log_4 \frac{1}{4}$
   e. $\log_2 \frac{1}{8}$
   f. $\log_9 \frac{1}{3}$

3. a. $\ln (e^{10})$
   b. $\ln \sqrt{e}$
   c. $\ln \sqrt[3]{e^4}$
   d. $\ln 1$
   e. $\ln (\ln (e^e))$
   f. $\ln \left(\dfrac{1}{e^3} \right)$

4. a. $\ln (e^{-5})$
   b. $\ln e$
   c. $\ln \sqrt[3]{e}$
   d. $\ln \sqrt{e^5}$
   e. $\ln \left(\dfrac{1}{e} \right)$
   f. $\ln (\ln e)$

> ✎ **FOR HELP GETTING STARTED**
> with Exercises 1–4, see Examples 1–4.

5–14. Use the properties of natural logarithms to simplify each function.

5. $f(x) = \ln (9x) - \ln 9$

6. $f(x) = \ln \left(\dfrac{x}{2} \right) + \ln 2$

7. $f(x) = \ln (x^3) - \ln x$
  8. $f(x) = \ln (4x) - \ln 4$

9. $f(x) = \ln \left(\dfrac{x}{4} \right) + \ln 4$
  10. $f(x) = \ln (x^5) - 3 \ln x$

11. $f(x) = \ln (e^{5x}) - 2x - \ln 1$

12. $f(x) = \ln (e^{-2x}) + 3x + \ln 1$

13. $f(x) = 8x - e^{\ln x}$
  14. $f(x) = e^{\ln x} + \ln (e^{-x})$

> ✎ **FOR HELP GETTING STARTED**
> with Exercises 5–14, see Examples 5–8.

15–16. Find the domain and range of each function:

15. $f(x) = \ln (x^2 - 1)$
  16. $f(x) = \ln (1 - x^2)$

Applied Exercises

17. PERSONAL FINANCE: Interest An investment grows at 24% compounded monthly. How many years will it take to:
   a. double?
   b. increase by 50%?

18. PERSONAL FINANCE: Interest An investment grows at 36% compounded monthly. How many years will it take to:
   a. double?
   b. increase by 50%?

19. PERSONAL FINANCE: Interest A bank offers 7% compounded continuously. How soon will a deposit:
   a. triple?
   b. increase by 25%?

20. PERSONAL FINANCE: Interest A bank offers 6% compounded continuously. How soon will a deposit:
   a. quadruple?
   b. increase by 75%?

21. PERSONAL FINANCE: Depreciation An automobile depreciates by 30% per year. How soon will it be worth only half its original value? [*Hint:* Depreciation is like interest but at a negative rate *r*.]

22. GENERAL: Population Growth From 2010 to 2012, the population of North Dakota (the fastest growing state) increased by 4% annually. Assuming that this trend continues, in how many years will the population double?
Source: U.S. Census Bureau

23. ECONOMICS: Energy Output The world's output of primary energy (petroleum, natural gas, coal, hydroelectricity, and nuclear electricity) is increasing at the rate of 2.4% annually. How long will it take to increase by 50%?
Source: Worldwatch

24. PERSONAL FINANCE: Art Appreciation In 2010, Pablo Picasso's painting *Nude, Green Leaves and Bust* sold for $106.5 million, shattering the record for an auctioned painting, having increased in value by 16% annually. At this rate, how long will it take for the painting to double in value? [*Note:* Picasso painted it in less than a day.]
Source: New York Times

25–26. BUSINESS: Advertising After *t* days of advertisements for a new laundry detergent, the proportion of shoppers in a town who have seen the ads is $1 - e^{-0.03t}$. How long must the ads run to reach:

25. 90% of the shoppers?
  26. 99% of the shoppers?

27. BEHAVIORAL SCIENCE: Forgetting The proportion of students in a psychology experiment who could remember an eight-digit number correctly for *t* minutes was $0.9 - 0.2 \ln t$ (for $t > 1$). Find the proportion that remembered the number for 5 minutes.

28. SOCIAL SCIENCE: Diffusion of Information by Mass Media Election returns are broadcast in a town of 1 million people, and the number of people who have heard the news within *t* hours is $1{,}000{,}000 \, (1 - e^{-0.4t})$. How long will it take for 900,000 people to hear the news?

29. BEHAVIORAL SCIENCE: Learning After *t* weeks of practice, a typing student can type $100(1 - e^{-0.4t})$ words per minute (wpm). How soon will the student type 80 wpm?

30. BIOMEDICAL: Drug Dose If the original concentration of a drug in a patient's bloodstream is *c* (milligrams per liter), *t* hours later the concentration will be $C(t) = ce^{-kt}$, where *k* is the absorption constant. If the original concentration of the asthma medication theophylline is $c = 20$ and the absorption constant is $k = 0.23$, when should the drug be readministered so that the concentration does not fall below the minimum effective concentration of 5?

Source: T. R. Harrison, ed., Principles of Internal Medicine

31–32. GENERAL: Carbon-14 Dating The proportion of carbon-14 still present in a sample after t years is $e^{-0.00012t}$.

Source: A. Dickin, Radiogenic Isotope Geology

31. Estimate the age of the cave paintings discovered in the Ardéche region of France if the carbon with which they were drawn contains only 2.3% of its original carbon-14. They are the oldest known paintings in the world.

32. Estimate the age of the Shroud of Turin, believed by many to be the burial cloth of Christ (see the Application Preview on page 245), from the fact that its linen fibers contained 92.3% of their original carbon-14.

33–34. GENERAL: Potassium-40 Dating The radioactive isotope potassium-40 is used to date very old remains. The proportion of potassium-40 that remains after t million years is $e^{-0.00054t}$. Use this function to estimate the age of the following fossils.

Source: A. Dickin, Radiogenic Isotope Geology

33. **DATING OLDER WOMEN** The most complete skeleton of an early human ancestor ever found, dubbed *Ardi*, was discovered in Ethiopia. Use the formula to estimate Ardi's age if rocks near the remains contained 99.76% of their original potassium-40.

34. **LUCY** Use the formula to estimate the age of the partial skeleton of *Australopithecus afarensis* (known as "Lucy") that was found in Ethiopia if rocks near it had 99.763% of their original potassium-40.

Robert Preston Photography/Alamy

35. **ENVIRONMENTAL SCIENCE:** Radioactive Waste Hospitals use radioactive tracers in many medical tests. After the tracer is used, it must be stored as radioactive waste until its radioactivity has decreased enough for it to be disposed of as ordinary chemical waste. For the radioactive isotope iodine-131, the proportion of radioactivity remaining after t days is $e^{-0.087t}$. How soon will the

proportion of radioactivity decrease to 0.001 so that it can be disposed of as ordinary chemical waste?

Source: New York Times

36. **ENVIRONMENTAL SCIENCE:** Rain Forests It has been estimated that the world's tropical rain forests are disappearing at the rate of 1.8% per year. If this rate continues, how soon will the rain forests be reduced to 50% of their present size? (Rain forests not only generate much of the oxygen that we breathe but also contain plants with unique medical properties, such as the rosy periwinkle which has revolutionized the treatment of leukemia.)

Source: Science 253

37–38. SOCIAL SCIENCE: Education and Income According to a study, each additional year of education increases one's income by 16%. Therefore, with x extra years of education, your income will be multiplied by a factor of 1.16^x.

Source: American Economic Review 84

37. How many additional years of education are required to *double* your income? That is, find the x that satisfies $1.16^x = 2$.

38. How many additional years of education are required to increase your income by 50%? That is, find the x that satisfies $1.16^x = 1.5$.

39–47. Solve the following exercises on a graphing calculator by graphing an appropriate exponential function (using x for ease of entry) together with a constant function and using INTERSECT to find where they meet. You will have to choose an appropriate window.

39. **GENERAL:** Inflation At 2% inflation, prices increase by 2% compounded annually. How soon will prices:
    a. double?   b. triple?

40. **GENERAL:** Inflation At 5% inflation, prices increase by 5% compounded annually. How soon will prices:
    a. double?   b. triple?

41. **PERSONAL FINANCE:** Interest A bank account grows at 6% compounded quarterly. How many years will it take to:
    a. double?   b. increase by 50%?

42. **PERSONAL FINANCE:** Interest A bank account grows at 7% compounded continuously. How many years will it take to:
    a. double?   b. increase by 25%?

43. **BUSINESS:** Advertising After a sale has been advertised for t days, the proportion of shoppers in a city who have seen the ad is $1 - e^{-0.08t}$. How long must the ad run to reach:
    a. 50% of the shoppers?
    b. 60% of the shoppers?

44. **BIOMEDICAL:** Drug Dosage If the original concentration of a drug in a patient's bloodstream is 5 (milligrams per milliliter), and if the absorption constant is 0.15, then t hours later the concentration will be $5e^{-0.15t}$. When

should the drug be readministered so that the concentration does not fall below the minimum effective concentration of 2.7?

45. GENERAL: Carbon-14 Dating In 1991 two hikers in the Italian Alps found the frozen but well-preserved body of the most ancient human ever found, dubbed "Iceman." Estimate the age of Iceman if his grass cape contained 53% of its original carbon-14. (Use the carbon-14 decay function stated in Exercises 31–32.)

46. GENERAL: Potassium-40 Dating Estimate the age of the oldest known dinosaur, a dog-sized creature called *Herrerasaurus* found in Argentina in 1988, if volcanic material found with it contained 88.4% of its original potassium-40. (Use the potassium-40 decay function given in Exercises 33–34.)

47. ENVIRONMENTAL SCIENCE: Nuclear Waste More than 70 years after the beginning of the nuclear age, we do not have a safe or permanent way to dispose of long-lived radioactive waste. Among the most hazardous radioactive waste is irradiated fuel from nuclear power plants, totaling more than 800,000 tons in 2010

and growing by 11.3% annually. At this rate, how long will it take for this amount to double?

Source: Worldwatch

48. BUSINESS: Satellite Radio Between 2009 and 2010, the number of satellite radio subscribers in the United States increased by 9%. At that rate, when will the number increase by 50%?

Source: Serius XM Radio

49. BUSINESS: Bonds DSG International recently sold five-year $1000 bonds with an annual yield of 9.25%. After how much time could they be sold for twice their original price? Give your answer in years and months.

Source: Bloomberg BusinessWeek

50. BUSINESS: Bonds General Electric recently sold $1000 bonds maturing in 30 years with an annual yield of 4.125%. After how much time could they be sold for twice their original price? Give your answer in years and months.

Source: Wall Street Journal

Conceptual Exercises

51–54. In each pair of equations, one is true and one is false. Choose the correct one.

51. $\ln 1 = 0$ or $\ln 0 = 1$.

52. $\ln (x + y) = \ln x \cdot \ln y$ or $\ln (x \cdot y) = \ln x + \ln y$.

53. $\dfrac{\ln x}{\ln y} = \ln (x - y)$ or $\ln \dfrac{x}{y} = \ln x - \ln y$.

54. $(\ln x)^n = n \cdot \ln x$ or $\ln (x^n) = n \cdot \ln x$.

55. If $y = 10^x$, then $x =$?

56. If $y = e^x$, then $x =$?

57. Why can't we define logs of negative numbers, such as $\ln(-2)$? [*Hint:* If $\ln(-2) = x$, what is the equivalent exponential statement? What is the sign of e^x?]

58. Why can't we define the logarithm of zero? [*Hint:* If $\ln 0 = x$, what is the equivalent exponential statement? What is the sign of e^x?]

59. Which type of compounding would give the shortest doubling time for a fixed interest rate: *daily, continuous,* or *annual*? Which would give the longest?

60. Recall that the concentration of a drug in the bloodstream after t hours is ce^{-kt}, where k is called the "absorbtion constant." If one drug has a larger absorbtion constant than another, will it require *more* or *less* time between doses? (Assume that both drugs have the same value of c.)

Explorations and Excursions   The following problems extend and augment the material presented in the text.

61. BIOMEDICAL: Half-Life of a Drug The time required for the amount of a drug in one's body to decrease by half is called the "half-life" of the drug.

  a. For a drug with absorption constant k, derive the following formula for its half-life:

$$\left(\begin{matrix} \text{Half-} \\ \text{life} \end{matrix} \right) = \frac{\ln 2}{k}$$

  [*Hint:* Solve the equation $e^{-kt} = \frac{1}{2}$ for t and use the properties of logarithms.]

  b. Find the half-life of the cardioregulator digoxin if its absorption constant is $k = 0.018$ and time is measured in hours.

  Source: T. R. Harrison, ed., *Principles of Internal Medicine*

62. BIOMEDICAL: Population Growth The *Gompertz growth curve* models the size $N(t)$ of a population at time $t \geq 0$ as $N(t) = Ke^{-ae^{-bt}}$ where K and b are positive constants. Show that if $N_0 = N(0)$ is the *initial population* at time $t = 0$, then $a = \ln(K/N_0)$.

  Source: F. Brauer and C. Castillo-Chávez, *Mathematical Models in Population Biology and Epidemiology*

63. BIOMEDICAL: Heterozygosity The frequency $f(x)$ of heterozygotes (averaged across many populations) after x generations is given by $f(x) = H_0\left(1 - \dfrac{1}{2N}\right)^x$ where H_0 is the initial heterozygosity and N is the number of individuals in the population. For a population of size $N = 500$, how many generations are needed to reduce the frequency by 6%?

Source: R. Frankham, J. Ballou, and D. Briscoe, Introduction to Conservation Genetics

64. BIOMEDICAL: Ricker Recruitment The population dynamics of many fish (such as salmon) can be described by the *Ricker curve* $y = axe^{-bx}$ for $x \geq 0$ where $a > 1$ and $b > 0$ are constants, x is the size of the parental stock, and y is the number of recruits (offspring). Determine the size of the equilibrium population for which $y = x$.

*Source: J. Fish. Res. Board Can. **11***

65–66. BUSINESS: Rule of 72 If an amount is invested at interest rate r compounded continuously, the doubling time (the time in which it will double in value) is found by solving the equation $Pe^{rt} = 2P$. The solution (by the usual method of canceling the P and taking logs) is $t = \dfrac{\ln 2}{r} \approx \dfrac{0.69}{r}$. For *annual* compounding, the doubling time should be somewhat longer, and may be estimated by replacing 69 by 72.

> **Rule of 72**
>
> For r% interest compounded annually, the doubling time is approximately $\dfrac{72}{r}$ years.

For example, to estimate the doubling time for an investment at 8% compounded annually we would divide 72 by 8, giving $\frac{72}{8} = 9$ years. The 72, however, is only a rough "upward adjustment" of 69, and the rule is most accurate for interest rates around 9%. For each interest rate:

a. Use the rule of 72 to estimate the doubling time for annual compounding.

b. Use the compound interest formula $P(1 + r)^t$ to find the actual doubling time for annual compounding.

65. 6%   (*See instructions above.*)

66. 1% (This shows that for interest rates very different from 9% the rule of 72 is less accurate.)

67. ECONOMICS: Interest-Growth Times Find a formula for the time required for an investment to grow to k times its original size if it grows at interest rate r compounded annually.

68. ECONOMICS: Interest-Growth Times Find a formula for the time required for an investment to grow to k times its original size if it grows at interest rate r compounded continuously.

4.3   Differentiation of Logarithmic and Exponential Functions

Introduction

You may have noticed that Sections 4.1 and 4.2 contained no calculus. Their purpose was to develop the properties of logarithmic and exponential functions. In this section we differentiate these new functions and use their derivatives for graphing, optimization, and finding rates of change. We emphasize *natural* (base e) logs and exponentials, since most applications use these exclusively. Verifications of the differentiation rules are given at the end of the section.

Derivatives of Logarithmic Functions

The rule for differentiating the natural logarithm function is as follows:

> **Derivative of   ln *x***
>
> $$\frac{d}{dx} \ln x = \frac{1}{x}$$
>
> The derivative of $\ln x$ is 1 over x

EXAMPLE 1 DIFFERENTIATING A LOGARITHMIC FUNCTION

Differentiate $f(x) = x^3 \ln x$.

Solution

The function is a *product*, x^3 times $\ln x$, so we use the Product Rule.

$$\frac{d}{dx}(x^3 \ln x) = 3x^2 \ln x + x^3 \frac{1}{x} = 3x^2 \ln x + x^2$$

Derivative of the first   Second left alone   First left alone   Derivative of $\ln x$ From $x^3 \frac{1}{x} = x^2$

PRACTICE PROBLEM 1

Differentiate $f(x) = \dfrac{\ln x}{x}$.

Solution on page 283 >

LOOKING BACK

The Chain Rule was stated on page 141.

The preceding rule, together with the Chain Rule, shows how to differentiate the natural logarithm of a *function*. For any differentiable function $f(x)$ that is positive:

Derivative of $\ln f(x)$

$$\frac{d}{dx} \ln f(x) = \frac{f'(x)}{f(x)}$$

The derivative of the natural log of a function is the derivative of the function over the function

Notice that the right-hand side does not involve logarithms at all.

EXAMPLE 2 DIFFERENTIATING A LOGARITHMIC FUNCTION

$$\frac{d}{dx} \ln (x^2 + 1) = \frac{2x}{x^2 + 1}$$

← Derivative of $x^2 + 1$
← Original function (without the ln)

As we observed, the answer does not involve logarithms.

PRACTICE PROBLEM 2

Find $\dfrac{d}{dx} \ln (x^3 - 5x + 1)$.

Solution on page 283 >

EXAMPLE 3   DIFFERENTIATING A LOGARITHMIC FUNCTION

Find the derivative of $f(x) = \ln (x^4 - 1)^3$.

Solution

We need the rule for differentiating the natural logarithm of a function, together with the Generalized Power Rule [for differentiating $(x^4 - 1)^3$].

$$\frac{d}{dx} \ln (x^4 - 1)^3 = \frac{\dfrac{d}{dx}(x^4 - 1)^3}{(x^4 - 1)^3} \qquad \text{Using } \frac{d}{dx} \ln f = \frac{f'}{f}$$

$$= \frac{3(x^4 - 1)^2 4x^3}{(x^4 - 1)^3} \qquad \text{Using the Generalized Power Rule}$$

$$= \frac{12x^3}{x^4 - 1} \qquad \text{Dividing top and bottom by } (x^4 - 1)^2$$

Alternative Solution   It is easier if we simplify first, using Property 8 of logarithms (see the inside back cover) to bring down the exponent 3:

$$\ln (x^4 - 1)^3 = 3 \ln (x^4 - 1) \qquad \text{Using } \ln (M^P) = P \cdot \ln M$$

Now we differentiate the simplified expression:

$$\frac{d}{dx} 3 \ln (x^4 - 1) = 3 \frac{4x^3}{x^4 - 1} = \frac{12x^3}{x^4 - 1} \qquad \text{Same answer as before}$$

FOR MORE HELP

with simplifying expressions, see the Algebra Review appendix, pages B13–B14

LOOKING BACK

The properties of logarithms were stated on pages 262–263.

Moral: Changing $\ln (\cdots)^n$ to $n \ln (\cdots)$ simplifies differentiation.

Derivatives of Exponential Functions

The rule for differentiating the exponential function e^x is as follows:

Derivative of e^x

$$\frac{d}{dx} e^x = e^x \qquad \text{The derivative of } e^x \text{ is simply } e^x$$

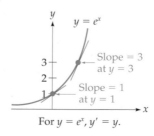

For $y = e^x$, $y' = y$.

This shows the rather surprising fact that e^x is its own derivative. Stated another way, the function e^x is unchanged by the operation of differentiation.

This rule can be interpreted graphically: If $y = e^x$, then $y' = e^x$, so that $y = y'$. This means that on the graph of $y = e^x$, the slope y' always equals the y-coordinate, as shown in the graph on the left. Since y' and y'' both equal e^x, they are always positive and the graph is always increasing and concave upwards.

Graphing Calculator Exploration

e^x or $\frac{d}{dx} e^x$?

a. Define $y_1 = e^x$ and y_2 as the derivative of y_1 (using NDERIV) and graph them together on the window $[-3, 3]$ by $[-1, 10]$.

b. Why does the screen show only one curve?

c. Use TRACE to compare the values of y_1 and y_2 at some chosen x-value. Do the y-values agree *exactly*? If not, explain the slight discrepancy. [*Hint:* Is NDERIV really the derivative?]

EXAMPLE 4   **FINDING A DERIVATIVE INVOLVING e^x**

Find $\dfrac{d}{dx}\left(\dfrac{e^x}{x}\right)$.

Solution

Since the function is a quotient, we use the Quotient Rule:

Derivative of e^x
Derivative of x

$$\frac{d}{dx}\left(\frac{e^x}{x}\right) = \frac{x \cdot e^x - 1 \cdot e^x}{x^2} = \frac{xe^x - e^x}{x^2}$$

EXAMPLE 5   **EVALUATING A DERIVATIVE INVOLVING e^x**

If $f(x) = x^2 e^x$, find $f'(1)$.

Solution

$$f'(x) = 2xe^x + x^2 e^x \qquad \text{Using the Product Rule on } x^2 \cdot e^x$$

$$f'(1) = 2(1)e^1 + (1)^2 e^1 \qquad \text{Substituting } x = 1$$

$$= 2e + e = 3e \qquad \text{Simplifying}$$

In these problems we leave our answers in their "exact" forms, leaving e as e. Later, in applied problems, we will approximate our answers using $e \approx 2.718$ or a calculator.

PRACTICE PROBLEM 3

If $f(x) = xe^x$, find $f'(1)$.

Solution on page 283 >

The rule for differentiating e^x, together with the Chain Rule, shows how to differentiate $e^{f(x)}$. For any differentiable function $f(x)$:

Derivative of $e^{f(x)}$

$$\frac{d}{dx}e^{f(x)} = e^{f(x)} \cdot f'(x)$$

The derivative of e to a function is e to the function times the derivative of the function

That is, to differentiate $e^{f(x)}$ we simply "copy" the original $e^{f(x)}$ and then multiply by the derivative of the exponent.

EXAMPLE 6   DIFFERENTIATING AN EXPONENTIAL FUNCTION

$$\frac{d}{dx}e^{x^4+1} = e^{x^4+1}(4x^3) = 4x^3e^{x^4+1} \qquad \text{Reversing the order}$$

$\underbrace{}_{\text{Copied}}$ Derivative of the exponent

EXAMPLE 7   DIFFERENTIATING AN EXPONENTIAL FUNCTION

Find $\dfrac{d}{dx}e^{x^2/2}$.

Solution

The exponent $x^2/2$ should first be rewritten as $\frac{1}{2}x^2$, a constant times x to a power, since then its derivative is easily seen to be x.

$$\frac{d}{dx}e^{x^2/2} = \frac{d}{dx}e^{\frac{1}{2}x^2} = e^{\frac{1}{2}x^2}(x) = xe^{\frac{1}{2}x^2} = xe^{x^2/2}$$

Rewriting the exponent in its original form

Derivative of the exponent

PRACTICE PROBLEM 4

Find $\dfrac{d}{dx}e^{1+x^3/3}$

Solution on page 283 >

The formulas for differentiating natural logarithmic and exponential functions are summarized as follows, with $f(x)$ written simply as f.

Logarithmic Formulas	*Exponential Formulas*	
$\dfrac{d}{dx}\ln x = \dfrac{1}{x}$	$\dfrac{d}{dx}e^x = e^x$	Top formulas apply only to $\ln x$ and e^x
$\dfrac{d}{dx}\ln f = \dfrac{f'}{f}$	$\dfrac{d}{dx}e^f = e^f \cdot f'$	Bottom formulas apply to $\ln$ and e of a *function*

Be Careful   Do *not* take the derivative of e^x by the Power Rule,

$$\frac{d}{dx} x^n = nx^{n-1}$$

The Power Rule applies to x^n, *a variable to a constant power*, while e^x is *a constant to a variable power*. The two types of functions are quite different, as their graphs show.

The graph of x^2
(a variable to a
constant power)

The graph of e^x
(a constant to a
variable power)

Each type of function has its own differentiation formula.

$$\frac{d}{dx} x^n = nx^{n-1}$$

For a variable
x to a constant
power n

$$\frac{d}{dx} e^x = e^x$$

For the constant
e to a variable
power x

EXAMPLE 8   **DIFFERENTIATING A LOGARITHMIC AND EXPONENTIAL FUNCTION**

Find the derivative of $\ln (1 + e^x)$.

Solution

$$\frac{d}{dx} \ln(1 + e^x) = \frac{\dfrac{d}{dx}(1 + e^x)}{1 + e^x} = \frac{e^x}{1 + e^x}$$

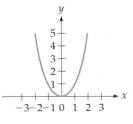

Using $\dfrac{d}{dx} \ln f = \dfrac{f'}{f}$

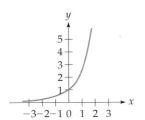

Working out
the numerator

Functions of the form e^{kx} (for constant k) arise in many applications. The derivative of e^{kx} is as follows:

$$\frac{d}{dx} e^{kx} = e^{kx} \cdot k = ke^{kx}$$ Using $\dfrac{d}{dx} e^f = e^f \cdot f'$

Derivative of the exponent

This result is so useful that we record it as a separate formula.

Derivative of e^{kx}

$$\frac{d}{dx} e^{kx} = ke^{kx}$$ For any constant k

This formula says that the rate of change (the derivative) of e^{kx} is proportional to itself. That is, the function $y = e^{kx}$ satisfies the *differential equation*

$$y' = ky$$

We noted this earlier when we observed that in exponential growth a quantity *grows in proportion to itself* (as in populations and savings accounts).

These differentiation formulas enable us to find instantaneous rates of change of logarithmic and exponential functions. In many applications the variable stands for time, so we use t instead of x.

EXAMPLE 9   FINDING A RATE OF IMPROVEMENT OF A SKILL

After t weeks of practice a pole vaulter can vault

$$H(t) = 15 - 11e^{-0.1t}$$

feet. Find the rate of change of the athlete's jumps after

a. 0 weeks (at the beginning of training)

b. 12 weeks

Solution

We differentiate to find the rate of change:

$$H'(t) = -11(-0.1)e^{-0.1t} \quad = \quad 1.1e^{-0.1t}$$

Differentiating $15 - 11e^{-0.1t}$

Using $\dfrac{d}{dt}\,e^{kt} = ke^{kt}$ Simplifying

a. For the rate of change after 0 weeks:

$$H'(0) = 1.1e^{-0.1(0)} = 1.1e^0 = 1.1$$

$H'(t) = 1.1e^{-0.1t}$ with $t = 0$

b. After 12 weeks:

$$H'(12) = 1.1e^{-0.1(12)}$$

$$= 1.1e^{-1.2} \approx 1.1(0.30) = 0.33$$

$H'(t) = 1.1e^{-0.1t}$ with $t = 12$

Using a calculator

At first, the vaults increased by 1.1 feet per week. After 12 weeks, the gain was only 0.33 foot (about 4 inches) per week.

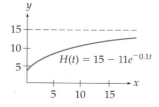

This result is typical of learning a new skill: Early improvement is rapid, later improvement is slower. This trend is called *diminishing returns* and may be seen in the leveling off of the polevault heights in the graph on the left.

Maximizing Consumer Expenditure

The amount of a commodity that consumers will buy depends on the price of the commodity. For a commodity whose price is p, let the consumer demand be given by a function $D(p)$. Multiplying the number of units $D(p)$ by the price p gives the total *consumer expenditure* for the commodity.

Consumer Demand and Expenditure

Let $D(p)$ be the consumer demand at price p. Then the consumer expenditure is

$$E(p) = p \cdot D(p)$$

EXAMPLE 10   MAXIMIZING CONSUMER EXPENDITURE

If consumer demand for a commodity is $D(p) = 10{,}000e^{-0.02p}$ units per week, where p is the selling price, find the price that maximizes consumer expenditure.

Solution

Using the preceding formula for consumer expenditure,

$$E(p) = p \cdot 10{,}000e^{-0.02p} = 10{,}000pe^{-0.02p} \qquad E(p) = p \cdot D(p)$$

To maximize $E(p)$ we differentiate:

$$E'(p) = \underbrace{10{,}000e^{-0.02p}}_{\substack{\text{Derivative} \\ \text{of } 10{,}000p}} + \underbrace{10{,}000p(-0.02)e^{-0.02p}}_{\substack{\text{Derivative} \\ \text{of } e^{-0.02p}}}$$

Using the Product Rule to differentiate $E(p) = 10{,}000p \cdot e^{-0.02p}$

$$= 10{,}000e^{-0.02p} - 200pe^{-0.02p} \qquad \text{Simplifying}$$

$$= 200e^{-0.02p}(50 - p) \qquad \text{Factoring}$$

$$\text{CN: } p = 50$$

Critical number from $(50 - p)$ (since e to a power is never zero)

We calculate E'' for the second-derivative test:

$$E''(p) = 200(-0.02)e^{-0.02p}(50 - p) + 200e^{-0.02p}(-1)$$

From $E'(p) = 200e^{-0.02}p \cdot (50 - p)$ using the Product Rule

$$= -4e^{-0.02p}(50 - p) - 200e^{-0.02p} \qquad \text{Simplifying}$$

At the critical number $p = 50$,

$$E''(50) = -4e^{-0.02(50)}(50 - 50) - 200e^{-0.02(50)} \qquad \text{Substituting } p = 50$$

$$= -200e^{-1} = \frac{-200}{e} \qquad \text{Simplifying}$$

E'' is negative, so the expenditure $E(p)$ is maximized at $p = 50$:

Consumer expenditure is maximized at price $50.

Graphing Calculator Exploration

Y₁=10000Xe^(-0.02X)

Maximum
X=49.999992   Y=183939.72

Interpret as $x = 50$

Use a graphing calculator to verify the answer to the previous Example by graphing $E(x) = 10{,}000xe^{-0.02x}$ (using x instead of p) on the window [0, 150] by [0, 200,000] and using MAXIMUM.

How did we choose the graphing window? We first found the critical number "by hand" (setting the derivative equal to zero and finding, in this example, 50) and chose x-values around it. For the y-values, we evaluated the function at the critical number (using CALCULATE, finding in this example $y \approx 184{,}000$) and chose y-values including it. Notice how graphing calculators and calculus are most effective when used together.

Graphing Logarithmic and Exponential Functions

To graph logarithmic and exponential functions using a graphing calculator, we first find critical points and possible inflection points and then graph the function on a window including these points. If graphing "by hand," make sign diagrams for the first and second derivatives and then sketch the graph, as in Chapter 3.

EXAMPLE 11   **GRAPHING AN EXPONENTIAL FUNCTION**

Graph $f(x) = e^{-x^2/2}$.

Solution

As before, we write the function as $f(x) = e^{-\frac{1}{2}x^2}$. The derivative is

$$f'(x) = e^{-\frac{1}{2}x^2}(-x) = -xe^{-\frac{1}{2}x^2} \qquad \text{Using } \frac{d}{dx}e^f = e^f \cdot f'$$

Derivative of the exponent

CN: $x = 0$ Critical number is 0

$\quad\;\; y = 1$ From $y = e^{-\frac{1}{2}x^2}$ evaluated at $x = 0$

The second derivative is

$$f''(x) = (-1)e^{-\frac{1}{2}x^2} - xe^{-\frac{1}{2}x^2}(-x) \qquad \begin{array}{l}\text{From } f'(x) = -x \cdot e^{-\frac{1}{2}x^2}\\ \text{using the Product Rule}\end{array}$$

$$= -e^{-\frac{1}{2}x^2} + x^2 e^{-\frac{1}{2}x^2} \qquad \text{Simplifying}$$

$$= e^{-\frac{1}{2}x^2}(-1 + x^2) \qquad \text{Factoring}$$

$$= (x^2 - 1)e^{-\frac{1}{2}x^2} \qquad \text{Rearranging}$$

$$= (x + 1)(x - 1)e^{-\frac{1}{2}x^2} \qquad \text{Factoring}$$

$$x = \pm 1 \qquad \text{Where } f'' = 0$$

$$y = e^{-\frac{1}{2}} \approx 0.6 \qquad \begin{array}{l}\text{From } y = e^{-\frac{1}{2}x^2}\\ \text{evaluated at } x = \pm 1\end{array}$$

Based on these values, we choose the graphing window as follows. For the x-values we choose [−3, 3] (to include 0 and ±1 and beyond), and for the y-values we choose [−1, 2] (to include 1 and 0.6 and above and below). This window gives the graph on the left. (Many other windows would be just as good, and after seeing the graph you might want to adjust the window.)

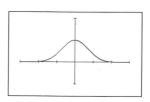

$f(x) = e^{-x^2/2}$ on [−3, 3] by [−1, 2]

This function, multiplied by the constant $1/\sqrt{2\pi}$, is the famous "bell-shaped curve" of statistics, used for predictions of many things from the IQs of newborn babies to stock prices.

Solutions TO PRACTICE PROBLEMS

Derivative of $\ln x$
Derivative of x

1. $f'(x) = \dfrac{d}{dx}\left(\dfrac{\ln x}{x}\right) = \dfrac{x\left(\frac{1}{x}\right) - 1 \ln x}{x^2}$

$\qquad\qquad = \dfrac{1 - \ln x}{x^2}$

2. $\dfrac{d}{dx}\ln(x^3 - 5x + 1) = \dfrac{3x^2 - 5}{x^3 - 5x + 1}$

3. $f'(x) = e^x + xe^x$

$\quad f'(1) = e^1 + 1e^1 = 2e$ (which is approximately $2 \cdot 2.718 = 5.436$)

4. $\dfrac{d}{dx}e^{1 + x^3/3} = \dfrac{d}{dx}e^{1 + \frac{1}{3}x^3} = e^{1 + \frac{1}{3}x^3}(x^2) = x^2 e^{1 + x^3/3}$

4.3   Section Summary

In this section we developed the formulas for differentiating logarithmic and exponential functions:

$$\frac{d}{dx}\ln x = \frac{1}{x} \qquad\qquad \frac{d}{dx}e^x = e^x$$

$$\frac{d}{dx}\ln f(x) = \frac{f'(x)}{f(x)} \qquad\qquad \frac{d}{dx}e^{f(x)} = e^{f(x)} \cdot f'(x)$$

The last of these has a useful special case when the exponent is of the form $f(x) = kx$ for a constant k:

$$\frac{d}{dx}e^{kx} = ke^{kx}$$

We used these formulas together with our other differentiation formulas (the Product, Quotient, and Chain Rules) to find rates of change and to optimize functions.

Verification of the Differentiation Formulas

Verification of the Power Rule for Arbitrary Powers

On page 108 we proved the Power Rule $\dfrac{d}{dx}x^n = nx^{n-1}$ for positive integer exponents, and on page 226 we extended it to rational exponents. We can now show that the Power Rule holds for *all* exponents. We begin by using one of the "inverse" properties of logarithms, $x = e^{\ln x}$ (for $x > 0$), but with x replaced by x^n:

$$f(x) = x^n = e^{\ln x^n} = e^{n \ln x}$$

Using the property that logs bring down exponents

Differentiating:

$$f'(x) = e^{n \ln x}\left(n\frac{1}{x}\right)$$

$\underbrace{\phantom{e^{n\ln x}}}_{x^n}$ $\underbrace{\phantom{\left(n\frac{1}{x}\right)}}$ Derivative of the exponent

Differentiating $f(x) = e^{n \ln x}$ by the $e^{f(x)}$ formula

$$= x^n \cdot \frac{1}{x} \cdot n = nx^{n-1}$$

$\underbrace{}_{x^{n-1}}$

Replacing $e^{n \ln x}$ by x^n, reordering, and combining x's

This is what we wanted to show—that the derivative of $f(x) = x^n$ is $f'(x) = nx^{n-1}$ for *all* exponents n. (The equation $x^n = e^{n \ln x}$ can be taken as a *definition* of x to a power.)

Verification of the Differentiation Formulas for $\ln x$ and $\ln f(x)$

The formula for the derivative of the natural logarithm function comes from applying the definition of the derivative to $f(x) = \ln x$.

$$f'(x) = \lim_{h \to 0} \frac{f(x+h) - f(x)}{h} = \lim_{h \to 0} \frac{\ln(x+h) - \ln x}{h}$$

Definition of $f'(x)$

$$= \lim_{h \to 0} \frac{1}{h}[\ln(x+h) - \ln x]$$

Dividing by h is equivalent to multiplying by $1/h$

$$= \lim_{h \to 0} \frac{1}{h}\ln\left(\frac{x+h}{x}\right) = \lim_{h \to 0} \frac{1}{h}\ln\left(1 + \frac{h}{x}\right)$$

Using Property 7 of logarithms (inside back cover), and simplifying

$$= \lim_{h \to 0} \frac{1}{x}\frac{x}{h}\ln\left(1 + \frac{h}{x}\right) = \lim_{h \to 0} \frac{1}{x}\ln\left(1 + \frac{h}{x}\right)^{x/h}$$

Dividing and multiplying by x, and then using Property 8 of logarithms

$$= \lim_{h \to 0} \frac{1}{x}\ln\underbrace{\left(1 + \frac{1}{n}\right)^n}_{\substack{\text{Approaches} \\ e \text{ as } n \to \infty}}$$

Defining $n = x/h$, so that $h \to 0$ implies $n \to \infty$ (for $h > 0$)

$$= \frac{1}{x}\ln e = \frac{1}{x}$$

Since $\ln e = 1$ (the same conclusion follows if $h < 0$)

This is the result that we wanted to show—that the derivative of the natural logarithm function $f(x) = \ln x$ is $f'(x) = \frac{1}{x}$.

For the rule to differentiate the natural logarithm of a (positive) *function*, we begin with

$$\frac{d}{dx}f(g(x)) = f'(g(x))g'(x)$$

Chain Rule (from page 141) for differentiable functions f and g

$$\frac{d}{dx}\ln(g(x)) = \frac{1}{g(x)}g'(x) = \frac{g'(x)}{g(x)}$$

Taking $f(x) = \ln x$, so $f'(x) = \frac{1}{x}$

Replacing g by f, this is exactly the formula we wanted to show:

$$\frac{d}{dx}\ln f(x) = \frac{f'(x)}{f(x)}$$

Verification of the Differentiation Formulas for e^x and $e^{f(x)}$

To derive the rule for differentiating e^x we begin with

$$\ln e^x = x \qquad\qquad \text{Property 3 of natural logarithms}$$

and differentiate both sides:

$$\frac{\dfrac{d}{dx}e^x}{e^x} = 1 \qquad\qquad \text{Using } \frac{d}{dx}\ln f = \frac{f'}{f} \text{ on the left side}$$

Multiplying each side by e^x gives

$$\frac{d}{dx}e^x = e^x$$

This is the rule for differentiating e^x. This rule together with the Chain Rule gives the rule for differentiating $e^{f(x)}$, just as before.

4.3 Exercises

1–44. Find the derivative of each function.

1. $f(x) = x^2 \ln x$

2. $f(x) = \dfrac{\ln x}{x^3}$

3. $f(x) = \ln x^2$

4. $f(x) = \ln(x^3 + 1)$

5. $f(x) = \ln \sqrt{x}$

6. $f(x) = \sqrt{\ln x}$

7. $f(x) = \ln(x^2 + 1)^3$

8. $f(x) = \ln(x^4 + 1)^2$

9. $f(x) = \ln(-x)$

10. $f(x) = \ln(5x)$

11. $f(x) = \dfrac{e^x}{x^2}$

12. $f(x) = x^3 e^x$

13. $f(x) = e^{x^3 + 2x}$

14. $f(x) = 2e^{7x}$

15. $f(x) = e^{x^3/3}$

16. $f(x) = \ln(e^x - 2x)$

17. $f(x) = x - e^{-x}$

18. $f(x) = x \ln x - x$

19. $f(x) = \ln e^{2x}$

20. $f(x) = \ln e^x$

21. $f(x) = e^{1 + e^x}$

22. $f(x) = \ln(e^x + e^{-x})$

23. $f(x) = x^e$

24. $f(x) = ex$

25. $f(x) = e^3$

26. $f(x) = \sqrt{e}$

27. $f(x) = \ln(x^4 + 1) - 4e^{x/2} - x$

28. $f(x) = x^2 e^x - 2\ln x + (x^2 + 1)^3$

29. $f(x) = x^2 \ln x - \frac{1}{2}x^2 + e^{x^2} + 5$

30. $f(x) = e^{-2x} - x \ln x + x - 7$

31. $f(x) = e^x \ln x^2$

32. $f(x) = e^x \ln(x + 1)$

33. $f(x) = \ln \dfrac{1}{x^2}$

34. $f(x) = \ln \dfrac{1}{e^{x^2}}$

35. $f(t) = (e^{2t} + 1)^3$

36. $f(t) = \sqrt{e^{2t} + 4}$

37. $f(t) = \sqrt{t^2 + 2\ln t}$

38. $f(t) = (2t + \ln t)^3$

39. $f(z) = \dfrac{e^z}{z^2 - 1}$

40. $f(z) = \dfrac{e^z}{1 + e^z}$

41. $f(z) = \dfrac{10}{1 + e^{-2z}}$

42. $f(z) = \dfrac{12}{1 + 2e^{-z}}$

43. $f(x) = \dfrac{e^x + e^{-x}}{e^x - e^{-x}}$

44. $f(x) = \dfrac{e^x - e^{-x}}{e^x + e^{-x}}$

45–50. For each function, find the indicated expressions.

45. $f(x) = \dfrac{\ln x}{x^5}$, find   **a.** $f'(x)$ **b.** $f'(1)$

46. $f(x) = x^4 \ln x$,   find   **a.** $f'(x)$ **b.** $f'(1)$

47. $f(x) = \ln(x^4 + 48)$, find   **a.** $f'(x)$ **b.** $f'(2)$

48. $f(x) = x^2 \ln x - x^2$, find   **a.** $f'(x)$ **b.** $f'(e)$

49. $f(x) = \ln(e^x - 3x)$, find   **a.** $f'(x)$ **b.** $f'(0)$

50. $f(x) = \ln(e^x + e^{-x})$, find   **a.** $f'(x)$ **b.** $f'(0)$

51–54. Find the equation for the tangent line to the curve $y = f(x)$ at the given x-value.

51. $f(x) = x \ln(x - 1)$ at $x = 2$

52. $f(x) = \dfrac{x^2}{1 + \ln x}$ at $x = 1$

53. $f(x) = e^{x^2 - 1}$ at $x = 1$

54. $f(x) = x^2 e^{x+1}$ at $x = -1$

55–58. For each function:

a. Find $f'(x)$.

b. Evaluate the given expression and approximate it to three decimal places.

55. $f(x) = 5x \ln x$, find and approximate $f'(2)$.

56. $f(x) = e^{x^2/2}$, find and approximate $f'(2)$.

57. $f(x) = \dfrac{e^x}{x}$, find and approximate $f'(3)$.

58. $f(x) = \ln(e^x - 1)$, find and approximate $f'(3)$.

59–60. Find the *second* derivative of each function.

59. $f(x) = e^{-x^5/5}$ **60.** $f(x) = e^{-x^6/6}$

61–62. By calculating the first few derivatives, find a formula for the nth derivative of each function (k is a constant).

61. $f(x) = e^{kx}$ **62.** $f(x) = e^{-kx}$

63–68. Find the differential of each function and evaluate it at the given values of x and dx.

63. $y = e^{-2x}$ at $x = 3$ and $dx = 0.5$

64. $y = e^{-\frac{1}{2}x^2}$ at $x = 2$ and $dx = 0.25$

65. $y = xe^x$ at $x = 1$ and $dx = 0.1$

66. $y = \ln x^3$ at $x = 4$ and $dx = 0.2$

67. $y = x^2 \ln x$ at $x = e$ and $dx = 0.01$

68. $y = \ln(x^2 + 1)$ at $x = 3$ and $dx = 0.25$

69–72. Use your graphing calculator to graph each function on a window that includes all relative extreme points and inflection points, and give the coordinates of these points (rounded to two decimal places). [*Hint:* Use NDERIV once or twice with ZERO.] (Answers may vary depending on the graphing window chosen.)

69. $f(x) = e^{-2x^2}$ **70.** $f(x) = 1 - e^{-x^2/2}$

71. $f(x) = \ln(1 + x^2)$ **72.** $f(x) = e^x + e^{-x}$

73–76. Use your graphing calculator to graph each function on the indicated interval, and give the coordinates of all relative extreme points and inflection points (rounded to two decimal places). [*Hint:* Use NDERIV once or twice together with ZERO.] (Answers may vary depending on the graphing window chosen.)

73. $f(x) = \dfrac{x^2}{e^x}$ for $-1 \le x \le 8$

74. $f(x) = \dfrac{x}{e^x}$ for $-1 \le x \le 5$

75. $f(x) = x \ln|x|$ for $-2 \le x \le 2$

[*Hint for Exercises 75–76:* $|x|$ is sometimes entered as ABS (x).]

76. $f(x) = x^2 \ln|x|$ for $-2 \le x \le 2$

77–78. Use implicit differentiation to find dy/dx.

77. $y^2 - ye^x = 12$ **78.** $y^2 - x \ln y = 10$

Applied Exercises

79. PERSONAL FINANCE: Earnings and Calculus A study found that one's earnings are affected by the mathematics courses one has taken. In particular, compared to someone making $40,000 who had taken *no* calculus, a comparable person who had taken x years of calculus would be earning $\$40{,}000e^{0.195x}$. Find the rate of change of this function at $x = 1$ and interpret your answer.

Source: Review of Economics and Statistics **86**

80. ENVIRONMENTAL SCIENCE: Solar Cells Cumulative world production of solar cells for generating electricity is predicted to be $58e^{0.43t}$ thousand megawatts, where t is the number of years since 2013. Find the rate of change of this quantity at the value of t corresponding to 2023 and interpret your answer. [*Note:* Three thousand megawatts is enough to power a million homes.]

Source: Earth Policy Institute

81. PERSONAL FINANCE: Compound Interest A sum of $1000 at 5% interest compounded continuously will grow to $V(t) = 1000e^{0.05t}$ dollars in t years. Find the rate of growth after:

a. 0 years (the time of the original deposit).

b. 10 years.

82. PERSONAL FINANCE: Depreciation A $10,000 automobile depreciates so that its value after t years is $V(t) = 10{,}000e^{-0.35t}$ dollars. Find the instantaneous rate of change of its value:

a. when it is new ($t = 0$).

b. after 2 years.

83. GENERAL: Population The United States population (in millions) is predicted to be $P(t) = 317e^{0.01t}$, where t is the number of years after 2013. Find the instantaneous rate of change of the population in the year 2023.

Source: U.S. Census Bureau

84. BEHAVIORAL SCIENCE: Ebbinghaus Memory Model According to the *Ebbinghaus model of memory*, if one is shown a list of items, the percentage of items remembered t time units later is $P(t) = (100 - a)e^{-bt} + a$, where a and b are constants. For $a = 25$ and $b = 0.2$, this function becomes $P(t) = 75e^{-0.2t} + 25$. Find the instantaneous rate of change of this percentage:

a. at the beginning of the test ($t = 0$).

b. after 3 time units.

Source: Gorfein and Hoffman, Memory and Learning

85. BIOMEDICAL: Drug Dosage A patient receives an injection of 1.2 milligrams of a drug, and the amount remaining in the bloodstream t hours later is $A(t) = 1.2e^{-0.05t}$. Find the instantaneous rate of change of this amount:

a. just after the injection (at time $t = 0$).
b. after 2 hours.

86. GENERAL: Temperature A covered cup of coffee at 200 degrees, if left in a 70-degree room, will cool to $T(t) = 70 + 130e^{-2.5t}$ degrees in t hours. Find the rate of change of the temperature:

a. at time $t = 0$.   **b.** after 1 hour.

87. BUSINESS: Sales The weekly sales (in thousands) of a new product are predicted to be $S(x) = 1000 - 900e^{-0.1x}$ after x weeks. Find the rate of change of sales after:

a. 1 week.   **b.** 10 weeks.

88. SOCIAL SCIENCE: Diffusion of Information by Mass Media The number of people in a town of 50,000 who have heard an important news bulletin within t hours of its first broadcast is $N(t) = 50{,}000(1 - e^{-0.4t})$. Find the rate of change of the number of informed people:

a. at time $t = 0$.   **b.** after 8 hours.

89–90. ECONOMICS: Consumer Expenditure If consumer demand for a commodity is given by the function below (where p is the selling price in dollars), find the price that maximizes consumer expenditure.

89. $D(p) = 5000e^{-0.01p}$ **90.** $D(p) = 8000e^{-0.05p}$

91–92. BUSINESS: Maximizing Revenue Each of the following functions is a company's price function, where p is the price (in dollars) at which quantity x (in thousands) will be sold.

a. Find the revenue function $R(x)$. [*Hint:* Revenue is price times quantity, $p \cdot x$.]
b. Find the quantity and price that will maximize revenue.

91. $p = 400e^{-0.20x}$ **92.** $p = 4 - \ln x$

93. GENERAL: Temperature A mug of beer chilled to 40 degrees, if left in a 70-degree room, will warm to a temperature of $T(t) = 70 - 30e^{-3.5t}$ degrees in t hours.

a. Find $T(0.25)$ and $T'(0.25)$ and interpret your answers.
b. Find $T(1)$ and $T'(1)$ and interpret your answers.

94. SOCIAL SCIENCE: Diffusion of Information by Mass Media The number of people in a city of 200,000 who have heard a weather bulletin within t hours of its first broadcast is $N(t) = 200{,}000(1 - e^{-0.5t})$.

a. Find $N(0.5)$ and $N'(0.5)$ and interpret your answers.
b. Find $N(3)$ and $N'(3)$ and interpret your answers.

95. BIOMEDICAL: Diabetes The percentage of the U.S. population with diabetes is

$$D(x) = 0.96 + 0.011e^x + \ln x$$

where x is the number of decades since 1950. Find $D'(5)$ and interpret your answer.

Source: Centers for Disease Control

96. BIOMEDICAL: DWI and Crash Risk The crash risk of an intoxicated driver relative to a similar driver with zero blood alcohol is

$$R(x) = \frac{195}{1 + 1650e^{-35.5x}}$$

where x is the blood alcohol level as a percent $(0.01 \le x \le 0.25)$. For example, $R(0.16) = 29.4$ means that a driver with blood alcohol level 0.16% is 29.4 times more likely to be involved in an accident than a similar driver who is not impaired. Find $R'(0.16)$ and interpret your answer. [*Note:* $x \ge 0.08$ defines "driving while intoxicated."]

Source: Dunlap and Associates, Inc.

97–98. ATHLETICS: World's Record 100-Meter Run In 1987 Carl Lewis set a new world's record of 9.93 seconds for the 100-meter run. The distance that he ran in the first x seconds was

$$11.274[x - 1.06(1 - e^{-x/1.06})] \text{ meters}$$

for $0 \le x \le 9.93$. Enter this function as y_1, and define y_2 as its derivative (using NDERIV), so that y_2 gives his velocity after x seconds. Graph them on the window $[0, 9.93]$ by $[0, 100]$.

Source: SIAM Review **35**

97. Trace along the velocity curve to verify that Lewis's maximum speed was about 11.27 meters per second. Find how quickly he reached a speed of 10 meters per second, which is 95% of his maximum speed.

98. Define y_3 as the derivative of y_2 (using NDERIV) so that y_3 gives his acceleration after x seconds, and graph y_2 and y_3 on the window $[0, 9.93]$ by $[0, 20]$. Evaluate both y_2 and y_3 at $x = 0.1$ and also at $x = 9.93$ (using CALCULATE). Interpret your answers.

99. ATHLETICS: How Fast Do Old Men Slow Down? The fastest times for the marathon (26.2 miles) for male runners aged 35 to 80 are approximated by the function

$$f(x) = \begin{cases} 106.2e^{0.0063x} & \text{if } x \le 58.2 \\ 850.4e^{0.000614x^2 - 0.0652x} & \text{if } x > 58.2 \end{cases}$$

in minutes, where x is the age of the runner.

Source: Review of Economics and Statistics **LXXVI**

a. Graph this function on the window $[35, 80]$ by $[0, 240]$.
[*Hint:* On some graphing calculators, enter $y_1 = (106.2e^{0.0063x})\ (x \le 58.2) + (850.4e^{0.000614x^2 - 0.0652x})\ (x > 58.2)$.]
b. Find $f(35)$ and $f'(35)$ and interpret these numbers. [*Hint:* Use NDERIV or dy/dx.]
c. Find $f(80)$ and $f'(80)$ and interpret these numbers.

Conceptual Exercises

100. Why can't we differentiate e^x by the *Power Rule*?

101–107. Choose the correct answer:

101. $\dfrac{d}{dx}e^x =$ **a.** xe^{x-1} **b.** e^x **c.** 0

102. $\dfrac{d}{dx}e^{f(x)} =$ **a.** $e^{f(x)}f'(x)$ **b.** $e^{f'(x)}f(x)$ **c.** $e^{f'(x)}f'(x)$

103. $\dfrac{d}{dx}e^5 =$ **a.** $5e^4$ **b.** e^5 **c.** 0

104. $\dfrac{d}{dx}x^e =$ **a.** ex^{e-1} **b.** x^e **c.** 0

105. $\dfrac{d}{dx}\ln x =$ **a.** $\dfrac{x}{1}$ **b.** $\dfrac{1}{x}$ **c.** 0

106. $\dfrac{d}{dx}\ln f(x) =$ **a.** $\dfrac{f(x)}{f'(x)}$ **b.** $\dfrac{f'(x)}{f(x)}$ **c.** $\dfrac{f'(x)}{f'(x)}$

107. $\dfrac{d}{dx}\ln 5 =$ **a.** $\dfrac{5}{1}$ **b.** $\dfrac{1}{5}$ **c.** 0

108. True or False: If a function involves natural logarithms, then its derivative will involve natural logarithms. Explain.

109. Explain why it is obvious, without any calculation, that $\dfrac{d}{dx}e^{\ln x} = 1$.

110. Explain why it is obvious, without any calculation, that $\dfrac{d}{dx}\ln e^x = 1$.

Explorations and Excursions   The following problems extend and augment the material presented in the text.

111. BIOMEDICAL: Population Growth The *Gompertz growth curve* models the size $N(t)$ of a population at time $t \geq 0$ as $N(t) = Ke^{-ae^{-bt}}$ where K and b are positive constants. Show that $\dfrac{dN}{dt} = bN \ln\left(\dfrac{K}{N}\right)$ and interpret this derivative to make statements about the population growth when $N < K$ and when $N > K$.

Source: S. Brauer and C. Castillo-Chávez, *Mathematical Models in Population Biology and Epidemiology*

112. BIOMEDICAL: Ricker Recruitment The population dynamics of many fish (such as salmon) can be described by the *Ricker curve* $y = axe^{-bx}$ for $x \geq 0$ where $a > 1$ and $b > 0$ are constants, x is the size of the parental stock, and y is the number of recruits (offspring). Determine the size of the parental stock that maximizes the number of recruits.

Source: J. Fish. Res. Board Can. **11**

113. BIOMEDICAL: Reynolds Number An important characteristic of blood flow is the *Reynolds number*. As the Reynolds number increases, blood flows less smoothly. For blood flowing through certain arteries, the Reynolds number is

$$R(r) = a \ln r - br$$

where a and b are positive constants and r is the radius of the artery. Find the radius r that maximizes the Reynolds number R. (Your answer will involve the constants a and b.)

114. BIOMEDICAL: Drug Concentration If a drug is injected intramuscularly, the concentration of the drug in the bloodstream after t hours will be

$$A(t) = \frac{c}{b-a}(e^{-at} - e^{-bt})$$

If the constants are $a = 0.4$, $b = 0.6$, and $c = 0.1$, find the time of maximum concentration.

Time of maximum concentration

115. CHANGE OF BASE FORMULA FOR LOGARITHMS: Derive the formula

$$\log_a x = \frac{\ln x}{\ln a} \quad \text{(for } a > 0 \text{ and } x > 0)$$

which expresses logarithms to *any* base a in terms of *natural* logarithms, as follows:

a. Define $y = \log_a x$, so that $x = a^y$, and take the natural logarithms of both sides of the last equation and obtain $\ln x = y \ln a$.

b. Solve the last equation for y to obtain

$$y = \frac{\ln x}{\ln a}$$ and then use the original definition

of y to obtain the stated change of base formula.

116. CHANGE OF BASE FORMULA FOR EXPONENTS Derive the formula

$$a^x = e^{(\ln a)x} \quad \text{(for } a > 0)$$

by identifying the property of natural logarithms (from the inside back cover) that justifies the first equality in the following sequence:

$$a^x = (e^{\ln a})^x = e^{(\ln a)x}$$

117–118. Use the formula derived in Exercise 116 to write each expression in the form e^{bx} for some number b.

117. 2.23^x **118.** 0.63^x

[*Note:* The graphing calculator operation ExpReg gives expressions like those above (see pages 59–60), and Exercises 117–118 show how to express such results using e.]

Exponential and Logarithmic Functions to Other Bases

The rules for differentiating exponential functions with (positive) base a are as follows:

Derivatives of a^x and $a^{f(x)}$

$$\frac{d}{dx}\, a^x = (\ln a)\, a^x$$

$$\frac{d}{dx}\, a^{f(x)} = (\ln a) a^{f(x)} f'(x) \qquad\qquad \text{For a differentiable function } f$$

For example,

$$\frac{d}{dx}\, 2^x = (\ln 2) 2^x$$

$$\frac{d}{dx}\, 5^{3x^2+1} = (\ln 5) 5^{3x^2+1}(6x) = 6(\ln 5)x\, 5^{3x^2+1}$$

These formulas are more complicated than the corresponding base e formulas (page 278), which is why e is called the "natural" base: It makes the derivative formulas simplest. These formulas reduce to the natural (base e) formulas if $a = e$.

119–120. Use the preceding formulas to find the derivative of each function.

119. a. $f(x) = 10^x$ **b.** $f(x) = 3^{x^2+1}$ **c.** $f(x) = 2^{3x}$ **120. a.** $f(x) = 5^x$ **b.** $f(x) = 2^{x^2-1}$ **c.** $f(x) = 3^{4x}$

     d. $f(x) = 5^{3x^2}$ **e.** $f(x) = 2^{4-x}$ **d.** $f(x) = 9^{5x^2}$ **e.** $f(x) = 10^{1-x}$

The rules for differentiating logarithmic functions with (positive) base a are as follows:

Derivatives of $\log_a x$ and $\log_a f(x)$

$$\frac{d}{dx}\, \log_a x = \frac{1}{(\ln a)x}$$

$$\frac{d}{dx}\, \log_a f(x) = \frac{f'(x)}{(\ln a)f(x)} \qquad\qquad \text{For a differentiable function } f > 0$$

For example,

$$\frac{d}{dx}\, \log_5 x = \frac{1}{(\ln 5)x}$$

$$\frac{d}{dx}\, \log_2 (x^3 + 1) = \frac{3x^2}{(\ln 2)(x^3 + 1)}$$

These formulas are more complicated than the corresponding base e formulas (page 278), and again the simplicity of the base e formulas is why e is called the "natural" base. As before, these formulas reduce to the natural (base e) formulas if $a = e$.

121–122. Use the formulas on the previous page to find the derivative of each function.

121. a. $\log_2 x$ **b.** $\log_{10}(x^2 - 1)$ **c.** $\log_3(x^4 - 2x)$ **122. a.** $\log_3 x$ **b.** $\log_2(x^2 + 1)$ **c.** $\log_{10}(x^3 - 4x)$

4.4   Two Applications to Economics: Relative Rates and Elasticity of Demand

Introduction

In this section we define **relative rates of change** and see how they are used in economics. (*Relative* rates are not the same as the *related* rates discussed in Section 3.6.) We then define the very important economic concept of **elasticity of demand.**

Relative Versus Absolute Rates

The derivative of a function gives its rate of change. For example, if $f(t)$ is the cost of a pair of shoes at time t years, then $f'(t)$ is the rate of change of cost (in dollars per year). That is, $f' = 3$ would mean that the price of shoes is increasing at the rate of $3 per year. Similarly, if $g(t)$ is the price of a new automobile at time t years, then $g' = 300$ would mean that automobile prices are increasing at the rate of $300 per year.

Does this mean that car prices are rising 100 times as fast as shoe prices? In absolute terms, yes. However, this does not take into account the enormous price difference between automobiles and shoes.

Relative Rates of Change

If shoe prices are increasing at the rate of $3 per year, and if the current price of a pair of shoes is $60, the *relative* rate of increase is $\frac{3}{60} = \frac{1}{20} = 0.05$, which means that shoe prices are increasing at the relative rate of 5% per year. Similarly, if the price of an average automobile is $15,000, then an increase of $300 relative to this price is $\frac{300}{15,000} = \frac{1}{50} = 0.02$, for a *relative* rate of 2% per year. Therefore, in a *relative* sense (that is, as a fraction of the current price), car prices are increasing *less* rapidly than shoe prices.

In general, if $f(t)$ is the price of an item at time t, then the rate of change is $f'(t)$, and the *relative* rate of change is $f'(t)/f(t)$, the derivative divided by the function. We will sometimes call the derivative $f'(x)$ the "absolute" rate of change to distinguish it from the relative rate of change $f'(x)/f(x)$.

Relative rates are often more meaningful than absolute rates. For example, it is easier to grasp the fact that the gross domestic product is growing at the relative rate of 3% a year than that it is growing at the absolute rate of $400,000,000,000 per year.

The expression $f'(x)/f(x)$ is the derivative of the natural logarithm of $f(x)$:

$$\frac{d}{dx}\ln f(x) = \frac{f'(x)}{f(x)}$$

This provides an alternative expression for the relative rate of change, in terms of logarithms:

Relative Rate of Change

$$\left(\begin{array}{c}\text{Relative rate of}\\\text{change of } f(t)\end{array}\right) = \frac{d}{dt}\ln f(t) = \frac{f'(t)}{f(t)}$$

For a differentiable function $f > 0$

We use the variable t since it often stands for time. Both formulas in the box give the same result, and are sometimes called the *logarithmic derivative*, since it may be found by first taking the logarithm and then the derivative.

The relative rate of change, being a ratio or a percent, does not depend on the units (dollars or euros, pounds or kilos) and so is called a *dimensionless quantity*. Therefore, relative rates can be compared between different products, and even between different nations. This is in contrast to *absolute* rates of change (that is, derivatives), which *do* depend on the units (for example, dollars per year).

EXAMPLE 1   FINDING A RELATIVE RATE OF CHANGE

If the gross domestic product t years from now is predicted to be $G(t) = 8.2e^{\sqrt{t}}$ trillion dollars, find the relative rate of change 25 years from now.

We give two solutions, showing the use of both formulas.

Solution $\left(\text{using the } \dfrac{d}{dt}\ln f(t) \text{ formula}\right)$

First we simplify:

$$\ln G(t) = \ln 8.2e^{\sqrt{t}} \qquad\qquad \text{Taking natural logs}$$

$$= \ln 8.2 + \ln e^{\sqrt{t}} \qquad\qquad \begin{array}{l}\text{Log of a product is}\\\text{the sum of the logs}\end{array}$$

$$= \ln 8.2 + \sqrt{t} \qquad\qquad \begin{array}{l}\ln e^{\sqrt{t}} = \sqrt{t} \text{ by Property 3}\\\text{of logs (see inside back cover)}\end{array}$$

$$= \ln 8.2 + t^{1/2} \qquad\qquad \text{In exponent form}$$

Then we differentiate:

$$\frac{d}{dt}(\ln 8.2 + t^{1/2}) = 0 + \frac{1}{2}t^{-1/2} = \frac{1}{2}t^{-1/2} \qquad \begin{array}{l}\ln 8.2 \text{ is a constant, so}\\\text{its derivative is zero}\end{array}$$

Finally, we evaluate at the given time $t = 25$:

$$\frac{1}{2}(25)^{-1/2} = \frac{1}{2}\frac{1}{\sqrt{25}} = \frac{1}{2}\frac{1}{5} = \frac{1}{10} = 0.10 \qquad \tfrac{1}{2}t^{-1/2} \text{ evaluated at } t = 25$$

Therefore, in 25 years the gross domestic product will be increasing at the relative rate of 0.10, or 10%, per year.

Alternative Solution $\left(\text{using the } \dfrac{f'(t)}{f(t)} \text{ formula}\right)$

$$G(t) = 8.2e^{\sqrt{t}} = 8.2e^{t^{1/2}} \qquad\qquad \text{Writing } G(t) \text{ with fractional exponents}$$

$$G'(t) = 8.2e^{t^{1/2}}\left(\frac{1}{2}t^{-1/2}\right) \qquad\qquad \text{Differentiating}$$

Derivative of the exponent

Therefore, the relative rate of change $\dfrac{G'(t)}{G(t)}$ is

$$\frac{G'(t)}{G(t)} = \frac{8.2e^{t^{1/2}}\left(\dfrac{1}{2}t^{-1/2}\right)}{8.2e^{t^{1/2}}} = \frac{1}{2}t^{-1/2}$$ Same result as with the first formula

At $t = 25$,

$$\frac{1}{2}(25)^{-1/2} = \frac{1}{2}\frac{1}{\sqrt{25}} = \frac{1}{2}\frac{1}{5} = \frac{1}{10} = 0.10$$ Again the same

Therefore, the relative growth rate is 10%, just as we found before.

Both formulas give the same answer, so you should use the one that is easier to apply in your particular problem. The $\dfrac{d}{dt}\ln f(t)$ formula sometimes allows simplification before the differentiation, while the $\dfrac{f'(t)}{f(t)}$ formula often involves simplification afterward.

PRACTICE PROBLEM 1

An investor estimates that if a piece of land is held for t years, it will be worth $f(t) = 300 + t^2$ thousand dollars. Find the relative rate of change at time $t = 10$ years. [*Hint:* Use the $\dfrac{f'(t)}{f(t)}$ formula.]

Solution on page 297 >

Graphing Calculator Exploration

Enter the function from Practice Problem 1 into a graphing calculator as $y_1 = 300 + x^2$, and then turn *off* the function so that it will not graph. Then graph $y_2 = \dfrac{d}{dx}\ln y_1$ and $y_3 = \dfrac{y_1'}{y_1}$ (using NDERIV) together on the window [0, 20] by [0, 0.1]. Why do you get only one curve for the two functions?

Elasticity of Demand

Farmers are aware of the paradox that an abundant harvest usually brings *lower* total revenue than a poor harvest. The reason is simply that the larger quantities in an abundant harvest result in lower prices, which in turn cause increased demand, but the demand does not increase enough to compensate for the lower prices.

Revenue is price times quantity, $R = p \cdot q$, and when one of these rises, the other falls. For this reason, we shall use only positive numbers for the percent changes and ignore the directions of the changes. The question is whether the rise in one is enough to compensate for the fall in the other. For example, if a 1% price decrease brings a 2% quantity increase, revenue will rise, but if the 1% price decrease brings only a $\frac{1}{2}$% quantity increase, revenue will fall. The concept of *elasticity of demand* was invented to analyze such problems. (Economists use the word *demand* instead of *quantity*.)

Roughly speaking, we may think of elasticity as *the percentage change in demand divided by the percentage change in price:*

Understanding Elasticity of Demand

$$E = \frac{\text{Percent change in demand}}{\text{Percent change in price}}$$

Brief Examples

If a 1% change in price brings a 2% change in demand:

$$E = \frac{2\%}{1\%} = 2$$

If a 1% change in price brings only a $\frac{1}{2}$% change in demand:

$$E = \frac{\frac{1}{2}\%}{1\%} = \frac{1}{2}$$

We classify demand as elastic or inelastic depending on whether elasticity is greater than 1 or less than 1:

Demand is *elastic* if $E > 1$

Demand is *inelastic* if $E < 1$

Demand is *unit-elastic* if $E = 1$

In the first Brief Example above, $E = 2$, so demand was *elastic*, while in the second example, $E = \frac{1}{2}$, so demand was *inelastic*.

PRACTICE PROBLEM 2

Suppose that Starbooks, purveyors of lattes and literature, raise prices by 5% and demand falls by 15%. Find the elasticity and identify the demand as elastic or inelastic.

Solution on page 297 >

Good or Service	Elasticity
Clothing	0.20
Housing	0.30
Gasoline	0.43
Movies	0.87
Automobiles	1.87
Restaurant meals	2.27
Fresh fruit	3.02

Source: Houthaker and Taylor, *Consumer Demand in the United States, Review of Economics and Statistics* **62**

Intuitively, we may think of elasticity of demand as measuring how *responsive* demand is to price changes: *Elastic* means *responsive* and *inelastic* means *unresponsive*. That is, for elastic demand, a price cut will bring a large increase in demand, so total revenue will rise. On the other hand, for inelastic demand, a price cut will bring only a slight increase in demand, so total revenue will *fall*.

Economists calculate elasticity of demand for many products, and some typical elasticities are shown in the table on the left. Notice that for *necessities* (clothing, food), demand is inelastic since consumers need them even if prices rise, while for luxuries (restaurant meals) demand is elastic since consumers can cut back or find substitutes in response to price increases.

The preceding formula for elasticity was useful for understanding the concept but is not entirely satisfactory since it does not reflect how elasticity may change as price changes. In fact, the elasticities in the table are only approximately correct since prices are always changing. To see how elasticity changes as prices change (which will be useful in applications), we need a more precise definition. First, we must define the *demand function,* $x = D(p)$, which gives the *demand* for an item if the price of the item is p.

Demand Function

The demand function

$$x = D(p)$$

gives the quantity x of an item that will be demanded by consumers if the price is p.

Law of downward-sloping demand

Since, in general, demand falls as prices rise, the slope of the demand function is negative, as shown on the left. This is known as the *law of downward-sloping demand.*

Calculating Elasticity of Demand

To analyze changes in demand compared to changes in price, we calculate the relative rate of change of demand divided by the relative rate of change of price. Using the derivative-of-the-logarithm formula,

$$\frac{\left(\begin{array}{c}\text{Relative rate of}\\\text{change of demand}\end{array}\right)}{\left(\begin{array}{c}\text{Relative rate of}\\\text{change of price}\end{array}\right)} = \frac{\dfrac{d}{dp}\ln D(p)}{\dfrac{d}{dp}\ln p} = \frac{\dfrac{D'(p)}{D(p)}}{\dfrac{1}{p}} = \underbrace{\frac{pD'(p)}{D(p)}}_{\text{Simplified}}$$

Because most demand functions are downward-sloping, the derivative $D'(p)$ is generally negative. Economists prefer to work with positive numbers, so the *elasticity of demand* is taken to be the negative of this quantity (in order to make it positive).*

Elasticity of Demand

For a demand function $D(p)$, the elasticity of demand is

$$E(p) = \frac{-p \cdot D'(p)}{D(p)}$$

Demand is elastic if $E(p) > 1$ and inelastic if $E(p) < 1$.

Elasticity, being composed of *relative* rates of change, does not depend on the units of the demand function, and so is *dimensionless*. Therefore, elasticities can be compared between different products, and even between different countries.

EXAMPLE 2   FINDING ELASTICITY OF DEMAND FOR COMMUTER BUS SERVICE

A bus line estimates the demand function for its daily commuter tickets to be $D(p) = 81 - p^2$ (in thousands of tickets), where p is the price in dollars ($0 \le p \le 9$). Find the elasticity of demand when the price is:

a. $3   b. $6

*Some economists omit the negative sign.

Solution

$$E(p) = \frac{-pD'(p)}{D(p)}$$ 　　　Definition of elasticity

$$= \frac{-p(-2p)}{81 - p^2}$$ 　　　Substituting $D(p) = 81 - p^2$ so $D'(p) = -2p$

$$= \frac{2p^2}{81 - p^2}$$ 　　　Simplifying

a. Evaluating at $p = 3$ gives

$$E(3) = \frac{2(3)^2}{81 - (3)^2} = \frac{18}{81 - 9} = \frac{18}{72} = \frac{1}{4}$$ 　　$E(p) = \frac{2p^2}{81 - p^2}$ with $p = 3$

Interpretation: The elasticity is less than 1, so demand for tickets is *inelastic* at a price of \$3. This means that a small price change (up or down from this level) will cause only a *slight* change in demand. More precisely, elasticity of $\frac{1}{4}$ means that a 1% price change will cause only about a $\frac{1}{4}$% change in demand.

b. At the price of \$6, the elasticity of demand is

$$E(6) = \frac{2(6)^2}{81 - (6)^2} = \frac{72}{81 - 36} = \frac{8}{5} = 1.6$$ 　　$E(p) = \frac{2p^2}{81 - p^2}$ with $p = 6$

Interpretation: The elasticity is greater than 1, so demand is *elastic* at a price of \$6. This means that a small change in price (up or down from this level) will cause a relatively *large* change in demand. In particular, an elasticity of 1.6 means that a price change of 1% will cause about a 1.6% change in demand.

The changes in demand are, of course, in the opposite direction from the changes in price. That is, if prices are *raised* by 1% (from the \$6 level), demand will *fall* by 1.6%, while if prices are *lowered* by 1%, demand will *rise* by 1.6%. In the future we will assume that the *direction* of the change is clear, and say simply that a 1% change in price will cause about a 1.6% change in demand.

PRACTICE PROBLEM 3

For the demand function $D(p) = 90 - p$, find the elasticity of demand $E(p)$ and evaluate it at $p = 30$ and $p = 75$. (Be sure to complete this Practice Problem, because the results will be used shortly.)

Solution on page 297 >

Graphing Calculator Exploration

For any demand function y_1 (written in terms of x), define y_2 as the elasticity function for y_1 by defining $y_2 = -x \cdot y_1' / y_1$ (using NDERIV). Try entering the demand function from Example 2 (as shown on the left) or Practice Problem 3 in y_1 and then evaluating y_2 at appropriate numbers to check the answers found there.

Using Elasticity to Increase Revenue

In Example 2 we found that at a price of $3, demand is inelastic ($E = \frac{1}{4} < 1$), and so demand responds only *weakly* to price changes. Therefore, to increase revenue the company should *raise* prices, since the higher prices will drive away only a relatively small number of customers. On the other hand, at a price of $6, demand is elastic ($E = 1.6 > 1$), and so demand is very responsive to price changes. In this case, to increase revenue the company should *lower* prices, since this will attract more than enough new customers to compensate for the price decrease. In general:

Elasticity and Revenue

To increase revenue:

If demand is elastic ($E > 1$), you should *lower* prices.

If demand is inelastic ($E < 1$), you should *raise* prices.

This should be intuitively clear: In simplest terms, you should *lower* prices if demand *will* change and *raise* prices if demand *won't* change. This statement shows why elasticity of demand is important to any company that cuts prices in an attempt to boost revenue, or to any utility that raises prices in order to increase revenue. Elasticity shows whether the strategy will succeed or fail.

 Spreadsheet Exploration

The following spreadsheet* is based on the demand function $D(p) = 90 - p$ from Practice Problem 3 on the previous page and shows the elasticity of demand along with the demand and revenue for various prices.

D6				= =$A6/(90-$A6)
	A	B	C	D
1	Price	Demand	Revenue	Elasticity
2				
3	30	60	1800	0.500
4	35	55	1925	0.636
5	40	50	2000	0.800
6	45	45	2025	1.000
7	50	40	2000	1.250
8	55	35	1925	1.571
9	60	30	1800	2.000
10	65	25	1625	2.600
11	70	20	1400	3.500
12	75	15	1125	5.000

Inelastic:
Revenue increases
as price increases (rows 3–5)

Elastic:
Revenue decreases
as price increases (rows 7–12)

Notice that where elasticity is less than 1 (the first three rows), revenue rises as the price increases, but where elasticity is greater than 1 (the last six rows), revenue falls as the price increases.

*To obtain this and other Spreadsheet Explorations, go to www.cengagebrain.com, search for this textbook and then scroll down to access the "free materials" tab.

The borderline case, $E = 1$ (called *unit-elasticity*), is where revenue cannot be raised, which will be the case if revenue is at its maximum. Therefore, elasticity must be unitary when revenue is maximized.

> At maximum revenue, elasticity of demand must equal 1.

We could use this fact as a basis for a new method for maximizing revenue, but instead, we will stick with our earlier (and easier) method of maximizing functions by finding critical numbers.

PRACTICE PROBLEM 4

According to the preceding spreadsheet, at what price is revenue maximized, and what is the elasticity of demand at that price?

Solution below   >

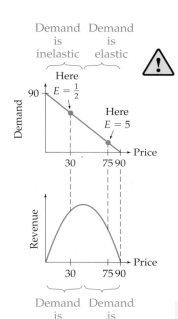

Be Careful Do not confuse elasticity of demand with the slope of the demand curve; the two ideas are quite different. For example, in Practice Problem 3 on page 295, we found that the linear demand function $D(p) = 90 - p$ (which has slope -1 all along it) has elasticity $\frac{1}{2}$ at one point and elasticity 5 at another, as shown on the left. In fact, the elasticities vary from 0 (at the top) to approaching infinity (at the bottom).

The second graph shows the revenue function for this demand function

$$R = p(90 - p) = 90p - p^2$$

From $R = p \cdot D(p)$
with $D(p) = 90 - p$

Notice that for values of p to the left of center, where demand is inelastic, increasing p will increase revenue, while for values of p to the right of center, where demand is elastic, you should *lower* p to increase revenue, just as stated in the box on the opposite page.

Solutions TO PRACTICE PROBLEMS

1. $\dfrac{f'(t)}{f(t)} = \dfrac{2t}{300 + t^2}$

    At $t = 10$, $\dfrac{2 \cdot 10}{300 + 10^2} = \dfrac{20}{400} = \dfrac{1}{20} = 0.05$ or 5%

2. $E = \dfrac{15\%}{5\%} = 3$,   so demand is elastic.

3. $E(p) = \dfrac{-pD'(p)}{D(p)} = \dfrac{-p(-1)}{90 - p} = \dfrac{p}{90 - p}$

    At $p = 30$, $E(30) = \dfrac{30}{90 - 30} = \dfrac{30}{60} = \dfrac{1}{2} = 0.5$ (demand is inelastic)

    At $p = 75$, $E(75) = \dfrac{75}{90 - 75} = \dfrac{75}{15} = 5$ (demand is elastic)

4. At $p = 45$.   Elasticity is 1.

4.4   Section Summary

The *relative rate of change* of a positive differentiable function $f(t)$ is found from either of the two equivalent formulas

$$\left(\begin{array}{c}\text{Relative rate of}\\\text{change of } f(t)\end{array}\right) = \frac{d}{dt}\ln f(t) = \frac{f'(t)}{f(t)}$$

It measures how a quantity changes *ralative to its current size*. We use t for the independent variable since it typically stands for *time*.

An important economic application of relative rates involves the *demand function* for a product, which is the amount $D(p)$ that will be demanded (purchased) by consumers if the price is p. The *elasticity of demand* for the product is defined as the negative of *the relative rate of change of demand divided by the relative rate of change of price*, which is equal to

$$E(p) = \frac{-p \cdot D'(p)}{D(p)}$$

The negative sign makes it *positive* since $D'(p)$ will be negative because of the *law of downward-sloping demand*. Demand at price p is said to be *elastic* if $E(p) > 1$, *inelastic* if $E(p) < 1$, and *unit-elastic* if $E(p) = 1$.

Intuitively, elasticity measures the *responsiveness* of demand to price changes: *Elastic* demand means that a price change will cause a relatively *large* change in demand, while *inelastic* demand means that the price change will cause a relatively *small* change in demand. Therefore to increase revenue, you should *lower* the price if demand is *elastic*, and *raise* the price if demand is *inelastic*.

Verification of the Relationship Between Elasticity and Revenue

We may verify the relationship between elasticity of demand and revenue as follows: revenue is price p times quantity x.

$$R = px = p \cdot D(p) \qquad \text{Using } x = D(p)$$

Differentiating will show how revenue responds to price changes.

$$R'(p) = D(p) + pD'(p) \qquad \begin{array}{l}\text{Using the Product Rule}\\\text{on } p \cdot D(p)\end{array}$$

$$= D(p)\left[1 + \frac{pD'(p)}{D(p)}\right] \qquad \text{Factoring out } D(p)$$

$$= D(p)\left[1 - \frac{-pD'(p)}{D(p)}\right] \qquad \begin{array}{l}\text{Replacing the plus sign}\\\text{by two minus signs}\end{array}$$

$$\qquad\qquad\qquad\qquad \begin{array}{l}\text{This is the definition}\\\text{of elasticity } E(p)\end{array}$$

$$= D(p)[1 - E(p)] \qquad \text{Replacing } \frac{-pD'(p)}{D(p)} \text{ by } E(p)$$

If demand is *elastic*, $E > 1$, then the quantity in brackets is negative, and so the derivative $R'(p)$ is negative, showing that revenue *decreases* as price increases. Therefore, to increase revenue, one should *lower* prices. On the other hand, if the demand is *inelastic*, $E < 1$, then the quantity in brackets is positive, and so the derivative $R'(p)$ is positive, showing that revenue *increases* as price increases. In this case, to increase revenue, one should *raise* prices. This proves the statements on page 296 in the "Elasticity and Revenue" box.

4.4   Exercises

Exercises on Relative Rates

1–10. For each function:

a. Find the relative rate of change.

b. Evaluate the relative rate of change at the given value(s) of t.

  1. $f(t) = t^2$, $t = 1$ and $t = 10$

  2. $f(t) = t^3$, $t = 1$ and $t = 10$

3. $f(t) = 100e^{0.2t}$, $t = 5$ **4.** $f(t) = 100e^{-0.5t}$, $t = 4$

5. $f(t) = e^{t^2}$, $t = 10$ **6.** $f(t) = e^{t^3}$, $t = 5$

7. $f(t) = e^{-t^2}$, $t = 10$ **8.** $f(t) = e^{-t^3}$, $t = 5$

9. $f(t) = 25\sqrt{t - 1}$, $t = 6$

10. $f(t) = 100\sqrt[3]{t + 2}$, $t = 8$

Applied Exercises on Relative Rates

11. BUSINESS: Wireless Communication Because of the popularity and power of smartphones, demand for wireless communication is growing rapidly. Monthly global data traffic is predicted to be $4.5e^{0.507x}$ peta-bytes, where x is the number of years since 2015. Find the relative growth rate in any year x.

Source: Standard Poor's

12. BUSINESS: 3D Systems Stock Price During 2012 and 2013, the price of 3D Systems stock was growing rapidly and was given approximately by $15e^{0.805x}$, where x is the number of years since 2011. Find the relative growth rate of this stock during that period (3D Systems manufactures 3D printers.)

Source: BigCharts.com

13–14. ECONOMICS: National Debt If the national debt of a country (in trillions of dollars) t years from now is given by the indicated function, find the relative rate of change of the debt 10 years from now.

13. $N(t) = 0.5 + 1.1e^{0.01t}$ **14.** $N(t) = 0.4 + 1.2e^{0.01t}$

 15–16. GENERAL: Population The population (in millions) of a city t years from now is given by the indicated function.

a. Find the relative rate of change of the population 8 years from now.

b. Will the relative rate of change ever reach 1.5%?

15. $P(t) = 4 + 1.3e^{0.04t}$ **16.** $P(t) = 6 + 1.7e^{0.05t}$

Exercises on Elasticity of Demand

17–28. For each demand function $D(p)$:

a. Find the elasticity of demand $E(p)$.

b. Determine whether the demand is elastic, inelastic, or unit-elastic at the given price p.

17. $D(p) = 200 - 5p$, $p = 10$

18. $D(p) = 60 - 8p$, $p = 5$

19. $D(p) = 300 - p^2$, $p = 10$

20. $D(p) = 100 - p^2$, $p = 5$

21. $D(p) = \dfrac{300}{p}$, $p = 4$

22. $D(p) = \dfrac{500}{p}$, $p = 2$

23. $D(p) = \sqrt{175 - 3p}$, $p = 50$

24. $D(p) = \sqrt{100 - 2p}$, $p = 20$

25. $D(p) = \dfrac{100}{p^2}$, $p = 40$

26. $D(p) = \dfrac{600}{p^3}$, $p = 25$

27. $D(p) = 4000e^{-0.01p}$, $p = 200$

28. $D(p) = 6000e^{-0.05p}$, $p = 100$

Applied Exercises on Elasticity of Demand

29. BUSINESS: Automobile Sales An automobile dealer is selling cars at a price of $12,000. The demand function is $D(p) = 2(15 - 0.001p)^2$, where p is the price of a car. Should the dealer raise or lower the price to increase revenue?

30. BUSINESS: Liquor Sales A liquor distributor wants to increase its revenues by discounting its bestselling liquor. If the demand function for this liquor is $D(p) = 60 - 3p$, where p is the price per bottle, and if the current price is $15, will the discount succeed?

31. BUSINESS: City Bus Revenues The manager of a city bus line estimates the demand function to be $D(p) = 150{,}000\sqrt{1.75 - p}$, where p is the fare in dollars. The bus line currently charges a fare of $1.25, and it plans to raise the fare to increase its revenues. Will this strategy succeed?

32. BUSINESS: Newspaper Sales The demand function for a newspaper is $D(p) = 80{,}000\sqrt{75 - p}$, where p is the price in cents. The publisher currently charges 50 cents, and it plans to raise the price to increase revenues. Will this strategy succeed?

33. GENERAL: Electricity Rates An electrical utility asks the Federal Regulatory Commission for permission to raise rates to increase revenues. The utility's demand function is

$$D(p) = \frac{120}{10 + p}$$

where p is the price (in cents) of a kilowatt-hour of electricity. If the utility currently charges 6 cents per kilowatt-hour, should the commission grant the request?

34. ECONOMICS: Oil Prices A Middle Eastern oil-producing country estimates that the demand for oil (in millions of barrels per day) is $D(p) = 9.5e^{-0.04p}$, where p is the price of a barrel of oil. To raise its revenues, should it raise or lower its price from its current level of $120 per barrel?

35. ECONOMICS: Oil Prices A European oil-producing country estimates that the demand for its oil (in millions of barrels per day) is $D(p) = 3.5e^{-0.06p}$, where p is the price of a barrel of oil. To raise its revenues, should it raise or lower its price from its current level of $120 per barrel?

36–37. BUSINESS: Liquor and Beer The demand functions for distilled spirits and for beer are given below, where p is the retail price and $D(p)$ is the demand in gallons per capita. For each demand function, find the elasticity of demand for any price p. [*Note:* You will find, in each case,

that demand is inelastic. This means that taxation, which acts like a price increase, is an ineffective way of discouraging liquor consumption, but is an effective way of raising revenue.]

Source: Journal of Consumer Research **12**

36. $D(p) = 3.509p^{-0.859}$ (for distilled spirits)

37. $D(p) = 7.881p^{-0.112}$ (for beer)

38–39. BUSINESS: Automobile Sales The demand function for automobiles in a dealership is given below, where p is the selling price.

a. Use the method described in the Graphing Calculator Exploration on page 295 to find the elasticity of demand at a price of $12,000.

b. Should the dealer raise or lower the price from this level to increase revenue?

c. Find the price at which elasticity equals 1. [*Hint:* Use INTERSECT.]

38. $D(p) = 3\sqrt{20 - 0.001p}$ **39.** $D(p) = \dfrac{200}{8 + e^{0.0001p}}$

40. ECONOMICS: Elasticities for Linear Demand For the linear demand function $D(p) = 90 - p$ used in Practice Problem 3 and graphed on page 297:

a. Find a formula for the elasticity of demand at *any* value of p. [*Hint:* Use the formula on page 294 with the given demand function.]

b. Show that the elasticity is zero at the *top* of the line (that is, for $p = 0$).

c. Show that the elasticity approaches infinity as p approaches the *bottom* of the line.

d. Show that the elasticity is $\frac{1}{2}$ at the midpoint of the line.

e. Show that statements (b), (c), and (d) are true for *any* linear demand function $D(p) = a - bp$ for any positive constants a and b. [*Hint:* Find an expression for the elasticity at any value of p, as you did in part (a).]

41. ECONOMICS: Cigarette Sales In 1998 the State of New Jersey raised cigarette taxes, with the result that the price of a pack of cigarettes rose from $2.40 to $2.80. Demand then fell from 52 to 47.5 million packs. Use these data to estimate the elasticity of demand for cigarettes. [*Hint:* Find the percentage change in price and the percentage change in demand and use the formula in the "Understanding Elasticity of Demand" box on page 293.

Source: P. Samuelson and W. Nordhaus, Economics

Conceptual Exercises

42–45. For each function, calculate "in your head" the relative rate of change.

42. $f(x) = e^x$ **43.** $f(x) = e^{1.2x}$

44. $f(x) = x$ **45.** $f(x) = x^n$

46–47. For each demand function, calculate "in your head" the elasticity.

46. $D(p) = e^{-p}$ **47.** $D(p) = e^{-2p}$

48. If demand is inelastic and you raise prices, would you expect demand to increase or decrease, and strongly or weakly?

49. If demand is elastic and you lower prices, would you expect demand to increase or decrease, and strongly or weakly?

50. For which of the following two items would you expect demand to be elastic and for which inelastic: *heating oil, olive oil.*

51. For which of the following two items would you expect demand to be elastic and for which inelastic: *cigarettes, jewelry.*

Explorations and Excursions   The following problems extend and augment the material presented in the text.

52. CONSTANT ELASTICITY
  a. Show that for a demand function of the form $D(p) = c/p^n$, where c and n are positive constants, the elasticity is constant.
  b. What type of demand function has elasticity equal to 1 for every value of p?

53. LINEAR ELASTICITY Show that for a demand function of the form $D(p) = ae^{-cp}$, where a and c are positive constants, the elasticity of demand is $E(p) = cp$.

54–55. ELASTICITY OF SUPPLY A supply function $S(p)$ gives the total amount of a product that producers

are willing to supply at a given price p. The *elasticity of supply* is defined as

$$E_s(p) = \frac{p \cdot S'(p)}{S(p)}$$

Elasticity of supply measures the relative increase in supply resulting from a small relative increase in price. It is less useful than elasticity of demand, however, since it is not related to total revenue.

54. Use the preceding formula to find the elasticity of supply for a supply function of the form $S(p) = ae^{cp}$, where a and c are positive constants.

55. Use the preceding formula to find the elasticity of supply for a supply function of the form $S(p) = ap^n$, where a and n are positive constants.

4	Chapter Summary with Hints and Suggestions

Reading the text and doing the exercises in this chapter have helped you to master the following concepts and skills, which are listed by section (in case you need to review them) and are keyed to particular Review Exercises. Answers for all Review Exercises are given at the back of the book, and full solutions can be found in the Student Solutions Manual.

4.1   Exponential Functions

• Find the value of money invested at compound interest. *(Review Exercise 1.)*

$$P\left(1 + \frac{r}{m}\right)^{mt} \qquad Pe^{rt}$$

• Determine which of two banks gives better interest. *(Review Exercise 2.)*

• Depreciate an asset. *(Review Exercise 3.)*

$$P(1 + r)^t \qquad \text{(for depreciation, } r \text{ is negative)}$$

• Determine which of two drugs provides more medication. *(Review Exercise 4.)*

• Predict the world's largest city. *(Review Exercise 5.)*

• Predict computer memory capacity. *(Review Exercise 6.)*

4.2   Logarithmic Functions

• Find doubling (and other) times for compound interest. *(Review Exercises 7–8.)*

$$\ln M^P = P \cdot \ln M \qquad \ln e^x = x$$

- Date a fossil by potassium-40. *(Review Exercises 9–10.)*

- Predict the spread of information, or oil demand. *(Review Exercises 11–12.)*

- Find the time required for a given percentage change in a bank deposit or in advertising results. *(Review Exercises 13–14.)*

4.3   Differentiation of Logarithmic and Exponential Functions

- Find the derivative of a logarithmic or exponential function. *(Review Exercises 15–30.)*

$$\frac{d}{dx}\ln x = \frac{1}{x} \qquad \frac{d}{dx}e^x = e^x \qquad \frac{d}{dx}\ln f = \frac{f'}{f}$$

$$\frac{d}{dx}e^f = e^f \cdot f' \qquad \frac{d}{dx}e^{kx} = ke^{kx}$$

- Find the tangent line to a curve at a given point. *(Review Exercises 31–32.)*

- Graph an exponential or a logarithmic function. *(Review Exercises 33–34.)*

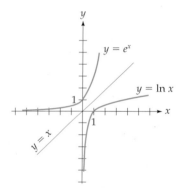

- Find the rate of change of sales, amount of medication, learning, temperature, or diffusion of information. *(Review Exercises 35–39.)*

- Find when a company maximizes its present value. *(Review Exercise 40.)*

- Maximize a company's revenue. *(Review Exercises 41–42.)*

- Maximize consumer expenditure for a product. *(Review Exercise 43.)*

- Graph an exponential or logarithmic function. *(Review Exercises 44–45.)*

- Maximize consumer expenditure or maximize revenue. *(Review Exercises 46–47.)*

4.4   Two Applications to Economics: Relative Rates and Elasticity of Demand

- Find the relative rate of change of a country's gross domestic product. *(Review Exercises 48–49.)*

$$\left(\begin{array}{c}\text{Relative}\\ \text{rate}\end{array}\right) = \frac{d}{dt}\ln f = \frac{f'}{f}$$

- Find the elasticity of demand for a product, and its consequences. *(Review Exercises 51–52.)*

$$\left(\begin{array}{c}\text{Elasticity}\\ \text{of demand}\end{array}\right) = \frac{-p \cdot D'(p)}{D(p)}$$

- Find the relative rate of change of population. *(Review Exercise 53.)*

- Find the elasticity of demand for a product, how it affects revenue, and what price gives unit-elasticity. *(Review Exercise 54.)*

Hints and Suggestions

- *(Overview)* Exponential and logarithmic functions should be thought of as just other types of functions, like polynomials, but having their own differentiation rules. In fact, in a sense they are more "natural" than polynomials because they give natural growth rates.

- A graphing calculator helps by drawing graphs of exponential and logarithmic functions, and finding intersection points (for example, where one population overtakes another). It is also useful for checking derivatives and graphically verifying maximum and minimum values.

- Interest rates are always *annual* rates unless stated otherwise. To use $P(1 + r/m)^{mt}$ for finding value under compound interest, r is the *annual interest rate*, m is the number of *compounding periods*, and t is the *number of years*.

- When do you use Pe^{rt} and when do you use $P(1 + r/m)^{mt}$? Use Pe^{rt} if the word "continuous" occurs, and use $P(1 + r/m)^{mt}$ if it does not. $P(1 + r)^t$ with negative r gives depreciation. *Dividing P by e^{rt} or $(1 + r/m)^{mt}$ gives present value.*

- The Power Rule $\frac{d}{dx}x^n = nx^{n-1}$ is for differentiating a *variable to a constant power,* and $\frac{d}{dx}e^x = e^x$ is for differentiating the *constant e to a variable power.*

| **4** | **Review Exercises and Chapter Test** | ◯ indicates a Chapter Test exercise. |

4.1 Exponential Functions

① **PERSONAL FINANCE:** Interest Find the value of $10,000 invested for 8 years at 8% interest if the compounding is done:
a. quarterly. **b.** continuously.

2. **PERSONAL FINANCE:** Interest One bank offers 6% compounded quarterly and a second offers 5.98% compounded continuously. Where should you take your money?

③ **BUSINESS:** Depreciation An $800,000 computer loses 20% of its value each year.

a. Give a formula for its value after t years.
b. Find its value after 4 years.

4. **BIOMEDICAL:** Drug Concentration If the concentration of a drug in a patient's bloodstream is c (milligrams per milliliter), then t hours later the concentration will be $C(t) = ce^{-kt}$, where k is a constant (the "elimination constant"). Two drugs are being compared: drug A with initial concentration $c = 2$ and elimination constant $k = 0.2$, and drug B with initial concentration $c = 3$ and elimination constant $k = 0.25$. Which drug will have the greater concentration 4 hours later?

⑤ **GENERAL:** Population The largest city in the world is Tokyo, with São Paulo (Brazil) smaller but growing faster. The population of Tokyo will be $28.5e^{0.0034x}$ and the population of São Paulo will be $19.4e^{0.011x}$ (both in millions) x years after 2010. Graph both functions on a calculator with the window [0, 90] by [0, 50]. When will São Paulo overtake Tokyo as the world's largest city (assuming that these rates continue to hold)?

Source: Census Bureau

6. **GENERAL:** Moore's Law of Computer Memory The amount of information that can be stored on a computer chip can be measured in megabits (a "bit" is a binary digit, 0 or 1, and a "megabit" is a million bits). The first 1-megabit chips became available in 1987, and 4-megabit chips became available in 1990. This quadrupling of capacity every 3 years is expected to continue, so that chip capacity will be $C(t) = 4^{t/3}$ megabits where t is the number of years after 1987. Use this formula (known as Moore's Law, after Gordon Moore, a founder of the Intel Corporation) to predict chip capacity in the year 2017. [*Hint:* What value of t corresponds to 2017?]

4.2 Logarithmic Functions

⑦ **PERSONAL FINANCE:** Interest Find how soon an investment at 10% interest compounded semiannually will:
a. double in value. **b.** increase by 50%.

⑧ **PERSONAL FINANCE:** Interest Find how soon an investment at 7% interest compounded continuously will:
a. double in value. **b.** increase by 50%.

9–10. GENERAL: Fossils In the following exercises, use the fact that the proportion of potassium-40 remaining after t million years is $e^{-0.00054t}$.

9. In 1984 in the Wind River Basin of Wyoming, scientists discovered a fossil of a small, three-toed horse, an ancestor of the modern horse. Estimate the age of this fossil if rocks near it contained 97.3% of their original potassium-40.

10. Estimate the age of a skull found in 1959 in Tanzania (dubbed "Nutcracker Man" because of its huge jawbone) if rocks near it contained 99.9% of their original potassium-40.

⑪ **SOCIAL SCIENCE:** Diffusion of Information by Mass Media In a city of a million people, election results broadcast over radio and television will reach $N(t) = 1,000,000(1 - e^{-0.3t})$ people within t hours. Find when the news will have reached 500,000 people.

12. **ECONOMICS:** Oil Demand The demand for oil in the United States is increasing by 3% per year. Assuming that this rate continues, how soon will demand increase by 50%?

⑬ **PERSONAL FINANCE:** Interest If a bank offers 6.5% interest, in how many years will a deposit increase by 50% if the compounding is done:
a. quarterly? **b.** continuously?

14. **BUSINESS:** Advertising After the opening of a new store has been advertised for t days, the proportion of people in a city who have seen the ad is $p(t) = 1 - e^{-0.032t}$. How long must the ad run to reach:

a. 30% of the people?
b. 40% of the people?

4.3 Differentiation of Logarithmic and Exponential Functions

15–30. Find the derivative of each function.
⑮ $f(x) = \ln 2x$ **16.** $f(x) = \ln(x^2 - 1)^2$
17. $f(x) = \ln(1 - x)$ **18.** $f(x) = \ln\sqrt{x^2 + 1}$
⑲ $f(x) = \ln\sqrt[3]{x}$ **20.** $f(x) = \ln e^x$
㉑ $f(x) = \ln x^2$ **22.** $f(x) = x\ln x - x$
㉓ $f(x) = e^{-x^2}$ **24.** $f(x) = e^{1-x}$
25. $f(x) = \ln e^{x^2}$ **26.** $f(x) = e^{x^2 \ln x - x^2/2}$
㉗ $f(x) = 5x^2 + 2x\ln x + 1$
28. $f(x) = 2x^3 + 3x\ln x - 1$
29. $f(x) = 2x^3 - 3xe^{2x}$ **30.** $f(x) = 4x - 2x^2e^{2x}$

31–32. Find the equation for the tangent line to the curve $y = f(x)$ at the given x-value.

31. $f(x) = [x + \ln(x + 2)]^2$ at $x = -1$

32. $f(x) = \dfrac{x^2 - 3}{e^{x-2}}$ at $x = 2$

33–34. Graph each function, showing all relative extreme points and inflection points.

33. $f(x) = \ln(x^2 + 4)$ **34.** $f(x) = 16e^{-x^2/8}$

35. BUSINESS: Sales The weekly sales (in thousands) of a new product after x weeks of advertising is $S(x) = 2000 - 1500e^{-0.1x}$. Find the rate of change of sales after:

   a. 1 week.   **b.** 10 weeks.

36. BIOMEDICAL: Drug Dosage A patient receives an injection of 1.5 milligrams of a drug, and the amount remaining in the bloodstream t hours later is $A(t) = 1.5e^{-0.08t}$. Find the instantaneous rate of change of this amount:

   a. immediately after the injection (time $t = 0$).
   b. after 5 hours.

37. BEHAVIORAL SCIENCE: Learning In a test of short-term memory, the percent of subjects who remember an eight-digit number for at least t seconds is $P(t) = 100 - 200 \ln(t + 1)$. Find the rate of change of this percent after 5 seconds.

38. GENERAL: Temperature A thermos bottle that is chilled to 35 degrees Fahrenheit and then left in a 70-degree room will warm to a temperature of $T(t) = 70 - 35e^{-0.1t}$ degrees after t hours. Find the rate of change of the temperature:

   a. at time $t = 0$.   **b.** after 5 hours.

39. SOCIAL SCIENCE: Diffusion of Information by Mass Media The number of people in a town of 30,000 who have heard an important news bulletin within t hours of its first broadcast is $N(t) = 30,000(1 - e^{-0.3t})$. Find the instantaneous rate of change of the number of informed people after:

   a. 1 hour.   **b.** 8 hours.

40. BUSINESS: Maximizing Present Value A new company is growing so that its value t years from now will be $50t^2$ dollars. Therefore, its present value (at the rate of 8% compounded continuously) is $V(t) = 50t^2 e^{-0.08t}$ dollars (for $t > 0$). Find the number of years that maximizes the present value.

41–42. BUSINESS: Maximizing Revenue The given function is a company's price function, where x is the quantity (in thousands) that will be sold at price p dollars.

a. Find the revenue function $R(x)$.
   [*Hint:* Revenue is price times quantity, $p \cdot x$.]
b. Find the quantity and price that will maximize revenue.

41. $p = 200e^{-0.25x}$ **42.** $p = 5 - \ln x$

43. ECONOMICS: Maximizing Consumer Expenditure Consumer demand for a commodity is estimated to be $D(p) = 25{,}000e^{-0.02p}$ units per month, where p is the selling price in dollars. Find the selling price that maximizes consumer expenditure.

44–45. Use your graphing calculator to graph each function on a window that includes all relative extreme points and inflection points, and give the coordinates of these points (rounded to two decimal places). [*Hint:* Use NDERIV once or twice with ZERO.] (Answers may vary depending on the graphing window chosen.)

44. $f(x) = \dfrac{x^4}{e^x}$ **45.** $f(x) = x^3 \ln|x|$

46. ECONOMICS: Consumer Expenditure If consumer demand for a commodity is $D(p) = 200e^{-0.0013p}$ (where p is the selling price in dollars), find the price that maximizes consumer expenditure.

47. BUSINESS: Maximizing Revenue A manufacturer finds that the price function for autofocus cameras is $p(x) = 20e^{-0.0077x}$, where p is the price in dollars at which quantity x (in thousands) will be sold. Find the quantity x that maximizes revenue.

4.4 Two Applications to Economics: Relative Rates and Elasticity of Demand

48–49. ECONOMICS: Relative Rate of Change The gross domestic product of a developing country is forecast to be $G(t) = 5 + 2e^{0.01t}$ million dollars t years from now. Find the relative rate of change:

48. 20 years from now.   **49.** 10 years from now.

50. ECONOMICS: Elasticity of Demand A South American country exports coffee and estimates the demand function to be $D(p) = 63 - 2p^2$. If the country wants to raise revenues to improve its balance of payments, should it raise or lower the price from the present level of $3 per pound?

51. ECONOMICS: Elasticity of Demand A South African country exports gold and estimates the demand function to be $D(p) = 400\sqrt{2600 - p}$. If the country wants to raise revenues to improve its balance of payments, should it raise or lower the price from the present level of $1600 per ounce?

52. ECONOMICS: Elasticity of Demand The demand function for cigarettes is of the form $D(p) = 1.2p^{-0.44}$, where p is the price of a pack of cigarettes and $D(p)$ is the demand measured in packs per day per capita. Find the elasticity of demand. [*Note:* You will find that demand is inelastic. This means that taxation, which acts like a price increase, is an ineffective way of discouraging smoking but is an effective way of raising revenue.]

Source: American Economic Review 70

(53) **SOCIAL SCIENCES:** Relative Rate of Change The population of a city x years from now is projected to be $P(x) = 3.25 + 0.04x + 0.002x^3$ million people (for $0 \le x \le 10$).   Find the relative rate of change 9 years from now.

54. **ECONOMICS:** Elasticity of Demand A boat dealer finds that the demand function for outboard motor boats near a large lake is $D(p) = 200 - 20p + p^2 - 0.03p^3$

(for $0 \le p \le 15$),   where p is the selling price in thousands of dollars.

a. Use a graphing calculator to find the elasticity of demand at a price of $10,000. [*Hint:* What value of p corresponds to $10,000?]

b. Should the dealer raise or lower the price from this level to increase revenue?

c. Find the price at which elasticity equals 1.

Integration and Its Applications

5

© sgm/Shutterstock.com

What You'll Explore

Many quantities grow by the continuous accumulation of small changes, the way a bank account grows by accumulating small amounts of interest. In this chapter you will learn how to add such continuous small changes using a technique called *integration,* and then use it in applications such as calculating total revenue, measuring the accumulation of knowledge, and even estimating the "tar" that accumulates in the filter when smoking a cigarette.

Cigarette Smoking

Most cigarettes today have filters to absorb some of the toxic material or "tar" in the smoke before it is inhaled. The tobacco near the filter acts like an additional filter, absorbing tar until it is itself smoked, at which time it releases all of its accumulated toxins. A typical cigarette consists of 8 centimeters of tobacco followed by a filter.

As the cigarette is smoked, tar is typically inhaled at the rate of $r(x) = 300e^{0.025x} - 240e^{0.02x}$ milligrams (mg) of tar per centimeter (cm) of tobacco, where x is the distance along the cigarette.* The amount of tar inhaled from any particular segment of the cigarette is the area under the graph of this function over that interval, which can be calculated by a process called *definite integration*, as explained in this chapter.

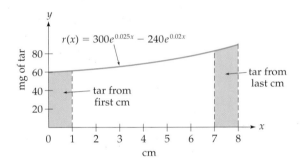

The results of integrating over the first and last centimeter of the cigarette are as follows (omitting the details—see Exercise 96 on page 343). The numbers represent milligrams of tar from the beginning and end of the cigarette.

$$\begin{pmatrix} \text{Tar from} \\ \text{first cm} \end{pmatrix} = \int_0^1 (300e^{0.025x} - 240e^{0.02x})\, dx \approx 61$$

Integrate from 0 to 1 for the first centimeter

$$\begin{pmatrix} \text{Tar from} \\ \text{last cm} \end{pmatrix} = \int_7^8 (300e^{0.025x} - 240e^{0.02x})\, dx \approx 83$$

Integrate from 7 to 8 for the last centimeter

Notice that the last centimeter releases significantly more tar (about 36% more) than the first centimeter.

Moral: The Surgeon General has determined that smoking is hazardous to your health, and the last puffs are 36% more hazardous than the first.

This is just one of the many applications of integration discussed in this chapter.

*The actual rate depends upon the type of cigarette and the proportion of time that it is smoked rather than left to burn in the air. For further information, see Helen Marcus-Roberts and Maynard Thompson (Eds.), *Modules in Applied Mathematics*, Vol. 4, *Life Science Models*, Springer-Verlag, pp. 238–249.

5.1 Antiderivatives and Indefinite Integrals

Introduction

We have been studying differentiation and its uses. We now consider the reverse process, *anti***differentiation,** which, for a given derivative, essentially recovers the original function. Antidifferentiation has many uses. For example, differentiation turns a cost function into a marginal cost function, and so antidifferentiation turns marginal cost back into cost. Later we will use antidifferentiation for other purposes, such as finding areas.

Antiderivatives and Indefinite Integrals

We begin with a simple example of antidifferentiation. Since the derivative of x^2 is $2x$, we may reverse the order and say:

> an antiderivative of $2x$ is x^2 Since the derivative of x^2 is $2x$

There are, however, other antiderivatives of $2x$. Each of the following is an antiderivative of $2x$:

$$x^2 + 1 \qquad x^2 - 17 \qquad x^2 + e \qquad\qquad \text{Since the derivative of each is } 2x$$

Clearly, we may add *any* constant to x^2 and the derivative will still be $2x$. There are no other possibilities, and so the *most general* antiderivative of $2x$ is $x^2 + C$ where C is any constant. The most general antiderivative is called the **indefinite integral** and is written with the function between an **integral sign** $\int$ and a *dx:*

$$\int 2x\,dx = x^2 + C$$

The indefinite integral of $2x$ is $x^2 + C$
(because the derivative of $x^2 + C$ is $2x$)

Integral sign ↑ ↑ Integrand ↑ Arbitrary constant

The function to be integrated (here $2x$) is called the **integrand.** The dx reminds us that the variable of integration is x. The constant C is called an **arbitrary constant** because it may take any value, positive, negative, or zero.

An indefinite integral (sometimes called simply an **integral**) can be checked by differentiation: The derivative of the answer must equal the integrand (as is the case with $x^2 + C$ and $2x$).

Indefinite Integral

$$\int f(x)\,dx = g(x) + C \qquad\qquad \text{The integral of } f(x) \text{ is } g(x) + C$$

if and only if

$$g'(x) = f(x) \qquad\qquad \text{The derivative of } g(x) \text{ is } f(x)$$

 LOOKING BACK

The power rule (for *differentiation*) was introduced on page 100.

Integration Rules

There are several "rules" that simplify integration. The first, which is one of the most useful rules in all of calculus, shows how to integrate x to a constant power. It comes from "reversing" the Power Rule for differentiation.

Power Rule for Integration

	Brief Example
$$\int x^n dx = \frac{1}{n+1}x^{n+1} + C \qquad (n \neq -1)$$	$$\int x^3\, dx = \frac{1}{4}x^4 + C$$

In words: To integrate x to a power, add 1 to the power and multiply by 1 over the new power (for any power other than -1).

 Be Careful To avoid mistakes, you should always check your answer by differentiating it to be sure you get back the original integrand. In this case, differentiating the answer $\frac{1}{4}x^4 + C$ *does* give the original integrand x^3, so the answer is correct.

PRACTICE PROBLEM 1

Find $\int x^2\, dx$ and check your answer by differentiation.

Solution on page 315 >

The proof of the Power Rule for Integration consists simply of differentiating the right-hand side.

$$\frac{d}{dx}\left(\frac{1}{n+1}x^{n+1} + C\right) = \frac{1}{n+1}(n+1)x^n = x^n \qquad\qquad (n+1)\text{'s cancel}$$

The power $n+1$ The power   Simplified
brought down   decreased by 1

Since the derivative is the integrand x^n, the Power Rule for Integration is correct.

To integrate functions like $\sqrt{x}$ and $\dfrac{1}{x^2}$ we first express them as powers.

EXAMPLE 1   EXPRESSING AS A POWER BEFORE INTEGRATING

$$\int \sqrt{x}\, dx = \int x^{1/2}\, dx = \frac{2}{3}x^{3/2} + C \qquad \text{Using the Power Rule for Integration with } n = 1/2$$

$$n = \frac{1}{2} \qquad\qquad n+1 = \frac{1}{2}+1 = \frac{3}{2}$$

$$\frac{1}{n+1} = \frac{1}{3/2} = \frac{2}{3}$$

$$= \frac{2}{3}\sqrt{x^3} + C \qquad\qquad \text{Returning to radical form}$$

Take Note

If the original problem was in radical form, we express the answer in radical form.

Note that in the answer, the multiple $\frac{2}{3}$ and the exponent $\frac{3}{2}$ are reciprocals of each other. Can you see, from the Power Rule, why this will always be so?

Take Note

Or if the original problem was in fractional form, we express the answer in fractional form.

EXAMPLE 2   **EXPRESSING AS A POWER BEFORE INTEGRATING**

$$\int \frac{1}{x^2}\,dx = \int x^{-2}\,dx = \frac{1}{-1}x^{-1} + C = -x^{-1} + C = -\frac{1}{x} + C$$

$n = -2$

$n + 1 = -1$

Simplified answer   Returning to fractional form

PRACTICE PROBLEM 2

Find $\int \dfrac{dx}{x^3}.$ $\left[\textit{Hint: } \text{Equivalent to } \int \dfrac{1}{x^3}\,dx. \right]$ Solution on page 315 >

EXAMPLE 3   **INTEGRATING 1**

$$\int 1\,dx = \int x^0\,dx = \frac{1}{1}x^1 + C = x + C$$ Using $x^0 = 1$

$n = 0$ $n + 1 = 1$

This result is so useful that it should be memorized.

$$\int 1\,dx = x + C$$ The integral of 1 is $x + C$

EXAMPLE 4   **INTEGRATING WITH OTHER VARIABLES**

a. $\displaystyle\int t^3\,dt = \frac{1}{4}t^4 + C$

Using the Power Rule ($n = 3$) with dt since the variable is t ("integrating with respect to t")

b. $\displaystyle\int u^{-4}\,du = \frac{1}{-3}u^{-3} + C = -\frac{1}{3}u^{-3} + C$

Using the Power Rule ($n = -4$) with du since the variable is u ("integrating with respect to u")

PRACTICE PROBLEM 3

Find $\displaystyle\int z^{-1/2}\,dz.$ Solution on page 315 >

   Be Careful   Notice what happens if we try to integrate x^{-1}:

$$\int x^{-1}\,dx = \frac{1}{0}x^0 + C$$ Undefined because of the $\frac{1}{0}$

$n = -1$ $n + 1 = 0$

The Power Rule for Integration fails for the exponent -1 because it leads to the undefined expression $\frac{1}{0}$. For this reason, the Power Rule for Integration includes the restriction "$n \neq -1$." It can integrate any power of x *except* x^{-1}. (In the next Section we will see how to integrate x^{-1}.)

The Constant Multiple and Sum Rules for differentiation lead immediately to analogous rules for simplifying integrals. The first rule says that a constant may be moved across the integral sign.

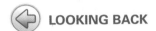

LOOKING BACK

These differentiation rules were introduced on pages 101 and 103.

Constant Multiple Rule for Integration

For any constant k,

$$\int k \cdot f(x)\, dx = k \int f(x)\, dx$$

The integral of a constant times a function is the constant times the integral of the function

---

EXAMPLE 5   **USING THE CONSTANT MULTIPLE RULE**

$$\int 6x^2\, dx = 6 \int x^2\, dx$$

Using the Constant Multiple Rule to move the 6 across the integral sign

$$= 6 \cdot \frac{1}{3}x^3 + C$$

Using the Power Rule with $n = 2$

$$= 2x^3 + C$$

Simplifying

EXAMPLE 6   **INTEGRATING A CONSTANT**

$$\int 7\, dx = 7 \int 1\, dx = 7x + C$$

Moving the constant outside   The integral of 1 is x (plus C)

This leads to a very useful general rule. For any constant k,

---

Integral of a Constant

$$\int k\, dx = kx + C$$

The integral of a constant is the constant times x (plus C)

The second rule says that the sum of two functions may be integrated one at a time.

Sum Rule for Integration

$$\int [f(x) + g(x)]\, dx = \int f(x)\, dx + \int g(x)\, dx$$

The integral of a sum is the sum of the integrals

EXAMPLE 7	**USING THE SUM RULE**

$$\int (x^2 + x^3)\, dx = \int x^2\, dx + \int x^3\, dx \qquad \text{Using the Sum Rule to break the integral into two integrals}$$

$$= \frac{1}{3}x^3 + \frac{1}{4}x^4 + C \qquad \text{Using the Power Rule on each (one ``+C'' is enough)}$$

The Sum Rule can be extended to integrate the sum or difference of *any* number of terms, writing only one +C at the end (since any number of arbitrary constants can be added together to give just one).

EXAMPLE 8	**INTEGRATING A SUM OF POWERS**

$$\int (6x^2 - 3x^{-2} + 5)\, dx$$

$$= 6\int x^2\, dx - 3\int x^{-2}\, dx + 5\int 1\, dx \qquad \text{Breaking up the integral and moving constants outside}$$

$$= 6 \cdot \frac{1}{3}x^3 - 3 \cdot \frac{1}{-1}x^{-1} + 5x + C \qquad \text{Integrating each separately}$$

From integrating the 1

$$= 2x^3 + 3x^{-1} + 5x + C \qquad \text{Simplifying}$$

EXAMPLE 9	**REWRITING BEFORE INTEGRATING**

$$\int \left(\frac{3\sqrt{x}}{2} - \frac{2}{\sqrt{x}} \right) dx = \int \left(\frac{3}{2}x^{1/2} - 2x^{-1/2} \right) dx \qquad \text{Writing as powers of } x$$

$$= \frac{3}{2} \cdot \frac{2}{3}x^{3/2} - 2 \cdot \frac{2}{1}x^{1/2} + C \qquad \text{Integrating each term separately}$$

$$= x^{3/2} - 4x^{1/2} + C \qquad \text{Simplifying}$$

$$= \sqrt{x^3} - 4\sqrt{x} + C \qquad \text{Returning to radical form}$$

PRACTICE PROBLEM 4

Find $\displaystyle\int \left(\sqrt[3]{w} - \frac{4}{w^3} \right) dw.$

Solution on page 316 >

Some integrals are so simple that they can be integrated "at sight."

EXAMPLE 10	**INTEGRATING "AT SIGHT"**

a. $\displaystyle\int 4x^3\, dx = x^4 + C \qquad$ By remembering that $4x^3$ is the derivative of x^4

b. $\displaystyle\int 7x^6\, dx = x^7 + C \qquad$ By remembering that $7x^6$ is the derivative of x^7

Integrate "at sight" by noticing that each integrand is of the form nx^{n-1} and integrating to x^n without working through the Power Rule.

a. $\int 5x^4\,dx$ **b.** $\int 3x^2\,dx$ Solutions on page 316 >

Algebraic Simplification of Integrals

Sometimes an integrand needs to be multiplied out or otherwise rewritten before it can be integrated.

EXAMPLE 11 EXPANDING BEFORE INTEGRATING

Find $\int x^2(x+6)^2\,dx$.

Solution

$$\int x^2(x+6)^2\,dx = \int x^2\underbrace{(x^2+12x+36)}_{(x+6)^2}\,dx \qquad \text{``Squaring out'' the } (x+6)^2$$

$$= \int (x^4 + 12x^3 + 36x^2)\,dx \qquad \text{Multiplying out}$$

$$= \frac{1}{5}x^5 + 12\cdot\frac{1}{4}x^4 + 36\cdot\frac{1}{3}x^3 + C \qquad \text{Integrating each term separately}$$

$$= \frac{1}{5}x^5 + 3x^4 + 12x^3 + C \qquad \text{Simplifying}$$

PRACTICE PROBLEM 6

Find $\int \dfrac{6t^2 - t}{t}\,dt$. [*Hint:* First simplify the integrand.]

Solution on page 316 >

Since differentiation turns a cost function into a marginal cost function, integration turns a marginal cost function back into a cost function. To evaluate the constant, however, we need the fixed costs.

EXAMPLE 12 RECOVERING COST FROM MARGINAL COST

A company's marginal cost function is $MC(x) = 6\sqrt{x}$ and the fixed cost is $1000. Find the cost function.

Solution

We integrate the marginal cost to find the cost function.

$$C(x) = \int MC(x)\,dx = \int 6\sqrt{x}\,dx = 6\int x^{1/2}\,dx \qquad \text{Integrating}$$

$$= 6\cdot\frac{2}{3}x^{3/2} + K = 4x^{3/2} + K = 4\sqrt{x^3} + K \qquad \begin{array}{l}\text{From the Power Rule}\\\text{(using } K \text{ to avoid}\\\text{confusion with } C \text{ for cost)}\end{array}$$

It remains to find the value of the constant K. The cost function evaluated at $x = 0$ always gives the fixed cost (because when nothing is produced, only the fixed cost remains):

$$\underbrace{C(0) = K}_{\text{Fixed cost}}$$

Evaluating $C(x) = 4\sqrt{x^3} + K$ at $x = 0$ gives K

Therefore, K equals the fixed cost, which is given as 1000, so $K = 1000$. The completed cost function is obtained by replacing K by 1000.

$$C(x) = 4\sqrt{x^3} + 1000$$

$C(x) = 4\sqrt{x^3} + K$ with $K = 1000$

 Be Careful In the preceding Example, the arbitrary constant turned out to be the fixed cost. This will not always be so, as we will see in the next section.

We found the cost function in two steps: integrating the marginal cost and then evaluating the arbitrary constant. Other problems that involve recovering a quantity from its *marginal* (or rate of change) together with a particular value are solved in the same way:

To Recover a Quantity from its Rate of Change

1. Integrate the *marginal* or *rate of change*.
2. Use the given fixed value to evaluate the arbitrary constant.

For example, given the rate of growth of an economy and its present size, we can find the size of the economy at any time in the future.

 EXAMPLE 13   **RECOVERING A QUANTITY FROM ITS RATE OF CHANGE**

The gross domestic product (GDP) of a country is \$78 billion and growing at the rate of $4.4t^{-1/3}$ billion dollars per year after t years. Find a formula for the GDP after t years. Then use your formula to find the GDP after 8 years.

Solution

$$G(t) = \int 4.4t^{-1/3}\,dt = 4.4 \int t^{-1/3}\,dt$$

Integrating the rate of change

$$= 4.4 \left(\frac{3}{2}\right) t^{2/3} + C = 6.6t^{2/3} + C$$

Using the Power Rule and then simplifying

As before, evaluating $G(t) = 6.6t^{2/3} + C$ at $t = 0$ gives $G(0) = C$, which shows that the constant C is the GDP at time $t = 0$, which is given as \$78 billion. Therefore

$$G(t) = 6.6t^{2/3} + 78$$

$G(t) = 6.6t^{2/3} + C$ with $C = 78$

For the GDP after 8 years, we evaluate $G(t)$ at $t = 8$:

$$G(8) = 6.6 \cdot \underbrace{8^{2/3}}_{4} + 78 = 6.6(4) + 78 = 104.4$$

$G(t) = 6.6t^{2/3} + 78$ at $t = 8$

Therefore, the GDP after 8 years is \$104.4 billion.

Geometrical Interpretation of the Arbitrary Constant

Notice that the $+C$, which at first may have seemed like a pointless formality, enabled us to give the cost function its correct fixed cost in Example 12 and to give the GDP its correct current value in Example 13.

The arbitrary constant can be interpreted *geometrically*. In $\int 2x\,dx = x^2 + C$, the solution $x^2 + C$ is actually a whole *collection* of curves, one for each value of C. The following graph shows three of these curves, $x^2 + 2$, x^2, and $x^2 - 2$, corresponding to C being 2, 0, and -2.

$y = x^2 + C$ with
tangent lines at $x = 1$

LOOKING BACK

The $+C$ amounts to a *vertical shift* of the curve, as we saw on page 56.

Notice that each curve has the same slope at $x = 1$ (their tangent lines are parallel). In fact, all three curves have the same slope at *any* given x-value (we call these *parallel* curves). This is because each comes from integrating $2x$, so each function has *derivative* $2x$, giving each the same slope. This is the geometric meaning of integrating a function: The result is a collection of parallel curves that all have the same slope (given by the original function) at any particular x-value.

Integrals as Continuous Sums

We have seen that differentiation is a kind of "breaking down," changing total cost into marginal (per unit) cost. Integration, the reverse process, is therefore a kind of "adding up," recovering the total from its parts. For example, integration recovers the total cost from marginal (per unit) cost, and integration adds up the growth of an economy over several years to give its total size.

To put this another way, ordinary addition is for summing "chunks," such as $2 plus $3 equals $5, while integration is for summing *continuous* change, such as the slow but steady growth of an economy over time. In other words:

> Integration is continuous summation.

Solutions TO PRACTICE PROBLEMS

1. $\int x^2\,dx = \dfrac{1}{3}x^3 + C$ Check: $\dfrac{d}{dx}\left(\dfrac{1}{3}x^3 + C\right) = x^2$

2. $\int \dfrac{1}{x^3}\,dx = \int x^{-3}\,dx = -\tfrac{1}{2}x^{-2} + C = -\dfrac{1}{2}\dfrac{1}{x^2} + C = -\dfrac{1}{2x^2} + C$

3. $\int z^{-1/2}\,dz = \dfrac{1}{1/2}z^{1/2} + C = 2z^{1/2} + C$

4. $\displaystyle\int (w^{1/3} - 4w^{-3})\, dw = \int w^{1/3}\, dw - 4\int w^{-3}\, dw$

$$= \tfrac{3}{4}w^{4/3} - 4\left(-\tfrac{1}{2}\right)w^{-2} + C$$

$$= \tfrac{3}{4}w^{4/3} + 2w^{-2} + C = \frac{3}{4}\sqrt[3]{w^4} + \frac{2}{w^2} + C$$

5. a. $\displaystyle\int 5x^4\, dx = x^5 + C$ **b.** $\displaystyle\int 3x^2\, dx = x^3 + C$

6. $\displaystyle\int \frac{6t^2 - t}{t}\, dt = \int \frac{t(6t - 1)}{t}\, dt$

$$= \int (6t - 1)\, dt$$

$$= 6\int t\, dt - \int 1\, dt = 6 \cdot \tfrac{1}{2}t^2 - t + C$$

$$= 3t^2 - t + C$$

5.1   Section Summary

In this section we have discussed both the *techniques* of integration and the *meanings* or *uses* of integration.

On the *technical* side, we defined indefinite integration as the reverse process of differentiation. The Power Rule enables us to integrate powers.

$$\int x^n\, dx = \frac{1}{n+1}x^{n+1} + C \qquad n \neq -1$$

(Increasing the exponent by 1 should seem quite reasonable: Differentiation *lowers* the exponent by 1, so integration, the reverse process, should *raise* it by 1.) The Constant Multiple and Sum Rules "extended" integration to more complicated functions such as polynomials. Sometimes algebra is necessary to express a function in power form.

 Be Careful Don't forget the C.

As for the *uses* of integration, it recovers cost, revenue, and profit from their marginals, and in general it recovers any quantity from its rate of change (together with a particular value). The inverse relationship between integration and differentiation is shown in the following diagram.

5.1   Exercises

1–40. Find each indefinite integral.

1. $\displaystyle\int x^4\,dx$ **2.** $\displaystyle\int x^7\,dx$ **3.** $\displaystyle\int x^{2/3}\,dx$

4. $\displaystyle\int x^{3/2}\,dx$ **5.** $\displaystyle\int \sqrt{u}\,du$ **6.** $\displaystyle\int \sqrt[3]{u}\,du$

7. $\displaystyle\int \frac{dw}{w^4}$ **8.** $\displaystyle\int \frac{dw}{w^2}$ **9.** $\displaystyle\int \frac{dz}{\sqrt{z}}$

10. $\displaystyle\int \frac{dz}{\sqrt[3]{z}}$ **11.** $\displaystyle\int 6x^5\,dx$ **12.** $\displaystyle\int 9x^8\,dx$

13. $\displaystyle\int (8x-5)\,dx$ **14.** $\displaystyle\int (2-4x)\,dx$

15. $\displaystyle\int (8x^3-3x^2+2)\,dx$ **16.** $\displaystyle\int (12x^3+3x^2-5)\,dx$

17. $\displaystyle\int \left(6\sqrt{x}+\frac{1}{\sqrt[3]{x}}\right)dx$ **18.** $\displaystyle\int \left(3\sqrt{x}+\frac{1}{\sqrt{x}}\right)dx$

19. $\displaystyle\int \left(16\sqrt[3]{x^5}-\frac{16}{\sqrt[3]{x^5}}\right)dx$ **20.** $\displaystyle\int \left(14\sqrt[4]{x^3}-\frac{3}{\sqrt[4]{x^3}}\right)dx$

21. $\displaystyle\int \left(10\sqrt[3]{t^2}+\frac{1}{\sqrt[3]{t^2}}\right)dt$ **22.** $\displaystyle\int \left(21\sqrt{t^5}+\frac{6}{\sqrt{t^5}}\right)dt$

23. $\displaystyle\int \left(\frac{4}{z^3}+\frac{1}{\sqrt{z}}\right)dz$ **24.** $\displaystyle\int \left(\frac{1}{z^2}+\frac{1}{\sqrt[3]{z}}\right)dz$

25. $\displaystyle\int (x-1)^2\,dx$ **26.** $\displaystyle\int (x+2)^2\,dx$

27. $\displaystyle\int 12x^2\,(x-1)\,dx$ **28.** $\displaystyle\int x^2\,(8x+3)\,dx$

29. $\displaystyle\int (1+10w)\sqrt{w}\,dw$ **30.** $\displaystyle\int (1-7w)\sqrt[3]{w}\,dw$

31. $\displaystyle\int \frac{6x^3-6x^2+x}{x}\,dx$ **32.** $\displaystyle\int \frac{4x^4+4x^2-x}{x}\,dx$

33. $\displaystyle\int (x-2)(x+4)\,dx$ **34.** $\displaystyle\int (x+5)(x-3)\,dx$

35. $\displaystyle\int (r-1)(r+1)\,dr$ **36.** $\displaystyle\int (3s+1)(3s-1)\,ds$

37. $\displaystyle\int \frac{x^2-1}{x+1}\,dx$ **38.** $\displaystyle\int \frac{x^2-1}{x-1}\,dx$

39. $\displaystyle\int (t+1)^3\,dt$ **40.** $\displaystyle\int (t-1)^3\,dt$

41. a. Verify that $\int x^2\,dx = \frac{1}{3}x^3 + C$.
   b. Graph the five functions $\frac{1}{3}x^3-2$, $\frac{1}{3}x^3-1$, $\frac{1}{3}x^3$, $\frac{1}{3}x^3+1$, and $\frac{1}{3}x^3+2$ (the solutions for five different values of C) on the window $[-3,3]$ by $[-5,5]$. Use TRACE to see how the constant shifts the curve vertically.
   c. Find the slopes (using NDERIV or dy/dx) of several of the curves at a particular x-value and check that in each case the slope is the square of the x-value. This verifies that the derivative of each curve is x^2, and so each is an integral of x^2.

42. a. Graph the five functions $\ln x-2$, $\ln x-1$, $\ln x$, $\ln x+1$, and $\ln x+2$ on the window $[0,4]$ by $[-3,3]$.
   b. Find the slope (using NDERIV or dy/dx) of several of the curves at a particular x-value and check that in each case the slope is the reciprocal of the x-value. This suggests that the derivative of each function is $1/x$.
   c. Based on part (b), conjecture what is the indefinite integral of the function $1/x$ (for $x>0$).

Applied Exercises

43. BUSINESS: Cost A company's marginal cost function is $MC = 20x^{3/2} - 15x^{2/3} + 1$, where x is the number of units, and fixed costs are $4000. Find the cost function.

44. BUSINESS: Cost A company's marginal cost function is $MC = 21x^{4/3} - 6x^{1/2} + 50$, where x is the number of units, and fixed costs are $3000. Find the cost function.

45. BUSINESS: Revenue A company's marginal revenue function is $MR = 120 - 12x^{1/3}$, where x is the number of units. Find the revenue function. (Evaluate C so that revenue is zero when nothing is produced.)

46. BUSINESS: Revenue A company's marginal revenue function is $MR = 100 - 10x^{1/4}$, where x is the number of units. Find the revenue function. (Evaluate C so that revenue is zero when nothing is produced.)

47. GENERAL: Velocity A Porsche 997 Turbo Cabriolet can accelerate from a standing start to a speed of $v(t) = -0.24t^2 + 18t$ feet per second after t seconds (for $0 \le t < 40$).

Source: Porsche

   a. Find a formula for the distance that it will travel from its starting point in the first t seconds. [*Hint:* Integrate velocity to find distance, and then use the fact that distance is 0 at time $t = 0$.]
   b. Use the formula that you found in part (a) to find the distance that the car will travel in the first 10 seconds.

48. GENERAL: Velocity A Mercedes-Benz Brabus Rocket can accelerate from a standing start to a speed of $v(t) = -0.15t^2 + 14t$ feet per second after t seconds (for $0 \le t < 40$).   *(continues)*

Source: Mercedes-Benz

a. Find a formula for the distance that it will travel from its starting point in the first t seconds. [*Hint:* Integrate velocity to find distance, and then use the fact that distance is 0 at time $t = 0.$]

b. Use the formula that you found in part (a) to find the distance that the car will travel in the first 10 seconds.

49. GENERAL: Learning A person can memorize words at the rate of $3/\sqrt{t}$ words per minute.

a. Find a formula for the total number of words that can be memorized in t minutes. [*Hint:* Evaluate C so that 0 words have been memorized at time $t = 0.$]

b. Use the formula that you found in part (a) to find the total number of words that can be memorized in 25 minutes.

50. BIOMEDICAL: Temperature A patient's temperature is 104 degrees Fahrenheit and is changing at the rate of $t^2 - 3t$ degrees per hour, where t is the number of hours since taking a fever-reducing medication $(0 \le t \le 3)$.

a. Find a formula for the patient's temperature after t hours. [*Hint:* Evaluate the constant C so that the temperature is 104 at time $t = 0.$]

b. Use the formula that you found in part (a) to find the patient's temperature after 3 hours.

51. ENVIRONMENTAL SCIENCE: Pollution A chemical plant is adding pollution to a lake at the rate of $40\sqrt{t^3}$ tons per year, where t is the number of years that the plant has been in operation.

a. Find a formula for the total amount of pollution that will enter the lake in the first t years of the plant's operation. [*Hint:* Evaluate C so that no pollution has been added at time $t = 0.$]

b. Use the formula that you found in part (a) to find how much pollution will enter the lake in the first 4 years of the plant's operation.

c. If all life in the lake will cease when 400 tons of pollution have entered the lake, will the lake "live" beyond 4 years?

52. BUSINESS: Appreciation A $20,000 art collection is increasing in value at the rate of $300\sqrt{t}$ dollars per year after t years.   (*continues*)

a. Find a formula for its value after t years. [*Hint:* Evaluate C so that its value at time $t = 0$ is $20,000.]

b. Use the formula that you found in part (a) to find its value after 25 years.

53. GENERAL: Young-Adult Population The U.S. Census Bureau estimates that the number of young adults (ages 18 to 24) in the United States to be 31,000 (in thousands) in 2010 and changing at the rate of $-x^2 + 50x - 300$ thousand per year, where x is the number of years since 2010. Find a formula for the size of this population at any time x and use your formula to predict the size of the young-adult population in 2040. [*Hint:* Keep all calculations in units of *thousands*.]

Source: Statistical Abstract of the United States

54. ENVIRONMENTAL SCIENCE: Global Temperatures In 2013, the average surface temperature of the earth was 58 (degrees Fahrenheit) and, according to one study, increasing at the rate of $0.0154t^{0.4}$ degrees per year, where t is the number of years since 2013. Find a formula for the temperature at any time t and use your formula to predict the temperature in 2100. [*Note:* Such global warming is expected to raise sea levels and cause widespread coastal flooding.]

Sources: Intergovernmental Panel on Climate Change, Worldwatch

55. BUSINESS: Revenue In 2013, Amazon's annual revenue was 74 billion dollars and growing at the rate of $3.2x + 17.4$ billion dollars per year, where x is the number of years since 2013. Find a formula to predict Amazon's annual revenue at any time x and use your formula to predict their revenue in 2020.

Source: New York Times

56. BUSINESS: Revenue In 2013, annual revenues at 3D Systems were 474 million dollars and growing at the rate of $28.4x + 120$ million dollars per year, where x stands for the number of years since 2013. Find a formula for 3D Systems' revenues at any time x and use your formula to predict their revenues in 2020.

Source: Standard & Poor's

Review Exercise

This exercise will be important in the next section.

57. Find $\dfrac{d}{dx} \ln(-x)$.

Conceptual Exercises

58. What power of x *cannot* be integrated by the Power Rule?

59. What is the integral of *zero*? $\int 0 \, dx = ?$

60. One student tells another: "Once you've got one anti-derivative, you've got them all." What did the student mean?

61. If $f(x)$ and $g(x)$ have the same derivative, how are $f(x)$ and $g(x)$ related?

62. Show that the integral of a product is *not* the product of the integrals by carrying out the following steps:

  a. Find the integral of the product $\int x \cdot x \, dx$ by evaluating $\int x^2 \, dx$.

  b. Find the corresponding product of the integrals $\int x \, dx \cdot \int x \, dx$.

  c. Do the answers for parts (a) and (b) agree?

63. You can move a *constant* across the integral sign, but can you do the same with a *variable*? Find out by carrying out the following steps:

  a. Find $\int x \cdot x \, dx$ by evaluating $\int x^2 \, dx$.

  b. Find $x \cdot \int x \, dx$.

  c. Do the answers for parts (a) and (b) agree?

64. Integrating x^2, you get $\frac{1}{3}x^3 + C$, and then differentiating $\frac{1}{3}x^3 + C$ gives back exactly the x^2 you began with. What if you begin with x^2 but *first* differentiate and *then* integrate—do you get back exactly what you began with?

65. Show that the integral of a quotient is *not* the quotient of the integrals by carrying out the following steps:

  a. Find the integral of the quotient $\int \dfrac{x}{x} \, dx$ by evaluating $\int 1 \, dx$.

  b. Find the corresponding quotient of the integrals $\dfrac{\int x \, dx}{\int x \, dx}$.

  c. Do the answers for parts (a) and (b) agree?

66. Supply the most appropriate single word: Integrating the *rate of change* of a population gives the _____.

67. Supply the most appropriate single word: Integrating the *marginal cost* gives the _____.

68. For each of the following casual, everyday sentences, fill in the blank with either the word *differentiate* or the word *integrate*, whichever is more appropriate.

  a. To find the value of an investment, we _____ how fast it is growing.

  b. To find how quickly it is getting warmer, we _____ temperature.

  c. To find distance, we _____ speed.

  d. To find acceleration, we _____ speed.

69. For each of the following casual, everyday sentences, fill in the blank with one of the words *differentiate* or *integrate*, whichever is more appropriate.

  a. To find marginal cost, we _____ cost.

  b. To find profit, we _____ marginal profit.

  c. To find how fast a stock price is rising, we _____ price.

  d. To find a population, we _____ how fast it is growing.

5.2	**Integration Using Logarithmic and Exponential Functions**

Introduction

In the previous section we defined integration as the reverse of differentiation, and we introduced several integration formulas. In this section we develop integration formulas involving logarithmic and exponential functions. One of these formulas will answer a question that we could not answer earlier—namely, how to integrate x^{-1}, the only power not covered by the Power Rule.

The Integral $\int e^{ax} \, dx$

On page 279 we saw that to *differentiate* e^{ax}, we *multiply* by a to get ae^{ax}. Therefore, to *integrate* e^{ax}, the reverse process, we must *divide* by a. That is, for any $a \neq 0$:

Integrating an Exponential Function

$$\int e^{ax} \, dx = \frac{1}{a} e^{ax} + C$$

Brief Example

$$\int e^{2x} \, dx = \frac{1}{2} e^{2x} + C$$

In words: The integral of e to a constant times x is 1 over the constant times the original function (plus C).

As always, we may check the answer by differentiation.

$$\frac{d}{dx}\left(\frac{1}{2}e^{2x} + C\right) = \frac{1}{2}\cdot 2 \cdot e^{2x} = e^{2x}$$ Using $\dfrac{d}{dx}e^{ax} = ae^{ax}$ and canceling the 2's

The result is the integrand e^{2x}, so the integration is correct.

The proof of this rule consists simply of differentiating the right-hand side.

$$\frac{d}{dx}\left(\frac{1}{a}e^{ax} + C\right) = \frac{1}{a}ae^{ax} = e^{ax}$$ Using $\dfrac{d}{dx}e^{ax} = ae^{ax}$ and canceling the a's

The result is the integrand, so the integration formula is correct.

⚠ **Be Careful**   To use the formula in the preceding box, the exponent of e cannot be more complicated than a constant times x. In Section 5.6 we will integrate more general exponential expressions.

EXAMPLE 1   **INTEGRATING EXPONENTIAL FUNCTIONS**

a. $\displaystyle\int e^{\frac{1}{2}x}\,dx = 2e^{\frac{1}{2}x} + C$
$\quad\underset{1/a}{\uparrow}$
$\qquad\qquad\qquad\qquad\qquad a = \dfrac{1}{2} \quad\text{so}\quad \dfrac{1}{a} = \dfrac{1}{\frac{1}{2}} = 2$

b. $\displaystyle\int 6e^{-3x}\,dx = 6\int e^{-3x}\,dx = 6\left(-\frac{1}{3}\right)\cdot e^{-3x} + C = -2e^{-3x} + C$
$\qquad\qquad\qquad\quad\underset{a=-3}{\uparrow}\qquad\underset{\frac{1}{a}=-\frac{1}{3}}{\uparrow}$

c. $\displaystyle\int e^{x}\,dx = 1e^{x} + C = e^{x} + C$
$\qquad\qquad\qquad\qquad\qquad\qquad a = 1, \quad\text{so}\quad \dfrac{1}{a} = 1$

Each of these answers may be checked by differentiation. The last one says that e^{x} is the integral of itself, just as e^{x} is the derivative of itself.

LOOKING BACK

We showed that e^{x} is its own derivative on page 276.

$$\int e^{x}\,dx = e^{x} + C$$ The integral of e^{x} is e^{x} (plus C)

PRACTICE PROBLEM 1

Find: **a.** $\displaystyle\int 12e^{4x}\,dx$ **b.** $\displaystyle\int e^{x/3}\,dx$

[*Hint:* For part (b), write $e^{x/3}$ as $e^{\frac{1}{3}x}$, then integrate, and rewrite again.]

Solutions on page 326 >

Be Careful When using these new integration formulas to solve applied problems, be careful to evaluate the constant C correctly. It will not always be equal to the initial value of the function.

EXAMPLE 2   FINDING TOTAL FLU CASES FROM THE RATE OF CHANGE

An influenza epidemic hits a large city and spreads at the rate of $12e^{0.2t}$ new cases per day, where t is the number of days since the epidemic began. The epidemic began with 4 cases.

a. Find a formula for the total number of flu cases in the first t days of the epidemic.

b. Use your formula to find the number of cases during the first 30 days.

Solution

a. To find the total number of cases, we integrate the growth rate $12e^{0.2t}$.

$$\underbrace{f(t)}_{\substack{\text{Total cases} \\ \text{in first } t \text{ days}}} = \int 12e^{0.2t}\, dt = 12 \int e^{0.2t}\, dt \qquad \text{Taking out the constant}$$

$$= 12\,\underset{\underset{5}{\big\uparrow}}{\frac{1}{0.2}}\, e^{0.2t} + C = 60e^{0.2t} + C \qquad \text{Using the } \int e^{ax}\,dx \text{ formula}$$

Evaluating $f(t)$ at $t = 0$ must give the initial number of cases:

$$\underbrace{f(0)}_{\substack{\text{Initial} \\ \text{number} \\ \text{of cases}}} = 60e^{0.2(0)} + C = \underbrace{60}_{e^0 = 1} + C \qquad \begin{array}{l} f(t) = 60e^{0.2t} + C \text{ evaluated} \\ \text{at } t = 0 \text{ (the beginning of} \\ \text{the epidemic)} \end{array}$$

This initial number must equal the given initial number, 4:

$$60 + C = 4 \qquad \begin{array}{l} \text{Initial number from the formula set} \\ \text{equal to the given initial number} \end{array}$$

$$C = -56 \qquad \text{Solving for } C$$

Replacing C by -56 gives the total number of flu cases within t days:

$$\underbrace{f(t)}_{\substack{\text{Number of cases} \\ \text{in first } t \text{ days}}} = \underbrace{60e^{0.2t} - 56}_{\substack{\text{Answer to} \\ \text{part (a)}}} \qquad f(t) = 60e^{0.2t} + C \text{ with } C = -56$$

b. To find the number within 30 days, we evaluate at $t = 30$:

$$f(30) = 60e^{0.2(30)} - 56 \qquad f(t) = 60e^{0.2t} + C \text{ at } t = 30$$

$$= 60e^6 - 56 \approx 24{,}150 \qquad \text{Using a calculator}$$

Therefore, within 30 days the epidemic will have grown to more than 24,000 cases.

Evaluating the Constant *C*

Notice that we did *not* simply replace the constant *C* by the initial number of cases, 4. (This would have given the wrong initial number of cases.) Instead, we evaluated the function at the initial time $t = 0$ and set it equal to the initial number of cases:

$$\underbrace{60e^0 + C}_{\substack{f(t) \text{ evaluated} \\ \text{at } t = 0}} = \underbrace{4}_{\substack{\text{Given} \\ \text{initial value}}}$$

We then solved to find the correct value of *C*, $C = -56$, which we then substituted into the formula. In general:

To evaluate the constant *C*

1. Evaluate the integral at the given number (usually $t = 0$) and set the result equal to the stated initial value.
2. Solve for *C*.
3. Write the answer with *C* replaced by its correct value.

The Integral $\int \dfrac{1}{x} \, dx$

The differentiation formula $\dfrac{d}{dx} \ln x = \dfrac{1}{x}$ can be read "backward" as an integration formula:

$$\int \frac{1}{x} \, dx = \ln x + C \qquad\qquad \text{The integral of 1 over } x \text{ is the} \\ \text{natural log of } x \text{ (plus } C\text{)}$$

This formula, however, is restricted to $x > 0$, for only then is $\ln x$ defined. For $x < 0$ we can differentiate $\ln(-x)$, giving

$$\frac{d}{dx} \ln(-x) = \frac{-1}{-x} = \frac{1}{x} \qquad\qquad \text{Using } \frac{d}{dx} \ln f = \frac{f'}{f} \text{ and simplifying}$$

This result says that for $x < 0$, the integral of $1/x$ is $\ln(-x)$. The negative sign in $\ln(-x)$ serves only to make the already negative *x* positive, and this could be accomplished just as well with absolute value bars.

$$\int \frac{1}{x} \, dx = \ln|x| + C \qquad\qquad \text{The integral of 1 over } x \text{ is the natural} \\ \text{logarithm of the absolute value of } x$$

This formula holds for negative *and* positive values of *x*, since in both cases $\ln|x|$ is defined. The integral can be written in three different ways, all of which have the same answer.

$$\int \frac{1}{x} \, dx = \int \frac{dx}{x} = \int x^{-1} \, dx = \ln|x| + C$$

EXAMPLE 3   INTEGRATING USING THE LN RULE

$$\int \frac{5}{2x} \, dx = \int \frac{5}{2} \frac{1}{x} \, dx = \frac{5}{2} \int \frac{1}{x} \, dx = \frac{5}{2} \ln|x| + C$$

$$\underset{\substack{\uparrow \\ \text{Taking out} \\ \text{the constant}}}{} \qquad\qquad \underset{\substack{\text{Using the} \\ \text{ln formula}}}{}$$

EXAMPLE 4   **INTEGRATING NEGATIVE POWERS**

$$\int (x^{-1} + x^{-2})\, dx = \int x^{-1}\, dx + \int x^{-2}\, dx = \ln|x| - x^{-1} + C$$

From the
natural log
formula

From the
Power Rule
with $n = -2$

PRACTICE PROBLEM 2

Find $\displaystyle\int \frac{3}{4x}\, dx.$

Solution on page 326 >

EXAMPLE 5   **FINDING TOTAL SALES FROM THE SALES RATE**

An electronics dealer estimates that during month t of a sale, a discontinued computer will sell at a rate of approximately $25/t$ per month, where $t = 1$ corresponds to the beginning of the sale, at which time none have been sold. Find a formula for the total number of computers that will be sold up to month t. Will the store's inventory of 64 computers be sold by month $t = 12$?

Solution

To find the total sales, we integrate the sales rate $\dfrac{25}{t}$:

$$\underbrace{S(t)}_{\substack{\text{Total sales in}\\\text{first } t \text{ months}}} = \int \frac{25}{t}\, dt = 25 \int \frac{1}{t}\, dt = 25 \ln t + C \qquad \text{Omitting absolute}\atop \text{values, since } t > 0$$

To evaluate C, we evaluate at the given starting time $t = 1$:

$$\underbrace{S(1)}_{\substack{\text{Initial number}\\\text{of sales}}} = 25 \underbrace{\ln 1}_{0} + C = C$$

The initial number of sales must be zero, giving $C = 0$, and substituting this into the sales function gives

$$S(t) = 25 \underset{\text{Total number sold up to month } t}{\ln t} \qquad\qquad S(t) = 25 \ln t + C \text{ with } C = 0$$

To find the number sold up to month 12, we evaluate at $t = 12$:

$$S(12) = 25 \ln 12 \approx 62 \qquad\qquad \text{Using a calculator}$$

Therefore, all but two of the 64 computers will be sold by month 12.

In the preceding Example the initial time (the beginning of the sale) was given as $t = 1$ rather than the more usual $t = 0$. The initial time will be clear from the problem.

LOOKING BACK

We first found a total from its rate of change on pages 313–314.

Consumption of Natural Resources

Just as the world population grows exponentially, so does the world's annual consumption of natural resources. We can estimate the total consumption at any time in the future by integrating the rate of consumption, and from this predict when the known reserves will be exhausted.

EXAMPLE 6	FINDING A FORMULA FOR TOTAL CONSUMPTION FROM THE RATE

The annual world consumption of silver is predicted to be $22.3e^{0.01t}$ thousand metric tons per year, where t is the number of years since 2014. Find a formula for the total silver consumption within t years of 2014 and estimate when the known world reserves of 540 thousand metric tons will be exhausted.

Source: U.S. Geological Survey

Solution

To find the total consumption, we integrate the rate $22.3e^{0.01t}$:

$$C(t) = \int 22.3e^{0.01t}\, dt = 22.3 \int e^{0.01t}\, dt \qquad \text{Taking out the constant}$$

Total silver consumed in first t years after 2014

$$= 22.3 \, \frac{1}{0.01} e^{0.01t} + C = 2230e^{0.01t} + C \qquad \text{Using the } \int e^{ax}\,dx \text{ formula}$$

100

The total consumed in the first zero years must be zero, so $C(0) = 0$.

$$C(0) = 2230e^0 + C = 2230 + C \qquad C(t) = 2230e^{0.01t} + C \text{ with } t = 0$$

0   Must equal zero

From $2230 + C = 0$ we find $C = -2230$. We substitute this into the formula for $C(t)$:

$$C(t) = 2230e^{0.01t} - 2230 \qquad C(t) = 2230\, e^{0.01t} + C \text{ with } C = -2230$$

Formula for total consumption within first t years since 2014

To predict when the total world reserves of 540 thousand metric tons will be exhausted, we set this function equal to 540 and solve for t:

$$2230e^{0.01t} - 2230 = 540 \qquad \text{Setting } C(t) \text{ equal to 540}$$

$$2230e^{0.01t} = 2770 \qquad \text{Adding 2230}$$

$$e^{0.01t} = \frac{2770}{2230} \approx 1.242 \qquad \text{Dividing by 2230}$$

$$\underbrace{\ln e^{0.01t}}_{0.01t} = \underbrace{\ln 1.242}_{0.217} \qquad \begin{array}{l}\text{Taking natural logs} \\ \text{and using } \ln e^x = x \\ \\ \text{Rounded}\end{array}$$

$$t \approx \frac{0.217}{0.01} = 21.7 \qquad \text{Solving } 0.01t = 0.217 \text{ for } t$$

Therefore, the known world supply of silver will be exhausted in about 22 years after 2014, which means in about the year 2036 (assuming that consumption continues at the predicted rate).

Realistically, silver will not suddenly disappear at this date, but its price will rise significantly as this date approaches because of scarcity. The theoretical depletion date is an indicator as to when such price rises will occur, unless other resources are found.

Spreadsheet Exploration

Since integration is continuous summation, instead of integrating we could add up the annual silver consumption for each year to see when the total will reach the known reserves of 540 thousand metric tons. The following spreadsheet* shows this annual consumption for each year (using the formula $22.3e^{0.01t}$ from the preceding Example) in column **B** continuing into column **F**, with the cumulative totals shown in column **C** continuing into column **G**.

B2   ✕ ✓ ◯   *fx* =22.3*EXP(0.01*A2)

	A	B	C	D	E	F	G
1	Year	Annual Consumption	Cumulative Consumption		Year	Annual Consumption	Cumulative Consumption
2	1	22.52411873	22.52411873		13	25.39587295	311.13781118
3	2	22.75048988	45.27460861		14	25.65110571	336.78891689
4	3	22.97913611	68.25374472		15	25.90890361	362.69782051
5	4	23.21008026	91.46382498		16	26.16929242	388.86711293
6	5	23.44334545	114.90717043		17	26.43229818	415.29941111
7	6	23.67895499	138.58612542		18	26.69794720	441.99735831
8	7	23.91693244	162.50305786		19	26.96626603	468.96362434
9	8	24.15730161	186.66035947		20	27.23728151	496.20090585
10	9	24.40008653	211.06044600		21	27.51102074	523.71192658
11	10	24.64531147	235.70575747		22	27.78751109	551.49943767
12	11	24.89300097	260.59875844		23	28.06678022	579.56621790
13	12	25.14317979	285.74193823		24	28.34885605	607.91507395

Known reserves
exhausted

Since the cumulative consumption reaches 540 thousand metric tons in year 22, we conclude that the known world reserves of silver will be exhausted sometime in 2036, in agreement with Example 6.

    The integration method of Example 6 has two advantages over addition: It is often easier (especially for a large number of years) and it includes changes in the amount *during* each year rather than just at the end of each year.

*To obtain this and other Spreadsheet Explorations, go to www.cengagebrain.com, search for this textbook and then scroll down to access the "free materials" tab.

Incidentally, silver is used mainly in batteries, electronics, and jewelry production.

Power Rule for Integration, Revisited

Our new integration formula shows how to integrate x^{-1}, the only power not covered by the Power Rule for Integration (see page 309). Therefore, we can now write one "combined" formula for the integral $\int x^n \, dx$ for *any power n*.

Integrals of Powers of *x*

$$\int x^n \, dx = \begin{cases} \dfrac{1}{n+1} x^{n+1} + C & \text{if } n \neq -1 \\[2mm] \ln|x| + C & \text{if } n = -1 \end{cases}$$

Use the Power Rule if *n* is *other* than -1

Use the ln formula if *n* equals -1

It is a curious fact that every power of *x* integrates to another power of *x*, with the single exception of x^{-1}, which integrates to an entirely different kind of function, the natural logarithm.

Solutions TO PRACTICE PROBLEMS

1. a. $\displaystyle\int 12e^{4x} \, dx = 12 \int e^{4x} \, dx = 12 \cdot \tfrac{1}{4} e^{4x} + C = 3e^{4x} + C$

b. $\displaystyle\int e^{x/3} \, dx = \int e^{\frac{1}{3}x} \, dx = 3e^{\frac{1}{3}x} + C = 3e^{x/3} + C$

2. $\displaystyle\int \frac{3}{4x} \, dx = \frac{3}{4} \int \frac{1}{x} \, dx = \frac{3}{4} \ln|x| + C$

5.2   Section Summary

We have three integration formulas:

$$\int x^n \, dx = \frac{1}{n+1} x^{n+1} + C \qquad\qquad n \neq -1$$

$$\int e^{ax} \, dx = \frac{1}{a} e^{ax} + C \qquad\qquad a \neq 0$$

$$\int \frac{1}{x} \, dx = \int \frac{dx}{x} = \int x^{-1} \, dx = \ln|x| + C$$

What is the difference between the bottom formula *with* and *without* the absolute value bars?

$$\int \frac{1}{x} \, dx = \ln|x| + C \qquad \text{versus} \qquad \int \frac{1}{x} \, dx = \ln x + C$$

Both are correct, but the second holds only for *positive* x-values (so that $\ln x$ is defined; see page 261), while the first holds for *negative* as well as positive x-values. We will sometimes use the second when we know that *x* is positive.

To evaluate *C* in an application, we set the function (evaluated at the given number) equal to the stated initial value and solve for *C*.

5.2 Exercises

1–34. Find each indefinite integral.

1. $\int e^{3x}\,dx$

2. $\int e^{4x}\,dx$

3. $\int e^{x/4}\,dx$

4. $\int e^{x/3}\,dx$

5. $\int e^{0.05x}\,dx$

6. $\int e^{0.02x}\,dx$

7. $\int e^{-2y}\,dy$

8. $\int e^{-3y}\,dy$

9. $\int e^{-0.5x}\,dx$

10. $\int e^{-0.4x}\,dx$

11. $\int 6e^{2x/3}\,dx$

12. $\int 24e^{-2u/3}\,du$

13. $\int -5x^{-1}\,dx$

14. $\int -\frac{1}{2}x^{-1}\,dx$

15. $\int \frac{3\,dx}{x}$

16. $\int \frac{dx}{2x}$

17. $\int \frac{3}{2v}\,dv$

18. $\int \frac{2}{3v}\,dv$

19. $\int (4e^{2x} - 6x)\,dx$

20. $\int (6e^{3x} + 4x)\,dx$

21. $\int \left(e^{3x} - \frac{3}{x}\right) dx$

22. $\int \left(e^{2x} - \frac{2}{x}\right) dx$

23. $\int (3e^{0.5t} - 2t^{-1})\,dt$

24. $\int (5e^{0.5t} - 4t^{-1})\,dt$

25. $\int (x^2 + x + 1 + x^{-1} + x^{-2})\,dx$

26. $\int (x^{-2} - x^{-1} + 1 - x + x^2)\,dx$

27. $\int (5e^{0.02t} - 2e^{0.01t})\,dt$

28. $\int (3e^{0.05t} - 2e^{0.04t})\,dt$

29. $\int \frac{e^w - w}{2}\,dw$

30. $\int \frac{e^w + w^2}{3}\,dw$

31. $\int \frac{z^3 + z}{z^2}\,dz$

32. $\int \frac{z^2 + 1}{z}\,dz$

33. $\int \frac{xe^x + 1}{x}\,dx$

34. $\int \frac{1}{x}(1 - xe^x)\,dx$

35–40. Find each indefinite integral. [*Hint:* Use some algebra first.]

35. $\int \frac{(x + 1)^2}{x}\,dx$

36. $\int \frac{(x - 1)^2}{x}\,dx$

37. $\int \frac{(t - 1)(t + 3)}{t^2}\,dt$

38. $\int \frac{(t + 2)(t - 4)}{t^2}\,dt$

39. $\int \frac{(x - 2)^3}{x}\,dx$

40. $\int \frac{(x + 2)^3}{x}\,dx$

Applied Exercises

41–42. BIOMEDICAL: Epidemics A flu epidemic hits a college community, beginning with five cases on day $t = 0$. The rate of growth of the epidemic (new cases per day) is given by the following function $r(t)$, where t is the number of days since the epidemic began.

a. Find a formula for the total number of cases of flu in the first t days.

b. Use your answer to part (a) to find the total number of cases in the first 20 days.

41. $r(t) = 18e^{0.05t}$

42. $r(t) = 20e^{0.04t}$

43. BUSINESS: Sales In an effort to reduce its inventory, a warehouse runs a sale on its least popular Blu-ray discs. The sales rate (discs sold per day) on day t of the sale is predicted to be $50/t$ (for $t \geq 1$), where $t = 1$ corresponds to the beginning of the sale, at which time none of the inventory of 200 discs had been sold.

a. Find a formula for the total number of discs sold up to day t.

b. Will the store have sold its inventory of 200 discs by day $t = 30$?

44. BUSINESS: Sales In an effort to make room for new inventory, a college bookstore runs a sale on its least popular mathematics books. The sales rate (books sold per day) on day t of the sale is predicted to be $60/t$ (for $t \geq 1$), where $t = 1$ corresponds to the beginning of the sale, at which time none of the inventory of 350 books had been sold.

a. Find a formula for the number of books sold up to day t.

b. Will the store have sold its inventory of 350 books by day $t = 30$?

45. ENVIRONMENTAL SCIENCE: Consumption of Natural Resources World consumption of tin is running at the rate of $342e^{0.02t}$ thousand metric tons per year, where t is measured in years and $t = 0$ corresponds to 2014.

a. Find a formula for the total amount of tin that will be consumed within t years of 2014.

b. When will the known world resources of 4900 thousand metric tons of tin be exhausted? [Tin is used mainly for coating steel (a "tin" can is actually a steel can with a thin protective coating of tin to prevent rust).]

Source: U.S. Geological Survey

46. ENVIRONMENTAL SCIENCE: Consumption of Natural Resources World consumption of copper is running at the rate of $18.8e^{0.04t}$ million metric per year, where t is measured in years and $t = 0$ corresponds to 2014.

   a. Find a formula for the total amount of copper that will be used within t years of 2014.

   b. When will the known world resources of 680 million metric tons of copper be exhausted?

   Source: U.S. Geological Survey

47. GENERAL: Cost of Maintaining a Home The cost of maintaining a home generally increases as the home becomes older. Suppose that the maintenance costs increase at the rate of $1800e^{0.05x}$ (dollars per year) when the home is x years old.

   a. Find a formula for the total maintenance cost during the first x years. (Total maintenance should be zero at $x = 0$.)

   b. Use your answer to part (a) to find the total maintenance cost during the first 5 years.

48. BIOMEDICAL: Cell Growth A culture of bacteria is growing at the rate of $20e^{0.8t}$ cells per day, where t is the number of days since the culture was started. Suppose that the culture began with 50 cells.

   a. Find a formula for the total number of cells in the culture after t days.

   b. If the culture is to be stopped when the population reaches 500, when will this occur?

49. GENERAL: Making Ice An ice cube tray filled with tap water is placed in the freezer, and the temperature of the water is changing at the rate of $-12e^{-0.2t}$ degrees Fahrenheit per hour after t hours. The original temperature of the tap water was 70 degrees.

   a. Find a formula for the temperature of water that has been in the freezer for t hours.

   b. When will the ice be ready? (Water freezes at 32 degrees.)

50. SOCIAL SCIENCE: Divorces The divorce rate in the United States (divorces per year) has been declining in recent years. The number of divorces per year is predicted to be $0.94e^{-0.02t}$ million, where t is the number of years since 2014.

   a. Find a formula for the total number of divorces within t years of 2014.

   b. Use your formula to find the total number of divorces from 2014 to 2020.

   Source: National Center for Health Statistics

51. BUSINESS: Total Savings A factory installs new equipment that is expected to generate savings at the rate of $800e^{-0.2t}$ dollars per year, where t is the number of years that the equipment has been in operation.

   a. Find a formula for the total savings that the equipment will generate during its first t years.

   b. If the equipment originally cost $2000, when will it "pay for itself"?

52. BUSINESS: Total Savings A company installs a new computer that is expected to generate savings at the rate of $20,000e^{-0.02t}$ dollars per year, where t is the number of years that the computer has been in operation.

   a. Find a formula for the total savings that the computer will generate during its first t years.

   b. If the computer originally cost $250,000, when will it "pay for itself"?

53. BUSINESS: Value of an Investment A real estate investment, originally worth $5000, grows continuously at the rate of $400e^{0.05t}$ dollars per year, where t is the number of years since the investment was made.

   a. Find a formula for the value of the investment after t years.

   b. Use your formula to find the value of the investment after 10 years.

54. BUSINESS: Value of an Investment A biotechnology investment, originally worth $20,000, grows continuously at the rate of $1000e^{0.10t}$ dollars per year, where t is the number of years since the investment was made.

   a. Find a formula for the value of the investment after t years.

   b. Use your formula to find the value of the investment after 7 years.

55. GENERAL: Consumption of Natural Resources World consumption of lead is running at the rate of $5e^{0.05t}$ million metric tons per year, where t is measured in years and $t = 0$ corresponds to 2014.

   a. Find a formula for the total amount of lead that will be consumed within t years of 2014.

    b. Use a graphing calculator to find when the world's known resources of 89 million metric tons of lead will be exhausted. [*Hint:* Use INTERSECT.] Lead has many uses, from batteries to shields against radioactivity.

   Source: U.S. Geological Survey

56. GENERAL: Total Savings A homeowner installs a solar water heater that is expected to generate savings at the rate of $70e^{0.03t}$ dollars per year, where t is the number of years since it was installed.

   a. Find a formula for the total savings within the first t years of operation.

    b. Use a graphing calculator to find when the heater will "pay for itself" if it cost $800. [*Hint:* Use INTERSECT.]

57. BUSINESS: Satellite Radio Subscriptions to satellite radio have been growing rapidly since they are now included in many new cars. The rate of increase has been approximately $2.15e^{0.086t}$ million subscriptions per year, where t is the number of years since 2014.

   a. Use this rate to find a formula to predict the total increase in subscriptions within t years of 2014.

   b. Use your formula to find the total increase from 2014 to 2020.

   Source: Sirius XM Radio

58. BUSINESS: Bond Growth The value of a recently issued General Electric bond increases in value at the rate of $40e^{0.04t}$ dollars per year, where $t = 0$ represents 2013.

a. Find a formula for the total increase in the value of the stock within t years of 2013.

b. Use your formula to find the total increase from 2013 to 2028.

Source: Wall Street Journal

Conceptual Exercises

59–63. Choose the correct answer.

59. $\int e^x \, dx = ?$

   a. $\frac{1}{x+1} e^{x+1} + C$ b. $e^x + C$ c. $e^x x + C$

60. $\int x^{-1} \, dx = ?$

   a. $\ln|x| + C$ b. $\frac{1}{0} x^0 + C$ c. $x^{-1}x + C$

61. $\int x^e \, dx = ?$

   a. $\frac{1}{e+1} x^{e+1} + C$ b. $x^e + C$ c. $x^e x + C$

62. $\int e^3 \, dx = ?$

   a. $\frac{1}{4}e^4 + C$ b. $e^3 + C$ c. $e^3 x + C$

63. $\int e^{-1} \, dx = ?$

   a. $\ln|e| + C$ b. $\frac{1}{0} e^0 + C$ c. $e^{-1}x + C$

64. Are there any powers of x that we still cannot integrate?

65. Find a formula for $\int e^{ax+b} \, dx$ where a and b are constants.

66. Does it follow from the box on page 320 that
$$\int e^{x^5} \, dx = e^{x^5} + C?$$

67. Does it follow from the lower box on page 322 that
$$\int \frac{1}{x^5} \, dx = \ln|x^5| + C?$$

68. Which one of these formulas is correct?

   a. $\int \ln x \, dx = \frac{1}{|x|} + C$

   b. $\int \ln|x| \, dx = \frac{1}{x} + C$

   c. $\int \frac{1}{x} \, dx = \ln|x| + C$

   d. $\int \frac{1}{\ln x} \, dx = |x| + C$

5.3   Definite Integrals and Areas

Introduction

We begin this section by calculating areas under curves, leading to a definition of the **definite integral** of a function. The **Fundamental Theorem of Integral Calculus** then provides an easier way to calculate definite integrals using *indefinite* integrals. Finally, we will illustrate the wide variety of applications of definite integrals.

Area Under a Curve

The diagram below shows a continuous nonnegative function. We want to calculate the *area* under the curve and above the x-axis between the vertical lines $x = a$ and $x = b$.

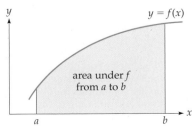

We begin by *approximating* the area by rectangles. In the graph on the left below, the area under the curve is approximated by *five* rectangles with equal bases and with heights equal to the height of the curve at the left-hand edge of the rectangle. (These are called *left* rectangles.) Five rectangles, however, do not give a very accurate approximation for the area under the curve: They underestimate the actual area by the small white spaces just above the rectangles.

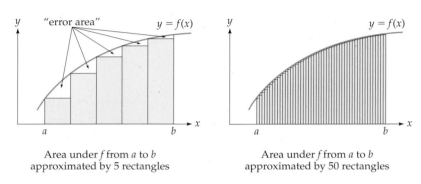

Area under f from a to b
approximated by 5 rectangles

Area under f from a to b
approximated by 50 rectangles

In the second graph, this same area is approximated by *fifty* rectangles, giving a much better approximation: the white *"error area"* between the curve and the rectangles is so small as to be almost invisible.

These diagrams suggest that more rectangles give a better approximation. In fact, the *exact* area under the curve is defined as the *limit* of the approximations as the number of rectangles *approaches infinity*. The following Example shows how to carry out such an approximation for the area under a given curve, and afterward we will find the *exact* area by letting the number of rectangles approach infinity.*

EXAMPLE 1   **APPROXIMATING AREA BY RECTANGLES**

Approximate the area under the curve $f(x) = x^2$ from 1 to 2 by five rectangles. Use rectangles with equal bases and with heights equal to the height of the curve at the left-hand edge of the rectangles.

Solution

For five rectangles, we divide the distance from $a = 1$ to $b = 2$ into five equal parts, so each rectangle has width (or change in x):

$$\Delta x = \frac{2-1}{5} = \frac{1}{5} = 0.2 \qquad \text{For } n \text{ rectangles,} \quad \Delta x = \frac{b-a}{n}$$

Along the x-axis beginning at $x = 1$ we mark successive points with spacing $\Delta x = 0.2$, giving points 1, 1.2, 1.4, 1.6, 1.8, and 2, as shown below. Above each of the resulting subintervals we draw a rectangle whose height

---

*This idea of dividing an area into many small parts and adding up their contributions to find a total was recommended by Tolstoy as the proper way to study history: "Only by taking infinitesimally small units for observation (the differentia of history, that is, the individual tendencies of men) and attaining to the art of integrating them (that is, finding the sum of these infinitesimals) can we hope to arrive at the laws of history." (From *War and Peace*, book 11, chapter 1) is the height of the curve at the left-hand edge of that rectangle. The curve is $y = x^2$, so these heights are the squares of the corresponding x-values.

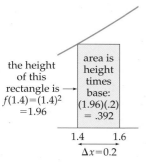

the height of this rectangle is $f(1.4) = (1.4)^2 = 1.96$

area is height times base: $(1.96)(.2) = .392$

1.4   1.6
$\Delta x = 0.2$

Middle rectangle

$(1)^2$ $(1.2)^2$ $(1.4)^2$ $(1.6)^2$ $(1.8)^2$ $y = x^2$

1   1.2   1.4   1.6   1.8   2
$\Delta x = .2$ $\Delta x = .2$ $\Delta x = .2$
$\Delta x = .2$ $\Delta x = .2$

An enlarged view of the middle rectangle is shown on the left. The sum of the areas of the five rectangles, height times base $\Delta x = 0.2$ for each rectangle, is

1st rectangle	2nd rectangle	3rd rectangle	4th rectangle	5th rectangle	Height times base for each rectangle

$$(1)^2 \cdot (0.2) + (1.2)^2 \cdot (0.2) + (1.4)^2 \cdot (0.2) + (1.6)^2 \cdot (0.2) + (1.8)^2 \cdot (0.2)$$

$$= 0.2 + 0.288 + 0.392 + 0.512 + 0.648 = 2.04$$

Multiplying out and summing

Therefore, the area under the curve is approximately 2.04 square units.

As we saw earlier, using only five rectangles does not give a very accurate approximation for the true area under the curve. For greater accuracy we use more rectangles, calculating the area in the same way. The following table gives the "rectangular approximation" for the area under the curve in Example 1 for larger numbers of rectangles, with answers rounded to three decimal places. Such results are easily found using any of the Riemann Sum programs for graphing calculators that are easily available on the Internet, as explained in the following Graphing Calculator Exploration.

Number of Rectangles	Sum of Areas of Rectangles	
5	2.04	← Found in Example 1
10	2.185	
100	2.318	The sum of the
1000	2.332	areas is approaching
10,000	2.333	$2\frac{1}{3}$

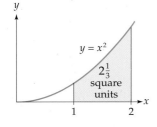

$y = x^2$

$2\frac{1}{3}$ square units

1   2

The areas in the right-hand column are approaching $2.333\ldots = 2\frac{1}{3}$, which is the *exact* area under the curve (as we will verify later). Therefore, the area under the curve $f(x) = x^2$ from 1 to 2 is $2\frac{1}{3}$ square units, as shown on the left.

Areas are given in "square units," meaning that if the units on the graph are inches, feet, or some other units, then the area is in *square* inches, *square* feet, or, in general, some other *square* units.

Graphing Calculator Exploration

5 *midpoint* rectangles for $f(x) = x^2$ on $[1, 2]$

N	Midpoint Riemann Sum
5	2.33
10	2.3325
25	2.3332
50	2.3333
100	2.333325
250	2.333332
500	2.333333

number of rectangles   sum of the areas

In the previous Example the rectangles touched the curve at the *left-hand* edge of the rectangle (and are called *left* rectangles). Similarly, we may use *midpoint* rectangles (which touch the curve at the *midpoint* of the rectangle) or *right* rectangles (which touch the curve at their right-hand edge). Graphing calculator programs that draw these (and other) types of rectangles for any function and calculate the sum of the areas are available on the Internet from sites such as *ticalc.org* (search for *Riemann*). Typical screens are shown on the left. The first screen shows five *midpoint* rectangles for $f(x) = x^2$ on the interval $[1, 2]$. The second screen shows the sum of the areas for various numbers of midpoint rectangles.

Notice from the second screen that the areas approach $2\frac{1}{3}$ very quickly, especially when compared with the previous table for *left* rectangles. Do you see why midpoint rectangles are more accurate?

Definite Integral

Approximating the area under a nonnegative function f by n rectangles means multiplying heights $f(x)$ by widths Δx and adding, obtaining

$$f(x_1) \cdot \Delta x + f(x_2) \cdot \Delta x + f(x_3) \cdot \Delta x + \cdots + f(x_n) \cdot \Delta x$$

The general procedure is as follows:

Area Under *f* from *a* to *b* Approximated by *n* Left Rectangles

1. Calculate the rectangle width $\Delta x = \dfrac{b - a}{n}$.

2. Find x-values $x_1, x_2, \ldots, x_n$ by successive additions of Δx beginning with $x_1 = a$.

3. Calculate the sum:

$$f(x_1) \cdot \Delta x + f(x_2) \cdot \Delta x + f(x_3) \cdot \Delta x + \cdots + f(x_n) \cdot \Delta x$$

The sum in Step 3 is called a **Riemann sum,** after the great German mathematician Georg Bernhard Riemann (1826–1866).* The *limit* of the Riemann sum as the number n of rectangles approaches infinity gives the **area under the curve,** and is called the **definite integral of the function** f **from** a **to** b, written $\displaystyle\int_a^b f(x)\, dx$. Formally:

---

*Actually, Riemann sums are slightly more general, allowing the subintervals to have different widths, with each width multiplied by the function evaluated at *any* x-value within that subinterval. For a continuous function, any Riemann sum will approach the same limiting value as the rectangles become arbitrarily narrow, so we may restrict our attention to the particular Riemann sums defined above (sometimes called "left Riemann sums").

Definite Integral

Let f be a continuous function on an interval $[a, b]$. The definite integral of f from a to b is defined as

$$\int_a^b f(x)\, dx = \lim_{n \to \infty}\left[f(x_1) \cdot \Delta x + f(x_2) \cdot \Delta x + \cdots + f(x_n) \cdot \Delta x \right]$$

where $\Delta x = \dfrac{b - a}{n}$, and $x_1, x_2, \ldots, x_n$ are x-values beginning with $x_1 = a$ and obtained by successive additions of Δx. If f is nonnegative on $[a, b]$, then the definite integral gives the *area under the curve from a to b*.

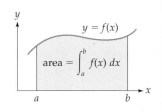

$$\text{area} = \int_a^b f(x)\, dx$$

The numbers a and b are called the **lower** and **upper limits of integration.**

Fundamental Theorem of Integral Calculus

A function followed by a vertical bar $\Big|_a^b$ with numbers a and b means evaluate the function at the *upper* number b and then subtract the evaluation at the *lower* number a.

$$F(x)\Big|_a^b = \underbrace{F(b)}_{\substack{\text{Evaluation at} \\ \text{upper number}}} - \underbrace{F(a)}_{\substack{\text{Evaluation at} \\ \text{lower number}}}$$

LOOKING BACK

We first used the vertical "evaluation bar" on page 104, but here we use it with *two* numbers.

EXAMPLE 2   USING THE EVALUATION NOTATION

$$\underset{\substack{\text{Evaluation} \\ \text{notation}}}{x^2\Big|_3^5} = \underset{\substack{x^2 \\ \text{at } x = 5}}{(5)^2} - \underset{\substack{x^2 \\ \text{at } x = 3}}{(3)^2} = \underset{\text{Simplifying}}{25 - 9} = 16$$

PRACTICE PROBLEM 1

Evaluate $\sqrt{x}\,\Big|_4^{25}$

Solution on page 338 >

The following **Fundamental Theorem of Integral Calculus** shows how to evaluate definite integrals by using *indefinite* integrals. A geometric and intuitive justification of the theorem is given at the end of this section.

Fundamental Theorem of Integral Calculus

For a continuous function f on an interval $[a, b]$,

$$\int_a^b f(x)\, dx = F(b) - F(a)$$

The right-hand side may be written $F(x)\Big|_a^b$

where F is any antiderivative of f.

The theorem is "fundamental" in that it establishes a deep and unexpected connection between definite integrals (limits of Riemann sums) and antiderivatives. It says that definite integrals can be evaluated in two simple steps:

1. Find an *indefinite* integral of the function (omitting the $+C$).
2. *Evaluate* the result at b and *subtract* the evaluation at a.

EXAMPLE 3   **FINDING A DEFINITE INTEGRAL BY THE FUNDAMENTAL THEOREM**

Find $\displaystyle\int_1^2 x^2\, dx$.

Solution

Because x^2 is continuous on $[1, 2]$, we can use the Fundamental Theorem.

$$\int_1^2 x^2\, dx = \frac{1}{3}x^3 \Big|_1^2 = \frac{1}{3}2^3 - \frac{1}{3}1^3 = \frac{8}{3} - \frac{1}{3} = \frac{7}{3}$$

Integrating x^2 Evaluating $\frac{1}{3}x^3$ at $x = 2$ and at $x = 1$ Subtracting

---

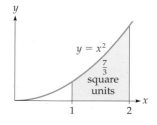

$y = x^2$

$\frac{7}{3}$ square units

Since definite integrals of nonnegative functions give areas, this result means that the area under $f(x) = x^2$ from $x = 1$ to $x = 2$ is $\frac{7}{3}$ square units.

On pages 330–331 we found this same area by the much more laborious process of calculating Riemann sums and taking the limit as the number of rectangles approached infinity, and our answer here, $\frac{7}{3}$, agrees with the answer there, $2\frac{1}{3}$ square units. Whenever possible, we will calculate definite integrals and find areas in this much simpler way, using the Fundamental Theorem of Integral Calculus.

Graphing Calculator Exploration

a. Graph $y_1 = x^2$ on the window $[0, 2]$ by $[0, 4]$.

b. Verify the result of Example 3 by having your graphing calculator find the definite integral of x^2 from 1 to 2. [*Hint:* Use a command like FnInt or $\int f(x)dx$.]

PRACTICE PROBLEM 2

Find the area under $y = x^3$ from 0 to 2 by evaluating $\displaystyle\int_0^2 x^3\, dx$.

Solution on page 338 >

EXAMPLE 4   FINDING THE AREA UNDER A CURVE

Find the area under $y = e^{2x}$ from $x = 0$ to $x = 1$.

Solution

Because e^{2x} is nonnegative and continuous on $[0, 1]$, the area is given by the definite integral:

$$\int_0^1 e^{2x}\, dx = \underbrace{\left. \frac{1}{2} e^{2x} \right|_0^1}_{\substack{\text{Indefinite integral} \\ \text{(using the formula} \\ \text{for } \int e^{ax}\, dx)}} = \underbrace{\frac{1}{2} e^2 - \frac{1}{2} e^0}_{\substack{\text{Evaluating} \\ \text{at } x = 1 \text{ and} \\ \text{at } x = 0}} = \underbrace{\frac{1}{2} e^2 - \frac{1}{2}}_{\substack{\text{Simplifying} \\ \text{(using } e^0 = 1)}}$$

Therefore, the area is $\frac{1}{2} e^2 - \frac{1}{2}$ square units.

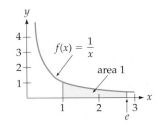

We leave the answer in this "exact" form (in terms of the number e). In an application we would use a calculator to approximate this answer as 3.19 square units.

 Be Careful Notice the difference between *definite* and *indefinite* integrals: *indefinite* integrals (the subject of the previous two sections) are *functions plus arbitrary constants*, while *definite* integrals are particular *numbers* (hence the word *definite*). The Fundamental Theorem of Integral Calculus shows how to use *indefinite* integrals to evaluate *definite* integrals.

EXAMPLE 5   FINDING THE AREA UNDER A CURVE

Find the area under $f(x) = \dfrac{1}{x}$ from $x = 1$ to $x = e$.

Solution

Because $\dfrac{1}{x}$ is nonnegative and continuous on $[1, e]$, the area is:

$$\int_1^e \frac{1}{x}\, dx = \left. \ln x \right|_1^e = \underbrace{\ln e}_{1} - \underbrace{\ln 1}_{0} = 1$$

Therefore, the area is 1 square unit.

From this Example we can give an alternate definition of the number e: e is the number such that the definite integral of $1/x$ from 1 to e is 1.

 Graphing Calculator Exploration

Use a graphing calculator to evaluate the definite integral

$$\int_0^1 e^{-x^2}\, dx.$$

This definite integral, which is important in probability and statistics, cannot be evaluated by the Fundamental Theorem because the function e^{-x^2} has no simple antiderivative. Your calculator may find the answer by approximating the area under the curve by modified Riemann sums.

Definite integrals have many of the properties of indefinite integrals. These properties follow from interpreting definite integrals as limits of Riemann sums.

Properties of Definite Integrals

$$\int_a^b c \cdot f(x)\, dx = c \int_a^b f(x)\, dx$$

A constant may be moved across the integral sign

$$\int_a^b [f(x) \pm g(x)]\, dx = \int_a^b f(x)\, dx \pm \int_a^b g(x)\, dx$$

⌞ Read both upper signs or both lower signs ⌟

The integral of a sum is the sum of the integrals (and similarly for differences)

EXAMPLE 6   USING THE PROPERTIES OF DEFINITE INTEGRALS

Find the area under $y = 24 - 6x^2$ from -1 to 1.

Solution

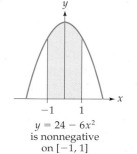

$$\int_{-1}^1 (24 - 6x^2)\, dx = \left(24x - 6 \cdot \frac{1}{3} x^3\right)\Bigg|_{-1}^1$$
$$= (24 - 2) - (-24 + 2)$$
$$= 44$$

$y = 24 - 6x^2$
is nonnegative
on $[-1, 1]$

Therefore, the area is 44 square units.

Total Cost of a Succession of Units

Given a marginal cost function, to find the *total* cost of producing, say, units 100 to 400, we could proceed as follows: *Integrate* the marginal cost to find the total cost, *evaluate* at 400 to find the total cost up to unit 400, and *subtract* the evaluation at 100 to leave just the cost of units 100 to 400. However, these steps of integrating, evaluating, and subtracting are just the steps in evaluating a definite integral by the Fundamental Theorem. Therefore, the cost of a succession of units is equal to the definite integral of the marginal cost function.

Cost of a Succession of Units

For a marginal cost function $MC(x)$:

$$\left(\begin{array}{c}\text{Total cost of} \\ \text{units } a \text{ to } b\end{array}\right) = \int_a^b MC(x)\, dx$$

EXAMPLE 7   FINDING THE COST OF A SUCCESSION OF UNITS

A company's marginal cost function is $MC(x) = \dfrac{75}{\sqrt{x}}$ where x is the number of units. Find the total cost of producing units 100 to 400.

Solution

$$\begin{pmatrix} \text{Total cost of} \\ \text{units 100 to 400} \end{pmatrix} = \int_{100}^{400} \frac{75}{\sqrt{x}} \, dx = 75 \int_{100}^{400} x^{-1/2} \, dx \qquad \text{Integrating} \atop \text{marginal cost}$$

$$= (75 \cdot 2 \cdot x^{1/2}) \Big|_{100}^{400}$$

$$= 150 \cdot (400)^{1/2} - 150 \cdot (100)^{1/2}$$

$$= 150 \cdot 20 - 150 \cdot 10 = 1500$$

The cost of producing units 100 to 400 is \$1500.

Similarly, integrating *any* rate from *a* to *b* gives the *total accumulation* at that rate between *a* and *b*.

Total Accumulation at a Given Rate

$$\begin{pmatrix} \text{Total accumulation at} \\ \text{rate } f \text{ from } a \text{ to } b \end{pmatrix} = \int_a^b f(x) \, dx$$

Take Note

On page 314 we used *indefinite* integrals to find total accumulations, and here we use *definite* integrals. What's the difference? Use *definite* integrals if you want the accumulations between two fixed numbers, and *indefinite* integrals if you want a general formula.

The following diagrams illustrate this idea. In each case, the *curve* represents a *rate*, and the *area under the curve*, given by the definite integral, gives the *total accumulation* at that rate.

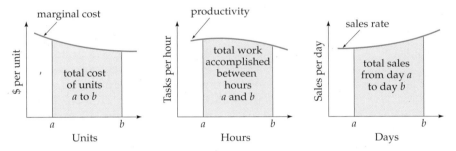

In repetitive tasks, a person's productivity usually increases with practice until it is slowed by monotony.

EXAMPLE 8   FINDING TOTAL PRODUCTIVITY FROM A RATE

A technician can test computer chips at the rate of $-3t^2 + 18t + 15$ chips per hour (for $0 \le t \le 6$), where t is the number of hours after 9:00 a.m. How many chips can be tested between 10:00 a.m. and 1:00 p.m.?

Solution

The total work accomplished is the integral of this rate from $t = 1$ (10 a.m.) to $t = 4$ (1 p.m.):

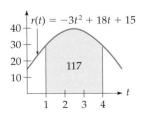

$$\int_1^4 (-3t^2 + 18t + 15) \, dt = \left(-t^3 + 18 \cdot \frac{1}{2} t^2 + 15t \right) \Big|_1^4$$

$$= (-64 + 144 + 60) - (-1 + 9 + 15) = 117$$

That is, between 10 a.m. and 1 p.m., 117 chips can be tested.

Integration Notation

The symbol Σ (the Greek letter S) is used in mathematics to indicate a *sum*, and so the Riemann sum can be written $\sum_{1}^{n} f(x_k)\,\Delta x$. The fact that the Riemann sum approaches the definite integral can be expressed as:

$$\sum_{1}^{n} f(x_k)\,\Delta x \longrightarrow \int_{a}^{b} f(x)\,dx$$

Σ becomes $\int$
Δ becomes d
as $n \to \infty$.

That is, the n approaching infinity changes the Σ (a Greek S) into an integral sign $\int$ (a "stretched out" S), and the Δ (a Greek D) into a d. In other words, the integral notation reminds us that a definite integral represents a *sum of rectangles* of height $f(x)$ and width dx.

Functions Taking Positive and Negative Values

Riemann sums can be calculated for *any* continuous function, not just nonnegative functions. The following diagram illustrates a Riemann sum for a continuous function that takes positive *and* negative values.

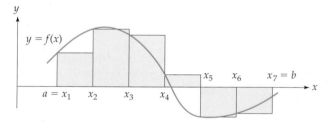

The two rightmost rectangles, where f is *negative*, will make a *negative* contribution to the sum. Taking the limit, the definite integral where the function lies *below* the x-axis will give the *negative* of the area between the curve and the x-axis. Therefore, the definite integral of such a function from a to b will give the *signed area* between the curve and the x-axis: the area *above* the axis minus the area *below* the axis, shown as A_{up} *minus* A_{down} in the diagram below.

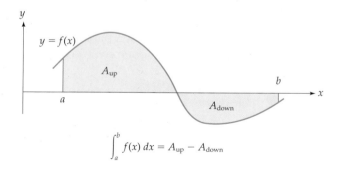

$$\int_{a}^{b} f(x)\,dx = A_{\text{up}} - A_{\text{down}}$$

Solutions TO PRACTICE PROBLEMS

1. $\left.\sqrt{x}\,\right|_{4}^{25} = \sqrt{25} - \sqrt{4} = 5 - 2 = 3$

2. $\displaystyle\int_{0}^{2} x^3\,dx = \left.\tfrac{1}{4}x^4\,\right|_{0}^{2} = \tfrac{1}{4}\cdot 16 - \tfrac{1}{4}\cdot 0 = 4$ square units

5.3   Section Summary

We began by approximating the area under a curve from a to b by Riemann sums (sums of rectangles). We then defined the *definite integral* $\int_a^b f(x)\,dx$ as the *limit* of the Riemann sum as the number of rectangles approaches infinity. The Fundamental Theorem of Integral Calculus showed how to evaluate definite integrals much more simply by evaluating *indefinite* integrals:

$$\int_a^b f(x)\,dx = F(x)\Big|_a^b \qquad \text{\textit{F} is any antiderivative of \textit{f}}$$

A *definite* integral is a *number*, but an *indefinite* integral is a function plus an arbitrary constant.

As for the *uses* of definite integrals, the definite integral of a nonnegative function gives the *area under the curve*:

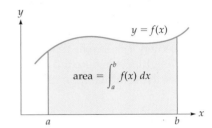

The definite integral of a *rate* gives the *total accumulation* at that rate:

$$\int_a^b f(x)\,dx = \left(\begin{array}{c}\text{Total accumulation at}\\ \text{rate } f \text{ from } a \text{ to } b\end{array}\right)$$

Verification of the Fundamental Theorem of Integral Calculus

The following is a geometric and intuitive justification of the Fundamental Theorem of Integral Calculus.

For a continuous and nonnegative function f on an interval $[a, b]$, we define a new function $A(x)$ as the *area under f from a to x*.

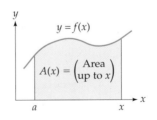

$$A(x) = \left(\begin{array}{c}\text{Area under } f\\ \text{from } a \text{ to } x\end{array}\right)$$

Therefore,

$$A(b) = \left(\begin{array}{c}\text{Area under } f\\ \text{from } a \text{ to } b\end{array}\right) \qquad \text{Replacing \textit{x} by \textit{b} gives the entire area from \textit{a} to \textit{b}}$$

and

$$A(a) = 0 \qquad \text{The area "from \textit{a} to \textit{a}" must be zero}$$

Subtracting these last two expressions, $A(b) - A(a)$, gives the total area from a to b [since $A(b)$ is the total area, and $A(a)$ is zero]. This same area can be expressed as the definite integral of f from a to b, leading to the equation

$$\int_a^b f(x)\,dx = A(b) - A(a) \qquad \text{Area under \textit{f} from \textit{a} to \textit{b} expressed in two ways}$$

If we can show that $A(x)$ is an antiderivative of $f(x)$, then the equation above will show that the definite integral can be found by an antiderivative evaluated at the upper and lower limits and subtracted, which will verify the Fundamental

Theorem. To show that $A(x)$ is an antiderivative of $f(x)$, we show that $A'(x) = f(x)$, differentiating $A(x)$ by the definition of the derivative:

$$A'(x) = \lim_{h \to 0} \frac{A(x + h) - A(x)}{h}$$

In the numerator, $A(x + h)$ is the area under the curve up to $x + h$ and $A(x)$ is the area up to x, so subtracting them, $A(x + h) - A(x)$, leaves just the area from x to $x + h$, as shown below.

| Area up to $x + h$ | minus | Area up to x | equals | Area from x to $x + h$ |

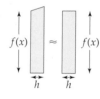

When h is small, this last area can be approximated by a rectangle of base h and height $f(x)$, where the approximation becomes exact as h approaches zero. Therefore, in the limit we may replace $A(x + h) - A(x)$ by the area of the rectangle, $h \cdot f(x)$:

$$A'(x) = \lim_{h \to 0} \frac{\overbrace{A(x + h) - A(x)}^{\approx\, h \cdot f(x)}}{h} = \lim_{h \to 0} \frac{h \cdot f(x)}{h} = f(x)$$

This equation says that $A'(x) = f(x)$, showing that $A(x)$ *is* an antiderivative of $f(x)$. This completes the verification of the Fundamental Theorem of Integral Calculus.

5.3 Exercises

1–6. Find the sum of the areas of the shaded rectangles under each graph. Round to two decimal places. [*Hint:* The width of each rectangle is the difference between the x-values at its base. The height of each rectangle is the height of the curve at the left edge of the rectangle.]

1.

2.

3.
4.

5.
6.

7–12. For each function:

i. *Approximate* the area under the curve from a to b by calculating a Riemann sum with the given number of rectangles. Use the method described in Example 1 on pages 330–331, rounding to three decimal places.

ii. Find the *exact* area under the curve from a to b by evaluating an appropriate definite integral using the Fundamental Theorem.

7. $f(x) = 2x$ from $a = 1$ to $b = 2$.
For part (i), use 5 rectangles.

8. $f(x) = x^2 + 1$ from $a = 0$ to $b = 1$.
For part (i), use 5 rectangles.

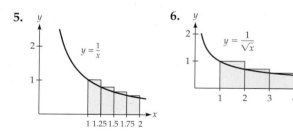

9. $f(x) = \sqrt{x}$ from $a = 1$ to $b = 4$.
For part (i), use 6 rectangles.

10. $f(x) = e^x$ from $a = -1$ to $b = 1$.
For part (i), use 8 rectangles.

11. $f(x) = \dfrac{1}{x}$ from $a = 1$ to $b = 2$.

For part (i), use 10 rectangles.

12. $f(x) = \dfrac{1}{\sqrt{x}}$ from $a = 1$ to $b = 4$.

For part (i), use 6 rectangles.

 13–18. Use a graphing calculator *Riemann Sum* program from the Internet (see page 332) to find the following Riemann sums.

i. Calculate the Riemann sum for each function for the following values of n: 10, 100, and 1000. Use left, right, or midpoint rectangles, making a table of the answers, rounded to three decimal places.

ii. Find the *exact* value of the area under the curve by evaluating an appropriate definite integral using the Fundamental Theorem. The values of the Riemann sums from part (i) should approach this number.

13. $f(x) = 2x$ from $a = 1$ to $b = 2$

14. $f(x) = x^2 + 1$ from $a = 0$ to $b = 1$

15. $f(x) = \sqrt{x}$ from $a = 1$ to $b = 4$

16. $f(x) = e^x$ from $a = -1$ to $b = 1$

17. $f(x) = \dfrac{1}{x}$ from $a = 1$ to $b = 2$

18. $f(x) = \dfrac{1}{\sqrt{x}}$ from $a = 1$ to $b = 4$

19–40. Use a definite integral to find the area under each curve between the given x-values. For Exercises 19–24, also make a sketch of the curve showing the region.

19. $f(x) = x^2$ from $x = 0$ to $x = 3$

20. $f(x) = x$ from $x = 0$ to $x = 4$

21. $f(x) = 4 - x$ from $x = 0$ to $x = 4$

22. $f(x) = 1 - x^2$ from $x = -1$ to $x = 1$

23. $f(x) = \dfrac{1}{x}$ from $x = 1$ to $x = 2$

24. $f(x) = e^x$ from $x = 0$ to $x = 1$

25. $f(x) = 8x^3$ from $x = 1$ to $x = 3$

26. $f(x) = 6x^2$ from $x = 2$ to $x = 3$

27. $f(x) = 6x^2 + 4x - 1$ from $x = 1$ to $x = 2$

28. $f(x) = 27 - 3x^2$ from $x = 1$ to $x = 3$

29. $f(x) = \dfrac{1}{\sqrt{x}}$ from $x = 4$ to $x = 9$

30. $f(x) = \dfrac{1}{x^2}$ from $x = 1$ to $x = 3$

31. $f(x) = 8 - 4\sqrt[3]{x}$ from $x = 0$ to $x = 8$

32. $f(x) = 9 - 3\sqrt{x}$ from $x = 0$ to $x = 9$

33. $f(x) = \dfrac{1}{x}$ from $x = 1$ to $x = 5$

34. $f(x) = \dfrac{1}{x}$ from $x = e$ to $x = e^3$

35. $f(x) = x^{-1} + x^2$ from $x = 1$ to $x = 2$

36. $f(x) = 6e^{2x}$ from $x = 0$ to $x = 2$

37. $f(x) = 2e^x$ from $x = 0$ to $x = \ln 3$

38. $f(x) = e^{-x}$ from $x = 0$ to $x = 1$

39. $f(x) = e^{x/2}$ from $x = 0$ to $x = 2$

40. $f(x) = e^{x/3}$ from $x = 0$ to $x = 3$

41–46. For each function:

a. Integrate ("by hand") to find the area under the curve between the given x-values.

b. Verify your answer to part (a) by having your calculator graph the function and find the area (using a command like FnInt or $\int f(x)dx$).

41. $f(x) = 12 - 3x^2$ from $x = 1$ to $x = 2$

42. $f(x) = 9x^2 - 6x + 1$ from $x = 1$ to $x = 2$

43. $f(x) = \dfrac{1}{x^3}$ from $x = 1$ to $x = 4$

44. $f(x) = \dfrac{1}{\sqrt[3]{x}}$ from $x = 8$ to $x = 27$

45. $f(x) = 2x + 1 + x^{-1}$ from $x = 1$ to $x = 2$

46. $f(x) = e^x$ from $x = 0$ to $x = 3$

47–66. Evaluate each definite integral.

47. $\displaystyle\int_0^1 (x^{99} + x^9 + 1)\, dx$

48. $\displaystyle\int_2^4 (1 + x^{-2})\, dx$

49. $\displaystyle\int_1^2 (6t^2 - 2t^{-2})\, dt$

50. $\displaystyle\int_{-2}^2 (3w^2 - 2w)\, dw$

51. $\displaystyle\int_1^4 \dfrac{1}{y^2}\, dy$

52. $\displaystyle\int_1^4 \dfrac{1}{\sqrt{z}}\, dz$

53. $\displaystyle\int_1^e \dfrac{dx}{x}$

54. $\displaystyle\int_1^{e^2} \dfrac{3}{x}\, dx$

55. $\displaystyle\int_1^3 (9x^2 + x^{-1})\, dx$

56. $\displaystyle\int_1^2 (x^{-1} - 4x^2)\, dx$

57. $\displaystyle\int_{-2}^{-1} 3x^{-1}\, dx$

58. $\displaystyle\int_{-3}^{-1} (1 + x^{-1})\, dx$

59. $\displaystyle\int_0^1 12e^{3x}\, dx$

60. $\displaystyle\int_0^2 3e^{x/2}\, dx$

61. $\displaystyle\int_{-1}^1 5e^{-x}\, dx$

62. $\displaystyle\int_0^1 (6x^2 - 4e^{2x})\, dx$

63. $\displaystyle\int_{\ln 2}^{\ln 3} e^x\, dx$

64. $\displaystyle\int_0^{\ln 5} e^x\, dx$

65. $\displaystyle\int_1^2 \dfrac{(x+1)^2}{x}\, dx$

66. $\displaystyle\int_1^2 \dfrac{(x+1)^2}{x^2}\, dx$

67–72. Use a graphing calculator to evaluate each definite integral, rounding answers to three decimal places. [*Hint:* Use a command like FnInt or ∫f(x)dx.]

67. $\displaystyle\int_0^2 \frac{1}{x^2+1}\,dx$ **68.** $\displaystyle\int_{-1}^1 \sqrt{x^4+1}\,dx$

69. $\displaystyle\int_{-1}^1 e^{x^2}\,dx$ **70.** $\displaystyle\int_{-2}^2 e^{(-1/2)x^2}\,dx$

71. $\displaystyle\int_0^4 \sqrt{x}\,e^x\,dx$ **72.** $\displaystyle\int_1^4 x^x\,dx$

73. OMITTING THE C IN DEFINITE INTEGRALS

a. Evaluate the definite integral $\displaystyle\int_0^3 x^2\,dx$.

b. Evaluate the same definite integral by completing the following calculation, in which the antiderivative includes a constant C.

$$\int_0^3 x^2\,dx = \left(\frac{1}{3}x^3 + C\right)\Big|_0^3 = \cdots$$

[The constant C should cancel out, giving the same answer as in part (a).]

c. Explain why the constant will cancel out of *any* definite integral. (We therefore omit the constant in definite integrals. However, be sure to keep the $+C$ in *indefinite* integrals.)

74. Show that for any number $a > 0$,

$$\int_1^a \frac{1}{x}\,dx = \ln a$$

This equation is often used as a *definition* of natural logarithms, defining $\ln a$ as the area under the curve $y = 1/x$ between 1 and a. [*Hint:* $\ln 1 = 0$.]

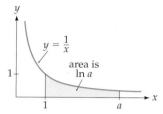

75–76. GEOMETRIC INTEGRATION Find $\displaystyle\int_0^4 f(x)\,dx$ for the function $f(x)$ graphed below. [*Hint:* No calculus is necessary—just as we used integrals to find areas, we can use areas to find integrals. The curves shown are quarter circles.]

75.

76.

Applied Exercises

77. GENERAL: Electrical Consumption On a hot summer afternoon, a city's electricity consumption is $-3t^2 + 18t + 10$ units per hour, where t is the number of hours after noon ($0 \le t \le 6$). Find the total consumption of electricity between the hours of 1 and 5 p.m.

78. BIOMEDICAL: Height An average child of age x years grows at the rate of $6x^{-1/2}$ inches per year (for $2 \le x \le 16$). Find the total height gain from age 4 to age 9.

Source: National Center for Health Statistics

79–80. GENERAL: Repetitive Tasks After t hours of work, a bank clerk can process checks at the rate of $r(t)$ checks per hour for the function $r(t)$ given below. How many checks will the clerk process during the first three hours (time 0 to time 3)?

79. $r(t) = -t^2 + 90t + 5$ **80.** $r(t) = -t^2 + 60t + 9$

81–82. BUSINESS: Cost A company's marginal cost function is $MC(x)$ (given below), where x is the number of units. Find the total cost of the first hundred units ($x = 0$ to $x = 100$).

81. $MC(x) = 6e^{-0.02x}$ **82.** $MC(x) = 8e^{-0.01x}$

83. BUSINESS: Money Stock Measure From 1964 to 2014, the money stock measure "M1" (currency, traveler's checks, demand deposits, and other checkable deposits) was growing at the rate of approximately $94e^{0.56x}$ billion dollars per decade, where x is the number of decades since 1964. Find the total increase in M1 from 1964 to 2014.

Source: Federal Reserve Board

84. BUSINESS: Money Stock Measure From 1984 to 2014, the money stock measure "M2" (M1 plus retail money market mutual funds, savings, and small time deposits) was growing at the rate of approximately $1130e^{0.54x}$ billion dollars per decade, where x is the number of decades since 1984. Find the total increase in M2 from 1984 to 2014.

Source: Federal Reserve Board

85. GENERAL: Price Increase The price of a double-dip ice cream cone is increasing at the rate of $15e^{0.05t}$ cents per year, where t is measured in years and $t = 0$ corresponds to 2014. Find the total change in price between the years 2014 and 2024.

86. BUSINESS: Sales An automobile dealer estimates that the newest model car will sell at the rate of $30/t$ cars per month, where t is measured in months and $t = 1$ corresponds to the beginning of January. Find the number of cars that will be sold from the beginning of January to the beginning of May.

87. BUSINESS: Tin Consumption World consumption of tin is running at the rate of $342e^{0.02t}$ thousand metric tons per year, where t is measured in years and $t = 0$ corresponds to 2014. Find the total consumption of tin from 2014 to 2024.

Source: U.S. Geological Survey

88. SOCIOLOGY: Marriages The marriage rate (marriages per year) in the United States has been declining recently, with about $1.97e^{-0.0102t}$ million marriages per year, where t is the number of years since 2014. Assuming that this rate continues, find the total number of marriages in the United States from 2014 to 2024.

Source: National Center for Health Statistics

89. BEHAVIORAL SCIENCE: Learning A student can memorize words at the rate of $6e^{-t/5}$ words per minute after t minutes. Find the total number of words that the student can memorize in the first 10 minutes.

90. BIOMEDICAL: Epidemics An epidemic is spreading at the rate of $12e^{0.2t}$ new cases per day, where t is the number of days since the epidemic began. Find the total number of new cases in the first 10 days of the epidemic.

91. GENERAL: Area
  a. Use your graphing calculator to find the area between 0 and 1 under the following curves: $y = x$, $y = x^2$, $y = x^3$, and $y = x^4$.
  b. Based on your answers to part (a), conjecture a formula for the area under $y = x^n$ between 0 and 1 for any value of $n > 0$.
  c. Prove your conjecture by evaluating an appropriate definite integral "by hand."

92. GENERAL: Dam Construction Ever since the Johnstown Dam burst in 1889 killing 2200 people, dam construction has become increasingly scientific.
  a. Find the area of the cross section of the dam shown in the next column and then multiply by its 574-foot length to determine the amount of concrete needed to build this dam. All dimensions are in feet. [*Hint:* Find the cross-sectional area by integrating and using area formulas.]
  b. If a mixing truck carries 300 cubic feet of concrete, about how many truckloads would be needed to build the dam?

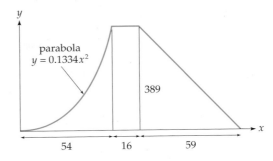

parabola $y = 0.1334x^2$

389

54   16   59

93. BIOMEDICAL: Drug Absorption An oral medication is absorbed into the bloodstream at the rate of $5e^{-0.04t}$ milliliters per minute, where t is the number of minutes since the medication was taken. Find the total amount of medication absorbed within the first 30 minutes.

94. BIOMEDICAL: Aortic Volume The rate of change of the volume of blood in the aorta t seconds after the beginning of the cardiac cycle is $-kP_0e^{-mt}$ milliliters per second, where k, P_0, and m are constants (depending, respectively, on the elasticity of the aorta, the initial aortic pressure, and various characteristics of the cardiac cycle). Find the total change in volume from time 0 to time T (the end of the cardiac cycle). (Your answer will involve the constants k, P_0, m, and T.)

Source: R. E. Klabunde, Cardiovascular Physiology Concepts

95. BUSINESS: Sales A dealer predicts that new cars will sell at the rate of $8xe^{-0.1x}$ sales per week in week x. Find the total sales in the first half year (week 0 to week 26).

96. GENERAL: Cigarette Smoking Reread, if necessary, the Application Preview on page 307.
  a. Evaluate the definite integrals
$$\int_0^1 (300e^{0.025x} - 240e^{0.02x})\, dx$$
and
$$\int_7^8 (300e^{0.025x} - 240e^{0.02x})\, dx$$
to verify the answers given there for the amount of tar inhaled from the first and last centimeters of the cigarette.
  b. Evaluate the definite integral
$$\int_0^8 (300e^{0.025x} - 240e^{0.02x})\, dx$$
to find the amount of tar inhaled from smoking the entire cigarette.

Source: H. Marcus-Roberts and M. Thompson, eds., Modules in Applied Mathematics, Vol. 4, Life Science Models

97. GENERAL: Population A resort community swells at the rate of $100e^{0.4\sqrt{x}}$ new arrivals per day on day x of its "high season." Find the total number of arrivals in the first two weeks (day 0 to day 14).

98. BEHAVIORAL SCIENCE: Repetitive Tasks After t hours of work, a medical technician can carry out T-cell counts at the rate of $2t^2e^{-t/4}$ tests per hour. How many tests will the technician process during the first eight hours (time 0 to time 8)?

Conceptual Exercises

99. Suppose that you have a positive, increasing function and you approximate the area under it by a Riemann sum with left rectangles. Will the Riemann sum *over*estimate or *under*estimate the actual area? [*Hint:* Make a sketch.]

100. Suppose that you have a positive, decreasing function and you approximate the area under it by a Riemann sum with left rectangles. Will the Riemann sum *over*estimate or *under*estimate the actual area? [*Hint:* Make a sketch.]

101. Suppose that you have a positive, increasing, concave up function and you approximate the area under it by a Riemann sum with *midpoint* rectangles. Will the Riemann sum *over*estimate or *under*estimate the actual area? [*Hint:* Make a sketch.]

102. Suppose that you have a positive, increasing, concave down function and you approximate the area under it by a Riemann sum with *midpoint* rectangles. Will the Riemann sum *over*estimate or *under*estimate the actual area? [*Hint:* Make a sketch.]

103. A friend says that definite and indefinite integrals are exactly the same except that one has numbers plugged in. A second friend disagrees, saying that the essential ideas are entirely different. Who is right?

104. Suppose that you have a positive function and you approximate the area under it using Riemann sums with midpoint rectangles. Explain why, if the function is *linear*, you will always get the *exact* area, no matter how many (or few) rectangles you use. [*Hint:* Make a sketch.]

105. Evaluate $\displaystyle\int_{1}^{1} \frac{x^{43}e^{-17x} + 219\sqrt[3]{x^2}}{\ln \sqrt[29]{6x^3 - x^{-11}} - \pi^3}\, dx.$

[*Hint:* No work necessary.]

106. Find a formula for $\displaystyle\int_{a}^{b} c\, dx.$ [*Hint:* No calculation necessary—just think of a graph.]

107. Finding $\displaystyle\int_{-1}^{1} \frac{1}{x^2}\, dx$ by the Fundamental Theorem of Integral Calculus gives an answer of -2, as you should check. However, shouldn't the area under a positive function be positive? Explain.

108. How will $\displaystyle\int_{a}^{b} f(x)\, dx$ and $\displaystyle\int_{b}^{a} f(x)\, dx$ differ?

[*Hint:* Assume that they can be evaluated by the Fundamental Theorem of Integral Calculus, and think how they will differ at the "evaluate and subtract" step.]

Explorations and Excursions   The following problems extend and augment the material presented in the text.

109. ECONOMICS: Pareto's Law   The economist Vilfredo Pareto (1848–1923) estimated that the number of people who have an income between A and B dollars ($A < B$) is given by a definite integral of the form

$$N = \int_{A}^{B} ax^{-b}\, dx \qquad (b \neq 1)$$

where a and b are constants. Evaluate this integral.

Source: A. Reynolds, *Income and Wealth*

110. BIOMEDICAL: Poiseuille's Law   According to *Poiseuille's law,* the speed of blood in a blood vessel is given by $V = \dfrac{p}{4Lv}(R^2 - r^2),$

where R is the radius of the blood vessel, r is the distance of the blood from the center of the blood vessel, and p, L, and v are constants determined by the pressure and viscosity of the blood and the length of the vessel. The total blood flow is then given by

$$\left(\begin{array}{c}\text{Total} \\ \text{blood flow}\end{array}\right) = \int_{0}^{R} 2\pi \frac{p}{4Lv}(R^2 - r^2)r\, dr$$

Find the total blood flow by finding this integral (p, L, v, and R are constants).

Source: E. Batschelet, *Introduction to Mathematics for Life Scientists*

111–112. BUSINESS: Capital Value of an Asset   The *capital value* of an asset (such as an oil well) that produces a continuous stream of income is the sum of the present value of all future earnings from the asset. Therefore, the capital value of an asset that produces income at the rate of $r(t)$ dollars per year (at a continuous interest rate i) is

$$\left(\begin{array}{c}\text{Capital} \\ \text{value}\end{array}\right) = \int_{0}^{T} r(t)e^{-it}\, dt$$

where T is the expected life (in years) of the asset.

Source: T. Lee, *Income and Value Measurement*

111. Use the formula in the preceding instructions to find the capital value (at interest rate $i = 0.06$) of an oil well that produces income at the constant rate of $r(t) = 240{,}000$ dollars per year for 10 years.

112. Use the formula in the preceding instructions to find the capital value (at interest rate $i = 0.05$) of a uranium mine that produces income at the rate of $r(t) = 560{,}000t^{1/2}$ dollars per year for 20 years.

5.4 Further Applications of Definite Integrals: Average Value and Area Between Curves

Introduction

In this section we will use definite integrals for two important purposes: finding average values of functions and finding areas between curves. We will find the average U.S. population over a decade, the average time to accomplish a task, and the average pollution level in a lake. Averages eliminate fluctuations, reducing a collection of numbers to a single "representative" number. We will use areas between curves to find quantities from trade deficits to lives saved by seat belts.

Average Value of a Function

The average value of n numbers is found by adding the numbers and dividing by n. For example,

$$\left(\begin{array}{c}\text{Average of}\\a, b, \text{and } c\end{array}\right) = \frac{1}{3}(a + b + c)$$

How can we find the average value of a *function* on an interval? For example, if a function gives the temperature over a 24-hour period, how can we calculate the *average temperature*? We could, of course, just take the temperature at every hour and then average these 24 values, but this would ignore the temperature at all of the intermediate times. Intuitively, the average should represent a "leveling off" of the curve to a uniform height, the horizontal line shown on the left.

This leveling should use the "hills" to fill in the "valleys," maintaining the same total area under the curve. Therefore, the area under the horizontal line (a rectangle with base $b - a$ and height up to the line) must equal the area under the curve (the definite integral of the function).

$$\underbrace{(b - a)\left(\begin{array}{c}\text{Average}\\\text{height}\end{array}\right)}_{\text{Area under line}} = \underbrace{\int_a^b f(x)\, dx}_{\text{Area under curve}} \qquad \begin{array}{l}\text{Equating the two ways of}\\\text{finding the area}\end{array}$$

Therefore:

$$\left(\begin{array}{c}\text{Average}\\\text{height}\end{array}\right) = \frac{1}{b - a}\int_a^b f(x)\, dx \qquad \text{Dividing by } (b - a)$$

This formula gives the average (or "mean") value of a continuous function on an interval.

Average Value of a Function

$$\left(\begin{array}{c}\text{Average value}\\\text{of } f \text{ on } [a, b]\end{array}\right) = \frac{1}{b - a}\int_a^b f(x)\, dx$$

Average value is the definite integral of the function divided by the length of the interval

---

Finding the average value of a function by integrating and dividing by $b - a$ is analogous to averaging n numbers by adding and dividing by n (since integrals are continuous sums). It also has the advantage of taking account of *all* of the values of the function in the interval. A derivation of this formula by Riemann sums is given at the end of this section.

| EXAMPLE 1 | FINDING THE AVERAGE VALUE OF A FUNCTION |

Find the average value of $f(x) = \sqrt{x}$ from $x = 0$ to $x = 9$.

Solution

$$\left(\begin{array}{c}\text{Average}\\\text{value}\end{array}\right) = \frac{1}{9-0}\int_0^9 \sqrt{x}\,dx = \frac{1}{9}\int_0^9 x^{1/2}\,dx \qquad \begin{array}{l}\text{Integral divided}\\\text{by the length of}\\\text{the interval}\end{array}$$

$$= \frac{1}{9}\frac{2}{3}x^{3/2}\Big|_0^9 = \frac{2}{27}9^{3/2} - \frac{2}{27}0^{3/2} \qquad \begin{array}{l}\text{Integrating and}\\\text{evaluating}\end{array}$$

$$= \frac{2}{27}\left(\sqrt{9}\right)^3 - 0 = \frac{2}{27}27 = 2 \qquad \text{Simplifying}$$

The average value of $f(x) = \sqrt{x}$ over the interval $[0, 9]$ is 2.

| EXAMPLE 2 | FINDING AVERAGE POPULATION |

The population of the United States is predicted to be $P(t) = 310e^{0.0073t}$ million people, where t is the number of years since 2010. Predict the average population between the years 2020 and 2030.

Source: U.S. Census Bureau

Solution

We integrate from $t = 10$ (year 2020) to $t = 20$ (year 2030).

$$\left(\begin{array}{c}\text{Average}\\\text{value}\end{array}\right) = \frac{1}{20-10}\int_{10}^{20} 310e^{0.0073t}\,dt \qquad \begin{array}{l}\text{Integral divided}\\\text{by the length of}\\\text{the interval}\end{array}$$

$$= \frac{310}{10}\int_{10}^{20} e^{0.0073t}\,dt$$

$$= 31\frac{1}{0.0073}e^{0.0073t}\Big|_{10}^{20} \qquad \begin{array}{l}\text{Integrating by the}\\\int e^{ax}\,dx \text{ formula}\end{array}$$

$$\approx 4247e^{0.146} - 4247e^{0.073} \approx 346 \qquad \begin{array}{l}\text{Evaluating, using}\\\text{a calculator}\end{array}$$

The average population of the United States during the third decade of the twenty-first century will be about 346 million people.

Be Careful When finding an average value, it is a common mistake to forget to divide by $b - a$. But that would be like finding the average of three numbers by adding them up and forgetting to divide by 3.

PRACTICE PROBLEM 1

Find the average value of $f(x) = 3x^2$ from $x = 0$ to $x = 2$.

Solution on page 351 >

Area Between Curves: Integrating "Upper Minus Lower"

We know that definite integrals give areas under curves. To calculate the area *between* two curves, we take the area under the *upper* curve and subtract the area under the *lower* curve.

In terms of integrals:

$$\int_a^b f(x)\,dx \qquad - \qquad \int_a^b g(x)\,dx \qquad = \qquad \int_a^b [f(x) - g(x)]\,dx$$

↑   ↑
Upper   Lower
curve   curve

Therefore, the area between the curves can be written as a single integral:

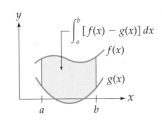

Area Between Curves

The area between two continuous curves $f(x) \geq g(x)$ on $[a, b]$ is

$$\begin{pmatrix} \text{Area between} \\ f \text{ and } g \text{ on } [a, b] \end{pmatrix} = \int_a^b [f(x) - g(x)]\,dx$$

Integrate "upper minus lower"

Finding area by integrating "upper minus lower" works regardless of whether one or both curves dip below the x-axis.

EXAMPLE 3   FINDING THE AREA BETWEEN CURVES

Find the area between $y = 3x^2 + 4$ and $y = 2x - 1$ from $x = -1$ to $x = 2$.

Solution

The area is shown in the diagram on the left. We integrate "upper minus lower" between the given x-values.

$$\int_{-1}^{2} [\underbrace{(3x^2 + 4)}_{\text{Upper}} - \underbrace{(2x - 1)}_{\text{Lower}}]\,dx$$

$$= \int_{-1}^{2} (3x^2 + 4 - 2x + 1)\,dx$$

$$= \int_{-1}^{2} (3x^2 - 2x + 5)\,dx \qquad\qquad \text{Simplifying}$$

$$= (x^3 - x^2 + 5x)\Big|_{-1}^{2} \qquad\qquad \text{Integrating}$$

$$= (8 - 4 + 10) - (-1 - 1 - 5) \qquad\qquad \text{Evaluating}$$

$$= 21 \text{ square units}$$

Be Careful When doing other problems, you may need to make a similar rough sketch to see which curve is "upper" and which is "lower."

PRACTICE PROBLEM 2

Find the area between
$y = 2x^2 + 1$ and $y = -x^2 - 1$
from $x = -1$ to $x = 1$.

Solution on page 351 >

If the two curves represent *rates* (one unit per another unit), then the area between the curves gives the *total accumulation* at the upper rate minus the lower rate.

 EXAMPLE 4   FINDING SALES FROM EXTRA ADVERTISING

A company marketing high-definition television sets expects to sell them at the rate of $2e^{0.05t}$ thousand sets per month, where t is the number of months since they became available. However, with additional advertising using a sports celebrity, they should sell at the rate of $3e^{0.1t}$ thousand sets per month. How many additional sales would result from the celebrity endorsement during the first year?

Solution

We integrate the difference of the rates from month $t = 0$ (the beginning of the first year) to month $t = 12$ (the end of the year):

$$\int_0^{12} (3e^{0.1t} - 2e^{0.05t})\, dt = \left(3\,\frac{1}{0.1}\,e^{0.1t} - 2\,\frac{1}{0.05}\,e^{0.05t} \right)\Bigg|_0^{12}$$

$$\underbrace{\phantom{3\,\frac{1}{0.1}}}_{30} \qquad \underbrace{\phantom{2\,\frac{1}{0.05}}}_{40}$$

$$= \underbrace{(30e^{1.2} - 40e^{0.6})}_{\text{Evaluating at } t = 12} - \underbrace{(30e^0 - 40e^0)}_{\text{Evaluating at } t = 0}$$

$$\underbrace{\approx 26.7 - (-10) = 36.7}_{\text{Using a calculator}} \qquad \text{In thousands}$$

The celebrity endorsement should result in 36.7 thousand, or 36,700 additional sales during the first year. (The profits from these additional sales must then be compared to the cost of the celebrity endorsement to decide whether it is worthwhile.)

Area Between Curves That Cross

At a point where two curves cross, the upper curve becomes the lower curve and the lower becomes the upper. The area between them must then be calculated by two (or more) integrals, upper minus lower on *each* interval.

| EXAMPLE 5 | FINDING THE AREA BETWEEN CURVES THAT CROSS |

Find the area between the curves $y = 12 - 3x^2$ and $y = 4x + 5$ from $x = 0$ to $x = 3$.

Solution

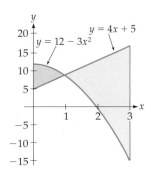

A sketch shows that the curves *do* cross. To find the intersection point, we set the functions equal to each other and solve.

$$4x + 5 = 12 - 3x^2 \qquad \text{Equating the two functions}$$

$$3x^2 + 4x - 7 = 0 \qquad \text{Combining all terms on the left}$$

$$(3x + 7)(x - 1) = 0 \qquad \text{Factoring}$$

$$x = 1 \qquad \begin{array}{l}\text{The other solution, } x = -7/3, \\ \text{is not in the interval } [0, 3]\end{array}$$

The curves cross at $x = 1$, so we must integrate separately over the intervals $[0, 1]$ and $[1, 3]$. The diagram shows which curve is upper and which is lower on each interval.

$$\underbrace{\int_0^1 [\underbrace{(12 - 3x^2)}_{\text{Upper}} - \underbrace{(4x + 5)}_{\text{Lower}}] \, dx} + \underbrace{\int_1^3 [\underbrace{(4x + 5)}_{\text{Upper}} - \underbrace{(12 - 3x^2)}_{\text{Lower}}] \, dx} \quad \begin{array}{l}\text{Integrating upper} \\ \text{minus lower on} \\ \text{each interval}\end{array}$$

$$= \int_0^1 (-3x^2 - 4x + 7) \, dx + \int_1^3 (3x^2 + 4x - 7) \, dx \qquad \begin{array}{l}\text{Simplifying the} \\ \text{integrands}\end{array}$$

$$= (-x^3 - 2x^2 + 7x) \Big|_0^1 + (x^3 + 2x^2 - 7x) \Big|_1^3 \qquad \begin{array}{l}\text{Integrating and} \\ \text{simplifying}\end{array}$$

$$= -1 - 2 + 7 + 27 + 18 - 21 - (1 + 2 - 7) = 32 \qquad \text{Evaluating}$$

Therefore, the area between the curves is 32 square units.

Graphing Calculator Exploration

Signed area —

Area (with absolute values, interpret as 32)

In Example 5 we used *two* integrals, since "upper" and "lower" switched at $x = 1$.

a. To see what happens if you integrate *without* regard to upper and lower, enter $y_1 = 12 - 3x$ and $y_2 = 4x + 5$ and graph them on the window $[0, 3]$ by $[-20, 20]$. Have your calculator find the definite integral of $y_1 - y_2$ on the interval $[0, 3]$. (Use a command like FnInt or $\int f(x) \, dx$. You will get a negative answer, which cannot be correct for an area.)

b. Explain why the answer was negative. [*Hint:* Look at the graph.]

c. Finally, obtain the correct answer for the area by returning to the calculation in part (a) and integrating the *absolute value* of the difference, $|y_1 - y_2|$ [on some calculators, entered as ABS($y_1 - y_2$)]

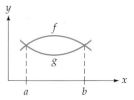

The *x*-values *a* and *b* are where the curves meet.

Areas Bounded by Curves

It is sometimes useful to find the *area bounded by two curves,* without being told the starting and ending *x*-values. In such problems the curves completely enclose an area, and the *x*-values for the upper and lower limits of integration are found by setting the functions equal to each other and solving.

EXAMPLE 6   FINDING AN AREA BOUNDED BY CURVES

Find the area bounded by the curves

$$y = 3x^2 - 12 \quad \text{and} \quad y = 12 - 3x^2$$

Solution

The *x*-values for the upper and lower limits of integration are not given, so we find them by setting the functions equal to each other and solving.

$3x^2 - 12 = 12 - 3x^2$	Setting the functions equal to each other
$6x^2 - 24 = 0$	Combining everything on one side
$6(x^2 - 4) = 0$	Factoring
$6(x + 2)(x - 2) = 0$	Factoring further
$x = -2 \quad \text{and} \quad x = 2$	Solving

The smaller of these, $x = -2$, is the lower limit of integration and the larger, $x = 2$, is the upper limit. To determine which function is "upper" and which is "lower," we choose a "test value" between $x = -2$ and $x = 2$ ($x = 0$ will do), which we substitute into each function to see which is larger. Evaluating each of the original functions at the test point $x = 0$ yields

$$3x^2 - 12 = 3(0)^2 - 12 = -12 \quad \text{(Smaller)} \qquad \begin{matrix} 3x^2 - 12 \\ \text{at } x = 0 \end{matrix}$$

$$12 - 3x^2 = 12 - 3(0)^2 = 12 \quad \text{(Larger)} \qquad \begin{matrix} 12 - 3x^2 \\ \text{at } x = 0 \end{matrix}$$

Therefore, $y = 12 - 3x^2$ is the "upper" function (since it gives a higher *y*-value) and $y = 3x^2 - 12$ is the "lower" function. We then integrate upper minus lower between the *x*-values found earlier.

$$\int_{-2}^{2} [\underbrace{12 - 3x^2}_{\text{Upper}} - (\underbrace{3x^2 - 12}_{\text{Lower}})] \, dx = \int_{-2}^{2} (24 - 6x^2) \, dx \qquad \text{Simplifying}$$

$$= (24x - 2x^3) \Big|_{-2}^{2} \qquad \text{Integrating}$$

$$= (48 - 16) - (-48 + 16) = 64 \qquad \text{Evaluating}$$

The area bounded by the two curves is 64 square units.

The two curves $y = 12 - 3x^2$ and $y = 3x^2 - 12$ are shown in the graph on the left. Notice that we were able to calculate the area between them without having to graph them.

PRACTICE PROBLEM 3

Find the area bounded by $y = 2x^2 - 1$ and $y = 2 - x^2$.

Solution on next page >

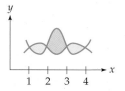

Integrate *upper minus lower* on three intervals.

For curves that intersect at *more* than two points, several integrals may be needed, integrating "upper minus lower" on each interval, as in Example 5. Test points in each interval will determine the upper and lower curves on that interval.

Solutions TO PRACTICE PROBLEMS

1. $\frac{1}{2}\int_0^2 3x^2\,dx = \frac{1}{2}\cdot x^3\Big|_0^2 = \frac{1}{2}\cdot 2^3 - \frac{1}{2}\cdot 0^3 = \frac{1}{2}\cdot 8 = 4$

2. $\int_{-1}^1 [(2x^2+1) - (-x^2-1)]\,dx = \int_{-1}^1 (2x^2+1+x^2+1)\,dx$

$= \int_{-1}^1 (3x^2+2)\,dx = (x^3+2x)\Big|_{-1}^1 = (1+2) - (-1-2) = 6\text{ square units}$

3. $2x^2 - 1 = 2 - x^2$

$3x^2 - 3 = 0$

$3(x^2 - 1) = 0$

$3(x+1)(x-1) = 0$

$x = 1 \qquad \text{and} \qquad x = -1.$

Test value $x = 0$ shows that $2 - x^2$ is "upper" and $2x^2 - 1$ is "lower."

$\int_{-1}^1 [(2-x^2) - (2x^2-1)]\,dx = \int_{-1}^1 (3-3x^2)\,dx = (3x-x^3)\Big|_{-1}^1$

$= (3-1) - (-3+1) = 4\text{ square units}$

5.4   **Section Summary**

The average value of a continuous function over an interval is defined as the definite integral of the function divided by the length of the interval:

$$\left(\begin{array}{c}\text{Average value}\\ \text{of } f \text{ on } [a,b]\end{array}\right) = \frac{1}{b-a}\int_a^b f(x)\,dx$$

To find the area between two curves:

1. If the x-values are not given, set the functions equal to each other and solve for the points of intersection.
2. Use a test point within each interval to determine which curve is "upper" and which is "lower."
3. Integrate "upper minus lower" on each interval.

If a curve lies *below* the x-axis, as in the diagram on the left, then the "upper" curve is the x-axis ($y = 0$) and the "lower" curve is $y = f(x)$, and so integrating "upper minus lower" results in integrating the *negative* of the function.

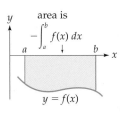

area is $-\int_a^b f(x)\,dx$

$y = f(x)$

$$\int_a^b [\underbrace{0}_{\text{Upper}} - \underbrace{f(x)}_{\text{Lower}}]\,dx = \int_a^b [-f(x)]\,dx = -\int_a^b f(x)\,dx$$

The case of area below the x-axis need not be remembered separately if you simply remember one idea:

> To find area, always integrate "upper minus lower" over each interval.

Definite integration provides a very powerful method for finding areas, taking us far beyond the few formulas (for rectangles, triangles, and circles) that we knew before studying calculus.

Derivation of the Average Value Formula from Riemann Sums

The formula for the average value of a function (page 345) can be derived using Riemann sums: For a continuous function f we could define a "sample average" by "sampling" the function at n points and averaging the results. From the interval $[a, b]$, we choose n numbers $x_1, x_2, \dots, x_n$ from successive subintervals of length $\Delta x = \dfrac{b - a}{n}$, sum the resulting values of the function, and divide by n. This gives a "sample average" of the following form:

$$\frac{1}{n}[f(x_1) + f(x_2) + \cdots + f(x_n)] \qquad \text{Sum of } n \text{ values divided by } n$$

$$= \frac{1}{b - a} \cdot \underbrace{\frac{b - a}{n}}_{\Delta x}[f(x_1) + f(x_2) + \cdots + f(x_n)] \qquad \text{Dividing and multiplying by } b - a$$

$$= \frac{1}{b - a}\underbrace{[f(x_1) + f(x_2) + \cdots + f(x_n)] \cdot \Delta x} \qquad \text{Moving } \Delta x = \frac{b - a}{n} \text{ to the right}$$

$$\text{This is a Riemann sum for } \int_a^b f(x)\, dx$$

To get a more "representative" average, we increase the number n of sample points. Letting n approach infinity makes the above Riemann sum approach the definite integral $\int_a^b f(x)\, dx$, leading to the definition of the average value of a function:

$$\left(\begin{matrix} \text{Average value} \\ \text{of } f \text{ on } [a, b] \end{matrix} \right) = \frac{1}{b - a} \int_a^b f(x)\, dx$$

5.4 Exercises

Exercises on Average Value

1–26. Find the average value of each function over the given interval.

1. $f(x) = x^2$ on $[0, 3]$

2. $f(x) = x^3$ on $[0, 2]$

3. $f(x) = 3\sqrt{x}$ on $[0, 4]$

4. $f(x) = \sqrt[3]{x}$ on $[0, 8]$

5. $f(x) = \dfrac{1}{x^2}$ on $[1, 5]$

6. $f(x) = \dfrac{1}{x^2}$ on $[1, 3]$

7. $f(x) = 2x + 1$ on $[0, 4]$

8. $f(x) = 4x - 1$ on $[0, 10]$

9. $f(x) = 36 - x^2$ on $[-2, 2]$

10. $f(x) = 9 - x^2$ on $[-3, 3]$

11. $f(z) = 3z^2 - 2z$ on $[-1, 2]$

12. $f(z) = 4z - 3z^2$ on $[-2, 2]$

13. $f(x) = 3$ on $[10, 50]$

14. $f(x) = 2$ on $[5, 100]$

15. $f(x) = e^{x/2}$ on $[0, 2]$

16. $f(x) = e^{-2x}$ on $[0, 1]$

17. $f(t) = e^{0.01t}$ on $[0, 10]$

18. $f(t) = e^{-0.1t}$ on $[0, 10]$

19. $f(x) = \dfrac{1}{x}$ on $[1, 2]$

20. $f(x) = \dfrac{1}{x}$ on $[1, 10]$

21. $f(x) = x^n$ on $[0, 1]$, where n is a constant $(n > 0)$

22. $f(x) = e^{kx}$ on $[0, 1]$, where k is a constant ($k \neq 0$)

23. $f(x) = ax + b$ on $[0, 2]$, where a and b are constants

24. $f(x) = \dfrac{1}{x}$ on $[1, c]$, where c is a constant ($c > 1$)

 25. $f(x) = e^{-x^4}$ on $[-1, 1]$

 26. $f(x) = \sqrt{1 + x^4}$ on $[-2, 2]$

Applied Exercises on Average Value

27–28. BUSINESS: Sales A store's sales on day x are given by the function $S(x)$ below. Find the average sales during the first 3 days (day 0 to day 3).

27. $S(x) = 200x + 6x^2$ **28.** $S(x) = 400x + 3x^2$

29. GENERAL: Temperature The temperature at time t hours is $T(t) = -0.3t^2 + 4t + 60$ (for $0 \leq t \leq 12$). Find the average temperature between time 0 and time 10.

30. BEHAVIORAL SCIENCE: Practice After x practice sessions, a person can accomplish a task in $f(x) = 12x^{-1/2}$ minutes. Find the average time required from the end of session 1 to the end of session 9.

31. ENVIRONMENTAL SCIENCE: Pollution The amount of pollution in a lake x years after the closing of a chemical plant is $P(x) = 100/x$ tons (for $x \geq 1$). Find the average amount of pollution between 1 and 10 years after the closing.

32. GENERAL: Population The population of the United States is predicted to be $P(t) = 310e^{0.0073t}$ million, where t is the number of years after the year 2010.

Predict the average population between the years 2010 and 2060.

Source: U.S. Census Bureau

33. BUSINESS: Compound Interest A deposit of $1000 at 5% interest compounded continuously will grow to $V(t) = 1000e^{0.05t}$ dollars after t years. Find the average value during the first 40 years (that is, from time 0 to time 40).

34. BIOMEDICAL: Bacteria A colony of bacteria is of size $S(t) = 300e^{0.1t}$ after t hours. Find the average size during the first 12 hours (that is, from time 0 to time 12).

35. BUSINESS: Average Revenue Revenues at 3D Systems are predicted to be $14.2x^2 + 120x + 474$ million dollars per year, where x is the number of years since 2013. Predict the average annual revenue from 2013 to 2023.

Source: Standard & Poor's

36. BUSINESS: Average Revenue Amazon's annual revenue is predicted to be $1.6x^2 + 17.4x + 74$ billion dollars, where x is the number of years since 2013. Predict Amazon's average revenue between 2013 and 2023.

Source: New York Times

Exercises on Area

37. Find the area between the curve $y = x^2 + 1$ and the line $y = 2x - 1$ (shown below) from $x = 0$ to $x = 3$.

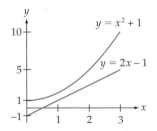

38. Find the area between the curve $y = x^2 + 3$ and the line $y = 2x$ (shown below) from $x = 0$ to $x = 3$.

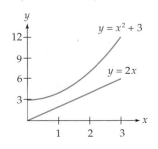

39. Find the area between the curves $y = e^x$ and $y = e^{2x}$ (shown below) from $x = 0$ to $x = 2$. (Leave the answer in its exact form.)

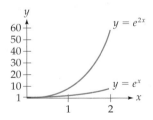

40. Find the area between the curves $y = e^x$ and $y = e^{-x}$ (shown below) from $x = 0$ to $x = 1$. (Leave the answer in its exact form.)

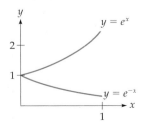

41–44. Sketch each parabola and line on the same graph and find the area between them from $x = 0$ to $x = 3$.

41. $y = x^2 + 4$ and $y = 2x + 1$

42. $y = x^2 + 5$ and $y = 2x + 3$

43. $y = 3x^2 - 3$ and $y = 2x + 5$

44. $y = 3x^2 - 12$ and $y = 2x - 11$

45–60. Find the area bounded by the given curves.

45. $y = x^2 - 1$ and $y = 2 - 2x^2$

46. $y = x^2 - 4$ and $y = 8 - 2x^2$

47. $y = 6x^2 - 10x - 8$ and $y = 3x^2 + 8x - 23$

48. $y = 3x^2 - x - 1$ and $y = 5x + 8$

49. $y = 3x^2 - 12x$ and $y = 0$

50. $y = x^2 - 6x$ and $y = 0$

51. $y = 3x^2$ and $y = 12$

52. $y = x^2$ and $y = 4$

53. $y = x^2$ and $y = x^3$

54. $y = x^3$ and $y = x^4$

55. $y = 4x^3 + 3$ and $y = 4x + 3$

56. $y = x^3$ and $y = 4x$

57. $y = 7x^3 - 36x$ and $y = 3x^3 + 64x$

58. $y = x^n$ and $y = x^{n-1}$ (for $n > 1$)

 59. $y = e^x$ and $y = x + 3$

[*Hint for Exercises 59–60:* Use INTERSECT to find the intersection points for the upper and lower limits of integration.]

60. $y = \ln x$ and $y = x - 2$

Applied Exercises on Area

61. GENERAL: Population In 2000 the birthrate in Africa increased from $17e^{0.02t}$ million births per year to $22e^{0.02t}$ million births per year, where t is the number of years since 2000. Find the total increase in population that will result from this higher birthrate between 2000 ($t = 0$) and 2050 ($t = 50$).

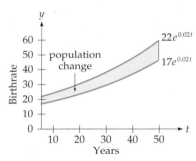

Source: New York Times

62. BUSINESS: Profit from Expansion A company expects profits of $60e^{0.02t}$ thousand dollars per month, but predicts that if it builds a new and larger factory, its profits will be $80e^{0.04t}$ thousand dollars per month, where t is the number of months from now. Find the extra profits resulting from the new factory during the first two years ($t = 0$ to $t = 24$). If the new factory will cost $1,000,000, will this cost be paid off during the first two years?

63. BUSINESS: Net Savings A factory installs new machinery that saves $S(x) = 1200 - 20x$ dollars per year, where x is the number of years since installation. However, the cost of maintaining the new machinery is $C(x) = 100x$ dollars per year.

a. Find the year x at which the maintenance cost $C(x)$ will equal the savings $S(x)$. (At this time, the new machinery should be replaced.)

b. Find the accumulated net savings [savings $S(x)$ minus cost $C(x)$] during the period from $t = 0$ to the replacement time found in part (a).

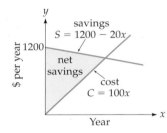

64. GENERAL: Design A graphic design consists of a white square containing the blue shape shown below. Find the area of the blue interior.

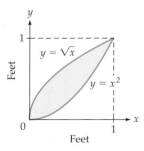

65. ECONOMICS: Balance of Trade A country's imports are $I(t) = 30e^{0.2t}$ and its exports are $E(t) = 25e^{0.1t}$, both in billions of dollars per year, where t is measured in years and $t = 0$ corresponds to the beginning of 2000. Find the country's accumulated trade deficit (imports minus exports) for the 10 years beginning with 2000. (*See diagram on next page.*)

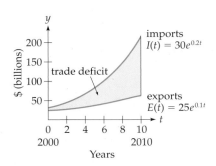

66. ECONOMICS: Cost of Labor Contracts An employer offers to pay workers at the rate of $30,000e^{0.04t}$ dollars per year, while the union demands payment at the rate of $30,000e^{0.08t}$ dollars per year, where $t = 0$ corresponds to the beginning of the contract. Find the accumulated difference in pay between these two rates over the 10-year life of the contract.

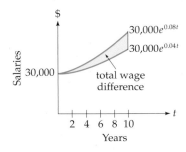

67–68. BUSINESS: Cumulative Profit A company's marginal revenue function is $MR(x) = 700x^{-1}$ and its marginal cost function is $MC(x) = 500x^{-1}$ (both in thousands of dollars), where x is the number of units $(x > 1)$. Find the total profit from

67. $x = 100$ to $x = 200$

68. $x = 200$ to $x = 300$

69. GENERAL: Lives Saved by Seat Belts Seat belt use in the United States has risen to 86%, but nonusers still risk needless expense and serious injury. The upper curve in the following graph represent an estimate of fatalities per year if seat belts were *not* used, and the lower curve is a prediction of actual fatalities per year *with* seat belt use, both in thousands (x represents years after 2010). Therefore, the area between the curves represents lives saved by seat belts. Find the area between the curves from 0 to 20, giving an estimate of the number of lives saved by seat belts during the years 2010–2030.

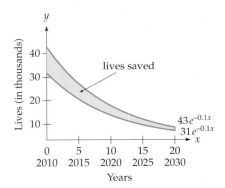

Source: National Highway Traffic Safety Administration

Review Exercises

These exercises review material that will be helpful in Section 5.6.

70–74. Find the derivative of each function.

70. e^{x^3+6x}

71. e^{x^2+5x}

72. $\ln(x^3 + 6x)$

73. $\ln(x^2 + 5x)$

74. $\sqrt{x^3 + 1}$

Conceptual Exercises

75. In the graph on page 346, the lowest and highest y-values are 0 and 3, the midpoint of which is $1\frac{1}{2}$. Yet the average of the *function* was found to be 2. Looking at the shape of the curve, can you explain why the average of the function is *larger* than the "middle" y-value?

76. Generalizing your observation in the preceding Exercise, what can you say about the average value of a function compared to its "middle" y-value if the curve is concave *down*? What if the curve is concave *up*?

77. If a linear function passes through two points (x_1, y_1) and (x_2, y_2), what is the average value of the function on the interval from x_1 to x_2?

78. If the values of a function on an interval are always positive, can the average value of the function over that interval be negative?

79. If the values of a function on an interval are always greater than 7, what can you say about the average value of the function on that interval?

80. Can the average value of a function on an interval be larger than the *maximum* value of the function on that interval? Can it be smaller than the *minimum* value on that interval?

81. If the average value of $f(x)$ on an interval is a number c, what will be the average value of the function $-f(x)$ on that interval?

82. If two curves cross twice, you can find the area contained by them by evaluating *one* definite integral (integrating "upper minus lower"). What if the curves cross *three* times—how many integrations of "upper

minus lower" would you need? What if the curves cross *ten* times?

83. Suppose that a company found its sales rate (in sales per day) if it *did* advertise, and also its (lower) sales rate if it did *not* advertise. If you integrated "upper minus lower" over a month, describe the meaning of the number that you would find.

84. Suppose that a company found its rate of revenue (dollars per day) and its (lower) rate of costs (also in dollars per day). If you integrated "upper minus lower" over a month, describe the meaning of the number that you would find.

5.5 Two Applications to Economics: Consumers' Surplus and Income Distribution

Introduction

In this section we discuss several important economic concepts—**consumers' surplus, producers' surplus,** and the **Gini index of income distribution**—each of which is defined as the area between two curves.

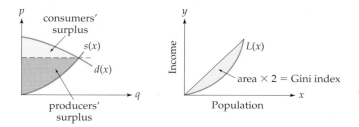

Consumers' Surplus

Imagine that you really liked pizza and were willing to pay $12 for a pizza pie. If, in fact, a pizza costs only $8, then you have, in some sense, "saved" $4, the $12 that you were willing to pay minus the $8 market price. If one were to add up this difference for all pizzas sold in a given period of time (the price that each consumer was willing to pay minus the price actually paid), the total savings would be called the *consumers' surplus* for that product. The consumers' surplus measures the benefit that consumers derive from an economy in which competition keeps prices low.

Demand Functions

Price and quantity are inversely related: if the price of an item rises, the quantity sold generally falls, and vice versa. Through market research, economists can determine the relationship between price and quantity for an item. This relationship can often be expressed as a **demand function** (or **demand curve**) $d(x)$, so called because it gives the price at which exactly x units will be demanded.

Price

demand
curve
$d(x)$

Quantity

Demand Function

The demand function $d(x)$ for a product gives the price at which exactly x units will be sold.

$$d(x) = \left(\begin{array}{c} \text{Price when} \\ \text{demand is } x \end{array} \right)$$

On page 193 we called $d(x)$ the *price function*.

Mathematical Definition of Consumers' Surplus

p

Price

consumers'
surplus

market
price

demand
curve

Quantity

The demand curve gives the price that consumers are *willing* to pay, and the *market price* is what they *do* pay, so the amount by which the demand curve is above the market price measures the benefit or "surplus" to consumers. We add up all of these benefits by integrating, so the area between the demand curve and the market price line gives the *total benefit* that consumers derive from being able to buy at the market price. This total benefit (the shaded area in the diagram on the left) is called the *consumers' surplus.*

Consumers' Surplus

For a demand function $d(x)$ and demand level A, the market price B is the demand function evaluated at $x = A$, so that $B = d(A)$. The consumers' surplus is the area between the demand curve and the market price.

$$\left(\begin{array}{c} \text{Consumers'} \\ \text{surplus} \end{array} \right) = \int_0^A [d(x) - B]\, dx$$

Demand   Market
function   price

$\$$

consumers'
surplus

Price

market price
$B = d(A)$

$d(x)$

A

Quantity

$\$$

Price

1100
900
700
500
300
100

$d(x) = 1100 - 10x$

20 40 60 80 100 x

Electricity consumption

EXAMPLE 1   FINDING CONSUMERS' SURPLUS FOR ELECTRICITY

If the demand function for electricity is $d(x) = 1100 - 10x$ dollars (where x is in millions of kilowatt-hours, $0 \le x \le 100$), find the consumers' surplus at the demand level $x = 80$.

Solution

First we find the market price, the demand function $d(x)$ evaluated at the demand level $x = 80$.

$$\left(\begin{array}{c} \text{Market} \\ \text{price } B \end{array} \right) = d(80) = 1100 - 10 \cdot 80 = 300$$

$d(x) = 1100 - 10x$
at $x = 80$

Electricity consumption

The consumers' surplus is then the area between the demand curve and the market price line.

$$\left(\begin{array}{c}\text{Consumers'}\\\text{surplus}\end{array}\right) = \int_0^{80} (\underbrace{1100 - 10x}_{\substack{\text{Demand}\\\text{function}}} - \underbrace{300}_{\substack{\text{Market}\\\text{price}}})\, dx$$

$$= \int_0^{80} (800 - 10x)\, dx = (800x - 5x^2)\Big|_0^{80} \quad \begin{array}{l}\text{Simplifying and}\\\text{integrating}\end{array}$$

$$= (64{,}000 - 32{,}000) - (0) = 32{,}000$$

Therefore, the consumers' surplus for electricity is $32,000.

How Consumers' Surplus Is Used

In Example 1, at demand level $x = 80$, the consumers' surplus was $32,000. If electricity usage were to increase to $x = 90$, the market price would then drop to $d(90) = 1100 - 10 \cdot 90 = 200$. We could then calculate the consumers' surplus at this higher demand level (and would find that the answer is $40,500). Therefore, a price decrease from $300 to $200 would mean that consumers would benefit by an additional $40,500 - $32,000 = $8500. This benefit would then be compared to the cost of a new generator to decide whether the expenditure would be worthwhile.

Producers' Surplus

Just as the consumers' surplus measures the total benefit to consumers, the *producers' surplus* measures the total benefit that producers derive from being able to sell at the market price. Returning to our pizza example, if a pizza producer might be just willing to remain in business if the price of pizzas dropped to $5, the fact that pizzas can be sold for $8 gives the producer a "benefit" of $3. The sum of all such benefits is the *producers' surplus* for the product.

Supply Functions

Clearly, as the price of an item rises, so does the quantity that producers are willing to supply at that price. The relationship between the price of an item and the quantity that producers are willing to supply at that price can be expressed as a **supply function** (or **supply curve**) $s(x)$.

Supply Function

Quantity

The supply function $s(x)$ for a product gives the price at which exactly x units will be supplied.

$$s(x) = \left(\begin{array}{c}\text{Price when}\\\text{supply is } x\end{array}\right)$$

Mathematical Definition of Producers' Surplus

As before, we integrate to find the total benefit, but now "upper" is the market price and "lower" is the supply curve $s(x)$.

Producers' Surplus

For a supply function $s(x)$ and demand level A, the market price is $B = s(A)$. The producers' surplus is the area between the market price and the supply curve.

$$\begin{pmatrix}\text{Producers'} \\ \text{surplus}\end{pmatrix} = \int_0^A [B - s(x)]\, dx$$

Market   Supply
price   function

EXAMPLE 2   FINDING PRODUCERS' SURPLUS

For the supply function $s(x) = 0.09x^2$ dollars and the demand level $x = 200$, find the producers' surplus.

Solution

First we find the market price, the supply function $s(x)$ evaluated at the demand level $x = 200$.

$$\begin{pmatrix}\text{Market} \\ \text{price } B\end{pmatrix} = s(200) = 0.09(200)^2 = 3600 \qquad \begin{array}{l}s(x) = 0.09x^2 \\ \text{at } x = 200\end{array}$$

The producers' surplus is then the area between the market price line and the supply curve.

$$\begin{pmatrix}\text{Producers'} \\ \text{surplus}\end{pmatrix} = \int_0^{200} (3600 - 0.09x^2)\, dx = (3600x - 0.03x^3)\Big|_0^{200}$$

Market Supply
price   function

$$= (720{,}000 - 240{,}000) - (0) = 480{,}000$$

Therefore, the producers' surplus is $480,000.

Consumers' Surplus and Producers' Surplus

Until now, we have discussed demand and supply curves separately. If we graph them together, the point at which they intersect is called the *market equilibrium*. It is important because it is where supply and demand will be in balance: The amount *supplied* at that price will exactly equal the amount *demanded* at that price. The demand x at this equilibrium point is called the **market demand.** The consumers' surplus and the producers' surplus can then be shown together on the same graph. These two areas together give a numerical measure of the total benefit that consumers and producers derive from competition, showing that both consumers and producers benefit from an open market.

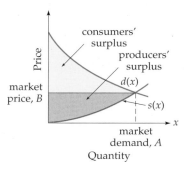

Gini Index of Income Distribution

In any society, some people make more money than others. To measure the "gap" between the rich and the poor, economists calculate the proportion of the total income that is earned by the lowest 20% of the population, and then the proportion that is earned by the lowest 40% of the population, and so on. This information (for the United States in the year 2012) is given in the table below (with percentages written as decimals) and is graphed on the right.

Proportion of Population	Proportion of Income
0.20	0.03
0.40	0.12
0.60	0.26
0.80	0.49
1.00	1.00

Source: U.S. Census Bureau

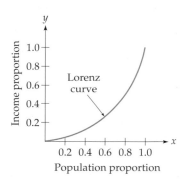

For example, the lowest 20% of the population earns only 3% of the total income, the lowest 40% earns only 12% of the total income, and so on. The curve is known as the *Lorenz curve*.*

Lorenz Curve

The Lorenz curve $L(x)$ gives the proportion of total income earned by the lowest proportion x of the population.

Gini Index

The Lorenz curve may be compared with two extreme cases of income distribution.

1. *Absolute equality of income* means that everyone earns exactly the same income, and so the lowest 10% of the population earns exactly 10% of the total income, the lowest 20% earns exactly 20% of the income, and so on. This gives the Lorenz curve $y = x$ shown below on the left.

---

*After Max Otto Lorenz (1880–1962), American statistician and economist.

Absolute equality of income

Absolute inequality of income

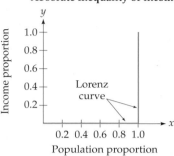

2. *Absolute inequality of income* means that nobody earns any income except one person, who earns all the income. This gives the Lorenz curve shown above on the right.

To measure how the actual distribution differs from absolute equality, we calculate the area between the actual distribution and the line of absolute equality $y = x$. Since this area can be at most $\frac{1}{2}$ (the area of the entire lower triangle), economists multiply the area by 2 to get a number between 0 (absolute equality) and 1 (absolute inequality). This measure is called the *Gini index*.* Note that a higher Gini index means greater *in*equality (greater deviation from the line of absolute equality).

Gini Index

For a Lorenz curve $L(x)$, the Gini index is

$$\left(\begin{array}{c}\text{Gini}\\\text{index}\end{array}\right) = 2 \int_0^1 [x - L(x)]\, dx$$

The Gini index varies from 0 (absolute equality) to 1 (absolute inequality).

EXAMPLE 3   **FINDING THE GINI INDEX**

The Lorenz curve for income distribution in the United States in 2012 was approximately $L(x) = x^{2.78}$. Find the Gini index.

Solution

First we calculate the area between the curve of absolute equality $y = x$ and the Lorenz curve $y = x^{2.78}$.

$$\int_0^1 (x - x^{2.78})\, dx = \left(\frac{1}{2} x^2 - \frac{1}{3.78} x^{3.78}\right)\Big|_0^1$$

$$= \frac{1}{2} \cdot 1^2 - \frac{1}{3.78} \cdot 1^{3.78} - 0$$

$$\approx 0.5 - 0.265 = 0.235$$

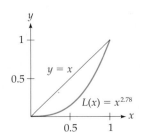

*After Corrado Gini (1884–1965), Italian statistician and sociologist.

Some Gini indices for incomes	
South Africa	0.63
Brazil	0.55
China	0.47
Britain	0.34
Germany	0.28
Sweden	0.25

Source: World Bank

Multiplying by 2 gives the Gini index.

$$\left(\frac{\text{Gini}}{\text{index}} \right) = 0.47 \qquad \text{Rounding to 2 decimal places.}$$

Source: U.S. Census Bureau

PRACTICE PROBLEM 1

In 1990, the Gini index for income was 0.43 and in 2000 it was 0.45. Since 1990, has income distribution become more equal or less equal?

Solution on next page >

Graphing Calculator Exploration

— Lorenz exponent

In Example 3, how was the Lorenz function of the form x^n found? It was found by a method called "least squares" (discussed in Section 7.4), which minimizes the squared differences between the income proportions (from the table on page 360) and the curve x^n at the x-values 0.2, 0.4, 0.6, and 0.8. This amounts to minimizing the function

$$(0.2^x - 0.03)^2 + (0.4^x - 0.12)^2 + (0.6^x - 0.26)^2 + (0.8^x - 0.49)^2$$

(Notice that here we are using x for the *exponent*.)

a. Graph this function on your calculator on the window [0, 5] by [0, 0.2].

b. Find the x that minimizes the function. The resulting value is the exponent. Your answer should agree with the exponent in Example 3.

The graph on the left, on the window [0, 1] by [0, 1], shows that the function $x^{2.78}$ fits the points from the table on page 360 very well. (Other types of functions besides x^n could also be used to fit these data.)

The Gini index for *total* wealth can be calculated similarly.

PRACTICE PROBLEM 2

The Lorenz curve for total wealth in the United States for 2006 was approximately $L(x) = x^{9.0}$. Calculate the Gini index.   Solution on next page >

Source: United Nations University

Practice Problem 2 shows that the Gini index for wealth is greater than the Gini index for income. That is, total wealth in the United States is distributed more unequally than total income. One reason for this is that we have an income tax but no "wealth" tax.

How the Gini Index Is Used

The Gini index is the most commonly used index of inequality and has been used to evaluate the fairness of new income tax systems. A *progressive* system taxes higher incomes at higher rates and so should tend to reduce income inequality. This can be measured by seeing how much the Gini index decreases after the new tax. Single Gini numbers have no particular meaning—they are meaningful only in comparison, with the higher Gini index indicating greater inequality. Gini indices have also been used in ecology to measure biodiversity and in education to measure equality of access to universities.

Solutions TO PRACTICE PROBLEMS

1. Less equal

2. $\displaystyle\int_0^1 (x - x^{9.0})\,dx = \left(\frac{1}{2}x^2 - \frac{1}{10}x^{10}\right)\Big|_0^1 = \frac{1}{2} - \frac{1}{10} - 0 = 0.4$

    Gini index for wealth is $2 \times 0.4 = 0.8$.

5.5 Exercises

1–8. For each demand function $d(x)$ and demand level x, find the consumers' surplus.

1. $d(x) = 4000 - 12x, \quad x = 100$

2. $d(x) = 500 - x, \quad x = 400$

3. $d(x) = 300 - \frac{1}{2}x, \quad x = 200$

4. $d(x) = 200 - \frac{1}{2}x, \quad x = 300$

5. $d(x) = 350 - 0.09x^2, \quad x = 50$

6. $d(x) = 840 - 0.06x^2, \quad x = 100$

 7. $d(x) = 200e^{-0.01x}, \quad x = 100$

 8. $d(x) = 400e^{-0.02x}, \quad x = 75$

9–12. For each supply function $s(x)$ and demand level x, find the producers' surplus.

9. $s(x) = 0.02x, \quad x = 100$

10. $s(x) = 0.4x, \quad x = 200$

11. $s(x) = 0.03x^2, \quad x = 200$

12. $s(x) = 0.06x^2, \quad x = 50$

13–18. For each demand function $d(x)$ and supply function $s(x)$:

a. Find the market demand (the positive value of x at which the demand function intersects the supply function).

b. Find the consumers' surplus at the market demand found in part (a).

c. Find the producers' surplus at the market demand found in part (a).

13. $d(x) = 300 - 0.4x, \quad s(x) = 0.2x$

14. $d(x) = 120 - 0.16x, \quad s(x) = 0.08x$

15. $d(x) = 300 - 0.03x^2, \quad s(x) = 0.09x^2$

16. $d(x) = 360 - 0.03x^2, \quad s(x) = 0.006x^2$

 17. $d(x) = 300e^{-0.01x}, \quad s(x) = 100 - 100e^{-0.02x}$

 18. $d(x) = 400e^{-0.01x}, \quad s(x) = 0.01x^{2.1}$

19–30. Find the Gini index for the given Lorenz curve.

Source: United Nations University-World Income Inequality Database

19. $L(x) = x^{3.2}$ (the Lorenz curve for U.S. income in 1929)

20. $L(x) = x^3$ (the Lorenz curve for U.S. income in 1935)

21. $L(x) = x^{2.1}$ (the Lorenz curve for income in Sweden in 1990)

22. $L(x) = x^{15.3}$ (the Lorenz curve for wealth in Great Britain in 1990)

23. $L(x) = 0.4x + 0.6x^2$ 24. $L(x) = 0.2x + 0.8x^3$

25. $L(x) = x^n$ (for $n > 1$)

26. $L(x) = \frac{1}{2}x + \frac{1}{2}x^n$ (for $n > 1$)

 27. $L(x) = \dfrac{e^{x^2} - 1}{e - 1}$

 28. $L(x) = 1 - \sqrt{1 - x}$

 29. $L(x) = \dfrac{x + x^2 + x^3}{3}$

 30. $L(x) = 0.62x^{7.15} + 0.38x^{9.47}$

 31–32. The following tables give the distribution of family income in the United States: Exercise 31 is for the year 1977 and Exercise 32 is for 1989. Use the procedure described in the Graphing Calculator Exploration on the previous page to find the Lorenz function of the form x^n for the data. Then find the Gini index. If you do both problems, did family income become more concentrated or less concentrated from 1977 to 1989?

Source: Congressional Budget Office

31.

Proportion (Lowest) of Families	Proportion of Income (1977)
0.20	0.06
0.40	0.18
0.60	0.34
0.80	0.57

32.

Proportion (Lowest) of Families	Proportion of Income (1989)
0.20	0.04
0.40	0.14
0.60	0.29
0.80	0.51

Review Exercises

These exercises review material that will be helpful in Section 5.6.

33–38. Find the derivative of each function.

33. $(x^5 - 3x^3 + x - 1)^4$ **34.** $(x^4 - 2x^2 - x + 1)^5$

35. $\ln(x^4 + 1)$ **36.** $\ln(x^3 - 1)$

37. e^{x^3} **38.** e^{x^4}

Conceptual Exercises

39. Suppose that for a demand function $d(x)$ we have $d(0) = 1000$. Describe in everyday language what this means about the number 1000.

40. Suppose that for a demand function $d(x)$ we have $d(1000) = 0$. Describe in everyday language what this means about the number 1000.

41. If $d(x)$ is the demand function for a product, what would it mean about the product if $d(0) = 0$?

42. Should demand curves slope *upward* or *downward*? Why?

43. Should supply curves slope *upward* or *downward*? Why?

44. For a Lorenz curve $L(x)$, what must be the values of $L(0)$ and $L(1)$?

45. Can a Lorenz curve look like this? Explain.

46. What would it mean about the distribution of income if the Lorenz curve looked like this?

47. In the country of Equalia, the Gini index for income is, by law, fixed at 0. State one good thing and one bad thing about living in Equalia.

48. Does a higher Gini index mean more *equality* or more *inequality*?

5.6   Integration by Substitution

Introduction

The Chain Rule (page 141) greatly expanded the range of functions that we could differentiate. In this section we will learn a related technique for integration, called the *substitution method*, which will greatly expand the range of functions that we can integrate. First, however, we must define **differentials.**

Differentials

One of the notations for the derivative of a function $f(x)$ is df/dx. Although written as a fraction, df/dx was not defined as the quotient of two quantities df and dx, but as a single object, the *derivative*. We will now define df and dx separately (they are called *differentials*) so that their quotient $df \div dx$ is equal to the derivative df/dx. We begin with

$$\frac{df}{dx} = f'$$ Since df/dx and f' are both notations for the derivative

$$df = f' \, dx$$ Multiplying each side by dx

This leads to a definition for the differential df.

Differential

For a differentiable function $f(x)$, the differential df is

$$df = f'(x)\, dx \qquad\qquad \textit{df is the derivative times } dx$$

Be Careful df does not mean d times f, but means the derivative of the function f times dx. What is dx? We have used it as the notation at the end of integrals, but it really represents a small change in x, just like the Δx that was the base of the rectangles in Riemann sums. The reason for finding differentials will be clear shortly.

LOOKING BACK

See page 331, where we used Δx for the base of rectangles.

EXAMPLE 1	FINDING DIFFERENTIALS

Function $f(x)$	Differential df	
$f(x) = x^2$	$df = 2x\, dx$	
$f(x) = \ln x$	$df = \dfrac{1}{x}\, dx$	Each differential is the derivative of the function times dx
$f(x) = e^{x^2}$	$df = e^{x^2}(2x)\, dx$	
$f(x) = x^4 - 5x + 2$	$df = \underbrace{(4x^3 - 5)}_{f'(x)}\, dx$	

PRACTICE PROBLEM 1

For $f(x) = x^3 - 4x - 2,$ find the differential df.   Solution on page 373 >

The differential formula $df = f'\, dx$ is easy to remember because dividing both sides by dx gives

$$\frac{df}{dx} = f'$$

which simply says "the derivative equals the derivative." We may use other letters besides f and x.

EXAMPLE 2	CALCULATING DIFFERENTIALS IN OTHER VARIABLES

Function	Differential	
$u = x^3 + 1$	$du = 3x^2\, dx$	Differentials end with d followed by the variable
$u = e^{2t} + 1$	$du = 2e^{2t}\, dt$	

PRACTICE PROBLEM 2

For $u = e^{-5t},$ find the differential du.   Solution on page 373 >

Substitution Method

Using differential notation, we can state three very useful integration formulas.

Substitution Formulas

(A) $\displaystyle\int u^n \, du = \frac{1}{n+1} u^{n+1} + C$ $\qquad n \neq -1$

(B) $\displaystyle\int e^u \, du = e^u + C$

(C) $\displaystyle\int \frac{du}{u} = \int \frac{1}{u} \, du = \int u^{-1} \, du = \ln|u| + C$

 LOOKING AHEAD

Exercises 61–63 on page 375 show how to prove these formulas.

These formulas are easy to remember because they are exactly the formulas that we learned earlier (see pages 320 and 326) except that here we use the letter u to stand for a *function*. The du is the **differential** of the function. Each of these formulas may be justified by differentiating the right-hand side. A few examples will illustrate their use.

EXAMPLE 3   INTEGRATING BY SUBSTITUTION

Find $\displaystyle\int (x^2 + 1)^3 \, 2x \, dx$.

Solution

The integral involves a function to a power:

$$\int (x^2 + 1)^3 \, 2x \, dx$$ $(x^2 + 1)^3$ is a function to a power

$\downarrow$

as does formula (A): $\displaystyle\int u^n \, du$ u^n is a function to a power in

$$\int u^n \, du = \frac{1}{n+1} u^{n+1} + C$$

To make the integral "fit" the formula we take $u = x^2 + 1$ and $n = 3$.

$$\int \underbrace{(x^2 + 1)^3}_{u^3} \, 2x \, dx$$ $u = x^2 + 1$ and $n = 3$

For $u = x^2 + 1$ the differential is $du = 2x \, dx$, which is exactly the remaining part of the integral. We then write the integral with each x-expression replaced by its equivalent u-expression.

$$\int \underbrace{(x^2 + 1)^3}_{u^3} \, \underbrace{2x \, dx}_{du} = \int \underbrace{u^3 \, du}_{\text{Written in terms of } u}$$ Using $u = x^2 + 1$ and $du = 2x \, dx$

The last integral we solve by formula (A):

$$\int u^3 \, du = \frac{1}{4} u^4 + C \qquad\qquad \int u^n \, du = \frac{1}{n+1} u^{n+1} + C$$

$$\text{[formula (A)] with } n = 3$$

Finally, we substitute back to the original variable x, using our relationship $u = x^2 + 1$, to get the answer:

$$\frac{1}{4} (x^2 + 1)^4 + C \qquad\qquad \frac{1}{4} u^4 + C \text{ with } u = x^2 + 1$$

The procedure is not as complicated as it might seem. All these steps may be written together as follows:

$$\int \underbrace{(x^2 + 1)^3}_{u^3} \underbrace{2x \, dx}_{du} \;=\; \int u^3 \, du \;=\; \frac{1}{4} u^4 + C \;=\; \frac{1}{4} (x^2 + 1)^4 + C$$

Choosing $u = x^2 + 1$ therefore $du = 2x \, dx$	Substituting $u^3 = (x^2 + 1)^3$ $du = 2x \, dx$	Integrating by formula (A) with $n = 3$	Substituting back to x using $u = x^2 + 1$

We may check this answer by differentiation (using the Generalized Power Rule or Chain Rule).

$$\frac{d}{dx} \left[\underbrace{\frac{1}{4} (x^2 + 1)^4 + C}_{\text{Answer}} \right] = \frac{1}{4} \cdot 4(x^2 + 1)^3 \underbrace{2x}_{\substack{\text{Derivative} \\ \text{of the inside}}} = \underbrace{(x^2 + 1)^3 \, 2x}_{\text{Integrand}}$$

Since the result of the differentiation agrees with the original integrand, the integration is correct.

Multiplying Inside and Outside by Constants

If the integral does not exactly match the form $\int u^n \, du$, we may sometimes still solve the integral by multiplying by constants.

EXAMPLE 4   INSERTING CONSTANTS BEFORE SUBSTITUTING

Find $\displaystyle\int (x^2 + 1)^3 x \, dx.$ Same as Example 3 but without the 2

Solution

As before, we use formula (A) with $u = x^2 + 1$, which gives $du = 2x \, dx$. But the integral has only an $x \, dx$, not $2x \, dx$, which would allow us to substitute du.

$$\int \underbrace{(x^2 + 1)^3}_{u^3} \underbrace{x \, dx}_{\substack{\text{not } du = 2x \, dx \\ \text{because there is no 2}}} \qquad\qquad u = x^2 + 1, \text{ so } du = 2x \, dx$$

Therefore, the integral is *not* in the form $\int u^3 \, du$. (The integral must fit the formula exactly: *Everything* in the integral must be accounted for either by the u^n or by the du.) However, we may multiply inside the integral by 2 as long as we compensate by also multiplying by $\frac{1}{2}$, and the $\frac{1}{2}$ may be written *outside* the integral (since constants may be moved across the integral sign), leading to the solution:

Multiplying by $\frac{1}{2}$ and 2

$$\frac{1}{2} \int \underbrace{(x^2+1)^3}_{\substack{u^3 \\ u = x^2+1 \\ du = 2x\,dx}} \underbrace{2x\,dx}_{du} = \underset{\text{Substituting}}{\frac{1}{2}\int u^3\,du} = \underset{\substack{\text{Integrating} \\ \text{by formula (A)}}}{\frac{1}{2}\cdot\frac{1}{4}u^4 + C} = \underset{\substack{\text{Substituting back} \\ \text{to } x \text{ using} \\ u = x^2+1}}{\frac{1}{8}(x^2+1)^4 + C}$$

This method of multiplying inside and outside by a constant is very useful and may be used with the other substitution formulas as well.

EXAMPLE 5   USING OTHER SUBSTITUTION FORMULAS

Find $\displaystyle\int e^{x^3} x^2 \, dx$.

Solution

The integral involves e to a function:

$$\int e^{x^3} x^2 \, dx \qquad\qquad e^{x^3} \text{ is } e \text{ to a function}$$

$$\updownarrow$$

as does formula (B) on page 366:

$$\int e^u \, du = e^u + C$$

Matching exponents of e gives $u = x^3$, and the differential is $du = 3x^2 \, dx$. The differential requires a 3, which is not in the integral, so we multiply inside by 3 and outside by $\frac{1}{3}$.

$$\int e^{x^3} x^2 \, dx = \frac{1}{3} \int \underbrace{e^{x^3}}_{e^u} \underbrace{3x^2 \, dx}_{du} = \underset{\text{Substituting}}{\frac{1}{3} \int e^u \, du} = \underset{\substack{\text{Integrating} \\ \text{using} \\ \text{formula (B)}}}{\frac{1}{3} e^u + C} = \underset{\substack{\text{Substituting} \\ \text{back to } x \\ \text{using } u = x^3}}{\frac{1}{3} e^{x^3} + C}$$

$u = x^3$
$du = 3x^2 \, dx$ Multiplying by $\frac{1}{3}$ and 3

Why did the 1/3 remain in the answer but the 3 disappeared? The 3 became part of the du ($du = 3x^2 \, dx$), which was then "used up" in the integration along with the integral sign.

EXAMPLE 6   RECOVERING COST FROM MARGINAL COST

A company's marginal cost function is $MC(x) = \dfrac{x^3}{x^4 + 1}$ and fixed costs are $1000. Find the cost function.

Solution   Cost is the integral of marginal cost.

$$C(x) = \int \frac{x^3 \, dx}{x^4 + 1}$$

The differential of the denominator is $4x^3 \, dx$, which except for the 4 is just the numerator. This suggests formula (C), $\displaystyle\int \frac{du}{u} = \ln|u| + C$ with $u = x^4 + 1$. We multiply inside by 4 (to complete the $du = 4x^3 \, dx$ in the numerator) and outside by $\frac{1}{4}$.

$$\int \frac{x^3 \, dx}{x^4 + 1} = \frac{1}{4} \int \frac{4x^3 \, dx}{x^4 + 1} = \frac{1}{4} \int \frac{du}{u} = \frac{1}{4} \ln|u| + C = \frac{1}{4} \ln(x^4 + 1) + C$$

$u = x^4 + 1$ $du = 4x^3 \, dx$	Multiplying by 4 and $\frac{1}{4}$	Substituting	Integrating by formula (C)	Substituting back to x

We dropped the absolute value bars because $x^4 + 1$ is positive. To evaluate the constant C, we set the cost function (evaluated at $x = 0$) equal to the given fixed cost.

$$\frac{1}{4} \underbrace{\ln(1)}_{0} + C = 1000$$

$\ln(x^4 + 1) + C$ at $x = 0$ set equal to 1000

This gives $C = 1000$. Therefore, the company's cost function is

$$C(x) = \frac{1}{4} \ln(x^4 + 1) + 1000$$

$C(x) = \frac{1}{4}\ln(x^4 + 1) + C$ with $C = 1000$

Which Formula to Use

The three formulas apply to three different types of integrals.

(A) $\displaystyle\int u^n \, du = \frac{1}{n + 1} u^{n+1} + C \quad (n \neq -1)$ Integrates a *function to a constant power* (except −1) times the differential of the function

(B) $\displaystyle\int e^u \, du = e^u + C$ Integrates *e to a power* times the differential of the exponent

(C) $\displaystyle\int \frac{du}{u} = \int u^{-1} \, du = \ln|u| + C$ Integrates a *fraction whose top is the differential of the bottom,* or equivalently, *a function to the power −1* times the differential of the function

To solve an integral by substitution, choose the formula whose left-hand side has the same form as the given integral.

PRACTICE PROBLEM 3

For each of the following integrals, choose the most appropriate formula: (A), (B), or (C). (Do not solve the integral.)

a. $\displaystyle\int e^{5x^2-1}x\,dx$ **b.** $\displaystyle\int \frac{x\,dx}{x^2+1}$ **c.** $\displaystyle\int (x^4-12)^4 x^3\,dx$

d. $\displaystyle\int (x^4-12)^{-1}x^3\,dx$

Solutions on page 373 >

Be Careful   Only constants can be adjusted. That is, we may multiply inside and outside only by constants, not variables, since only constants can be moved across the integral sign. Therefore, the *du* in a problem must already be "complete" except for adjusting the constant. Otherwise, the problem cannot be solved by a substitution. For example, the following integral cannot be found by a substitution.

$$\int \underbrace{e^{x^3}}_{e^u}\,\underbrace{x}_{\uparrow}\,dx$$

$$\text{---}\,not\,du\ (\text{does not have an } x^2)$$

$$u = x^3$$
$$du = 3x^2\,dx$$

PRACTICE PROBLEM 4

Which of these integrals can be found by a substitution? [*Hint:* See whether only a constant is needed to complete the *du*.]

a. $\displaystyle\int (x^3+1)^3 x^3\,dx$ **b.** $\displaystyle\int e^{x^2}\,dx$

Solutions on page 373 >

EXAMPLE 7   INTEGRATING BY SUBSTITUTION

Find $\displaystyle\int \sqrt{x^3-3x}\,(x^2-1)\,dx$.

Solution

Since $\sqrt{x^3-3x} = (x^3-3x)^{1/2}$ is a *function to a power*, we use the formula for $\int u^n\,du$ with $u = x^3-3x$. Comparing the differential $du = (3x^2-3)\,dx = 3(x^2-1)\,dx$ with the problem shows that we need to multiply by 3.

$$\int (x^3-3x)^{1/2}(x^2-1)\,dx = \frac{1}{3}\int \underbrace{(x^3-3x)^{1/2}}_{u^{1/2}}\underbrace{3(x^2-1)\,dx}_{du}$$

$$\quad\quad u = x^3-3x$$
$$\quad\quad du = (3x^2-3)\,dx$$
$$\quad\quad\quad = 3(x^2-1)\,dx$$

Multiplying by 3 and $\frac{1}{3}$

$$= \frac{1}{3}\int u^{1/2}\,du = \frac{1}{3}\cdot\frac{2}{3}u^{3/2} + C = \frac{2}{9}(x^3-3x)^{3/2} + C$$

Substituting   Integrating by   Substituting
                     formula (A)   back to x

EXAMPLE 8   INTEGRATING BY SUBSTITUTION

Evaluate $\displaystyle\int e^{\sqrt{x}}\, x^{-1/2}\, dx.$

Solution

The integral involves $e^{\sqrt{x}}$, e to a function, so we use the formula $\int e^u\, du = e^u + C$ with $u = x^{1/2}$.

$$\int e^{x^{1/2}} x^{-1/2}\, dx = 2 \int \underbrace{e^{x^{1/2}}}_{e^u}\, \underbrace{\frac{1}{2}\, x^{-1/2}\, dx}_{du} \qquad \text{Multiplying by } \tfrac{1}{2} \text{ and } 2$$

$$u = x^{1/2}$$
$$du = \frac{1}{2}\, x^{-1/2}\, dx$$

$$= 2 \int e^u\, du \;\;=\;\; 2e^u + C \;\;=\;\; 2e^{x^{1/2}} + C$$

$$\qquad\qquad \underset{\text{Substituting}}{} \qquad \underset{\text{Integrating}}{} \qquad \underset{\substack{\text{Substituting}\\ \text{back to } x}}{}$$

Evaluating Definite Integrals by Substitution

Sometimes a *definite* integral requires a substitution. In such cases changing from x to u also requires changing the limits of integration from x-values to u-values, using the substitution formula for u.

EXAMPLE 9   EVALUATING A DEFINITE INTEGRAL BY SUBSTITUTION

Evaluate $\displaystyle\int_4^5 \frac{dx}{3 - x}.$

Solution

The differential of the denominator $3 - x$ is $-1 \cdot dx$, which except for the -1 is just the numerator. This suggests formula (C), $\displaystyle\int \frac{du}{u} = \ln|u| + C$ with $u = 3 - x$ (from equating the denominators). We multiply inside and outside by -1.

$$u = 3 - 5 = -2$$

$$\int_4^5 \frac{dx}{3 - x} = -\int_4^5 \frac{-dx}{3 - x} = -\int_{-1}^{-2} \frac{du}{u}$$

$$u = 3 - 4 = -1$$

New upper and lower limits of integration for u are found by evaluating $u = 3 - x$ at the old x limits

$$u = 3 - x$$
$$du = -dx$$

$$= -\ln|u| \Big|_{-1}^{-2} = -\ln|-2| - (-\ln|-1|) = -\ln 2 + \ln 1 = -\ln 2$$

$\overline{\int f(X)dx} = ^-.6931472$
[4,5]

Why is the answer to Example 9 *negative*? Graph $f(x) = \dfrac{1}{3 - x}$ on the window [3, 6] by [− 2, 2] to see why. Have your calculator find the definite integral from 4 to 5. How would you change the integral so that it gives the *area* between the curve and the *x*-axis?

 LOOKING BACK

Using definite integrals to find total accumulations was summarized on page 337 and to find average values on page 345.

Definite integrals are used to find areas, total accumulations, and average values, and any of these uses may require a substitution.

EXAMPLE 10   FINDING TOTAL POLLUTION FROM A RATE

Pollution is being discharged into a lake at the rate of $r(t) = 400te^{t^2}$ tons per year, where t is the number of years since measurements began. Find the total amount of pollutant discharged into the lake during the first 2 years.

Solution

The total accumulation is the definite integral from $t = 0$ (the beginning) to $t = 2$ (2 years later). Since the integral involves e to a function, we use the formula for $\int e^u \, du$ with $u = t^2$ (by equating exponents).

$$\int_0^2 400\, t\underbrace{e^{t^2}}_{e^u}\, dt \;=\; 400 \cdot \frac{1}{2} \int_0^2 2t\, e^{t^2}\, dt = 200 \int_0^4 e^u \, du$$

$$\begin{array}{c} u = 2^2 = 4 \\ \downarrow \end{array}$$

Taking out the constant

$u = t^2$
$du = 2t\, dt$

du

$u = 0^2 = 0$

Changing the limits to u-values using $u = t^2$

$$= 200 e^u \Big|_0^4 = 200e^4 - 200e^0 \approx 10{,}720$$

Therefore, during the first 2 years 10,720 tons of pollutant were discharged into the lake.

$400te^{t^2}$ 10,720

Notice that the du does not need to be all together, but can be in several separate pieces, as long as it is all *there*.

EXAMPLE 11   FINDING AVERAGE WATER DEPTH

After x months the water level in a newly built reservoir is $L(x) = 40x(x^2 + 9)^{-1/2}$ feet. Find the average depth during the first 4 months.

Solution

The average value is the definite integral from $x = 0$ to $x = 4$ (the end of month 4) divided by the length of the interval.

$$u = 4^2 + 9 = 25$$

$$\frac{1}{4}\int_0^4 40x(x^2 + 9)^{-1/2}\, dx = \frac{1}{4}\cdot 40 \cdot \frac{1}{2}\int_0^4 \underbrace{2x(x^2+9)^{-1/2}}\, dx = 5\int_9^{25} u^{-1/2}\, du$$

Changing the limits to u-values using $u = x^2 + 9$

$u^{-1/2}$

$u = x^2 + 9$
$du = 2x\, dx$

$u^{-1/2}$

du

$u = 0^2 + 9 = 9$

$$= 5 \cdot 2u^{1/2}\Big|_9^{25} = 10(25)^{1/2} - 10(9)^{1/2} = 10\cdot 5 - 10\cdot 3 = 20$$

That is, the average depth of the reservoir over the last 4 months was 20 feet, as shown on the left.

$L(x) = 40x(x^2 + 9)^{-1/2}$

average

Months

Solutions TO PRACTICE PROBLEMS

1. $df = (3x^2 - 4)\, dx$

2. $du = -5e^{-5t}\, dt$

3. **a.** (B)   **b.** (C)   **c.** (A)   **d.** (C)

4. Neither.

   a. Try formula (A) with $u = x^3 + 1$, so $du = 3x^2\, dx$. The problem has an x^3 for the differential instead of the needed x^2.

   b. Try formula (B) with $u = x^2$, so $du = 2x\, dx$. The problem does not have the x that is needed for the differential.

5.6   Section Summary

The substitution method does for integration what the Chain Rule does for differentiation—it allows the use of simpler calculus formulas on complicated mathematical expressions by seeing them in "blocks," here the u and the du. In fact, if you use the substitution method to integrate, you should expect to use the Chain Rule to differentiate when you check your answer.

The three substitution formulas on page 366 are also listed on the inside back cover. Most of the work in using these formulas is making a problem "fit" the left-hand side of one of the formulas (choosing the u and adjusting constants to complete the du). Once a problem fits a left-hand side, the right-hand side immediately gives the answer (except for substituting back to the original variable).

Note that the du now plays a very important role: The du must be correct if the answer is to be correct. For example, the formula $\int e^u\, du = e^u + C$ should not be thought of as the formula for integrating e^u, but as the formula for integrating $e^u\, du$. The du is just as important as the e^u.

5.6 Exercises

1–8. Find each indefinite integral. [Integration formulas (A), (B), and (C) are on the inside back cover, numbered 5–7]

1. $\int (x^2 + 1)^9 \, 2x \, dx$ $\begin{bmatrix} Hint:\text{ Use } u = x^2 + 1 \\ \text{and formula 5.} \end{bmatrix}$

2. $\int (x^3 + 1)^4 \, 3x^2 \, dx$ $\begin{bmatrix} Hint:\text{ Use } u = x^3 + 1 \\ \text{and formula 5.} \end{bmatrix}$

3. $\int (x^2 + 1)^9 x \, dx$ $\begin{bmatrix} Hint:\text{ Use } u = x^2 + 1 \\ \text{and formula 5.} \end{bmatrix}$

4. $\int (x^3 + 1)^4 x^2 \, dx$ $\begin{bmatrix} Hint:\text{ Use } u = x^3 + 1 \\ \text{and formula 5.} \end{bmatrix}$

5. $\int e^{x^5} x^4 \, dx$ $\begin{bmatrix} Hint:\text{ Use } u = x^5 \\ \text{and formula 7.} \end{bmatrix}$

6. $\int e^{x^4} x^3 \, dx$ $\begin{bmatrix} Hint:\text{ Use } u = x^4 \\ \text{and formula 7.} \end{bmatrix}$

7. $\int \dfrac{x^5 \, dx}{x^6 + 1}$ $\begin{bmatrix} Hint:\text{ Use } u = x^6 + 1 \\ \text{and formula 6.} \end{bmatrix}$

8. $\int \dfrac{x^4 \, dx}{x^5 + 1}$ $\begin{bmatrix} Hint:\text{ Use } u = x^5 + 1 \\ \text{and formula 6.} \end{bmatrix}$

9–12. Show that each integral *cannot* be found by our substitution formulas.

9. $\int \sqrt{x^3 + 1} \, x \, dx$

10. $\int \sqrt{x^5 + 9} \, x^2 \, dx$

11. $\int e^{x^4} x^5 \, dx$

12. $\int e^{x^3} x^4 \, dx$

13–46. Find each indefinite integral by the substitution method or state that it cannot be found by our substitution formulas.

13. $\int (x^4 - 16)^5 x^3 \, dx$

14. $\int (x^5 - 25)^6 x^4 \, dx$

15. $\int e^{-x^2} x \, dx$

16. $\int e^{-x^4} x^3 \, dx$

17. $\int e^{3x} \, dx$

18. $\int e^{5x} \, dx$

19. $\int e^{x^2} x^2 \, dx$

20. $\int e^{x^3} x \, dx$

21. $\int \dfrac{dx}{1 + 5x}$

22. $\int \dfrac{dx}{1 + 3x}$

23. $\int (x^2 + 1)^9 5x \, dx$

24. $\int (x^2 - 4)^6 3x \, dx$

25. $\int \sqrt[4]{z^4 + 16} \, z^3 \, dz$

26. $\int \sqrt[3]{z^3 - 8} \, z^2 \, dz$

27. $\int \sqrt[4]{x^4 + 16} \, x^2 \, dx$

28. $\int \sqrt[3]{x^3 - 8} \, x \, dx$

29. $\int (2y^2 + 4y)^5 (y + 1) \, dy$

30. $\int (3y^2 - 6y)^3 (y - 1) \, dy$

31. $\int e^{x^2 + 2x + 5}(x + 1) \, dx$

32. $\int e^{x^3 - 3x + 7}(x^2 - 1) \, dx$

33. $\int \dfrac{x^3 + x^2}{3x^4 + 4x^3} \, dx$

34. $\int \dfrac{x^2 - x}{2x^3 - 3x^2} \, dx$

35. $\int \dfrac{x^3 + x^2}{(3x^4 + 4x^3)^2} \, dx$

36. $\int \dfrac{x^2 - x}{(2x^3 - 3x^2)^3} \, dx$

37. $\int \dfrac{x}{1 - x^2} \, dx$

38. $\int \dfrac{1}{1 - x} \, dx$

39. $\int (2x - 3)^7 \, dx$

40. $\int (5x + 9)^9 \, dx$

41. $\int \dfrac{e^{2x}}{e^{2x} + 1} \, dx$

42. $\int \dfrac{e^{3x}}{e^{3x} - 1} \, dx$

43. $\int \dfrac{\ln x}{x} \, dx$ [*Hint:* Let $u = \ln x$.]

44. $\int \dfrac{(\ln x)^2}{x} \, dx$ [*Hint:* Let $u = \ln x$.]

45. $\int \dfrac{e^{\sqrt{x}}}{\sqrt{x}} \, dx$ [*Hint:* Let $u = \sqrt{x}$.]

46. $\int \dfrac{e^{1/x}}{x^2} \, dx$ $\begin{bmatrix} Hint:\text{ Let } u = \dfrac{1}{x}. \end{bmatrix}$

47–50. Find each integral. [*Hint:* Try some algebra.]

47. $\int (x + 1)x^2 \, dx$

48. $\int (x + 4)(x - 2) \, dx$

49. $\int (x + 1)^2 x^3 \, dx$

50. $\int (x - 1)^2 \sqrt{x} \, dx$

51–60. For each definite integral:
a. Evaluate it "by hand."
b. Check your answer by using a graphing calculator.

51. $\int_0^3 e^{x^2} x \, dx$

52. $\int_0^2 e^{x^3} x^2 \, dx$

53. $\int_0^1 \dfrac{x}{x^2 + 1} \, dx$

54. $\int_2^3 \dfrac{x^2}{x^3 - 7} \, dx$

55. $\int_0^4 \sqrt{x^2 + 9} \, x \, dx$

56. $\int_0^3 \sqrt{x^2 + 16} \, x \, dx$

57. $\int_2^3 \dfrac{dx}{1 - x}$

58. $\int_3^4 \dfrac{dx}{2 - x}$

59. $\int_1^8 \dfrac{e^{\sqrt[3]{x}}}{\sqrt[3]{x^2}}\,dx$ **60.** $\int_1^4 \dfrac{e^{\sqrt{x}}}{\sqrt{x}}\,dx$

61. Prove the integration formula

$$\int u^n\,du = \dfrac{1}{n+1}u^{n+1} + C \quad (n \neq -1)$$

as follows.

  a. Differentiate the right-hand side of the formula with respect to x (remembering that u is a function of x).
  b. Verify that the result of part (a) agrees with the integrand in the formula (after replacing du in the formula by $u'\,dx$).

62. Prove the integration formula $\int e^u\,du = e^u + C$ by following the steps in Exercise 61.

63. Prove the integration formula $\displaystyle\int \dfrac{du}{u} = \ln u + C$
  $(u > 0)$ by following the steps in Exercise 61. (Absolute value bars come from applying the same argument to $-u$ for $u < 0$.)

64. Find $\int (x+1)\,dx$:
  a. By using the formula for $\int u^n\,du$ with $n = 1$.
  b. By dropping the parentheses and integrating directly.
  c. Can you reconcile the two seemingly different answers? [*Hint:* Think of the arbitrary constant.]

Applied Exercises

65. BUSINESS: Cost A company's marginal cost function is $MC(x) = \dfrac{1}{2x+1}$ and its fixed costs are 50. Find the cost function.

66. BUSINESS: Cost A company's marginal cost function is $MC(x) = \dfrac{1}{\sqrt{2x+25}}$ and its fixed costs are 100. Find the cost function.

67. GENERAL: Average Value The population of a city is expected to be $P(x) = x(x^2 + 36)^{-1/2}$ million people after x years. Find the average population between year $x = 0$ and year $x = 8$.

68. GENERAL: Area Find the area between the curve $y = xe^{x^2}$ and the x-axis from $x = 1$ to $x = 3$. (Leave the answer in its exact form.)

69. BUSINESS: Average Sales A company's sales (in millions) during week x are given by $S(x) = \dfrac{1}{x+1}$.   Find the average sales from week $x = 1$ to week $x = 4$.

70. BEHAVIORAL SCIENCE: Repeated Tasks A subject can perform a task at the rate of $\sqrt{2t+1}$ tasks per minute at time t minutes. Find the total number of tasks performed from time $t = 0$ to time $t = 12$.

71. BIOMEDICAL: Cholesterol An experimental drug lowers a patient's blood serum cholesterol at the rate of $t\sqrt{25 - t^2}$ units per day, where t is the number of days since the drug was administered $(0 \leq t \leq 5)$. Find the total change during the first 3 days.

72. BUSINESS: Total Sales During an automobile sale, cars are selling at the rate of $\dfrac{12}{x+1}$ cars per day, where x

is the number of days since the sale began. How many cars will be sold during the first 7 days of the sale?

73. BUSINESS: Total Sales A real estate office is selling condominiums at the rate of $100e^{-x/4}$ per week after x weeks. How many condominiums will be sold during the first 8 weeks?

74. BUSINESS: Revenue An aircraft company estimates its marginal revenue function for helicopters to be $MR(x) = (x + 40)\sqrt{x^2 + 80x}$ thousand dollars, where x is the number of helicopters sold. Find the total revenue from the sale of the first 10 helicopters.

75–76. ENVIRONMENTAL SCIENCE: Pollution A factory is discharging pollution into a lake at the rate of $r(t)$ tons per year given below, where t is the number of years that the factory has been in operation. Find the total amount of pollution discharged during the first 3 years of operation.

75. $r(t) = \dfrac{t}{t^2 + 1}$ **76.** $r(t) = t\sqrt{t^2 + 16}$

77. BIOMEDICAL: Weight Gain An average young female in the United States gains weight at the rate of $14(x - 10)^{-1/2}$ pounds per year, where x is her age $(10 \leq x \leq 19)$.   Find the total weight gain from age 11 to 19.
Source: National Center for Health Statistics

78. BIOMEDICAL: Weight Gain An average young male in the United States gains weight at the rate of $18(x - 10)^{-1/2}$ pounds per year, where x is his age $(11 \leq x \leq 20)$.   Find the total weight gain from age 11 to 19.
Source: National Center for Health Statistics

Conceptual Exercises

79. A friend says that finding differentials is as easy as finding derivatives—you just multiply the derivative by dx. Is your friend right?

80. A friend says that if you can move *numbers* across the integral sign, you can do the same for *variables* since variables stand for numbers, and in this way you can always "fix" the differential du to be what you want. Is your friend right?

81. Explain why the substitution $u = x$ is useless for finding an integral by the substitution method. Give an example.

82. What is wrong with the following use of the substitution $u = x^2$?

$$\int e^{x^2}\, dx = \int e^u\, du = e^u + C = e^{x^2} + C$$

83. What is wrong with the following use of the substitution $u = x^2 - 1$?

$$\int \frac{1}{x^2 - 1}\, dx = \int \frac{1}{u}\, du = \ln|u| + C = \ln|x^2 - 1| + C$$

84. Al and Betty carry out the substitution method for definite integrals slightly differently. Both use the same substitution to change the original "x" integral into a "u" integral, which they then integrate.

Then Al changes the "x" limits of integration into their corresponding "u" values, substitutes, and subtracts to get the final answer.

Betty uses the substitution to return the integrand to its "x" form and uses the original limits of integration to get the final answer.

Which way is better?

Explorations and Excursions   The following problems extend and augment the material presented in the text.

Further Use of the Substitution Method

85–94. The substitution method can be used to find integrals that do not fit our formulas. For example, observe how we find the following integral using the substitution $u = x + 4$ which implies that $x = u - 4$ and so $dx = du$.

$$\int (x - 2)(x + 4)^8\, dx = \int (u - 4 - 2)u^8\, du$$

$$= \int (u - 6)u^8\, du$$

$$= \int (u^9 - 6u^8)\, du$$

$$= \frac{1}{10} u^{10} - \frac{2}{3} u^9 + C$$

$$= \frac{1}{10}(x + 4)^{10} - \frac{2}{3}(x + 4)^9 + C$$

It is often best to choose u to be the quantity that is raised to a power. The following integrals may be found as explained on the left (as well as by the methods of Section 6.1).

85. $\displaystyle\int (x + 1)(x - 5)^4\, dx$ **86.** $\displaystyle\int (x - 2)(x + 4)^5\, dx$

87. $\displaystyle\int x(x - 2)^6\, dx$ **88.** $\displaystyle\int x(x + 4)^7\, dx$

89. $\displaystyle\int \frac{x - 1}{x - 2}\, dx$ **90.** $\displaystyle\int \frac{x - 4}{x - 5}\, dx$

91. $\displaystyle\int x\sqrt[3]{x - 4}\, dx$ **92.** $\displaystyle\int (x - 1)\sqrt{x + 2}\, dx$

93. $\displaystyle\int \frac{x}{\sqrt{x + 2}}\, dx$ **94.** $\displaystyle\int \frac{x}{\sqrt[3]{x + 1}}\, dx$

---

5   Chapter Summary with Hints and Suggestions

Reading the text and doing the exercises in this chapter have helped you to master the following concepts and skills, which are listed by section (in case you need to review them) and are keyed to particular Review Exercises. Answers for all Review Exercises are given at the back of the book, and full solutions can be found in the Student Solutions Manual.

5.1   Antiderivatives and Indefinite Integrals

- Find an indefinite integral using the Power Rule.   *(Review Exercises 1–8.)*

$$\int x^n\, dx = \frac{1}{n + 1} x^{n+1} + C   (n \neq -1)$$

- Solve an applied problem involving integration. *(Review Exercises 9–10.)*

5.2   Integration Using Logarithmic and Exponential Functions

- Find an indefinite integral involving e^x or $\dfrac{1}{x}$. *(Review Exercises 11–18.)*

$$\int e^{ax}\, dx = \frac{1}{a} e^{ax} + C$$

$$\int \frac{1}{x}\, dx = \int x^{-1}\, dx = \ln|x| + C$$

- Solve an applied problem involving integration.
  (*Review Exercises 19–22.*)

5.3   Definite Integrals and Areas

- Evaluate a definite integral. (*Review Exercises 23–30.*)

$$\int_a^b f(x)\,dx = F(b) - F(a)$$

- Find the area under a curve. (*Review Exercises 31–36.*)

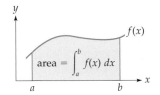

- Graph a function and find the area under it.
  (*Review Exercises 37–38.*)

- Solve an applied problem using definite
  integration.   (*Review Exercises 39–44.*)

- Approximate the area under a curve by rectangles
  (Riemann sum).   (*Review Exercises 45–46.*)

$$\int_a^b f(x)\,dx = \lim_{n \to \infty} [f(x_1) \cdot \Delta x + \cdots + f(x_n) \cdot \Delta x]$$

- Use a Riemann sum program to find Riemann
  sums.   (*Review Exercises 47–48.*)

5.4   Further Applications of Definite Integrals: Average Value and Area Between Curves

- Find the area bounded by two curves.
  (*Review Exercises 49–56.*)

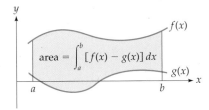

- Find the average value of a function on an
  interval.   (*Review Exercises 57–60.*)

$$\left(\begin{array}{c} \text{Average of } f \\ \text{from } a \text{ to } b \end{array} \right) = \frac{1}{b-a} \int_a^b f(x)\,dx$$

- Solve an applied problem involving average
  value.   (*Review Exercises 61–64.*)

- Solve an applied problem involving area between
  curves.   (*Review Exercises 65–66.*)

5.5   Two Applications to Economics: Consumers' Surplus and Income Distribution

- Find the consumers' surplus for a product.
  (*Review Exercises 67–70.*)

$$\left(\begin{array}{c} \text{Consumers'} \\ \text{surplus} \end{array} \right) = \int_0^A [d(x) - B]\,dx$$

- Find the Gini index of income distribution.
  (*Review Exercises 71–74.*)

$$\left(\begin{array}{c} \text{Gini} \\ \text{index} \end{array} \right) = 2 \cdot \int_0^1 [x - L(x)]\,dx$$

5.6   Integration by Substitution

- Use a substitution to find an integral.
  (*Review Exercises 75–90.*)

$$\int u^n\,du = \frac{1}{n+1} u^{n+1} + C \qquad (n \neq -1)$$

$$\int e^u\,du = e^u + C \qquad \int \frac{1}{u}\,du = \ln|u| + C$$

- Use a substitution to find a definite integral.
  (*Review Exercises 91–98.*)

- Use a substitution to find the area under a
  curve.   (*Review Exercises 99–100.*)

- Use a substitution to find the average value of a
  function.   (*Review Exercises 101–102.*)

- Use a substitution to solve an applied problem.
  (*Review Exercises 103–104.*)

Hints and Suggestions

- An *indefinite* integral is a function plus a constant,
  while a *definite* integral is a *number* (the signed area
  between the curve and the *x*-axis). The Fundamental
  Theorem of Integral Calculus shows how to evaluate
  the second in terms of the first.

- To integrate any power of *x except* x^{-1}, use the Power
  Rule; for x^{-1}, use the ln rule.

- *Indefinite* integrals have a $+ C$. *Definite* integrals do not.

- Differentiation *breaks things down into parts*—for example, turning cost into marginal cost (cost per unit).
  Integration *combines back into a whole*—for example,
  combining all of the per-unit costs back into a total cost.
  In fact, the word "integrate" means "make whole."

- To find the area between two curves, integrate *upper*
  minus *lower.*

- For average values, don't forget the $\frac{1}{b-a}$.

- The substitution method can only "fix up" the *constant*; the variable part must already be correct (or else
  the function cannot be integrated by this technique).

- A graphing calculator helps by graphing functions,
  evaluating definite integrals, and finding areas under
  curves and total accumulations. With an appropriate
  program it can calculate Riemann sums with many
  rectangles.

5   Review Exercises and Chapter Test   ○ indicates a Chapter Test exercise.

5.1   Antiderivatives and Indefinite Integrals

1–8. Find each indefinite integral.

① $\int (24x^2 - 8x + 1)\, dx$ **2.** $\int (12x^3 + 6x - 3)\, dx$

3. $\int \left(6\sqrt{x} - 5\right) dx$ **4.** $\int \left(8\sqrt[3]{x} - 2\right) dx$

⑤ $\int \left(10\sqrt[3]{x^2} - 4x\right) dx$ **6.** $\int \left(5\sqrt{x^3} - 6x\right) dx$

7. $\int (x + 4)(x - 4)\, dx$ **8.** $\int \dfrac{3x^3 + 2x^2 + 4x}{x}\, dx$

⑨ **BUSINESS: Cost** A company's marginal cost function is $MC(x) = x^{-1/2} + 4$, where x is the number of units. If fixed costs are 20,000, find the company's cost function.

10. GENERAL: Population The population of a town is now 40,000 and t years from now will be growing at the rate of $300\sqrt{t}$ people per year.

    a. Find a formula for the population of the town t years from now.

    b. Use your formula to find the population of the town 16 years from now.

5.2   Integration Using Logarithmic and Exponential Functions

11–18. Find each indefinite integral.

11. $\int e^{x/2}\, dx$ **12.** $\int e^{-2x}\, dx$

13. $\int 4x^{-1}\, dx$ **14.** $\int \dfrac{2}{x}\, dx$

⑮ $\int \left(6e^{3x} - \dfrac{6}{x}\right) dx$ **16.** $\int (x - x^{-1})\, dx$

17. $\int (9x^2 + 2x^{-1} + 6e^{3x})\, dx$

18. $\int \left(\dfrac{1}{x^2} + \dfrac{1}{x} + e^{-x}\right) dx$

19. ENVIRONMENTAL SCIENCE: Consumption of Natural Resources World consumption of antimony (used in flame retardants and batteries) is running at the rate of $146e^{0.02t}$ thousand metric tons per year, where t is the number of years since 2014.

    a. Find a formula for the total amount of antimony consumed within t years of 2014.

    b. If consumption continues at this rate, when will the known resources of 1800 thousand metric tons of antimony be exhausted?

Source: U.S. Geological Survey

20. PERSONAL FINANCE: Total Savings A homeowner installs a solar heating system, which is expected to generate savings at the rate of $200e^{0.1t}$ dollars per year, where t is the number of years since the system was installed.

    a. Find a formula for the total savings in the first t years.

    b. If the system originally cost \$1500, when will it "pay for itself"?

㉑ **ENVIRONMENTAL SCIENCE: Consumption of Natural Resources** World consumption of zinc is running at the rate of $15e^{0.06t}$ million metric tons per year, where t is the number of years since 2014.

    a. Find a formula for the total amount of zinc consumed within t years of 2014.

    b. If consumption continues at this rate, when will the known resources of 250 million metric tons of zinc be exhausted? (Zinc is used to make protective coatings for iron and steel.)

Source: U.S. Geological Survey

22. BUSINESS: Profit A company's profit is growing at the rate of $200x^{-1}$ thousand dollars per month after x months, for $x \geq 1$.

    a. Find a formula for the total growth in the profit from month 1 to month x.

    b. When will the total growth in profit reach 600 thousand dollars?

5.3   Definite Integrals and Areas

23–30. Evaluate each definite integral.

㉓ $\int_{1}^{9} \left(x - \dfrac{1}{\sqrt{x}}\right) dx$ **24.** $\int_{2}^{5} (3x^2 - 4x + 5)\, dx$

25. $\int_{1}^{e^4} \dfrac{dx}{x}$ **26.** $\int_{1}^{5} \dfrac{dx}{x}$

㉗ $\int_{0}^{2} e^{-x}\, dx$ **28.** $\int_{0}^{2} e^{x/2}\, dx$

29. $\int_{0}^{100} (e^{0.05x} - e^{0.01x})\, dx$ **30.** $\int_{0}^{10} (e^{0.04x} - e^{0.02x})\, dx$

31–36. For each function:

a. Find the area under the curve between the given x-values.

b. Verify your answer to part (a) by using a graphing calculator to find the area.

㉛ $f(x) = 6x^2 - 1$, $x = 1$ to $x = 2$

32. $f(x) = 9 - x^2$, $x = -3$ to $x = 3$

33. $f(x) = 12e^{2x}$, $x = 0$ to $x = 3$

34. $f(x) = e^{x/2}$, $x = 0$ to $x = 4$

35. $f(x) = \dfrac{1}{x}$, $x = 1$ to $x = 100$

36. $f(x) = x^{-1}$, $x = 1$ to $x = 1000$

37–38. Use a calculator to graph each function and find the area under it between the given x-values.

37. $f(x) = \dfrac{10}{x^4 + 1}$ from $x = -2$ to $x = 2$

38. $f(x) = e^{x^4}$ from $x = -1$ to $x = 1$

39. **BIOMEDICAL:** Weight Gain An average young person in the United States gains weight at the rate of $16(x - 10)^{-1/2}$ pounds per year, where x is the person's age $(11 \le x \le 19)$. Find the total weight gain from age 11 to 19.

Source: National Center for Health Statistics

40. **BEHAVIORAL SCIENCE:** Learning A student can memorize foreign vocabulary words at the rate of $2/\sqrt[3]{t}$ words per minute, where t is the number of minutes since the studying began. Find the number of words that can be memorized from time $t = 1$ to time $t = 8$.

41. **ENVIRONMENTAL SCIENCE:** Global Warming The temperature of the earth is rising, as a result of the "greenhouse effect," in which carbon dioxide prevents the escape of heat from the atmosphere. According to one study, the temperature is rising at the rate of $0.014t^{0.4}$ degrees Fahrenheit per year. Find the total rise in temperature over the next 100 years.

Source: Intergovernmental Panel on Climate Change

42. **BUSINESS:** Cost A company's marginal cost function is $MC(x) = x^{-1/2} + 4$ dollars, where x is the number of units. Find the total cost of the first 400 units (units $x = 0$ to $x = 400$).

 43. **BUSINESS:** Cost A company's marginal cost function is $MC(x) = 22e^{-\sqrt{x}/5}$ dollars, where x is the number of units. Find the total cost of the first hundred units (units $x = 0$ to $x = 100$).

 44. **BEHAVIORAL SCIENCE:** Repetitive Tasks A proofreader can read $15xe^{-0.25x}$ pages per hour, where x is the number of hours worked. Find the total number of pages that can be proofread in 8 hours.

Exercises on Riemann Sums

45. **a.** Approximate the area under the curve $f(x) = x^2$ from 0 to 2 using 10 left rectangles with equal bases.
  b. Find the *exact* area under the curve between the given x-values by evaluating an appropriate definite integral using the Fundamental Theorem.

46. **a.** Approximate the area under the curve $f(x) = \sqrt{x}$ from 0 to 4 using 10 left rectangles with equal bases. (Round calculations to three decimal places.)
  b. Find the *exact* area under the curve between the given x-values by evaluating an appropriate definite integral using the Fundamental Theorem.

 47–48. Use a graphing calculator Riemann sum program from the Internet (see page 332) to find the following Riemann sums.

a. Calculate the Riemann sum for each function below for the following values of n: 10, 100, 1000. Use left, right, or midpoint rectangles, making a table of the answers, keeping three decimal places of accuracy.

b. Find the *exact* value of the area under the curve in each exercise below by evaluating an appropriate definite integral using the Fundamental Theorem. Your answers from part (a) should approach the number found from this integral.

47. $f(x) = e^x$ from $a = -2$ to $b = 2$

48. $f(x) = \dfrac{1}{x}$ from $a = 1$ to $b = 4$

5.4 Further Applications of Definite Integrals: Average Value and Area Between Curves

49–56. Find the area bounded by each pair of curves.

49. $y = x^2 + 3x$ and $y = 3x + 1$

50. $y = 12x - 3x^2$ and $y = 6x - 24$

51. $y = x^2$ and $y = x$

52. $y = x^4$ and $y = x$

53. $y = 4x^3$ and $y = 12x^2 - 8x$

54. $y = x^3 + x^2$ and $y = x^2 + x$

55. $y = e^x$ and $y = x + 5$

56. $y = \ln x$ and $y = \dfrac{x^2}{10}$

57–60. Find the average value of the function on the given interval.

57. $f(x) = \dfrac{1}{x}$ on $[1, 4]$

58. $f(x) = 6\sqrt{x}$ on $[1, 4]$

 59. $f(x) = \sqrt{x^3 + 1}$ on $[0, 5]$

 60. $f(x) = \ln(e^{x^2} + 10)$ on $[-2, 2]$

 61. **GENERAL:** Average Population The population of the United States is predicted to be $P(t) = 310e^{0.0073t}$ million, where t is the number of years after 2010. Find the average population between the years 2010 and 2110.

Source: U.S. Census Bureau

62. **PERSONAL FINANCE:** Compound Interest A deposit of \$3000 in a bank paying 6% interest compounded continuously will grow to $V(t) = 3000e^{0.06t}$ dollars after t years. Find the average value during the first 20 years $(t = 0$ to $t = 20)$.

 63. **PERSONAL FINANCE:** Real Estate The value of a suburban plot of land being considered for rezoning is assessed at $4.3e^{0.01x^2}$ hundred thousand dollars x years from now. Find the average value over the next 10 years (year 0 to year 10).

64. **PERSONAL FINANCE:** Stock Price The price of a share of stock is expected to be $28e^{0.01x^{1.2}}$ dollars where *x* is the number of weeks from now. Find the average price over the next year (week 0 to week 52).

65. GENERAL: Art An artist wants to paint the interior of the shape shown below on the side of a building. How much area (in square meters) will the artist need to paint?

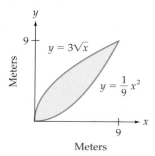

66. ECONOMICS: Balance of Trade A country's annual exports will be $E(t) = 40e^{0.2t}$ and its imports will be $I(t) = 20e^{0.1t}$ (both in billions of dollars per year), where t is the number of years from now. Find the accumulated trade surplus (exports minus imports) over the next 10 years.

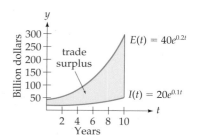

5.5   Two Applications to Economics: Consumers' Surplus and Income Distribution

67–70. ECONOMICS: Consumers' Surplus For each demand function $d(x)$ and demand level x, find the consumers' surplus.

⑥⑦ $d(x) = 8000 - 24x, \quad x = 200$

68. $d(x) = 1800 - 0.03x^2, \quad x = 200$

69. $d(x) = 300e^{-0.2\sqrt{x}}, \quad x = 120$

70. $d(x) = \dfrac{100}{1 + \sqrt{x}}, \quad x = 100$

71–74. ECONOMICS: Gini Index For each Lorenz curve, find the Gini index.

⑦① $L(x) = x^{3.5}$ **72.** $L(x) = x^{2.5}$

73. $L(x) = \dfrac{x}{2 - x}$ **74.** $L(x) = \dfrac{e^{x^4} - 1}{e - 1}$

5.6   Integration by Substitution

75–90. Find each integral or state that it cannot be evaluated by our substitution formulas.

75. $\displaystyle\int x^2 \sqrt[3]{x^3 - 1}\, dx$ **76.** $\displaystyle\int x^3 \sqrt{x^4 - 1}\, dx$

77. $\displaystyle\int x \sqrt[3]{x^3 - 1}\, dx$ **78.** $\displaystyle\int x^2 \sqrt{x^4 - 1}\, dx$

⑦⑨ $\displaystyle\int \dfrac{dx}{9 - 3x}$ **80.** $\displaystyle\int \dfrac{dx}{1 - 2x}$

⑧① $\displaystyle\int \dfrac{dx}{(9 - 3x)^2}$ **82.** $\displaystyle\int \dfrac{dx}{(1 - 2x)^2}$

83. $\displaystyle\int \dfrac{x^2}{\sqrt[3]{8 + x^3}}\, dx$ **84.** $\displaystyle\int \dfrac{x}{\sqrt{9 + x^2}}\, dx$

85. $\displaystyle\int \dfrac{w + 3}{(w^2 + 6w - 1)^2}\, dw$ **86.** $\displaystyle\int \dfrac{t - 2}{(t^2 - 4t + 1)^2}\, dt$

87. $\displaystyle\int \dfrac{\left(1 + \sqrt{x}\right)^2}{\sqrt{x}}\, dx$ **88.** $\displaystyle\int \dfrac{\left(1 + \sqrt[3]{x}\right)^2}{\sqrt[3]{x^2}}\, dx$

89. $\displaystyle\int \dfrac{e^x}{e^x - 1}\, dx$ **90.** $\displaystyle\int \dfrac{1}{x \ln x}\, dx$

91–98. For each definite integral:

a. Evaluate it ("by hand") or state that it cannot be evaluated by our substitution formulas.

b. Verify your answer to part (a) by using a graphing calculator.

⑨① $\displaystyle\int_0^3 x \sqrt{x^2 + 16}\, dx$ **92.** $\displaystyle\int_0^4 \dfrac{dz}{\sqrt{2z + 1}}$

93. $\displaystyle\int_0^4 \dfrac{w}{\sqrt{25 - w^2}}\, dw$ **94.** $\displaystyle\int_1^2 \dfrac{x + 1}{(x^2 + 2x - 2)^2}\, dx$

95. $\displaystyle\int_3^9 \dfrac{dx}{x - 2}$ **96.** $\displaystyle\int_4^5 \dfrac{dx}{x - 6}$

97. $\displaystyle\int_0^1 x^3 e^{x^4}\, dx$ **98.** $\displaystyle\int_0^1 x^4 e^{x^5}\, dx$

99–100. Find the area under the given curve between the given x-values.

99. $y = \dfrac{x^2 + 6x}{\sqrt[3]{x^3 + 9x^2 + 17}}$ from $x = 1$ to $x = 3$

100. $y = \dfrac{x + 6}{\sqrt{x^2 + 12x + 4}}$ from $x = 0$ to $x = 3$

101–102. Find the average value of the function on the given interval.

101. $f(x) = xe^{-x^2}$ on $[0, 2]$

102. $f(x) = \dfrac{x}{x^2 - 3}$ on $[2, 4]$

103. BUSINESS: Cost A company's marginal cost function is $MC(x) = \dfrac{1}{\sqrt{2x + 9}}$ and fixed costs are 100. Find the cost function.

104. BIOMEDICAL: Temperature An experimental drug changes a patient's temperature at the rate of $\dfrac{3x^2}{x^3 + 1}$ degrees per milligram of the drug, where x is the amount of the drug administered. Find the total change in temperature resulting from the first 3 milligrams of the drug. [*Note:* The rate of change of temperature with respect to dosage is called the "drug sensitivity."]

Integration Techniques

6

imagebroker.net/SuperStock

What You'll Explore

In the previous chapter we defined definite integrals in terms of *areas*, and then used integrals to find average values and total accumulation of quantities that grow at predictable rates, such as populations, profits, and pollution. In this chapter you will expand the range of functions that you can integrate, finding present values of income streams and the amounts of money needed to generate fixed incomes in perpetuity. You will also see how to estimate definite integrals that cannot be found exactly.

Improper Integrals and Eternal Recognition

Suppose that after you become rich and famous, you decide to commission a statue of yourself for your hometown. Your town, however, will accept this self-less gesture only if you pay for the perpetual upkeep of the statue by establishing a fund that will generate $2000 annually for every year in the future. Before deciding whether or not to accept this condition, you of course want to know how much it will cost. (Interestingly, we will see that a fund that will generate income forever does *not* require an infinite amount of money.) On page 252 we found that to realize a yield of $2000 t years from now requires only its *present value* deposited now in a bank. At an interest rate of, say, 5% compounded continuously, $2000 in t years requires a deposit now of

$$\left(\begin{array}{l}\text{Present value of \$2000 at 5\%}\\\text{compounded continuously}\end{array}\right) = 2000e^{-0.05t} \qquad \text{Amount times } e^{-0.05t}$$

Therefore, the size of the fund needed to generate $2000 annually *forever* is found by summing (integrating) this present value over the infinite time interval from zero to infinity (∞).

$$\left(\begin{array}{l}\text{Size of fund to yield an}\\\text{annual \$2000 forever}\end{array}\right) = \int_0^\infty 2000e^{-0.05t}\, dt \qquad \text{Integrating out to infinity}$$

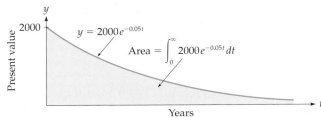

On pages 408–409 we will learn how to evaluate such "improper integrals," finding that the value of this integral is $40,000, which shows that $40,000 deposited in a bank at 5% compounded continuously will generate $2000 every year *forever*. That is, $40,000 would pay for the perpetual upkeep of your statue, buying you (or at least your likeness) a kind of immortality.

Incidentally, to generate $2000 annually for only the first *hundred* years would

require $\displaystyle\int_0^{100} 2000e^{-0.05t}\, dt \approx \$39{,}730$ (integrating from 0 to 100). This amount

is only $270 less than the amount needed to generate the same sum *forever*. This small additional cost shows that the short term is expensive, but eternity is cheap.

6.1 Integration by Parts

Introduction

In this chapter we introduce further techniques for finding integrals: **integration by parts, integration by tables,** and **numerical integration.** We also discuss **improper integrals** (integrals over infinite intervals).

You may have felt that integration is "harder" than differentiation. One reason is that integration is an *inverse* process, carried out by *reversing* differentiation, and inverse processes are generally more difficult. Another reason is that, while we have the product and quotient rules for *differentiating* complicated expressions, there are no product and quotient rules for integration. The method of integration by parts, explained in this section, is in some sense a "Product Rule for integration" in that it comes from interpreting the Product Rule as an integration formula.

Integration by Parts

For two differentiable functions $u(x)$ and $v(x)$, hereafter denoted simply u and v, the Product Rule is

LOOKING BACK

On page 114 we stated the product rule in terms of functions *f* and *g*.

$$(uv)' = u'v + uv'$$

The derivative of a product is the derivative of the first times the second, plus the first times the derivative of the second

If we integrate both sides of this equation, integrating the left side "undoes" the differentiation.

$$uv = \int u'v \, dx + \int uv' \, dx$$

Using differential notation, $du = u' \, dx, \quad dv = v' \, dx$

$$uv = \int v \, du + \int u \, dv$$

Formula above in differential notation

Solving this equation for the second integral $\int u \, dv$ gives

$$\int u \, dv = uv - \int v \, du$$

This formula is the basis for a technique called "integration by parts."

Integration by Parts

For differentiable functions u and v,

$$\int u \, dv = uv - \int v \, du$$

We use this formula to solve integrals by a "double substitution," substituting u for part of the given integral and dv for the rest, and then expressing the integral in the form $uv - \int v \, du$. The point is to choose the u and the dv so that the resulting integral $\int v \, du$ is *simpler* than the original integral $\int u \, dv$. A few examples will make the method clear.

EXAMPLE 1   INTEGRATING BY PARTS

Use integration by parts to find $\int xe^x \, dx$.

Solution

$$\int \underbrace{x}_{u}\ \underbrace{e^x\ dx}_{dv}$$

Original integral $\int u\ dv$

We choose $u = x$
and $dv = e^x\ dx$

$$\left[\begin{array}{cc} u = x & dv = e^x\ dx \\ \downarrow & \downarrow \\ du = 1\ dx = dx & v = \int e^x\ dx = e^x \end{array}\right]$$

Differentiating u to find du
and integrating dv to find
v (omitting the C)

$$= \underbrace{x\ e^x}_{u\ v} - \int \underbrace{e^x\ dx}_{v\ du}$$

$u{\cdot}v - \int v\ du$ with $u = x$,
$v = e^x$, and $du = dx$

$$= xe^x - e^x + C$$
$$\underset{\text{From } \int e^x\ dx}{\uparrow}$$

Finding the new integral
$\int e^x\ dx = e^x$ to give the
final answer (with $+C$)

The procedure is not as complicated as it might seem. All the steps in using the formula $\int u\ dv = uv - \int v\ du$ may be written together as follows:

$$\int \underset{u\ dv}{xe^x\ dx} \qquad = \underset{u\ v}{xe^x} - \underset{v\ du}{\int e^x\ dx} = xe^x - e^x + C$$
$$\underset{\text{From } \int e^x\ dx}{}$$

$$\left[\begin{array}{cc} u = x & dv = e^x\ dx \\ du = dx & v = \int e^x\ dx = e^x \end{array}\right]$$

We can check this answer by differentiation.

$$\overset{\overbrace{\qquad\text{Cancel}\qquad}}{\frac{d}{dx}(xe^x - e^x + C) = e^x + \underbrace{xe^x - e^x}_{} = xe^x}$$

Differentiating xe^x
by the Product Rule

Agrees with the orig-
inal integrand, so the
integration is correct

Remarks on the Integration by Parts Procedure

i. The differentials du and dv include the dx.

ii. We omit the constant C when we integrate dv to get v because one $+C$ at the end is enough (See Exercise 74).

iii. The integration by parts formula does not give a "final answer" but rather expresses the given integral as $uv - \int v\ du$, a product $u \cdot v$ (already integrated) and a new integral $\int v\ du$. That is, integration by parts "exchanges" the original integral $\int u\ dv$ for another integral $\int v\ du$. The hope is that the second integral will be simpler than the first. In our example we "exchanged" $\int xe^x\ dx$ for the simpler $\int e^x\ dx$, which could be integrated immediately by formula 3 (inside back cover).

iv. Integration by parts should be used only if the simpler formulas 1 through 7 (inside back cover) fail to solve the integral.

How to Choose the *u* and the *dv*

In Example 1, the choice of $\begin{cases} u = x \\ dv = e^x\,dx \end{cases}$ "exchanged" the original integral $\int xe^x\,dx$ for the simpler integral $\int e^x\,dx$. If we had instead chosen $\begin{cases} u = e^x \\ dv = x\,dx \end{cases}$ we would have "exchanged" the original integral for $\int x^2e^x\,dx$ (as you may check), which is *more* difficult than the original (because of the x^2). Therefore, the first choice was the "right" choice in that it led to a solution. Generally, one choice for *u* and *dv* will be "best," and finding it may involve some trial and error. While there is no foolproof rule for finding the best *u* and *dv*, the following guidelines often help.

Guidelines for Choosing *u* and *dv*

1. Choose *dv* to be the most complicated part of the integral that can be integrated easily.
2. Choose *u* so that *u′* is simpler than *u*.

EXAMPLE 2   INTEGRATING BY PARTS

Find $\displaystyle\int x^2 \ln x\,dx$.

Solution

None of the easier formulas (1 through 7 on the inside back cover) will solve the integral, as you may easily check. Therefore, we try integration by parts. The integrand is a product, x^2 times $\ln x$. The guidelines say to choose *dv* to be the most complicated part that can be easily integrated. We can integrate x^2 but not $\ln x$ (we know how to *differentiate* $\ln x$, but not how to *integrate* it), so we choose $dv = x^2\,dx$, and therefore $u = \ln x$.

$$\int \underbrace{x^2}_{\substack{u \\ dv}} \ln x\,dx = (\ln x)\underbrace{\tfrac{1}{3}x^3}_{u} \underbrace{}_{v} - \int \underbrace{\tfrac{1}{3}x^3}_{v}\,\underbrace{\tfrac{1}{x}\,dx}_{du} = \frac{1}{3}x^3 \ln x - \frac{1}{3}\int x^2\,dx$$

$$\begin{bmatrix} u = \ln x & dv = x^2\,dx \\[2mm] du = \dfrac{1}{x}\,dx & v = \int x^2\,dx = \tfrac{1}{3}x^3 \end{bmatrix}$$

Moving the $\frac{1}{3}$ outside and simplifying $x^3\frac{1}{x}$ to x^2

$$= \frac{1}{3}x^3 \ln x - \frac{1}{3}\frac{1}{3}x^3 + C = \frac{1}{3}x^3 \ln x - \frac{1}{9}x^3 + C$$

We may check this answer by differentiation.

$$\frac{d}{dx}\left(\frac{1}{3}x^3 \ln x - \frac{1}{9}x^3 + C\right) = x^2 \ln x + \underbrace{\frac{1}{3}x^3\frac{1}{x} - \frac{1}{3}x^2}$$

Differentiating $\frac{1}{3}x^3 \ln x$
by the Product Rule

$$= x^2 \ln x + \frac{1}{3}x^2 - \frac{1}{3}x^2 = x^2 \ln x$$

Agrees with the original integrand, so the integration is correct

From $x^3 \frac{1}{x}$ Cancel

PRACTICE PROBLEM 1

Use integration by parts to find $\displaystyle\int x^3 \ln x \, dx$.

Solution on page 390 >

Integration by parts is also useful for integrating products of linear functions to powers.

EXAMPLE 3   INTEGRATING BY PARTS

Use integration by parts to find $\displaystyle\int (x - 2)(x + 4)^8 \, dx$.

Solution

The guidelines recommend that dv be the most complicated part that can be integrated. Both $x - 2$ and $(x + 4)^8$ can be integrated. For example,

$$\int (x + 4)^8 \, dx = \frac{1}{9}(x + 4)^9 + C$$

By the substitution method with $u = x + 4$ (omitting the details)

Since $(x + 4)^8$ is more complicated than $(x - 2)$, we take $dv = (x + 4)^8 \, dx$.

$$\underbrace{\int \underbrace{(x - 2)}_{u} \underbrace{(x + 4)^8 \, dx}_{dv}} = \underbrace{(x - 2)}_{u} \underbrace{\frac{1}{9}(x + 4)^9}_{v} - \int \underbrace{\frac{1}{9}(x + 4)^9}_{v} \underbrace{dx}_{du}$$

$$\begin{bmatrix} u = x - 2 & dv = (x + 4)^8 \, dx \\ du = dx & v = \int (x + 4)^8 \, dx \\ & = \frac{1}{9}(x + 4)^9 \end{bmatrix}$$

$$= \frac{1}{9}(x - 2)(x + 4)^9 - \frac{1}{9}\int (x + 4)^9 \, dx$$

Taking out the $\frac{1}{9}$

$$= \frac{1}{9}(x - 2)(x + 4)^9 - \frac{1}{9}\frac{1}{10}(x + 4)^{10} + C$$

Integrating by the substitution method

$$= \frac{1}{9}(x - 2)(x + 4)^9 - \frac{1}{90}(x + 4)^{10} + C$$

Again we could check this answer by differentiation. (Do you see how the Product Rule, applied to the first part of this answer, will give a piece that will cancel with the derivative of the second part?) This integral can also be found by the substitution method (see page 376).

PRACTICE PROBLEM 2

Use integration by parts to find $\displaystyle\int (x+1)(x-1)^3\, dx$.   Solution on page 390 >

Present Value of a Continuous Stream of Income

If a business or some other asset generates income continuously at the rate $C(t)$ dollars per year, where t is the number of years from now, then $C(t)$ is called a **continuous stream of income.*** On page 252 we saw that to find the *present value* of an amount of money (the amount now that will later yield the stated amount) under continuous compounding we multiply by e^{-rt}, where r is the interest rate and t is the number of years. (We will refer to a rate with continuous compounding as a "continuous interest rate.") Therefore, the present value of the continuous stream $C(t)$ is found by multiplying by e^{-rt} and summing (integrating) over the time period.

 LOOKING BACK

Recall that integrals are *continuous sums*, as we saw on page 315.

Present Value of a Continuous Stream of Income

The present value of the continuous stream of income $C(t)$ dollars per year, where t is the number of years from now, for T years at continuous interest rate r is

$$\begin{pmatrix} \text{Present} \\ \text{value} \end{pmatrix} = \int_0^T C(t)e^{-rt}\, dt$$

 EXAMPLE 4   **FINDING THE PRESENT VALUE OF A CONTINUOUS STREAM OF INCOME**

A business generates income at the rate of $2t$ million dollars per year, where t is the number of years from now. Find the present value of this continuous stream for the next 5 years at the continuous interest rate of 10%.

Solution

$$\begin{pmatrix} \text{Present} \\ \text{value} \end{pmatrix} = \int_0^5 2te^{-0.1t}\, dt$$

Multiplying $C(t) = 2t$ by $e^{-0.1t}$ (since 10% = 0.1) and integrating from 0 to 5 years

This is a *definite* integral, but we will ignore the limits of integration until after we have found the *indefinite* integral. None of the formulas 1 through 7 on the inside back cover will find this integral, so we try integration by parts with $u = 2t$ and $dv = e^{-0.1t}\, dt$. (Do you see why the guidelines on page 386 suggest this choice?)

$$\int \underbrace{2t}_{u}\, \underbrace{e^{-0.1t}\, dt}_{dv} = \underbrace{(2t)}_{u}\underbrace{(-10e^{-0.1t})}_{v} - \int \underbrace{(-10e^{-0.1t})}_{v}\, \underbrace{2\, dt}_{du}$$

$$\begin{bmatrix} u = 2t & dv = e^{-0.1t}\, dt \\ du = 2dt & v = \int e^{-0.1t}\, dt \\ & = -10e^{-0.1t} \end{bmatrix}$$

**C(t)* must be continuous, meaning that the income is being paid continuously rather than in "lump-sum" payments. However, even lump-sum payments can be approximated by a continuous stream if the payments are frequent enough or last long enough.

$$= -20te^{-0.1t} + 20 \int e^{-0.1t} \, dt$$

$$= -20te^{-0.1t} + 20(-10)e^{-0.1t} + C$$

$$= -20te^{-0.1t} - 200e^{-0.1t} + C$$

For the *definite* integral, we evaluate this from 0 to 5:

$$(-20te^{-0.1t} - 200e^{-0.1t}) \Big|_0^5 = \underbrace{(-20 \cdot 5e^{-0.5} - 200e^{-0.5})}_{\text{Evaluation at } t = 5} - (\overbrace{-20 \cdot 0e^0}^{0} - \overbrace{200e^0}^{200})$$

Evaluation at $t = 0$

$$= -300e^{-0.5} + 200 \approx 18$$

In millions of dollars (using a calculator)

The present value of the stream of income over 5 years is approximately $18 million.

This answer means that $18 million at 10% interest compounded continuously would generate the continuous stream $C(t) = 2t$ million dollars for 5 years. This method is often used to determine the fair value of a business or some other asset, since it gives the present value of all future income.

Graphing Calculator Exploration

on [0, 5] by [0, 7]

a. Verify the answer to Example 4 by graphing $2xe^{-0.1x}$ and finding the area under the curve from 0 to 5.

b. Can you explain why the curve increases less steeply farther to the right? [*Hint:* Think of the present value of money to be paid in the more distant future.]

Why is it necessary to learn integration by parts if integrals like the one in Example 4 can be evaluated on graphing calculators? One answer is that you should have a *variety* of ways to approach a problem—sometimes geometrically, sometimes analytically, and sometimes numerically (using a calculator). These various approaches mutually support one another; for example, you can solve a problem one way and check it another way. Furthermore, not all applications involve *definite* integrals. Example 4 might have asked for a *formula* for the present value up to any time *t*, and such formulas may be found by integrating "by hand" but not from most graphing calculators.

 Be Careful You should use integration by parts *only* if the "easier" formulas (1 through 7 on the inside back cover) fail to solve the integral.

PRACTICE PROBLEM 3

Which of the following integrals require integration by parts, and which can be found by the substitution formula $\int e^u\, du = e^u + C$? (Do not solve the integrals.)

a. $\displaystyle\int xe^x\, dx$ **b.** $\displaystyle\int xe^{x^2}\, dx$ Solution below >

Solutions TO PRACTICE PROBLEMS

1. $\displaystyle\int x^3 \ln x\, dx = (\ln x)\left(\frac{1}{4}x^4\right) - \int \frac{1}{4}x^4\frac{1}{x}\, dx = (\ln x)\left(\frac{1}{4}x^4\right) - \frac{1}{4}\int x^3\, dx$

$$\left[\begin{array}{ll} u = \ln x & dv = x^3\, dx \\ du = \dfrac{1}{x}\, dx & v = \displaystyle\int x^3\, dx = \dfrac{1}{4}x^4 \end{array}\right]$$

$$= \frac{1}{4}x^4 \ln x - \frac{1}{16}x^4 + C$$

2. $\displaystyle\int (x + 1)(x - 1)^3\, dx = (x + 1)\frac{1}{4}(x - 1)^4 - \int \frac{1}{4}(x - 1)^4\, dx$

$$\left[\begin{array}{ll} u = x + 1 & dv = (x - 1)^3\, dx \\ du = dx & v = \int(x - 1)^3\, dx \\ & = \frac{1}{4}(x - 1)^4 \end{array}\right]$$

$$= \frac{1}{4}(x + 1)(x - 1)^4 - \frac{1}{4}\cdot\frac{1}{5}(x - 1)^5 + C$$

$$= \frac{1}{4}(x + 1)(x - 1)^4 - \frac{1}{20}(x - 1)^5 + C$$

3. a. Requires integration by parts
   b. Can be solved by the substitution $u = x^2$

6.1   Section Summary

The integration by parts formula

$$\int u\, dv = uv - \int v\, du$$

is simply the integration version of the Product Rule. In fact, if you use integration by parts to find an integral, you should expect to use the Product Rule when you check your answer by differentiation. The guidelines on page 386 lead to the following suggestions for choosing u and dv in some commonly occurring integrals.

For Integrals of the Form	Choose	
$\displaystyle\int x^n e^{ax}\, dx$	$u = x^n$	$dv = e^{ax}\, dx$
$\displaystyle\int x^n \ln x\, dx$	$u = \ln x$	$dv = x^n\, dx$
$\displaystyle\int (x + a)(x + b)^n\, dx$	$u = x + a$	$dv = (x + b)^n\, dx$

Integration by parts can be useful in any situation involving integrals, such as recovering total cost from marginal cost, calculating areas, average values (the definite integral divided by the length of the interval), continuous accumulations, or present values of continuous income streams.

6.1 Exercises

1–8. Integration by parts often involves finding integrals like the following when integrating dv to find v. Find the following integrals *without* using integration by parts (using formulas 1 through 7 on the inside back cover). Be ready to find similar integrals during the integration by parts procedure.

1. $\int e^{2x}\,dx$

2. $\int x^5\,dx$

3. $\int (x+2)\,dx$

4. $\int (x-1)\,dx$

5. $\int \sqrt{x}\,dx$

6. $\int e^{-0.5t}\,dt$

7. $\int (x+3)^4\,dx$

8. $\int (x-5)^6\,dx$

9–36. Use integration by parts to find each integral.

9. $\int xe^{2x}\,dx$

10. $\int xe^{3x}\,dx$

11. $\int x^5 \ln x\,dx$

12. $\int x^4 \ln x\,dx$

13. $\int (x+2)e^x\,dx$ [*Hint:* Take $u = x+2$.]

14. $\int (x-1)e^x\,dx$ [*Hint:* Take $u = x-1$.]

15. $\int \sqrt{x}\ln x\,dx$

16. $\int \sqrt[3]{x}\ln x\,dx$

17. $\int (x-3)(x+4)^5\,dx$ **18.** $\int (x+2)(x-5)^5\,dx$

19. $\int te^{-0.5t}\,dt$

20. $\int te^{-0.2t}\,dt$

21. $\int \dfrac{\ln t}{t^2}\,dt$

22. $\int \dfrac{\ln t}{\sqrt{t}}\,dt$

23. $\int s(2s+1)^4\,ds$

24. $\int \dfrac{x+1}{e^{3x}}\,dx$

25. $\int \dfrac{x}{e^{2x}}\,dx$

26. $\int \dfrac{\ln(x+1)}{\sqrt{x+1}}\,dx$

27. $\int \dfrac{x}{\sqrt{x+1}}\,dx$

28. $\int x\sqrt{x+1}\,dx$

29. $\int xe^{ax}\,dx$ $(a \neq 0)$

30. $\int (x+b)e^{ax}\,dx$ $(a \neq 0)$

31. $\int x^n \ln ax\,dx$ $(a \neq 0,\ n \neq -1)$

32. $\int (x+a)^n \ln(x+a)\,dx$ $(n \neq -1)$

33. $\int \ln x\,dx$ [*Hint:* Take $u = \ln x$, $dv = dx$.]

34. $\int \ln x^2\,dx$ [*Hint:* Take $u = \ln x^2$, $dv = dx$.]

35. $\int x^3 e^{x^2}\,dx$ [*Hint:* Take $u = x^2$, $dv = xe^{x^2}$,

and use a substitution to find v from dv.]

36. $\int x^3(x^2-1)^6\,dx$

[*Hint:* Take $u = x^2$, $dv = x(x^2-1)^6\,dx$, and use a substitution to find v from dv.]

37–38. Find each integral by integration by parts or a substitution, as appropriate.

37. a. $\int xe^{x^2}\,dx$

b. $\int \dfrac{(\ln x)^3}{x}\,dx$

c. $\int x^2 \ln 2x\,dx$

d. $\int \dfrac{e^x}{e^x+4}\,dx$

38. a. $\int \sqrt{\ln x}\,\dfrac{1}{x}\,dx$

b. $\int x^2 e^{x^3}\,dx$

c. $\int x^7 \ln 3x\,dx$

d. $\int xe^{4x}\,dx$

39–46. Evaluate each definite integral using integration by parts. (Leave answers in exact form.)

39. $\displaystyle\int_0^2 xe^x \, dx$

40. $\displaystyle\int_0^3 xe^x \, dx$

41. $\displaystyle\int_1^3 x^2 \ln x \, dx$

42. $\displaystyle\int_1^2 x \ln x \, dx$

43. $\displaystyle\int_0^2 z(z-2)^4 \, dz$

44. $\displaystyle\int_0^4 z(z-4)^6 \, dz$

45. $\displaystyle\int_0^{\ln 4} te^t \, dt$

46. $\displaystyle\int_1^e \ln x \, dx$

47–48. Find in two different ways and check that your answers agree.

47. $\displaystyle\int x(x-2)^5 \, dx$

    a. Use integration by parts.

    b. Use the substitution $u = x - 2$ (so x is replaced by $u + 2$) and then multiply out the integrand.

48. $\displaystyle\int x(x+4)^6 \, dx$

    a. Use integration by parts.

    b. Use the substitution $u = x + 4$ (so x is replaced by $u - 4$) and then multiply out the integrand.

49–50. Derive each formula by using integration by parts on the left-hand side. (Assume $n > 0$.)

49. $\displaystyle\int x^n e^x \, dx = x^n e^x - n \int x^{n-1} e^x \, dx$

50. $\displaystyle\int (\ln x)^n \, dx = x(\ln x)^n - n \int (\ln x)^{n-1} \, dx$

51. Use the formula in Exercise 49 to find the integral $\int x^2 e^x \, dx$.
[*Hint:* Apply the formula twice.]

52. Use the formula in Exercise 50 to find the integral $\int (\ln x)^2 \, dx$.
[*Hint:* Apply the formula twice.]

Applied Exercises

53. BUSINESS: Revenue If a company's marginal revenue function is $MR(x) = xe^{-x/4}$, find the revenue function. [*Hint:* Evaluate the constant C so that revenue is 0 at $x = 0$.]

54. BUSINESS: Cost A company's marginal cost function is $MC(x) = xe^{-x/2}$ and fixed costs are 200. Find the cost function. [*Hint:* Evaluate the constant C so that the cost is 200 at $x = 0$.]

55. BUSINESS: Present Value of a Continuous Stream of Income An electronics company generates a continuous stream of income of $4t$ million dollars per year, where t is the number of years that the company has been in operation. Find the present value of this stream of income over the first 10 years at a continuous interest rate of 10%.

56–57. For each exercise:

a. Solve *without* using a graphing calculator.

b. Verify your answer to part (a) using a graphing calculator.

56. BUSINESS: Present Value of a Continuous Stream of Income An oil well generates a continuous stream of income of $60t$ thousand dollars per year, where t is the number of years that the rig has been in operation. Find the present value of this stream of income over the first 20 years at a continuous interest rate of 5%.

57. BIOMEDICAL: Drug Dosage A drug taken orally is absorbed into the bloodstream at the rate of $te^{-0.5t}$ milligrams per hour, where t is the number of hours since the drug was taken. Find the total amount of the drug absorbed during the first 5 hours.

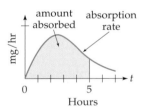

58. ENVIRONMENTAL SCIENCE: Pollution Contamination is leaking from an underground waste-disposal tank at the rate of $t \ln t$ thousand gallons per month, where t is the number of months since the leak began. Find the total leakage from the end of month 1 to the end of month 4.

59. GENERAL: Area Find the area under the curve $y = x \ln x$ and above the x-axis from $x = 1$ to $x = 2$.

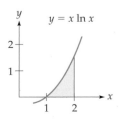

60. BUSINESS: Tablet Computers When tablet computers were first introduced, sales grew at the rate of $21xe^{0.1x}$ million per year. Find the cumulative sales over the first five years.

Source: Morgan Stanley Research

61. BUSINESS: Cell Phones When cell phones were first introduced, sales grew at the rate of $11xe^{0.2x}$ million per year. Find the cumulative sales during the first four years.

Source: Morgan Stanley Research

62. BUSINESS: Product Recognition A company begins advertising a new product and finds that after t weeks the product is gaining customer recognition at the rate of $t^2 \ln t$ thousand customers per week (for $t \geq 1$). Find the total gain in recognition from the end of week 1 to the end of week 6.

63. ATHLETICS: Fluid Absorption Runners and other athletes know that their ability to absorb water when exercising varies over time as their electrolyte levels change. Under certain circumstances, the absorption rate of water through the gastrointestinal tract will be $f(x) = 2xe^{-\frac{1}{2}x}$ liters per hour, where t is the number of hours of exercise. Find the number of liters of water absorbed during the first 5 hours of exercise.

Source: J. Keener and J. Sneyd, *Mathematical Physiology*

64. GENERAL: Internet Host Computers The number of Internet host computers (computers connected directly to the Internet, for networks, bulletin boards, or online services) has been growing at the rate of $f(x) = xe^{0.1x}$ million per year, where x is the number of years since 1990. Find the total number of Internet host computers that will have been added from 1990 to 2020.

Source: Worldwatch

Conceptual Exercises

65. Choose the correct formula:

a. $\int u\, dv = uv + \int v\, du$ \qquad or

b. $\int u\, dv = uv - \int v\, du$

66. Choose the correct formula:

a. $\int u\, dv = uv - \int v\, du$ \qquad or

b. $\int u\, dv = \int v\, du - uv$

67–69. Each equation follows from the integration by parts formula by replacing u by $f(x)$ and v by a particular function. What is the function v?

(See instruction in previous column.)

67. $\int f(x)e^x\, dx = f(x)e^x - \int e^x f'(x)\, dx$

68. $\int f(x)\frac{1}{x}\, dx = f(x)\ln x - \int \ln x\, f'(x)\, dx$

69. $\int f(x)\, dx = f(x)x - \int x f'(x)\, dx$

70–72. According to the guidelines in the text, what is the correct choice for u in each of the following integrals? (Do not carry out the solution, just state what u should be.)

70. $\int xe^{3x}\, dx$ \qquad **71.** $\int x^7 \ln x\, dx$

72. $\int (x+1)^5(x+2)\, dx$

Explorations and Excursions

The following problems extend and augment the material presented in the text.

Constants of Integration

73. a. Find the integral $\int x^{-1}\, dx$ by integration by parts (using $u = x^{-1}$ and $dv = dx$), obtaining

$$\int x^{-1}\, dx = x^{-1}x - \int (-x^{-2})x\, dx$$

which gives

$$\int x^{-1}\, dx = 1 + \int x^{-1}\, dx$$

b. Subtract the integral from both sides of this last equation, obtaining $0 = 1$. Explain this apparent contradiction.

74. We omit the constant of integration when we integrate dv to get v. Including the constant C in this step simply replaces v by $v + C$, giving the formula

$$\int u\, dv = u(v + C) - \int (v + C)\, du$$

Multiplying out the parentheses and expanding the last integral into two gives

$$\int u\, dv = uv + Cu - \int v\, du - C\int du$$

Show that the second and fourth terms on the right cancel, giving the "old" integration by parts formula $\int u\, dv = uv - \int v\, du$. This shows that including the constant in the dv to v step gives the same formula. *One constant of integration at the end is enough.*

Repeated Integration by Parts

Sometimes an integral requires two or more integrations by parts. As an example, we apply integration by parts to the integral $\int x^2 e^x \, dx$.

$$\int \underbrace{x^2}_{u} \underbrace{e^x \, dx}_{dv} = \underbrace{x^2 e^x}_{u \ v} - \int \underbrace{e^x \, 2x \, dx}_{v \ \ du} = x^2 e^x - 2\int xe^x \, dx$$

$$\begin{bmatrix} u = x^2 & dv = e^x \, dx \\ du = 2x \, dx & v = \int e^x \, dx = e^x \end{bmatrix}$$

The new integral $\int xe^x \, dx$ is solved by a second integration by parts. Continuing with the previous solution, we choose new u and du:

$$= x^2 e^x - 2\left(\int \underbrace{x}_{u} \underbrace{e^x \, dx}_{dv} \right) \qquad \begin{bmatrix} u = x & dv = e^x \, dx \\ du = dx & v = e^x \end{bmatrix}$$

$$= x^2 e^x - 2\left(\underbrace{xe^x}_{uv} - \int \underbrace{e^x \, dx}_{v \ du} \right)$$

$$= x^2 e^x - 2(xe^x - e^x) + C$$

$$= x^2 e^x - 2xe^x + 2e^x + C$$

75–80. After reading the preceding explanation, find each integral by repeated integration by parts.

75. $\displaystyle\int x^2 e^{-x} \, dx$

76. $\displaystyle\int x^2 e^{2x} \, dx$

77. $\displaystyle\int (x+1)^2 e^x \, dx$

78. $\displaystyle\int (\ln x)^2 \, dx$

79. $\displaystyle\int x^2 (\ln x)^2 \, dx$

80. $\displaystyle\int x^3 e^x \, dx$

81–82. For each definite integral:

a. Evaluate it by integration by parts. (Give answer in its *exact* form.)

b. Verify your answer to part (a) using a graphing calculator.

81. $\displaystyle\int_0^2 x^2 e^x \, dx$

82. $\displaystyle\int_1^5 (\ln x)^2 \, dx$

Repeated Integration by Parts Using a Table

The solution to a repeated integration by parts problem can be organized in a table. As an example, we solve $\int x^2 e^{3x} \, dx$. We begin by choosing

$$u = x^2 \qquad dv = v' \, dx = e^{3x} \, dx$$

We then make a table consisting of the following three columns:

Alternating Signs	$u = x^2$ and Its Derivatives	$v' = e^{3x}$ and Its Antiderivatives
+	x^2	e^{3x}
−	$2x$	$\dfrac{1}{3} e^{3x}$
+	2	$\dfrac{1}{9} e^{3x}$
−	0	$\dfrac{1}{27} e^{3x}$

Using the formula for $\int e^{ax} \, dx$

Stop when you get to 0

Finally, the solution is found by adding the *signed* products of the diagonals shown in the table:

$$\int x^2 e^{3x} \, dx = \frac{1}{3} x^2 e^{3x} - \frac{2}{9} xe^{3x} + \frac{2}{27} e^{3x} + C$$

83–88. After reading the preceding explanation, find each integral by repeated integration by parts using a table.

83. $\displaystyle\int x^2 e^{-x} \, dx$

84. $\displaystyle\int x^2 e^{2x} \, dx$

85. $\displaystyle\int x^3 e^{2x} \, dx$

86. $\displaystyle\int x^3 e^{-x} \, dx$

87. $\displaystyle\int (x-1)^3 e^{3x} \, dx$

88. $\displaystyle\int (x+1)^2 (x+2)^5 \, dx$

6.2   Integration Using Tables

Introduction

There are many techniques of integration, and only some of the most useful ones will be discussed in this book. Many of the advanced techniques lead to integration formulas, which can then be collected into a "table of integrals." In this section we will see how to find integrals by choosing an appropriate formula from such a table.*

On the inside back cover is a short table of integrals that we shall use. The formulas are grouped according to the type of integrand (for example, "Forms Involving $x^2 - a^2$ and $a^2 - x^2$"). Look at the table now (formulas 9 through 23) to see how it is organized.

Using Integral Tables

Given a particular integral, we first look for a formula that fits it exactly.

EXAMPLE 1   INTEGRATING USING AN INTEGRAL TABLE

Find $\displaystyle\int \frac{1}{x^2 - 4}\,dx$.

Solution

The denominator $x^2 - 4$ is of the form $x^2 - a^2$ (with $a = 2$), so we look in the table of integrals under "Forms Involving $x^2 - a^2$ and $a^2 - x^2$." Formula 15,

$$\int \frac{1}{x^2 - a^2}\,dx = \frac{1}{2a}\ln\left|\frac{x - a}{x + a}\right| + C \qquad \text{Formula 15}$$

with $a = 2$ becomes our answer:

$$\int \frac{1}{x^2 - 4}\,dx = \frac{1}{4}\ln\left|\frac{x - 2}{x + 2}\right| + C \qquad \begin{array}{l}\text{Formula 15 with } a = 2\\ \text{substituted on both sides}\end{array}$$

 LOOKING BACK

See page 315 for using integrals to find continuous sums.

Note that the expression $x^2 - a^2$ does not require that the last number be a "perfect square." For example, $x^2 - 3$ can be written $x^2 - a^2$ with $a = \sqrt{3}$.

Integral tables are useful in many applications, such as integrating a rate to find the total accumulation.

EXAMPLE 2   FINDING TOTAL SALES FROM THE SALES RATE

A company's sales rate is $\dfrac{x}{\sqrt{x + 9}}$ sales per week after x weeks. Find a formula for the total sales after x weeks.

---

*Another way is to use a more advanced graphing calculator (see page 400) or a computer software package such as Maple, Mathcad, or Mathematica.

$$\left(\begin{matrix}\text{Sales} \\ \text{rate}\end{matrix}\right) = \frac{x}{\sqrt{x+9}}$$

Sales

total sales $S(x)$

Weeks

Solution

To find the *total* sales $S(x)$ we integrate the *rate* of sales.

$$S(x) = \int \frac{x}{\sqrt{x+9}}\,dx$$

In the table on the inside back cover under "Forms Involving $\sqrt{ax+b}$" we find

$$\int \frac{x}{\sqrt{ax+b}}\,dx = \frac{2ax-4b}{3a^2}\sqrt{ax+b} + C \qquad \text{Formula 13}$$

This formula with $a = 1$ and $b = 9$ gives the integral

$$S(x) = \int \frac{x}{\sqrt{x+9}}\,dx = \frac{2x-36}{3}\sqrt{x+9} + C \qquad \begin{matrix}\text{Formula 13 with}\\ a=1 \text{ and } b=9\end{matrix}$$

$$= \left(\frac{2}{3}x - 12\right)\sqrt{x+9} + C \qquad \text{Simplifying}$$

To evaluate the constant C, we use the fact that total sales at time $x = 0$ must be zero: $S(0) = 0$.

$$(-12)\sqrt{9} + C = 0 \qquad \begin{matrix}\left(\frac{2}{3}x-12\right)\sqrt{x+9}+C\\ \text{at } x=0 \text{ set equal to zero}\end{matrix}$$

$$-36 + C = 0 \qquad \text{Simplifying}$$

Therefore, $C = 36$. Substituting this into $S(x)$ gives the formula for the total sales in the first x weeks.

$$S(x) = \left(\frac{2}{3}x - 12\right)\sqrt{x+9} + 36 \qquad \begin{matrix}S(x)=\left(\frac{2}{3}x-12\right)\sqrt{x+9}+C\\ \text{with } C=36\end{matrix}$$

Modern methods of biotechnology are being used to develop many new products, including powerful antibiotics, disease-resistant crops, and bacteria that literally "eat" oil spills. These "gene splicing" techniques require evaluating definite integrals such as the following.

EXAMPLE 3   **GENETIC ENGINEERING**

Under certain circumstances, the number of generations of bacteria needed to increase the frequency of a gene from 0.2 to 0.5 is

$$n = 2.5 \int_{0.2}^{0.5} \frac{1}{q^2(1-q)}\,dq$$

Find n (rounded to the nearest integer).

Solution

Formula 12 (inside back cover) integrates a similar-looking fraction.

$$\int \frac{1}{x^2(ax+b)}\,dx = -\frac{1}{b}\left(\frac{1}{x} + \frac{a}{b}\ln\left|\frac{x}{ax+b}\right|\right) + C \qquad \text{Formula 12}$$

To make $(ax + b)$ into $(1 - x)$, we take $a = -1$ and $b = 1$, so the left-hand side of the formula becomes

$$\int \frac{1}{x^2(-x+1)}\,dx \qquad \text{or} \qquad \int \frac{1}{x^2(1-x)}\,dx \qquad \begin{matrix}\text{From formula 12}\\ \text{with } a=-1,\ b=1\end{matrix}$$

Except for replacing x by q, this is the same as our integral. Therefore, the indefinite integral is found by formula 12 with $a = -1$ and $b = 1$ (which we express in the variable q).

$$\int \frac{1}{q^2(1-q)} \, dq = -\left(\frac{1}{q} - \ln\left|\frac{q}{1-q}\right|\right) + C \qquad \text{Formula 12 with } a = -1,$$
$$b = 1, \text{ and } x \text{ replaced by } q$$

For the *definite* integral from 0.2 to 0.5, we evaluate and subtract.

$$\underbrace{-\left(\frac{1}{0.5} - \ln\left|\frac{0.5}{1-0.5}\right|\right)}_{\text{Evaluation at } q = 0.5} - \underbrace{\left[-\left(\frac{1}{0.2} - \ln\left|\frac{0.2}{1-0.2}\right|\right)\right]}_{\text{Evaluation at } q = 0.2}$$

$$= -\underbrace{(2 - \ln 1)}_{0} + \left(5 - \underbrace{\ln\frac{0.2}{0.8}}_{\ln 0.25}\right) = -2 + 5 - \ln 0.25 \underbrace{\approx 4.39}_{\substack{\text{Using a} \\ \text{calculator}}}$$

We multiply this by the 2.5 in front of the original integral.

$$(2.5)(4.39) \approx 10.98$$

Therefore, 11 generations are needed to raise the gene frequency from 0.2 to 0.5.

 Graphing Calculator Exploration

on [0, 1] by [0, 100]
(shaded from 0.2 to 0.5)

a. Verify the answer to Example 3 by finding the definite integral of $y = \dfrac{2.5}{x^2(1-x)}$ from 0.2 to 0.5.

b. From the graph of the function shown on the left, which would require more generations: increasing the gene frequency from 0.1 to 0.2 or from 0.5 to 0.6?

c. Can you think of a reason for this? [*Hint:* Genes reproduce from other similar genes.]

Sometimes a substitution is needed to transform a formula to fit a given integral. In such cases both the x and the dx must be transformed. A few examples will make the method clear.

EXAMPLE 4   **USING A TABLE WITH A SUBSTITUTION**

Find $\displaystyle\int \frac{x}{\sqrt{x^4 + 1}} \, dx.$

Solution

The table of integrals on the inside back cover has no formula involving x^4. However, $x^4 = (x^2)^2$, so a formula involving x^2, along with a substitution, might work. Formula 18 looks promising:

$$\int \frac{1}{\sqrt{x^2 \pm a^2}} \, dx = \ln|x + \sqrt{x^2 \pm a^2}| + C \qquad \begin{array}{l}\text{The } \pm \text{ means: use either} \\ \text{the upper sign or the} \\ \text{lower sign on both sides}\end{array}$$

$$\int \frac{1}{\sqrt{x^2 + 1}} \, dx = \ln|x + \sqrt{x^2 + 1}| + C \qquad \begin{array}{l}\text{Formula 18 with } a = 1 \\ \text{and the upper sign}\end{array}$$

With the substitution

$$x = z^2$$
$$dx = 2z\,dz \qquad \text{Differential of } x = z^2$$

this becomes

$$\int \frac{1}{\sqrt{z^4 + 1}}\, 2z\,dz = \ln|z^2 + \sqrt{z^4 + 1}| + C \qquad \text{Previous formula with } x = z^2 \text{ and } dx = 2z\,dz$$

Dividing by 2 and replacing z by x gives the integral that we wanted:

$$\int \frac{x}{\sqrt{x^4 + 1}}\, dx = \frac{1}{2}\ln\left(x^2 + \sqrt{x^4 + 1}\right) + C \qquad \text{Dropping the absolute value bars since } x^2 + \sqrt{x^4 + 1} \text{ is positive}$$

Given a particular integral, how do we choose a formula?

How to Choose a Formula

Find a formula that matches the *most complicated part* of the integral, making appropriate substitutions to change the formula into the given integral.

For instance, in Example 4 we matched the $\sqrt{x^4 + 1}$ in the given integral with the $\sqrt{x^2 \pm a^2}$ in the formula, and the rest of the integral followed from the differential.

 Be Careful As indicated in the box, it is usually simpler to change the *formula* into the given *integral* rather than the given integral into the formula. Also, most mistakes come from incorrectly changing the dx into the new differential (using the chosen substitution), so this step should be done with particular care.

PRACTICE PROBLEM

Find $\displaystyle\int \frac{t}{9t^4 - 1}\, dt.$ [*Hint:* Use formula 15 with $x = 3t^2$.]

Solution on page 400 >

EXAMPLE 5   USING A TABLE WITH A SUBSTITUTION

Find $\displaystyle\int \frac{e^{-2t}}{e^{-t} + 1}\, dt.$

Solution

Looking in the table of integrals under "Forms Involving e^{ax} and $\ln x$," none of the formulas looks anything like this integral. However, replacing e^{-t} by x

would make the denominator of our integral into $x + 1$, so formula 9 might help. This formula with $a = 1$ and $b = 1$ is

$$\int \frac{x}{x + 1}\, dx = x - \ln|x + 1| + C$$

Formula 9 with $a = 1$ and $b = 1$

With the substitution

$$x = e^{-t}$$
$$dx = -e^{-t}\, dt$$

Differential of $x = e^{-t}$

formula 9 becomes

$$\int \frac{e^{-t}}{e^{-t} + 1}\,(-e^{-t})\, dt = e^{-t} - \ln|e^{-t} + 1| + C$$

or

$$-\int \frac{e^{-2t}}{e^{-t} + 1}\, dt = e^{-t} - \ln(e^{-t} + 1) + C$$

Except for the negative sign, this is the given integral. Multiplying through by -1 gives the final answer.

$$\int \frac{e^{-2t}}{e^{-t} + 1}\, dt = -e^{-t} + \ln(e^{-t} + 1) + C$$

Reduction Formulas

Sometimes we must apply a formula several times to simplify an integral in stages.

EXAMPLE 6   USING A REDUCTION FORMULA

Find $\displaystyle\int x^3 e^{-x}\, dx$.

Solution

In the integral table, we find formula 21.

$$\int x^n e^{ax}\, dx = \frac{1}{a} x^n e^{ax} - \frac{n}{a} \int x^{n-1} e^{ax}\, dx$$

Formula 21 with $n = 3$ and $a = -1$, the left side fits our integral

The right-hand side of this formula involves a new integral, but with a *lower* power of x. We will apply formula 21 several times, each time reducing the power of x until we eliminate it completely. Applying formula 21 with $n = 3$ and $a = -1$,

$$\int x^3 e^{-x}\, dx = -x^3 e^{-x} + 3 \int x^2 e^{-x}\, dx$$

After one application

The power has been reduced

$$= -x^3 e^{-x} + 3\left(-x^2 e^{-x} + 2 \int x^1 e^{-x}\, dx \right)$$

Applying formula 21 again (now with $n = 2$) to the last integral above

$$= -x^3 e^{-x} - 3x^2 e^{-x} + 6 \int x^1 e^{-x}\, dx$$

Multiplying out

$$= -x^3e^{-x} - 3x^2e^{-x} + 6\left(-xe^{-x} + \underbrace{\int x^0e^{-x}\,dx}_{1}\right)$$

Using formula 21 a third time (now with $n = 1$)

$$= -x^3e^{-x} - 3x^2e^{-x} - 6xe^{-x} + 6\int e^{-x}\,dx$$

Now solve this last integral by the formula $\int e^{ax}\,dx = \frac{1}{a}e^{ax} + C$

$$= -x^3e^{-x} - 3x^2e^{-x} - 6xe^{-x} - 6e^{-x} + C$$

The solution, after three applications of formula 21

$$= -e^{-x}(x^3 + 3x^2 + 6x + 6) + C$$

Factoring

We used formula 21 three times, reducing the x^3 in steps, first down to x^2, then to x^1, and finally to $x^0 = 1$, at which point we could solve the integral easily. If the power of x in the integral had been higher, more applications of formula 21 would have been necessary. Formulas such as 21 and 22 are called **reduction formulas,** since they express an integral in terms of a similar integral but with a smaller power of x.

 Graphing Calculator Exploration

Some advanced graphing calculators can find *indefinite* integrals. For example, the Texas Instruments *TI-89* graphing calculator finds the integrals in Examples 5 and 6 as follows.

For Example 5   For Example 6

With some algebra, you can verify that these answers agree with those found in Examples 5 and 6, even though they do not look the same. (Notice that this calculator omits the arbitrary constant of integration.)

Solution   TO PRACTICE PROBLEM

Formula 15 with the substitution $x = 3t^2$, $dx = 6t\,dt$, and $a = 1$ becomes

$$\int \frac{1}{9t^4 - 1}\, 6t\, dt = \frac{1}{2}\ln\left|\frac{3t^2 - 1}{3t^2 + 1}\right| + C$$

Dividing each side by 6 gives the answer:

$$\int \frac{t}{9t^4 - 1}\, dt = \frac{1}{12}\ln\left|\frac{3t^2 - 1}{3t^2 + 1}\right| + C$$

Section Summary

To find a formula that "fits" a given integral, we look for the formula whose left-hand side most closely matches the most complicated part of the integral. Then we choose constants (and possibly a substitution) to make the formula fit exactly. Although we have been using a very brief table, the technique is the same with a more extensive table. Many integral tables have been published, some of which are book-length, containing several thousand formulas.*

6.2 Exercises

1–6. For each integral, state the number of the integration formula (from the inside back cover) and the values of the constants a and b so that the formula fits the integral. (Do not evaluate the integral.)

1. $\displaystyle\int \frac{1}{x^2(5x-1)}\,dx$ **2.** $\displaystyle\int \frac{x}{2x-3}\,dx$

3. $\displaystyle\int \frac{1}{x\sqrt{-x+7}}\,dx$ **4.** $\displaystyle\int \frac{x}{\sqrt{-2x+1}}\,dx$

5. $\displaystyle\int \frac{x}{1-x}\,dx$ **6.** $\displaystyle\int \frac{1}{x\sqrt{1-4x}}\,dx$

7–42. Find each integral by using the integral table on the inside back cover.

7. $\displaystyle\int \frac{1}{9-x^2}\,dx$

[*Hint:* Use formula 16 with $a=3$.]

8. $\displaystyle\int \frac{1}{x^2-25}\,dx$

[*Hint:* Use formula 15 with $a=5$.]

9. $\displaystyle\int \frac{1}{x^2(2x+1)}\,dx$

[*Hint:* Use formula 12 with $a=2,\ b=1$.]

10. $\displaystyle\int \frac{x}{x+2}\,dx$

[*Hint:* Use formula 9 with $a=1,\ b=2$.]

11. $\displaystyle\int \frac{x}{1-x}\,dx$ [*Hint:* Use formula 9.]

12. $\displaystyle\int \frac{x}{\sqrt{1-x}}\,dx$ [*Hint:* Use formula 13.]

13. $\displaystyle\int \frac{1}{(2x+1)(x+1)}\,dx$

14. $\displaystyle\int \frac{x}{(x+1)(x+2)}\,dx$

15. $\displaystyle\int \sqrt{x^2-4}\,dx$ **16.** $\displaystyle\int \frac{1}{\sqrt{x^2-1}}\,dx$

17. $\displaystyle\int \frac{1}{z\sqrt{1-z^2}}\,dz$ **18.** $\displaystyle\int \frac{\sqrt{4+z^2}}{z}\,dz$

19. $\displaystyle\int x^3 e^{2x}\,dx$ **20.** $\displaystyle\int x^{99}\ln x\,dx$

21. $\displaystyle\int x^{-101}\ln x\,dx$ **22.** $\displaystyle\int (\ln x)^2\,dx$

23. $\displaystyle\int \frac{1}{x(x+3)}\,dx$ **24.** $\displaystyle\int \frac{1}{x(x-3)}\,dx$

25. $\displaystyle\int \frac{z}{z^4-4}\,dz$ **26.** $\displaystyle\int \frac{z}{9-z^4}\,dz$

27. $\displaystyle\int \sqrt{9x^2+16}\,dx$ **28.** $\displaystyle\int \frac{1}{\sqrt{16x^2-9}}\,dx$

29. $\displaystyle\int \frac{1}{\sqrt{4-e^{2t}}}\,dt$ **30.** $\displaystyle\int \frac{e^t}{9-e^{2t}}\,dt$

31. $\displaystyle\int \frac{e^t}{e^{2t}-1}\,dt$ **32.** $\displaystyle\int \frac{e^{2t}}{1-e^t}\,dt$

33. $\displaystyle\int \frac{x^3}{\sqrt{x^8-1}}\,dx$ **34.** $\displaystyle\int x^2\sqrt{x^6+1}\,dx$

35. $\displaystyle\int \frac{1}{x\sqrt{x^3+1}}\,dx$ **36.** $\displaystyle\int \frac{\sqrt{1-x^6}}{x}\,dx$

37. $\displaystyle\int \frac{e^t}{(e^t-1)(e^t+1)}\,dt$ **38.** $\displaystyle\int \frac{e^{2t}}{(e^t-1)(e^t+1)}\,dt$

39. $\displaystyle\int xe^{x/2}\,dx$ **40.** $\displaystyle\int \frac{x}{e^x}\,dx$

41. $\displaystyle\int \frac{1}{e^{-x}+4}\,dx$ **42.** $\displaystyle\int \frac{1}{\sqrt{e^{-x}+4}}\,dx$

43–48. For each definite integral:
a. Evaluate it using the table of integrals on the inside back cover. (Leave answers in *exact* form.)
b. Use a graphing calculator to verify your answer to part (a).

43. $\displaystyle\int_4^5 \sqrt{x^2-16}\,dx$ **44.** $\displaystyle\int_0^4 \frac{1}{\sqrt{x^2+9}}\,dx$

*A useful table of integrals containing more than 400 formulas is found in *CRC Standard Mathematical Tables and Formulae*, CRC Press.

(See instructions on previous page.)

45. $\displaystyle\int_2^3 \frac{1}{x^2 - 1}\, dx$ **46.** $\displaystyle\int_2^4 \frac{1}{1 - x^2}\, dx$

47. $\displaystyle\int_3^5 \frac{\sqrt{25 - x^2}}{x}\, dx$ **48.** $\displaystyle\int_3^4 \frac{1}{x\sqrt{25 - x^2}}\, dx$

49–56. Find each integral by whatever means are necessary (either substitution or tables).

49. $\displaystyle\int \frac{1}{2x + 6}\, dx$ **50.** $\displaystyle\int \frac{x}{x^2 - 4}\, dx$

51. $\displaystyle\int \frac{x}{2x + 6}\, dx$ **52.** $\displaystyle\int \frac{1}{4 - x^2}\, dx$

53. $\displaystyle\int x\sqrt{1 - x^2}\, dx$ **54.** $\displaystyle\int \frac{x}{\sqrt{1 - x^2}}\, dx$

55. $\displaystyle\int \frac{\sqrt{1 - x^2}}{x}\, dx$ **56.** $\displaystyle\int \frac{1}{\sqrt{x^2 - 1}}\, dx$

57–62. Find each integral. [*Hint:* Separate each integral into two integrals, using the fact that the numerator is a sum or difference, and find the two integrals by two different formulas.]

57. $\displaystyle\int \frac{x - 1}{(3x + 1)(x + 1)}\, dx$ **58.** $\displaystyle\int \frac{x - 1}{x^2(x + 1)}\, dx$

59. $\displaystyle\int \frac{x + 1}{x\sqrt{1 + x^2}}\, dx$ **60.** $\displaystyle\int \frac{x - 1}{x\sqrt{x^2 + 4}}\, dx$

61. $\displaystyle\int \frac{x + 1}{x - 1}\, dx$ [*Hint:* After separating into two integrals, find one by a formula and the other by a substitution.]

62. $\displaystyle\int \frac{x + 1}{\sqrt{x^2 + 1}}\, dx$ [*Hint:* After separating into two integrals, find one by a formula and the other by a substitution.]

Applied Exercises

63. BUSINESS: Total Sales A company's sales rate is $x^2 e^{-x}$ million sales per month after x months. Find a formula for the total sales in the first x months. [*Hint:* Integrate the sales rate to find the total sales and determine the constant C so that total sales are zero at time $x = 0$.]

64. GENERAL: Population The population of a city is expected to grow at the rate of $x/\sqrt{x + 9}$ thousand people per year after x years. Find the total change in population from year 0 to year 27.

65. BIOMEDICAL: Gene Frequency Under certain circumstances, the number of generations necessary to increase the frequency of a gene from 0.1 to 0.3 is

$$n = 3 \int_{0.1}^{0.3} \frac{1}{q^2(1 - q)}\, dq$$

Find n (rounded to the nearest integer).

66. BEHAVIORAL SCIENCE: Response Rate A subject in a psychology experiment gives responses at the rate of $t/\sqrt{t + 1}$ correct answers per minute after t minutes.

  a. Find the total number of correct responses from time $t = 0$ to time $t = 15$.

  b. Verify your answer to part (a) using a graphing calculator.

67. BUSINESS: Cost The marginal cost function for a computer chip manufacturer is $MC(x) = 1/\sqrt{x^2 + 1}$, and fixed costs are \$2000. Find the cost function.

68. SOCIAL SCIENCE: Employment An urban job placement center estimates that the number of residents seeking employment t years from now will be $t/(2t + 4)$ million people.

  a. Find the average number of job seekers during the period $t = 0$ to $t = 10$.

  b. Verify your answer to part (a) using a graphing calculator.

Conceptual Exercises

69–72. Which formula (from formulas 9–23 on the inside back cover) and what substitution (of the form $x = f(t)$) would you use to find the given integral? (Do not carry out the solution, just indicate a formula and a substitution.)

69. $\displaystyle\int \frac{e^{100t}}{e^{50t} + 7}\, dt$ **70.** $\displaystyle\int \frac{e^{1000t}}{(e^{500t} + 5)(e^{500t} + 7)}\, dt$

71. $\displaystyle\int \frac{t^{49}}{\sqrt{t^{100} + 1}}\, dt$ **72.** $\displaystyle\int \frac{\sqrt{1 + t^{100}}}{t}\, dt$

73. How many uses of formula 21 would you need to find $\displaystyle\int t^{476} e^t\, dt$?

74. How many uses of formula 22 would you need to find $\int (\ln t)^{623} \, dt$?

75. State (without carrying them out) two different *methods* to find $\int \ln t \, dt$.

76. State (without carrying them out) two different *methods* to find $\int t e^t \, dt$.

77. Which *two* formulas can find $\int \dfrac{1}{t^2 - 1} \, dt$?

78. Which *two* formulas can find $\int \dfrac{1}{1 - t^2} \, dt$?

6.3 Improper Integrals

Introduction

In this section we define integrals over intervals that are infinite in length. Such **improper integrals** have many applications, such as the permanent endowments discussed in the Application Preview on page 383 at the beginning of this chapter.

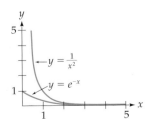

LOOKING BACK

We introduced limits as $x \to \pm\infty$ on pages 78–79.

Limits as x Approaches ± ∞

Recall that the notation $x \to \infty$ ("x approaches infinity") means that x takes on arbitrarily large values.

$x \to \infty$ and $x \to -\infty$		
$x \to \infty$	means:	x takes values arbitrarily far to the *right* on the number line.
$x \to -\infty$	means:	x takes values arbitrarily far to the *left* on the number line.

Evaluating limits as x approaches positive or negative infinity is simply a matter of thinking about large and small numbers. The reciprocal of a large number is a small number. For example,

$$\frac{1}{1{,}000{,}000} = 0.000001$$

One over a million is one one-millionth

Therefore,

$$\lim_{x \to \infty} \frac{1}{x^2} = 0$$

As the denominator approaches infinity (with the numerator constant), the value approaches zero

$$\lim_{x \to \infty} e^{-x} = \lim_{x \to \infty} \frac{1}{e^x} = 0$$

Geometrically, these statements mean that the curves approach the x-axis asymptotically as $x \to \infty$, as shown on the left. These examples illustrate the following general rules.

Limits as $x \to \infty$

			Brief Examples
$\lim_{x \to \infty} \dfrac{1}{x^n} = 0$	$(n > 0)$	As x approaches infinity, 1 over x to a *positive* power approaches zero	$\lim_{x \to \infty} \dfrac{1}{x^2} = 0$
$\lim_{x \to \infty} e^{-ax} = 0$	$(a > 0)$	As x approaches infinity, e to a *negative* number times x approaches zero	$\lim_{x \to \infty} e^{-3x} = 0$

EXAMPLE 1   EVALUATING LIMITS

 LOOKING AHEAD

We will use these (or very similar) limits later in this section.

a. $\lim_{b \to \infty} \left(-\dfrac{1}{b} + 1 \right) = 1$ Because the $\dfrac{1}{b}$ approaches zero

b. $\lim_{b \to \infty} \left(-10e^{-0.05b} + 40 \right) = 40$ Because the $e^{-0.05b}$ approaches zero

Similar rules hold for x approaching *negative* infinity.

Limits as $x \to -\infty$

			Brief Examples
$\lim_{x \to -\infty} \dfrac{1}{x^n} = 0$	(for integer $n > 0$)	As x approaches *negative* infinity, 1 over x to a positive integer approaches zero	$\lim_{x \to -\infty} \dfrac{1}{x^3} = 0$
$\lim_{x \to -\infty} e^{ax} = 0$	$(a > 0)$	As x approaches negative infinity, e to a positive number times x approaches zero (because the exponent is approaching $-\infty$)	$\lim_{x \to -\infty} e^{2x} = 0$

EXAMPLE 2   EVALUATING LIMITS

a. $\lim_{a \to -\infty} \dfrac{1}{a^2} = 0$ Using the first rule in the box above

b. $\lim_{a \to -\infty} e^{0.5a} = 0$ Since the exponent is approaching $-\infty$

Both approach 0

$y = e^{0.5a}$

Recall that for a limit to exist, it must be a *number*, that is, a *finite quantitiy*. There-fore, if a function becomes arbitrarily large, we say its limit *does not exist* or is *un-defined*, which we indicate by writing "$=\infty$" after the limit. That is, the notation $\lim_{x \to \infty} f(x) = \infty$ does not mean that the limit exists, but rather that the limit does *not* exist because the function takes on arbitrarily large values as $x \to \infty$. The follow-ing are some examples of limits that are undefined.

Limits as $x \to \infty$ That Do Not Exist

			Brief Example
$\lim\limits_{x \to \infty} x^n = \infty$	$(n > 0)$	As x approaches infinity, x to a positive power grows arbitrarily large	$\lim\limits_{x \to \infty} x^4 = \infty$
$\lim\limits_{x \to \infty} e^{ax} = \infty$	$(a > 0)$	As x approaches infinity, e to a positive number times x grows arbitrarily large	$\lim\limits_{x \to \infty} e^{3x} = \infty$
$\lim\limits_{x \to \infty} \ln x = \infty$		As x approaches infinity, the natural logarithm of x grows arbitrarily large	

EXAMPLE 3   FINDING WHETHER A LIMIT EXISTS

a. $\lim\limits_{b \to \infty} b^3 = \infty$ (does not exist).   Because b^3 becomes arbitrarily large as b approaches infinity

b. $\lim\limits_{b \to \infty} 2\sqrt{b} - 2 = \infty$ (does not exist).   Because $\sqrt{b}$ becomes arbitrarily large as b approaches infinity

LOOKING AHEAD

We will use the limit in part (b) later in this section.

PRACTICE PROBLEM 1

Evaluate the following limits (if they exist).

a. $\lim\limits_{b \to \infty} \left(4 - \dfrac{1}{b^2} \right)$ **b.** $\lim\limits_{b \to \infty} \left(\sqrt[3]{b} + 3 \right)$ Solution on page 411 >

Improper Integrals

A definite integral over an interval of infinite length is an *improper integral*. As a first example, we evaluate the improper integral $\displaystyle\int_1^\infty \dfrac{1}{x^2}\, dx,$ which gives the area under the curve $y = \dfrac{1}{x^2}$ from $x = 1$ extending arbitrarily far to the right as shown in the graph.

EXAMPLE 4   EVALUATING AN IMPROPER INTEGRAL

Evaluate $\displaystyle\int_1^\infty \dfrac{1}{x^2}\, dx.$

Solution

To integrate to infinity, we first integrate over a *finite* interval, from 1 to some number b (think of b as some very large number), and then take the limit as b approaches ∞. First integrate from 1 to b.

$$\int_1^b \frac{1}{x^2}\, dx = \int_1^b x^{-2}\, dx = \left. (-x^{-1}) \right|_1^b = \left. \left(-\frac{1}{x} \right) \right|_1^b \qquad \text{Using the Power Rule}$$

$$= -\frac{1}{b} - \left(-\frac{1}{1}\right) = -\frac{1}{b} + 1$$ Evaluating and simplifying

$$\underbrace{\phantom{-\frac{1}{b}}}_{\text{at } x = b} \quad \underbrace{\phantom{-\frac{1}{1}}}_{\text{at } x = 1}$$

Then take the limit of this answer as $b \to \infty$.

$$\lim_{b \to \infty}\left(-\frac{1}{b} + 1\right) = 1$$ Limit as $b \to \infty$ (the $\frac{1}{b}$ approaches zero)

This gives the answer:

$$\int_1^\infty \frac{1}{x^2}\, dx = 1$$ Integral from 1 to ∞ equals 1

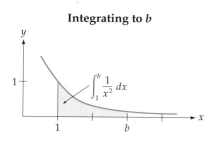

(left margin figure) $y = \frac{1}{x^2}$, area is 1 square unit

Since the limit exists, we say that the improper integral is **convergent.** Geometrically, this procedure amounts to finding the area under the curve from 1 to some number b, shown on the left below, and then letting $b \to \infty$ to find the area arbitrarily far to the right.

Integrating to b   **Integrating to ∞**

Improper Integrals: Integrating to ∞

If f is continuous and nonnegative for $x \geq a$, we define

$$\int_a^\infty f(x)\, dx = \lim_{b \to \infty} \int_a^b f(x)\, dx$$

provided that the limit exists. The improper integral is said to be *convergent* if the limit exists, and *divergent* if the limit does not exist.

It is possible to define improper integrals for functions that take negative values, and even for discontinuous functions, but we shall not do so in this book, since most applications involve functions that are positive and continuous.

PRACTICE PROBLEM 2

Evaluate $\displaystyle\int_2^\infty \frac{1}{x^2}\, dx$.

Solution on page 411 >

EXAMPLE 5   FINDING WHETHER AN INTEGRAL DIVERGES

Evaluate $\displaystyle\int_{1}^{\infty} \frac{1}{\sqrt{x}}\, dx.$

Solution

Integrating up to b:

$$\int_{1}^{b} \frac{1}{\sqrt{x}}\, dx = \int_{1}^{b} x^{-1/2}\, dx = \underbrace{2 \cdot x^{1/2}\Big|_{1}^{b}}_{\substack{\text{Integrating by}\\\text{the Power Rule}}} = \underbrace{2\sqrt{b} - 2\sqrt{1}}_{\text{Evaluating}} = 2\sqrt{b} - 2$$

LOOKING BACK

We found this limit in Example 3 on page 405.

Letting b approach infinity:

$$\lim_{b \to \infty}\left(2\sqrt{b} - 2\right) = \infty \qquad \begin{array}{l}\text{Because } \sqrt{b} \text{ becomes}\\ \text{arbitrarily large } b \to \infty\end{array}$$

Therefore,

$$\int_{1}^{\infty} \frac{1}{\sqrt{x}}\, dx \quad \text{is } divergent \qquad \text{The integral cannot be evaluated}$$

Notice from Examples 4 and 5 that $\displaystyle\int_{1}^{\infty} \frac{1}{x^2}\, dx$ is *convergent* (its value is 1), but $\displaystyle\int_{1}^{\infty} \frac{1}{\sqrt{x}}\, dx$ is *divergent* (it does not have a finite value), as illustrated in the following diagrams.

Intuitively, the areas differ because the curve $\dfrac{1}{x^2}$ approaches the x-axis more rapidly than does the curve $\dfrac{1}{\sqrt{x}}$ for large values of x, and so has a "smaller" area under it.

📶 Spreadsheet Exploration

The following spreadsheet* allows us to numerically "see" that one of these integrals converges while the other diverges. From Example 4 on pages 405–406 we know that $\displaystyle\int_{1}^{\infty} \frac{1}{x^2}\, dx = \lim_{b \to \infty}\left(1 - \frac{1}{b}\right),$ while from Example 5 above we have

---

*To obtain this and other Spreadsheet Explorations, go to www.cengagebrain.com, search for this textbook and then scroll down to access the "free materials" tab.

$\int_1^\infty \dfrac{1}{\sqrt{x}}\, dx = \lim_{b \to \infty}\left(2\sqrt{b} - 2\right).$ The increasingly large values of b shown in column A

are used to evaluate $1 - \dfrac{1}{b}$ in column B and $2\sqrt{b} - 2$ in column C.

C7		▼	= =2*$A7^(1/2)-2
	A	B	C
1	b	Integral of $x^{\wedge}(-2)$	Integral of $x^{\wedge}(-1/2)$
2	1	0	0.000
3	5	0.8	2.472
4	10	0.9	4.325
5	50	0.98	12.142
6	100	0.99	18.000
7	500	0.998	42.721
8	1000	0.999	61.246
9	5000	0.9998	139.421
10	10000	0.9999	198.000
11	50000	0.99998	445.214
12	100000	0.99999	630.456
13	500000	0.999998	1412.214

b 1   areas up to b ∞

Is it clear which of these integrals (areas) is converging and which is diverging?

We may combine the two steps of integrating up to b and letting $b \to \infty$ into a single line. We show how to do this by evaluating the integral from Example 4 again, but more briefly.

$$\int_1^\infty x^{-2}\, dx = \lim_{b \to \infty} \int_1^b x^{-2}\, dx = \lim_{b \to \infty} \underbrace{\left(-x^{-1}\right)\Big|_1^b}_{\substack{\text{Integrating;} \\ \text{then use} \\ x^{-1} = \frac{1}{x}}} = \lim_{b \to \infty}\left[\underbrace{-\frac{1}{b}}_{\substack{\text{Approaches} \\ 0}} - \left(-\frac{1}{1}\right)\right] = 1$$

$\uparrow$ Same answer as before

Permanent Endowments

Funds that generate steady income forever are called *permanent endowments*.

EXAMPLE 6   **FINDING THE SIZE OF A PERMANENT ENDOWMENT**

In the Application Preview on page 383 we found that the size of the fund necessary to generate $2000 annually forever (at 5% interest compounded continuously) is $\int_0^\infty 2000e^{-0.05t}\, dt.$ Find the size of this permanent endowment by evaluating the integral.

Solution

$$\int_0^\infty \underbrace{2000}_{\substack{\text{Annual}\\\text{income}}} e^{-0.05t}\,dt = \lim_{b\to\infty}\left(\underbrace{2000}_{\substack{\text{Moved}\\\text{outside}}}\int_0^b e^{-0.05t}\,dt\right)$$

Continuous interest rate

$$= \lim_{b\to\infty}\left[2000(-20)e^{-0.05t}\,\Big|_0^b\right] \qquad \begin{array}{l}\text{Integrating by}\\ \int e^{ax}\,dx = \dfrac{1}{a}e^{ax}\end{array}$$

$$= \lim_{b\to\infty}\big(\underbrace{-40{,}000e^{-0.05b}}_{\substack{\text{Approaches}\\ 0}} + 40{,}000\underbrace{e^0}_{1}\big) = 40{,}000$$

Therefore, the size of the permanent endowment that will pay the $2000 annual maintenance forever is $40,000. (The area represented by this number is shown on the left.)

$y = 2000e^{-0.05t}$

40,000

Permanent endowments are used to estimate the ultimate cost of anything that requires continuous long-term funding, from buildings to government agencies to toxic waste sites.

Finding Improper Integrals Using Substitutions

Solving an improper integral may require a substitution. In such cases we apply the substitution not only to the integrand, but also to the differential and the upper and lower limits of integration.

EXAMPLE 7 **FINDING AN IMPROPER INTEGRAL USING A SUBSTITUTION**

Evaluate $\displaystyle\int_2^\infty \frac{x}{(x^2+1)^2}\,dx.$

Solution

We use the substitution $u = x^2 + 1$, so $du = 2x\,dx$, requiring multiplication by 2 and by $\frac{1}{2}$. Notice how the substitution changes the limits.

as $x\to\infty,\ x^2+1\to\infty$

$$\int_2^\infty \frac{x}{(x^2+1)^2}\,dx = \frac{1}{2}\int_2^\infty \frac{2x}{(x^2+1)^2}\,dx = \frac{1}{2}\int_5^\infty \frac{du}{u^2} = \frac{1}{2}\lim_{b\to\infty}\int_5^b u^{-2}\,du$$

$$\begin{bmatrix} u = x^2 + 1\\ du = 2x\,dx \end{bmatrix}$$

at $x = 2,\quad x^2 + 1 = 5$

$$= \frac{1}{2} \lim_{b \to \infty} \left[-u^{-1} \right] \Big|_5^b = \frac{1}{2} \lim_{b \to \infty} \left[-\frac{1}{b} - \left(-\frac{1}{5} \right) \right] = \frac{1}{2} \cdot \frac{1}{5} = \frac{1}{10}$$

$\underbrace{}_{\text{Integrating}}$ $\underbrace{}_{\text{Approaches 0}}$

To integrate over an interval that extends arbitrarily far to the *left*, we again integrate over a finite interval and then take the limit.

Improper Integrals: Integrating to $-\infty$

If f is continuous and nonnegative for $x \leq b$, we define

$$\int_{-\infty}^b f(x)\, dx = \lim_{a \to -\infty} \int_a^b f(x)\, dx$$

provided that the limit exists. The improper integral is *convergent* if the limit exists, and *divergent* if the limit does not exist.

To integrate over the *entire* x-axis, from $-\infty$ to ∞, we use two integrals, one from $-\infty$ to 0, and the other from 0 to ∞, and then add the results.

Improper Integrals: Integrating from $-\infty$ to ∞

If f is continuous and nonnegative for *all* values of x, we define

$$\int_{-\infty}^{\infty} f(x)\, dx = \lim_{a \to -\infty} \int_a^0 f(x)\, dx + \lim_{b \to \infty} \int_0^b f(x)\, dx$$

The improper integral is *convergent* if both limits exist, and *divergent* if either limit does not exist.

EXAMPLE 8   INTEGRATING TO $-\infty$

Evaluate $\displaystyle\int_{-\infty}^3 4e^{2x}\, dx$.

Solution

$$\int_{-\infty}^{3} 4e^{2x}\, dx = \lim_{a \to -\infty} \int_{a}^{3} 4e^{2x}\, dx$$

$$= \lim_{a \to -\infty} \left[4 \cdot \frac{1}{2} \cdot e^{2x} \Big|_{a}^{3} \right] \qquad \text{Integrating}$$

$$= \lim_{a \to -\infty} (2e^{6} - 2e^{2a}) = 2e^{6} \qquad \text{Convergent}$$

$$\underbrace{\qquad\qquad}_{\substack{\text{Approaches 0} \\ \text{as } a \to -\infty}}$$

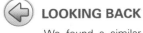

LOOKING BACK
We found a similar limit in Example 2b on page 404.

PRACTICE PROBLEM 3

Evaluate the improper integral $\displaystyle\int_{-\infty}^{1} 12e^{3x}\, dx.$ **Solution below >**

Solutions TO PRACTICE PROBLEMS

1. a. $\displaystyle\lim_{b \to \infty} \left(4 - \frac{1}{b^2} \right) = 4$ Because $\dfrac{1}{b^2}$ approaches 0

b. $\displaystyle\lim_{b \to \infty} \left(\sqrt[3]{b} + 3 \right) = \infty$ Because $\sqrt[3]{b}$ gets arbitrarily large

2. $\displaystyle\int_{2}^{b} x^{-2}\, dx = (-x^{-1}) \Big|_{2}^{b} = -\frac{1}{b} - \left(-\frac{1}{2} \right) = \frac{1}{2} - \frac{1}{b}$

$\displaystyle\lim_{b \to \infty} \left(\frac{1}{2} - \frac{1}{b} \right) = \frac{1}{2}$

Therefore, $\displaystyle\int_{2}^{\infty} \frac{1}{x^2}\, dx = \frac{1}{2}$

3. $\displaystyle\int_{-\infty}^{1} 12e^{3x}\, dx = \lim_{a \to -\infty} \int_{a}^{1} 12e^{3x}\, dx = \lim_{a \to -\infty} \left[12 \cdot \frac{1}{3} e^{3x} \Big|_{a}^{1} \right] = \lim_{a \to -\infty} (4e^3 - 4e^{3a}) = 4e^3$

6.3   Section Summary

A definite integral in which one or both limits of integration are infinite is called an "improper" integral. The improper integral of a continuous nonnegative function is defined as the *limit* of the integral over a finite interval. The integral is *convergent* if the limit exists, and *divergent* otherwise. This idea of dealing with the infinite by "dropping back to the finite and then taking the limit" is a standard technique in mathematics.

Several particular limits are helpful in evaluating improper integrals.

Approaching Infinity	*Approaching Negative Infinity*
$\displaystyle\lim_{x \to \infty} \frac{1}{x^n} = 0 \quad (n > 0)$	$\displaystyle\lim_{x \to -\infty} \frac{1}{x^n} = 0 \quad (\text{for integer } n > 0)$
$\displaystyle\lim_{x \to \infty} e^{-ax} = 0 \quad (a > 0)$	$\displaystyle\lim_{x \to -\infty} e^{ax} = 0 \quad (a > 0)$

The two notations $\lim\limits_{x \to \infty} f(x) = \infty$ and $\lim\limits_{x \to \infty} f(x) = -\infty$ both mean that the limit does not exist, and furthermore they explain *why* it is undefined: in the first case because the function takes on values *greater* than every number, and in the second because the function takes on values *less* than every number.

Improper integrals give continuous sums over infinite intervals. For example, the total future output of an oil well can be found by integrating the production rate out to infinity, and the value of an asset that lasts indefinitely (such as land) can be found by integrating the present value of the future income out to infinity. Even when infinite duration is unrealistic, ∞ is used to represent "long-term behavior."

6.3 Exercises

1–16. Evaluate each limit (or state that it does not exist using ∞ and $-\infty$ where appropriate).

1. $\lim\limits_{x \to \infty} \dfrac{1}{x^2}$

2. $\lim\limits_{b \to \infty} \left(\dfrac{1}{\sqrt{b}} - 8 \right)$

3. $\lim\limits_{b \to \infty} (1 - 2e^{-5b})$

4. $\lim\limits_{b \to \infty} (3e^{3b} - 4)$

5. $\lim\limits_{x \to \infty} (2 - e^{x/2})$

6. $\lim\limits_{x \to \infty} (1 - e^{-x/3})$

7. $\lim\limits_{b \to \infty} (3 + \ln b)$

8. $\lim\limits_{b \to \infty} (2 - \ln b^2)$

9. $\lim\limits_{x \to -\infty} \dfrac{1}{x^3}$

10. $\lim\limits_{x \to -\infty} \dfrac{1}{x^2}$

11. $\lim\limits_{a \to -\infty} e^{2a}$

12. $\lim\limits_{a \to -\infty} e^{-3a}$

13. $\lim\limits_{a \to -\infty} e^{-2a}$

14. $\lim\limits_{a \to -\infty} e^{\frac{1}{2}a}$

15. $\lim\limits_{x \to -\infty} \dfrac{1}{(x^2 + 1)^2}$

16. $\lim\limits_{x \to -\infty} \dfrac{1}{e^x + e^{-x}}$

17–40. Evaluate each improper integral or state that it is divergent.

17. $\displaystyle\int_1^\infty \dfrac{1}{x^3}\, dx$

18. $\displaystyle\int_1^\infty \dfrac{1}{\sqrt[3]{x^4}}\, dx$

19. $\displaystyle\int_2^\infty 3x^{-4}\, dx$

20. $\displaystyle\int_0^\infty e^{-t}\, dt$

21. $\displaystyle\int_2^\infty \dfrac{1}{x}\, dx$

22. $\displaystyle\int_1^\infty \dfrac{1}{x^{0.99}}\, dx$

23. $\displaystyle\int_1^\infty \dfrac{1}{x^{1.01}}\, dx$

24. $\displaystyle\int_{10}^\infty e^{-x/5}\, dx$

25. $\displaystyle\int_0^\infty e^{-0.05t}\, dt$

26. $\displaystyle\int_0^\infty e^{0.01t}\, dt$

27. $\displaystyle\int_5^\infty \dfrac{1}{(x-4)^3}\, dx$

28. $\displaystyle\int_0^\infty \dfrac{x}{(x^2+1)^2}\, dx$

29. $\displaystyle\int_0^\infty \dfrac{x}{x^2 + 1}\, dx$

30. $\displaystyle\int_0^\infty \dfrac{x^2}{x^3 + 1}\, dx$

31. $\displaystyle\int_0^\infty x^2 e^{-x^3}\, dx$

32. $\displaystyle\int_e^\infty (\ln x)^{-2} \dfrac{1}{x}\, dx$

33. $\displaystyle\int_{-\infty}^0 e^{3x}\, dx$

34. $\displaystyle\int_{-\infty}^0 \dfrac{x^4}{(x^5 - 1)^2}\, dx$

35. $\displaystyle\int_{-\infty}^1 \dfrac{1}{2 - x}\, dx$

36. $\displaystyle\int_{-\infty}^0 \dfrac{1}{1 - x}\, dx$

37. $\displaystyle\int_{-\infty}^\infty \dfrac{e^x}{(1 + e^x)^2}\, dx$

38. $\displaystyle\int_{-\infty}^\infty \dfrac{e^{-x}}{(1 + e^{-x})^3}\, dx$

39. $\displaystyle\int_{-\infty}^\infty \dfrac{e^x}{1 + e^x}\, dx$

40. $\displaystyle\int_{-\infty}^\infty \dfrac{e^{-x}}{1 + e^{-x}}\, dx$

41. Use a graphing calculator to estimate the improper integrals $\displaystyle\int_0^\infty e^{\sqrt{x}}\, dx$ and $\displaystyle\int_0^\infty e^{-x^2}\, dx$ (if they converge) as follows:

  a. Define y_1 to be the definite integral (using FnInt) of $e^{\sqrt{x}}$ from 0 to x.

  b. Define y_2 to be the definite integral of e^{-x^2} from 0 to x.

  c. y_1 and y_2 then give the *areas* under these curves out to any number x. Make a TABLE of values of y_1 and y_2 for x-values such as 1, 10, 100, and 500. Which integral converges (and to what number, approximated to five decimal places) and which diverges?

42. Use a graphing calculator to estimate the improper integrals $\displaystyle\int_0^\infty \dfrac{1}{x^2 + 1}\, dx$ and $\displaystyle\int_0^\infty \dfrac{1}{\sqrt{x} + 1}\, dx$ (if they converge) as follows:

  a. Define y_1 to be the definite integral (using FnInt) of $\dfrac{1}{x^2 + 1}$ from 0 to x.

  b. Define y_2 to be the definite integral of $\dfrac{1}{\sqrt{x} + 1}$ from 0 to x.

c. y_1 and y_2 then give the *areas* under these curves out to any number x. Make a TABLE of values of y_1 and y_2 for x-values such as 1, 10, 100, 500, and

10,000. Which integral converges (and to what number, approximated to five decimal places) and which diverges?

Applied Exercises

43. GENERAL: Permanent Endowments Find the size of the permanent endowment needed to generate an annual $12,000 forever at a continuous interest rate of 6%.

44. GENERAL: Permanent Endowments Show that the size of the permanent endowment needed to generate an annual C dollars forever at interest rate r compounded continuously is C/r dollars.

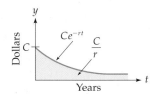

45. GENERAL: Permanent Endowments

a. Find the size of the permanent endowment needed to generate an annual $1000 forever at a continuous interest rate of 10%.

b. At this same interest rate, the size of the fund needed to generate an annual $1000 for precisely 100 years is $\int_0^{100} 1000e^{-0.1t}\, dt$. Evaluate this integral (it is not an improper integral), approximating your answer using a calculator.

c. Notice that the cost for the first 100 years is almost the same as the cost forever. This illustrates again the principle that in endowments, the short term is expensive, but eternity is cheap.

46–48. BUSINESS: Capital Value of an Asset The *capital value of an asset* is defined as the present value of all future earnings. For an asset that may last indefinitely (such as real estate or a corporation), the capital value is

$$\begin{pmatrix} \text{Capital} \\ \text{value} \end{pmatrix} = \int_0^\infty C(t)e^{-rt}\, dt$$

where $C(t)$ is the income per year and r is the continuous interest rate. Find the capital value of a piece of property that will generate an annual income of $C(t)$, for the function $C(t)$ given below, at a continuous interest rate of 5%.

46. $C(t) = 8000$ dollars

47. $C(t) = 50\sqrt{t}$ thousand dollars

48. $C(t) = 59t^{0.1}$ thousand dollars

49. BUSINESS: Oil Well Output An oil well is expected to produce oil at the rate of $50e^{-0.05t}$ thousand barrels per month indefinitely, where t is the number of months that the well has been in operation. Find the total output over the lifetime of the well by integrating this rate from 0 to ∞. [*Note:* The owner will shut down the well when production falls too low, but it is convenient to estimate the total output as if production continued forever.]

50. GENERAL: Duration of Telephone Calls The proportion of telephone calls that last longer than t minutes is approximately $\int_t^\infty 0.3e^{-0.3s}\, ds$. Use this formula to find the proportion of telephone calls that last longer than 4 minutes.

Source: J. F. Haynes and T. V. J. Ganesh Baba, Modeling and Analysis of Telecommunications Networks

51. AREA Find the area between the curve $y = 1/x^{3/2}$ and the x-axis from $x = 1$ to ∞.

52. AREA Find the area between the curve $y = e^{-4x}$ and the x-axis from $x = 0$ to ∞.

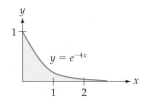

53. AREA Find the area between the curve $y = e^{-ax}$ (for $a > 0$) and the x-axis from $x = 0$ to ∞.

54. AREA Find the area between the curve $y = 1/x^n$ (for $n > 1$) and the x-axis from $x = 1$ to ∞.

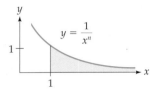

55. BEHAVIORAL SCIENCE: Mazes In a psychology experiment, rats were placed in a T-maze, and the proportion of rats who required more than t seconds to reach the end was $\int_t^\infty 0.05e^{-0.05s}\,ds$. Use this formula to find the proportion of rats who required more than 10 seconds.

56. SOCIOLOGY: Prison Terms If the proportion of prison terms that are longer than t years is given by the improper integral $\int_t^\infty 0.2e^{-0.2s}\,ds$, find the proportion of prison terms that are longer than 5 years.

57. BUSINESS: Product Reliability The proportion of light bulbs that last longer than t hours is predicted to be $\int_t^\infty 0.001e^{-0.001s}\,ds$. Use this formula to find the proportion of light bulbs that will last longer than 1200 hours.

58. BUSINESS: Warranties When a company sells a product with a lifetime guarantee, the number of items returned for repair under the guarantee usually decreases with time. A company estimates that the annual rate of returns after t years will be $800e^{-0.2t}$. Find the total number of returns by summing (integrating) this rate from 0 to ∞.

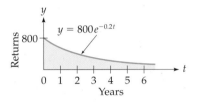

59. BUSINESS: Sales A publisher estimates that a book will sell at the rate of $16,000e^{-0.8t}$ books per year t years from now. Find the total number of books that will be sold by summing (integrating) this rate from 0 to ∞.

60. BIOMEDICAL: Drug Absorption To determine how much of a drug is absorbed into the body, researchers measure the difference between the dosage D and the amount of the drug excreted from the body. The total amount excreted is found by integrating the excretion rate $r(t)$ from 0 to ∞. Therefore, the amount of the drug absorbed by the body is

$$D - \int_0^\infty r(t)\,dt.$$

If the initial dose is $D = 200$ milligrams (mg), and the excretion rate is $r(t) = 40e^{-0.5t}$ mg per hour, find the amount of the drug absorbed by the body.

Conceptual Exercises

61–64. Which of the two limits exists?

61. a. $\displaystyle\lim_{x \to \infty} x^3$ **b.** $\displaystyle\lim_{x \to \infty} \frac{1}{x^3}$

62. a. $\displaystyle\lim_{x \to -\infty} x^3$ **b.** $\displaystyle\lim_{x \to -\infty} \frac{1}{x^3}$

63. a. $\displaystyle\lim_{x \to -\infty} e^{3x}$ **b.** $\displaystyle\lim_{x \to -\infty} e^{-3x}$

64. a. $\displaystyle\lim_{x \to \infty} e^{3x}$ **b.** $\displaystyle\lim_{x \to \infty} e^{-3x}$

65. True or False: If $f(x)$ is continuous, nonnegative, and $\displaystyle\lim_{x \to \infty} f(x) = 0$, then $\displaystyle\int_1^\infty f(x)\,dx$ converges.

66. True or False: If $f(x)$ is continuous, nonnegative, and $\displaystyle\lim_{x \to \infty} f(x) = 2$, then $\displaystyle\int_1^\infty f(x)\,dx$ diverges.

67. Does $\displaystyle\int_0^\infty 5\,dx$ converge or diverge?

68. Does $\displaystyle\int_0^\infty \frac{1}{1000}\,dx$ converge or diverge?

69. True or False: If $f(-x) = f(x)$ for every x and if $\int_0^\infty f(x)\,dx = 7$, then $\int_{-\infty}^0 f(x)\,dx = 7$.

70. True or False: If $\int_{-\infty}^\infty f(x)\,dx$ converges, then $\int_0^\infty f(x)\,dx$ converges.

Explorations and Excursions   The following problems extend and augment the material presented in the text.

71. BUSINESS: Preferred Stock Since preferred stock can remain outstanding indefinitely, the present value per share is the limit of the present value of an annuity* paying that share's dividend D at interest rate r:

$$\begin{pmatrix} \text{Present} \\ \text{value} \end{pmatrix} = \lim_{t \to \infty} D\left(\frac{1 - (1 + r)^{-t}}{r} \right)$$

Find this limit in terms of D and r.

72. BIOMEDICAL: Population Growth The *Gompertz growth curve* models the size $N(t)$ of a population at time $t \geq 0$ as

$$N(t) = Ke^{-ae^{-bt}}$$

where K and b are positive constants. Find $\lim_{t \to \infty} N(t)$.

Source: F. Brauer and C. Castillo-Chávez, *Mathematical Models in Population Biology and Epidemiology*

*See Section 2.4 of *FINITE* by the same authors and publisher for a derivation of this formula.

73–74. GENERAL: Permanent Endowments The formula for integrating the exponential function a^{bx} is

$$\int a^{bx}\,dx = \frac{1}{b \ln a} a^{bx} + C \quad \text{for constants } a > 0 \text{ and } b, \text{ as}$$

may be verified by using the differentiation formulas on page 289.

73. Use the formula above to find the size of the permanent endowment needed to generate an annual $2000 forever at 5% interest compounded annually.

[*Hint:* Find $\int_0^\infty 2000 \cdot 1.05^{-x}\,dx$.] Compare your answer with that found in Example 6 (pages 408–409) for the same interest rate but compounded continuously.

74. Use the formula above to find the size of the permanent endowment needed to generate an annual $12,000 forever at 6% interest compounded annually.

[*Hint:* Find $\int_0^\infty 12{,}000 \cdot 1.06^{-x}\,dx$.] Compare your answer with that found in Exercise 43 (page 413) for the same interest rate but compounded continuously.

6.4   Numerical Integration

Introduction

In spite of the many techniques of integration, there are still integrals that cannot be found by *any* method (as finite combinations of elementary functions). One example is the integral $\int e^{-x^2}\,dx$, which is closely related to the famous "bell-shaped curve" of probability and statistics. For *definite* integrals, however, it is always possible to *approximate* the actual value by interpreting it as the area under a curve and using a process called **numerical integration.** We will discuss two of the most useful methods of numerical integration, based on approximating areas by **trapezoids** and by **parabolas** (known as **Simpson's Rule**). Both methods can be programmed on a calculator or computer. In fact, when you evaluate a definite integral on a graphing calculator using a command like FnInt, the calculator is using a numerical integration procedure similar to those described here. This section explains the mathematics behind such operations.

Rectangular Approximation and Riemann Sums

Area under $y = f(x)$ from a to b approximated by three rectangles

In Section 5.3 we approximated the area under a curve by rectangles, and we called the sum of the areas of the rectangles a *Riemann sum.* However, these rectangles underestimate the true area under the curve by the white "error area" just above the rectangles shown on the left. For greater accuracy we could increase the number of rectangles (as we did in Section 5.3), but this would involve more calculation and consequently more roundoff errors. Instead, we will replace the rectangles by shapes that fit the curve more closely.

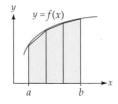

Area under $y = f(x)$ from
a to b approximated
by three trapezoids

Trapezoidal Approximation

We modify the approximating rectangles by allowing their tops to slant with the curve, as shown on the left, giving a much better "fit" to the curve. Such shapes, in which two sides are parallel, are called *trapezoids*. Clearly, approximating the area under the curve by trapezoids is more accurate than approximating it by rectangles: the white "error area" is much smaller.

trapezoidal error   rectangular error

Area under a curve approximated by a
trapezoid (left) and by a rectangle (right).

The area of a trapezoid is the average of the two heights times the width.

$$\left(\begin{matrix}\text{Area of a} \\ \text{trapezoid}\end{matrix}\right) = \underbrace{\frac{h_1 + h_2}{2}}_{\substack{\text{Average} \\ \text{height}}} \cdot \underbrace{\Delta x}_{\text{Width}}$$

If we use trapezoids that all have the same width Δx, we may add up all of the average heights first and *then* multiply by Δx. Furthermore, averaging the heights means dividing each height by 2 since $\dfrac{h_1 + h_2}{2} = \dfrac{h_1}{2} + \dfrac{h_2}{2}$. However, a side *between* two trapezoids will be counted twice, once for the trapezoid on the left and again for the one on the right, thereby canceling the division by 2. Therefore, in adding up the heights, only the two outside heights, at a and b, should be divided by 2. This leads to the following procedure for trapezoidal approximation.

Trapezoidal Approximation

To approximate $\displaystyle\int_a^b f(x)\, dx$ by using n trapezoids:

1. Calculate $\Delta x = \dfrac{b - a}{n}$.   Trapezoid width

2. Find numbers $x_1, x_2, \ldots, x_{n+1}$ starting with $x_1 = a$ and successively adding Δx, ending with $x_{n+1} = b$.

3. The approximation for the integral is

$$\int_a^b f(x)\, dx \approx \left[\frac{1}{2}f(x_1) + f(x_2) + \cdots + f(x_n) + \frac{1}{2}f(x_{n+1}) \right] \Delta x$$

This last formula calculates the total area of the n trapezoids shown on the next page.

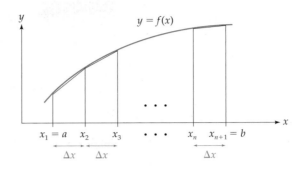

It is easiest to carry out the calculation in a table.

EXAMPLE 1   APPROXIMATING AN INTEGRAL USING TRAPEZOIDS

Approximate $\displaystyle\int_1^2 x^2\,dx$ using four trapezoids.

Solution

The limits of integration are $a = 1$ and $b = 2$, and we are using $n = 4$ trapezoids. The method consists of the following six steps.

1. Calculate the trapezoid width $\Delta x = \dfrac{b-a}{n} = \dfrac{2-1}{4} = 0.25.$

2. List the x-values
   a through b with
   spacing Δx.

3. Apply $f(x)$ to
   each x-value.

x	$f(x) = x^2$	
1	$(1)^2 = \cancel{1}$	0.5
1.25	$(1.25)^2 \approx 1.56$	
1.5	$(1.5)^2 = 2.25$	
1.75	$(1.75)^2 \approx 3.06$	
2.0	$(2)^2 = \cancel{4}$	2

Initial point a — add Δx — add Δx — add Δx — add Δx — Final point b

4. Take half of first
   and last entries.

5. Sum last column.

6. Multiply by Δx.

$$9.37 \cdot (0.25) \approx 2.34$$

Final
answer

Therefore, the estimate using four trapezoids is $\displaystyle\int_1^2 x^2\,dx \approx 2.34.$

Earlier (see page 334) we evaluated this integral exactly, obtaining $\displaystyle\int_1^2 x^2\,dx = \tfrac{7}{3} \approx 2.33$, and we can use this result to assess the accuracy of the trapezoidal method: Our approximation of 2.34 is very accurate in spite of the fact that we used only four trapezoids. The relative error (the actual error, 0.01, divided by the actual value, 7/3) is $\dfrac{0.01}{7/3} \approx 0.004.$ Expressed as a percentage, this is 0.4% (four tenths of one percent), which is also remarkably small. Notice also that the trapezoidal approximation of 2.34 using four trapezoids is far more accurate than the Riemann sum of 2.04 that we found on pages 330–331 using five (left) rectangles.

Error in Trapezoidal Approximation

The maximum error in trapezoidal approximation obeys the following formula.

Trapezoidal Error

For the trapezoidal approximation of $\displaystyle\int_a^b f(x)\,dx$ with n trapezoids,

$$\text{(Error)} \le \frac{(b-a)^3}{12n^2} \max_{a \le x \le b} |f''(x)|$$

This formula is very difficult to use because it involves maximizing the absolute value of the second derivative. We will not make further use of it except to observe that the n^2 in the denominator means that doubling the number of trapezoids reduces the maximum error by a factor of *four.*

Trapezoidal approximation is most easily carried out on a calculator or a computer, as shown in the following Graphing Calculator Exploration.

Graphing Calculator Exploration

Approximating $\displaystyle\int_1^2 x^2\,dx$ by four trapezoids

Graphing calculator programs that calculate the trapezoidal approximation of a definite integral and, for small values of n, draw the approximating trapezoids are available on the Internet. See, for example, *ticalc.org* and search for *Trapezoidal.* Typical *screens* are shown on the left. The first screen shows the approximation of $\displaystyle\int_1^2 x^2\,dx$ using four trapezoids, giving a value 2.34375 that agrees with the 2.34 found in the previous Example. Even for only four trapezoids, the curve $y = x^2$ is almost indistinguishable from the tops of the trapezoids, the difference appearing only as a slight thickening of the line where they separate slightly.

N	Trapezoid Apprxmtion
5	2.34
10	2.335
25	2.3336
50	2.3334
100	2.33335
250	2.333336
500	2.333334

For 500 trapezoids, the approximation is about $2\frac{1}{3}$

The second screen shows the trapezoidal approximations as the number n of trapezoids increases from 5 to 500. The approximations do approach the *exact* value $2\frac{1}{3}$ found in Example 3 on page 334 by evaluating the definite integral.

Try using a command like FnInt to find the area under $y = x^2$ from $x = 1$ to $x = 2$. How do the answers compare? (FnInt uses a method similar to those described in this section but is faster since it is built into the calculator.)

The above error formula is of limited usefulness on a computer since maximizing the second derivative is itself subject to error. What is done instead is to calculate the approximations for larger and larger values of n until successive approximations agree to the desired degree of accuracy. For example, the last two results in the table above give answers agreeing to five decimal places, indicating an error of less than 0.00001. While this procedure does not *guarantee* this accuracy, it is often accepted in practice.

Simpson's Rule (Parabolic Approximation)

For even greater accuracy, we could increase n (resulting in more calculations and more roundoff errors) or we could change the trapezoids, replacing the tops by *curves* chosen to fit the given curve more closely. Replacing the tops of the trapezoids by *parabolas* that pass through three points of the given curve leads to an even more accurate method of approximation, called Simpson's Rule.* The following diagram shows such a curve and its approximation by two parabolas.

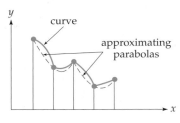

A curve approximated by two
parabolas

Since each parabola spans two intervals, the number of intervals must be even. The area under the approximating parabolas is easily found by integration. The procedure is described as follows and is illustrated in Example 2. A justification of Simpson's Rule is given in Exercise 47.

Simpson's Rule (Parabolic Approximation)

To approximate $\displaystyle\int_a^b f(x)\,dx$ by Simpson's Rule using n intervals:

1. Calculate $\Delta x = \dfrac{b-a}{n}$.   *n must be even*

2. Find numbers $x_1, x_2, x_3, \ldots, x_{n+1}$ starting with $x_1 = a$ and successively adding Δx, ending with $x_{n+1} = b$.

3. The approximation for the integral is

$$\int_a^b f(x)\,dx \approx [\,\underbrace{f(x_1) + 4f(x_2) + 2f(x_3) + \cdots + 4f(x_n)}_{\text{Alternating 4's and 2's}} + f(x_{n+1})\,]\,\frac{\Delta x}{3}$$

> **LOOKING BACK**
>
> Compare this formula to the one for trapezoidal approximation on page 416. The only differences are the numbers that multiply the f's and that Δx is divided by 3 at the end.

The function values are multiplied by "weights," which, written out by themselves, are

$$1 \quad 4 \quad 2 \quad 4 \quad 2 \quad 4 \quad \cdots \quad 4 \quad 2 \quad 4 \quad 1$$

$\uparrow$ Initial 1   Alternating 4's and 2's, beginning and ending with 4 $\uparrow$ Final 1

EXAMPLE 2   APPROXIMATING AN INTEGRAL USING SIMPSON'S RULE

Approximate $\displaystyle\int_3^5 \frac{1}{x}\,dx$ by Simpson's Rule with $n = 4$.   *n = 4 means 2 parabolas*

Solution

The method consists of the following six steps:

1. Calculate $\Delta x = \dfrac{b-a}{n} = \dfrac{5-3}{4} = 0.5$ *n must be even*

---

*Named after Thomas Simpson (1701–1761), an early user, but not the discoverer, of the formula.

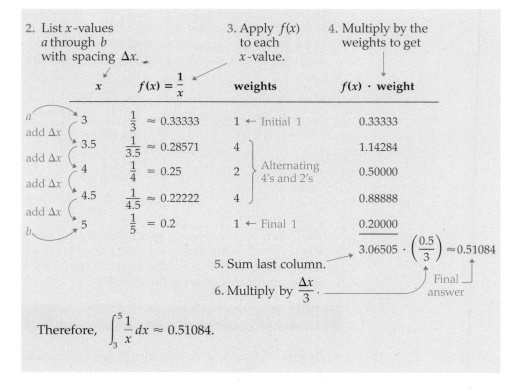

2. List x-values a through b with spacing Δx.

3. Apply $f(x)$ to each x-value.

4. Multiply by the weights to get

x	$f(x) = \dfrac{1}{x}$	weights	$f(x) \cdot$ **weight**
a 3	$\dfrac{1}{3} \approx 0.33333$	1 ← Initial 1	0.33333
add Δx 3.5	$\dfrac{1}{3.5} \approx 0.28571$	4	1.14284
add Δx 4	$\dfrac{1}{4} = 0.25$	2 $\bigg\}$ Alternating 4's and 2's	0.50000
add Δx 4.5	$\dfrac{1}{4.5} \approx 0.22222$	4	0.88888
add Δx b. 5	$\dfrac{1}{5} = 0.2$	1 ← Final 1	0.20000

$$3.06505 \cdot \left(\frac{0.5}{3}\right) \approx 0.51084$$

5. Sum last column.

6. Multiply by $\dfrac{\Delta x}{3}$.

Final answer

Therefore, $\displaystyle\int_3^5 \frac{1}{x}\, dx \approx 0.51084$.

This integral, too, can be found exactly. The answer is $\ln 5 - \ln 3 \approx 0.51083$, for an error of only 0.00001. The *relative* error (actual error divided by actual value) is

$$\frac{0.00001}{0.51083} \approx 0.00002 = 0.002\%$$

(two thousandths of one percent) which is also extremely small.

 Be Careful Remember that in Simpson's Rule n must be *even*. Also in Simpson's Rule you multiply by $\dfrac{\Delta x}{3}$ while in trapezoidal approximation you multiply by just Δx.

 ## IQ Distribution

Although it is increasingly clear that human intelligence cannot be measured by a single number, IQ tests are still widely used. IQ stands for Intelligence Quotient, and is defined as mental age divided by chronological age, multiplied by 100. The average American IQ is 100, and the distribution of IQs follows the famous "bell-shaped curve" so often used in statistics.*

| | 70 | 85 | 100 | 115 | 130 |

*For those familiar with statistics, IQ scores are normally distributed with mean 100 and standard deviation 15.

The proportion of Americans with IQs between two numbers A and B (with $A < B$) is given by the following integral:

$$\begin{pmatrix} \text{Proportion of Americans} \\ \text{with IQs between A and B} \end{pmatrix} \approx \frac{1}{\sqrt{2\pi}} \int_{(A-100)/15}^{(B-100)/15} e^{-x^2/2}\, dx$$

For example, the proportion of Americans who have IQs between 115 and 145 is found by substituting $A = 115$ and $B = 145$ into the lower and upper limits in the formula above:

$$\begin{pmatrix} \text{Proportion of IQs} \\ \text{between 115 and 145} \end{pmatrix} = \frac{1}{\sqrt{2\pi}} \int_{1}^{3} e^{-x^2/2}\, dx$$

$\longleftarrow 3$ from $\dfrac{B-100}{15} = \dfrac{145-100}{15} = \dfrac{45}{15} = 3$

$\longleftarrow 1$ from $\dfrac{A-100}{15} = \dfrac{115-100}{15} = \dfrac{15}{15} = 1$

This integral cannot be found "by hand" (as a finite combination of elementary functions). It can be approximated by Simpson's Rule, just as in Example 2, but it is easier to carry out the calculation on a graphing calculator.

Graphing Calculator Exploration

Approximating $\dfrac{1}{\sqrt{2\pi}} \int_{1}^{3} e^{-x^2/2}\, dx$ using four intervals

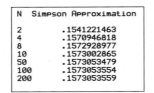

N	Simpson Approximation
2	.1541221463
4	.1570946818
8	.1572928977
10	.1573002865
50	.1573053479
100	.1573053554
200	.1573053559

For 100 and 200 intervals the approximations agree to 9 decimal places

Graphing calculator programs that use Simpson's Rule to approximate definite integrals and, for small values of n, draw the approximating parabolas are available on the Internet. See, for example, *ticalc. org* and search for *Simpson*. Typical screens are shown on the left. The first screen shows the approximation of $\dfrac{1}{\sqrt{2\pi}} \int_{1}^{3} e^{-x^2/2}\, dx$ using $n = 4$ intervals (2 parabolas), giving approximately 0.157. Simpson's Rule is so accurate that the curve is almost indistinguishable from the approximating parabolas, the difference again appearing only as a slight thickening of the line.

The second screen shows that Simpson's Rule converges extremely rapidly.

Try using a command like FnInt to find the area under $y = 1/\sqrt{2\pi}\, e^{-x^2/2}$ from $x = 1$ to $x = 3$. How do the answers compare?

Converting the answer 0.157 to a percentage, about 16% of all Americans have IQs in the range 115 to 145.

The Error in Simpson's Rule

The maximum error in Simpson's Rule obeys the following formula.

LOOKING BACK

Compare this error formula to the one for trapezoidal approximation on page 418. The n^4 in the denominator will make this error much smaller even for $n = 10$ (compared to the n^2 in the trapezoidal case).

Error in Simpson's Rule

In approximating $\displaystyle\int_a^b f(x)\, dx$ by Simpson's Rule with n intervals,

$$(\text{Error}) \leq \frac{(b-a)^5}{180n^4} \max_{a \leq x \leq b} |f^{(4)}(x)|$$

This formula is difficult to use because it involves maximizing the absolute value of the fourth derivative. It does, however, show that Simpson's Rule is *exact* for cubics (third-degree polynomials), since cubics have zero fourth derivative, and that doubling the value of n reduces the maximum error by a factor of 16 (because of the n^4).

6.4 Section Summary

Some definite integrals cannot be evaluated exactly because it is impossible (or very difficult) to find an antiderivative. However, any definite integral can be *approximated* by numerical integration, and two of the most useful methods are trapezoidal approximation and Simpson's Rule (parabolic approximation). Each method involves choosing a number n of intervals and calculating a "weighted average" of function values at the endpoints of these intervals. Higher values of n generally give greater accuracy but also involve more calculation (and therefore more roundoff errors). In practice, Simpson's Rule is generally the method of choice, since it gives greater accuracy for only slightly more effort. Trapezoidal approximation, however, has the advantage of having a simpler formula for its error. It is particularly appropriate to seek an approximation if the "exact" answer involves logarithms or exponentials, since these functions will probably be approximated anyway in the evaluation step.

It is curious that in Chapter 5 we used integrals to evaluate areas, and here we are using areas to evaluate integrals. This is typical of mathematics, in which any equivalence (such as definite integrals and areas) is exploited in both directions.

6.4 Exercises

Exercises on Trapezoidal Approximation

1–4. For each definite integral:

a. Approximate it "by hand," using trapezoidal approximation with $n = 4$ trapezoids. Round calculations to three decimal places.

b. Evaluate the integral exactly using antiderivatives, rounding to three decimal places.

c. Find the actual error (the difference between the actual value and the approximation).

d. Find the relative error (the actual error divided by the actual value, expressed as a percent).

1. $\displaystyle\int_1^3 x^2\, dx$ **2.** $\displaystyle\int_1^2 x^3\, dx$

3. $\displaystyle\int_2^4 \frac{1}{x}\, dx$ **4.** $\displaystyle\int_1^3 \frac{1}{x}\, dx$

5–8. Approximate each integral using trapezoidal approximation "by hand" with the given value of n. Round all calculations to three decimal places.

5. $\displaystyle\int_0^1 \sqrt{1 + x^2}\, dx, \quad n = 3$

6. $\displaystyle\int_0^1 \sqrt{1 + x^3}\, dx, \quad n = 3$

7. $\displaystyle\int_0^1 e^{-x^2}\, dx, \quad n = 4$ **8.** $\displaystyle\int_0^1 e^{x^2}\, dx, \quad n = 4$

9–14. Use a graphing calculator *trapezoidal approximation* program from the Internet (see page 418) to approximate

each integral. Use the following values for the numbers of intervals: 10, 50, 100, 200, 500. Then give an estimate for the value of the definite integral, keeping as many decimal places as the last two approximations agree (when rounded).

9. $\displaystyle\int_1^2 \sqrt{\ln x}\, dx$ **10.** $\displaystyle\int_0^1 \ln(x^2 + 1)\, dx$

11. $\displaystyle\int_{-1}^1 \sqrt{16 + 9x^2}\, dx$ **12.** $\displaystyle\int_{-1}^1 \sqrt{25 - 9x^2}\, dx$

13. $\displaystyle\int_{-1}^1 e^{x^2}\, dx$ **14.** $\displaystyle\int_0^4 \sqrt{1 + x^4}\, dx$

Applied Exercises on Trapezoidal Approximation

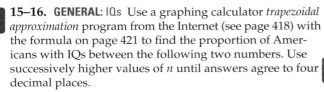

15–16. GENERAL: IQs Use a graphing calculator *trapezoidal approximation* program from the Internet (see page 418) with the formula on page 421 to find the proportion of Americans with IQs between the following two numbers. Use successively higher values of n until answers agree to four decimal places.

15. 100 and 130   **16.** 130 and 145

17. GENERAL: World Oil Consumption In recent decades, world consumption of oil has been running at the rate of $f(x) = \sqrt{12 + 0.025x^2}$ billion tons per year, where x is the number of years since 2000. The total amount of oil consumed from 2010 to 2020 is then given by the integral $\displaystyle\int_{10}^{20} \sqrt{12 + 0.025x^2}\, dx$. Estimate

this amount by approximating the integral using trapezoidal approximation with $n = 10$ trapezoids.
Source: Worldwatch

18. BUSINESS: Investment Growth An investment grows at a rate of $3.2\, e^{\sqrt{t}}$ thousand dollars per year, where t is the number of years since the beginning of the investment. Use a graphing calculator *trapezoidal approximation* program from the Internet (see page 418) to estimate the total growth of the investment during the first 3 years (year 0 to year 3). Use successively higher values of n until answers agree to two decimal places.

Exercises on Simpson's Rule

19–26. Estimate each definite integral "by hand," using Simpson's Rule with $n = 4$. Round all calculations to three decimal places. Exercises 19–26 correspond to Exercises 1–8, in which the same integrals were estimated using trapezoids. If you did the corresponding exercise, compare your Simpson's Rule answer with your trapezoidal answer.

19. $\displaystyle\int_1^3 x^2\, dx$ **20.** $\displaystyle\int_1^2 x^3\, dx$

21. $\displaystyle\int_2^4 \frac{1}{x}\, dx$ **22.** $\displaystyle\int_1^3 \frac{1}{x}\, dx$

23. $\displaystyle\int_0^1 \sqrt{1 + x^2}\, dx$ **24.** $\displaystyle\int_0^1 \sqrt{1 + x^3}\, dx$

25. $\displaystyle\int_0^1 e^{-x^2}\, dx$ **26.** $\displaystyle\int_0^1 e^{x^2}\, dx$

27–32. Use a graphing calculator *Simpson's Rule* program from the Internet (see page 421) to approximate each integral. Use the following values for the numbers of intervals: 10, 20, 50, 100, 200. Then give an estimate for the value of the definite integral, keeping as many decimal places as the last two approximations agree to (when rounded). Exercises 27– 32 correspond to Exercises 9–14 in which the same integrals were estimated using trapezoids. If you did the corresponding exercise, compare your Simpson's Rule answer with your trapezoidal answer.

27. $\displaystyle\int_1^2 \sqrt{\ln x}\, dx$ **28.** $\displaystyle\int_0^1 \ln(x^2 + 1)\, dx$

29. $\displaystyle\int_{-1}^1 \sqrt{16 + 9x^2}\, dx$ **30.** $\displaystyle\int_{-1}^1 \sqrt{25 - 9x^2}\, dx$

31. $\displaystyle\int_{-1}^1 e^{x^2}\, dx$ **32.** $\displaystyle\int_0^4 \sqrt{1 + x^4}\, dx$

33–34. APPROXIMATION OF IMPROPER INTEGRALS

For each improper integral:

a. Make it a "proper" integral by using the substitution $x = \dfrac{1}{t}$ and simplifying.

b. Approximate the proper integral using Simpson's Rule (either "by hand" or using a program) with $n = 4$ intervals, rounding your answer to three decimal places.

33. $\displaystyle\int_{1}^{\infty} \frac{1}{x^3 + 1}\, dx$ 34. $\displaystyle\int_{1}^{\infty} \frac{x}{x^3 + 1}\, dx$

Applied Exercises on Simpson's Rule

 35. GENERAL: Suspension Bridges The cable of a suspension bridge hangs in a parabolic curve. The equation of the cable shown below is $y = \dfrac{x^2}{2000}$. Its length in feet is given by the integral

$$\int_{-400}^{400} \sqrt{1 + \left(\frac{x}{1000}\right)^2}\, dx$$

Approximate this integral using Simpson's Rule, using successively higher values of n until answers agree to the nearest whole number.

36. APPROXIMATION OF π The number π is the ratio of the circumference of a circle to its diameter (since $C = \pi D$). By Exercise 78 on page 581,

$$\int_{0}^{1} \frac{4}{x^2 + 1}\, dx = \pi$$

Find π by approximating this integral using Simpson's Rule, using successively higher values of n until answers agree to four decimal places.

 37. ECONOMICS: Gross World Product The gross world product (GWP), the total value of all finished goods and services produced worldwide, is predicted to be $f(x) = \sqrt{4096 + 100x^2}$ trillion dollars per year, where x is the number of years since 2010. The total value of all finished goods and services produced during the years 2010 to 2020 is then given by the integral

$$\int_{0}^{10} \sqrt{4096 + 100x^2}\, dx.$$ Estimate this total GWP by approximating the integral using Simpson's Rule with $n = 10$.

Source: International Monetary Fund

Conceptual Exercises

38. Explain why trapezoidal approximation is better than Riemann sums for approximating definite integrals.

39. Explain why the trapezoidal method is *exact* for linear functions. [*Hint:* Either use the error formula on page 418 or think of the graph.]

40. True or False: In trapezoidal approximation, the number of trapezoids must be even.

41. If you *triple* the number of trapezoids in trapezoidal approximation, by what factor will the maximum error be reduced?

42. If you *triple* the size of n in Simpson's Rule, by what factor will the maximum error be reduced?

43. Use the *idea* of trapezoidal approximation to approximate the area of the shape shown below.

44. Use the *idea* of Simpson's Rule to approximate the area of the shape shown above.

45. Since Simpson's Rule is *exact* for cubics (third-degree polynomials), if you use Simpson's Rule to approximate the integral of a cubic, what value of n should you use?

46. What are the "weights" for Simpson's Rule when $n = 2$?

Explorations and Excursions
The following problems extend and augment the material presented in the text.

47. JUSTIFICATION OF SIMPSON'S RULE Justify Simpson's Rule by carrying out the following steps, which lead to the formula for Simpson's Rule (page 419) in a simple case.

i. Observe that if the three points shown in the following diagram lie on the parabola $f(x) = ax^2 + bx + c$, then the following three equations hold:

$$a(-d)^2 + b(-d) + c = y_1 \qquad \text{Since} \quad f(-d) = y_1$$

$$a(0)^2 + b(0) + c = y_2 \qquad \text{Since} \quad f(0) = y_2$$

$$a(d)^2 + b(d) + c = y_3 \qquad \text{Since} \quad f(d) = y_3$$

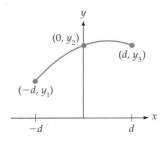

ii. Simplify these three equations to obtain

$$ad^2 - bd + c = y_1$$

$$c = y_2$$

$$ad^2 + bd + c = y_3$$

iii. Add the first and last equation plus four times the middle equation to obtain

$$2ad^2 + 6c = y_1 + 4y_2 + y_3$$

You will use this equation in Step v.

iv. Evaluate the integral $\displaystyle\int_{-d}^{d} (ax^2 + bx + c)\, dx$ and simplify to show that the area under the parabola from $-d$ to d is

$$\text{Area} = \frac{2}{3}ad^3 + 2cd = \left(2ad^2 + 6c\right)\frac{d}{3}$$

v. Use the equation found in Step iii to write this area as

$$\text{Area} = (y_1 + 4y_2 + y_3)\frac{d}{3}$$

vi. Use $y_1 = f(-d)$, $y_2 = f(0)$, $y_3 = f(d)$ and the fact that the spacing Δx is equal to d to rewrite this last equation as

$$\text{Area} = [f(-d) + 4f(0) + f(d)]\frac{\Delta x}{3}$$

This is exactly the formula for Simpson's Rule using one parabola ($n = 2$). For several parabolas placed next to each other, the function values *between* two neighboring parabolas are added twice (once for each side), and so should have weight 2. Therefore, successive function values are multiplied by the weights given on page 419:

$$1 \quad 4 \quad 2 \quad 4 \quad 2 \quad 4 \quad \ldots \quad 4 \quad 1$$

48. While Simpson's Rule is generally more accurate than trapezoidal approximation, show that this is *not always* the case by considering the function $f(x) = 9x^2 - 5x^4$ on the interval $[-1, 1]$ as follows:

a. Find the *exact* area under the curve by integration.

b. Use trapezoidal approximation with two trapezoids to approximate the area.

c. Use Simpson's Rule with $n = 2$ to approximate the area.

d. Which approximation method gave greater accuracy?

| 6 | **Chapter Summary with Hints and Suggestions** |

Reading the text and doing the exercises in this chapter have helped you to master the following concepts and skills, which are listed by section (in case you need to review them) and are keyed to particular Review Exercises. Answers for all Review Exercises are given at the back of the book, and full solutions can be found in the Student Solutions Manual.

6.1 Integration by Parts

- Find an integral using integration by parts. *(Review Exercises 1–14.)*

$$\int u \, dv = uv - \int v \, du$$

- Find an integral by whatever technique is necessary. *(Review Exercises 15–22.)*

- Solve an applied problem using integration by parts. *(Review Exercises 23–24.)*

$$\left(\begin{matrix} \text{Present} \\ \text{value} \end{matrix}\right) = \int_0^T C(t)e^{-rt} \, dt$$

$$\left(\begin{matrix} \text{Total} \\ \text{accumulation} \end{matrix}\right) = \int_0^T r(t) \, dt$$

6.2 Integration Using Tables

- Find an integral using a table of integrals. *(Review Exercises 25–36.)*

- Solve an applied problem using an integral table. *(Review Exercises 37–38.)*

6.3 Improper Integrals

- Evaluate an improper integral (if it is convergent). *(Review Exercises 39–56.)*

$$\int_a^\infty f(x) \, dx = \lim_{b \to \infty} \int_a^b f(x) \, dx$$

- Solve an applied problem involving an improper integral. *(Review Exercises 57–60.)*

- Predict whether an improper integral converges, and then check by evaluating it. *(Review Exercises 61–62.)*

6.4 Numerical Integration

- Approximate an integral using trapezoidal approximation "by hand." *(Review Exercises 63–68.)*

$$\int_a^b f(x) \, dx \approx \left[\frac{1}{2}f(x_1) + f(x_2) + \cdots + f(x_n) + \frac{1}{2}f(x_{n+1}) \right] \cdot \Delta x$$

- Use a program to approximate an integral by trapezoidal approximation. *(Review Exercises 69–74.)*

- Approximate an integral using Simpson's Rule "by hand." *(Review Exercises 75–80.)*

$$\int_a^b f(x) \, dx \approx [f(x_1) + 4f(x_2) + 2f(x_3) + \cdots + 4f(x_n) + f(x_{n+1})] \cdot \frac{\Delta x}{3}$$

- Use a program to approximate an integral by Simpson's Rule. *(Review Exercises 81–86.)*

- Approximate an improper integral using trapezoidal approximation. *(Review Exercises 87–88.)*

Hints and Suggestions

- The unifying idea of this chapter is extensions of the concept of integration.

- Integration by parts takes one integral and gives another integral that, it is hoped, is simpler than the original integral. The formula is simply an integration version of the Product Rule. When using it, try to choose *dv* (including the *dx*) to be the most complicated part of the integrand that you can integrate, and, if possible, choose the *u* to be something that simplifies when differentiated.

- There are tables of integrals that are much longer than the one on the inside back cover. Longer tables, however, require much more time to search for the "right" formula. Other techniques (such as a substitution, integration by parts, or use of a formula more than once) may be used with an integral table.

- To find an integral, try the following methods. First try the "basic" formulas 1 through 4 on the inside back cover. Then try a substitution (formulas 5 through 7). If these methods fail, try integration by parts or an integral table. Remember that some integrals *cannot* be integrated (in terms of elementary functions). A *definite* integral can always be approximated by numerical methods.

- Before "evaluating" an improper integral, be sure that the integrand is defined over the interval, and that the integral is convergent. If the integral diverges, then it has no value and we simply state that it is divergent.

- Numerical integration involves approximating the area under a curve using geometric figures such as trapezoids or parabolas (Simpson's Rule). In practice, the calculations are usually carried out on a calculator or computer, but doing some "by hand" helps to make the method clear.

- A graphing calculator is very helpful for approximating definite integrals by trapezoidal approximation or Simpson's Rule for large values of n. Graphing

calculators also have their own built-in numerical procedures for approximating integrals when you use FnInt.

6 Review Exercises and Chapter Test   ◯ indicates a Chapter Test exercise.

6.1 Integration by Parts

1–14. Find each integral using integration by parts.

1. $\displaystyle\int xe^{2x}\,dx$

2. $\displaystyle\int xe^{-x}\,dx$

③ $\displaystyle\int x^8 \ln x\,dx$

4. $\displaystyle\int \sqrt[4]{x}\ln x\,dx$

⑤ $\displaystyle\int (x-2)(x+1)^5\,dx$

6. $\displaystyle\int (x+3)(x-1)^4\,dx$

7. $\displaystyle\int \frac{\ln t}{\sqrt{t}}\,dt$

8. $\displaystyle\int x^7 e^{x^4}\,dx$

9. $\displaystyle\int x^2 e^x\,dx$

10. $\displaystyle\int (\ln x)^2\,dx$

11. $\displaystyle\int x(x+a)^n\,dx$ (for constants a and $n > 0$)

12. $\displaystyle\int x(1-x)^n\,dx$ (for constant $n > 0$)

⑬ $\displaystyle\int_0^5 xe^x\,dx$

14. $\displaystyle\int_1^e x\ln x\,dx$

15–22. Find each integral by a substitution or by integration by parts, as appropriate.

15. $\displaystyle\int \frac{dx}{1-x}$

16. $\displaystyle\int xe^{-x^2}\,dx$

17. $\displaystyle\int x^3 \ln 2x\,dx$

18. $\displaystyle\int \frac{dx}{(1-x)^2}$

19. $\displaystyle\int \frac{\ln x}{x}\,dx$

20. $\displaystyle\int \frac{e^{2x}}{e^{2x}+1}\,dx$

21. $\displaystyle\int \frac{e^{\sqrt{x}}}{\sqrt{x}}\,dx$

22. $\displaystyle\int (e^{2x}+1)^3 e^{2x}\,dx$

23. BUSINESS: Present Value of a Continuous Stream of Income A company generates a continuous stream of income of $25t$ million dollars per year, where t is the number of years that the company has been in operation.

  a. Find the present value of this stream for the first 10 years at 5% interest compounded continuously. (Do not use a graphing calculator.)

  b. Verify your answer to part (a) using FnInt on a graphing calculator.

24. ENVIRONMENTAL SCIENCE: Pollution Radioactive waste is leaking out of cement storage vessels at the rate

of $te^{0.2t}$ hundred gallons per month, where t is the number of months since the leak began.

  a. Find the total leakage during the first 3 months. (Do not use a graphing calculator.)

  b. Verify your answer to part (a) using FnInt on a graphing calculator.

6.2 Integration Using Tables

25–36. Use the integral table on the inside back cover to find each integral.

㉕ $\displaystyle\int \frac{1}{25-x^2}\,dx$

26. $\displaystyle\int \frac{1}{x^2-4}\,dx$

㉗ $\displaystyle\int \frac{x}{(x-1)(x-2)}\,dx$

28. $\displaystyle\int \frac{1}{(x-1)(x-2)}\,dx$

29. $\displaystyle\int \frac{1}{x\sqrt{x+1}}\,dx$

30. $\displaystyle\int \frac{x}{\sqrt{x+1}}\,dx$

31. $\displaystyle\int \frac{1}{\sqrt{x^2+9}}\,dx$

32. $\displaystyle\int \frac{1}{\sqrt{x^2+16}}\,dx$

㉝ $\displaystyle\int \frac{z^3}{\sqrt{z^2+1}}\,dz$

34. $\displaystyle\int \frac{e^{2t}}{e^t+2}\,dt$

35. $\displaystyle\int x^2 e^{2x}\,dx$

36. $\displaystyle\int (\ln x)^4\,dx$

37. BUSINESS: Cost A company's marginal cost function is $MC(x) = \dfrac{1}{(2x+1)(x+1)}$ and fixed costs are 1000 (all in dollars). Find the company's cost function.

38. GENERAL: Population The population of a town is growing at the rate of $\sqrt{t^2+1600}$ people per year, where t is the number of years from now.

  a. Find the total increase in population during the first 30 years. (Do not use a graphing calculator.)

  b. Verify your answer to part (a) using FnInt on a graphing calculator.

6.3 Improper Integrals

39–56. Find the value of each improper integral or state that it is divergent.

㊴ $\displaystyle\int_1^\infty \frac{1}{x^5}\,dx$

40. $\displaystyle\int_1^\infty \frac{1}{x^6}\,dx$

㊶ $\displaystyle\int_1^\infty \frac{1}{\sqrt[5]{x}}\,dx$

42. $\displaystyle\int_1^\infty \frac{1}{\sqrt[6]{x}}\,dx$

43. $\displaystyle\int_0^\infty e^{-2x}\,dx$

44. $\displaystyle\int_4^\infty e^{-0.5x}\,dx$

45. $\displaystyle\int_0^\infty e^{2x}\,dx$

46. $\displaystyle\int_4^\infty e^{0.5x}\,dx$

47. $\displaystyle\int_0^\infty e^{-t/5}\,dt$

48. $\displaystyle\int_{100}^\infty e^{-t/10}\,dt$

49. $\displaystyle\int_0^\infty \frac{x^3}{(x^4+1)^2}\,dx$

50. $\displaystyle\int_0^\infty \frac{x^4}{(x^5+1)^2}\,dx$

51. $\displaystyle\int_{-\infty}^0 e^{2t}\,dt$

52. $\displaystyle\int_{-\infty}^0 e^{4t}\,dt$

53. $\displaystyle\int_{-\infty}^4 \frac{1}{(5-x)^2}\,dx$

54. $\displaystyle\int_{-\infty}^8 \frac{1}{(9-x)^2}\,dx$

55. $\displaystyle\int_{-\infty}^\infty \frac{e^{-x}}{(1+e^{-x})^4}\,dx$

56. $\displaystyle\int_{-\infty}^\infty \frac{e^{-x}}{(1+e^{-x})^3}\,dx$

57. **GENERAL: Permanent Endowments** Find the size of the permanent endowment needed to generate an annual $6000 forever at an interest rate of 10% compounded continuously.

58. **GENERAL: Automobile Age** Insurance records indicate that the proportion of cars on the road that are more than x years old is approximated by the integral $\int_x^\infty 0.21e^{-0.21t}\,dt$. Find the proportion of cars that are more than 5 years old.

59. **BUSINESS: Book Sales** A publisher estimates that the demand for a certain book will be $12e^{-0.05t}$ thousand copies per year, where t is the number of years since the book's publication. Find the total number of books that will be sold from the publication date onward.

60. **ENVIRONMENTAL SCIENCES: Consumption of Natural Resources** Because of environmental concerns and available substitutes, mercury usage has been declining over recent years. Annual U.S. consumption is estimated to be $270e^{-0.02t}$ metric tons of mercury per year, where t is the number of years since 2010. Assuming that this rate continues, find the total amount of mercury that will be used in the United States from 2010 on.

Source: U.S. Geological Survey

 61–62. For each improper integral, use a graphing calculator to evaluate it (or to show that it diverges) as follows:

a. Define y_1 to be the definite integral (using FnInt) of the integrand from 1 to x.

b. Make a TABLE of values of y_1 for x-values such as 1, 10, 100, and 1000. Does the integral converge (and if so, to what number) or does it diverge?

c. Verify your answers to part (b) by evaluating the improper integral "by hand."

61. $\displaystyle\int_1^\infty \frac{1}{x^3}\,dx$

62. $\displaystyle\int_1^\infty \frac{1}{\sqrt[3]{x}}\,dx$

6.4 Numerical Integration

63–68. Estimate each integral using trapezoidal approximation with the given value of n. (Round all calculations to three decimal places.)

63. $\displaystyle\int_0^1 \sqrt{1+x^4}\,dx,\ \ n=3$

64. $\displaystyle\int_0^1 \sqrt{1+x^5}\,dx,\ \ n=3$

65. $\displaystyle\int_0^1 e^{x^2/2}\,dx,\ \ n=4$

66. $\displaystyle\int_0^1 e^{-x^2/2}\,dx,\ \ n=4$

67. $\displaystyle\int_{-1}^1 \ln(1+x^2)\,dx,\ \ n=4$

68. $\displaystyle\int_{-1}^1 \ln(x^3+2)\,dx,\ \ n=4$

69–74. Use a graphing calculator *trapezoidal approximation* program from the Internet (See page 418) to approximate each integral. Use successively higher values of n until the results agree to three decimal places (rounded).

69. $\displaystyle\int_0^1 \sqrt{1+x^4}\,dx$

70. $\displaystyle\int_0^1 \sqrt{1+x^5}\,dx$

71. $\displaystyle\int_0^1 e^{x^2/2}\,dx$

72. $\displaystyle\int_0^1 e^{-x^2/2}\,dx$

73. $\displaystyle\int_{-1}^1 \ln(1+x^2)\,dx$

74. $\displaystyle\int_{-1}^1 \ln(x^3+2)\,dx$

75–80. Estimate each integral using *Simpson's Rule* with the given value of n. (Round all calculations to four decimal places.)

75. $\displaystyle\int_0^1 \sqrt{1+x^4}\,dx,\ \ n=4$

76. $\displaystyle\int_0^1 \sqrt{1+x^5}\,dx,\ \ n=4$

77. $\displaystyle\int_0^1 e^{x^2/2}\,dx,\ \ n=4$

78. $\displaystyle\int_0^1 e^{-x^2/2}\,dx,\ \ n=4$

79. $\displaystyle\int_{-1}^1 \ln(1+x^2)\,dx,\ \ n=4$

80. $\displaystyle\int_{-1}^1 \ln(x^3+2)\,dx,\ \ n=4$

81–86. Use a graphing calculator *Simpson's Rule* approximation program from the Internet (see page 421) to approximate each integral. Use successively higher values of *n* until the rounded results agree to six decimal places.

81. $\displaystyle\int_0^1 \sqrt{1 + x^4}\, dx$

82. $\displaystyle\int_0^1 \sqrt{1 + x^5}\, dx$

83. $\displaystyle\int_0^1 e^{x^2/2}\, dx$

84. $\displaystyle\int_0^1 e^{-x^2/2}\, dx$

85. $\displaystyle\int_{-1}^1 \ln(1 + x^2)\, dx$

86. $\displaystyle\int_{-1}^1 \ln(x^3 + 2)\, dx$

87–88. For each improper integral:

a. Make it a "proper" integral by using the substitution $x = \dfrac{1}{t}$ and simplifying.

b. Approximate the proper integral using trapezoidal approximation with $n = 4$. Keep three decimal places.

87. $\displaystyle\int_1^\infty \frac{1}{x^2 + 1}\, dx$

88. $\displaystyle\int_1^\infty \frac{x^2}{x^4 + 1}\, dx$

Calculus of Several Variables

7

John Howard/Photodisc/Getty Images

What You'll Explore

In this chapter you will apply calculus to functions of two or more variables. For example, a company's output depends on the available capital and labor, and its profit depends on the quantities of the products that it produces. You will use derivatives to find marginal profits (now there will be more than one) and to maximize functions, and integrals to find volumes and to sum continuous changes, now in several variables. Along the way, you will derive the method of least squares, the most widely used method for fitting lines and curves to data.

APPLICATION PREVIEW

Gender Pay Gap

Women's wages have always lagged behind men's. The reasons for this are not entirely clear (possibly discrimination or different career choices made by men and women), but what *is* clear is that women's wages have been steadily catching up. In 1980 women working full-time made only 64% of men's salaries, but by 2010 that percentage had risen to 81%. Can we estimate when in the future men's and women's salaries might actually be equal? The mathematical technique of *linear regression* or *least squares,* used extensively in Chapter 1, fits a line to data points. The following graph shows the least squares lines for men's wages and for women's wages.

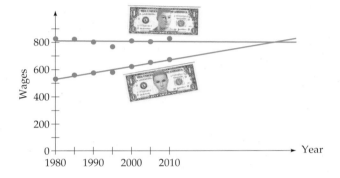

Year	Men's Salaries	Women's Salaries
1980	828	532
1985	824	561
1990	803	577
1995	769	581
2000	812	624
2005	806	654
2010	829	675

Median Weekly Earnings of Full-Time and Salary Workers Ages 15 and Older, in 2010 Dollars

Source: U.S. Bureau of Labor Statistics

By finding the intersection point of these lines, we may predict that men's and women's salaries will be approximately equal in mid-2037. While in Chapter 1 you may have carried out linear regression by pressing buttons on a graphing calculator, in this chapter we will see how the least squares method *works* and that it is based on the same calculus techniques that we used for optimization in Chapter 3.

Introduction

Many quantities depend on *several* variables. For example, the "windchill" factor announced by the weather bureau during the winter depends on two variables: temperature and wind speed. The cost of a loan depends on *three* variables: amount, duration, and interest rate.

In this chapter we define functions of two or more variables and learn how to differentiate and integrate them. We use derivatives for calculating rates of change and optimizing functions, and integrals for finding volumes, continuous sums, and average values.

Graphing calculators will be less useful in this chapter because of their small screens and limited computing power, but computer-drawn pictures of three-dimensional surfaces will be very useful.

Functions of Two Variables

A function f that depends on *two* variables, x and y, is written $f(x, y)$ (read: f of x and y). The **domain** of the function is the set of all ordered pairs (x, y) for which it is defined. The **range** is the set of all resulting values of the function. Formally:

LOOKING BACK

Note the similarity of this to the definition of a function of *one* variable on page 33.

Function of Two Variables

A function f of two variables is a rule that assigns to each ordered pair (x, y) in a set exactly one number $f(x, y)$.

The set of allowable ordered pairs (x, y) is called the *domain*.
The set of all values $f(x, y)$ for (x, y) in the domain is called the *range*.

If the domain is not stated, it will always be taken to be the largest set of ordered pairs for which the function is defined (the **natural domain**).

EXAMPLE 1   FINDING THE DOMAIN OF A FUNCTION

For $f(x, y) = \dfrac{\sqrt{x}}{y^2}$ find

a. the domain

b. $f(9, -1)$

Solution

a. $\{(x, y) \mid x \geq 0, y \neq 0\}$

> For $\sqrt{x}/y^2$, x cannot be negative (because of the $\sqrt{\ }$), and y cannot be zero

b. $f(9, -1) = \dfrac{\sqrt{9}}{(-1)^2} = \dfrac{3}{1} = 3$

> $f(x, y) = \sqrt{x}/y^2$
> with $x = 9$ and $y = -1$

EXAMPLE 2	FINDING THE DOMAIN OF A FUNCTION INVOLVING LOGARITHMS AND EXPONENTIALS

For $g(u, v) = e^{uv} - \ln u$, find   **a.** the domain   **b.** $g(1, 2)$

Solution

a. $\{(u, v) \mid u > 0\}$

> u must be positive so that its logarithm is defined

b. $g(1, 2) = e^{1 \cdot 2} - \underbrace{\ln 1}_{0} = e^2 - 0 = e^2$

> $g(u, v) = e^{uv} - \ln u$
> with $u = 1$ and $v = 2$

PRACTICE PROBLEM 1

For $f(x, y) = \dfrac{\ln x}{e^{\sqrt{y}}}$, find   **a.** the domain   **b.** $f(e, 4)$

Solutions on page 440 >

Functions of two variables are used in many applications.

EXAMPLE 3	FINDING A COMPANY'S COST FUNCTION

A company manufactures three-speed and ten-speed bicycles. It costs $100 to make each three-speed bicycle, it costs $150 to make each ten-speed bicycle, and fixed costs are $2500. Find the cost function, and use it to find the cost of producing 15 three-speed bicycles and 20 ten-speed bicycles.

Solution

Let

$$x = \text{the number of three-speed bicycles}$$
$$y = \text{the number of ten-speed bicycles}$$

The cost function is

$$C(x, y) = 100x + 150y + 2500$$

Unit cost   Quantity   Unit cost   Quantity   Fixed costs

The cost of producing 15 three-speed bicycles and 20 ten-speed bicycles is found by evaluating $C(x, y)$ at $x = 15$ and $y = 20$:

$$C(15, 20) = 100 \cdot 15 + 150 \cdot 20 + 2500$$
$$= 1500 + 3000 + 2500 = 7000$$

Producing 15 three-speed and 20 ten-speed bicycles costs $7000.

The variables x and y in the preceding Example stand for numbers of bicycles and so should take only integer values. Instead, however, we will allow x and y to be "continuous" variables, and round to integers at the end if necessary.

Some other "everyday" examples of functions of two variables are:

$$A(l, w) = lw$$
Area of a rectangle of length l and width w

$$f(w, v) = kwv^2$$
Length of the skid marks for a car of weight w and velocity v skidding to a stop (k is a constant depending on the road surface)

Cobb–Douglas Production Functions

A function used to model the output of a company or a nation is called a **production function,** and the most famous is the Cobb–Douglas production function*

$$P(L, K) = aL^bK^{1-b}$$
For constants $a > 0$ and $0 < b < 1$

This function expresses the total production P as a function of L, the number of units of labor, and K, the number of units of capital. (Labor is measured in work-hours, and capital means *invested* capital, including the cost of buildings, equipment, and raw materials.)

| EXAMPLE 4 | EVALUATING A COBB–DOUGLAS PRODUCTION FUNCTION |

Cobb and Douglas modeled the output of the American economy by the function $P(L, K) = 1.01L^{0.75}K^{0.25}$. Find $P(150, 220)$.

Solution

$$P(150, 220) = 1.01(150)^{0.75}(220)^{0.25}$$
$P(L, K) = 1.01L^{0.75}K^{0.25}$ with $L = 150$ and $K = 220$

$$\approx 167$$
Using a calculator

That is, 150 units of labor and 220 units of capital should result in approximately 167 units of production.

Source: American Economic Review **18**

Graphing Calculator Exploration

The windchill index announced by the weather bureau during the winter to measure the combined effect of wind and cold is calculated from the formula below, where x is wind speed (miles per hour) and y is temperature (degrees Fahrenheit).

$$W(x, y) = 35.74 + 0.6215y - 35.75x^{0.16} + 0.4275yx^{0.16}$$

a. Enter this function into your graphing calculator but with y (temperature) replaced by 32 so that it becomes a function of just one variable, x (wind speed). Then graph the function on the window [0, 45] by [0, 40]. The graph shows how the perceived temperature decreases as wind speed increases.

b. Evaluate this function at $x = 15$ to find the perceived temperature resulting from a 15-mph wind when the actual temperature is 32 degrees.

---

*First used by Charles Cobb (1875–1949) and Paul Douglas (1892–1976) in a landmark study of the American economy published in 1928.

c. Notice that the graph drops more steeply for low wind speeds than for high wind speeds. What does this mean about the effect of an extra 5 miles per hour of wind on a calm day as opposed to a windy day? (Exercises 38 and 39 continue this analysis.)

Source: National Weather Service

Functions of Three or More Variables

Functions of three (or more) variables are defined analogously. Some examples are:

$$V(l, w, h) = lwh$$
Volume of a rectangular solid of length l, width w, and height h

$$W(P, r, t) = Pe^{rt}$$
Worth of P dollars invested at a continuous interest rate r for t years

$$f(w, x, y, z) = \frac{w + x + y + z}{4}$$
Average of four numbers

EXAMPLE 5   FINDING THE DOMAIN OF A FUNCTION OF THREE VARIABLES

For $f(x, y, z) = \dfrac{\sqrt{x}}{y} + \ln \dfrac{1}{z}$, find

a. the domain

b. $f(4, -1, 1)$

Solution

a. In $f(x, y, z) = \dfrac{\sqrt{x}}{y} + \ln \dfrac{1}{z}$ we must have $x \geq 0$ (because of the square root), $y \neq 0$ (since it is a denominator), and $z > 0$ (so that $1/z$ has a logarithm). Therefore, the domain is

$$\{(x, y, z) \mid x \geq 0, y \neq 0, z > 0\}$$

b. $f(4, -1, 1) = \dfrac{\sqrt{4}}{-1} + \ln \dfrac{1}{1} = \dfrac{2}{-1} + \underbrace{\ln 1}_{0} = -2$

EXAMPLE 6   FINDING THE VOLUME AND AREA OF A DIVIDED BOX

An open-top box is to have a center divider, as shown in the diagram. Find formulas for the volume V of the box and for the total amount of material M needed to construct the box.

Solution

The volume is length times width times height.

$$V = xyz$$

The box consists of a bottom, a front and back, two sides, and a divider, whose areas are shown in the diagram. Therefore, the total amount of material (the area) is

$$M = xy + 2xz + 3yz$$

$$\underset{\text{Bottom}}{\diagup} \quad \underset{\text{Back}}{\mid} \quad \underset{\text{Sides and}}{\diagdown}$$
$$\text{and front} \quad \text{divider}$$

bottom area xy

side area yz (3 of these)

front area xz (2 of these)

PRACTICE PROBLEM 2

Find a formula for the total amount of material M needed to construct an open-top box with three parallel dividers. Use the variables shown in the diagram.

Solution on page 440 >

Graph of a Function of Two Variables

Graphing a function of two variables requires a **three-dimensional coordinate system.** We draw three perpendicular axes as shown on the right.* We will usually draw only the positive half of each axis, although each axis extends infinitely far in the negative direction as well. The plane at the base is called the *x-y plane.*

A point in a three-dimensional coordinate system is specified by three coordinates, giving its distances from the origin in the x, y, and z directions. For example, the point

The three-dimensional ("right-handed") coordinate system

$$(2, 3, 4)$$

z-coordinate
y-coordinate
x-coordinate

The point (2, 3, 4)

*This is called a "right-handed" coordinate system because the x, y, and z axes correspond to the first two fingers and thumb of the right hand.

is plotted by starting at the origin, moving 2 units in the x direction, 3 units in the y direction, and then 4 units in the (vertical) z direction.

To graph a function $f(x, y)$ of two variables, we choose values for x and y, let z stand for the *function* values $z = f(x, y)$, and plot the points (x, y, z).

EXAMPLE 7   GRAPHING A FUNCTION OF TWO VARIABLES

To graph $f(x, y) = 18 - x^2 - y^2$, we set z equal to the function.

$$z = 18 - x^2 - y^2 \qquad\qquad z \text{ replaces } f(x, y)$$

Then we choose values for x and y. Choosing $x = 1$ and $y = 2$ gives

$$z = 18 - 1^2 - 2^2 = 13 \qquad \begin{array}{l} z = 18 - x^2 - y^2 \text{ with} \\ x = 1 \text{ and } y = 2 \end{array}$$

for the point

$$(1, 2, 13) \qquad\qquad \begin{array}{l}\text{The chosen } x = 1, \ y = 2, \\ \text{and the calculated } z \end{array}$$

Choosing $x = 2$ and $y = 3$ gives

$$z = 18 - 2^2 - 3^2 = 5 \qquad \begin{array}{l} z = 18 - x^2 - y^2 \text{ with} \\ x = 2 \text{ and } y = 3 \end{array}$$

for the point

$$(2, 3, 5) \qquad\qquad \begin{array}{l}\text{The chosen } x = 2, \ y = 3, \\ \text{and the calculated } z \end{array}$$

These points $(1, 2, 13)$ and $(2, 3, 5)$ are plotted on the graph on the left below. The completed graph of the function is shown on the right.

The points $(1, 2, 13)$ and $(2, 3, 5)$ of the function $f(x, y) = 18 - x^2 - y^2$

The graph of $f(x, y) = 18 - x^2 - y^2$

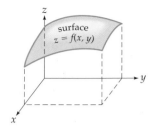

The graph of a function of two variables is a surface whose height above the point (x, y) in the x-y plane is $z = f(x, y)$.

In general, the graph of a function of *two* variables is a *surface* above or below the x-y plane, just as the graph of a function of *one* variable is a *curve* above or below the x-axis.

Graphing functions of two variables involves drawing three-dimensional graphs, which is very difficult. Graphing functions of *more* than two variables requires *more* than three dimensions and is impossible. For this reason we will not graph functions of several variables. We will, however, often speak of a function of two variables as representing a *surface* in three-dimensional space.

Spreadsheet Exploration

The following spreadsheet* graph of $f(x, y) = 18 - x^2 - y^2$ from the previous Example is a chart showing the values of the function that were calculated for values of x and y between -5 and 5.

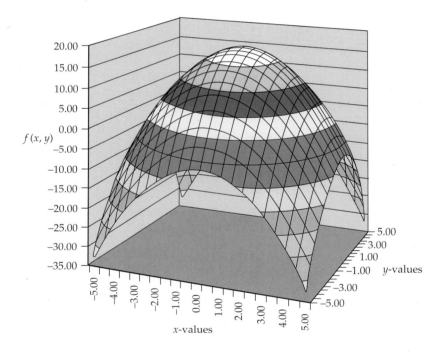

*To obtain this and other Spreadsheet Explorations, go to www.cengagebrain.com, search for this textbook and then scroll down to access the "free materials" tab.

Just as before, useful graphs can be constructed based on a few important points. For functions of one variable, these were maximum, minimum, and inflection points. For functions of two variables, these will be maximum and minimum points and a new type, *saddle points*.

Relative Extreme Points and Saddle Points

Certain points on a surface of such a graph are of special importance.

Relative Maximum Point

A point (a, b, c) on a surface $z = f(x, y)$ is a *relative maximum point* if $f(a, b) \geq f(x, y)$ for all (x, y) in some region surrounding (a, b).

Relative Minimum Point

A point (a, b, c) on a surface $z = f(x, y)$ is a *relative minimum point* if $f(a, b) \leq f(x, y)$ for all (x, y) in some region surrounding (a, b).

As before, the term **relative extreme point** means a point that is either a relative maximum or a relative minimum point. A surface can have any number of relative extreme points, even none.

A surface with two relative extreme points: one relative maximum and one relative minimum.

The point shown on the left is called a **saddle point** (so named because the diagram resembles a saddle).

A saddle point is a point that is the highest point along one curve of the surface and the lowest point along another curve. A saddle point is *not* a relative extreme point.

If we think of a surface $z = f(x, y)$ as a landscape, then relative maximum and minimum points correspond to "hilltops" and "valley bottoms," and a saddle point corresponds to a "mountain pass" between two peaks. The first three graphs below show points of each of these types.

Gallery of Surfaces

The following are the graphs of a few functions of two variables.

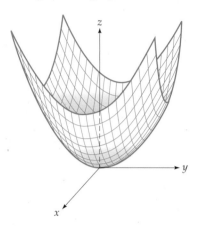

The surface $f(x, y) = x^2 + y^2$ has a relative minimum point at the origin.

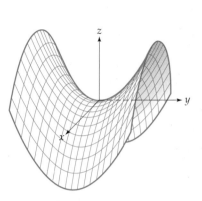

The surface $f(x, y) = y^2 - x^2$ has a saddle point at the origin.

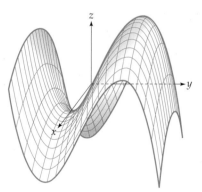

The surface $f(x, y) = 12y + 6x - x^2 - y^3$ has a saddle point and a relative maximum point.

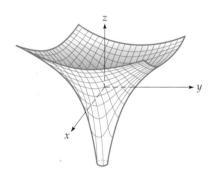

The surface $f(x, y) = \ln(x^2 + y^2)$ has no relative extreme points. It is undefined at $(0, 0)$.

Solutions TO PRACTICE PROBLEMS

1. a. $\{(x, y) \mid x > 0, y \geq 0\}$

   b. $f(e, 4) = \dfrac{\ln e}{e^{\sqrt{4}}} = \dfrac{1}{e^2} = e^{-2}$

2. $M = xy + 2xz + 5yz$

7.1 Section Summary

Just as for a function of one variable, a *function of several variables* gives exactly one value for each point in its domain. For a function of *two* variables, denoted $f(x, y)$, the values $z = f(x, y)$ determine a *surface* above or below the *x-y* plane. The surface may have *relative maximum points* and *relative minimum points* ("hilltops" and "valley bottoms") or even *saddle points* (high points along one curve and low points along another), as shown on the previous pages.

7.1 Exercises

1–8. For each function, find the domain.

1. $f(x, y) = \dfrac{1}{xy}$

2. $f(x, y) = \dfrac{\sqrt{x}}{\sqrt{y}}$

3. $f(x, y) = \dfrac{1}{x - y}$

4. $f(x, y) = \dfrac{\sqrt[3]{x}}{\sqrt[3]{y}}$

5. $f(x, y) = \dfrac{\ln x}{y}$

6. $f(x, y) = \dfrac{x}{\ln y}$

7. $f(x, y, z) = \dfrac{e^{1/y} \ln z}{x}$

8. $f(x, y, z) = \dfrac{\sqrt{x} \ln y}{z}$

> ✏ **FOR HELP GETTING STARTED**
> on Exercises 1–8, see Examples 1 and 2 on pages 432–433.

9–22. For each function, evaluate the given expression.

9. $f(x, y) = \sqrt{99 - x^2 - y^2}$, find $f(3, -9)$

10. $f(x, y) = \sqrt{75 - x^2 - y^2}$, find $f(5, -1)$

11. $g(x, y) = \ln(x^2 + y^4)$, find $g(0, e)$

12. $g(x, y) = \ln(x^3 - y^2)$, find $g(e, 0)$

13. $w(u, v) = \dfrac{1 + 2u + 3v}{uv}$, find $w(-1, 1)$

14. $w(u, v) = \dfrac{2u + 4u}{v - u}$, find $w(1, -1)$

15. $h(x, y) = e^{xy + y^2 - 2}$, find $h(1, -2)$

16. $h(x, y) = e^{x^2 - xy - 4}$, find $h(1, -2)$

17. $f(x, y) = xe^y - ye^x$, find $f(1, -1)$

18. $f(x, y) = xe^y + ye^x$, find $f(-1, 1)$

19. $f(x, y, z) = xe^y + ye^z + ze^x$, find $f(1, -1, 1)$

20. $f(x, y, z) = xe^y + ye^z + ze^x$, find $f(-1, 1, -1)$

21. $f(x, y, z) = z \ln \sqrt{xy}$, find $f(-1, -1, 5)$

22. $f(x, y, z) = z \sqrt{x} \ln y$, find $f(4, e, -1)$

Applied Exercises

23. PERSONAL FINANCE: Stock Yield The *yield* of a stock is defined as $Y(d, p) = \dfrac{d}{p}$ where d is the dividend per share and p is the price of a share of stock. Find the yield of a stock that sells for \$140 and offers a dividend of \$2.20.

24. PERSONAL FINANCE: Price-Earnings Ratio The price-earnings ratio of a stock is defined as

$R(P, E) = \dfrac{P}{E}$ where P is the price of a share of stock and E is its earnings. Find the price-earnings ratio of a stock that is selling for \$140 with earnings of \$1.70.

25. GENERAL: Scuba Diving The maximum duration of a scuba dive (in minutes) can be estimated from the formula

$$T(v, d) = \dfrac{33v}{d + 33}$$

where v is the volume of air (at sea-level pressure) in the tank and d is the depth of the dive. Find $T(90, 33)$.

Source: U.S. Navy Diving Manual, Vol. 2

26. BIOMEDICAL: Body Area The surface area (in square feet) of a person of weight w pounds and height h feet is approximated by the function $A(w, h) = 0.55 \, w^{0.425} h^{0.725}$. Use this function to estimate the surface area of a person who weighs 160 pounds and who is 6 feet tall. (Such estimates are important in certain medical procedures.)

Source: Archives of Internal Medicine 17

27. ECONOMICS: Cobb–Douglas Functions A company's production is estimated to be $P(L, K) = 2L^{0.6} K^{0.4}$. Find $P(320, 150)$.

Source: Journal of Political Economy 84

28. ECONOMICS: Cobb–Douglas Functions At one time the production of American manufacturing was estimated to be $P = 2.39L^{0.76}K^{0.24}$. Find $P(2500, 450)$.

Source: American Economic Review **31**

29–30. ECONOMICS: Isoquant Curves An *isoquant curve* (*iso* means "same" and *quant* is short for "quantity") is a "level curve" of a Cobb–Douglas function having a constant production level. Such a curve is also called an *indifference curve* because the production level is the same for every (L, K) pair of labor and capital values on the curve, so the consumers should be indifferent to the labor–capital choice of the manufacturer.

For each Cobb–Douglas production function and production level, verify that the given (L, K) pairs all result in the same production level.

29. $P = 1.5L^{1/2}K^{1/2}$
  $(400, 8100)$, $(900, 3600)$, $(3600, 900)$, $(8100, 400)$

30. $P = 1.25L^{1/3}K^{2/3}$
  $(30, 240)$, $(270, 80)$, $(750, 48)$, $(4320, 20)$

31. ECONOMICS: Cobb–Douglas Functions
Show that the Cobb–Douglas production function $P(L, K) = aL^bK^{1-b}$ satisfies the equation $P(2L, 2K) = 2 \cdot P(L, K)$. This shows that doubling the amounts of labor and capital doubles production, a property called *returns to scale.*

32. ECONOMICS: Cobb–Douglas Functions Show that the Cobb–Douglas function $P(L, K) = aL^bK^{1-b}$ with $0 < b < 1$ satisfies

  $$P(2L, K) < 2P(L, K) \quad \text{and} \quad P(L, 2K) < 2P(L, K)$$

This shows that doubling the amounts of either labor or capital alone results in *less* than double production, a property called *diminishing returns.*

33. GENERAL: Telephone Calls For two cities with populations x and y that are d miles apart, the number of telephone calls per hour between them can be estimated by the function of three variables

  $$f(x, y, d) = \frac{3xy}{d^2}$$

(This is called the *gravity model.*) Use the gravity model to estimate the number of calls between two cities of populations 40,000 and 60,000 that are 600 miles apart.

Source: about.com

34. ENVIRONMENTAL SCIENCE: Tag and Recapture Estimates Ecologists estimate the size of animal populations by capturing and tagging a few animals, and then releasing them. After the first group has mixed with the population, a second group of animals is captured, and the number of tagged animals in this group is counted. If originally T animals were tagged, and the second group is of size S and contains t tagged animals, then the population is estimated by the function of three variables

  $$P(T, S, t) = \frac{TS}{t}$$

Estimate the size of a deer population if 100 deer were tagged, and then a second group of 250 contained 20 tagged deer.

Source: J. Blower, L. Cook, and J. Bishop, Estimating the Size of Animal Populations

35. BUSINESS: Cost Function It costs an appliance company $210 to manufacture each washer and $180 to manufacture each dryer, and fixed costs are $4000. Find the company's cost function $C(x, y)$, using x and y for the numbers of washers and dryers, respectively.

36–37. GENERAL: Box Design For each open-top box shown below, find formulas for:

a. the volume.
b. the total amount of material (the area).

36.

37.

38. GENERAL: Windchill Enter the formula for windchill from the Graphing Calculator Exploration on page 434 and again replace y (temperature) by 32 so that it becomes a function of one variable, x (wind speed).

a. Graph the function on the window $[0, 45]$ by $[0, 40]$.
b. Use NDERIV or dy/dx to find the slope of this curve at $x = 5$. Interpret the answer.
c. Repeat part (b) but for $x = 20$, interpreting the answer.
d. What do the answers indicate about the windchill effect of additional wind on a calm day as opposed to on an already windy day?

Source: National Weather Service

39. GENERAL: Windchill Enter the formula for windchill index from the Graphing Calculator Exploration on page 434, but with y (temperature) replaced by *several* temperatures: 20, 30, 40, and 50 degrees (using y_1, y_2, y_3, and y_4).

a. Graph the functions on the window $[0, 45]$ by $[-10, 60]$.
b. Notice that the lower temperature curves slope downward more steeply than the others. What does this mean about the effect of wind on a colder day?   *(continues)*

c. Use NDERIV or dy/dx to find the slope of the lowest and the highest curves at $x = 10$. Interpret the answers.

d. Do your answers to part (c) support your conclusion in part (b)?

Source: National Weather Service

40. BIOMEDICAL: Oxygen Consumption The oxygen consumption of a well-insulated nonsweating mammal can be estimated from the formula $f(t_b, t_a, w) = 2.5(t_b - t_a)w^{-0.67}$, where t_b is the animal's body temperature, t_a is the air temperature (both in degrees Celsius), and w is the animal's weight (in kilograms). Find the oxygen consumption of a 40-kilogram animal whose body temperature is 35 degrees when the air temperature is 5 degrees.

Source: D. J. Clow and N. S. Urquhart, Mathematics in Biology

41. ATHLETICS: Pythagorean Baseball Standings Many professional baseball teams (including the Cincinnati Reds and the Boston Red Sox) use Bill James's formula to estimate their probability of winning a league pennant:

$$\binom{\text{Probability}}{\text{of winning}} = \frac{(\text{Runs scored})^2}{(\text{Runs scored})^2 + (\text{Runs allowed})^2}$$

This formula, whose form is reminiscent of the Pythagorean theorem, is considered more accurate than just the proportion of games won because it takes into consideration the scores of the games. Find this probability for a team that has scored 400 runs and allowed 300 runs.

Source: B. James, Baseball Abstract 1983

42. ATHLETICS: Olympics A country's population and wealth certainly contribute to its success in the Olympics. The following formula, based on the country's population p and per capita gross domestic product d, has proved accurate in predicting the proportion of Olympic medals that a country will win:

$$\binom{\text{Proportion}}{\text{of medals}} = 0.0062 \ln p + 0.0064 \ln d - 0.0652$$

Estimate the proportion of Olympic medals that the United States will win based on a population of 308,746,000 and a per capita gross domestic product of \$47,123.

Source: Review of Economics and Statistics 86

Conceptual Exercises

43. Which is a *right-handed* coordinate system?

a.
b.

44. Fill in the blanks with the correct numbers: To graph a function $f(x)$ of *one* variable we need ____ axes, and to graph a function $f(x, y)$ of *two* variables we need ____ axes.

45–46. The surface graphed below has:

45. How many relative extreme points?

46. How many saddle points?

47. Which function has a relative maximum point and which has a relative minimum point?
  a. $f(x, y) = x^2 + y^2 - 1$ b. $g(x, y) = 1 - x^2 - y^2$

48. True or False: Every function $f(x, y)$ of two variables can be written as the sum of two functions of one variable, $g(x) + h(y)$.

49. True or False: A function f of two variables is a rule such that to each number $f(x, y)$ in the domain there corresponds one and only one ordered pair (x, y).

50–54. For each function, state whether it satisfies:
  a. $f(-x, -y) = f(x, y)$ for all x and y,
  b. $f(-x, -y) = -f(x, y)$ for all x and y, or
  c. neither of these conditions.

50. $f(x, y) = x - y$ **51.** $f(x, y) = x^2 - y^2$

52. $f(x, y) = x^2 - y^3$ **53.** $f(x, y) = x^2 y^3$

54. $f(x, y) = x^2 y^4$

7.2   # Partial Derivatives

Introduction

Functions of several variables have *several* derivatives, called **partial derivatives**, one for each variable. In this section you will learn how to calculate and interpret these derivatives.

Partial Derivatives

Before defining partial derivatives, we review the rules governing derivatives and constants.

$$\frac{d}{dx} c = 0$$

For a constant *standing alone*, the derivative is zero.

$$\frac{d}{dx}(cx^3) = c \cdot 3x^2$$

For a constant *multiplying* a function, the constant is carried along.

Carry along the constant · Derivative of x^3

LOOKING BACK

This follows from the Constant Multiple Rule (page 101).

These rules will be very useful in this section.

A function $f(x, y)$ has two *partial derivatives*, one with respect to x and the other with respect to y.

Partial Derivatives

$$\frac{\partial}{\partial x} f(x, y) = \lim_{h \to 0} \frac{f(x + h, y) - f(x, y)}{h}$$

Partial derivative of f with respect to x (x is changed by h; y is held constant)

$$\frac{\partial}{\partial y} f(x, y) = \lim_{h \to 0} \frac{f(x, y + h) - f(x, y)}{h}$$

Partial derivative of f with respect to y (y is changed by h; x is held constant)

(provided that the limits exist).

Partial derivatives are written with a "curly" ∂, $\partial/\partial x$ instead of d/dx, and are often called "partials." (The Greek letter ∂ is a lowercase delta, equivalent to our letter d, and we have already used the capital delta, Δ, to denote change.)

Rewriting the first formula, but without the y's, gives

$$\lim_{h \to 0} \frac{f(x + h) - f(x)}{h}$$

$$\lim_{h \to 0} \frac{f(x + h, y) - f(x, y)}{h}$$
but omitting the y's

LOOKING AHEAD

The word *partial* is used to indicate that only one variable is changing. On page 494 we will define the *total differential*, which includes *both* partials.

which is just the "ordinary derivative" from page 92. Therefore, the partial derivative with respect to x is just the ordinary derivative with respect to x with y held constant. Similarly, the partial with respect to y is just the ordinary derivative with respect to y, but now with x held constant.

$$\frac{\partial}{\partial x} f(x, y) = \left(\begin{array}{l}\text{Derivative of } f \text{ with respect}\\ \text{to } x \text{ with } y \text{ held constant}\end{array}\right)$$

$$\frac{\partial}{\partial y} f(x, y) = \left(\begin{array}{l}\text{Derivative of } f \text{ with respect}\\ \text{to } y \text{ with } x \text{ held constant}\end{array}\right)$$

EXAMPLE 1   FINDING A PARTIAL DERIVATIVE WITH RESPECT TO X

Find $\dfrac{\partial}{\partial x} x^3 y^4$.

Solution

The $\partial/\partial x$ means differentiate with respect to x, holding y (and therefore y^4) constant. We therefore differentiate the x^3 and carry along the "constant" y^4:

Derivative of x^3

$$\frac{\partial}{\partial x} x^3 y^4 = 3x^2 y^4$$

Carry along the "constant" y^4

$\partial/\partial x$ means differentiate with respect to x, treating y (and therefore y^4) as a constant

EXAMPLE 2   FINDING A PARTIAL WITH RESPECT TO Y

Find $\dfrac{\partial}{\partial y} x^3 y^4$.

Solution

$$\frac{\partial}{\partial y} x^3 y^4 = x^3 4y^3 \qquad = \qquad 4x^3 y^3$$

Carry along the "constant" x^3 — Derivative of y^4 — Writing the 4 first

$\partial/\partial y$ means differentiate with respect to y, treating x (and therefore x^3) as a constant

PRACTICE PROBLEM 1

Find **a.** $\dfrac{\partial}{\partial x} x^4 y^2$ **b.** $\dfrac{\partial}{\partial y} x^4 y^2$

Solutions on page 453 >

EXAMPLE 3   FINDING A PARTIAL DERIVATIVE

Find $\dfrac{\partial}{\partial x} y^4$.

Solution

$$\frac{\partial}{\partial x} y^4 = 0 \qquad \text{The derivative of a constant is zero}$$
$$\text{(since } \partial/\partial x \text{ means hold } y \text{ constant)}$$

Partial with respect to x ⟶   ⟵ Function of y alone

PRACTICE PROBLEM 2

Find $\dfrac{\partial}{\partial y} x^2$.

Solution on page 453 >

EXAMPLE 4   **FINDING A PARTIAL OF A POLYNOMIAL IN TWO VARIABLES**

Find $\dfrac{\partial}{\partial x}(2x^4 - 3x^3 y^3 - y^2 + 4x + 1)$.

Solution

$$\frac{\partial}{\partial x}(2x^4 - 3x^3 y^3 - y^2 + 4x + 1) = 8x^3 - 9x^2 y^3 + 4$$

Differentiating with respect to x, so each y is held constant

$$\frac{\partial}{\partial x} y^2 = 0$$

PRACTICE PROBLEM 3

Find $\dfrac{\partial}{\partial y}(2x^4 - 3x^3 y^3 - y^2 + 4x + 1)$.

Solution on page 453 >

Subscript Notation for Partial Derivatives

Partial derivatives are often denoted by subscripts: a subscript x means the partial with respect to x, and a subscript y means the partial with respect to y.*

Subscript Notation		*∂ Notation*	
$f_x(x, y)$	$=$	$\dfrac{\partial}{\partial x} f(x, y)$	f_x means the partial of f with respect to x
$f_y(x, y)$	$=$	$\dfrac{\partial}{\partial y} f(x, y)$	f_y means the partial of f with respect to y

---

*Sometimes subscripts 1 and 2 are used to indicate partial derivatives with respect to the first and second variables: $f_1(x, y)$ means $f_x(x, y)$ and $f_2(x, y)$ means $f_y(x, y)$. We will not use this notation in this book.

| EXAMPLE 5 | USING SUBSCRIPT NOTATION |

Find $f_x(x, y)$ if $f(x, y) = 5x^4 - 2x^2y^3 - 4y^2$.

Solution

$$f_x(x, y) = 20x^3 - 4xy^3$$

Differentiating with respect to x, holding y constant

| EXAMPLE 6 | FINDING A PARTIAL INVOLVING LOGS AND EXPONENTIALS |

Find both partials of $f = e^x \ln y$.

Solution

$$f_x = e^x \ln y$$

The derivative of e^x is e^x (times the "constant" $\ln y$)

$$f_y = e^x \frac{1}{y}$$

The derivative of $\ln y$ is $\frac{1}{y}$ (times the "constant" e^x)

| EXAMPLE 7 | FINDING A PARTIAL OF A FUNCTION TO A POWER |

Find f_y if $f = (xy^2 + 1)^4$.

Solution

$$f_y = 4(xy^2 + 1)^3(x2y)$$

Partial of the inside with respect to y

Using the Generalized Power Rule (the derivative of f^n is $nf^{n-1}f'$, but with f' meaning a *partial*)

$$= 8xy(xy^2 + 1)^3$$

Simplifying

| EXAMPLE 8 | FINDING A PARTIAL OF A QUOTIENT |

Find $\dfrac{\partial g}{\partial x}$ if $g = \dfrac{xy}{x^2 + y^2}$.

Solution

Partial of the top with respect to x
Partial of the bottom with respect to x

$$\frac{\partial g}{\partial x} = \frac{(x^2 + y^2)y - 2x \cdot xy}{(x^2 + y^2)^2}$$

Using the Quotient Rule

Bottom squared

$$= \frac{x^2y + y^3 - 2x^2y}{(x^2 + y^2)^2} = \frac{y^3 - x^2y}{(x^2 + y^2)^2}$$

Simplifying

EXAMPLE 9	FINDING A PARTIAL OF THE LOGARITHM OF A FUNCTION

Find $f_x(x, y)$ if $f(x, y) = \ln(x^2 + y^2)$.

Solution

$$f_x(x, y) = \frac{2x}{x^2 + y^2}$$

    Partial of the bottom with respect to x

Derivative of $\ln f$ is $\dfrac{f'}{f}$

An expression such as $f_x(2, 5)$, which involves both differentiation and evaluation, means *first differentiate and then evaluate.**

EXAMPLE 10	EVALUATING A PARTIAL DERIVATIVE

Find $f_y(1, 3)$ if $f(x, y) = e^{x^2 + y^2}$.

Solution

$$f_y(x, y) = e^{x^2 + y^2}(2y)$$

    Partial of the exponent with respect to y

Derivative of e^f is $e^f \cdot f'$

$$f_y(1, 3) = e^{1^2 + 3^2}(2 \cdot 3)$$

$$= 6e^{10}$$

Evaluating at $x = 1$ and $y = 3$

Simplifying

PRACTICE PROBLEM 4

Find $f_y(1, 2)$ if $f(x, y) = e^{x^3 + y^3}$.

Solution on page 453 >

Partial Derivatives in Three or More Variables

Partial derivatives in three or more variables are defined similarly. That is, the partial derivative of $f(x, y, z)$ with respect to any one variable is the "ordinary" derivative with respect to that variable, holding all other variables constant.

EXAMPLE 11	FINDING A PARTIAL OF A FUNCTION OF THREE VARIABLES

$$\frac{\partial}{\partial x}(x^3 y^4 z^5) = 3x^2 y^4 z^5$$

$\partial/\partial x$ means differentiate with respect to x, holding y and z constant

Hold constant   Derivative of x^3

---

*$f_x(2, 5)$ may be written $\dfrac{\partial f}{\partial x}(2, 5)$ or $\dfrac{\partial f}{\partial x}\bigg|_{(2, 5)}$ again meaning first differentiate, then evaluate.

PRACTICE PROBLEM 5

Find $\dfrac{\partial}{\partial y}(x^3y^4z^5)$.

Solution on page 453 >

EXAMPLE 12	**EVALUATING A PARTIAL IN THREE VARIABLES**

Find $f_z(1, 1, 1)$ if $f(x, y, z) = e^{x^2+y^2+z^2}$.

Solution

$$f_z(x, y, z) = e^{x^2+y^2+z^2}(2z)$$ Partial with respect to z

$$= 2ze^{x^2+y^2+z^2}$$ Writing the $2z$ first

$$f_z(1, 1, 1) = 2 \cdot 1 \cdot e^{1^2+1^2+1^2} = 2e^3$$ Evaluating

Interpreting Partial Derivatives as Rates of Change

Since partials are just "ordinary" derivatives with the other variable(s) held constant, they give *instantaneous rates of change* with respect to one variable at a time.

Partials as Rates of Change

$$f_x(x, y) = \left(\begin{matrix}\text{Instantaneous rate of change of } f \text{ with} \\ \text{respect to } x \text{ when } y \text{ is held constant}\end{matrix}\right)$$

$$f_y(x, y) = \left(\begin{matrix}\text{Instantaneous rate of change of } f \text{ with} \\ \text{respect to } y \text{ when } x \text{ is held constant}\end{matrix}\right)$$

This is why they are called *partial* derivatives: Not all the variables are changed at once, only a "partial" change is made.

Cobb–Douglas Production Functions

Recall that a Cobb–Douglas production function $P(L, K) = aL^bK^{1-b}$ expresses production P as a function of L (units of labor) and K (units of capital). Each partial therefore gives the rate of increase of production with respect to one of these variables while the other is held constant.

EXAMPLE 13	**INTERPRETING THE PARTIALS OF A COBB–DOUGLAS PRODUCTION FUNCTION**

Find and interpret $P_L(120, 200)$ and $P_K(120, 200)$ for the Cobb–Douglas function $P(L, K) = 20L^{0.6}K^{0.4}$.

Solution

$$P_L = 12L^{-0.4}K^{0.4}$$ Partial with respect to L
(the 12 is 20 times 0.6)

$$P_L(120, 200) = 12(120)^{-0.4}(200)^{0.4} \approx 14.7$$ Substituting $L = 120$ and $K = 200$,   and evaluating using a calculator

Interpretation: $P_L = 14.7$ means that production increases by about 14.7 units for each additional unit of labor (when $L = 120$ and $K = 200$). This is called the *marginal productivity of labor.*

$$P_K = 8L^{0.6}K^{-0.6}$$

Partial with respect to K (the 8 is 20 times 0.4)

$$P_K(120, 200) = 8(120)^{0.6}(200)^{-0.6} \approx 5.9$$

Substituting $L = 120$ and $K = 200$, and evaluating using a calculator

Interpretation: $P_K = 5.9$ means that production increases by about 5.9 units for each additional unit of capital (when $L = 120$ and $K = 200$). This is called the *marginal productivity of capital.*

These numbers show that to increase production, additional units of labor are more than twice as effective as additional units of capital (at the levels $L = 120$ and $K = 200$).

Marginal Analysis: Partials in Business and Economics

Just as before, partial derivatives give the marginals, but now for one product at a time.

Interpreting Partials as Marginals

LOOKING BACK

We first saw these ideas when we defined marginal cost as the derivative of cost (for just one variable) on page 105. We then used them to develop marginal analysis (again for one variable) on pages 234–235.

Let $C(x, y)$ be the (total) cost function for x units of product A and y units of product B. Then

$$C_x(x, y) = \left(\begin{array}{l} \text{Marginal cost function for product A when} \\ \text{production of product B is held constant} \end{array} \right)$$

$$C_y(x, y) = \left(\begin{array}{l} \text{Marginal cost function for product B when} \\ \text{production of product A is held constant} \end{array} \right)$$

Similar statements hold, of course, for revenue and profit functions: The partials give the marginals for one variable at a time when the other variables are held constant.

EXAMPLE 14   **INTERPRETING PARTIALS OF A PROFIT FUNCTION**

A company's profit from producing x radios and y televisions per day is $P(x, y) = 4x^{3/2} + 6y^{3/2} + xy$. Find the marginal profit functions. Then find and interpret $P_y(25, 36)$.

Solution

$$P_x(x, y) = 6x^{1/2} + y$$

Marginal profit for radios when television production is held constant

$$P_y(x, y) = 9y^{1/2} + x$$

Marginal profit for televisions when radio production is held constant

$$P_y(25, 36) = 9(36)^{1/2} + 25 = 79$$
$$\underbrace{\hphantom{9(36)^{1/2}}}_{6}$$

Evaluating P_y at $x = 25$ and $y = 36$

Interpretation: Profit increases by about \$79 per additional television (when producing 25 radios and 36 televisions per day).

LOOKING AHEAD

These ideas will be used for further development of marginal analysis on pages 497–498.

The use of partials to find changes in revenue, cost, and profit resulting from one additional unit is called *marginal analysis.*

Interpreting Partials Geometrically

A function $f(x, y)$ represents a surface in three-dimensional space, and the partial derivatives are the **slopes** along the surface in different directions: $\partial f / \partial x$ gives the slope of the surface "in the x-direction," and $\partial f / \partial y$ gives the slope of the surface "in the y-direction" at the point (x, y).

To put this colloquially, if you were walking on the surface $z = f(x, y)$, then $\partial f / \partial x$ would be the steepness of the surface *in the x-direction,* and $\partial f / \partial y$ would be the steepness of the surface *in the y-direction* from the point (x, y).

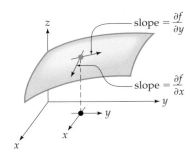

Partial derivatives are slopes.

For example, the following graph shows gridlines in the x-direction (roughly up and down the page, with the positive x-direction being down) and in the y-direction (roughly across the page, with the positive y-direction being to the right).

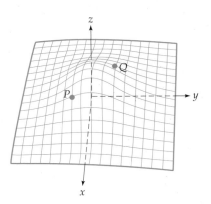

Therefore, at the point P on the graph the partials would have the following signs:

$$\frac{\partial f}{\partial x} < 0$$

Walking from P in the positive x-direction would mean walking downhill

$$\frac{\partial f}{\partial y} > 0$$

Walking from P in the positive y-direction would mean walking uphill

PRACTICE PROBLEM 6

On the preceding graph at the point Q,

a. Is $\dfrac{\partial f}{\partial x}$ positive or negative?

b. Is $\dfrac{\partial f}{\partial y}$ positive or negative?

[*Hint:* Leaving Q in those directions, would you be walking uphill or downhill?]

Solutions on next page >

Higher-Order Partial Derivatives

We can differentiate a function more than once to obtain *higher-order* partials.

Second-Order Partials

Subscript Notation		∂ Notation	In Words
f_{xx}	$=$	$\dfrac{\partial^2}{\partial x^2} f$	Differentiate twice with respect to x
f_{yy}	$=$	$\dfrac{\partial^2}{\partial y^2} f$	Differentiate twice with respect to y
f_{xy}	$=$	$\dfrac{\partial^2}{\partial y\, \partial x} f$	Differentiate first with respect to x, then with respect to y
f_{yx}	$=$	$\dfrac{\partial^2}{\partial x\, \partial y} f$	Differentiate first with respect to y, then with respect to x

Each notation means differentiate first with respect to the variable *closest* to f.

Calculating a "second partial" such as f_{xy} is a two-step process: First calculate f_x, and then differentiate the *result* with respect to y.

EXAMPLE 15   FINDING SECOND-ORDER PARTIALS

Find all four second-order partials of $f(x, y) = x^4 + 2x^2y^2 + x^3y + y^4$.

Solution

First we calculate

$$f_x = 4x^3 + 4xy^2 + 3x^2y \qquad \text{Partial of } f \text{ with respect to } x$$

Then from this we find f_{xx} and f_{xy}:

$$f_{xx} = 12x^2 + 4y^2 + 6xy$$
Differentiating $f_x = 4x^3 + 4xy^2 + 3x^2y$
with respect to x

$$f_{xy} = 8xy + 3x^2$$
Differentiating $f_x = 4x^3 + 4xy^2 + 3x^2y$
with respect to y

Now, returning to the original function $f = x^4 + 2x^2y^2 + x^3y + y^4$, we calculate

$$f_y = 4x^2y + x^3 + 4y^3$$
Partial of f with respect to y

Then, from this,

$$f_{yy} = 4x^2 + 12y^2$$
Differentiating $f_y = 4x^2y + x^3 + 4y^3$
with respect to y

$$f_{yx} = 8xy + 3x^2$$
Differentiating $f_y = 4x^2y + x^3 + 4y^3$
with respect to x

Notice that these "mixed partials" are equal:

$$f_{xy} = 8xy + 3x^2$$
$$f_{yx} = 8xy + 3x^2$$

Equal

That is, $f_{xy} = f_{yx}$, so reversing the order of differentiation (first x, then y, or first y, then x) makes no difference. This is not true for all functions, but it is true for all the functions that we will encounter in this book, and it is also true for all functions that you are likely to encounter in applications.*

PRACTICE PROBLEM 7

For the function $f(x, y) = x^3 - 3x^2y^4 + y^3$,   find:

a. f_x **b.** f_{xy} **c.** f_y **d.** f_{yx}

Solutions below >

Solutions TO PRACTICE PROBLEMS

1 a. $\dfrac{\partial}{\partial x} x^4y^2 = 4x^3y^2$ **b.** $\dfrac{\partial}{\partial y} x^4y^2 = x^4 2y = 2x^4y$

2. $\dfrac{\partial}{\partial y} x^2 = 0$

3. $\dfrac{\partial}{\partial y}(2x^4 - 3x^3y^3 - y^2 + 4x + 1) = 0 - 3x^3 3y^2 - 2y + 0 = -9x^3y^2 - 2y$

4. $f_y(x, y) = e^{x^3 + y^3}(3y^2) = 3y^2 e^{x^3 + y^3}$

  $f_y(1, 2) = (3 \cdot 4)e^{1^3 + 2^3} = 12e^9$

5. $\dfrac{\partial}{\partial y}(x^3y^4z^5) = x^3 4y^3 z^5 = 4x^3y^3z^5$

6. a. Positive   **b.** Negative

7. a. $f_x = 3x^2 - 6xy^4$ **b.** $f_{xy} = -6x4y^3 = -24xy^3$

  c. $f_y = -3x^2 4y^3 + 3y^2 = -12x^2y^3 + 3y^2$ **d.** $f_{yx} = -24xy^3$

*$f_{xy} = f_{yx}$ if these partials are continuous. A more detailed statement can be found in a book on advanced calculus.

7.2   Section Summary

For a function $f(x, y)$ the partial derivatives are:

$$\frac{\partial}{\partial x} f(x, y) = f_x(x, y) = \lim_{h \to 0} \frac{f(x + h, y) - f(x, y)}{h} \qquad \text{Partial derivative of } f \text{ with respect to } x$$

$$\frac{\partial}{\partial y} f(x, y) = f_y(x, y) = \lim_{h \to 0} \frac{f(x, y + h) - f(x, y)}{h} \qquad \text{Partial derivative of } f \text{ with respect to } y$$

provided that the limits exist. The partial derivative with respect to either variable is found by applying the usual differentiation rules to that variable and treating the other variable as a constant. Partial derivatives (or "partials") can be interpreted as instantaneous rates of change:

$$f_x(x, y) = \left(\begin{array}{c} \text{Instantaneous rate of change of } f \\ \text{with respect to } x \text{ when } y \text{ is held constant} \end{array} \right)$$

$$f_y(x, y) = \left(\begin{array}{c} \text{Instantaneous rate of change of } f \\ \text{with respect to } y \text{ when } x \text{ is held constant} \end{array} \right)$$

Partials can also be interpreted as marginals for one product while production of the other is held constant (see page 450). Geometrically, a function $z = f(x, y)$ of two variables represents a *surface* in three-dimensional space, and the partial $\partial f / \partial x$ gives the *slope* or *steepness* along the surface *in the x-direction*, and the partial $\partial f / \partial y$ gives the slope or steepness *in the y-direction*, as shown in the diagram on page 451.

7.2 Exercises

1–16. For each function, find the partials
a. $f_x(x, y)$ and **b.** $f_y(x, y)$.

> **FOR HELP GETTING STARTED**
>
> on Exercises 1–5, see Examples 4 and 5 on pages 446–447.

1. $f(x, y) = x^3 + 3x^2y^2 - 2y^3 - x + y$

2. $f(x, y) = 2x^4 - 7x^3y^2 - xy + 1$

3. $f(x, y) = 12x^{1/2}y^{1/3} + 8$

4. $f(x, y) = x^{-1}y + xy^{-2}$

5. $f(x, y) = 100x^{0.05}y^{0.02}$ **6.** $f(x, y) = \dfrac{x}{y}$

7. $f(x, y) = (x + y)^{-1}$

8. $f(x, y) = (x^2 + xy + 1)^4$

9. $f(x, y) = \ln(x^3 + y^3)$ **10.** $f(x, y) = x^2e^y$

11. $f(x, y) = 2x^3e^{-5y}$ **12.** $f(x, y) = e^{x+y}$

13. $f(x, y) = e^{xy}$ **14.** $f(x, y) = \ln(xy^3)$

15. $f(x, y) = \ln\sqrt{x^2 + y^2}$ **16.** $f(x, y) = \dfrac{xy}{x + y}$

17–20. For each function, find **a.** $\dfrac{\partial w}{\partial u}$ and **b.** $\dfrac{\partial w}{\partial v}$.

17. $w = (uv - 1)^3$ **18.** $w = (u - v)^3$

19. $w = e^{(u^2 - v^2)/2}$ **20.** $w = \ln(u^2 + v^2)$

21–26. For each function, evaluate the stated partials.

21. $f(x, y) = 4x^3 - 3x^2y^2 - 2y^2$, find $f_x(-1, 1)$ and $f_y(-1, 1)$

22. $f(x, y) = 2x^4 - 5x^2y^3 - 4y$, find $f_x(1, -1)$ and $f_y(1, -1)$

23. $f(x, y) = e^{x^2+y^2}$, find $f_x(0, 1)$ and $f_y(0, 1)$

24. $g(x, y) = (xy - 1)^5$, find $g_x(1, 0)$ and $g_y(1, 0)$

25. $h(x, y) = x^2y - \ln(x + y)$, find $h_x(1, 1)$

26. $f(x, y) = \sqrt{x^2 + y^2}$, find $f_y(8, -6)$

27–32. For each function, find the second-order partials
a. f_{xx}, **b.** f_{xy}, **c.** f_{yx}, and **d.** f_{yy}.

27. $f(x, y) = 5x^3 - 2x^2y^3 + 3y^4$

28. $f(x, y) = 4x^2 - 3x^3y^2 + 5y^5$

29. $f(x, y) = 9x^{1/3}y^{2/3} - 4xy^3$

30. $f(x, y) = 32x^{1/4}y^{3/4} - 5x^3y$

31. $f(x, y) = ye^x - x \ln y$

32. $f(x, y) = y \ln x + xe^y$

33–34. For each function, calculate the third-order partials
a. f_{xxy}, **b.** f_{xyx}, and **c.** f_{yxx}.

33. $f(x, y) = x^4y^3 - e^{2x}$ **34.** $f(x, y) = x^3y^4 - e^{2y}$

35–40. For each function of three variables, find the partials **a.** f_x, **b.** f_y, and **c.** f_z.

35. $f = xy^2z^3$

36. $f = x^2y^3z^4$

37. $f = (x^2 + y^2 + z^2)^4$

38. $f = (xyz + 1)^3$

39. $f = e^{x^2+y^2+z^2}$

40. $f = \ln(x^2 - y^3 + z^4)$

41–44. For each function, evaluate the stated partial.

41. $f = 3x^2y - 2xz^2$, find $f_x(2, -1, 1)$

42. $f = 2yz^3 - 3x^2z$, find $f_z(2, -1, 1)$

43. $f = e^{x^2+2y^2+3z^2}$, find $f_y(-1, 1, -1)$

44. $f = e^{2x^3+3y^3+4z^3}$, find $f_y(1, -1, 1)$

Applied Exercises

45–46. BUSINESS: Marginal Profit An electronics company's profit $P(x, y)$ from making x DVD players and y CD players per day is given below.

a. Find the marginal profit function for DVD players.
b. Evaluate your answer to part (a) at $x = 200$ and $y = 300$ and interpret the result.
c. Find the marginal profit function for CD players.
d. Evaluate your answer to part (c) at $x = 200$ and $y = 100$ and interpret the result.

45. $P(x, y) = 2x^2 - 3xy + 3y^2 + 150x + 75y + 200$

46. $P(x, y) = 3x^2 - 4xy + 4y^2 + 80x + 100y + 200$

47–48. BUSINESS: Cobb–Douglas Production Functions A company's production is given by the Cobb–Douglas function $P(L, K)$ below, where L is the number of units of labor and K is the number of units of capital.

a. Find $P_L(27, 125)$ and interpret this number.
b. Find $P_K(27, 125)$ and interpret this number.
c. From your answers to parts (a) and (b), which will increase production more: an additional unit of labor or an additional unit of capital?

47. $P(L, K) = 270L^{1/3}K^{2/3}$

48. $P(L, K) = 225L^{2/3}K^{1/3}$

49. BUSINESS: Sales A store's TV sales depend on x, the price of the televisions, and y, the amount spent on advertising, according to the function $S(x, y) = 200 - 0.1x + 0.2y^2$. Find and interpret the marginals S_x and S_y.

50. ECONOMICS: Value of an MBA A study found that a businessperson with a master's degree in business administration (MBA) earned an average salary of $S(x, y) = 48,340 + 4930x + 3840y$ dollars in 2005, where x is the number of years of work experience before the MBA, and y is the number of years of work experience after the MBA. Find and interpret the marginals S_x and S_y.

51. SOCIOLOGY: Status A study found that a person's status in a community depends on the person's

income and education according to the function $S(x, y) = 7x^{1/3}y^{1/2}$, where x is income (in thousands of dollars) and y is years of education beyond high school.

a. Find $S_x(27, 4)$ and interpret this number.
b. Find $S_y(27, 4)$ and interpret this number.

52. BIOMEDICAL: Blood Flow The resistance of blood flowing through an artery of radius r and length L (both in centimeters) is $R(r, L) = 0.08Lr^{-4}$.

a. Find $R_r(0.5, 4)$ and interpret this number.
b. Find $R_L(0.5, 4)$ and interpret this number.

Source: E. Batschelet, *Introduction to Mathematics for Life Scientists*

53. GENERAL: Highway Safety The length in feet of the skid marks from a truck of weight w (tons) traveling at velocity v (miles per hour) skidding to a stop on a dry road is $S(w, v) = 0.027wv^2$.

a. Find $S_w(4, 60)$ and interpret this number.
b. Find $S_v(4, 60)$ and interpret this number.

Source: National Highway Traffic Safety Administration

54. GENERAL: Windchill Temperature The windchill temperature announced by the weather bureau during the cold weather measures how cold it "feels" for a given temperature and wind speed. The formula is $C(t, w) = 35.74 + 0.6215t - 35.75w^{0.16} + 0.4275tw^{0.16}$ where t is the temperature (in degrees Fahrenheit) and w is the wind speed (in miles per hour). Find and interpret $C_w(30, 20)$.

Source: National Weather Service

55. ATHLETICS: Pythagorean Baseball Standings The Pythagorean standing of a baseball team is given by Bill James's formula $f(x, y) = x^2/(x^2 + y^2)$ where x is the number of runs scored and y is the number of runs allowed.

a. Find $f_x(400, 300)$, giving the rate of change of the standings per additional run scored (for 400 runs scored and 300 allowed).
b. Multiply your answer to part (a) by 20 to find the change that would result from 20 additional runs

scored, and then multiply this result by 160 (the approximate number of games per season) to estimate the number of games that would be won by those 20 additional runs.

Source: B. James, Baseball Abstract 1983

56. ATHLETICS: Olympics A country's share of medals at an Olympic games can be estimated from the formula $f(x, y) = 0.0062 \ln x + 0.0064 \ln y - 0.0652$, where x is the population and y is the per capita gross domestic product (PCGDP) of the country.

a. Find $f_y(x, y)$ and evaluate it at $y = 1000$ to find the rate of change in the proportion of medals per extra dollar when PCGDP is \$1000.

b. Multiply your answer to part (a) by 500 to find the change in the proportion that would result from an additional \$500 in PCGDP, and then multiply this result by 920 (the number of medals at a typical Olympic games) to estimate the number of additional Olympic medals that would be won.

Source: Review of Economics and Statistics 86

Conceptual Exercises

57. Find $\dfrac{\partial}{\partial x} e^{\sqrt{1 + 5y^{-23}}} \ln (1 - 17y^{97}/z^{47})$

58. Find $\dfrac{\partial}{\partial y} 108^{17x^2}/(x^{23} - 2/z^8)^{13}$

59. You are walking on a surface $z = f(x, y)$, and for each unit that you walk in the x-direction you rise 3 units and for each unit that you walk in the y-direction you fall 2 units. Find the partial derivatives of $f(x, y)$.

60. Your company makes two products, called X and Y, in quantities x and y. If your profit function satisfies $\dfrac{\partial P}{\partial x} = 12$ and $\dfrac{\partial P}{\partial y} = -8$ (both in dollars), how much money are you making on each of your two products?

61. Look at the first graph of a surface on page 451. Is $\dfrac{\partial f}{\partial x}$ positive or negative at the point shown? Same question for $\dfrac{\partial f}{\partial y}$.

62–65. A *linear function of two variables* is of the form $f(x, y) = ax + by + c$ where a, b, and c are constants. Find the linear function of two variables satisfying the following conditions.

62. $\dfrac{\partial f}{\partial x} = 3$, $\dfrac{\partial f}{\partial y} = -2$, and $f(0, 0) = 5$

63. $\dfrac{\partial f}{\partial x} = -1$, $\dfrac{\partial f}{\partial y} = 1$, and $f(0, 0) = 0$

64. $\dfrac{\partial f}{\partial x} = 0$, $\dfrac{\partial f}{\partial y} = 0$, and $f(100, 100) = 100$

65. $\dfrac{\partial f}{\partial x} = \pi$, $\dfrac{\partial f}{\partial y} = e$, and $f(0, 0) = \ln 2$

66. Find all second partial derivatives of a linear function of two variables.

67–68. SOCIAL SCIENCE: Procrastination A ten-year study of procrastination found that if you have a task to do, your desire to complete the task (denoted D) is given by $D = \dfrac{E \cdot V}{T \cdot P}$, where E is the expectation of success, V is the value of completing the task, T is the time needed to complete the task, and P is your tendency to procrastinate, all of which are positive quantities.

Source: Scientific American

67. Find the signs of $\dfrac{\partial D}{\partial E}$ and $\dfrac{\partial D}{\partial T}$ and interpret these signs.

68. Find the signs of $\dfrac{\partial D}{\partial V}$ and $\dfrac{\partial D}{\partial P}$ and interpret these signs.

Explorations and Excursions
The following problems extend and augment the material presented in the text.

Competitive and Complementary Commodities

69. ECONOMICS: Competitive Commodities Certain commodities (such as butter and margarine) are called "competitive" or "substitute" commodities because one can substitute for the other. If $B(b, m)$ gives the daily sales of butter as a function of b, the price of butter, and m, the price of margarine:

a. Give an interpretation of $B_b(b, m)$.

b. Would you expect $B_b(b, m)$ to be positive or negative? Explain.

c. Give an interpretation of $B_m(b, m)$.

d. Would you expect $B_m(b, m)$ to be positive or negative? Explain.

70. ECONOMICS: Complementary Commodities Certain commodities (such as washing machines and clothes dryers) are called "complementary"commodities because they are often used together. If $D(d, w)$ gives the monthly sales of dryers as a function of d, the price of dryers, and w, the price of washers:

a. Give an interpretation of $D_d(d, w)$.

b. Would you expect $D_d(d, w)$ to be positive or negative? Explain.

c. Give an interpretation of $D_w(d, w)$.

d. Would you expect $D_w(d, w)$ to be positive or negative? Explain.

Harmonic Functions

71–74. A function $f(x, y)$ that has continuous second derivatives and that satisfies $f_{xx} + f_{yy} = 0$ is called a *harmonic function*. Harmonic functions have many interesting properties, including the fact that their value at the center of any circle is the average of their values around the circumference of the circle. So if you stretch the edges of a flexible rubber sheet, the shape will be a harmonic function. Find whether each of the following functions is harmonic.

71. $f(x, y) = x^2 - y^2$ **72.** $f(x, y) = xy$

73. $f(x, y) = x^2 + y^2$ **74.** $f(x, y) = x^3 - 3xy^2$

Finding Partial Derivatives from the Definition

75–78. Use the *definition* of partial derivatives on page 444 to find both $\dfrac{\partial}{\partial x} f(x, y)$ and $\dfrac{\partial}{\partial y} f(x, y)$ for each of the following functions. [*Hint:* The method is the same as that on page 92 but applied to just *one* variable at a time.]

75. $f(x, y) = 2x^2 + 3y^2 - 3x + 7y + 4$

76. $f(x, y) = 3x^2 - 2y^2 + 5x - 4y - 9$

77. $f(x, y) = 5x^2 - 3x + 7y + 4$

78. $f(x, y) = 4y^2 + 2x - 5y + 12$

7.3   **Optimizing Functions of Several Variables**

Introduction

The graph of a function of two variables is a *surface*, with **relative maximum** and **minimum points** ("hilltops" and "valley bottoms") and **saddle points** ("mountain passes"). In this section we will see how to maximize and minimize such functions by finding **critical points** and using a two-variable version of the second-derivative test.

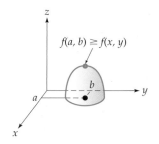

f has a relative *maximum* value at (a, b).

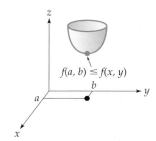

f has a relative *minimum* value at (a, b).

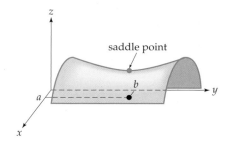

f has a saddle point at (a, b) (neither a maximum nor a minimum).

Such optimization techniques have many applications, such as maximizing profit for a company that makes several products. For simplicity, we will consider only functions whose first and second partials are defined everywhere.

Critical Points

At the very top of a smooth hill, the slope or steepness in any direction must be zero. That is, a straight stick would balance horizontally at the top. Since the partials f_x and f_y are the slopes in the x- and y-directions, these partials must both be

A critical point
(both partials are zero)

zero at a relative maximum point, and similarly for a relative minimum point or a saddle point. A point (a, b) at which both partials are zero is called a *critical point* of the function.*

Critical Point

(a, b) is a critical point of $f(x, y)$ if

$$f_x(a, b) = 0 \quad \text{and} \quad f_y(a, b) = 0$$

Both first partials are zero

Relative maximum and minimum values can occur only at critical points.

EXAMPLE 1   FINDING CRITICAL POINTS

Find all critical points of

$$f(x, y) = 3x^2 + y^2 + 3xy + 3x + y + 6$$

Solution

We want all points at which both partials are zero.

$$f_x = 6x + 3y + 3 \quad \text{and} \quad f_y = 2y + 3x + 1 \qquad \text{Partials}$$

$$6x + 3y + 3 = 0 \left.\vphantom{\begin{matrix}a\\b\end{matrix}}\right\} \text{Partials set equal to zero}$$
$$3x + 2y + 1 = 0 \leftarrow$$

Reordered so the x- and y-terms line up

To solve these equations simultaneously, we multiply the second by -2 so that the x-terms drop out when we add.

$$6x + 3y + 3 = 0 \qquad\qquad \text{First equation}$$
$$\underline{-6x - 4y - 2 = 0} \qquad\quad \text{Second equation times } -2$$
$$-y + 1 = 0 \qquad\qquad \text{Adding (x drops out)}$$
$$y = 1 \qquad\qquad\quad \text{From solving } -y + 1 = 0$$

Substituting $y = 1$ into either equation gives x:

$$6x + 3 + 3 = 0 \qquad\qquad \text{Substituting } y = 1 \text{ into } 6x + 3y + 3 = 0$$
$$6x = -6 \qquad\qquad\qquad \text{Simplifying}$$
$$x = -1 \qquad\qquad\qquad \text{Solving}$$

These x- and y-values give one critical point.

$$\text{CP: } (-1, 1) \qquad\qquad \text{From } x = -1, \; y = 1$$

Second-Derivative Test for Functions $f(x, y)$: The D-Test

To determine whether $f(x, y)$ has a relative maximum, a relative minimum, or a saddle point at a critical point, we use the following **D-test,** which is a generalization of the second-derivative test on page 184.

*We use the term "critical point" for the *pair* of values (a, b) at which the two partials are zero. The corresponding point on the graph has *three* coordinates $(a, b, f(a, b))$.

D-Test

If (a, b) is a critical point of the function f and for D defined by

$$D = f_{xx}(a, b) \cdot f_{yy}(a, b) - [f_{xy}(a, b)]^2$$

More briefly,
$D = f_{xx}f_{yy} - (f_{xy})^2$

then f at the point (a, b) has a:

  i. relative *maximum* if $D > 0$ and $f_{xx}(a, b) < 0$

  ii. relative *minimum* if $D > 0$ and $f_{xx}(a, b) > 0$

Different signs for $f_{xx}(a, b)$

  iii. *saddle point* if $D < 0$.

 Be Careful Be sure to apply the *D*-test correctly:

1. The *D*-test is used only *after* finding the critical points. The test is then applied to each critical point, one at a time.

2. $D > 0$ appears only in parts i and ii, and so guarantees that the function has either a relative maximum or a relative minimum. You then use the "old" second-derivative test (checking the sign of f_{xx}) to determine which one occurs (maximum or minimum).

3. $D < 0$ means a *saddle point*, regardless of the sign of f_{xx}. (A saddle point is neither a maximum nor a minimum.)

4. $D = 0$ means that the *D*-test is *inconclusive*—the function may have a maximum, a minimum, or a saddle point at the critical point.

EXAMPLE 2   **FINDING RELATIVE EXTREME VALUES OF A POLYNOMIAL**

Find the relative extreme values of

$$f(x, y) = 3x^2 + y^2 + 3xy + 3x + y + 6$$

Solution

We find critical points by setting the two partials equal to zero and solving. But we did this for the same function in Example 1, finding one critical point, $(-1, 1)$. For the *D*-test, we calculate the second partials.

$$f_{xx} = 6 \qquad\qquad \text{From } f_x = 6x + 3y + 3$$

$$f_{yy} = 2 \qquad\qquad \text{From } f_y = 2y + 3x + 1$$

$$f_{xy} = 3 \qquad\qquad \text{From } f_x = 6x + 3y + 3$$

Calculating D:

$$D = 6 \cdot 2 - (3)^2 = 12 - 9 = 3 \qquad D = f_{xx}f_{yy} - (f_{xy})^2$$

        f_{xx} f_{yy} f_{xy} Positive

D is positive and f_{xx} is positive (since $f_{xx} = 6$), so f has a *relative minimum* (part ii of the *D*-test) at the critical point $(-1, 1)$. (If f_{xx} had been negative, there would have been a relative *maximum*.) The relative minimum *value* is found by evaluating $f(x, y)$ at $(-1, 1)$.

$$f(-1, 1) = 3 + 1 - 3 - 3 + 1 + 6$$

$f = 3x^2 + y^2 + 3xy + 3x + y + 6$
evaluated at $x = -1, \ y = 1$

$$= 5$$

Relative minimum value: $f = 5$ at $x = -1, \ y = 1$.

The graph of this function on the left shows this relative minimum point.

$f(x, y) = 3x^2 + y^2 + 3xy + 3x + y + 6$

PRACTICE PROBLEM

Suppose that a function $f(x, y)$ at a particular critical point had second partials with values:

$$f_{xx} = -4$$
$$f_{yy} = -6$$
$$f_{xy} = 3$$

What could you conclude about the graph of the function at that critical point?

What if instead the values were:

$$f_{xx} = 2$$
$$f_{yy} = 4$$
$$f_{xy} = -3$$

Solution on page 466 >

EXAMPLE 3	FINDING RELATIVE EXTREME VALUES OF AN EXPONENTIAL FUNCTION

Find the relative extreme values of $f(x, y) = e^{x^2 - y^2}$.

Solution

$$\left. \begin{array}{l} f_x = e^{x^2-y^2}(2x) \\ f_y = e^{x^2-y^2}(-2y) \end{array} \right\} \text{Partials}$$

$$\left. \begin{array}{l} e^{x^2-y^2}(2x) = 0 \\ e^{x^2-y^2}(-2y) = 0 \end{array} \right\} \text{Partials set equal to zero}$$

CP: $(0, 0)$ Since $\begin{cases} e^{x^2-y^2}(2x) = 0 \\ e^{x^2-y^2}(-2y) = 0 \end{cases}$ only at $x = 0, y = 0$

For the D-test we calculate the second partials:

$$f_{xx} = e^{x^2-y^2}(2x)(2x) + e^{x^2-y^2}(2)$$

From $f_x = e^{x^2-y^2}(2x)$ using the Product Rule

$$= 4x^2 e^{x^2-y^2} + 2e^{x^2-y^2}$$

Simplifying

$$f_{yy} = e^{x^2-y^2}(-2y)(-2y) + e^{x^2-y^2}(-2)$$

From $f_y = e^{x^2-y^2}(-2y)$ using the Product Rule

$$= 4y^2 e^{x^2-y^2} - 2e^{x^2-y^2}$$

Simplifying

$$f_{xy} = e^{x^2-y^2}(2x)(-2y) = -4xye^{x^2-y^2}$$

From $f_x = e^{x^2-y^2}(2x)$ treating x as a constant

Evaluating at the critical point $(0, 0)$:

$$f_{xx}(0, 0) = 0e^0 + 2e^0 = 0 + 2 = 2 \qquad f_{xx} = 4x^2 e^{x^2 - y^2} + 2e^{x^2 - y^2} \text{ at } (0, 0)$$

$$f_{yy}(0, 0) = 0e^0 - 2e^0 = 0 - 2 = -2 \qquad f_{yy} = 4y^2 e^{x^2 - y^2} - 2e^{x^2 - y^2} \text{ at } (0, 0)$$

$$f_{xy}(0, 0) = 0e^0 = 0 \qquad f_{xy} = -4xye^{x^2 - y^2} \text{ at } (0, 0)$$

Therefore D is

$$D = (2)(-2) - (0)^2 = -4 - 0 = -4 \qquad D = f_{xx}f_{yy} - (f_{xy})^2$$

$\begin{array}{ccc} / & \backslash & \backslash \\ f_{xx} & f_{yy} & f_{xy} \end{array}$ $\qquad\qquad$ Negative

Since D is negative, the function has a *saddle point* (part iii of the D-test) at the critical point $(0, 0)$.

f has no relative extreme values
(it has a saddle point at $x = 0$, $y = 0$).

The graph of this function on the left shows this saddle point.

$f(x, y) = e^{x^2 - y^2}$

Maximizing Profit

If a company makes too few of its products, the resulting lost sales will lower the company's profits. On the other hand, making too many will "flood the market" and depress prices, again resulting in lower profits. Therefore, any realistic profit function must have a maximum at some "intermediate" point (x, y), which therefore must be a *relative* maximum point. Many applied problems are solved in this way: by knowing that the *absolute* extreme values (the highest and lowest values on the entire domain) must exist, and finding them as *relative* extreme points.

Suppose that a company produces two products, A and B, and that the two price functions are

$$p(x) = \left(\begin{array}{c} \text{Price at which exactly } x \text{ units} \\ \text{of product A will be sold} \end{array} \right)$$

and

$$q(y) = \left(\begin{array}{c} \text{Price at which exactly } y \text{ units} \\ \text{of product B will be sold} \end{array} \right)$$

If $C(x, y)$ is the (total) cost function, then the company's profit will be

$$\underbrace{P(x, y)}_{\text{Profit}} = \underbrace{p(x) \cdot x}_{\substack{\text{Price times} \\ \text{quantity for} \\ \text{product A}}} + \underbrace{q(y) \cdot y}_{\substack{\text{Price times} \\ \text{quantity for} \\ \text{product B}}} - \underbrace{C(x, y)}_{\text{Cost}}$$

Revenue for each product (price times quantity) minus the cost function

EXAMPLE 4   **MAXIMIZING PROFIT FOR A COMPANY**

Universal Motors makes compact and midsized cars. The price function for compacts is $p = 17 - 2x$ (for $0 \le x \le 8$), and the price function for midsized cars is $q = 20 - y$ (for $0 \le y \le 20$), both in thousands of dollars, where x and y are, respectively, the numbers of compact and midsized cars produced per hour. If the company's cost function is

$$C(x, y) = 15x + 16y - 2xy + 5$$

thousand dollars, find how many of each car should be produced and the prices that should be charged in order to maximize profit. Also find the maximum profit.

Solution

The profit function is

$$P(x, y) = (17 - 2x)x + (20 - y)y - (15x + 16y - 2xy + 5)$$

Price Quantity   Price Quantity   Cost

For compacts   For midsized

$$= 17x - 2x^2 + 20y - y^2 - 15x - 16y + 2xy - 5 \qquad \text{Multiplying out}$$

$$= -2x^2 - y^2 + 2xy + 2x + 4y - 5 \qquad \text{Simplifying}$$

We maximize $P(x, y)$ in the usual way:

$$P_x = -4x + 2y + 2$$
$$P_y = -2y + 2x + 4$$
$\Big\}$ Partials

$-4x + 2y + 2 = 0$	Partials set equal to zero
$2x - 2y + 4 = 0$	Rearranged to line up x's and y's
$-2x \qquad + 6 = 0$	Adding (the y's cancel)
$x = 3$	From solving $-2x + 6 = 0$
$y = 5$	From substituting $x = 3$ into either equation (omitting the details)

These two values give one critical point.

$$\text{CP: } (3, 5)$$

For the D-test we calculate the second partials:

$$P_{xx} = -4 \qquad P_{xy} = 2 \qquad P_{yy} = -2 \qquad \begin{array}{l}\text{From } P_x = -4x + 2y + 2 \\ \text{and } P_y = -2y + 2x + 4\end{array}$$

$$D = (-4)(-2) - (2)^2 = 4 \qquad D = P_{xx}P_{yy} - (P_{xy})^2$$

D is positive and $P_{xx} = -4$ is negative, so profit is indeed *maximized* at $x = 3$ and $y = 5$. To find the prices, we evaluate the price functions:

$$p = 17 - 2 \cdot 3 = 11 \quad \text{(thousand dollars)} \qquad \begin{array}{l}p = 17 - 2x \\ \text{evaluated at } x = 3\end{array}$$

$$q = 20 - 5 = 15 \quad \text{(thousand dollars)} \qquad \begin{array}{l}q = 20 - y \\ \text{evaluated at } y = 5\end{array}$$

The profit comes from the profit function:

$$P(3, 5) = -2 \cdot 3^2 - 5^2 + 2 \cdot 3 \cdot 5 + 2 \cdot 3 + 4 \cdot 5 - 5$$

$P = -2x^2 - y^2 + 2xy + 2x + 4y - 5$
evaluated at $x = 3$, $y = 5$

$$= 8 \text{ (thousand dollars)}$$

Profit is maximized when the company produces 3 compacts per hour, selling them for \$11,000 each, and 5 midsized cars per hour, selling them for \$15,000 each. The maximum profit will be \$8000 per hour.

The graph of the profit function on the left shows this maximum point.

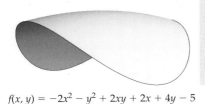

$f(x, y) = -2x^2 - y^2 + 2xy + 2x + 4y - 5$

The *D*-test ensures that you have *maximized* profit rather than minimized it.*

Some functions have *more* than one critical point.

EXAMPLE 5   FINDING RELATIVE EXTREME VALUES

Find the relative extreme values of

$$f(x, y) = x^2 + y^3 - 6x - 12y$$

Solution

$$\begin{array}{ll} 2x - 6 = 0 & \qquad f_x = 0 \\ 3y^2 - 12 = 0 & \qquad f_y = 0 \end{array}$$

The first gives

$$x = 3 \qquad\qquad \text{Solving } 2x - 6 = 0$$

and the second gives

$$3y^2 = 12 \qquad \begin{array}{l}\text{Adding 12 to each}\\ \text{side of } 3y^2 - 12 = 0\end{array}$$

$$y^2 = 4 \qquad \text{Dividing by 3}$$

$$y = \pm 2 \qquad \text{Taking square roots}$$

From $x = 3$ and $y = \pm 2$ we get *two* critical points:

$$\text{CP: } (3, 2) \quad \text{and} \quad (3, -2)$$

Calculating the second partials and substituting them into D:

$$D = (2)(6y) - (0)^2 = 12y \qquad \begin{array}{l} D = (f_{xx})(f_{yy}) - (f_{xy})^2 \text{ with}\\ f_{xx} = 2, \ f_{yy} = 6y, \ f_{xy} = 0 \end{array}$$

*How are the price and cost functions in such problems found? Price functions may be constructed by the methods used on pages 203–205 or by more sophisticated techniques of market research. Cost functions may be found simply as the sum of the unit cost times the number of units for each product, or by regression techniques based on the least squares method described in the following section for constructing functions from data.

We apply the D-test to the critical points one at a time.

At $(3, 2)$: $D = 12 \cdot 2 > 0$ $D = 12y$ evaluated at $(3, 2)$

$f_{xx} = 2 > 0$

relative minimum at $x = 3$, $y = 2$ Since D and f_{xx} are both positive

At $(3, -2)$: $D = 12 \cdot (-2) < 0$ $D = 12y$ evaluated at $(3, -2)$

saddle point at $x = 3$, $y = -2$ Since D is negative

Answer:

Relative minimum value: $f = -25$ at $\begin{cases} x = 3 \\ y = 2 \end{cases}$ $f = -25$ from $f = x^2 + y^3 - 6x - 12y$ at $(3, 2)$
(saddle point at $x = 3$, $y = -2$)

The graph of this function on the left shows the relative minimum and saddle points.

$f(x, y) = x^2 + y^3 - 6x - 12y$

Competition and Collusion

In 1838, Antoine Cournot* published the following comparison of a **monopoly** (a market with only one supplier) and a **duopoly** (a market with two suppliers).

Monopoly.   Imagine that you are selling spring water from your own spring (or any product whose cost of production is negligible). Since you are the only supplier in town, you have a "monopoly." Suppose that your price function is $p = 6 - 0.01x$, where p is the price in dollars at which you will sell precisely x gallons $(0 \le x \le 600)$.   Your revenue is then

$$R(x) = (6 - 0.01x)x = 6x - 0.01x^2$$ Price $(6 - 0.01x)$ times quantity x

You maximize revenue by setting its derivative equal to zero:

$$R'(x) = 6 - 0.02x = 0$$

$$x = \frac{6}{0.02} = 300$$ Solving $6 - 0.02x = 0$

Therefore, you should sell 300 gallons per day. (The second-derivative test will verify that revenue is maximized.) The price will be

$$p = 6 - 0.01 \cdot 300 = 6 - 3 = 3$$ $p = 6 - 0.01x$ evaluated at $x = 300$

or $3 dollars per gallon. Your maximum revenue will be

$$R(300) = 6 \cdot 300 - 0.01(300)^2 = \$900$$ $R(x) = 6x - 0.01x^2$ evaluated at $x = 300$

Duopoly.   Suppose now that your neighbor opens a competing spring water business. (A market such as this with two suppliers is called a "duopoly.") Now both of you must share the same market. If your neighbor sells y gallons per day (and you sell x), you must both sell at price

$$p = 6 - 0.01(x + y) = 6 - 0.01x - 0.01y$$ Price function $p = 6 - 0.01x$ with x replaced by the combined quantity $x + y$

*French mathematician, philosopher, and economist (1801–1887).

Each of you calculates revenue as price times quantity:

$$\begin{pmatrix} \text{Your} \\ \text{revenue} \end{pmatrix} = p \cdot x = (6 - 0.01x - 0.01y)x = 6x - 0.01x^2 - 0.01xy$$

$$\begin{pmatrix} \text{Neighbor's} \\ \text{revenue} \end{pmatrix} = p \cdot y = (6 - 0.01x - 0.01y)y = 6y - 0.01xy - 0.01y^2$$

You each want to maximize revenue, so you set the partials equal to zero:

$$6 - 0.02x - 0.01y = 0$$

Partial of $6x - 0.01x^2 - 0.01xy$ with respect to x, set equal to zero

$$6 - 0.01x - 0.02y = 0$$

Partial of $6y - 0.01xy - 0.01y^2$ with respect to y, set equal to zero

These are easily solved by multiplying one of them by 2 and subtracting the other. The solution (omitting the details) is

$$x = 200 \qquad y = 200$$

so each of you sells 200 gallons per day. The selling price for both will be

$$p = 6 - 0.01(200 + 200) = 6 - 4 = 2$$

$p = 6 - 0.01(x + y)$ evaluated at $x = 200$, $y = 200$

or \$2 per gallon. Revenue is price (\$2) times quantity (200), resulting in \$400 for each of you.

Comparison of the Monopoly and the Duopoly

The two systems may be compared as follows:

	Monopoly	**Duopoly**
Quantity:	300 gallons	200 gallons each, 400 total
Price:	\$3 per gallon	\$2 per gallon
Revenue:	\$900	\$400 each, \$800 total

Notice that the duopoly produces *more* than the monopoly (400 gallons versus only 300 in the monopoly), and does so at a *lower price* (\$2 versus \$3 in the monopoly). Cournot therefore concluded that consumers benefit more from a duopoly than from a monopoly.

However, the smart duopolist will notice that the revenue of \$400 is less than half of the monopoly revenue of \$900. Therefore, it benefits each duopolist to co-operate and share the market as a single monopoly. This is called *collusion*. It is for this reason that markets with only a few suppliers tend toward collusion rather than competition. (In the United States, the Federal Trade Commission restricts collusion in major industries.)

Optimizing Functions of Three or More Variables

Functions of *more* than two variables are optimized in the same way: by finding critical points (where all of the first partials are zero). For example, to optimize a function $f(x, y, z)$ we would set the three partials equal to zero and solve:

$$f_x = 0$$

$$f_y = 0$$

$$f_z = 0$$

The second-derivative test for functions of three or more variables is very complicated, and we shall not discuss it.

Solutions TO PRACTICE PROBLEM

Relative maximum (since $D = f_{xx}f_{yy} - (f_{xy})^2 = 15 > 0$ and $f_{xx} = -4 < 0$)

Saddle point (since $D = f_{xx}f_{yy} - (f_{xy})^2 = -1 < 0$)

7.3   Section Summary

To optimize a function $f(x, y)$ of two variables, first find all critical points, the points (a, b) where both first partials are zero.

$$f_x(a, b) = 0$$
$$f_y(a, b) = 0$$

Then apply the D-test (page 459) to each critical point to determine whether the function has a relative maximum, a relative minimum, or a saddle point at that critical point.

7.3 Exercises

1–20. Find the relative extreme values of each function.

1. $f(x, y) = x^2 + 2y^2 + 2xy + 2x + 4y + 7$

2. $f(x, y) = 2x^2 + y^2 + 2xy + 4x + 2y + 5$

3. $f(x, y) = 2x^2 + 3y^2 + 2xy + 4x - 8y$

4. $f(x, y) = 3x^2 + 2y^2 + 2xy + 8x - 4y$

5. $f(x, y) = 3xy - 2x^2 - 2y^2 + 14x - 7y - 5$

6. $f(x, y) = 2xy - 2x^2 - 3y^2 + 4x - 12y + 5$

7. $f(x, y) = xy + 4x - 2y + 1$

8. $f(x, y) = 5xy - 2x^2 - 3y^2 + 5x - 7y + 10$

9. $f(x, y) = 3x - 2y - 6$

10. $f(x, y) = 5x - 4y + 5$

11. $f(x, y) = e^{(x^2 + y^2)/2}$ **12.** $f(x, y) = e^{5(x^2 + y^2)}$

13. $f(x, y) = \ln(x^2 + y^2 + 1)$

14. $f(x, y) = \ln(2x^2 + 3y^2 + 1)$

15. $f(x, y) = -x^3 - y^2 + 3x - 2y$

16. $f(x, y) = x^3 - y^2 - 3x + 6y$

17. $f(x, y) = y^3 - x^2 - 2x - 12y$

18. $f(x, y) = -x^2 - y^3 - 6x + 3y + 4$

19. $f(x, y) = x^3 - 2xy + 4y$

20. $f(x, y) = y^3 - 2xy - 4x$

Applied Exercises

21. BUSINESS: Maximum Profit A company manufactures two products. The price function for product A is $p = 12 - \frac{1}{2}x$ (for $0 \le x \le 24$), and for product B is $q = 20 - y$ (for $0 \le y \le 20$), both in thousands of dollars, where x and y are the amounts of products A and B, respectively. If the cost function is

$$C(x, y) = 9x + 16y - xy + 7$$

thousands of dollars, find the quantities and the prices of the two products that maximize profit. Also find the maximum profit.

22. BUSINESS: Maximum Profit A company manufactures two products. The price function for product A is $p = 16 - x$ (for $0 \le x \le 16$), and for product B is $q = 19 - \frac{1}{2}y$ (for $0 \le y \le 38$), both in thousands of dollars, where x and y are the amounts of product A and B, respectively. If the cost function is

$$C(x, y) = 10x + 12y - xy + 6$$

thousand dollars, find the quantities and the prices of the two products that maximize profit. Also find the maximum profit.

23. **BUSINESS:** Price Discrimination An automobile manufacturer sells cars in America and Europe, charging different prices in the two markets. The price function for cars sold in America is $p = 20 - 0.2x$ thousand dollars (for $0 \le x \le 100$), and the price function for cars sold in Europe is $q = 16 - 0.1y$ thousand dollars (for $0 \le y \le 160$), where x and y are the numbers of cars sold per day in America and Europe, respectively. The company's cost function is

$$C = 20 + 4(x + y) \text{ thousand dollars}$$

   a. Find the company's profit function. [*Hint:* Profit is revenue from America plus revenue from Europe minus costs, where each revenue is price times quantity.]
   b. Find how many cars should be sold in each market to maximize profit. Also find the price for each market.

24. **BIOMEDICAL:** Drug Dosage In a laboratory test the combined antibiotic effect of x milligrams of medicine A and y milligrams of medicine B is given by the function

$$f(x, y) = xy - 2x^2 - y^2 + 110x + 60y$$

   (for $0 \le x \le 55, 0 \le y \le 60$). Find the amounts of the two medicines that maximize the antibiotic effect.

25. **PSYCHOLOGY:** Practice and Rest A subject in a psychology experiment who practices a skill for x hours and then rests for y hours achieves a test score of

$$f(x, y) = xy - x^2 - y^2 + 11x - 4y + 120$$

   (for $0 \le x \le 10, 0 \le y \le 4$). Find the numbers of hours of practice and rest that maximize the subject's score.

26. **SOCIOLOGY:** Absenteeism The number of office workers near a beach resort who call in "sick" on a warm summer day is

$$f(x, y) = xy - x^2 - y^2 + 110x + 50y - 5200$$

   where x is the air temperature ($70 \le x \le 100$) and y is the water temperature ($60 \le y \le 80$). Find the air and water temperatures that maximize the number of absentees.

27–28. **ECONOMICS:** Competition and Collusion Compare the outputs of a monopoly and a duopoly by repeating the analysis on pages 464–465 for the following price function. That is:

   a. For a monopoly, calculate the quantity x that maximizes your revenue. Also calculate the price p and the revenue R.
   b. For the duopoly, calculate the quantities x and y that maximize revenue for each duopolist. Calculate the price p and the two revenues.
   c. Are more goods produced under a monopoly or a duopoly?
   d. Is the price lower under a monopoly or a duopoly?

27. $p = 12 - 0.005x$ dollars ($0 \le x \le 2400$)

28. $p = a - bx$ dollars (for positive numbers a and b with $0 \le x \le a/b$)

29. **BUSINESS:** Maximum Profit An automobile dealer can sell 8 sedans per day at a price of $20,000 and 4 SUVs (sport utility vehicles) per day at a price of $25,000. She estimates that for each $400 decrease in price of the sedans she can sell two more per day, and for each $600 decrease in price for the SUVs she can sell one more. If each sedan costs her $16,800 and each SUV costs her $19,000, and fixed costs are $1000 per day, what price should she charge for the sedans and the SUVs to maximize profit? How many of each type will she sell at these prices? [*Hint:* Let x be the number of $400 price decreases for sedans and y be the number of $600 price decreases for SUVs, and use the method of Examples 1 and 2 on pages 203–204 for each type of car.]

30. **BUSINESS:** Maximum Revenue An airline flying to a Midwest destination can sell 20 coach-class tickets per day at a price of $250 and six business-class tickets per day at a price of $750. It finds that for each $10 decrease in the price of the coach ticket, it will sell four more per day, and for each $50 decrease in the business-class price, it will sell two more per day. What prices should the airline charge for the coach- and business-class tickets to maximize revenue? How many of each type will be sold at these prices? [*Hint:* Let x be the number of $10 price decreases for coach tickets and y be the number of $50 price decreases for business-class tickets, and use the method of Examples 1 and 2 on pages 203–204 for each type of ticket.]

Optimizing Functions of Three Variables

31. **BUSINESS:** Price Discrimination An automobile manufacturer sells cars in America, Europe, and Asia, charging a different price in each of the three markets. The price function for cars sold in America is $p = 20 - 0.2x$ (for $0 \le x \le 100$), the price function for cars sold in Europe is $q = 16 - 0.1y$ (for $0 \le y \le 160$), and the price function for cars sold in Asia is $r = 12 - 0.1z$ (for $0 \le z \le 120$), all in thousands of dollars, where x, y, and z are the numbers of cars sold in America, Europe, and Asia, respectively. The company's cost function is $C = 22 + 4(x + y + z)$ thousand dollars.

   a. Find the company's profit function $P(x, y, z)$. [*Hint:* The profit will be revenue from America plus revenue from Europe plus revenue from Asia minus costs, where each revenue is price times quantity.]
   b. Find how many cars should be sold in each market to maximize profit. [*Hint:* Set the three partials P_x, P_y, and P_z equal to zero and solve. Assuming that the maximum exists, it must occur at this point.]

32. **ECONOMICS: Competition and Collusion** Suppose that in the discussion of competition and collusion (pages 464–465), *two* of your neighbors began selling spring water. Use the price function $p = 36 - 0.01x$ (for $0 \le x \le 3600$) and repeat the analysis, but now comparing a monopoly with competition among *three* suppliers (a "triopoly"). That is:

    a. For a monopoly, calculate the quantity x that maximizes your revenue. Also calculate the price p and the revenue R.

    b. For a triopoly, find the quantities x, y, and z for the three suppliers that maximize revenue for each. Also calculate the price p and the three revenues. [*Hint:* Find the three revenue functions, one for each supplier, and maxmimize each with respect to that supplier's variable.]

    c. Are more goods produced under a monopoly or under a triopoly?

    d. Is the price lower under a monopoly or under a triopoly?

Exercises with More Than One Critical Point

33–38. Find the relative extreme values of each function.

33. $f(x, y) = x^3 + y^3 - 3xy$

34. $f(x, y) = x^5 + y^5 - 5xy$

35. $f(x, y) = 12xy - x^3 - 6y^2$

36. $f(x, y) = 6xy - x^3 - 3y^2$

37. $f(x, y) = 2x^4 + y^2 - 12xy$

38. $f(x, y) = 16xy - x^4 - 2y^2$

Conceptual Exercises

39. Maximum and minimum points occurred with functions of one variable as well as with functions of two variables. What *new* type of point becomes possible with functions of *two* variables?

40. True or False: If $f_x = 0$ and $f_y = 0$ at a point, then that point must be a relative maximum or minimum point.

41. True or False: If a function has no critical points, then it has no relative extreme points.

42. True or False: If a function has no relative extreme points, then it has no critical points.

43. If a function has a critical point at which $D > 0$, what can you conclude?

44. If a function has a critical point at which $D < 0$, what can you conclude?

45. If a function has a critical point at which $f_{xx} > 0$, what can you conclude?

46. In the D-test we did not discuss what happens if $D > 0$ and $f_{xx} = 0$. Why not? [*Hint:* Can this case occur?]

47. True or False: At a critical point, $D = 0$ means that the point is *not* a relative maximum or minimum point.

48. True or False: If f_{xx} and f_{yy} have different signs at a critical point, then that point is a saddle point.

7.4 Least Squares

Introduction

You may have wondered how the mathematical models in this book were developed. For example, how were the functions that predict populations found from census data, and how were the constants a and b in a Cobb–Douglas production function found from production data? The problem of finding a function that fits a collection of data can be viewed geometrically as the problem of fitting a curve to a collection of points. The simplest case of this problem is fitting a straight line to the points, and the most widely used method for doing this, and the method used by computers and calculators, is called **least squares.** You may have found regression lines and curves in Chapter 1 by pressing buttons on your calculator, and you have certainly used many functions in this book that were developed in this way. Although using a calculator or computer is both reasonable and practical, it is important to understand the method, and that is what this section explains.

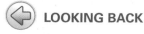 **LOOKING BACK**

Least squares was first mentioned in the mile run Application Preview on page 3. It was used throughout Chapter 1, where it was called *linear regression.*

A First Example

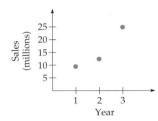

The graph on the left shows a company's annual sales (in millions) over a three-year period. How can we fit a straight line to these three points? Clearly, these points do not lie exactly on a line, and so rather than an "exact" fit, we want the line $y = ax + b$ that fits these three points most closely.

Let d_1, d_2, and d_3 stand for the vertical distances between the three points and the line $y = ax + b$. The line that minimizes the sum of the squares of these vertical deviations is called the *least squares line* or the *regression line*.* Squaring the deviations ensures that none are negative, so that a deviation below the line does not "cancel" one above the line.

EXAMPLE 1   MINIMIZING THE SQUARED DEVIATIONS

The table below gives a company's sales (in millions) in year x. Find the least squares line for the data. Then use the line to predict sales in year 4.

x	y
1	10
2	12
3	25

Solution

The vertical deviations are found by calculating the heights of the line $y = ax + b$ at each x-value minus the y-values from the table. These differences are then squared and summed.

$$S = (a \cdot 1 + b - 10)^2 + (a \cdot 2 + b - 12)^2 + (a \cdot 3 + b - 25)^2$$

Height of the line $y = ax + b$ at $x = 1$	y-value of the point at $x = 1$	Height of the line $y = ax + b$ at $x = 2$	y-value of the point at $x = 2$	Height of the line $y = ax + b$ at $x = 3$	y-value of the point at $x = 3$

This sum S depends on a and b, the numbers that determine the line $y = ax + b$. To minimize S, we set its partials with respect to a and b equal to zero:

---

*The word "regression" comes from an early use of this technique to determine whether unusually tall parents have unusually tall children. It seems that tall parents do have tall offspring, but not quite as tall, with successive generations exhibiting a "regression" toward the average height of the population.

$$\frac{\partial S}{\partial a} = 2(a + b - 10) + 2(2a + b - 12) \cdot 2 + 2(3a + b - 25) \cdot 3 \qquad \text{Differentiating each part of } S \text{ by the Generalized Power Rule}$$

$$= 2a + 2b - 20 + 8a + 4b - 48 + 18a + 6b - 150 \qquad \text{Multiplying out}$$

$$= 28a + 12b - 218 \qquad \text{Combining terms}$$

$$\frac{\partial S}{\partial b} = 2(a + b - 10) + 2(2a + b - 12) + 2(3a + b - 25) \qquad \text{Differentiating each part of } S \text{ by the Generalized Power Rule}$$

$$= 2a + 2b - 20 + 4a + 2b - 24 + 6a + 2b - 50 \qquad \text{Multiplying out}$$

$$= 12a + 6b - 94 \qquad \text{Combining terms}$$

We set the two partials equal to zero and solve:

$$\left. \begin{array}{l} 28a + 12b - 218 = 0 \\ 12a + 6b - 94 = 0 \end{array} \right\} \text{ Partials set equal to zero}$$

$$28a + 12b - 218 = 0 \qquad \text{First equation}$$

$$\underline{-24a - 12b + 188 = 0} \qquad \text{Second multiplied by } -2$$

$$4a \qquad\quad - 30 = 0 \qquad \text{Adding (the } b\text{'s drop out)}$$

$$a = \frac{30}{4} = 7.5 \qquad \text{Solving } 4a - 30 = 0$$

$$b = \frac{4}{6} \approx 0.67 \qquad \begin{array}{l}\text{From substituting } a = 7.5 \text{ into} \\ 12a + 6b - 94 = 0 \text{ and solving for } b\end{array}$$

These values for a and b give the least squares line. (The D-test would show that S has indeed been minimized.)

$$y = 7.5x + 0.67$$
$$\uparrow$$
$$y = ax + b \text{ with}$$
$$a = 7.5 \text{ and } b = 0.67$$

least squares line
$y = 7.5x + 0.67$

To predict the sales in year 4, we evaluate the least squares line at $x = 4$:

$$y = 7.5 \cdot 4 + 0.67 = 30.67 \qquad \begin{array}{l} y = 7.5x + 0.67 \\ \text{evaluated at } x = 4 \end{array}$$

Prediction: 30.67 million sales in year 4.

The slope of the line is 7.5, meaning that the linear trend in the company's sales is a growth of 7.5 million sales per year.

Graphing Calculator Exploration

on [0, 4] by [0, 30]

Verify that your graphing calculator gives the same least squares line $y = 7.5x + 0.67$ that we found in Example 1 by entering the data from the table on page 469 into your calculator and finding the linear regression line. The r that your calculator might find is a measure of "goodness of fit," with values closer to 1 indicating a better fit.

Least Squares Line for *n* Points

Example 1 used only three points, which is too few for most realistic applications. If we carry out the same steps for *n* points, we would obtain the following formulas (in which Σ stands for sum).

Least Squares Line

For data

x	y
x_1	y_1
x_2	y_2
.	.
.	.
x_n	y_n

calculate

$$a = \frac{n \sum xy - (\sum x)(\sum y)}{n \sum x^2 - (\sum x)^2}$$

$$b = \frac{1}{n}\left(\sum y - a \sum x\right)$$

n = number of data points
$\sum x$ = sum of x's
$\sum y$ = sum of y's
$\sum xy$ = sum of products $x \cdot y$
$\sum x^2$ = sum of squares of x's

The least squares line is then $y = ax + b$

 Be Careful $\sum x^2$ is *not* the same as $(\sum x)^2$.

From now on we will find least squares lines by using these formulas, a derivation of which is given on pages 475–476. While the derivation may appear complicated, it is just the procedure of Example 1 carried out for an arbitrary number of points.

EXAMPLE 2   **FINDING THE LEAST SQUARES LINE**

A study compared cigarette smoking with the mortality rate for lung cancer in several countries. Find the least squares line that fits these data. Then use the line to predict lung cancer deaths if per capita cigarette consumption is 600 cigarettes per month (a pack a day).

	Cigarette Consumption (per capita)	Lung Cancer Deaths (per million)
Norway	250	90
Sweden	300	120
Denmark	350	170
Canada	500	150

Source: www.deathsfromsmoking.net

Solution

The procedure for calculating a and b consists of six steps, beginning with the following table.

1. List the x- and y-values   2. Multiply $x \cdot y$ 3. Square each x

x	y	xy	x^2
250	90	22,500	62,500
300	120	36,000	90,000
350	170	59,500	122,500
500	150	75,000	250,000
1400	530	193,000	525,000
$\sum x$	$\sum y$	$\sum xy$	$\sum x^2$

← 4. Sum each column

5. Calculate a and b using the formulas above

$$a = \frac{(4)(193,000) - (1400)(530)}{(4)(525,000) - 1400^2}$$

$$= \frac{30,000}{140,000} \approx 0.21$$

$$b = \frac{1}{4}[530 - 0.21(1400)] = 59$$

n = number of points

$$a = \frac{n \sum xy - (\sum x)(\sum y)}{n \sum x^2 - (\sum x)^2}$$

$$b = \frac{1}{n}\left(\sum y - a \sum x\right)$$

6. The least squares line is $y = ax + b$ with the a and b values above

$$y = 0.21x + 59 \qquad y = ax + b \text{ with } a = 0.21 \text{ and } b = 59$$

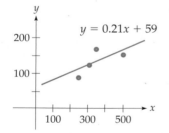

$y = 0.21x + 59$

This line is graphed on the left and shows that lung cancer mortality increases with cigarette smoking.

To predict the lung cancer deaths if per capita cigarette consumption reaches 600, we evaluate the least squares line at $x = 600$.

$$y = 0.21 \cdot 600 + 59 = 185 \qquad y = 0.21x + 59 \text{ with } x = 600$$

Predicted annual mortality from a pack a day: 185 deaths per million.

Criticism of Least Squares

Least squares is the most widely used method for fitting lines to points, but it does have one weakness: The vertical deviations from the line are squared, so one large deviation, when squared, can have an unexpectedly large influence on the line. For example, the graph on the right shows four points and their least squares line. The line fits the points quite closely.

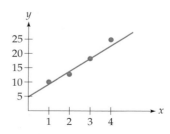

This graph adds a fifth point, one quite out of line with the others, and shows the least squares line for the five points. The added point has an enormous effect on the line, causing it to slope downward even though all of the other points suggest an upward slope. In actual applications, one should inspect the points for such "outliers" before calculating the least squares line.

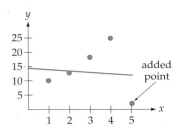

Fitting Exponential Curves by Least Squares

It is not always appropriate to fit a straight line to a set of points. Sometimes a collection of points will suggest a *curve* rather than a line, such as one of the following exponential curves.

Take Note

This technique is widely used to create exponential growth models from data, both in this book and elsewhere.

Least squares can be used to fit an exponential curve of the form $y = Be^{Ax}$ to a collection of points as follows. Taking natural logs of both sides of $y = Be^{Ax}$ gives

$$\ln y = \ln(Be^{Ax}) = \ln B + \ln e^{Ax} = \ln B + Ax \qquad \text{Using the properties of natural logarithms}$$

If we introduce a new variable $Y = \ln y$, and also let $b = \ln B$, then $\ln y = \ln B + Ax$ becomes linear:

$$Y = b + Ax \qquad\qquad \text{Or equivalently, } Y = Ax + b$$

Therefore, we fit a straight line to the *logarithms* of the y-values. (Now we are minimizing not the squared deviations, but the squared deviations of the logarithms.) The procedure consists of the eight steps shown in the following Example.

| EXAMPLE 3 | FITTING AN EXPONENTIAL CURVE BY LEAST SQUARES |

The world population since 1850 is shown in the table below. Fit an exponential curve to these data and predict the world population in the year 2050.

Year	Population (billions)
1850	1.26
1900	1.65
1950	2.52
2000	6.08

Source: U.S. Census Bureau

Solution

We number the years 1 through 4 (since they are evenly spaced).

1. List the x- and y-values
2. Take ln of the y-values (call them *capital Y*)
3. Multiply $x \cdot Y$ (capital Y)
4. Square each x

x	y	$Y = \ln y$	xY	x^2
1	1.26	0.231	0.231	1
2	1.65	0.501	1.002	4
3	2.52	0.924	2.772	9
4	6.08	1.805	7.220	16
10		3.461	11.225	30
$\parallel$		$\parallel$	$\parallel$	$\parallel$
Σx		ΣY	ΣxY	Σx^2

5. Add each column (except y)
6. Calculate A and b (n = number of points)

$$A = \frac{4(11.225) - 10(3.461)}{4(30) - 10^2} = \frac{10.290}{20} \approx 0.515 \qquad A = \frac{n \sum xY - (\sum x)(\sum Y)}{n \sum x^2 - (\sum x)^2}$$

$$b = \frac{1}{4}(3.461 - 0.515 \cdot 10) \approx -0.422 \qquad b = \frac{1}{n}(\sum Y - A \sum x)$$

$$B = e^b = e^{-0.422} \approx 0.656$$

7. Calculate $B = e^b$ (since $b = \ln B$)

The exponential curve is

$$y = 0.656 e^{0.515x}$$

8. The curve is $y = Be^{Ax}$ with the A and B found above

The population in 2050 is this function evaluated at $x = 5$:

$$y = 0.656 e^{0.515 \cdot 5} \approx 8.614$$

Therefore, the world population in the year 2050 is predicted to be about 8.6 billion.

Graphing Calculator Exploration

on $[0, 5]$ by $[0, 9]$

Verify that your graphing calculator gives the same curve $y = 0.656 e^{0.515x}$ that we found in Example 3 by entering the x- and y-values from the previous table into your calculator and finding the exponential regression curve. [You may find $y = 0.656 \cdot 1.673^x$, which is the same curve, since $\ln 1.673 \approx 0.515$ so $1.673^x \approx e^{0.515x}$.]

Section Summary

The technique of least squares can be used to determine the equation of a line that "best fits" a collection of points in the sense that it minimizes the sum of the squared vertical deviations. The method is described on pages 471–472. Least squares can also be used to fit a *curve* to a collection of points. The procedure for fitting an exponential curve $y = Be^{Ax}$ is described on pages 473–474. The method is widely used in forecasting and for detecting underlying trends in data.

To determine whether to fit a line or a curve to a collection of points, you should first graph the points. For example, the points in the left-hand graph below suggest a line, while those in the right-hand graph suggest an exponential curve.

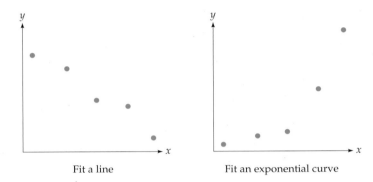

Fit a line   Fit an exponential curve

Derivation of the Formula for the Least Squares Line

The formulas for a and b in the least squares line come from minimizing the squared vertical deviations, just as in Example 1, but now for the n points (x_1, y_1), (x_2, y_2), ..., (x_n, y_n).

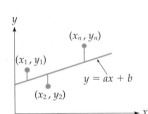

$$S = (ax_1 + b - y_1)^2 + (ax_2 + b - y_2)^2 + \cdots + (ax_n + b - y_n)^2$$

The partials are:

$$\frac{\partial S}{\partial a} = 2(ax_1 + b - y_1)x_1 + 2(ax_2 + b - y_2)x_2 + \cdots + 2(ax_n + b - y_n)x_n$$

$$= 2ax_1^2 + 2bx_1 - 2x_1y_1 + 2ax_2^2 + 2bx_2 - 2x_2y_2 + \cdots \qquad \text{Multiplying out}$$
$$+ 2ax_n^2 + 2bx_n - 2x_ny_n$$

$$= 2a(x_1^2 + x_2^2 + \cdots + x_n^2) + 2b(x_1 + x_2 + \cdots + x_n) \qquad \text{Regrouping}$$
$$- 2(x_1y_1 + x_2y_2 + \cdots + x_ny_n)$$

$$= 2a \sum x^2 + 2b \sum x - 2 \sum xy \qquad \text{Using } \Sigma \text{ for sum}$$

$$\frac{\partial S}{\partial b} = 2(ax_1 + b - y_1) + 2(ax_2 + b - y_2) + \cdots + 2(ax_n + b - y_n)$$

$$= 2ax_1 + 2b - 2y_1 + 2ax_2 + 2b - 2y_2 + \cdots + 2ax_n + 2b - 2y_n \quad \text{Multiplying out}$$

$$= 2a(x_1 + x_2 + \cdots + x_n) + 2b(1 + 1 + \cdots + 1) \qquad \text{Regrouping}$$
$$- 2(y_1 + y_2 + \cdots + y_n)$$

$$= 2a \sum x + 2bn - 2 \sum y \qquad \text{Using } \Sigma \text{ for sum}$$

We set the partials equal to zero:

$$a \sum x^2 + b \sum x - \sum xy = 0$$

$$a \sum x + bn - \sum y = 0$$

Dividing each by 2

To solve for a we multiply the first of these equations by n, the second by $\sum x$, and subtract (so that the b drops out):

$$an \sum x^2 + bn \sum x \ - n \sum xy = 0$$

$$a\left(\sum x\right)^2 + bn \sum x - \left(\sum x\right)\left(\sum y\right) = 0$$

Be Careful: $\sum xy$ is not the same as $(\sum x)(\sum y)$

$$a\left(n \sum x^2 - \left(\sum x\right)^2\right) - \left(n \sum xy - \left(\sum x\right)\left(\sum y\right)\right) = 0$$

Subtracting the last two equations

Solving this last equation for a gives:

$$a = \frac{n \sum xy - \left(\sum x\right)\left(\sum y\right)}{n \sum x^2 - \left(\sum x\right)^2}$$

Formula for a

For b we obtain:

$$b = \frac{1}{n}\left(\sum y - a \sum x\right)$$

$a \sum x + nb - \sum y = 0$ solved for b

These are the formulas in the box on page 471.

7.4 Exercises

Exercises on Least Squares Lines

1–8. Find the least squares line for each table of points.

1.

x	y
1	2
2	5
3	9

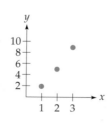

2.

x	y
1	2
2	4
3	7

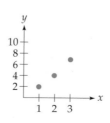

3.

x	y
1	6
3	4
6	2

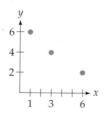

4.

x	y
1	9
4	6
5	1

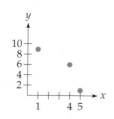

5.

x	y
0	7
1	10
2	10
3	15

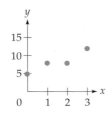

6.

x	y
0	5
1	8
2	8
3	12

7.

x	y
−1	10
0	8
1	5
3	0
5	−2

8.

x	y
−2	12
0	10
2	6
4	0
5	−3

Applied Exercises on Least Squares Lines

9. BUSINESS: Sales A company's annual sales are shown in the following table. Find the least squares line. Use your line to predict the sales in the next year $(x = 5)$.

Year	1	2	3	4
Sales (millions)	7	10	11	14

10. GENERAL: Automobile Costs The following table gives the cost per mile of operating a compact car, depending upon the number of miles driven per year. Find the least squares line for these data. Use your answer to predict the cost per mile for a car driven 25,000 miles annually $(x = 5)$.

Annual Mileage	x	Cost per Mile (cents)
5000	1	50
10,000	2	35
15,000	3	27
20,000	4	25

11. SOCIOLOGY: Crime A sociologist finds the following data for the number of felony arrests per year in a town. Find the least squares line. Then use it to predict the number of felony arrests in the next year.

Year	1	2	3	4
Arrests	120	110	90	100

12. GENERAL: Farming A farmer's wheat yield (bushels per acre) depends on the amount of fertilizer (hundreds of pounds per acre) according to the following table. Find the least squares line. Then use the line to predict the yield using 3 hundred pounds of fertilizer per acre.

Fertilizer	1.0	1.5	2.0	2.5
Yield	30	35	38	40

13. ATHLETICS: The Disappearance of the .400 Hitter Between 1901 and 1930, baseball boasted several .400 hitters (Lajoie, Cobb, Jackson, Sisler, Heilmann, Hornsby, and Terry), but only one since then (Ted Williams in 1941). The decline of the "heavy hitter" is evidenced by the following data, showing the highest batting average for each 20-year period (National or American League). Find the least squares line for these data and use it to predict the highest batting average for the period 2001–2020 $(x = 6)$.

Source: S. J. Gould, *Triumph and Tragedy in Mudville*

	x	Highest Average
1901–1920	1	.422
1921–1940	2	.424
1941–1960	3	.406
1961–1980	4	.390
1981–2000	5	.394

14. ECONOMICS: Consumer Price Index The consumer price index (CPI) is shown in the following table. Fit a least squares line to the data. Then use the line to predict the CPI in the year 2020 $(x = 7)$.

Source: U.S. Bureau of Labor Statistics

	x	CPI
1990	1	130.7
1995	2	152.4
2000	3	172.2
2005	4	195.3
2010	5	218.1

15–16. GENERAL: Percentage of Smokers The following tables show the percentage of smokers among the adult population in the United States for every five years since 1990 (the first table is males, the second females). Find the least squares lines for these data, and use your answer to predict the percentage of that sex who will smoke in the year 2020 $(x = 7)$.

Source: U.S. Department of Health and Human Services

15.

	x	Percent Males
1990	1	28
1995	2	27
2000	3	25
2005	4	23
2010	5	23

16.

	x	Percent Females
1990	1	23
1995	2	23
2000	3	21
2005	4	18
2010	5	18

17. BIOMEDICAL: Smoking and Longevity The following data show the life expectancy of a 25-year-old male based on the number of cigarettes smoked daily. Find the least squares line for these data. The slope of the line estimates the years lost per extra cigarette per day.

Cigarettes Smoked Daily	Life Expectancy
0	73.6
5	69.0
15	68.1
30	67.4
40	65.3

18. ENVIRONMENTAL SCIENCES: Pollution and Absenteeism The following table shows the relationship between the sulfur dust content of the air (in micrograms per cubic meter) and the number of female absentees in industry. (Only absences of at least seven days were counted.) Find the least squares line for these data. Use your answer to predict absences in a city with a sulfur dust content of 25.

	Sulfur	Absences per 1000 Employees
Cincinnati	7	19
Indianapolis	13	44
Woodbridge	14	53
Camden	17	61
Harrison	20	88

Exercises on Fitting Exponential Curves

19–26. Use least squares to find the exponential curve $y = Be^{Ax}$ for the following tables of points.

19.

x	y
1	2
2	4
3	7

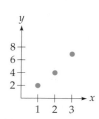

20.

x	y
1	3
2	6
3	11

21.

x	y
1	10
3	5
6	1

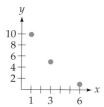

22.

x	y
1	12
4	3
5	2

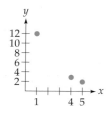

23.

x	y
0	1
1	2
2	5
3	10

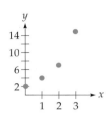

24.

x	y
0	2
1	4
2	7
3	15

25.

x	y
−1	20
0	18
1	15
3	4
5	1

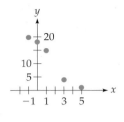

26.

x	y
−2	20
0	12
2	9
4	6
5	5

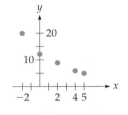

Applied Exercises on Fitting Exponential Curves

27. POLITICAL SCIENCE: Cost of a Congressional Victory The following table shows average amounts spent by winners of seats in the House of Representatives in recent election years. Fit an exponential curve to these data and use it to predict the cost of a House seat in the year 2020.

Source: Center for Responsive Politics

	x	Cost (million $)
2008	1	1.373
2010	2	1.440
2012	3	1.567

28. GENERAL: Drunk Driving The following table shows how a driver's blood-alcohol level (% grams per dekaliter) affects the probability of being in a collision. A collision factor of 3 means that the probability of a collision is 3 times as large as normal. Fit an exponential curve to the data. Then use your curve to estimate the collision factor for a blood-alcohol level of 15.

Blood-Alcohol Level	Collision Factor
0	1
6	1.1
8	3
10	6

29. BUSINESS: Super Bowl Tickets The following table gives the maximum price of a Super Bowl ticket in 2014 and at preceding decades. Fit an exponential

curve to the data and then use it to predict the minimum cost of a ticket in 2024.

Source: National Football League

Year	x	Price
1994	1	175
2004	2	500
2014	3	2,600

30. BUSINESS: Movie Prices The following table shows the national average theater admission prices for recent decades. Fit an exponential curve to these data and use your answer to predict the cost of a movie ticket in the year 2020.

Source: Entertainment Weekly

	x	Price
1980	1	2.69
1990	2	7.22
2000	3	5.39
2010	4	7.85

31. GENERAL: Aging World Population The number of children in the world (age 14 years and younger) has always exceeded the number of elderly (age 60 years and older), but changes in fertility and mortality are increasing the elderly population much faster than the children's population. The following table gives these populations (in billions) for past years and predicted for 2025. Fit an *exponential curve* to the elderly data and a *line* to the children's data (it is roughly linear). Then evaluate each function at $x = 4$ to see which population will be larger at that time. What year does $x = 4$ correspond to? This aging of the world population has enormous economic consequences.

Source: U.N. Department of Economic and Social Affairs

Year	x	Elderly	Children
1985	1	0.4	1.6
2005	2	0.7	1.8
2025	3	1.2	1.9

32. GENERAL: Cost of College Education   The following table shows the cost (tuitions, fees, and room and board, in

2013 dollars) of one year at a public four-year college. Fit an exponential curve to the data and then use it to predict the cost for the 2023–2024 academic year.

Source: The College Board

	x	Cost ($1000)
1998–99	1	11.1
2003–04	2	13.4
2008–09	3	15.3
2013–14	4	18.4

Conceptual Exercises

33. True or False: The least squares line will always pass through at least *one* of the data points.

34. True or False: If there are just *two* data points, the least squares line will be the line that passes through them. (Assume that the x-coordinates of the points are different.)

35. True or False: If the data points all lie on a line, then the least squares line for the data will *be* that line. (Assume that the line is not vertical.)

36. Suppose that the least squares line for a set of data points is $y = ax + b$. If you *doubled* each y-value, what would be the new least squares line? [*Hint:* How has the line been changed?]

37. Suppose that the least squares line for a set of data points is $y = ax + b$. If you *added 5* to each y-value, what would be the new least squares line? [*Hint:* How has the line been changed?]

38. What do you get if you find the least squares line for just *one* data point? [*Hint:* Try it for the point (1, 2).]

39. What goes wrong if you try to find the least squares line for just *two* data points and they have the *same* x-coordinate? [*Hint:* Try it for the points (1, 2) and (1, 5).]

40. What goes wrong if you try to fit an exponential curve to data to just *one* data point? [*Hint:* Try it for the point (1, 2).]

41. What goes wrong if you try to fit an exponential curve to data and one of the points has a y-coordinate that is 0 or a negative number? [*Hint:* Look at the table on page 474.]

42. Suppose when you fit an exponential curve to a set of data points you obtain the equation $y = Be^{Ax}$. If you *doubled* each y-value, what would be the new exponential curve? [*Hint:* How has the curve been changed?]

7.5   Lagrange Multipliers and Constrained Optimization

Introduction

In Section 7.3 we optimized functions of several variables. Some problems, however, involve maximizing or minimizing a function subject to a **constraint.** For example, a company might want to maximize production subject to the constraint of staying within its budget, or a soft drink distributor might want to design the least expensive aluminum can subject to the constraint that it hold exactly 12 ounces of soda. In this section we solve such "constrained optimization" problems by the method of **Lagrange multipliers,** invented by the French mathematician Joseph Louis Lagrange (1736–1813).

Constraints

If a constraint can be written as an equation, such as

$$x^2 + y^2 = 100$$

then by moving all the terms to the left-hand side, it can be written with zero on the right-hand side:

$$\underbrace{x^2 + y^2 - 100}_{g(x,\, y)} = 0$$

In general, any equation can be written with all terms moved to the left-hand side, and we will write all constraints in the form

$$g(x, y) = 0$$

PRACTICE PROBLEM 1

Write the constraint $2y^3 = 3x - 1$ in the form $g(x, y) = 0$.

Solution on page 489 >

First Example of Lagrange Multipliers

We illustrate the method of Lagrange multipliers by an example. The method requires a new variable, and it is customary to use λ ("lambda," the Greek letter l, in honor of Lagrange).

EXAMPLE 1   MAXIMIZING THE AREA OF AN ENCLOSURE

A farmer wants to build a rectangular enclosure along an existing stone wall. If the side along the wall needs no fence, find the dimensions of the largest enclosure that can be made using only 400 feet of fence.

Solution

We want the enclosure of largest *area*. Let

$$x = \text{width (perpendicular to the wall)}$$

$$y = \text{length (parallel to the wall)}$$

Two widths and one length must be made from the 400 feet of fence, so the constraint is

$$2x + y = 400$$

Therefore, the problem becomes

maximize $A = xy$ Area is length times width

subject to $\underbrace{2x + y - 400}_{g(x,\, y)} = 0$ Constraint $2x + y = 400$ written with zero on the right

We write a new function $F(x, y, \lambda)$, called the **Lagrange function,** which consists of the function to be maximized plus λ times the constraint function:

$$F(x, y, \lambda) = xy + \lambda(2x + y - 400)$$

$$F(x, y, \lambda) = \begin{pmatrix} \text{Function to} \\ \text{be optimized} \end{pmatrix}$$
$$+ \lambda \begin{pmatrix} \text{Constraint} \\ \text{function} \end{pmatrix}$$

$$= xy + \lambda 2x + \lambda y - \lambda 400 \qquad \text{Multiplied out}$$

The Lagrange function $F(x, y, \lambda)$ is a function of *three* variables, x, y, and λ, and we begin as usual by setting its partials with respect to each variable equal to zero.

$$F_x = y + 2\lambda \qquad = 0 \qquad \text{Partial of } xy + \lambda 2x + \lambda y - \lambda 400 \text{ with respect to } x$$

$$F_y = x + \lambda \qquad = 0 \qquad \text{Partial of } xy + \lambda 2x + \lambda y - \lambda 400 \text{ with respect to } y$$

$$F_\lambda = 2x + y - 400 = 0 \qquad \text{Partial of } xy + \lambda 2x + \lambda y - \lambda 400 \text{ with respect to } \lambda$$

Constraint $g = 0$

We solve the first two of these equations for λ:

$$\lambda = -\frac{1}{2}y \qquad\qquad \text{Solving } y + 2\lambda = 0 \text{ for } \lambda$$

$$\lambda = -x \qquad\qquad \text{Solving } x + \lambda = 0 \text{ for } \lambda$$

Then we set these two expressions for λ equal to each other:

$$-\frac{1}{2}y = -x \qquad\qquad \text{Equating } \lambda = -\frac{1}{2}y \text{ and } \lambda = -x$$

$$y = 2x \qquad\qquad \text{Multiplying by } -2$$

We use $y = 2x$ to eliminate the y in the equation $2x + y - 400 = 0$ (which came from the third partial $F_\lambda = 0$):

$$2x + 2x - 400 = 0 \qquad 2x + y - 400 = 0 \text{ with } y \text{ replaced by } 2x$$

$$4x = 400 \qquad \text{Simplifying}$$

$$x = 100 \qquad \text{Dividing by 4}$$

$$y = 200 \qquad \text{From } y = 2x \text{ with } x = 100$$

The largest possible enclosure has width 100 feet (perpendicular to the wall) and length 200 feet (parallel to the wall).

wall

100'   100'

200'

Method of Lagrange Multipliers

In general, the function to be maximized or minimized is called the **objective function,** because the "objective" of the whole procedure is to optimize it. In Example 1 the objective function was the area function $A = xy$, which was to be maximized. The variable λ is called the **Lagrange multiplier.** The entire method may be summarized as follows. A justification of Lagrange's method is given on pages 489–490.

Lagrange Multipliers

To optimize $f(x, y)$ subject to $g(x, y) = 0$:

Objective function and constraint

1. Write $F(x, y, \lambda) = f(x, y) + \lambda g(x, y)$.

   Objective function plus λ times the constraint function

2. Set the partials of F equal to zero:
   $$F_x = 0 \qquad F_y = 0 \qquad F_\lambda = 0$$
   and solve for the critical points.

3. The solution to the original problem (if it exists) will occur at one of these critical points.

It is important to realize that Lagrange's method only finds *critical points*—it does not tell whether the function is maximized, minimized, or neither at the critical point. (The *D*-test, which involves calculating $D = f_{xx}f_{yy} - (f_{xy})^2$, is for *unconstrained* optimization and cannot be used with constraints.) In each problem we must *know* that the maximum or minimum (whichever is requested) *does* exist, and it then follows that it must occur at a critical point (found by Lagrange multipliers).

How do we know that the maximum or minimum does exist? Most reasonable applied problems *do* have solutions. In Example 1, for instance, making the width or length too small would make the area (length times width) small, so the area must be largest for some "intermediate" length and width. Therefore, the problem *does* have a solution, which is what Lagrange multipliers find.

We could have solved Example 1 *without* Lagrange multipliers by solving the constraint equation $2x + y = 400$ for y and using it to eliminate the y in the objective function, as we did in Sections 3.3 and 3.4. However, Lagrange's method has two advantages: It eliminates the need to solve the constraint equation (which can sometimes be difficult), and it is symmetric in the variables, thereby allowing you to solve for whichever variable is easier to find.

Hints for Solving the Partial Equations

Solving the partial derivative equations $F_x = 0$, $F_y = 0$, and $F_\lambda = 0$ can sometimes be difficult. The following strategy (used in each example in this section) may be helpful.

1. Solve each of $F_x = 0$ and $F_y = 0$ for λ.
2. Set the two expressions for λ equal to each other.
3. Solve the equation resulting from step 2 together with $F_\lambda = 0$ to find x and y.

EXAMPLE 2	DESIGNING THE MOST EFFICIENT CONTAINER

A container company wants to design an aluminum can that requires the least amount of aluminum but that contains exactly 12 fluid ounces (21.3 cubic inches). Find the radius and height of the can.

Solution

Minimizing the amount of aluminum means minimizing the surface area (top plus bottom plus sides) of the cylindrical can. If we let r and h stand for the radius and height (in inches) of the can, then the diagram shows that the area is:

$$A = 2\pi r^2 + 2\pi rh$$

Top and bottom area   Side area

$A = \pi r^2$

$2\pi r$

side area $= 2\pi rh$
(unrolled)

circumference $2\pi r$

The volume is

$$V = \pi r^2 h$$

Therefore, the problem becomes

minimize	$A = 2\pi r^2 + 2\pi rh$	Area = (top and bottom) + side
subject to	$\pi r^2 h = 21.3$	Volume must equal 21.3 in.[3]

The Lagrange function is

$$F = 2\pi r^2 + 2\pi rh + \lambda(\pi r^2 h - 21.3)$$ Objective function A plus λ times the constraint

$$F_r = 4\pi r + 2\pi h + \lambda 2\pi rh = 0$$
$$F_h = 2\pi r + \lambda\pi r^2 \qquad\quad = 0$$
$$F_\lambda = \pi r^2 h - 21.3 \qquad\quad = 0$$

Partials set equal to zero

Solving, we have

$$\lambda = -\frac{4\pi r + 2\pi h}{2\pi rh} = -\frac{2r + h}{rh}$$ $4\pi r + 2\pi h + \lambda 2\pi rh = 0$ solved for λ

$$\lambda = -\frac{2\pi r}{\pi r^2} = -\frac{2}{r}$$ $2\pi r + \lambda\pi rh^2 = 0$ solved for λ

Equating λ's yields

$$\frac{2r + h}{rh} = \frac{2}{r}$$ Equating the two expressions for λ and multiplying each side by -1

$$2r + h = 2h$$ Multiplying each side by rh

$$2r = h$$ Subtracting h from each side

Therefore,

$$\pi r^2(2r) = 21.3 \qquad \begin{array}{l} \pi r^2 h - 21.3 = 0 \quad \text{(the third partial} \\ \text{equation) with } h = 2r \end{array}$$

$$2\pi r^3 = 21.3 \qquad \text{Simplifying}$$

$$r^3 = \frac{21.3}{2\pi} \approx 3.39 \qquad \text{Dividing by } 2\pi \text{ (using a calculator)}$$

$$r \approx \sqrt[3]{3.39} \approx 1.5 \qquad \text{Taking cube roots}$$

$$h = 3 \qquad \text{From } h = 2r \text{ with } r = 1.5$$

The most economical 12-fluid-ounce can has radius $r = 1.5$ inches and height $h = 3$ inches.

Notice that the height (3 inches) is twice the radius (1.5 inches), *so the height equals the diameter.* This shows that the most efficient can (least area for given volume) has a "squarish" shape.

It is interesting to consider why so few cans are shaped like this. The most common 12-ounce can for soft drinks is about twice as tall as it is across, requiring about 67% more aluminum than the most efficient can. This results in an enormous waste for the millions of cans manufactured each year. It seems that beverage companies prefer taller cans because they have more area for advertising, and they are easier to hold. Some products, however, are sold in "efficient" cans, with the height equal to the diameter.

PRACTICE PROBLEM 2

Which of the cans pictured below is most "efficiently" proportioned?

cola   paint   motor oil   tuna

Solution on page 489 >

With Lagrange multipliers we can maximize a company's output subject to a budget constraint.

EXAMPLE 3   MAXIMIZING PRODUCTION

A company's output is given by the Cobb–Douglas production function $P = 840L^{2/3}K^{1/3}$, where L and K are the numbers of units of labor and capital that are used. To stay within its budget of \$2520, these amounts must satisfy $35L + 140K = 2520$ (known as the *isocost line*). Find the amounts of labor and capital that maximize production. What is the maximum number of items that can be produced?

Solution

The problem can be stated:

maximize $P = 840L^{2/3}K^{1/3}$

subject to $35L + 140K - 2520 = 0$ Constraint with zero on the right

$F(L, K, \lambda) = 840L^{2/3}K^{1/3} + \lambda(35L + 140K - 2520)$ Lagrange function

$$\left. \begin{aligned} F_L &= 560L^{-1/3}K^{1/3} + 35\lambda = 0 \\ F_K &= 280L^{2/3}K^{-2/3} + 140\lambda = 0 \\ F_\lambda &= 35L + 140K - 2520 = 0 \end{aligned} \right\}$$
Partials set equal to zero

$\lambda = -16L^{-1/3}K^{1/3}$ Solving the first equation for λ

$\lambda = -2L^{2/3}K^{-2/3}$ Solving the second equation for λ

$-16L^{-1/3}K^{1/3} = -2L^{2/3}K^{-2/3}$ Equating the two expressions for λ

$8L^{-1/3}K^{1/3} = L^{2/3}K^{-2/3}$ Dividing each side by -2

$8K = L$ Multiplying by $L^{1/3}K^{2/3}$ on each side

$35(8K) + 140K - 2520 = 0$ Substituting $L = 8K$ into $35L + 140K - 2520 = 0$

$420K = 2520$ Simplifying

$K = 6$ Solving for K

$L = 48$ From $L = 8K$ with $K = 6$

The company should use 48 units of labor and 6 units of capital. The number of items that can then be produced is

$P = 840(48)^{2/3}(6)^{1/3}$ Substituting $L = 48$, $K = 6$ into $P = 840L^{2/3}K^{1/3}$

$= 20{,}160$ Using a calculator

The company can produce a maximum of 20,160 items.

Meaning of the Lagrange Multiplier

The Lagrange multiplier λ has a useful interpretation. If we call the units of the objective function "objective units" and the units of the constraint function "constraint units," then λ (or, more precisely, its absolute value) has the following interpretation, which will be justified in the next section (see pages 501–502).

Interpretation of λ

$$|\lambda| = \begin{pmatrix} \text{Number of additional objective units} \\ \text{for each additional constraint unit} \end{pmatrix}$$

As an illustration, in the preceding Example we maximized production (units) subject to a budget constraint (dollars), so $|\lambda|$ gives *the number of additional units produced per additional dollar,* or, in the language of economics, *the marginal productivity of money.* We can calculate the value of λ from either of the expressions for it.

$$\lambda = -16L^{-1/3}K^{1/3} = -16(48)^{-1/3}(6)^{1/3} = -8$$

From Example 3   Substituting $L = 48$, $K = 6$ Using a calculator   From the previous page

Therefore* $|\lambda| = 8$, meaning that production increases by about 8 units for each additional dollar in the budget. For example, an additional \$100 would result in about 800 extra units of production.

PRACTICE PROBLEM 3

In Example 1 we maximized the area of an enclosure subject to the constraint of having only 400 feet of fence.

a. Interpret the meaning of $|\lambda|$ in Example 1. [*Hint:* The objective function is area, and the constraint units are feet of fence.]
b. Calculate $|\lambda|$ in Example 1. [*Hint:* Use either expression for λ on page 482.]
c. Use this number to approximate the additional area that could be enclosed by an additional 5 feet of fence.

Solutions on page 489 >

Geometry of Constrained Optimization

A constrained optimization problem can be visualized as a surface (the objective function), with a curve along it determined by the constraint. The *constrained* maximum is the highest point along the *curve*, while the *unconstrained* maximum is the highest point on the *entire surface.*

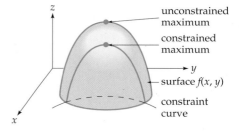

If there are several critical points, we evaluate the objective function at each of them. The maximum and minimum values (which we assume to exist) are the largest and smallest of the resulting values of the function. In this way we can find both the maximum and minimum values in a constrained optimization problem.

EXAMPLE 4   FINDING BOTH EXTREME VALUES

Maximize *and* minimize $f(x, y) = 4xy$
subject to the constraint $x^2 + y^2 = 50$

Solution

$$F(x, y, \lambda) = 4xy + \lambda(x^2 + y^2 - 50)$$ Lagrange function

$$F_x = 4y + \lambda 2x \quad = 0$$

$$F_y = 4x + \lambda 2y \quad = 0$$ Partials set equal to zero

$$F_\lambda = x^2 + y^2 - 50 = 0$$

---

*The negative sign in λ comes from our defining the Lagrange function as $F = f + \lambda g$. We could equally well have defined it as $F = f - \lambda g$, in which case λ would have been positive.

$$\lambda = -\frac{4y}{2x} = -\frac{2y}{x}$$

Solving $4y + \lambda 2x = 0$ for λ (and simplifying)

$$\lambda = -\frac{4x}{2y} = -\frac{2x}{y}$$

Solving $4x + \lambda 2y = 0$ for λ (and simplifying)

$$\frac{2y}{x} = \frac{2x}{y}$$

Equating the two λ's (multiplying by -1)

$$2y^2 = 2x^2$$

Multiplying both sides by xy (or "cross-multiplying")

$$y^2 = x^2$$

Dividing by 2

$$y = \pm x$$

Taking square roots ($+$ and $-$)

$$x^2 + x^2 - 50 = 0$$

$x^2 + y^2 - 50 = 0$ (the third partial equation) with $y = \pm x$

$$2x^2 - 50 = 0$$

Simplifying

$$2(x^2 - 25) = 0$$

Factoring

$$\underbrace{(x + 5)(x - 5)}$$

$$x = \pm 5$$

Solving

$$y = \pm 5$$

From $y = \pm x$

Thus there are *four* critical points (all possible combinations of $x = \pm 5$ and $y = \pm 5$).   We evaluate the objective function $f(x, y) = 4xy$ at each of them.

Critical point	$f(x, y) = 4xy$
(5, 5)	100
(−5, −5)	100
(5, −5)	−100
(−5, 5)	−100

$\left.\begin{array}{l}\end{array}\right\}$ 100 is the highest value of $f = 4xy$, occurring at (5, 5) and at (−5, −5)

$\left.\begin{array}{l}\end{array}\right\}$ −100 is the lowest value of $f = 4xy$, occurring at (−5, 5) and at (5, −5)

Maximum value of f is 100, occurring at $\begin{cases} x = 5 \\ y = 5 \end{cases}$ and $\begin{cases} x = -5 \\ y = -5 \end{cases}$

Minimum value of f is −100, occurring at $\begin{cases} x = -5 \\ y = 5 \end{cases}$ and $\begin{cases} x = 5 \\ y = -5 \end{cases}$

 Be Careful   When solving an equation such as $y^2 = x^2$ in the previous Example, be sure to consider all possible combinations of $+$ and $-$ signs so that you find *all* of the critical points.

Constrained Optimization of Functions of Three or More Variables

To optimize a function f of any number of variables subject to a constraint $g = 0$, we proceed just as before, writing the Lagrange function $F = f + \lambda g$ and setting the partial with respect to each variable equal to zero. For example, to maximize $f(x, y, z)$ subject to $g(x, y, z) = 0$, we would write

$$F(x, y, z, \lambda) = f(x, y, z) + \lambda g(x, y, z)$$ Lagrange function

and solve the partial equations

$$F_x = 0$$
$$F_y = 0$$
$$F_z = 0$$
$$F_\lambda = 0$$

Solutions TO PRACTICE PROBLEMS

1. $2y^3 - 3x + 1 = 0$ Moving all terms to the left-hand side

2. The paint can

3. a. The approximate additional area for each additional foot of fence

   b. $|\lambda| = 100$ (using $\lambda = -x$ with $x = 100$)

   c. Approximately 500 more square feet of area (from $5 \cdot 100$)

7.5   Section Summary

To optimize (maximize or minimize) an *objective* function $f(x, y)$ subject to a *constraint* $g(x, y) = 0$, we form the *Lagrange function* $F(x, y, \lambda) = f(x, y) + \lambda g(x, y)$ and set its partials equal to zero:

$$f_x(x, y) + \lambda g_x(x, y) = 0 \qquad\qquad F_x = 0$$
$$f_y(x, y) + \lambda g_y(x, y) = 0 \qquad\qquad F_y = 0$$
$$g(x, y) = 0 \qquad\qquad F_\lambda = 0$$

We then find all solutions (x, y) of these equations. The maximum or minimum (if they exist) of $f(x, y)$ subject to $g(x, y) = 0$ will occur at one of these solutions.

   The absolute value of the *Lagrange multiplier* λ found from solving these equations gives the approximate change in the *objective* function that would result from changing the *constraint* function by one unit.

Justification of Lagrange's Method

The following is a geometric justification of Lagrange's method for maximizing an objective function $f(x, y)$ subject to a constraint $g(x, y) = 0$. In the graph below, each of the parallel curves represents a curve along which the objective function f takes a constant value, with successively higher curves corresponding to higher values of the constant. Maximizing f subject to $g = 0$ means finding the highest of these "objective" curves that still meets the constraint curve.

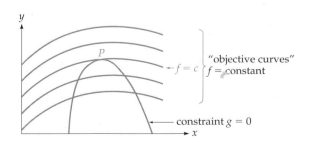

Observe that at the point where the highest objective curve meets the constraint curve (the point P), *the two curves have the same slope.* In Exercise 45 on page 504 we will show that the slope of the curve $f = c$ is $-f_x/f_y$, and the slope of the curve $g = 0$ is $-g_x/g_y$. Equating these slopes gives

$$-\frac{f_x}{f_y} = -\frac{g_x}{g_y} \quad \text{or, equivalently,} \quad -\frac{f_x}{g_x} = -\frac{f_y}{g_y} \qquad \text{Multiplying by } f_y \text{ and dividing by } g_x$$

The two sides of this last equation being equal is equivalent to equating each side to a number λ:

$$\lambda = -\frac{f_x}{g_x} \quad \text{and} \quad \lambda = -\frac{f_y}{g_y}$$

With a little algebra (clearing fractions, and moving everything to one side), these equations become

$$f_x + \lambda g_x = 0 \quad \text{and} \quad f_y + \lambda g_y = 0$$

However, these two equations, together with the constraint $g = 0$, are precisely the equations obtained from Lagrange's method by setting the partials of $F = f + \lambda g$ equal to zero. This completes the justification of Lagrange's method, and also exhibits its simplicity and power by showing all that is accomplished by the familiar steps of setting the partials equal to zero.

7.5 Exercises

1–10. Use Lagrange multipliers to maximize each function $f(x, y)$ subject to the constraint. (The maximum values *do* exist.)

1. $f(x, y) = 3xy, \quad x + 3y = 12$

2. $f(x, y) = 2xy, \quad 2x + y = 20$

3. $f(x, y) = 6xy, \quad 2x + 3y = 24$

4. $f(x, y) = 3xy, \quad 3x + 2y = 60$

5. $f(x, y) = xy - 2x^2 - y^2, \quad x + y = 8$

6. $f(x, y) = 12xy - 3y^2 - x^2, \quad x + y = 16$

7. $f(x, y) = x^2 - y^2 + 3, \quad 2x + y = 3$

8. $f(x, y) = y^2 - x^2 - 5, \quad x + 2y = 9$

9. $f(x, y) = \ln(xy), \quad x + y = 2e$

10. $f(x, y) = e^{xy}, \quad x + 2y = 8$

11–20. Use Lagrange multipliers to minimize each function $f(x, y)$ subject to the constraint. (The minimum values *do* exist.)

11. $f(x, y) = x^2 + y^2, \quad 2x + y = 15$

12. $f(x, y) = x^2 + y^2, \quad x + 2y = 30$

13. $f(x, y) = xy, \quad y = x + 8$

14. $f(x, y) = xy, \quad y = x + 6$

15. $f(x, y) = x^2 + y^2, \quad 2x + 3y = 26$

16. $f(x, y) = 5x^2 + 6y^2 - xy, \quad x + 2y = 24$

17. $f(x, y) = \ln(x^2 + y^2), \quad 2x + y = 25$

18. $f(x, y) = \sqrt{x^2 + y^2 + 5}, \quad 2x + y = 10$

19. $f(x, y) = e^{x^2 + y^2}, \quad x + 2y = 10$

20. $f(x, y) = 2x + y, \quad 2 \ln x + \ln y = 12$

21–24. Use Lagrange multipliers to maximize *and* minimize each function subject to the constraint. (The maximum and minimum values *do* exist.)

21. $f(x, y) = 2xy, \quad x^2 + y^2 = 8$

22. $f(x, y) = 2xy, \quad x^2 + y^2 = 18$

23. $f(x, y) = x + 2y, \quad 2x^2 + y^2 = 72$

24. $f(x, y) = 12x + 30y, \quad x^2 + 5y^2 = 81$

Applied Exercises

25–42. Solve each using Lagrange multipliers. (The stated extreme values *do* exist.)

25. GENERAL: Parking Lot Design A parking lot, divided into two equal parts, is to be constructed against a building, as shown in the diagram. Only 6000 feet of fence are to be used, and the side along the building needs no fence.

a. What are the dimensions of the largest area that can be so enclosed?

b. Evaluate and give an interpretation for $|\lambda|$.

parking lot

26. GENERAL: Fences   Three adjacent rectangular lots are to be fenced in, as shown in the diagram using 12,000 feet of fence. What is the largest total area that can be so enclosed?

27–28.   GENERAL: Container Design   A cylindrical tank without a top is to be constructed with the least amount of material (bottom plus side area). Find the dimensions if the volume is to be:

open top

h

27. 160 cubic feet.

28. 120 cubic feet.

29. GENERAL: Postal Regulations   The U.S. Postal Service will accept a package if its length plus its girth is not more than 84 inches. Find the dimensions and volume of the largest package with a square end that can be mailed.

Source: U.S. Postal Service

length

square

girth (all the way around)

30. GENERAL: Postal Regulations   Solve Exercise 29, but now for a package with a round end, so that the package is a cylinder rather than a rectangular solid. Compare the volume with that of Exercise 29.

length

girth

31–33.   BUSINESS: Maximum Production   For each Cobb–Douglas production function P and isocost line (budget constraint, in dollars), find the amounts of labor L and capital K that maximize production, and also find the maximum production. Then evaluate and give an interpretation for $|\lambda|$ and use it to answer the question.

31. a. Maximize $P = 280L^{3/4}K^{1/4}$ with budget constraint $525L + 0.28K = 16{,}800$.
   b. Evaluate and give an interpretation for $|\lambda|$.
   c. Approximate the increase in production if the budget is increased by $300.

32. a. Maximize $P = 2000L^{3/5}K^{2/5}$ with budget constraint $15L + 320K = 8000$.
   b. Evaluate and give an interpretation for $|\lambda|$.
   c. Approximate the increase in production if the budget is increased by $100.

33. a. Maximize $P = 180L^{1/2}K^{1/2}$ with budget constraint $45L + 20K = 2520$.
   b. Evaluate and give an interpretation for $|\lambda|$.
   c. Approximate the increase in production if the budget is increased by $50.

34. BUSINESS: Production Possibilities   A company manufactures two products, in quantities x and y. Because of limited materials and capital, the quantities produced must satisfy the equation $2x^2 + 5y^2 = 32{,}500$. (This curve is called a *production possibilities curve.*) If the company's profit function is $P = 4x + 5y$ dollars, how many of each product should be made to maximize profit? Also find the maximum profit.

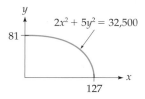

35. GENERAL: Package Design   A metal box with a square base is to have a volume of 45 cubic inches. If the top and bottom cost 50 cents per square inch and the sides cost 30 cents per square inch, find the dimensions that minimize the cost. [*Hint:* The cost of the box is the area of each part (top, bottom, and sides) times the cost per square inch for that part. Minimize this subject to the volume constraint.]

36. GENERAL: Building Design A one-story building is to have 8000 square feet of floor space. The front of the building is to be made of brick, which costs $120 per linear foot, and the back and sides are to be made of cinderblock, which costs only $80 per linear foot.

  a. Find the length and width that minimize the cost of the building. [*Hint:* The cost of the building is the length of the front, back, and sides, each times the cost per foot for that part. Minimize this subject to the area constraint.]
  b. Evaluate and give an interpretation for $|\lambda|$.

$80 per ft

$80 per ft   8000 sq. ft $80 per ft

$120 per ft

Functions of Three Variables
(The stated extreme values *do* exist.)

37. Minimize $f(x, y, z) = x^2 + y^2 + z^2$
    subject to $2x + y - z = 12$.

38. Minimize $f(x, y, z) = x^2 + y^2 + z^2$
    subject to $x - y + 2z = 6$.

39. Maximize $f(x, y, z) = x + y + z$
    subject to $x^2 + y^2 + z^2 = 12$.

40. Maximize $f(x, y, z) = xyz$
    subject to $x^2 + y^2 + z^2 = 12$.

41. GENERAL: Building Design A one-story storage building is to have a volume of 2000 cubic feet. The roof costs $32 per square foot, the walls $10 per square foot, and the floor $8 per square foot. Find the dimensions that minimize the cost of the building.

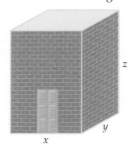

42. GENERAL: Container Design An open-top box with two parallel partitions, as in the diagram, is to have volume 64 cubic inches. Find the dimensions that require the least amount of material.

Conceptual Exercises

43. True or False: If $f(x, y)$ is the objective function and $g(x, y)$ is the constraint function, then the Lagrange function is

$$F(x, y, \lambda) = g(x, y) + \lambda f(x, y)$$

44. True or False:

$$|\lambda| = \left(\begin{array}{c} \text{Number of additional constraint units} \\ \text{for each additional objective unit} \end{array} \right)$$

45. True or False: If the Lagrange function has no critical values, then the constrained optimization problem has no solution. (Assume for simplicity that the Lagrange function is defined for all values of its variables.)

46. Suppose you know that a constrained maximum problem has a solution. If the Lagrange function has one critical point, then what conclusion can you draw?

47. Suppose you know that a constrained maximum problem and the corresponding constrained minimum problem both have solutions. If the Lagrange function has two critical points, what conclusion can you draw?

48. If the maximum value of a function is a number a, and the maximum value of the function *subject to a constraint* is a number b, then what can you say about the relationship between the numbers a and b?

49. If the minimum value of a function is a number a, and the minimum value of the function *subject to a constraint* is a number b, then what can you say about the relationship between the numbers a and b?

50. Explain why we could have defined the Lagrange function to be $F = f - \lambda g$ (instead of $F = f + \lambda g$) and still obtain the same solutions to constrained optimization problems.

Explorations and Excursions   The following problems extend and augment the material presented in the text.

51. BUSINESS: Maximum Production Show that for the Cobb–Douglas production function $P = aL^bK^{1-b}$ restricted to the isocost line $wL + rK = C^*$ (where L and K are the amounts of labor and capital with prices w and r, and C^* is a given total cost), the maximum possible production is

$$P^* = a\left(\frac{b}{w}\right)^b\left(\frac{1-b}{r}\right)^{1-b}C^*$$

when the labor and capital are

$$L^* = \left(\frac{b}{w}\right)C^* \quad \text{and} \quad K^* = \left(\frac{1-b}{r}\right)C^*$$

and λ is

$$\lambda^* = -a\left(\frac{b}{w}\right)^b\left(\frac{1-b}{r}\right)^{1-b}$$

52. BUSINESS: Minimum Cost Show that for the total cost function $C = wL + rK$ restricted to an isoquant of the Cobb–Douglas production function $aL^bK^{1-b} = P^*$ (where L and K are the amounts of labor and capital with prices w and r, and P^* is a given level of production), the minimum possible cost is

$$C^* = \frac{1}{a}\left(\frac{b}{w}\right)^{-b}\left(\frac{1-b}{r}\right)^{-(1-b)}P^*$$

when the labor and capital are

$$L^* = \frac{1}{a}\left(\frac{1-b}{r}\right)^{b-1}\left(\frac{w}{b}\right)^{b-1}P^*$$

$$K^* = \frac{1}{a}\left(\frac{1-b}{r}\right)^b\left(\frac{w}{b}\right)^b P^*$$

and λ is

$$\lambda^* = -\frac{1}{a}\left(\frac{b}{w}\right)^{-b}\left(\frac{1-b}{r}\right)^{-(1-b)}$$

53. BUSINESS: Maximum Production and Minimum Cost Show that the formulas for P^*, C^*, L^*, and K^* in Exercises 51 and 52 are the same. Explain why the λ for the minimum problem of Exercise 52 is the reciprocal of the λ in the maximum problem of Exercise 51.

54. BUSINESS: Marginal Rate of Substitution Show that at the maximum production found in Exercise 51,

$$MRTS = \frac{w}{r}$$

[*Hint:* Use the formula for *MRTS* from Exercise 50 on page 228.]

55. BUSINESS: Marginal Rate of Substitution Verify the formula from Exercise 54 for Example 3 on pages 485–486.

56. BUSINESS: Marginal Productivities Find the partial derivatives P_L and P_K of the Cobb–Douglas production function $P = aL^bK^{1-b}$ and then use the results of Exercise 51 to show that at maximum production,

$$\frac{P_L}{P_K} = \frac{w}{r}$$

57. BUSINESS: Marginal Productivities Verify the formula from Exercise 56 for Example 3 on pages 485–486.

58. BUSINESS: Least Cost Rule Deduce the *least cost rule* that when a company produces output in the least costly way,

$$\frac{\left(\begin{array}{c}\text{marginal productivity}\\\text{of labor}\end{array}\right)}{\text{cost of labor}} = \frac{\left(\begin{array}{c}\text{marginal productivity}\\\text{of capital}\end{array}\right)}{\text{cost of capital}}$$

7.6   Total Differentials, Approximate Changes, and Marginal Analysis

Introduction

In this section we define the **total differential** of a function of several variables, and use it to approximate the change in the function resulting from changes in the independent variables. In addition to giving several applications, we will justify the interpretation of the Lagrange variable λ stated on page 486.

This section parallels the developments of Section 3.7, *Differentials, Approximations, and Marginal Analysis*, where we defined and used differentials for functions of a *single* variable. You may read this section even if you did not read Section 3.7.

Total Differential of a Function of Two Variables

For a function $f(x)$ of one variable, we defined (see page 231 or 365) the differential to be the *derivative multiplied by dx*, $dy = f'(x) \cdot dx$. For a function $f(x, y)$ of *two* variables, we define the total differential analogously as the partial derivatives multiplied by dx and dy and added.*

Total Differential of $f(x, y)$

For a function $f(x, y)$, the total differential df is

$$df = f_x(x, y) \cdot dx + f_y(x, y) \cdot dy$$

More briefly,
$$df = f_x \cdot dx + f_y \cdot dy$$

Take Note

This can be interpreted as saying: The total change in f is the change in the x-direction plus the change in the y-direction.

The total differential can also be written with the partials in the ∂ notation:

$$df = \frac{\partial f}{\partial x} \cdot dx + \frac{\partial f}{\partial y} \cdot dy$$

Total differential in ∂ notation

EXAMPLE 1   FINDING A TOTAL DIFFERENTIAL

Find the total differential of $f(x, y) = 5x^3 - 4xy^{-1} + 3y^4$.

Solution

The partials are

$$f_x = 15x^2 - 4y^{-1}$$

Partial of $5x^3 - 4xy^{-1} + 3y^4$ with respect to x

$$f_y = 4xy^{-2} + 12y^3$$

Partial of $5x^3 - 4xy^{-1} + 3y^4$ with respect to y

Then the total differential is

$$df = \underbrace{(15x^2 - 4y^{-1})}_{f_x(x,\,y)} \cdot dx + \underbrace{(4xy^{-2} + 12y^3)}_{f_y(x,\,y)} \cdot dy$$

$df = f_x\,dx + f_y\,dy$ with the above partials

If dx and dy are considered as new variables, then the total differential df is a function of the *four* variables x, y, dx, and dy. The total differential of $z = f(x, y)$ can be denoted dz or df.

EXAMPLE 2   FINDING THE TOTAL DIFFERENTIAL OF A LOGARITHMIC FUNCTION

Find the total differential of $z = \ln(x^2 + y^3)$.

---

*Technically, the partials f_x and f_y must be continuous. However, since we have not defined continuity for functions of two variables, we will not discuss this requirement further, except to say that it is satisfied by all functions in this section and by most functions encountered in applications.

Solution

The partials are

Partials of the denominator

$$z_x = \frac{2x}{x^2 + y^3} \qquad z_y = \frac{3y^2}{x^2 + y^3} \qquad \text{Derivative of } \ln f \text{ is } \frac{f'}{f}$$

Therefore, dz is

$$dz = \frac{2x}{x^2 + y^3} \cdot dx + \frac{3y^2}{x^2 + y^3} \cdot dy \qquad \text{Partials times } dx \text{ and } dy$$

PRACTICE PROBLEM

Find the total differential of $g(x, y) = x^2 e^{5y}$.

Solution on page 501 >

Approximating Changes by Total Differentials

For a function $f(x, y)$, changing the value of x by Δx and y by Δy generally changes the value of the function. The change Δf in the function is found by evaluating f at the "changed" values and subtracting f evaluated at the original values.

Change in f(x, y)

$$\Delta f = f(x + \Delta x, y + \Delta y) - f(x, y)$$

f at the "new" values f at the "old" values

Be Careful   For *independent* variables (such as x and y) we use "Δ" and "d" interchangeably to denote changes. That is,

$$\Delta x = dx \quad \text{and} \quad \Delta y = dy \qquad \text{For independent variables,} \\ \text{"}\Delta\text{" = "}d\text{"}$$

However, for *dependent* variables, "Δ" and "d" have different meanings: Δ indicates the *actual* change, and d indicates the total differential.

For some functions, calculating the actual change Δf can be complicated. The total differential provides a simple **linear approximation** for the actual change. The partials f_x and f_y give the rate of change of f per unit change in x and y, respectively. Changing x by Δx units changes f by approximately $f_x \cdot \Delta x$ units, and changing y by Δy units changes f by approximately $f_y \cdot \Delta y$ units. Therefore, changing *both* x and y should change f by approximately the *sum* of these changes. In symbols:

Differential Approximation Formula

$$\underbrace{f(x + \Delta x, y + \Delta y) - f(x, y)}_{\text{Change in } f} \approx \underbrace{f_x \cdot \Delta x + f_y \cdot \Delta y}_{\substack{\text{Total differential of } f \\ \text{(since } \Delta x = dx \text{ and } \Delta y = dy)}}$$

Written more compactly,

$$\Delta f \approx df$$

$$\Delta f = f(x + \Delta x, y + \Delta y) - f(x, y)$$
$$df = f_x(x, y) \cdot dx + f_y(x, y) \cdot dy$$

The partials in the first formula are evaluated at (x, y), and the approximation improves as Δx and Δy approach zero. (See page 499 for a geometric explanation of the approximation $\Delta f \approx df$.)

EXAMPLE 3 **APPROXIMATING AN ACTUAL CHANGE BY A DIFFERENTIAL**

For $f(x, y) = x^2 + 4xy + y^3$ and values $x = 3$, $y = 2$, $\Delta x = 0.2$, and $\Delta y = -0.1$, find **a.** Δf **b.** df

Solution

a. From the given values,

$$x + \Delta x = 3 + 0.2 = 3.2 \qquad \text{and} \qquad y + \Delta y = 2 - 0.1 = 1.9$$

The change Δf is

$$\Delta f = f(3.2, 1.9) - f(3, 2) \qquad\qquad \Delta f = f(x + \Delta x, y + \Delta y) - f(x, y)$$

$$= \overbrace{3.2^2 + 4 \cdot (3.2) \cdot (1.9) + 1.9^3}^{f(3.2,\, 1.9)} - \overbrace{(3^2 + 4 \cdot 3 \cdot 2 + 2^3)}^{f(3,\, 2)} \qquad \text{Using } f(x, y) = x^2 + 4xy + y^3$$

$$= 10.24 + 24.32 + 6.859 - (9 + 24 + 8) \qquad \text{Evaluating}$$

$$= 41.419 - 41 = 0.419 \qquad \begin{array}{l} \text{Change in } f \text{ is} \\ \Delta f = 0.419 \end{array}$$

b. The total differential df is

$$df = \overbrace{(2x + 4y)}^{f_x(x,\, y)} \cdot dx + \overbrace{(4x + 3y^2)}^{f_y(x,\, y)} \cdot dy \qquad \begin{array}{l} \text{Partials of} \\ x^2 + 4xy + y^3 \\ \text{times } dx \text{ and } dy \end{array}$$

$$= (2 \cdot 3 + 4 \cdot 2) \cdot (0.2) + (4 \cdot 3 + 3 \cdot 2^2) \cdot (-0.1) \qquad \begin{array}{l} \text{Evaluating at } x = 3, \\ y = 2, \ dx = 0.2, \\ \text{and } dy = -0.1 \end{array}$$

$$= 2.8 - 2.4 = 0.4 \qquad \begin{array}{l} \text{Total differential} \\ \text{is } df = 0.4 \end{array}$$

We found $\begin{cases} \Delta f = 0.419 \\ df = 0.4 \end{cases}$

The total differential $df = 0.4$ provides a reasonably accurate approximation for the actual change $\Delta f = 0.419$. The approximation would be even more accurate for smaller values of Δx and Δy.

Why should we bother calculating an *approximation df* when with a calculator we can easily find the *exact* change Δf? The total differential *df* has the advantage of *linearity*: If we were to *double* the changes Δx and Δy in the independent variables, then the differential would also double, from 0.4 to 0.8; if we were to *halve* the changes Δx and Δy, then the differential would also be halved, from 0.4 to 0.2. The *actual* change Δf admits no such simple modification—whenever Δx and Δy change it has to be recalculated "from scratch" using the formula $\Delta f = f(x + \Delta x, y + \Delta y) - f(x, y)$. The linearity of the total differential *df* is an advantage when you are trying to understand how various changes in x and y would affect the value of *f*.

Marginal Analysis

In business and economics, the term *marginal analysis* means estimating the changes in cost, revenue, or profit that result from small changes in production. The total differential of a cost, revenue, or profit function provides exactly these estimates when several goods are produced.

Approximate Changes

For cost $C(x, y)$, revenue $R(x, y)$, and profit $P(x, y)$ functions of x and y units of goods,

$$\Delta C \approx C_x dx + C_y dy$$

$$\Delta R \approx R_x dx + R_y dy$$

$$\Delta P \approx P_x dx + P_y dy$$

with these approximations becoming more accurate for $\Delta x = dx$ and $\Delta y = dy$ closer to zero.

EXAMPLE 4   ESTIMATING ADDITIONAL PROFIT

The American Farm Machinery Company finds that if it manufactures x economy tractors and y heavy-duty tractors per month, then its profit (in thousands of dollars) will be $P(x, y) = 3x^{4/3} + 0.05xy + 4y$. If the company now manufactures 125 economy tractors and 100 heavy-duty tractors per month, find an approximation for the additional profit from manufacturing 3 more economy tractors and 2 more heavy-duty tractors per month.

Solution

We want an estimate for the change in the profit $P(x, y)$ when production increases above the levels $x = 125$ and $y = 100$ by amounts $\Delta x = 3$ and $\Delta y = 2$. The partials are

$$P_x = 4x^{1/3} + 0.05y \qquad P_y = 0.05x + 4$$

Partials of $P = 3x^{4/3} + 0.05xy + 4y$ with respect to x and y

The total differential is

$$dP = (4x^{1/3} + 0.05y) \cdot dx + (0.05x + 4) \cdot dy$$

$dP = P_x \cdot dx + P_y \cdot dy$

$$= (4 \cdot \underbrace{125^{1/3}}_{\sqrt[3]{125}\, =\, 5} + 0.05 \cdot 100) \cdot 3 + (0.05 \cdot 125 + 4) \cdot 2$$

Substituting $x = 125$, $y = 100$, $\Delta x = 3$, and $\Delta y = 2$

$$= (20 + 5) \cdot 3 + (6.25 + 4) \cdot 2 = 75 + 20.5 = 95.5$$

In thousands of dollars

Therefore, producing 3 more economy tractors and 2 more heavy-duty tractors will generate about $95,500 in additional profit.

The actual change in the profit function, found by applying the Δf formula on page 495, is

$$\Delta P = P(125 + 3, 100 + 2) - P(125, 100) \approx 96.04,$$

so the approximation of 95.5 is indeed quite accurate.

The *linearity* of the total differential means that if sometime later the company wanted to estimate the additional profit from *doubling* these changes—making 6 more economy tractors and 4 more heavy-duty tractors—they would need only to double the estimate of $95,500 to immediately obtain $191,000.

Estimating Errors

No physical measurement can ever be made with perfect accuracy. If you can estimate the maximum error in a measurement, then you can use differentials to estimate the resulting error in a calculation. The measurement errors may be expressed as *percentage* errors or *actual* numbers (as in Example 6, to be discussed shortly). The following Example estimates the percentage error in calculating the volume of a cylinder. Such calculations have applications from predicting variations in soft-drink cans to ensuring safety margins for artificial arteries.

| EXAMPLE 5 | ESTIMATING THE ERROR IN CALCULATING VOLUME |

A cylinder is measured to have radius r and height h, but these measurements may be in error by up to 1%. Estimate the resulting percentage error in calculating the volume of the cylinder.

Solution

The height and radius being in error by 1% means that

$$\Delta r = 0.01r$$

For each, the change is 1% of the value

$$\Delta h = 0.01h$$

The volume of a cylinder is $V = \pi r^2 h$, and the total differential is

$$dV = 2\pi rh \cdot dr + \pi r^2 \cdot dh \qquad \text{Partials of } V = \pi r^2 h \text{ times } dr \text{ and } dh$$

$$= 2\pi rh \cdot 0.01r + \pi r^2 \cdot 0.01h \qquad \text{Substituting } dr = \Delta r = 0.01r \text{ and } dh = \Delta h = 0.01h$$

$$= \pi r^2 h(\underbrace{2 \cdot 0.01 + 0.01}_{0.03}) \qquad \text{Factoring out } \pi r^2 h$$

$$= 0.03\underbrace{\pi r^2 h}_{} = 0.03 \cdot V \qquad \text{Change is 3\% of volume}$$

Volume V of the cylinder

This result, $dV = 0.03 \cdot V$, means that the volume may be in error by as much as 3% if the radius and height are "off" by 1%.

Geometric Visualization of *df* and Δ*f*

A function $f(x, y)$ represents a *surface* in three-dimensional space, and the change Δf represents the change in height when moving from one point to another *along the surface*, as shown below.

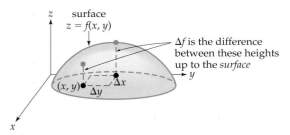

The total differential *df* represents the change in height when moving from one point to another *along the plane* that best fits the surface at the first point, as shown below. This plane is called the *tangent plane* to the surface at the (blue) point.

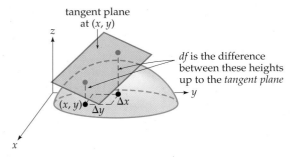

Since the tangent plane fits the surface closely near the point (x, y), changes *df* in height along the *tangent plane* should be very close to changes Δf in height along the *surface*, which is why the total differential *df* closely approximates the actual change Δf for small changes in x and y.

Total Differential of a Function of Three Variables

The total differential may be generalized to apply to functions of three (or more) variables, multiplying each partial by "*d*" of that variable and adding.

Total Differential of $f(x, y, z)$

For a function $f(x, y, z)$, the total differential df is

$$df = f_x(x, y, z) \cdot dx + f_y(x, y, z) \cdot dy + f_z(x, y, z) \cdot dz$$

Partials times $dx, dy,$ and dz

The total differential of a function $f(x, y, z)$ gives an estimate for the change Δf when the variables are changed by amounts Δx, Δy, and Δz:

$$\underbrace{f(x + \Delta x, y + \Delta y, z + \Delta z) - f(x, y, z)}_{\text{Change in } f} \approx \underbrace{\frac{\partial f}{\partial x} \cdot \Delta x + \frac{\partial f}{\partial y} \cdot \Delta y + \frac{\partial f}{\partial z} \cdot \Delta z}_{\substack{\text{Total differential of } f \\ (\text{since } \Delta x = dx, \\ \Delta y = dy, \ \Delta z = dz)}}$$

or, more briefly,

$$\Delta f \approx df$$

The partials in the formula above are evaluated at (x, y, z), and the approximation is increasingly accurate for smaller values of Δx, Δy, and Δz.

EXAMPLE 6   ESTIMATING THE ERROR IN CALCULATING VOLUME

A rectangular box is measured to be 30 inches long, 24 inches wide, and 10 inches high. If the maximum errors in measuring the length, width, and height of the box are, respectively, 0.3, 0.2, and 0.1 inch, estimate the maximum error in calculating its volume.

10 ± 0.1

24 ± 0.2

30 ± 0.3

Solution

The volume of the box is length times width times height, $V = x \cdot y \cdot z$. We want to estimate the change in volume resulting from changing the dimensions $x = 30$, $y = 24$, and $z = 10$ by the amounts $\Delta x = 0.3$, $\Delta y = 0.2$, and $\Delta z = 0.1$. The total differential is

$$df = \overbrace{y \cdot z \cdot dx}^{V_x} + \overbrace{x \cdot z \cdot dy}^{V_y} + \overbrace{x \cdot y \cdot dz}^{V_z}$$

Partials of $V = x \cdot y \cdot z$ times $dx, dy,$ and dz

$$= 24 \cdot 10 \cdot 0.3 + 30 \cdot 10 \cdot 0.2 + 30 \cdot 24 \cdot 0.1$$

Substituting $x = 30$, $y = 24$, $z = 10$, $dx = 0.3$, $dy = 0.2$, and $dz = 0.1$

$$= 72 + 60 + 72 = 204$$

Multiplying out and adding

That is, the maximum error in calculating the volume is approximately 204 cubic inches.

While an error of 204 cubic inches may seem large, the *relative* error (the error divided by the volume, expressed as a percentage) is only

$$\frac{204}{30 \cdot 24 \cdot 10} = \frac{204}{7200} \approx 0.028 = 2.8\%$$

Relative error is 2.8%

Solution TO PRACTICE PROBLEM

$$g_x = 2xe^{5y} \qquad g_y = x^2 5e^{5y}$$ Partials

$$dg = 2xe^{5y} \cdot dx + 5x^2 e^{5y} \cdot dy$$ Total differential

---

7.6 Section Summary

For a function $f(x, y)$, the total differential df is defined as the partials multiplied by dx and dy and added:

$$df = f_x(x, y) \cdot dx + f_y(x, y) \cdot dy \qquad\qquad df = \frac{\partial f}{\partial x} dx + \frac{\partial f}{\partial y} dy$$

The actual change in the function when x and y change by amounts Δx and Δy is

$$\Delta f = f(x + \Delta x, y + \Delta y) - f(x, y)$$

For small values of $\Delta x = dx$ and $\Delta y = dy$, the actual change Δf can be approximated by the total differential df:

$$f(x + \Delta x, y + \Delta y) - f(x, y) \approx f_x(x, y) \cdot dx + f_y(x, y) \cdot dy \qquad \Delta f \approx df$$

The actual change Δf may depend on Δx and Δy in very complicated ways, but the total differential df is *linear* in $\Delta x = dx$ and $\Delta y = dy$, and is therefore easier to calculate.

Justification of the Interpretation of the Lagrange Multiplier λ

In the preceding Section we used Lagrange multipliers to optimize an objective function $f(x, y)$ subject to a constraint $g(x, y) = 0$. We did so by setting the partials of $F(x, y, \lambda) = f(x, y) + \lambda g(x, y)$ equal to zero, and we interpreted the absolute value of the "Lagrange multiplier" $|\lambda|$ as *the number of additional objective units per additional constraint unit*. This interpretation may be justified as follows. Setting the partials of F with respect to x and y equal to zero gives

$$\begin{cases} f_x + \lambda g_x = 0 \\ f_y + \lambda g_y = 0 \end{cases} \quad \text{or, equivalently,} \quad \begin{cases} f_x = -\lambda g_x \\ f_y = -\lambda g_y \end{cases}$$

If we increase x and y by amounts Δx and Δy, the resulting change in the objective function $f(x, y)$ can be approximated by the total differential:

$$\underbrace{f(x + \Delta x, y + \Delta y) - f(x, y)}_{\Delta f} \approx f_x \Delta x + f_y \Delta y \qquad \Delta f \approx df$$

$$= -\lambda g_x \Delta x - \lambda g_y \Delta y$$

Substituting
$f_x = -\lambda g_x$ and
$f_y = -\lambda g_y$

$$= -\lambda(g_x\,\Delta x + g_y\,\Delta y)$$

Factoring out $-\lambda$ leaves the total differential of g

$$\underbrace{}_{dg}$$

$$\approx -\lambda \cdot \Delta g$$

Replacing dg by Δg (since $dg \approx \Delta g$)

These equations, read from beginning to end, say that $\Delta f \approx -\lambda \cdot \Delta g$, or

$$\frac{\Delta f}{\Delta g} \approx -\lambda$$

Dividing by Δg

The left-hand side of this is the ratio of the change in the objective function f to the change in the constraint function g. Taking absolute values and letting Δx and Δy approach zero (to make the approximation exact) shows that $|\lambda|$ is *the number of objective units per additional constraint unit*, as stated on page 486.

7.6 Exercises

1–20. Find the total differential of each function.

1. $f(x, y) = x^2y^3$

2. $f(x, y) = x^4y^{-1}$

3. $f(x, y) = 6x^{1/2}y^{1/3} + 8$

4. $f(x, y) = 100x^{0.05}y^{0.02} - 7$

5. $g(x, y) = \dfrac{x}{y}$

6. $g(x, y) = \dfrac{x}{x + y}$

7. $g(x, y) = (x - y)^{-1}$

8. $g(x, y) = \sqrt{x^2 + y^2}$

9. $z = \ln(x^3 - y^2)$

10. $z = x^2 \ln y$

11. $z = xe^{2y}$

12. $z = e^{3x-2y}$

13. $w = 2x^3 + xy + y^2$

14. $w = 3x^2 - xy^{-1} + y^3$

15. $f(x, y, z) = 2x^2y^3z^4$

16. $f(x, y, z) = xy + yz + xz$

17. $f(x, y, z) = \ln(xyz)$

18. $f(x, y, z) = \ln(x^2 + y^2 + z^2)$

19. $f(x, y, z) = e^{xyz}$

20. $f(x, y, z) = e^{x^2+y^2+z^2}$

21–26. For the given function and values, find:
a. Δf **b.** df

21. $f(x, y) = x^2 + xy + y^3$,
$x = 4$, $\Delta x = dx = 0.2$,
$y = 2$, $\Delta y = dy = -0.1$

> **FOR HELP GETTING STARTED**
> on Exercises 21–26, see Example 3 on pages 496–497

22. $f(x, y) = x^3 + xy + y^3$,
$x = 5$, $\Delta x = dx = 0.01$,
$y = 3$, $\Delta y = dy = -0.01$

23. $f(x, y) = e^x + xy + \ln y$,
$x = 0$, $\Delta x = dx = 0.05$,
$y = 1$, $\Delta y = dy = 0.01$

24. $f(x, y) = \ln(x^2 + y^2)$, $x = 6$, $\Delta x = dx = 0.1$, $y = 8$, $\Delta y = dy = 0.2$

25. $f(x, y, z) = xy + z^2$, $x = 3$, $\Delta x = dx = 0.03$, $y = 2$, $\Delta y = dy = 0.02$, $z = 1$, $\Delta z = dz = 0.01$

26. $f(x, y, z) = x^2 + y^2 + z^2$, $x = 3$, $\Delta x = dx = 0.1$, $y = 4$, $\Delta y = dy = 0.1$, $z = 5$, $\Delta z = dz = 0.1$

Applied Exercises

27–37. Use total differentials to solve the following exercises.

27. GENERAL: Measurement Errors A rectangle is measured to be 150 feet by 100 feet, but each measurement may be "off" by half a foot. Estimate the error in calculating the area. Then estimate the error in calculating the area if each measurement is "off" by one foot.

28. GENERAL: Telephone Calls For two cities with populations x and y (in thousands) that are 500 miles apart, the number of telephone calls per day between them can be modeled by the function $12xy$. For two cities with populations 40 thousand and 60 thousand, estimate the number of additional telephone calls if each city grows by 1 thousand people. Then estimate the number of additional calls if instead each city were to grow by only 500 people.

29–30. BUSINESS:Profit An electronics company's profit in dollars from making x DVD players and y Blu-ray players per day is given by the following profit function $P(x, y)$. If the company currently produces 200 DVD players and 300 Blu-ray players, estimate the extra profit that would result from producing five more DVD players and four more Blu-ray players.

29. $P(x, y) = 2x^2 - 3xy + 3y^2$

30. $P(x, y) = 3x^2 - 4xy + 4y^2$

31. GENERAL: Highway Safety The emergency stopping distance in feet for a truck of weight w tons traveling at v miles per hour on a dry road is $S = 0.027wv^2$. For a truck that weighs 4 tons and is usually driven at 60 miles per hour, estimate the extra stopping distance if it has an extra half ton of load and is traveling 5 miles per hour faster than usual.

Source: National Highway Traffic Safety Administration

32. GENERAL: Scuba Diving The maximum duration (in minutes) of a scuba dive can be estimated by the formula $T = \dfrac{33v}{d + 33}$, where v is the volume of air in the tank (in cubic feet at sea-level pressure) and d is the depth (in feet) of the dive. For values $v = 100$ and $d = 67$, estimate the change in duration if an extra 20 cubic feet of air is added and the dive is 10 feet deeper.

Source: U.S. Navy Diving Manual, Vol. 2

33. GENERAL: Relative Error in Calculating Area A rectangle is measured to have length x and width y, but each measurement may be in error by 1%. Estimate the percentage error in calculating the area.

34. GENERAL: Relative Error in Calculating Volume A rectangular solid is measured to have length x, width y, and height z, but each measurement may be in error by 1%. Estimate the percentage error in calculating the volume.

35. BIOMEDICAL: Cardiac Output Medical researchers calculate the quantity of blood pumped through the lungs (in liters per minute) by the formula $C = \dfrac{x}{y - z}$, where x is the amount of oxygen absorbed by the lungs (in milliliters per minute), and y and z are, respectively, the concentrations of oxygen in the blood just after and just before passing through the lungs (in milliliters of oxygen per liter of blood). Typical measurements are $x = 250$, $y = 160$, and $z = 150$. Estimate the error in calculating the cardiac output C if each measurement may be "off" by 5 units.

Source: T. Aherns and K. Rutherford (eds.), Essentials of Oxygenation

36. GENERAL: Windchill The windchill index announced during the winter by the weather bureau measures how cold it "feels" for a given temperature t (in degrees Fahrenheit) and wind speed w (in miles per hour). It is calculated by the formula $C(t, w) = 35.74 + 0.6215t - 35.75w^{0.16} + 0.4275tw^{0.16}$. If the temperature is 30 degrees and the wind speed is 10 miles per hour, estimate the change in the windchill temperature if the wind speed increases by 4 miles per hour and the temperature drops by 5 degrees.

Source: National Weather Service

37. BIOMEDICAL: Blood Vessel Volume A section of an artery is measured to have length 12 centimeters and diameter 0.8 centimeter. If each measurement may be off by 0.1 centimeter, find the volume with an estimate of the error.

Conceptual Exercises

38. Let $f(x, y)$ be an arbitrary function. For each equation, state whether it is True or False:

   a. $dx = \Delta x$ **b.** $dy = \Delta y$ **c.** $df = \Delta f$

39. What is the total differential of a *constant* function?

40. What is the total differential of the *linear* function $f(x, y) = ax + by + c$ where a, b, and c are constants?

41. What is the most general function $f(x, y)$ that satisfies $df = dx + dy$?

42. What is the value of the total differential of a function evaluated at a *critical point* of the function? [*Hint:* See the definition of critical point on page 458.]

43. If the total differential of a function $f(x, y)$ simplifies to $df = f_x(x, y) \cdot dx$, what conclusion can you draw about the function $f(x, y)$?

44. In Example 6 on page 500 we found that the *actual* error was 204 and then that the *relative* error was the much smaller number $\frac{204}{7200} = 0.028$. Can the relative error ever be *greater* than the actual error?

Explorations and Excursions   The following problem extends and augments the material presented in the text.

45. THE SLOPE OF $f(x, y) = c$ On page 490 we used the fact that the slope in the x-y plane of the curve defined by $f(x, y) = c$ (for constant c) is given by the formula $-\dfrac{f_x}{f_y}$. Verify this formula by justifying the following five steps.

a. If $f(x, y) = c$ can be solved explicitly for a function $y = F(x)$, then we may write $f(x, F(x)) = c$. Justify: $f(x + \Delta x, F(x + \Delta x)) - f(x, F(x)) = 0$.

b. Justify: $f(x + \Delta x, F(x + \Delta x) - F(x) + F(x)) - f(x, F(x)) = 0$.

c. Defining ΔF by $\Delta F = F(x + \Delta x) - F(x)$, we may write the previous equation as

$$f(x + \Delta x, \Delta F + F(x)) - f(x, F(x)) = 0$$

Then, writing F for $F(x)$, this becomes

$$f(x + \Delta x, F + \Delta F) - f(x, F) = 0$$

Justify: $f_x \, \Delta x + f_y \, \Delta F \approx 0$.

d. Justify: $\dfrac{\Delta F}{\Delta x} \approx -\dfrac{f_x}{f_y}$.

e. Justify: $\dfrac{dF}{dx} = -\dfrac{f_x}{f_y}$.

This shows that the slope of $F(x)$, and therefore the slope of $f(x, y) = c$, is $-\dfrac{f_x}{f_y}$.

7.7   Multiple Integrals

Introduction

This section discusses *integration* of functions of several variables. We define the **double integral** of a function by considering the volume under a surface $z = f(x, y)$. We then evaluate double integrals by **iterated integrals,** that is, by repeated single integrations. Finally, we use double integrals to calculate volumes, average values, and total accumulations. We restrict our attention to continuous functions (surfaces that have no holes or breaks), since most functions encountered in applications satisfy this restriction.

Rectangular Regions, Volumes, and Double Integrals

The points (x, y) in the plane for x taking values between numbers a and b and y taking values between numbers c and d determine a **rectangular region** R.

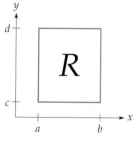

Rectangular region $R = \{(x,y) \mid a \le x \le b, c \le y \le d\}$

The graph below shows a nonnegative function $f(x, y)$ defined on a rectangular region R. We want to find the **volume** under the surface f and above the region R.

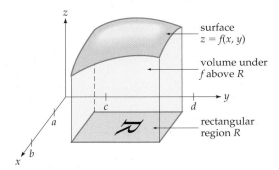

Volume under the surface $z = f(x, y)$
lying above a rectangular region R

We begin by *approximating* the volume by rectangular solids ("boxes") extending from R up to the surface. We divide R into small rectangles by drawing lines parallel to the x- and y-axes with spacing Δx and Δy, as shown below.* On each of these small rectangles we erect a rectangular solid with height $f(x_i, y_j)$, the height of the surface at some point (x_i, y_j) in the base rectangle.

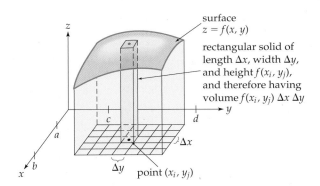

Volume under f over R showing one of the rectangular solids

The volume of the rectangular solid is $f(x_i, y_j) \cdot \Delta x \cdot \Delta y$ (height times length times width), and the sum of the volumes of all such rectangular solids approximates the volume under f above R:

$$\binom{\text{Volume under}}{f \text{ above } R} \approx \sum f(x_i, y_j) \cdot \Delta x \cdot \Delta y$$

Σ means sum over all base rectangles

The *limit* of this sum as both Δx and Δy approach zero (so that the base rectangles become smaller and more numerous) gives the *exact* volume, and is called the *double integral of f over R*, denoted $\iint\limits_{R} f(x, y)\, dx\, dy$.

---

*Technically, the parallel lines need not have equal spacing. However, we will be letting the spacings approach zero, and the final results are the same for equal or unequal spacing.

Double Integrals

The double integral of a continuous function $f(x, y)$ on a rectangular region R is

$$\iint\limits_{R} f(x, y)\, dx\, dy = \lim_{\Delta x, \Delta y \to 0} \sum f(x_i, y_j) \cdot \Delta x \cdot \Delta y$$

The sum is over all rectangles in R, each containing one (x_i, y_j)

If $f(x, y)$ is nonnegative on R, then the double integral gives the *volume under f over R.*

LOOKING BACK

Notice how similar this definition is to the one for a function of just one variable, on page 333 (upper box).

Iterated Integrals

Evaluating double integrals from the definition is difficult. Fortunately, double integrals can be evaluated by two separate "single" integrations, integrating with respect to one variable at a time while holding the other variable constant. (This is analogous to partial differentiation with respect to one variable, holding the other variable constant.) Such repeated integrals are called *iterated integrals* (*iterated* means "repeated"). A proof that double integrals can be evaluated as iterated integrals can be found in a book on advanced calculus.

EXAMPLE 1 **EVALUATING AN ITERATED INTEGRAL**

Evaluate the iterated integral $\displaystyle\int_0^1 \int_0^2 (3x^2 + 6xy^2)\, dx\, dy$.

Solution

The two separate integrations will be clearer if we use parentheses:

$$\int_0^1 \left(\int_0^2 (3x^2 + 6xy^2)\, dx \right) dy$$

An inner x-integral and an outer y-integral

The inner integral gives

$$\int_0^2 (3x^2 + 6xy^2)\, dx = \left(x^3 + 6 \cdot \frac{1}{2} x^2 y^2 \right)\Bigg|_{x=0}^{x=2}$$

dx means integrate with respect to x holding y constant   Integral of $3x^2$ Integral of x Held constant

$$= 2^3 + 3 \cdot 2^2 y^2 \quad - \quad 0 \quad = 8 + 12y^2$$

Evaluated at $x = 2$ And at $x = 0$ Simplified

We now apply the outer y-integral to this result:

$$\int_0^1 (8 + 12y^2)\, dy = (8y + 4y^3)\Bigg|_{y=0}^{y=1} = 8 + 4 \quad - \quad 0 \quad = \quad 12$$

Result of the inner integral dy means integrate with respect to y $12 \cdot \frac{1}{3}$ Evaluated at $y = 1$ And at $y = 0$ Final answer

Therefore:

$$\int_0^1 \int_0^2 (3x^2 + 6xy^2)\, dx\, dy = 12$$

The iterated integral
equals 12

Always solve an iterated integral "from the inside out."

$$\int_0^1 \left(\int_0^2 (3x^2 + 6xy^2)\, dx \right) dy$$

↑   ↑   ↗   ↑

Limits Limits First integrate Then with
for y for x with respect to x respect to y

In Example 1 we integrated first with respect to x and then with respect to y. The next Example shows that switching the order of integration gives the same answer, provided that we also switch the x and y limits of integration. That is,

$$\int_0^1 \int_0^2 (3x^2 + 6xy^2)\, dx\, dy$$

is equal to

$$\int_0^2 \int_0^1 (3x^2 + 6xy^2)\, dy\, dx$$

 Be Careful Be sure to switch the x and y limits of integration when you switch the order of integration.

EXAMPLE 2   REVERSING THE ORDER OF INTEGRATION

Evaluate $\displaystyle\int_0^2 \int_0^1 (3x^2 + 6xy^2)\, dy\, dx.$

Same as Example 1, but with the
order of integration reversed

Solution

First we evaluate the inner y-integral:

$$\int_0^1 (3x^2 + 6xy^2)\, dy = (3x^2y + 2xy^3)\Big|_{y=0}^{y=1} = 3x^2 + 2x - 0$$

↗   ↑ ↗   ↑

Integrate with x held $\frac{1}{3} \cdot 6$ Evaluated And at
respect to y constant   at $y = 1$ $y = 0$

$$= 3x^2 + 2x$$

Then we apply the outer x-integral to this expression:

$$\int_0^2 (3x^2 + 2x)\, dx = (x^3 + x^2)\Big|_{x=0}^{x=2} = 8 + 4 - 0 = 12$$

From inner   Evaluated And at Final
integration   at $x = 2$ $x = 0$ answer

Therefore:

$$\int_0^2 \int_0^1 (3x^2 + 6xy^2)\, dy\, dx = 12$$

Notice that Examples 1 and 2 (in which the order of integration was reversed) gave the same answer, 12. Reversing the order of integration *always* gives the same answer, provided that the function is continuous.

Reversing the Order of Integration

For a continuous $f(x, y)$

$$\int_c^d \int_a^b f(x, y)\, dx\, dy$$

is equal to

$$\int_a^b \int_c^d f(x, y)\, dy\, dx$$

Double Integrals and Volumes

Earlier, we defined double integrals over rectangular regions. Double integrals can be evaluated by *either* of two iterated integrals (integrating in either order). The limits of integration are taken directly from the region R.

Evaluating Double Integrals

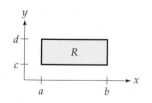

The double integral $\iint\limits_R f(x, y)\, dx\, dy$

over the region $R = \{\, (x, y) \mid a \le x \le b, c \le y \le d \,\}$

can be evaluated by finding *either* of the iterated integrals

$$\int_c^d \int_a^b f(x, y)\, dx\, dy \qquad \text{or} \qquad \int_a^b \int_c^d f(x, y)\, dy\, dx$$

The limits of integration come from the definition of the rectangle R.

EXAMPLE 3 EVALUATING A DOUBLE INTEGRAL

Evaluate $\iint\limits_R y^2 e^{-x}\, dx\, dy$ where $R = \{(x, y) \mid 0 \le x \le 2, -1 \le y \le 1\}$.

Solution

This double integral can be evaluated by finding either of the iterated integrals

$$\int_{-1}^{1} \int_{0}^{2} y^2 e^{-x} \, dx \, dy \quad \text{or} \quad \int_{0}^{2} \int_{-1}^{1} y^2 e^{-x} \, dy \, dx \qquad \text{Limits of integration come from } R$$

We find the second one, beginning with the inner integral.

$$\int_{-1}^{1} y^2 e^{-x} \, dy = \left(\frac{1}{3} y^3 e^{-x} \right) \Big|_{y=-1}^{y=1} = \underbrace{\frac{1}{3} e^{-x}}_{\text{Evaluated at } y=1} - \underbrace{\left(\frac{1}{3} (-1) e^{-x} \right)}_{\text{And at } y=-1}$$

Integrated ⌐ ⌐ Held constant

$$= \frac{1}{3} e^{-x} + \frac{1}{3} e^{-x} = \frac{2}{3} e^{-x}$$

Then we integrate this with respect to x:

$$\int_{0}^{2} \frac{2}{3} e^{-x} \, dx = -\frac{2}{3} e^{-x} \Big|_{x=0}^{x=2} = -\frac{2}{3} e^{-2} - \left(-\frac{2}{3} e^{0} \right) = -\frac{2}{3} e^{-2} + \frac{2}{3}$$

PRACTICE PROBLEM

Show that evaluating this same double integral by the iterated integral in the *other* order gives the same answer. That is, evaluate the iterated integral

$$\int_{-1}^{1} \int_{0}^{2} y^2 e^{-x} \, dx \, dy \qquad \text{Solution on page 513 >}$$

The volume under a surface can be found by evaluating a double integral (since this is how double integrals were defined).

Volume by Double Integrals

For a nonnegative continuous function $f(x, y)$, the volume under the surface $z = f(x, y)$ and above a rectangular region R in the x-y plane is

$$\binom{\text{Volume under}}{f \text{ above } R} = \iint\limits_{R} f(x, y) \, dx \, dy$$

If the surface lies *below* the x-y plane, this integral gives the *negative* of the volume.

EXAMPLE 4   FINDING THE VOLUME UNDER A SURFACE

A dome tent with closed sides is constructed according to the function shown below. To design ventilation and heating systems, it is necessary to know the volume under the tent. Find the volume under the tent on the indicated rectangle. All dimensions are in feet.

Solution

The volume is the integral of the function over the rectangle R:

$$\int_{-10}^{10} \int_{-20}^{20} (15 - 0.03x^2 - 0.03y^2)\, dx\, dy \qquad \text{Limits of integration come from } R$$

The inner integral is

$$\int_{-20}^{20} (15 - 0.03x^2 - 0.03y^2)\, dx = (15x - 0.01x^3 - 0.03y^2x)\Big|_{x=-20}^{x=20}$$

$$= \underbrace{300 - 80 - 0.6y^2}_{\text{Evaluated at } x = 20} - \underbrace{(-300 + 80 + 0.6y^2)}_{\text{Evaluated at } x = -20} = 440 - 1.2y^2$$

Integrating this with respect to y,

$$\int_{-10}^{10} (440 - 1.2y^2)\, dy = (440y - 0.4y^3)\Big|_{-10}^{10} \qquad \overset{\text{From integrating } 1.2y^2}{}$$

$$= 4400 - 400 - (-4400 + 400) = 8000$$

Therefore, the volume under the tent is 8000 cubic feet.

Average Value

On page 345 the average value of a function of *one* variable over an interval was defined as the definite integral of the function divided by the length of the interval. For similar reasons, the average value of a function $f(x, y)$ of *two* variables over a region is defined as the *double* integral divided by the *area* of the region.

Average Value

$$\begin{pmatrix} \text{Average value} \\ \text{of } f \text{ over } R \end{pmatrix} = \frac{1}{\text{Area of } R} \iint\limits_{R} f(x, y) \, dx \, dy$$

Double integral over the region divided by the area of the region

For a rectangular region R, the area of R is simply length times width.

EXAMPLE 5   **FINDING THE AVERAGE TEMPERATURE OVER A REGION**

The temperature x miles east and y miles north of a weather station is $T(x, y) = 60 + 2x - 4y$ degrees. Find the average temperature over the rectangular region R extending 2 miles north and south from the station and 5 miles east (as shown in the following diagram).

Solution

The area of the region R is $4 \cdot 5 = 20$ square miles (length times width). The average temperature is the double integral divided by 20:

$$\text{Average} = \frac{1}{20} \int_{-2}^{2} \int_{0}^{5} (60 + 2x - 4y) \, dx \, dy$$

The inner integral is

$$\int_{0}^{5} (60 + 2x - 4y) \, dx = (60x + x^2 - 4yx) \Big|_{x=0}^{x=5}$$

$$= \underbrace{300 + 25 - 20y}_{\text{Evaluated at } x = 5} - 0 = 325 - 20y$$

The (outer) integral of this expression is

$$\int_{-2}^{2} (325 - 20y) \, dy = (325y - 10y^2) \Big|_{y=-2}^{y=2} = 650 - 40 - (-650 - 40)$$

$$= 610 + 690 = 1300$$

Finally, for the average we divide by 20 (the area of the region):

$$\frac{1300}{20} = 65$$

The average temperature over the region is 65 degrees.

Integrating over More General Regions

We can also integrate over regions R that are bounded by curves, provided that the curves as well as the function being integrated are continuous.

Double Integrals over Regions Between Curves

Let R be the region bounded by a lower curve $y = g(x)$ and an upper curve $y = h(x)$ from $x = a$ to $x = b$, as shown below. Then the double integral of $f(x, y)$ over R is

$$\iint\limits_{R} f(x, y) \, dx \, dy = \int_a^b \int_{g(x)}^{h(x)} f(x, y) \, dy \, dx$$

If f is nonnegative, this double integral gives the volume under the surface $f(x, y)$ above R.

EXAMPLE 6   FINDING THE VOLUME UNDER A SURFACE

Find the volume under the surface $f(x, y) = 12xy$ and above the region R shown on the right.

Solution

The region R is bounded by the upper curve $h(x) = \sqrt{x}$ and the lower curve $g(x) = x^2$ from $x = 0$ to $x = 1$. From the box above, the volume is given by:

$$\text{Volume} = \int_0^1 \int_{x^2}^{\sqrt{x}} 12xy \, dy \, dx$$

We find the inner integral first:

$$\int_{x^2}^{\sqrt{x}} 12xy \, dy = 12x \frac{1}{2} y^2 \Big|_{y=x^2}^{y=\sqrt{x}} = 6xy^2 \Big|_{y=x^2}^{y=\sqrt{x}}$$

$$= \underbrace{6x(\sqrt{x})^2}_{\substack{\text{Evaluating} \\ \text{at } y = \sqrt{x}}} - \underbrace{6x(x^2)^2}_{\substack{\text{Evaluating} \\ \text{at } y = x^2}} = \underbrace{6x^2 - 6x^5}_{\text{Simplified}}$$

Now we integrate this result with respect to x:

$$\int_0^1 (6x^2 - 6x^5) \, dx = \left(6 \cdot \frac{1}{3} x^3 - x^6 \right) \Big|_0^1$$

$$= (2x^3 - x^6) \Big|_0^1 = 2 - 1 - (0) = 1$$

Therefore, the volume under the surface and above the region R is 1 cubic unit.

Solution   TO PRACTICE PROBLEM

$$\int_0^2 y^2 e^{-x}\, dx = -y^2 e^{-x}\Big|_{x=0}^{x=2} = -y^2 e^{-2} + y^2 e^0$$

$$= -y^2 e^{-2} + y^2$$

$$\int_{-1}^1 (-y^2 e^{-2} + y^2)\, dy = \left(-\frac{1}{3}y^3 e^{-2} + \frac{1}{3}y^3\right)\Big|_{-1}^1$$

$$= -\frac{1}{3}e^{-2} + \frac{1}{3} - \left(\frac{1}{3}e^{-2} - \frac{1}{3}\right) = -\frac{2}{3}e^{-2} + \frac{2}{3}$$ (same as before)

7.7   Section Summary

The double integral of a nonnegative function $f(x, y)$ over a region R gives the *volume* under the surface above R. (This is analogous to defining the definite integral of a function $f(x)$ of *one* variable as the *area* under the curve.)

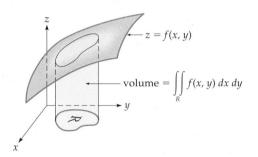

We evaluate a double integral by evaluating an iterated (repeated) integral:

$$\iint_R f(x, y)\, dx\, dy \quad \text{over} \quad R = \{(x, y) \mid a \le x \le b, c \le y \le d\}$$

is found by evaluating either of the two iterated integrals

$$\int_c^d \int_a^b f(x, y)\, dx\, dy \quad \text{or} \quad \int_a^b \int_c^d f(x, y)\, dy\, dx \qquad \text{Integrating in either order}$$

Note the distinction: *Double* integrals are written with R (which must be specified) under the double integral, and *iterated* integrals are written with upper and lower *limits* on each integral sign.

Double integrals do more than just find volumes; they give *continuous sums* (as illustrated in Exercises 43 and 44), and when divided by the area of the region they give the *average value of the function over the region.*

Exercises 47–50 discuss "triple" integrals of functions $f(x, y, z)$ of three variables. Triple integrals are evaluated by *iterated* integrals (but *three* of them), integrating successively with respect to one variable at a time, holding all others constant.

7.7 Exercises

1–6. Evaluate each (single) integral.

1. $\displaystyle\int_{1}^{x^2} 8xy^3\, dy$

2. $\displaystyle\int_{1}^{y^2} 10x^4\, dx$

3. $\displaystyle\int_{-y}^{y} 9x^2y\, dx$

4. $\displaystyle\int_{-x}^{x} 6xy^2\, dy$

5. $\displaystyle\int_{0}^{x} (6y - x)\, dy$

6. $\displaystyle\int_{0}^{y} (4x - y)\, dx$

7–28. Evaluate each iterated integral.

7. $\displaystyle\int_{0}^{2}\int_{0}^{1} 4xy\, dx\, dy$

8. $\displaystyle\int_{0}^{2}\int_{0}^{1} 8xy\, dy\, dx$

9. $\displaystyle\int_{0}^{2}\int_{0}^{1} x\, dy\, dx$

10. $\displaystyle\int_{0}^{4}\int_{0}^{3} y\, dx\, dy$

11. $\displaystyle\int_{0}^{1}\int_{0}^{2} x^3y^7\, dx\, dy$

12. $\displaystyle\int_{0}^{1}\int_{0}^{3} x^8y^2\, dy\, dx$

13. $\displaystyle\int_{1}^{3}\int_{0}^{2} (x + y)\, dy\, dx$

14. $\displaystyle\int_{1}^{2}\int_{0}^{4} (x - y)\, dx\, dy$

15. $\displaystyle\int_{-1}^{1}\int_{0}^{3} (x^2 - 2y^2)\, dx\, dy$

16. $\displaystyle\int_{-1}^{1}\int_{0}^{3} (2x^2 + y^2)\, dy\, dx$

17. $\displaystyle\int_{-3}^{3}\int_{0}^{3} y^2e^{-x}\, dy\, dx$

18. $\displaystyle\int_{-2}^{2}\int_{0}^{2} xe^{-y}\, dx\, dy$

19. $\displaystyle\int_{-2}^{2}\int_{-1}^{1} ye^{xy}\, dx\, dy$

20. $\displaystyle\int_{-1}^{1}\int_{-1}^{1} xe^{xy}\, dy\, dx$

21. $\displaystyle\int_{0}^{2}\int_{x}^{1} 12xy\, dy\, dx$

22. $\displaystyle\int_{0}^{1}\int_{y}^{1} 4xy\, dx\, dy$

23. $\displaystyle\int_{3}^{5}\int_{0}^{y} (2x - y)\, dx\, dy$

24. $\displaystyle\int_{2}^{4}\int_{0}^{x} (x - 2y)\, dy\, dx$

25. $\displaystyle\int_{-3}^{3}\int_{0}^{4x} (y - x)\, dy\, dx$

26. $\displaystyle\int_{-1}^{1}\int_{0}^{2y} (x + y)\, dx\, dy$

27. $\displaystyle\int_{0}^{1}\int_{-y}^{y} (x + y^2)\, dx\, dy$

28. $\displaystyle\int_{0}^{2}\int_{-x}^{x} (x^2 - y)\, dy\, dx$

29–32. For each double integral:
a. Write the *two* iterated integrals that are equal to it.
b. Evaluate *both* iterated integrals (the answers should agree).

29. $\displaystyle\iint_{R} 3xy^2\, dx\, dy$

with $R = \{(x, y) \mid 0 \le x \le 2, 1 \le y \le 3\}$

30. $\displaystyle\iint_{R} 6x^2y\, dx\, dy$

with $R = \{(x, y) \mid 0 \le x \le 1, 1 \le y \le 2\}$

31. $\displaystyle\iint_{R} ye^x\, dx\, dy$

with $R = \{(x, y) \mid -1 \le x \le 1, 0 \le y \le 2\}$

32. $\displaystyle\iint_{R} xe^y\, dx\, dy$

with $R = \{(x, y) \mid 0 \le x \le 1, -2 \le y \le 2\}$

33–36. Use integration to find the volume under each surface $f(x, y)$ above the region R.

33. $f(x, y) = x + y$
$R = \{(x, y) \mid 0 \le x \le 2, 0 \le y \le 2\}$

34. $f(x, y) = 8 - x - y$
$R = \{(x, y) \mid 0 \le x \le 4, 0 \le y \le 4\}$

35. $f(x, y) = 2 - x^2 - y^2$
$R = \{(x, y) \mid 0 \le x \le 1, 0 \le y \le 1\}$

36. $f(x, y) = x^2 + y^2$
$R = \{(x, y) \mid 0 \le x \le 2, 0 \le y \le 2\}$

37–40. Use integration to find the volume under each surface $f(x, y)$ above the region R.

37. $f(x, y) = 24xy$
over the region shown in the graph.

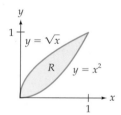

38. $f(x, y) = 3xy^2$
over the region shown in the graph.

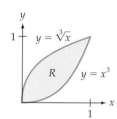

39. $f(x, y) = e^y$
over the region shown in the graph.

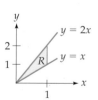

40. $f(x, y) = e^y$
over the region shown in the graph.

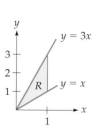

Applied Exercises

41. GENERAL: Average Temperature The temperature x miles east and y miles north of a weather station is given by the function $f(x, y) = 48 + 4x - 2y$. Find the average temperature over the region R shown below.

42. ENVIRONMENTAL SCIENCE: Average Air Pollution The air pollution near a chemical refinery is $f(x, y) = 20 + 6x^2y$ parts per million (ppm), where x and y are the numbers of miles east and north of the refinery. Find the average pollution level for the region R shown below.

43. GENERAL: Total Population The population density (people per square mile) x miles east and y miles north of the center of a city is $P(x, y) = 12{,}000e^{x-y}$. Find the total population of the region R shown in the following diagram. [*Hint:* Integrate the population density over the region R. This is an example of a double integral as a *sum*, giving a *total* population over a region.]

44. BUSINESS: Value of Mineral Deposit The value of an off-shore mineral deposit x miles east and y miles north of a certain point is $f(x, y) = 4x + 6y^2$ million dollars per square mile. Find the total value of the tract shown below. [*Hint:* Integrate the function over the region R. This is an example of a double integral as a *sum*, giving a *total* value over a region.]

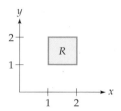

45–46. GENERAL: Volume of a Building To estimate heating and air conditioning costs, it is necessary to know the volume of a building.

45. A conference center has a curved roof whose height is $f(x, y) = 40 - 0.006x^2 + 0.003y^2$. The building sits on a rectangle extending from $x = -50$ to $x = 50$ and $y = -100$ to $y = 100$. Use integration to find the volume of the building. (All dimensions are in feet.)

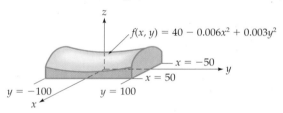

46. An airplane hangar has a curved roof whose height is $f(x, y) = 40 - 0.03x^2$. The building sits on a rectangle extending from $x = -20$ to $x = 20$ and $y = -100$ to $y = 100$. Use integration to find the volume of the building. (All dimensions are in feet.)

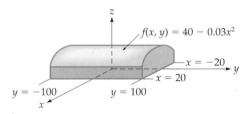

Exercises on Triple Integrals

47–50. Evaluate each triple iterated integral. [*Hint:* Integrate with respect to one variable at a time, treating the other variables as constants, working from the inside out.]

47. $\int_1^2 \int_0^3 \int_0^1 (2x + 4y - z^2)\, dx\, dy\, dz$

48. $\int_1^2 \int_0^3 \int_0^2 (6x - 2y + z^2)\, dx\, dy\, dz$

49. $\int_1^2 \int_0^2 \int_0^1 2xy^2z^3\, dx\, dy\, dz$

50. $\int_1^3 \int_0^1 \int_0^2 12x^3y^2z\, dx\, dy\, dz$

Conceptual Exercises

Assume, for simplicity, that all functions are continuous and all denominators are nonzero.

51. Which of the following is an *iterated* integral and which is a *double* integral?

   a. $\displaystyle\iint_R f(x, y)\, dx\, dy$ **b.** $\displaystyle\int_a^b \int_c^d f(x, y)\, dx\, dy$

52. If a surface lies partly above the *x-y* plane and partly below it on a region, what does the double integral represent?

53. For a *double* integral there are two possible orders of integration. How many possible orders of integration are there for a *triple* integral?

54. True or False:

$$\int_a^b \int_c^d f(x)\, g(y)\, dx\, dy = \int_c^d f(x)\, dx \cdot \int_a^b g(y)\, dy.$$

55. Can $\displaystyle\int_a^b \int_c^d e^{f(x) + g(y)}\, dx\, dy$ be written as the product of two integrals?

56. Can $\displaystyle\int_a^b \int_c^d \frac{f(x)}{g(y)}\, dx\, dy$ be written as the product of two integrals?

57. True or False: If $f(x, y) \le g(x, y)$ for all *x* and *y*,

   then $\displaystyle\int_a^b \int_c^d f(x, y)\, dx\, dy \le \int_a^b \int_c^d g(x, y)\, dx\, dy.$

58. How could you find the volume *between* two surfaces $f(x, y)$ and $g(x, y)$ over a region *R* by using one double integral? (Assume that surface *f* lies above surface *g*.)

59. If $f(x, y)$ is in *widgets per square blarg*, and *x* and *y* are in *blargs*, then what are the units of

$$\int_a^b \int_c^d f(x, y)\, dx\, dy\,?$$

60. If $\displaystyle\int_a^b \int_c^d f(x, y)\, dx\, dy$ is in *blargs*, and *x* and *y* are in *widgets*, then what are the units of $f(x, y)$?

---

7   Chapter Summary with Hints and Suggestions

Reading the text and doing the exercises in this chapter have helped you to master the following concepts and skills, which are listed by section (in case you need to review them) and are keyed to particular Review Exercises. Answers for all Review Exercises are given at the back of the book, and full solutions can be found in the Student Solutions Manual.

7.1   Functions of Several Variables

- Find the domain of a function of two variables. *(Review Exercises 1–4.)*

7.2   Partial Derivatives

- Find the first and second partials of a function of two variables. *(Review Exercises 5–12.)*

$$\frac{\partial}{\partial x} f(x, y) = f_x(x, y) = \lim_{h \to 0} \frac{f(x + h, y) - f(x, y)}{h}$$

$$\frac{\partial}{\partial y} f(x, y) = f_y(x, y) = \lim_{h \to 0} \frac{f(x, y + h) - f(x, y)}{h}$$

- Evaluate the first partials of a function of two variables. *(Review Exercises 13–16.)*

- Solve an applied problem involving partials, and interpret the answer. *(Review Exercises 17–18.)*

$$\frac{\partial}{\partial x} f(x, y) = \begin{pmatrix} \text{Rate of change of } f \text{ with respect} \\ \text{to } x \text{ when } y \text{ is held constant} \end{pmatrix}$$

$$\frac{\partial}{\partial y} f(x, y) = \begin{pmatrix} \text{Rate of change of } f \text{ with respect} \\ \text{to } y \text{ when } x \text{ is held constant} \end{pmatrix}$$

7.3   Optimizing Functions of Several Variables

- Find the relative extreme values of a function. *(Review Exercises 19–30.)*

$$\begin{cases} f_x = 0 \\ f_y = 0 \end{cases} \qquad D = f_{xx}f_{yy} - (f_{xy})^2$$

- Solve an applied problem by optimizing a function of two variables. *(Review Exercises 31–32.)*

7.4   Least Squares

- Find a least squares line "by hand." *(Review Exercises 33–34.)*

- Find the least squares line or curve for actual data, and use it to make a prediction. *(Review Exercises 35–36.)*

7.5 Lagrange Multipliers and Constrained Optimization

- Solve a constrained maximum or minimum problem using Lagrange multipliers. *(Review Exercises 37–42.)*

$$F(x, y, \lambda) = f(x, y) + \lambda g(x, y) \quad \begin{cases} F_x = 0 \\ F_y = 0 \\ F_\lambda = 0 \end{cases}$$

- Find *both* extreme values in a constrained optimization problem using Lagrange multipliers. *(Review Exercises 43–44.)*

- Solve an applied problem using Lagrange multipliers. *(Review Exercises 45–48.)*

7.6 Total Differentials, Approximate Changes, and Marginal Analysis

- Find the total differential of a function. *(Review Exercises 49–54.)*

$$df = \frac{\partial f}{\partial x}\, dx + \frac{\partial f}{\partial y}\, dy$$

- Solve an applied problem by using the total differential to estimate an actual change. *(Review Exercise 55.)*

$$\Delta f = f(x + \Delta x, y + \Delta y) - f(x, y) \qquad \Delta f \approx df$$

- Estimate the relative error in an area calculation. *(Review Exercise 56.)*

7.7 Multiple Integrals

- Evaluate an iterated integral. *(Review Exercises 57–60.)*

$$\int_c^d \int_a^b f(x, y)\, dx\, dy = \int_a^b \int_c^d f(x, y)\, dy\, dx$$

- Find the volume under a surface above a region using a double integral. *(Review Exercises 61–64.)*

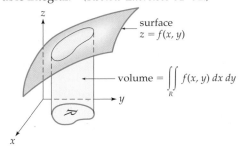

- Find the average population over a region using a double integral. *(Review Exercise 65.)*

- Find the total value of a region using a double integral. *(Review Exercise 66.)*

Hints and Suggestions

- The graph of a function of two variables is represented by a *surface* above (or below) the x-y plane. Such three-dimensional graphs are difficult to draw "by hand" but can be shown on some graphing calculators and computer screens. Such surfaces have maximum and minimum points (just as with functions of one variable), but also a new phenomenon, *saddle points* (see page 439).

- The graph of a function of *three* or more variables would require *four* or more dimensions, and so cannot be drawn.

- When finding partial derivatives, remember which is the variable of differentiation and then treat the other variable as a constant. Other than this, the differentiation formulas are the same.

- Partials give instantaneous rates of change with respect to one variable while the other is held constant. They also give *marginals* with respect to one product while production of the other is held constant.

- The D-test (page 459) applies only to critical points, where the first partials are zero. These critical points must be found first, and then the D-test is applied to each, one at a time.

- Least squares is carried out automatically by a graphing calculator or computer when it finds the linear regression line.

- Solving a constrained optimization problem by Lagrange multipliers is often easier than eliminating one of the variables, as was done in Sections 3.3 and 3.4.

- Multiple integrals give volume under a function if the function is nonnegative.

- When finding the average value of a function of two variables, don't forget to divide by the area of the region over which you are integrating.

7 | Review Exercises and Chapter Test   ◯ indicates a Chapter Test exercise.

7.1   Functions of Several Variables

1–4. For each function, state the domain.

1. $f(x, y) = \dfrac{\sqrt{x}}{\sqrt[3]{y}}$ **2.** $f(x, y) = \dfrac{\sqrt[3]{x}}{\sqrt{y}}$

③ $f(x, y) = e^{1/x} \ln y$ **4.** $f(x, y) = \dfrac{\ln y}{x}$

7.2   Partial Derivatives

5–12. For each function f, find:
a. f_x, **b.** f_y, **c.** f_{xy}, and **d.** f_{yx}.

5. $f(x, y) = 2x^5 - 3x^2y^3 + y^4 - 3x + 2y + 7$

6. $f(x, y) = 3x^4 + 5x^3y^2 - y^6 - 6x + y - 9$

⑦ $f(x, y) = 18x^{2/3}y^{1/3}$ **8.** $f(x, y) = \ln(x^2 + y^3)$

9. $f(x, y) = e^{x^3 - 2y^3}$ **10.** $f(x, y) = 3x^2e^{-5y}$

11. $f(x, y) = ye^{-x} - x \ln y$

12. $f(x, y) = x^2e^y + y \ln x$

13–16. For each function, calculate:
a. $f_x(1, -1)$ and **b.** $f_y(1, -1)$.

13. $f(x, y) = \dfrac{x + y}{x - y}$ **14.** $f(x, y) = \dfrac{x}{x^2 + y^2}$

⑮ $f(x, y) = (x^3 + y^2)^3$ **16.** $f(x, y) = (2xy - 1)^4$

⑰ BUSINESS: Marginal Productivity A company's production is given by the Cobb–Douglas function $P(L, K) = 160L^{3/4}K^{1/4}$, where L is the number of units of labor and K is the number of units of capital.
a. Find $P_L(81, 16)$ and interpret this number.
b. Find $P_K(81, 16)$ and interpret this number.
c. From your answers to parts (a) and (b), which will increase production more, an additional unit of labor or an additional unit of capital?

18. BUSINESS: Advertising A clothing designer's sales S depend on x, the amount spent on television advertising, and y, the amount spent on print advertising (both in thousands of dollars), according to the function

$$S(x, y) = 60x^2 + 90y^2 - 6xy + 200$$

Find $S_x(2, 3)$ and $S_y(2, 3)$, and interpret these numbers.

7.3   Optimizing Functions of Several Variables

19–30. For each function, find all relative extreme values.

⑲ $f(x, y) = 2x^2 - 2xy + y^2 - 4x + 6y - 3$

20. $f(x, y) = x^2 - 2xy + 2y^2 - 6x + 4y + 2$

21. $f(x, y) = 2xy - x^2 - 5y^2 + 2x - 10y + 3$

22. $f(x, y) = 2xy - 5x^2 - y^2 + 10x - 2y + 1$

23. $f(x, y) = 2xy + 6x - y + 1$

24. $f(x, y) = 4xy - 4x + 2y - 4$

25. $f(x, y) = e^{-(x^2 + y^2)}$ **26.** $f(x, y) = e^{2(x^2 + y^2)}$

27. $f(x, y) = \ln(5x^2 + 2y^2 + 1)$

28. $f(x, y) = \ln(4x^2 + 3y^2 + 10)$

29. $f(x, y) = x^3 - y^2 - 12x - 6y$

30. $f(x, y) = y^2 - x^3 + 12x - 4y$

31. BUSINESS: Maximum Profit A boatyard builds 18-foot and 22-foot sailboats. Each 18-foot boat costs \$3000 to build, each 22-foot boat costs \$5000 to build, and the company's fixed costs are \$6000. The price function for the 18-foot boats is $p = 7000 - 20x$, and that for the 22-foot boats is $q = 8000 - 30y$ (both in dollars), where x and y are the numbers of 18-foot and 22-foot boats, respectively.
a. Find the company's cost function $C(x, y)$.
b. Find the company's revenue function $R(x, y)$.
c. Find the company's profit function $P(x, y)$.
d. Find the quantities and prices that maximize profit. Also find the maximum profit.

32. BUSINESS: Price Discrimination A company sells farm equipment in America and Europe, charging different prices in the two markets. The price function for harvesters sold in America is $p = 80 - 0.2x$, and the price function for harvesters sold in Europe is $q = 64 - 0.1y$ (both in thousands of dollars), where x and y are the numbers sold per day in America and Europe, respectively. The company's cost function is $C = 100 + 12(x + y)$ thousand dollars.
a. Find the company's profit function.
b. Find how many harvesters should be sold in each market to maximize profit. Also find the price for each market.

7.4   Least Squares

33–34. For each table of points:
a. Find the least squares line "by hand."
b. Check your answer using a graphing calculator.

㉝

x	y
1	−1
3	6
4	6
5	10

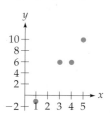

34.

x	y
1	7
2	4
4	2
5	−1

35. BUSINESS: Aging Markets The growth of the elderly population (over-65) is outpacing all other segments of the American market. Find the least squares line for the following data for the over-65 population, and use it to predict the size of the over-65 population in the year 2020 ($x = 6$).

Source: U.S. Census Bureau

	x	Population (millions)
1970	1	20.1
1980	2	25.5
1990	3	31.2
2000	4	35.0
2010	5	40.2

36 SOCIAL SCIENCES: Violent Crime Crime, and in particular violent crime, has been decreasing in the United States. The following table gives the number of violent crimes per 1000 people in the United States for recent years. Fit an exponential curve to the data and then use it to predict the number of violent crimes per 1000 people in 2023.

Source: FBI Uniform Crime Reports

Year	x	Crimes
1998	1	5.7
2003	2	4.8
2008	3	4.6
2013	4	3.8

7.5 Lagrange Multipliers and Constrained Optimization

37–42. Use Lagrange multipliers to optimize each function subject to the given constraint. (The stated extreme values *do* exist.)

37. Maximize $f(x, y) = 6x^2 - y^2 + 4$
subject to $3x + y = 12$.

38. Maximize $f(x, y) = 4xy - x^2 - y^2$
subject to $x + 2y = 26$.

39. Minimize $f(x, y) = 2x^2 + 3y^2 - 2xy$
subject to $2x + y = 18$.

40. Minimize $f(x, y) = 12xy - 1$
subject to $y - x = 6$.

41 Minimize $f(x, y) = e^{x^2 + y^2}$
subject to $x + 2y = 15$.

42. Maximize $f(x, y) = e^{-x^2 - y^2}$
subject to $2x + y = 5$.

43–44. Use Lagrange multipliers to find the maximum *and* minimum values of each function subject to the given constraint. (Both extreme values *do* exist.)

43. $f(x, y) = 6x - 18y$ subject to $x^2 + y^2 = 40$

44. $f(x, y) = 4xy$ subject to $x^2 + y^2 = 32$

45 BUSINESS: Maximum Profit A company's profit is $P = 300x^{2/3}y^{1/3}$, where x and y are, respectively, the amounts spent on production and advertising. The company has a total of $60,000 to spend.
a. Use Lagrange multipliers to find the amounts for production and advertising that maximize profit.
b. Evaluate and give an interpretation for $|\lambda|$.

46. BIOMEDICAL: Nutrition A nursing home uses two vitamin supplements, and the nutritional value of x ounces of the first together with y ounces of the second is $4x + 2xy + 8y$. The first costs $2 per ounce, the second costs $1 per ounce, and the nursing home can spend only $8 per patient per day.
a. Use Lagrange multipliers to find how much of each supplement should be used to maximize the nutritional value subject to the budget constraint.
b. Evaluate and give an interpretation for $|\lambda|$.

47. BUSINESS: Maximum Production A company's output is given by the Cobb–Douglas production function $P = 400L^{3/5}K^{2/5}$, where L and K are the number of units of labor and capital. Each unit of labor costs $50 and each unit of capital costs $60, and $1500 is available to pay for labor and capital.
a. How many units of labor and of capital should be used to maximize production?
b. Evaluate and give an interpretation for $|\lambda|$.

48. GENERAL: Container Design An open-top box with a square base and two perpendicular dividers, as shown in the diagram, is to have a volume of 576 cubic inches. Use Lagrange multipliers to find the dimensions that require the least amount of material.

7.6 Total Differentials, Approximate Changes, and Marginal Analysis

49–54. Find the total differential of each function.

49 $f(x, y) = 3x^2 + 2xy + y^2$ **50.** $f(x, y) = x^2 + xy - 3y^2$

51 $g(x, y) = \ln(xy)$ **52.** $g(x, y) = \ln(x^3 + y^3)$

53. $z = e^{x-y}$ **54.** $z = e^{xy}$

55. BUSINESS: Sales A clothing designer's sales S depend on x, the amount spent on television advertising, and y, the amount spent on print advertising (all in thousands of dollars) according to the formula $S(x, y) = 60x^2 - 6xy + 90y^2 + 200$. If the company now spends 2 thousand dollars on television advertising and 3 thousand dollars on print advertising, use the total differential to estimate the change in sales if television advertising is increased by \$500 and print advertising is decreased by \$500. [*Hint:* Δx and Δy must be in thousands of dollars, just as x and y are.] Then estimate the change in sales if each of the changes in advertising were halved.

56. GENERAL: Relative Error in Calculating Area   A triangular piece of real estate is measured to have length x feet and altitude y feet, but each measurement may be in error by 1%. Estimate the percentage error in calculating the area by using a total differential.

7.7   Multiple Integrals

57–60. Evaluate each iterated integral.

57 $\displaystyle\int_0^4 \int_{-1}^1 2xe^{2y}\, dy\, dx$ **58.** $\displaystyle\int_{-1}^1 \int_0^3 (x^2 - 4y^2)\, dx\, dy$

59. $\displaystyle\int_{-1}^1 \int_{-y}^y (x + y)\, dx\, dy$ **60.** $\displaystyle\int_{-2}^2 \int_{-x}^x (x + y)\, dy\, dx$

61–64. Find the volume under the surface $f(x, y)$ above the region R.

61. $f(x, y) = 8 - x - y$
$R = \{(x, y)\,|\, 0 \le x \le 2, 0 \le y \le 4\}$

62. $f(x, y) = 6 - x - y$
$R = \{(x, y)\,|\, 0 \le x \le 4, 0 \le y \le 2\}$

63. $f(x, y) = 12xy^3$ over the region shown in the graph.

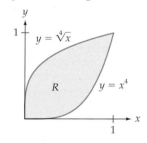

64. $f(x, y) = 15xy^4$ over the region shown in the graph.

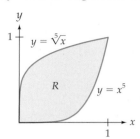

65 GENERAL: Average Population of a Region   The population per square mile x miles east and y miles north of the center of a city is $P(x, y) = 12{,}000 + 100x - 200y$. Find the *average* population over the region shown below.

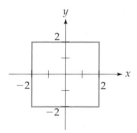

66. GENERAL: Total Real Estate Value   The value of land x blocks east and y blocks north of the center of a town is $V(x, y) = 40 - 4x - 2y$ hundred thousand dollars per block. Find the total value of the parcel of land shown in the graph above.

<table>
<tr><td>**1–7**</td><td>**Cumulative Review for Chapters 1–7**</td></tr>
</table>

The following exercises review some of the basic techniques that you learned in Chapters 1–7. Answers to all of these cumulative review exercises are given in the answer section at the back of the book. A graphing calculator is suggested but not required.

1. Draw the graph of the function

$$f(x) = \begin{cases} 2x - 1 & \text{if } x \le 3 \\ 7 - x & \text{if } x > 3 \end{cases}$$

2. Simplify $\left(\dfrac{1}{8}\right)^{-2/3}$.

3. Use the definition of the derivative

$$f'(x) = \lim_{h \to 0} \frac{f(x + h) - f(x)}{h}$$

to find the derivative of $f(x) = \dfrac{1}{x}$.

4. If $f(x) = 12\sqrt[3]{x^2} - 4$, find $f'(8)$.

5. A camera store finds that if it sells disposable cameras at a price of p dollars each, it will sell
$$S(p) = \frac{800}{p + 8}$$ of them per week. Find $S'(12)$ and interpret the answer.

6. Find $\dfrac{d}{dx}[x^2 + (2x + 1)^4]^3$.

7. Make sign diagrams for the first and second derivatives and graph the function $f(x) = x^3 + 9x^2 - 48x - 148$, showing all relative extreme points and inflection points.

8. Make a sign diagram for the first derivative and graph the function $f(x) = \dfrac{1}{x^2 - 4x}$, showing all asymptotes and relative extreme points.

9. A homeowner wants to use 80 feet of fence to make a rectangular enclosure along an existing stone wall. If the side along the existing wall needs no fence, what are the dimensions of the enclosure that has the largest possible area?

10. An open-top box with a square base is to have a volume of 108 cubic feet. Find the dimensions of the box of this type that can be made with the least amount of material.

11. A spherical balloon is being inflated at the constant rate of 128 cubic feet per minute. Find how fast the radius is increasing at the moment when the radius is 4 feet.

12. A sum of $1000 is deposited in a bank account paying 6% interest. Find the value of the account after 3 years if the interest is compounded:
    a. quarterly.   b. continuously.

13. In t years the population of a county is predicted to be $P(t) = 12{,}000e^{0.02t}$, and the population of a neighboring county is predicted to be $Q(t) = 9000e^{0.04t}$. In how many years will the second county overtake the first in population?

14. A sum is deposited in a bank account paying 6% interest compounded monthly. How many years will it take for the value to increase by 50%?

15. Make sign diagrams for the first and second derivatives and graph the function $f(x) = e^{-x^2/2}$, showing all relative extreme points and inflection points.

16. Find $\displaystyle\int (12x^2 - 4x + 1)\,dx$.

17. Pollution is being discharged into a lake at the rate of $18e^{0.02t}$ million gallons per year, where t is the number of years from now. Find a formula for the total amount of pollution that will be discharged into the lake during the next t years.

18. Find the area bounded by $y = 20 - x^2$ and $y = 8 - 4x$.

19. Find the average value of $f(x) = 12\sqrt{x}$ over the interval $[0, 4]$.

20. Find: a. $\displaystyle\int \frac{x^2\,dx}{x^3 + 1}$ b. $\displaystyle\int \frac{e^{\sqrt{x}}\,dx}{\sqrt{x}}$

21. Find by integration by parts: $\displaystyle\int xe^{4x}\,dx$

22. Use the integral table on the inside back cover to find $\displaystyle\int \frac{\sqrt{4 - x^2}}{x}\,dx$. Check your answer by differentiating and then simplifying.

23. Evaluate $\displaystyle\int_1^\infty \frac{1}{x^3}\,dx$

24. Use trapezoidal approximation with $n = 4$ trapezoids to approximate $\int_0^1 \sqrt{x^2 + 1}\,dx$. (If you use a graphing calculator, compare your answer with the value obtained by using FnInt.)

25. Use Simpson's Rule with $n = 4$ to approximate $\int_0^1 \sqrt{x^2 + 1}\,dx$. (If you use a graphing calculator, compare your answer with the value obtained by using FnInt.)

26. For the function $f(x, y) = \dfrac{3\sqrt{x} + \ln y}{x - y}$, find:
    a. the domain.
    b. $f(4, 1)$.

27. Find the first partial derivatives of the function $f(x, y) = x \ln y + ye^{2x}$.

28. Find the relative extreme values of $f(x, y) = 2x^2 - 2xy + y^2 + 4x - 6y + 12$.

29. Find the least squares line for the following points:

x	y
1	-3
3	1
5	3
7	8

30. Use Lagrange multipliers to find the minimum value of $f(x, y) = 3x^2 + 2y^2 - 2xy$ subject to the constraint $x + 2y = 18$.

31. Find the total differential of $f(x, y) = 2x^2 + xy - 3y^2 + 4$.

32. Find the volume under the surface $f(x, y) = 12 - x - 2y$ over the rectangle
$$R = \{(x, y) \mid 0 \le x \le 2, 0 \le y \le 3\}.$$

Trigonometric Functions

8

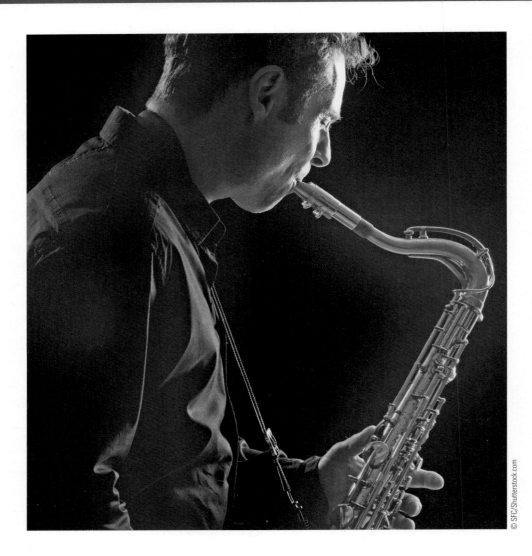

© SFC/Shutterstock.com

What You'll Explore

Many quantities, from sound waves to seasonal temperatures, behave in a *cyclic* fashion, repeating their values over time. In this chapter you will use trigonometric functions to model cyclic behaviour, predicting daily temperatures and seasonal profits. You will use derivatives to find rates of income growth, and integrals to give cumulative sales. Along the way, you will find the best angle for hitting a home run, shooting a basket, or kicking a fieldgoal.

8.1 Triangles, Angles, and Radian Measure

8.2 Sine and Cosine Functions

8.3 Derivatives of Sine and Cosine Functions

8.4 Integrals of Sine and Cosine Functions

8.5 Other Trigonometric Functions

Tones, Temperature, and Trigonometry

This chapter introduces the sine and cosine functions, "wavy" curves that repeat themselves at regular intervals. Such functions have many applications. For example, musical sounds are caused by vibrations in the air, which can be picked up by a microphone and displayed on an oscilloscope. The following graph shows the sound wave from a saxophone playing a G#. This curve can be expressed as the sum of modified sine and cosine curves.*

Saxophone G#

The next graph shows the daily temperature range in New York City for the years 2012 and 2013 (the very jagged line), and the normal daily high and low temperatures (the smoother curves), which may be closely approximated by sine and cosine curves.

Source: Climatestations.com

The following graph shows the rising level of carbon dioxide in the atmosphere as measured at the Mauna Loa Observatory in Hawaii. It is known as the *Keeling curve*, after its originator, C. David Keeling. Carbon dioxide contributes to the "greenhouse effect," which causes global warming. This curve is a sine curve (the seasonal fluctuations) added to an exponential function (the upward trend).

*See Dorothea Bone, "Music and the Circular Functions," UMAP unit 588, COMAP, Lexington, Mass., and John R. Pierce, *The Science of Musical Sound* (Scientific American Books, W. H. Freeman & Co.).

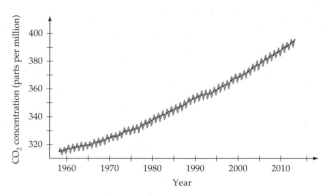

Source: National Oceanic and Atmospheric Administration

Triangles, Angles, and Radian Measure

Introduction

The graph on the previous page shows the annual cycle of temperatures, varying "periodically" from winter cold to summer heat and back to winter cold. In this chapter we will study **trigonometric functions** and use them to model this and other periodic behavior, from business cycles to home heating costs. It is assumed that you have studied trigonometry at some time in the past. The first two sections review trigonometry from the beginning, but selectively rather than comprehensively, covering only the topics that will be needed for applications. Later sections discuss differentiation and integration of trigonometric functions. We begin with triangles, angles, and radian measure.

Right Triangles

A triangle with a right (90°) angle is called a **right triangle.** The lengths of the sides of a right triangle are related by the Pythagorean Theorem.*

Pythagorean Theorem

In a right triangle, the square of the hypotenuse equals the sum of the squares of the other two sides.

$$a^2 + b^2 = c^2$$

---

*This theorem was first attributed to the school of the Greek philosopher Pythagoras (sixth century B.C.), although the Babylonian tablet "Plimpton 322" shows that it was known several centuries earlier.

A proof of the Pythagorean Theorem is outlined in Exercise 37. We first use the theorem as follows: Given an equilateral triangle of side 2, cutting it in half, as shown on the left, leaves a right triangle with hypotenuse 2 and side 1. The third side, b, can then be found from the Pythagorean Theorem as follows.

$$1^2 + b^2 = 2^2$$ $a^2 + b^2 = c^2$ with $a = 1$ and $c = 2$

$$1 + b^2 = 4$$ Simplifying

$$b^2 = 3$$ Subtracting 1

Therefore, the third side has length $b = \sqrt{3}$,   leading to the following result.

30°–60°–90° Triangle

A triangle with angles 30°, 60°, 90° has sides in proportion to 1, 2, $\sqrt{3}$.

We say "in proportion to" because the numbers 1, 2, $\sqrt{3}$ can be doubled, tripled, halved, or multiplied by *any* positive number, to obtain other 30°−60°−90° triangles. Note that the smallest side (1) goes opposite the smallest angle (30°), and the largest side (2) goes opposite the largest angle (90°).

We now consider a triangle with angles 45°, 45°, and 90°, as shown on the left. The two 45° angles make the triangle isosceles, so it has two equal sides, which we take to be of length 1. We find the length of the hypotenuse from the Pythagorean Theorem, as follows.

$$1^2 + 1^2 = c^2$$ $a^2 + b^2 = c^2$ with $a = b = 1$

$$2 = c^2$$ Simplifying

Therefore, the hypotenuse has length $c = \sqrt{2}$,   giving the following result.

45°–45°–90° Triangle

A triangle with angles 45°, 45°, 90° has sides in proportion to 1, 1, $\sqrt{2}$.

Again, the largest side $\left(\sqrt{2}\right)$ goes opposite the largest angle (90°). These two particular right triangles will be used extensively in the next section.

Angles and Radian Measure

Trigonometry requires a precise definition of the word **angle.** An *angle* is formed when a line segment is rotated around one of its endpoints (the **vertex**), from an **initial side** to a **terminal side.** We use an arrow to indicate the rotation.

An angle in the *x*-*y*-plane is in **standard position** if its vertex is at the origin and its initial side lies along the positive *x*-axis. An angle is **positive** if its rotation is *counterclockwise*, and **negative** if its rotation is *clockwise*. The first three angles in the following diagram are positive angles, and the last is a negative angle.

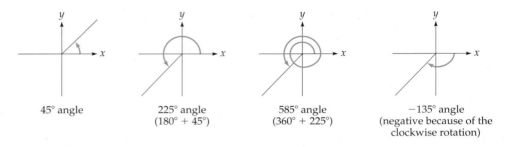

45° angle	225° angle	585° angle	−135° angle
	(180° + 45°)	(360° + 225°)	(negative because of the clockwise rotation)

 Be Careful It is not enough to specify the initial and terminal sides of an angle—the rotation must also be specified. For example, the second, third, and fourth angles in the above diagrams have the same initial and terminal sides but are different because they have different rotations.

Angles that have the same initial and terminal sides (even if formed by different rotations) are called **coterminal angles.** Any two coterminal angles differ by a multiple of 360°.

The choice of 360 for the number of degrees in a full circle dates back to the ancient Babylonians, whose numbering system was based on 60, the "sexagesimal" system. We still use their system for time: 60 seconds in a minute and 60 minutes in an hour.

In calculus we use another method for measuring angles, called **radian measure.** The radian measure of an angle is found by placing the angle at the center of a circle of radius 1 and measuring the length of the arc of the circle cut off by the angle: The length of the arc gives the measure of the angle.

Radian Measure

LOOKING AHEAD

Measuring angles in this way, although counterintuitive, will lead to the simplest formulas for derivatives and integrals on pages 547 and 559.

An angle of *x* radians is an angle that cuts off an arc of length *x* on a circle of radius 1 centered at the vertex.

As usual, a positive angle means counterclockwise rotation. The following diagrams show angles of $\frac{1}{2}$ radian, 1 radian, and 2 radians.

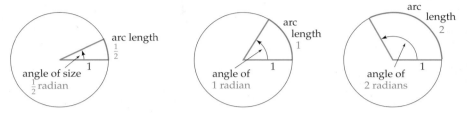

Radian measure of angle equals the arc length cut off on a unit circle.

An angle whose vertex is at the center of a circle is called a **central angle**.

The circumference formula $C = 2\pi r$ with $r = 1$ shows that the circumference of the unit circle is 2π. Therefore, a full circle of 360° is equivalent to 2π radians.

$$360° = 2\pi \text{ radians}$$

Dividing each side by 2 shows that $180° = \pi$ radians, and dividing again shows that $90° = \pi/2$ radians. The degree and radian measures of the following **quadrant angles** should be memorized.

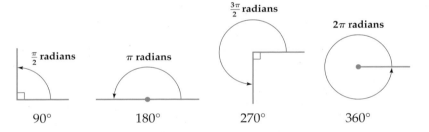

Converting Between Degrees and Radians

Since degrees and radians are proportional, we can convert between them using the following relationship.

$$\frac{\text{Degrees}}{180} = \frac{\text{Radians}}{\pi}$$

This equation may be solved for either degrees or radians, resulting in the two formulas

$$\text{Radians} = \frac{\pi}{180} \cdot \text{degrees}$$

To change from degrees to radians, multiply by $\dfrac{\pi}{180}$

and

$$\text{Degrees} = \frac{180}{\pi} \cdot \text{radians}$$

To change from radians to degrees, multiply by $\dfrac{180}{\pi}$

EXAMPLE 1   **CONVERTING DEGREES TO RADIANS**

Find the radian measure of each angle: **a.** $45°$ **b.** $150°$ **c.** $-60°$

Solution

To convert to radians, multiply degrees by $\dfrac{\pi}{180}$.

a. $45° = \overset{1}{\cancel{45}} \cdot \dfrac{\pi}{\underset{4}{\cancel{180}}} = \dfrac{\pi}{4}$ radians   Multiplying by $\dfrac{\pi}{180}$

b. $150° = \overset{5}{\cancel{150}} \cdot \dfrac{\pi}{\underset{6}{\cancel{180}}} = \dfrac{5\pi}{6}$ radians

c. $-60° = \overset{-1}{\cancel{-60}} \cdot \dfrac{\pi}{\underset{3}{\cancel{180}}} = -\dfrac{\pi}{3}$ radians

We leave π in our answers to keep them exact. For an approximation, we could replace π by 3.14 or use a calculator.

EXAMPLE 2   **CONVERTING RADIANS TO DEGREES**

Find the degree measure of each angle:

a. $\dfrac{\pi}{6}$ radians   **b.** $\dfrac{3\pi}{4}$ radians   **c.** $-\dfrac{\pi}{2}$ radians   **d.** 1 radian

Solution

To convert to degrees, multiply radians by $\dfrac{180}{\pi}$.

a. $\dfrac{\pi}{6}$ radians $= \dfrac{\cancel{\pi}}{\underset{1}{\cancel{6}}} \cdot \dfrac{\overset{30}{\cancel{180}}}{\cancel{\pi}} = 30°$ Multiplying by $\dfrac{180}{\pi}$

b. $\dfrac{3\pi}{4}$ radians $= \dfrac{3\cancel{\pi}}{\underset{1}{\cancel{4}}} \cdot \dfrac{\overset{45}{\cancel{180}}}{\cancel{\pi}} = 3 \cdot 45° = 135°$

c. $-\dfrac{\pi}{2}$ radians $= -\dfrac{\cancel{\pi}}{\underset{1}{\cancel{2}}} \cdot \dfrac{\overset{90}{\cancel{180}}}{\cancel{\pi}} = -90°$

d. 1 radian $= 1 \cdot \dfrac{180}{\pi} \approx 57°$

1 radian $\approx 57°$

Graphing Calculator Exploration

For calculator in *radian* mode

```
( -π/2)ʳ
                              -90
```

For calculator in *degree* mode

a. You can use a graphing calculator to convert from degrees to radians. For example, to check the answer to part (a) of Example 1, set your calculator to RADIANS (on some calculators use MODE), enter 45 degrees with the ° symbol (use ANGLE), and press ENTER. Since your calculator gives *decimal* answers (using an approximation for π), you can check the decimal by evaluating π/4.

b. You can also convert from radians to degrees. For example, to check the answer to part (c) of Example 2, set your calculator for DEGREES (using MODE), enter −π/2 radians with the ʳ symbol (using ANGLE). The result will be in degrees.

Be sure to reset your calculator for RADIANS (using MODE) since that is what you will be using almost exclusively in this chapter.

PRACTICE PROBLEM

a. Find the radian measure of an angle of 60°.

b. Find the degree measure of an angle of $\dfrac{2\pi}{3}$ radians.

Solutions on page 532 >

The angles 30°, 45°, and 60° occur so frequently that their radian measures should be memorized.

These angles may be added to the quadrant angles to find corresponding angles in other quadrants. For example, the largest angle shown in the diagram on the right has measure

$$\pi + \frac{\pi}{4} = \frac{5\pi}{4}$$

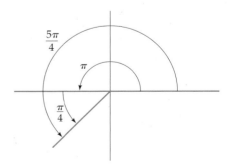

Other angles resulting from such additions are shown in the following diagram.

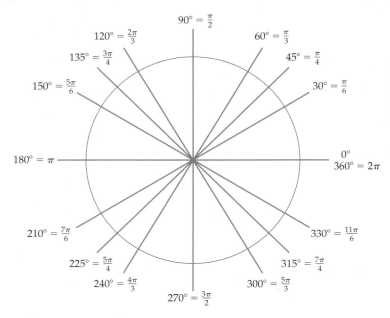

Alternative Definition of Radian Measure

The circumference formula $C = 2\pi r$ shows that the circumference C of a circle is proportional to the radius r. Therefore, we may define radian measure using a circle of *any* radius r, as long as arc length is measured in "radius" units, which means dividing the actual arc length by the radius r.

Alternative Definition of Radian Measure

The central angle that cuts off an arc of length s on a circle of radius r has radian measure s/r:

$$\left(\begin{array}{c} \text{Radian} \\ \text{measure} \end{array}\right) = \frac{s}{r} \quad \leftarrow \text{arc length} \\ \qquad\quad \leftarrow \text{radius}$$

Since radian measure is a *quotient,* arc length over radius, any units (inches, feet, etc.) will cancel out. For this reason, radians are called *dimensionless* units.

Arc Length Along a Circle

Some applications involve finding the length of the arc of a circle that is cut off by a central angle. For positive angles, the definition of radian measure is

$$\text{Radians} = \frac{\text{Arc length}}{\text{Radius}}$$

Solving this equation for arc length (multiplying each side by "Radius") gives a formula for the arc length along any circle.

Arc Length

The arc length cut off by any central angle is

$$\left(\begin{matrix}\text{Arc} \\ \text{length}\end{matrix}\right) = (\text{Radians}) \cdot (\text{Radius})$$

Arc length is Radians times Radius

Speed of Rotation of the Earth

How fast are you moving if you "stand still" on the equator? The earth makes one full rotation (2π radians) every 24 hours, so during 1 hour it rotates $\dfrac{2\pi}{24} = \dfrac{\pi}{12}$ radian. Since the earth's radius is approximately 4000 miles, during 1 hour a point on the equator will move a distance of

$$\left(\begin{matrix}\text{Arc} \\ \text{length}\end{matrix}\right) = \frac{\pi}{12} \cdot 4000 \approx 1047 \text{ miles}$$

Radians times radius

That is, standing on the equator you are actually rotating around the center of the earth at a speed of over 1000 miles per hour. You don't feel it because everything around you is moving with you.

Graphing Calculator Exploration

(For those familiar with the cosine function)

What is your speed (resulting from the earth's rotation) if you live somewhere other than the equator? It depends on your latitude. If you are at latitude x, then your speed is $1047 \cos x$.

a. Set your calculator for DEGREES and graph $y_1 = 1047 \cos x$ on the window [0, 90] by [0, 1100]. Observe how the speed varies from 1047 miles per hour at the equator (0° latitude) to 0 miles per hour at the North Pole (90° latitude).

b. Find the latitude where you are and evaluate $y_1 = 1047 \cos x$ at that number to find your speed of rotation.

Afterwards, be sure to reset your calculator to RADIANS.

EXAMPLE 3	INTERCITY DISTANCES D.C.

The cities of Washington, D.C., and Moscow determine a central angle of 1.22 radians (at the center of the earth). Find the distance between these cities measured along the surface of the earth. (Assume that the earth is a sphere of radius 4000 miles.)

Solution

Since arc length is radians times radius,

$$\begin{pmatrix} \text{Distance from} \\ \text{Washington, D.C.,} \\ \text{to Moscow} \end{pmatrix} = 1.22 \cdot \underbrace{4000}_{\text{radius}} = 4880$$

$$\underbrace{}_{\text{radians}}$$

Therefore, an airplane flying between Washington, D.C., and Moscow must fly approximately 4880 miles (since the altitude of the airplane is negligible compared to the radius of the earth).

Solutions TO PRACTICE PROBLEM

a. $60° = \overset{1}{\cancel{60}} \cdot \dfrac{\pi}{\underset{3}{\cancel{180}}} = \dfrac{\pi}{3}$ radians

b. $\dfrac{2\pi}{3}$ radians $= \dfrac{2\cancel{\pi}}{\underset{1}{\cancel{3}}} \cdot \dfrac{\overset{60}{\cancel{180}}}{\cancel{\pi}} = 120°$

8.1 Section Summary

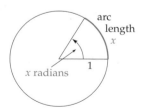

We began by finding the sides of two particular triangles: a $30°-60°-90°$ triangle (sides proportional to 1, 2, $\sqrt{3}$) and a $45°-45°-90°$ triangle (sides proportional to 1, 1, $\sqrt{2}$). We then defined the *radian measure* of an angle as the arc length cut off on a unit circle centered at the vertex of the angle.

Alternatively, using a circle of *any* radius, the radian measure is the arc length divided by the radius. To put this colloquially, if we were walking along the circumference of a circle, then walking a distance of x radii determines an angle of x radians. The diagram on page 530 shows the measures of many common angles in both degrees and radians.

To convert between degrees and radians, we use the relationship

$$\frac{\text{Degrees}}{180} = \frac{\text{Radians}}{\pi}$$

The arc length of a circle intercepted by a central angle measured in radians can be calculated from the formula

$$\begin{pmatrix} \text{Arc} \\ \text{length} \end{pmatrix} = (\text{Radians}) \cdot (\text{Radius})$$

8.1   Exercises

1–8. Find the degree and radian measure of each angle shown below.

1.

2.

3.

(1, 1)

4.

(−1, 1)

5.

6.

(2, 2)

7.

(1, −1)

8.

9–14. For each angle:

 i. Find the radian measure (without using a calculator).
 ii. Check your answers using a graphing calculator.

9. a. 30°   **b.** 225°   **c.** −180°

10. a. 15°   **b.** 210°   **c.** −90°

11. a. 60°   **b.** 315°   **c.** −120°

12. a. 45°   **b.** 240°   **c.** −150°

13. a. 540°   **b.** 135°   **c.** 1°

14. a. 750°   **b.** 150°   **c.** 10°

15–20. For each angle:

 i. Find the degree measure (without using a calculator).
 ii. Check your answers using a graphing calculator.

15. a. $\dfrac{\pi}{4}$ **b.** $\dfrac{2\pi}{3}$ **c.** $-\dfrac{5\pi}{6}$

16. a. $\dfrac{\pi}{6}$ **b.** $\dfrac{3\pi}{2}$ **c.** $-\dfrac{5\pi}{4}$

17. a. $\dfrac{\pi}{3}$ **b.** $\dfrac{5\pi}{4}$ **c.** $-\dfrac{7\pi}{6}$

18. a. $\dfrac{3\pi}{4}$ **b.** $\dfrac{4\pi}{3}$ **c.** $-\dfrac{7\pi}{4}$

19. a. $\dfrac{9\pi}{4}$ **b.** -3π **c.** 5

20. a. $\dfrac{7\pi}{3}$ **b.** -5π **c.** 3

Applied Exercises

21. GENERAL: Arc Length A 27-inch-long pendulum swings through an angle of $\dfrac{\pi}{9}$ radians. Find the length of arc through which it swings. (Give your answer in terms of π, and then approximate it using $\pi \approx 3.14$.)

27"

$\dfrac{\pi}{9}$

← arc →

22. GENERAL: Building Swaying Modern skyscrapers are designed to bend in high winds. As a general rule, a 100 mile per hour wind can cause a tall building to bend by 0.002 radian. Find the length of arc through which the 2717-foot-tall Burj Khalifa in Dubai (the tallest building in the world) would bend in such a wind. [*Note:* Because of back-and-forth motion, the swaying might be twice this great.]

23. GENERAL: Airplanes An airliner is flying in a circular holding pattern with a radius of 12 miles. Find the central angle (in radians) that the airliner will have flown through by flying 18 miles along this circular path.

24. GENERAL: Distance Find the distance along the surface of the earth between Los Angeles and Honolulu if the two cities determine a central angle of 0.64 radian. (Assume that the earth is a sphere of radius 4000 miles.)

25. GENERAL: Distance Find the distance along the surface of the earth between New York City and Paris if the two cities determine a central angle of 0.9 radian. (Assume that the earth is a sphere of radius 4000 miles.)

26. GENERAL: Moon Diameter Estimate the diameter of the moon if it is 240,000 miles from the earth and it forms an angle of 0.009 radian when viewed from the earth. [*Hint:* For very small angles, the diameter can be approximated by arc length.]

0.009 radian

240,000 miles   moon

Conceptual Exercises

27. True or False: All 30°–60°–90° triangles have sides of length 1, 2, $\sqrt{3}$.

28. True or False: Positive angles are measured *clockwise*.

29. True or False: 45 radians = $\dfrac{\pi}{4}$ degrees.

30. Which angles are in standard position?

a.

b.

c.

d.

31. What is the size of the smallest angle coterminal with an angle of 1000π radians?

32. If the shortest side of a 45°−45°−90° triangle has length 1000 kilometers, then what is the length of the longest side?

33. Fill in the blank: $\dfrac{180}{\pi}$ degrees = _____ radians.

34. Fill in the blank: $\dfrac{\pi}{180}$ radians = _____ degrees.

35. If a central angle of 0.123 radian cuts off an arc length of 5 on a circle, then what central angle cuts off an arc length of 10?

36. If you walk along the circumference of a circle a distance of 1 diameter, what is the size of the corresponding central angle (in radians)?

Explorations and Excursions   The following problem extends and augments the material presented in the text.

37. PROOF OF THE PYTHAGOREAN THEOREM Use the diagram on the right to prove the Pythagorean Theorem for the right triangle with sides a and b and hypotenuse c as follows:

   i. Calculate the area of the outer square by multiplying length times width. [*Hint:* Length and width are each $a + b$.]

   ii. Calculate the area of the outer square again, but now as the inner square plus four right triangles, each of area $\frac{1}{2}ab$.

   iii. Set the two expressions for the area equal to each other and simplify, obtaining $a^2 + b^2 = c^2$.

8.2 Sine and Cosine Functions

Introduction

In this section we define the **sine** and **cosine** functions and use them to model quantities that fluctuate periodically.

Sine and Cosine Functions

We use the Greek letter θ ("theta") to represent an angle (always measured in *radians* unless otherwise indicated). For an angle θ in standard position, we choose a point (x, y) on its terminal side and define the sine and cosine functions as follows.

$\sin \theta$ and $\cos \theta$

Let (x, y) be a point on the terminal side of an angle of θ radians in standard position, and let r be its distance from the origin. We define $\sin \theta$ and $\cos \theta$ as follows:

$$\sin \theta = \frac{y}{r}$$

$$\cos \theta = \frac{x}{r}$$

It does not matter which point on the terminal side (other than the origin) we choose; any two points (x, y) and (x', y') determine similar triangles, so the ratios of the sides will be the same.

If we choose the point (x, y) to be a distance $r = 1$ from the origin, then the definitions of $\sin \theta$ and $\cos \theta$ simplify to

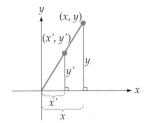

$$\sin \theta = y \qquad\qquad \sin \theta = \frac{y}{r} \text{ with } r = 1$$

$$\cos \theta = x \qquad\qquad \cos \theta = \frac{x}{r} \text{ with } r = 1$$

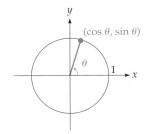

To put this another way, a point on the circle of radius 1 centered at the origin can be represented as $(\cos \theta, \sin \theta)$, where θ is the angle at the origin, as shown in the diagram on the left.

EXAMPLE 1   **FINDING VALUES OF SINE AND COSINE**

Evaluate $\sin \theta$ and $\cos \theta$ for the following values of θ: $0, \dfrac{\pi}{2}, \pi, \text{ and } \dfrac{3\pi}{2}$.

Solution

The angles are shown below with a point (x, y) on the terminal side 1 unit from the origin. The sine and cosine are then just the y- and x-coordinates (since $\sin \theta = y$ and $\cos \theta = x$ when $r = 1$).

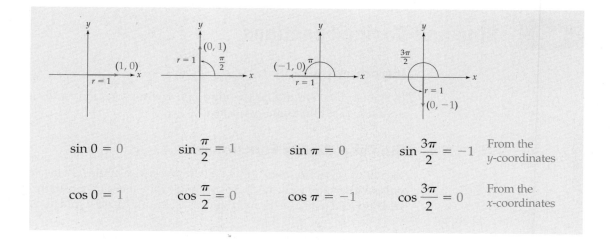

$$\sin 0 = 0 \qquad \sin \frac{\pi}{2} = 1 \qquad \sin \pi = 0 \qquad \sin \frac{3\pi}{2} = -1 \quad \text{From the } y\text{-coordinates}$$

$$\cos 0 = 1 \qquad \cos \frac{\pi}{2} = 0 \qquad \cos \pi = -1 \qquad \cos \frac{3\pi}{2} = 0 \quad \text{From the } x\text{-coordinates}$$

EXAMPLE 2   **FINDING VALUES OF SINE AND COSINE**

Find $\sin \dfrac{\pi}{6}$ and $\cos \dfrac{\pi}{6}$.

Solution

Draw an angle of $\frac{\pi}{6}$ radians (30°) in standard position.

Complete the sides of the 30°-60°-90° triangle with sides 1, 2, $\sqrt{3}$.

Write the coordinates of a point based on the triangle sides.

From these values of x, y, and r we find the sine and cosine:

$$\sin \frac{\pi}{6} = \frac{1}{2} \qquad\qquad\qquad \sin \theta = \frac{y}{r}$$

$$\cos \frac{\pi}{6} = \frac{\sqrt{3}}{2} \qquad\qquad\qquad \cos \theta = \frac{x}{r}$$

PRACTICE PROBLEM

Find $\sin \dfrac{\pi}{4}$ and $\cos \dfrac{\pi}{4}$.

Solution on page 543 >

Similar calculations lead to sines and cosines of the following angles. You should either memorize this table or be able to calculate these values by remembering the sides of $30° - 60° - 90°$ and $45° - 45° - 90°$ triangles.

θ	0	$\dfrac{\pi}{6}$	$\dfrac{\pi}{4}$	$\dfrac{\pi}{3}$	$\dfrac{\pi}{2}$
$\sin \theta$	0	$\dfrac{1}{2}$	$\dfrac{\sqrt{2}}{2}$	$\dfrac{\sqrt{3}}{2}$	1
$\cos \theta$	1	$\dfrac{\sqrt{3}}{2}$	$\dfrac{\sqrt{2}}{2}$	$\dfrac{1}{2}$	0

Notice the patterns:

$$\frac{\sqrt{0}}{2} \quad \frac{\sqrt{1}}{2} \quad \frac{\sqrt{2}}{2} \quad \frac{\sqrt{3}}{2} \quad \frac{\sqrt{4}}{2}$$

$$\frac{\sqrt{4}}{2} \quad \frac{\sqrt{3}}{2} \quad \frac{\sqrt{2}}{2} \quad \frac{\sqrt{1}}{2} \quad \frac{\sqrt{0}}{2}$$

| EXAMPLE 3 | FINDING SINE AND COSINE IN OTHER QUADRANTS |

Find $\sin \dfrac{2\pi}{3}$ and $\cos \dfrac{2\pi}{3}$.

Solution

Draw an angle of $\dfrac{2\pi}{3}$ radians (120°).
Note the remaining 60° angle.

Draw a perpendicular to an axis to make a 30°-60°-90° triangle with sides $1, 2, \sqrt{3}$

Write the coordinates of a point based on the triangle sides (*with the correct signs*).

From these values of x, y, and r:

$$\sin \frac{2\pi}{3} = \frac{\sqrt{3}}{2} \qquad\qquad \sin\theta = \frac{y}{r}$$

$$\cos \frac{2\pi}{3} = \frac{-1}{2} = -\frac{1}{2} \qquad\qquad \cos\theta = \frac{x}{r}$$

 Graphing Calculator Exploration

```
sin(2π/3)
               .8660254038
√3/2
               .8660254038
cos(2π/3)
                      -.5
```
agree

Your calculator should find a value for $\sin \dfrac{2\pi}{3}$ that agrees with $\dfrac{\sqrt{3}}{2}$ to as many decimal places as the calculator shows, and a value for $\cos \dfrac{2\pi}{3}$ that agrees with the result found above.

⚠ **Be Careful** If you get incorrect answers, make sure that your calculator is (and remains) set for *radians*.

| EXAMPLE 4 | FINDING SINE AND COSINE OF A NEGATIVE NUMBER |

Find $\sin\left(-\dfrac{\pi}{4}\right)$ and $\cos\left(-\dfrac{\pi}{4}\right)$.

Solution

Using only one graph:

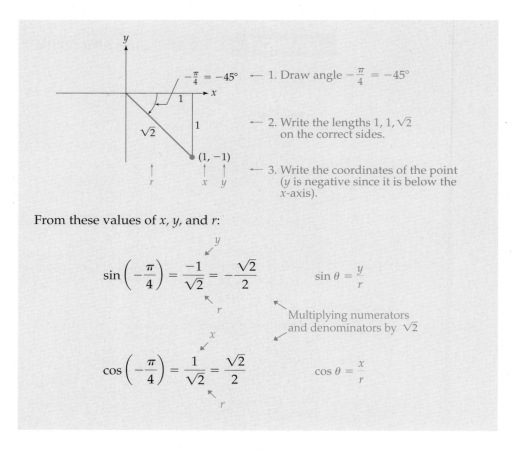

From these values of x, y, and r:

$$\sin\left(-\frac{\pi}{4}\right) = \frac{-1}{\sqrt{2}} = -\frac{\sqrt{2}}{2} \qquad \sin\theta = \frac{y}{r}$$

Multiplying numerators and denominators by $\sqrt{2}$

$$\cos\left(-\frac{\pi}{4}\right) = \frac{1}{\sqrt{2}} = \frac{\sqrt{2}}{2} \qquad \cos\theta = \frac{x}{r}$$

Sine and Cosine of an Acute Angle

An *acute* angle is an angle between 0 and $\pi/2$ radians. For an acute angle θ, $\sin\theta$ and $\cos\theta$ can be expressed as ratios of sides of a right triangle (called hyp, opp, and adj for the *hyp*otenuse, *opp*osite side, and *adj*acent side).

LOOKING BACK

These agree with our earlier definitions on page 535 by using

$opp = y$
$adj = x$
$hyp = r$

Those definitions are more general than these because they apply to *any* triangle, not just acute ones.

Right Triangle Formulas for $\sin\theta$ and $\cos\theta$

$$\sin\theta = \frac{\text{opp}}{\text{hyp}}$$

$$\cos\theta = \frac{\text{adj}}{\text{hyp}}$$

EXAMPLE 5   CALCULATING DISTANCES

How long a ladder is needed to reach a roof 20 feet above the ground if the ladder makes a 60° angle with the ground?

Solution

The ladder and building form the hypotenuse and opposite side of the right triangle shown above, suggesting that we use $\sin \theta = \dfrac{\text{opp}}{\text{hyp}}$ with $\theta = \dfrac{\pi}{3}$.

$$\sin \frac{\pi}{3} = \frac{20}{x} \qquad\qquad \sin \theta = \frac{\text{opp}}{\text{hyp}}$$

$$\frac{\sqrt{3}}{2} = \frac{20}{x} \qquad\qquad \text{Using } \sin \frac{\pi}{3} = \frac{\sqrt{3}}{2}$$

$$\sqrt{3}\, x = 40 \qquad\qquad \text{Cross-multiplying}$$

$$x = \frac{40}{\sqrt{3}} \approx 23.09 \qquad \text{Solving for } x, \text{ then using a calculator}$$

Therefore, the ladder must be about 24 feet long.

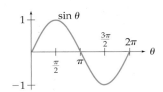

Graphs of the Sine and Cosine Functions

On page 535, we saw that the coordinates of a point on a circle of radius 1 centered at the origin can be represented as $(\cos\theta, \sin\theta)$, where θ is the angle at the origin. Therefore, as θ increases, $\sin\theta$ behaves just like the y-coordinate of the point (x, y) as it moves around the circle—first increasing from 0 to 1, then decreasing to 0 to -1, then increasing back to 0. Plotting θ on the horizontal axis, we obtain the graph on the left.

This pattern of values repeats itself each time the angle completes another full circle of 2π radians, giving the following graph of $\sin t$ (where we have replaced θ by t).

Graph of   sin t

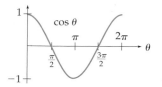

Similarly, $\cos\theta$ behaves just like the *x*-coordinate of the point (x, y) as it moves around the circle—first decreasing from 1 to 0 to -1, then increasing back to 1. Again, this cycle repeats itself each time the angle completes another full circle of 2π radians, giving the following graph of $\cos t$.

Graph of $\cos t$

(These graphs could also have been found by the more tedious process of plotting points.) $\mathrm{Sin}\ t$ and $\cos t$ are said to be **periodic with period 2π**, meaning that they repeat their values every 2π units. We will now think of $\sin t$ and $\cos t$ as *ordinary functions,* unrelated to angles, functions that oscillate between $+1$ and -1 in a "wavy" pattern. We will use $\sin t$ and $\cos t$ to model quantities that undergo regular fluctuations, such as seasonal sales.

Modified Sine and Cosine Curves—Changing Amplitude and Period

The **amplitude** of a sine or cosine curve is half the distance between its highest and lowest values (sometimes called its *half-height*). Therefore, $\sin t$ and $\cos t$ have amplitude 1. Multiplying by constants, we may create similar "wavy" functions with any amplitude and period. For positive numbers a and b, the function $a \sin bt$ oscillates between the values a and $-a$, and so has amplitude a. The period of $a \sin bt$ is found by setting $bt = 2\pi$, giving period $t = \dfrac{2\pi}{b}$. Similar statements hold for the function $a \cos bt$.

Modified Sine and Cosine Functions

$$\left.\begin{array}{l} y = a \sin bt \\ y = a \cos bt \end{array}\right\} \text{ have amplitude } a \text{ and period } \frac{2\pi}{b}.$$

EXAMPLE 6	GRAPHING A MODIFIED SINE CURVE

Sketch the graph of $f(t) = 3 \sin 2t$.

Solution

The amplitude is 3 (since the sine is multiplied by 3), and the period is π (the sine's period 2π divided by the 2). Therefore, the graph is a sine curve, but between heights 3 and -3, with a complete cycle occurring between 0 and π, as shown below.

Graphing Calculator Exploration

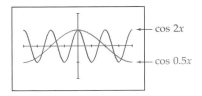

Graph $y_1 = \cos 2x$ and $y_2 = \cos 0.5x$ on the window $[-2\pi, 2\pi]$ by $[-2, 2]$. (Be sure that your calculator is set for radians.) Notice how the number in front of the x changes the period: A larger number makes the curve "wiggle" faster.

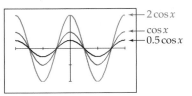

Graph $y_1 = \cos x$, $y_2 = 2\cos x$, and $y_3 = 0.5 \cos x$ on the same window. Notice how the number in front of the *function* changes the amplitude: a larger number makes the amplitude larger.

Predict what the curve $y = 2\cos 2x$ would look like. Then check your prediction by graphing the curve. What about $y = -2\cos x$? What about $y = -2\sin 2x$?

Seasonal Sales

The sine and cosine functions with modified amplitude and period are often used to model quantities that fluctuate in a regular pattern, such as seasonal sales.

EXAMPLE 7	PREDICTING SEASONAL REVENUE

A travel agency specializing in Caribbean vacations models its daily revenue (in dollars) on day n of the year by

$$R(n) = 600 + 500 \cos\left(\frac{2\pi n}{365}\right)$$

Find the company's revenue on March 1 (day $n = 60$).

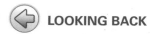 **LOOKING BACK**

Notice that the 600 is a *vertical shift*, as defined on page 56, so the graph never reaches zero.

Solution

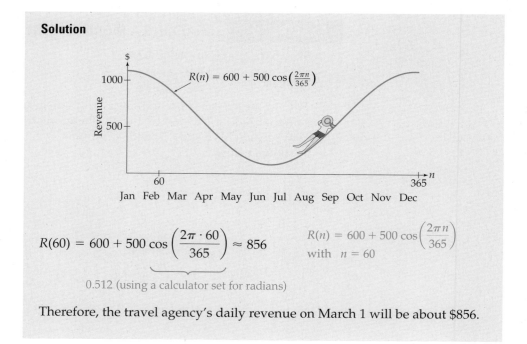

$$R(60) = 600 + 500 \cos\left(\frac{2\pi \cdot 60}{365}\right) \approx 856$$

$$R(n) = 600 + 500 \cos\left(\frac{2\pi n}{365}\right)$$
with $n = 60$

$\underbrace{}$
0.512 (using a calculator set for radians)

Therefore, the travel agency's daily revenue on March 1 will be about $856.

Trigonometric Identities

LOOKING BACK

See pages 524 and 534 for a statement and proof of the Pythagorean Theorem.

A formula that holds for *all* values of the variable is called an *identity*. There are many trigonometric identities, and we will mention only a few. The first is called the *Pythagorean identity*, since it comes directly from the Pythagorean Theorem.

$$\sin^2 t + \cos^2 t = 1$$

$\sin^2 t$ means $(\sin t)^2$
$\cos^2 t$ means $(\cos t)^2$

To prove this identity, recall that $\sin t = \dfrac{y}{r}$ and $\cos t = \dfrac{x}{r}$, so

$$\sin^2 t + \cos^2 t = \left(\frac{y}{r}\right)^2 + \left(\frac{x}{r}\right)^2 = \frac{x^2 + y^2}{r^2} = \frac{r^2}{r^2} = 1$$ Since $x^2 + y^2 = r^2$

This and other useful identities are listed below, and the remaining proofs are developed in Exercises 52–54.

Trigonometric Identities

$$\sin^2 t + \cos^2 t = 1$$ Pythagorean identity

$$\sin(t + 2\pi) = \sin t$$
$$\cos(t + 2\pi) = \cos t$$ Periodicity identities

$$\sin(-t) = -\sin t$$
$$\cos(-t) = \cos t$$ Negative angle identities

$$\sin t = \cos\left(\frac{\pi}{2} - t\right)$$

$$\cos t = \sin\left(\frac{\pi}{2} - t\right)$$ Complementary angle identities

$$\sin(s \pm t) = (\sin s) \cdot (\cos t) \pm (\cos s) \cdot (\sin t)$$
$$\cos(s \pm t) = (\cos s) \cdot (\cos t) \mp (\sin s) \cdot (\sin t)$$ Sum and difference identities

In the last two identities, use the *upper* signs throughout the equation, or the *lower* signs throughout.

Solution   TO PRACTICE PROBLEM

$$\sin \frac{\pi}{4} = \frac{1}{\sqrt{2}} = \frac{\sqrt{2}}{2}$$

$$\cos \frac{\pi}{4} = \frac{1}{\sqrt{2}} = \frac{\sqrt{2}}{2}$$

8.2	**Section Summary**

The sine and cosine functions are defined in terms of the coordinates (x, y) of a point on a unit circle forming an angle of θ radians:

$$\sin \theta = y \quad \text{and} \quad \cos \theta = x$$

For acute angles, these functions can also be interpreted in terms of the sides of a right triangle:

$$\sin \theta = \frac{\text{opp}}{\text{hyp}} \quad \text{and} \quad \cos \theta = \frac{\text{adj}}{\text{hyp}}$$

For a few particular angles (such as 30°, 45°, and 60°) we evaluated these functions *exactly*, while for other angles we used a calculator. We then graphed these functions and used them to model quantities that undergo regular fluctuations. The trigonometric identities will be used in the exercises and in the following section.

8.2 Exercises

1–10. Find the sine and cosine of each angle θ. When r is not given, calculate it from the Pythagorean Theorem.

FOR HELP GETTING STARTED

on Exercises 1–10, see Examples 2 and 3 on pages 536–537.

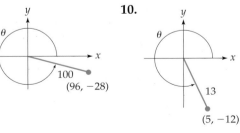

11–16. Evaluate *without* using a calculator, leaving answers in exact form.

11. a. $\sin \dfrac{\pi}{3}$ **b.** $\cos \dfrac{\pi}{3}$

 c. $\sin \dfrac{\pi}{4}$ **d.** $\cos \dfrac{\pi}{4}$

12. a. $\sin \dfrac{3\pi}{4}$ **b.** $\cos \dfrac{3\pi}{4}$

 c. $\sin \dfrac{5\pi}{3}$ **d.** $\cos \dfrac{5\pi}{3}$

13. a. $\sin \dfrac{5\pi}{4}$ **b.** $\cos \dfrac{5\pi}{4}$

 c. $\sin \dfrac{5\pi}{6}$ **d.** $\cos \dfrac{5\pi}{6}$

14. a. $\sin \dfrac{11\pi}{6}$ **b.** $\cos \dfrac{11\pi}{6}$

 c. $\sin \dfrac{7\pi}{4}$ **d.** $\cos \dfrac{7\pi}{4}$

15. a. $\sin \left(-\dfrac{\pi}{3}\right)$ **b.** $\cos \left(-\dfrac{\pi}{3}\right)$

 c. $\sin \left(-\dfrac{\pi}{4}\right)$ **d.** $\cos \left(-\dfrac{\pi}{4}\right)$

16. a. $\sin \dfrac{5\pi}{2}$ **b.** $\cos \dfrac{5\pi}{2}$

 c. $\sin 3\pi$ **d.** $\cos 3\pi$

17–18. Approximate using a calculator (set for *radians*). Round answers to two decimal places.

17. a. $\sin \dfrac{\pi}{10}$ **b.** $\cos \dfrac{\pi}{10}$

 c. $\sin 1$ **d.** $\cos 1$

18. a. $\sin \dfrac{\pi}{5}$ **b.** $\cos \dfrac{\pi}{5}$

 c. $\sin 4$ **d.** $\cos 4$

19–26. Sketch the graph of each function showing the amplitude and period.

19. $y = 4 \sin t$ **20.** $y = 6 \cos t$

21. $y = \sin 4t$ **22.** $y = \cos 6t$

23. $y = 3 \cos 4t$ **24.** $y = 5 \sin 4t$

25. $y = -\cos 2t$ **26.** $y = -\sin 2t$

27–28. Use the sum and difference identities on page 542 to find the following sines and cosines. Do *not* use a calculator.

27. a. $\sin \dfrac{5\pi}{12}$ **b.** $\cos \dfrac{5\pi}{12}$

 [*Hint:* Take $s = \dfrac{\pi}{4}$ and $t = \dfrac{\pi}{6}$ with upper signs.]

28. a. $\sin \dfrac{7\pi}{12}$ **b.** $\cos \dfrac{7\pi}{12}$

 [*Hint:* Take $s = \dfrac{\pi}{4}$ and $t = \dfrac{\pi}{3}$ with upper signs.]

Applied Exercises

29. GENERAL: Ladder Height A 30-foot ladder leaning against a wall makes a 70° angle with the ground. How high up the wall does the ladder reach?

30. GENERAL: Kite Flying A straight kite string makes an angle of 40° with the ground, as shown in the diagram. If 250 feet of string has been let out, how high is the kite?

31. GENERAL: Distance A boat sailing due west toward a point 200 feet directly north of you makes an angle of 48°, as shown in the diagram. Find the distance between you and the boat.

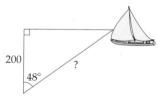

32. GENERAL: Wheelchair Ramp According to the American National Standards Institute, a wheelchair ramp should make an angle of 4.75° with the horizontal. Find the length of such a ramp that rises 2 feet.

33. BUSINESS: Seasonal Sales An ice cream parlor estimates that the number of ice cream cones that it will sell on day n of the year is

$$S(n) = 700 - 600 \cos \frac{2\pi n}{365}$$

Find the number of ice cream cones that will be sold on August 1 (day 213).

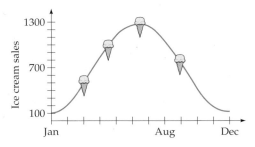

34. BUSINESS: Seasonal Sales A store estimates that the number of ski parkas that it will sell on day n of the year is

$$S(n) = 27 + 25 \cos \frac{2\pi n}{365}$$

Find the number of ski parkas that will be sold on February 1 (day 32).

35. BIOMEDICAL: Breathing When you breathe normally, you inhale and exhale about 0.8 liter of air approximately every 4 seconds, with about 4.5 liters remaining in your lungs. Under these conditions, the total amount of air in your lungs t seconds after you last inhaled is

$$L(t) = 4.9 + 0.4 \cos \frac{\pi t}{2} \quad \text{liters}$$

Find (*without* using a calculator) the amount of air in your lungs at time:

a. $t = 0$ (just after inhaling).
b. $t = 2$ (just after exhaling).

Source: E. Batschelet, Introduction to Mathematics for Life Scientists

 36. COSINE AND SINE ON THE UNIT CIRCLE
  a. Set your calculator for PARAMETRIC equations (on some calculators, use MODE), and enter $x_1 = \cos T$, $y_1 = \sin T$. (Be sure your calculator is set for *radians*.)
  b. GRAPH the curve on the window with T in $[0, 2\pi]$, x in $[-2, 2]$ and y in $[-1.3, 1.3]$.   (*continues*)

  c. Use TRACE to move a point around the circle, with X and Y showing the cosine and sine of the angle. Observe how the sine and cosine change as the point moves around the circle.
  d. Use CALCULATE to check the sine and cosine at T-values such as $\frac{\pi}{6}$, $\frac{\pi}{4}$, and $\frac{\pi}{3}$. (Reset your calculator to FUNCTION mode after this exercise.)

37. MODIFIED SINE CURVES
  a. Graph the curves $y_1 = \sin x$, $y_2 = \sin 2x$, and $y_3 = \sin 0.5x$ on the window $[-2\pi, 2\pi]$ by $[-2, 2]$. (Be sure your calculator is set for *radians*.) Notice how a number in front of x changes the *period*: a larger number makes the curve "wiggle faster."
  b. Now graph $y_1 = \sin x$, $y_2 = 2 \sin x$, and $y_3 = 0.5 \sin x$ on the same window. Notice how a number in front of the function changes the *amplitude*: a larger number makes the amplitude larger.
  c. Predict what the curve $y = 2 \sin 2x$ would look like. Then check your answer by graphing it. What about $y = -2 \sin x$? What about $y = -2 \sin 2x$?

38. THREE SIMILAR CURVES Graph $y_1 = 1 - x^2$, $y_2 = \sqrt{1 - x^2}$, and $y_3 = \cos \frac{\pi x}{2}$ on the window $[-1, 1]$ by $[0, 1.3]$. Which curve is a parabola, which is a semicircle, and which is a trigonometric function? (This shows that trigonometric curves should be drawn more "pointed" than circles and slightly more pointed than parabolas.)

39. GENERAL: Temperature The mean temperature in Fairbanks, Alaska, on day x of the year is approximately

$$T(x) = 36 + 22 \sin \frac{2\pi (x - 102)}{365}$$

degrees Fahrenheit.

  a. Graph this function on the window $[0, 365]$ by $[0, 60]$. (Be sure your calculator is set for *radians*.)
  b. Find the temperature on January 10 (day 10).
  c. Find the temperature on July 12 (day 193).

Source: Alaska Climate Research Center

40. VERIFICATION OF $\sin^2 x + \cos^2 x = 1$

Graph $y_1 = (\sin x)^2$, $y_2 = (\cos x)^2$, and $y_3 = y_1 + y_2$ on the window $[-2\pi, 2\pi]$ by $[-0.5, 1.5]$. Why does the fact that y_3 is a horizontal line at height 1 verify the identity $\sin^2 x + \cos^2 x = 1$?

41. GENERAL: Greenhouse Gases The "greenhouse effect," which may be increasing global temperatures and, through the melting of polar ice, raising sea levels, is caused by increasing levels of carbon dioxide and other atmospheric gases. The carbon dioxide ("Keeling") curve graphed on page 524 is

approximated by $C(x) = 319e^{0.0041x} - 2.5 \cos(2\pi x)$ ppm (parts per million), where x is the number of years since January 1, 1960.

a. Graph this function on the window [50, 60] by [385, 415].

Notice how the sine curve (seasonal variation) follows an upward trend.

b. Use this formula to predict the carbon dioxide level on January 1, 2018.

Source: National Oceanic and Atmospheric Administration

Conceptual Exercises

42. True or False: An angle of θ radians in standard position intersects a circle of radius 1 at the point $(\sin \theta, \cos \theta)$.

43. What is the shortest horizontal shift of a cosine curve that will turn it into a sine curve?

44. A modified sine curve is shifted up so that it varies between heights 4 and 10. What is the amplitude of this sine curve?

45. What is the maximum height of $f(x) = 8 + 3 \sin x$? What is its *minimum* height?

46. What is the period of $\cos(2\pi x)$?

47. Evaluate at sight: **a.** $\sin 4000\pi$ **b.** $\cos 4000\pi$

48. Evaluate at sight: **a.** $\sin 100.5\pi$ **b.** $\cos 100.5\pi$

49. True or False: $\sin(x - y) = -\sin(y - x)$.

50. True or False: $\cos(x - y) = \cos(y - x)$.

51. Evaluate $\sin^2(e^{3-\pi}/103^{7\pi/8}) + \cos^2(e^{3-\pi}/103^{7\pi/8})$

Explorations and Excursions   The following problems extend and augment the material presented in the text.

52. PROOFS OF NEGATIVE ANGLE IDENTITIES

As illustrated in the following diagram, changing the sign of an angle (from t to $-t$) changes the sign of the y-coordinate of a point on the terminal side, but not the sign of the x-coordinate. Use this fact together with the definitions of $\sin t$ and $\cos t$ on page 535 to prove the negative angle identities:
$\sin(-t) = -\sin t$ and $\cos(-t) = \cos t$.

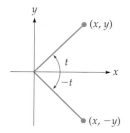

53. PROOFS OF COMPLEMENTARY ANGLE IDENTITIES:

Prove the identities $\sin t = \cos\left(\dfrac{\pi}{2} - t\right)$ and $\cos t = \sin\left(\dfrac{\pi}{2} - t\right)$ for an acute angle t as follows.

a. Show that if one acute angle of a right triangle is of size t radians, then the other acute angle is of size $\dfrac{\pi}{2} - t$ radians.

b. Observe that the "opposite" side for angle t is the "adjacent" side for angle $\dfrac{\pi}{2} - t$, and vice versa. Use these facts with the expressions for $\sin t$ and $\cos t$ on page 538 to prove the identities above.

54. PROOFS OF THE SUM AND DIFFERENCE IDENTITIES

Prove the identity

$$\sin(s + t) = (\sin s)(\cos t) + (\cos s)(\sin t)$$

for acute angles whose sum is also acute by using the diagram on the right as follows (the notation AB means the length of the line segment between points A and B, and similarly for any other points).

a. Show that $AB = \sin(s + t)$.

b. Show that $AC = \sin t$.

c. Show that $CD = \cos t$.

d. Show that $CE = \dfrac{\dfrac{CE}{DE}}{\dfrac{CD}{DE}} \cdot CD = \dfrac{\sin s}{\cos s} \cdot \cos t$.

e. Show that

$$AB = (\cos s)(AC + CE)$$

$$= (\cos s)\left(\sin t + \frac{\sin s}{\cos s} \cos t \right)$$

f. From (a) and (e), conclude that

$$\sin(s + t) = (\sin s)(\cos t) + (\cos s)(\sin t)$$

[*Note:* The identity

$$\cos(s + t) = (\cos s)(\cos t) - (\sin s)(\sin t)$$

will be proved in Exercise 95 on page 559. The identities for $\sin(s - t)$ and $\cos(s - t)$ are derived by changing t to $-t$ in the identities for $\sin(s + t)$ and $\cos(s + t)$ and then using the negative angle identities.]

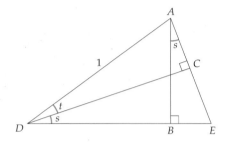

8.3   Derivatives of Sine and Cosine Functions

Introduction

The last two sections contained no calculus, their purpose being to introduce the sine and cosine functions. In this section we will differentiate these functions, and use these derivatives to optimize functions and calculate rates of change.

Derivatives of the Sine and Cosine Functions

The derivatives of the sine and cosine functions are given by the following formulas.

 LOOKING AHEAD

These formulas are as simple as they are because we use *radians*. For the more complicated forms if we used *degrees*, see page 553.

Derivatives of $\sin t$ and $\cos t$	
$\dfrac{d}{dt} \sin t = \cos t$	The derivative of sine is cosine (for t in radians)
$\dfrac{d}{dt} \cos t = -\sin t$	The derivative of cosine is negative sine

To see graphically that the derivative (the slope) of $\sin t$ is $\cos t$,   observe in the following diagram that, at any particular t-value, the *slope* of the *sine* function is the same as the *value* (height) of the *cosine* function.

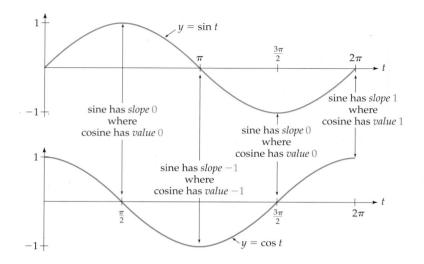

Similarly, to see that the derivative of $\cos t$ is $-\sin t,$ observe in the diagram above that the *slope* of the *cosine* curve is the *negative* of the *value* of the *sine* curve. Justifications of these two formulas are given at the end of this section. The following examples show how to use these formulas.

EXAMPLE 1   **DIFFERENTIATING A QUOTIENT INVOLVING SINE**

Differentiate $f(t) = \dfrac{\sin t}{t}.$

Solution

$\qquad\qquad$ Derivative of $\sin t$
$\qquad\qquad$ Derivative of t

$$\frac{d}{dt}\frac{\sin t}{t} = \frac{t\cdot\cos t - 1\cdot\sin t}{t^2} = \frac{t\cos t - \sin t}{t^2}\qquad\text{Using the Quotient Rule}$$

EXAMPLE 2   **DIFFERENTIATING A PRODUCT INVOLVING COSINE**

Differentiate $g(t) = t^3\cos t.$

Solution

$$\frac{d}{dt}(t^3\cdot\cos t) = 3t^2\cdot\cos t + t^3\cdot(-\sin t) = 3t^2\cos t - t^3\sin t$$

Using the Product Rule

$\qquad$ Derivative $\quad$ Second $\qquad$ First $\quad$ Derivative
$\qquad$ of the first $\quad$ left alone $\quad$ left alone $\quad$ of $\cos t$

PRACTICE PROBLEM 1

Differentiate $f(t) = t \sin t$.

Solution on page 554 >

These rules for differentiating sine and cosine, combined with the Chain Rule, show how to differentiate the sine and cosine of a *function*. For a differentiable function $f(t)$:

Derivative of $\sin f(t)$ and $\cos f(t)$

$$\frac{d}{dt} \sin f(t) = \cos f(t) \cdot \frac{df}{dt}$$

The derivative of the sine of a function is the cosine of the function times the derivative of the function

$$\frac{d}{dt} \cos f(t) = -\sin f(t) \cdot \frac{df}{dt}$$

The derivative of the cosine of a function is negative the sine of the function times the derivative of the function

EXAMPLE 3   DIFFERENTIATING THE COSINE OF A FUNCTION

Differentiate $g(t) = \cos(t^2 + 1)$.

Solution

$$\frac{d}{dt} \cos(\underbrace{t^2 + 1}_{f(t)}) = \underbrace{-\sin(t^2 + 1)}_{-\sin f(t)} \cdot \underbrace{2t}_{\frac{df}{dt}}$$

Using $\dfrac{d}{dt} \cos f = -\sin f \cdot \dfrac{df}{dt}$

$$= -2t \sin(t^2 + 1)$$

EXAMPLE 4   FINDING A SECOND DERIVATIVE

Find the *second* derivative of $g(t) = \sin 4t$.

Solution

$$g'(t) = \frac{d}{dt}(\sin 4t) = \cos 4t \cdot (4) = 4 \cos 4t$$

Using $\dfrac{d}{dt} \sin f = \cos f \cdot \dfrac{df}{dt}$

Derivative of $4t$

$$g''(t) = \frac{d}{dt}(4 \cos 4t) = 4[-\sin 4t \cdot (4)]$$

$\dfrac{d}{dt} \cos f = -\sin f \cdot \dfrac{df}{dt}$

$$= -16 \sin 4t$$

LOOKING AHEAD

Writing $\frac{t}{2}$ as $\frac{1}{2}t$ makes it obvious that the derivative is $\frac{1}{2}$. We will use this "trick" several times in this and the following sections.

PRACTICE PROBLEM 2

Differentiate $f(t) = \sin\frac{t}{2}$. $\left[\textit{Hint: Write } \frac{t}{2} \text{ as } \frac{1}{2}t, \text{ a constant times the variable.}\right]$

Solution on page 554 >

EXAMPLE 5   **EVALUATING A DERIVATIVE**

For $f(x) = \dfrac{\sin x}{\cos x}$, find $f'\left(\dfrac{\pi}{3}\right)$.

Solution

First differentiating:

Derivative of $\sin x$
Derivative of $\cos x$

$$f'(x) = \frac{\cos x \cdot \cos x - (-\sin x) \cdot \sin x}{\cos^2 x}$$ Using the Quotient Rule

$$= \frac{\cos^2 x + \sin^2 x}{\cos^2 x} = \frac{1}{\cos^2 x}$$ Using the identity $\sin^2 x + \cos^2 x = 1$

Evaluating at the given $\dfrac{\pi}{3}$:

$$f'\left(\frac{\pi}{3}\right) = \frac{1}{\cos^2\dfrac{\pi}{3}} = \frac{1}{\left(\cos\dfrac{\pi}{3}\right)^2} = \frac{1}{\left(\dfrac{1}{2}\right)^2} = \frac{1}{\left(\dfrac{1}{4}\right)} = 4$$ Using $\cos\dfrac{\pi}{3} = \dfrac{1}{2}$

EXAMPLE 6   **DIFFERENTIATING AN EXPONENTIAL TRIGONOMETRIC FUNCTION**

If $f(z) = e^{\sin z}$, find $f'(\pi)$.

Solution

$$f'(z) = e^{\sin z} \cos z$$ Using $\dfrac{d}{dt}e^f = e^f \cdot \dfrac{df}{dt}$

$$f'(\pi) = \underbrace{e^{\sin \pi}}_{0} \underbrace{\cos \pi}_{-1} = \underbrace{e^0(-1)}_{1} = -1$$ Evaluating at π

PRACTICE PROBLEM 3

Differentiate $f(x) = \ln(1 + \cos x)$. Solution on page 554 >

EXAMPLE 7   **DIFFERENTIATING A TRIGONOMETRIC FUNCTION**

Differentiate $f(t) = \sin^2 t^3$.

Solution

We write $\sin^2 t^3$ as $(\sin t^3)^2$ and then differentiate using the Generalized Power Rule.

$$\frac{d}{dt}(\sin t^3)^2 = 2(\sin t^3)^1 \cdot \underbrace{\cos t^3 \cdot 3t^2}_{\substack{\text{Derivative of} \\ \sin t^3}} = 6t^2 \sin t^3 \cos t^3$$

Growth of Sales

In the preceding section we used sine and cosine functions to model the fluctuations in sales of seasonal goods. The *derivative* of such a sales function then gives the *rate of change of sales,* showing how rapidly sales are increasing or decreasing.

EXAMPLE 8   **FINDING GROWTH OF SEASONAL SALES**

A department store predicts that the number of bathing suits that it will sell during week t of the year (with $t = 1$ meaning the first week of January) is given by the formula

$$S(t) = 400 - 400 \cos \frac{2\pi t}{52}$$

Weekly sales of bathing suits

Find the rate of change of the sales during the first week of May (week $t = 18$).

Solution

The rate of change means the derivative, and the differentiation is easier if we write $\dfrac{2\pi t}{52}$ as $\dfrac{2\pi}{52} t$, a constant times the variable.

$$S'(t) = \frac{d}{dt}\left(400 - 400 \cos \frac{2\pi}{52} t\right)$$

$$= 400 \sin \frac{2\pi}{52} t \cdot \left(\frac{2\pi}{52}\right) \qquad \text{Using } \frac{d}{dt}\cos f = -\sin f \cdot \frac{df}{dt}$$

$$= 400 \frac{2\pi}{52} \sin \frac{2\pi}{52} t \qquad \text{Rearranging}$$

Evaluating at week $t = 18$,

$$S'(18) = 400 \underbrace{\frac{2\pi}{52}}_{48.3} \underbrace{\sin\left(\frac{2\pi}{52} 18\right)}_{0.823} \approx 39.8 \qquad \text{Using a calculator set for radians}$$

Interpretation: Sales will be increasing by about 40 per week.

**Graphing
Calculator
Exploration**

Verify the answer to the previous Example as follows:

a. Graph the function $y_1 = 400 - 400 \cos(2\pi x/52)$
on the window [0, 52] by [0, 1000].

b. Find the derivative at $x = 18$ (using dy/dx or
NDERIV). Your answer should agree with that
found in the preceding Example.

As before, we optimize a function by setting its first derivative equal to zero and
solving, then using the second-derivative test to determine whether we have
maximized or minimized.

EXAMPLE 9   **MAXIMIZING VOLUME**

A water trough consists of two boards, each 10 feet long and 1 foot wide,
attached along one edge at an angle of θ radians, and closed at both ends,
as shown below. Find the angle θ that maximizes the volume of the trough.

Solution

The volume will be maximized if we maximize
the cross-sectional area (shown at the right).

Turning this triangle on its side gives a triangle of base 1 and height $\sin \theta$

$$\left(\text{since }\ \sin \theta = \frac{\text{opp}}{\text{hyp}} = \frac{h}{1} = h\right).$$

The area of the triangle is

$$A = \frac{1}{2} \cdot 1 \cdot \sin \theta = \frac{1}{2} \sin \theta \qquad \text{One-half base times height}$$

Differentiating,

$$A' = \frac{1}{2} \cos \theta = 0 \qquad \begin{array}{l}\text{To maximize, set the}\\ \text{derivative equal to zero}\end{array}$$

The only angle between 0 and π whose cosine is zero (think of the graph
of cos x) is

$$\theta = \frac{\pi}{2} \qquad \text{Solution to }\ \cos \theta = 0$$

The second-derivative test gives

$$A'' = -\frac{1}{2}\sin\theta \qquad\qquad \text{Differentiating } A' = \tfrac{1}{2}\cos\theta$$

$$= -\frac{1}{2}\sin\frac{\pi}{2} = -\frac{1}{2} \qquad\qquad \text{Evaluating at } \frac{\pi}{2}$$

Negative

Since A'' is negative, the area is *maximized* at $\theta = \dfrac{\pi}{2}$, or 90°.

Therefore, joining the boards at an angle of 90° will maximize the volume of the trough.

Why We Use Radians Instead of Degrees

We can now explain why we use radians instead of degrees in calculus. On page 527 we saw that multiplying by $\dfrac{\pi}{180}$ converts degrees into radians. Therefore, the sine and cosine functions, written as functions of t *degrees,* are $\sin\left(\dfrac{\pi}{180}t\right)$ and $\cos\left(\dfrac{\pi}{180}t\right)$. The derivatives of these "degree" sine and cosine functions are then

$$\frac{d}{dt}\sin\left(\frac{\pi}{180}t\right) = \cos\left(\frac{\pi}{180}t\right)\cdot\frac{\pi}{180} \qquad \text{Using } \frac{d}{dt}\sin f = \cos f \cdot \frac{df}{dt}$$

and

$$\frac{d}{dt}\cos\left(\frac{\pi}{180}t\right) = -\sin\left(\frac{\pi}{180}t\right)\cdot\frac{\pi}{180} \qquad \text{Using } \frac{d}{dt}\cos f = -\sin f \cdot \frac{df}{dt}$$

Using the notations $\sin t°$ and $\cos t°$ for the "degree" sine and cosine functions, these derivative formulas become:

$$\frac{d}{dt}\sin t° = \frac{\pi}{180}\cos t° \qquad\qquad\qquad \frac{d}{dt}\sin t = \cos t$$

and *radian formulas*

$$\frac{d}{dt}\cos t° = -\frac{\pi}{180}\sin t° \qquad\qquad\qquad \frac{d}{dt}\cos t = -\sin t$$

It is the extra factor $\pi/180$ in these formulas that makes them more complicated than the simpler *radian* formulas on page 547. This is why we use radians: they simplify the formulas for differentiation and (as we shall see on page 559) integration.

Solutions TO PRACTICE PROBLEMS

1. $\dfrac{d}{dt}(t \cdot \sin t) = 1 \cdot \sin t + t \cdot \cos t = \sin t + t \cos t$

2. $\dfrac{d}{dt} \sin \dfrac{1}{2}t = \left(\cos \dfrac{1}{2}t\right)\left(\dfrac{1}{2}\right) = \dfrac{1}{2}\cos \dfrac{t}{2}$

3. $\dfrac{d}{dx} \ln(1 + \cos x) = \dfrac{-\sin x}{1 + \cos x}$

8.3   Section Summary

The derivatives of the sine and cosine functions are given by the following formulas.

$$\frac{d}{dt} \sin t = \cos t \qquad\qquad \frac{d}{dt} \sin f(t) = \cos f(t) \cdot \frac{df}{dt}$$

$$\frac{d}{dt} \cos t = -\sin t \qquad\qquad \frac{d}{dt} \cos f(t) = -\sin f(t) \cdot \frac{df}{dt}$$

In the formulas on the right, $f(t)$ is any differentiable function.

We used derivatives to find the rate of change of daily sales (showing how rapidly sales are changing), and to optimize functions (maximizing volume) by setting the derivative equal to zero.

Justification of the Differentiation Formulas for $\sin t$ and $\cos t$

The differentiation formulas for the sine and cosine functions come from the definition of the derivative. Before we can show how this is done, we need to establish the following two limits.

$$\lim_{h \to 0} \frac{\sin h}{h} = 1 \qquad\qquad \frac{\sin h}{h} \text{ approaches 1 as } h \text{ approaches 0}$$

and

$$\lim_{h \to 0} \frac{\cos h - 1}{h} = 0 \qquad\qquad \frac{\cos h - 1}{h} \text{ approaches 0 as } h \text{ approaches 0}$$

Instead of presenting formal proofs of these limits, we justify them by using a graphing calculator to calculate values of the functions as the variable takes values approaching zero.

Graphing Calculator Exploration

Find the two preceding limits as follows, using the TABLE feature of a graphing calculator (using x instead of h for ease of entry).

a. Enter $y_1 = (\sin x)/x$ and $y_2 = (\cos x - 1)/x$.

b. Set the TABLE feature to ASK you for the x-values.

c. Enter x-values approaching zero, such as 1, .1, .01, .001, .0001, ... or their negatives. You should get a table such as the one shown below. Observe that the values of $y_1 = (\sin x)/x$ approach 1 and the values of $y_2 = (\cos x - 1)/x$ approach zero.

1 E−4 means
$1 \times 10^{-4} = 0.0001$

$\leftarrow -5 \times 10^{-4} = -0.0005$
$\leftarrow -5 \times 10^{-5} = -0.00005$

$(\sin x)/x$ approaches 1

$(\cos x - 1)/x$ approaches 0

The derivative of $f(t) = \sin t$ is then found as follows.

$$f'(t) = \lim_{h \to 0} \frac{f(t+h) - f(t)}{h} = \lim_{h \to 0} \frac{\sin(t+h) - \sin t}{h}$$

Definition of $f'(t)$, with $f(t) = \sin t$

$$= \lim_{h \to 0} \frac{(\sin t)(\cos h) + (\cos t)(\sin h) - \sin t}{h}$$

Using the identity $\sin(s + t) = (\sin s)(\cos t) + (\cos s)(\sin t)$

$$= \lim_{h \to 0} \left(\frac{(\sin t)(\cos h) - \sin t}{h} + \frac{(\cos t)(\sin h)}{h} \right)$$

Rearranging the numerator, and then separating into two fractions

$$= \lim_{h \to 0} \left(\sin t \, \frac{\cos h - 1}{h} + \cos t \, \frac{\sin h}{h} \right)$$

Factoring sin t from the first numerator, then moving sin t and cos t out front

$$= (\sin t) \left(\lim_{h \to 0} \frac{\cos h - 1}{h} \right) + (\cos t) \left(\lim_{h \to 0} \frac{\sin h}{h} \right)$$

Separating the limit into two limits, then taking sin t and cos t outside of the h limits

$\underbrace{\qquad\qquad}_{0} \qquad\qquad \underbrace{\qquad\qquad}_{1}$

$$= (\sin t) \cdot (0) + (\cos t) \cdot (1) = \cos t$$

Using the two limits from the previous page

$\underbrace{\qquad}_{0} \qquad \underbrace{\qquad}_{\cos t}$

This is the result that we wanted, showing that the derivative of $f(t) = \sin t$ is $f'(t) = \cos t$. While the derivative formula for the cosine can be derived similarly (see Exercise 92), it can be found even more easily by using the identities $\cos t = \sin\left(\dfrac{\pi}{2} - t\right)$ and $\sin t = \cos\left(\dfrac{\pi}{2} - t\right)$ from page 542:

$$\frac{d}{dt}\cos t = \frac{d}{dt}\sin\left(\frac{\pi}{2} - t\right) = \cos\left(\frac{\pi}{2} - t\right)\cdot(-1) = -\sin t$$

$$\underbrace{}_{\substack{\text{Using}\\ \cos t = \sin\left(\frac{\pi}{2} - t\right)}} \qquad \underbrace{}_{\substack{\text{Using}\\ \frac{d}{dt}\sin f = \cos f\cdot f'}} \qquad \underbrace{}_{\substack{\text{Using}\\ \sin t = \cos\left(\frac{\pi}{2} - t\right)}}$$

8.3 Exercises

1–60. Differentiate each function.

1. $f(t) = t^3 \sin t$

2. $f(t) = t^2 \cos t$

3. $f(t) = \dfrac{\cos t}{t}$

4. $f(t) = (1 + \sin t)^4$

5. $f(t) = \sin(t^3 + 1)$

6. $f(t) = \cos(t^3 + t + 1)$

7. $f(t) = \sin 5t$

8. $f(t) = \cos 4t$

9. $f(t) = 2 \cos 3t$

10. $f(t) = 6 \sin 2t$

11. $f(t) = \cos \dfrac{\pi t}{180}$

12. $f(t) = \sin \dfrac{\pi t}{200}$

13. a. $f(t) = t^2 + \sin \pi t$
   b. Find $f'(0)$

14. a. $f(t) = t \cos \pi t$
   b. Find $f'(0)$

15. a. $f(t) = 4 \cos \dfrac{t}{2}$
   b. Find $f'\left(\dfrac{\pi}{3}\right)$

16. a. $f(t) = 48 \sin \dfrac{t}{12}$
   b. Find $f'(4\pi)$

17. a. $f(t) = \sin(\pi - t)$
   b. Find $f'\left(\dfrac{\pi}{2}\right)$

18. a. $f(t) = \cos(2\pi - t)$
   b. Find $f'(\pi)$

19. $f(t) = \cos(t + \pi)^3$

20. $f(t) = \sin^3(\pi + t)$

21. $f(t) = \cos^3(t + \pi)$

22. $f(t) = \sin(\pi + t)^3$

23. $f(t) = \sin(\pi - t)^2$

24. $f(t) = \sin^2(\pi - t)$

25. $f(t) = \sin^3 t^2$

26. $f(t) = \cos^4 t^2$

27. $f(x) = \cos \sqrt{x + 1}$

28. $f(x) = \sin \sqrt[3]{x}$

29. $f(x) = 4 \sin x + \sin 4x + \sin^4 x$

30. $f(x) = 4 \cos x + \cos 4x + \cos^4 x$

31. $f(x) = \sin e^x$

32. $f(x) = e^{\cos x}$

33. $f(z) = e^{z + \cos z}$

34. $f(z) = \cos(z + e^z)$

35. $f(z) = (1 - \cos z)^4$

36. $f(z) = (\sin z + \cos z)^2$

37. a. $f(z) = \ln(\sin z)$
   b. Find $f'\left(\dfrac{\pi}{2}\right)$

38. a. $f(z) = \cos(\ln z)$
   b. Find $f'(1)$

39. $f(z) = \sin(z + \ln z)$

40. $f(z) = \ln(z + \sin z)$

41. $f(x) = 3 \sin^2 x - 4 \cos^2 x$

42. $f(x) = 4 \cos^2 x + 5 \sin^2 x$

43. $f(x) = \dfrac{1}{x} \cos x^2$

44. $f(x) = x^2 \sin \dfrac{1}{x}$

45. $f(x) = e^{-x} \sin e^x$

46. $f(x) = e^x \cos e^{-x}$

47. a. $f(x) = \dfrac{\cos x}{\sin x}$
   b. Find $f'\left(\dfrac{\pi}{3}\right)$

48. a. $f(x) = \dfrac{\sin 2x}{\cos 2x}$
   b. Find $f'\left(\dfrac{\pi}{8}\right)$

49. $f(t) = \sin^2 t - \cos^2 t$

50. $f(t) = \cos^2 t + \sin^2 t$

51. a. $f(t) = \dfrac{1 + \sin t}{\cos t}$
   b. Find $f'(\pi)$

52. a. $f(t) = \dfrac{1 - \cos t}{\sin t}$
   b. Find $f'\left(\dfrac{\pi}{2}\right)$

53. $f(x) = \sin(\cos x)$

54. $f(x) = \cos(\sin x)$

55. $f(x) = \left(\cos \dfrac{x}{2} - 1\right)^2$

56. $f(x) = 2x \cos \dfrac{x}{2}$

57. $f(x) = x^2 \sin x + 2x \cos x - 2 \sin x$

58. $f(x) = x^2 \cos x - 2x \sin x - 2 \cos x$

59. $f(t) = \sin^2(t^2 + 1)$

60. $f(t) = \cos^2 \sqrt{t}$

61–64. Find the equation for the tangent line to the curve at the given point.

61. $f(x) = \sin x + \cos x$ at $x = 0$

62. $f(x) = \sin(\sin x)$ at $x = 0$

63. $f(x) = \ln(2e^{\sin x} - 1)$ at $x = \pi$

64. $f(x) = \sin 3x \cos 2x$ at $x = \dfrac{\pi}{3}$

65–70. Find the *second* derivative of each function.

65. $f(t) = t \sin t$

66. $f(t) = \sin at + \cos bt$ (*a* and *b* are constants)

67. $f(x) = e^{\sin x}$

68. $f(x) = e^{\cos x}$

69. $f(z) = z \sin z + \cos z$

70. $f(z) = \sin z - z \cos z$

Applied Exercises

71–72. BUSINESS: Seasonal Sales A manufacturer estimates that the number of air conditioners sold during week *t* of the year will be

$$S(t) = 500 - 500 \cos \frac{\pi t}{26}$$

Find the rate of change of sales at

71. $t = 19$ (mid-May)

72. $t = 37$ (mid-September)

73. GENERAL: Maximizing Volume of a Tent A tent made by hanging a square sheet over a ridge pole at angle θ as shown will enclose a volume of

$$V(\theta) = 1000 \sin \theta$$

cubic feet. Find the angle θ that maximizes the volume $V(\theta)$.

74. BUSINESS: Maximum Temperature The temperature in a chemical refining tower after *t* hours is $T(t) = 100 + 60(\sin t + \cos t)$ degrees Fahrenheit (for $0 \le t \le \pi$). Find when the temperature is the greatest.

75. ENVIRONMENTAL SCIENCE: Honeycombs A honeycomb consists of cells like the one shown in the next column, where θ is the angle between the upper faces and the axis of the cell. Minimizing the surface of the area is equivalent to minimizing the function

$$S(\theta) = \frac{\sqrt{3} - \cos \theta}{\sin \theta}$$

Minimize this function by setting its derivative equal to zero and solving, deriving the relationship $\cos \theta = \dfrac{1}{\sqrt{3}}$. The acute angle that solves this equation is approximately 54.7°, which is the actual angle used by honeybees.

Sources: D'A. W. Thompson, *On Growth and Form* and E. Batschelet, *Introduction to Mathematics for Life Scientists*

76. GENERAL: Maximizing Area

a. Find a formula for the area of a right triangle (shown below) with hypotenuse 1 and one acute angle of *x* radians $(0 < x < \pi/2)$.

b. Find the angle *x* that maximizes the area. [*Hint:* Either graph the area function and use MAXIMUM, or use NDERIV to find where its derivative equals zero.]

c. Convert *x* into degrees by multiplying by $\dfrac{180}{\pi}$. Did you expect to obtain this answer?

77. ATHLETICS: Maximizing Distance of a Thrown Ball A ball thrown with an initial velocity of 60 miles per hour at an angle of inclination with the ground of *x* radians, as shown on the next page, will travel a horizontal distance of

$$D(x) = 242(\sin x)(\cos x)$$

feet (neglecting air resistance).

a. Find the angle *x* that maximizes this distance [*Hint:* Either graph the distance function and use MAXIMUM, or use NDERIV to find where its derivative equals zero.]

b. Convert x into degrees by multiplying by $\dfrac{180}{\pi}$.
[*Note:* The angle that you find holds for all velocities.*]

$\overset{\longleftarrow}{\underset{\longrightarrow}{}}$ 242 sin x cos x

78. GENERAL: Temperature The mean temperature in Fairbanks, Alaska, on day x of the year is approximately

$$T(x) = 36 + 22\sin\left(\frac{2\pi(x - 102)}{365}\right)$$

degrees Fahrenheit.

a. Graph its derivative (using NDERIV) on the window $[0, 365]$ by $[-0.5, 0.5]$.

b. Find the rate of change of temperature on March 15 (day 74) and on September 12 (day 255) and interpret these answers.

c. Find the days on which the temperature will be rising the fastest.

Source: Alaska Climate Research Center

79. BIOMEDICAL: Blood Pressure A person's blood pressure varies periodically according to the formula $p(t) = 90 + 15\sin(2.5\pi t)$, where t is the number of seconds since the beginning of a cardiac cycle.

a. Graph the function on the window $[0, 1.6]$ by $[0, 120]$.

b. When is blood pressure the *highest* for $0 \le t \le 1.6$, and what is the maximum blood pressure?

(*continues*)

c. When is blood pressure the *lowest* for $0 \le t \le 1.6$ and what is the minimum blood pressure?

80. ATHLETICS: Basketball To make a basket, a basketball player must launch the ball with just the right angle and initial velocity for it to pass through the hoop. It can be shown from Newton's laws of motion that the initial velocity v and angle θ must satisfy the equation

$$v = \sqrt{\frac{16d}{\cos^2\theta\left(\dfrac{\sin\theta}{\cos\theta} - \dfrac{h}{d}\right)}}$$

where the constants d and h are as shown in the following diagram. To minimize the chances of the ball bouncing off the rim, coaches recommend shooting with the smallest possible velocity, called the "soft shot." Assume that for a free throw the constants are $d = 14$ and $h = 3$, and graph the resulting function on the window $[0, \pi/2]$ by $[0, 100]$. Then find the "soft shot" angle θ that minimizes the function.*

Source: American Journal of Physics **49**

Conceptual Exercises

81. Find $\dfrac{d}{dx}\sin(7\pi^{23} - e^{99})$

[*Hint:* No work necessary.]

82. Find $\dfrac{d}{dx}\cos(e^{-44} - \ln 99\pi)$

[*Hint:* No work necessary.]

83. Find $\dfrac{d}{dx}(\sin^2 x + \cos^2 x)$

[*Hint:* No work necessary.]

84. What is the smallest positive value of n such that $\dfrac{d^n}{dx^n}\sin x = \sin x$?

85. What is the smallest positive value of n such that $\dfrac{d^n}{dx^n}\cos x = \cos x$?

86. What is the 1000th derivative of $5\sin x$?

87. What is the 1000th derivative of $\cos(x + 973)$?

88. What is the maximum angle (in degrees) that the sine or cosine curve makes with the horizontal?

89. Find $\lim\limits_{h \to 0}\dfrac{h}{\sin h}$ [*Hint:* No work necessary if you use a similar known limit.]

90. Find $\lim\limits_{h \to 0}\dfrac{1 - \cos h}{h}$ [*Hint:* No work necessary if you use a similar known limit.]

*The result of this exercise applies to a *thrown ball*. For a ball hit by a bat, the best angle is 35°. This is because a batted ball can be given a backspin of 2000 rpm or more, giving it lift through what is called the Magnus effect. See Robert K. Adair, *The Physics of Baseball* (Harper & Row).

*In general, the angle for the soft shot is

$$\theta = 45° + \frac{1}{2}\left(\begin{array}{c}\text{Angle of elevation}\\\text{from ball to rim}\end{array}\right),\quad \text{found by setting the derivative}$$

of v equal to zero and solving for θ.

Explorations and Excursions   The following problems extend and augment the material presented in the text.

91. DERIVATIVE FORMULA FOR cos *x* Derive the formula for the derivative of cos *x* (assuming that the derivative exists) from the formula for the derivative of sin *x* as follows.

a. Differentiate both sides of the identity $\sin^2 x + \cos^2 x = 1$, obtaining

$$2 \sin x \cos x + 2 \cos x \frac{d}{dx} \cos x = 0$$

b. Solve this equation for $\frac{d}{dx} \cos x$ to obtain

$$\frac{d}{dx} \cos x = -\sin x.$$

92. Use the definition of the derivative to show that the derivative of $f(t) = \cos t$ is $f'(t) = -\sin t$. [*Hint:* The steps are similar to those on page 555, but using the identity $\cos(t + h) = (\cos t)(\cos h) - (\sin t)(\sin h)$.]

93. Show that the formulas on the right side of the box on page 554 reduce to the formulas on the left side in the case that the function *f* is simply $f(t) = t$.

94. Why do the functions $\sin^2 x$ and $-\cos^2 x$ have the same derivative?

95. PROOF OF cos(*s* + *t*) IDENTITY Prove the identity

$$\cos(s + t) = (\cos s)(\cos t) - (\sin s)(\sin t)$$

by differentiating the identity from page 542,

$$\sin(s + t) = (\sin s)(\cos t) + (\cos s)(\sin t)$$

with respect to *s*, holding *t* constant.

 96. DIFFERENTIATION FORMULAS FOR sin *x* AND cos *x*

a. Verify the formula $\frac{d}{dx} \sin x = \cos x$ by defining $y_1 = \sin x$ and y_2 as its derivative (using NDERIV). Then graph y_1 and y_2 on the window $[-2\pi, 2\pi]$ by $[-2, 2]$, observing that y_2 appears to be the cosine curve.

b. Then change y_1 to cos *x* to verify that $\frac{d}{dx} \cos x = -\sin x$ by observing that y_2 appears to be the negative of the sine curve.

Note that NDERIV finds derivatives *numerically*, using the difference quotient with a small value of *h*, so this is an independent verification of the formulas.

Integrals of Sine and Cosine Functions

Introduction

In this section we will integrate sine and cosine functions and use their integrals to find *areas under curves, total accumulations,* and *average values.*

Integrals of Sine and Cosine

The integrals of the sine and cosine functions are given by the following formulas, which are proved by differentiating their right-hand sides.

Integrals of sin *t* and cos *t*	
$$\int \sin t \, dt = -\cos t + C$$	The integral of sine is negative cosine (plus a constant)
$$\int \cos t \, dt = \sin t + C$$	The integral of cosine is sine (plus a constant)

These formulas are easily remembered. The derivative of sine is cosine, so the integral of cosine is sine, which is just the second formula. The first formula is similar, but with a negative sign (from the differentiation formula for cosine).

EXAMPLE 1 FINDING AREA BY INTEGRATION

Find the area under one arch of the sine curve $y = \sin t$.

Solution

For the area under one arch of $\sin t$, we integrate from 0 to π:

$$\text{Area} = \int_0^\pi \sin t \, dt = -\cos t \Big|_0^\pi = -\cos \pi - (-\cos 0)$$

Using
$\int \sin t \, dt = -\cos t + C$ Evaluating at π And at 0

$$= -(-1) - (-1) = 1 + 1 = 2$$

The area is 2 square units.

Graphing Calculator Exploration

Use a graphing calculator to verify the answer to the preceding Example as follows:

a. Graph $y_1 = \sin x$ on the window $[0, \pi]$ by $[0, 1]$.

b. Integrate this function from 0 to π using $\int f(x) \, dx$. Your answer should agree with that found in the preceding Example.

The substitution method (introduced on page 366) greatly expands the class of functions that we can integrate. As before, we write the integration formulas with x replaced by u, where u stands for a differentiable function.

$$\int \sin u \, du = -\cos u + C$$

$$\int \cos u \, du = \sin u + C$$

To evaluate a particular integral, we represent part of the integrand by u, calculate its differential du, and multiply inside and outside by constants (if necessary) to complete the du. The right-hand side of the formula then gives the solution, after substituting back to the original variable.

EXAMPLE 2 USING THE SUBSTITUTION METHOD

Find $\displaystyle\int x^2 \sin x^3 \, dx$.

Solution

The integrand involves $\sin x^3$, the sine of a *function*, suggesting the $\int \sin u\, du$ formula with $u = x^3$. The differential is $du = 3x^2\, dx$ (the derivative times dx).

$$\int \underbrace{x^2 \sin x^3}_{\sin u}\, dx = \frac{1}{3}\int \underbrace{\sin x^3}_{\sin u}\, \underbrace{3x^2\, dx}_{du} = \frac{1}{3}\int \underbrace{\sin u\, du}_{\text{Substituting}}$$

$$\begin{array}{cc} u = x^3 & \text{Multiplying by 3 and } \frac{1}{3} \\ du = 3x^2\, dx & \end{array}$$

$$= \underbrace{\frac{1}{3}(-\cos u) + C}_{\substack{\text{Integrating by} \\ \int \sin u\, du = -\cos u + C}} = \underbrace{-\frac{1}{3}\cos x^3 + C}_{\substack{\text{Substituting back} \\ \text{to } x \text{ using } u = x^3}}$$

We can check this answer by differentiation:

$$\underbrace{\frac{d}{dx}\left(-\frac{1}{3}\cos x^3 + C\right)}_{\text{Answer}} = \underbrace{-\frac{1}{3}(-\sin x^3)(3x^2)}_{\text{Using } \frac{d}{dx}\cos f = \sin f \cdot \frac{df}{dt}} = \underbrace{x^2 \sin x^3}_{\text{Simplifying}}$$

It checks: the derivative of the answer is the integrand

PRACTICE PROBLEM 1

Find $\displaystyle\int \cos \frac{2t}{3}\, dt$.   [*Hint:* Write $\dfrac{2t}{3}$ as $\dfrac{2}{3}t$, a constant times the variable.]

Solution on page 565 >

Other Substitution Formulas

An integral may require one of the earlier integration formulas (from page 366), but now with u being a trigonometric function.

(A) $\displaystyle\int u^n\, du = \frac{1}{n+1}u^{n+1} + C$ $(n \neq -1)$

(B) $\displaystyle\int e^u\, du = e^u + C$

(C) $\displaystyle\int u^{-1}\, du = \int \frac{du}{u} = \ln|u| + C$

EXAMPLE 3 **USING ANOTHER SUBSTITUTION FORMULA**

Find $\displaystyle\int \sin^5 2t \, \cos 2t \, dt$.

Solution

The integrand includes $\sin^5 2t = (\sin 2t)^5$, a function to a power, suggesting the $\int u^n \, du$ formula with $u = \sin 2t$.

$$\int \underbrace{(\sin 2t)^5}_{u^5} \cos 2t \, dt = \frac{1}{2} \int \underbrace{(\sin 2t)^5}_{u^5} \underbrace{2\cos 2t \, dt}_{du} = \frac{1}{2} \int \underbrace{u^5 \, du}_{\text{Substituting}}$$

$u = \sin 2t$

$du = \cos 2t \cdot 2dt$

Multiplying by 2 and $\frac{1}{2}$

$$= \underbrace{\frac{1}{2}\frac{1}{6} u^6 + C}_{\substack{\text{Integrating} \\ \text{by formula (A)}}} = \underbrace{\frac{1}{12} \sin^6 2t + C}_{\substack{\text{Substituting back} \\ \text{using } u = \sin 2t}}$$

PRACTICE PROBLEM 2

Find $\displaystyle\int \cos^2 x \, \sin x \, dx$.

Solution on page 565 >

EXAMPLE 4 **EVALUATING A DEFINITE INTEGRAL**

Evaluate the definite integral $\displaystyle\int_0^{\pi/2} e^{\sin z} \cos z \, dz$.

Solution

The integrand involves $e^{\sin z}$, e to a function, suggesting the $\int e^u \, du$ formula with $u = \sin z$. With definite integrals, we must also change the upper and lower limits of integration to their corresponding u-values.

$$u = \sin \frac{\pi}{2} = 1$$

$$\int_0^{\pi/2} \underbrace{e^{\sin z}}_{e^u} \underbrace{\cos z \, dz}_{du} = \underbrace{\int_0^1 e^u \, du}_{\text{Substituting}} = \underbrace{e^u \Big|_0^1}_{\substack{\text{Integrating} \\ \text{by formula (B)}}} = \underbrace{e^1 - e^0}_{\text{Evaluating}} = \underbrace{e - 1}_{\substack{\text{Final} \\ \text{answer}}}$$

$u = \sin z$

$du = \cos z \, dz$

$u = \sin 0 = 0$

We leave the answer in its *exact* form, $e - 1$, although we could use $e \approx 2.718$ to obtain the approximate answer 1.718.

| EXAMPLE 5 | INTEGRATING A QUOTIENT OF TRIGONOMETRIC FUNCTIONS |

Find $\displaystyle\int \frac{1 + \cos t}{t + \sin t}\, dt.$

Solution

The derivative of the denominator is the numerator: $\dfrac{d}{dt}(t + \sin t) = 1 + \cos t,$

suggesting the $\displaystyle\int \frac{du}{u}$ formula with $u = t + \sin t.$

$$\int \frac{1 + \cos t}{t + \sin t}\, dt = \int \frac{du}{u} = \ln|u| + C = \ln|t + \sin t| + C$$

Substituting   Integrating by formula (C)   Substituting back using $u = t + \sin t$

$u = t + \sin t$
$du = (1 + \cos t)\, dt$

PRACTICE PROBLEM 3

Evaluate $\displaystyle\int \frac{\sin t}{\cos t}\, dt.$

Solution on page 565 >

Many applications involve the *modified* trigonometric functions sin *at* and cos *at*, with a *scale constant a* that changes the period (see page 540). These functions are easily integrated by the substitution method, and the integrals occur so often that we list them as separate formulas.

$$\int \sin at\, dt = -\frac{1}{a}\cos at + C$$

$$\int \cos at\, dt = \frac{1}{a}\sin at + C$$

For constant $a \neq 0$

For example, Practice Problem 1 on page 561 is evaluated by the second formula in one step:

$$\int \cos \frac{2}{3} t\, dt = \frac{3}{2} \sin \frac{2}{3} t + C$$

Using the second formula above

a $\frac{1}{a}$

Recall that definite integration represents *continuous summation*, summing a rate to give the *total accumulation* at that rate (see page 337). For example, integrating a sales rate (such as sales per week) from *a* to *b* gives the *total sales* from week *a* to week *b*.

EXAMPLE 6   **SUMMING TOTAL SALES OF A SEASONAL PRODUCT**

A clothing store estimates that the number of overcoats sold per week will be

$$S(t) = 30 + 30 \cos \frac{\pi t}{26} \qquad \text{Coats per week}$$

where t is the number of weeks into the winter. Find the total number of overcoats that will be sold during the first 8 weeks of winter.

Solution

To find the total for the first 8 weeks, we integrate the rate from 0 to 8:

$$\int_0^8 \left(30 + 30 \cos \frac{\pi}{26} t \right) dt = \left(30t + 30 \frac{26}{\pi} \sin \frac{\pi}{26} t \right)\Big|_0^8$$

Writing $\frac{\pi t}{26}$ as $\frac{\pi}{26} t$

Using $\int \cos at\, dt = \frac{1}{a} \sin at + C$

$$= \underbrace{\left(\overbrace{30 \cdot 8}^{240} + \overbrace{30 \frac{26}{\pi} \sin \frac{8\pi}{26}}^{\approx 204} \right)}_{\text{Evaluating at } t = 8} - \underbrace{0}_{\substack{\text{And at} \\ t = 0}} \approx 444$$

Approximately 444 overcoats will be sold during the first 8 weeks.

Integrals also give average values when multiplied by $\dfrac{1}{b - a}$ (see page 345).

EXAMPLE 7   **FINDING AVERAGE TEMPERATURE**

The mean daily temperature in Fairbanks, Alaska, on day x of the year is

$$T(x) = 36 + 22 \sin \frac{2\pi (x - 102)}{365} \qquad \text{Degrees Fahrenheit}$$

Find the average summer temperature, June 21 to September 23 (day 172 to day 266).

Source: Alaska Climate Research Center

Solution

The Average Value of a Function formula (page 345) gives

$$\frac{1}{266 - 172} \int_{172}^{266} \left(36 + 22 \sin \frac{2\pi (x - 102)}{365} \right) dx \qquad \begin{array}{l}\text{The average is the} \\ \text{integral from } a \text{ to } b \\ \text{divided by } (b - a).\end{array}$$

We could calculate this "by hand," but instead we use a graphing calculator, graphing the function $T(x)$ on the window $[0, 365]$ by $[0, 70]$ and using the $\int f(x)dx$ command to integrate from 172 to 266.

Dividing the result by $(266 - 172)$ gives the average summer temperature in Fairbanks:

$$\frac{5055}{266 - 172} \approx 53.8$$

About 54 degrees!

Solutions TO PRACTICE PROBLEMS

1. $\displaystyle \int \cos \tfrac{2}{3}t \, dt = \tfrac{3}{2} \int \underbrace{\cos \tfrac{2}{3}t}_{\cos u} \, \tfrac{2}{3} dt = \frac{3}{2} \int \underbrace{\cos u}_{\cos u} \, \underbrace{du}_{du} = \frac{3}{2} \sin u + C = \frac{3}{2} \sin \frac{2t}{3} + C$

$u = \tfrac{2}{3}t$
$du = \tfrac{2}{3} dt$

2. $\displaystyle \int \underbrace{\cos^2 x}_{u^2} \sin x \, dx = -\int \underbrace{\cos^2 x}_{u^2} \underbrace{(-\sin x) \, dx}_{du} = -\int u^2 \, du = -\tfrac{1}{3}u^3 + C = -\tfrac{1}{3} \cos^3 x + C$

$u = \cos x$
$du = -\sin x \, dx$

3. $\displaystyle \int \frac{\sin t}{\cos t} \, dt = -\int \frac{-\sin t}{\cos t} \, dt = -\int \frac{du}{u} = -\ln|u| + C = -\ln|\cos t| + C$

$u = \cos t$
$du = -\sin t \, dt$

8.4   Section Summary

The Differentiation Formulas for the sine and cosine functions (page 547) led immediately to the integration formulas

$$\int \sin t \, dt = -\cos t + C$$

$$\int \cos t \, dt = \sin t + C$$

These two formulas extend immediately to their "u forms" by the substitution method (see page 560) and to a "scale constant" a (see page 563). We use integration for *areas*, for *continuous summations*, for *total accumulations*, and for *average values*.

8.4 Exercises

1–28. Find each integral.

1. $\displaystyle \int (\cos t - \sin t) \, dt$

2. $\displaystyle \int (\sin t + \cos t) \, dt$

5. $\displaystyle \int (t + 1) \cos(t^2 + 2t) \, dt$

6. $\displaystyle \int t^2 \sin(t^3 + 1) \, dt$

3. $\displaystyle \int (1 + \sin x) \, dx$

4. $\displaystyle \int (\cos x - 1) \, dx$

7. $\displaystyle \int \sin \pi t \, dt$

8. $\displaystyle \int \cos 2t \, dt$

9. $\displaystyle\int \cos \frac{\pi t}{2}\, dt$

10. $\displaystyle\int \sin \frac{2t}{5}\, dt$

11. $\displaystyle\int \sin(\pi - t)\, dt$

12. $\displaystyle\int \cos(1 - t)\, dt$

13. $\displaystyle\int \cos \frac{2\pi (t + 20)}{365}\, dt$

14. $\displaystyle\int \sin \frac{\pi (t + 3)}{26}\, dt$

15. $\displaystyle\int \sin^2 t \cos t\, dt$

16. $\displaystyle\int \cos^3 t \sin t\, dt$

17. $\displaystyle\int \frac{\sin t}{\cos^3 t}\, dt$

18. $\displaystyle\int \sqrt{\sin t}\, \cos t\, dt$

19. $\displaystyle\int e^{1 + \sin x} \cos x\, dx$

20. $\displaystyle\int e^{2x} \cos e^{2x}\, dx$

21. $\displaystyle\int \frac{\sin t}{\cos t + 3}\, dt$

22. $\displaystyle\int \frac{\cos t}{1 + \sin t}\, dt$

23. $\displaystyle\int \frac{\sin w}{\sqrt{1 - \cos w}}\, dw$

24. $\displaystyle\int \sqrt[3]{1 + \sin w}\, \cos w\, dw$

25. $\displaystyle\int \frac{\sin y \cos y}{\sqrt{1 + \sin^2 y}}\, dy$

26. $\displaystyle\int \frac{\cos \sqrt{z}}{\sqrt{z}}\, dz$

27. $\displaystyle\int \left(x^2 - \frac{1}{x} + \cos 2x \right) dx$

28. $\displaystyle\int \left(e^{2x} + \sqrt{x} - \sin 3x \right) dx$

29–36. For each definite integral:

a. Evaluate it "by hand," leaving the answer in *exact* form.

b. Check your answer to part (a) using a graphing calculator.

29. $\displaystyle\int_0^{1/2} \sin \pi t\, dt$

30. $\displaystyle\int_0^{\pi/4} \cos 2t\, dt$

31. $\displaystyle\int_0^{\pi} (t + \cos t)\, dt$

32. $\displaystyle\int_0^{\pi} (1 - \sin t)\, dt$

33. $\displaystyle\int_0^{1} (t + \sin \pi t)\, dt$

34. $\displaystyle\int_0^{1} (t^2 - \cos \pi t)\, dt$

35. $\displaystyle\int_{\pi/4}^{3\pi/4} \cos \left(t + \frac{\pi}{4} \right) dt$

36. $\displaystyle\int_{\pi/2}^{\pi} e^{\sin t} \cos t\, dt$

37. Find the area under the curve $y = \cos \dfrac{x}{2}$ and above the x-axis from $x = 0$ to $x = \pi$.

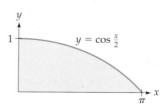

38. Find the area under the curve $y = x + \sin x$ and above the x-axis from $x = 0$ to $x = \pi$. Leave your answer in its *exact* form. If you are using a graphing calculator, verify your answer numerically using FnInt or $\int f(x)dx$.

39. Find the area under $y = \sin x + \cos x$ and above the x-axis from $x = 0$ to $x = \pi/2$. Leave your answer in its *exact* form. If you are using a graphing calculator, verify your answer numerically using FnInt or $\int f(x)dx$.

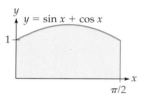

40. Find the area under the curve $y = x \sin x^2$ and above the x-axis from $x = 0$ to $x = \sqrt{\pi}$. Leave your answer in its *exact* form. If you are using a graphing calculator, verify your answer numerically using FnInt or $\int f(x)dx$.

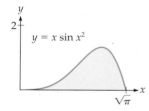

41. Prove the integration formulas in the box on page 563 by differentiating the right-hand sides to obtain the integrands.

42. Find the integral $\int \sin t \cos t\, dt$ in two ways:

a. Using the substitution method with $u = \sin t$.

b. Using the substitution method with $u = \cos t$.

c. Can you reconcile the two seemingly different answers? [*Hint:* Try using a trigonometric identity.]

Applied Exercises

43–44. BUSINESS: Seasonal Sales A ski shop predicts sales of

$$S(t) = 10 + 9 \cos \frac{\pi t}{26}$$

thousand dollars during week t of the year, with $t = 0$ corresponding to the beginning of January.

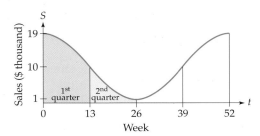

Predict the company's total sales during:

43. the first quarter of the year $(t = 0$ to $t = 13)$.

44. the second quarter of the year $(t = 13$ to $t = 26)$.

45. ENVIRONMENTAL SCIENCE: Pollution A chemical refinery discharges $P(t) = 5 + 3 \sin \dfrac{\pi t}{6}$ tons of pollution each month, where t is the number of months that the plant has been in operation. For the first 18 months $(t = 0$ to $t = 18)$, find:

 a. the *total* amount of pollutant discharged.
 b. the *average* monthly amount of pollutant discharged.

46. BUSINESS: Summing an Income Stream An investment generates a stream of income at the rate of $S(t) = 50 + 10 \sin \dfrac{\pi t}{6}$ thousand dollars per month, where t is the number of months since the investment was made. For the first 6 months $(t = 0$ to $t = 6)$, find:

 a. the *total* income from the income stream.
 b. the *average* monthly income from the income stream.

47. ENVIRONMENTAL: Bird Counts The number of migratory birds flying north past a check-point is $300 - 300 \cos \dfrac{\pi t}{6}$ per month, where t is the number of months since the beginning of the year. Find the total number for the year.

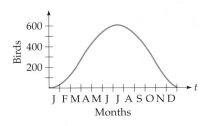

48. BIOMEDICAL: Blood Flow Blood flows through an artery at the rate of $30 - 30 \cos(2.4\pi t)$ cubic centimeters

per second, where t is the number of seconds after the beginning of the cardiac cycle. Find the total amount of blood flowing through the artery in the first 5/6 second (one complete cardiac cycle).

49. PERSONAL FINANCE: Home Heating Costs Suppose that the temperature outside of a house on day t of the year is $T(t) = 60 - 30 \cos \dfrac{2\pi t}{365}$ degrees Fahrenheit. If it costs \$1 to raise the temperature of the house by $1°$ for 1 day, then setting the thermostat at $75°$ will result in an annual heating bill equal to the area of the shaded region shown below.

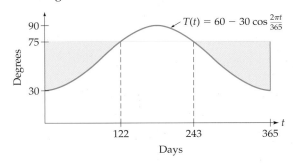

This area is given by the integral below (with the two equal areas calculated as twice the first). Use a graphing calculator to evaluate this integral to find the annual heating cost.

$$\left(\begin{array}{c} \text{Heating} \\ \text{costs at } 75° \end{array} \right) = 2 \int_0^{122} \left[75 - \left(60 - 30 \cos \frac{2\pi t}{365} \right) \right] dt$$

50. PERSONAL FINANCE: Home Heating Costs (*continuation*) If the thermostat is lowered to $68°$, the annual heating cost is equal to the area shown below.

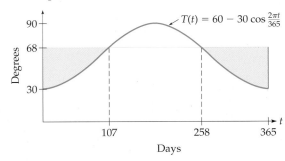

The annual heating cost is given by:

$$\left(\begin{array}{c} \text{Heating} \\ \text{costs at } 68° \end{array} \right) = 2 \int_0^{107} \left[68 - \left(60 - 30 \cos \frac{2\pi t}{365} \right) \right] dt$$

 a. Use a graphing calculator to evaluate this integral to find the new heating cost.
 b. Compare your answers from part (a) with your answer from Exercise 49 to find how much is saved by lowering the thermostat.

51. BUSINESS: Hotel Occupancy A Caribbean hotel estimates that it will have an occupancy of $200 + 100 \cos \dfrac{\pi t}{26}$ guests during week t of the year. Find the *average* weekly occupancy during the first 10 weeks of the year ($t = 0$ to $t = 10$).

52. BUSINESS: Average Demand A motorbike manufacturer estimates demand to be $400 + 300 \sin \dfrac{\pi t}{26}$ motorbikes per week, where $t = 0$ corresponds to the beginning of the year. Find the *average* weekly demand over the first 20 weeks ($t = 0$ to $t = 20$).

53. GENERAL: Temperature in New York The mean temperature in New York City on day x of the year is

$$54 + 24 \sin \frac{2\pi(x - 101)}{365}$$

degrees Fahrenheit. Use a graphing calculator to find:

a. The average temperature during January and February (days 0 to 59).

b. The average temperature during July and August (days 181 to 243).

Source: National Weather Service

54. GENERAL: Temperature in Minneapolis/St. Paul The mean temperature in Minneapolis/St. Paul on day x of the year is

$$45 + 32 \sin \frac{2\pi(x - 101)}{365}$$

degrees Fahrenheit. Use a graphing calculator to find:

a. The average temperature during January and February (days 0 to 59).

b. The average temperature during July and August (days 181 to 243).

Source: National Weather Service

Conceptual Exercises

55–60. Evaluate each of the following definite integrals by thinking of the graphs of the functions, without any calculation.

55. $\displaystyle\int_0^{2\pi} \sin t \, dt$

56. $\displaystyle\int_0^1 \sin 2\pi t \, dt$

57. $\displaystyle\int_0^{1000} \cos \pi \, dt$

58. $\displaystyle\int_0^{100} \cos 2\pi \, dt$

59. $\displaystyle\int_{2.345}^{2.345 + 1000\pi} \cos t \, dt$

60. $\displaystyle\int_{-1000\pi}^{1000\pi} \sin t \, dt$

61. Give an approximate value for the average value of $9 + \cos x$ over a very long interval.

62. Give an approximate value for the average value of $\sin 4x - 2$ over a very long interval.

63. By just looking at the temperature function in Exercise 53, state the average annual temperature in New York City.

64. By just looking at the temperature function in Exercise 54, state the average annual temperature in Minneapolis/St. Paul.

8.5   Other Trigonometric Functions

Introduction

In this section we define the **tangent, cotangent, secant,** and **cosecant** functions, differentiate and integrate them, and discuss some of their applications.

The Tangent, Cotangent, Secant, and Cosecant Functions

The functions $\tan \theta$, $\cot \theta$, $\sec \theta$, and $\csc \theta$ are defined as follows:

tan θ,   cot θ,   sec θ,   and   csc θ

Let (x, y) be any point on the terminal side of an angle of θ radians in standard position, and let r be its distance from the origin. We define:

$$\tan \theta = \frac{\sin \theta}{\cos \theta} = \frac{y}{x}$$

$$\cot \theta = \frac{\cos \theta}{\sin \theta} = \frac{x}{y}$$

$$\sec \theta = \frac{1}{\cos \theta} = \frac{r}{x}$$

$$\csc \theta = \frac{1}{\sin \theta} = \frac{r}{y}$$

The domain of each function is the set of numbers such that the denominator is not zero. Unlike the sine and cosine functions, these four functions can take very large (positive) values and very small (negative) values. Notice from the above definitions that tan θ and cot θ are reciprocals of each other, so that at values where both are defined:

$$\cot \theta = \frac{1}{\tan \theta}$$

EXAMPLE 1   EVALUATING TRIGONOMETRIC FUNCTIONS AT A QUADRANT ANGLE

Find $\tan \pi$, $\cot \pi$, $\sec \pi$,   and $\csc \pi$.

Solution

An angle of π radians is shown on the left, together with a point (x, y) on its terminal side. From these x, y, and r values, we calculate the values of the trigonometric functions:

$$\tan \pi = \frac{0}{-1} = 0 \qquad\qquad \tan \theta = \frac{y}{x}$$

$$\cot \pi = \frac{-1}{0} \qquad \text{Undefined!} \qquad \cot \theta = \frac{x}{y}$$

$$\sec \pi = \frac{1}{-1} = -1 \qquad\qquad \sec \theta = \frac{r}{x}$$

$$\csc \pi = \frac{1}{0} \qquad \text{Undefined!} \qquad \csc \theta = \frac{r}{y}$$

x y
↓ ↓
$(-3, 4)$

| EXAMPLE 2 | EVALUATING TRIGONOMETRIC FUNCTIONS FROM A GRAPH |

Find $\tan \theta$, $\cot \theta$, $\sec \theta$, and $\csc \theta$ for the angle θ shown on the left.

Solution

The distance from the point $(-3, 4)$ to the origin is found from the Pythagorean Theorem:

$$r = \sqrt{(-3)^2 + 4^2} = \sqrt{9 + 16} = \sqrt{25} = 5$$

From these x, y, and r values we calculate

$$\tan \theta = \frac{4}{-3} = -\frac{4}{3} \qquad\qquad \tan \theta = \frac{y}{x}$$

$$\cot \theta = \frac{-3}{4} = -\frac{3}{4} \qquad\qquad \cot \theta = \frac{x}{y}$$

$$\sec \theta = \frac{5}{-3} = -\frac{5}{3} \qquad\qquad \sec \theta = \frac{r}{x}$$

$$\csc \theta = \frac{5}{4} \qquad\qquad\qquad \csc \theta = \frac{r}{y}$$

| EXAMPLE 3 | EVALUATING TRIGONOMETRIC FUNCTIONS |

Find $\tan \dfrac{\pi}{3}$, $\cot \dfrac{\pi}{3}$, $\sec \dfrac{\pi}{3}$, and $\csc \dfrac{\pi}{3}$.

Solution

Recall that $\sin \dfrac{\pi}{3} = \dfrac{\sqrt{3}}{2}$ and $\cos \dfrac{\pi}{3} = \dfrac{1}{2}$.   See page 536
Therefore:

$$\tan \frac{\pi}{3} = \frac{\sin \dfrac{\pi}{3}}{\cos \dfrac{\pi}{3}} = \frac{\dfrac{\sqrt{3}}{2}}{\dfrac{1}{2}} = \sqrt{3} \qquad\qquad \text{Since } \tan \theta = \frac{\sin \theta}{\cos \theta}$$

$$\cot \frac{\pi}{3} = \frac{1}{\tan \dfrac{\pi}{3}} = \frac{1}{\sqrt{3}} = \frac{\sqrt{3}}{3} \qquad\qquad \text{Multiplying numerator and denominator by } \sqrt{3}$$

$$\sec \frac{\pi}{3} = \frac{1}{\cos \dfrac{\pi}{3}} = \frac{1}{\dfrac{1}{2}} = 2 \qquad\qquad \text{Since } \sec \theta = \frac{1}{\cos \theta}$$

$$\csc \frac{\pi}{3} = \frac{1}{\sin \dfrac{\pi}{3}} = \frac{1}{\dfrac{\sqrt{3}}{2}} = \frac{2}{\sqrt{3}} = \frac{2\sqrt{3}}{3} \qquad\qquad \text{Multiplying numerator and denominator by } \sqrt{3}$$

Using the [SIN], [COS], and [TAN] keys on your calculator, together with the [1/x] or [x^{-1}] key for reciprocals, you can evaluate trigonometric functions at other values.

EXAMPLE 4   APPROXIMATING USING A CALCULATOR

Use a calculator to approximate $\cot \dfrac{\pi}{9}$.

Solution

$$\cot \frac{\pi}{9} = \frac{1}{\tan \dfrac{\pi}{9}} \approx 2.747 \qquad \text{Using a calculator (set for radians)}$$

```
1/tan(π/9)
            2.747477419
```

PRACTICE PROBLEM 1

Use a calculator to find $\csc \dfrac{\pi}{7}$.

Solution on page 577 >

For *acute* angles θ, these trigonometric functions can be expressed as ratios of sides of a right triangle. We simply take the earlier definitions and replace $x, y,$ and r by adj, opp, and hyp.

Right Triangle Formulas for $\tan \theta$, $\cot \theta$, $\sec \theta$,   and $\csc \theta$

  LOOKING BACK

These agree with our earlier definitions on page 569 by using

$opp = y$
$adj = x$
$hyp = r$

Those definitions are more general than these because they apply to *any* triangle, not just acute ones.

$$\tan \theta = \frac{\text{opp}}{\text{adj}}$$

$$\cot \theta = \frac{\text{adj}}{\text{opp}}$$

$$\sec \theta = \frac{\text{hyp}}{\text{adj}}$$

$$\csc \theta = \frac{\text{hyp}}{\text{opp}}$$

EXAMPLE 5   ESTIMATING HEIGHT

Estimate the height of a tree if its top makes an angle of 41° with the ground when seen from 50 feet away.

Solution

Since the height and the known distance (50 feet) are the opposite and adjacent sides of a right triangle, we use the tangent (opp/adj) of the angle. Letting h = height of tree:

$$\tan 41° = \frac{h}{50} \qquad \tan \theta = \frac{\text{opp}}{\text{adj}}$$

Solving for h:

$$h = 50 \tan 41° \approx 50 \cdot 0.869 \approx 43.5 \qquad \tan 41° \approx 0.869 \text{ (from a calculator set for } degrees)$$

The tree is about 43.5 feet high.

If an object is thrown with initial velocity v (feet per second) and with an initial angle θ to the horizontal, then its height y when it has gone a horizontal distance x from its starting point is given by the following formula:*

$$y = x \tan \theta - \left(\frac{4x}{v} \sec \theta\right)^2$$

 EXAMPLE 6   **HOME RUNS**

A baseball is hit at an angle of 35° with the horizontal with an initial velocity of 120 feet per second. Will it make a home run by clearing the outfield fence 400 feet away if the fence is 10 feet higher than the initial height of the ball?

Solution

The formula with the given data becomes

$$y = 400 \tan 35° - \left(\frac{4 \cdot 400}{120} \sec 35°\right)^2$$

Previous formula with distance $x = 400$, angle $\theta = 35°$, and initial velocity $v = 120$

$\underbrace{\qquad}_{0.700}$ $\underbrace{\qquad}_{1.22}$ Using a calculator set for *degrees*

$$\approx 15.1$$

The ball will be 15 feet higher, easily clearing the fence and making a home run.

 Graphing Calculator Exploration

What is the best angle to get the baseball as high as possible over the outfield fence (as before, neglecting air resistance and backspin)?

Set your calculator for DEGREES and enter the expression for y from the preceding Example into your graphing calculator as y_1 but with 35° replaced by x. Then graph y_1 on the window [0, 90] by [0, 50] and use MAXIMUM to find the best angle x. (Afterward, reset your calculator for RADIANS.)

---

*Neglecting air resistance and backspin. See Philippe de Mezieres, *The Mathematics of Projectiles in Sports* (Australian Academy of Sciences).

Graphs of Tangent, Cotangent, Secant, and Cosecant Functions

The graphs of tan t, cot t, sec t, and csc t (for t in *radians*) are shown below.

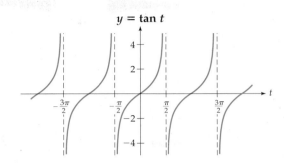

Period: π
Undefined at:
$\pm\frac{\pi}{2}, \pm\frac{3\pi}{2}, \dots$

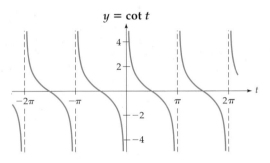

Period: π
Undefined at:
$0, \pm\pi, \pm2\pi, \dots$

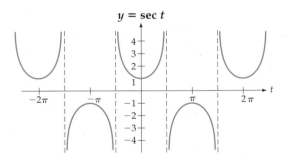

Period: 2π
Undefined at:
$\pm\frac{\pi}{2}, \pm\frac{3\pi}{2}, \dots$

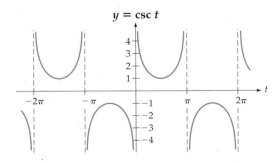

Period: 2π
Undefined at:
$0, \pm\pi, \pm2\pi, \dots$

Verify the preceding graphs by setting the window of your graphing calculator to $[-2\pi, 2\pi]$ by $[-4, 4]$, and graphing successively:

$$\tan x, \quad \cot x \ \left(\text{as } \frac{1}{\tan x}\right), \quad \sec x \ \left(\text{as } \frac{1}{\cos x}\right), \quad \text{and} \quad \csc x \ \left(\text{as } \frac{1}{\sin x}\right)$$

Be Careful Each of these functions has x-values where it is *undefined*. At these values, your graphing calculator screen may show "false lines," which should be ignored.

Derivatives of Trigonometric Functions

Derivatives of the tangent, cotangent, secant, and cosecant functions (always in *radians*) are given by the formulas in the first column in the following box. Combining these formulas with the Chain Rule gives the formulas in the second column, in which $f(t)$ stands for a differentiable function.

Derivatives of $\tan t$, $\cot t$, $\sec t$, and $\csc t$

$$\frac{d}{dt}\tan t = \sec^2 t \qquad\qquad \frac{d}{dt}\tan f(t) = \sec^2 f(t) \cdot \frac{df}{dt}$$

$$\frac{d}{dt}\cot t = -\csc^2 t \qquad\qquad \frac{d}{dt}\cot f(t) = -\csc^2 f(t) \cdot \frac{df}{dt}$$

$$\frac{d}{dt}\sec t = \sec t \cdot \tan t \qquad\qquad \frac{d}{dt}\sec f(t) = \sec f(t) \cdot \tan f(t) \cdot \frac{df}{dt}$$

$$\frac{d}{dt}\csc t = -\csc t \cdot \cot t \qquad\qquad \frac{d}{dt}\csc f(t) = -\csc f(t) \cdot \cot f(t) \cdot \frac{df}{dt}$$

Take Note

Notice the pattern: Adding a "co" to each function and adding a negative sign gives a new formula. Do you see how applying this to the "tan x" formula gives the "cot x" formula?

EXAMPLE 7 **PROVING A FORMULA**

Prove the first formula: $\dfrac{d}{dt}\tan t = \sec^2 t$.

Solution

$$\frac{d}{dt}\tan t = \frac{d}{dt}\left(\frac{\sin t}{\cos t}\right) \qquad\qquad\qquad \text{Since } \tan t = \frac{\sin t}{\cos t}$$

Derivative of sin t

Derivative of cos t

$$= \frac{\cos t \cos t - (-\sin t)\sin t}{\cos^2 t} \qquad\qquad \text{Using the Quotient Rule}$$

$$= \frac{\cos^2 t + \sin^2 t}{\cos^2 t} = \frac{1}{\cos^2 t} \qquad\qquad \begin{array}{l}\text{Simplifying and using} \\ \sin^2 t + \cos^2 t = 1\end{array}$$

$$= \sec^2 t \qquad\qquad\qquad\qquad\qquad \text{Since } \frac{1}{\cos t} = \sec t$$

This proves the first formula. The other three formulas are proved similarly (see Exercises 68–70).

EXAMPLE 8   **DIFFERENTIATING THE TANGENT OF A FUNCTION**

Find $\dfrac{d}{dt}\tan(t^3 + 1)$

Solution

$$\frac{d}{dt}\tan(t^3 + 1) = \underbrace{\sec^2(t^3 + 1) \cdot (3t^2)}_{\substack{\text{Derivative} \\ \text{of } t^3 + 1}} \qquad \text{Using } \frac{d}{dt}\tan f = \sec^2 f \cdot \frac{df}{dt}$$

$$= 3t^2 \sec^2(t^3 + 1)$$

PRACTICE PROBLEM 2

Find $\dfrac{d}{dt}\sec \pi t.$

Solution on page 577 >

Integrals of Trigonometric Functions

The differentiation formulas in the preceding box, when read in reverse, become the following integration formulas.

Integrals of Trigonometric Functions

$$\int \sec^2 t \; dt = \tan t + C \qquad\qquad \text{Since } \frac{d}{dt}\tan t = \sec^2 t$$

$$\int \csc^2 t \; dt = -\cot t + C \qquad\qquad \text{Since } \frac{d}{dt}(-\cot t) = \csc^2 t$$

$$\int \sec t \tan t \; dt = \sec t + C \qquad\qquad \text{Since } \frac{d}{dt}\sec t = \sec t \tan t$$

$$\int \csc t \cot t \; dt = -\csc t + C \qquad\qquad \text{Since } \frac{d}{dt}(-\csc t) = \csc t \cot t$$

Take Note

The pattern for the box on the previous page works here too, reducing what you need to memorize by half.

Replacing the variable t by u gives analogous formulas for the substitution method.

EXAMPLE 9   **INTEGRATING USING A SUBSTITUTION**

Find $\displaystyle\int \sec^2 5t \, dt$.

Solution

We use the first formula in the preceding box, but in its "u" form.

$$\int \underbrace{\sec^2 5t}_{u} \, dt = \frac{1}{5} \int \underbrace{\sec^2 5t}_{u} \underbrace{5 \, dt}_{du} = \frac{1}{5} \int \underbrace{\sec^2 u \, du}_{\text{Substituting}}$$

$$u = 5t$$
$$du = 5 \, dt$$
Multiplying and dividing by 5

$$= \underbrace{\frac{1}{5} \tan u + C}_{\substack{\text{Integrating by} \\ \int \sec^2 u \, du = \tan u + C}} = \underbrace{\frac{1}{5} \tan 5t + C}_{\substack{\text{Substituting back} \\ \text{using } u = 5t}}$$

EXAMPLE 10   **INTEGRATING USING A SUBSTITUTION**

Find $\displaystyle\int \tan^3 t \sec^2 t \, dt$.

Solution

The derivative of $\tan t$ is $\sec^2 t$, so the integral is of the form $\int u^3 du$.

$$\int \underbrace{\tan^3 t}_{u^3} \underbrace{\sec^2 t \, dt}_{du} = \underbrace{\int u^3 \, du}_{\text{Substituting}} = \underbrace{\frac{1}{4} u^4 + C}_{\substack{\text{Integrating by} \\ \text{the Power Rule}}} = \underbrace{\frac{1}{4} \tan^4 t + C}_{\substack{\text{Substituting back} \\ \text{using } u = \tan t}}$$

$$u = \tan t$$
$$du = \sec^2 t \, dt$$

Exercises 73–77 show how to derive the following and other integration formulas.

Integrals of $\tan t$, $\cot t$, $\sec t$ and $\csc t$

$$\int \tan t \, dt = -\ln|\cos t| + C$$

$$\int \cot t \, dt = \ln|\sin t| + C$$

$$\int \sec t \, dt = \ln|\sec t + \tan t| + C$$

$$\int \csc t \, dt = \ln|\csc t - \cot t| + C$$

**Graphing
Calculator
Exploration**

One of the difficulties of trigonometry is that the trigonometric identities allow results to be written in many different forms. Some advanced graphing calculators can find trigonometric integrals. For example, the Texas Instruments *TI-89* graphing calculator finds the integral $\int \sec t \, dt$ as follows.

Show that this answer agrees with the third formula in the box on the previous page. [*Hint:* Begin with the answer in the screen, multiply numerator and denominator by $\sin t + 1$ and simplify.]

For a quick reference, all of the integration formulas used in this book are summarized on the inside back cover.

EXAMPLE 11   FINDING AREA

Find the area under $y = \tan x$ from $x = 0$ to $x = \dfrac{\pi}{4}$.

Solution

Since $\tan x$ is nonnegative between $x = 0$ and $x = \dfrac{\pi}{4}$, we integrate from 0 to $\dfrac{\pi}{4}$:

$$\int_0^{\pi/4} \tan t \, dt = -\ln|\cos t| \, \Big|_0^{\pi/4} = -\ln \left| \cos \frac{\pi}{4} \right| + \ln|\cos 0|$$

$$= -\ln \frac{\sqrt{2}}{2} = -\ln \frac{1}{\sqrt{2}} = \frac{1}{2}\ln 2$$

The area is $\dfrac{1}{2}\ln 2 \approx 0.347$ square unit.

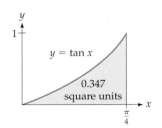

Solutions TO PRACTICE PROBLEMS

1. $\csc \dfrac{\pi}{7} \approx 2.305$ From entering $1/\sin(\pi/7)$

2. $\dfrac{d}{dt}\sec \pi t = (\sec \pi t)(\tan \pi t)\, \pi = \pi \sec \pi t \tan \pi t$

<table>
<tr><td>8.5</td><td>**Section Summary**</td></tr>
</table>

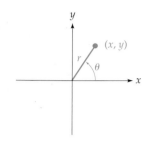

In this section we defined the remaining four trigonometric functions: tangent, cotangent, secant, and cosecant. If (x, y) is any point on the terminal side of an angle of θ radians in standard position, and r is the distance from the origin, then:

$$\tan \theta = \frac{\sin \theta}{\cos \theta} = \frac{y}{x} \qquad\qquad \sec \theta = \frac{1}{\cos \theta} = \frac{r}{x}$$

$$\cot \theta = \frac{\cos \theta}{\sin \theta} = \frac{x}{y} \qquad\qquad \csc \theta = \frac{1}{\sin \theta} = \frac{r}{y}$$

Tangent and cotangent are particularly useful in applications since they give the ratio of the sides of a right triangle with acute angle θ. The graphs of these functions are shown on page 573. We often regard these as functions of a variable t, without regard to angles. The derivative and corresponding integral formulas are given below.

$$\frac{d}{dt} \tan t = \sec^2 t \qquad\qquad \int \sec^2 t \, dt = \tan t + C$$

$$\frac{d}{dt} \cot t = -\csc^2 t \qquad\qquad \int \csc^2 t \, dt = -\cot t + C$$

$$\frac{d}{dt} \sec t = \sec t \cdot \tan t \qquad\qquad \int \sec t \cdot \tan t \, dt = \sec t + C$$

$$\frac{d}{dt} \csc t = -\csc t \cdot \cot t \qquad\qquad \int \csc t \cdot \cot t \, dt = -\csc t + C$$

Some other integral formulas are given on page 576.

8.5 Exercises

1–4. For each angle θ, find:

a. $\tan \theta$ **b.** $\cot \theta$ **c.** $\sec \theta$ **d.** $\csc \theta$.

1.

2.

3.

4.

6. a. $\cot \dfrac{\pi}{6}$ **b.** $\sec \dfrac{\pi}{6}$

7. a. $\tan \dfrac{3\pi}{2}$ **b.** $\csc \dfrac{3\pi}{2}$

8. a. $\cot \dfrac{3\pi}{2}$ **b.** $\sec \dfrac{3\pi}{2}$

9. a. $\tan \dfrac{\pi}{4}$ **b.** $\cot \dfrac{\pi}{4}$

10. a. $\tan \dfrac{3\pi}{4}$ **b.** $\sec \dfrac{3\pi}{4}$

11. a. $\tan \dfrac{7\pi}{6}$ **b.** $\csc \dfrac{7\pi}{6}$

12. a. $\tan \dfrac{5\pi}{3}$ **b.** $\cot \dfrac{5\pi}{3}$

5–12. Evaluate without using a calculator.

5. a. $\tan \dfrac{\pi}{6}$ **b.** $\csc \dfrac{\pi}{6}$

13–16. Use a calculator to approximate each value.

13. $\cot \dfrac{\pi}{5}$ **14.** $\sec \dfrac{\pi}{7}$ **15.** $\sec \dfrac{5\pi}{12}$ **16.** $\csc \dfrac{9\pi}{52}$

Applied Exercises

17. GENERAL: Flagpole Height Estimate the height of a flagpole if its top makes an angle of 62° with the ground when seen from 30 feet away.

18. GENERAL: River Width Estimate the width of a river if two points on opposite shores make a 35° angle for someone standing 40 feet down the shore.

19. GENERAL: Slope and Angle of Inclination Show that the slope of a line is the *tangent* of the "angle of inclination" between the line and the *x*-axis.

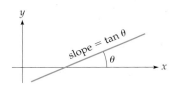

20. ATHLETICS: Baseball A baseball is hit so that it leaves the bat with an initial velocity of 115 feet per second and at an angle of 50° with the horizontal. Will it make a home run by clearing the outfield fence 400 feet away if the fence is 8 feet higher than the initial height of the ball? [*Hint:* Use the formula on page 572.]

21. ATHLETICS: Football A football is kicked from the ground with an initial velocity of 60 feet per second and at an angle of 40° with the horizontal. Will it make a field goal by clearing the 10-foot-high goal post that is 90 feet away? [*Hint:* Use the formula on page 572.]

22. ATHLETICS: Baseball *(continuation of 20)* At what angle should the ball be hit (with initial velocity 115 feet per second) so as to maximize its height above the outfield fence? Set your calculator for degrees and enter the height function that you used in Exercise 20 as y_1, but with 50° replaced by x. Then graph y_1 on an appropriate window and use MAXIMUM to maximize it and find the angle. (Afterward, reset your calculator for *radians*.)

23. ATHLETICS: Football *(continuation of 21)* At what angle should the football be kicked (with initial velocity 60 feet per second) so as to maximize its height over the goal post? Set your calculator for degrees and enter the height function that you used in Exercise 21 as y_1, but with 40° replaced by x. Then graph y_1 on an appropriate window and use MAXIMUM to maximize it and find the angle. (Afterward, reset your calculator for *radians*.)

24. Graph $y_1 = \csc^2 x$ on the window $[-2\pi, 2\pi]$ by $[-4, 4]$. What is its period? Where is it undefined? (This graph will be used in Exercise 68.)

25. Graph $y_1 = \sec x \tan x$ on the window $[-2\pi, 2\pi]$ by $[-4, 4]$. What is its period? Where is it undefined? (This graph will be used in Exercise 69.)

26. Graph $y_1 = \csc x \cot x$ on the window $[-2\pi, 2\pi]$ by $[-4, 4]$. What is its period? Where is it undefined? (This graph will be used in Exercise 70.)

27. Derive the identity $\tan^2 t + 1 = \sec^2 t$. [*Hint:* Divide the identity $\sin^2 t + \cos^2 t = 1$ by $\cos^2 t$. This identity will be used in Exercise 78.]

28. Derive the identity $1 + \cot^2 t = \csc^2 t$. [*Hint:* Divide the identity $\sin^2 t + \cos^2 t = 1$ by $\sin^2 t$.]

29–36. Differentiate each function.

29. $f(t) = t \cot t$

30. $f(t) = t^2 \sec t$

31. $f(t) = \tan(t^2 - 1)$

32. $f(t) = \cot(2t - 1)$

33. $f(x) = \sec(\pi x + 1)$

34. $f(x) = \tan\left(\dfrac{x}{2} - \pi\right)$

35. $f(z) = \cot^2 z^3$

36. $f(z) = \csc^4 \sqrt{z}$

37–40. Find the equation for the tangent line to the curve at the given point.

37. $f(x) = \tan(x^2 + \frac{\pi}{4} - 1)$ at $x = 1$

38. $f(x) = \tan x \sec^2 x$ at $x = \frac{\pi}{4}$

39. $f(x) = e^{\cot x - 1}$ at $x = \frac{\pi}{4}$

40. $f(x) = \ln(\csc x - 1)$ at $x = \frac{\pi}{6}$

41. GENERAL: Length of Shadow When the sun is x degrees above the horizon, a 6-foot-tall person will cast a shadow of length $S(x) = 6 \cot \dfrac{\pi x}{180}$ feet. Calculate $S'(10)$ and interpret the resulting number.

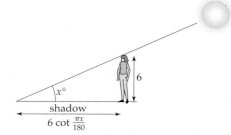

shadow
$6 \cot \frac{\pi x}{180}$

42. GENERAL: Ladder Reach A sufficiently long extension ladder whose base is 4 feet from a wall and that makes an angle of x degrees with the ground will reach to a height of $H(x) = 4 \tan \dfrac{\pi x}{180}$ feet. Find $H'(75)$ and interpret the resulting number.

$4 \tan \frac{\pi x}{180}$

43–52. Find each integral.

43. $\displaystyle \int t^2 \sec^2(t^3 + 1)\, dt$ **44.** $\displaystyle \int \csc^2 \pi t\, dt$

45. $\displaystyle \int \csc \pi t \cot \pi t\, dt$

46. $\displaystyle \int t \sec(t^2 + 1) \tan(t^2 + 1)\, dt$

47. $\displaystyle \int \tan(1 - t)\, dt$ **48.** $\displaystyle \int \cot \dfrac{t}{2}\, dt$

49. $\displaystyle \int x^4 \csc x^5\, dx$ **50.** $\displaystyle \int x^{-1/2} \sec x^{1/2}\, dx$

51. $\displaystyle \int \cot^4 x \csc^2 x\, dx$ **52.** $\displaystyle \int \sqrt{\tan x}\, \sec^2 x\, dx$

53–54. Evaluate each definite integral.

53. $\displaystyle \int_{\pi/4}^{\pi/2} \cot t\, dt$ **54.** $\displaystyle \int_{0}^{\pi/4} \sec^2 t\, dt$

55. a. Find (without using a graphing calculator) the area under the curve $y = \sec x$ and above the x-axis from $x = 0$ to $x = \pi/4$.

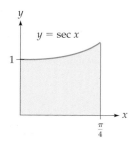

b. Use a graphing calculator to verify your answer (using FnInt, ∫f(x)dx, or a similar command).

56. a. Find (without using a graphing calculator) the area under the curve $y = \sec x \tan x$ and above the x-axis between $x = 0$ and $x = \pi/4$.

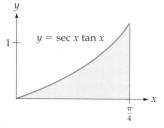

b. Use a graphing calculator to verify your answer (using FnInt, ∫f(x)dx, or a similar command).

Conceptual Exercises

57. True or False: All six trigonometric functions have period 2π.

58. True or False: All six trigonometric functions can be expressed in terms of the sine and cosine functions.

59. Which trigonometric function is the same as $\dfrac{1}{\sec \theta}$ (where both are defined)?

60. Which trigonometric function is the same as $\dfrac{1}{\csc \theta}$ (where both are defined)?

61. Find all values of t that satisfy $\sec t = \frac{1}{2}$. [*Hint:* Look at the graph of the secant function.]

62. Find $\dfrac{d}{dt}(\sec^2 t - \tan^2 t)$. [*Hint:* No work necessary—see Exercise 27.]

63. Find $\dfrac{d}{dt}(\csc^2 t - \cot^2 t)$. [*Hint:* No work necessary—see Exercise 28.]

64. What does the formula $\dfrac{d}{dx}\tan t = \sec^2 t$ tell you about the sign of the slope of the tangent function? Does the graph of the tangent function verify your observation?

65. What does the formula $\dfrac{d}{dx}\cot t = -\csc^2 t$ tell you about the sign of the slope of the cotangent function?

Does the graph of the cotangent function verify your observation?

66. Find $\displaystyle\int_{-\pi/4}^{\pi/4} \tan t \, dt$ [*Hint:* No work necessary—look at the graph of the tangent function.]

Explorations and Excursions   The following problems extend and augment the material presented in the text.

More on Derivatives and Integrals of Trigonometric Functions

67. Why do $\tan^2 t$ and $\sec^2 t$ have the same derivative?

68. a. Derive the formula $\dfrac{d}{dt}\cot t = -\csc^2 t$ by writing $\cot t = \dfrac{\cos t}{\sin t}$ and differentiating by the Quotient Rule.

  b. Verify this formula on a graphing calculator by entering $y_1 = \cot x$ [entered as $(\tan x)^{-1}$], graphing its derivative (using NDERIV), and observing that the result is the negative of the graph of $\csc^2 x$ found in Exercise 24.

69. a. Derive the formula $\dfrac{d}{dt}\sec t = \sec t \tan t$ by writing $\sec t = (\cos t)^{-1}$ and differentiating by the Generalized Power Rule.

  b. Verify this formula on a graphing calculator by entering $y_1 = \sec x$ [entered as $(\cos x)^{-1}$], graphing its derivative (using NDERIV), and observing that the result is the same as the graph of $\sec x \tan x$ found in Exercise 25.

70. a. Derive the formula $\dfrac{d}{dt}\csc t = -\csc t \cot t$ by writing $\csc t = (\sin t)^{-1}$ and differentiating by the Generalized Power Rule.

  b. Verify this formula on a graphing calculator by entering $y_1 = \csc x$ [entered as $(\sin x)^{-1}$], graphing its derivative (using NDERIV), and observing that the result is the negative of the graph of $\csc x \cot x$ found in Exercise 26.

71. Find the integral $\displaystyle\int \tan t \sec^2 t \, dt$ in two ways:
  a. Using the substitution method with $u = \tan t$.
  b. Using the substitution method with $u = \sec t$.
  c. Can you reconcile the two seemingly different answers?

72. Find the integral $\displaystyle\int \cot t \csc^2 t \, dt$ in two ways:
  a. Using the substitution method with $u = \cot t$.
  b. Using the substitution method with $u = \csc t$.
  c. Can you reconcile the two seemingly different answers?

73. Verify the formula $\displaystyle\int \tan t \, dt = -\ln |\cos t| + C$ in the box on page 576.
  [*Hint:* Use $\tan t = \dfrac{\sin t}{\cos t}$ and integrate by the substitution method with $u = \cos t$.]

74. Verify the formula $\displaystyle\int \cot t \, dt = \ln |\sin t| + C$ in the box on page 576. [*Hint:* Use $\cot t = \dfrac{\cos t}{\sin t}$ and integrate by the substitution method with $u = \sin t$.]

75. Verify the formulas $\displaystyle\int \tan t \, dt = \ln |\sec t| + C$ and $\displaystyle\int \cot t \, dt = -\ln |\csc t| + C$ which may serve as alternates to the first two formulas in the box on page 576. [*Hint:* Replace sec t and csc t by their definitions in terms of cos t and sin t and use the properties of naturals logarithms to show that these formulas agree with those on page 576.

76. Verify the formula $\displaystyle\int \sec t \, dt = \ln |\sec t + \tan t| + C$ from page 576. [*Hint:* Multiply the integrand by $\dfrac{\sec t + \tan t}{\sec t + \tan t}$ and integrate by the substitution method.]

77. Verify the formula $\displaystyle\int \csc t \, dt = \ln |\csc t - \cot t| + C$ from page 576. [*Hint:* Multiply the integrand by $\dfrac{\csc t - \cot t}{\csc t - \cot t}$ and integrate by the substitution method.]

78. Show that $\displaystyle\int_0^1 \dfrac{1}{x^2 + 1} \, dx = \dfrac{\pi}{4}$ by making the substitution $x = \tan t$ so that $x^2 + 1 = \tan^2 t + 1 = \sec^2 t$ (see Exercise 27), $dx = \sec^2 t \, dt$ and the limits of integration $x = 0$ to $x = 1$ become $t = 0$ to $t = \pi/4$.

8 Chapter Summary with Hints and Suggestions

Reading the text and doing the exercises in this chapter have helped you master the following concepts and skills, which are listed by section (in case you need to review them) and are keyed to particular Review Exercises. Answers for all Review Execises are given at the back of the book, and full solutions can be found in the Student Solutions Manual.

8.1 Triangles, Angles, and Radian Measure

- Convert angle measures between degrees and radians. *(Review Exercises 1–2.)*

$$\frac{\text{Degrees}}{180} = \frac{\text{Radians}}{\pi}$$

- Calculate the length of a section of a circle. *(Review Exercises 3–4.)*

$$\left(\begin{array}{c}\text{Arc} \\ \text{length}\end{array}\right) = (\text{Radians})(\text{Radius})$$

8.2 Sine and Cosine Functions

- Find the exact value of a sine or cosine function. *(Review Exercises 5–6.)*

$$\sin t = \frac{y}{r}$$

$$\cos t = \frac{x}{r}$$

- Sketch the graph of a modified sine or cosine function. *(Review Exercises 7–8.)*

- Solve an applied problem involving a trigonometric function. *(Review Exercises 9–12.)*

8.3 Derivatives of Sine and Cosine Functions

- Find the derivative of a sine or cosine function. *(Review Exercises 13–20.)*

$$\frac{d}{dt}\sin t = \cos t \qquad \frac{d}{dt}\cos t = -\sin t$$

- Find the tangent line to a curve at a given point. *(Review Exercise 21)*

- Solve an applied problem by differentiating a sine or cosine function. *(Review Exercises 22–27.)*

8.4 Integrals of Sine and Cosine Functions

- Find the indefinite integral of a sine or cosine function. *(Review Exercises 28–37.)*

$$\int \sin t \, dt = -\cos t + C$$

$$\int \cos t \, dt = \sin t + C$$

- Find the definite integral of a sine or cosine function. *(Review Exercises 38–39.)*

- Find the average value of a sine or cosine function on an interval. *(Review Exercises 40–41.)*

- Find an area involving a sine or cosine function. *(Review Exercises 42–43.)*

- Solve an applied problem by integrating a sine or cosine function. *(Review Exercises 44–45.)*

8.5 Other Trigonometric Functions

- Find the exact value of one of the other trigonometric functions. *(Review Exercises 46–47.)*

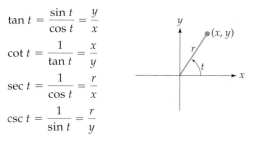

$$\tan t = \frac{\sin t}{\cos t} = \frac{y}{x}$$

$$\cot t = \frac{1}{\tan t} = \frac{x}{y}$$

$$\sec t = \frac{1}{\cos t} = \frac{r}{x}$$

$$\csc t = \frac{1}{\sin t} = \frac{r}{y}$$

- Solve an applied problem involving one of the other trigonometric functions. *(Review Exercises 48–49.)*

- Find the derivative of one of these other functions. *(Review Exercises 50–55.)*

$$\frac{d}{dt}\tan t = \sec^2 t$$

$$\frac{d}{dt}\cot t = -\csc^2 t$$

$$\frac{d}{dt}\sec t = \sec t \tan t$$

$$\frac{d}{dt}\csc t = -\csc t \cot t$$

- Find the tangent line to a curve at a given point. *(Review Exercise 56.)*

- Find the indefinite integral of one of the other trigonometric functions. *(Review Exercises 57–62.)*

- Find the definite integral of one of the other trigonometric functions. *(Review Exercises 63–64.)*

- Find the area under a trigonometric curve. *(Review Exercises 65–66.)*

Sections 8.3, 8.4, and 8.5

- Solve a more involved graphing, differentiation, integration, or numerical integration problem involving a trigonometric function. *(Review Exercises 67–86.)*

Hints and Suggestions

- We measure angles in *radians* because it simplifies the differentiation and integration formulas. All equivalent measures can be found by using the equation $180° = \pi$ radians.

- Think of π as a natural unit for radian measure, just as it occurs naturally in the formulas for the circumference and area of a circle.

- The "triangle" definitions such as $\sin t = \frac{\text{opp}}{\text{hyp}}$

    hold only for *acute* angles, while the "coordinate"

    definitions such as $\sin t = \frac{y}{r}$ hold for *all* angles.

- We can find $\sin t$ and $\cos t$ *exactly* for some

    angles, such as $\frac{\pi}{6}, \frac{\pi}{4}$, and $\frac{\pi}{3}$, using $30°-60°-90°$

and $45°-45°-90°$ triangles. For other angles we approximate the value using a calculator.

- Try not to think of $\sin t$ and $\cos t$ as functions of *angles,* but as ordinary "numerical" functions, so that you can use them to model any quantity that varies periodically, such as a business cycle. Each has period 2π, but $\sin t$ begins at the origin and $\cos t$ begins one unit up.

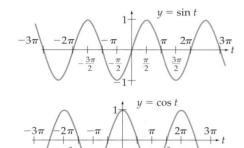

- The uses of differentiation remain the same (slopes, rates of change, marginals, and velocities) and the uses of integration remain the same (areas, total accumulations, and average values), but we can now apply these ideas to *periodic* behavior.

- The derivative formulas for the other trigonometric functions can be derived from those for $\sin t$ and $\cos t$ by using the Quotient Rule.

- A graphing calculator helps by graphing trigonometric functions and finding derivatives (for optimization) and definite integrals (for areas, total accumulations, and average values).

---

8   Review Exercises and Chapter Test   ◯ indicate a Chapter Test exercise.

8.1   Triangles, Angles, and Radian Measure

① **a.** Convert $150°$ into radians.

 b. Convert $\frac{\pi}{9}$ radians into degrees.

2. a. Convert $120°$ into radians.

 b. Convert $\frac{\pi}{5}$ radians into degrees.

3–4. GENERAL: Largest Clock The largest single-faced clock in the world is the Colgate clock on the Hudson River in Jersey City, New Jersey, with a minute hand that is 26 feet long. Find the arc length through which the tip of the minute hand travels between the following times.

[*Hint:* Find the angle in radians. Give an approximate answer using $\pi \approx 3.14$.]

 3. Between noon and 12:40 p.m.

 4. Between 3 p.m. and 3:30 p.m.

8.2   Sine and Cosine Functions

5–6. Evaluate *without* using a calculator (leaving answers in *exact* form).

⑤ **a.** $\sin \dfrac{2\pi}{3}$ **b.** $\cos \dfrac{2\pi}{3}$

 c. $\sin \dfrac{5\pi}{4}$ **d.** $\cos \dfrac{5\pi}{4}$

(See instructions on previous page.)

6. a. $\sin \dfrac{7\pi}{6}$ **b.** $\cos \dfrac{7\pi}{6}$

  c. $\sin \dfrac{3\pi}{4}$ **d.** $\cos \dfrac{3\pi}{4}$

7–8. Sketch the graph of each function. (If you use a graphing calculator, make a hand-drawn sketch showing the amplitude and period.)

7 $y = 7 \cos 2x$ **8.** $y = 6 \sin 2x$

9 **GENERAL: Building Construction** A roof is to make an angle of 30° with the horizontal, and rise 12 feet in the middle. Find the length of the rafters.

10. GENERAL: Water Depth A 40-foot anchor line makes an angle of 30° with the surface. How deep is the water?

11. GENERAL: Daylight The number of hours of daylight in Seattle, Washington, on day x of the year is approximated by the function

$$D(x) = 12.25 + 3.75 \sin \frac{2\pi(x - 81)}{365}$$

a. Graph this function on the window [0, 365] by [0, 20]. (Be sure your calculator is set for *radians*.)
b. Find the number of hours of daylight on May 1 (day 121).
c. Find the number of hours of daylight on December 1 (day 335).

Source: Ecological Modeling 80

12. GENERAL: Daylight *(continuation)* For the "daylight" function in Exercise 11, find the days with the largest and smallest amounts of daylight. [*Hint:* Use MAXIMUM and MINIMUM.]

8.3 Derivatives of Sine and Cosine Functions

13–20. Find the derivative of each function.

13. $f(t) = 2 \cos(4t - 1)$ **14.** $f(t) = \sin(2 - 3t)$

15 $f(x) = x^3 \sin x + 3x^2 \cos x - 6x \sin x - 6 \cos x$

16. $f(t) = \dfrac{1 + \sin t}{1 + \cos t}$

17. $f(t) = \sin^3 t^2 - \cos \dfrac{\pi}{3}$ **18.** $f(x) = \sin e^{x^2}$

19 $f(t) = \sqrt{\sin t}$ **20.** $f(t) = \sin \sqrt{t}$

21 Find the equation for the tangent line to the curve $y = e^{\cos x}$ at $x = \frac{\pi}{2}$.

22 **BUSINESS: Seasonal Sales** A manufacturer estimates that the number of overcoats sold during week t of the year is

$$S(t) = 250 + 250 \cos \frac{\pi t}{26}$$

a. Graph this function on the window [0, 52] by [−100, 600].
b. Find the rate of change of sales at week $t = 37$ (mid September).

23. BUSINESS: Seasonal Sales *(continuation)* For the sales function in Exercise 22, find when the rate of increase in sales is the greatest. [*Hint:* Use NDERIV and MAXIMUM.]

24. GENERAL: Temperature The normal mean temperature in St. Louis, Missouri, on day x of the year is approximately

$$T(x) = 50 - 26 \cos \frac{2\pi(x - 15)}{365}$$

degrees Fahrenheit.

a. Graph this function on the window [0, 365] by [0, 90].
b. Find the rate of change of temperature on October 15 (day 288).

Source: National Climatic Data Center

25. GENERAL: Temperature *(continuation)* For the temperature function in Exercise 24, find when the rate of increase is the greatest. [*Hint:* Use NDERIV and MAXIMUM.]

(26) **GENERAL: Maximizing Volume** An enclosure made by leaning a sheet of plywood against a wall at an angle θ with the ground will have a volume of $V(\theta) = 128 \sin \theta \cos \theta$ cubic feet. Find the angle θ that maximizes the volume $V(\theta)$.

27. BIOMEDICAL: Vascular Branching A surgeon wants to attach a new artery to an existing one at the angle θ that minimizes resistance of blood flow.

Under certain conditions, minimizing the resistance is equivalent to minimizing the function

$$f(\theta) = \frac{r^{-4} - R^{-4} \cos \theta}{\sin \theta}$$

Derive the relationship $\cos \theta = r^4/R^4$ by setting the derivative f' equal to zero and solving.

Source: E. Batschelet, *Introduction to Mathematics for Life Scientists*

8.4 Integrals of Sine and Cosine Functions

28–37. Find each integral.

(28) $\displaystyle\int \cos \frac{\pi t}{5} \, dt$

29. $\displaystyle\int t \sin(t^2 + 1) \, dt$

30. $\displaystyle\int \sin(2\pi - t) \, dt$

31. $\displaystyle\int e^{-x} \sin e^{-x} \, dx$

32. $\displaystyle\int e^{\cos x} \sin x \, dx$

33. $\displaystyle\int \frac{\cos t}{\sqrt{\sin t}} \, dt$

(34) $\displaystyle\int \frac{\cos 3t}{1 + \sin 3t} \, dt$

35. $\displaystyle\int (x^3 - e^{-x} - \cos \pi x) \, dx$

36. $\displaystyle\int \frac{\cos(\ln y)}{y} \, dy$

37. $\displaystyle\int \frac{\sin \frac{1}{z}}{z^2} \, dz$

38–39. For each definite integral:

a. Evaluate it "by hand."

b. Check your answer to part (a) by evaluating using a graphing calculator.

38. $\displaystyle\int_{\pi/6}^{\pi/2} \sin 2t \, dt$

39. $\displaystyle\int_{\pi/6}^{\pi} \sin^2 t \cos t \, dt$

(40) Find the average value of $f(t) = \sin \pi t$ on the interval $[0, 3]$. (Give an exact and also an approximate answer.)

41. Find the average value of $f(t) = 200 + 100 \cos \pi t$ on the interval $[0, 3]$.

(42) Find the area under the curve $y = \sin^2 x \cos x$ and above the x-axis from $x = 0$ to $x = \dfrac{\pi}{2}$.

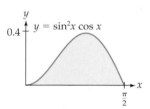

43. Find the area between the curves $y = \cos x$ and $y = \sin x$ from $x = 0$ to $x = \dfrac{\pi}{4}$, as shown below. (Give the answer in *exact* form and also approximate it.)

(44) **BUSINESS: Total Sales** A boatyard estimates that sales during week x of the year will be

$$S(x) = 25 - 20 \cos \frac{\pi x}{26}$$

thousand dollars, where $x = 0$ corresponds to the beginning of the year.

a. Graph the sales function on $[0, 52]$ by $[-10, 50]$.
b. Find the total sales during the first half of the year ($x = 0$ to $x = 26$).

45. GENERAL: Average Temperature The temperature at time x hours in a warehouse is given by the formula

$$T(x) = 60 - 10 \sin \frac{\pi x}{12}$$

degrees Fahrenheit.

a. Graph the function on $[0, 24]$ by $[0, 70]$.
b. Find the average temperature during the first 6 hours.

8.5   Other Trigonometric Functions

46–47. Evaluate *without* using a calculator (leaving answers in *exact* form).

(46) a. $\tan \dfrac{\pi}{3}$ b. $\cot \dfrac{\pi}{3}$

c. $\sec \dfrac{\pi}{3}$ d. $\csc \dfrac{\pi}{3}$

47. a. $\tan \dfrac{3\pi}{4}$ b. $\cot \dfrac{3\pi}{4}$

c. $\sec \dfrac{3\pi}{4}$ d. $\csc \dfrac{3\pi}{4}$

48. GENERAL: Calculating Height Estimate the height of a monument if its top, from 600 feet away, makes an angle of $40°$ with the ground.

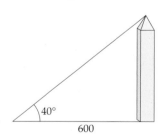

49. GENERAL: Calculating Distance Estimate the distance from a boat to an 80-foot-tall lighthouse if the top of the lighthouse makes an angle of $20°$ with the water.

50–54. Differentiate each function.

(50) $f(x) = x \tan x^2$ **51.** $f(x) = x^2 \cot x$

52. $f(t) = \csc(\pi - t)$ **53.** $f(t) = \sec \sqrt{t}$

54. $f(t) = e^{\tan \pi t}$

55. Why do $\cot^2 t$ and $\csc^2 t$ have the same derivative?

56. Find the equation for the tangent line to the curve $y = \ln(2 \tan x - 1)$ at $x = \dfrac{\pi}{4}$.

57–62. Find each integral.

(57) $\displaystyle\int \sec^2 \frac{\pi t}{2} \, dt$ **58.** $\displaystyle\int x \tan x^2 \, dx$

59. $\displaystyle\int t^2 \sec t^3 \tan t^3 \, dt$

60. $\displaystyle\int \csc(\pi - t) \cot(\pi - t) \, dt$

61. $\displaystyle\int \sqrt{\cot x} \csc^2 x \, dx$ **62.** $\displaystyle\int \frac{\sec^2 x}{\sqrt{\tan x}} \, dx$

63–64. For each definite integral:

a. Evaluate it "by hand."

b. Check your answer to part (a) by evaluating using a graphing calculator.

63. $\displaystyle\int_0^{\pi/4} \sec t \, dt$ **64.** $\displaystyle\int_{\pi/4}^{\pi/2} \csc^2 t \, dt$

(65) a. **AREA UNDER CURVE** Find "by hand" the area under the curve $y = \cot x$ and above the x-axis from $x = \dfrac{\pi}{4}$ to $x = \dfrac{\pi}{2}$.

b. Verify your answer to part (a) using a graphing calculator.

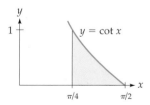

66. a. **AREA UNDER CURVE** Find "by hand" the area under the curve $y = \csc x$ and above the x-axis from $x = \dfrac{\pi}{4}$ to $x = \dfrac{\pi}{2}$.

b. Verify your answer to part (a) using a graphing calculator.

Additional Topics

67–68. GRAPHING TRIGONOMETRIC FUNCTIONS

Graph each function for $0 \le x \le 2\pi$ (using an appropriate range of y-values). Make a hand-drawn sketch, showing all relative extreme points.

67. $f(x) = \sin^2 x$ **68.** $f(x) = \sin x + \cos x$

69–70. IMPLICIT DIFFERENTIATION For each equation:

a. Find dy/dx by implicit differentiation.
b. Evaluate dy/dx at the given point.

69. $\sin 2x + \cos y = 0$

Evaluate dy/dx at $\left(\dfrac{\pi}{2}, \dfrac{\pi}{2} \right)$.

70. $\sin y - \cos 4x = 0$

Evaluate dy/dx at $\left(\dfrac{\pi}{8}, 0 \right)$.

71. IMPLICIT DIFFERENTIATION: Radar Tracking A helicopter, rising vertically at the rate of 10 feet per second, is tracked by a radar station 200 feet away, as shown in the diagram. How fast is the angle θ of the radar beam changing at the moment when the angle is $\dfrac{\pi}{3}$?

$$\left[Hint: \quad \tan \theta = \frac{y}{200}. \right]$$

$$\frac{dy}{dt} = 10$$

72. IMPLICIT DIFFERENTIATION: Radar Tracking
A plane is flying at a constant altitude of 5 miles toward a radar station at the rate of 450 miles per hour, as shown in the diagram in the next column. How fast is the angle θ of the radar beam changing at the moment when the angle is $\dfrac{\pi}{6}$?

$$\left[Hint: \quad \cot \theta = \frac{x}{5}. \right]$$

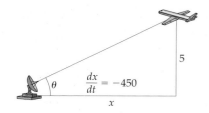

$$\frac{dx}{dt} = -450$$

73–76. FUNCTIONS OF SEVERAL VARIABLES For each function, find the partials:

a. $f_x(x, y)$ and   **b.** $f_y(x, y)$.

73. $f(x, y) = \sin 2x - y \cos x$

74. $f(x, y) = \ln \cos(x - y)$

75. $f(x, y) = (\sin x)e^{\cos y}$

76. $f(x, y) = \sin(x^3 y)$

77–82. INTEGRATION BY PARTS Evaluate each integral by integration by parts.

77. $\displaystyle \int t \sin t \, dt$ **78.** $\displaystyle \int t \cos t \, dt$

79. $\displaystyle \int x \sec^2 x \, dx$ **80.** $\displaystyle \int x \sec x \tan x \, dx$

81. $\displaystyle \int e^t \sin t \, dt$ **82.** $\displaystyle \int e^t \cos t \, dt$

[*Hint for Exercises* 81 *and* 82: Integrate by parts twice and solve the resulting equation for the original integral.]

83–84. NUMERICAL INTEGRATION Approximate each integral using trapezoidal approximation with $n = 10$, $n = 100$, and $n = 500$ trapezoids. Give the most accurate approximation that your answers will support.

83. $\displaystyle \int_0^{\pi} e^{\sin x} \, dx$ **84.** $\displaystyle \int_0^{\pi/2} \sqrt{\sin x} \, dx$

85–86. NUMERICAL INTEGRATION Approximate each integral using Simpson's Rule with $n = 10$, $n = 100$, and $n = 500$. Give the most accurate approximation that your answers will support.

85. $\displaystyle \int_0^{\pi} e^{\sin x} \, dx$ **86.** $\displaystyle \int_0^{\pi/2} \sqrt{\sin x} \, dx$

Differential Equations

9

© Zadorozhnyi Viktor/Shutterstock.com

What You'll Explore

Differential equations provide a very powerful method for modeling almost any quantity based on its rate of growth. In this chapter you will model the value of assets, the spread of epidemics, the concentration of air pollution, and the growth of personal wealth, to name only a few. You will learn several of the most widely used methods for solving these differential equations, and even a method for approximating their solutions if they cannot be found exactly.

9.1 Separation of Variables

9.2 Further Applications of Differential Equations: Three Models of Growth

9.3 First-Order Linear Differential Equations

9.4 Approximate Solutions of Differential Equations: Euler's Method

APPLICATION PREVIEW

Personal Wealth and Differential Equations

Like most people, you probably have (or soon will have) income that comes from two sources: salary and investments (such as interest on savings). From this income, you pay "necessary" expenses (housing and food), spend some of the remainder on "luxuries" (vacations and entertainment), and save the rest (increasing your investments). Given some simplifying assumptions, we can construct a mathematical model for your wealth at any time t. Suppose that after paying for necessities, you spend a fixed percentage (typically 75%) of the remainder on luxuries, and that your investments generate income at a fixed rate. Let

s = your salary

$W(t)$ = your wealth (savings and investments), which is a function of time t

r = rate of interest on your wealth

n = amount spent on necessities

p = proportion of your income after necessities that you spend on luxuries

A model of your wealth is then given by the following "differential equation," which states that *income* minus *outflow* is what you can add to savings:

$$\underbrace{\frac{dW}{dt}}_{\substack{\text{Change} \\ \text{in wealth}}} = \overbrace{\underbrace{s}_{\text{Salary}} + \underbrace{r \cdot W}_{\substack{\text{Interest} \\ \text{on wealth}}}}^{\text{Income}} - \overbrace{(\underbrace{n}_{\text{Necessities}} + \underbrace{p(s + r \cdot W - n))}_{\substack{\text{Luxuries (a fixed} \\ \text{proportion of income} \\ \text{after necessities)}}}}^{\text{Outflow}}$$

With a little algebra, this "differential equation" can be written

$$\frac{dW}{dt} = (1 - p)(s - n + r \cdot W)$$

The *solution* to this differential equation (assuming zero initial wealth) can be found by the methods of this chapter, giving your "wealth" function shown below and graphed on the left:

$$W(t) = \frac{s - n}{r}(e^{(1 - p)rt} - 1)$$

Given numbers for salary, necessities, interest rate, and proportion spent on luxuries, this solution could be used to predict when your wealth will reach a particular value (for example, enough to buy a house). It could also be used to show how your wealth would change if you lowered your "luxury proportion" from, say, 75% to 70% (see Exercise 71 on page 603).

This is but one illustration of the wide variety of applications of differential equations, which are discussed in this and the following sections.

9.1   Separation of Variables

Introduction

A **differential equation** is simply an equation involving derivatives. In this section we will solve differential equations from business and economics, biology and medicine, and other fields, using a technique called *separation of variables*. We will denote the quantity being studied as $y(x)$ or sometimes just y, denoting its derivative by either y' or $\dfrac{dy}{dx}$.

Differential Equation $y' = f(x)$

We have actually been solving differential equations since the beginning of Chapter 5. For example, the differential equation

$$y' = 2x$$

saying that the derivative of a function is $2x$, is solved simply by integrating:

$$y = \int 2x\,dx = x^2 + C \qquad\qquad y = x^2 + C \text{ satisfies } y' = 2x$$

In general, a differential equation of the form

$$y' = f(x)$$

is solved by integrating:

$$y = \int f(x)\,dx$$

Therefore, whenever we integrate a marginal cost function to find cost, or integrate a rate to find the total accumulation, we are solving a differential equation of the form $y' = f(x)$.

General and Particular Solutions

The solution of the differential equation $y' = 2x$ is $y = x^2 + C$, with an arbitrary constant C. We call $y = x^2 + C$ the **general solution** because taking all possible values of the constant C gives *all* solutions of the differential equation. If we take C to be a particular number, we get a **particular solution**. Some particular solutions of the differential equation $y' = 2x$ are

$$y = x^2 + 2 \qquad \text{(taking } C = 2)$$
$$y = x^2 \qquad \text{(taking } C = 0)$$
$$y = x^2 - 2 \qquad \text{(taking } C = -2)$$

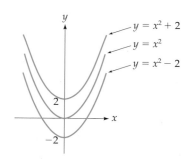

Three particular solutions from the general solution $y = x^2 + C$.

The different values of the arbitrary constant C give a "family" of curves, and the general solution $y = x^2 + C$ may be thought of as the entire family.

The solution of a differential equation is a *function.* The *general* solution contains an arbitrary constant, and a *particular* solution has this constant replaced by a particular number.

Verifying Solutions

Verifying that a function is a solution of a differential equation is simply a matter of calculating the necessary derivatives, substituting them into the differential equation, and checking that the two sides of the equation are equal.

EXAMPLE 1   **VERIFYING A SOLUTION OF A DIFFERENTIAL EQUATION**

Verify that $y = e^{2x} + e^{-x} - 1$ is a solution of the differential equation

$$y'' - y' - 2y = 2$$

Solution

The differential equation involves y, y', and y'', so we calculate

$y = e^{2x} + e^{-x} - 1$	Given function
$y' = 2e^{2x} - e^{-x}$	Derivative
$y'' = 4e^{2x} + e^{-x}$	Second derivative

Then we substitute these into the differential equation.

$$y'' - y' - 2y = 2$$

$$(4e^{2x} + e^{-x}) - (2e^{2x} - e^{-x}) - 2(e^{2x} + e^{-x} - 1) \stackrel{?}{=} 2 \qquad \stackrel{?}{=} \text{ means the equation may not be true}$$

$$4e^{2x} + e^{-x} - 2e^{2x} + e^{-x} - 2e^{2x} - 2e^{-x} + 2 \stackrel{?}{=} 2 \qquad \text{Expanding}$$

$$\cancel{4e^{2x}} + \cancel{e^{-x}} - \cancel{2e^{2x}} + \cancel{e^{-x}} - \cancel{2e^{2x}} - \cancel{2e^{-x}} + 2 \stackrel{?}{=} 2 \qquad \text{Canceling}$$

$$2 = 2 \quad \checkmark \qquad \text{It checks!}$$

Since the equation is satisfied, the given function is indeed a solution of the differential equation.

If the two sides had *not* turned out to be exactly the same, the given function y would *not* have been a solution of the differential equation.

PRACTICE PROBLEM 1

Verify that $y = e^{-x} + e^{3x}$ is a solution of the differential equation

$$y'' - 2y' - 3y = 0$$

Solution on page 599 >

The differential equations in Example 1 and Practice Problem 1 are called *second-order* differential equations because they involve second derivatives but no higher-order derivatives. From here on we will restrict our attention to **first-order** differential equations—that is, differential equations involving only the *first* derivative. Many first-order differential equations can be solved by a method called **separation of variables**.

Separation of Variables

A differential equation is said to be **separable** if the variables can be "separated" by moving every x and dx to one side of the equation and every y and dy to the other side. We may then solve the differential equation by integrating both sides. Several examples will make the method clear.

EXAMPLE 2   **FINDING A GENERAL SOLUTION**

Find the general solution of the differential equation $\dfrac{dy}{dx} = 2xy^2$.

Solution

$$dy = 2xy^2\, dx$$

Multiplying both sides of the differential equation by dx

$$\frac{dy}{y^2} = 2x\, dx$$

Dividing each side by y^2 ($y \neq 0$); the variables are now separated: every y on one side, every x on the other

$$y^{-2}\, dy = 2x\, dx$$

In power form

$$\int y^{-2}\, dy = \int 2x\, dx$$

Integrating both sides

$$-y^{-1} = x^2 + C$$

Using the Power Rule (writing one C for both integrations)

$$\frac{1}{y} = -x^2 - C$$

Writing y^{-1} as $\dfrac{1}{y}$ and multiplying by -1

$$y = \frac{1}{-x^2 - C}$$

Taking reciprocals of both sides

This is the general solution of the differential equation. The solution may be left in this form, but if we replace the arbitrary constant C by $-c$, another constant but with the opposite sign, this solution may be written

$$y = \frac{1}{-x^2 + c}$$

or, slightly shorter,

$$y = \frac{1}{c - x^2}$$

Reversing the order, giving the general solution of the differential equation

---

The differential equation $y' = 2xy^2$ has another solution: the function that is identically zero, $y(x) \equiv 0$. This is known as a **singular solution** and cannot be obtained from the general solution. We will not consider singular solutions further in this book except to say that $y = 0$ will generally be a solution whenever separating the variables results in dy being divided by a positive power of y, as in the previous Example.

Separable Differential Equations

A first-order differential equation is *separable* if it can be written in the following form for some functions $f(x)$ and $g(y)$:

$$\frac{dy}{dx} = \frac{f(x)}{g(y)} \qquad \text{for } g(y) \neq 0$$

It is solved by separating variables and integrating:

$$\int g(y)\, dy = \int f(x)\, dx \qquad \text{Multiplying by } dx \text{ and } g(y) \text{ and integrating}$$

The preceding Example asked for the *general* solution of a differential equation. Sometimes we will be given a differential equation together with some additional information that selects a *particular* solution from the general solution, information that determines the value of the arbitrary constant in the general solution. This additional information is called an **initial condition.**

EXAMPLE 3   FINDING A PARTICULAR SOLUTION

Solve the differential equation $y' = \dfrac{6x}{y^2}$ with the initial condition $y(1) = 2$.

Solution

First we find the general solution by separating the variables.

$$\frac{dy}{dx} = \frac{6x}{y^2} \qquad \text{Replacing } y' \text{ by } \frac{dy}{dx}$$

$$y^2\, dy = 6x\, dx \qquad \begin{array}{l} \text{Multiplying both sides} \\ \text{by } dx \text{ and } y^2 \text{ (the} \\ \text{variables are separated)} \end{array}$$

$$\int y^2\, dy = \int 6x\, dx \qquad \text{Integrating both sides}$$

$$\frac{1}{3} y^3 = 3x^2 + C \qquad \text{Using the Power Rule}$$

$$y^3 = 9x^2 + \underbrace{3C}_{c} \qquad \text{Multiplying by 3}$$
$$ \quad \begin{array}{l}\text{(3 times a constant is} \\ \text{just another constant)}\end{array}$$

$$y^3 = 9x^2 + c \qquad \text{Replacing } 3C \text{ by } c$$

$$y = \sqrt[3]{9x^2 + c} \qquad \text{Taking cube roots}$$

This is the general solution, with arbitrary constant c. The initial condition $y(1) = 2$ says that $y = 2$ when $x = 1$. We substitute these values into the general solution $y = \sqrt[3]{9x^2 + c}$ and solve for c.

$$2 = \sqrt[3]{9 + c} \qquad y = \sqrt[3]{9x^2 + c} \text{ with } x = 1 \text{ and } y = 2$$

$$8 = 9 + c \qquad \text{Cubing each side}$$

$$-1 = c \qquad \text{Solving for } c \text{ gives } c = -1$$

Therefore, we replace c by -1 in the general solution to obtain the particular solution:

$$y = \sqrt[3]{9x^2 - 1} \qquad y = \sqrt[3]{9x^2 + c} \text{ with } c = -1$$

This solution $y = \sqrt[3]{9x^2 - 1}$ with $x = 1$ gives $y = \sqrt[3]{9 - 1} = \sqrt[3]{8} = 2,$ so the initial condition $y(1) = 2$ is indeed satisfied. We could also verify that the differential equation is satisfied by substituting this solution into it.

In general, solving a differential equation with an initial condition just means finding the general solution and then using the initial condition to evaluate the constant. Several particular solutions of the differential equation $y' = 6x/y^2$ are shown below, with the solution $y = \sqrt[3]{9x^2 - 1}$ shown in color.

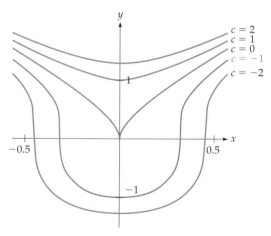

The solution $y = \sqrt[3]{9x^2 + c}$
for various values of c.

Slope field for $\dfrac{dy}{dx} = \dfrac{6x}{y^2}$ on
$[-0.6, 0.6]$ by $[-1.5, 1.5]$

Graphing Calculator Exploration

A graphing calculator can help you to see what a differential equation is saying. The differential equation in the preceding Example,

$$\frac{dy}{dx} = \frac{6x}{y^2} \qquad\qquad \text{Slope equals } \frac{6x}{y^2}$$

says that the slope at any point (x, y) in the plane is given by the formula $6x/y^2$. Graphing calculator programs that draw *slope fields*, many small line segments with the slopes specified by the differential equation, are easily available on the Internet, from sites such as ticalc.org (search for *slope field*). Typical screens are shown on the left. The first screen shows the slope field of the above differential equation, and below it is the same slope field with the particular solution found in the preceding Example.

The same slope field with
the solution $y = \sqrt[3]{9x^2 - 1}$

Such programs show the slope field of any differential equation that can be written in the form $y' = f(x,y)$.

Look at the above graph of the five solutions and see how each of them follows the prescribed slopes in the slope field. In fact, starting at *any* point in the plane you could draw a curve following the indicated slopes, thereby geometrically constructing a *solution curve*.

PRACTICE PROBLEM 2

Solve the differential equation
and initial condition

$$\begin{cases} y' = \dfrac{6x^2}{y^4} \\ y(0) = 2 \end{cases}$$

Solution on page 599 >

Recall that to solve for y in the logarithmic equation

$$\ln y = f(x)$$

we exponentiate both sides and simplify

$$y = e^{f(x)}$$

Using $e^{\ln y} = y$ on the left side

This idea will be useful in the next Example.

EXAMPLE 4   **FINDING A PARTICULAR SOLUTION**

Solve the differential equation and initial condition $\begin{cases} \dfrac{dy}{dx} = xy \\ y(0) = 2 \end{cases}$

Solution

$$dy = xy\, dx$$

Multiplying by dx

$$\frac{dy}{y} = x\, dx$$

Dividing by y ($y \neq 0$). The variables are separated

$$\int \frac{dy}{y} = \int x\, dx$$

Integrating

$$\ln y = \frac{1}{2} x^2 + C$$

Evaluating the integrals (assuming $y > 0$)

$$y = e^{\frac{1}{2}x^2 + C}$$

Solving for y by exponentiating

$$y = e^{\frac{1}{2}x^2} \cdot \underbrace{e^C}_{c}$$

e to a sum can be expressed as a product

e^C is a positive constant c

$$y = ce^{x^2/2}$$

Replacing e^C by c and $e^{\frac{1}{2}x^2}$ by $e^{x^2/2}$

This is the general solution. To satisfy the initial condition $y(0) = 2$, we substitute $y = 2$ and $x = 0$ and solve for c.

$$2 = ce^0$$

$y = ce^{x^2/2}$ with $y = 2$ and $x = 0$

$$2 = c$$

Since $e^0 = 1$

Substituting $c = 2$ into the general solution gives the particular solution

$$y = 2e^{x^2/2}$$

$y = ce^{x^2/2}$ with $c = 2$

We wrote the solution of the integral $\displaystyle\int \frac{dy}{y}$ as $\ln y$, without absolute value bars, thereby assuming that $y > 0$. *Keeping* the absolute values would have given $|y| = e^{\frac{1}{2}x^2 + C} = ce^{\frac{1}{2}x^2}$ so $y = \pm ce^{\frac{1}{2}x^2}$ where c is a positive constant, showing that the general solution is $y = ce^{x^2/2}$, just as before, but now where c is *any* (positive or negative) constant. We will make similar simplifying assumptions in similar situations.

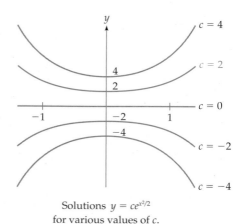

Solutions $y = ce^{x^2/2}$
for various values of c.

$c = 4$

$c = 2$

$c = 0$

$c = -2$

$c = -4$

Note that the constant that was *added* in the integration step became a constant *multiplier* in the solution $y = ce^{x^2/2}$. In general, the constant may appear *anywhere* in the solution. The diagram on the left shows the solutions $y = ce^{x^2/2}$ for various values of c, with the particular solution $y = 2e^{x^2/2}$ shown in color.

Graphing Calculator Exploration

The slope field for the differential equation in the preceding Example, found by entering $y_1 = xy$, setting the window to $[-2, 2]$ by $[-6, 6]$, and running a slope field program, is shown on the left below.

Slope field for $\dfrac{dy}{dx} = xy$

With the solution $y = 2e^{x^2/2}$

The screen on the right shows the same slope field with the particular solution found in the Example. Do you see how the other solutions graphed above follow these slopes?

PRACTICE PROBLEM 3

Solve the differential equation and initial condition

$$\begin{cases} \dfrac{dy}{dx} = x^2 y \\ y(0) = 5 \end{cases}$$

Solution on page 600 >

EXAMPLE 5   FINDING A GENERAL SOLUTION

Find the general solution of the differential equation

$$yy' - x = 0$$

Solution

$$y\frac{dy}{dx} = x$$ Replacing y' by dy/dx and moving the x to the other side

$$y\,dy = x\,dx$$ Multiplying by dx

$$\int y\,dy = \int x\,dx$$ Integrating

$$\frac{1}{2}y^2 = \frac{1}{2}x^2 + C \qquad \qquad \text{Using the Power Rule}$$

$$y^2 = x^2 + \underbrace{2C}_{c} \qquad \qquad \text{Multiplying by 2}$$

To solve for y we take the square root of each side. However, there are *two* square roots, one positive and one negative.

$$y = \sqrt{x^2 + c} \qquad \text{and} \qquad y = -\sqrt{x^2 + c}$$

These two solutions together are the general solution:

$$y = \pm\sqrt{x^2 + c}$$

For some differential equations, the integration step requires a substitution.

EXAMPLE 6   FINDING A PARTICULAR SOLUTION USING A SUBSTITUTION

Solve the differential equation and initial condition $\begin{cases} y' = xy - x \\ y(0) = 4 \end{cases}$

Solution

$$\frac{dy}{dx} = xy - x \qquad \qquad \text{Replacing } y' \text{ by } \frac{dy}{dx}$$

$$\frac{dy}{dx} = x(y - 1) \qquad \qquad \text{Factoring (to separate variables)}$$

$$\frac{dy}{y - 1} = x\, dx \qquad \qquad \text{Dividing by } y - 1 \text{ and multiplying by } dx$$

$$\int \frac{dy}{y - 1} = \int x\, dx \qquad \qquad \text{Integrating}$$

$$\begin{array}{l} u = y - 1 \\ du = dy \end{array} \qquad \qquad \text{Using a substitution}$$

$$\int \frac{du}{u} = \int x\, dx \qquad \qquad \text{Substituting}$$

$$\ln u = \frac{1}{2}x^2 + C \qquad \qquad \text{Integrating (assuming } u > 0)$$

$$\ln(y - 1) = \frac{1}{2}x^2 + C \qquad \qquad \text{Substituting back to } y \text{ using } u = y - 1$$

$$y - 1 = e^{\frac{1}{2}x^2 + C} \qquad \qquad \text{Solving for } y - 1 \text{ by exponentiating}$$

$$y - 1 = e^{\frac{1}{2}x^2} \cdot \underbrace{e^C}_{c} = c \cdot e^{x^2/2} \qquad \qquad \text{Replacing } e^C \text{ by } c \text{ and } e^{\frac{1}{2}x^2} \text{ by } e^{x^2/2}$$

$$y = ce^{x^2/2} + 1 \qquad \qquad \text{Adding 1 to each side}$$

This is the general solution. To satisfy the initial condition $y(0) = 4$, we substitute $y = 4$ and $x = 0$.

$$4 = ce^0 + 1$$

$y = ce^{x^2/2} + 1$ with
$y = 4$ and $x = 0$

$$4 = c + 1$$

Since $e^0 = 1$

This gives $c = 3$. Therefore, the particular solution is

$$y = 3e^{x^2/2} + 1$$

$y = ce^{x^2/2} + 1$
with $c = 3$

Accumulation of Wealth

The examples so far have *given* us a differential equation to solve. In this application we will first *derive* a differential equation to represent a situation and then solve it.

EXAMPLE 7   PREDICTING WEALTH

Suppose that you have saved $5000, and that you expect to save an additional $3000 during each year. If you deposit these savings in a bank account paying 5% interest compounded continuously, find a formula for your bank balance after t years.

Solution

Let $y(t)$ stand for your bank balance (in thousands of dollars) after t years. Each year $y(t)$ grows by 3 (thousand dollars) plus 5% interest. This growth can be modeled by a differential equation:

$$y' = 3 + 0.05y$$

y increases by 3
plus 5% of itself

Rate of   is   3   plus   5% of y
change of y

Before continuing, be sure that you understand how this differential equation models the changes due to savings and interest.

We solve it by separating variables.

$$\frac{dy}{dt} = 3 + 0.05y$$

Replacing y' by $\dfrac{dy}{dt}$

$$\int \frac{dy}{3 + 0.05y} = \int dt$$

Dividing by $3 + 0.05y$,
multiplying by dt, and
then integrating

$u = 3 + 0.05y$
$du = 0.05dy$

Using a substitution

$$20 \int \frac{du}{u} = \int dt$$

Substituting (the 20 comes
from $\dfrac{1}{0.05} = 20$)

$$20 \ln u = t + C$$

Integrating (assuming $u > 0$)

$$\ln(3 + 0.05y) = 0.05t + \underbrace{0.05C}_{c}$$

Substituting $u = 3 + 0.05y$
and dividing by 20

$$\ln(3 + 0.05y) = 0.05t + c \qquad \text{Replacing } 0.05C \text{ by } c$$

$$3 + 0.05y = e^{0.05t + c} = e^{0.05t}\underbrace{e^c}_{k} = ke^{0.05t} \qquad \text{Exponentiating and then simplifying constants}$$

$$0.05y = ke^{0.05t} - 3 \qquad \text{Subtracting 3}$$

$$y = \underbrace{20k}_{b}e^{0.05t} - 60 \qquad \text{Dividing by 0.05}$$

(Always simplify arbitrary constants)

$$y = be^{0.05t} - 60 \qquad \text{With arbitrary constant } b$$

You began at time $t = 0$ with 5 thousand dollars, which gives the initial condition $y(0) = 5$. We substitute $y = 5$ and $t = 0$.

$$5 = be^0 - 60 \qquad \begin{array}{l} y = be^{0.05t} - 60 \text{ with} \\ y = 5 \text{ and } t = 0 \end{array}$$

$$5 = b - 60 \qquad \text{Since } e^0 = 1$$

Therefore, $b = 65$, which we substitute into the general solution, obtaining a formula for your accumulated wealth after t years.

$$y = 65e^{0.05t} - 60 \qquad y = be^{0.05t} - 60 \text{ with } b = 65$$

For example, to find your wealth after 10 years, we evaluate the solution of the differential equation at $t = 10$.

$$y = 65e^{0.5} - 60 \approx 47.167 \qquad y = 65e^{0.05t} - 60 \text{ with } t = 10$$

Therefore, after 10 years, you will have \$47,167 in the bank.* This wealth function is graphed on the left.

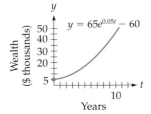

Solutions TO PRACTICE PROBLEMS

1. $y = e^{-x} + e^{3x}$

$y' = -e^{-x} + 3e^{3x}$

$y'' = e^{-x} + 9e^{3x}$

Substituting these expressions into the differential equation:

$$(e^{-x} + 9e^{3x}) - 2(-e^{-x} + 3e^{3x}) - 3(e^{-x} + e^{3x}) \overset{?}{=} 0$$

$$\cancel{e^{-x}} + \cancel{9e^{3x}} + \cancel{2e^{-x}} - \cancel{6e^{3x}} - \cancel{3e^{-x}} - \cancel{3e^{3x}} \overset{?}{=} 0$$

$$0 = 0 \ \checkmark \qquad \text{It checks}$$

2. $\dfrac{dy}{dx} = \dfrac{6x^2}{y^4}$

$$y^4\,dy = 6x^2\,dx$$

$$\int y^4\,dy = \int 6x^2\,dx$$

*How would this amount change if we were to replace *continuous* compounding with the more common *daily* compounding? At 5% interest with equal monthly deposits (in the middle of each month) adding up to the stated \$3000 per year, the total after 10 years would be \$47,166, differing from our figure by just \$1 after 10 years.

$$\tfrac{1}{5}y^5 = 2x^3 + C$$
$$y^5 = 10x^3 + 5C = 10x^3 + c$$
$$y = \sqrt[5]{10x^3 + c}$$

The initial condition gives

$$2 = \sqrt[5]{0 + c}$$
$$2 = \sqrt[5]{c}$$
$$32 = c \qquad\qquad\qquad \text{Raising each side to the fifth power}$$

Solution: $y = \sqrt[5]{10x^3 + 32}$

3. $\dfrac{dy}{dx} = x^2 y$

$$\int \frac{dy}{y} = \int x^2\, dx$$

$$\ln y = \tfrac{1}{3}x^3 + C$$

$$y = e^{\frac{1}{3}x^3 + C} = e^{\frac{1}{3}x^3}e^C = ce^{x^3/3}$$

The initial condition gives

$$5 = ce^0 = c$$

Solution: $y = 5e^{x^3/3}$

9.1   Section Summary

We solved separable differential equations by separating the variables (moving every y to one side and every x to the other, with dx and dy in the numerators) and integrating both sides. The *general* solution of a differential equation includes an arbitrary constant, while a *particular* solution results from evaluating the constant, usually by applying an initial condition. The *slope field* of a differential equation allows you to construct geometrically a solution from any point by following the slopes away from the point.

We also saw how to *derive* a differential equation from information about how a quantity changes. For example, we can "read" the differential equation as follows.

The applied exercises show how differential equations lead to important formulas in a wide variety of fields.

9.1   Exercises

1–4. Verify that the function y satisfies the given differential equation.

1. $y = e^{2x} - 3e^x + 2$
$y'' - 3y' + 2y = 4$

2. $y = e^{5x} - 4e^x + 1$
$y'' - 6y' + 5y = 5$

3. $y = ke^{ax} - \dfrac{b}{a}$ (for constants a, b, and k)
$y' = ay + b$

4. $y = ax^2 + bx$ (for constants a and b)
$y' = \dfrac{y}{x} + ax$

5–36. Find the general solution of each differential equation or state that the differential equation is not separable. If the exercise says "and check," verify that your answer is a solution.

5. $y^2 y' = 4x$ **6.** $y^4 y' = 8x$

7. $y' = x + y$ **8.** $y' = xy - 1$

9. $y' = 6x^2 y$ and check

10. $y' = 12x^3 y$ and check

11. $y' = \dfrac{y}{x}$ and check

12. $y' = \dfrac{y^2}{x^2}$ and check

13. $yy' = 4x$ **14.** $yy' = 6x^2$

15. $y' = e^{xy}$ **16.** $y' = e^x + y$

17. $y' = 9x^2$ **18.** $y' = 6e^{-2x}$

19. $y' = \dfrac{x}{x^2 + 1}$ **20.** $y' = xy^2$

21. $y' = x^3 y$ **22.** $y' = \dfrac{x}{y}$

23. $y' = x^m y^n$ (for $m > 0$, $n \neq 1$)

24. $y' = x^m y$ (for $m > 0$)

25. $y' = 2\sqrt{y}$ **26.** $y' = 5 + y$

27. $xy' = x^2 + y^2$ **28.** $y' = \sqrt{x + y}$

29. $y' = xy + x$ **30.** $y' = x - 2xy$

31. $y' = ye^x - e^x$ **32.** $y' = ye^x - y$

33. $y' = ay^2$ (for constant $a > 0$)

34. $y' = axy$ (for constant a)

35. $y' = ay + b$ (for constants a and b)

36. $y' = (ay + b)^2$ (for constants $a \neq 0$ and b)

37–50. Solve each differential equation and initial condition and verify that your answer satisfies both the differential equation and the initial condition.

37. $\begin{cases} y^2 y' = 2x \\ y(0) = 2 \end{cases}$ **38.** $\begin{cases} y^4 y' = 3x^2 \\ y(0) = 1 \end{cases}$

39. $\begin{cases} y' = xy \\ y(0) = -1 \end{cases}$ **40.** $\begin{cases} y' = y^2 \\ y(2) = -1 \end{cases}$

41. $\begin{cases} y' = 2xy^2 \\ y(0) = 1 \end{cases}$ **42.** $\begin{cases} y' = 2xy^4 \\ y(0) = 1 \end{cases}$

43. $\begin{cases} y' = \dfrac{y}{x} \\ y(1) = 3 \end{cases}$ **44.** $\begin{cases} y' = \dfrac{2y}{x} \\ y(1) = 2 \end{cases}$

45. $\begin{cases} y' = 2\sqrt{y} \\ y(1) = 4 \end{cases}$ **46.** $\begin{cases} y' = \sqrt{y}\,e^x - \sqrt{y} \\ y(0) = 1 \end{cases}$

47. $\begin{cases} y' = y^2 e^x + y^2 \\ y(0) = 1 \end{cases}$ **48.** $\begin{cases} y' = xy - 5x \\ y(0) = 4 \end{cases}$

49. $\begin{cases} y' = ax^2 y \\ y(0) = 2 \end{cases}$ (for constant $a > 0$)

50. $\begin{cases} y' = axy \\ y(0) = 4 \end{cases}$ (for constant $a > 0$)

Applied Exercises

51. BUSINESS: Elasticity For a demand function $D(p)$, the elasticity of demand (see page 294) is defined as $E = \dfrac{-pD'}{D}$. Find demand functions $D(p)$ that have *constant elasticity* by solving the differential equation $\dfrac{-pD'}{D} = k$, where k is a constant.

52. BIOMEDICAL: Cell Growth A cell receives nutrients through its surface, and its surface area is proportional to the two-thirds power of its weight. Therefore, if $w(t)$ is the cell's weight at time t, then $w(t)$ satisfies $w' = aw^{2/3}$, where a is a positive constant. Solve this differential equation with the initial condition $w(0) = 1$ (initial weight 1 unit).

Source: J. M. Smith, Mathematical Ideas in Biology

53–54. BUSINESS: Continuous Annuities An annuity is a fund into which one makes equal payments at regular intervals. If the fund earns interest at rate r compounded continuously, and deposits are made continuously at the rate of d dollars per year (a *continuous annuity*), then the value $y(t)$ of the fund after t years satisfies the differential equation $y' = d + ry$. (Do you see why?)

Source: S. Broverman, Mathematics of Investment and Credit

53. Solve the differential equation in the preceding instructions for the continuous annuity $y(t)$ with deposit rate $d = \$1000$ and continuous interest rate $r = 0.05$, subject to the initial condition $y(0) = 0$ (zero initial value).

54. Solve the differential equation in the preceding instructions for the continuous annuity $y(t)$, where d and r are unknown constants, subject to the initial condition $y(0) = 0$ (zero initial value).

55. GENERAL: Time of a Murder A medical examiner called to the scene of a murder will usually take the temperature of the body. A corpse cools at a rate proportional to the difference between its temperature and the temperature of the room. If $y(t)$ is the temperature (in degrees Fahrenheit) of the body t hours after the murder, and if the room temperature is $70°$, then y satisfies

$$y' = -0.32(y - 70)$$

$$y(0) = 98.6 \text{ (body temperature initially } 98.6°)$$

a. Solve this differential equation and initial condition.

b. Use your answer to part (a) to estimate how long ago the murder took place if the temperature of the body when it was discovered was 80°. [*Hint:* Find the value of *t* that makes your solution equal 80°.]

Source: D. Lomen and D. Lovelock, *Exploring Differential Equations via Graphics and Data*

56. BIOMEDICAL: Glucose Levels Hospital patients are often given glucose (blood sugar) through a tube connected to a bottle suspended over their beds. Suppose that this "drip" supplies glucose at the rate of 25 mg per minute, and each minute 10% of the accumulated glucose is consumed by the body. Then the amount $y(t)$ of glucose (in excess of the normal level) in the body after *t* minutes satisfies

$$y' = 25 - 0.1y \quad \text{(Do you see why?)}$$
$$y(0) = 0 \quad \text{(zero excess glucose at } t = 0)$$

Solve this differential equation and initial condition.

57. GENERAL: Friendships Suppose that you meet 30 new people each year, but each year you forget 20% of all of the people that you know. If $y(t)$ is the total number of people who you remember after *t* years, then *y* satisfies the differential equation $y' = 30 - 0.2y$. (Do you see why?) Solve this differential equation subject to the condition $y(0) = 0$ (you knew no one at birth).

58. ENVIRONMENTAL SCIENCE: Pollution For more than 75 years the Flexfast Rubber Company in Massachusetts discharged toxic toluene solvents into the ground at a rate of 5 tons per year. Each year approximately 10% of the accumulated pollutants evaporated into the air. If $y(t)$ is the total accumulation of pollution in the ground after *t* years, then *y* satisfies

$$y' = 5 - 0.1y \quad \text{(Do you see why?)}$$
$$y(0) = 0 \quad \text{(initial accumulation zero)}$$

Solve this differential equation and initial condition to find a formula for the accumulated pollutant after *t* years.

Source: Greater Boston Legal Services

59. PERSONAL FINANCE: Wealth Accumulation Suppose that you now have $6000, you expect to save an additional $3000 during each year, and all of this is deposited in a bank paying 10% interest compounded continuously. Let $y(t)$ be your bank balance (in thousands of dollars) *t* years from now.

a. Write a differential equation that expresses the fact that your balance will grow by 3 (thousand dollars) and also by 10% of itself. [*Hint:* See Example 7.]

b. Write an initial condition to say that at time zero the balance is 6 (thousand dollars).

c. Solve your differential equation and initial condition.

d. Use your solution to find your bank balance $t = 25$ years from now.

60. PERSONAL FINANCE: Wealth Accumulation Suppose that you now have $2000, you expect to save an additional $6000 during each year, and all of this is deposited in a bank paying 4% interest compounded continuously.

Let $y(t)$ be your bank balance (in thousands of dollars) *t* years from now.

a. Write a differential equation that expresses the fact that your balance will grow by 6 (thousand dollars) and also by 4% of itself. [*Hint:* See Example 7.]

b. Write an initial condition to say that at time zero the balance is 2 (thousand dollars).

c. Solve your differential equation and initial condition.

d. Use your solution to find your bank balance $t = 20$ years from now.

61. BUSINESS: Sales Your company has developed a new product, and your marketing department has predicted how it will sell. Let $y(t)$ be the (monthly) sales of the product after *t* months.

a. Write a differential equation that says that the rate of growth of the sales will be four times the one-half power of the sales.

b. Write an initial condition that says that at time $t = 0$ sales were 10,000.

c. Solve this differential equation and initial value.

d. Use your solution to predict the sales at time $t = 12$ months.

62. BUSINESS: Sales Your company has developed a new product, and your marketing department has predicted how it will sell. Let $y(t)$ be the (monthly) sales of the product after *t* months.

a. Write a differential equation that says that the rate of growth of the sales will be six times the two-thirds power of the sales.

b. Write an initial condition that says that at time $t = 0$ sales were 1000.

c. Solve this differential equation and initial value.

d. Use your solution to predict the sales at time $t = 12$ months.

63. BIOMEDICAL: Bacterial Colony Let $y(t)$ be the size of a colony of bacteria after *t* hours.

a. Write a differential equation that says that the rate of growth of the colony is equal to eight times the three-fourths power of its present size.

b. Write an initial condition that says that at time zero the colony is of size 10,000.

c. Solve the differential equation and initial condition.

d. Use your solution to find the size of the colony at time $t = 6$ hours.

64. BUSINESS: Value of a Building Let $y(t)$ be the value of a commercial building (in millions of dollars) after *t* years.

a. Write a differential equation that says that the rate of growth of the value of the building is equal to two times the one-half power of its present value.

b. Write an initial condition that says that at time zero the value of the building is 9 million dollars.

c. Solve the differential equation and initial condition.

d. Use your solution to find the value of the building at time $t = 5$ years.

65–67. The following exercises require the use of a slope field program.

For each differential equation and initial condition:

a. Use a graphing calculator *slope field* program to graph the slope field for the differential equation on the window $[-5, 5]$ by $[-5, 5]$.
b. Sketch the slope field on a piece of paper and draw a solution curve that follows the slopes and that passes through the point (0, 2).
c. Solve the differential equation and initial condition.
d. Use your slope field program to graph the slope field and the solution that you found in part (c). How good was the sketch that you made in part (b) compared with the solution graphed in part (d)?

65. $\begin{cases} \dfrac{dy}{dx} = \dfrac{6x^2}{y^4} \\ y(0) = 2 \end{cases}$ **66.** $\begin{cases} \dfrac{dy}{dx} = \dfrac{x^2}{y^2} \\ y(0) = 2 \end{cases}$ **67.** $\begin{cases} \dfrac{dy}{dx} = \dfrac{4x}{y^3} \\ y(0) = 2 \end{cases}$

 68–70. The following exercises require the use of a slope field program.

For each differential equation:

a. Use a graphing calculator *slope field* program to graph the slope field for the differential equation on the window $[-5, 5]$ by $[-5, 5]$.

b. Sketch the slope field on a piece of paper and draw a solution curve that follows the slopes and that passes through the given point.

68. $\dfrac{dy}{dx} = x - y^2$

point: (0, 1)

69. $\dfrac{dy}{dx} = \dfrac{x}{y^2 + 1}$

point: (0, −1)

70. $\dfrac{dy}{dx} = x \ln(y^2 + 1)$

point: (0, –2)

71. PERSONAL FINANCE: Wealth Accumulation Reread the Application Preview on page 589.

a. Make up some reasonable numbers of s (annual salary) and n (annual cost of necessities) and substitute them together with $r = 0.06$ and $p = 0.75$ into the formula for $W(t)$. Then evaluate $W(20)$ to predict your wealth in 20 years.
b. Change p to 0.70 to see how your wealth $W(20)$ would change if you lived more frugally.
c. Give yourself a raise by increasing s by 20% to see how that affects $W(20)$.
d. Suppose that you take out a mortgage to buy a house, thereby increasing n by 15%. How does $W(20)$ change?

Conceptual Exercises

72–75. The following are differential equations *stated in words*. Find the general solution of each.

72. The derivative of a function at each point is 6.

73. The derivative of a function at each point is −2.

74. The derivative of a function at each point is 0.

75. The derivative of a function at each point is itself.

76–79. Determine whether each differential equation is separable. (Do not solve it, just find whether it's separable.)

76. $y' = x + y$

77. $y' = \ln(xy)$

78. $y' = \ln(x + y)$

79. $y' = e^{x+y}$

80. Think of the slope field for the differential equation $\dfrac{dy}{dx} = \dfrac{6x}{y^2}$. What is the sign of the slope in quadrant I (where x and y are both positive)? What is the sign of the slope in each of the other three quadrants? Check your answers by looking at the slope field on page 594.

81. Think of the slope field for the differential equation $\dfrac{dy}{dx} = xy$. What is the sign of the slope in quadrant I (where x and y are both positive)? What is the sign of the slope in each of the other three quadrants? Check your answers by looking at the slope field on page 596.

Explorations and Excursions   The following problems extend and augment the material presented in the text.

82. BIOMEDICAL: Fick's Law Fick's Law governs the diffusion of a solute across a cell membrane. According to Fick's Law, the concentration $y(t)$ of the solute inside the cell at time t satisfies

$\dfrac{dy}{dt} = \dfrac{kA}{V}(C_0 - y)$, where k is the diffusion constant,

A is the area of the cell membrane, V is the volume of the cell, and C_0 is the concentration outside the cell.

a. Find the general solution of this differential equation. (Your solution will involve the constants k, A, V, and C_0.)

b. Find the particular solution that satisfies the initial condition $y(0) = y_0$, where y_0 is the initial concentration inside the cell.

83. **BIOMEDICAL: Heart Function** In the *reservoir model,* the heart is viewed as a balloon that swells as it fills with blood (during a period called the *systole*), and then at time t_0 it shuts a valve and contracts to force the blood out (the *diastole*). Let $p(t)$ represent the pressure in the heart at time t.

a. During the diastole, which lasts from t_0 to time T, $p(t)$ satisfies the differential equation

$$\frac{dp}{dt} = -\frac{K}{R}p$$

Find the general solution $p(t)$ of this differential equation. (K and R are positive constants determined, respectively, by the strength of the heart and the resistance of the arteries. The differential equation states that as the heart contracts, the pressure decreases (dp/dt is negative) in proportion to itself.)

b. Find the particular solution that satisfies the condition $p(t_0) = p_0$. (p_0 is a constant representing the pressure at the transition time t_0.)

c. During the systole, which lasts from time 0 to time t_0, the pressure $p(t)$ satisfies the differential equation

$$\frac{dp}{dt} = KI_0 - \frac{K}{R}p$$

Find the general solution of this differential equation. (I_0 is a positive constant representing the constant rate of blood flow into the heart while it is expanding.) [*Hint:* Use the same u-substitution technique that was used in Example 7.]

d. Find the particular solution that satisfies the condition $p(t_0) = p_0$.

e. In parts (b) and (d) you found the formulas for the pressure $p(t)$ during the diastole $(t_0 \le t \le T)$ and the systole $(0 \le t \le t_0)$. Since the heart behaves in a cyclic fashion, these functions must satisfy $p(T) = p(0)$. Equate the solutions at these times (use the correct formula for each time) to derive the important relationship

$$R = \frac{p_0}{I_0} \frac{1 - e^{-KT/R}}{1 - e^{-Kt_0/R}}$$

9.2 **Further Applications of Differential Equations: Three Models of Growth**

Introduction

This section continues our study of differential equations, but with a different approach. Instead of solving individual differential equations, we will solve three important *classes* of differential equations (for **unlimited growth, limited growth,** and **logistic growth**) and remember their solutions. This will enable us to solve many problems by identifying the appropriate differential equation and then immediately writing the solution. In this section we begin *to think in terms of differential equations.*

Proportional Quantities

The circumference of a circle is proportional to its diameter.

We say that one quantity is *proportional* to another quantity if the first quantity is a *constant multiple* of the second. That is, y is proportional to x if $y = ax$ for some "proportionality constant" a. For example, the formula $C = \pi D$ for the circumference of a circle shows that the circumference C is proportional to the diameter D.

Unlimited Growth

In many situations the growth of a quantity is proportional to its present size. For example, a population of cells will grow in proportion to its present size, and a bank account earns interest in proportion to its current value. If a quantity y grows so that its rate of growth y' is proportional to its present size y, then y satisfies the differential equation

$$y' = ay$$

Rate of growth | is proportional to | current size

We solve this differential equation by separating variables.

$$\frac{dy}{dt} = ay$$

Replacing y' by $\frac{dy}{dt}$

$$\int \frac{dy}{y} = \int a\, dt$$

Dividing by y ($y \neq 0$), multiplying by dt, and integrating

$$\ln y = at + C$$

Integrating ($y > 0$)

$$y = e^{at + C} = e^{at}e^{C} = ce^{at}$$

Solving for y by exponentiating and then replacing e^{C} by c

$$\underset{\quad \uparrow\, c}{}$$

At time $t = 0$ this becomes

$$y(0) = ce^{0} = c$$

$y(t) = ce^{at}$ with $t = 0$

Summarizing:

Unlimited Growth

The differential equation
with initial condition
is solved by

$$y' = ay$$
$$y(0) = c$$
$$y = ce^{at}$$

Such growth, where the rate of growth is proportional to the present size, is called *unlimited growth* because the solution y grows arbitrarily large. Given this result, whenever we encounter a differential equation of the form $y' = ay$, we can immediately write the solution $y = ce^{at}$, where c is the initial value.

EXAMPLE 1   PREDICTING ART APPRECIATION (UNLIMITED GROWTH)

An art collection, initially worth \$20,000, continuously grows in value at the rate of 5% a year. Express this growth as a differential equation and find a formula for the value of the collection after t years. Then estimate the value of the art collection after 10 years.

Solution

Growing continuously at the rate of 5% means that the value $y(t)$ grows by 5% of itself:

$$y' = 0.05y$$

Rate of   is   5% of the
growth   current value

This differential equation is of the form $y' = ay$ (unlimited growth) with $a = 0.05$ and initial value 20,000, so we may immediately write its solution:

$$y(t) = 20,000e^{0.05t}$$

$y = ce^{at}$ with $a = 0.05$
and $c = 20,000$

This formula gives the value of the art collection after t years. To find the value after 10 years, we evaluate at $t = 10$:

$$y(10) = 20{,}000e^{0.05 \cdot 10}$$

$$= 20{,}000e^{0.5} \approx 32{,}974$$

$y(t) = 20{,}000e^{0.05t}$ with $t = 10$

Using a calculator

In 10 years the art collection will be worth \$32,974.

The solution ce^{at} is the same as the continuous compounding formula Pe^{rt} (except for different letters). On page 255 we derived the formula Pe^{rt} rather laboriously, using the discrete interest formula $P(1 + r/m)^{mt}$, replacing m by rn, and taking the limit as $n \to \infty$. The present derivation, using differential equations, is much simpler and shows that ce^{at} applies to *any* situation governed by the differential equation $y' = ay$.

Limited Growth

No real population can undergo unlimited growth for very long. Restrictions of food and space would soon slow its growth. If a quantity $y(t)$ cannot grow larger than a certain fixed maximum size M, and if its growth rate y' is proportional to how far it is from its upper limit, then y satisfies

$$y' = a(M - y) \qquad\qquad a > 0$$

Rate of   is propor-   distance below
growth   tional to   upper bound M

We solve this by separating variables.

$$\frac{dy}{dt} = a(M - y)$$

Replacing y' by $\dfrac{dy}{dt}$

$$\int \frac{dy}{M - y} = \int a\,dt$$

Dividing by $M - y$, multiplying by dt, and integrating

$$u = M - y$$
$$du = -dy$$

Using a substitution

$$-\int \frac{du}{u} = \int a\,dt$$

Substituting

$$-\ln u = at + C$$

Integrating $(u > 0)$

$$\ln(M - y) = -at - C$$

Multiplying by -1 and replacing u by $(M - y)$

$$M - y = e^{-at-C} = e^{-at}e^{-C} = ce^{-at}$$

Solving for $M - y$ by exponentiating and simplifying

$$y = M - ce^{-at}$$

Subtracting M and multiplying by -1

We impose the initial condition $y(0) = 0$ (size zero at time $t = 0$).

$$0 = M - ce^0 = M - c$$

$y = M - ce^{-at}$ with $y = 0$ and $t = 0$

Therefore, $c = M$, which gives the solution

$$y = M - Me^{-at} = M(1 - e^{-at})$$

$y = M - ce^{-at}$ with $c = M$

Summarizing:

Limited Growth

The differential equation
$$y' = a(M - y)$$
with initial condition
$$y(0) = 0$$
is solved by
$$y = M(1 - e^{-at})$$

This type of growth, in which the rate of growth is proportional to the distance below an upper limit M, is called *limited growth* because the solution $y = M(1 - e^{-at})$ asymptotically approaches the limit M as $t \to \infty$. Given this result, whenever we encounter a differential equation of the form
$$y' = a(M - y)$$
with initial value zero, we may immediately write the solution
$$y = M(1 - e^{-at})$$

Diffusion of Information by Mass Media

If a news bulletin is repeatedly broadcast over radio and television, the news spreads quickly at first, but later more slowly when most people have already heard it. Sociologists often assume that the rate at which news spreads is proportional to the number who have not yet heard the news. Let M be the population of a city, and let $y(t)$ be the number of people who have heard the news within t time units. Then y satisfies the differential equation

$$y' \quad = \quad a(M - y)$$

Rate of   is   propor-   number who have
growth   tional to   not heard the news

We recognize this as the differential equation for limited growth, whose solution is $y = M(1 - e^{-at})$. It remains only to determine the values of the constants M and a.

EXAMPLE 2   PREDICTING SPREAD OF INFORMATION (LIMITED GROWTH)

An important news bulletin is broadcast to a town of 50,000 people, and after 2 hours 30,000 people have heard the news. Find a formula for the number of people who have heard the bulletin within t hours. Then find how many people will have heard the news within 6 hours.

Solution

If $y(t)$ is the number of people who have heard the news within t hours, then y satisfies

$$y' = a(50{,}000 - y)$$

$y' = a(M - y)$ with $M = 50{,}000$

Number who have
not heard the news

This is the differential equation for limited growth with $M = 50{,}000$, so the solution (from the preceding box) is

$$y = 50{,}000(1 - e^{-at})$$

$y = M(1 - e^{-at})$ with $M = 50{,}000$

To find the value of the constant a, we use the given information that 30,000 people have heard the news within 2 hours.

$$30{,}000 = 50{,}000(1 - e^{-a \cdot 2})$$ $y = 50{,}000(1 - e^{-at})$ with $y = 30{,}000$
and $t = 2$

$$0.6 = 1 - e^{-2a}$$ Dividing each side by 50,000

$$0.4 = e^{-2a}$$ Subtracting 1 and then multiplying by -1

$$-0.916 \approx -2a$$ Taking natural logs (using $\ln e^x = x$ on the right)

$$a \approx 0.46$$ Dividing by -2

Therefore, the number of people who have heard the news within t hours is

$$y(t) = 50{,}000(1 - e^{-0.46t})$$ $y = 50{,}000(1 - e^{-at})$ with $a = 0.46$

as shown in the graph. To find the number who have heard the news within 6 hours, we evaluate this solution at $t = 6$.

$$y(6) = 50{,}000(1 - e^{-0.46 \cdot 6}) = 50{,}000(1 - e^{-2.76}) \approx 46{,}835$$ Using a calculator

Within 6 hours about 46,800 people have heard the news.

We found the value of the constant a by substituting the given data into the solution, simplifying, and taking logs. Similar steps will be required in many other problems, and in the future we shall omit the details.

Learning Theory

Psychologists have found that there seems to be an upper limit to the number of meaningless words that a person can memorize, and that memorizing becomes increasingly difficult approaching that bound. If M is this upper limit and $y(t)$ is the number of words that can be memorized in t minutes, then the situation is modeled by the differential equation for limited growth.

$$y' = a(M - y)$$

Rate of   is propor-   upper limit M minus
increase   tional to   number already memorized

This can be interpreted as saying that the rate at which new words can be memorized is proportional to the "unused memory capacity."

EXAMPLE 3   **PREDICTING MEMORIZATION (LIMITED GROWTH)**

Suppose that a person can memorize at most 100 meaningless words, and that after 15 minutes 10 words have been memorized. How long will it take to memorize 50 words?

Solution

The number $y(t)$ of words that can be memorized in t minutes satisfies

$$y' = a(100 - y)$$

$y' = a(M - y)$ with $M = 100$

This is the differential equation for limited growth, so the solution is

$$y = 100(1 - e^{-at})$$

$y = M(1 - e^{-at})$ with $M = 100$

To evaluate the constant a, we use the given information that 10 words have been memorized in 15 minutes.

$$10 = 100(1 - e^{-a \cdot 15})$$

$y = 100(1 - e^{-at})$ with $y = 10$ and $t = 15$

Solving this equation for the constant a (omitting the details, which are the same as in Example 2) gives $a = 0.007$.

$$y(t) = 100(1 - e^{-0.007t})$$

$y = 100(1 - e^{-at})$ with $a = 0.007$

$y = 100(1 - e^{-0.007t})$

Words memorized

100
80
60
40
20

60 120 180 240

Minutes

The slope (the rate at which additional words can be memorized) decreases near the upper limit.

This solution (graphed on the left) gives the number of words that can be memorized in t minutes. To find how long it takes to memorize 50 words, we set this solution equal to 50 and solve for t.

$$50 = 100(1 - e^{-0.007t})$$

$y = 100(1 - e^{-0.007t})$ with $y = 50$

Solving for t (again the details are similar to those in Example 2) gives $t = 99$ minutes. Therefore, 50 words can be memorized in about 1 hour and 39 minutes.

Logistic Growth

Some quantities grow in proportion to both their present size *and* their distance from an upper limit M.

$$y' \;=\; ay(M - y)$$

Rate of growth   is   propor-tional to   present size   upper limit M minus present size

This differential equation can be solved by separation of variables (see Exercise 65) to give the solution

$$y = \frac{M}{1 + ce^{-aMt}}$$

for constants $a > 0$ and $c > -1$. This function is called the *logistic* function,* governing *logistic growth*.

*Or sometimes the *Verhulst model* or the *Beverton-Holt model*.

Logistic Growth

The differential equation

$$y' = ay(M - y)$$

with initial condition

$$y(0) = \frac{M}{1 + c}$$

is solved by

$$y = \frac{M}{1 + ce^{-aMt}}$$

The upper and lower curves in the box represent solutions whose initial values are, respectively, greater than or less than M as determined by whether the constant c is negative or positive. As before, this result enables us to solve a differential equation "at sight," leaving only the evaluation of constants.

The lower curve in the box that rises to M is called a *sigmoidal* or *S-shaped curve* and is used to model growth that begins slowly, then becomes more rapid, and finally slows again near the upper limit. Many different quantities grow according to sigmoidal curves.

Quinoa imports

Source: U.S. Customs

Wikipedia articles

Source: en.wikipedia.org

Music downloads

Source: NielsonSoundScan

Environmental Science

For an animal environment (such as a lake or a forest), the population that it can support will have an upper limit, called the *carrying capacity of the environment*. Ecologists often assume that an animal population grows in proportion to both its present size and its distance below the carrying capacity of the environment.

$$y' \;=\; ay(M - y)$$

Rate of growth   is   propor-tional to   present size   carrying capacity minus present size

Since this is the logistic differential equation, we know that the solution is the logistic function

$$y = \frac{M}{1 + ce^{-aMt}}$$

| EXAMPLE 4 | **PREDICTING AN ANIMAL POPULATION (LOGISTIC GROWTH)** |

Ecologists estimate that an artificial lake can support a maximum of 2500 fish. The lake is initially stocked with 500 fish, and after 6 months the fish population is estimated to be 1500. Find a formula for the number of fish in the lake after t months, and estimate the fish population at the end of the first year.

Solution

Letting $y(t)$ stand for the number of fish in the lake after t months, the situation is modeled by the logistic differential equation.

$$y' = ay(2500 - y) \qquad \begin{array}{l} y' = ay(M - y) \\ \text{with } M = 2500 \end{array}$$

Rate of growth | is | propor-tional to | present size | carrying capacity minus present size

The solution is the logistic function with $M = 2500$.

$$y = \frac{2500}{1 + ce^{-a2500t}} \qquad y = \frac{M}{1 + ce^{-aMt}} \text{ with } M = 2500$$

$$= \frac{2500}{1 + ce^{-bt}} \qquad \text{Replacing } a \cdot 2500 \text{ by another constant } b$$

To evaluate the constants c and b, we use the fact that the lake was originally stocked with 500 fish.

$$500 = \frac{2500}{1 + ce^0} \qquad y = \frac{2500}{1 + ce^{-bt}} \text{ with } y = 500, \ t = 0$$

$$500 = \frac{2500}{1 + c} \qquad \text{Simplifying}$$

Solving this for c (omitting the details—the first step is to multiply both sides by $1 + c$) gives $c = 4$, so the logistic function becomes

$$y = \frac{2500}{1 + 4e^{-bt}} \qquad y = \frac{2500}{1 + ce^{-bt}} \text{ with } c = 4$$

To evaluate b, we substitute the information that the population is $y = 1500$ at $t = 6$.

$$1500 = \frac{2500}{1 + 4e^{-b \cdot 6}} \qquad y = \frac{2500}{1 + 4e^{-bt}} \text{ with } y = 1500, \ t = 6$$

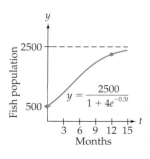

$$y = \frac{2500}{1 + 4e^{-0.3t}}$$

Solving for b (again omitting the details—the first step is to multiply both sides by $1 + 4e^{-b \cdot 6}$) gives $b = 0.30$, so the logistic function becomes

$$y(t) = \frac{2500}{1 + 4e^{-0.3t}}$$

$$y = \frac{2500}{1 + 4e^{-bt}} \quad \text{with} \quad b = 0.3$$

This is the formula for the population after t months. To find the population after a year, we evaluate at $t = 12$.

$$y(12) = \frac{2500}{1 + 4e^{-0.3 \cdot 12}} = \frac{2500}{1 + 4e^{-3.6}} \approx 2254 \qquad \text{Using a calculator}$$

Therefore, the population at the end of the first year is 2254, which is about 90% of the carrying capacity of the lake.

Epidemics

Many epidemics spread at a rate proportional to both the number of people already infected (the "carriers") and also the number who have yet to catch the disease (the "susceptibles"). If $y(t)$ is the number of infected people at time t from a population of size M, then $y(t)$ satisfies the logistic differential equation

$$y' = ay(M - y)$$

Rate of   is propor-   number   number
growth   tional to infected   susceptible

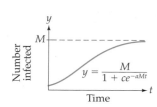

$$y = \frac{M}{1 + ce^{-aMt}}$$

Therefore, the size of the infected population is given by the logistic function

$$y = \frac{M}{1 + ce^{-aMt}}$$

The constants are evaluated just as in Example 4, using the (initial) number of cases reported at time $t = 0$, and also the number of cases at some later time.

Spread of Rumors

Sociologists have found that rumors spread at a rate proportional to the number who have heard the rumor (the "informed") and the number who have not heard the rumor (the "uninformed"). Therefore, in a population of size M, the number $y(t)$ who have heard the rumor within t time units satisfies the logistic differential equation

$$y' = ay(M - y)$$

Rate of   is propor-   number   number
growth   tional to informed   uninformed

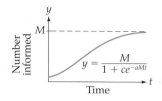

The solution $y(t)$ is then the logistic function

$$y = \frac{M}{1 + ce^{-aMt}}$$

It remains only to evaluate the constants. The spread of a rumor is analogous to the spread of a disease, with an "informed" person being one who has been "infected."

Limited and Logistic Growth of Sales

The sales of a product whose total sales will approach an upper limit (market saturation) can be modeled by either the limited or the logistic equation. Which do you use when? For a product advertised over mass media, sales will at first grow rapidly, indicating a *limited* model. For a product becoming known only through "word of mouth," sales will at first grow slowly, indicating a *logistic* model.

PRACTICE PROBLEM

The graphs below show the total sales through day t for two different products, A and B. Which of these products was advertised and which became known only by "word of mouth"? State an appropriate differential equation for each curve.

Solution below >

Solution TO PRACTICE PROBLEM

Product A, whose growth begins slowly, was not advertised, and the differential equation is logistic: $y' = ay(M - y)$. Product B, whose growth begins rapidly, was advertised, and the differential equation is limited: $y' = a(M - y)$.

9.2 Section Summary

The unlimited, limited, and logistic growth models are summarized in the following table. If one of these differential equations governs a particular situation, we can write the solution immediately, evaluating the constants from the given data.

Three Models of Growth

Type	Differential Equation	Solution	Graph	Examples
Unlimited Growth is proportional to present size.	$y' = ay$	$y = ce^{at}$		Investments Bank accounts Unlimited populations
Limited Growth (starting at 0) is proportional to maximum size M minus present size.	$y' = a(M - y)$	$y = M(1 - e^{-at})$		Information spread by mass media Memorizing random information Total sales (advertised)
Logistic Growth is proportional to present size and to maximum size M minus present size.	$y' = ay(M - y)$	$y = \dfrac{M}{1 + ce^{-aMt}}$		Confined populations Epidemics Rumors Total sales (unadvertised)

To decide which (if any) of the three models applies in a given situation, think of whether the growth is proportional to *size*, to *unused capacity*, or to *both* (as shown in the chart). Notice that the differential equation gives much more insight into how the growth occurs than does the solution. This is what we meant at the beginning of the section by "thinking in terms of differential equations."

Graphing Calculator Exploration

Use a graphing calculator *slope field* program to graph the slope fields for the following differential equations (one at a time) on the window [0, 5] by [0, 5]:

a. $y' = y$ (unlimited)   **b.** $y' = 3 - y$ (limited)   **c.** $y' = y(3 - y)$ (logistic)

Do you see how the slope field gives a "picture" of the differential equation?

$y' = y$

$y' = 3 - y$

$y' = y(3 - y)$

9.2 Exercises

1. Verify that $y(t) = ce^{at}$ solves the differential equation for unlimited growth, $y' = ay$, with initial condition $y(0) = c$.

2. Verify that $y(t) = M(1 - e^{-at})$ solves the differential equation for limited growth, $y' = a(M - y)$, with initial condition $y(0) = 0$.

3–12. Determine the type of each differential equation: *unlimited* growth, *limited* growth, *logistic* growth, or *none* of these. (Do not solve, just identify the type.)

3. $y' = 0.02y$

4. $y' = 5(100 - y)$

5. $y' = 30(0.5 - y)$

6. $y' = 0.4y(0.01 - y)$

7. $y' = 2y^2(0.5 - y)$

8. $y' = 6y$

9. $y' = y(6 - y)$

10. $y' = 0.01(100 - y^2)$

11. $y' = 4y(0.04 - y)$

12. $y' = 4500(1 - y)$

13–34. Find the solution $y(t)$ by recognizing each differential equation as determining unlimited, limited, or logistic growth, and then finding the constants.

13. $y' = 6y$
  $y(0) = 1.5$

14. $y' = 0.25y$
  $y(0) = 4$

15. $y' = -y$
  $y(0) = 100$

16. $y' = \dfrac{y}{2}$
  $y(0) = 8$

17. $y' = -0.45y$
  $y(0) = -1$

18. $y' = 0$
  $y(0) = 5$

19. $y' = 2(100 - y)$
  $y(0) = 0$

20. $y' = 48(2 - y)$
  $y(0) = 0$

21. $y' = 0.05(0.25 - y)$
  $y(0) = 0$

22. $y' = \frac{2}{3}(1 - y)$
  $y(0) = 0$

23. $y' = 80 - 2y$
  $y(0) = 0$

24. $y' = 27 - 3y$
  $y(0) = 0$

[*Hint for Exercises 23–26:* Use factoring to write the differential equation in the form $y' = a(M - y)$.]

25. $y' = 2 - 0.01y$
  $y(0) = 0$

26. $y' = 6 - 8y$
  $y(0) = 0$

27. $y' = 5y(100 - y)$
  $y(0) = 10$

28. $y' = y(1 - y)$
  $y(0) = \frac{1}{2}$

29. $y' = 0.25y(0.5 - y)$
  $y(0) = 0.1$

30. $y' = \frac{1}{3}y(\frac{1}{2} - y)$
  $y(0) = \frac{1}{6}$

31. $y' = 3y(10 - y)$
  $y(0) = 20$

32. $y' = y(2 - y)$
  $y(0) = 4$

33. $y' = 6y - 2y^2$
  $y(0) = 1$

34. $y' = 3y - 6y^2$
  $y(0) = \frac{1}{6}$

[*Hint for Exercises 33–34:* Use factoring to write the differential equation in the form $y' = ay(M - y)$.]

Applied Exercises

35–48. Write the differential equation (unlimited, limited, or logistic) that applies to the situation described. Then use its solution to solve the problem.

35. PERSONAL FINANCE: Stamp Appreciation The value of a stamp collection, initially worth $1500, grows continuously at the rate of 8% per year. Find a formula for its value after t years.

36. PERSONAL FINANCE: Home Appreciation The value of a home, originally worth $25,000, grows continuously at the rate of 6% per year. Find a formula for its value after t years.

37. BUSINESS: Total Sales A manufacturer estimates that he can sell a maximum of 100,000 Blu-ray players in a city. His total sales grow at a rate proportional to the distance below this upper limit. If after 5 months total sales are 10,000, find a formula for the total sales after t months. Then use your answer to estimate the total sales at the end of the first year.

38. BUSINESS: Product Recognition Let $p(t)$ be the number of people in a city who have heard of a new product after t weeks of advertising. The city is of size 1,000,000, and $p(t)$ grows at a rate proportional to the number of people in the city who have *not* heard of the product. If after 8 weeks 250,000 people have heard of the product, find a formula for $p(t)$. Use your formula to estimate the number of people who will have heard of the product after 20 weeks of advertising.

39. GENERAL: Fund-Raising In a drive to raise $5000, fund-raisers estimate that the rate of contributions is proportional to the distance from the goal. If $1000 was raised in 1 week, find a formula for the amount raised in t weeks. How many weeks will it take to raise $4000?

40. GENERAL: Learning A person can memorize at most 40 two-digit numbers. If that person can memorize 15 numbers in the first 20 minutes, find a formula for the number that can be memorized in t minutes. Use your answer to estimate how long the person will take to memorize 30 numbers.

41. BUSINESS: Sales An electronics dealer estimates the maximum market for GPS navigation systems in a city to be 10,000. Total sales are proportional to both the number already sold and the size of the remaining market. If 100 systems have been sold at time $t = 0$ and after 6 months 2000 have been sold, find a formula for the total sales after t months. Use your answer to estimate the total sales at the end of the first year.

42. BIOMEDICAL: Epidemics During a flu epidemic in a city of 1,000,000, a flu vaccine sells in proportion to both the number of people already inoculated and the number not yet inoculated. If 100 doses have been sold at time $t = 0$ and after 4 weeks 2000 doses have been sold, find a formula for the total number of doses sold within t weeks. Use your formula to predict the sales after 10 weeks.

43. **SOCIOLOGY: Rumors** One person at an airport starts a rumor that a plane has been hijacked, and within 10 minutes 200 people have heard the rumor. If there are 800 people in the airport, find a formula for the number who have heard the rumor within t minutes. Use your answer to estimate how many will have heard the rumor within 15 minutes.

44. **BIOMEDICAL: Epidemics** A flu epidemic on a college campus of 4000 students begins with 12 cases, and after 1 week has grown to 100 cases. Find a formula for the size of the epidemic after t weeks. Use your answer to estimate the size of the epidemic after 2 weeks.

45. **ENVIRONMENTAL SCIENCE: Deer Population** A wildlife refuge is initially stocked with 100 deer, and can hold at most 800 deer. If 2 years later the deer population is 160, find a formula for the deer population after t years. Use your answer to estimate when the deer population will reach 400.

46. **POLITICAL SCIENCE: Voting** Suppose that a bill in the U.S. Senate gains votes in proportion to the number of votes that it already has and to the number of votes that it does not have. If it begins with one vote (from its sponsor) and after 3 days it has 30 votes, find a formula for the number of votes that it will have after t days. (*Note:* The number of votes in the Senate is 100.) When will the bill have "majority support" of 51 votes?

47. **BUSINESS: Personal Computer Sales** Retailers estimate the upper limit for per capita sales of personal computers (sales per 1000 inhabitants) to be 360 annually and find that sales grow in proportion to both current sales and the difference between sales and the upper limit. In 2005 sales were 289, and in 2010 were 340. Find a formula for the annual sales t years after 2005. Use your answer to predict sales in 2020.

Source: Consumer USA

48. **BUSINESS: Digital Camera Sales** Manufacturers estimate the upper limit for sales of digital cameras to be 65 million annually and find that sales increase in proportion to both current sales and the difference between the sales and the upper limit. In 2005 sales were 8 million, and in 2010 were 24 million. Find a formula for the annual sales (in millions) t years after 2005. Use your answer to predict sales in 2020.

Source: Consumer USA

49–52. Solve by recognizing that the differential equation is one of the three types whose solutions we know.

49. **BIOMEDICAL: Drug Absorption** A drug injected into a vein is absorbed by the body at a rate proportional to the amount remaining in the blood. For a certain drug, the amount $y(t)$ remaining in the blood after t hours

satisfies $y' = -0.15y$ with $y(0) = 5$ mg. Find $y(t)$ and use your answer to estimate the amount present after 2 hours.

50. **BUSINESS: Stock Value** One model for the growth of the value of stock in a corporation assumes that the stock has a limiting "market value" L, and that the value $v(t)$ of the stock on day t satisfies the differential equation $v' = a(L - v)$ for some constant a. Find a formula for the value $v(t)$ of a stock whose market value is $L = 40$ if on day $t = 10$ it was selling for $v = 30$.

51. **GENERAL: Dam Sediment** A hydroelectric dam generates electricity by forcing water through turbines. Sediment accumulating behind the dam, however, will reduce the flow and eventually require dredging. Let $y(t)$ be the amount of sediment (in thousands of tons) accumulated in t years. If sediment flows in from the river at the constant rate of 20 thousand tons annually, but each year 10% of the accumulated sediment passes through the turbines, then the amount of sediment remaining satisfies the differential equation $y' = 20 - 0.1y$.
   a. By factoring the right-hand side, write this differential equation in the form $y' = a(M - y)$. Note the value of M, the maximum amount of sediment that will accumulate.
   b. Solve this (factored) differential equation together with the initial condition $y(0) = 0$ (no sediment until the dam was built).
   c. Use your solution to find when the accumulated sediment will reach 95% of the value of M found in step (a). This is when dredging is required.

52. **BIOMEDICAL: Glucose Levels** Solve Exercise 56 on page 602 by factoring the right-hand side of the differential equation to write it in the form $y' = a(M - y)$.

53. **GENERAL: Raindrops** *(Requires a slope field program)* Why do larger sized raindrops fall faster than smaller ones? It depends on the resistance they encounter as they fall through the air. For large raindrops, the resistance to gravity's acceleration is proportional to the *square* of the velocity, while for small droplets, the resistance is proportional to the *first power* of the velocity. More precisely, their velocities obey the following differential equations, with each differential equation leading to a different *terminal velocity* for the raindrop:

i. $\dfrac{dv}{dt} = 32.2 - 0.1115v^2$	Downpour droplets, about 0.05 inch in diameter
ii. $\dfrac{dv}{dt} = 32.2 - 52.6v$	Drizzle droplets, about 0.003 inch in diameter
iii. $\dfrac{dv}{dt} = 32.2 - 5260v$	Fog droplets, about 0.0003 inch in diameter

(The 32.2 represents gravitational acceleration, and the other constant is determined experimentally.)

a. Use a slope field program to graph the slope field of differential equation (i) on the window [0, 3] by [0, 20] (using x and y instead of t and v). From the slope field, must the solution curves rising from the bottom level off at a particular y-value? Estimate the value. This number is the terminal velocity (in feet per second) for a downpour droplet.

b. Do the same for differential equation (ii), but on the window [0, 0.1] by [0, 1]. What is the terminal velocity for a drizzle droplet?

c. Do the same for differential equation (iii), but on the window [0, 0.001] by [0, 0.01]. What is the terminal velocity for a fog droplet?

d. At this speed [from part (c)], how long would it take a fog droplet to fall 1 foot? This shows why fog clears so slowly.

Source: Journal of Meteorology 6

 54. GENERAL: Automobile Fatalities A study used a logistics curve to predict the number of automobile fatalities

based on the total spending on alcohol advertising in the United States in a year.

a. Enter the data from the following table into your graphing calculator. (Incidentally, the last line gives the situation at the time of the study: $1 billion of alcohol advertising and 20,000 alcohol-related deaths annually.)

b. Use the STAT then CALC then LOGISTIC operations of your calculator to fit a logistics curve to the data.

c. Use the logistics function to predict the number of alcohol-related deaths if alcohol advertising were eliminated ($x = 0$).

Alcohol Advertising ($ millions)	Alcohol-related Fatalities
325	12,500
850	18,700
1000	20,000

Source: Review of Economics and Statistics 79

Conceptual Exercises

55. Each of the following is algebraically equivalent to the differential equations for one of our three types of growth. Which is which?

a. $\dfrac{y'}{M - y} = a$ **b.** $\dfrac{y'}{y} = a$ **c.** $\dfrac{y'}{y} = a(M - y)$

56. Which of our three types of growth would be most likely to apply to each situation?
a. An investment that grows by $\frac{1}{2}$% each year
b. A new technology, such as high-definition TV or, long ago, the automobile
c. A new "fad" that gets a lot of news coverage

57. Each of the following is an example of one of the three types of growth functions (see page 614). Which is which?

a. $y = (1 + e^{-t})^{-1}$ **b.** $y = \frac{1}{2}e^{t/5}$
c. $y = 15 - 15e^{-t}$

58. What is the value of each growth function at $t = 0$?

a. $y = \dfrac{0.03}{1 + 2e^{-3t}}$ **b.** $y = 2.5(1 - e^{-t/2})$

c. $y = \frac{1}{2}e^{0.04t}$

59. What is the limiting behavior of each growth function as $t \to \infty$?

a. $y = \dfrac{0.03}{1 + 2e^{-3t}}$ **b.** $y = 2.5(1 - e^{-t/2})$

c. $y = \frac{1}{2}e^{0.04t}$

60. Make up a logistic growth function that begins *higher* than M and descends to M, as does the upper curve in the box on page 610. [*Hint:* Begin with $y = \dfrac{10}{1 + ce^{-t}}$ so that $y(0) = \dfrac{10}{1 + c}$ and choose a value for c that will make $y(0) > 10$.]

Explorations and Excursions The following problems extend and augment the material presented in the text.

Other Growth Models

Solve each differential equation by separation of variables.

61. BIOMEDICAL: Individual Birthrate If $y(t)$ is the size of a population at time t, then $\dfrac{y'}{y}$ is the population growth rate divided by the size of the population, and is called the *individual birthrate*. Suppose that the

individual birthrate is proportional to the size of the population, $\dfrac{y'}{y} = ay$ for some constant a. Find a formula for the size of the population after t years.

62. BIOMEDICAL: Gompertz Curve Another differential equation that is used to model the growth of a population $y(t)$ is $y' = bye^{-at}$, where a and b are constants. Solve this differential equation.

Source: F. Brauer and C. Castillo-Chávez, Mathematical Models in Population Biology and Epidemiology

63. BIOMEDICAL: Allometry Solve the differential equation of allometric growth: $y' = \dfrac{ay}{x}$ (where a is a constant).

This differential equation governs the relative growth rates of different parts of the same animal.

64. GENERAL: Population Growth Suppose that a population $y(t)$ in a certain environment grows in proportion to the square of the difference between the carrying capacity M and the present population, that is, $y' = a(M - y)^2$, where a is a constant. Solve this differential equation.

Logistic Growth Function

65. Solve the logistic differential equation $y' = ay(M - y)$ as follows:
   a. Separate variables to obtain
   $$\frac{dy}{y(M - y)} = a\,dt$$

b. Integrate, using on the left-hand side the integration formula
$$\int \frac{dy}{y(M - y)} = \frac{1}{M} \ln\!\left(\frac{y}{M - y}\right)$$
(which may be checked by differentiation).

c. Exponentiate to solve for $\dfrac{y}{(M - y)}$ and then solve for y.

d. Show that the solution can be expressed as
$$y = \frac{M}{1 + ce^{-aMt}}$$

66. Find the inflection point of the logistic curve
$$f(x) = \frac{M}{1 + ce^{-aMx}}$$
and show that it occurs at midheight between $y = 0$ and the upper limit $y = M$. [*Hint:* Do you already know $f'(x)$?]

9.3   First-Order Linear Differential Equations

Introduction

Not all differential equations can be solved by separation of variables. For example, the differential equation

$$y' = x + y$$

cannot be solved by separating variables (as you should check for yourself). But it can be solved by the method of *integrating factors,* which is the subject of this section. The method involves a clever use of the Product Rule, which we now review.

Using the Product Rule Backwards

When we apply the Product Rule to an expression containing an unknown function y, we must write the derivative of the unknown function y as y'. For example:

$$\frac{d}{dx}(x^2 y) = 2xy + x^2 y'$$

Both are functions of x | Derivative of the first | Second left alone | First left alone | Derivative of the second

Using the Product Rule
$$\frac{d}{dx}(f \cdot g) = f' \cdot g + f \cdot g'$$

Now we will reverse this process: beginning with the two terms on the right (even if written in a different order), we will express them in the form of the derivative on the left.

EXAMPLE 1 USING THE PRODUCT RULE BACKWARDS

Express $x^5 y' + 5x^4 y$ as the derivative of a product.

Solution

$$\underbrace{x^5 y' + 5x^4 y}_{\substack{\text{Original} \\ \text{expression}}} \quad = \quad \underbrace{\frac{d}{dx}(x^5 y)}_{\substack{\text{Written as the} \\ \text{derivative of a product}}}$$

Recognizing
$\frac{d}{dx}(x^5 y) = 5x^4 y + x^5 y'$
in the reverse order

PRACTICE PROBLEM 1

Express $e^{2x} y' + 2e^{2x} y$ as the derivative of a product.

Solution on page 626 >

First-Order Linear Differential Equations and Integrating Factors

This technique will be useful for solving a type of differential equation called **first-order linear,** which is defined as follows:

First-Order Linear Differential Equations

A differential equation is *first-order linear* if it can be written in the form

$$y' + p(x) \cdot y = q(x)$$

y' and y on the left, with functions of x as shown

where $p(x)$ and $q(x)$ are functions of x.

The previous equation is called the **standard form** of a first-order linear differential equation.* We solve it by multiplying both sides by an **integrating factor,** enabling us to write the left-hand side as the derivative of a product. A few examples will demonstrate the method.

EXAMPLE 2 SOLVING A FIRST-ORDER LINEAR DIFFERENTIAL EQUATION

Solve the differential equation $y' + \dfrac{2}{x} y = x.$

Solution

This *is* in standard form: $y' + \underbrace{\dfrac{2}{x}}_{p(x)} \cdot y = \underbrace{x}_{q(x)}$

$y' + p(x)y = q(x)$ with
$p(x) = \dfrac{2}{x}$ and $q(x) = x$

---

*Technically, "first-order" means that only the *first* derivative of y appears, and "linear" means that the equation is linear in y' and y. If $p(x)$ and $q(x)$ are continuous on a closed interval, then the differential equation is guaranteed to have a solution on that interval.

We multiply through by x^2. (The choice of x^2 will be explained shortly.)

$$x^2 y' + x^2 \frac{2}{x} y = x^2 x \qquad\qquad y' + \frac{2}{x}y = x \quad \text{times } x^2$$

$$\underbrace{\phantom{x^2\frac{2}{x}}}_{2x} \quad \underbrace{}_{x^3}$$

$$\underbrace{x^2 y' + 2xy}_{\frac{d}{dx}(x^2 y)} = x^3 \qquad\qquad\qquad \text{Simplifying}$$

$$\qquad\qquad\qquad\qquad\qquad\qquad \text{The left-hand side is now} \\ \text{the derivative of } x^2 y$$

Therefore,

$$\frac{d}{dx}(x^2 y) = x^3$$

We integrate both sides, on the left side simply dropping the d/dx, and on the right side using the Power Rule:

$$\underbrace{x^2 y = \frac{1}{4}x^4 + C}_{} \qquad\qquad\qquad \text{Integrating}$$

$$\qquad \uparrow\!\!\text{— From } \int x^3 dx = \frac{1}{4}x^4 + C$$

Finally, we solve for the unknown function y by dividing both sides by x^2, obtaining the general solution of the differential equation:

$$y = \frac{1}{4}x^2 + \frac{C}{x^2} \qquad\qquad x^2 y = \tfrac{1}{4}x^4 + C \quad \text{divided by } x^2$$

PRACTICE PROBLEM 2

Check this solution by verifying that it solves the original differential equation.

Solution on page 626 >

How to Choose the Integrating Factor

In Example 2, the x^2 that we multiplied by is called an "integrating factor" because it enabled us to write the left-hand side as a derivative, which could then be integrated. How do we choose the integrating factor? There is a simple formula.

Integrating Factor

For the first-order linear differential equation

$$y' + p(x) \cdot y = q(x)$$

the integrating factor is

$$I(x) = e^{\int p(x)dx} \qquad\qquad \begin{array}{l} e \text{ to the integral of the} \\ \text{function multiplying } y \end{array}$$

Take Note

In this form, $p(x)$ multiplies the y. Then you integrate and exponentiate it to get the integrating factor.

We omit the constant of integration in calculating the integrating factor, since it has no effect on the solution (as is shown in Exercise 57).

EXAMPLE 3   FINDING AN INTEGRATING FACTOR AND SOLVING

Solve the differential equation $y' + y = e^{3x}$.

Solution   The equation is already in standard form: $y' + y = e^{3x}$

with $\begin{cases} p(x) = 1 \\ q(x) = e^{3x} \end{cases}$ The integrating factor is

$$\underset{\substack{\uparrow \\ p(x)=1}}{} \quad \overset{\overbrace{}^{q(x)}}{}$$

$$I(x) = e^{\int 1\,dx} = e^{x} \qquad\qquad I(x) = e^{\int p(x)\,dx}$$

Multiplying by $I(x) = e^{x}$,

$$\underbrace{e^{x}y' + e^{x}y}_{\frac{d}{dx}(e^{x}y)} = \underbrace{e^{x}e^{3x}}_{e^{4x}} \qquad\qquad \begin{array}{l} y' + y = e^{3x} \\ \text{multiplied by } e^{x} \end{array}$$

$$\frac{d}{dx}(e^{x}y) = e^{4x} \qquad\qquad \begin{array}{l}\text{Writing the left-hand} \\ \text{side as the derivative} \\ \text{of a product}\end{array}$$

$$\underbrace{e^{x}y = \frac{1}{4}e^{4x} + C}_{} \qquad\qquad \begin{array}{l}\text{Integrating both sides} \\ \left(\text{dropping the } \dfrac{d}{dx}\right)\end{array}$$

$$\overset{\uparrow}{\quad}\text{From}\ \ \int e^{4x}\,dx = \frac{1}{4}e^{4x} + C$$

$$y = \frac{1}{4}e^{3x} + Ce^{-x} \qquad\qquad \begin{array}{l}\text{Solving for } y \text{ by dividing} \\ \text{both sides by } e^{x}. \text{ This is} \\ \text{the general solution.}\end{array}$$

FOR MORE HELP

with natural logarithms, see
pages 261–264.

Note that in the solution the constant of integration C (from integrating the right-hand side) does not stand "by itself," but is multiplied by e^{-x}. In general, the constant C may appear *anywhere* in the solution.

Finding integrating factors may involve using the properties of logarithms. For example, the integrating factor for the differential equation $y' + \dfrac{2}{x}y = x$ in Example 2 (pages 619–620) is found as follows:

$$I(t) = e^{\int p(x)\,dx} = e^{\int \frac{2}{x}\,dx} = e^{2\int \frac{1}{x}\,dx} \quad = \quad e^{2\ln x}$$

$$\underset{p(x)}{\uparrow} \quad \underset{\substack{\text{Taking out} \\ \text{the constant}}}{\underbrace{}} \qquad \underset{\substack{\text{Integrating} \\ 1/x \text{ gives } \ln x}}{\underbrace{}}$$

$$= \ \underbrace{e^{\ln x^{2}}}_{\substack{\text{Using} \\ 2\ln x = \ln x^{2}}} \ = \ \underbrace{x^{2}}_{\substack{\text{Using} \\ e^{\ln z} = z}}$$

You should study the steps in this calculation carefully, since they occur frequently with integrating factors. The last two steps used natural logarithm properties 8 and 4 (see the inside back cover). We omitted the absolute value bars when we integrated $1/x$ to get $\ln x$, since in most applications the independent variable is positive, often representing time. Furthermore, in this particular case, including the absolute value bars leads to the same integrating factor as *without* the absolute value bars, since the x is squared.

PRACTICE PROBLEM 3

Find the integrating factor for $y' + \dfrac{4}{x}y = x$.

Solution on page 626 >

As before, an initial condition is used to evaluate the arbitrary constant.

EXAMPLE 4   **FINDING A PARTICULAR SOLUTION**

Solve the differential equation and initial condition $\begin{cases} xy' - 3y = x^2 \\ y(1) = 5 \end{cases}$.

Solution

$$y' - \frac{3}{x}y = x$$

Dividing by x to write the equation in standard form with $p(x) = -3/x$

↦ Try to skip to here ↧

$$I(x) = e^{\int \frac{-3}{x}dx} = e^{-3\int \frac{1}{x}dx} = e^{-3\ln x} = e^{\ln x^{-3}} = x^{-3}$$

Integrating factor is x^{-3}

$$x^{-3}y' - x^{-3}\frac{3}{x}y = x^{-3}x$$

$y' - \dfrac{3}{x}y = x$ multiplied by x^{-3}

$$x^{-3}y' - 3x^{-4}y = x^{-2}$$

Simplifying

$$\frac{d}{dx}(x^{-3}y) = x^{-2}$$

Expressing the left side as the derivative of a product

$$x^{-3}y = -x^{-1} + C$$

Integrating both sides

$\underbrace{\phantom{x^{-3}y = -x^{-1}+C}}$
⌞ From $\int x^{-2}dx = -x^{-1} + C$

$$y = -x^2 + Cx^3$$

Solving for y by multiplying by x^3

$$5 = -1 + C$$

Substituting $y = 5$ and $x = 1$ (from the given $y(1) = 5$)

Therefore $C = 6$, giving

$$y = -x^2 + 6x^3$$

$y = -x^2 + Cx^3$ with $C = 6$ (the particular solution)

The general procedure for solving first-order linear differential equations is as follows:

Solving First-Order Linear Differential Equations

1. Write the differential equation in standard form $y' + p(x) \cdot y = q(x)$.

2. Find the integrating factor $I(x) = e^{\int p(x)\,dx}$ (omitting the constant of integration).

3. Multiply the equation through by the integrating factor $I(x)$.

4. Express the left-hand side as the derivative of a product, $\dfrac{d}{dx}[I(x) \cdot y]$.

5. Integrate both sides (dropping the d/dx on the left side, and writing the constant of integration on the right side).

6. Solve the resulting equation for y. If there is an initial condition, use it to evaluate the arbitrary constant of integration.

This general procedure works because multiplying by the integrating factor changes the left-hand side of the differential equation into

$$e^{\int p(x)\,dx}\,y' + e^{\int p(x)\,dx}\,p(x) \cdot y \quad \text{which equals} \quad \frac{d}{dx}\left(e^{\int p(x)\,dx}\,y\right)$$

This last expression can then be integrated and the equation solved for y. The next Example shows that integrating the right-hand side may require a substitution or some other integration technique.

EXAMPLE 5   USING AN INTEGRATING FACTOR AND A SUBSTITUTION

Solve the differential equation and initial condition $\begin{cases} y' = 6x - 2xy \\ y(0) = 2 \end{cases}$.

Solution

$$y' + 2xy = 6x \qquad\qquad \text{Standard form, with } p(x) = 2x$$

$$I(x) = e^{\int 2x\,dx} = e^{x^2} \qquad\qquad \text{Integrating factor}$$

$$e^{x^2}y' + e^{x^2}2xy = e^{x^2}6x \qquad\qquad y' + 2xy = 6x \text{ times } e^{x^2}$$

$$\frac{d}{dx}(e^{x^2}y) = e^{x^2}6x \qquad\qquad \begin{array}{l}\text{Expressing the left side as}\\\text{the derivative of a product}\end{array}$$

$$e^{x^2}y = \int e^{x^2}6x\,dx = 3\underbrace{\int \underbrace{e^{x^2}}_{e^u}\,\underbrace{2x\,dx}_{du}}_{\substack{\text{Using}\\ \int e^u\,du = e^u + C}} = 3e^{x^2} + C \qquad \begin{array}{l}\text{Integrating both sides}\\\text{(using the substitution}\\\text{method on } \int e^{x^2}6x\,dx)\end{array}$$

$$y = 3 + Ce^{-x^2} \qquad\qquad \begin{array}{l}\text{Solving for } y \text{ by multiplying}\\ e^{x^2}y = 3e^{x^2} + C \text{ by } e^{-x^2}\end{array}$$

$$2 = 3 + C \qquad\qquad \begin{array}{l}\text{Substituting } y = 2\\\text{and } x = 0 \text{ (from}\\\text{the given } y(0) = 2)\end{array}$$

Therefore, $C = -1$, giving

$$y = 3 - e^{-x^2} \qquad\qquad \begin{array}{l}y = 3 + Ce^{-x^2} \text{ with}\\ C = -1 \text{ (the particular}\\\text{solution)}\end{array}$$

Radon Pollution

Radon is a colorless, odorless gas formed by the natural decay of uranium in the earth. When dispersed in the atmosphere it is harmless, but if allowed to concentrate in a room or basement, it becomes a serious health hazard. Indoor radon pollution is second only to smoking as a certified cause of lung cancer, killing an estimated 7000 to 30,000 Americans each year.*

EXAMPLE 6   VENTING RADON POLLUTION

A 1000-cubic-foot room has a radon level of 500 pCi (picocuries) per cubic foot. A ventilation system is installed, which each hour brings in 100 cubic feet of outside air (containing 4 pCi per cubic foot), while an equal volume of air leaves the room. Assuming that the air in the room mixes thoroughly, find how soon the radon level will fall to 112 pCi per cubic foot, the safety level set by the Environmental Protection Agency (EPA).

Solution

Let

$$y(t) = \begin{pmatrix} \text{Amount of radon in} \\ \text{room after } t \text{ hours} \end{pmatrix}$$

so

$$y(0) = 500,000 \qquad \text{Initial amount (500 per cubic foot times 1000 cubic feet)}$$

The concentration of radon *in the incoming air* is given as 4, and the concentration *in the room* at any time is $y/1000$ [the amount (y) of radon in the room divided by the volume (1000) of the room]. Multiplying these concentrations by 100, the volume of air flowing in and out, gives the rates at which radon enters and leaves the room each hour. We show these rates in an inflow-outflow diagram for a 1-hour time period.

$$\textbf{IN} \xrightarrow[\substack{= 400}]{\substack{4 \text{ per cu ft} \\ \times\ 100 \text{ cu ft}}} \left[y(t) = \left\{ \begin{array}{c} \text{Total radon} \\ \text{in room} \end{array} \right\} \right] \xrightarrow[\substack{= \frac{y}{10} = 0.1y}]{\substack{\frac{y}{1000} \text{ per cu ft} \\ \times\ 100 \text{ cu ft}}} \textbf{OUT}$$

(originally 500,000)

volume: 1000 cu ft

This leads to the following differential equation:

$$y' \;=\; 400 \;-\; 0.1y \qquad\qquad \begin{pmatrix} \text{Rate of} \\ \text{change} \end{pmatrix} = \begin{pmatrix} \text{Amount} \\ \text{added} \end{pmatrix} - \begin{pmatrix} \text{Amount} \\ \text{removed} \end{pmatrix}$$

Rate of   is   400   0.1y
change of y added   removed

---

*See Bernard Cohen, *Radon, a Homeowner's Guide to Detection and Control* (Consumer Reports Books). His book measures radon per *liter*, while we measure it per *cubic foot*. To change our numbers to "per liter," multiply them by 0.0353.

$$y' + 0.1y = 400 \qquad \text{In standard form}$$

$$I(t) = e^{\int 0.1\, dt} = e^{0.1t} \qquad \text{Integrating factor}$$

$$e^{0.1t}y' + 0.1e^{0.1t}y = 400e^{0.1t} \qquad y' + 0.1y = 400 \text{ times } e^{0.1t}$$

$$\frac{d}{dt}\left(e^{0.1t}y\right) = 400e^{0.1t} \qquad \begin{array}{l}\text{Left-hand side as the}\\ \text{derivative of a product}\end{array}$$

$$e^{0.1t}y = 400 \cdot 10e^{0.1t} + C \qquad \text{Integrating}$$

From $\displaystyle\int e^{0.1t}\,dt = \frac{1}{0.1}e^{0.1t} + C$

$$y = 4000 + Ce^{-0.1t} \qquad \begin{array}{l}\text{Solving for } y \text{ by}\\ \text{multiplying by } e^{-0.1t}\end{array}$$

$$500{,}000 = 4000 + C \qquad \begin{array}{l}\text{Substituting } y = 500{,}000\\ \text{and } t = 0\\ \text{(from } y(0) = 500{,}000)\end{array}$$

Therefore, $C = 496{,}000$, giving

$$y = 4000 + 496{,}000e^{-0.1t} \qquad \begin{array}{l}y = 4000 + Ce^{-0.1t}\\ \text{with } C = 496{,}000\end{array}$$

Amount of radon after t hours

To find when the EPA limit of 112 pCi per cubic foot is reached, we solve

$$\underbrace{4000 + 496{,}000e^{-0.1t}}_{\substack{\text{Total amount of radon}\\ \text{(according to formula)}}} = \underbrace{112{,}000}_{\substack{\text{Total radon at EPA safety level}\\ \text{(112 per cubic foot times 1000 cubic feet)}}}$$

We could solve this equation for t as we did in the preceding section, or we could use a graphing calculator (see the following Graphing Calculator Exploration). Either way, we would find that the radon concentration will fall to the EPA level in about 15.2 hours.

The amount of radon remaining in the room after t hours is shown by the following graph.

Graph $y_1 = 4000 + 496{,}000e^{-0.1x}$ and $y_2 = 112{,}000$ on the window $[0, 24]$ by $[0, 500{,}000]$. Use INTERSECT to find where they meet.

The "long-run" amount of radon in the room is found by taking the limit as t approaches infinity:

$$\underset{t \to \infty}{\lim} \; (4000 + \underbrace{496{,}000e^{-0.1t}}_{\substack{\text{Approaches } 0 \\ \text{as } t \to \infty}}) = 4000 \qquad\qquad 4000 \text{ pCI}$$

This same number can be found as follows: outside air (with 4 pCi per cubic foot) filling the 1000-cubic-foot room will eventually result in $4 \cdot 1000 = 4000$ pCi of radon. This agreement serves as a check of the solution.

    Example 6 is a typical "mixture problem," in which two mixtures (here, outside and inside air), each having a different concentration of a particular substance (radon), are combined and you want to find the concentration at some later time. The same technique works for *any* mixture, whether it is air pollution in a room or chemicals in a container. It is generally easiest to let $y(t)$ stand for the *total amount* of the substance in the room (or container) at any time, so that the *concentration* can be expressed $\dfrac{y}{\text{Total volume}}$. Making an *inflow-outflow* diagram for the gain and loss per time unit leads to a differential equation, which you then solve. (This differential equation could also have been solved by separation of variables.)

Solutions TO PRACTICE PROBLEMS

1. $e^{2x}y' + 2e^{2x}y = \dfrac{d}{dx}(e^{2x}y)$ Using $\dfrac{d}{dx}e^{2x} = 2e^{2x}$

2. $y = \dfrac{1}{4}x^2 + \dfrac{C}{x^2} = \dfrac{1}{4}x^2 + Cx^{-2}$ The solution

$y' = \dfrac{1}{2}x - 2Cx^{-3} = \dfrac{1}{2}x - \dfrac{2C}{x^3}$

Substituting these expressions into the differential equation:

$$\left(\dfrac{1}{2}x - \dfrac{2C}{x^3}\right) + \dfrac{2}{x}\left(\dfrac{1}{4}x^2 + \dfrac{C}{x^2}\right) \overset{?}{=} x$$

$$\dfrac{1}{2}x - \dfrac{2\cancel{C}}{\cancel{x^3}} + \dfrac{1}{2}x + \dfrac{2\cancel{C}}{\cancel{x^3}} \overset{?}{=} x$$

$$x = x \;\checkmark \qquad\qquad\qquad \text{It checks}$$

$\longrightarrow$ Try to skip to here $\longrightarrow$

3. $I(x) = e^{\int \frac{4}{x}dx} = e^{4\int \frac{1}{x}dx} = e^{4\ln x} = e^{\ln x^4} = x^4$

9.3   ## Section Summary

A first-order linear differential equation can be written in the standard form

$$y' + p(x) \cdot y = q(x) \qquad \text{\textit{p} and \textit{q} are functions of \textit{x} alone}$$

We solve first-order linear differential equations by multiplying both sides by the *integrating factor* $I(x) = e^{\int p(x)\,dx}$, expressing the left-hand side as the derivative of a product, integrating both sides, and then solving for y.

9.3 Exercises

1–4. Solve and check each first-order linear differential equation.

1. $y' + 2y = 8$

2. $y' - y = 2$

3. $y' - 2y = e^{-2x}$

4. $y' + 3y = e^{3x}$

5–20. Solve each first-order linear differential equation.

5. $y' + \dfrac{5}{x}y = 24x^2$

6. $y' + \dfrac{4}{x}y = 12x$

7. $xy' - y = x^2$

8. $xy' - y = x$

9. $y' + 3x^2y = 9x^2$

10. $y' - 4x^3y = 8x^3$

11. $y' - 2xy = 0$

12. $y' + xy = 0$

13. $(x + 1)y' + y = 2x$

14. $y' + \dfrac{2x}{x^2 + 1}y = 3$

15. $y' - \dfrac{2}{x}y = 6x^3 - 9x^2$

16. $xy' - y = x^3 e^{x^2}$

17. $y' = x + y$

[Hint for Exercises 17 and 18: Use integration by parts or an integral table.]

18. $xy' + y = x^2 \ln x$

19. $y' + (\cos x)y = \cos x$

20. $y' - (\sin x)y = \sin x$

21–28. Solve each differential equation with the given initial condition.

21. $y' + 3y = 12e^x$
$y(0) = 5$

22. $y' + 4y = e^{-3x}$
$y(0) = 4$

23. $xy' + 2y = 14x^5$
$y(1) = 1$

24. $xy' + 4y = 10x$
$y(1) = 0$

25. $xy' = 2y + x^2$
$y(1) = 3$

26. $xy' = 3y + x^4$
$y(1) = 7$

27. $y' + 2xy = 4x$
$y(0) = 0$

28. $y' - 3x^2y = 6x^2$
$y(0) = 1$

29–30. Solve each differential equation in two ways: first by using an integrating factor, and then by separation of variables (as in Section 9.1).

29. $y' = y + 1$

30. $y' = xy$

31–32. Each exercise below has one differential equation that can be solved by an integrating factor and one that can be solved by separation of variables (as in Section 9.1). Solve each by the appropriate technique.

31. a. $y' + xy^2 = 0$
b. $y' = y + x^2 e^x$

32. a. $yy' = x$
b. $y' + y = e^{-x}$

Applied Exercises

33. BUSINESS: Value of a Company A company grows in value by 10% each year, and also gains 20% of a growing market estimated at $100e^{0.1t}$ million dollars, where t is the number of years that the company has been in business. Therefore, the value $y(t)$ of the company (in millions of dollars) satisfies

$$y' = 0.1y + 20e^{0.1t}$$

$y(0) = 5$ \qquad (Initial value $5 million)

a. Solve this differential equation and initial condition to find a formula for the value of the company after t years.

b. Use your solution to find the value of the company after 2 years.

34. ECONOMICS: Exports A country's cumulative exports $y(t)$ (in millions of dollars) grow in proportion to its *average* size y/t over the last t years, plus a fixed growth rate (10), and so satisfy

$$y' = \frac{1}{t}y + 10 \qquad \text{(for } t \geq 1\text{)}$$

$y(1) = 8$ \qquad ($8 million after 1 year)

a. Solve this differential equation and initial condition to find the country's cumulative exports after t years.

b. Use your solution to find the country's cumulative exports after 5 years.

35. BIOMEDICAL: AIDS The total number $y(t)$ of cases of AIDS (acquired immunodeficiency syndrome) diagnosed in a certain region after t years satisfies

$$y' - \frac{3}{t}y = 0 \qquad (\text{for} \quad t \geq 1)$$
$$y(1) = 125 \qquad (\text{125 cases after 1 year})$$

a. Solve this differential equation and initial condition. (Your solution will show that AIDS does not spread logistically, as do most epidemics, but like a power. This means that AIDS will spread more slowly, which seems to result from its being transmitted at different rates within different subpopulations.)

b. Use your solution to predict the number of AIDS cases in the region by the year $t = 15$.

Source: Centers for Disease Control

36. ENVIRONMENTAL SCIENCE: Global Warming The burning of coal and oil is increasing the amount of carbon dioxide in the atmosphere, which is expected to trap more solar radiation and increase global temperatures (the "greenhouse effect"), possibly raising the sea level by melting polar ice and flooding coastal regions. The current 3300 billion tons of carbon dioxide in the atmosphere is growing by about 50 billion tons per year, while about 1% of the accumulated carbon dioxide is removed by natural processes. Therefore, the amount $y(t)$ of carbon dioxide (in billions of tons) t years from now satisfies

$$y' = -0.01y + 50$$
$$y(0) = 3300 \qquad (\text{Initial amount})$$

a. Solve this differential equation and initial condition.

b. Find when the amount of carbon dioxide will reach 4000 billion tons, at which time global temperatures are expected to have risen by 3 degrees Fahrenheit.

c. Use your solution to find the long-run amount of carbon dioxide in the atmosphere.

Source: Worldwatch

37–38. BUSINESS: Ponzi Schemes A Ponzi scheme is an investment fraud that promises high returns but in which funds, instead of being invested, are merely used to pay returns to the investors and profits to the fund manager. To avoid running out of money, new investors must be brought in at an increasing rate to provide funds to pay off existing investors. Eventually the scheme must collapse in debt when not enough new investors can be found.*

37. Suppose that each of 10 investors deposits $100,000 into a fund and is promised a 20% annual return. However, the $1,000,000 collected is used to pay each investor the required $20,000 return, with the

*They are named after Charles Ponzi, who in 1920 collected $15 million from 40,000 investors before being imprisoned and deported. Beginning in the 1980's, Bernard Madoff ran a Ponzi scheme that left debts of $65 billion. He is currently serving a 150-year prison sentence.

remaining $800,000 kept by the fund manager. Let $y(t)$ be the total number of investors needed after t years so that incoming funds will be enough to pay the existing investors plus the manager's $800,000. Representing dollar amounts in thousands, we have

$$\left(\begin{array}{c}\text{Annual} \\ \text{inflow}\end{array}\right) = 100 \cdot \frac{dy}{dt}$$

$$\left(\begin{array}{c}\text{Annual} \\ \text{outflow}\end{array}\right) = 20 \cdot y + 800$$

For inflow to equal outflow, we must have the differential equation and initial condition

$$\begin{cases} 100 \cdot \dfrac{dy}{dt} = 20y + 800 \\ y(0) = 10 \end{cases}$$

a. Solve this differential equation and initial condition.

b. Use your solution to find how many investors would be needed after 10 years, after 20 years, after 30 years, and after 50 years.

38. Suppose that each of 20 investors deposits $100,000 into a fund and is promised a 25% annual return. However, the $2,000,000 collected is used to pay each investor the required $25,000 return, with the remaining $1,500,000 kept by the fund manager. Let $y(t)$ be the total number of investors needed after t years so that incoming funds will be enough to pay the existing investors plus the manager's $1,500,000. Representing dollar amounts in thousands, we have

$$\left(\begin{array}{c}\text{Annual} \\ \text{inflow}\end{array}\right) = 100 \cdot \frac{dy}{dt}$$

$$\left(\begin{array}{c}\text{Annual} \\ \text{outflow}\end{array}\right) = 25 \cdot y + 1500$$

For inflow to equal outflow, we must have the differential equation and initial condition

$$\begin{cases} 100 \cdot \dfrac{dy}{dt} = 25y + 1500 \\ y(0) = 20 \end{cases}$$

a. Solve this differential equation and initial condition.

b. Use your solution to find how many investors would be needed after 10 years, after 20 years, after 30 years, and after 50 years.

39. ENVIRONMENTAL SCIENCE: Algae Bloom An algae bloom in a lake is a sudden growth of algae that consumes nutrients and blocks sunlight, killing other life in the water. Suppose that the number of tons $y(t)$ of algae after t weeks satisfies

$$y' = ty + t$$
$$y(0) = 2 \qquad (\text{Initial tonnage})$$

a. Solve this differential equation and initial condition.

b. Use your solution to find the amount of algae after 2 weeks.

c. Graph the solution on a graphing calculator and find when the algae bloom will reach 40 tons.

40. BIOMEDICAL: Drug Absorption A patient's ability to absorb a drug sometimes changes with time, and the dosage must therefore be adjusted. Suppose that the number of milligrams $y(t)$ of a drug remaining in the patient's bloodstream after t hours satisfies

$$y' = -\frac{1}{t}y + t \qquad (\text{for } t \geq 1)$$

$$y(3) = 5 \qquad \text{(5 mg after 3 hours)}$$

Solve this differential equation and initial condition to find the amount remaining in the bloodstream after t hours.

41. ENVIRONMENTAL SCIENCE: Radon Pollution A 10,000-cubic-foot-room has an initial radon level of 800 pCi (picocuries) per cubic foot. A ventilation system is installed that each hour brings in 500 cubic feet of outside air (containing 5 pCi per cubic foot), while an equal volume of air leaves the room. Assume that the air in the room mixes thoroughly.

   a. Find a differential equation and initial condition that govern the total amount $y(t)$ of radon in the room after t hours.
   b. Solve this differential equation and initial condition.
    c. Graph the solution on a graphing calculator and find how soon the radon level will fall to the EPA safety level of 112 pCi per cubic foot.

42. ENVIRONMENTAL SCIENCE: Smoke Pollution A 5000-cubic-foot room has 400 smoke particles per cubic foot. A ventilation system is turned on that each minute brings in 50 cubic feet of outside air (containing 10 smoke particles per cubic foot), while an equal volume of air leaves the room. Assume that the air in the room mixes thoroughly.

   a. Find a differential equation and initial condition that govern the total number $y(t)$ of smoke particles in the room after t minutes.
   b. Solve this differential equation and initial condition.
    c. Graph the solution on a graphing calculator and find how soon the smoke level will fall to 100 smoke particles per cubic foot.

43. ENVIRONMENTAL SCIENCE: Smoke Pollution A 12,000-cubic-foot room has 500 smoke particles per cubic foot. A ventilation system is turned on that each minute brings in 600 cubic feet of smoke-free air, while an equal volume of air leaves the room. Also, during each minute, smokers in the room add a total of 10,000 particles of smoke to the room. Assume that the air in the room mixes thoroughly.

   a. Find a differential equation and initial condition that govern the total number $y(t)$ of smoke particles in the room after t minutes.
   b. Solve this differential equation and initial condition.
   c. Find how soon the smoke level will fall to 100 smoke particles per cubic foot.

44. ENVIRONMENTAL SCIENCE: Water Purity A 500-gallon tank is filled with water containing 0.2 ounce of impurities per gallon. Each hour, 200 gallons of water (containing 0.01 ounce of impurities per gallon) is added and mixed into the tank, while an equal volume of water is removed.

   a. Write a differential equation and initial condition that describe the total amount $y(t)$ of impurities in the tank after t hours.
   b. Solve this differential equation and initial condition.
   c. Use your solution to find when the impurities will reach 0.05 ounce per gallon, at which time the water may be used for drinking.
   d. Use your solution to find the "long-run" amount of impurities in the tank.

45. ENVIRONMENTAL SCIENCE: Water Purity The water in a 100,000-gallon reservoir contains 0.1 gram of pesticide per gallon. Each hour, 2000 gallons of water (containing 0.01 gram of pesticide per gallon) is added and mixed into the reservoir, and an equal volume of water is drained off.

   a. Write a differential equation and initial condition that describe the amount $y(t)$ of pesticide in the reservoir after t hours.
   b. Solve this differential equation and initial condition.
    c. Graph your solution on a graphing calculator and find when the amount of pesticide will reach 0.02 gram per gallon, at which time the water is safe to drink.
   d. Use your solution to find the "long-run" amount of pesticide in the reservoir.

46. PERSONAL FINANCE: Banking You deposit $8000 into a bank account paying 5% interest compounded continuously, and you withdraw funds continuously at the rate of $1000 per year. Therefore, the amount $y(t)$ in the account after t years satisfies

$$y' = 0.05y - 1000$$

$$y(0) = 8000$$

   a. Solve this differential equation and initial value.
   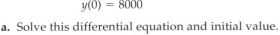 b. Graph your solution on a graphing calculator and find how long it will take until the account is empty.

47. BIOMEDICAL: Dieting A person's weight $w(t)$ after t days of eating c calories per day can be modeled by the following differential equation

$$w' + 0.005w = \frac{c}{3500}$$

where the 0.005 represents the proportional weight loss per day when eating nothing, and 3500 is the conversion rate for calories into pounds.

   a. If a person initially weighing 170 pounds goes on a diet of 2100 calories per day, find a formula for the person's weight after t days.
   b. Use your solution to find when the person will have lost 15 pounds.   *(continues)*

c. Find the "limiting weight" that will be a approached if the person continues on this diet indefinitely.

48. BIOMEDICAL: Dieting (*continuation*) Use the differential equation in Exercise 47 to find the number of calories a person weighing 170 pounds should eat each day in order to lose 6 pounds in 21 days.

Conceptual Exercises

49. Is a differential equation of the form
$u(x)y' + v(x)y = w(x)$ first-order linear? If so, what is its standard form?

50. Can we solve *any* first-order linear differential equation by the method of this section?[*Hint:* Can you always carry out every step?]

51. Which of the differential equations for the three types of growth summarized on page 614 are first-order linear?

52. Can a *separable* differential equation also be first-order linear? [*Hint:* See the definition of *separable* on page 592 and compare it to the first-order linear standard form. For what types of function $g(y)$ can these be the same?]

53. If you tried to solve the "unlimited growth" differential equation (see page 614) by the method of this section, what would be the integrating factor?

54. If you tried to solve the "limited growth" differential equation (see page 614) by the method of this section, what would be the integrating factor?

55. The air in a 100,000-cubic-foot room contains 483 pCi of radon per cubic foot. A ventilation system is turned on that each hour will bring in 7250 cubic feet of fresh air containing 8 pCi of radon per cubic foot, while an equal amount of air leaves the room. Assuming that the air mixes thoroughly, in the long run, what will be the total amount of radon in the room? [*Hint:* No calculation necessary.]

56. Explain why we cannot use an integrating factor $I(x)$ to solve the *nonlinear* differential equation $y' + p(x)y^2 = q(x)$. [*Hint:* Can we obtain the left-hand side by differentiating an expression of the form $I(x)y$ or $I(x)y^2$?]

Explorations and Excursions   The following problems extend and augment the material presented in the text.

57. OMITTING THE CONSTANT Show that the constant of integration in the integrating factor can be omitted. [*Hint:* Show that multiplying by the integrating factor $e^{\int p(x)\,dx+C}$ just multiplies the differential equation by a nonzero constant, which does not change the solution.]

58. SOLVING BY A FORMULA Derive the formula

$$y(x) = \frac{1}{I(x)} \int I(x)q(x)\,dx \quad \text{for the solution of}$$

the first-order linear differential equation
$y' + p(x)y = q(x)$, where $I(x) = e^{\int p(x)\,dx}$.
(*Note:* Rather than memorizing this formula, it is better to understand the method, which generalizes to an even larger class of differential equations.)

59–60. BIOMEDICAL: Drug Dynamics When you swallow a pill, the medication passes through your stomach lining into your bloodstream, where some is absorbed by the cells of your body and the rest continues to circulate for future absorption. The amount $y(t)$ of medication remaining in the bloodstream after t hours can be modeled by the differential equation

$$\frac{dy}{dt} = abe^{-bt} - cy$$

for constants a, b, and c (respectively the dosage of the pill, the dissolution constant of the pill, and the absorption

constant of the medication). For the given values of the constants:

a. Substitute the constants into the stated differential equation.
b. Solve the differential equation (with the initial condition of having no medicine in the bloodstream at time $t = 0$) to find a formula for the amount of medicine in the bloodstream at any time t (hours).
c. Use your solution to find the amount of medicine in the bloodstream at time $t = 2$ hours.
d. Graph your solution on a graphing calculator and find when the amount of medication in the bloodstream is maximized.

Source: J. Keener and J. Sneyd, *Mathematical Physiology*

59. $a = 10\,\text{mg}$, $b = 3$, $c = 0.2$

60. $a = 30\,\text{mg}$, $b = 5$, $c = 0.3$

61. BIOMEDICAL: Drug Dynamics Read the explanation for the Exercises 59–60. Then solve the stated differential equation with the initial condition $y(0) = 0$ (initially no medication in the bloodstream). Show that your solution can be written in the form

$$y(t) = \frac{ab}{b - c}(e^{-ct} - e^{-bt}).$$

Approximate Solutions of Differential Equations: Euler's Method

Introduction

There are many techniques for solving differential equations besides those described in this book. Despite all the techniques, however, there are many differential equations that cannot be solved by *any* method (as finite combinations of elementary functions). Even if a differential equation *can* be solved, its solution may be too complicated to be useful. In such cases we must settle for an **approximate solution** that is *close* to the actual solution, and one way of finding such an approximation is *Euler's method*. Leonhard Euler (1707–1783; his name is pronounced "Oiler") was a prolific Swiss mathematician in whose honor the constant e was named.

Euler's Method

A differential equation together with an initial condition is called an **initial value problem.** For an initial value problem of the form

$$y' = g(x, y)$$

$$y(a) = y_0$$

Initial value $y = y_0$
at $x = a$

Euler's method finds an approximation to the solution on a closed interval beginning at $x = a$.* Euler observed that if the solution $y(x)$ passes through a point (x, y) in the plane, then at that point the solution has slope $g(x, y)$, since $y' = g(x, y)$. This idea is used to construct a **polygonal approximation** (consisting of connected line segments) that is tangent to the true solution at the initial point, as shown in the following graph.

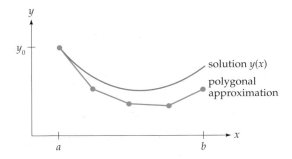

The actual (but generally unknown) solution
$y(x)$ in red and below it a polygonal approximation
with $n = 4$ segments.

To construct the Euler approximation on an interval $[a, b]$ using n segments, we divide $[a, b]$ into n subintervals by equally spaced points $a = x_0 < x_1 < x_2 < \ldots < x_n = b$ with spacing (step size) $h = \dfrac{b - a}{n}$ as shown in the following diagram.

---

*Technically, if g and $\dfrac{\partial g}{\partial y}$ are continuous for all x in an interval $[a, b]$ and all values of y, then the initial value problem has a solution for x-values in some subinterval beginning at a. Less stringent conditions can also be given.

Beginning with $x_0 = a$, successive x-values are found by adding h to the previous x-value until reaching $x_n = b$. The segments are constructed on these subintervals as follows.

The initial condition $y(a) = y_0$ gives the initial point (x_0, y_0) (since $x_0 = a$), and the differential equation gives the slope at this point as $y' = g(x_0, y_0)$. The first segment is constructed from this point, (x_0, y_0), with this slope, $g(x_0, y_0)$, extending a step size h to the right, as illustrated below.

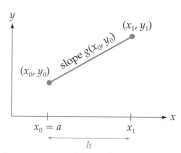

The first segment of the Euler approximation.

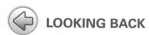

LOOKING BACK

The point-slope formula (page 9) is $y - y_1 = m(x - x_1)$.

The equation of this segment comes from the point-slope form for a line, with the given point and slope:

$$y - y_0 = g(x_0, y_0) \cdot (x - x_0)$$

point: (x_0, y_0)
slope: $m = g(x_0, y_0)$

or

$$y = y_0 + g(x_0, y_0) \cdot (x - x_0)$$

Adding y_0 to each side
(for $x_0 \le x \le x_1$)

The y-coordinate y_1 at the end of the first segment is found by evaluating at $x = x_1$:

$$y_1 = y_0 + g(x_0, y_0) \cdot \underbrace{(x_1 - x_0)}_{h} = y_0 + g(x_0, y_0) \cdot h$$

Formula above
at $x = x_1$

The second segment is constructed similarly, beginning at the previous end point (x_1, y_1) and with slope $g(x_1, y_1)$ [from $y' = g(x, y)$ evaluated at this same point], as illustrated in the following diagram.

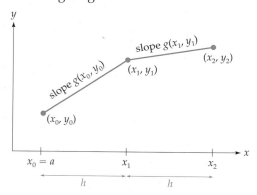

First two segments of the Euler approximation.
The slope of each segment is $g(x, y)$ evaluated
at the left-hand endpoint of the segment.

Just as before, this leads to a formula for the second segment:

$$y = y_1 + g(x_1, y_1) \cdot (x - x_1) \qquad \text{For} \quad x_1 \le x \le x_2$$

This process is then repeated, with each new segment beginning at the previous endpoint and having slope $y' = g(x, y)$ evaluated at this left endpoint. The process continues until it reaches the final x-value $x_n = b$. An example will illustrate the method.

EXAMPLE 1　FINDING AN APPROXIMATE SOLUTION

For the initial value problem $\begin{cases} y' = 3x - y \\ y(0) = 4 \end{cases}$, find the Euler approximation

to the solution on the interval $[0, 1.5]$ using $n = 3$ segments. State the approximation to the solution at $x = 1.5$.

Solution

The function g is

$$g(x, y) = 3x - y$$

Right-hand side of the differential equation

The step size (between successive x-values) is

$$h = \frac{1.5 - 0}{3} = \frac{1.5}{3} = 0.5$$

$h = \dfrac{b - a}{n}$ with $a = 0$, $b = 1.5$, and $n = 3$

The initial point is

$$x_0 = 0, \quad y_0 = 4$$

From the given $y(0) = 4$

Segment 1

Slope　$g(x_0, y_0) = g(0, 4) = 3 \cdot 0 - 4 = -4$

$g(x, y) = 3x - y$ with $x = 0$ and $y = 4$

Endpoint $x_1 = 0.5$, $y_1 = 4 + (-4) \cdot (0.5) = 2$

$x_1 = x_0 + h$, $y_1 = y_0 + g(x_0, y_0) \cdot h$

Segment 2

Slope　$g(x_1, y_1) = g(0.5, 2) = 3 \cdot (0.5) - 2 = -0.5$

$g(x, y) = 3x - y$ with $x = 0.5$ and $y = 2$

Endpoint $x_2 = 1$, $y_2 = 2 + (-0.5) \cdot (0.5) = 1.75$

$x_2 = x_1 + h$, $y_2 = y_1 + g(x_1, y_1) \cdot h$

Segment 3

Slope　$g(x_2, y_2) = g(1, 1.75) = 3 \cdot 1 - 1.75 = 1.25$

$g(x, y) = 3x - y$ with $x = 1$ and $y = 1.75$

Endpoint $x_3 = 1.5$, $y_3 = 1.75 + (1.25) \cdot (0.5) = 2.375$

$x_3 = x_2 + h$, $y_3 = y_2 + g(x_2, y_2) \cdot h$

The points $(0, 4)$, $(0.5, 2)$, $(1, 1.75)$, and $(1.5, 2.375)$ determine the polygonal approximation of the solution. The y-values are the approximations of the solution at the x-values. In particular, the approximation at the endpoint $x = 1.5$ is

$$y(1.5) \approx y_3 = 2.375$$

The polygonal approximation of the solution to this problem is graphed below.

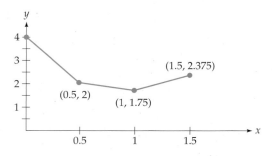

Euler approximation to the solution

of $\begin{cases} y' = 3x - y \\ y(0) = 4 \end{cases}$ with $n = 3$ segments.

 **Graphing
Calculator
Exploration**

Graphing calculator programs that carry out the Euler approximation for any number of segments, draw the polygonal approximation, and show the coordinates of the points, are easily available on the Internet, from sites such as `ticalc.org` (search for *Euler*). A typical screen is shown on the left. It shows the approximation for the initial value problem in the preceding Example along with the coordinates of the final point.

Accuracy of the Euler Approximation

In Example 1 we used only $n = 3$ segments, so we should not expect great accuracy. Small values of n like 3 are only for demonstration purposes. More realistic values of n are in the hundreds or thousands, with the calculations carried out on a programmable calculator or computer. The following diagram shows the Euler approximation for $n = 3$ that we found, and also the approximations for $n = 6$ and $n = 100$, together with the actual solution (found by using an integrating factor).

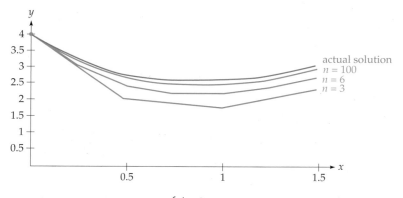

Actual solution to $\begin{cases} y' = 3x - y \\ y(0) = 4 \end{cases}$ (in red) and

the Euler approximations using 3, 6, and 100 segments.

The accuracy for $n = 100$ segments may be seen numerically from the following table.

x-Values	Euler Approximation ($n = 100$)	Actual y-Values
0	4	4
0.5	2.73	2.75
1	2.56	2.58
1.5	3.04	3.06

very close

Inaccuracies in Euler's approximation come from two sources. First, the *actual* solution changes slope *continuously,* while the approximation changes slope only at the segment endpoints. Second, the points at which the new slopes are calculated are not generally on the actual solution (except for the initial point), and so do not give the true slope of the solution curve. For greater accuracy one may increase n (but this involves more calculation and therefore more roundoff errors), or use a more advanced method (such as the Runge-Kutta method).

PRACTICE PROBLEM

For the initial value problem $\begin{cases} y' = 7x - 2y \\ y(3) = 6 \end{cases}$, find:

a. the initial point (x_0, y_0)

b. the slope of the solution at the initial point

Solutions below >

The following differential equation cannot be solved by either of the "exact" methods of this chapter, so we use an *Euler's method* graphing calculator program.

 Graphing Calculator Exploration

 Let $y(x)$ be the value (in millions of dollars) of an electronics company x years after it was founded. The company is expected to grow, but more slowly as inefficiency increases with its size. Suppose that its value $y(x)$ satisfies

$$y' = 2.6x - e^{0.3y}$$
$$y(0) = 1 \qquad \text{(Starting value is \$1 million)}$$

Use Euler's method with $n = 50$ to estimate the value of the company after 5 years.

The screen on the left shows the graph from a typical *Euler's method* program on the interval [0, 5] beginning with $y(0) = 1$ and using 50 line segments, along with the coordinates of the final point, which gives the value of the company.

The value of the company will be \$8.35 million.

Solutions TO PRACTICE PROBLEM

a. $(3, 6)$

b. Slope: $y' = g(3, 6) = 7 \cdot 3 - 2 \cdot 6 = 21 - 12 = 9$ Since $g(x, y) = 7x - 2y$

For an initial value problem of the form $\begin{cases} y' = g(x, y) \\ y(a) = y_0 \end{cases}$, Euler's method gives an *approximation* of the solution on an interval $[a, b]$. The approximation consists of n connected line segments, each on an interval of length $h = (b - a)/n$, beginning from the initial point at $x_0 = a$, with slope calculated from $y' = g(x, y)$ evaluated at the left-hand endpoint of the segment. The endpoints of these segments, after the initial point (x_0, y_0) with $x_0 = a$, are calculated successively from the formulas

$$\left. \begin{array}{l} x_i = x_{i-1} + h \\ y_i = y_{i-1} + g(x_{i-1}, y_{i-1}) \cdot h \end{array} \right\} \quad i = 1, 2, \ldots, n$$

The accuracy generally increases as the number of segments n increases, but at the cost of more calculation. While the calculations may be carried out "by hand" (as in Example 1), they are easily done on a graphing calculator or computer.

9.4 Exercises

1. If the solution to $y' = 5x - 4y$ passes through the point $(4, 3)$, what is the slope of the solution at that point?

2. If the solution to $y' = x + 4y$ passes through the point $(3, 1)$, what is the slope of the solution at that point?

3. For the initial value problem $\begin{cases} y' = 4xy \\ y(1) = 3 \end{cases}$ state the initial point (x_0, y_0) and calculate the slope of the solution at this point.

4. For the initial value problem $\begin{cases} y' = x/y \\ y(6) = 2 \end{cases}$ state the initial point (x_0, y_0) and calculate the slope of the solution at this point.

5–10. For each initial value problem, calculate the Euler approximation for the solution on the interval $[0, 1]$ using $n = 4$ segments. Draw the graph of your approximation. (Carry out the calculations "by hand" with the aid of a calculator, rounding to two decimal places. Answers may differ slightly, depending on when you do the rounding.)

5. $y' = 3x - 2y$
$y(0) = 2$

6. $y' = x + 2y$
$y(0) = 1$

7. $y' = 4xy$
$y(0) = 1$

8. $y' = 8x^2 - y$
$y(0) = 2$

9. $y' + 2y = e^{4x}$
$y(0) = 2$

10. $y' + e^y = 8x$
$y(0) = 0$

 11–16. For each initial value problem:

a. Use an *Euler's method* graphing calculator program to find the estimate for $y(2)$. Use the interval $[0, 2]$ with $n = 50$ segments.

b. Solve the differential equation and initial condition *exactly* by separating variables or using an integrating factor.

c. Evaluate the solution that you found in part (b) at $x = 2$. Compare this *actual* value of $y(2)$ with the *estimate* of $y(2)$ that you found in part (a).

11. $y' = \dfrac{x}{y}$
$y(0) = 1$

12. $y' = -xy$
$y(0) = 1$

13. $\dfrac{dy}{dx} = 0.2y$
$y(0) = 1$

14. $\dfrac{dy}{dx} = -y$
$y(0) = 1$

15. $y' + y = 2e^x$
$y(0) = 5$

16. $y' + y = e^{-x}$
$y(0) = 0$

17–22. For each initial value problem, use an *Euler's method* graphing calculator program to find the approximate solution at the stated x-value, using 50 segments. [*Hint:* Use an interval that begins at the initial x-value and ends at the stated x-value.]

17. $y' = xe^{-y}$
$y(1) = 0.5$

Approximate the solution at $x = 3$

18. $y' = e^{x/y}$
$y(1) = 0.4$

Approximate the solution at $x = 3$

19. $y' = \dfrac{x+1}{y+1}$
$y(1) = \dfrac{1}{2}$

Approximate the solution at $x = 2.5$

20. $y' = 1 + \dfrac{y}{x}$
$y(2) = -1$

Approximate the solution at $x = 3.5$

21. $\dfrac{dy}{dx} = \sqrt{x + y}$

$y(3) = 1$

Approximate the
solution at $x = 3.8$

22. $\dfrac{dy}{dx} = (x - y)^2$

$y(2) = 0$

Approximate the
solution at $x = 2.8$

23–24. Use an *Euler's method* graphing calculator program to find the approximate solution at the stated x-value, using the given numbers of segments.

23. $y' = e^{x/y}$

$y(0) = 1$

Approximate the solution at $x = 1$ using:

a. $n = 10$

b. $n = 100$

c. $n = 500$

24. $y' = x^2 + y^2$

$y(0) = 0$

Approximate the solution at $x = 2$ using:

a. $n = 10$

b. $n = 100$

c. $n = 500$

Applied Exercises

25. BUSINESS: Sales Growth Sales of a popular new computer game are growing so that the number $y(t)$ (in thousands) sold in the first t months satisfies

$$y' = 2e^{0.05y} + 2$$

$$y(0) = 0$$

Use Euler's method with $n = 3$ to estimate $y(3)$, the number sold in the first 3 months. (Carry out the calculations "by hand" with the aid of a calculator, rounding to two decimal places.)

26. BUSINESS: Value of a Company A biotechnology company is growing so that its value $y(t)$ (in millions of dollars) after t years satisfies

$$y' = \frac{1}{5} e^{\sqrt{y}} + 1$$

$$y(0) = 1 \qquad \text{(Initially worth \$1 million)}$$

Use Euler's method with $n = 4$ to estimate $y(2)$, the company's value after 2 years. (Carry out the calculations "by hand" with the aid of a calculator, rounding to two decimal places.)

27. GENERAL: Dam Sediment Sediment is flowing into a lake behind a hydroelectric dam at the rate of 1000 tons per year, while 10% of the upper layer of accumulated sediment is flushed through the turbines each year. The amount $y(t)$ of sediment after t years satisfies

$$\frac{dy}{dt} = 1000 - 0.1y^{2/3}$$

$$y(0) = 0 \qquad \text{(Initially no sediment)}$$

Use an *Euler's method* graphing calculator program with $n = 50$ to estimate $y(2)$, the amount of sediment after 2 years.

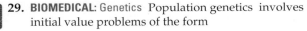

28. ENVIRONMENTAL SCIENCE: Crop Parasite A crop parasite is spreading so that the affected area $y(t)$ (in square miles) in t years is growing according to

$$\frac{dy}{dt} = \sqrt{y^2 + 100}$$

$$y(0) = 5 \qquad \text{(Initially affects 5 square miles)}$$

Use an *Euler's method* graphing calculator program with $n = 50$ to estimate the area affected by the parasite after 1 year.

29. BIOMEDICAL: Genetics Population genetics involves initial value problems of the form

$$q' = -0.01q^2(1 - q)$$

$$q(0) = 0.9$$

in which $q(t)$ is the frequency of a gene after t generations of evolution. Use an *Euler's method* graphing calculator program with $n = 100$ to approximate $q(200)$, the gene frequency after 200 generations.

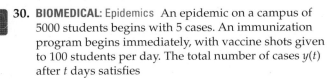

30. BIOMEDICAL: Epidemics An epidemic on a campus of 5000 students begins with 5 cases. An immunization program begins immediately, with vaccine shots given to 100 students per day. The total number of cases $y(t)$ after t days satisfies

$$y' = 0.0001y(5000 - y - 100t)$$

$$y(0) = 5$$

a. Use an *Euler's method* graphing calculator program with $n = 50$ to estimate $y(4)$, the total number of cases after 4 days.

b. Find the answer to part (a) if 300 students (instead of 100) are vaccinated each day.

Conceptual Exercises

31. When using Euler's method to solve $y' = g(x, y)$, suppose that the function $g(x, y)$ *increases* as x increases (for every fixed value of y). Will the approximation *under*estimate or *over*estimate the exact solution? [*Hint:* Think of the graphs of the true solution and the approximation.]

32. Suppose that you use Euler's method to solve $y' = g(x, y)$ with initial condition $y(0) = 5$ on the interval [0, 2] with just *one* segment. What will be the value of $y(2)$? [*Hint:* No calculation needed—your answer will involve g.]

33. When will a differential equation of the form $y' = g(x, y)$ be *separable*? That is, what must be the form of $g(x, y)$? [*Hint:* See page 592.]

34. When will a differential equation of the form $y' = g(x, y)$ represent *logistic growth*? That is, what must be the form of $g(x, y)$? [*Hint:* See page 610.]

35. When will a differential equation of the form $y' = g(x, y)$ be *first-order linear*? That is, what must be the form of $g(x, y)$? [*Hint:* See page 619.]

36. What happens with Euler's method when the function $g(x, y)$ is a *constant*? To be specific, suppose that you use Euler's method to solve $y' = 7$ with initial condition $y(0) = 1$ on the interval [0, 2]. What will be your approximation for $y(2)$, and how accurate will it be? [*Hint:* What will be the slope of each segment? Since the segments are attached together, how many segments do you really need?]

37. In Euler's method applied to $y' = g(x, y)$, the slope of each segment is chosen to be g evaluated at the *left-hand endpoint* of the segment. If g is in fact a function of x alone, can you think of a better way to choose the slope of the segment?

38. If the function $g(x, y)$ in Euler's method turns out to be a function of x alone, then approximating the solution to $y' = g(x)$ amounts to approximating the integral of a function on an interval. Long ago (in Section 5.3) we had another way of approximating integrals (areas), using Riemann sums with left rectangles. Is there a connection?

---

9   Chapter Summary with Hints and Suggestions

Reading the text and doing the exercises in this chapter have helped you to master the following concepts and skills, which are listed by section (in case you need to review them) and are keyed to particular Review Exercises. Answers for all Review Exercises are given at the back of the book, and full solutions can be found in the Student Solutions Manual.

9.1   Separation of Variables

- Find the general solution of a differential equation by separation of variables. (*Review Exercises 1–10.*)

- Find a particular solution of a differential equation with an initial condition. (*Review Exercises 11–14.*)

- Solve an applied problem involving a differential equation. (*Review Exercises 15–18.*)

- Use a program to graph the slope field of a differential equation, and sketch the solution through a given point. (*Review Exercises 19–20.*)

9.2   Further Applications of Differential Equations: Three Models of Growth

- Choose an appropriate differential equation for an applied problem, and use it to solve the problem. (*Review Exercises 21–28.*)

9.3   First-Order Linear Differential Equations

- Find the general solution of a first-order linear differential equation. (*Review Exercises 29–34.*)

$$y' + p(x) \cdot y = q(x) \qquad I(x) = e^{\int p(x)\, dx}$$

- Find a particular solution of a first-order linear differential equation. (*Review Exercises 35–38.*)

- Solve an applied problem involving a first-order linear differential equation. (*Review Exercises 39–42.*)

9.4   Approximate Solutions of Differential Equations: Euler's Method

- Find an Euler approximation for the solution of a differential equation (see the diagram below and the formulas on the next page). Then find the *exact* solution and compare the results. (*Review Exercises 43–44.*)

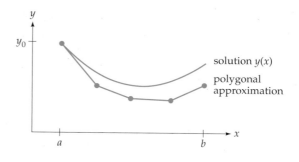

$$\begin{cases} y' = g(x, y) \\ y(a) = y_0 \end{cases} \qquad \begin{cases} x_i = x_{i-1} + h \\ y_i = y_{i-1} + g(x_{i-1}, y_{i-1}) \cdot h \end{cases}$$
$$\text{for } i = 1, 2, \ldots, n$$

- Find an Euler approximation for the solution of a differential equation. (*Review Exercises 45–46.*)

- Solve an applied problem by finding an Euler approximation for the solution of a differential equation. (*Review Exercises 47–48.*)

Hints and Suggestions

- A differential equation is an equation involving derivatives (rates of change). A solution involving an arbitrary constant is called a *general solution,* while a solution with the arbitrary constant replaced by a number is called a *particular solution.* The constant is determined by an *initial condition,* specifying the value of the solution at a particular point.

- Solving a differential equation by separation of variables involves moving the x's and y's to opposite sides of the equation and integrating both sides. Many useful differential equations can be solved by this technique, but many cannot. In fact, many differential equations cannot be solved by *any* method.

- A graphing calculator with a slope field program can show a "picture" of a differential equation of the form $\dfrac{dy}{dx} = f(x, y)$, drawing little slanted dashes with the correct slopes at many points of the plane. A solution can then be drawn through a given point following the indicated slopes.

- With practice, you can *guess* the integrating factor that makes a first-order linear differential equation integrable. If not, use the formula on page 620.

- Euler's method is easily carried out on a graphing calculator or a computer.

---

9   Review Exercises and Chapter Test

◯ indicates a Chapter Test exercise.

9.1   Separation of Variables

1–10. Find the general solution of each differential equation.

(1) $y^2 y' = x^2$ **2.** $y' = x^2 y$ **3.** $y' = \dfrac{x^3}{x^4 + 1}$

4. $y' = xe^{-x^2}$ **5.** $y' = y^2$ **6.** $y' = y^3$

7. $y' = 1 - y$ **8.** $y' = \dfrac{1}{y}$

9. $y' = xy - y$ **10.** $y' = x^2 + x^2 y$

11–14. Solve each differential equation and initial condition.

11. $y^2 y' = 3x^2$ **12.** $y' = \dfrac{y}{x^2}$

     $y(0) = 1$ $y(1) = 1$

(13) $y' = \dfrac{y}{x^3}$ **14.** $y' = \sqrt[3]{y}$

     $y(1) = 1$ $y(1) = 0$

(15) PERSONAL FINANCE: Wealth Accumulation Suppose that you now have $10,000 and that you expect to save an additional $4000 during each year, and all of this is deposited in a bank paying 5% interest compounded continuously. Let $y(t)$ be your bank balance (in thousands of dollars) after t years.

   a. Write a differential equation and initial condition to model your bank balance.
   b. Solve your differential equation and initial condition.
   c. Use your solution to find your bank balance after 10 years.

16. ENVIRONMENTAL SCIENCE: Pollution A town discharges 4 tons of pollutant annually into a once-pristine lake, and each year bacterial action removes 25% of the accumulated pollution.

   a. Write a differential equation and initial condition for the amount of pollution in the lake.
   b. Solve your differential equation to find a formula for the amount of pollution in the lake after t years.

17. BIOMEDICAL: Fever Thermometers How long should you keep a thermometer in your mouth to take your temperature? *Newton's Law of Cooling* says that the thermometer reading rises at a rate proportional to the difference between your actual temperature and the present reading. For a fever of 106 degrees Fahrenheit, the thermometer reading $y(t)$ after t minutes in your mouth satisfies $y' = 2.3(106 - y)$ with $y(0) = 70$ (initially at room temperature). (The constant 2.3 is typical for household thermometers.) Solve this differential equation and initial condition.

18. BIOMEDICAL: Fever Thermometers (*continuation*) Use your solution $y(t)$ in Exercise 17 to calculate $y(1)$, $y(2)$, and $y(3)$, the thermometer readings after 1, 2, and 3 minutes. Do you see why 3 minutes is the usually recommended time for keeping the thermometer in your mouth?

19–20. For each differential equation and initial condition:

   a. Use a graphing calculator *slope field* program to graph the slope field for the differential equation on the window $[-5, 5]$ by $[-5, 5]$.
   b. Sketch the slope field on a piece of paper and draw a solution curve that follows the slopes and that passes through the point $(0, -2)$.

(continues)

c. Solve the differential equation and initial condition.

d. Use the program to graph the slope field and the solution that you found in part (c). How good was the sketch that you made in part (b) compared with the solution graphed in part (d)?

(19) $\begin{cases} \dfrac{dy}{dx} = \dfrac{x^2}{y^2} \\ y(0) = -2 \end{cases}$ **20.** $\begin{cases} \dfrac{dy}{dx} = \dfrac{x}{y^2} \\ y(0) = -2 \end{cases}$

9.2 Further Applications of Differential Equations: Three Models of Growth

For each situation, write an appropriate differential equation (unlimited, limited, or logistic). Then find its solution and solve the problem.

21. BUSINESS: Movie Prices Since the year 2000 the price of a movie ticket has been growing continuously at the rate of about 4% per year. If in 2010 the nationwide average price was $7.85, estimate the price in the year 2018.

Source: Entertainment Weekly

22. BUSINESS: Computer Expenditure The amount spent by U.S. businesses on computers and related hardware is increasing continuously at the rate of 12% per year. If in 2008 the amount was $132 billion, estimate the amount in the year 2018.

Source: Census Bureau

(23) **BIOMEDICAL: Epidemics** A virus spreads through a university community of 8000 people at a rate proportional to both the number already infected and the number not yet infected. If it begins with 10 cases and grows in a week to 150 cases, estimate the size of the epidemic after 2 weeks.

24. SOCIAL SCIENCE: Rumors A rumor spreads through a school of 500 students at a rate proportional to both the number who have heard and the number who have not heard the rumor. If the rumor began with 2 students and within a day had spread to 75, how many students will have heard the rumor within 2 days?

(25) **BUSINESS: Total Sales** A manufacturer estimates that he can sell a maximum of 10,000 DVD players in a city. His total sales grow at a rate proportional to how far they are below this upper limit. If after 7 months the total sales are 3000, find a formula for the total sales after *t* months. Then use your answer to estimate the total sales at the end of the first year.

26. BEHAVIORAL SCIENCE: Learning Suppose that the maximum rate at which a mail carrier can sort letters is 60 letters per minute, and that she learns at a rate proportional to her distance from this upper limit. If after 2 weeks on the route she can sort 25 letters per minute, how many weeks will it take her to sort 50 letters per minute?

27. BUSINESS: Advertising A new product is advertised extensively on television to a city of 500,000 people, and the number of people who have seen the ads increases at a rate proportional to the number who have not yet seen the ads. If within 2 weeks 200,000 have seen the ads, how long must the product be advertised to reach 400,000 people?

28. BUSINESS: Sales A company estimates the maximum market for fax (facsimile transmission) machines in a city to be 40,000. Sales are growing in proportion to both the number already sold and the size of the remaining market. If 1000 fax machines have been sold at time $t = 0$, and after 1 year 4000 have been sold, estimate how long it will take for 20,000 to be sold.

9.3 First-Order Linear Differential Equations

29–34. Solve each first-order linear differential equation.

(29) $xy' - 5y = 4x^7$ **30.** $xy' + \dfrac{1}{2}y = 3x$

31. $y' + xy = x$ **32.** $y' - 2xy = -4x$

33. $xy' + y = xe^x$ **34.** $xy' + y = x \ln x$

[*Hint for Exercises 33 and 34:* Use integration by parts or an integral table.]

35–38. Solve each first-order linear differential equation and initial condition.

(35) $y' - xy = 0$ **36.** $y' + 4xy = 0$
     $y(0) = 3$ $y(0) = 2$

37. $xy' + 2y = 6x$ **38.** $xy' - y = 4$
     $y(1) = 0$ $y(1) = 6$

39. BUSINESS: Investment Fund An investment fund grows in value by 5% each year, and also attracts new capital at the rate of $10e^{0.05t}$ thousand dollars per year, where *t* is the number of years since the fund was established. Therefore, its value $y(t)$ (in thousands of dollars) satisfies

$$y' = 0.05y + 10e^{0.05t}$$

$y(0) = 100$ (Initial value $100 thousand)

a. Solve this differential equation and initial condition.

b. Use your answer to part (a) to find the value of the fund after 5 years.

 c. Graph the solution you found in part (a) on a graphing calculator, using an appropriate window. Find when the fund will reach a quarter of a million dollars. [*Hint:* Use INTERSECT with $y_2 = 250$.]

40. GENERAL: Population Growth with Immigration A country's population has a natural growth rate of 2%, and also accepts immigrants amounting to 1% of a neighboring population of $10e^{0.01t}$ million at time t years. Therefore, the population $p(t)$ (in millions) satisfies

$$p' = 0.02p + 0.1e^{0.01t}$$

$$p(0) = 100 \qquad \text{(Initial population)}$$

a. Solve this differential equation and initial condition.

b. Use your answer to part (a) to estimate the population at time $t = 10$ years.

c. Graph the solution you found in part (a) on a graphing calculator, using an appropriate window. Find when the population will reach 150 million. [*Hint:* Use INTERSECT with $y_2 = 150$.]

41 **ENVIRONMENTAL SCIENCE:** Air Quality The air in a stuffy 10,000-cubic-foot lecture hall is only 10% oxygen. Each minute, 500 cubic feet of air are exchanged with the outside (containing 20% oxygen). Assume that the air in the room mixes thoroughly.

a. Write a differential equation and initial condition to describe the total amount $y(t)$ of oxygen in the room after t minutes.

b. Solve the differential equation and initial condition.

c. Use your solution to find the amount of oxygen after 15 minutes of ventilation.

d. Use your solution to find the long-run oxygen content of the room.

e. Graph the solution you found in part (b) on a graphing calculator, using an appropriate window. Find when the oxygen level will reach 19%. [*Hint:* Use INTERSECT with y_2 equal to the amount of oxygen for the 19% level.]

42. **ENVIRONMENTAL SCIENCE:** Bird Population The natural growth rate of a population of bald eagles in a nature sanctuary is 20% per year, and two more eagles are added to the population each year. Therefore, the population $p(t)$ t years from now satisfies

$$p' = 0.2p + 2$$

$$p(0) = 5 \qquad \text{(Initial population)}$$

a. Solve this differential equation and initial condition.

b. Use your solution to predict the eagle population after 2 years.

c. Graph the solution you found in part (b) on a graphing calculator, using an appropriate window. Find when the eagle population will reach 24.

9.4 Approximate Solutions of Differential Equations: Euler's Method

43–44. For each initial value problem:

a. Calculate the Euler approximation on the interval [0, 1] using $n = 4$ segments. Carry out the calculations "by hand" with the aid of a calculator, rounding

calculations to two decimal places. Draw the graph of your approximation.

b. Solve the differential equation and initial condition *exactly* by separating variables or using an integrating factor.

c. Evaluate at $x = 1$ the solution that you found in part (b). Compare this actual value of $y(1)$ (rounded to two decimal places) with the *estimate* for $y(1)$ that you found in part (a). [*Note:* Expect low accuracy using $n = 4$.]

43. $y' = 2 - 2y$ **44.** $y' = \dfrac{4x}{y}$

$y(0) = 2$ $y(0) = 1$

45–46. For each initial value problem, use an *Euler's method* graphing calculator program to find an estimate for the solution at the stated x-value, using 50 segments.

45 $y' = \dfrac{1}{y^2 + 1}, \quad y(0) = 1$

Approximate the solution at $x = 2$

46. $y' = ye^{-x}, \quad y(1) = 5$

Approximate the solution at $x = 2.5$

47 **BUSINESS:** Advertising If a new product becomes known *both* by word of mouth *and* by advertising, then $p(t)$, the proportion of people who have heard of the new product after t weeks, satisfies a differential equation and initial condition of the form

$$p' = 0.1(1 - p) + 0.1p(1 - p)$$

$$p(0) = 0.3 \quad \text{(Initially 30\% know the product)}$$

a. Show that the differential equation simplifies to $p' = 0.1(1 - p^2)$

b. Use an *Euler's method* graphing calculator program with 90 segments to estimate $p(4)$, the proportion of people who have heard of the product after 4 weeks.

48. **ENVIRONMENTAL SCIENCE:** Insect Population A colony of flying insects occupying a spherical nest grows so that the number $y(t)$ of insects after t weeks satisfies

$$y' = 0.2y - 0.1\sqrt[3]{y}$$

$$y(0) = 100 \qquad \text{(Initially 100 insects)}$$

Use an *Euler's method* graphing calculator program with 80 segments to estimate the population after 2 weeks.

Sequences and Series

10

rubberball/Getty Images

What You'll Explore

Series are *sums*, and this chapter will concentrate on *infinite* series, sums of infinitely many terms. You will use infinite series to calculate long-term effects of tax cuts, government expenditures, and drug dosages, and to design *perpetuities* that generate a fixed income forever. You will also find remarkably simple formulas for functions such as sin x, cos x, and e^x, and a method for approximating solutions to equations that cannot be solved exactly .

10.1 Geometric Series

10.2 Taylor Polynomials

10.3 Taylor Series

10.4 Newton's Method

Tax Cuts, Government Spending, and Multiplier Effects

Tax cuts and government spending have been much in the news recently, connecting economics, politics, and social philosophy. How could we assess the effect of, say, a $30 billion government expenditure or tax cut on the economy? If divided equally (a highly unlikely event), it would give just $100 to each American. On a personal level, what would you do with your $100? You might spend most of it, say, 80%, amounting to $100 \cdot 0.8 = \$80$, and save the rest. (Economists call this 80% the *marginal propensity to consume,* or *MPC*.) The amount you spend would go to other people, who might similarly spend 80% of their share (in dollars, $100 \cdot 0.8 \cdot 0.8$ or $100 \cdot 0.8^2$), paying others who would in turn spend 80% (or $100 \cdot 0.8^3$), and so on in an endless chain.

Therefore, the original $100 would have the effect of generating a total expenditure throughout the economy of

$$100 + 100 \cdot 0.8 + 100 \cdot 0.8^2 + 100 \cdot 0.8^3 + \cdots$$

Such "infinitely long sums" are called *infinite series* and are the subject of this chapter. This particular infinite series sums to 500, *five times* the original $100. This 5 is called the *multiplier,* since any initial amount (in place of the 100) would similarly be multiplied by 5 (see page 650). Scaling up to the national level, the $30 billion expenditure or tax cut would also be multiplied by 5, thereby generating expenditures of $150 billion throughout the economy.

A few caveats about the effects of cutting taxes: A tax cut deprives the government of exactly the same amount of money, and the resulting decrease in governmental spending is subject to a roughly similar multiplier, with a countervailing effect on the economy. Also, wealthy people (who can afford to save) have a lower *MPC* than poorer people (who must spend all they have), with the result that a tax cut for the wealthy will have a much smaller multiplier effect in the economy than one for the lower economic classes. For the same reason, spending on social programs will have a much larger multiplier effect on the economy. We will further analyze the multiplier effect in the first section of this chapter.

10.1   Geometric Series

Introduction

The decimal expansion of $\frac{1}{3}$ is the "infinitely long decimal" $0.333\ldots$, which represents an "infinitely long sum"

$$0.3333\ldots = \frac{3}{10} + \frac{3}{100} + \frac{3}{1000} + \frac{3}{10000} + \cdots$$

Such sums are called **infinite series** and are the subject of this chapter. This first section discusses *geometric* series, first finite and then infinite.

Finite Geometric Series

A **finite series** is a sum of numbers in a particular order. The numbers are called the **terms** of the series. A **finite** *geometric* **series** is a finite series in which each successive term is obtained from the previous term by multiplying by a fixed number, denoted r (for **ratio**).

Finite Geometric Series

$$a \; + \; ar \; + \; ar^2 \; + \; ar^3 \; + \cdots + \; ar^{n-1}$$

  ↑   ↑   ↑   ↑   ↑

First   Second   Third   Fourth nth
term   term   term   term   term

$a =$ first term
$r =$ ratio
$n =$ number of terms

 Be Careful   Observe that the *number* of terms is always one more than the final exponent of r (so, the nth term has exponent $n - 1$).

Some examples of finite geometric series are:

$$2 + 4 + 8 + 16 + 32$$

First term is $a = 2$
Ratio is $r = 2$
Number of terms is $n = 5$

$$3 - 30 + 300 - 3000$$

$a = 3, \; r = -10, \; n = 4$

$$5 + \frac{5}{2} + \frac{5}{2^2} + \frac{5}{2^3} + \frac{5}{2^4} + \cdots + \frac{5}{2^{100}}$$

$a = 5, \; r = \frac{1}{2}, \; n = 101$
(n is one more than the final power of r)

A series is a sum, and so it may be evaluated by adding the terms. For the first two examples above, the sums are easily found to be 62 and -2727 (as you may check). For the last example, adding up the 101 terms would be extremely tedious. Shortly we will develop a simple *formula* for summing such a series.

    Not all series are geometric. The following series is *not* geometric, since successive terms are found by adding 1 to the previous denominator, *not* by multiplying by a fixed ratio.

$$\frac{1}{1} + \frac{1}{2} + \frac{1}{3} + \frac{1}{4} + \frac{1}{5} + \frac{1}{6}$$

PRACTICE PROBLEM 1

For each series below, if it is geometric, find the first term a, the ratio r, and the number of terms n. If the series is not geometric, say so.

a. $1 - \pi + \pi^2 - \pi^3 + \pi^4 - \pi^5$

b. $2 + \dfrac{2}{10} + \dfrac{2}{10^2} + \dfrac{2}{10^3} + \cdots + \dfrac{2}{10^8}$

c. $0.01 + 0.02 + 0.03 + 0.04 + 0.05$ Solutions on page 651 >

To find a formula for the *sum* of a finite geometric series, we denote the sum by S_n:

$$S_n = a + ar + ar^2 + \cdots + ar^{n-1}$$

From this we subtract r times S_n:

$S_n \quad = a + ar + ar^2 + \cdots + ar^{n-1}$		S_n (sum of first n terms)
$\underline{rS_n \quad = \qquad ar + ar^2 + \cdots + ar^{n-1} + ar^n}$		S_n multiplied by r
$S_n - rS_n = a \qquad\qquad\qquad\qquad\qquad - ar^n$		Subtracting (most terms cancel)
$S_n(1 - r) = a(1 - r^n)$		Factoring each side of $S_n - rS_n = a - ar^n$
$S_n = a\,\dfrac{1 - r^n}{1 - r}$		Dividing each side by $1 - r$ (for $r \neq 1$)

This formula is very useful for summing a finite geometric series.

Sum of a Finite Geometric Series

$$a + ar + ar^2 + \cdots + ar^{n-1} = a\,\frac{1 - r^n}{1 - r} \qquad\qquad (r \neq 1)$$

Memorization hint: The power n on the right is the first power *left out* of the series on the left.

EXAMPLE 1   SUMMING CUMULATIVE FINES

In the summer of 1988, the city of Yonkers, New York, was fined for refusing to build low-income housing. The fine was \$100 for the first day, \$200 for the second day, \$400 for the third day, and so on, with the fines doubling each day until the City Council approved the housing. Find the total fine for the 14 days until the housing was approved.

Source: New York Times

Solution

$$\left(\begin{array}{c}\text{Total}\\\text{fine}\end{array}\right) = \underbrace{100}_{\text{1st day}} + \underbrace{2\cdot 100}_{\text{2nd day}} + \underbrace{4\cdot 100}_{\text{3rd day}} + \cdots + \underbrace{2^{13}\cdot 100}_{\text{14th day}}$$

Geometric series (final power is 1 less than the final day)

The sum formula gives

$$\left(\begin{array}{c}\text{Total}\\\text{fine}\end{array}\right) = 100\,\frac{1 - 2^{14}}{1 - 2} = 100\,\frac{1 - \underbrace{16{,}384}_{}}{-1} = 1{,}638{,}300$$

$$\underbrace{\phantom{1 - 16{,}384}}_{16{,}383}$$

$a\,\dfrac{1 - r^n}{1 - r}$ with $a = 100$, $r = 2$, and $n = 14$

The total fine was \$1,638,300.

LOOKING BACK

Continuous annuities were analyzed by differential equations in Exercises 53–54 on page 601. Most annuities, however, are of the type discussed here.

Sum the finite geometric series $3 + 3 \cdot 2 + 3 \cdot 2^2 + \cdots + 3 \cdot 2^5$.

Solution on page 651 >

Annuities

An **annuity** is a series of equal payments at regular intervals. The total value or *amount* of an annuity, including interest, may be found from the formula for the sum of a geometric series.

| EXAMPLE 2 | FINDING THE AMOUNT OF AN ANNUITY |

At the end of each month for 5 years you deposit $200 in a bank at 6% interest compounded monthly. Find the amount of this annuity.

Solution

At 6% interest compounded monthly, you receive $\frac{1}{2}\%$ per month, so payments are multiplied by $(1 + 0.005)$ or (1.005) for each month on deposit. In 5 years there will be $5 \cdot 12 = 60$ monthly deposits, beginning with the first $200 (at the end of the first month, and so earning interest for 59 months) and ending with the final $200 (at the end of the last month, and so earning *no* interest). We sum these amounts *in reverse order*:

$$200 + 200(1.005) + 200(1.005)^2 + \cdots + 200(1.005)^{59}$$

Final payment · · · First payment

Geometric series with $a = 200$, $r = 1.005$, and $n = 60$ (final power plus 1)

The formula gives

$$\left(\begin{array}{c}\text{Amount}\\ \text{of annuity}\end{array}\right) = 200\,\frac{1 - 1.005^{60}}{1 - 1.005} \approx 13{,}954$$

Using a calculator

$a\,\dfrac{1 - r^n}{1 - r}$ with the stated values of a, r, and n

The amount of the annuity is $13,954.

Note that five years of monthly payments of $200 adds up to only $60 \cdot 200 = \$12{,}000$. The remainder here is interest.

Infinite Series

An **infinite series** is an expression written as a sum of infinitely many terms.*

$$a_1 + a_2 + a_3 + a_4 + \cdots$$

The following are examples of infinite series:

$$1 + 2 + 3 + 4 + 5 + \cdots$$

$$1 + \frac{1}{2} + \frac{1}{4} + \frac{1}{8} + \cdots + \frac{1}{2^n} + \cdots$$

$\dfrac{1}{2^n}$ represents a "typical" term

*We will sometimes begin with a term a_0, writing the series as $a_0 + a_1 + a_2 + \cdots$.   In Section 10.3 we will consider infinite series whose terms are *functions*.

We cannot sum an infinite series simply by adding the terms, because adding infinitely many numbers one-by-one would be impossible. Instead, we sum an infinite series by first finding the **partial sum** S_n of the first n terms, and then taking the *limit* as n approaches infinity. The partial sums of the infinite series $a_1 + a_2 + a_3 + \cdots$ are defined as follows:

$$S_1 = a_1 \qquad\qquad \text{First partial sum is the first term}$$

$$S_2 = a_1 + a_2 \qquad\qquad \text{Second partial sum is the sum of the first \textit{two} terms}$$

$$S_3 = a_1 + a_2 + a_3 \qquad\qquad \text{Third partial sum is the sum of the first \textit{three} terms}$$
$$\vdots$$
$$S_n = a_1 + a_2 + a_3 + \cdots + a_n \qquad\qquad \textit{n}\text{th partial sum is the sum of the first \textit{n} terms}$$
$$\vdots$$

 LOOKING BACK

The idea of dealing with the infinite by *dropping back to the finite and taking the limit* should be familiar to you—we used this same idea with improper integrals (see page 411).

If the *limit* of this sequence $S_1, S_2, S_3, \dots, S_n, \dots$ as n approaches infinity *exists* and equals a number S, then the series is said to **converge** to S, and S is the **sum of the series.**

Convergence of an Infinite Series

For an infinite series $a_1 + a_2 + a_3 + \cdots$, if the partial sums S_n approach a limit S,

$$\lim_{n \to \infty} S_n = S$$

then the series is *convergent*, and *converges* to S:

$$S = a_1 + a_2 + a_3 + \cdots$$

An infinite series that does *not* converge is said to **diverge**, or be a **divergent series**, and therefore has no sum.

Infinite Geometric Series

An infinite series in which each successive term is a fixed multiple of the previous term is called an **infinite geometric series.**

Infinite Geometric Series

$$a + ar + ar^2 + ar^3 + \cdots \qquad\qquad \begin{aligned} a &= \text{first term} \\ r &= \text{ratio} \end{aligned}$$

The partial sums of an infinite geometric series are *finite* geometric series, for which we may use our formula:

$$S_n = a\,\frac{1 - r^n}{1 - r} \qquad\qquad \text{Partial sum of an infinite geometric series (from page 645)}$$

The sum of an *infinite* geometric series is then the *limit* of S_n as n approaches infinity. If the ratio r satisfies $|r| < 1$, then r^n will approach zero as $n \to \infty$.

$$\lim_{n \to \infty} r^n = 0 \qquad\qquad \text{for } |r| < 1$$

Graphing Calculator Exploration

Verify this last statement by choosing an r satisfying $|r| < 1$ and raising it to a high power. Try a few different values of r and n.

$(1/2)^{25}$	
2.980232239ᴇ-8	← means 0.0000000298 . . .
$(-9/10)^{200}$	
7.05507911ᴇ-10	← means 0.0000000007055 . . .

In each case the result is very small, and the limit is zero.

The sum S of the infinite series is then found by taking the limit of the partial sums, and for $|r| < 1$,

Limit is zero

$$S = \lim_{n \to \infty} S_n = \lim_{n \to \infty} \underbrace{\left(a\,\frac{1 - r^n}{1 - r} \right)}_{S_n} = a\,\frac{1 - 0}{1 - r} = a\,\frac{1}{1 - r} = \frac{a}{1 - r}$$

The last expression gives a formula for the sum of a convergent infinite geometric series.

Take Note

This *infinite series* formula is simpler than the *finite series* formula on page 645 in that is omits the r^n. Surprisingly, it is often the case that the infinite is simpler than the finite.

Sum of an Infinite Geometric Series

An infinite geometric series with $|r| < 1$ is convergent, and its sum is:

$$a + ar + ar^2 + ar^3 + \cdots = \frac{a}{1 - r} \qquad \text{for } |r| < 1$$

An infinite geometric series with $|r| \geq 1$ is divergent.

The divergence of infinite geometric series with $|r| \geq 1$ will be shown in Exercise 84.

EXAMPLE 3   SUMMING GEOMETRIC SERIES

a. $\dfrac{1}{2} + \dfrac{1}{4} + \dfrac{1}{8} + \dfrac{1}{16} + \cdots = \dfrac{\dfrac{1}{2}}{1 - \dfrac{1}{2}} = \dfrac{\dfrac{1}{2}}{\dfrac{1}{2}} = 1$ \qquad $\dfrac{a}{1 - r}$ with first term $a = \frac{1}{2}$ and ratio $r = \frac{1}{2}$

b. $8 - \dfrac{8}{3} + \dfrac{8}{9} - \dfrac{8}{27} + \cdots = \dfrac{8}{1 - \left(-\dfrac{1}{3}\right)} = \dfrac{8}{\dfrac{4}{3}}$ \qquad $\dfrac{a}{1 - r}$ with $a = 8$ and $r = -\frac{1}{3}$

$$= 8 \cdot \dfrac{3}{4} = 6$$

c. $1 + \dfrac{3}{2} + \dfrac{9}{4} + \dfrac{27}{8} + \cdots$ \qquad *diverges.* \qquad $r = \dfrac{3}{2}$ so $|r| > 1$

PRACTICE PROBLEM 3

Find the sum of the series $2 + 2\left(\dfrac{3}{5}\right) + 2\left(\dfrac{3}{5}\right)^2 + 2\left(\dfrac{3}{5}\right)^3 + \cdots$

Solution on page 651 >

For medications taken over an extended period of time, the amount in the blood-stream at any time is the sum of small contributions from all past doses. This amount can be approximated by an infinite series.

EXAMPLE 4   **FINDING LONG-TERM DRUG CONCENTRATION**

A patient takes a daily dose of 150 milligrams (mg) of a drug, and each day 70% of the accumulated drug in the bloodstream is absorbed by the body. Find the long-term maximum and minimum amounts of the drug in the bloodstream.

Solution

One day after taking a dose, 70% of that dose will have been absorbed, so 30% of it will remain. After a second day, only 30% of that 30% will remain, with proportional reductions on subsequent days. The maximum amount in the bloodstream will occur just after a new dose, and it will be the most recent 150 mg, plus 30% of the previous 150 mg, plus $(0.30)^2$ of the dose previous to that, and so on, giving the infinite series

$$\left(\begin{array}{c}\text{Maximum} \\ \text{amount}\end{array}\right) = 150 + 150(0.3) + 150(0.3)^2 + \cdots$$

$$= \frac{150}{1 - 0.3} = \frac{150}{0.7} \approx 214 \qquad \text{Using } \frac{a}{1-r} \text{ with } a = 150 \text{ and } r = 0.3$$

Therefore, the long-term amount in the bloodstream just after a dose will be 214 mg. Just before the next dose, 70% of this amount will have been absorbed, leaving 30%, or $0.30 \cdot 214 \approx 64$ mg.

The long-term amount in the bloodstream will vary between 214 mg and 64 mg.

These maximum and minimum amounts are important in calculating dosages that avoid toxicity and maintain efficacy. Notice that we used an *infinite* series even though the number of past doses is certainly finite. The *approximation of the finite by the infinite* is permissible because the contribution from earlier doses rapidly becomes negligible. It also has the advantage that the infinite formula is *simpler* than the finite formula.

Graphing Calculator Exploration

```
sum(seq(150*0.3^X,X,0,4,1))
                     213.765
```

Show that the sum of the first five terms of the series in the preceding Example is already very close to the sum of the *infinite* series. [*Hint:* Sum a sequence of terms of the form $150 \cdot (0.3)^x$ with x going from 0 to 4 by increments of 1, as shown on the left.]

The long-term amount for medication given *continuously* (intravenously) may be found using differential equations (see Exercise 56 on page 602).

Multiplier Effect in Economics

As explained in the Application Preview on page 643, the **multiplier effect** shows how the effects of an expenditure spread throughout an economy. Suppose that you pay someone $1000. The recipient may spend a part of it (say 80%, or $800) and save the rest. The $800 spent goes to other recipients, who in turn spend a part (say 80%) for purchases of their own and save the rest. If this continues, with each recipient spending 80% of what is received, the result will be an infinite series of decreasing expenditures throughout the economy. We will see that the sum of all of these expenditures is a *multiple* of the original expenditure (hence the term "multiplier effect"). The percentage spent (here 80%) is called the "marginal propensity to consume" or *MPC*.

| EXAMPLE 5 | FINDING THE MULTIPLIER |

If the marginal propensity to consume is 80%, find the total of all of the expenditures generated by an initial expenditure of $1000, and find the multiplier.

Solution

The sum of all of the expenditures, beginning with the original $1000, is

$$1000 + 1000(0.8) + 1000(0.8)^2 + \cdots = \frac{1000}{1 - 0.8}$$

$$\frac{a}{1 - r} \text{ with}$$
$$a = 1000 \text{ and}$$
$$r = 0.8$$

$$= \frac{1000}{0.2} = 5000$$

The original expenditure of $1000 generated a total of $5000 of expenditures in the economy, so the multiplier is 5.

Notice that the $1000 can be replaced by *any* initial expenditure, showing that the multiplier is independent of the initial amount, depending only on the *MPC*. The multiplier is used by economists to assess the effects of government spending and tax increases and cuts.

Graphing Calculator Exploration

How many terms in the series in the preceding Example are required for the sum to reach 4500? To reach 4900? [*Hint*: Use the same operations as in the previous Graphing Calculator Exploration.]

Summation Notation

A series may be written very compactly using the symbol Σ (the Greek capital letter "sigma" corresponding to our S). Σ is followed by a formula for the terms, using an "index variable" that takes the values between those indicated at the bottom and top of the Σ (similar to the notation for definite integrals). An infinity symbol (∞) at the top indicates an *infinite* series.

Terms written out	Sigma notation	
$2 + 4 + 8 + \cdots + 2^{10} \quad =$	$\displaystyle\sum_{i=1}^{10} 2^i$	← Last value of index i is 10 ← Formula for terms ← First value of index i is 1
$\dfrac{1}{3} + \dfrac{1}{4} + \dfrac{1}{5} + \dfrac{1}{6} + \cdots \quad =$	$\displaystyle\sum_{k=3}^{\infty} \dfrac{1}{k}$	← ∞ means an *infinite series* ← Formula for terms ← First value of index k is 3

Read: the sum of $1/k$ from k equals 3 to infinity

Solutions TO PRACTICE PROBLEMS

1. a. $a = 1, \quad r = -\pi, \quad n = 6$

  b. $a = 2, \quad r = \frac{1}{10}, \quad n = 9$

  c. Not geometric

2. $3 + 3 \cdot 2 + 3 \cdot 2^2 + \cdots + 3 \cdot 2^5 = 3\,\dfrac{1 - 2^6}{1 - 2} = 3\,\dfrac{1 - 64}{-1} = 3 \cdot 63 = 189$

3. $2 + 2\left(\dfrac{3}{5}\right) + 2\left(\dfrac{3}{5}\right)^2 + 2\left(\dfrac{3}{5}\right)^3 + \cdots = \dfrac{2}{1 - \dfrac{3}{5}} = \dfrac{2}{\dfrac{2}{5}} = 2 \cdot \dfrac{5}{2} = 5$

10.1   Section Summary

A series is *geometric* if multiplying any term by a fixed number gives the next term. The sum of a *finite* geometric series is given by the following formula:

First power omitted

$$\sum_{k=0}^{n-1} ar^k = a + ar + ar^2 + \cdots + ar^{n-1} = a\,\frac{1 - r^n}{1 - r} \qquad \text{for } r \neq 1$$

In sigma notation

Terms written out (n = number of terms)

An *infinite* geometric series converges if the ratio r satisfies $|r| < 1$, and diverges if $|r| \geq 1$. The sum of a convergent infinite geometric series is given by the following formula:

$$\sum_{k=0}^{\infty} ar^k = a + ar + ar^2 + ar^3 + \cdots = \frac{a}{1 - r} \qquad \text{for } |r| < 1$$

In sigma notation

Terms written out

Notice that the "infinite" formula is simpler than the "finite" formula. These formulas for the sums of *geometric* series are particularly important since there are no such formulas for most (nongeometric) series.

10.1 Exercises

1–6. Write out each finite series.

1. $\displaystyle\sum_{i=1}^{5} \frac{1}{i+1}$

2. $\displaystyle\sum_{i=1}^{6} \frac{i}{i+1}$

3. $\displaystyle\sum_{k=1}^{4} \left(-\frac{1}{3}\right)^k$

4. $\displaystyle\sum_{k=1}^{5} \frac{(-1)^k}{k^2}$

5. $\displaystyle\sum_{n=1}^{6} \frac{1-(-1)^n}{2}$

6. $\displaystyle\sum_{n=1}^{4} (-n)^n$

7–12. Write each infinite series in sigma notation, beginning with $\sum_{i=1}^{\infty}$.

7. $\dfrac{1}{3} + \dfrac{1}{9} + \dfrac{1}{27} + \dfrac{1}{81} + \cdots$

8. $\dfrac{1}{7} + \dfrac{2}{7} + \dfrac{3}{7} + \dfrac{4}{7} + \dfrac{5}{7} + \cdots$

9. $-2 + 4 - 8 + 16 - \cdots$

10. $-3 + 9 - 27 + 81 - \cdots$

11. $-1 + 2 - 3 + 4 - \cdots$

12. $-1 + 4 - 9 + 16 - \cdots$

13–18. Find the sum of each finite geometric series.

13. $1 + 2 + 2^2 + 2^3 + \cdots + 2^9$

14. $1 + 2 + 2^2 + 2^3 + \cdots + 2^{10}$

15. $3 + 3\cdot 4 + 3\cdot 4^2 + \cdots + 3\cdot 4^5$

16. $2 + 2\cdot 3 + 2\cdot 3^2 + \cdots + 2\cdot 3^9$

17. $3 - 3\cdot 2 + 3\cdot 2^2 - 3\cdot 2^3 + \cdots + 3\cdot 2^6$

18. $4 - 4\cdot 3 + 4\cdot 3^2 - 4\cdot 3^3 + \cdots + 4\cdot 3^6$

19–40. Determine whether each infinite geometric series converges or diverges. If it converges, find its sum.

19. $4 + \dfrac{4}{5} + \dfrac{4}{25} + \dfrac{4}{125} + \cdots$

20. $5 + \dfrac{5}{6} + \dfrac{5}{36} + \dfrac{5}{216} + \cdots$

21. $2 - \dfrac{2}{3} + \dfrac{2}{9} - \dfrac{2}{27} + \dfrac{2}{81} - \cdots$

22. $27 - \dfrac{27}{2} + \dfrac{27}{4} - \dfrac{27}{8} + \dfrac{27}{16} - \cdots$

23. $\dfrac{1}{3} + \dfrac{2}{3} + \dfrac{4}{3} + \dfrac{8}{3} + \dfrac{16}{3} + \cdots$

24. $\dfrac{1}{100} - \dfrac{3}{100} + \dfrac{9}{100} - \dfrac{27}{100} + \dfrac{81}{100} - \cdots$

25. $\dfrac{3}{2} + \dfrac{3^2}{2^3} + \dfrac{3^3}{2^5} + \dfrac{3^4}{2^7} + \dfrac{3^5}{2^9} + \cdots$

26. $\dfrac{1}{3} + \dfrac{1}{3^3} + \dfrac{1}{3^5} + \dfrac{1}{3^7} + \dfrac{1}{3^9} + \cdots$

27. $8 + 6 + \dfrac{9}{2} + \dfrac{27}{8} + \cdots$

28. $9 + 6 + 4 + \dfrac{8}{3} + \cdots$

29. $\displaystyle\sum_{i=1}^{\infty} \frac{100}{5^i}$

30. $\displaystyle\sum_{i=1}^{\infty} 30\cdot 4^{-i}$

31. $\displaystyle\sum_{j=1}^{\infty} \frac{3^j}{7\cdot 2^j}$

32. $\displaystyle\sum_{j=1}^{\infty} \frac{5\cdot 2^j}{3^j}$

33. $\displaystyle\sum_{j=0}^{\infty} \left(-\frac{1}{2}\right)^j$

34. $\displaystyle\sum_{n=0}^{\infty} 3^{-n}$

35. $\displaystyle\sum_{k=1}^{\infty} (1.01)^k$

36. $\displaystyle\sum_{n=1}^{\infty} \pi^n$

37. $\displaystyle\sum_{k=0}^{\infty} (0.99)^k$

38. $\displaystyle\sum_{n=1}^{\infty} \pi^{-n}$

39. $\displaystyle\sum_{j=0}^{\infty} \frac{3^j}{4^{j+1}}$

40. $\displaystyle\sum_{i=0}^{\infty} \frac{(-1)^i}{1000}$

41–48. Find the value of each repeating decimal. [*Hint:* Write each as an infinite series. For example,

$$0.\overline{25} = 0.252525\ldots = \frac{25}{100} + \frac{25}{100^2} + \frac{25}{100^3} + \cdots$$

The bar indicates the repeating part.]

41. $0.\overline{36} = 0.363636\ldots$

42. $0.\overline{63} = 0.636363\ldots$

43. $2.\overline{54}$ $(= 2 + 0.\overline{54})$

44. $1.\overline{24}$ $(= 1 + 0.\overline{24})$

45. $0.\overline{027}$

46. $0.\overline{9}$

47. $0.\overline{012345679}$

48. $0.\overline{987654320}$

(For Exercises 47 and 48, after finding the fraction, try dividing the numerator and denominator by 12345679.)

Applied Exercises on Finite Geometric Series

49. a. GENERAL: Yonkers Fine Suppose that the fine in Example 1 (beginning at $100 and doubling each day; see page 645) had continued for a total of 18 days. Find the total fine.

   b. (Graphing calculator with series operations helpful) How many days would it have taken for the fine to reach a billion dollars?

Source: New York Times

50. PERSONAL FINANCE: Annuity A star baseball player expects his career to last 10 years, and for his retirement deposits $4000 at the end of each month in a bank account earning 3% interest compounded monthly. Find the amount of this annuity at the end of 10 years.

51. PERSONAL FINANCE: Annuity For your retirement "nest egg," you deposit $300 at the end of each month into a bank account paying 6% interest compounded monthly. Find the amount of this annuity at the end of your 40-year working career.

52. PERSONAL FINANCE: Present Value of an Annuity
A retirement plan pays $1000 at retirement and every month thereafter for a total of 12 years. Find the sum of the present values of these payments (at an interest rate of 6% compounded monthly) by summing the series

$$1000 + \frac{1000}{1.005} + \frac{1000}{(1.005)^2} + \cdots + \frac{1000}{(1.005)^{143}}$$

53. GENERAL: Million Dollar Lottery Most state lottery jackpots are paid out over time (often 20 years), so the "real" cost to the state is the sum of the present values of the payments. Find the cost of a "million dollar" lottery by summing the present values of 240 monthly payments of $4167, beginning now, if the interest rate is 6% compounded monthly. [*Hint:* The present value of $4167 in k months is $4167/(1 + 0.005)^k$.]

54. BUSINESS: Sinking Fund A sinking fund is an annuity designed to reach a given value at a given time in the future (often to pay off a debt or to buy new equipment). A company's $100,000 printing press is expected to last 8 years. If equal payments are to be made at the end of each quarter into an account paying 8% compounded quarterly, find the size of the quarterly payments needed to yield $100,000 at the end of 8 years. [*Hint:* Solve

$$x + x(1.02) + x(1.02)^2 + \cdots + x(1.02)^{31} = 100,000.]$$

55. a. GENERAL: Saving Pennies You make a New Year's resolution to save pennies, promising to save on successive days 1¢, 2¢, 4¢, 8¢, and so on, doubling the number each day throughout January. How much will you have saved by the end of January?

  b. (Graphing calculator with series operations helpful) How many days will it take for the savings to reach a million dollars?

56. GENERAL: Chessboards It is said that the game of chess was invented by the Grand Vizier Sissa Ben Dahir for the Indian King Shirham. When the king offered him any reward, Sissa asked for one grain of wheat to be placed on the first square of the chessboard, two on the second, four on the third, and so on, doubling each time for each of the 64 squares. Estimate the total number of grains of wheat required. (It is also said that the king, upon learning that the total was many times the world production of wheat, executed Sissa for making a fool out of him.)

  57. GENERAL: Cycling (Graphing calculator with series operations helpful) You plan to cycle across the United States, cycling 10 miles the first day, 10% further the second day (so 11 miles), increasing each day's distance by 10%. On which day will you reach the opposite coast 3000 miles away?

  58. PERSONAL FINANCE: Payments (Graphing calculator with series operations helpful) A person gives you $1, waits 1 minute, gives you another dollar, waits 2 minutes, gives you another dollar, waits four minutes, etc., doubling the wait each time. How much money will you have at the end of the first year?

Applied Exercises on Infinite Geometric Series

59. a. BIOMEDICAL: Drug Dosage A patient has been taking a daily dose of 12 units of insulin for an extended period of time. If 80% of the total amount in the bloodstream is absorbed by the body during each day, find the long-term maximum and minimum amounts of insulin in the bloodstream.

  b. (Graphing calculator with series operations helpful) How many daily doses of insulin does it take for the maximum amount in the blood to reach 14.9 units?

60. BIOMEDICAL: Drug Dosage A patient has been taking a daily dose of D milligrams of a drug for an extended period of time, and a proportion p $(0 < p < 1)$ of the total amount in the bloodstream is absorbed by the body during each day. Find the long-term maximum and minimum amounts in the bloodstream.

61. ECONOMICS: Multiplier Effect In the United States, the marginal propensity to consume (*MPC*) is approximately 90%. Find the total economic effect of a $50 billion tax cut. Also find the multiplier.
Source: P. Samuelson and W. Nordhaus, *Economics*

62. ECONOMICS: Multiplier Effect After several years of stock market gains, in 2000 the marginal propensity to consume (*MPC*) peaked at 96%. Find the total economic effect that a $40 billion tax cut would have had at that time. Also find the multiplier.
Source: economics.about.com

63. ECONOMICS: Multiplier Effect The Department of Education decides to increase grants to college students by $300 million. Find the total economic effect of this increase if the marginal propensity to consume (*MPC*) is 95%. Also find the multiplier.

64. ECONOMICS: Multiplier Effect A town decides to spend $1,500,000 to build a new library. If the marginal propensity to consume (*MPC*) is 84%, find the total economic effect of this expenditure. Also find the multiplier.

65. ECONOMICS: Multiplier Effect In Japan, the marginal propensity to consume (*MPC*) is approximately 75%. Find the multiplier.
Source: economics.about.com

66. ECONOMICS: Multiplier Effect If the marginal propensity to *consume* (MPC) is 70%, then the marginal propensity to *save* (MPS) is 30%. In general, $MPS = 1 - MPC$. Show that for a given MPC, the multiplier is $\dfrac{1}{MPS}$.

67–68. BUSINESS: Perpetuities A *perpetuity* is an annuity that continues forever. The *capital value* of a perpetuity is the present value of all (future) payments. (Perpetuities are also called permanent endowments, and the continuous case was treated on pages 408–409 using improper integrals.)

67. The capital value of a perpetuity that pays $300 now and every quarter from now on at an interest rate of 6% compounded quarterly is

$$300 + \frac{300}{1 + 0.015} + \frac{300}{(1 + 0.015)^2} + \cdots$$

Find this value.

68. The capital value of a perpetuity that pays $1000 annually, beginning now, at an interest rate of 8% compounded annually is

$$1000 + \frac{1000}{1 + 0.08} + \frac{1000}{(1 + 0.08)^2} + \frac{1000}{(1 + 0.08)^3} + \cdots$$

Find this value.

69. a. GENERAL: Bouncing Ball A ball dropped from a height of 6 feet bounces to two-thirds of its former height with each bounce. Find the total vertical distance that the ball travels.

 b. (Graphing calculator with series operations helpful) At which bounce will the total vertical distance traveled by the ball first exceed 29 feet?

70. GENERAL: Coupons A 45¢ candy bar comes with a coupon worth 10% toward the next candy bar (which includes a coupon for 10% of the next, and so on). Therefore, the "real" value of a single 45¢ candy bar is

$$45 + 45(0.10) + 45(0.10)^2 + 45(0.10)^3 + \cdots$$

Find this value.

71–72. BIOMEDICAL: Long-Term Population A population (of cells or people) is such that each year a number a of individuals are added (call them "immigrants"), and a proportion p of the individuals who have been there die. Therefore, the proportion that survives is $(1 - p)$, so that just after an immigration the population will consist of new immigrants plus $a(1 - p)$ from the previous year's immigration plus $a(1 - p)^2$ from the immigration before that, and so on. In the long run, the size of the population just after an immigration will then be the sum $a + a(1 - p) + a(1 - p)^2 + a(1 - p)^3 + \cdots$.

71. If the number of immigrants each year is 800 and the survival proportion is 0.95, find the long-run size of the population just after an immigration. Then find the long-run size just *before* an immigration.

72. Find the long-run size of the population just after an immigration for *any* number of immigrants a and *any* survival proportion $(1 - p)$. Then find the long-run size just *before* an immigration.

Conceptual Exercises

73. How many terms are in the following finite geometric series? $2 + 2 \cdot 3 + 2 \cdot 3^2 + \cdots + 2 \cdot 3^{5280}$.

74. What is the value of the first term a_1 in the following finite geometric series? $a_1 + 1 + \dfrac{1}{\pi}$.

75. True or False: If the partial sums S_n of an infinite series all satisfy $S_n \leq 10$, then the sum S also satisfies $S \leq 10$.

76. True or False: If the partial sums S_n of an infinite series all satisfy $S_n < 1$, then the sum S also satisfies $S < 1$.

77. What is the ratio of the following infinite geometric series? $e + \dfrac{1}{e} + \cdots$. Does it converge?

78. What is the 501st term in the following infinite geometric series? $\pi^{-500} + \pi^{-499} + \pi^{-498} + \cdots$. Does it converge?

79. Does the infinite geometric series $1{,}000{,}000 + 100{,}000 + 10{,}000 + \cdots$ converge or diverge?

80. Does the infinite geometric series $\dfrac{1}{1{,}000{,}000} + \dfrac{1}{100{,}000} + \dfrac{1}{10{,}000} + \cdots$ converge or diverge?

81. Applying the sum formula to the following infinite geometric series gives the result that a series of *positive* terms adds up to a *negative* number. What's wrong?

$$1 + 2 + 4 + 8 + \cdots \overset{?}{=} \frac{1}{1 - 2} = \frac{1}{-1} = -1$$

82. Applying the sum formula to the following infinite geometric series gives the result that a series of *integer* terms adds up to a *fraction*. What's wrong?

$$1 - 2 + 4 - 8 + \cdots \overset{?}{=} \frac{1}{1 - (-2)} = \frac{1}{3}$$

Explorations and Excursions   The following problems extend and augment the material presented in the text.

83. APPROXIMATING THE FINITE BY THE INFINITE For the patient in Example 4 (page 649), the amount of the drug in the bloodstream just after the nth dose is

$$150 + 150(0.3) + 150(0.3)^2 + \cdots + 150(0.3)^{n-1}$$

Use a graphing calculator to sum this finite geometric series for the following values of n: 1, 2, 3, 4, 5, 6, and 7.

a. For which dose does the cumulative amount first rise above 200 milligrams?

b. For which dose does the cumulative amount first rise above 213 milligrams?

84. DIVERGENCE OF GEOMETRIC SERIES Show that the geometric series

$$a + ar + ar^2 + ar^3 + \cdots$$

diverges if $|r| \geq 1$, as follows (for $a \neq 0$):

a. If $r = 1$, the series becomes $a + a + a + \cdots$. Show that this series diverges by showing that the sum of the first n terms is $S_n = n \cdot a$, which does not approach a limit as $n \to \infty$ (since $a \neq 0$).

b. If $r = -1$, then the series becomes $a - a + a - \cdots$. Show that this series diverges by showing that the partial sums $S_1, S_2, S_3, S_4, \ldots$ take

values $a, 0, a, 0, \ldots$, and so do not approach any limit.

c. If $|r| > 1$, use the fact that r^n does not approach a limit as $n \to \infty$ to show that the partial sum formula $a \dfrac{1 - r^n}{1 - r}$ does not approach a limit as $n \to \infty$.

85. Consider *Grandi's series* $1 - 1 + 1 - 1 + \cdots$

a. Show that applying the formula $\dfrac{a}{1 - r}$ gives $\dfrac{1}{2}$.

b. Can this formula legitimately be applied to the series?

c. Calculate the partial sums of this series. What would be the average of these partial sums *in the long run*? [*Note:* Such an average is called a *Cesàro sum*.]

Source: Mathematics Magazine 56

86. Derive the formula for the sum S of a geometric series as follows:

$$S = a + ar + ar^2 + ar^3 + \cdots$$
$$= a + r\underbrace{(a + ar + ar^2 + \cdots)}_{S} = a + rS$$

Now solve the equation $S = a + rS$ for S. (Note that this derivation *assumes* that the series converges.)

10.2   Taylor Polynomials

Introduction

Often in mathematics we approximate complicated objects by simpler objects, such as approximating a curve by its tangent line. In this section we will approximate functions such as e^x and $\sin x$ by *polynomials*, called **Taylor polynomials,** named after the English mathematician Brook Taylor (1685–1731), a student of Isaac Newton.

Factorial Notation

We begin by defining **factorial** notation. For any positive integer n, $n!$ (read: "n factorial") is defined as the product of all integers from n down to 1.

Factorial Notation	
$n! = n \cdot (n - 1) \cdot \cdots \cdot 2 \cdot 1$	$n!$ is the product of all integers from n down to 1
$0! = 1$	Zero factorial is defined as 1

For example,

$$4! = 4 \cdot 3 \cdot 2 \cdot 1 = 24$$

! (factorial) may be found in the MATH/PROB menu on your calculator

A justification for defining 0! as 1 is given in Exercise 30.

First Taylor Polynomial at $x = 0$

We begin by finding the polynomial of degree 1 that best approximates a given function $f(x)$ near $x = 0$. A polynomial of degree 1 is a *linear* function $y = mx + b$, and it is clear from the following diagram that the best linear approximation to $f(x)$ at $x = 0$ will be the *tangent line* at $x = 0$. The tangent line has the same slope and y-intercept as the curve at $x = 0$, so that $m = f'(0)$ and $b = f(0)$. Substituting these into $y = mx + b$ gives $y = f'(0)x + f(0)$, or, writing it in reverse order, $y = f(0) + f'(0)x$.

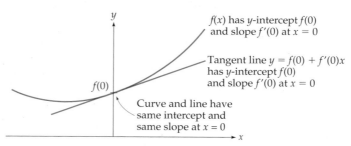

$f(x)$ has y-intercept $f(0)$
and slope $f'(0)$ at $x = 0$

Tangent line $y = f(0) + f'(0)x$
has y-intercept $f(0)$
and slope $f'(0)$ at $x = 0$

$f(0)$

Curve and line have
same intercept and
same slope at $x = 0$

$f(x)$ and its tangent line at $x = 0$

LOOKING BACK

In Exercise 75 on page 113 we proved that the tangent line is the best linear approximation.

The tangent line is called the *best linear approximation* or the *first Taylor polynomial at $x = 0$* for the function, and is denoted $p_1(x)$.*

First Taylor Polynomial at $x = 0$

The first Taylor polynomial at $x = 0$ of $f(x)$ is

$$p_1(x) = f(0) + f'(0) \cdot x \qquad \qquad \text{Tangent line at } x = 0$$

EXAMPLE 1   **FINDING A FIRST TAYLOR POLYNOMIAL**

Find the first Taylor polynomial at $x = 0$ for $f(x) = e^x$.

Solution

The derivative of $f(x) = e^x$ is e^x, so $f(0) = f'(0) = e^0 = 1$. Therefore, the first Taylor polynomial is

$$p_1(x) = 1 + x$$

$p_1(x) = f(0) + f'(0) \cdot x$
with $f(0) = 1$ and $f'(0) = 1$

$f(x) = e^x$

$p_1(x) = 1 + x$
(first Taylor polynomial)

$f(x) = e^x$ and its first Taylor polynomial at $x = 0$

---

*Taylor polynomials at $x = 0$ are sometimes called Maclaurin polynomials, although there is little historical justification for this terminology, and we will not use it.

Higher Taylor Polynomials at $x = 0$

This idea of finding the best *linear* approximation at $x = 0$ by matching y-intercepts and slopes (derivatives) at $x = 0$ can be extended to finding the best *quadratic* approximation by also matching the "curl" or concavity (second derivative). In fact, we may find the best *polynomial* approximation by matching higher-order derivatives of the polynomial to those of the function. For example, to approximate $f(x)$ by a *third*-degree polynomial $p(x) = a_0 + a_1 x + a_2 x^2 + a_3 x^3$ at $x = 0$, we match derivatives up to the *third* order at $x = 0$, as is done in the right-hand column of the following table.

Polynomial $p(x)$ and Its Derivatives	$p(x)$ and Its Derivatives Evaluated at $x = 0$	$f(x)$ and Its Derivatives Evaluated at $x = 0$	Equating the Last Two Columns
$p(x) = a_0 + a_1 x + a_2 x^2 + a_3 x^3$	a_0	$f(0)$	$a_0 = f(0)$
$p'(x) = a_1 + 2a_2 x + 3a_3 x^2$	a_1	$f'(0)$	$a_1 = f'(0)$
$p''(x) = 2a_2 + 3 \cdot 2a_3 x$	$2 \cdot a_2$	$f''(0)$	$2! a_2 = f''(0)$
$p'''(x) = 3 \cdot 2a_3$	$3 \cdot 2a_3$	$f'''(0)$	$3! a_3 = f'''(0)$

The coefficients a_0, a_1, a_2, and a_3 of the polynomial are found by solving the equations in the extreme right column, giving

$$a_0 = f(0), \qquad a_1 = f'(0), \qquad a_2 = \frac{f''(0)}{2!}, \qquad a_3 = \frac{f'''(0)}{3!} \qquad a_n = \frac{f^{(n)}(0)}{n!}$$

Further coefficients follow the same pattern, and substituting them into the polynomial $p(x) = a_0 + a_1 x + a_2 x^2 + a_3 x^3 + \cdots + a_n x^n$ gives the nth **Taylor polynomial** of an n-times differentiable function.

Taylor Polynomials at $x = 0$

The nth Taylor polynomial at $x = 0$ of $f(x)$ is

$$p_n(x) = f(0) + \frac{f'(0)}{1!} x + \frac{f''(0)}{2!} x^2 + \frac{f'''(0)}{3!} x^3 + \cdots + \frac{f^{(n)}(0)}{n!} x^n$$

EXAMPLE 2   FINDING TAYLOR POLYNOMIALS

Find the first three Taylor polynomials at $x = 0$ for $f(x) = e^x$.

Solution

All derivatives of e^x are e^x, so $f(0) = f'(0) = f''(0) = f'''(0) = e^0 = 1$. Therefore, the first, second, and third Taylor polynomials for e^x at $x = 0$ are

$p_1(x) = 1 + x$ 　　　　　　　Found in Example 1

$p_2(x) = 1 + x + \dfrac{1}{2!} x^2$ 　　　$f(0) + \dfrac{f'(0)}{1!} x + \dfrac{f''(0)}{2!} x^2$
　　　　　　　　　　　　　with $f(0) = f'(0) = f''(0) = 1$

$p_3(x) = 1 + x + \dfrac{1}{2!} x^2 + \dfrac{1}{3!} x^3$ 　　$f(0) + \dfrac{f'(0)}{1!} x + \dfrac{f''(0)}{2!} x^2 + \dfrac{f'''(0)}{3!} x^3$
　　　　　　　　　　　　　with $f(0) = f'(0) = f''(0) = f'''(0) = 1$

The following diagram shows e^x and these approximating Taylor polynomials.

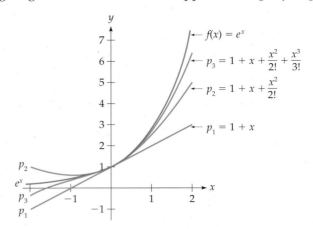

The original function $f(x) = e^x$ (red) and
its first three Taylor polynomials at $x = 0$.
Which polynomial fits e^x most closely?

This graph shows two general properties of Taylor polynomials at $x = 0$:

1. Higher-order Taylor polynomials generally approximate functions more closely.

2. Each approximation is more accurate closer to $x = 0$ (and is *exact at* $x = 0$).

PRACTICE PROBLEM 1

Based on the pattern of the first three Taylor polynomials found in Example 2, find the fifth Taylor polynomial at $x = 0$ of e^x.

Solution on page 664 >

**Graphing
Calculator
Exploration**

Graph the fifth Taylor polynomial (shown in blue) found in Practice Problem 1 together with e^x (shown in red) on the window $[-3.5, 3.5]$ by $[-5, 25]$. Observe where the curves are close and where they diverge. Adjust the window to see how they diverge on a wider interval.

Try adding another term, $\dfrac{x^6}{6!} = \dfrac{x^6}{720}$, to graph the *sixth* Taylor polynomial.

Error in Taylor Approximation at $x = 0$

An approximation is useful only if we can estimate how far "off" it is. The *error* or **remainder** in approximating $f(x)$ by $p_n(x)$ is defined as

$$|R_n(x)| = |f(x) - p_n(x)|$$

the vertical distance between the curves at that x-value. If $f(x)$ is $n + 1$ times differentiable between 0 and x, the error can be estimated by the following formula (which is shown in Exercises 44–50).

Error in Taylor Approximation at $x = 0$

For the nth Taylor polynomial $p_n(x)$ at $x = 0$ of $f(x)$, the error at x, $|R_n(x)| = |f(x) - p_n(x)|$, satisfies

$$|R_n(x)| \le \frac{M}{(n+1)!} |x|^{n+1}$$

Resembles the $n + 1$st

term $\dfrac{f^{(n+1)}(x)}{(n+1)!} x^{n+1}$

where M is any number such that $|f^{(n+1)}(t)| \le M$ for all t between 0 and x.

EXAMPLE 3   APPROXIMATING BY A TAYLOR POLYNOMIAL

Approximate $e^{0.5}$ using the third Taylor polynomial for e^x, and estimate the error.

Solution

Using the third Taylor polynomial (found in Example 2),

$$p_3(0.5) = 1 + 0.5 + \frac{0.5^2}{2!} + \frac{0.5^3}{3!} \approx 1.6458$$

$p_3(x) = 1 + x + \dfrac{x^2}{2!} + \dfrac{x^3}{3!}$

evaluated at $x = 0.5$

Estimate for $e^{0.5}$

The error formula for the third Taylor polynomial requires a number M such that $|f^{(4)}(t)| \le M$ for $0 \le t \le 0.5$. Since $|f^{(4)}(t)| = e^t$ is an increasing function, it is maximized at its right-hand endpoint $t = 0.5$, so

$$|f^{(4)}(t)| = e^t \le e^{0.5} < e^1 < 3$$

Since $e \approx 2.7 < 3$

We therefore take $M = 3$ (although smaller estimates could be found). With this value of M, the error formula with $n = 3$, $M = 3$, and $x = 0.5$ becomes

$$|R_3(0.5)| \le \frac{3}{4!} |0.5|^4 = \frac{3}{24}(0.0625) \approx 0.00781 \approx 0.008$$

Error estimate

Therefore,

$$e^{0.5} \approx 1.646 \quad \text{with an error less than } 0.008$$

We always round the error estimate *up* to the first nonzero digit (for example, an error of 0.0072 would be rounded up to 0.008) so as not to claim more accuracy than is justified. We then round the approximation to the same number of decimal places.

Why Bother to Calculate $e^{0.5}$?

You may wonder why we approximate $e^{0.5}$ when a calculator or computer will do so at the press of a few keys. How do you think computers actually evaluate such functions? They do so by using (modified) preprogrammed Taylor polynomials.*

*Most computers use "economized" Taylor series that make the error more uniform over an interval. Most calculators, on the other hand, use a system called CORDIC, for COordinate Rotation DIgital Computer.

Therefore, this section explains what happens when you press the keys. Furthermore, at some time you may need to evaluate a new or more complicated function, and Taylor polynomials enable you to do this in terms of "easier" polynomials.

Not only can we find the accuracy of a particular approximation, but we can also determine the Taylor polynomial needed to *guarantee* a given accuracy.

| EXAMPLE 4 | APPROXIMATING WITHIN A GIVEN ACCURACY |

Find the Taylor polynomial at $x = 0$ that approximates e^x with an error less than 0.005 on the interval $-1 \le x \le 1$.

Solution

We want the value of n that makes the error $\dfrac{M}{(n + 1)!} |x|^{n+1}$ less than 0.005. This error formula with $M = 3$ (since $|e^1| < 3$, as in Example 3) and $|x| = 1$ becomes $\dfrac{3}{(n + 1)!}$. We evaluate this formula at $n = 1, 2, 3, \dots$ until its value first falls below 0.005.

n	$\dfrac{3}{(n + 1)!}$	
1	$\dfrac{3}{2!} = \dfrac{3}{2} = 1.5$	
2	$\dfrac{3}{3!} = \dfrac{3}{6} = 0.5$	We could have skipped some early values of n that clearly would not give small enough errors
3	$\dfrac{3}{4!} = \dfrac{3}{24} = 0.125$	
4	$\dfrac{3}{5!} = \dfrac{3}{120} = 0.025$	
5	$\dfrac{3}{6!} = \dfrac{3}{720} \approx 0.004$	< 0.005 for the first time

The first value of n for which the error is small enough (less than 0.005) is $n = 5$, so the fifth Taylor polynomial (found in Practice Problem 1) provides the desired accuracy:

$$e^x \approx 1 + x + \frac{x^2}{2!} + \frac{x^3}{3!} + \frac{x^4}{4!} + \frac{x^5}{5!}$$

For $-1 \le x \le 1$, with error less than 0.005

Graphing Calculator Exploration

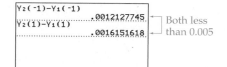

Both less than 0.005

(You should still have e^x and its fifth Taylor polynomial at $x = 0$ in your calculator.) Verify that e^x and the Taylor polynomial are indeed within 0.005 of each other between $x = -1$ and $x = 1$. How close are they actually? (Remember that the error formula generally *overstates* the error.)

EXAMPLE 5   FINDING AN APPROXIMATION IN AN APPLICATION

A company that produces seasonal goods can sell sin 1.1 thousand units on a certain day. Approximate sin 1.1 by using the fifth Taylor polynomial at $x = 0$ for sin x and estimate the error.

Solution

The function sin x and its derivatives (from page 547) at $x = 0$ are

$$f(x) = \sin x \qquad\qquad f(0) = 0$$
$$f'(x) = \cos x \qquad\qquad f'(0) = 1$$
$$f''(x) = -\sin x \qquad\qquad f''(0) = 0$$
$$f'''(x) = -\cos x \qquad\qquad f'''(0) = -1$$
$$f^{(4)}(x) = \sin x \qquad\qquad f^{(4)}(0) = 0$$
$$f^{(5)}(x) = \cos x \qquad\qquad f^{(5)}(0) = 1$$

sin x and its derivatives evaluated at $x = 0$

Therefore, the fifth Taylor polynomial is

$$p_5(x) = 0 + \frac{1}{1!}x + 0 - \frac{1}{3!}x^3 + 0 + \frac{1}{5!}x^5$$

$f(0) + \dfrac{f'(0)}{1!}x + \cdots + \dfrac{f^{(5)}(0)}{5!}x^5$

with the preceding values for $f(0), f'(0), \dots, f^{(5)}(0)$

$$= x - \frac{x^3}{3!} + \frac{x^5}{5!}$$

Simplifying (all even powers vanish)

Evaluating,

$$\sin 1.1 \approx 1.1 - \frac{(1.1)^3}{3!} + \frac{(1.1)^5}{5!} \approx 0.89159$$

Evaluating at $x = 1.1$

Therefore, sin 1.1 ≈ 0.89159. For the error formula, we need an M such that $|f^{(6)}(t)| \le M$ for all t between 0 and 1.1. For the function sin x, we have $|f^{(6)}(x)| = |-\sin x| \le 1 = M$. With this M, the error formula becomes

$$|R_5(x)| \le \frac{1}{6!}|1.1|^6 \approx 0.0024 \approx 0.003$$

$\dfrac{M}{(n+1)!}|x|^{n+1}$ with

$M = 1, \quad n = 5, \quad x = 1.1$

└── Rounded *up*

Therefore,

$$\sin 1.1 \approx 0.892 \quad \text{with an error less than } 0.003$$

These are production numbers in *thousands*, so the company should produce 892 units, with a possible error of 3 units. (To be safe, it might choose to produce 895 units.)

PRACTICE PROBLEM 2

As we saw in the preceding Example, the Taylor polynomial for sin x at $x = 0$ consists only of *odd* powers of x. Therefore, the sixth Taylor polynomial will be the same as the fifth, so we can use the error formula for the sixth Taylor polynomial to estimate the accuracy of the approximation that we found. Use the error formula for $|R_6(x)|$ to estimate the accuracy of the approximation sin 1.1 ≈ 0.89159 from Example 5.

Solution on page 664 >

The function $\sin x$ and its fifth Taylor polynomial are shown in the following graph. Notice that the approximation is quite accurate for x-values near 0.

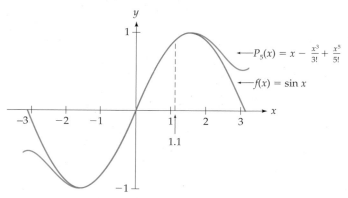

Sin x and its fifth Taylor polynomial at $x = 0$

Taylor Polynomials at $x = a$

We have been using Taylor polynomials centered *at* $x = 0$, meaning that they approximate $f(x)$ most accurately near $x = 0$. There are analogous formulas for Taylor polynomials centered at *any* number $x = a$ (provided that f is n times differentiable at a).

LOOKING BACK

Like the "$x = 0$" version on page 657, but with *a* instead of 0.

Taylor Polynomials at $x = a$

The nth Taylor polynomial at $x = a$ of $f(x)$ is

$$p_n(x) = f(a) + \frac{f'(a)}{1!}(x - a) + \frac{f''(a)}{2!}(x - a)^2 + \cdots + \frac{f^{(n)}(a)}{n!}(x - a)^n$$

Notice that for Taylor polynomials at $x = a$:

1. The polynomial is in powers of $(x - a)$ rather than powers of x.
2. The coefficients involve derivatives at $x = a$.
3. The polynomials approximate $f(x)$ more accurately nearer $x = a$ (and are *exact at* $x = a$).
4. If $a = 0$, we obtain the Taylor polynomials at $x = 0$.

EXAMPLE 6   FINDING A TAYLOR POLYNOMIAL AT $x = 4$

Find the second Taylor polynomial for $f(x) = \sqrt{x}$ at $x = 4$, and use it to approximate $\sqrt{4.2}$.

Solution

The function $f(x) = \sqrt{x}$ and its derivatives at $x = 4$ are

$$f(x) = x^{1/2} \qquad\qquad f(4) = 4^{1/2} = 2$$

$$f'(x) = \frac{1}{2}x^{-1/2} \qquad f'(4) = \frac{1}{2}4^{-1/2} = \frac{1}{2}\cdot\frac{1}{2} = \frac{1}{4}$$

$$f''(x) = -\frac{1}{4}x^{-3/2} \qquad f''(4) = -\frac{1}{4}4^{-3/2} = -\frac{1}{4}\cdot\frac{1}{8} = -\frac{1}{32}$$

The second Taylor polynomial at $x = 4$ is

$$p_2(x) = 2 + \frac{\frac{1}{4}}{1!}(x-4) - \frac{\frac{1}{32}}{2!}(x-4)^2$$

$p_2(x) = f(a) + \frac{f'(a)}{1!}(x-a) + \frac{f''(a)}{2!}(x-a)^2$ with $a = 4$ and the calculated values of $f(4), f'(4),$ and $f''(4)$

$$= 2 + \frac{x-4}{4} - \frac{(x-4)^2}{64}$$

Simplifying

Evaluating,

$$p_2(4.2) = 2 + \frac{0.2}{4} - \frac{(0.2)^2}{64} = 2.049375$$

Evaluating at $x = 4.2$

Approximation for $\sqrt{4.2}$

$$\sqrt{4.2} \approx 2.0494$$

Rounding to four decimal places

Actually, $\sqrt{4.2} \approx 2.04939$, so the estimate above is correct in all four decimal places. The function $f(x) = \sqrt{x}$ and its second Taylor polynomial at $x = 4$ are graphed below, showing that the approximation is very good near $x = 4$.

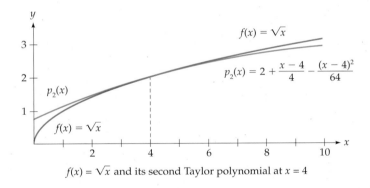

$f(x) = \sqrt{x}$ and its second Taylor polynomial at $x = 4$

Graphing Calculator Exploration

Some advanced graphing calculators can find Taylor polynomials. For example, the Texas Instruments *TI-89* graphing calculator finds the second Taylor polynomial at $x = 4$, for the function $f(x) = \sqrt{x}$ as follows.

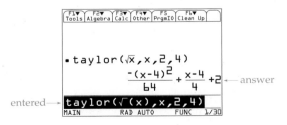

This answer agrees with that in the preceding Example, except that it is written in reverse order.

If $f(x)$ is $n + 1$ times differentiable between a and x, the error in approximating $f(x)$ by $p_n(x)$ can be estimated by the following formula (which is shown in Exercises 44–49).

Error in Taylor Approximation at $x = a$

For the nth Taylor approximation $p_n(x)$ at $x = a$ of $f(x)$, the error at x, $|R_n(x)| = |f(x) - p_n(x)|$, satisfies

$$|R_n(x)| \le \frac{M}{(n+1)!} |x - a|^{n+1}$$

where M is any number such that $|f^{(n+1)}(t)| \le M$ for all t between a and x.

LOOKING BACK

Again, like the "$x = 0$" version (page 659) but with a instead of 0.

Solutions TO PRACTICE PROBLEMS

1. $1 + x + \dfrac{1}{2!} x^2 + \dfrac{1}{3!} x^3 + \dfrac{1}{4!} x^4 + \dfrac{1}{5!} x^5,$ or $1 + x + \dfrac{x^2}{2!} + \dfrac{x^3}{3!} + \dfrac{x^4}{4!} + \dfrac{x^5}{5!}$

2. $|R_6(x)| \le \dfrac{1}{7!} |1.1|^7 = \dfrac{1.9487}{5040} \approx 0.0004$

Therefore, $\sin 1.1 \approx 0.8916$ with an error of less than 0.0004. (The actual value of $\sin 1.1$ is 0.8912....)

10.2   Section Summary

Taylor polynomials approximate a curve near a single chosen point $x = a$. Any sufficiently differentiable function can be approximated by a polynomial, using the formulas

$$\sum_{k=0}^{n} \frac{f^{(k)}(0)}{k!} x^k \qquad \begin{array}{l} n\text{th Taylor polynomial at } x = 0 \\ [\text{first term simplifies to } f(0)] \end{array}$$

and

$$\sum_{k=0}^{n} \frac{f^{(k)}(a)}{k!} (x - a)^k \qquad \begin{array}{l} n\text{th Taylor polynomial at } x = a \\ [\text{first term simplifies to } f(a)] \end{array}$$

Taylor polynomials can also be found from other Taylor polynomials by replacing the variable by a polynomial expression. The accuracy of the approximation can be found from the error formulas on page 659 and above.

10.2 Exercises

1–8. Find the third Taylor polynomial at $x = 0$ of each function.

1. $f(x) = e^{2x}$

2. $f(x) = \ln(1 - x)$

3. $f(x) = \sqrt{x + 1}$

4. $f(x) = \sqrt{2x + 1}$

5. $f(x) = 7 + x - 3x^2$

6. $f(x) = 2 - x + 3x^2 - x^3$

7. $f(x) = e^{x^2}$

8. $f(x) = \ln(\cos x)$

📝 **FOR HELP GETTING STARTED**
on Exercises 1–8, see Example 2 on page 657.

9–12. For each function:

a. Find the fourth Taylor polynomial at $x = 0$.

b. Graph the original function and the Taylor polynomial together on the indicated window.

9. $f(x) = \ln(x + 1)$
(for (b), use the window $[-2, 2]$ by $[-2, 2]$)

10. $f(x) = e^{-2x}$
(for (b), use the window $[-2, 2]$ by $[-2, 8]$)

11. $f(x) = \cos x$
(for (b), use the window $[-\pi, \pi]$ by $[-2, 2]$)

12. $f(x) = \cos(x + \pi)$
for (b), use the window $[-\pi, \pi]$ by $[-2, 2]$)

13. Find the fourth Taylor polynomial for e^{x^2} by taking the second Taylor polynomial for e^x (page 657) and replacing x by x^2.

14. Find the third Taylor polynomial for e^{-x} by taking the third Taylor polynomial for e^x (page 657) and replacing x by $-x$.

15. Find the fifth Taylor polynomial for $\sin 2x$ by taking the fifth Taylor polynomial for $\sin x$ (page 661) and replacing x by $2x$.

16. Find the tenth Taylor polynomial for $\sin x^2$ by taking the fifth Taylor polynomial for $\sin x$ (page 661) and replacing x by x^2.

17. **a.** For $f(x) = e^x$, find the fourth Taylor polynomial at $x = 0$.
b. Use your answer in part (a) to approximate $e^{-1/2}$.
c. Estimate the error by the error formula (page 659). [*Hint:* First calculate $f^{(5)}(x)$ and find an M such that $|f^{(5)}(x)| \le M$ for $-\frac{1}{2} \le x \le 0$, using the fact that an increasing function is maximized at its right-hand endpoint.]
d. Restate the approximation and error, rounded appropriately.
e. Graph $f(x) = e^x$ and its Taylor polynomial on $[-3, 3]$ by $[0, 12]$.

18. **a.** For $f(x) = \ln(x + 1)$, find the fifth Taylor polynomial at $x = 0$.
b. Use your answer in part (a) to approximate $\ln \frac{3}{2}$.
c. Estimate the error by the error formula (page 659). [*Hint:* Calculate $f^{(6)}(x)$ and find an M such that $|f^{(6)}(x)| \le M$ for $0 \le x \le \frac{1}{2}$, using the fact that a

decreasing function is maximized at its left-hand endpoint.]
d. Restate the approximation and error, rounded appropriately.
e. Graph $f(x) = \ln(x + 1)$ and its Taylor polynomial on $[-2, 2]$ by $[-2, 2]$.

19. **a.** Find the Taylor polynomial at $x = 0$ that approximates $f(x) = \cos x$ on the interval $-1 \le x \le 1$ with error less than 0.0002.
b. Graph $f(x) = \cos x$ and the Taylor polynomial on $[-2\pi, 2\pi]$ by $[-2, 2]$. Use TRACE to estimate the maximum difference between the two curves for $-1 \le x \le 1$.

20. For the function
$$f(x) = 1 + 2x + \frac{9}{2!}x^2 + \frac{5}{3!}x^3 - \frac{7}{4!}x^4$$
find $f'''(0)$ and $f^{(4)}(0)$. [*Hint:* No calculation is necessary.]

21. **a.** What linear function best approximates $\sqrt[3]{1 + x}$ near $x = 0$?
b. Use the error formula to find an interval $[0, x]$ on which the error is less than 0.01.
c. Graph $\sqrt[3]{1 + x}$ and the linear approximation on $[-2, 2]$ by $[-2, 2]$ and use TRACE to verify that the difference is less than 0.01 for the x-values found in part (b).

22. **a.** What quadratic function best approximates $\sqrt[3]{1 + x}$ near $x = 0$?
b. Use the error formula to estimate the maximum error on $[-0.1, 0.1]$.
c. Graph $\sqrt[3]{1 + x}$ and the quadratic approximation on $[-2, 2]$ by $[-2, 2]$ and use TRACE to verify that on $[-0.1, 0.1]$ the difference is less than the error estimate found in part (b).

Taylor Polynomials at $x = a$

23. **a.** For $f(x) = \sqrt{x}$, find the third Taylor polynomial at $x = 1$.
b. Graph $\sqrt{x}$ and its Taylor polynomial on $[-1, 3]$ by $[-1, 2]$.

24. **a.** For $f(x) = 3x^2 - 4x + 5$, find the second Taylor polynomial at $x = 2$.
b. Multiply out the Taylor polynomial found in part (a) and show that it is equal to the original polynomial.

25. **a.** For $f(x) = \sin x$, find the third Taylor polynomial at $x = \pi$.
b. Graph $\sin x$ and its Taylor polynomial on $[0, 2\pi]$ by $[-2, 2]$.

26. **a.** For $f(x) = \cos x$, find the fourth Taylor polynomial at $x = \pi$.
b. Graph $\cos x$ and its Taylor polynomial on $[0, 2\pi]$ by $[-2, 2]$.

27. **a.** For $f(x) = \sqrt[3]{x}$, find the second Taylor polynomial at $x = 1$.
b. Use your answer from part (a) to approximate $\sqrt[3]{1.3}$.
c. Estimate the error by the error formula (page 664). [*Hint:* Find an M such that $|f^{(3)}(x)| \le M$ for $1 \le x \le 1.3$, using the fact that a decreasing function is maximized at its left-hand endpoint.]
d. Restate the approximation and error, rounded appropriately.

28. **a.** For $f(x) = \sqrt[4]{x}$, find the second Taylor polynomial at $x = 1$.
b. Use your answer from part (a) to approximate $\sqrt[4]{1.2}$.
c. Estimate the error by the error formula (page 664). [*Hint:* Find an M such that $|f^{(3)}(x)| \le M$ for $1 \le x \le 1.2$, using the fact that a decreasing function is maximized at its left-hand endpoint.]

(continues)

d. Restate the approximation and error, rounded appropriately.

29. a. For $f(x) = \ln x$, find the sixth Taylor polynomial at $x = 1$.
b. Use your answer in part (a) to approximate $\ln \frac{3}{2}$.
c. Estimate the error by the error formula (page 664). [*Hint:* Find an M such that $|f^{(7)}(x)| \leq M$ for $1 \leq x \leq 1.5$, using the fact that a decreasing function is maximized at its left-hand endpoint.]

d. Restate the approximation and error, rounded appropriately.

30. ZERO FACTORIAL The relationship $(n + 1)! = (n + 1)n!$, solved for $n!$, gives

$$n! = \frac{(n + 1)!}{n + 1}$$

Evaluate this at $n = 0$ to see the reason for defining $0! = 1$.

Applied Exercises

31. BUSINESS: Advertising Five days after the end of an advertising campaign, a company's daily sales (as a proportion of its peak sales) will be $e^{-0.25}$. Approximate $e^{-0.25}$ by using the second Taylor polynomial at $x = 0$ for e^x (rounding to two decimal places). Check your answer by using the e^x key on your calculator.

32. ENVIRONMENTAL SCIENCES: Deer Population A population of deer in a wildlife refuge varies with the season and is predicted to be $300 + 50 \cos 0.52$ on the first day of summer. Approximate this number by using the second Taylor polynomial at

$x = 0$ for $\cos x$. Then check your answer using the $\boxed{\cos}$ key on your calculator (set for radians).

33. ECONOMICS: Supply and Demand The demand for a new toy is predicted to be $4 \cos x$, and the supply is $x + 3$ (both in thousands of units), where x is the number of years $(0 \leq x \leq 2)$. Find when supply will equal demand by solving the equation

$$4 \cos x = x + 3$$

as follows. Replace $\cos x$ by its second Taylor polynomial at $x = 0$ and solve the resulting quadratic equation for the (positive) value of x.

Conceptual Exercises

34. Find: **a.** $\dfrac{1000!}{999!}$ **b.** $0^{0!}$ **c.** $(0!)^0$

35–38. Find, if possible, the best linear approximation at $x = 0$ for each function. [*Hint:* No work necessary—just think about the graph.]

35. $f(x) = \sin x$ **36.** $f(x) = x$
37. $f(x) = \dfrac{1}{x}$ **38.** $f(x) = \ln x$

39. What is an upper bound for the absolute value of *any* derivative of $f(x) = \sin x$ on any interval?

40. Fill in the blank: If a function is *increasing* on a closed interval, then an upper bound for the function is its value at the _____-hand endpoint of the interval.

41. Fill in the blank: If a function is *decreasing* on a closed interval, then an upper bound for its value is its value at the _____-hand endpoint of the interval.

42. True or False: The nth Taylor approximation of a function agrees with the function for at least one value of x.

43. Suppose that $f(0) = 1$ and $f^{(n)}(0) = n!$ for $n = 1, 2, 3, \ldots$. What will be the nth Taylor polynomial for f at $x = 0$? [*Hint:* See page 657.] Use a formula from the preceding section to express this function in a different form.

Explorations and Excursions   The following problems extend and augment the material presented in the text.

Error Term for Taylor Approximation

44–50. Let a be a given number and suppose that the function f is $n + 1$ times differentiable between a and x. Since a and x are now fixed, we will use t for a variable taking values between them.

44. Show that the *Fundamental Theorem of Integral Calculus* on page 333 may be rewritten as

$$\int_a^x f'(t)\, dt = f(x) - f(a)$$

and therefore

$$f(x) = f(a) + \int_a^x f'(t)\, dt$$

45. Show that

$$\int_a^x f'(t) \cdot dt = (x - a)f'(a) + \int_a^x f''(t)(x - t)\, dt$$

by using integration by parts with $u = f'(t)$, so $du = f''(t)\, dt$, and $dv = dt$, so that we may choose to take $v = -(x - t)$ since x is a "constant" that is independent of the integration.

46. Combine the results of Exercises 44 and 45 to deduce that

$$f(x) = p_1(x) + \int_a^x f''(t)(x - t)\, dt$$

where $p_1(x)$ is the first Taylor polynomial of f at $x = a$.

47. Show that

$$\int_a^x f^{(k+1)}(t) \cdot (x - t)^k\, dt$$
$$= \frac{1}{k + 1}(x - a)^{k+1} f^{(k+1)}(a)$$
$$+ \frac{1}{k + 1} \int_a^x f^{(k+2)}(t) \cdot (x - t)^{k+1}\, dt$$

for $1 < k < n$ by using integration by parts with $u = f^{(k+1)}(t)$, so $du = f^{(k+2)}(t)\, dt$, and $dv = (x-t)^k\, dt$.

48. Combine Exercise 46 with repeated applications of Exercise 47 to deduce that

$$f(x) = p_n(x) + \int_a^x f^{(n+1)}(t) \cdot \frac{1}{n!}(x - t)^n\, dt$$

where $p_n(x)$ is the nth Taylor polynomial of f at $x = a$ and thus the "error" $R_n(x)$ between $f(x)$ and $p_n(x)$ is

$$R_n(x) = \int_a^x f^{(n+1)}(t) \cdot \frac{1}{n!}(x - t)^n\, dt$$

49. Show that if $|f^{(n+1)}(t)| \leq M$ for all t between a and x, then

$$|R_n(x)| \leq \frac{M}{(n + 1)!} |x - a|^{n+1}$$

This establishes the error estimate on page 664.

50. Use Exercise 49 with $a = 0$ to establish the error estimate on page 659.

10.3 Taylor Series

Introduction

In this section we return to infinite series, but with the terms of the series now being *functions*. We begin by discussing *power series*, then specializing to *Taylor series*, which may be thought of as "infinitely long" Taylor polynomials. Taylor series lead to remarkably simple infinite series for functions such as e^x, $\ln x$, and $\sin x$, and enable us to find difficult integrals involving such functions.

Power Series

In this (and the next) section we will consider infinite series whose terms involve a *variable*, such as

$$1 + x + x^2 + x^3 + \cdots$$

Replacing x by $\frac{1}{2}$ gives

$$1 + \frac{1}{2} + \frac{1}{4} + \frac{1}{8} + \cdots \qquad \text{From } 1 + \frac{1}{2} + \left(\frac{1}{2}\right)^2 + \left(\frac{1}{2}\right)^3 + \cdots$$

which is a convergent geometric series with ratio $r = \frac{1}{2}$. If instead we replace x by 1, we obtain

$$1 + 1 + 1 + 1 + \cdots \qquad \text{From } 1 + 1 + (1)^2 + (1)^3 + \cdots$$

which clearly diverges since the sum becomes arbitrarily large. In general, a series involving a *variable* may converge for some values of the variable and diverge for others. It will therefore be important to determine the values of the variable for which the series converges. The most useful type of series is called a

power series, which is similar to the series $1 + x + x^2 + x^3 + \cdots$ but with the terms multiplied by coefficients. That is, a power series is of the form

$$a_0 + a_1 x + a_2 x^2 + \cdots + a_n x^n + \cdots$$

in which the coefficients $a_0, a_1, a_2, \ldots, a_n, \ldots$ are (real) numbers. Every power series in x converges for $x = 0$ (where its value is clearly a_0). It also converges at all points within some radius R around $x = 0,$ with the understanding that R may be a positive number, or $R = 0$ (in which case the series converges only at the single number $x = 0$), or $R = \infty$ (in which case the series converges for *all* values of x). Formally:

Power Series in *x* and Radius of Convergence

For a power series in x, $a_0 + a_1 x + a_2 x^2 + \cdots + a_n x^n + \cdots$, exactly one of the following statements is true:

1. There is a positive number R (the "radius of convergence") such that the series converges for $|x| < R$ and diverges for $|x| > R$.
2. The series converges only at $x = 0$ (radius $R = 0$).
3. The series converges for *all* x (radius $R = \infty$).

These three cases can be shown graphically as follows.

Positive radius	Zero radius	Infinite radius

For a positive radius of convergence, the series may converge at *neither* endpoint, at just *one* endpoint, or at *both* endpoints. We will generally ignore these endpoints, finding only the radius of convergence by using the following **Ratio Test**.

Ratio Test

For an infinite series $c_0 + c_1 + c_2 + \cdots$ of nonzero terms such that the limit

$$r = \lim_{n \to \infty} \left| \frac{c_{n+1}}{c_n} \right| \quad \text{exists or is infinite:}$$

1. If $r < 1,$ the series *converges*.
2. If $r > 1$ or if the limit is infinite, the series *diverges*.

The Ratio Test may be understood by recalling a fact about geometric series (page 648): The series *converges* if the ratio r of successive terms satisfies $|r| < 1$. The Ratio Test says that even if a series is *not* geometric, as long as the *limiting* ratio of successive terms is less than 1 in absolute value, then the series converges. A proof of the Ratio Test is outlined in Exercises 58–59. Finding the radius of convergence of a power series by the Ratio Test involves four steps:

1. Find expressions for successive terms c_n and c_{n+1} (found from c_n by replacing each n by $n + 1$).

2. Write the quotient $\dfrac{c_{n+1}}{c_n}$ and simplify.

3. Calculate $r = \lim\limits_{n \to \infty} \left| \dfrac{c_{n+1}}{c_n} \right|$.

4. Solve the inequality $r < 1$ for $|x|$ to find the radius of convergence R.

EXAMPLE 1   FINDING THE RADIUS OF CONVERGENCE

Find the radius of convergence of the power series

$$\frac{1}{2}x + \frac{2}{4}x^2 + \frac{3}{8}x^3 + \cdots + \frac{n}{2^n}x^n + \cdots$$

Solution

Successive terms are $c_n = \dfrac{n}{2^n} x^n$ and $c_{n+1} = \dfrac{n+1}{2^{n+1}} x^{n+1}$, and their ratio is

$$\frac{c_{n+1}}{c_n} = \frac{\dfrac{n+1}{2^{n+1}} x^{n+1}}{\dfrac{n}{2^n} x^n} = \frac{n+1}{2^{n+1}} \cdot \frac{x^{n+1}}{x^n} \cdot \frac{2^n}{n} = \underbrace{\frac{n+1}{n}} \cdot \underbrace{\frac{2^n}{2^{n+1}}} \cdot x = \underbrace{\left(1 + \frac{1}{n}\right)} \cdot \underbrace{\frac{1}{2}} \cdot x \to \frac{x}{2}$$

with $\xrightarrow{\text{Inverting}}$ under x; $1 + \dfrac{1}{n}$; $\dfrac{1}{2}$; Approaches 1 as $n \to \infty$.

Therefore, $r = \lim\limits_{n \to \infty} \left| \dfrac{c_{n+1}}{c_n} \right| = \dfrac{|x|}{2}$. For convergence, this ratio must be less

than 1, $\dfrac{|x|}{2} < 1$, which is equivalent to $|x| < 2$. Therefore, the radius of convergence is $R = 2$, meaning that the series converges for $|x| < 2$ and diverges for $|x| > 2$.

Taylor Series

We now consider a special type of power series called a **Taylor series,** which may be thought of as an "infinitely long" Taylor polynomial. We begin with a function f with derivatives of all orders at $x = 0$.

Taylor Series at $x = 0$

LOOKING BACK

This is just like Taylor *polynomials* (page 657), but not stopping.

The Taylor series at $x = 0$ of f is

$$f(0) + \frac{f'(0)}{1!}x + \frac{f''(0)}{2!}x^2 + \frac{f'''(0)}{3!}x^3 + \cdots + \frac{f^{(n)}(0)}{n!}x^n + \cdots$$

Note that the coefficients are calculated just as in Taylor polynomials.

EXAMPLE 2	FINDING A TAYLOR SERIES

Find the Taylor series at $x = 0$ for e^x, and find its interval of convergence.

Solution

The Taylor series for e^x can be written down immediately from the Taylor polynomials for e^x in Example 2 on page 657 simply by letting the terms continue indefinitely:

$$1 + \frac{x}{1!} + \frac{x^2}{2!} + \frac{x^3}{3!} + \frac{x^4}{4!} + \cdots \qquad \text{Taylor series for } e^x$$

For the radius of convergence, we calculate the ratio of successive terms and use the Ratio Test:

$$\frac{\dfrac{x^{n+1}}{(n+1)!}}{\dfrac{x^n}{n!}} = \frac{x^{n+1}}{(n+1)!} \cdot \frac{n!}{x^n} = \frac{x^{n+1}}{x^n} \cdot \frac{n!}{(n+1)!}$$

$c_n = \dfrac{x^n}{n!}$ and

$c_{n+1} = \dfrac{x^{n+1}}{(n+1)!}$

 Inverting

x

$\dfrac{n \cdots 1}{(n+1) \cdot n \cdots 1} = \dfrac{1}{n+1}$

$$= x \cdot \frac{1}{n+1} \longrightarrow 0$$

Approaches 0 r
as $n \to \infty$

This ratio approaches 0, so $r = 0$ satisfying $|r| < 1$ for *any* fixed value of x. Therefore, by the Ratio Test, the series converges for *all* values of x (the radius of convergence is $R = \infty$).

⚠️ **Be Careful** Don't confuse r and R. *Lowercase r* comes from the ratio of terms (after taking absolute values) and so may involve x (as in Example 1). *Uppercase R* is the radius of convergence, which is simply a number or the symbol ∞. In extreme cases they are inversely related: if $r = 0$ then $R = \infty$, while if $r = \infty$ then $R = 0$.

We have shown that the series for e^x converges for every value of x, but we have not shown that it converges *to e^x* (it might instead converge to some other value). Proving convergence to the original function requires showing that the error $|R_n(x)|$ in the nth Taylor polynomial approaches zero as $n \to \infty$. Such proofs are difficult, and are relegated to the Exercises (see Exercises 56 and 57). For every Taylor series in this book, and for every Taylor series you are likely to encounter in applications, if it converges, it converges to its defining function. Based on this statement, we will write the series as *equal* to its defining function wherever the series converges. For example, the series for e^x *converges* for all x, and *equals e^x*:

Taylor Series at $x = 0$ for e^x

$$e^x = 1 + \frac{x}{1!} + \frac{x^2}{2!} + \frac{x^3}{3!} + \frac{x^4}{4!} + \cdots \qquad \text{for} \quad -\infty < x < \infty$$

This series is called the *Taylor series expansion of e^x*. Evaluating at $x = 1$ gives a remarkable series for the constant e:

$$e = 1 + \frac{1}{1!} + \frac{1}{2!} + \frac{1}{3!} + \frac{1}{4!} + \cdots$$

e is the sum of the reciprocals of factorials

Graphing Calculator Exploration

a. Sum the first seven terms of the series above. To how many decimal places does the result agree with $e \approx 2.71828$?

b. How many terms are required to get agreement to *five* decimal places?

EXAMPLE 3   FINDING THE TAYLOR SERIES FOR $\sin x$

Find the Taylor series at $x = 0$ for $\sin x$.

Solution

Differentiating $\sin x$ gives the following repeating pattern of positive and negative sines and cosines, which on the third line below are evaluated at $x = 0$.

f	f'	f''	f'''	$f^{(4)}$	$f^{(5)}$	$f^{(6)}$	$f^{(7)}$	$\cdots$
$\sin x$	$\cos x$	$-\sin x$	$-\cos x$	$\sin x$	$\cos x$	$-\sin x$	$-\cos x$	$\cdots$
0	1	0	-1	0	1	0	-1	$\cdots$

Substituting these into the Taylor series formula (page 669) gives

$$0 + \frac{1}{1!}x + \frac{0}{2!}x^2 + \frac{-1}{3!}x^3 + \frac{0}{4!}x^4 + \frac{1}{5!}x^5 + \frac{0}{6!}x^6 + \frac{-1}{7!}x^7 + \cdots$$

Omitting the zero terms and simplifying gives the following Taylor series for $\sin x$.

$$\sin x = x - \frac{x^3}{3!} + \frac{x^5}{5!} - \frac{x^7}{7!} + \cdots$$

Alternating signs, with only odd powers

This series converges for *all* x (as is shown in Exercise 53).*

Taylor Series at $x = 0$ for $\sin x$

$$\sin x = x - \frac{x^3}{3!} + \frac{x^5}{5!} - \frac{x^7}{7!} + \cdots \qquad \text{for} \quad -\infty < x < \infty$$

---

*This series gives another reason for expressing trigonometric functions in *radians* instead of degrees. The Taylor series for $\sin x$ for *x in degrees*, written $\sin x°$, is found by multiplying each x by $\frac{\pi}{180}$, obtaining

$$\sin x° = \frac{\pi}{180}x - \left(\frac{\pi}{180}\right)^3 \frac{x^3}{3!} + \left(\frac{\pi}{180}\right)^5 \frac{x^5}{5!} - \left(\frac{\pi}{180}\right)^7 \frac{x^7}{7!} + \cdots$$

This series lacks the simplicity of the radian version.

Operations on Taylor Series

In many ways, Taylor series behave like polynomials. For example, the Taylor series for $f(x)$ can be differentiated term by term to obtain the Taylor series for $f'(x)$ (but only for x *within* the interval of convergence, avoiding the endpoints). A similar statement holds for integration. A Taylor series can also be multiplied by a constant c or by a positive integer power x^n, or the variable may be replaced by an expression of the form cx^n. In each case, the resulting series will be the Taylor series for the function similarly modified, and the radius of convergence will remain the same (except in the case of replacing the variable, where a finite radius of convergence may change). Two series may be added or subtracted, giving the series for the sum or difference of the original functions, and the resulting series will have a radius of convergence equal to the *smaller* of the radii of the original series.

Any power series that converges to a function must be the *Taylor* series for that function. For example, according to the formula on page 648, the geometric series $1 + x + x^2 + x^3 + \cdots$ converges to $\dfrac{1}{1-x}$ for $|x| < 1$, and so must be the *Taylor* series for that function for $|x| < 1$.

Taylor Series at $x = 0$ for $\dfrac{1}{1-x}$

$$\frac{1}{1-x} = 1 + x + x^2 + x^3 + \cdots \qquad \text{for} \quad |x| < 1$$

EXAMPLE 4   **FINDING ONE TAYLOR SERIES FROM ANOTHER**

Find the Taylor series at $x = 0$ for $\cos x$.

Solution

The easiest way is to differentiate the series for $\sin x$ found in Example 3:

$$\sin x = x - \frac{1}{3!}x^3 + \frac{1}{5!}x^5 - \frac{1}{7!}x^7 + \cdots \qquad \text{for} \quad -\infty < x < \infty$$

$$\cos x = 1 - \frac{1}{3!}3x^2 + \frac{1}{5!}5x^4 - \frac{1}{7!}7x^6 + \cdots \qquad \text{Differentiating term by term by the Power Rule}$$

$$= 1 - \frac{1}{2!}x^2 + \frac{1}{4!}x^4 - \frac{1}{6!}x^6 + \cdots \qquad \text{Simplifying}$$

This gives the following Taylor series for $\cos x$.

$$\cos x = 1 - \frac{x^2}{2!} + \frac{x^4}{4!} - \frac{x^6}{6!} + \cdots \qquad \text{Alternating signs, with only } even \text{ powers}$$

This series converges for *all* x (since term-by-term differentiation does not change the radius of convergence).

Taylor Series at $x = 0$ for $\cos x$

$$\cos x = 1 - \frac{x^2}{2!} + \frac{x^4}{4!} - \frac{x^6}{6!} + \cdots \qquad \text{for} \quad -\infty < x < \infty$$

PRACTICE PROBLEM

Find the Taylor series at $x = 0$ for $\cos x^2$. [*Hint:* Use the preceding series with a substitution.]

Solution on page 676 >

By integrating entire series, we can construct infinite series for difficult integrals that we could not otherwise evaluate.

Baby Booms and Echoes

An increase in births (a "baby boom") is usually followed in later generations by smaller fluctuations (called "echoes" of the boom) as the children grow up and in turn have more children. The resulting fluctuations in the number of births can be modeled by a function such as $f(t) = \dfrac{\sin t}{t}$.

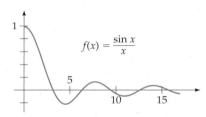

Source: National Center for Health Statistics

EXAMPLE 5   SUMMING EXCESS BIRTHS

A country is predicted to have an excess of $\dfrac{\sin t}{t}$ million births per year (above the usual level) in year t. During the first x years, this means an additional $\displaystyle\int_0^x \frac{\sin t}{t}\, dt$ million births. Find the Taylor series for this integral. Then use the series to estimate the number of extra births during the first two years.

Solution

There is no simple antiderivative for the function $\dfrac{\sin t}{t}$, so we express it as a Taylor series (dividing the series for $\sin t$ by t), which we then integrate.

$$\frac{\sin t}{t} = \frac{1}{t}\left(t - \frac{1}{3!}t^3 + \frac{1}{5!}t^5 - \frac{1}{7!}t^7 + \cdots \right) \qquad \text{Using the series for } \sin t$$

$$= 1 - \frac{1}{3!}t^2 + \frac{1}{5!}t^4 - \frac{1}{7!}t^6 + \cdots \qquad \text{Simplifying}$$

Then

$$\int_0^x \frac{\sin t}{t}\, dt = \int_0^x \left(1 - \frac{1}{3!}t^2 + \frac{1}{5!}t^4 - \frac{1}{7!}t^6 + \cdots \right) dt$$

$$= \left(t - \frac{1}{3!}\frac{1}{3}t^3 + \frac{1}{5!}\frac{1}{5}t^5 - \frac{1}{7!}\frac{1}{7}t^7 + \cdots \right)\Bigg|_0^x$$

Integrating term by term by the Power Rule

$$= x - \frac{x^3}{3\cdot 3!} + \frac{x^5}{5\cdot 5!} - \frac{x^7}{7\cdot 7!} + \cdots$$

Substituting $t = x$ (the evaluation at $t = 0$ is zero)

This infinite series gives the *exact* value of the integral $\int_0^x \frac{\sin t}{t}\, dt$. Evaluating at $x = 2$ and approximating by the first four terms gives

$$2 - \frac{2^3}{3\cdot 3!} + \frac{2^5}{5\cdot 5!} - \frac{2^7}{7\cdot 7!} \approx 1.605$$

Therefore, there will be about 1.605 million extra births during the first two years.

Although $\frac{\sin t}{t}$ is undefined at $t = 0$, its *limit* as $t \to 0$ is 1 (see pages 554–555), which is also the *value* of its Taylor series (at the bottom of the previous page) evaluated at $t = 0$.

Why Is Convergence So Important?

The radius of convergence of a Taylor series is important because only *within* that radius from the central point does the series *equal* the function. Beyond that radius the series will have no meaning, even though the *original function* may still be defined.

For example, the function $\frac{1}{1 - x}$ is defined for all x-values except $x = 1$, but its Taylor series $1 + x + x^2 + x^3 + \cdots$ converges only on the interval $(-1, 1)$, as we saw on page 672. Outside of this interval the *function* is still defined (its value at $x = 2$ is $\frac{1}{1-2} = \frac{1}{-1} = -1$), but its Taylor series at $x = 2$ is $1 + 2 + 2^2 + 2^3 + \cdots$, which does not converge and is meaningless. That is, using a Taylor series outside of the radius of convergence leads to incorrect results.

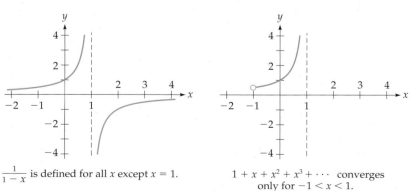

$\frac{1}{1-x}$ is defined for all x except $x = 1$.

$1 + x + x^2 + x^3 + \cdots$ converges only for $-1 < x < 1$.

Taylor Series at $x = a$

Just as with Taylor polynomials, we may define Taylor series in powers of $(x - a)$, shifting the "center point" from $x = 0$ to $x = a$. Let f be a function that has derivatives of all orders at $x = a$.

Taylor Series at $x = a$

LOOKING BACK

Like the "$x = 0$" version on page 669, but with a instead of 0.

The Taylor series at $x = a$ of f is

$$f(a) + \frac{f'(a)}{1!}(x - a) + \frac{f''(a)}{2!}(x - a)^2 + \cdots + \frac{f^{(n)}(a)}{n!}(x - a)^n + \cdots$$

Taylor *series* at $x = a$ are simply Taylor *polynomials* at $x = a$ with infinitely many terms. Taking $a = 0$ gives the Taylor series at $x = 0$, as defined earlier.

EXAMPLE 6   **FINDING A TAYLOR SERIES AT $x = 1$**

Find the Taylor series at $x = 1$ for $\ln x$.

Solution

Differentiating $f(x) = \ln x$ and evaluating at $x = 1$ gives

f	f'	f''	f'''	$f^{(4)}$	$\cdots$	
$\ln x$	x^{-1}	$-x^{-2}$	$2x^{-3}$	$-3!x^{-4}$	$\cdots$	$\ln x$ and its derivatives
0	1	-1	2	$-3!$	$\cdots$	Evaluated at $x = 1$

Therefore, the Taylor series at $x = 1$ for $\ln x$ is

$$(x - 1) - \frac{1}{2!}(x - 1)^2 + \frac{2}{3!}(x - 1)^3 - \frac{3!}{4!}(x - 1)^4 + \cdots \qquad \text{Taylor series with } a = 1 \text{ and the derivatives found above}$$

$$= (x - 1) - \frac{1}{2}(x - 1)^2 + \frac{1}{3}(x - 1)^3 - \frac{1}{4}(x - 1)^4 + \cdots \qquad \text{Simplifying}$$

The radius of convergence is found from the Ratio Test (page 668), calculating the ratio of successive terms (and now including absolute value bars because of the alternating signs):

$$\left| \frac{\frac{1}{n+1}(x-1)^{n+1}}{\frac{1}{n}(x-1)^n} \right| = \left| \frac{n}{n+1} \cdot \frac{(x-1)^{n+1}}{(x-1)^n} \right| \to |x - 1|$$

$$\underbrace{\quad}_{\substack{\text{Approaches 1} \\ \text{as } n \to \infty}} \quad \underbrace{\quad}_{(x-1)} \quad \underbrace{\quad}_{r}$$

For convergence, this ratio must be less than 1: $|x - 1| < 1$, which is equivalent to $0 < x < 2$. Therefore, the Taylor series at $x = 1$ for $\ln x$ is

$$\ln x = (x - 1) - \frac{1}{2}(x - 1)^2 + \frac{1}{3}(x - 1)^3 - \frac{1}{4}(x - 1)^4 + \cdots$$

for $0 < x < 2$.

This series converges at the endpoint $x = 2$ but *not* at the endpoint $x = 0$ (see Exercises 60 and 61).

Taylor Series at $x = 1$ for $\ln x$

$$\ln x = (x - 1) - \frac{(x - 1)^2}{2} + \frac{(x - 1)^3}{3} - \frac{(x - 1)^4}{4} + \cdots \qquad \text{for } 0 < x \le 2$$

Solution   TO PRACTICE PROBLEM

$$\cos x^2 = 1 - \frac{x^4}{2!} + \frac{x^8}{4!} - \frac{x^{12}}{6!} + \cdots \qquad\qquad \text{Replacing } x \text{ by } x^2$$

10.3   Section Summary

Every power series in x, $a_0 + a_1 x + a_2 x^2 + \cdots + a_n x^n + \cdots$, has a *radius of convergence R* (which may be zero, a positive number, or ∞), and the series converges for $|x| < R$ and diverges for $|x| > R$. We find R by using the Ratio Test, which may be thought of as testing whether the series is "asymptotically geometric."

A *Taylor series* is a type of power series whose coefficients are calculated from a given function f, using the same formulas as for Taylor polynomials:

$$f(x) = \sum_{n=0}^{\infty} \frac{f^{(n)}(0)}{n!} x^n \qquad\qquad \begin{array}{l} \text{Taylor series at } x = 0 \text{ for} \\ f(x) \text{ [the first term is } f(0)] \end{array}$$

$$f(x) = \sum_{n=0}^{\infty} \frac{f^{(n)}(a)}{n!} (x - a)^n \qquad\qquad \begin{array}{l} \text{Taylor series at } x = a \text{ for} \\ f(x) \text{ [the first term is } f(a)] \end{array}$$

A Taylor series can also be found by modifying an already existing Taylor series, as explained on page 672. We derived the following Taylor series:

$$\frac{1}{1 - x} = 1 + x + x^2 + x^3 + \cdots \qquad\qquad \text{for } -1 < x < 1$$

$$e^x = 1 + \frac{x}{1!} + \frac{x^2}{2!} + \frac{x^3}{3!} + \frac{x^4}{4!} + \cdots \qquad\qquad \text{for } -\infty < x < \infty$$

$$\sin x = x - \frac{x^3}{3!} + \frac{x^5}{5!} - \frac{x^7}{7!} + \cdots \qquad\qquad \text{for } -\infty < x < \infty$$

$$\cos x = 1 - \frac{x^2}{2!} + \frac{x^4}{4!} - \frac{x^6}{6!} + \cdots \qquad\qquad \text{for } -\infty < x < \infty$$

$$\ln x = (x - 1) - \frac{(x - 1)^2}{2} + \frac{(x - 1)^3}{3} - \frac{(x - 1)^4}{4} + \cdots \qquad \text{for } 0 < x \le 2$$

10.3 Exercises

1–10. Find the radius of convergence of each power series.

1. $1 - \dfrac{x}{2} + \dfrac{x^2}{2^2} - \dfrac{x^3}{2^3} + \cdots$

2. $1 + 2x + 2^2 x^2 + 2^3 x^3 + \cdots$

3. $\dfrac{x}{1 \cdot 2} + \dfrac{x^2}{2 \cdot 3} + \dfrac{x^3}{3 \cdot 4} + \dfrac{x^4}{4 \cdot 5} + \cdots$

4. $1 - \dfrac{x}{3} + \dfrac{x^2}{3^2} - \dfrac{x^3}{3^3} + \cdots + \dfrac{(-1)^n x^n}{3^n} + \cdots$

5. $\displaystyle\sum_{n=0}^{\infty} \frac{(-1)^n x^n}{n!}$

6. $\displaystyle\sum_{n=0}^{\infty} \frac{2^n x^n}{n!}$

7. $\displaystyle\sum_{n=0}^{\infty} \frac{n! x^n}{5^n}$

8. $\displaystyle\sum_{n=0}^{\infty} n! x^n$

9. $\displaystyle\sum_{n=0}^{\infty} \frac{x^{2n}}{n!}$

10. $\displaystyle\sum_{n=0}^{\infty} x^{n^2}$

11–16. Find the Taylor series at $x = 0$ for each function by calculating three or four derivatives and using the definition of Taylor series.

11. $\ln(1 + x)$

12. e^{2x}

13. $\cos 2x$

14. $\dfrac{1}{3 - x}$

15. $\sqrt{2x + 1}$

16. $2x^2 - 7x + 3$

17–18. Find the Taylor series at $x = 0$ for each function in *two* ways:

a. By calculating derivatives and using the definition of Taylor series

b. By modifying a known Taylor series

17. $e^{x/5}$

18. $\sin 2x$

19. Find the Taylor series at $x = \dfrac{\pi}{2}$ for $\sin x$.

20. Find the Taylor series at $x = 1$ for $\sqrt{x}$.

21–28. Find the Taylor series at $x = 0$ for each function by modifying one of the Taylor series from this section.

21. $\dfrac{x^2}{1 - x}$

22. $\dfrac{1}{1 + 5x}$

23. $\sin x^2$

24. e^{-x^2}

25. $\dfrac{e^x - 1}{x}$

26. $\sin x + \cos x$

[*Hint:* For Exercise 26, use two series from this section, writing the terms in order of increasing powers of x.]

27. $\dfrac{x - \sin x}{x^3}$

28. $\dfrac{1 - \cos x}{x^2}$

29. Find the Taylor series at $x = 0$ for $\ln(x + 1)$ by integrating both sides of

$$\frac{1}{1 + x} = 1 - x + x^2 - x^3 + \cdots \quad \text{for } |x| < 1$$

30. Find the Taylor series at $x = 0$ for $\dfrac{1}{(1 - x)^2}$ by differentiating both sides of

$$\frac{1}{1 - x} = 1 + x + x^2 + x^3 + \cdots \quad \text{for } |x| < 1$$

31. **GREGORY-LEIBNIZ SERIES FOR π.**

a. Find the Taylor series at $x = 0$ for $\dfrac{1}{1 + x^2}$ by modifying one of the series derived in this section.

b. Use the Ratio Test to find the radius of convergence of the series from part (a). Notice that although

the original function $\dfrac{1}{1 + x^2}$ is defined for *all* values of x, the series is defined only for a much narrower set of x-values.

c. Find a series for $\displaystyle\int_0^1 \frac{1}{1 + x^2}\, dx$ by integrating your Taylor series from part (a) and evaluating.

d. Exercise 78 on page 581 showed that this integral has value $\pi/4$. Combine this result with your answer to part (c) to show that

$$1 - \frac{1}{3} + \frac{1}{5} - \frac{1}{7} + \cdots = \frac{\pi}{4}$$

This is known as *Leibniz's formula* or the *Gregory-Leibniz series* (after James Gregory, 1638–1675).

32. **HYPERBOLIC SINE FUNCTION** Use the Taylor series at $x = 0$ for e^x to find the series for $\dfrac{e^x - e^{-x}}{2}$. (This function is called the *hyperbolic sine*, written $\sinh x$. Notice how similar its Taylor series is to that of $\sin x$.)

33. **HYPERBOLIC COSINE FUNCTION** Use the Taylor series at $x = 0$ for e^x to find the series for $\dfrac{e^x + e^{-x}}{2}$. (This function is called the *hyperbolic cosine*, written $\cosh x$. Note how similar its Taylor series is to that of $\cos x$. The graph of $\cosh x$ is called a "catenary." The St. Louis Gateway Arch was built in this shape: see Exercise 48 on page 258.)

34. **SERIES CALCULATION OF sin 1**

a. Use a calculator set for *radians* to find $\sin 1$ (rounded to five decimal places).

b. Estimate $\sin 1$ by using the first three terms of the Taylor series

$$\sin x = x - \frac{x^3}{3!} + \frac{x^5}{5!} - \cdots$$

evaluated at $x = 1$ (rounded to five decimal places). [*Note:* Computers use series similar to this to calculate values of trigonometric functions.]

35. **SERIES CALCULATION OF $e^{0.5}$**

a. Use a calculator to find $e^{0.5}$ (rounded to five decimal places).

b. Estimate $e^{0.5}$ by using the first four terms of the Taylor series

$$e^x = 1 + x + \frac{x^2}{2!} + \frac{x^3}{3!} + \frac{x^4}{4!} + \cdots$$

evaluated at $x = 0.5$ (rounded to five decimal places).

 c. (Requires sequence and series operations) Set your calculator to find the sum of the series in part (b) up to any number of terms. How many terms are required for the sum (rounded to five decimal places) to agree with the value found in part (a)?

36. **SERIES CALCULATION OF e^2** (Requires sequence and series operations)

a. Use your calculator to find e^2 rounded to six decimal places.

b. The Taylor series for e^x evaluated at $x = 2$ gives $e^2 = \sum_{n=0}^{\infty} \dfrac{2^n}{n!}$. Set your calculator to find the sum of this series up to any number of terms. How many terms are required for the sum (rounded to six decimal places) to agree with your answer to part (a)?

37. TAYLOR SERIES FOR $\sin x$ **AND** $\cos x$
Differentiate the series

$$\cos x = 1 - \frac{x^2}{2!} + \frac{x^4}{4!} - \frac{x^6}{6!} + \cdots$$

and check that the resulting series is the negative of the series for $\sin x$, showing (again) that

$$\frac{d}{dx} \cos x = -\sin x.$$

38. TAYLOR SERIES FOR e^x Differentiate the series

$$e^x = 1 + \frac{x}{1!} + \frac{x^2}{2!} + \frac{x^3}{3!} + \frac{x^4}{4!} + \cdots$$

showing (again) that e^x is its own derivative.

Applied Exercises

39. GENERAL: Bell-Shaped Curve The "normal distribution" or "bell-shaped curve" is used to predict many things, including IQ's, incomes, and manufacturing errors. The area under this curve from 0 up to any number x is given by the integral

$$\frac{1}{\sqrt{2\pi}} \int_0^x e^{-t^2/2} dt.$$

a. Find the Taylor series at 0 for $e^{-t^2/2}$. [*Hint:* Modify a known series.]
b. Integrate this series from 0 to x and multiply by 0.4 $\left(\text{which approximates } \dfrac{1}{\sqrt{2\pi}}\right)$, obtaining a Taylor series for $\dfrac{1}{\sqrt{2\pi}} \int_0^x e^{-t^2/2} \, dt.$

c. Estimate $\dfrac{1}{\sqrt{2\pi}} \int_0^1 e^{-t^2/2} \, dt$ by using the first three terms of the series found in part (b) evaluated at $x = 1$. Your answer finds, for example, the proportion of people with IQ's between 100 and 115 (see pages 420–421).

40. SOCIAL SCIENCE: Baby Booms If the birthrate (thousands of births per year) rises by an amount $\dfrac{1 - \cos t}{t}$ (above the usual level) in year t, the number of excess births during the first x years will be $\displaystyle\int_0^x \frac{1 - \cos t}{t} dt.$

a. Find the Taylor series at 0 for $\dfrac{1 - \cos t}{t}$.
[*Hint:* Modify a known series.]

b. Integrate this series from 0 to x, obtaining a Taylor series for the integral $\displaystyle\int_0^x \frac{1 - \cos t}{t} dt$

c. Estimate $\displaystyle\int_0^1 \frac{1 - \cos t}{t} dt$ by using the first three terms of the series found in part (b) evaluated at $x = 1$.

41. BUSINESS: Sales Fluctuations As the pace of change in modern society quickens, popular fashions may fluctuate increasingly rapidly. Suppose that sales (above a certain minimum level) for a fashion item are $\cos t^2$ in year t, so that extra sales during the first x years are $\int_0^x \cos t^2 \, dt$ (in thousands).

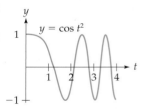

a. Find the Taylor series at 0 for $\cos t^2$.
[*Hint:* Modify a known series.]
b. Integrate this series from 0 to x, obtaining a Taylor series for the integral $\int_0^x \cos t^2 \, dt$.
c. Estimate $\int_0^1 \cos t^2 \, dt$ by using the first three terms of the series found in part (b) evaluated at $x = 1$.

42. GENERAL: Reactor Temperature Under emergency conditions, the temperature in a nuclear containment vessel is expected to rise at the rate of $200e^{t^2}$ degrees per hour, so that the temperature change in the first x hours will be $200\int_0^x e^{t^2} \, dt$ degrees. Estimate the temperature rise as follows:

a. Find the Taylor series at 0 for e^{t^2}.
[*Hint:* Modify a known series.]
b. Integrate this series from 0 to x and multiply by 200, obtaining a Taylor series for $200\int_0^x e^{t^2} \, dt$.
c. Estimate the temperature change in the first half hour by using the first three terms of the series found in part (b) evaluated at $x = \frac{1}{2}$.

Conceptual Exercises

43. Can a power series in x diverge at 3 but converge at 4?

44. What is the maximum number of terms in the Taylor series for a polynomial of degree 7?

45. Fill in the blank: If in the Ratio Test you find $r = \dfrac{|x|}{17}$, then the radius of convergence is _____.

46. Suppose that $f(0) = 1$ and $f^{(n)}(0) = n!$ for $n = 1, 2, 3, \ldots$. What will be the Taylor series for f at $x = 0$? [*Hint:* See page 669.] Do you know a formula from earlier in the chapter to express this function in a different form?

47–50. Find, by inspection, the sum of each series.

47. $1 - \dfrac{1}{1!} + \dfrac{1}{2!} - \dfrac{1}{3!} + \cdots$ [*Hint:* See page 670.]

48. $\pi - \dfrac{\pi^3}{3!} + \dfrac{\pi^5}{5!} - \dfrac{\pi^7}{7!} + \cdots$ [*Hint:* See page 671.]

49. $1 - \dfrac{\pi^2}{2!} + \dfrac{\pi^4}{4!} - \dfrac{\pi^6}{6!} + \cdots$ [*Hint:* See page 673.]

50. $1 - \dfrac{1}{2} + \dfrac{1}{3} - \dfrac{1}{4} + \cdots$ [*Hint:* See page 676.]

51. If a Taylor series at $x = a$ converges on the interval [3, 17] but on no larger interval, what is the value of a?

52. Fill in the blank: If a Taylor series at $x = 2$ converges at $x = 5$, then the radius of convergence is at least _____.

Explorations and Excursions   The following problems extend and augment the material presented in the text.

More on Convergence of Series

53. TAYLOR SERIES FOR $\sin x$ Use the Ratio Test to show that the Taylor series

$$\sin x = \sum_{n=0}^{\infty} \frac{(-1)^n x^{2n+1}}{(2n+1)!}$$ converges for all x.

54. TAYLOR SERIES AND POWER SERIES For a power series that converges to a function f,

$$f(x) = a_0 + a_1 x + a_2 x^2 + a_3 x^3 + \cdots$$

show that the coefficients must be $a_0 = f(0)$, $a_1 = f'(0)$, $a_2 = \dfrac{f''(0)}{2!}, \ldots$. Since these are exactly the Taylor series coefficients, this shows that a power series that converges to a function *is* the Taylor series for that function. [*Hint:* For a_0, just evaluate the given equation at 0. For $a_1, a_2, \ldots$, differentiate one or more times and then evaluate at $x = 0$. We are assuming the fact that a power series can be differentiated term by term inside its radius of convergence.]

55. nth TERM TEST For a convergent infinite series $S = c_1 + c_2 + c_3 + \cdots$, let S_n be the sum of the first n terms: $S_n = c_1 + c_2 + c_3 + \cdots + c_n$.

a. Show that $S_n - S_{n-1} = c_n$.

b. Take the limit of the equation in part (a) as $n \to \infty$ to show that $\lim_{n \to \infty} c_n = 0$. This proves the nth term test for infinite series:

   In a convergent infinite series, the nth term must approach zero as $n \to \infty$.

(*continues*)

Or equivalently:

   If the nth term of a series does *not* approach zero as $n \to \infty$, then the series diverges.

56. TAYLOR SERIES FOR e^x Show that the Taylor series for e^x, $\sum_{n=0}^{\infty} \dfrac{x^n}{n!}$, converges to e^x by showing that the error for the nth Taylor approximation approaches zero as follows:

a. Use the Ratio Test to show that the series

$$\sum_{n=0}^{\infty} \frac{|x|^n}{n!}$$ converges for all x.

b. Use the nth term test (Exercise 55) to conclude that $\dfrac{|x|^n}{n!} \to 0$ as $n \to \infty$.

c. Use part (b) to show that

$$|R_n(x)| \le \frac{e^{|x|}}{(n+1)!} |x|^{n+1} \to 0$$

as $n \to \infty$ for any fixed value of x.

57. TAYLOR SERIES FOR $\sin x$ Show that the Taylor series for $\sin x$, $\sum_{n=0}^{\infty} \dfrac{(-1)^n x^{2n+1}}{(2n+1)!}$, converges to $\sin x$ by showing that the error for the nth Taylor approximation approaches zero as follows:

a. Use the Ratio Test to show that the series

$$\sum_{n=0}^{\infty} \frac{|x|^n}{n!}$$ converges for all x.

b. Use the nth term test (Exercise 55) to

conclude that $\dfrac{|x|^{2n+1}}{(2n+1)!} \to 0$ as $n \to \infty$.

[*Hint:* If the nth term approaches zero, then so does the $2n + 1$st.]

c. Use part (b) to show that

$$|R_{2n}(x)| \le \frac{1}{(2n+1)!}|x|^{2n+1} \to 0$$

as $n \to \infty$ for any fixed value of x.

58. RATIO TEST Prove part 1 of the Ratio Test (page 668) for positive terms as follows:

a. If $\displaystyle\lim_{n \to \infty} \frac{c_{n+1}}{c_n} = r < 1$, and if s is a number between r and 1, so that $r < s < 1$, explain why there is an integer N such that $\dfrac{c_{n+1}}{c_n} \le s$ for all $n \ge N$.

b. Show that this last inequality implies that

$$c_{N+1} \le s \cdot c_N, \quad c_{N+2} \le s^2 \cdot c_N, \quad c_{N+3} \le s^3 \cdot c_N, \dots$$

c. Show that $\sum_{n=0}^{\infty} c_{N+n}$ converges by showing that

$$\sum_{n=0}^{\infty} c_{N+n} \le \sum_{n=0}^{\infty} s^n \cdot c_N = c_N \sum_{n=0}^{\infty} s^n = c_N \frac{1}{1-s}$$

It can be shown that if a series of positive terms remains below a (finite) number, then the series converges.

d. Lastly, show that $\sum_{n=0}^{\infty} c_n$ converges.

It can be shown that if a series of positive terms converges, then the series obtained by changing the signs of some or all of the terms will also converge.

59. Prove part 2 of the Ratio Test (page 668) as follows:

a. If $\displaystyle\lim_{n \to \infty} \left|\frac{c_{n+1}}{c_n}\right| = r > 1$, explain why there is an integer N such that $\left|\dfrac{c_{n+1}}{c_n}\right| \ge 1$ for all $n \ge N$.

b. Show that this last inequality implies that

$$|c_{N+1}| \ge |c_N|, \quad |c_{N+2}| \ge |c_N|, \quad |c_{N+3}| \ge |c_N|, \dots$$

c. Then show that these inequalities imply that $\displaystyle\lim_{n \to \infty} c_n \ne 0$, so $\sum_{n=0}^{\infty} c_n$ diverges by the nth term test (Exercise 55).

60. TAYLOR SERIES FOR ln x Show that the Taylor series for ln x,

$$(x-1) - \frac{(x-1)^2}{2} + \frac{(x-1)^3}{3} - \frac{(x-1)^4}{4} + \cdots$$

diverges at $x = 0$ by evaluating the series at 0 and explaining why it diverges.

61. TAYLOR SERIES FOR ln x Show that the Taylor series for ln x,

$$(x-1) - \frac{(x-1)^2}{2} + \frac{(x-1)^3}{3} - \frac{(x-1)^4}{4} + \cdots$$

converges at $x = 2$ by carrying out the following steps.

a. Evaluate the series at $x = 2$ to obtain

$$1 - \frac{1}{2} + \frac{1}{3} - \frac{1}{4} + \frac{1}{5} - \cdots$$

b. Group the terms in pairs to obtain

$$\left(1 - \frac{1}{2}\right) + \left(\frac{1}{3} - \frac{1}{4}\right) + \left(\frac{1}{5} - \frac{1}{6}\right) + \cdots \quad \text{and}$$

explain why this expresses the series as a sum of positive terms.

c. Then group the terms as

$$1 - \left(\frac{1}{2} - \frac{1}{3}\right) - \left(\frac{1}{4} - \frac{1}{5}\right) - \left(\frac{1}{6} - \frac{1}{7}\right) - \cdots \quad \text{and}$$

explain why this shows that the sum must be less than 1.

It can be shown that if a series of positive terms remains below a (finite) number, then the series converges.

62. USING SERIES TO FIND LIMITS* Carry out the following steps to prove that if $a > 0$, then $\displaystyle\lim_{x \to \infty} xe^{-ax} = 0$.

a. Show that $e^x > \dfrac{x^2}{2}$ for $x > 0$.

[*Hint:* Begin with the Taylor series

$$e^x = 1 + x + \frac{x^2}{2!} + \frac{x^3}{3!} + \cdots \quad \text{and drop all but the third}$$

term on the right.]

b. Replace x by ax to obtain $e^{ax} > \dfrac{a^2x^2}{2}$.

c. Take the reciprocal of each side to obtain $e^{-ax} < \dfrac{2}{a^2x^2}$.

d. Multiply each side by x to obtain $xe^{-ax} < \dfrac{2}{a^2x}$.

e. Let $x \to \infty$ to obtain the claimed result.

63. USING SERIES TO FIND LIMITS* Prove that if $a > 0$, then $\displaystyle\lim_{x \to \infty} x^2 e^{-ax} = 0$.

[*Hint:* Use steps similar to those in Exercise 62, but in step (a) keep only the fourth term, and in step (d) multiply by x^2.]

64. RADIUS OF CONVERGENCE Show that the power series $\sum_{n=0}^{\infty} a_n x^n$ has radius of convergence

$$R = \lim_{n \to \infty} \left|\frac{a_n}{a_{n+1}}\right|, \quad \text{provided that this limit exists.}$$

[*Hint:* Use the Ratio Test.]

* The results of Exercises 62 and 63 will be used to find the mean and variance of an exponential random variable on pages 723 and 729.

10.4 Newton's Method

Introduction

Many times we have set a function equal to zero and solved for x. For example, setting the first or second derivative equal to zero is the first step in finding relative extreme points and inflection points. Newton's method is a procedure for *approximating* the solutions to such equations in case the *exact* solutions are difficult or impossible to find. The method was invented by Isaac Newton, one of the originators of calculus, and has many applications, such as finding the internal rate of return (interest rate) on a loan.*

Zeros of a Function

Recall from page 39 that an x-value at which a function $f(x)$ equals zero is called a **zero** of the function, or equivalently, a **root** of the equation $f(x) = 0$. A zero of a function is not simply the number 0, but is a value that, when substituted into the function, gives the *result* zero. Graphically, the zeros of a function are the x-values at which the graph meets the x-axis.

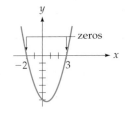

Zeros of $f(x) = x^2 - x - 6$ are the x-values where the curve crosses the x-axis.

EXAMPLE 1	FINDING ZEROS BY FACTORING

Find the zeros of $f(x) = x^2 - x - 6$.

Solution

$$f(x) = x^2 - x - 6 = (x - 3) \cdot (x + 2) = 0$$

Factoring and setting equal to zero

The zeros are

$$\begin{cases} x = 3 \\ x = -2 \end{cases}$$

From setting $(x - 3)$ and $(x + 2)$ equal to zero

These zeros are shown on the graph on the left.

Newton's Method

Only relatively simple functions can be factored. For more complicated functions, the zeros can only be *approximated,* and one of the most useful techniques is Newton's method. The method begins with an **initial approximation** $x = x_0$, which may be merely a "guess" (more on this later). The function is then replaced by its first Taylor polynomial at x_0 (the tangent line at x_0), since it is the best linear approximation for $f(x)$ near x_0. That is, instead of solving $f(x) = 0$, we set the *Taylor polynomial* equal to zero:

$$f(x_0) + f'(x_0) \cdot (x - x_0) = 0$$

First Taylor polynomial at $x = x_0$ set equal to zero

$$f'(x_0) \cdot (x - x_0) = -f(x_0)$$

Subtracting $f(x_0)$ from each side

*Newton's method is also called the Newton–Raphson method, after Joseph Raphson (1648–1715), a contemporary and friend of Issac Newton. Raphson first published the method, giving credit to Newton, who had described it in an unpublished manuscript many years earlier.

$$x - x_0 = -\frac{f(x_0)}{f'(x_0)}$$ Dividing each side by $f'(x_0)$
                                         [assuming $f'(x_0) \neq 0$)]

$$x = x_0 - \frac{f(x_0)}{f'(x_0)}$$ Solving for x by adding x_0
                                         to each side

This last formula gives the x-value (which we call x_1) where the first Taylor polynomial (the tangent line) crosses the x-axis. This x_1 should be a better approximation than the initial appoximation x_0 for the zero of the function $f(x)$, as shown below.

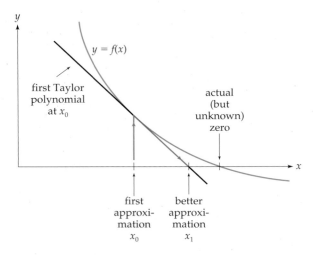

Initial approximation x_0 and the better approximation x_1,
which is closer to the actual zero.

The same formula can then be applied to the "better" approximation x_1 to obtain an even better approximation x_2, and then again to x_2 for a still better x_3, and so on:

$$x_1 = x_0 - \frac{f(x_0)}{f'(x_0)}$$

$$x_2 = x_1 - \frac{f(x_1)}{f'(x_1)}$$ Same formula applied to
                                          successive approximations

$$x_3 = x_2 - \frac{f(x_2)}{f'(x_2)}$$

$\vdots$

Each use of the formula is called an "iteration" (to "iterate" means to "repeat"). If the results of the iterations $x_1, x_2, x_3, \ldots$ approach a limit, we say that the iterations "converge" to this limit. In summary:

Newton's Method

To approximate a solution to $f(x) = 0$, choose an initial approximation x_0, and calculate $x_1, x_2, x_3, \ldots$ using

$$x_{n+1} = x_n - \frac{f(x_n)}{f'(x_n)} \qquad \text{for } n = 0, 1, 2, \ldots$$

If the numbers $x_0, x_1, x_2, \ldots$ converge, they converge to a solution of $f(x) = 0$.

EXAMPLE 2   APPROXIMATING BY NEWTON'S METHOD

Approximate $\sqrt{2}$ by using three iterations of Newton's method.

Solution

We approximate $\sqrt{2}$ by seeking a zero $f(x) = x^2 - 2$, which equals zero at $x = \sqrt{2}$. Since $\sqrt{2}$ lies between 1 and 2, we choose our initial approximation to be $x_0 = 1$ (we could just as well have chosen $x_0 = 2$ or $x_0 = 1.5$). Calculating $f(x)$ and $f'(x)$:

$$f(x) = x^2 - 2 \qquad \text{The function whose zero we want}$$

$$f'(x) = 2x \qquad \text{Its derivative}$$

The iteration formula is

$$x_{n+1} = x_n - \frac{x_n^2 - 2}{2x_n} \qquad x_{n+1} = x_n - \frac{f(x_n)}{f'(x_n)} \text{ with the } f \text{ and } f' \text{ above}$$

The approximations are

$$x_0 = 1 \qquad \text{Initial approximation}$$

$$x_1 = 1 - \frac{1^2 - 2}{2 \cdot 1} = 1.5 \qquad x_1 = x_0 - \frac{f(x_0)}{f'(x_0)} \text{ applied to } x_0 = 1$$

$$x_2 = 1.5 - \frac{1.5^2 - 2}{2 \cdot 1.5} \approx 1.4167 \qquad x_2 = x_1 - \frac{f(x_1)}{f'(x_1)} \text{ applied to } x_1 = 1.5$$

$$x_3 = 1.4167 - \frac{1.4167^2 - 2}{2 \cdot 1.4167} \approx 1.4142 \qquad \text{Iteration formula applied to } x_2 = 1.4167$$

The x_3 is our approximation for $\sqrt{2}$, the zero of $f(x) = x^2 - 2$:

$$\sqrt{2} \approx 1.4142$$

This estimate is correct to four decimal places (after only three iterations).

The approximations $x_0, x_1, x_2,$ and x_3 to $\sqrt{2}$ are shown in the following graph. The convergence is rapid enough that the second and third iterations require a magnified view.

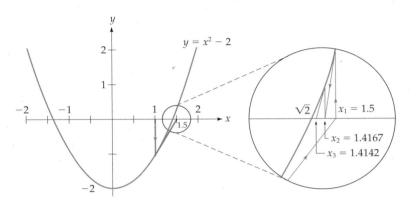

$f(x) = x^2 - 2$ and the iterations x_0, x_1, x_2, and x_3 approaching $\sqrt{2}$.

PRACTICE PROBLEM 1

Carry out one more iteration "by hand" with the formula above, using $x_3 = 1.4142$, keeping (if possible) nine decimal places of accuracy.

Solution on page 687 >

Observe that the number of correct digits in successive approximations roughly *doubled* with each iteration:

$x_0 = 1$	Initial "guess"
$x_1 = 1.5$	1 digit correct
$x_2 = 1.4167$	3 digits correct
$x_3 = 1.4142$	5 digits correct
$x_4 = 1.414213562$	10 digits correct (from Practice Problem 1)

How do you decide when to stop the iterations? Generally, you first choose a degree of accuracy, say 0.001, and stop when two successive approximations differ by less than your chosen accuracy.

How do you choose the function to use in the iterations? The number in Example 1, $\sqrt{2}$, satisfies $x^2 = 2$, or equivalently $x^2 - 2 = 0$, so we sought the zeros of the function $f(x) = x^2 - 2$. Other roots require different equations. For example, $\sqrt[3]{11}$ satisfies $x^3 = 11$, so we want to find a zero of $f(x) = x^3 - 11$, beginning with, perhaps, $x_0 = 2$.

How do you choose the initial estimate x_0? There is no general rule—you just try to find a number close to the number that you are seeking. The better the initial estimate, the faster the given accuracy will be achieved. If a function has *several* zeros, the one that the iterations converge to will depend upon the initial choice x_0. In Example 2, the equation $x^2 - 2 = 0$ actually has *two* solutions, $\sqrt{2}$ and $-\sqrt{2}$, corresponding to the two places where the curve crosses the x axis (see the graph above). The initial choice of $x_0 = 1$ led to an approximation for $\sqrt{2}$, but choosing $x_0 = -1$ would have led to an approximation for $-\sqrt{2}$.

PRACTICE PROBLEM 2

To approximate $\sqrt[5]{30}$, what function $f(x)$ would you use, and what would be a good initial estimate x_0?

Solution on page 687 >

The ZERO and SOLVE operations in a graphing calculator use Newton's method or one similar to it. The "initial guess" that these operations request is simply the initial x_0 of Newton's method.

EXAMPLE 3   **APPROXIMATING WITHIN A GIVEN TOLERANCE**

Approximate the solution to $e^x = 2 - 2x$, continuing until two successive iterations agree to three decimal places.

Solution

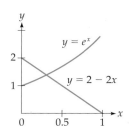

We find the initial estimate x_0 from a quick sketch of the graphs of e^x and $2 - 2x$, as shown on the left. They intersect at an x-value closer to 0 than to 1, so we choose $x_0 = 0$ as our initial estimate (other choices are also possible). We write the equation $e^x = 2 - 2x$ with zero on the right:

$$\underbrace{e^x - 2 + 2x}_{f(x)} = 0 \qquad\qquad \text{Definition of } f(x)$$

We could carry out the calculations "by hand," using the iteration formula

$$x_{n+1} = x_n - \frac{e^{x_n} - 2 + 2x_n}{e^{x_n} + 2} \qquad\qquad x_{n+1} = x_n - \frac{f(x_n)}{f'(x_n)}$$

beginning with $x_0 = 0$, or using a graphing calculator program from the Internet (see, for example, `ticalc.org`, and search for *Newton*), but instead we show a simple and direct method using a graphing calculator.

Graphing Calculator Exploration

Take Note

For $\rightarrow$ press STO▸

First enter the function whose zero you want to find into Y_1 and its derivative into Y_2. For the function in the preceding Example, these are shown on the left below.

Both round to 0.315

Then (if 0 is the *initial guess*) enter the two statements in the home screen $0 \rightarrow X$ and $X - Y_1/Y_2 \rightarrow X$ and repeatedly press ENTER to carry out successive iterations until two values agree to the desired accuracy. The results are shown on the right above, with the last two iterations agreeing to three (rounded) decimal places.

$$x \approx 0.315 \qquad\qquad \text{Approximation to the solution of } e^x = 2 - 2x$$

Check that pressing ENTER just twice more will result in agreement in all ten decimal places, giving $x \approx 0.3149230578$.

Internal Rate of Return on a Loan

By federal law, any loan agreement must disclose the rate of interest. If the amount of the loan and the repayments are specified, finding the implied interest rate (called the **internal rate of return**) can involve an equation whose roots can only be approximated.*

EXAMPLE 4   **FINDING THE INTERNAL RATE OF RETURN**

A bank loan of $2500 is to be repaid in three payments of $1000 at the end of the first, second, and third years. Find the internal rate of return.

Solution

Present value was first discussed on page 249.

LOOKING BACK

The internal rate of return is defined as the interest rate r for which the present values of the payments add up to the amount of the loan. To find the present value at the time of the loan, we divide the payment after n years by $(1 + r)^n$:

$$2500 = \frac{1000}{(1+r)^1} + \frac{1000}{(1+r)^2} + \frac{1000}{(1+r)^3}$$

Amount of loan | Present value of first payment | Present value of second payment | Present value of third payment

Dividing by 1000 and replacing $1 + r$ by x:

$$2.5 = \frac{1}{x} + \frac{1}{x^2} + \frac{1}{x^3}$$

$x = 1 + r$

$$2.5x^3 - x^2 - x - 1 = 0$$

$f(x)$

Multiplying by x^3 and moving all terms to the left

Definition of $f(x)$

Again, we could calculate the iterations by hand using the formula on page 682 or by a graphing calculator program, but instead we use the same method as before.

Graphing Calculator Exploration

We enter the function into Y_1 and its derivative into Y_2, as shown on the left.

Both round to 1.097

Take Note

To recover past entries, press **2nd** **ENTER** repeatedly.

For initial guess 1, we enter $1 \rightarrow X$ and $X - Y_1/Y_2 \rightarrow X$ and then repeatedly press **ENTER** until two values agree to the desired accuracy. Since we want a percentage (two decimal places), we continue until agreement to *three* decimal places

*See H. Paley, P. Colwell, and R. Cannaday, "Internal Rates of Return," UMAP Module 640, COMAP.

for greater precision. The iterations shown on the right above give $x \approx 1.097$. To find r, we subtract 1 (since we defined $x = 1 + r$), giving $r = 0.097$.

The internal rate of return is 9.7%   Converting 0.097 to a percent

You may check that again only two more pressings of [ENTER] will result in agreement of all ten decimal places.

To state this result in terms of bank interest, an account with $2500 at 9.7% would yield exactly the same amount at the end of three years as another account (also at 9.7%) with three $1000 payments made at the end of the first, second, and third years.

The Iterations May Not Converge

Although Newton's method often converges rapidly, there are functions for which it converges very slowly, or not at all. For example, if the derivative is *zero* at any approximation, then the iteration formula will involve division by zero, and so the process will stop. (This can often be remedied by choosing a different starting point.) The following two diagrams illustrate other problems: The first shows what can happen if there is an inflection point between two successive approximations, and the second shows the difficulties caused by a nearby critical point.

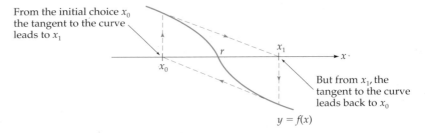

From the initial choice x_0 the tangent to the curve leads to x_1

But from x_1, the tangent to the curve leads back to x_0

$y = f(x)$

For this curve, the approximations will alternate between x_0 and x_1, never approaching the true root r in the center of the diagram.

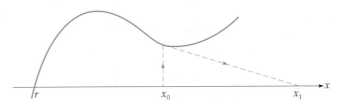

Starting at x_0, the tangent line to the curve leads to x_1 far to the *right*, instead of to the actual root r, which is to the left.

Although these diagrams show difficulties that can occur, Newton's method is generally very successful in applications, usually doubling the number of correct digits with each successive iteration.

Solutions TO PRACTICE PROBLEMS

1. $x_4 = 1.4142 - \dfrac{1.4142^2 - 2}{2 \cdot 1.4142} \approx 1.414213562$

2. Use $f(x) = x^5 - 30$ [or $f(x) = 30 - x^5$] and $x_0 = 2$ (although other values are possible).

10.4   Section Summary

Newton's method is a procedure for approximating the *zeros* of a function, or equivalently, the *roots* or *solutions* of $f(x) = 0$. An initial estimate x_0 is successively "improved" using the formula

$$x_{\text{new}} = x_{\text{old}} - \frac{f(x_{\text{old}})}{f'(x_{\text{old}})}$$

The convergence is often quite rapid, but is affected by such considerations as the accuracy of the initial estimate, roundoff errors, and how closely the function is approximated by its first Taylor polynomial (the tangent line).

Besides finding internal rates of return, the method is useful for solving many other equations, such as finding break-even points (where revenue = cost) or market equilibrium points (where supply = demand), as we will see in the exercises.

10.4   Exercises

1–4. Use Newton's method beginning with the given x_0 to find the first two approximations x_1 and x_2. Carry out the calculation "by hand" with the aid of a calculator, rounding to two decimal places.

1. $x^3 + x - 4 = 0$ **2.** $x^3 + 2x - 4 = 0$
   $x_0 = 1$ $x_0 = 1$

3. $e^x - 3x = 0$ **4.** $x - \cos x = 0$
   $x_0 = 0$ $x_0 = 0$

5–6. Use Newton's method to approximate each root, continuing until two successive iterations agree to three decimal places. Carry out the calculation "by hand" with the aid of a calculator, rounding to three decimal places.

5. $\sqrt{5}$ **6.** $\sqrt[3]{22}$

7–10. Use Newton's method to approximate the root of each equation, beginning with the given x_0 and continuing until two successive approximations agree to three decimal places. Carry out the calculation "by hand" with the aid of a calculator, rounding to three decimal places.

7. $x^3 + 3x - 8 = 0$ **8.** $x^3 - 3x + 3 = 0$
   $x_0 = 2$ $x_0 = -2$

9. $e^x + 3x + 2 = 0$ **10.** $e^x + 4x - 4 = 0$
   $x_0 = -1$ $x_0 = 0$

 11–12. Use a graphing calculator program for Newton's method to approximate each root, continuing until two successive iterations agree to nine decimal places.

11. $\sqrt[3]{130}$ **12.** $\sqrt{130}$

 13–20. Use a graphing calculator program for Newton's method to approximate the root of each equation

beginning with the given x_0 and continuing until two successive approximations agree to nine decimal places.

13. $x^5 + 2x - 1 = 0$ **14.** $x^5 + 5x - 4 = 0$
   $x_0 = 0$ $x_0 = 1$

15. $x + \ln x - 5 = 0$ **16.** $x^2 + \ln x - 3 = 0$
   $x_0 = 2$ $x_0 = 2$

17. $e^{x^2} + x - 3 = 0$ **18.** $e^{2x} + 3x = 0$
   $x_0 = 1$ $x_0 = 0$

19. $x + \sin x - 2 = 0$ **20.** $x^2 + \cos x - 4 = 0$
   $x_0 = 1$ $x_0 = 1$

 21–24. Use a graphing calculator to graph $f(x)$ and $g(x)$ together on a reasonable window and estimate the x-value where the curves meet. Then use Newton's method to approximate the solution of $f(x) = g(x)$, beginning with your estimate and continuing until two successive iterations agree to nine decimal places. (You may check your answer using INTERSECT, which uses a method similar to Newton's.)

21. $f(x) = x$, $g(x) = 2 \cos x$

22. $f(x) = \sin x$, $g(x) = 1 - x$

23. $f(x) = \ln x$, $g(x) = -x$

24. $f(x) = e^x$, $g(x) = 2 - x$

25. Use Newton's method (either "by hand" or using a graphing calculator) to solve $x^2 + 3 = 0$, beginning with $x_0 = 1$. Explain why it doesn't work.

26. Apply Newton's method (either "by hand" or using a graphing calculator) to the function $f(x) = x^{1/3}$ beginning with $x_0 = 1$. The following graph may help explain what goes wrong. [*Hint:* See the first diagram on page 687.] Does choosing another starting point help?

$f(x) = x^{1/3}$

27. What happens when Newton's method is applied to $f(x) = 2x - 4$? (Begin at any number other than 2.)

28. What happens when Newton's method is applied to a linear function $f(x) = mx + b$ with $m \neq 0$?

29. Use a graphing calculator program for Newton's method to calculate the first 20 or so iterations for the zero of $f(x) = x^{10}$, beginning with $x_0 = 1$. Notice how slowly the values converge to the actual zero, $x = 0$. Can you see why from the following graph?

How many iterations does it take to reach 0.001 (rounded)? Try using ZERO on your graphing calculator to find the root, starting at $x = 1$.

$f(x) = x^{10}$

30. SQUARE ROOTS BY ITERATION Show that finding $\sqrt{N}$ by applying Newton's method to the equation $x^2 - N = 0$ leads to the iteration formula

$$x_{n+1} = \frac{1}{2}\left(x_n + \frac{N}{x_n}\right).$$

This formula is known as the *Mechanic's Rule.*

Applied Exercises

31. ECONOMICS: Supply and Demand The supply and demand functions for a product are $S(p) = p$ and $D(p) = 10e^{-0.1p}$, where p is its price (in dollars). Find the "market equilibrium price" p at which $S(p) = D(p)$. Use Newton's method with two iterations, beginning with $x_0 = 4$. Carry out the calculation "by hand" with the aid of a calculator, rounding to two decimal places.

32. BUSINESS: Cost and Revenue The cost and revenue functions for a company are $C(x) = 45 + 2x$ and $R(x) = 20\sqrt{x}$ where x is the quantity (in thousands). Find the "break-even" quantity x at which $C(x) = R(x)$. Use Newton's method with two iterations, beginning with $x_0 = 10$. Carry out the calculation "by hand" with the aid of a calculator, rounding to one decimal place.

33. ECONOMICS: Supply and Demand The supply and demand functions for a product are $S(p) = 2e^{0.5p}$ and $D(p) = 10 - p$, where p is the price of the product (in dollars) and $0 < p < 10$. Find the "market equilibrium price" p at which $S(p) = D(p)$. Use a graphing calculator program for Newton's method, beginning with some initial guess between 0 and 10, and continuing until the iterations agree to two decimal places.

34. BUSINESS: Cost and Revenue The cost and revenue functions for a company are $C(x) = 50 + 40\sqrt{x}$ and $R(x) = 3x$, where x is the quantity ($x \geq 100$). Find the "break-even" quantity x at which $C(x) = R(x)$. Use a graphing calculator program for Newton's method, beginning with an initial guess of at least 100, and continuing until the iterations agree when rounded to the nearest unit.

35. ECONOMICS: Internal Rate of Return A $15,000 loan is to be repaid in four installments of $5000 each at the end of the first, second, third, and fourth years. Use a graphing calculator program for Newton's method to find the internal rate of return (to the nearest tenth of a percent).

36. ECONOMICS: Internal Rate of Return An $8000 loan is to be repaid in installments of $1000, $2000, $3000, and $4000 at the end of the first, second, third, and fourth years, respectively. Use a graphing calculator program for Newton's method to find the internal rate of return (to the nearest tenth of a percent).

37. BIOMEDICAL: Epidemics An epidemic is subject to two effects: linear growth, as infected people move into the area, and exponential decline, as existing cases are cured. Find when the two effects will be in equilibrium by solving $5t = 30e^{-0.2t}$, where t is the number of weeks since the epidemic began. Use a graphing calculator program for Newton's method, beginning with some appropriate initial value and continuing until the iterations agree to two decimal places.

38. SOCIAL SCIENCE: Immigration Immigration into a country is occurring at the rate of $2e^{0.1t}$ thousand people in year t, and the government sets an upper limit of $20 - t$ thousand people in year t (for $0 \leq t \leq 20$). Find when this limit will be reached by solving $2e^{0.1t} = 20 - t$. Use a graphing calculator program for Newton's method, beginning with some appropriate initial estimate and continuing until the iterations agree to two decimal places.

Conceptual Exercises

39. What is the zero of the linear function
$f(x) = mx + b$ with $m \neq 0$?

40. Find all zeros of: **a.** $\ln x$ **b.** e^x **c.** $\sin x$

41. If 13 is a zero of $f(x)$, then name a zero of
$g(x) = f(x + 17)$.

42. What happens in Newton's method if your initial
estimate is actually a *solution* to $f(x) = 0$?

43. What happens in Newton's method if the function is
linear with nonzero slope? Assume that you do not
begin *at* the solution.

44. How many iterations will Newton's method take if
applied to the absolute value function? (See page 53.)
Assume that you do not begin at 0.

45. Let f be a function that has one zero, has positive slope,
and is concave up. Will Newton's method make estim-
ates to the *right* or the *left* of the zero? Assume that you
do not begin *at* the solution. [*Hint:* Make a sketch.]

46. Let f be a function that has one zero, has negative
slope, and is concave down. Will Newton's method
make estimates to the *right* or the *left* of the zero?
Assume that you do not begin *at* the solution.
[*Hint:* Make a sketch.]

47. To estimate the solution to $f(x) = g(x)$, one person
might apply Newton's method to $f(x) - g(x)$ and
another might apply it to $g(x) - f(x)$. If both use the
same initial point, how will their iterations differ?

48. What is the internal rate of return if you borrow $100
and pay back $110 a year later?

| **10** | **Chapter Summary with Hints and Suggestions** |

Reading the text and doing the exercises in this chapter have helped you to master the following concepts and skills, which are listed by section (in case you need to review them) and are keyed to particular Review Exercises. Answers for all Review Exercises are given at the back of the book, and full solutions can be found in the Student Solutions Manual.

10.1 Geometric Series

- Determine whether an infinite geometric series
converges, and if so, find its sum.
(Review Exercises 1–8.)

$$a + ar + ar^2 + ar^3 + \cdots = \frac{a}{1 - r} \text{ for } |r| < 1$$

- Find the value of a repeating decimal.
(Review Exercises 9–10.)

- Find the value of an annuity. *(Review Exercises 11–12.)*

$$a + ar + ar^2 + \cdots + ar^{n-1} = a\frac{1 - r^n}{1 - r}$$

- Solve an applied problem using an infinite
series. *(Review Exercises 13–14.)*

10.2 Taylor Polynomials

- Find a Taylor polynomial for a function.
(Review Exercises 15–18.)

$$p_n(x) = f(0) + \frac{f'(0)}{1!}x + \frac{f''(0)}{2!}x^2 + \cdots + \frac{f^{(n)}(0)}{n!}x^n$$

- Find a Taylor polynomial for a function, use it to
approximate a value, and estimate the error.
(Review Exercises 19–20.)

$$|R_n(x)| \leq \frac{M}{(n + 1)!}|x|^{n+1} \text{ for } |f^{(n+1)}| \leq M$$

- Find the best linear approximation for a function.
(Review Exercise 21.)

- Find a Taylor polynomial that approximates a function within a given accuracy. *(Review Exercise 22.)*

- Find a Taylor polynomial at $x = a$ for function and
graph both. *(Review Exercises 23–24.)*

- Approximate a value using a Taylor polynomial and
estimate the error. *(Review Exercise 25.)*

- Solve an applied problem using a Taylor polynomial
and check using a calculator. *(Review Exercises 26–28.)*

10.3 Taylor Series

- Find the radius of convergence of a power
series. *(Review Exercises 29–32.)*

$$r = \lim_{n \to \infty} \left| \frac{c_{n+1}}{c_n} \right|$$

- Use one Taylor series to find another.
(Review Exercises 33–36.)

$$\frac{1}{1 - x} = 1 + x + x^2 + x^3 + \cdots \text{ on}(-1, 1)$$

$$e^x = 1 + \frac{x}{1!} + \frac{x^2}{2!} + \frac{x^3}{3!} + \frac{x^4}{4!} + \cdots \text{ on } (-\infty, \infty)$$

$$\sin x = x - \frac{x^3}{3!} + \frac{x^5}{5!} - \frac{x^7}{7!} + \cdots \text{ on } (-\infty, \infty)$$

$$\cos x = 1 - \frac{x^2}{2!} + \frac{x^4}{4!} - \frac{x^6}{6!} + \cdots \quad \text{on } (-\infty, \infty)$$

$$\ln x = (x - 1) - \frac{(x-1)^2}{2} + \frac{(x-1)^3}{3} - \cdots \quad \text{on } (0, 2]$$

- Find a Taylor series in two ways: from the definition using derivatives and by modifying a known series. *(Review Exercises 37–38.)*

$$f(0) + \frac{f'(0)}{1!}x + \frac{f''(0)}{2!}x^2 + \frac{f'''(0)}{3!}x^3 + \cdots$$

- Find a Taylor series at $x = a$. Compare it to the original function. *(Review Exercise 39.)*

$$f(a) + \frac{f'(a)}{1!}(x - a) + \frac{f''(a)}{2!}(x - a)^2 + \cdots$$

- Verify trigonometric identities using Taylor series. *(Review Exercise 40.)*

- Solve an applied problem by using a Taylor series. *(Review Exercises 41–42.)*

10.4 Newton's Method

- Use Newton's method "by hand" to approximate a root within a given accuracy. *(Review Exercises 43–46.)*

$$x_{\text{new}} = x_{\text{old}} - \frac{f(x_{\text{old}})}{f'(x_{\text{old}})}$$

- Use Newton's method with a calculator or computer to approximate a root within a given accuracy. *(Review Exercises 47–52.)*

- Solve an applied problem by using Newton's method. *(Review Exercises 53–55.)*

Hints and Suggestions

- In a finite geometric series, the number of terms is always one more than the final power of x.

- In mathematics, the terms *series* and *sequence* mean different things. Terms in a *series* are separated by + or − signs, and those in a *sequence* by *commas* (as in a sequence of approximations from Newton's method).

- Compare the formulas for the sum of finite and infinite geometric series: the *infinite* formula is simpler than the *finite* formula.

- For an infinite geometric series, be sure to check convergence ($|r| < 1$) before using the formula, or you may get an incorrect answer.

- The *first* Taylor polynomial is the best *linear* approximation, the *second* Taylor polynomial is the best *quadratic* approximation (at a point), and so on. Taylor polynomials are found by matching values and derivatives at a point.

- A Taylor polynomial at $x = a$ is *exact at* $x = a$, and less accurate further from $x = a$ (as judged by the error formula). The accuracy generally improves as the degree increases.

- The error formula for Taylor polynomials is easy to remember, since it closely resembles the first term left off.

- Taylor series are most easily found by modifying existing Taylor series.

- In practice, Newton's method for approximating the root of an equation generally converges very quickly, typically doubling the number of digits of accuracy with each iteration.

- A graphing calculator helps by summing many terms of a series, but it cannot sum an *infinite* series (unless you give it a formula). It can also *graph* a function together with several Taylor polynomials to show how close they are. It is also useful for calculating iterations of Newton's method.

---

10 **Review Exercises and Chapter Test** ◯ indicates a Chapter Test exercise.

10.1 Geometric Series

1–8. Determine whether each infinite geometric series converges or diverges. If it converges, find its sum.

①. $\frac{3}{10} + \frac{9}{100} + \frac{27}{1000} + \cdots$ **2.** $\frac{2}{5} - \frac{4}{25} + \frac{8}{125} - \cdots$

③. $\frac{3}{100} - \frac{9}{100} + \frac{27}{100} - \cdots$ **4.** $\frac{2}{3} - \frac{4}{9} + \frac{8}{27} - \cdots$

5. $10 + 8 + \frac{32}{5} + \frac{128}{25} + \cdots$ **6.** $\sum_{i=0}^{\infty} \left(-\frac{3}{4}\right)^i$

⑦. $\sum_{k=0}^{\infty} 6(-0.2)^k$ **8.** $\sum_{n=0}^{\infty} \frac{2^n}{3}$

9–10. Find the value of each repeating decimal.

9. $0.\overline{54} = 0.545454\ldots$ **10.** $0.\overline{72} = 0.727272\ldots$

11. PERSONAL FINANCE: Annuity A star basketball player preparing for his parents' retirement plans to deposit $6000 at the end of each month for 8 years into an account paying 6% compounded monthly. Find the amount of this annuity.

12. BUSINESS: Annuities Derive the formula

$$\text{Amount} = D\left[\frac{(1+i)^n - 1}{i}\right]$$

for the amount of an annuity consisting of n payments of D dollars each, with payments made at the end of each period, at compound interest rate i per period. (The formula in the brackets is denoted $S_{\overline{n}|i}$ (read "s angle n at i") and is calculated in tables.)

 13 **BIOMEDICAL:** Drug Dosage A patient has been taking daily doses of 0.26 milligram (mg) of digoxin for an extended period of time, and 65% of the drug in the bloodstream is absorbed by the body during each day.

   a. Find the maximum and minimum long-term amounts in the bloodstream.

   b. How many doses does it take for the cumulative amount in the bloodstream to reach 0.39 mg? To reach 0.399 mg?

14. ECONOMICS: Banking Reserves When a bank receives a deposit, it keeps only a small part (say 10%) as its "reserves," lending out the rest. These loans result in other purchases, with the sums eventually deposited in other banks, which in turn keep 10% reserves and lend out the other 90%. This process continues, with an original deposit of $1000 generating deposits totaling

$$1000 + 1000(0.90) + 1000(0.90)^2 + \cdots$$

   a. Sum this series and find the "multiplier" by which the original deposit is multiplied. This illustrates how the banking system multiplies money.

   b. (Graphing calculator with series operations helpful) How many terms of the preceding series are required for the sum to surpass $9000? To surpass $9900?

10.2 Taylor Polynomials

15–18. For each function:

a. Find the third Taylor polynomial at $x = 0$.
b. Graph the original function and the Taylor polynomial together on the indicated window.

15. $f(x) = \sqrt{x + 9}$
   (for (b), use window $[-15, 15]$ by $[-6, 6]$)

16. $f(x) = e^{3x}$
   (for (b), use window $[-1, 1]$ by $[-1, 8]$)

17. $f(x) = \ln(2x + 1)$
   (for (b), use window $[-1, 1]$ by $[-5, 5]$)

18. $f(x) = \sin^2 x$
   (for (b), use window $[-3, 3]$ by $[-2, 2]$)

19 **a.** For $f(x) = \sqrt{x + 1}$, find the third Taylor polynomial at $x = 0$.
   b. Use your answer to part (a) to approximate $\sqrt{2}$.
   c. Estimate the error by the error formula. [*Hint:* Calculate $f^{(4)}(x)$ and find an M such that $|f^{(4)}(x)| \le M$ for $0 \le x \le 1$.]
   d. Restate the approximation and error, rounded appropriately.

20. a. For $f(x) = \cos x$, find the fourth Taylor polynomial at $x = 0$.
   b. Use your answer to part (a) to approximate $\cos 1$.
   c. Estimate the error by the error formula. [*Hint:* Calculate $f^{(5)}(x)$ and find an M such that $|f^{(5)}(x)| \le M$ for $0 \le x \le 1$.] *(continues)*

   d. Restate the approximation and error, rounded appropriately.

21. What linear function best approximates $\sqrt[n]{1 + x}$ near $x = 0$?

22. Find the Taylor polynomial at $x = 0$ that approximates e^x on the interval $-1 \le x \le 1$ with error less than 0.0001.

23 **a.** For $f(x) = \sin x$, find the fourth Taylor polynomial at $x = \pi/2$.
   b. Graph $\sin x$ and its Taylor polynomial together on the window $[-2, 5]$ by $[-2, 2]$.

24. a. For $f(x) = x^{3/2}$, find the third Taylor polynomial at $x = 4$.
   b. Graph $x^{3/2}$ and its Taylor polynomial together on the window $[-2, 15]$ by $[-5, 60]$.

25. a. For $f(x) = x^{2/3}$, find the second Taylor polynomial at $x = 1$.
   b. Use your answer to part (a) to approximate $(1.2)^{2/3}$.
   c. Calculate $f^{(3)}(x)$ and find an M such that $|f^{(3)}(x)| \le M$ for $1 \le x \le 1.2$.
   d. Estimate the error by the error formula.
   e. Restate the approximation and error, rounded appropriately.

26. BIOMEDICAL: Drug Dosage Two hours after a patient takes a 1-gram dose of penicillin, the amount of penicillin in the patient's bloodstream will be $e^{-0.22}$ gram. Approximate $e^{-0.22}$ by using the second Taylor polynomial for e^x (rounded to two decimal places). Check your answer by using $\boxed{e^x}$ on a calculator.

27. GENERAL: Temperature The mean temperature in New York City on December 25 is predicted to be $55 - 24 \cos 0.45$. Approximate this number by using the second Taylor polynomial at $x = 0$ for $\cos x$. Then check your answer using the $\boxed{\cos}$ key on a calculator.

28. BUSINESS: Sales Just after introducing a new product, a company's weekly sales are predicted to be $200e^{0.1t^2}$ units.

   a. Find the fourth Taylor polynomial at $t = 0$ for this function. [*Hint:* Begin with the second Taylor polynomial for e^x and use substitution, then multiply by 200.]
   b. Use your Taylor polynomial to approximate the sales in week $t = 2$.
   c. Use a calculator to evaluate the original function at $t = 2$ and compare answers.

10.3 Taylor Series

29–32. Find the radius of convergence of each power series.

29 $\displaystyle\sum_{n=0}^{\infty} \frac{nx^n}{3^n}$ **30.** $\dfrac{x}{\sqrt{1}} + \dfrac{x^2}{\sqrt{2}} + \dfrac{x^3}{\sqrt{3}} + \dfrac{x^4}{\sqrt{4}} + \cdots$

31. $\displaystyle\sum_{n=0}^{\infty} \frac{n!x^n}{6^n}$ **32.** $\displaystyle\sum_{n=1}^{\infty} \frac{(-1)^n x^n}{(n - 1)!}$

33–36. For each function, find the Taylor series at $x = 0$ by modifying a known Taylor series.

33. $\dfrac{x^3}{1 - x^2}$ **34.** $x \cos 2x$

35. xe^{x^2} **36.** xe^{-x}

37–38. For each function, find the Taylor series at $x = 0$ in two ways:

a. By calculating derivatives and using the definition of Taylor series.

b. By modifying a known Taylor series

37. $\dfrac{1}{1 - 2x}$ **38.** $e^{-x/2}$

39. a. Find the Taylor series at $x = \pi/2$ for $\cos x$.

 b. Graph $\cos x$ and the sum of the first three nonzero terms of the Taylor series from part (a) on the window $[-3, 6]$ by $[-2, 2]$.

40. Verify the negative angle identities (page 542) $\sin(-x) = -\sin x$ and $\cos(-x) = \cos x$ by using the Taylor series for the sine and cosine functions.

41. **GENERAL: Commuter Traffic** The number of commuter cars passing through a toll plaza per hour is $\dfrac{1}{1 + t^2}$ thousand, where t is the number of hours after 9 a.m. Therefore, the total number of cars (in thousands) up to x hours after 9 a.m. is given by the integral $\displaystyle\int_0^x \dfrac{1}{1 + t^2}\, dt$.

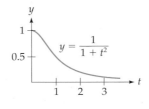

a. Find the Taylor series at $t = 0$ for $\dfrac{1}{1 + t^2}$ by modifying an existing series.

b. Integrate this series from 0 to x, thereby finding a Taylor series for $\displaystyle\int_0^x \dfrac{1}{1 + t^2}\, dt$.

c. Estimate the number of cars passing through the toll plaza during the first half hour after 9 a.m. by evaluating the first three terms of the series at $x = \frac{1}{2}$.

 d. Compare your answer to that found by using $\int f(x)\, dx$ on your graphing calculator to integrate the original function from 0 to 0.5.

42. **BIOMEDICAL: Gene Frequency** Under certain circumstances, the number of generations necessary to increase the frequency of a gene from 0.2 to 0.5 is

$$2.5 \int_{0.2}^{0.5} \dfrac{1}{x^2(1 - x)}\, dx$$

rounded to the nearest integer.

a. Use the Taylor series at $x = 0$ for $1/(1 - x)$ to find an infinite series for $1/(x^2(1 - x))$.

b. Integrate this series term by term to find a series expression for $\displaystyle\int \dfrac{1}{x^2(1 - x)}\, dx$.

c. Use the first five terms of this series to estimate the definite integral

$$\int_{0.2}^{0.5} \dfrac{1}{x^2(1 - x)}\, dx$$

d. How many generations are necessary to raise the gene frequency from 0.2 to 0.5? Compare this solution to that found for Example 3 on pages 396–397.

10.4 Newton's Method

43–44. Use Newton's method to approximate each root, continuing until two successive iterations agree to three decimal places. Carry out the calculations "by hand" with the aid of a calculator, rounding to three decimal places.

43. $\sqrt{26}$ **44.** $\sqrt[3]{30}$

45–46. Use Newton's method to approximate the root of each equation, beginning with the given x_0 and continuing until two successive iterations agree to three decimal places. Carry out the calculations "by hand" with the aid of a calculator, rounding to three decimal places.

45. $x^3 + 3x - 3 = 0$ **46.** $e^x + 10x - 4 = 0$
       $x_0 = 1$ $x_0 = 0$

 47–48. Use a graphing calculator program for Newton's method to approximate each root, continuing until two successive iterations agree to nine decimal places.

47. $\sqrt[3]{100}$ **48.** $\sqrt{10}$

 49–52. Use a graphing calculator program for Newton's method to approximate the root of each equation, beginning with the given x_0 and continuing until two successive approximations agree to nine decimal places.

49. $e^x - 5x - 3 = 0$ **50.** $e^x + 4x - 100 = 0$
       $x_0 = 1$ $x_0 = 3$

51. $x^3 = 5 \ln x + 10$ **52.** $x^4 = 2 + \sin x$
       $x_0 = 2$ $x_0 = 0$

 53. **ECONOMICS: Internal Rate of Return** For an investment of $20,000, you receive three payments of $9000 at the end of the first, second, and third years. Find the internal rate of return (to the nearest tenth of a percent).

54. **ECONOMICS: Supply and Demand** The supply and demand functions for a product are $S(p) = 20 + 3p$ and $D(p) = 50 - 8\sqrt{p}$ (for $0 \le p \le 25$), where p is the price (in dollars) of the product. Find the "market equilibrium price" p at which $S(p) = D(p)$. Use Newton's method, beginning with some reasonable initial guess and continuing until the iterations agree to two decimal places.

 55. **BUSINESS: Cost and Revenue** The cost and revenue functions for a company are $C(x) = 3x + 100$ and $R(x) = 40\sqrt{x}$, where x is the quantity (in thousands). Find the "break-even" quantity, the x at which $C(x) = R(x)$. Use Newton's method, beginning with $x = 10$ and continuing until the iterations agree to the nearest whole unit.

Probability

11

© Stephen VanHorn/Shutterstock.com

What You'll Explore

Probability affects our lives in many ways: Insurance rates are calculated from accident statistics, election are predicted from random samples, and airplanes are overbooked based on expected "no-shows." In this chapter, we will calculate the probability of a computer failure or of catching a bus, evaluate the expected value of an insurance policy or an investment, and even find the actual probability of a coincidence.

Coincidences

We often hear of "amazing" coincidences. For example, a woman named Evelyn Marie Adams won the New Jersey Lottery twice, an event that was widely reported to have a probability of 1 in 17 trillion.

Actually, the figure 1 in 17 trillion is misleading; such events are not all that unlikely. In fact, given enough tries, the most outrageous things are virtually certain to happen. For example, if a coincidence is defined as an event with a one-in-a-million chance of happening to you today, then in the United States, with about 300 million people, we should expect more than 300 coincidences each day, and more than 100,000 in a year.

Returning to the supposed 1-in-17-trillion double lottery winning, that figure is the right answer to the wrong question: What is the probability that a *preselected* person who buys just *two* tickets for separate lotteries will win on both? The more relevant question is: What is the probability that *some* person, among the many millions who buy lottery tickets (most buying multiple tickets), will win twice in a lifetime? It has been calculated* that such a double winning is likely to occur once in seven years, with the likelihood approaching certainty for longer periods.

Some knowledge of probability is necessary for an understanding of the world, if only to cast doubt on the misleading statements that one often hears. In the first section we lay the foundations for the probability discussed in later sections. For further information on probability, see the two books listed below.**

*See Persi Diaconis and Frederick Mosteller, "Methods for Studying Coincidences," *Journal of the American Statistical Association* **84**.

**For a readable introduction to probability, see Warren Weaver, *Lady Luck*, Dover Publications. For a more complete exposition, see William Feller, *An Introduction to Probability Theory and Its Applications*, vol. 1, 3rd ed., John Wiley.

11.1   Discrete Probability

Introduction

Probability measures the *likelihood* of an event; a 90% probability of rain means that rain is very likely, and should occur on about nine of ten similar days. We begin with **discrete probability,** in which probabilities are *added,* with later sections discussing **continuous probability,** in which probabilities are summed by *integration.*

In this section we will define the important concept of a (discrete) *random variable,* along with its mean, variance, and standard deviation. Then we will introduce the *Poisson random variable* and use it to model numbers of rare events, such as frequencies of hurricanes or the number of people living to age 100.

Elementary Events and Their Probabilities

Probability begins with an experiment, real or imagined, whose outcome depends on chance. In the experiment of rolling one die (the singular of dice), the possible outcomes are the numbers of dots 1, 2, 3, 4, 5, and 6, which are called **elementary events.** An **event** is a set consisting of elementary events. For example, the event of *rolling at least a 5* on a die would be represented

$$E_1 = \{5, 6\} \qquad \text{Rolling at least 5 means rolling 5 or 6}$$

The event of rolling an *even* number would be represented

$$E_2 = \{2, 4, 6\} \qquad \text{Rolling evens means rolling 2, 4, or 6}$$

In general:

Elementary Events, Events, and Sample Spaces

Given an experiment, the possible outcomes, exactly one of which must occur, are called *elementary events,* denoted $E_1, E_2, E_3, \ldots, E_n$. An *event* is a set of elementary events. The set of *all* elementary events is called the *sample space.*

If there are infinitely many elementary events, we denote them $E_1, E_2, E_3, \ldots$. Further examples of experiments and their sample spaces are given below.

Experiment	Sample Space	
	Elementary events	
Tossing a coin twice	{HH, HT, TH, TT}	"HT" means "heads then tails"
Counting defects in a manufacturing process	{0, 1, 2, 3, ... }	An infinite sample space

To each elementary event we assign a number that represents the **probability** of that event. In practice, we choose the number to be the long-run proportion of times that we expect the event to occur, but mathematically there are only two requirements: each probability must be between 0 and 1 (inclusive), and they must add to 1. Formally:

Probabilities of Elementary Events

The elementary events $E_1, E_2, \ldots, E_n$ are assigned probabilities $P(E_i) = p_i$ (for $i = 1, 2, \ldots, n$) such that for each i,

$$0 \le p_i \le 1 \qquad \text{Probabilities between 0 and 1}$$

and

$$p_1 + p_2 + \cdots + p_n = 1 \qquad \text{Probabilities sum to 1}$$

If there are infinitely many elementary events $E_1, E_2, \ldots$, then the sum of the probabilities will be an *infinite series* that must sum to 1. The condition that the probabilities sum to 1 can be interpreted as meaning that *something* must happen.

EXAMPLE 1   ASSIGNING PROBABILITIES TO ELEMENTARY EVENTS

For the experiment of rolling a die, the six possible faces 1, 2, 3, 4, 5, 6 are equally likely, each expected to occur about one-sixth of the time, so we assign probability $\frac{1}{6}$ to each.

$$P(1) = \frac{1}{6}, \qquad P(2) = \frac{1}{6}, \qquad P(3) = \frac{1}{6},$$

$$P(4) = \frac{1}{6}, \qquad P(5) = \frac{1}{6}, \qquad P(6) = \frac{1}{6}$$

P is read "the probability of"

EXAMPLE 2   ASSIGNING PROBABILITIES TO ELEMENTARY EVENTS

For the experiment of tossing a (fair) coin twice, the four elementary events HH, HT, TH, TT should each occur about one-quarter of the time, leading to the assignment

$$P(\text{HH}) = \frac{1}{4}, \qquad P(\text{HT}) = \frac{1}{4}, \qquad P(\text{TH}) = \frac{1}{4}, \qquad P(\text{TT}) = \frac{1}{4}$$

Be Careful In the above Example, you might be tempted to say that there are only *three* elementary events, namely, tossing 0, 1, or 2 heads. The difficulty is that 1 head can occur in two different ways, HT or TH, and so is twice as likely than an event like 0 heads that can occur in only one way, TT. The assignment of equal probabilities here is supported by experimental evidence from tossing coins.

In other situations, we may assign unequal probabilities to elementary events.

EXAMPLE 3   ASSIGNING UNEQUAL PROBABILITIES

In America, newborn babies are slightly more often male than female, the ratio being about 52 to 48. Therefore, for the experiment of observing the sex of a newborn baby, with elementary events M and F, we would assign probabilities

$$P(\text{M}) = 0.52 \quad \text{and} \quad P(\text{F}) = 0.48 \qquad \text{Probabilities need not be equal}$$

Source: U.S. Census Bureau

An *event* consists of *elementary events*, and the *probability* of the event is found by adding the probabilities of the constituent elementary events.

LOOKING AHEAD

In the *discrete* case, we *add* probabilities. In the *continuous* case (page 711), we will *integrate* probabilities.

Probability of an Event

The probability $P(E)$ of an event E is the sum of the probabilities of the elementary events in E.

EXAMPLE 4   FINDING THE PROBABILITY OF AN EVENT

For one roll of a die, find the probability of rolling an even number.

Solution

The event "rolling an even number" is made up of the elementary events 2, 4, and 6, each of which has probability $\frac{1}{6}$ (from Example 1). Therefore, the probability of rolling an even number is the *sum* of the probabilities of these elementary events:

$$P\left(\begin{array}{c}\text{Rolling an}\\\text{even number}\end{array}\right) = P(2) + P(4) + P(6) = \frac{1}{6} + \frac{1}{6} + \frac{1}{6} = \frac{3}{6} = \frac{1}{2}$$

$$\uparrow \qquad \uparrow \qquad \uparrow$$
Elementary events

Therefore, the probability of rolling "evens" is $\frac{1}{2}$.

EXAMPLE 5   FINDING THE PROBABILITY OF AN EVENT

For two tosses of a coin, find the probability of at least one head.

Solution

The event "at least one head" consists of the elementary events HH, HT, and TH, with probability $\frac{1}{4}$ each (from Example 2). Adding these probabilities gives

$$P\left(\begin{array}{c}\text{At least}\\\text{one head}\end{array}\right) = P(\text{HH}) + P(\text{HT}) + P(\text{TH}) = \frac{1}{4} + \frac{1}{4} + \frac{1}{4} = \frac{3}{4}$$

Therefore, the probability of rolling at least one head is $\frac{3}{4}$.

Random Variables

Often in probability we are interested in some numerical quantity, such as the number of heads in two tosses of a coin, or the sum of the faces when rolling two dice. Such quantities are called **random variables.** (Technically, a random variable is a *function defined on the sample space*, since for each elementary event it takes a numerical value.) If we calculate the probability that a random variable takes each of its possible values, the collection of these probabilities is called the *probability distribution* of the random variable.

Random Variables and Probability Distributions

A *random variable* is a numerical quantity whose value is determined by the outcome that occurs. Its *probability distribution* gives the probability that it takes each of its possible values.

An example should make matters clear.

EXAMPLE 6   FINDING THE PROBABILITY DISTRIBUTION OF A RANDOM VARIABLE

For two tosses of a (fair) coin, find the probability distribution of

$$X = \binom{\text{Number}}{\text{of heads}}.$$

Solution

The possible values of X are 0, 1, and 2 (the possible numbers of heads in two tosses). The probability that X takes any one of these values is calculated by finding the elementary events for which X takes that value and adding their probabilities.

$$P(X = 0) = \frac{1}{4}$$

$X = 0$ (no heads) only for TT, which occurs with probability $\frac{1}{4}$

$$P(X = 1) = \frac{1}{2}$$

$X = 1$ for HT and TH, which occur with probabilities $\frac{1}{4} + \frac{1}{4} = \frac{1}{2}$

$$P(X = 2) = \frac{1}{4}$$

$X = 2$ only for HH, which occurs with probability $\frac{1}{4}$

These three probabilities make up the probability distribution for X.

A probability distribution may be shown as a **bar graph,** with bars whose heights (and also areas) show the probability of each possible value. The probabilities adding to 1 means that the *total area* under the graph must be 1.

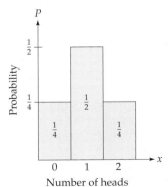

Probability distribution of $X = \binom{\text{Number}}{\text{of heads}}$ in two tosses

> **EXAMPLE 7**   **FINDING A PROBABILITY FOR A RANDOM VARIABLE**
>
> For two tosses of a coin and $X = \left(\begin{matrix} \text{Number} \\ \text{of heads} \end{matrix}\right)$, find $P(0 \le X \le 1)$.
>
> **Solution**
>
> $0 \le X \le 1$ means that $X = 0$ or $X = 1$, so we *add* these probabilities:
>
> $$P(0 \le X \le 1) = P(X = 0) + P(X = 1) = \frac{1}{4} + \frac{1}{2} = \frac{3}{4} \qquad \text{\small $\frac{1}{4}$ and $\frac{1}{2}$ from Example 6}$$
>
> The probability of tossing between 0 and 1 heads (inclusive) is $\frac{3}{4}$.

We interpret this probability as meaning that if you were to toss a pair of coins many times, in about three-quarters of the tosses you would get either no heads or one head.

The probability $P(0 \le X \le 1)$ is given by the first two bars in the graph of the probability distribution.

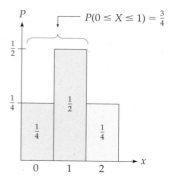

LOOKING AHEAD

Interpreting probabilities as areas under graphs is important since later we will find probabilities as areas under graphs by *integration*.

Expected Value

Suppose that a game of chance pays $8 with probability $\frac{1}{4}$ and $4 with probability $\frac{3}{4}$. If you were to play this game many times, what would be your *long-term average winnings per play*? We multiply each prize by its probability (after all, winning $8 with probability $\frac{1}{4}$ means winning $8 about every fourth time, for an *average* of $\frac{1}{4} \cdot 8 = \$2$ each time).

$$\left(\begin{matrix} \text{Average} \\ \text{winnings} \end{matrix}\right) = 8 \cdot \frac{1}{4} + 4 \cdot \frac{3}{4} = 2 + 3 = 5 \qquad \text{\small Each prize times its probability}$$

This answer, $5, is called the *expected value* of the game, meaning that *in the long run* your winnings should average about $5 per game. Therefore, $5 would be a *fair price* to charge for playing this game. For *any* random variable X, its **mean** or **expected value** is written $E(X)$ or μ ("mu," the Greek letter m), and is found by multiplying the possible values of X by their probabilities and adding.

> ### Expected Value of a Random Variable
>
> A random variable X taking values $x_1, x_2, \ldots, x_n$ with probabilities $p_1, p_2, \ldots, p_n$ has an *expected value* or *mean*
>
> $$\mu = E(X) = \sum_{i=1}^{n} x_i \cdot p_i \qquad \text{\small $x_1 p_1 + \cdots + x_n p_n$}$$

The expected value is a **weighted average,** with the values of X weighted by their probabilities. If a random variable has infinitely many possible values, then the sum will be an infinite series.*

EXAMPLE 8   **FINDING THE EXPECTED VALUE OF A RANDOM VARIABLE**

For two tosses of a coin and $X = \begin{pmatrix} \text{Number} \\ \text{of heads} \end{pmatrix},$ find $E(X)$.

Solution

Using the probabilities calculated in Example 6 on page 699:

$$E(X) = 2 \cdot \frac{1}{4} \quad + \quad 1 \cdot \frac{1}{2} \quad + \quad 0 \cdot \frac{1}{4} \qquad \text{Values times their probabilities, added}$$

$$\underset{P(X=2)}{\uparrow} \qquad \underset{P(X=1)}{\uparrow} \qquad \underset{P(X=0)}{\uparrow}$$

$$= \frac{1}{2} + \frac{1}{2} = 1$$

That is, in two tosses of a fair coin, the expected number of heads is $E(X) = 1$.

We interpret this expected value as meaning that repeating this experiment (tossing two coins) many times and keeping track of the numbers of heads, in the long run you should average about 1 head per repetition (just as you might have expected).

For this random variable, the expected value 1 is the "middle number" of its possible values. In general, this will happen whenever the probability distribution is symmetric around a central value.

EXAMPLE 9   **FINDING THE EXPECTED VALUE OF REAL ESTATE**

You are considering buying a piece of property whose value V depends on whether oil is found on it. You estimate that with probability 20% it will be worth $300,000 and with probability 80% it will be worth only $10,000. Find the expected value $E(V)$ of the property.

---

*In this case there is an added technicality that the series $\sum_{i=1}^{\infty} |x_i| \cdot p_i$ must converge.

Solution

$$E(V) = 300{,}000 \cdot 0.2 + 10{,}000 \cdot 0.8 \qquad \text{Values times their probabilities, added}$$

$$= 60{,}000 + 8000 = 68{,}000 \qquad \text{\$68,000}$$

This expected value of $68,000 could then be compared with the selling price to decide whether it is a "good buy."

Variance

For a random variable X, the possible values may be clustered closely around the mean, or they may be distributed widely on either side of the mean. The *spread* or *dispersion* of a random variable around its mean is measured by the **variance.**

Variance of a Random Variable

A random variable X taking values $x_1, x_2, \ldots, x_n$ with probabilities $p_1, p_2, \ldots, p_n$, and having mean μ, has *variance*

$$\mathrm{Var}(X) = \sum_{i=1}^{n} (x_i - \mu)^2 p_i \qquad (x_1 - \mu)^2 p_1 + \cdots + (x_n - \mu)^2 p_n$$

The variance is the differences between the possible values and the mean, squared, multiplied by their probabilities, and added. It is sometimes called the **mean square deviation.** The squaring of the differences $x_i - \mu$ ensures that a positive difference does not cancel a negative one. Obviously, finding the variance requires first finding the mean μ. (An alternative formula for the variance is developed in Exercise 47.)

EXAMPLE 10   FINDING THE VARIANCE OF A RANDOM VARIABLE

For two tosses of a fair coin and $X = \left(\begin{matrix}\text{Number} \\ \text{of heads}\end{matrix}\right)$, find $\mathrm{Var}(X)$.

Solution

In Example 6 we found the possible values of X and their probabilities, and in Example 8 we found the mean $\mu = 1$. From these,

$$\mathrm{Var}(X) = (0 - 1)^2 \cdot \frac{1}{4} + (1 - 1)^2 \cdot \frac{1}{2} + (2 - 1)^2 \cdot \frac{1}{4} \qquad \begin{matrix}\text{Values minus} \\ \mu = 1, \text{ squared,} \\ \text{and multiplied by} \\ \text{their probabilities}\end{matrix}$$

$$= \frac{1}{4} + 0 + \frac{1}{4} = \frac{1}{2} \qquad \text{Simplified}$$

A particular variance such as $\mathrm{Var}(X) = \frac{1}{2}$ has little intuitive meaning by itself. Variance may be understood *comparatively:* a variance of 2 would indicate a greater spread away from the mean than a variance of $\frac{1}{2}$.

If the units of the random variable are "dollars," then the variance has units "dollars squared," which is difficult to interpret. The **standard deviation** of X, written $\sigma(X)$, is the square root of the variance, and so has the same units as the random variable. [The symbol σ ("sigma") is the Greek letter s.]

Standard Deviation of a Random Variable

The standard deviation of a random variable is the square root of the variance:

$$\sigma(X) = \sqrt{\mathrm{Var}(X)}$$

The standard deviation is sometimes called the **root mean square deviation** of the random variable. Standard deviations will be used extensively in Section 11.4.

EXAMPLE 11   FINDING THE STANDARD DEVIATION OF A RANDOM VARIABLE

For two tosses of a fair coin and $X = \begin{pmatrix} \text{Number} \\ \text{of heads} \end{pmatrix}$, find $\sigma(X)$.

Solution

$$\sigma(X) = \sqrt{\mathrm{Var}(X)} = \sqrt{0.5} \approx 0.707$$

Using $\mathrm{Var}(X) = \frac{1}{2}$ from Example 10

Poisson Distribution

Until now we have defined random variables by imagining an experiment (rolling a die, tossing a coin) and finding the probability distribution. We now turn our attention to certain classic random variables that are useful in applications. The first of these is the **Poisson random variable,** named for the French mathematician Siméon Poisson (1781–1840). A Poisson random variable has an infinite number of possible values, 0, 1, 2, 3, … , and its probability distribution depends on a constant or *parameter* $a > 0$, which is also its mean.

Poisson Random Variable

For a Poisson random variable X with mean $a > 0$,

$$P(X = k) = e^{-a} \frac{a^k}{k!}$$

for $k = 0, 1, 2, \ldots$

EXAMPLE 12   FINDING PROBABILITIES FOR A POISSON RANDOM VARIABLE

For the Poisson random variable X with mean $a = 3$, find $P(X = 0)$, $P(X = 1)$, and $P(X = 2)$.

Solution

$$P(X = 0) = e^{-3}\frac{3^0}{0!} = e^{-3}\frac{1}{1} \approx 0.05 \qquad\qquad e^{-a}\frac{a^k}{k!} \quad \text{with} \quad a = 3 \\ \text{and} \quad k = 0$$

$$P(X = 1) = e^{-3}\frac{3^1}{1!} = e^{-3}\frac{3}{1} \approx 0.15 \qquad\qquad e^{-a}\frac{a^k}{k!} \quad \text{with} \quad a = 3 \\ \text{and} \quad k = 1$$

$$P(X = 2) = e^{-3}\frac{3^2}{2!} = e^{-3}\frac{9}{2} \approx 0.22 \qquad\qquad e^{-a}\frac{a^k}{k!} \quad \text{with} \quad a = 3 \\ \text{and} \quad k = 2$$

The probability distribution of a Poisson random variable with mean $a = 3$ is graphed below. The heights of the first three bars were calculated above.

Probability distribution for the
Poisson random variable with mean $a = 3$

To show that the probabilities for a Poisson random variable sum to 1, we factor e^{-a} out of the sum and recognize the remaining sum as the Taylor series for e^a (from page 670):

$$\underbrace{\sum_{k=0}^{\infty} e^{-a}\frac{a^k}{k!}}_{\substack{\text{Sum of the}\\\text{probabilities}}} = \underbrace{e^{-a}}_{\substack{\text{Factored}\\\text{out}}}\underbrace{\left(1 + \frac{a}{1} + \frac{a^2}{2!} + \frac{a^3}{3!} + \cdots\right)}_{\substack{\text{Taylor series for }e^a,\\\text{which equals }e^a}} = \underbrace{e^{-a}(e^a)}_{\substack{\text{Replacing}\\\text{the series}\\\text{by }e^a}} = \underbrace{e^0 = 1}_{\substack{\text{Adding}\\\text{exponents}}}$$

Therefore, the probabilities *do* sum to 1. Exercise 13 shows that a Poisson random variable X has mean $E(X) = a$. (The variance is also a, so the standard deviation is $\sqrt{a}$, although we will not use these facts.)

Poisson random variables are often used to model independent events that occur rarely, such as the number of defective items in a manufacturing process or people living to age 100.* The Poisson distribution is sometimes (although inaccurately) called the **law of rare events**.

---

*Two events are *independent* if the occurrence of one does not make the occurrence of the other more or less likely. For further examples that fit the Poisson distribution, see F. Thorndike, "Applications of Poisson's Probability Summation," *Bell System Technical Journal* **5**.

EXAMPLE 13   **HURRICANE PREDICTION**

During the years 1900 through 2013, 190 hurricanes reached the United States, for an average of 1.7 per year during those 113 years. Use the Poisson distribution with mean $a = 1.7$ to find the probability that three or fewer hurricanes will reach the United States during a particular year.

Source: National Oceanic and Atmospheric Administration

Solution

If $X = \left(\begin{array}{c}\text{Number of} \\ \text{hurricanes}\end{array}\right),$ then we want to find $P(X \le 3),$ where X is a Poisson random variable with $a = 1.7.$

$P(X \le 3)$

$= P(X = 0) + P(X = 1) + P(X = 2) + P(X = 3)$ $X \le 3$ means
                                                                                          $X = 0, 1, 2,$ or 3

$= e^{-1.7}\left(\dfrac{1.7^0}{0!} + \dfrac{1.7^1}{1!} + \dfrac{1.7^2}{2!} + \dfrac{1.7^3}{3!}\right)$ $e^{-a}\left(\dfrac{a^0}{0!} + \dfrac{a^1}{1!} + \dfrac{a^2}{2!} + \dfrac{a^3}{3!}\right)$
                                                                                          with $a = 1.7$

≈ 0.907 Using a calculator

Therefore, with probability about 91% at most three hurricanes will reach the United States in any given year.

![Graphing Calculator Exploration logo] **Graphing Calculator Exploration**

A graphing calculator with a command like POISSONCDF (in the probability DISTRibution menu) finds the probability that a Poisson random variable takes values *less than or equal to* a given number. To find the probability in the preceding Example, we would use the POISSONCDF command with the given mean (1.7) and the maximum number of occurrences (3). The screen on the left verifies the answer we found.

What is the number such that with probability 99% at most that many hurricanes will reach the United States in a given year? [*Hint:* Experiment with POISSONCDF, replacing 3 by larger numbers.] [Answer: 5]

Calculations such as this are used to determine insurance premiums and land values in hurricane-prone areas. The Poisson distribution with $a = 1.7$ is graphed below. The entire area is 1, and the darker portion shows the probability $P(X \le 3).$

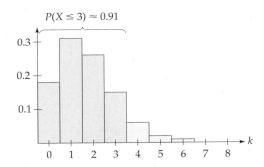

PRACTICE PROBLEM

In a shipment of 100 light bulbs, an average of 2 bulbs will be defective. Suppose that the number of defective bulbs in each shipment is modeled by a Poisson random variable.

a. What should be the value of the parameter a?

b. Find the probability that a given shipment has no defective bulbs.

Solutions below >

Solutions TO PRACTICE PROBLEM

a. $a = 2$

b. $P(X = 0) = e^{-2} \dfrac{2^0}{0!} = e^{-2} \approx 0.135$

11.1 Section Summary

For an experiment, the possible outcomes (exactly one of which must occur) are called *elementary events,* and collectively they make up the *sample space.* The elementary events are assigned nonnegative probabilities that sum to 1, usually based on long-run proportions. An *event* consists of one or more elementary events, and its probability is the sum of the probabilities of its constituent elementary events.

A *random variable* (represented by a capital letter, such as X, Y, or Z) takes values that depend on the elementary events. Its *probability distribution* consists of the probabilities that it takes these various values. The *mean* or *expected value* $E(X)$ gives an average value for X, and the *variance* $\text{Var}(X)$ and *standard deviation* $\sigma(X)$ measure the *deviation* of the values away from the mean.

$$E(X) = \mu = \sum_{i=1}^{n} x_i p_i$$

For possible values $x_1, x_2, \dots, x_n$ with probabilities $p_1, p_2, \dots, p_n$

$$\text{Var}(X) = \sum_{i=1}^{n} (x_i - \mu)^2 p_i$$

$$\sigma(X) = \sqrt{\text{Var}(X)}$$

A *Poisson* random variable with mean $a > 0$ has probability distribution

$$P(X = k) = e^{-a} \dfrac{a^k}{k!} \qquad \text{for } k = 0, 1, 2, \dots$$

Poisson random variables are used to model the frequencies of independent rare events, with the parameter a set equal to the average number of occurrences.

11.1 Exercises

1. For the event of rolling one die, find:

   a. $P\left(\begin{array}{c}\text{Rolling at}\\ \text{least a 3}\end{array}\right)$ **b.** $P\left(\begin{array}{c}\text{Rolling an}\\ \text{odd number}\end{array}\right)$

2. For the event of tossing a coin twice, find:

   a. $P\left(\begin{array}{c}\text{Tossing exactly}\\ \text{one head}\end{array}\right)$ **b.** $P\left(\begin{array}{c}\text{Tossing at most}\\ \text{one head}\end{array}\right)$

3–4. For the event of tossing a coin *three times,* find:

3. a. $P\left(\begin{array}{c}\text{Tossing exactly}\\ \text{one head}\end{array}\right)$ **b.** $P\left(\begin{array}{c}\text{Tossing exactly}\\ \text{two heads}\end{array}\right)$

4. a. $P\left(\begin{array}{c}\text{Tossing exactly}\\ \text{three heads}\end{array}\right)$ **b.** $P\left(\begin{array}{c}\text{Tossing more}\\ \text{heads than tails}\end{array}\right)$

5–6. Find the expected value, variance, and standard deviation of each random variable. Notice how the variance and standard deviation are affected by the "spread" of the values.

5. a. X takes values 9 and 11, each with probability $\frac{1}{2}$.
   b. Y takes values 0 and 20, each with probability $\frac{1}{2}$.

6. a. X takes values 19 and 21, each with probability $\frac{1}{2}$.
   b. Y takes values 0 and 40, each with probability $\frac{1}{2}$.

7–10. For each spinner, the arrow is spun and then stops, pointing in one of the numbered sectors, with probability proportional to the area of the sector. Let X stand for the number so chosen. Find:

   a. the probability distribution of X.
   b. $E(X)$.
   c. Var(X).

7.   **8.**

9.   **10.**

11. For a spinner such as that described in Exercises 7–10, what is the probability that the arrow ends up pointing

   a. in the top 30°?   **b.** in the top 10°?

c. exactly upward?

12. *(continuation)* If lines are drawn from the center of the spinner to 360 points evenly spaced around the circumference, what is the probability that the arrow ends up falling exactly on one of these lines?

13. Show that the mean of a Poisson random variable X with parameter a is $E(X) = a$.

14. A die is rolled, and X is defined as the number of dots on the face that lands upward. Find:
   a. $E(X)$ (Your answer shows that the mean need not be a possible value.)
   b. Var(X)

15. For a Poisson random variable X with mean 2, find (to three decimal places):
   a. $P(X = 0)$ **b.** $P(X = 1)$
   c. $P(X = 2)$ **d.** $P(X = 3)$

16. For a Poisson random variable X with mean 1.5, find (to three decimal places):
   a. $P(X = 0)$ **b.** $P(X = 1)$
   c. $P(X = 2)$ **d.** $P(X = 3)$

17. For a Poisson random variable X with mean 2, use POISSONCDF to find (to three decimal places):
   a. $P(X \le 1)$ **b.** $P(X \le 2)$
   c. $P(X \le 3)$ **d.** $P(X > 3)$
   [*Hint for (d):* Use your answer to (c).]

18. For a Poisson random variable X with mean 1.5, use POISSONCDF to find (to three decimal places):
   a. $P(X \le 1)$ **b.** $P(X \le 2)$
   c. $P(X \le 3)$ **d.** $P(X > 3)$
   [*Hint for (d):* Use your answer to (c).]

Applied Exercises

19. SOCIAL SCIENCE: Family Composition A young couple hopes to have four children.
   a. State the sample space for the sexes of the children listed in birth order, using B for boy and G for girl. (For example, GGBB means two girls then two boys.)
   b. If each possible outcome is equally likely, what is the probability of each?
   c. Find the probability distribution of the random variable $X = \begin{pmatrix} \text{Number} \\ \text{of girls} \end{pmatrix}$. Make a bar graph for the probability distribution.
   d. Find the probability $P(X \ge 1)$.
   e. Find $E(X)$ and Var(X).

20. BUSINESS: Expected Value of a Company A venture capitalist buys a biotechnology company, estimating that with probability 0.2 it will be worth $2,000,000, with probability 0.7 it will be worth $300,000, and with probability 0.1 it will be worthless. Find the expected value of the company.

21. BUSINESS: Price of an Insurance Policy An insurance company predicts that a policy it plans to offer will cost the company the following amounts in claims: $10,000 with probability $\frac{1}{10}$, $2000 with probability $\frac{1}{2}$, $1000 with probability $\frac{3}{10}$, and nothing at all with probability $\frac{1}{10}$. If it wants to sell the policy for $500 more than the expected cost of claims, what price should it set?

22. GENERAL: Lawsuits A lawyer in a court case estimates that with probability 0.25 she will win $100,000, with probability 0.4 she will win $20,000, and with probability 0.35 she will win nothing. Her client is offered a settlement of $45,000 to drop the suit. Based on expected value, should she advise her client to accept the settlement?

Applied Exercises Using Poisson Random Variables

23. BUSINESS: Quality Control An automobile assembly line produces 200 cars per day, and an average of 4 per day fail inspection. Use the Poisson distribution to find the probability that on a given day, 2 or fewer cars will fail inspection.

24. GENERAL: Misprints An 11-chapter book has 35 misprints, for an average of 3.2 per chapter. Use the Poisson distribution to find the probability that a given chapter has 2 or fewer misprints.

25. BIOMEDICAL: Bacilli In an examination of 1000 phagocytes (white blood cells), an average of 1.93 bacilli were found per phagocyte. Use the Poisson distribution to find the probability that a given phagocyte has 2 or fewer bacilli.

26. GENERAL: Spacial Poisson Distribution A paper company produces rolls of paper with an average of 0.2 flaw per square yard. Use the Poisson distribution to find the probability that a given square yard has no flaws.

27. GENERAL: Hurricanes An average of 1.7 hurricanes reach the United States each year. Use the Poisson distribution to find the probability that no hurricane reaches the United States in a given year.
Source: National Oceanic and Atmospheric Administration

28. SOCIAL SCIENCE: Centenarians In a group of 100 twenty-year-old Americans, an average of 2.15 will live to be 100 years old ("centenarians"). Use the Poisson distribution to find the probability that in such a group of 100, at least one will be a centenarian. [*Hint:* Find the probability that *none* will be a centenarian, and subtract it from 1.]
Source: National Vital Statistics Report

29–36. A graphing calculator with series operations is useful but not necessary.

29. BUSINESS: Quality Control In a box of 100 screws, an average of 2 are defective. The manufacturer offers a refund if 5 or more are defective. Use the Poisson distribution to find the probability of a refund. [*Hint:* First find $P(X < 5)$.]

30. GENERAL: Parking A college parking lot has an average of six empty parking places at any time. If eight cars arrive at the same time, use the Poisson distribution to find the probability that they can all be accommodated. [*Hint:* First find the probability that the cars *cannot* be accommodated.]

31. GENERAL: Traffic A certain mathematics professor driving to work encounters, on average, 2 traffic jams per month. Use the Poisson distribution to find the probability that in a given month he will encounter:
a. no traffic jams.
b. at least 1 traffic jam. [*Hint:* Use your answer to part (a).]
c. 3 or fewer traffic jams.

32. BIOMEDICAL: Blood Type People with blood type AB+ can receive blood from people with *any* blood type and so are called *universal recipients*. On the average, 3 people in 100 have this blood type. Use the Poisson distribution to find the probability that in a group of 100 people:
a. none will be universal recipients.
b. 3 or fewer will be universal recipients.
Source: U.S. Census Bureau

33. BIOMEDICAL: Blood Type People with blood type O− can give blood to people with *any* blood type and so are called *universal donors*. On average, 7 people in 100 have this blood type. Use the Poisson distribution to find the probability that in a group of 100 people:
a. no more than 3 will be universal donors.
b. no more than 5 will be universal donors.
Source: U.S. Census Bureau

34. SOCIAL SCIENCE: Crime On average in the United States, there will be 1.2 robberies per 1000 people in a year. Use the Poisson distribution to find the probability that in a neighborhood of 1000 residents there will be:
a. no robberies.
b. no more than 2 robberies.
Source: Statistical Abstract of the United States, 2013

35. SOCIAL SCIENCE: Crime On average in the United States, there will be 2.4 motor vehicle thefts per 1000 people in a year. Use the Poisson distribution to find the probability that in a neighborhood of 1000 residents there will be:

a. no automobile thefts.

b. no more than 5 automobile thefts.

Source: Statistical Abstract of the United States, 2013

36. **GENERAL:** Airplane Accidents During the years 1982 to 2013 there were, on average, 1.6 serious commercial airplane accidents per year. Use the Poisson distribution to find the probability that in any particular year the number of accidents will not exceed 3.

Source: National Transportation Safety Board

Conceptual Exercises

37. If a sample space has 275 equally likely elementary events, what is the probability of each?

38. If a sample space contains just two elementary events, and one occurs twice as often as the other, what are their probabilities?

39. Find a simple expression for $P(X \leq 5) - P(X < 5)$.

40. If $P(X \leq 10) = \frac{2}{3}$, then which of the following can you find and what is its value: $P(X \geq 10)$ or $P(X > 10)$?

41. True or False: If a random variable has mean μ, then one of its values must be μ.

42. Can the mean of a random variable be greater than each of its values?

43. If a random variable X has mean 5, then what is the mean of $X + 100$? Of $2X$?

44. If a random variable X has variance 5, then what is the variance of $2X$? If X has *standard deviation* 5, then what is the standard deviation of $2X$?

45. What can you say about a random variable whose variance is *zero*? To be specific, what can you say about X if $E(X) = 5$ and $\text{Var}(X) = 0$?

46. If X is a Poisson random variable and $P(X = 0) = e^{-7}$, what is its mean?

Explorations and Excursions   The following problem extends and augments the material presented in the text.

Alternative Formula for Variance

47. For a random variable X with mean μ that takes values $x_1, x_2, \ldots, x_n$ with probabilities $p_1, p_2, \ldots, p_n$, prove the following "alternative" formula for the variance

$$\text{Var}(X) = \sum_{i=1}^{n} x_i^2 \cdot p_i - \mu^2$$

by giving a justification for each numbered equality in the following derivation.

$$\text{Var}(X) \overset{1}{=} \sum_{i=1}^{n} (x_i - \mu)^2 p_i \overset{2}{=} \sum_{i=1}^{n} (x_i^2 - 2x_i\mu + \mu^2) p_i$$

$$\overset{3}{=} \sum_{i=1}^{n} x_i^2 p_i - \sum_{i=1}^{n} 2x_i\mu p_i + \sum_{i=1}^{n} \mu^2 p_i$$

$$\overset{4}{=} \sum_{i=1}^{n} x_i^2 p_i - 2\mu \sum_{i=1}^{n} x_i p_i + \mu^2 \sum_{i=1}^{n} p_i$$

$$\overset{5}{=} \sum_{i=1}^{n} x_i^2 p_i - 2\mu\mu + \mu^2 1 \overset{6}{=} \sum_{i=1}^{n} x_i^2 p_i - \mu^2$$

11.2   **Continuous Probability**

Discrete Random Variables

The random variables that we saw in the previous section took values that are separated from each other and so can be written out in a *list*, such as 0, 1, 2, 3, ... Such random variables are called *discrete* random variables.* For example, the Poisson random variable (see page 703) is a discrete random variable. The probability distribution of a discrete random variable may be shown as a bar graph over the possible values, with total area 1.

---

*Do not confuse the word "discrete," which means *separated* or *distinct*, with the word "discreet," which means *prudent* or *showing self-restraint*.

Continuous Random Variables

Other random variables take values that make up a *continuous interval* of the real line. For example, the proportion of time that you study today can be any number in the interval [0, 1], and the duration of a telephone call can be any number in the interval (0, ∞). For a **continuous random variable,** the probability distribution is given by a nonnegative function over the interval of possible values, with total area 1. Such a function is called a **probability density function.**

Probability Density Function

A probability density function on an interval [A, B] is a function f such that

1. $f(x) \geq 0$ for $A \leq x \leq B$

2. $\displaystyle\int_A^B f(x)\,dx = 1$

Total area is 1

A probability density function on [A, B]

The interval may be infinite, in which case $A = -\infty$,   or $B = \infty$,   or both, with the integral above being an *improper* integral. For simplicity, we define $f(x)$ to be zero for all x-values outside of the interval [A, B].

EXAMPLE 1   CONSTRUCTING A PROBABILITY DENSITY FUNCTION

Find the value of the constant a that makes $f(x) = ax(1 - x)$ a probability density function on the interval [0, 1].

Solution

To find the area under the curve, we integrate over the given interval:

$$\underbrace{a \int_0^1}_{\substack{\text{Moved} \\ \text{outside}}} x(1-x)\,dx = a \int_0^1 \underbrace{(x - x^2)}_{\substack{\text{Multiplied} \\ \text{out}}}\,dx = a \underbrace{\left(\frac{1}{2}x^2 - \frac{1}{3}x^3\right)\Big|_0^1}_{\substack{\text{Integrating by} \\ \text{the Power Rule}}}$$

$$= a \underbrace{\left(\frac{1}{2} - \frac{1}{3}\right.}_{\substack{\text{Evaluating} \\ \text{at 1}}} \underbrace{\left. - \; 0\right)}_{\substack{\text{and} \\ \text{at 0}}} = \underbrace{\frac{a}{6}}_{\substack{\text{Area under} \\ \text{the curve}}} \qquad \substack{\text{Must} \\ \text{equal 1}}$$

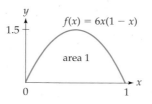

For this area $a/6$ to equal 1, we must have $a = 6$.   The probability density function with a replaced by 6 is then $f(x) = 6x(1 - x)$ on [0, 1], which is graphed on the left and is clearly nonnegative.*

---

*This probability density function is an example of a *beta density*, defined more generally as $f(x) = ax^{\alpha}(1 - x)^{\beta}$ for positive constants α and β and $0 \leq x \leq 1$, where a is a constant that makes the area under the curve equal to 1. Therefore, the probability density function in Example 1 is a beta density with constants $\alpha = 1$, $\beta = 1$, and $a = 6$. Beta distributions are of secondary importance compared to the other distributions discussed in this chapter.

For *discrete* random variables, probabilities were given by *areas of bars*. For *continuous* random variables, probabilities are given by *areas under curves*. More precisely, the probability that X takes values between two given numbers is found by integrating the probability density function between those two numbers.

Probabilities from Probability Density Functions

For a continuous random variable X with probability density function f,

$$P(a \leq X \leq b) = \int_a^b f(x)\, dx$$

$P(a \leq X \leq b)$ is the area under the probability density function from a to b.

EXAMPLE 2   **FINDING A PROBABILITY FOR A CONTINUOUS RANDOM VARIABLE**

Let X be defined as the number of inches of rain in a town during the first week of April in each year. Suppose that a meteorologist has used past weather records to determine that the probability density function for X is $f(x) = 6x(1 - x)$ on $[0, 1]$. Find the probability that the rainfall is between 0 and 0.4 inch.

Solution

To find $P(0 \leq X \leq 0.4)$ we integrate the probability density function from 0 to 0.4:

$$P(0 \leq X \leq 0.4) = \int_0^{0.4} 6x(1 - x)\, dx \qquad \text{Integrating the probability density function}$$

$$= \int_0^{0.4} (6x - 6x^2)\, dx \qquad \text{Multiplying out}$$

$$= (3x^2 - 2x^3)\Big|_0^{0.4} \qquad \text{Using the Power Rule}$$

$$= (0.48 - 0.128) - 0 = 0.352 \qquad \text{Evaluating}$$

Therefore, the probability that the rainfall will be between 0 and 0.4 inch is about 35%.

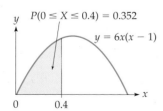

On pages 329–333 we defined definite integrals by summing areas of rectangles under curves. Therefore, finding continuous probabilities as areas under curves is a natural extension of finding discrete probabilities as sums of areas of rectangles.

PRACTICE PROBLEM

a. Find the value of a that makes $f(x) = ax^2$ a probability density function on [0, 1].

b. For X with this probability density function, find $P(0 \leq X \leq 0.5)$.

Solutions on page 717 >

One-Point Probabilities Are Zero

For a continuous random variable X, the probability that X takes any particular single value is zero: $P(X = a) = 0$. This is because $P(X = a)$ is found by integrating the probability density function "from a to a," which gives zero:

$$P(X = a) = \underbrace{P(a \leq X \leq a)}_{\substack{\text{Equivalent} \\ \text{to } X = a}} = \int_a^a f(x)\, dx = 0$$

Same upper and lower limits

area under a single point is zero

Therefore, adding one or both endpoints to the inequality in $P(a < X < b)$ does not change the probability:

$$P(a < X < b) = P(a \leq X \leq b) = \int_a^b f(x)\, dx \qquad \begin{array}{l}\text{Endpoints do not}\\ \text{change the probability}\end{array}$$

In continuous probability, what does *zero probability* mean? Zero probability does *not* mean that the event is impossible, only that it is "overwhelmingly unlikely." Think of choosing a number from the interval [0, 1] by some means such as throwing a dart at a line segment. The probability of choosing (hitting) exactly one single preselected number a is $P(X = a)$, which is zero, as we saw above. The probability being zero means only that predicting the exact point in advance is theoretically possible but overwhelmingly unlikely. In *discrete* probability, however, zero probability *does* mean that the event is impossible. Intuitively, the difference is that in continuous probability a random variable has so many more possible values that the probability must be spread over them much more "thinly."

Mean, Variance, and Standard Deviation

The mean and variance of a continuous random variable are defined by *integrals* that are analogous to the sums that we used for discrete probability.

Mean, Variance, and Standard Deviation

For a continuous random variable X with probability density function f on $[A, B]$,

$$\mu = E(X) = \int_A^B x f(x)\, dx$$

Mean or expected value is the integral of x times $f(x)$

$$\text{Var}(X) = \int_A^B (x - \mu)^2 f(x)\, dx$$

Variance is x minus the mean, squared, multiplied by $f(x)$, and integrated

$$\sigma(X) = \sqrt{\text{Var}(X)}$$

Standard deviation is the square root of the variance

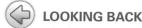 **LOOKING BACK**

Notice the similarity of these definitions to the ones in the boxes on pages 700–703.

If the interval is infinite, with $A = -\infty$, or $B = \infty$, or both, then the integrals are *improper* integrals.* As before, the mean $\mu = E(X)$ gives an *average* value, and the variance $\text{Var}(X)$ and standard deviation $\sigma(X)$ measure the *spread* or *dispersion* of the values away from this average. Also as before, finding the variance involves first finding the mean.

EXAMPLE 3   FINDING THE EXPECTED VALUE OF A CONTINUOUS RANDOM VARIABLE

For the random variable in Example 2, $X = \left(\begin{array}{c}\text{Inches of}\\\text{rainfall}\end{array}\right)$ with probability density function $f(x) = 6x(1 - x)$ on $[0, 1]$, find $E(X)$.

Solution

$$E(X) \quad = \quad \underbrace{\int_0^1 x \cdot 6x(1 - x)\, dx}_{\substack{\text{Integrating } x \text{ times the}\\\text{probability density function}}} \quad = \quad \underbrace{\int_0^1 (6x^2 - 6x^3)\, dx}_{\substack{\text{Multiplying}\\\text{out}}}$$

$$= \underbrace{\left(2x^3 - \frac{6}{4}x^4\right)\Big|_0^1}_{\substack{\text{Integrating by}\\\text{the Power Rule}}} = \underbrace{2 - \frac{3}{2}}_{\text{Evaluating}} = \underbrace{\frac{1}{2}}_{E(X)}$$

The expected value is $\mu = E(X) = \frac{1}{2}$ so that as a long-term average, there should be about half an inch of rain during that week. It is reasonable that the mean is the "middle value" of the interval $[0, 1]$, since this probability density function is symmetric about this number.

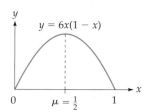

We could find the *variance* of the rainfall by evaluating the integral

$$\text{Var}(X) = \int_0^1 \left(x - \frac{1}{2}\right)^2 6x(1 - x)\, dx \qquad \int_A^B (x - \mu)^2 f(x)\, dx \quad \text{with} \quad \mu = \frac{1}{2}$$

There is, however, an easier formula, which is proved at the end of this section.

---

*In this case there is an added technicality that the improper integral $\int |x| f(x)\, dx$ over the infinite interval must converge.

Alternative Formula for Variance

LOOKING BACK

Again, notice the similarity to the *discrete* version (see Exercise 47 on page 709).

For a continuous random variable X with mean μ and probability density function f on $[A, B]$,

$$\text{Var}(X) = \int_A^B x^2 f(x)\, dx - \mu^2$$

Integrate x^2 times $f(x)$, then subtract the square of the mean

EXAMPLE 4   FINDING THE VARIANCE AND STANDARD DEVIATION

Find the variance and standard deviation of the random variable with probability density function $f(x) = 6x(1 - x)$ on $[0, 1]$.

Solution

In Example 3 we found that the mean of this random variable is $\mu = \frac{1}{2}$. To use the alternative formula for variance, we first integrate x^2 times the probability density function:

$$\int_0^1 x^2 \underbrace{6x(1 - x)}_{f(x)}\, dx = \int_0^1 (6x^3 - 6x^4)\, dx = \left(\frac{6}{4} x^4 - \frac{6}{5} x^5 \right) \Big|_0^1$$

$$= \frac{3}{2} - \frac{6}{5} = \frac{3}{10}$$

Then

$$\text{Var}(X) = \underbrace{\frac{3}{10}}_{\substack{\text{Preceding} \\ \text{integral}}} - \underbrace{\left(\frac{1}{2} \right)^2}_{\mu^2} = \frac{3}{10} - \frac{1}{4} = \frac{1}{20} = 0.05$$

$\int_A^B x^2 f(x)\, dx - \mu^2$ with $\mu = \frac{1}{2}$

$$\sigma(X) = \sqrt{0.05} \approx 0.224$$

Square root of the variance

Therefore, $\text{Var}(X) = 0.05$ and $\sigma(X) \approx 0.224$

 Graphing Calculator Exploration

Use a graphing calculator to verify that the earlier formula (on page 713) for variance gives the same answer as in the proceeding Example.

The variance and standard deviation will be used extensively in Section 11.4.

Cumulative Distribution Functions

Because we so often integrate a probability density function, it is useful to have a name for the integral of it up to an arbitrary number x. The result is called the **cumulative distribution function.**

Cumulative Distribution Function

For a continuous random variable X with probability density function f on $[A, B]$, the *cumulative distribution function* is

$$F(x) = P(X \le x) = \int_A^x f(t)\, dt$$

$F(x)$ is the probability that X is less than or equal to x, and is found by integrating the probability density function from A to x

Be Careful We use uppercase letters (such as F, G, and H) for cumulative distribution functions, and lowercase letters (such as f, g, and h) for probability density functions.

Graphically, the cumulative distribution function $F(x)$ is the *area* under the probability density function f up to x:

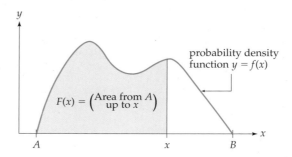

The curve $y = f(x)$ is the probability density function. The area under it up to x is the cumulative distribution function $F(x)$.

Since integrating f gives F, differentiating F gives f. (To be completely precise, this relationship holds at x-values where $f(x)$ is continuous.)

$$\frac{d}{dx} F(x) = f(x)$$

The derivative of the cumulative distribution function $F(x)$ is the probability density function $f(x)$.

Probabilities $P(a \le X \le b)$, which we found by integrating the probability density function, can also be found by *subtracting* values of the cumulative distribution function.

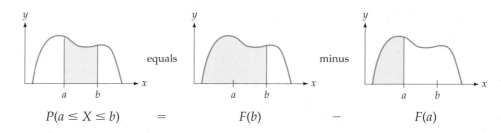

In words:

Probabilities from the Cumulative Distribution Function

For a continuous random variable X with cumulative distribution function F,

$$P(a \le X \le b) = F(b) - F(a)$$

Either or both $\le$ may be replaced by $<$

EXAMPLE 5   **FINDING AND USING A CUMULATIVE DISTRIBUTION FUNCTION**

For a random variable W with probability density function $f(x) = \frac{1}{9}x^2$ on $[0, 3]$:

a. Find the cumulative distribution function $F(x)$.

b. Use $F(x)$ to find $P(1 \le W \le 2)$.

Solution

a. The cumulative distribution function is the integral of the probability density function up to x. We use t to avoid confusion with the upper limit x.

$$F(x) = \int_0^x \frac{1}{9} t^2 \, dt = \left(\frac{1}{9} \frac{1}{3} t^3 \right) \Big|_0^x = \frac{1}{27} x^3 \qquad F(x) = \int_A^x f(t) \, dt$$

Therefore, the cumulative distribution function is $F(x) = \frac{1}{27} x^3$ on $[0, 3]$.

b. To find $P(1 \le W \le 2)$, we evaluate $F(x) = \frac{1}{27} x^3$ at $x = 1$ and $x = 2$ and subtract:

$$P(1 \le W \le 2) = F(2) - F(1) = \underbrace{\frac{1}{27} 2^3 - \frac{1}{27} 1^3}_{\substack{\text{Evaluating } F \\ \text{at 2 and 1}}} = \frac{8}{27} - \frac{1}{27} = \frac{7}{27}$$

Values of F subtracted

Therefore, $P(1 \le W \le 2) = \frac{7}{27}$.

Graphing Calculator Exploration

```
Y₁(2)-Y₁(1)
           .2592592593
Ans▶Frac
              7/27
```

To check the answer to the preceding Example:

a. Define y_1 as the definite integral of $\frac{1}{9}x^2$ from 0 to x by entering $y_1 = \text{FnInt}(x^2/9, x, 0, x)$.

b. Then calculate $y_1(2) - y_1(1)$.

In some applications it is easier to *first* find the cumulative distribution function and *then* find the probability density function by differentiation.

EXAMPLE 6   **FINDING AND USING A CUMULATIVE DISTRIBUTION FUNCTION**

A dart is thrown at a circular target of radius 12 inches, hitting a point at random. Let X be defined as

$$X = \left(\begin{array}{c} \text{Distance of dart} \\ \text{from center of circle} \end{array} \right)$$

Find: **a.** the cumulative distribution function $F(x)$

   b. the probability density function $f(x)$

   c. the mean $E(X)$

Solution

A point being chosen "at random" from a region means that the probability of the point being in any subregion is proportional to the *area* of that subregion. (After all, the dart should have twice the probability of landing in an area twice as large.) The dart landing within x inches of the center means that it lands inside a circle of radius x, and the probability of this, $P(X \le x)$, is the area of the inner circle (πx^2) divided by the area of the entire target ($\pi 12^2$):

a. $P(X \le x) = \dfrac{\pi x^2}{\pi 12^2} = \dfrac{x^2}{144}$

   Therefore, $F(x) = \dfrac{x^2}{144}$ for $0 \le x \le 12$.

b. $f(x) = \dfrac{d}{dx}\left(\dfrac{1}{144} x^2 \right) = \dfrac{1}{72} x$ *Differentiating $F(x)$ to find $f(x)$*

   Therefore, $f(x) = \dfrac{1}{72} x$ for $0 \le x \le 12$.

c. $E(X) = \displaystyle\int_0^{12} x \cdot \dfrac{1}{72} x \, dx = \dfrac{1}{72} \int_0^{12} x^2 \, dx = \dfrac{1}{72} \cdot \dfrac{1}{3} x^3 \Big|_0^{12} = \dfrac{12^3}{216} = 8$

The expected value $E(X) = 8$ means that if many darts are thrown, their average distance from the center will be about 8 inches.

Solutions TO PRACTICE PROBLEM

a. $a \displaystyle\int_0^1 x^2 \, dx = a \dfrac{1}{3} x^3 \Big|_0^1 = \dfrac{a}{3}$

   For the integral to equal 1, we must have $a = 3$.

b. $P(0 \le X \le 0.5) = \displaystyle\int_0^{0.5} 3x^2 \, dx = x^3 \Big|_0^{0.5} = (0.5)^3 = 0.125$

11.2   Section Summary

A random variable is *continuous* if its possible values form an *interval* [A, B], which may be infinite, and if its probabilities are found by integrating its *probability density function* (a nonnegative function whose integral over [A, B] is 1):

$$P(a \le X \le b) = \int_a^b f(x)\, dx$$

Either or both $\le$ may be replaced by $<$

The *cumulative distribution function* for a random variable is

$$F(x) = P(X \le x) = \int_A^x f(t)\, dt$$

A is the lowest possible value of the random variable

Differentiating the cumulative distribution function $F(x)$ gives the probability density function $f(x)$. Probabilities $P(a \le X \le b)$ may also be found by subtracting values of the cumulative distribution function:

$$P(a \le X \le b) = F(b) - F(a)$$

Either or both $\le$ may be replaced by $<$

The *mean* or *expected value* gives the *average* of the random variable, and the *variance* and *standard deviation* indicate how widely the values are spread around the mean:

$$\mu = E(X) = \int_A^B x\, f(x)\, dx$$

$$\text{Var}(X) = \int_A^B (x - \mu)^2 f(x)\, dx = \int_A^B x^2 f(x)\, dx - \mu^2$$

Both formulas give the same answer

$$\sigma(X) = \sqrt{\text{Var}(X)}$$

The height of the probability density function at a point does not, by itself, give a probability, but it can be interpreted in a *relative* sense: If the probability density function is twice as high at one x-value as at another, the random variable is about twice as likely to be near the first x-value as near the other (where by "near" we mean "within a small fixed distance of").

Proof of the Alternative Formula for Variance

To prove the alternative formula for variance stated on page 714, we begin with the *definition* of variance given on page 713:

$$\text{Var}(X) = \int_A^B (x - \mu)^2 f(x)\, dx = \int_A^B (x^2 - 2x\mu + \mu^2) f(x)\, dx$$

Expanding $(x - \mu)^2$

$$= \int_A^B x^2 f(x)\, dx - 2\mu \underbrace{\int_A^B x f(x)\, dx}_{\substack{\mu \\ \text{(by definition} \\ \text{of the mean)}}} + \mu^2 \underbrace{\int_A^B f(x)\, dx}_{\substack{1 \\ \text{(by property 2} \\ \text{on page 710)}}}$$

Separating into 3 integrals, and taking out constants

$$= \int_A^B x^2 f(x)\, dx \underbrace{- 2\mu^2 + \mu^2}_{-\mu^2} = \int_A^B x^2 f(x)\, dx - \mu^2$$

This last expression is the alternative formula for the variance

11.2 Exercises

1–10. Find the value of the constant a that makes each function a probability density function on the stated interval.

1. $ax^2(1 - x)$ on $[0, 1]$

2. $a\sqrt{x}$ on $[0, 1]$

3. ax^2 on $[0, 3]$

4. $ax(2 - x)$ on $[0, 2]$

5. $\dfrac{a}{x}$ on $[1, e]$

6. $a \sin x$ on $[0, \pi]$

7. axe^x on $[0, 1]$ [*Hint:* Integrate by parts.]

8. $a \ln x$ on $[1, e]$ [*Hint:* Integrate by parts using $u = \ln x$.]

 9. ae^{-x^2} on $[-1, 1]$ **10.** $\dfrac{a}{1 + x^2}$ on $[-1, 1]$

11–16. For each probability density function $f(x)$, find:
a. $E(X)$ **b.** $\text{Var}(X)$ **c.** $\sigma(X)$

11. $f(x) = 3x^2$ on $[0, 1]$

12. $f(x) = \tfrac{1}{2}x$ on $[0, 2]$

13. $f(x) = 12x^2(1 - x)$ on $[0, 1]$

14. $f(x) = 20x^3(1 - x)$ on $[0, 1]$

 15. $\dfrac{4}{\pi} \dfrac{1}{1 + x^2}$ on $[0, 1]$ **16.** $1.34e^{-x^2}$ on $[0, 1]$

17. If X has probability density function $f(x) = \tfrac{1}{12}x$ on $[1, 5]$, find:
a. the cumulative distribution function $F(x)$
b. $P(3 \leq X \leq 5)$

18. If X has probability density function $f(x) = \tfrac{1}{10}$ on $[0, 10]$, find:
a. the cumulative distribution function $F(x)$
b. $P(3 \leq X \leq 7)$

19. If X has probability density function $f(x) = \dfrac{2}{(1 + x)^2}$ on $[0, 1]$, find $P\left(X \leq \tfrac{1}{3}\right)$.

20. If X has probability density function $f(x) = \tfrac{1}{6}x^{-1/2}$ on $[1, 16]$, find $P(4 \leq X \leq 9)$.

21. If X has cumulative distribution function $F(x) = \tfrac{1}{3}x - \tfrac{1}{3}$ on $[1, 4]$, find $P(2 \leq X \leq 3)$.

22. If X has cumulative distribution function $F(x) = \tfrac{1}{8}x^{3/2}$ on $[0, 4]$, find $P(1 \leq X \leq 4)$.

23. If X has cumulative distribution function $F(x) = x^2$ on $[0, 1]$, find:
a. $P(0.5 \leq X \leq 1)$
b. the probability density function $f(x)$

24. If X has cumulative distribution function $F(x) = \tfrac{1}{2}\sqrt{x} - \tfrac{1}{2}$ on $[1, 9]$, find:
a. $P(4 \leq X \leq 9)$
b. the probability density function $f(x)$

25. For any cumulative distribution function $F(x)$ on the interval $[A, B]$, show that:
a. $F(A) = 0$ **b.** $F(B) = 1$

26. For any cumulative distribution function $F(x)$, show that if $a \leq b$, then $F(a) \leq F(b)$.

 27. If X has probability density function $f(x) = \dfrac{2}{\pi} \dfrac{1}{1 + x^2}$ on $[-1, 1]$, find $P\left(-\tfrac{1}{2} \leq X \leq \tfrac{1}{2}\right)$

 28. If X has probability density function $f(x) = 0.67e^{-x^2}$ on $[-1, 1]$, find $P\left(-\tfrac{1}{2} \leq X \leq \tfrac{1}{2}\right)$.

Applied Exercises

29. SOCIAL SCIENCE: Voter Turnout If the proportion of registered voters who actually vote in an election is a random variable X with probability density function $f(x) = 6x(1 - x)$, find $P(0.4 \leq X \leq 0.6)$.

30. BEHAVIORAL SCIENCE: Learning If the number of days required for a worker to learn a new skill is a random variable X with probability density function $f(x) = \tfrac{4}{27}x^2(3 - x)$ on $[0, 3]$, find the expected time $E(X)$.

31. BIOMEDICAL: Life Expectancy In a 4-year study, the number of years that a patient survives after an experimental medical procedure is a random variable X with probability density function $f(x) = \tfrac{3}{64}x^2$ on $[0, 4]$. Find:
a. the expected survival time $E(X)$
b. $P(2 \leq X \leq 4)$

32. GENERAL: Milk Freshness If the shelf life of a carton of milk (in days) is a random variable X with probability density function $f(x) = \tfrac{1}{32}x$ on $[0, 8]$, find:
a. the expected shelf life $E(X)$
b. the time at which the probability of spoilage is only 25% (that is, find b such that $P(X \leq b) = 0.25$)

33. ATHLETICS: Darts A dart hits a point at random on a circular dart board of radius 9 inches. For
$$X = \left(\begin{array}{c}\text{Distance of dart} \\ \text{from center}\end{array}\right), \quad \text{find:}$$
a. the cumulative distribution function $F(x)$
b. the probability density function $f(x)$
c. $E(X)$
d. $P(3 \leq X \leq 6)$ using the *probability density function*
e. $P(3 \leq X \leq 6)$ using the *cumulative distribution function*

34. BUSINESS: Product Failure A spherical ball bearing of radius 1 inch has a flaw located randomly somewhere within it. If the flaw is within half an inch of the center, the ball bearing will fracture when it is used. Find:

a. the probability that the ball bearing will fracture. [*Hint:* Use the volumes of the "inner" and entire spheres.]

b. the expected distance from the flaw to the center of the ball bearing.

35. BUSINESS: Profitability If the number of years until a new store makes a profit (or goes out of business) is a random variable X with probability density function $f(x) = 0.75x(2 - x)$ on [0, 2], find:

a. the expected number of years $E(X)$

b. the variance and standard deviation

c. $P(X \geq 1.5)$

36. BUSINESS: Newspaper Advertising If the proportion of a newspaper devoted to advertising is a random variable X with probability density function $f(x) = 20x^3(1 - x)$ on [0, 1], find:

a. the expected proportion $E(X)$

b. the variance and standard deviation

c. the probability that more than 80% is devoted to advertising

37. BIOMEDICAL: Weibull Aging Model The probability that the fruit fly *Drosophila melanogaster* lives longer than t days is given by the *Weibull survivorship function*

$$S(t) = e^{-(0.02t)^{3.3}}$$

a. Graph this function on the window [−5, 105] by [−0.5, 1.5].

b. What is the probability that a fly has a life span of more than 30 days?

Source: H. McCallum, *Population Parameters: Estimation for Ecological Models*

38. ATHLETICS: Baseball Based on an analysis of five years of 9-inning major league baseball games, the number of hits per team per game can be approximated by a random variable X with probability density function $f(x) = 3(0.05x)^2(1 - 0.05x)^3$ on [0, 20]. Find $P(10 \leq X \leq 20)$.

Source: J. Albert, J. Bennett, and J. Cochran, *Anthology of Statistics in Sports*

39. ATHLETICS: Baseball Based on an analysis of 1782 major league baseball games, the winning margin in a game can be approximated by a random variable X with probability density function $f(x) = 0.55e^{-0.37x}$ on [1, 11]. Find $P(1 \leq X \leq 3)$.

Source: J. Albert, J. Bennett, and J. Cochran, *Anthology of Statistics in Sports*

40–41. GENERAL: Richest Americans A study of the richest Americans (those with a net worth of at least $600 million) found that their net worth X (in millions of dollars) has probability density function $f(x) = 1250x^{-2.1}$ on [600, ∞].

Source: Contemporary Physics **46**

40. Find $P(600 \leq X \leq 1000)$ (that is, that the net worth of such a person is between $600 million and $1 billion).

41. Find $P(1000 \leq X \leq 5000)$ (that is, that the net worth of such a person is between $1 billion and $5 billion).

Conceptual Exercises

42. Can a cumulative distribution function F satisfy $F(2) > F(3)$?

43. Explain what it means for an event in *discrete* probability to have probability 0 as opposed to an event in *continuous* probability to have probability zero.

44. If X has probability density function $f(x)$, then what is the probability density function for $X + 5$? [*Hint:* Think of how the graph is shifted.]

45. If X has cumulative distribution function $F(x)$, then what is the cumulative distribution function for $X - 7$? [*Hint:* Think of how the graph is shifted.]

46. Fill in the blanks: If f is a probability density function on [A, B] and F is the cumulative distribution function, then $F(A) =$ _____ and $F(B) =$ _____.

47. In Example 6 (page 717), the dart could land between 0 and 12 inches from the center, but the mean was 8. Explain in intuitive terms why the mean was *more* than 6, halfway to the edge.

48. Suppose that a probability density function on the interval [A, B] has the shape shown below. Would you expect the mean to be closer to A or to B?

49. True or False: If a random variable X has probability density function f on [A, B], then $A \leq E(X) \leq B$.

50. If a continuous random variable X has mean 5, then what is the mean of $X + 100$? Of $2X$?

51. If a continuous random variable X has variance 5, then what is the variance of $2X$? If X has *standard deviation* 5, then what is the standard deviation of $2X$?

11.3   Uniform and Exponential Random Variables

Introduction

The preceding section introduced the essentials of continuous probability: continuous random variables, probability density functions, cumulative distribution functions, mean, variance, and standard deviation. However, little was said about how to *use* continuous random variables to model particular situations. Although a complete answer to this important question would require further courses in probability, statistics, and mathematical modeling, many situations can be modeled by just a few standard types of random variables. In this section we will study two of the most useful, *uniform* and *exponential* random variables, and in the next section we will study *normal* random variables.

Uniform Random Variables

Uniform probability density function on $[0, B]$

Suppose that you want to choose a point at random from an interval $[0, B]$. If all points are to be "equally likely," the probability density function must be constant over the interval $[0, B]$, and if the area under it is to be 1, its uniform height must be $1/B$, as shown on the left.

The resulting probability density function, $f(x) = 1/B$ for $0 \leq x \leq B$,   is called the **uniform probability density function** on $[0, B]$. A random variable with this probability density function is called a **uniform random variable** on $[0, B]$ and is said to be **uniformly distributed** between 0 and B. The mean of a uniform random variable on $[0, B]$ is easily calculated from the formula $E(X) = \int_A^B x\, f(x)\, dx$.

$$E(X) = \int_0^B x\,\frac{1}{B}\,dx = \frac{1}{B}\int_0^B x\,dx = \left(\frac{1}{B}\frac{1}{2}x^2\right)\Big|_0^B = \frac{1}{B}\frac{1}{2}B^2 - 0 = \frac{B}{2}$$

$f(x)$ Moved outside   Integrating by the Power Rule   Evaluating at B and at 0 $E(X)$

The mean $\mu = E(X) = B/2$ is the midpoint of the interval $[0, B]$, as should be reasonable. For the variance, we use the alternative formula from page 714, first integrating x^2 times the probability density function:

$$\int_0^B x^2\,\frac{1}{B}\,dx = \frac{1}{B}\int_0^B x^2\,dx = \frac{1}{B}\frac{1}{3}x^3\Big|_0^B = \frac{1}{B}\frac{1}{3}B^3 = \frac{B^2}{3}$$

Then

$$\mathrm{Var}(X) = \frac{B^2}{3} - \left(\frac{B}{2}\right)^2 = \frac{B^2}{3} - \frac{B^2}{4} = \frac{B^2}{12} \qquad\qquad \mathrm{Var}(X) = \int_A^B x^2 f(x)\, dx - \mu^2$$

Calculated μ above $\mathrm{Var}(X)$

Summarizing:

Uniform Random Variable

A uniform random variable X on the interval $[0, B]$ has probability density function

$$f(x) = \frac{1}{B} \quad \text{for } 0 \leq x \leq B$$

Its mean, variance, and standard deviation are

$$E(X) = \frac{B}{2} \qquad \text{Var}(X) = \frac{B^2}{12} \qquad \sigma(X) = \sqrt{\text{Var}(X)} = \frac{B}{2\sqrt{3}}$$

EXAMPLE 1   **FINDING THE MEAN AND VARIANCE OF A UNIFORM RANDOM VARIABLE**

A campus shuttle bus departs every 18 minutes. For a person arriving at a random time, find the mean and variance of the wait for the bus.

Solution

Arriving at a random time means that all waiting times between 0 and 18 minutes are equally likely, so $X = \begin{pmatrix} \text{Waiting} \\ \text{time} \end{pmatrix}$ is uniformly distributed on the interval $[0, 18]$. Therefore, the preceding formulas for the uniform distribution give

$$E(X) = \frac{18}{2} = 9 \qquad\qquad E(X) = \frac{B}{2} \text{ with } B = 18$$

$$\text{Var}(X) = \frac{18^2}{12} = 27 \qquad\qquad \text{Var}(X) = \frac{B^2}{12} \text{ with } B = 18$$

$$\sigma(X) = \sqrt{27} \approx 5.2 \qquad\qquad \text{Standard deviation}$$

The fact that the average wait is 9 minutes for buses that come every 18 minutes should be intuitively reasonable. The standard deviation of 5.2 minutes gives a rough indication of the expected deviation from this mean, although an exact interpretation of this number is difficult.

Exponential Random Variables

We next define **exponential random variables,** which are used to model waiting times that are potentially unlimited in duration, such as the time until an electronic component fails or the time between the arrival of telephone calls. An exponential random variable takes values in the interval $[0, \infty)$ and has probability density function $f(x) = ae^{-ax}$ for $x \geq 0$ with parameter $a > 0$, as shown on the left.

Exponential probability density function with parameter a.

To verify that $f(x)$ is a probability density function, we must show that the area under it is 1. The integral is improper, so we first integrate from 0 to B:

$$\int_0^B f(x)\, dx = a \int_0^B e^{-ax}\, dx = a\left(-\frac{1}{a}\right) e^{-ax} \Big|_0^B = -e^{-aB} + e^0 = 1 - e^{-aB}$$

Moved outside   Using formula 4 (inside back cover)   Evaluating   Reversing the order

Then we let $B \to \infty$:

$$\int_0^\infty f(x)\,dx = \lim_{B \to \infty}\,(1 - e^{-aB}) = 1 \qquad\qquad \text{Area is 1}$$

$$\underbrace{}_{\to 0}$$

To find the mean $E(X) = \displaystyle\int_0^\infty x\,f(x)\,dx,$ we use integration by parts:

$$\int \underbrace{x}_{u}\ \underbrace{ae^{-ax}\,dx}_{dv} = \underbrace{x}_{u}\underbrace{(-e^{-ax})}_{v} - \int \underbrace{-e^{-ax}}_{v}\ \underbrace{dx}_{du} = -xe^{-ax} - \underbrace{\frac{1}{a}e^{-ax}}_{\text{Using formula 4}}$$

Indefinite
integral

Using formula 4
(inside back cover)

$$u = x \qquad dv = ae^{-ax}\,dx$$
$$du = dx \qquad v = \int ae^{-ax}\,dx = -e^{-ax}$$

Evaluating from 0 to B:

$$\left(-xe^{-ax} - \frac{1}{a}e^{-ax}\right)\Bigg|_0^B = -Be^{-aB} - \frac{1}{a}e^{-aB} + \frac{1}{a}$$

Finally, taking the limit as $B \to \infty$ (using $\displaystyle\lim_{B \to \infty} Be^{-aB} = 0,$ from Exercise 62 on page 680),

$$E(X) = \lim_{B \to \infty}\left(\underbrace{-Be^{-aB}}_{\substack{\to 0 \\ \text{as } B \to \infty}} - \underbrace{\frac{1}{a}e^{-aB}}_{\substack{\to 0 \\ \text{as } B \to \infty}} + \frac{1}{a}\right) = \underbrace{\frac{1}{a}}_{E(X)}$$

A similar calculation (see Exercise 29) shows that $\text{Var}(X) = 1/a^2,$ and therefore $\sigma(X) = 1/a$.   These results are easy to remember: both the mean and standard deviation are $1/a$, the reciprocal of the parameter. Summarizing:

Exponential Random Variable

An exponential random variable X with parameter $a > 0$ takes values in $[0, \infty)$ and has probability density function

$$f(x) = ae^{-ax} \qquad \text{for } x \geq 0$$

Its mean, variance, and standard deviation are

$$E(X) = \frac{1}{a}$$

$$\text{Var}(X) = \frac{1}{a^2}$$

$$\sigma(X) = \frac{1}{a}$$

PRACTICE PROBLEM 1

Find the mean, variance, and standard deviation of the random variable with probability density function $f(x) = 2e^{-2x}$ for $x \geq 0$.
[*Hint:* Solve by inspection.]

Solution on page 727 >

Exponential random variables are used to model waiting times for events whose *non*occurrence up to any time t has no effect on the probability of the occurrence in any later time interval. (For example, your *not* winning the lottery last week has no effect on your chances this week, since each week's game is independent, so your time between winnings would be an exponential random variable.) In practice, the value of the parameter a is found by setting the theoretical mean $1/a$ equal to the observed average based on past experience.

EXAMPLE 2   FINDING PROBABILITIES OF COMPUTER FAILURE

A company's computer occasionally breaks down and, based on past records, the time X between failures averages 50 days. Model X by an exponential random variable and find:

a. $P(X \leq 20)$ **b.** $P(X > 20)$ **c.** $P(30 \leq X \leq 60)$

Solution

The value of the parameter a is found by setting the theoretical mean $1/a$ equal to the observed average, here 50:

$$\frac{1}{a} = 50 \quad \text{so} \quad a = \frac{1}{50} = 0.02 \qquad \text{Parameter} \quad a = 0.02$$

Therefore, the random variable X has probability density function

$$f(x) = 0.02e^{-0.02x} \quad \text{on } [0, \infty) \qquad f(x) = ae^{-ax} \quad \text{with} \quad a = 0.02$$

which is shown in the graph on the left.

For part (a), we integrate this function from 0 to 20 to find $P(X \leq 20)$:

$$0.02 \int_0^{20} e^{-0.02x}\,dx = 0.02\left(\frac{1}{-0.02}\right)e^{-0.02x}\Big|_0^{20} = -e^{-0.4} + e^0 \approx 0.33$$

Moved outside   Using formula 4 (inside back cover)   Evaluating   Using a calculator

Therefore, with probability about $\frac{1}{3}$ the next failure will occur within 20 days.

For part (b), $P(X > 20)$, we could integrate from 20 to ∞, but it is easier to find the probability of all *other* values, $P(X \leq 20)$, which we found in part (a) to be 0.33, and subtract it from 1:

$$P(X > 20) = 1 - P(X \leq 20) \approx 1 - 0.33 = 0.67$$

Therefore, with probability about $\frac{2}{3}$ the next failure will occur some time after 20 days.

For part (c), $P(30 \leq X \leq 60)$, we could evaluate the integral $\int_{30}^{60} 0.02e^{-0.02x}\,dx$ "by hand." Instead, we use a graphing calculator as shown on the left. The value 0.248 means that with probability about $\frac{1}{4}$ the next failure of the computer will occur some time between 30 and 60 days (1 to 2 months) after the preceding failure. Such calculations are used to determine the need for a backup computer system.

on [0, 100] by [0, 0.02]

EXAMPLE 3   **FINDING AND USING A CUMULATIVE DISTRIBUTION FUNCTION**

Find the cumulative distribution function for the probability density function in Example 2, and use it to find $P(30 \leq X \leq 60)$.

Solution

For the cumulative distribution function $F(x) = P(X \leq x)$,   we integrate the probability density function up to x:

$$F(x) = \int_0^x 0.02e^{-0.02t}\, dt = -e^{-0.02t}\Big|_0^x$$

$$= -e^{-0.02x} + 1 = 1 - e^{-0.02x}$$

Therefore, the cumulative distribution function is $F(x) = 1 - e^{-0.02x}$ for $0 \leq x < \infty$,   as shown on the left.

We then find $P(30 \leq X \leq 60)$ by evaluating $F(x)$ at 60 and at 30 and subtracting:

$$P(30 \leq X \leq 60) = F(60) - F(30) = \underbrace{1 - e^{-0.02(60)}}_{F(60)} - \underbrace{(1 - e^{-0.02(30)})}_{F(30)}$$

$$= -e^{-1.2} + e^{-0.6} \approx 0.248 \qquad \text{Same answer as in Example 2(c)}$$

Relationship Between Exponential and Poisson Random Variables

There is a useful relationship between the *Poisson* random variable (discussed on page 703) and the *exponential* random variables discussed here. If the time *intervals* between events are independent *exponential* random variables with parameter a, then the total *number* of events to occur up to time t will be a *Poisson* random variable with parameter at. For instance, in the computer failure situation described in Examples 2 and 3, the times between failures are independent exponential random variables with parameter 0.02, so the total number of failures within t days will be Poisson with parameter (mean) $0.02t$. Such ideas are used in designing service queues, such as waiting lines for customers in a bank or docking facilities for ships in a port.

Median Value of a Random Variable

The mean or expected value of a random variable gives a "representative" or "middle" value for the random variable. There is another possible choice for a "representative" value—the **median,** which is a number m such that the random variable is equally likely to be above m as below m. More precisely:

Median of a Random Variable

The median of a random variable X is a number m such that x is equally likely to be on either side of m.

$$P(X \leq m) = P(X \geq m) = \frac{1}{2}$$

half of the probability on either side

0.5   0.5

probability density function

median
m

EXAMPLE 4   **FINDING THE MEDIAN OF A CONTINUOUS RANDOM VARIABLE**

Find the median time between failures for the computer in Example 2.

Solution

The median is most easily found by setting the cumulative distribution function $F(x)$ equal to $\frac{1}{2}$ and solving for x. We could do this "by hand," solving $1 - e^{-0.02x} = 0.5$ (using the cumulative distribution function found in Example 3), but instead we use a graphing calculator.

Graphing Calculator Exploration

Define $y_1 = 1 - e^{-0.02x}$ and $y_2 = 0.5$, graph them on the window [0, 100] by [0, 1], and use INTERSECT to find where they meet.

The median waiting time is about 35 days, meaning that about half of the time the failure will occur *before* 35 days, and half of the time *after* 35 days.

Median Versus Mean

In the computer failure example, the median $m \approx 35$ was smaller than the mean $\mu = 50$ (calculated from the formula $\mu = \frac{1}{a} = \frac{1}{0.02} = 50$). Intuitively, the mean is larger because such waiting times can occasionally be very long, and these very large values increase the mean but not the median.

To clarify the difference between mean and median, think of the numbers 1, 2, 3, and 10. The mean (the average) is $\frac{1 + 2 + 3 + 10}{4} = \frac{16}{4} = 4$, but three of the four numbers are *below* this mean. For the median value, we could choose any number between 2 and 3 (such as 2.5), since two values are below and two are above this number. If we were to replace the 10 by an even larger number, this would increase the mean but not the median. In general, the mean is influenced by the values of the highest and lowest numbers, but the median is not.

PRACTICE PROBLEM 2

a. A news commentator recently read the statistic that the mean family income in the United States is about $40,000, and that 65% of families are below this level. He then said that this must be wrong, because only 50% of the families can be below the mean. Is he right?

b. In the 1995 major league baseball strike, it was revealed that the mean salary for major league baseball players was $1.2 million, while the median salary was only $500,000. What does this tell you about the distribution of salaries among the players?

Solutions below >

Solutions TO PRACTICE PROBLEMS

1. The probability density function is exponential with parameter $a = 2$, so

$$E(X) = \frac{1}{a} = \frac{1}{2} \qquad \text{Var}(X) = \frac{1}{a^2} = \frac{1}{4} \qquad \sigma(X) = \frac{1}{2}$$

2. a. He is wrong. It is the *median*, not the mean, that has 50% of the people below it and 50% above it.

b. Baseball salaries are distributed very unequally, with a few players having extraordinarily high salaries that "skew" the mean, but most having much lower salaries, half of them under $500,000.

11.3 Section Summary

A *uniform* random variable takes values in a finite interval, with all values equally likely. An *exponential* random variable takes values in the infinite interval $[0, \infty)$, with decreasing probabilities for higher values.

Random Variable	Mean $E(X)$	Variance $Var(X)$	Probability Density Function	
Uniform on $[0, B]$	$\dfrac{B}{2}$	$\dfrac{B^2}{12}$	$f(x) = \dfrac{1}{B}$ for $0 \le x \le B$	
Exponential (parameter $a > 0$)	$\dfrac{1}{a}$	$\dfrac{1}{a^2}$	$f(x) = ae^{-ax}$ for $0 \le x < \infty$	

How do we choose the constants for these random variables in a given application? For the uniform distribution, B is chosen as the largest possible value of the random variable. For the exponential distribution, the parameter a is determined by setting the theoretical mean $1/a$ equal to the *observed average value* based on past experience.

The *median* of a random variable X is a number m such that the random variable is equally likely to be above it as below it: $P(X \geq m) = P(X \leq m)$.   The median is most easily found by setting the cumulative distribution function equal to $\frac{1}{2}$ and solving.

11.3   Exercises

Exercises on Uniform Random Variables

1. If X is a uniform random variable on the interval $[0, 10]$, find:

  a. the probability density function $f(x)$

  b. $E(X)$ **c.** Var(X)

  d. $\sigma(X)$ **e.** $P(8 \leq X \leq 10)$

2. If X is a uniform random variable on the interval $[0, 0.01]$, find:

  a. the probability density function $f(x)$

  b. $E(X)$ **c.** Var(X)

  d. $\sigma(X)$ **e.** $P(X \geq 0.005)$

3. For a uniform random variable on the interval $[0, B]$, find:

  a. the cumulative distribution function $F(x)$

  b. the median

Applied Exercises for Uniform Random Variables

4. GENERAL: Waiting Time Buses arrive at a station every 12 minutes, so if you arrive at a random time, your wait is uniformly distributed on the interval $[0, 12]$. Find:

  a. $E(X)$ **b.** Var(X)

  c. $\sigma(X)$ **d.** $P(X \leq 3)$

5. GENERAL: Accidents A highway patrol station responds to emergencies within 30 miles of its location on a highway. Suppose that an accident occurs on this part of the highway and that the distance X between the accident and the station is uniformly distributed over the interval $[0, 30]$. Find:

  a. $E(X)$ **b.** Var(X)

  c. $\sigma(X)$ **d.** $P(X > 24)$

6. GENERAL: Waiting Time Your calculus teacher is often late to class, and the number of minutes you wait is uniformly distributed on the interval $[0, 10]$ (since you leave after 10 minutes). Find:

  a. $E(X)$ **b.** Var(X)

  c. $\sigma(X)$ **d.** $P(X \geq 9)$

7. Uniform Random Variable on $[A, B]$ For any two numbers A and B with $A < B$, a *uniform random variable on the interval $[A, B]$* has probability density function

$$f(x) = \frac{1}{B - A} \quad \text{for} \quad A \leq x \leq B$$

Find:

  a. the cumulative distribution function $F(x)$

  b. the mean $E(X)$

  c. the variance Var(X)

8. GENERAL: Roundoff Errors When rounding numbers to the nearest integer, the error X in the rounding is uniformly distributed on the interval $[-0.5, 0.5]$. Use Exercise 7 to find:

  a. $E(X)$ **b.** Var(X)

  c. $\sigma(X)$ **d.** $P(-0.25 \leq X \leq 0.25)$

Exercises on Exponential Random Variables

9–12. For the exponential random variables with the following densities, find:

a. $E(X)$ **b.** $\text{Var}(X)$ **c.** $\sigma(X)$
[*Hint:* Solve by inspection.]

9. $f(x) = 5e^{-5x}$ **10.** $f(x) = e^{-x}$

11. $f(x) = \frac{1}{2}e^{-x/2}$ **12.** $f(x) = 0.001e^{-0.001x}$

13–16. State the probability density function for an exponential random variable with:

13. Mean 3   **14.** Mean 0.01

15. Variance $\frac{1}{4}$ **16.** Standard deviation $\frac{1}{4}$

17. For an exponential random variable with parameter a, find:

a. the cumulative distribution function $F(x)$
b. the median

18. Find the probability that an exponential random variable is less than its mean.

Applied Exercises for Exponential Random Variables

19. BUSINESS: Employee Safety Records indicate that the average time between accidents on a factory floor is 20 days. If the time between accidents is an exponential random variable, find the probability that the time between accidents is less than a month (30 days).

20. GENERAL: Emergency Calls The number of minutes between calls to an emergency 911 center is exponentially distributed with mean 8. There is one emergency operator who takes 2 minutes to deal with each call. What is the probability that the next caller will get a busy signal?

21. BUSINESS: Quality Control The time until failure of a TV picture tube is an exponential random variable with mean 10 years. What is the probability that the tube will fail during the 2-year guarantee period?

22. POLITICAL SCIENCE: Supreme Court Vacancies The length of time between vacancies on the Supreme Court is exponentially distributed with mean 2 years. Find the probability that at least 4 years will elapse between vacancies.

Source: New York Times

23. GENERAL: Powerball Powerball is a multistate lottery with drawings twice a week, in which there may or may not be jackpot winners. On average, 4 weeks elapse between jackpot wins. Find the probability that the time between jackpot wins is:

a. no more than one week.
b. no more than six weeks.
(Note that here you are approximating a discrete-time event by a continuous-time model, as is often done in practice. Incidentally, the largest powerball jackpot was $590.5 million in 2013.)

Source: powerball.com

24. BUSINESS: Product Reliability The lifetime of a light bulb is exponentially distributed. What must the mean lifetime be if the manufacturer wants the bulb to last at least 750 hours with probability 0.90?

25–28. MEDIAN: Find the median of the random variable with the given probability density function.

25. $f(x) = 2x$ on $[0, 1]$ **26.** $f(x) = 3x^2$ on $[0, 1]$

27. $f(x) = \frac{1}{2}x$ on $[0, 2]$ **28.** $f(x) = \frac{1}{6}x^{-1/2}$ on $[0, 16]$

29. Find the variance of an exponential random variable with parameter a. [*Hint:* Use the alternative formula for the variance on page 714, integrate by parts, and then use the result of Exercise 63 on page 680 together with the mean as found on page 723.]

30. GENERAL: Telephone Calls You follow the telephone company's recommendation to let a number ring for ten rings (50 seconds) before hanging up. If the time it takes a government bureaucrat to answer his phone is exponentially distributed with mean 45 seconds, what is the probability that you will reach him?

31. GENERAL: Cash Machines The number of minutes between arrivals of customers at an automated teller machine is exponentially distributed with mean 7 minutes. If a customer's transactions take 3 minutes, what is the probability that the next arrival will have to wait?

32. BIOMEDICAL: ICU Admissions The number of hours between the arrivals of new patients in an intensive care unit is exponentially distributed with mean 8 hours. Find the median time between arrivals.

33. GENERAL: Airplane Maintenance The amount of flight time between failures of an airplane engine is exponentially distributed with mean 700 hours. If the engine is inspected every 100 hours of flight time, what is the probability that the engine will fail before it is inspected?

34. BUSINESS: Computer Failure The number of days between failures of a company's computer system is exponentially distributed with mean 10 days. What is the probability that the next failure will occur between 7 and 14 days after the last failure?

35. GENERAL: Light Bulbs The life of a light bulb is exponentially distributed with mean 1000 hours. What is
the probability that the light bulb will burn out sometime between 800 and 1200 hours?

36–37. BIOMEDICAL: Survival Time For a person who has
received treatment for a life-threatening disease, such as
cancer, the number of years of life after the treatment (the
survival time) can be modeled by an exponential random
variable.

36. Suppose that the average survival time for a group of
patients is 5 years. Find the probability that a randomly
selected patient survives for no more than 7 years.

37. The *survival function* $S(x)$ gives the probability that
a randomly selected patient will survive for at least
x years after treatment has ended. Show that
$S(x) = 1 - F(x)$,   where $F(x)$ is the cumulative
distribution function for the exponential random variable, and find a formula for $S(x)$ in terms of the exponential parameter a.

Conceptual Exercises

38. An angle is uniformly distributed between 0 and
180 degrees. What is its probability density function
expressed in *radians?*

39. If a uniform random variable (as described on
page 722) has mean 10, what is its probability density
function?

40. If a uniform random variable (as described on page 722)
has variance 3, what is its probability density function?

41. If X is uniformly distributed on $[0, 2]$, what is the *probability* density function for $3X$?

42. If X is an exponential random variable with mean $\frac{1}{5}$,
what is its probability density function?

43. Consider a uniform random variable (as described on
page 722) on the interval $[0, 1]$. What is the probability
density function of an *exponential* random variable that
has the same mean?

44. Fill in the blank: If X is an exponential random variable with parameter 4, then $2X$ is an exponential random variable with parameter ____. [*Hint:* What is the
mean of X? What is the mean of $2X$? Then what is the
parameter for $2X$?]

45. In a daily lottery that may or may not award a jackpot,
the intervals between jackpots are independent

exponential random variables with mean 10 days.
Then the total number of jackpots in the next 30 days
is a random variable. What kind of a random variable
is it? [*Hint:* See the middle paragraph on page 725.]

46. Consider the incomes of three people: you, your
math teacher, and Bill Gates (the richest man in the
United States). Which will be greater: the median or
the mean?

47. The number of children in an American family is a
random variable with mean 2. For this random variable,
which will be greater: the median or the mean?

48. Between 2004 and 2007, U.S. *mean* family income
increased by 8.5%, but *median* family income decreased
by 0.4%. How is it possible for the mean to rise and the
median to fall?

Source: Federal Reserve

49. A real estate reporter discussing home prices in
Duarte, California, recently wrote "So a single high-
priced listing (like the mammoth nine-bedroom, built
this year, that's selling for $19.8 million) is enough to
skew the median price skywards." Find the mistake in
this statement.

Source: realestate.yahoo.com

11.4 Normal Random Variables

Introduction

Studies show that heights of adults cluster around a mean of 65 inches, with
larger and smaller heights occurring with lower probabilities. Similarly, IQ's
cluster around a mean of 100, with values farther away from the mean occurring
with lower probabilities. Many other measurements exhibit similar clustering
about a central value.

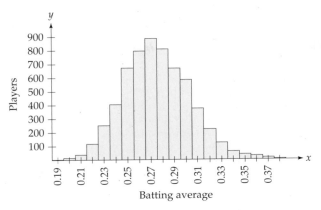

Annual batting averages of
major league players 1995–2004

Source: O'Reilly Media

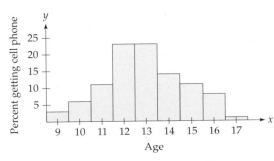

Age at which American children
get their first cell phone

Source: Pew Internet

Measurements that cluster around a mean μ, with values farther from the mean occurring with lower probabilities, are often modeled by a **normal random variable.** The probability density function of a normal random variable is often referred to as the *bell-shaped curve,* but more formally called the **Gaussian probability density function,** after the German mathematician Carl Friedrich Gauss (1777–1855). Although normal random variables are continuous, they are often used to model discrete quantities.

Normal Random Variables

The probability density function of a normal random variable depends on two parameters, denoted μ and σ, which are, respectively, the mean and standard deviation of the random variable.

Normal Random Variable

A normal random variable with mean μ and standard deviation $\sigma > 0$ has probability density function

$$f(x) = \frac{1}{\sigma \sqrt{2\pi}} e^{-\frac{1}{2}\left(\frac{x-\mu}{\sigma}\right)^2} \qquad \text{for} \quad -\infty < x < \infty$$

A normal random variable is **standard normal** if it has mean $\mu = 0$ and standard deviation $\sigma = 1$. A standard normal random variable will be denoted by the letter Z. Any normal random variable X can be **standardized** by subtracting the mean and dividing by the standard deviation:

$$Z = \frac{X - \mu}{\sigma} \qquad \begin{array}{l} \mu = \text{mean} \\ \sigma = \text{standard deviation} \end{array}$$

The probability density function for a *standard* normal random variable Z takes the following simpler form, which is graphed on the next page.

Take Note

This is the formula two boxes earlier with $\mu = 0$ and $\sigma = 1$.

Standard Normal Random Variable

A standard normal random variable Z has probability density function

$$f(x) = \frac{1}{\sqrt{2\pi}} e^{-\frac{1}{2}x^2} \qquad \text{for} \quad -\infty < x < \infty$$

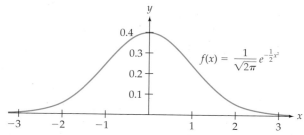

Normal probability density function with mean
$\mu = 0$ and standard deviation $\sigma = 1$

The area under the normal probability density function is 1, as is shown in Exercise 41 on pages 738–739. The mean μ gives the location of the central peak of this probability density function. The following graph shows three normal probability density functions with different means.

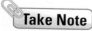

Take Note

Changing μ shifts the curve right or left.

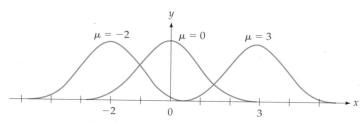

Three normal probability density
functions with different means μ

The standard deviation σ measures how the values *spread* away from the mean, so the peak of the probability density function will be higher and more pointed if σ is small, and lower and more rounded if σ is large.

Take Note

Changing σ makes the curve "taller and narrower" or "shorter and wider."

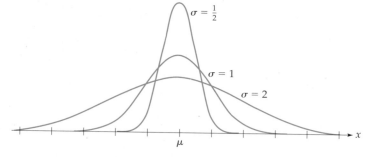

Three normal probability density
functions with different standard deviations σ

IMPORTANT NOTE For those *not* using graphing calculators, turn now to Appendix C: Normal Probabilities from Tables, at the back of the book.

Probabilities for Normal Random Variables

The probability that a normal random variable takes values between two numbers is the area under its probability density function between the two numbers.

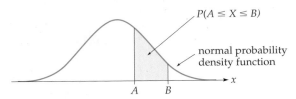

The integral of the normal probability density function cannot be expressed in terms of elementary functions, so we cannot find the area by integration. However, we can *approximate* the area by *numerical* integration, as in Section 6.4. The results can be listed in a table (as in Appendix C), or they can be found directly on a graphing calculator, using the probability DISTRibution menu and the NORMAL-CDF command. The NORMALCDF command finds the probability that a normal random variable takes values between any two given numbers.

EXAMPLE 1 **FINDING A STANDARD NORMAL PROBABILITY**

For a standard normal random variable Z, find $P(0 \leq Z \leq 1.24)$.

Solution

We use the NORMALCDF command with the given upper and lower limits 0 and 1.24.

Note that in the first line of the screen, the final two numbers, 0 and 1, are the mean and standard deviation (which are correct for *standard* normal). They can be omitted, as in the second entry, either way giving the same answer.

Therefore, $P(0 \leq Z \leq 1.24) \approx 0.3925$.

Graphing Calculator Exploration

Use the NORMALCDF command to find $P(0 \leq Z \leq 0.52)$ for a standard normal random variable Z. (Answer: 0.1985)

On some calculators, you can show such results graphically using the SHADENORM command:

Window: $[-3, 3]$ by $[0, 0.5]$. Use CLRDRAW afterward to clear the screen

⌐̅⌐ either command ⌐̅⌐
leads to

Note that in the first line of the screen on the left, the final number, 10, specifies the color.

The values of a normal random variable are said to be *normally distributed*.

EXAMPLE 2   FINDING THE PROBABILITY OF MEN'S HEIGHTS

The heights of American men are approximately normally distributed, with mean $\mu = 70$ inches and standard deviation $\sigma = 3.1$ inches. Find the proportion of men who are less than 6 feet tall.

Source: G. Salvendy, *Handbook of Human Factors and Ergonomics,* 3rd ed.

Solution

If $X =$ height, we want $P(X < 72)$.   72 inches = 6 feet

For the lower limit, we enter a sufficiently small number to include essentially *all* of the area to the left of the upper limit (here, 0 will do). We use the NORMALCDF command, entering the given mean and standard deviation after the upper and lower limits.

Therefore, approximately 74% of American men are under 6 feet tall.

Such calculations are used for designing automobiles and other consumer products.

Why did we need to enter the mean and standard davation in Example 2 but had the option to enter them or not in Example 1? Because the calculator will *assume* that $\mu = 0$ and $\sigma = 1$ (standard normal, which was the case in Example 1) unless told otherwise. How do we pick a lower limit to include essentially *all* of the area below the upper limit? As a "rule of thumb," choose a convenient number at least five standard deviations below the mean. In this Example, five standard deviations would be $5 \cdot 2.6 = 13$, and our lower limit of 0 was far more than this distance below the mean of 70. A similar rule of thumb applies when there is no upper limit: Go at least five standard deviations *above* the mean.

Graphing Calculator Exploration

The weights of American women are approximately normally distributed, with mean $\mu = 151.8$ pounds and standard deviation $\sigma = 38.8$ pounds. Find the proportion of women who weigh more than 140 pounds. (Answer: 62%)

Source: G. Salvendy, *Handbook of Human Factors and Ergonomics,* 3rd ed.

EXAMPLE 3	FINDING THE PROBABILITY OF ACCEPTABLE CHOLESTEROL LEVELS

Cholesterol is a fat-like substance that, in the bloodstream, can increase the risk of heart disease. Cholesterol levels below 200 (milligrams per deciliter of blood) are generally considered acceptable. A survey by the U.S. Department of Health found that cholesterol levels for one age group were approximately normally distributed, with mean $\mu = 222$ and standard deviation $\sigma = 28$. What proportion of the people in the study had acceptable cholesterol levels?

Source: Rogosin Institute

Solution

Letting $X = \left(\dfrac{\text{Cholesterol}}{\text{level}}\right)$, we want $P(X < 200)$. We again use 0 for the lower limit. (Do you see why it satisfies the "at least five standard deviations below the mean" rule?) The NORMALCDF command with these numbers gives the following result.

Therefore, only about 22% of the people in the study had acceptable cholesterol levels.

Inverse Probabilities for Normal Random Variables

We may *reverse* the process of finding probabilities, finding instead the value that results in a *given* probability. A graphing calculator command like INVNORM finds the number such that, with a given probability, a normal random variable is *below* that number.

EXAMPLE 4	FINDING THE HEIGHT FOR A GIVEN PROBABILITY

Women's heights are approximately normally distributed, with mean $\mu = 64.6$ inches and standard deviation $\sigma = 2.9$ inches. Find the height that marks the tallest 10%.

Source: G. Salvendy, *Handbook of Human Factors and Ergonomics,* 3rd ed.

Solution

The height that marks the tallest 10% means that 90% are below that height. The INVNORM command with the given numbers provides the following answer.

Therefore, a height of about 68.3 inches, or 5 feet 8 inches, marks the tallest 10% of women.

Measuring distance from the mean in *standard deviation units* leads to some useful probabilities for a normal random variable X.

With probability 68%, X is within 1 standard deviation of its mean.

With probability 95%, X is within 2 standard deviations of its mean.

With probability 99.7%, X is within 3 standard deviations of its mean.

These results are shown graphically as follows.

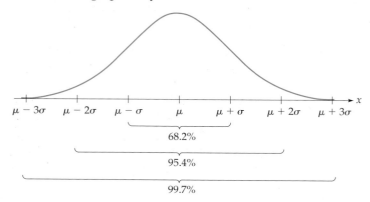

For example, from the fact that men's heights are normally distributed with $\mu = 70$ and $\sigma = 3.1$, we can conclude

About 68% of men have heights 70 ± 3.1 inches (that is, heights 66.9 inches to 73.1 inches).   Mean ± 1 standard deviation

About 95% of men have heights $70 \pm 2(3.1)$ inches (that is, heights 63.8 inches to 76.2 inches).   Mean ± 2 standard deviations

About 99.7% of men have heights $70 \pm 3(3.1)$ inches (that is, heights 60.7 inches to 79.3 inches).   Mean ± 3 standard deviations

11.4   Section Summary

Normal random variables are used to model values that cluster symmetrically around a mean, μ, with the probability decreasing as the distance from μ increases. The probability density function of a normal random variable with mean μ and standard deviation σ is

$$f(x) = \frac{1}{\sigma\sqrt{2\pi}}\, e^{-\frac{1}{2}\left(\frac{x-\mu}{\sigma}\right)^2} \qquad \text{for } -\infty < x < \infty$$

The mean μ gives the location of the central peak, and the standard deviation σ measures the *spread* of the values around this mean (larger σ meaning greater spread).

The probability density of a *standard* normal random variable ($\mu = 0$ and $\sigma = 1$) is

$$f(x) = \frac{1}{\sqrt{2\pi}}\, e^{-\frac{1}{2}x^2} \qquad \text{for } -\infty < x < \infty$$

Probabilities for a normal random variable can be found either in a *table* (by first standardizing, defining $Z = \dfrac{X - \mu}{\sigma}$ and then using the standard normal table, as explained in Appendix C: Normal Probabilities from Tables) or by using a *graphing calculator* with an operation like NORMALCDF. Similarly, we can find the *x*-value such that with given probability the random variable is below that value either by using tables or a graphing calculator operation like INVNORM.

11.4 Exercises

1–10. Find each probability for a standard normal random variable *Z*.

1. $P(0 \leq Z \leq 1.95)$

2. $P(-1.23 \leq Z \leq 0)$

3. $P(-2.55 \leq Z \leq 0.48)$

4. $P(-1.5 \leq Z \leq 1.05)$

5. $P(1.4 \leq Z \leq 2.8)$

6. $P(1 \leq Z \leq 2)$

7. $P(Z < -3.1)$

8. $P(Z > 3)$

9. $P(Z \leq 3.45)$

10. $P(Z \geq -3.5)$

11. If *X* is normal with mean 9 and standard deviation 2, find $P(10 < X < 11)$.

12. If *X* is normal with mean 3 and standard deviation $\frac{1}{2}$, find $P(2 \leq X \leq 3.5)$.

13. If *X* is normal with mean -5 and standard deviation 2, find $P(X \leq -6)$.

14. If *X* is normal with mean 0.05 and standard deviation 0.01, find $P(X > 0.04)$.

Applied Exercises

15. SOCIAL SCIENCE: Workload A motor vehicle office can process only 250 license applications per day. If the number of applications on a given day is normally distributed with mean 210 and standard deviation 30, what is the probability that the office can handle the day's applications? [*Hint:* Find $P(0 \leq X \leq 250)$.]

16. BIOMEDICAL: Hospital Stays A patient's medical insurance will pay for at most 20 days in the hospital. Past experience with similar cases shows that the length of hospital stay is normally distributed with mean 17 days and standard deviation 2 days. What is the probability that the insurance coverage will be enough? [*Hint:* Find $P(0 \leq X \leq 20)$.]

17. BUSINESS: Advertising The percentage of television viewers who watch a certain show is a normal random variable with mean 22% and standard deviation 3%. The producer guarantees to advertisers that the viewer percentage will be at least 20%, or else the advertisers will receive free air time. What is the probability that the producers will have to give free air time? [*Hint:* Find $P(X < 20)$.]

18. PSYCHOLOGY: Learning The time that a rat takes to "solve" a maze is normally distributed with mean 3 minutes and standard deviation 30 seconds. Find the probability that the rat will take less than 2 minutes.

19. BUSINESS: Orders The number of orders received by a company each day is normally distributed with mean 2000 and variance 160,000. If the orders exceed 2500, the company will have to hire extra help. Find the probability that extra help will be needed. [*Hint:* First find σ.]

20. BUSINESS: Airline Overbooking In order to avoid empty seats, airlines generally sell more tickets than there are seats. Suppose that the number of ticketed passengers who show up for a flight with 100 seats is a normal random variable with mean 97 and standard deviation 6. What is the probability that at least one person will have to be "bumped" (denied space on the plane)?

21. BIOMEDICAL: Weights Weights of men are approximately normally distributed with mean 190 pounds and standard deviation 40. The minimum and maximum weights in order to be an FBI Police Officer (providing security at FBI offices) are 117 and 238 pounds. What proportion of men fall in this range?

Sources: Federal Bureau of Investigation; G. Salvendy, Handbook of Human Factors and Ergonomics, 3rd ed.

22. GENERAL: SAT Scores Scores on the 2013 College Board SAT Test in Mathematics were approximately normally distributed with mean 514 and standard deviation 118. What score would put you in the top 20%?

Source: College Board

23. SOCIAL SCIENCE: IQ's Although intelligence cannot be measured on a single scale, IQ (Intelligence Quotient) scores are still used in some situations. IQ scores are normally distributed with mean 100 and standard deviation 15. Find the IQ score that places you in the top 2%.

24. BUSINESS: Product Reliability The number of years that an electrical generator will last is a normal random variable. Company A makes generators with a mean lifetime of 8 years and a standard deviation of 1 year. Company B makes generators with a mean lifetime of 7 years and a standard deviation of 2 years. Which company's generators have a higher probability of lasting at least 10 years?

25. **BUSINESS: Quality Control** A soft drink bottler fills 32-ounce bottles with a high-speed bottling machine. Assume that the amount of the soft drink put into the bottle is a normal random variable. The mean amount that the machine puts into the bottles is adjustable, but the standard deviation is 0.25 ounce. What should

be the mean amount such that with probability 98% a bottle contains at least 32 ounces?

26. **ECONOMICS: Portfolio Management** Modern portfolio theory generally assumes that the yield on a diversified portfolio of investments will be normally distributed around its historic mean. What is the probability that a diversified portfolio will yield at least 5% if the historic mean is 8% and standard deviation 1.5%?

Conceptual Exercises

27–30. If Z is a standard normal random variable, find the mean and standard deviation of the given random variable. No calculation is necessary.

27. $2Z$ **28.** $Z + 5$ **29.** $2Z + 5$ **30.** $2(Z + 5)$

31–36. Assume that X and Y are normal random variables. No calculation is necessary.

31. If X and Y both have mean 0 but $\sigma(X) = 2$ and $\sigma(Y) = 3$, then which is greater: $P(X > 4)$ or $P(Y > 4)$?

32. If X and Y both have standard deviation 1 but $E(X) = 2$ and $E(Y) = 3$, then which is greater: $P(X > 4)$ or $P(Y > 4)$?

33. If X and Y both represent the heights of people, but X is in feet and Y is in inches, which has the greater mean? Which has the greater standard deviation?

34. If X and Y both represent the temperature at noon tomorrow but X is in degrees Fahrenheit and Y is in degrees Celsius (or Centigrade), which has the greater mean? Which has the greater standard deviation? [*Hint:* The relationship is $X = \frac{9}{5}Y + 32$.]

35. If $P(X > 47.8) = \frac{1}{2}$, then what do you know about X?

36. If $\frac{1}{2}X + 2$ is *standard* normal, then what do you know about X?

Explorations and Excursions   The following problems extend and augment the material presented in the text.

Calculus of the Normal Distribution

37. Find the x-value at which the standard normal probability density function $f(x) = \dfrac{1}{\sqrt{2\pi}}e^{-x^2/2}$ is maximized. Also find the value of the function at this x-value.

38. Find the x-value at which the (nonstandard) normal probability density function $f(x) = \dfrac{1}{\sigma\sqrt{2\pi}}e^{-\frac{1}{2}\left(\frac{x-\mu}{\sigma}\right)^2}$ (for constants μ and σ) is maximized. Also find the maximum value of the function at this x-value.

39. Find the inflection points of the standard normal probability density function $f(x) = \dfrac{1}{\sqrt{2\pi}}e^{-x^2/2}$.

40. Find the x-coordinates of the inflection points of the (nonstandard) normal probability density function $f(x) = \dfrac{1}{\sigma\sqrt{2\pi}}e^{-\frac{1}{2}\left(\frac{x-\mu}{\sigma}\right)^2}$ (for constants μ and σ).

41. Verify that the area under the standard normal curve is 1 by carrying out the following steps. (*Requires Sections 6.3, 7.7, and 8.5.*)

a. Define $I = \displaystyle\int_{-\infty}^{\infty} e^{-\frac{1}{2}x^2}\,dx$ and show that
$$I = 2\int_0^{\infty} e^{-\frac{1}{2}x^2}\,dx. \quad [\textit{Hint: Use symmetry.}]$$

b. Show that $I^2 = 4\displaystyle\int_0^{\infty} e^{-\frac{1}{2}x^2}\,dx \cdot \int_0^{\infty} e^{-\frac{1}{2}y^2}\,dy.$
[*Hint:* I^2 consists of two copies of I expressed with different variables and multiplied together.]

c. Show that $I^2 = 4\displaystyle\int_0^{\infty}\left(\int_0^{\infty} e^{-\frac{1}{2}(x^2+y^2)}\,dy\right)dx.$
[*Hint:* Move one integral (which is a constant) inside the other and use some algebra.]

d. Show that $I^2 = 4\displaystyle\int_0^{\infty}\left(\int_0^{\infty} e^{-\frac{1}{2}(x^2+x^2s^2)}\,x\,ds\right)dx.$
[*Hint:* In the inner integral in step (c), use the substitution $y = x \cdot s$ so that $dy = x \cdot ds$]

e. Show that $I^2 = 4\displaystyle\int_0^{\infty}\left(\int_0^{\infty} e^{-\frac{1}{2}x^2(1+s^2)}\,x\,ds\right)dx$
$$= 4\int_0^{\infty}\left(\int_0^{\infty} e^{-\frac{1}{2}x^2(1+s^2)}\,x\,dx\right)ds.$$
[*Hint:* Use some algebra and then reverse the order of integration.]

f. Show that $I^2 = 4\displaystyle\int_0^{\infty} \dfrac{1}{1+s^2}\,ds$. [*Hint:* In the inner integral of the last part of step (e), use the $\int e^u\,du$ formula with $u = -\frac{1}{2}x^2(1+s^2)$ and $du = -x(1+s^2)\,dx$ and evaluate (by limits) at 0 and ∞ to show that the inner integral equals $\dfrac{1}{1+s^2}$.]

g. Define $s = \tan\theta$ so that $s = 0$ corresponds to $\theta = 0$, $s = \infty$ corresponds to $\theta = \pi/2$, $ds = \sec^2\theta\,d\theta$, and $1 + s^2 = 1 + \tan^2\theta = \sec^2\theta$. [See Exercise 27 on page 579.]

h. Show that

i. Show that $I = \sqrt{2\pi}$. [*Hint:* Since $I^2 = 2\pi$ from step (h).]

$$I^2 = 4\int_0^{\pi/2} \dfrac{1}{1+\tan^2\theta}\sec^2\theta\,d\theta$$

$$= 4\int_0^{\pi/2} \dfrac{1}{\sec^2\theta}\sec^2\theta\,d\theta = 4\int_0^{\pi/2} d\theta = 4\cdot\dfrac{\pi}{2}$$

$$= 2\pi$$

[*Hint:* In the integral from step (f), use the substitution from step (g).]

j. Show that $\dfrac{1}{\sqrt{2\pi}}\displaystyle\int_{-\infty}^{\infty} e^{-\frac{1}{2}x^2}\,dx = 1$, as was to be shown. [*Hint:* Use step (i) with the definition of I (from step (a)) and divide by $\sqrt{2\pi}$.]

---

11   Chapter Summary with Hints and Suggestions

Reading the text and doing the exercises in this chapter have helped you to master the following concepts and skills, which are listed by section (in case you need to review them) and are keyed to particular Review Exercises. Answers for all Review Exercises are given at the back of the book, and full solutions can be found in the Student Solutions Manual.

11.1 Discrete Probability

- Find $E(X)$, $\mathrm{Var}(X)$, and $\sigma(X)$ for a discrete random variable. *(Review Exercises 1–2.)*

$$\mu = E(X) = \sum_{i=1}^{n} x_i p_i$$

$$\mathrm{Var}(X) = \sum_{i=1}^{n} (x_i - \mu)^2 p_i = \sum_{i=1}^{n} x_i^2 p_i - \mu^2$$

$$\sigma(X) = \sqrt{\mathrm{Var}(X)}$$

- Solve an applied problem using expected value. *(Review Exercises 3–4.)*

- Solve an applied problem using a Poisson random variable. *(Review Exercises 5–10.)*

$$P(X = k) = e^{-a}\dfrac{a^k}{k!} \qquad k = 0, 1, 2, \ldots$$

$$E(X) = a$$

11.2 Continuous Probability

- Make a function into a probability density function. *(Review Exercises 11–16.)*

$$\int_A^B f(x)\,dx = 1$$

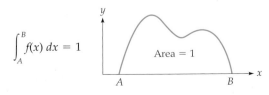

- Find a probability and the cumulative distribution function. *(Review Exercise 17.)*

$$P(a \le X \le b) = \int_a^b f(x)\,dx$$

$$F(x) = \int_A^x f(t)\,dt$$

- Find a probability and the probability density function. *(Review Exercise 18.)*

$$P(a \le X \le b) = F(b) - F(a) \qquad f(x) = \dfrac{d}{dx}F(x)$$

- Find $E(X)$, $\mathrm{Var}(X)$, and $\sigma(X)$ for a continuous random variable. *(Review Exercises 19–20.)*

$$\mu = E(X) = \int_A^B x f(x)\,dx$$

$$\mathrm{Var}(X) = \int_A^B (x - \mu)^2 f(x)\,dx$$

$$= \int_A^B x^2 f(x)\,dx - \mu^2$$

$$\sigma(X) = \sqrt{\mathrm{Var}(X)}$$

- Solve an applied problem involving a continuous random variable. *(Review Exercises 21–22.)*
- Find $E(X)$, $Var(X)$, and $\sigma(X)$ for a continuous random variable. *(Review Exercises 23–24.)*
- Solve an applied problem involving a continuous random variable. *(Review Exercises 25–26.)*

11.3 Uniform and Exponential Random Variables

- Find the probability density function, $E(X)$, $Var(X)$, $\sigma(X)$, and a probability for a uniform random variable. *(Review Exercises 27–30.)*

$$E(X) = \frac{B}{2} \qquad Var(X) = \frac{B^2}{12} \qquad \sigma(X) = \frac{B}{2\sqrt{3}}$$

- Solve an applied problem using an exponential random variable. *(Review Exercises 31–32.)*

$$E(X) = \frac{1}{a} \qquad Var(X) = \frac{1}{a^2} \qquad \sigma(X) = \frac{1}{a}$$

- Find the median of a continuous random variable. *(Review Exercises 33–34.)*

$$P(X \le m) = P(X \ge m) = \frac{1}{2}$$

- Solve an applied problem involving an exponential random variable. *(Review Exercises 35–36.)*

11.4 Normal Random Variables

- Find a probability for a standard normal random variable. *(Review Exercises 37–40.)*

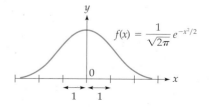

- Find a probability for a normal random variable with a given mean and variance. *(Review Exercises 41–42.)*

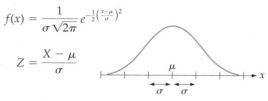

$$f(x) = \frac{1}{\sigma\sqrt{2\pi}} e^{-\frac{1}{2}\left(\frac{x-\mu}{\sigma}\right)^2}$$

$$Z = \frac{X - \mu}{\sigma}$$

- Solve an applied problem using a normal random variable. *(Review Exercises 43–46.)*

Hints and Suggestions

- In discrete probability, probabilities are found by *adding*. In continuous probability, they are found by *integrating* a probability density function (or by subtracting values of a cumulative distribution function).

- The *expected value* or *mean* of a random variable gives a "representative" value, while the *variance* and *standard deviation* measure the "spread" or "dispersion" of the values away from the mean.

- Poisson random variables are often used to model the number of occurrences of "rare events," with the parameter a set equal to the *average number of occurrences*.

- The alternative formula for the variance (page 714) is generally easier to use than the definition (page 713). The corresponding alternative variance formula for a *discrete* random variable is given in Exercise 47 on page 709.

- The mean and the median are both used as "representative" numbers for a random variable. The mean is influenced by the extreme values, but the median is not. For example, if X takes values $\{1, 2, 99\}$ with equal probability, the median would be 2, but the mean would be 34 (showing the influence of the 99).

- Uniform random variables are used to model waiting times for which all numbers in a *fixed interval* are *equally likely*. The parameter B is set equal to the largest possible value.

- Exponential random variables are used to model waiting times that can be *arbitrarily long*. The parameter a is found by setting the mean $1/a$ equal to the observed average value based on past experience.

- Normal random variables are used to model measurements that cluster symmetrically around the mean. About 68% of the probability is within one standard deviation of the mean, and about 95% is within two standard deviations of the mean.

- A graphing calculator helps by finding medians and integrating probability density functions to find probabilities, especially for the normal distribution.

11 Review Exercises and Chapter Test 11 ◯ indicates a Chapter Test exercise.

11.1 Discrete Probability

1–2. For each spinner, the arrow is spun and then stops, pointing in one of the numbered sectors. Let X stand for the number so chosen. If the probability of each number is proportional to the area of its sector, find:
a. $E(X)$ **b.** $\text{Var}(X)$ **c.** $\sigma(X)$

①   **2.**

3. BUSINESS: Land Value A builder estimates a parcel of land to be worth $2,000,000 if he can get a zoning variance, and only $50,000 if he cannot. He believes that the probability of getting the variance is 40%. What is the expected value of the land?

4. BUSINESS: Land Value A land speculator must choose between two pieces of property, whose values depend on whether they contain oil. Property A will be worth $1,200,000 with probability 10% (that is, if oil is found), and only $2000 with probability 90%. Property B will be worth $1,000,000 with probability 15%, and only $1000 with probability 85%. On the basis of expected value, which property should the speculator buy?

⑤ **BUSINESS:** Quality Control Computer flash drives are manufactured in batches of 1000, with an average of two defective drives per batch. Use the Poisson distribution to find the probability that a batch has two or fewer defective drives.

6. BUSINESS: Insurance According to a recent insurance company study, the number of major natural "catastrophes" (such as earthquakes and floods) resulting in at least $5 million of claims has recently averaged 2.63 per month. Use the Poisson distribution to estimate the probability that the number X in a given month satisfies $2 \leq X \leq 4$.
Source: Insurance Information Institute

7. GENERAL: Mail Delivery In an average hour you receive 4 emails. Use a Poisson random variable to find the probability that in a given hour you receive no emails.

8. GENERAL: Cookies Suppose that 100 cookies are to be made from dough containing 500 chocolate chips. Use a Poisson random variable to find the probability that a given cookie has no chips.

9. GENERAL: Earthquakes From 2000 through 2013 there were 84 large (magnitude 6 or more) earthquakes in the United States, for an average of 6.46 per year. Use a Poisson random variable to estimate the probability that in a given year the number of earthquakes will be:
a. 4 or fewer.
b. 6 or fewer.
c. 8 or fewer.
d. 9 or more.
Source: U.S. Geological Survey

10. BUSINESS: Quality Control In a shipment of 500 television sets, an average of two sets will be defective. The dealer will return the shipment if more than four sets are defective. Use the Poisson distribution to find the probability of a return. [*Hint:* First find $P(X \leq 4)$ and then subtract it from 1.]

11.2 Continuous Probability

11–16. Find the value of a that makes each function a probability density function on the given interval.

11. $ax^2(2 - x)$ on $[0, 2]$ **12.** $a\sqrt{9 - x}$ on $[0, 9]$

⑬ $\dfrac{a}{(1 + x)^2}$ on $[0, 1]$ **14.** $a \sin x$ on $[0, \pi]$

15. $a\sqrt{1 - x^2}$ on $[-1, 1]$ **16.** $ae^{-|x|}$ on $[-2, 2]$

⑰ For the probability density function $f(x) = \frac{1}{9} x^2$ on $[0, 3]$, find:
a. $P(1 \leq X \leq 2)$ using the probability density function
b. the cumulative distribution function $F(x)$
c. $P(1 \leq X \leq 2)$ using the cumulative distribution function

18. For the cumulative distribution function $F(x) = \frac{1}{4}\sqrt{x} - \frac{1}{4}$ on $[1, 25]$, find:
a. $P(4 \leq X \leq 9)$ using the cumulative distribution function
b. the probability density function $f(x)$
c. $P(4 \leq X \leq 9)$ using the probability density function

⑲ For the probability density function $f(x) = \frac{3}{2}(1 - x^2)$ on $[0, 1]$, find:
a. $E(X)$ **b.** $\text{Var}(X)$ **c.** $\sigma(X)$

20. For the probability density function $f(x) = \frac{2}{3} x^{-3/2}$ on $[1, 16]$, find:
a. $E(X)$ **b.** $\text{Var}(X)$ **c.** $\sigma(X)$

21. **GENERAL:** Electrical Demand The amount of electricity (in millions of kilowatt-hours) used by a city is a random variable X with probability density function $f(x) = \frac{1}{32} x$ on [0, 8]. Find:

a. the expected demand $E(X)$

b. the power level that, with probability 0.95, will be sufficient [that is, find b such that $P(X \le b) = 0.95$]

22. **BEHAVIORAL SCIENCE:** Learning In a test of memory, the proportion X of items recalled is a random variable with probability density function $f(x) = 12x(1 - x)^2$ on [0, 1]. Find:

a. $P(X \le \frac{1}{2})$ b. $E(X)$

23. For the probability density function

$$f(x) = \frac{2}{\pi} \sqrt{1 - x^2} \quad \text{on } [-1, 1], \quad \text{find:}$$

a. $E(X)$ b. $\text{Var}(X)$ c. $\sigma(X)$

24. For the probability density function $f(x) = 504x^3(1 - x)^5$ on [0, 1], find:

a. $E(X)$ b. $\text{Var}(X)$ c. $\sigma(X)$

25. **BUSINESS:** Manufacturing Flaws A flaw is randomly located in a computer disk. The x-coordinate of the flaw has probability density function

$$f(x) = \frac{1}{2\pi} \sqrt{4 - x^2} \quad \text{on } [-2, 2]. \text{ Find the prob-}$$

ability that the x-coordinate of the flaw lies between $x = -1$ and $x = 1$.

26. **BEHAVIORAL SCIENCE:** Time to Complete a Task The fraction of a minute that a person takes to complete a task has probability density function $f(x) = 105x^2(1 - x)^4$ for x in [0, 1]. Find the probability that the time lies between $x = 0.25$ and $x = 0.5$ minute.

11.3 Uniform and Exponential Random Variables

27. If X is a uniform random variable on the interval [0, 50], find:

a. the probability density function $f(x)$

b. $E(X)$ c. $\text{Var}(X)$ d. $\sigma(X)$

e. $P(15 \le X \le 45)$

28. **GENERAL:** Rainfall Rain is forecast, and you estimate that the actual number of inches X of rain is uniformly distributed on the interval [0, 2]. Find:

a. $E(X)$ b. $\text{Var}(X)$

c. $\sigma(X)$ d. $P(X < 0.5)$

29. For the exponential random variable with probability density function $f(x) = 0.01e^{-0.01x}$ on [0, ∞), find:

a. $E(X)$ b. $\text{Var}(X)$

30. State the probability density function for an exponential random variable with mean 4.

31. **GENERAL:** Airplane Flight Delays For flights from U.S. airports, the "taxi-out" time (the number of minutes between leaving the gate and takeoff) can be approximated by an exponential random variable with mean 20. Find the probability that for a given flight the taxi-out time is more than:

a. 1 hour (60 minutes) b. 3 hours (180 minutes)

Source: U.S. Bureau of Transportation Statistics

32. **BUSINESS:** Quality Control The "life" of a typical incandescent light bulb is exponentially distributed with mean 1000 hours. Find the probability that the bulb will burn out within the first 500 hours.

Source: General Electric

33–34. Find the median of the random variable with the given probability density function.

33. $f(x) = \frac{3}{8} x^2$ on [0, 2]

34. $f(x) = 3e^{-3x}$ on [0, ∞)

35. **GENERAL:** Fire Alarms The time between calls to a fire department is an exponential random variable with mean 4 hours. If the fire company is out of the firehouse for 1 hour on each call, what is the probability that the company is out when the next call comes in?

36. **GENERAL:** Ship Arrivals The time between the arrival of ships in a port is an exponential random variable with mean 5 hours. If each arrival requires 1 hour of docking time with the harbor's tugboat, what is the probability that the next arriving ship will have to wait?

11.4 Normal Random Variables

37–40. Find the indicated probability for a standard normal random variable Z.

37. $P(0.95 \le Z \le 2.54)$ **38.** $P(-1.11 \le Z \le 1.44)$

39. $P(Z < 2.5)$ **40.** $P(Z > 1)$

41. If X is normal with mean 12 and standard deviation 2, find $P(10 < X \le 15)$.

42. If X is normal with mean 1 and standard deviation 0.4, find $P(0 < X \le 1.8)$.

43. **BUSINESS:** Refunds The "life" of an automobile tire is normally distributed with mean 50,000 miles and standard deviation 5000. If the company guarantees its tires for 40,000 miles, what is the probability that a tire will outlast its guarantee?

44. **SOCIAL SCIENCE:** Smoking In a large-scale study carried out in the 1960s among male smokers 35 to 45 years of age, the number of cigarettes smoked daily was approximately normal with mean 28 and standard deviation 10. What proportion had a two-pack-a-day habit (40 or more cigarettes)?

Source: National Cancer Institute Monograph No. 19

45. **BIOMEDICAL:** Heights Heights of women are approximately normally distributed with mean 64.6 inches and standard deviation 2.9. The minimum and maximum heights in order to be an FBI Police

Officer (providing security at FBI offices) are 60 and 72 inches. What proportion of women fall in this range?

Sources: Federal Bureau of Investigation; G. Salvendy, Handbook of Human Factors and Ergonomics, 3rd ed.

46. GENERAL: SAT Scores Scores on the 2013 SAT Critical Reasoning Test were approximately normally distributed with mean 496 and standard deviation 115. What score will put you in the top 30%?

Source: College Board

1-11 Cumulative Review for Chapters 1–11

The following exercises review some of the basic concepts and techniques that you learned in Chapters 1–11. Answers to all of these Cumulative Review exercises are given in the answer section at the back of the book. A graphing calculator is suggested but not required.

1. a. Find an equation for the horizontal line through the point $(-2, 5)$.
  b. Find an equation for the vertical line through the point $(-2, 5)$.

2. Simplify: $\left(\frac{1}{4}\right)^{-3/2}$

3. Use the definition of the derivative,
$$f'(x) = \lim_{h \to 0} \frac{f(x+h) - f(x)}{h},$$ to find the derivative
of $f(x) = 3x^2 - 7x + 1$.

4. The temperature in an industrial refining tank after t hours is $T(t) = 12\sqrt{t^3} + 225$ degrees Fahrenheit. Find $T'(4)$ and $T''(4)$ and interpret these values.

5. Find $\dfrac{d}{dx} \sqrt[3]{x^3 + 8}$.

6. Find the derivative of $f(x) = (2x + 3)^3(3x + 2)^4$.

7. Make sign diagrams for the first and second derivatives and then graph the function $f(x) = x^3 - 12x^2 - 60x + 400$, showing all relative extreme points and inflection points.

8. Make sign diagrams for the first and second derivatives and graph the function $f(x) = \sqrt[3]{x} + 1$ showing all relative extreme points and inflection points.

9. Make a sign diagram for the first derivative and graph the function $f(x) = \dfrac{1}{x^2 - 4}$ showing all asymptotes and relative extreme points.

10. A gardener wants to enclose two identical rectangular pens along an existing wall. If the side along the wall needs no fence, what is the largest total area that can be enclosed using only 300 feet of fence?

11. An open-top box is to be made from a square sheet of metal 12 inches on each side by cutting a square from each corner and folding up the sides. Find the volume of the largest box that can be made in this way.

12. For $xy - 3y = 5$, use implicit differentiation to find $\dfrac{dy}{dx}$ at the point $(4, 5)$.

13. A circular oil slick is expanding such that its area is growing at the rate of 3 square miles per day. Find how fast the radius is growing at the moment when the radius is 2 miles.

14. A sum of $5000 is deposited in a bank account earning 6% interest. Find the value after 4 years if the interest is compounded:
  a. semiannually.   **b.** continuously.

15. An investment grows by 9% compounded continuously. How long will it take to increase by 75%?

16. How much must you deposit now in a bank account paying 8% compounded quarterly to have $100,000 in 25 years?

17. Find:
  a. $\dfrac{d}{dx}\left(\dfrac{e^{2x}}{x}\right)$ **b.** $\dfrac{d}{dx}\ln(x^3 + 1)$

18. Consumer demand for a product is $D(p) = 12{,}000e^{-0.04p}$ units, where p is the selling price in dollars. Find the price that maximizes consumer expenditure. [*Hint:* Expenditure is $E(p) = p \cdot D(p)$.]

19. The gross domestic product of a country after t years is $G(t) = 3 + 5e^{0.06t}$. Find the relative rate of change after 2 years.

20. Find:
  a. $\displaystyle\int (12x^3 - 4x + 5)\, dx$

  b. $\displaystyle\int 6e^{-2x}\, dx$

21. Evaluate $\displaystyle\int_1^e \frac{2}{x}\, dx$.

22. A company's marginal cost function is $MC(x) = 3x^{-1/2}$ thousand dollars, where x is the number of units. Find the total cost of producing units 25 through 100.

23. Radioactive waste is being deposited into a new underground storage vault at the rate of $120e^{0.15t}$ tons per year, where t is the number of years that the depository has been open. Find a formula for the total amount of radioactive waste deposited during the first t years.

24. Find the area bounded by $y = x^2 + 4$ and $y = 6x + 4$.

25. The population of a city is predicted to be $P(t) = 60e^{0.04t}$ thousand people, where t is the number of years from now. Find the average population over the next 10 years (year 0 to year 10).

26. Find:

a. $\displaystyle\int \frac{x\,dx}{\sqrt{x^2 + 1}}$ **b.** $\displaystyle\int_0^1 xe^{-x^2}\,dx$

27. Find by integration by parts: $\displaystyle\int x^2 \ln x\,dx$.

28. Use the integral table on the inside back cover to find $\displaystyle\int \frac{x}{\sqrt{x - 1}}\,dx$.

29. Evaluate $\displaystyle\int_0^\infty e^{-2x}\,dx$.

30. Use trapezoidal approximation with $n = 4$ trapezoids to approximate $\displaystyle\int_0^1 e^{\sqrt{x}}\,dx$. (If you are using a graphing calculator, compare your answer with that using FnInt or $\int f(x)dx$.)

31. Use Simpson's Rule with $n = 4$ to approximate $\displaystyle\int_0^1 e^{\sqrt{x}}\,dx$.

(If you are using a graphing calculator, compare your answer with that using FnInt or $\int f(x)dx$.)

32. Find the first partial derivatives of $f(x, y) = xe^y - y\ln(x^2 + 1)$.

33. Find the extreme values of $f(x, y) = 3x^2 + 2y^2 - 2xy - 8x - 4y + 15$.

34. Find the least squares line for the following points:

x	y
2	4
4	3
6	0
8	-1

35. Use Lagrange multipliers to find the maximum value of $f(x, y) = 100 - 2x^2 - 3y^2 + 2xy$ subject to the constraint $2x + y = 18$.

36. Find the total differential of $f(x, y) = 5x^2 - 2xy + 2y^3 - 12$.

37. Find the volume under the surface $f(x, y) = x^2 + y^2$ over the rectangle $R = \{(x, y) \mid 0 \le x \le 3, 0 \le y \le 3\}$.

38. a. Convert 135° into radians.

b. Convert $\dfrac{5\pi}{6}$ into degrees.

39. Evaluate *without* using a calculator.

a. $\sin \dfrac{3\pi}{4}$ **b.** $\cos \dfrac{3\pi}{4}$

c. $\sin \dfrac{5\pi}{6}$ **d.** $\cos \dfrac{5\pi}{6}$

40. Sketch the graph of $f(x) = 2\cos 4x$.

41. Find $\dfrac{d}{dx}(\sin x \cos x)$.

42. Find $\displaystyle\int e^{\cos x} \sin x\,dx$.

43. Find the area under $y = \cos x$ from $x = 0$ to $x = \pi/2$.

44. Evaluate *without* using a calculator.

a. $\tan \dfrac{5\pi}{6}$ **b.** $\cot \dfrac{5\pi}{6}$

c. $\sec \dfrac{3\pi}{4}$ **d.** $\csc \dfrac{3\pi}{4}$

45. Find:

a. $\dfrac{d}{dx}\tan(x^2 + 1)$

b. $\displaystyle\int \sec(2t + 1)\tan(2t + 1)\,dt$

46. a. Find the general solution to the differential equation $y' = x^3 y$.

b. Then find the particular solution that satisfies the initial condition $y(0) = 2$.

47. Solve the first-order linear differential equation and initial condition $\begin{cases} y' + 6xy = 0 \\ y(0) = 4 \end{cases}$

48. Find the Euler approximation on the interval $[0, 1]$ using $n = 4$ segments for the following initial value problem, and draw a graph of your approximation. $\begin{cases} y' = \dfrac{12x}{y} \\ y(0) = 2 \end{cases}$

49. Determine whether the geometric series converges, and if so, find its sum:

$$\tfrac{4}{5} + \tfrac{16}{25} + \tfrac{64}{125} + \cdots$$

50. Find the third Taylor polynomial at $x = 0$ for $f(x) = e^{2x}$.

51. Find the Taylor series at $x = 0$ for $f(x) = \sin x^2$ by modifying a known Taylor series. What is its radius of convergence?

52. Use Newton's method to approximate $\sqrt{105}$, continuing until two successive iterations agree to five decimal places.

53. A department store receives umbrellas in batches of 120, and, on average, four are defective. Use a Poisson random variable to find the probability that three or fewer are defective.

54. If X is a uniform random variable on the interval $[0, 18]$, find:

    a. $E(X)$ **b.** $\text{Var}(X)$

55. Suppose that the waiting time for an elevator in a tall building is exponentially distributed with mean 1 minute. Find the probability that your wait will be $1\frac{1}{2}$ minutes or less.

56. If a random variable X is normally distributed with mean 12 and standard deviation 2, find $P(11 \leq X \leq 13)$.

Appendix A

Graphing Calculator Basics

Although the (optional) Graphing Calculator Explorations may be carried out on most graphing calculators, the screens shown in this book are from the Texas Instruments TI-84 Plus C. (We occasionally show a screen from the TI-89 calculator, but for illustrative purposes only.) Any specific instructions we give are designed to work with the entire family of TI-83 and TI-84 calculators. To do the Graphing Calculator Explorations, you should be familiar with the terms described in the "Graphing Calculator Terminology" section below. To do the (optional) modeling Examples and Exercises in Chapter 1, you should be familiar with the regression commands described in the "Entering Data" section below.

In what follows, **Bold** words refer to standard terms that appear on the screen of your calculator. Particular keys to be pressed are indicated by key symbols. For example, you turn your calculator "on" by pressing the [ON] button and "off" using two keys [2nd] [ON].

Graphing Calculator Terminology

The *viewing window* is that part of the *x-y* plane shown in the screen of your graphing calculator. **Xmin** and **Xmax** are the smallest and largest *x*-values shown, and **Ymin** and **Ymax** are the smallest and largest *y*-values shown. **Xscl** and **Yscl** define the distance between tick marks on the *x*- and *y*-axes. These values can be set using the [WINDOW] command or changed automatically using [ZOOM] selections. The result of [ZOOM] [6] for "ZStandard" is shown below.

Xscl and Yscl are each set at 1, so the tick marks are 1 unit apart. The unit distances in the *x*- and *y*-directions on the screen may differ.

Viewing Window [−10, 10] by [−10, 10]

The viewing window is always [Xmin, Xmax] by [Ymin, Ymax] and it is best to set Xscl and Yscl so that there is a reasonable number of tick marks (generally 2 to 20) on each axis. The *x*- and *y*-axes will not be visible if the viewing window does not include the origin.

[ZOOM] [2] for "Zoom In" allows you to magnify part of the viewing window to see finer detail around a chosen point. [ZOOM] [3] "Zoom Out" does the opposite, like stepping back to see a larger portion of the plane but with less detail. These and other [ZOOM] commands change the viewing window.

[Y=] allows you to define one or more functions to be graphed in the viewing window, and then [GRAPH] plots them by calculating their *y*-values for *x*-values from Xmin to Xmax. Points with *y*-values not between Ymin and Ymax are not shown!

[2nd] [GRAPH] (TABLE F5) lists the values of your function in table form, just as you have probably done when graphing a curve by hand. Use [2nd] [WINDOW] (TBLSET F2) and choose "Ask" on the "Indpnt" line and "Auto" on the "Depend" line to have your calculator *ask* you for the *x*-values and *automatically* calculate the *y*-values.

[TRACE] allows you to move a cursor along a curve by using the left and right arrows with the *x*- and *y*-coordinates shown at the bottom of the screen. If you have graphed several curves, use the up and down arrows to change between them. ***Useful Hint***: To make the *x*-values in [TRACE] take simple values like 0.1, 0.2, and so on, use [ZOOM] [4] for "ZDecimal" and for larger windows, multiply both Xmin and Xmax (or Ymin and Ymax) by the *same* whole number.

[2nd] [TRACE] (CALC F4) displays a list of calculations that can be carried out on functions that you have graphed:

[2nd] [TRACE] (CALC F4) [1] for "value" evaluates your function at a specified *x*-value.

[2nd] [TRACE] (CALC F4) [2] for "zero" searches for an *x*-intercept of your curve by solving $f(x) = 0$ to find the root.

[2nd] [TRACE] (CALC F4) [3] for "minimum" and [2nd] [TRACE] (CALC F4) [4] for "maximum" find the minimum and maximum values of your curve between specified *x*-values. These can also be found directly using [MATH] [6] "fMin(" and [MATH] [7] "fMax(" without first graphing your function.

[2nd] [TRACE] (CALC F4) [5] for "intersect" finds the *x*-value where two curves meet.

[2nd] [TRACE] (CALC F4) [6] for "dy/dx" approximates the derivative of your function at a point. This can also be done directly using [MATH] [8] "nDeriv(" without first graphing your function.

[2nd] [TRACE] (CALC F4) [7] for "∫f(x)dx" approximates the definite integral of your function on an interval. This can also be done directly using [MATH] [9] "fnInt (" without first graphing your function.

The *order of operations* used by most calculators evaluates operations in the following order: first powers and roots, then operations like [LN] and [LOG], then multiplication and division, then addition and subtraction—left to right within each level. For example, 5^2x means $(5^2)x$, *not* $5^{(2x)}$. Also, $1/x+1$ means $(1/x)+1$, *not* $1/(x+1)$. See the instruction manual that came with your calculator for further information.

Be Careful! Some calculators evaluate $1/2x$ as $(1/2)x$ and some as $1/(2x)$. When in doubt, use parentheses to clarify the expression.

Entering Data

To ensure that your calculator is properly prepared to accept data, turn off all statistical plots by pressing 2nd Y= [STAT PLOT F1] 4 ENTER

clear all data lists by pressing 2nd + [MEM] 4 ENTER

set up the statistical data editor by pressing STAT 5 ENTER

and finally clear all functions by pressing Y= and repeatedly using CLEAR and the down-arrow key ▾ until all functions are cleared. You will need to carry out the preceding steps only once if you are using your calculator only with this book.

To enter the data shown in the following table (which is taken from Example 9 on pages 14–15), first put the *x*-values into list **L1** by pressing STAT 1 "Edit..." and entering each value followed by an ENTER. Then put the *y*-values into list **L2** by pressing the right-arrow key ▸ and entering each value followed by an ENTER.

x	0	1	2	3
y	50	140	320	420

Turn **Plot1** "on" by pressing 2nd Y= [STAT PLOT F1] ENTER ENTER and use the arrow keys and ENTER (along with 2nd 1 [L1 Y] for list **L1** and 2nd 2 [L2 Z] for list **L2**) to set **Plot1** as shown below. Press ZOOM 9 "ZoomStat" to plot these data on an appropriate viewing window. You may then use TRACE and the arrow keys to check the values of your data points.

When you are finished studying the data you have entered, press [2nd] [Y=] (STAT PLOT F1) [ENTER] [>] [ENTER] to turn **Plot1** "off" and then [2nd] [+] [4] [ENTER] (MEM) to clear all lists.

Graphing Calculator Programs

Several of the Graphing Calculator Explorations show screens drawn by programs that carry out extended calculations and use advanced graphing techniques. Many such programs are available on the Internet at `http://www.ticalc.org/` and can be downloaded to your laptop and then transferred to your calculator using a USB cable. You can also use [2nd] [X,T,θ,n] (LINK) to "link" your calculator with a friend's and share programs (and other calculator data). See the instruction manual that came with your calculator for further information.

Appendix B

Algebra Review

CONTENTS

Complete Solutions to the Diagnostic Test (page xxi)

1. The inequality symbols $<$ and $>$ *point* to the smaller quantity, so $\frac{1}{2} < -3$ is *false* because the positive number $\frac{1}{2}$ is *not* smaller than negative number -3. Using the real number line to visualize the real numbers, smaller values are to the left and larger values are to the right:

See pages 4–5 for more about the real number line and inequalities.

 Be Careful Inequalities indicate the relative *left/right position* of values on the number line, while the absolute value is the *distance* from the number to zero. So $\left|\frac{1}{2}\right| < |-3|$ because $\frac{1}{2}$ is *closer to zero* than is -3. See pages 52–53 for more about the absolute value function.

2. The numbers in the set $\{x \mid -4 < x \le 5\}$ form an interval from -4 to 5 excluding the left endpoint -4 but including the right endpoint 5. We say that the interval is *open* on the left and *closed* on the right. Interval notation uses parentheses "(" and ")" to indicate open endpoints, and square brackets "[" and "]" to indicate closed endpoints. So this interval is written $(-4, 5]$. See pages 5–6 for more about intervals.

3. The slope of the line through $(x_1, y_1) = (6, -7)$ and $(x_2, y_2) = (9, 8)$ is

$$m = \frac{\Delta y}{\Delta x} = \frac{y_2 - y_1}{x_2 - x_1} = \frac{8 - (-7)}{9 - 6} = \frac{15}{3} = 5$$

See page 7 for more about slope, and pages B5–B6 later in this Appendix for more about order of operations and evaluating expressions.

4. The line $y = 3x + 4$ is written in slope-intercept form $y = mx + b$, so the slope is $m = 3$. Since

$$m = \frac{\Delta y}{\Delta x}$$

we have

$$3 = \frac{\Delta y}{2} \quad \text{giving} \quad \Delta y = 3 \cdot 2 = 6$$

See page 7 for more about slope, and page 9 for more about the slope-intercept form of the equation for a line.

5. The line $y = 2x - 1$ has slope $m = 2$ and y-intercept $(0, b) = (0, -1)$. Since the slope is positive, the line slopes upward, eliminating (d). Since $b = -1$ is negative, the y-intercept is below the x-axis, eliminating (c). Since the slope is the amount the line rises when x increases by 1, starting from $(0, -1)$ in either of (a) or (b) and moving *right* 1 and *up* 2, we see (a) is correct while (b) is not.

See pages 7 and 9 for more about slopes, y-intercepts, slope-intercept equations, and graphs of lines.

6. The statement is *true* because

$$\left(\frac{\sqrt{x}}{y}\right)^{-2} = \left(\frac{y}{\sqrt{x}}\right)^2 = \frac{y^2}{(\sqrt{x})^2} = \frac{y^2}{x}$$

where in the first step we used the fact that a negative exponent means the *reciprocal*. See pages 22–23 for more about exponents and roots.

7. Since $9x^2 - 6x - 1$ does not have a simple factorization, we solve the equation $9x^2 - 6x - 1 = 0$ by the Quadratic Formula

$$x = \frac{-b \pm \sqrt{b^2 - 4ac}}{2a}$$

Using $a = 9$, $b = -6$, and $c = -1$, we obtain

$$x = \frac{-(-6) \pm \sqrt{(-6)^2 - 4(9)(-1)}}{2(9)} = \frac{6 \pm \sqrt{72}}{18} = \frac{6 \pm 6\sqrt{2}}{18} = \frac{1 \pm \sqrt{2}}{3}$$

See page 39 for more about quadratic functions and the Quadratic Formula, and pages B5 and B7 later in this Appendix for more about order of operations and evaluating expressions.

8. Multiplying out to remove the parentheses,

$$x(8 - x) - (3x + 7) = 8x - x^2 - 3x - 7$$

Reordering and combining *like* terms (that is, terms with the same power of x) gives

$$-x^2 + (8x - 3x) - 7 = -x^2 + 5x - 7$$

See pages 48–49 for more about working with polynomials, and page B5 later in this Appendix for more about the distributive law.

9. The domain of the rational function $f(x) = \dfrac{x^2 - 3x + 2}{x^3 + x^2 - 6x}$ is the set of numbers for which the denominator $x^3 + x^2 - 6x$ is not zero, so we factor:

$$x^3 + x^2 - 6x = x(x^2 + x - 6) = x(x + 3)(x - 2)$$

Since the denominator is zero when x is 0, -3, or 2, the domain is

$$\{x \mid x \neq -3, x \neq 0, x \neq 2\}$$

See pages 49–50 for more about rational functions, and pages B11–B12 later in this Appendix for more about factoring.

 Be Careful When finding the domain of a rational function, do not begin by "simplifying" the function. If you first factor the numerator as $(x - 1)(x - 2)$ and cancel the $(x - 2)$'s, you would overlook the fact that $x = 2$ makes the original denominator zero. In fact, "canceling" the $(x - 2)$'s requires assuming that $x \neq 2$, or else you are dividing by zero, which is not allowed.

10. For $f(x) = x^2 - 5x$, the difference quotient is

$$\frac{f(x + h) - f(x)}{h} = \frac{(x + h)^2 - 5(x + h) - (x^2 - 5x)}{h}$$

$$= \frac{x^2 + 2hx + h^2 - 5x - 5h - x^2 + 5x}{h}$$

$$= \frac{2hx + h^2 - 5h}{h} = 2x + h - 5$$

See pages 57–58 for more about difference quotients, and pages B8 and B11 later in this Appendix for more about expanding and simplifying expressions.

Real Number Laws

The laws for real numbers are the basic facts of addition and multiplication, with multiplication sometimes indicated by a dot $\cdot$ for clarity but often just indicated by adjacent symbols, so that ab means $a \cdot b$.

Laws for Real Numbers

Let a, b, and c represent real numbers.

Law	for addition	for multiplication
closure	$a + b$ is a real number	ab is a real number
commutative	$a + b = b + a$	$ab = ba$
associative	$(a + b) + c = a + (b + c)$	$(ab)c = a(bc)$
identity element	$a + 0 = a$	$a \cdot 1 = a$
inverse	$a + (-a) = 0$	$a \cdot \frac{1}{a} = 1$
		(provided $a \neq 0$)

Zero-product law: If $ab = 0$, then either $a = 0$ or $b = 0$ or both.

Distributive law of multiplication over addition: $a(b + c) = ab + ac$

Brief Examples

For instance, choosing $a = 3$, $b = 4$, and $c = 5$, these laws simply state:

closure	$3 + 4 = 7$ is a real number	$3 \cdot 4 = 12$ is a real number
commutative	$3 + 4 = 4 + 3$	$3 \cdot 4 = 4 \cdot 3$
associative	$(3 + 4) + 5 = 3 + (4 + 5)$	$(3 \cdot 4) \cdot 5 = 3 \cdot (4 \cdot 5)$
	(that is, $7 + 5 = 3 + 9$)	(that is, $12 \cdot 5 = 3 \cdot 20$)
identity element	$3 + 0 = 3$	$3 \cdot 1 = 3$
inverse	$3 + (-3) = 0$	$3 \cdot \frac{1}{3} = 1$

Zero-product law: If $(x - 3)(x - 4) = 0$, then $x = 3$ or $x = 4$

Distributive law of multiplication over addition: $3(4 + 5) = 3 \cdot 4 + 3 \cdot 5$
(that is, $3 \cdot 9 = 12 + 15$)

Negation may also be thought of as multiplication by -1, so that $-a = (-1) \cdot a$. Then, $-(-a) = (-1)(-1) \cdot a = a$, which shows that $(-1)^2 = 1$, so *a negative times a negative is a positive.*

From these results, we have

$$a(b - c) = a(b + (-c)) = ab + a(-c) = ab - ac$$

so the Distributive Law also applies to *subtraction*:

$$a(b - c) = ab - ac$$

The number $\frac{1}{a}$ or $1/a$ is called the *reciprocal* of a, and *division* means multiplication by the reciprocal. Notice that the reciprocal of a reciprocal is just the original number: $1/\frac{1}{a} = a$ since $\frac{1}{a} \cdot a = 1$. For instance, the reciprocal of $\frac{1}{3}$ is $1/\frac{1}{3} = 3$. Therefore, for example, $4/\frac{1}{3} = 4 \cdot 3 = 12$.

Since multiplication by zero always results in 0, there is no number that when multiplied by zero gives 1. Therefore, zero itself does not have a reciprocal. This is why "division by zero" is impossible.

Try It Yourself! Find the zeros of each factored polynomial. [Remember: finding the zeros of a polynomial does not mean *plugging in zero*, but means finding the values of x that make the polynomial *equal* zero.]

$$(x + 3)(x - 4) \qquad (x - 1)(x + 1) \qquad x(x - 2) \qquad (x - 3)(x - 3)$$

$$x = -3,\ x = 4 \qquad\qquad x = -1,\ x = 1 \qquad\qquad x = 0,\ x = 2 \qquad\qquad x = 3$$

The Distributive Law explains how to *expand* or *multiply out* $a(b + c)$ to remove the parentheses, as well as how to *factor* or *pull out the common muiltiple* from $ab + ac$.

Try It Yourself! Multiply out each expression by using the Distributive Law.

$$x(y + z) \qquad 2(x + y) \qquad 3(x - 2) \qquad u(2 - v)$$

$$xy + xz \qquad\quad 2x + 2y \qquad\quad 3x - 6 \qquad\quad 2u - uv$$

Try It Yourself! Factor each expression by using the Distributive Law to pull out the common multiple.

$$pq + pr \qquad 3u - 3v \qquad 12x + 8y \qquad xy + yz$$

$$p(q + r) \qquad 3(u - v) \qquad 4(3x + 2y) \qquad y(x + z)$$

Simplifying an Expression: Using the Distributive Law

Expressions containing parentheses can usually be simplified by using the Distributive Law to multiply out any "$a(b + c)$" parts and then collecting *like* terms (that is, terms with x to the same power). For instance, to simplify the expression $2x(x - 3) - 4(5 - x)$, we use the Distributive Law for subtraction to proceed as follows:

$$2x(x - 3) - 4(5 - x) = 2xx - 2x3 - 4 \cdot 5 + 4x$$

$$= 2x^2 - 6x - 20 + 4x$$

$$= 2x^2 - 2x - 20$$

Try It Yourself! Simplify each expression.

$$x(3 + x) + 2(x + 1) \qquad 3u(4 - u) - 5(2u + 6) \qquad p(p + q) - q(p - q)$$

$$x^2 + 5x + 2 \qquad\qquad -3u^2 + 2u - 30 \qquad\qquad p^2 + q^2$$

Order of Operations

To simplify a complicated expression, you have to know which part to do first, then second, and so on. Calculations should be done in the following order.

Order of Operations	
First	Any expression in parentheses or other symbol of inclusion (brackets, division bars, and so on), starting with the innermost part.
Second	Multiplications (and divisions) indicated by an exponent.
Third	Multiplications (and divisions) in order from left to right.
Fourth	Additions (and subtractions) in order from left to right.

Brief Examples

$2 + 3 \cdot 4 = 14$	does *not* mean	$(2 + 3) \cdot 4 = 20$
$\dfrac{10 + 20}{5} = 6$	does *not* mean	$10 + 20/5 = 14$
$\sqrt{9 + 7} = 4$	does *not* mean	$\sqrt{9} + 7 = 10$
$\sqrt{3^2 + 4^2} = 5$	does *not* mean	$\sqrt{3^2} + \sqrt{4^2} = 3 + 4 = 7$
$(3 + 4)^2 = 49$	does *not* mean	$3^2 + 4^2 = 25$
$(2 \cdot 3)^2 = 36$	does *not* mean	$2 \cdot 3^2 = 18$

Sometimes reading written mathematics "out loud" can clarify what needs to be done. For instance, the first example above might be spoken aloud as "two plus [pause] three times four is fourteen but two plus three [pause] times four is twenty."

Try It Yourself! Read aloud the next two lines of the above Brief Examples. Be sure your pauses are sufficient to convey your meaning to someone listening over the telephone who is unable to see the written mathematics.

"The square root of [pause] nine plus seven is four but the square root of nine [pause] plus seven is ten."

"Ten plus twenty [pause] divided by 5 is 6 but ten plus twenty [pause] divided by five is fourteen."

The same order of operations applies to algebraic expressions, so that

$$9x^2 \text{ does } not \text{ mean } (9x)^2$$

$$\frac{1 + x}{x} \text{ cannot be "simplified" to } \frac{1 + \cancel{x}}{\cancel{x}} = \frac{1 + 1}{1} = 2$$

$$\sqrt{x^2 + y^2} \text{ cannot be "simplified" to } x + y$$

Expression Evaluation: The Slope Formula

The slope m of the line through the points (x_1, y_1) and (x_2, y_2) is

$$m = \frac{y_2 - y_1}{x_2 - x_1} = \frac{\Delta y}{\Delta x} \qquad \text{The change in } y \text{ over} \\ \text{the change in } x$$

The Order of Operations says that the differences in the numerator and denominator must be calculated *before* the division. For instance, the slope of the line through the points $(1, -2)$ and $(3, 4)$ is

$$m = \frac{4 - (-2)}{3 - 1} = \frac{6}{2} = 3$$

Try It Yourself! Find the slope of the line through each pair of points.

   $(1, 3)$ and $(2, 4)$ $(3, 4)$ and $(1, -2)$ $(4, 2)$ and $(1, 3)$ $(3, 1)$ and $(2, -4)$

   1 3 $-\dfrac{1}{3}$ 5

Expression Evaluation: The Quadratic Formula

Applying the Order of Operations to the Quadratic Formula

$$x = \frac{-b \pm \sqrt{b^2 - 4ac}}{2a}$$

we must find the numerator and denominator *before* dividing, and the two parts of the numerator separated by the $\pm$ sign must to be found before adding and subtracting them. And, before taking the square root, we must evaluate the "discriminant" $b^2 - 4ac$.

For instance, to solve

$$4x^2 + 11x - 3 = 0$$

$a^2 + bx + c = 0$ with
$a = 4,\ b = 11,\ c = -3$

we begin by replacing the letters a, b, c in the quadratic formula by the respective values $4, 11, -3$ and then following the Order of Operations:

$$x = \frac{-(11) \pm \sqrt{11^2 - 4(4)(-3)}}{2(4)}$$

$$= \frac{-11 \pm \sqrt{121 - (-48)}}{8} = \frac{-11 \pm \sqrt{121 + 48}}{8}$$

$$= \frac{-11 \pm \sqrt{169}}{8} = \frac{-11 \pm 13}{8}$$

$$= \frac{-11 + 13}{8} \quad \text{or} \quad \frac{-11 - 13}{8}$$

$$= \frac{2}{8} \quad \text{or} \quad \frac{-24}{8} = \frac{1}{4} \text{ or } -3$$

That is,

$$4x^2 + 11x - 3 = 0 \quad \text{when} \quad x = \frac{1}{4} \text{ or } x = -3$$

To check that we have calculated correctly, we evaluate $f(x) = 4x^2 + 11x - 3$ at these values:

$$f(\tfrac{1}{4}) = 4(\tfrac{1}{4})^2 + 11(\tfrac{1}{4}) - 3 = 4 \cdot \tfrac{1}{16} + 11 \cdot \tfrac{1}{4} - 3 = \tfrac{1}{4} + \tfrac{11}{4} - 3 = \tfrac{12}{4} - 3 = 0$$

and

$$f(-3) = 4(-3)^2 + 11(-3) - 3 = 4 \cdot 9 - 33 - 3 = 36 - 36 = 0$$

Both calculations give 0, verifying the two solutions from the Quadratic Formula.

Try It Yourself! Use the Quadratic Formula to solve each quadratic equation and check your answers by evaluating at your solutions.

$$2x^2 + 5x - 12 = 0 \qquad 3x^2 + 4x - 4 = 0 \qquad -5x^2 + 22x - 21 = 0$$

$x = \frac{3}{2}$ or $x = -4$ $\qquad$ $x = \frac{2}{3}$ or $x = -2$ $\qquad$ $x = \frac{7}{5}$ or $x = 3$

Simplifying Square Roots

Of course, not every discriminant will have a simple square root. You should simplify those that are more complicated by factoring out any perfect squares and moving them outside of the square root. For instance, since $63 = 9 \cdot 7$,

$$\sqrt{63} = \sqrt{9 \cdot 7} = \sqrt{9} \cdot \sqrt{7} = 3\sqrt{7}$$

When you move the 9 outside of the square root, be sure to write 3 and not 9.

Try It Yourself! Simplify each square root by removing all square factors.

$$\sqrt{18} \qquad \sqrt{75} \qquad \sqrt{32} \qquad \sqrt{500} \qquad \sqrt{72}$$
$$3\sqrt{2} \qquad 5\sqrt{3} \qquad 4\sqrt{2} \qquad 10\sqrt{5} \qquad 6\sqrt{2}$$

Try It Yourself! Use the Quadratic Formula to solve each quadratic equation and be sure to simplify any square roots in your final answers.

$$25x^2 - 20x + 1 = 0 \qquad 9x^2 - 24x + 14 = 0 \qquad 2x^2 + 10x + 9 = 0$$
$$x = \frac{2 \pm \sqrt{3}}{5} \qquad\qquad x = \frac{4 \pm \sqrt{2}}{3} \qquad\qquad x = \frac{-5 \pm \sqrt{7}}{2}$$

Expanding an Expression: Squaring a Binomial

A *binomial* is a sum of two terms, such as $x + y$, $a - b$, $z + 5$ or the number $\frac{1}{3} + \frac{1}{3}\sqrt{2}$. How do you *square* such an expression? In simplest terms, this amounts to finding

$$(x + y)^2$$

Be Careful $(x + y)^2$ is *not* the same as $x^2 + y^2$, as we saw on page B6.

To expand $(x + y)^2$, we start with the fact that squaring means *multiply by itself*, so that

$$(x + y)^2 = (x + y)(x + y)$$

and then work out the product using the *Distributive Law* repeatedly:

$$(x + y)(x + y) = x \cdot (x + y) + y \cdot (x + y) \qquad \text{Using the Distributive Law}$$
$$= x \cdot x + x \cdot y + y \cdot x + y \cdot y \qquad \text{And again twice more}$$
$$= x^2 + xy + xy + y^2 \qquad\qquad \text{Multiplication is communtative}$$
$$= x^2 + 2xy + y^2 \qquad\qquad \text{Adding like terms}$$

Therefore , we have the following result , which will be very useful:

Square of a Binomial

$$(x + y)^2 = x^2 + 2xy + y^2$$

Replacing y by $-y$, and using the fact that $(-y)^2 = y^2$, we also have the formula:

$$(x - y)^2 = x^2 - 2xy + y^2$$

Notice that only the *middle* term changes sign when squaring a difference rather than a sum.

Expression Evaluation: Using the Square of a Binomial

In the solution on pages B2–B3 of Problem 7 of the Diagnostic Test, we found that the equation $9x^2 - 6x - 1 = 0$ was satisfied by $x = \frac{1 \pm \sqrt{2}}{3} = \frac{1}{3} \pm \frac{\sqrt{2}}{3}$. We can now verify that these numbers are correct by evaluating $f(x) = 9x^2 - 6x - 1$ at each of them.

$$f\left(\tfrac{1}{3} + \tfrac{\sqrt{2}}{3}\right) = 9\left(\tfrac{1}{3} + \tfrac{\sqrt{2}}{3}\right)^2 - 6\left(\tfrac{1}{3} + \tfrac{\sqrt{2}}{3}\right) - 1 \qquad \text{$f(x)$ at $x = \tfrac{1}{3} + \tfrac{\sqrt{2}}{3}$}$$

$$= 9\left(\tfrac{1}{9} + \tfrac{2\sqrt{2}}{9} + \tfrac{2}{9}\right) - 6\left(\tfrac{1}{3} + \tfrac{\sqrt{2}}{3}\right) - 1 \qquad \text{Squaring}$$

$$= 1 + 2\sqrt{2} + 2 - 2 - 2\sqrt{2} - 1 \qquad \text{Multiplying out}$$

$$= 0 \qquad \text{It checks!}$$

and

$$f\left(\tfrac{1}{3} - \tfrac{\sqrt{2}}{3}\right) = 9\left(\tfrac{1}{3} - \tfrac{\sqrt{2}}{3}\right)^2 - 6\left(\tfrac{1}{3} - \tfrac{\sqrt{2}}{3}\right) - 1 \qquad \text{$f(x)$ at $x = \tfrac{1}{3} - \tfrac{\sqrt{2}}{3}$}$$

$$= 9\left(\tfrac{1}{9} - \tfrac{2\sqrt{2}}{9} + \tfrac{2}{9}\right) - 6\left(\tfrac{1}{3} - \tfrac{\sqrt{2}}{3}\right) - 1 \qquad \text{Squaring}$$

$$= 1 - 2\sqrt{2} + 2 - 2 + 2\sqrt{2} - 1 \qquad \text{Multiplying out}$$

$$= 0 \qquad \text{It checks!}$$

Both calculations give 0, verifying the two solutions from the Quadratic Formula.

Try It Yourself! Use the "Square of a Binomial" formulas to evaluate the quadratic function at the given values to verify that they are zeros of the function.

$$f(x) = 25x^2 - 20x + 1 \quad \text{at} \quad x = \tfrac{2}{5} + \tfrac{\sqrt{3}}{5} \quad \text{and} \quad x = \tfrac{2}{5} - \tfrac{\sqrt{3}}{5}$$

Expanding an Expression: More Binomial Products

Just as we used the Distributive Law several times to expand $(x + y)^2$, we can now expand the general binomial product $(x + y)(u + v)$:

$$(x + y)(u + v) = x(u + v) + y(u + v) \qquad \text{Using the Distributive Law}$$

$$= xu + xv + yu + yv \qquad \text{and twice more}$$

This gives the formula:

Product of Two Binomials

$$(x + y)(u + v) = xu + xv + yu + yv$$

Notice that the expression on the right consists of *all possible products of one term from the first parenthesis and one from the second*. You may remember the choices from the two parentheses by the letters FOIL, for *First, Inner, Outer, Last*.

Using this formula with $u = x, v = y$ gives our earlier formula for $(x + y)^2$. Other choices for u and v give the following results.

Some Useful Binomial Products

	Brief Examples
$(x + a)^2 = x^2 + 2ax + a^2$	$(x + 5)^2 = x^2 + 10x + 25$
$(x - a)^2 = x^2 - 2ax + a^2$	$(x - 5)^2 = x^2 - 10x + 25$
$(x + a)(x - a) = x^2 - a^2$	$(x + 5)(x - 5) = x^2 - 25$
$(x + a)(x + b) = x^2 + (a + b)x + ab$	$(x + 5)(x + 6) = x^2 + 11x + 30$
$(x + a)(x - b) = x^2 + (a - b)x - ab$	$(x + 5)(x - 6) = x^2 - x - 30$
	$(x - 5)(x + 6) = x^2 + x - 30$
$(x - a)(x - b) = x^2 - (a + b)x + ab$	$(x - 5)(x - 6) = x^2 - 11x + 30$

Try It Yourself! Find each binomial product.

$$(x + 2)(x - 2) \qquad (x + 2)(x + 3) \qquad (x + 3)(x - 1) \qquad (x - 3)(x + 1)$$

$$x^2 - 4 \qquad\qquad x^2 + 5x + 6 \qquad\quad x^2 + 2x - 3 \qquad\quad x^2 - 2x - 3$$

It is important to remember that x and a in these formulas may be replaced by *any* variables, numbers, or expressions. That is, $(x + y)^2 = x^2 + 2xy + y^2$ has the same meaning as $(p + q)^2 = p^2 + 2pq + q^2$. As another example, we can immediately state that

$$(\sqrt{x} + 3)^2 = x + 6\sqrt{x} + 9$$

$(p + q)^2 = p^2 + 2pq + q^2$
with $p = \sqrt{x}$ and $q = 3$

Similarly, our "Product of Two Binomials" formula on the previous page allows us to immediately state that

$$(x^3 + 4)(x^2 + 5) = x^5 + 5x^3 + 4x^2 + 20$$

Try It Yourself! Find each binomial product.

$$(\sqrt{x} + 2)(\sqrt{x} - 2) \qquad (\sqrt{x} + 2)(\sqrt{x} + 3) \qquad (x^4 + 3)(x^4 - 3) \qquad (x^4 + 3)(x^3 - 3)$$

$$x - 4 \qquad\qquad x + 5\sqrt{x} + 6 \qquad\qquad x^8 - 9 \qquad\quad x^7 - 3x^4 + 3x^3 - 9$$

Expanding an Expression: Higher Powers of a Binomial

The cube of a binomial is the product of three copies of the binomial, and we may multiply out two of them first and then multiply the result by the other:

$$(x + y)^3 = (x + y)(x + y)(x + y)$$

$$= (x + y)(x + y)^2$$

$$= (x + y)(x^2 + 2xy + y^2)$$

$$= x(x^2 + 2xy + y^2) + y(x^2 + 2xy + y^2)$$

$$= x^3 + 2x^2y + xy^2 + x^2y + 2xy^2 + y^3$$

$$= x^3 + 3x^2y + 3xy^2 + y^3 \qquad \text{Combining like terms}$$

Therefore, we have:

Cube of a Binomial

$$(x + y)^3 = x^3 + 3x^2y + 3xy^2 + y^3$$

Since this result is true in general, it is true in particular for whatever values we might choose for x and y. For instance, replacing y by h, we find

$$(x + h)^3 = x^3 + 3x^2h + 3xh^2 + h^3$$

Try It Yourself! Find each cube.

$$(x + 2)^3 \qquad\qquad (x^2 - 1)^3$$

$$x^3 + 6x^2 + 12x + 8 \qquad\qquad x^6 - 3x^4 + 3x^2 - 1$$

Since our derivation of $(x + y)^3$ was really just the expansion of $(x + y)$ times our formula for the square of a binomial, the formula for $(x + y)^4$ follows from multiplying the above answer by $(x + y)$.

$$(x + y)^4 = (x + y) \cdot (x + y)^3 = (x + y) \cdot (x^3 + 3x^2y + 3xy^2 + y^3)$$

Try It Yourself! Expand this product to show that

$$(x + y)^4 = x^4 + 4x^3y + 6x^2y^2 + 4xy^3 + y^4$$

Simplifying an Expression: Multiplying Out and Collecting Terms

Expressions containing parentheses can usually be *simplified* by expanding the various parts and then collecting "like" terms that contain the same power. For instance:

$$(x - 3)(x + 4) - 5(6 - x) = x^2 + \underbrace{(-3 + 4)}_{1}x \underbrace{- 12 - 30}_{-42} + 5x$$

$$= x^2 + x - 42 + 5x$$

$$= x^2 + 6x - 42$$

Try It Yourself! Simplify the expression.

$$(x + 4)(x - 4) + 3x(x - 2)$$

$$4x^2 - 6x - 16$$

Factoring

We have already carried out the simplest form of factoring by using the Distributive Law to *pull out* common terms, and then used factoring to find the zeros of a factored polynomial.

   Much of factoring amounts to *reversing* the steps we have done in multiplying binomials. From our review of binomial products, we know:

$$x^2 + 5x + 6 = (x + 2)(x + 3)$$
$$x^2 + 2x - 3 = (x + 3)(x - 1)$$
$$x^2 + 10x + 25 = (x + 5)(x + 5) = (x + 5)^2$$

If there is a common factor, take it out first, and then factor the remaining expression.

$$2x^2 - 10x + 12 = 2(x^2 - 5x + 6) = 2(x - 2)(x - 3)$$
$$x^5 - 2x^4 - 3x^3 = x^3(x^2 - 2x - 3) = x^3(x - 3)(x + 1)$$
$$3x^3 - 30x^2 + 75x = 3x(x^2 - 10x + 25) = 3x(x - 5)^2$$

Each of these factorizations can be verified by multiplying it back out and collecting like terms.

Try It Yourself! Factor each expression.

| $x^2 - 4$ | $5x^2 - 45$ | $x^2 - 7x + 12$ | $x^3 + 2x^2 - 8x$ |

(ʒ − x)(ʒ + x)   (ɛ + x)(ɛ − x)ϛ   (ᔭ − x)(ɛ − x)   (ʒ − x)(ᔭ + x)x

Try It Yourself! Solve each equation by factoring.

$3x^2 - 18x + 15 = 0$ $12x^2 - 96x + 144 = 0$ $3x^2 - 18x + 24 = 0$ $4x^3 - 12x^2 = 0$

ϛ = x ʻI = x   9 = x ʻʒ = x   ᔭ = x ʻʒ = x   ɛ = x ʻ0 = x

Be Careful Just because part of an expression is a multiplication does *not* mean that it is in factored form. For instance,

$$(x + 4)(x - 2) - 7$$

might look *mostly* factored, but it isn't, and −4 and 2 are *not* its zeros. The only way to find the zeros is to simplify this expression and then factor:

$$= x^2 - 2x + 4x - 8 - 7 = x^2 + 2x - 15 = (x + 5)(x - 3)$$

The zeros are −5 and 3.

Try It Yourself! Factor each expression and find the zeros.

$(x + 5)(x - 1) + 8$ $(x + 3)(x - 2) - 6$ $2x(x + 1) - 4$ $(x + 2)^2 - (8x + 9)$

I− ʻɛ−   ɛ ʻᔭ−   I ʻʒ−   ϛ ʻI−

(I + x)(ɛ + x)   (ɛ − x)(ᔭ + x)   (I − x)(ʒ + x)ʒ   (ϛ − x)(I + x)

The Myth of Cancellation

There is no mathematical operation called "cancellation" — the word is just a catch-all term for some other valid mathematical operation such as *adding two terms that sum to zero*, or perhaps *dividing top and bottom of a fraction by the same quantity*. For instance, "canceling the *x*" could have the following *correct* meanings:

$$x + y + 2z - x$$ Their sum is zero

$$x + y = 2 + x$$ Subtract x from both sides

$$\frac{xyz}{vwx}$$ Divide a common factor from numerator and denominator

$$\frac{y}{vx} = \frac{z}{wx}$$ Multiply both sides by x

The following are *incorrect* uses of cancelation:

$$x + y - (2z - x) \qquad x + y = 2 - x \qquad \frac{x + yz}{vwx} \qquad \frac{y}{vx} = \frac{xz}{w}$$

 Be Careful If you're going to "cancel" something, be sure you can give the *real* reason why it is *correct* to do so.

Try It Yourself! Identify each cancellation as "valid" or "invalid."

$$\frac{x(y + 2) - (x + y)}{4z} \qquad \frac{z + 2}{}$$

$$(w + 2)(w - 3) - 6 = (w - 3)(w + 4)$$

$$\frac{2h + h^2}{h} = 2 + h^2$$

all invalid

Simplifying an Expression: Fractions in Lowest Terms

A fraction is in *lowest terms* if the numerator and the denominator have no factor in common. When a fraction is *not* in lowest terms, the factorizations of both the numerator and the denominator will have a common term. For instance, the fraction

$$\frac{35}{10}$$

is not in lowest terms because the numerator is 5×7 and the denominator is 2×5 and thus they have a factor of 5 in common. To say that the 5's cancel is a *correct* use of that word, and we may write

$$\frac{35}{10} = \frac{5 \times 7}{2 \times 5} = \frac{\overset{1}{5} \times 7}{2 \times \underset{1}{5}} = \frac{7}{2}$$

Note that when we cancel in this sense, we are *dividing a quantity by itself*, which leaves a 1. We will sometimes write the 1s, as above, but when we don't, as below, the 1s are understood. This same kind of cancellation applies to fractions with algebraic expressions:

$$\frac{x^2 + 2x - 35}{x^2 - 7x + 10} = \frac{(x - 5)(x + 7)}{(x - 2)(x - 5)} = \frac{x + 7}{x - 2}$$

 Be Careful If the original fraction represents a *function*, then the cancellation of the common factor includes the *assumption* that the common term is *not* zero, since division by zero is not possible. For example, the functions on the extreme right and left above are not precisely equal, since the one on the right is defined at $x = 5$ and the one on the left isn't. However, the two functions are the same at all *other* x-values, so we will continue to write an equals sign between them, but remembering when necessary that they are not exactly equal.

Simplifying fractions with more complicated numerators and denominators in lowest terms proceeds in the same manner. For instance,

$$\frac{(4x^3 + 2x)(x^2 + 1) - (x^4 + x^2)(2x)}{(x^2 + 1)^2} = \frac{(4x^3 + 2x)(x^2 + 1) - x^2(x^2 + 1)(2x)}{(x^2 + 1)^2}$$

$$= \frac{\cancel{(x^2 + 1)}((4x^3 + 2x) - x^2(2x))}{\cancel{(x^2 + 1)}(x^2 + 1)}$$

$$= \frac{4x^3 + 2x - 2x^3}{x^2 + 1} = \frac{2x^3 + 2x}{x^2 + 1}$$

$$= \frac{2x\cancel{(x^2 + 1)}}{\cancel{x^2 + 1}} = 2x$$

Try It Yourself! Express each fraction in lowest terms.

$$\frac{x^2 + 3x + 2}{x^2 + 5x + 6} \qquad\qquad \frac{x^2 + 4x + 3}{(x + 1)(x + 6) + 6} \qquad\qquad \frac{(x + 3)^2 + x + 1}{(x - 1)^2 - 9}$$

$$\frac{x + 1}{x + 3} \qquad\qquad\qquad\qquad \frac{x + 1}{x + 4} \qquad\qquad\qquad\qquad \frac{x + 5}{x - 4}$$

Final Note: If you have understood the material presented in this Appendix and successfully caried out the *Try It Yourself!* exercises, you are ready to start learning calculus.

Normal Probabilities from Tables

Introduction

This Appendix is for readers of Section 11.4 who do not use graphing calculators. You should read Section 11.4 through "Normal Random Variables," which ends on the bottom of page 732, then read this Appendix, and then return to the top of page 736.

Standard Normal Random Variables

A normal random variable with mean $\mu = 0$ and standard deviation $\sigma = 1$ is called a *standard* normal random variable. Its probability density function

$$f(x) = \frac{1}{\sqrt{2\pi}} e^{-\frac{1}{2}x^2} \qquad \text{for } -\infty < x < \infty$$

Probability density function for a standard normal random variable

was graphed on page 732. Throughout this section, the letter Z will always stand for a *standard normal random variable*. The probability that a standard normal random variable Z lies between two numbers is the area under the probability density function between the two numbers.

Probabilities for Standard Normal Random Variables

For a standard normal random variable Z,

$$P(a \leq Z \leq b) = \int_a^b \frac{1}{\sqrt{2\pi}} e^{-\frac{1}{2}x^2}\, dx$$

The integral of the normal probability density function cannot be expressed in terms of elementary functions, so we cannot find the area by integration. However, we can *approximate* the area by *numerical* integration as in Section 6.4, and then list the results in a table. One such table, on page C7 at the end of this Appendix, gives the areas under this curve from 0 to any positive number x. The first two examples show how to use the table.

EXAMPLE 1   FINDING A PROBABILITY FOR A STANDARD NORMAL RANDOM VARIABLE

Find $P(0 \leq Z \leq 1.24)$.

Solution

The table on page C7 gives the probability that a standard normal random variable Z is between 0 and any higher number. For the given number 1.24, we find the *row* headed 1.2 and the *column* headed .04 (the second decimal place), which intersect at the table value of .3925:

From the table at the end of the Appendix: the row headed 1.2 and the column headed .04 lead to the table value .3925

Therefore:

$$P(0 \leq Z \leq 1.24) = .3925$$

Also written 0.3925

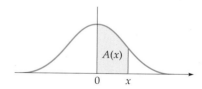

PRACTICE PROBLEM 1

Find $P(0 \leq Z \leq 0.52)$

Solution on page C6 >

We will use the notation

$$A(x) = \left(\begin{array}{c} \text{Area under the standard} \\ \text{normal probability density} \\ \text{function from 0 to } x \end{array} \right)$$

That is, $A(x)$ is the number found in the standard normal table in the row and column determined by x. For instance, in Example 1 we found $A(1.24) = 0.3925$, and in Practice Problem 1 you found $A(0.52) = 0.1985$.

    The probabilitiy that Z is between two numbers can be found by adding and subtracting values from the table, using the symmetry of the curve about $x = 0$ and the fact that the area under each half of the curve is 0.5 (since the total area is 1). The following Example illustrates the technique, with the calculation shown graphically as well.

EXAMPLE 2   FINDING PROBABILITIES FROM A TABLE

Find: **a.** $P(-0.75 \leq Z \leq 0)$ **b.** $P(Z \leq 0.65)$

   c. $P(-1.2 \leq Z \leq 0.68)$ **d.** $P(0.65 \leq Z \leq 1.45)$

   e. $P(Z > 0.95)$

Solution

a. $P(-0.75 \leq Z \leq 0)$ $= A(0.75) = 0.2734$ From the table on page C7

  $=$ Using the symmetry of the curve about $x = 0$

b. $P(Z \leq 0.65)$ $= \frac{1}{2} + A(0.65) = 0.5 + 0.2422 = 0.7422$

  $=$ Using the fact that the area to the left of $x = 0$ is $\frac{1}{2}$

c. $P(-1.2 \leq Z \leq 0.68)$ $= A(1.2) + A(0.68) = 0.3849 + 0.2518 = 0.6367$

  $=$ Using symmetry

d. $P(0.65 \leq Z \leq 1.45)$ $= A(1.45) - A(0.65) = 0.4265 - 0.2422 = 0.1843$

  found from   Area up to 1.45 minus area up to 0.65

e. $P(Z > 0.95)$ $= \frac{1}{2} - A(0.95) = 0.5 - 0.3289 = 0.1711$

found from   Area beyond 0.95 is $\frac{1}{2}$ minus area from 0 to 0.95

PRACTICE PROBLEM 2

Find: **a.** $P(0.5 \leq Z \leq 1.25)$

   b. $P(-2.1 \leq Z \leq 0.75)$ Solutions on page C6 >

Probabilities for Nonstandard Random Variables

How do we find probabilities for a *non*standard normal random variable with mean μ and standard deviation σ? Instead of developing tables for all possible means and standard deviations, we use a process called *standardization* that changes a nonstandard random variable into one with mean 0 and standard deviation 1, so we can use the standard normal table. Standardizing is achieved by defining a new random variable by subtracting the mean μ and dividing by the standard deviation σ.

Standardizing Normal Random Variables

A normal random variable X with mean μ and standard deviation σ is *standardized* by defining

$$Z = \frac{X - \mu}{\sigma}$$

Subtracting the mean and dividing by the standard deviation

Z is then a normal random variable with mean 0 and standard deviation 1.

The values of a normal random variable are said to be *normally distributed*.

EXAMPLE 3   STANDARDIZING TO FIND A PROBABILITY

The heights of American men are approximately normally distributed, with mean $\mu = 70$ inches and standard deviation $\sigma = 3.1$ inches. Find the proportion of American men who are less than 6 feet tall.

Source: G. Salvendy, *Handbook of Human Factors and Ergonomics,* 3rd ed.

Solution

If $X = $ height, we want $P(X < 72)$
Standardizing:

72 inches = 6 feet

$$X < 72 \quad \text{is equivalent to} \quad \underbrace{\frac{X - \mu}{\sigma}}_{Z} < \underbrace{\frac{72 - 70}{3.1}}_{0.65}$$

Subtracting mean $\mu = 70$ and dividing by $\sigma = 3.1$

Therefore:

$$P(X < 72) = P(Z < 0.65) = 0.7422$$

$P(Z < 0.65) = 0.7422$ from Example 2b

Normal   Standard normal

That is, approximately 74% of American men are under 6 feet tall.

Such calculations are used for designing automobiles and other consumer products.

PRACTICE PROBLEM 3

The weights of American women are approximately normally distributed, with mean 151.8 pounds and standard deviation 38.8 pounds. Find the proportion of women who weigh more than 140 pounds.

Source: G. Salvendy, *Handbook of Human Factors and Ergonomics,* 3rd ed.

Solution on next page >

EXAMPLE 4	FINDING THE PROBABILITY OF ACCEPTABLE CHOLESTEROL LEVELS

Cholesterol is a fat-like substance that, in the bloodstream, can increase the risk of heart disease. Cholesterol levels below 200 (milligrams per deciliter of blood) are generally considered acceptable. A survey by the U.S. Department of Health found that cholesterol levels for one age group were approximately normally distributed with mean $\mu = 222$ and standard deviation $\sigma = 28$. What proportion of people in the study had acceptable cholesterol levels?

Source: Rogosin Institute

Solution

Letting $X = \left(\begin{matrix} \text{Cholesterol} \\ \text{level} \end{matrix}\right)$, we want $P(X \le 200)$. The inequality

$$X \le 200$$

with X standardized becomes

$$\underbrace{\frac{X - \mu}{\sigma}}_{Z} \le \underbrace{\frac{200 - 222}{28}}_{-0.79} \qquad \begin{array}{l}\text{Using } \mu = 222 \\ \text{and } \sigma = 28\end{array}$$

Therefore:

$$\underbrace{P(X \le 200)}_{\substack{\text{Normal}}} = \underbrace{P(Z \le -0.79)}_{\substack{\text{Standard} \\ \text{normal}}} = \underbrace{\tfrac{1}{2} - A(0.79)}_{\substack{\text{Using the} \\ \text{graph below,} \\ \text{as in} \\ \text{Example 2}}} = \underbrace{0.5000 - 0.2852}_{\substack{\text{From the} \\ \text{standard} \\ \text{normal} \\ \text{table}}} = \underbrace{0.2148}_{\substack{\text{Answer}}}$$

Therefore, only about 21% of the people in the study had acceptable cholesterol levels.

Inverse Probabilities for Normal Random Variables

The standard normal table can also be used "in reverse" to find the values that result in a given probability.

EXAMPLE 5   **FINDING THE HEIGHT FOR A GIVEN PROBABILITY**

Women's heights are approximately normally distributed, with mean $\mu = 64.6$ inches and standard deviation $\sigma = 2.9$ inches. Find the height that marks the tallest 10%.

Source: G. Salvendy, Handbook of Human Factors and Ergonomics, 3rd ed.

Solution

We begin by finding the value of the *standard* normal random variable Z that marks the top $10\% = 0.10$ (or, equivalently, the number z with area 0.40 between 0 and z, as shown on the right).

To find this z-value, we look for .40 *inside* the standard normal table. The value that comes closest to .40 is .3997, and the "outside" value that gives it is 1.28. Therefore:

$$z = 1.28$$

z-value of the top 10% (using z to represent a value of the random variable Z)

$$\frac{x - 64.6}{2.9} = 1.28$$

Replacing z by $\frac{x - \mu}{\sigma}$ with the given μ and σ

$$x - 64.6 = 3.712$$

Multiplying by 2.9

$$x = 68.312$$

Adding 64.6

Therefore, a height of about 68.3 inches, or 5 feet 8 inches, marks the tallest 10% of women.

You should now return to Section 11.4 at the top of page 736.

Solutions TO PRACTICE PROBLEMS

1. $P(0 \leq Z \leq 0.52) = 0.1985$

2. a. $P(0.5 \leq Z \leq 1.25) = A(1.25) - A(0.5) = 0.3944 - 0.1915 = 0.2029$

b. $P(-2.1 \leq Z \leq 0.75) = A(2.1) + A(0.75) = 0.4821 + 0.2734 = 0.7555$

3. $P(X > 140) = P\left(\frac{X - \mu}{\sigma} > \frac{140 - 151.8}{38.8}\right) = P(Z > -0.30) = \frac{1}{2} + A(0.30)$

$$= 0.5 + 0.1179 = 0.6179$$

About 62% of women weigh more than 140 pounds.

**Area Under the Standard
Normal Density from 0 to x**

x	.00	.01	.02	.03	.04	.05	.06	.07	.08	.09
0.0	.0000	.0040	.0080	.0120	.0160	.0199	.0239	.0279	.0319	.0359
0.1	.0398	.0438	.0478	.0517	.0557	.0596	.0636	.0675	.0714	.0754
0.2	.0793	.0832	.0871	.0910	.0948	.0987	.1026	.1064	.1103	.1141
0.3	.1179	.1217	.1255	.1293	.1331	.1368	.1406	.1443	.1480	.1517
0.4	.1554	.1591	.1628	.1664	.1700	.1736	.1772	.1808	.1844	.1879
0.5	.1915	.1950	.1985	.2019	.2054	.2088	.2123	.2157	.2190	.2224
0.6	.2258	.2291	.2324	.2357	.2389	.2422	.2454	.2486	.2518	.2549
0.7	.2580	.2612	.2642	.2673	.2704	.2734	.2764	.2794	.2823	.2852
0.8	.2881	.2910	.2939	.2967	.2996	.3023	.3051	.3078	.3106	.3133
0.9	.3159	.3186	.3212	.3238	.3264	.3289	.3315	.3340	.3365	.3389
1.0	.3413	.3438	.3461	.3485	.3508	.3531	.3554	.3577	.3599	.3621
1.1	.3643	.3665	.3686	.3708	.3729	.3749	.3770	.3790	.3810	.3820
1.2	.3849	.3869	.3888	.3907	.3925	.3944	.3962	.3980	.3997	.4015
1.3	.4032	.4049	.4066	.4082	.4099	.4115	.4131	.4147	.4162	.4177
1.4	.4192	.4207	.4222	.4236	.4251	.4265	.4279	.4292	.4306	.4319
1.5	.4332	.4345	.4357	.4370	.4382	.4394	.4406	.4418	.4429	.4441
1.6	.4452	.4463	.4474	.4484	.4495	.4505	.4515	.4525	.4535	.4545
1.7	.4554	.4564	.4573	.4582	.4591	.4599	.4608	.4616	.4625	.4633
1.8	.4641	.4649	.4656	.4664	.4671	.4678	.4686	.4693	.4699	.4706
1.9	.4713	.4719	.4726	.4732	.4738	.4744	.4750	.4756	.4761	.4767
2.0	.4772	.4778	.4783	.4788	.4793	.4798	.4803	.4808	.4812	.4817
2.1	.4821	.4826	.4830	.4834	.4838	.4842	.4846	.4850	.4854	.4857
2.2	.4861	.4864	.4868	.4871	.4875	.4878	.4881	.4884	.4887	.4890
2.3	.4893	.4896	.4898	.4901	.4904	.4906	.4909	.4911	.4913	.4916
2.4	.4918	.4920	.4922	.4925	.4927	.4929	.4931	.4932	.4934	.4936
2.5	.4938	.4940	.4941	.4943	.4945	.4946	.4948	.4949	.4951	.4952
2.6	.4953	.4955	.4956	.4957	.4959	.4960	.4961	.4962	.4963	.4964
2.7	.4965	.4966	.4967	.4968	.4969	.4970	.4971	.4972	.4973	.4974
2.8	.4974	.4975	.4976	.4977	.4977	.4978	.4979	.4979	.4980	.4981
2.9	.4981	.4982	.4982	.4983	.4984	.4984	.4985	.4985	.4986	.4986
3.0	.4987	.4987	.4987	.4988	.4988	.4989	.4989	.4989	.4990	.4990
3.1	.4990	.4991	.4991	.4991	.4992	.4992	.4992	.4992	.4993	.4993
3.2	.4993	.4993	.4994	.4994	.4994	.4994	.4994	.4995	.4995	.4995
3.3	.4995	.4995	.4995	.4996	.4996	.4996	.4996	.4996	.4996	.4997
3.4	.4997	.4997	.4997	.4997	.4997	.4997	.4997	.4997	.4997	.4998
3.5	.4998	.4998	.4998	.4998	.4998	.4998	.4998	.4998	.4998	.4998

Answers to Selected Exercises

Exercises 1.1 page 16

1. $\{x \mid 0 \le x < 6\}$ **3.** $\{x \mid x \le 2\}$ **5. a.** Increase by 15 units.

b. Decrease by 10 units. **7.** $m = -2$ **9.** $m = \frac{1}{3}$ **11.** $m = 0$ **13.** Slope is undefined.

15. $m = 3$, $(0, -4)$ **17.** $m = -\frac{1}{2}$, $(0, 0)$ **19.** $m = 0$, $(0, 4)$

  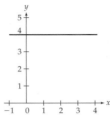

21. Slope and y-intercept do not exist. **23.** $m = \frac{2}{3}$, $(0, -4)$ **25.** $m = -1$, $(0, 0)$

27. $m = 1$, $(0, 0)$ **29.** $m = \frac{1}{3}$, $\left(0, \frac{2}{3}\right)$ **31.** $m = \frac{2}{3}$, $(0, -1)$

33. $y = -2.25x + 3$ **35.** $y = 5x + 3$ **37.** $y = -4$ **39.** $x = 1.5$ **41.** $y = -2x + 13$

43. $y = -1$ **45. a.** $y = \frac{3}{4}x - 7$ **b.** $y = -\frac{4}{3}x + 18$ **47.** $y = -2x + 1$ **49.** $y = \frac{3}{2}x - 2$

51. $y = -x + 5$, $y = -x - 5$, $y = x + 5$, $y = x - 5$

53. Substituting $(0, b)$ into $y - y_1 = m(x - x_1)$ gives $y - b = m(x - 0)$, or $y = mx + b$.

55. a.

on $[-5, 5]$ by $[-5, 5]$

b.

on $[-5, 5]$ by $[-5, 5]$

D1

57. Low: [0, 8); average: [8, 20); high: [20, 40); critical [40, ∞)

59. a. 3 minutes 34.72 seconds   **b.** In about 2033

61. a. $y = 38x + 32$ **b.** Sales are increasing by 38 million units per year.   **c.** 412 million units

63. a. $y = \frac{9}{5}x + 32$ **b.** 68°

65. a. $V = 50,000 - 2200t$ **67. a.** $10L + 5K = 1000$

b. \$39,000

c.

on [0, 20] by [0, 50,000]

b. Every pair gives 1000.

69. a.

71. a.

b. Men: 30.3 years; women: 27.8 years   **b.** 28%   **c.** 18%

c. Men: 32.1 years; women: 29.2 years

73. a. $y = -0.094x + 1.58$ **b.** Cigarette consumption is declining by 94 cigarettes (from .094 thousand, so about 5 packs) per person per year.   **c.** About 360 cigarettes

75. a. $y = 2.13x + 65.35$ **b.** The male life expectancy is increasing by 2.13 years per decade, which is 0.213 year (or about 2.6 months) per year.   **c.** 79.2 years

77. a. $y = -0.864x + 75.46$ **b.** Future longevity decreases by 0.864 year (or about 10.4 months) per year.
c. 53.9 years   **d.** It would not make sense to use the regression line to predict future longevity at age 90 because the line predicts −2.3 years of life remaining.   **79.** False. Infinity is not a number.

81. $m = \dfrac{y_2 - y_1}{x_2 - x_1}$ for any two points (x_1, y_1) and (x_2, y_2) on the line   or $m = \left(\begin{array}{c}\text{Amount that the line rises}\\\text{when } x \text{ increases by 1}\end{array}\right)$

83. False. For a vertical line the slope is *undefined*.   **85.** True   **87.** False. It should be $\dfrac{y_2 - y_1}{x_2 - x_1}$.
89. $\frac{4}{3}$, or $-\frac{4}{3}$ if the ladder slopes downward
91. $(-b/m, 0)$, $m \neq 0$ **93.** Smaller populations increase toward the carrying capacity.

Exercises 1.2 page 29

1. 64   **3.** $\frac{1}{16}$ **5.** 8   **7.** $\frac{8}{5}$ **9.** $\frac{1}{32}$ **11.** $\frac{8}{27}$ **13.** 1   **15.** $\frac{4}{9}$ **17.** 5   **19.** 125   **21.** 8

23. 4   **25.** −32   **27.** $\frac{125}{216}$ **29.** $\frac{9}{25}$ **31.** $\frac{1}{4}$ **33.** $\frac{1}{2}$ **35.** $\frac{1}{8}$ **37.** $\frac{1}{4}$ **39.** $-\frac{1}{2}$ **41.** $\frac{1}{4}$ **43.** $\frac{4}{5}$

45. $\frac{64}{125}$ **47.** −243   **49.** 2.14   **51.** 274.37   **53.** 0.977 (rounded)   **55.** 2.720 (rounded)

57. $4x^{-5}$ **59.** $2x^{-4/3}$ **61.** $3x^{-3/2}$ **63.** $3x^{-2}$ **65.** $5x^{3/2}$ **67.** $4x^{-4/3}$ **69.** $3x^{-1/2}$ **71.** x^{10} **73.** z^{27} **75.** x^8

77. $27x^6y^{15}z^3$ **79.** w^5 **81.** y^5/x **83.** $27y^4$ **85.** $u^2v^2w^2$ **87.** 25.6 feet   **89.** Costs will be multiplied by 2.3.

91. a. 2.77 percent   **b.** 0.12 percent   **93.** 125 beats per minute   **95.** About 42.6 thousand work-hours, or 42,600 work-hours, rounded to the nearest hundred hours.   **97. a.** About 32 times more ground motion **b.** About 12.6 times more ground motion   **99.** $K = 200$ **101.** About 312 mph

103.

$x \approx 18.2$.   Therefore, the land area must be increased by a factor of more than 18 to double the number of species.

on [0, 100] by [0, 4]

105. 60 mph   **107. a.** $y = 79.9x^{0.138}$ (rounded)   **b.** \$110 billion   **109. a.** $y = 41.6x^{0.0388}$ (rounded)   **b.** \$45.7 million

111. 3, since $\sqrt{9}$ means the *principal* square root. (To get ± 3 you would have to write $\pm \sqrt{9}$.)

113. False. $\dfrac{2^6}{2^2} = \dfrac{64}{4} = 16$, while $2^{6/2} = 2^3 = 8$. (The correct statement is $\dfrac{x^m}{x^n} = x^{m-n}$.)

115. All nonnegative values of x.   **117.** All values of x except 0.

Exercises 1.3 page 44

1. Yes   **3.** No   **5.** No   **7.** No   **9.** Domain: $\{x \mid x \le 0 \text{ or } x \ge 1\}$; Range: $\{y \mid y \ge -1\}$

11. a. $f(10) = 3$ **b.** $\{x \mid x \ge 1\}$ **c.** $\{y \mid y \ge 0\}$ **13. a.** $h(-5) = -1$ **b.** $\{z \mid z \ne -4\}$ **c.** $\{y \mid y \ne 0\}$

15. a. $h(81) = 3$ **b.** $\{x \mid x \ge 0\}$ **c.** $\{y \mid y \ge 0\}$ **17. a.** $f(-8) = 4$ **b.** $\mathbb{R}$ **c.** $\{y \mid y \ge 0\}$

19. a. $f(0) = 2$ **b.** $\{x \mid -2 \le x \le 2\}$ **c.** $\{y \mid 0 \le y \le 2\}$

21. a. $f(-25) = 5$ **b.** $\{x \mid x \le 0\}$ **c.** $\{y \mid y \ge 0\}$

23.   **25.**   **27.**   **29.**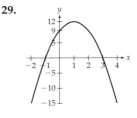

31. a. (20, 100)   **33. a.** (−40, −200)

b.   **b.**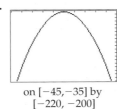

on [15, 25] by [100, 120]   on [−45, −35] by [−220, −200]

35. $x = 7$, $x = -1$ **37.** $x = 3$, $x = -5$
39. $x = 4$, $x = 5$ **41.** $x = 0$, $x = 10$
43. $x = 5$, $x = -5$ **45.** $x = -3$
47. $x = 1$, $x = 2$ **49.** No solutions
51. No solutions   **53.** $x = -4$, $x = 5$
55. $x = 4$, $x = 5$ **57.** $x = -3$
59. No (real) solutions

61. $x = 1.14$, $x = -2.64$ **63. a.** The slopes are all 2, but the y-intercepts differ.   **b.** $y = 2x - 8$

65. $C(x) = 4x + 20$ **67.** $P(x) = 15x + 500$ **69. a.** 17.7 lbs/in.^{2} **b.** 15,765 lbs/in.^{2} **71.** 132 feet

73. a. 400 cells   **b.** 5200 cells   **75.** About 230 mph   **77.** 2.92 seconds

79. a. Break even at 40 units and at 200 units.   **b.** Profit maximized at 120 units. Max profit is \$12,800 per week.

81. a. Break even at 20 units and at 80 units.   **b.** Profit maximized at 50 units. Max profit is \$1800 per day.

83. $v = \dfrac{c}{w + a} - b$

85. a. 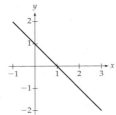   **87. a.** $f(x) = 100x - x^2$ or $-x^2 + 100x$ **b.** \$50
89. a. $y = 0.434x^2 - 3.26x + 11.6$ **b.** 15.7%

b. About 40% (from 39.8)
c. About 60% (from 60.4)

91. Yes—many parabolas cross the x-axis twice. No—that would violate the vertical line test.

93. $f(5) = 9$ (halfway between 7 and 11)   **95.** x is in prendles and y is in blargs.

97. No—that would violate the vertical line test. [*Note: A parabola* is a geometric shape and so *may* open sideways, but a quadratic function, being a *function*, must pass the vertical line test.]

Exercises 1.4 page 61

1. Domain: $\{x \mid x > 0 \text{ or } x < -4\}$; Range: $\{y \mid y > 0 \text{ or } y < -2\}$ **3. a.** $f(-3) = 1$ **b.** $\{x \mid x \neq -4\}$
c. $\{y \mid y \neq 0\}$ **5. a.** $f(-1) = -\frac{1}{2}$ **b.** $\{x \mid x \neq 1\}$ **c.** $\{y \mid y \leq 0 \text{ or } y \geq 4\}$ **7. a.** $f(2) = 1$
b. $\{x \mid x \neq 0, x \neq -4\}$ **c.** $\{y \mid y > 0 \text{ or } y \leq -3\}$ **9. a.** $g\left(-\frac{1}{2}\right) = \frac{1}{2}$ **b.** $\mathbb{R}$ **c.** $\{y \mid y > 0\}$

11. $x = 0, \quad x = -3, \quad x = 1$ **13.** $x = 0, \quad x = 2, \quad x = -2$ **15.** $x = 0, \quad x = 3$ **17.** $x = 0, \quad x = 5$

19. $x = 0, \quad x = 3$ **21.** $x = 0, \quad x = 1$ **23.** $x \approx -1.79, \quad x = 0, \quad x \approx 2.79$

25.   **27.**   **29.**   **31.**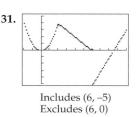

Includes $(6, -5)$
Excludes $(6, 0)$

33. Polynomial   **35.** Exponential   **37.** Polynomial   **39.** Rational   **41.** Piecewise linear
43. Polynomial   **45.** None (not a polynomial because of the fractional exponent)

47. a. y_4 **b.** y_1 **49. a.** $(7x - 1)^5$ **b.** $7x^5 - 1$ **c.** x^{25} **51. a.** $\dfrac{1}{x^2 + 1}$ **b.** $\left(\dfrac{1}{x}\right)^2 + 1$ **c.** x

c.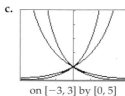

on $[-3, 3]$ by $[0, 5]$

53. a. $\sqrt{x^3 - x^2} - 1$ **b.** $(\sqrt{x} - 1)^3 - (\sqrt{x} - 1)^2$ **c.** $\sqrt{\sqrt{x} - 1} - 1$
55. a. $\left(\dfrac{x^3 - 1}{x^3 + 1}\right)^2 - \dfrac{x^3 - 1}{x^3 + 1}$ **b.** $\dfrac{(x^2 - x)^3 - 1}{(x^2 - x)^3 + 1}$ **c.** $(x^2 - x)^2 - (x^2 - x)$

d. $(0, 1)$, because $a^0 = 1$ for any constant $a \neq 0$.

57. a. $f(g(x)) = x$ **b.** $g(f(x)) = x$ **59.** $5x^2 + 10xh + 5h^2$ or $5(x^2 + 2xh + h^2)$

61. $2x^2 + 4xh + 2h^2 - 5x - 5h + 1$ **63.** $10x + 5h$ or $5(2x + h)$ **65.** $4x + 2h - 5$ **67.** $14x + 7h - 3$

69. $3x^2 + 3xh + h^2$ **71.** $\dfrac{-2}{(x + h)x}$ **73.** $\dfrac{-2x - h}{x^2(x + h)^2}$ or $\dfrac{-2x - h}{x^2(x^2 + 2xh + h^2)}$ or $\dfrac{-2x - h}{x^4 + 2x^3h + x^2h^2}$

75. a. 2.70481   **b.** 2.71815   **c.** 2.71828   **d.** Yes, 2.71828

77. Shifted left 3 units and up 6 units   **79.** About 680 million people

on $[-10, 10]$ by $[-10, 10]$

81. a. $300   **b.** $500   **c.** $2000

d.

83. a. $f\left(\frac{2}{3}\right) = 7, \quad f\left(1\frac{1}{3}\right) = 14, \quad f(4) = 29,$ and $f(10) = 53$ **85.** $(8, 3)$

b.

87. $R(v(t)) = 2(60 + 3t)^{0.3}, \quad R(v(10)) \approx 7.714$ million dollars

89. a. About 1 million cells   **b.** No

91. a. $y = 9 \cdot 0.547^x$ **b.** About 0.24 weeks (using $x = 6$), or less than 2 days

93. One will have "missing points" at the excluded x-values.

95. A slope of 1 corresponds to a tax rate of 100%, which means that every dollar earned would be paid as the tax.

97. $f(f(x)) = x + 2a$ **99.** $f(x + 10)$ is shifted 10 units to the *left*.

101. False. $f(x + h) = (x + h)^2 = x^2 + 2xh + h^2$

103. a.

$y = \text{INT}(x)$ on
$[-5, 5]$ by $[-5, 5]$

Note that each line segment in this graph includes its left endpoint but excludes its right endpoint, and so should be drawn like •——∘.

 b. Domain: $\mathbb{R}$; range: { ... , $-3, -2, -1, 0, 1, 2, 3, ...$ }—that is, the set of all integers

105. a. $f(g(x)) = acx + ad + b$ **b.** Yes

Chapter 1 Review Exercises and Chapter Test page 66

1. $\{x \mid 2 < x \le 5\}$ **2.** $\{x \mid -2 \le x < 0\}$

3. $\{x \mid x \ge 100\}$ **4.** $\{x \mid x \le 6\}$

5. Hurricane: $[74, \infty)$; storm: $[55, 74)$; gale: $[38, 55)$; small craft warning: $[21, 38)$

6. a. $(0, \infty)$ **b.** $(-\infty, 0)$ **c.** $[0, \infty)$ **d.** $(-\infty, 0]$ **7.** $y = 2x - 5$ **8.** $y = -3x + 3$ **9.** $x = 2$

10. $y = 3$ **11.** $y = -2x + 1$ **12.** $y = 2x - 13$ **13.** $y = 2x - 1$ **14.** $y = -\frac{1}{2}x + 1$

15. a. $V = 25{,}000 - 3000t$ **b.** \$13,000   **16. a.** $V = 78{,}000 - 5000t$ **b.** \$38,000

17. a. $y = 5.04x - 2.45$ **b.** The number is increasing by about 5 each year.   **c.** About 58 thousand

18. 36   **19.** $\frac{3}{4}$ **20.** 8   **21.** 10   **22.** $\frac{1}{27}$ **23.** $\frac{1}{1000}$ **24.** $\frac{9}{4}$ **25.** $\frac{64}{27}$ **26.** 13.97   **27.** 112.32

28. a. About 42 lbs   **b.** About 1630 lbs   **29. a.** About 127 lbs   **b.** About 7185 lbs

30. a. $y = 4.55x^{0.643}$ (rounded)   **b.** About \$22.5 billion

31. a. $f(11) = 2$ **b.** $\{x \mid x \ge 7\}$ **c.** $\{y \mid y \ge 0\}$ **32. a.** $g(-1) = \frac{1}{2}$ **b.** $\{t \mid t \ne -3\}$ **c.** $\{y \mid y \ne 0\}$

33. a. $h(16) = \frac{1}{8}$ **b.** $\{w \mid w > 0\}$ **c.** $\{y \mid y > 0\}$ **34. a.** $w(8) = \frac{1}{16}$ **b.** $\{z \mid z \ne 0\}$ **c.** $\{y \mid y > 0\}$

35. Yes   **36.** No

37. **38.** **39.** **40.**

41. $x = 0,\quad x = -3$ **42.** $x = 5,\quad x = -1$ **43.** $x = -2,\quad x = 1$ **44.** $x = 1,\quad x = -1$

45. a. Vertex $(5, -50)$ **46. a.** Vertex: $(-7, -64)$ **47.** $C(x) = 45 + 0.25x$

 b.   **b.**   **48.** $I(t) = 800t$

49. $T(x) = 70 - \dfrac{x}{300}$

on $[2, 8]$ by $[-50, -40]$ on $[-10, -4]$ by $[-64, -54]$

50. $C(t) = 0.45t + 20.3$; in about 2020 (from 10.4 years after 2010).

51. a. Break even at 15 and 65 units   **b.** Profit maximized at 40 units. Max profit is $1250 per week.

52. a. Break even at 150 and 450 units.   **b.** Profit maximized at 300 units. Max profit is $67,500 per month.

53. a. $y = 1.675x^2 + 0.435x + 21.6$ (rounded)   **b.** About $268 billion

54. a. $f(-1) = 1$ **b.** $\{x \mid x \neq 0 \text{ and } x \neq 2\}$ **c.** $\{y \mid y > 0 \text{ or } y \leq -3\}$

55. a. $f(-8) = \frac{1}{2}$ **b.** $\{x \mid x \neq 0 \text{ and } x \neq -4\}$ **c.** $\{y \mid y > 0 \text{ or } y \leq -4\}$

56. a. $g\left(\frac{3}{2}\right) = 27$ **b.** $\mathbb{R}$ **c.** $\{y \mid y > 0\}$ **57. a.** $g\left(\frac{5}{3}\right) = 32$ **b.** $\mathbb{R}$ **c.** $\{y \mid y > 0\}$

58. $x = 0$, $x = 1$, $x = -3$ **59.** $x = 0$, $x = 2$, $x = -4$ **60.** $x = 0$, $x = 5$ **61.** $x = 0$, $x = 2$

62.   **63.**   **64.**   **65.**

66. a. $f(g(x)) = \left(\frac{1}{x}\right)^2 + 1 = \frac{1}{x^2} + 1$ **b.** $g(f(x)) = \frac{1}{x^2 + 1}$

67. a. $f(g(x)) = \sqrt{5x - 4}$ **b.** $g(f(x)) = 5\sqrt{x} - 4$ **68. a.** $f(g(x)) = \frac{x^3 + 1}{x^3 - 1}$ **b.** $g(f(x)) = \left(\frac{x + 1}{x - 1}\right)^3$

69. a. $f(g(x)) = 2^{x^2}$ **b.** $g(f(x)) = (2^x)^2 = 2^{2x}$ **70.** $4x + 2h - 3$ **71.** $\dfrac{-5}{(x + h)x}$

72. $A(p(t)) = 2(18 + 2t)^{0.15}$, $A(p(4)) \approx \$3.26$ million

73. a. $x = -1$, $x = 0$, $x = 3$ **74. a.** $x = -3$, $x = 0$, $x = 1$ **75. a.** $y = 6.52 \cdot 0.761^x$ (rounded)

b.   **b.**   **b.** 2.2 crimes per 100,000

on $[-5, 5]$ by $[-5, 5]$ on $[-5, 5]$ by $[-5, 5]$

Exercises 2.1 page 83

1.

x	$5x - 7$	x	$5x - 7$
1.9	2.5	2.1	3.5
1.99	2.95	2.01	3.05
1.999	2.995	2.001	3.005

a. $\lim\limits_{x \to 2^-} (5x - 7) = 3$ **b.** $\lim\limits_{x \to 2^+} (5x - 7) = 3$ **c.** $\lim\limits_{x \to 2} (5x - 7) = 3$

3.

x	$\dfrac{x^3 - 1}{x - 1}$	x	$\dfrac{x^3 - 1}{x - 1}$
0.9	2.71	1.1	3.31
0.99	2.97	1.01	3.03
0.999	2.997	1.001	3.003

a. $\lim\limits_{x \to 1^-} \left(\dfrac{x^3 - 1}{x - 1}\right) = 3$ **b.** $\lim\limits_{x \to 1^+} \left(\dfrac{x^3 - 1}{x - 1}\right) = 3$ **c.** $\lim\limits_{x \to 1} \left(\dfrac{x^3 - 1}{x - 1}\right) = 3$

5. 7.389 (rounded)   **7.** -0.25 **9.** 1   **11.** 2   **13.** 8   **15.** 2   **17.** $\sqrt{2}$ **19.** 6   **21.** $5x^3$

23. 4   **25.** $\frac{1}{2}$ **27.** -9 **29.** $2x$ **31.** $4x^2$ **33. a.** 1   **b.** 3   **c.** Does not exist

35. a. -1 **b.** -1 **c.** -1 **37. a.** -1 **b.** 2   **c.** Does not exist

39. a. -2 **b.** -2 **c.** -2 **41. a.** 0   **b.** 0   **c.** 0

43. a. -1 **b.** 1   **c.** Does not exist

45. $\lim\limits_{x \to -\infty} f(x) = 0$; $\lim\limits_{x \to 3^-} f(x) = \infty$ and $\lim\limits_{x \to 3^+} f(x) = -\infty$, so that $\lim\limits_{x \to 3} f(x)$ does not exist; and $\lim\limits_{x \to \infty} f(x) = 0$.

47. $\lim\limits_{x \to -\infty} f(x) = 0$; $\lim\limits_{x \to 0^-} f(x) = \infty$ and $\lim\limits_{x \to 0^+} f(x) = \infty$, so that $\lim\limits_{x \to 0} f(x) = \infty$; and $\lim\limits_{x \to \infty} f(x) = 0$.

49. $\lim\limits_{x \to -\infty} f(x) = 2$; $\lim\limits_{x \to 1^-} f(x) = -\infty$ and $\lim\limits_{x \to 1^+} f(x) = \infty$, so that $\lim\limits_{x \to 1} f(x)$ does not exist; and $\lim\limits_{x \to \infty} f(x) = 2$.

51. $\lim\limits_{x \to -\infty} f(x) = 1$; $\lim\limits_{x \to -2^-} f(x) = \infty$ and $\lim\limits_{x \to -2^+} f(x) = \infty$, so that $\lim\limits_{x \to 2} f(x) = \infty$; and $\lim\limits_{x \to \infty} f(x) = 1$.

53. Continuous   **55.** Discontinuous, (3) is violated   **57.** Discontinuous, (1) is violated

59. Discontinuous, (2) is violated

61. a.   **63. a.**   **65.** Continuous

67. Discontinuous at $x = 1$

b. $\lim\limits_{x \to 3^-} f(x) = 3$, $\lim\limits_{x \to 3^+} f(x) = 3$ **b.** $\lim\limits_{x \to 3^-} f(x) = 3$, $\lim\limits_{x \to 3^+} f(x) = 4$

c. Yes   **c.** No, (2) is violated

69. Discontinuous at $x = -1$, $x = 0$,   and $x = 1$ **71.** Discontinuous at $x = 4$

73. Continuous   **75.** Continuous   **77.** Discontinuous at $x = 6$

79. The two functions are *not* equal to each other, since at $x = 1$ one is defined and the other is not (see page 50).

81. 1.11 (dollars)   **83.** 100

85. As x approaches c, the function is approaching $\lim\limits_{x \to c} f(x)$ even if the value of the function *at c* is different, so the limit is where it's "going."

87. False. The value of the function *at 2* has nothing to do with the limit as x *approaches* 2.

89. False. *Both* one-sided limits would have to exist and agree to guarantee that the limit exists.

91. False. On the left side the limit exists and equals 2 (as we saw in Example 4 on pages 75–76), but on the right side the denominator of the fraction is zero. Therefore, one side of the equation is defined and the other is not.

93. True. The third requirement for continuity at $x = 2$ is that the *limit* and the *value* at 2 must agree, so if one is 7 the other must be 7.   **95.** None of the limits exist.

Exercises 2.2 page 96

1. At P_1: positive slope   **3.** At P_1: positive slope   **5.** At P_1: slope is 3
At P_2: negative slope   At P_2: negative slope   At P_2: slope is $-\frac{1}{2}$
At P_3: zero slope   At P_3: zero slope

7. Your graph should look roughly like the following:

9. a. 5   **b.** 4   **c.** 3.5   **d.** 3.1   **e.** 3.01   **f.** 3
11. a. 13   **b.** 11   **c.** 10   **d.** 9.2   **e.** 9.02   **f.** 9
13. a. 5   **b.** 5   **c.** 5   **d.** 5   **e.** 5   **f.** 5
15. a. 0.2247   **b.** 0.2361   **c.** 0.2426   **d.** 0.2485   **e.** 0.2498   **f.** 0.25
17. 3   **19.** 5   **21.** 9   **23.** $\frac{1}{4}$ **25.** $f'(x) = 2x - 3$ **27.** $f'(x) = -2x$
29. $f'(x) = 9$ **31.** $f'(x) = \frac{1}{2}$ **33.** $f'(x) = 0$ **35.** $f'(x) = 2ax + b$
37. $f'(x) = 5x^4$ **39.** $f'(x) = \dfrac{-2}{x^2}$ **41.** $f'(x) = \dfrac{1}{2\sqrt{x}}$ **43.** $f'(x) = 3x^2 + 2x$

45. a. $y = x + 1$ **b.**

on window
$[-10, 10]$ by $[-10, 10]$

47. a.   **b.**

49. a. $f'(x) = 3$ **b.** The graph of $f(x) = 3x - 4$ is a straight line with slope 3.

51. a. $f'(x) = 0$ **b.** The graph of $f(x) = 5$ is a horizontal straight line with slope 0.

53. a. $f'(x) = m$ **b.** The graph of $f(x) = mx + b$ is a straight line with slope m.

55. a. $f'(x) = 2x - 8$ **b.** Decreasing at the rate of 4 degrees per minute (since $f'(2) = -4$)
   c. Increasing at the rate of 2 degrees per minute (since $f'(5) = 2$)

57. a. $f'(x) = 4x - 1$
   b. $f'(5) = 19$.   *Interpretation:* When 5 words have been memorized, the memorization time is increasing at the rate of 19 seconds per word.

59. a. $f'(x) = x - 3.7$
   b. -2.7.   *Interpretation:* In 1940 the percentage of immigrants was decreasing by 2.7 percentage points per decade (so about 0.27 of a percentage point per year).
   c. Increasing by 4.3 percentage points per decade (so almost half a percentage point per year).

61. The *average* rate of change requires *two* x-values and is the change in the function-values divided by the change in the x-values. The *instantaneous* rate of change is at a *single* x-value, and is found from the formula

$$f'(x) = \lim_{h \to 0} \frac{f(x + h) - f(x)}{h}.$$

63. Substituting $h = 0$ would make the denominator zero, and we can't divide by zero. That's why we need to do some algebra on the difference quotient—to cancel the terms that are zero so that afterward we can evaluate by direct substitution.

65. The units of x are *blargs,* and the units of *f* are *prendles.*

67. The patient's health is deteriorating during the first day (temperature is rising above normal). The patient's health is improving during the second day (temperature is falling back to normal).

Exercises 2.3 page 110

1. $4x^3$ **3.** $500x^{499}$ **5.** $\frac{1}{2}x^{-1/2}$ **7.** $2x^3$ **9.** $2w^{-2/3}$ **11.** $-6x^{-3}$ **13.** $8x - 3$

15. $-\dfrac{1}{2}x^{-3/2} = -\dfrac{1}{2\sqrt{x^3}}$ **17.** $-2x^{-4/3} = -\dfrac{2}{\sqrt[3]{x^4}}$ **19.** $2\pi r$ **21.** $\frac{1}{2}x^2 + x + 1$ **23.** $\frac{1}{2}x^{-1/2} + x^{-2}$

25. $4x^{-1/3} + 4x^{-4/3}$ **27.** $-5x^{-3/2} - 3x^{2/3}$ **29.** $1 + 2x$ **31.** 80   **33.** 3   **35.** 27   **37.** 1

39. a. $y = 4x - 7$ **41. a.** $y = 2x - 6$

b.   **b.**

43. For $y_1 = 5$ and window $[-10, 10]$ by $[-10, 10]$, your calculator screen should look like the following:

45. a. -2.6 **b.** -0.29
Interpretation: Near full employment, inflation greatly decreases with small increases in unemployment but with higher unemployment, inflation hardly decreases with increases in unemployment.

47. a. 8; when purchasing 8 items, the cost of the last item is about \$8.
  b. 4; when purchasing 64 items, the cost of the last item is about \$4.

49. 4.01, which is very close to 4

51. a. 300; in 2030 this population group will be increasing by 300 thousand per year.
  b. -300; in 2010 this population group was decreasing by 300 thousand per year.

53. Increasing by about 8000 people per additional day

55. 21; in 2020 the number of 3D movies will be increasing by 21 per year.

57. Increasing by about 6 phrases per hour

59. a. $MU(x) = 50x^{-1/2}$ **b.** 50   **c.** 0.05

61. a. $f(12) \approx 40$. *Interpretation:* The probability of a high school graduate quitting smoking is about 40%.
    $f'(12) \approx 1.8$. *Interpretation:* The probability of quitting increases by about 1.8% for each additional year of education.
  b. $f(16) \approx 60$. *Interpretation:* The probability of a college graduate quitting smoking is about 60%.
    $f'(16) \approx 8.5$. *Interpretation:* The probability of quitting increases by about 8.5% for each additional year of education.

63. a. \$43,250
  b. 9500; in 2020–21 private college tuition will be increasing by \$9500 every five years.
  c. \$1900

65. $f(x) = 2$ will have a graph that is a horizontal line at height 2, and the slope (the derivative) of a horizontal line is zero. A function that stays constant will have a rate of change of zero, so its derivative (instantaneous rate of change) will be zero.

67. If f has a particular rate of change, then the rate of change of $2 \cdot f(x)$ will be *twice* as large, and the rate of change of $c \cdot f(x)$ will be c times as large, so the derivative of $c \cdot f(x)$ will be $c \cdot f'(x)$, which is just the Constant Multiple Rule.

69. Since $-f$ slopes down by the same amount that f slopes up, the slope of $-f$ should be the *negative* of the slope of f. The constant multiple rule with $c = -1$ also says that the slope of $-f$ will be the negative of the slope of f.

71. Evaluating first would give a constant, and the derivative of a constant is zero, so evaluating and *then* differentiating would always give zero, regardless of the function and number. This supports the idea that we should always differentiate and *then* evaluate to obtain anything meaningful.

73. Each additional year of education increases life expectancy by 1.7 years.

75. a–d. The result follows from the indicated steps.
  e. Since the tangent line is the line with the smallest possible differences between it and the curve at and near the point, it is the best linear approximation to the function at that point.

Exercises 2.4 page 124

1. $10x^9$ **3.** $9x^8 + 4x^3$ **5.** $5x^4 + 2x$ **7.** $15x^2 - 1$ **9.** $9\sqrt{x} + \dfrac{1}{\sqrt{x}}$ **11.** $4x^3$ **13.** $9x^2 + 8x + 1$

15. $4x^3 + 9x^2 - 2x$ **17.** $-6x^2 + 4x - 1$ **19.** 1   **21.** $36t + 8t^{1/3}$ (after simplification)

23. $7z^6 - 1$ (after simplification)   **25.** $6\sqrt{z} + \dfrac{1}{\sqrt{z}} + 10$ **27.** $6x^5$ **29.** $-\dfrac{3}{x^4}$ or $-3x^{-4}$ **31.** $\dfrac{x^4 - 3}{x^4}$

33. $-\dfrac{2}{(x-1)^2}$ **35.** $\dfrac{5}{(2+x)^2}$ (after simplification)   **37.** $\dfrac{4t}{(t^2+1)^2}$ **39.** $\dfrac{2s^3 + 3s^2 + 1}{(s+1)^2}$ (after simplification)

41. $\dfrac{x^2 + 2x - 5}{(x+1)^2}$ (after simplification)   **43.** $\dfrac{2x^5 + 4x^3}{(x^2+1)^2}$ (after simplification)   **45.** $-\dfrac{t^2 + 4t + 5}{(t^2 + t - 3)^2}$

47. $y = 3x^{-1}$, $\dfrac{dy}{dx} = -3x^{-2}$, $\dfrac{dy}{dx} = -\dfrac{3}{x^2}$ **49.** $y = \dfrac{3}{8}x^4$, $\dfrac{dy}{dx} = \dfrac{3}{2}x^3$, $\dfrac{dy}{dx} = \dfrac{3x^3}{2}$

51. $y = \dfrac{1}{3}x^2 - \dfrac{5}{3}x$, $\dfrac{dy}{dx} = \dfrac{2}{3}x - \dfrac{5}{3}$, $\dfrac{dy}{dx} = \dfrac{2x - 5}{3}$ **53.** $3x^2\,\dfrac{x^2 + 1}{x + 1} + (x^3 + 2)\,\dfrac{x^2 + 2x - 1}{(x + 1)^2}$

55. $\dfrac{3x^6 + 13x^4 + 18x^2 - 2x}{(x^2 + 2)^2}$ **57.** $\dfrac{x^{-1/2}}{(x^{1/2} + 1)^2} = \dfrac{1}{\sqrt{x}(\sqrt{x} + 1)^2}$ **59.** $MAR(x) = \dfrac{xR'(x) - R(x)}{x^2}$

61. a. $C'(x) = \dfrac{100}{(100 - x)^2}$ **b.** Increasing by 4¢ per additional percentage of purity

    c. Increasing by 25¢ per additional percentage of purity

63. b. Rates of change are 4 and 25

65. a. $8 + \dfrac{45}{x}$ **b.** $-\dfrac{45}{x^2}$ **c.** $-\dfrac{1}{20}$, so the average cost is decreasing by 5¢ per additional clock.

67. Increasing at the rate of 7 degrees per hour   **69. b.** 7   **c.** About 104.5 degrees

71. a. $65,468, $59,425 (from 65.468, 59.425)

    b. $-151, so in 2020 the per capita national debt will be shrinking by $151 per year;
    $-2231, so in 2025 the per capita national debt will be shrinking by $2231 per year.

73. $-1.24, 0.83$. Interest rates were declining by 1.24 percent per year in 2010 and rising by 0.83 percent per year in 2011.

75. 13.99, 8.14. Median weekly earnings were rising by $13.99 per year in 2009 and by $8.14 per year in 2013.

77. a. $y' = -2x^{-3} = -\dfrac{2}{x^3}$ (which is undefined at $x = 0$)

    b. Your calculator should give "Error" but may, incorrectly, give "0."

79. False. The product rule gives the correct right-hand side.   **81.** True

83. The right-hand side multiplies out to $g \cdot f' + f \cdot g'$, which agrees with the product rule.

85. False. This would be equivalent to saying that the derivative (instantaneous rate of change) of a product is the product of the derivatives. The *Product Rule* gives the correct way of finding the derivative of a product.

87. $\dfrac{d}{dx}(fgh) = f'(gh) + f(gh)' = f'gh + fg'h + fgh'$ **89.** $2f(x)f'(x)$

91. $\dfrac{dy}{dx} = \dfrac{R}{\left(1 + \dfrac{R - 1}{K}x\right)^2} > 0$. As the population increases, the number of offspring increases.

Exercises 2.5 page 136

1. a. $4x^3 - 6x^2 - 6x + 5$ **b.** $12x^2 - 12x - 6$ **c.** $24x - 12$ **d.** 24

3. a. $1 + x + \dfrac{1}{2}x^2 + \dfrac{1}{6}x^3 + \dfrac{1}{24}x^4$ **b.** $1 + x + \dfrac{1}{2}x^2 + \dfrac{1}{6}x^3$ **c.** $1 + x + \dfrac{1}{2}x^2$ **d.** $1 + x$

5. a. $\dfrac{5}{2}x^{3/2}$ **b.** $\dfrac{15}{4}x^{1/2}$ **c.** $\dfrac{15}{8}x^{-1/2}$ **d.** $-\dfrac{15}{16}x^{-3/2}$ **7. a.** $-\dfrac{2}{x^3}$ or $-2x^{-3}$ **b.** $-\dfrac{2}{27}$

9. a. $\dfrac{1}{x^3} = x^{-3}$ **b.** $\dfrac{1}{27}$ **11. a.** x^{-4} **b.** $\dfrac{1}{81}$ **13.** $12x^2 + 2$ **15.** $12x^{-7/3}$

17. $\dfrac{2x - 2}{(x^2 - 2x + 1)^2} = \dfrac{2}{(x - 1)^3}$ **19.** 2π **21.** 90   **23.** -720 **25.** 3

27. $20x^3 - 12x^2 + 6x - 2$ **29.** $\dfrac{2x^5 - 4x^3 - 6x}{(x^2 + 1)^4} = \dfrac{2x^3 - 6x}{(x^2 + 1)^3}$ **31.** $\dfrac{-32x - 16}{(4x^2 + 4x + 1)^2} = \dfrac{-16}{(2x + 1)^3}$

33. a. 54 mph   **b.** -42 mph or 42 mph south   **c.** 24 mi/hr^2 **35.** 310 ft/sec, 61 ft/sec^2

37. a. About 13.03 seconds   **b.** At about 417 feet per second. [*Note:* This is about 284 miles per hour. More realistically, the ball would be slowed by air resistance, going no faster than its *terminal velocity*. For a steel ball the size of a marble, this would be about 200 mph.]   **c.** -32 feet per second per second

39. a. $-32t + 1280$ **b.** 40 seconds   **c.** 25,600 feet

41. $D'(8) = 24$: After 8 years the debt is growing by \$24 billion per year.
$D''(8) = 1$: After 8 years the debt will be growing increasingly rapidly, with the rate of growth growing by about \$1 billion per year.

43. $L(10) = 93$; *Interpretation:* By 2100 sea levels may have risen by 93 centimeters (about 3 feet). $L'(10) = 12.6$; *Interpretation:* In 2100 sea levels will be rising by about 12.6 centimeters per decade, or about 1.26 cm (about half an inch) per year. $L''(10) = 1.06$. *Interpretation:* The rise in the sea level will be speeding up (by about 1 cm per decade per decade).

45. a.

on [0, 50] by [0, 40]

Approximately 22° and 18°

b. Each successive 1-mph increase in wind speed lowers the windchill index, but less so as wind speed rises.
c. $y_2(15) \approx -0.4°$ and $y_2(30) \approx -0.2°$. *Interpretation:* At a wind speed of 15 mph, each additional mph decreases the windchill index by about 0.4°, while at a wind speed of 30 mph, each additional mph decreases the windchill index by only about 0.2°.

47. a. Negative   **b.** Negative   **49. a.** Negative   **b.** Positive   **51.** True   **53. a.** iii   **b.** i   **c.** ii

55. 100!   **57.** $\dfrac{d^2}{dx^2}(fg) = \dfrac{d}{dx}(f'g + fg') = f''g + f'g' + f'g' + fg'' = f''g + 2f'g' + fg''$

Exercises 2.6 page 147

Note: For Exercises 1 through 9, there are other possible correct answers.

1. $f(x) = \sqrt{x}, g(x) = x^2 - 3x + 1$ **3.** $f(x) = x^{-3}, g(x) = x^2 - x$ **5.** $f(x) = \dfrac{x+1}{x-1}, g(x) = x^3$

7. $f(x) = x^4, g(x) = \dfrac{x+1}{x-1}$ **9.** $f(x) = \sqrt{x} + 5, g(x) = x^2 - 9$ **11.** $6x(x^2 + 1)^2$

13. $4(2x^2 - 7x + 3)^3(4x - 7)$ **15.** $4(3z^2 - 5z + 2)^3(6z - 5)$ **17.** $\frac{1}{2}(x^4 - 5x + 1)^{-1/2}(4x^3 - 5)$

19. $\frac{1}{3}(9z - 1)^{-2/3}(9) = 3(9z - 1)^{-2/3}$ **21.** $-8x(4 - x^2)^3$ **23.** $-12w^2(w^3 - 1)^{-5}$

25. $4x^3 - 4(1 - x)^3$ **27.** $-\frac{1}{2}(3x^2 - 5x + 1)^{-3/2}(6x - 5)$

29. $-\frac{2}{3}(9x + 1)^{-5/3}(9) = -6(9x + 1)^{-5/3}$ **31.** $-\frac{2}{3}(2x^2 - 3x + 1)^{-5/3}(4x - 3)$

33. $3[(x^2 + 1)^3 + x]^2[6x(x^2 + 1)^2 + 1]$ **35.** $6x(2x + 1)^5 + 30x^2(2x + 1)^4 = 6x(2x + 1)^4(7x + 1)$

37. $6(2x + 1)^2(2x - 1)^4 + 8(2x + 1)^3(2x - 1)^3 = 2(2x + 1)^2(2x - 1)^3(14x + 1)$

39. $2(3z^2 - z + 1)^4 + 8z(3z^2 - z + 1)^3(6z - 1) = 2(3z^2 - z + 1)^3(27z^2 - 5z + 1)$

41. $-6\dfrac{(x + 1)^2}{(x - 1)^4}$ **43.** $2x(1 + x^2)^{1/2} + x^3(1 + x^2)^{-1/2}$ **45.** $\dfrac{1}{4}x^{-1/2}(1 + x^{1/2})^{-1/2}$

47. $y = -8x - 7$ **49.** $y = -2x + 5$ **51. a.** $4x(x^2 + 1)$ **b.** $4x^3 + 4x$

53. a. $-\dfrac{3}{(3x + 1)^2}$ **b.** $-3(3x + 1)^{-2}$ **55.** $20(x^2 + 1)^9 + 360x^2(x^2 + 1)^8$

57. $MC(x) = 4x(4x^2 + 900)^{-1/2}$ so $MC(20) = \frac{8}{5} = 1.60$ **59.** $x = 27$

61. $S'(25) \approx 2.1$. *Interpretation:* At income level \$25,000, status increases by about 2.1 units for each additional \$1000 of income.

63. $R'(50) = 32\frac{1}{3}$ **65.** 26 mg   **67.** $P'(2) = 0.24$; pollution is increasing by about 0.24 ppm per year.

69. b. 0.05184   **c.** 34.7, so about 35 years

71. False. There should not be a prime of the first g on the right-hand side.

73. The Generalized Power Rule is a special case of the Chain Rule, when the *outer* function is just a *power* of the variable.

75. True.

77. False. The *outer* function, $\sqrt{x}$, was not differentiated. The correct right-hand side is $\frac{1}{2}[g(x)]^{-1/2} \cdot g'(x)$.

79. No. Since instantaneous rates of change are derivatives, this would be saying that $\frac{d}{dx}[f(x)]^2 = [f'(x)]^2$ where $f(x)$ is the length of a side. The Chain Rule gives the correct derivative of $[f(x)]^2$.

81. $\dfrac{d}{dx} L(g(x)) = L'(g(x))\, g'(x) = \dfrac{1}{g(x)} g'(x) = \dfrac{g'(x)}{g(x)}$

83. The result follows from the hint.

85. Solve for $F(h)$ to find $F(h) = \dfrac{f(x+h) - f(x)}{h}$ and use the continuity of F at 0 to obtain

$F(0) = \lim\limits_{h \to 0} F(h) = \lim\limits_{h \to 0} \dfrac{f(x+h) - f(x)}{h}$, which is the definition of the derivative on page 92.

Exercises 2.7 page 155

1. $-2, 0,$ and 2   **3.** -3 and 3

5. $\lim\limits_{h \to 0} \dfrac{f(x+h) - f(x)}{h}$ simplifies to $\lim\limits_{h \to 0} \dfrac{|2h|}{h}$, which gives $\begin{cases} 2 & \text{for } h > 0 \\ -2 & \text{for } h < 0 \end{cases}$ so the limit

(and therefore the derivative at $x = 0$) does not exist.

7. $\lim\limits_{h \to 0} \dfrac{f(x+h) - f(x)}{h}$ simplifies to $\lim\limits_{h \to 0} \dfrac{h^{2/5}}{h} = \lim\limits_{h \to 0} \dfrac{1}{h^{3/5}}$ which does not exist. Therefore, the derivative

at $x = 0$ does not exist.

9. If you got a numerical answer, it is wrong, since the function is undefined at $x = 0$, so the derivative at $x = 0$ does not exist. (For an explanation, see the Graphing Calculator Exploration on page 152.)

11. a. The formula comes from substituting the given function and x-value into the definition of the derivative and simplifying.

   b. 6.3, 251.2, 10000 (rounded)   **c.** No, no   **d.**

on $[-1, 1]$ by $[-1, 1]$

13. One explanation: At a corner point, a proposed "tangent line" can tip back and forth (see page 95), and so there is no well-defined slope.

15. At a discontinuity, the values of the function take a sudden *jump,* and so a (steady) rate of change cannot be defined.

17. True.   **19.** One possibility is:

Chapter 2 Review Exercises and Chapter Test page 157

1.

x	$4x + 2$
1.9	9.6
1.99	9.96
1.999	9.996

x	$4x + 2$
2.1	10.4
2.01	10.04
2.001	10.004

a. $\lim_{x \to 2^-} (4x + 2) = 10$ **b.** $\lim_{x \to 2^+} (4x + 2) = 10$ **c.** $\lim_{x \to 2} (4x + 2) = 10$

2.

x	$\dfrac{\sqrt{x+1}-1}{x}$
-0.1	0.513
-0.01	0.501
-0.001	0.500

x	$\dfrac{\sqrt{x+1}-1}{x}$
0.1	0.488
0.01	0.499
0.001	0.500

a. $\lim_{x \to 0^-} \dfrac{\sqrt{x+1}-1}{x} = 0.5$ **b.** $\lim_{x \to 0^-} \dfrac{\sqrt{x+1}-1}{x} = 0.5$ **c.** $\lim_{x \to 0} \dfrac{\sqrt{x+1}-1}{x} = 0.5$

3. a. 3   **b.** -2 **c.** Does not exist   **4. a.** -1 **b.** -1 **c.** -1 **5.** 5   **6.** π **7.** 4

8. 2   **9.** $\frac{1}{2}$ **10.** -3 **11.** $2x^2$ **12.** $-x^2$

13. $\lim_{x \to -\infty} f(x) = 0$; $\lim_{x \to -1^-} f(x) = \infty$ and $\lim_{x \to -1^+} f(x) = \infty$,   so that $\lim_{x \to -1} f(x) = \infty$; and $\lim_{x \to \infty} f(x) = 0$.

14. $\lim_{x \to -\infty} f(x) = 3$; $\lim_{x \to 2^-} f(x) = -\infty$ and $\lim_{x \to 2^+} f(x) = \infty$, so that $\lim_{x \to 2} f(x)$ does not exist; and $\lim_{x \to \infty} f(x) = 3$.

15. Continuous   **16.** Continuous   **17.** Discontinuous at $x = -1$ **18.** Continuous

19. Discontinuous at $x = 0$ and at $x = -1$ **20.** Discontinuous at $x = 3$ and at $x = -3$

21. Discontinuous at $x = 5$ **22.** Continuous   **23.** $4x + 3$ **24.** $6x + 2$ **25.** $-\dfrac{3}{x^2} = -3x^{-2}$

26. $\dfrac{2}{\sqrt{x}}$ **27.** $10x^{2/3} + 2x^{-3/2}$ **28.** $10x^{3/2} + 2x^{-4/3}$ **29.** -16 **30.** -9 **31.** 1   **32.** $\frac{1}{2}$ **33.** $y = x + 1$

34. a. 20. *Interpretation:* Costs are increasing by about $20 per additional license.
　　b. 10. *Interpretation:* Costs are increasing by about $10 per additional license.

35. $f'(10) \approx -2.3$ (thousand hours). *Interpretation:* After 10 planes, the construction time is decreasing by about 2300 hours for each additional plane built.

36. a. $\dfrac{dA}{dr} = 2\pi r$ **b.** As the radius increases, the area "grows by a circumference."

37. a. $V' = \frac{4}{3}\pi r^2 \cdot 3 = 4\pi r^2$ **b.** As radius increases, volume "grows by a surface area."

38. $40x^3 + 6$ (after simplification)   **39.** $15x^4 - 2x$ **40.** $2x(x^2 - 5) + (x^2 + 5)2x = 4x^3$ **41.** $4x^3$

42. $(4x^3 + 2x)(x^5 - x^3 + x) + (x^4 + x^2 + 1)(5x^4 - 3x^2 + 1) = 9x^8 + 5x^4 + 1$

43. $(5x^4 + 3x^2 + 1)(x^4 - x^2 + 1) + (x^5 + x^3 + x)(4x^3 - 2x) = 9x^8 + 5x^4 + 1$

44. $\dfrac{2}{(x+1)^2}$ **45.** $\dfrac{-2}{(x-1)^2}$ **46.** $\dfrac{-10x^4}{(x^5-1)^2}$ **47.** $\dfrac{12x^5}{(x^6+1)^2}$

48. a. $-\dfrac{1}{x^2}$ **b.** $-x^{-2}$ (after simplification)   **c.** By simplifying to $f(x) = 2 + x^{-1}$, the derivative is $-x^{-2}$.

49. $y = -\dfrac{1}{2}x - \dfrac{3}{2}$

50. $S'(6) = -10$, so at a price of $6 each, sales will decrease by about 10 for each dollar increase in price.

51. a. $AP(x) = \dfrac{6x - 200}{x}$ **b.** $MAP(x) = \dfrac{200}{x^2}$ **c.** $MAP(10) = 2$, so average profit is increasing by about $2 per additional unit.

52. a. $7.5 + \dfrac{50}{x}$ **b.** $-\dfrac{50}{x^2}$ **c.** $-\dfrac{1}{50}$, so the average cost is decreasing by 2¢ per additional mouse.

53. $9x^{-1/2} + 2x^{-5/3}$ **54.** $-4x^{-4/3} - 3x^{-1/2}$ **55.** $2x^{-4}$ **56.** $6x^{-5}$ **57.** -24 **58.** 60

59. 480   **60.** $\frac{3}{8}$ **61.** 15   **62.** 70

63. $P(10) = 200$, $P'(10) = 20$, $P''(10) = 9$. *Interpretation:* 10 years from now the population will be 200 thousand, growing at the rate of 20 thousand per year, and the growth will be accelerating.

64. Velocity 2500 ft/sec; acceleration 150 ft/sec^2 **65. a.** 347.25 feet

66. $3(4z^2 - 3z + 1)^2(8z - 3)$ **67.** $4(3z^2 - 5z - 1)^3(6z - 5)$ **68.** $-5(100 - x)^4$ **69.** $-4(1000 - x)^3$

70. $\frac{1}{2}(x^2 - x + 2)^{-1/2}(2x - 1)$ **71.** $\frac{1}{2}(x^2 - 5x - 1)^{-1/2}(2x - 5)$ **72.** $2(6z - 1)^{-2/3}$ **73.** $(3z + 1)^{-2/3}$

74. $-2(5x + 1)^{-7/5}$ **75.** $-6(10x + 1)^{-8/5}$ **76.** $2x(2x - 1)^4 + 8x^2(2x - 1)^3 = 2x(2x - 1)^3(6x - 1)$

77. $5(x^3 - 2)^4 + 60x^3(x^3 - 2)^3 = 5(x^3 - 2)^3(13x^3 - 2)$ **78.** $3x^2(x^3 + 1)^{1/3} + x^5(x^3 + 1)^{-2/3}$

79. $4x^3(x^2 + 1)^{1/2} + x^5(x^2 + 1)^{-1/2}$ **80.** $3[(2x^2 + 1)^4 + x^4]^2[16x(2x^2 + 1)^3 + 4x^3]$

81. $2[(3x^2 - 1)^3 + x^3][18x(3x^2 - 1)^2 + 3x^2]$ **82.** $\frac{1}{2}[(x^2 + 1)^4 - x^4]^{-1/2}[8x(x^2 + 1)^3 - 4x^3]$

83. $[(x^3 + 1)^2 + x^2]^{-1/2}[3x^2(x^3 + 1) + x]$ **84.** $12(3x + 1)^3(4x + 1)^3 + 12(3x + 1)^4(4x + 1)^2$

85. $6x(x^2 + 1)^2(x^2 - 1)^4 + 8x(x^2 + 1)^3(x^2 - 1)^3$ **86.** $\dfrac{-20}{x^2}\left(\dfrac{x + 5}{x}\right)^3 = \dfrac{-20(x + 5)^3}{x^5}$

87. $-20\left(\dfrac{x + 4}{x}\right)^4 \dfrac{1}{x^2} = -20\dfrac{(x + 4)^4}{x^6}$ **88.** $y = \dfrac{1}{3}x + \dfrac{4}{3}$

89. $20(2w^2 - 4)^4 + 320w^2(2w^2 - 4)^3$ **90.** $24(3w^2 + 1)^3 + 432w^2(3w^2 + 1)^2$

91. $6z(z + 1)^3 + 18z^2(z + 1)^2 + 6z^3(z + 1)$ **92.** $12z^2(z + 1)^4 + 32z^3(z + 1)^3 + 12z^4(z + 1)^2$

93. a. $6x^2(x^3 - 1)$ **b.** $6x^5 - 6x^2$ **94. a.** $\dfrac{-3x^2}{(x^3 + 1)^2}$ **b.** $-(x^3 + 1)^{-2}(3x^2)$

95. $P'(5) = 3$. *Interpretation:* When producing 5 tons, profit increases by about 3 thousand dollars for each additional ton.

96. $V'(8) = 17.496$. *Interpretation:* Value increased by about $17.50 for each additional percentage of interest.

97. a. $P(5) - P(4) \approx 2.73$, $P(6) - P(5) \approx 3.23$, both of which are near 3   **b.** At about $x = 7.6$

98. $x \approx 16$ **99.** 0.08

100. $N'(96) = -250$. *Interpretation:* At age 96, the number of survivors is decreasing by about 250 people per year.

101. $x = -3, x = 1, x = 3$ **102.** $x = 2, x = -2$ **103.** $x = 0, x = 3.5$ **104.** $x = 0, x = 3$

105. $\displaystyle\lim_{h \to 0} \dfrac{f(x + h) - f(x)}{h}$ simplifies to $\displaystyle\lim_{h \to 0} \dfrac{|5h|}{h}$, which gives $\begin{cases} 5 & \text{if } h > 0 \\ -5 & \text{if } h < 0 \end{cases}$ and so the limit

(and therefore the derivative at $x = 0$) does not exist.

106. $\displaystyle\lim_{h \to 0} \dfrac{f(x + h) - f(x)}{h}$ simplifies to $\displaystyle\lim_{h \to 0} \dfrac{h^{3/5}}{h} = \lim_{h \to 0} \dfrac{1}{h^{2/5}}$ which does not exist. Therefore, the derivative at $x = 0$ does not exist.

Exercises 3.1 page 173

1. a. $(-\infty, -2)$ and $(0, \infty)$ **b.** $(-2, 0)$ **3.** All but 3 (where the function is undefined)   **5.** 4 and -4

7. -1 and 5   **9.** $0, -4,$ and 1   **11.** 3   **13.** No CNs   **15.** $\frac{1}{3}$ and -1

17.

Open intervals of increase: $(-\infty, -1)$ and $(3, \infty)$; open intervals of decrease: $(-1, 3)$

19.

Open intervals of increase: $(-4, 0)$ and $(1, \infty)$; open intervals of decrease: $(-\infty, -4)$ and $(0, 1)$

21.

Increase: $(-\infty, 0)$ and $(1, 2)$; decrease: $(0, 1)$ and $(2, \infty)$

23.

Increase: $(0, 1)$ and $(1, \infty)$; decrease: $(-\infty, 0)$

25.

Increase: $(1, \infty)$; decrease: $(-\infty, 1)$

27.

Increase: $(-2, 0)$ and $(2, \infty)$; decrease: $(-\infty, -2)$ and $(0, 2)$

29.

$f' < 0$	$f' = 0$	$f' > 0$	$f' = 0$	$f' < 0$	$f' = 0$	$f' > 0$
	$x = 0$		$x = 2$		$x = 4$	
↘	→	↗	→	↘	→	↗
	rel min		rel max		rel min	
	$(0, 0)$		$(2, 16)$		$(4, 0)$	

Increase: $(0, 2)$ and $(4, \infty)$; decrease: $(-\infty, 0)$ and $(2, 4)$

31.

$f' > 0$	$f' = 0$	$f' < 0$	$f' = 0$	$f' > 0$	$f' = 0$	$f' > 0$
	$x = 0$		$x = 2$		$x = 5$	
↗	→	↘	→	↗	→	↗
	rel max		rel min		neither	
	$(0, 0)$		$(2, -108)$		$(5, 0)$	

Increase: $(-\infty, 0)$, $(2, 5)$, and $(5, \infty)$; decrease: $(0, 2)$

33.

$f' < 0$	f und	$f' < 0$
	$x = -3$	
↘		↘

35.

$f' < 0$	f und	$f' < 0$
	$x = 2$	
↘		↘

37.

$f' > 0$	f und	$f' > 0$
	$x = -2$	
↗		↗

39.

$f' > 0$	f und	$f' < 0$
	$x = 2$	
↗		↘

41.

$f' > 0$	f und	$f' > 0$	$f' = 0$	$f' < 0$	f und	$f' < 0$
	$x = -1$		$x = 1$		$x = 3$	
↗		↗	→	↘		↘
			rel max			
			$(1, -3)$			

43.

$f' > 0$	$f' = 0$	$f' < 0$
	$x = 0$	
↗	→	↘
	rel max	
	$(0, 2)$	

45.

$f' < 0$	$f' = 0$	$f' > 0$
	$x = 0$	
↘	→	↗
	rel min	
	$(0, 0)$	

47.

$f' < 0$	$f' = 0$	$f' > 0$	$f' = 0$	$f' < 0$
	$x = -1$		$x = 1$	
↘	→	↗	→	↘
	rel min		rel max	
	$(-1, -1)$		$(1, 1)$	

49.

$f' > 0$	f und	$f' > 0$	$f' = 0$	$f' < 0$	f und	$f' < 0$
	$x = -1$		$x = 0$		$x = 1$	
↗		↗	→	↘		↘
			rel max			
			$(0, -3)$			

51.

$f' < 0$	f und	$f' < 0$	$f' = 0$	$f' < 0$	f und	$f' < 0$
	$x = -3$		$x = 0$		$x = 3$	
↘		↘	→	↘		↘

53.

$f' > 0$	f und	$f' > 0$	$f' = 0$	$f' < 0$	f und	$f' < 0$
	$x = -1$		$x = 1$		$x = 3$	
↗		↗	→	↘		↘
			rel max			
			$(1, -1)$			

55.

$f' > 0$	f und	$f' > 0$	$f' = 0$	$f' < 0$	f und	$f' < 0$
	$x = -1$		$x = 0$		$x = 1$	
↗		↗	→	↘		↘
			rel max			
			$(0, 0)$			

57.

$f' < 0$	f und	$f' < 0$
	$x = -2$	
↘		↘

59.

$f' < 0$	f und	$f' > 0$	$f' = 0$	$f' < 0$	f und	$f' < 0$
	$x = 0$		$x = 2$		$x = 3$	
↘		↗	→	↘		↘
			rel max			
			$(2, -1)$			

61.

$f' > 0$ $f' = 0$ $f' < 0$ $f' = 0$ $f' > 0$ $f' = 0$ $f' < 0$

$x = -1$ $x = 0$ $x = 1$

↗   ↘   →   ↗   →   ↘

rel max   rel min   rel max
$(-1, 1)$ $(0, 0)$ $(1, 1)$

63. $f'(x) = 2ax + b = 0$ at $x = \dfrac{-b}{2a}$

65.

67. i. $f(0) = 0$, giving $d = 0$ **ii.** $f'(x) = 3ax^2 + 2bx + c$ so $f'(0) = c$, giving $c = 0$
iii. $f(100) = 5$ means $a \cdot 100^3 + b \cdot 100^2 = 5$, and $f'(100) = 0$ means $3a \cdot 100^2 + 2b \cdot 100 = 0$, giving
$a = -0.00001$ and $b = 0.0015$. Therefore, $f(x) = -0.00001x^3 + 0.0015x^2$.

iv.

69. a.

on [0, 100] by [0, 100]

b. The derivative would
be positive, but rather
flat at first, and then
rising steeply.

c.

on [0, 99.99] by [0, 100]

71. a. $\displaystyle\lim_{x \to \infty} AC(x) = \lim_{x \to \infty} 3 + \frac{50}{x} = 3$ and $MC(x) = (3x + 50)' = 3$

b.

73. False. The original function must also be *defined* at the critical number.
75. False. For example, see the graph on page 164.
77. Decreasing on $(-\infty, 1)$ and increasing on $(1, \infty)$. [*Remember:* Wherever the
derivative is positive (above the x-axis) the function will be increasing.]

79. In a rational function, the denominator of the derivative is the *square* of the denominator of the original
function (from the Quotient Rule). Therefore, both denominators will be zero at the same x-values, making both the
function and its derivative undefined at the same x-values.

81. True. See the diagrams on pages 168.

83. If $p(c) = 0$ and $q(c) = 0$, they may have a common factor that can be canceled out so that $\displaystyle\lim_{x \to c} \frac{p(x)}{q(x)}$ may
exist. See Example 4 on pages 75–76 and the following Graphing Calculator Exploration.

85. d.

on [0, 10] by [0, 10]

Exercises 3.2 page 186

1. Point 2   **3.** Points 3 and 5   **5.** Points 4 and 6

7. a.

$f' > 0$ $f' = 0$ $f' < 0$ $f' = 0$ $f' > 0$

$x = -3$ $x = 1$

↗   →   ↘   →   ↗

rel max   rel min
$(-3, 32)$ $(1, 0)$

b.

$f'' < 0$ $f'' = 0$ $f'' > 0$

$x = -1$

con dn   con up
          IP $(-1, 16)$

c.

9. a.

$$f' > 0 \qquad f' = 0 \qquad f' > 0$$
$$x = 1$$

↗   →   ↗

neither
(1, 5)

b.

$$f'' < 0 \qquad f'' = 0 \qquad f'' > 0$$
$$x = 1$$
con dn   con up
IP (1, 5)

c.

11. a.

$$f' < 0 \quad f' = 0 \quad f' > 0 \quad f' = 0 \quad f' > 0$$
$$x = 0 \qquad\qquad x = 3$$

↘   →   ↗   →   ↗

rel min   neither
(0, 2)   (3, 29)

b.

$$f'' > 0 \quad f'' = 0 \quad f'' < 0 \quad f'' = 0 \quad f'' > 0$$
$$x = 1 \qquad\qquad x = 3$$
con up   con dn   con up
IP (1, 13)   IP (3, 29)

c.

13. a.

$$f' < 0 \quad f' = 0 \quad f' > 0 \quad f' = 0 \quad f' > 0$$
$$x = -3 \qquad\qquad x = 0$$

↘   →   ↗   →   ↗

rel min   neither
(−3, −12)   (0, 15)

b.

$$f'' > 0 \quad f'' = 0 \quad f'' < 0 \quad f'' = 0 \quad f'' > 0$$
$$x = -2 \qquad\qquad x = 0$$
con up   con dn   con up
IP (−2, −1)   IP (0, 15)

c.

15. a.

$$f' < 0 \quad f' = 0 \quad f' > 0 \quad f' = 0 \quad f' < 0$$
$$x = 0 \qquad\qquad x = 4$$

↘   →   ↗   →   ↘

rel min   rel max
(0, 0)   (4, 256)

b.

$$f'' > 0 \quad f'' = 0 \quad f'' > 0 \quad f'' = 0 \quad f'' < 0$$
$$x = 0 \qquad\qquad x = 3$$
con up   con up   con dn
IP (3, 162)

c.

17. a.

$$\underline{\quad f' > 0 \qquad f' = 0 \qquad f' > 0 \quad}$$
$$x = -2$$

↗   neither   ↗
$(-2, 0)$

b.

$$\underline{\quad f'' < 0 \qquad f'' = 0 \qquad f'' > 0 \quad}$$
$$x = -2$$

con dn   con up
IP $(-2, 0)$

c.

19. a.

$$\underline{\quad f' > 0 \qquad f' = 0 \qquad f' < 0 \qquad f' = 0 \qquad f' > 0 \quad}$$
$$\qquad\qquad x = 1 \qquad\qquad\qquad x = 3$$

↗   rel max   ↘   rel min   ↗
$(1, 4)$ $(3, 0)$

b.

$$\underline{\quad f'' < 0 \qquad f'' = 0 \qquad f'' > 0 \quad}$$
$$x = 2$$

con dn   con up
IP $(2, 2)$

c.

21. a.

$$\underline{\quad f' > 0 \qquad f' \text{ und} \qquad f' > 0 \quad}$$
$$x = 0$$

↗   neither   ↗
$(0, 0)$

b.

$$\underline{\quad f'' > 0 \qquad f'' \text{ und} \qquad f'' < 0 \quad}$$
$$x = 0$$

con up   con dn
IP $(0, 0)$

c.

23. a.

$$\underline{\quad f' < 0 \qquad f' \text{ und} \qquad f' > 0 \quad}$$
$$x = 0$$

↘   rel min   ↗
$(0, 2)$

b.

$$\underline{\quad f'' < 0 \qquad f'' \text{ und} \qquad f'' < 0 \quad}$$
$$x = 0$$

con dn   con dn

c.

25. a.

$$\underline{\quad f' \text{ und} \qquad f' > 0 \quad}$$
$$x = 0$$

↗

b.

$$\underline{\quad f'' \text{ und} \qquad f'' < 0 \quad}$$
$$x = 0$$

con dn

c.

27. a.

$$\underline{\quad f' < 0 \qquad f' \text{ und} \qquad f' > 0 \quad}$$
$$x = 1$$

↘   rel min   ↗
$(1, 0)$

b.

$$\underline{\quad f'' < 0 \qquad f'' \text{ und} \qquad f'' < 0 \quad}$$
$$x = 1$$

con dn   con dn

c.

29. a.

$$\underline{\quad f' < 0 \qquad f \text{ und} \qquad f' < 0 \qquad f \text{ und} \qquad f' < 0 \quad}$$
$$\qquad\quad x = -1 \qquad\qquad\quad x = 1$$

↘   ↘   ↘

b.

$$\underline{\quad f'' < 0 \qquad f \text{ und} \qquad f'' > 0 \qquad f'' = 0 \qquad f'' < 0 \qquad f \text{ und} \qquad f'' > 0 \quad}$$
$$\qquad\quad x = -1 \qquad\qquad\qquad x = 0 \qquad\qquad\qquad x = 1$$

con dn   con up   con dn   con up
IP $(0, 0)$

c.

31.

f' und	$f' > 0$
$x = 0$	↗

on [0, 10] by [0, 4]

33.

f' und	$f' < 0$
$x = 0$	↘

on [0, 10] by [0, 4]

35.

$f' < 0$	f' und	$f' > 0$	$f' = 0$	$f' < 0$
	$x = 0$		$x = 1$	
↘		↗	→	↘
	rel min		rel max	
	(0, 0)		(1, 3)	

on [−2, 10] by [−10, 10]

37.

$f' > 0$	f' und	$f' < 0$	$f' = 0$	$f' > 0$
	$x = 0$		$x = 1$	
↗		↘	→	↗
	rel max		rel min	
	(0, 0)		(1, −2)	

on [−2, 10] by [−10, 10]

39.

$f' > 0$	$f' = 0$	$f' < 0$	f' und	$f' > 0$	$f' = 0$	$f' < 0$
	$x = -1$		$x = 0$		$x = 1$	
↗	→	↘		↗	→	↘
	rel max		rel min		rel max	
	(−1, 2)		(0, 0)		(1, 2)	

on [−5, 5] by [−10, 5]

41. Relative maximum at $x = 1$; relative minimum at $x = 3$
43. Relative minimum at $x = 0$ and at $x = 2$; relative maximum at $x = 1$
45. Relative maximum at $x = -3$; relative minimum at $x = 3$
47. −0.77, 0, and 0.77

49.   **51.**   **53.**   **55.**

57. a.

$f' > 0$	$f' = 0$	$f' < 0$	$f' = 0$	$f' > 0$

$x = 1$ $x = 5$

↗   →   ↘   →   ↗

rel max   rel min
(1, 32)   (5, 0)

$f'' < 0$	$f'' = 0$	$f'' > 0$

$x = 3$

con dn   con up

IP (3, 16)

b.

c. The decline in revenue first begins to slow at $x = 3$ years.

59. a.

$f' = 0$	$f' < 0$	$f' = 0$	$f' > 0$

0          $x = 3$

↘   →   ↗

rel min
(3, 85)

$f'' = 0$	$f'' < 0$	$f'' = 0$	$f'' > 0$

0          $x = 2$

con dn   con up

IP (2, 96)

b.

c. The decline in temperature first begins to slow at $x = 2$ hours.

61.

f' und	$f' > 0$

$x = 0$

rel min   ↗
(0, 0)

f'' und	$f'' < 0$

$x = 0$

con dn

63. a.

S' und	$S' > 0$

$i = 0$ ↗

S'' und	$S'' < 0$

$i = 0$

con dn

b. Concave down. Status increases more slowly at higher income levels.

65. a.

$S' > 0$	$S' = 0$	$S' < 0$	$S' = 0$	$S' > 0$

$x = 2$ $x = 4$

↗   →   ↘   →   ↗

rel max   rel min
(2, 260)   (4, 252)

$S'' < 0$	$S'' = 0$	$S'' > 0$

$x = 3$

con dn   con up

IP (3, 256)

b.

c. The inflection point in 2008 is the first sign that the decline in semiconductor sales is beginning to slow.

67. $800,000 (from $x = 8$).   **69.** $4,000,000 (from $x = 4$).

71. (50, 2.5). The curve is concave up (slope increasing) before $x = 50$ and concave down (slope decreasing) after $x = 50$. Therefore, the slope is maximized at $x = 50$.

73. Concave up for $a > 0$ and concave down for $a < 0$.

75. False. The sign of the second derivative must actually *change* (*positive* on one side and *negative* on the other) for the point to be an inflection point.

77. Yes. The second derivative is positive (above the x-axis) on one side and negative (below the x-axis) on the other, so the concavity *does* change.

79. The inflection point was the first sign of the coming downturn. Alternative answer: The sales were growing fastest at the inflection point.

81. True

83. $f''(x) = 2a$, therefore: For $a > 0$, $f'' > 0$, so f is concave up.
For $a < 0$, $f'' < 0$, so f is concave down.

85. Where the curve is concave *up*, it lies *above* its tangent line, and where it is concave *down*, it lies *below* its tangent line, so *at* an inflection point it must cross its tangent line.

Exercises 3.3 page 199

1. Max f is 12 (at $x = 1$), min f is -8 (at $x = -1$). **3.** Max f is 16 (at $x = -2$), min f is -16 (at $x = 2$).

5. Max f is 9 (at $x = 1$), min f is 0 (at $x = 0$ and at $x = -2$).

7. Max f is 20 (at $x = 1$), min f is -12 (at $x = -3$). **9.** Max f is 81 (at $x = 3$), min f is -16 (at $x = 2$).

11. Max f is 4 (at $x = 2$), min f is -50 (at $x = 5$). **13.** Max f is 5 (at $x = 0$), min f is 0 (at $x = 5$).

15. Max f is 4 (at $x = -1$ and at $x = 2$), min f is 0 (at $x = 0$ and at $x = 3$).

17. Max f is 1 (at $x = 0$), min f is 0 (at $x = -1$ and at $x = 1$).

19. Max f is $\frac{1}{2}$ (at $x = 1$), min f is $-\frac{1}{2}$ (at $x = -1$). **21. a.** The number is $\frac{1}{2}$. **b.** The number is 3.

23. a. Both at endpoints **b.** One at a critical number (the maximum) and one at an endpoint (the minimum)
c. Both at critical numbers **d.** Yes; for example, [2, 10]

25. On the 20th day **27.** 31 mph **29.** 52 **31.** 2/3 **33.** 36 years **35.** 12 miles from A toward B

37. Produce 40 per day, price = $400, max profit = $6500

39. Rent 40 per day, price = $120, max revenue = $4800

41. 400 feet along the building and 200 feet perpendicular to the building

43. Each is 200 yards parallel to the river and 150 yards perpendicular to the river

45. 3 inches high with a base 12 inches by 12 inches; volume: 432 cubic inches

47. The numbers are 25 and 25. **49.** $r = 110/\pi \approx 35$ yards, $x = 110$ yards

51. a. At time 10 hours; 1,500,000 bacteria (since $N(t)$ is in thousands)
b. At time 5 hours; growing by 75,000 bacteria per hour (inflection point)

53. Remove a square of side $x \approx 0.96$ inch; volume ≈ 15 square inches

55. A function can have at most *one* absolute maximum value (the highest value it attains), but it can have several *relative* maximum values (values that are high *relative* to their neighboring values).

57. False. Think of a function like the first graph on page 190.

59. For example **61.** f has a relative minimum at $x = 5$. **63.** True **65.** $r = 2$ cm

There are many others.

Exercises 3.4 page 209

1. Price: $14,400; sell 16 cars per day (from $x = 2$ price reductions)

3. Ticket price: $150; number sold: 450 (from $x = 5$ price reductions)

5. Rent the cars for $90, and expect to rent 54 cars (from $x = 2$ price increases)

7. 25 trees per acre (from $x = 5$ extra trees per acre) **9.** Base: 2 feet by 2 feet; height: 1 foot

11. Base: 14 inches by 14 inches; height: 28 inches; volume: 5488 cubic inches

13. 50 feet along the driveway and 100 feet perpendicular to the driveway; cost: $800

15. 6.4% **17.** 16 years

19. [*Hint:* If area is A (a constant) and one side is x, then show that the perimeter is $P = 2x + 2\dfrac{A}{x}$, which is minimized at $x = \sqrt{A}$. Then show that this means the rectangle is a square.]

21. The page should be 8 inches wide and 12 inches tall.

23. e.   **f.** Price: \$325; quantity: 35 bicycles   **g.** Price: \$350; quantity: 40 bicycles

25. Revenue is price times quantity.

27. \$65. (*Negative* three price reductions means three price *increases* of \$5, so \$50 goes up to \$65.)

29. A linear cost function will have a positive slope (the unit cost), and so will be minimized at its left-hand endpoint. At this point, where $x = 0$, nothing is produced, which is not a realistic solution for a business situation.

31. Producing more will generally drive the price down, which will eventually reduce profits.

33. No. Revenue and profit will generally be maximized at different points, since one has price subtracted and the other does not. A careful look at the graphs on page 41 will show that revenue and profit are maximized at different x-values.

35. $R' = 2cpx - 3cx^2 = xc(2p - 3x)$, which is zero when $x = \frac{2}{3}p$. The second-derivative test will show that R is maximized.

Exercises 3.5 page 217

1. Lot size: 400 boxes; 10 orders during the year   **3.** Lot size: 500 bottles; 20 orders during the year

5. Lot size: 40 cars per order; 20 orders during the year

7. Produce 1000 games per run; 2 production runs during the year

9. Produce 40,000 circuit boards per run; 25 runs for the year

11. Population: 20,000; yield: 40,000 (from $p = 200$)   **13.** Population: 75,000; yield: 2250 (from $p = 75$)

15. Population: 26,000 metric tons; yield: 84,500 metric tons (from $p = 26$)

17. Population: 625,000; yield: 625,000 (from $p = 625$)

19. Population: 100,000 metric tons; yield: 50,000 metric tons (from $p = 1000$)

21. Because of the larger number of items on hand, storage costs will increase and reorder costs will decrease.

23. Lot size should decrease.

25. If $f(p) < p$ for all values of p, next year's population will always be smaller than this year's population, so the animal will eventually become extinct.

27. Animal populations, such as fish in the sea, can be very difficult to estimate. Also, one year's population depends on factors other than last year's population, such as weather, predators, disease, and changing environmental conditions.

Exercises 3.6 page 226

1. $\dfrac{dy}{dx} = \dfrac{2x}{3y^2}$ **3.** $\dfrac{dy}{dx} = \dfrac{3x^2}{2y}$ **5.** $\dfrac{dy}{dx} = \dfrac{3x^2 + 2}{4y^3}$ **7.** $\dfrac{dy}{dx} = -\dfrac{x+1}{y+1}$ **9.** $\dfrac{dy}{dx} = -\dfrac{2y}{x}$ (after simplification)

11. $\dfrac{dy}{dx} = \dfrac{-y+1}{x}$ **13.** $\dfrac{dy}{dx} = -\dfrac{y-1}{2x}$ (after simplification)   **15.** $\dfrac{dy}{dx} = \dfrac{1}{3y^2 - 2y + 1}$

17. $\dfrac{dy}{dx} = -\dfrac{y^2}{x^2}$ (after simplification)   **19.** $\dfrac{dy}{dx} = \dfrac{3x^2}{2(y-2)}$ **21.** $\dfrac{dy}{dx} = 2$ **23.** $\dfrac{dy}{dx} = -3$ **25.** $\dfrac{dy}{dx} = -1$

27. $\dfrac{dy}{dx} = -4$ **29.** $\dfrac{dp}{dx} = -\dfrac{2}{2p+1}$ **31.** $\dfrac{dp}{dx} = \dfrac{1}{24p+4}$ **33.** $\dfrac{dp}{dx} = -\dfrac{p}{3x}$ (after simplification)

35. $\dfrac{dp}{dx} = -\dfrac{p+5}{x+2}$ **37.** $y = 4x + 9$ **39.** $y = -x - 1$

41. $\dfrac{dp}{dx} = -\dfrac{1}{2}$. *Interpretation:* The rate of change of price with respect to quantity is $-1/2$, so price decreases by about \$0.50 or 50¢ when quantity increases by 1.

43. $\dfrac{dx}{dp} = -1$, so sales will decrease by 1 for each \$1 increase in price.

45. $\dfrac{dh}{dp} = \dfrac{1}{25} = 0.04$, so a person will work for an extra $\frac{1}{25}$ hour (2.4 minutes) for an extra \$1 in pay (at the given pay rate).

47. a. $\dfrac{ds}{dr} = \dfrac{3r^2}{2s} = 8$ **b.** $\dfrac{dr}{ds} = \dfrac{2s}{3r^2} = \dfrac{1}{8}$ **c.** $\dfrac{ds}{dr} = 8$ means that the rate of change of sales with respect to

research expenditures is 8, so that increasing research by $1 million will increase sales by about $8 million (at these levels of r and s).

$\dfrac{dr}{ds} = \dfrac{1}{8}$ means that the rate of change of research expenditures with respect to sales is $\dfrac{1}{8}$, so that

increasing sales by $1 million will increase research by about $\dfrac{1}{8}$ million dollars (at these levels of r and s).

49. $\left|\dfrac{dK}{dL}\right| = 6.75$. At a production level of 144 units using 8 units of labor and 27 units of capital, each additional unit of labor reduces the capital requirement by 6.75 units.

51. $3x^2 \dfrac{dx}{dt} + 2y \dfrac{dy}{dt} = 0$ **53.** $2x \dfrac{dx}{dt} y + x^2 \dfrac{dy}{dt} = 0$ **55.** $6x \dfrac{dx}{dt} - 7 \dfrac{dx}{dt} y - 7x \dfrac{dy}{dt} = 0$

57. $2x \dfrac{dx}{dt} + \dfrac{dx}{dt} y + x \dfrac{dy}{dt} = 2y \dfrac{dy}{dt}$ **59.** Decreasing by $72\pi \approx 226$ in³/hr

61. Increasing by $32\pi \approx 101$ cm³/week   **63.** Growing by $16,000 per day

65. Increasing by 400 cases per year   **67.** Slowing by $\frac{1}{2}$ mm/sec per year   **69.** Yes (65.8 mph)

71. $\dfrac{dx}{dt} = 2$, so sales are increasing at the rate of 2 per week.   **73.** $y^3 - x^2 + 1 = 0$

75. $\dfrac{dy}{dx}$ would be the distance driven per gallon of gas, or miles per gallon. $\dfrac{dx}{dy}$ would be the amount of gas used per mile of driving, or gallons per mile (which, you would hope, is a small fraction).

77. No. The point $(10, 10)$ does not satisfy the original equation and so is not on the curve. Substituting $(10, 10)$ gives a meaningless result.

79. $\dfrac{dP}{dt} = -1500$, where P is the population and t is time in years.

81. $\dfrac{dR}{dt} = 2 \dfrac{dP}{dt}$, where R stands for revenue, P for profit, and t for time in years.

Exercises 3.7 page 236

1. $dy = (2x - 4)\, dx$, 0.5   **3.** $dy = \dfrac{-2}{(x - 1)^2}\, dx$, 0.3   **5.** $dy = \dfrac{1 - 6x - x^2}{(x^2 + 1)^2}\, dx$, -0.15

7. $dy = -0.25$, $\Delta y = -0.24$ so dy is a very good approximation for Δy.

9. $dy = -0.1$, $\Delta y = -0.091$ so dy is a very good approximation for Δy.

11. $dy = -0.5$, $\Delta y = -0.48$ so dy is a very good approximation for Δy.

13. $(x + \Delta x)^2 \approx x^2 + 2x\, dx$, 140

15. $\dfrac{1}{x + \Delta x} \approx \dfrac{1}{x} - \dfrac{1}{x^2}\, dx$, 0.16

17. $7 + \frac{1}{14} \approx 7.0714$

19. $2 + \frac{1}{12} \approx 2.0833$

21. -2.1%. At "low" unemployment, increasing unemployment by just 0.5% significantly lowers the inflation rate by 2.1%.

23. $dV = \$47.34$, $\Delta V = \$48.02$, for a difference of just 68 cents.

25. $dR = -6$, $\Delta R = -5.2$, which are fairly close.

27. $3.33, $2153.33   **29.** ± 3 in³, $\pm 0.3\%$ **31.** ± 1.48 mph, $\pm 2.5\%$

33. $100.75. At this level of production, the cost increases about $1.25 per additional hundred items.

35. $4820. At this level of sales, the revenue increases about $1 per additional hundred items. The price will need to be lowered by about 5 cents.

37. $4950. At this level of production and sales, the profit increases by about 5 hundred dollars per additional hundred items.

39. A linear function.

41. $\Delta y \approx dy = a \cdot bx^{b-1} \, dx$ with $\Delta x = \frac{1}{100}x = dx$ becomes

$\Delta y \approx (a \cdot bx^{b-1}) \cdot (\frac{x}{100}) = a(x^{b-1} \cdot x) \cdot \frac{b}{100} = (ax^b) \cdot \frac{b}{100} = y \cdot \frac{b}{100}$ which is a $b\%$ change in y.

43. 32% increase

45. Only if $du = g'(x) \, dx$ is not zero.

Chapter 3 Review Exercises and Chapter Test page 240

1.

2.

3.

4.

5.

6.

7.

8.

9.

10.

11.

12.

13.

14.

15.

16.

17.

18.

19.

20.

21. Max f is 220 (at $x = 5$), min f is -4 (at $x = 1$).   **22.** Max f is 130 (at $x = 5$), min f is -32 (at $x = 2$).

23. Max f is 64 (at $x = 0$), min f is -64 (at $x = 4$).

24. Max f is 6401 (at $x = 10$), min f is 1 (at $x = 0$ and $x = 2$).

25. Max h is 4 (at $x = 9$), min h is 0 (at $x = 1$).

26. Max f is 10 (at $x = 0$), min f is 0 (at $x = 10$ and $x = -10$).

27. Max g is 25 (at $w = 3$ and $w = -3$), min g is 0 (at $w = 2$ and $w = -2$).

28. Max g is 16 (at $x = 4$), min g is 0 (at $x = 0$ and $x = 8$).

29. Max f is $\frac{1}{2}$ (at $x = 1$), min f is $-\frac{1}{2}$ (at $x = -1$).   **30.** Max f is $\frac{1}{4}$ (at $x = 2$), min f is $-\frac{1}{4}$ (at $x = -2$).

31. $v = 2c$, which means that the tugboat should travel through the water at twice the speed of the current.

32. $v = \sqrt[4]{\dfrac{aw^2}{3b}}$ **33.** 3600 square feet   **34.** 1800 square feet

35. 15 cubits (gilded side) by 135 cubits   **36.** Base: 10 inches by 10 inches; height: 5 inches
37. Radius: 2 inches; height: 4 inches   **38.** Radius: 2 inches; height: 2 inches

39. Price: $2400 each; quantity: 9 per week   **40.** 5 weeks   **41.** $x = \frac{3}{4}$ mile
42. a. $t \approx 0.5 = 50\%$ **b.** $2   **43.** Radius ≈ 1.2 inches; height ≈ 4.8 inches
44. $x \approx 1.59$ inches; volume ≈ 33.07 cubic inches   **45.** $x \approx 1.13$ inches; volume ≈ 12.13 cubic inches
46. 600 per run; $1\frac{1}{2}$ runs per year (or 3 runs in 2 years)   **47.** Lot size: 50; 10 orders during the year
48. Population: 150,000 ($p = 150$); yield: 450,000
49. Population: $p = 900$ (thousand); yield: 900 (thousand)

50. $\dfrac{dy}{dx} = \dfrac{-6x - 4y}{4x + y}$ **51.** $\dfrac{dy}{dx} = \dfrac{-y^2}{2xy - 1}$ (after simplification)   **52.** $\dfrac{dy}{dx} = \dfrac{-2y^2 + 6xy}{4xy - 3x^2}$

53. $\dfrac{dy}{dx} = \dfrac{y^{1/2}}{x^{1/2}}$ (after simplification)   **54.** $\dfrac{dy}{dx} = -1$ **55.** $\dfrac{dy}{dx} = \dfrac{1}{7}$ **56.** $\dfrac{dy}{dx} = -\dfrac{1}{6}$

57. $\dfrac{dy}{dx} = 1$ **58.** $y = 2x - 2$ **59.** $y = x + 3$

60. 600 in^3/hr   **61.** Increasing by $4200 per day   **62.** Increasing by $45,000 per month
63. a. Decreasing by about 0.31 cm^3/min   **b.** Decreasing by about 0.05 cm^3/min

64. $dy = \left(1 - \dfrac{1}{2\sqrt{x}}\right) dx, 0.015$

65. $dy = 0.0300$, $\Delta y = 0.0301$ so dy is a very good approximation for Δy.
66. $6 - \frac{1}{12} \approx 5.92$
67. ± 1.2 in^2, $\pm 0.2\%$
68. $401. At this level of production, the cost increases about $10 per additional thousand items.

Cumulative Review for Chapters 1–3 page 243

1. $y = -\frac{1}{2}x + 1$ **2.** $\frac{5}{2}$ **3.** 20.086
4. a.
    b. 4   **5.** $f'(x) = 4x - 5$ **6.** $f'(x) = 12x^{1/2} + 6x^{-3}$
    c. 1
    d. Does not exist   **7.** $f'(x) = 9x^8 + 10x^4 - 8x^3$ **8.** $f'(x) = \dfrac{11}{(3x - 2)^2}$
    e. Discontinuous at $x = 3$ **9.** $y = \frac{3}{4}x - 2$

10. $P'(8) = 1200$, so in 8 years the population will be increasing by 1200 people per year.
    $P''(8) = -50$, so in 8 years the rate of growth will be slowing by 50 people per year.

11. $\dfrac{2x}{\sqrt{2x^2 - 5}}$ **12.** $12(3x + 1)^3(4x + 1)^3 + 12(3x + 1)^4(4x + 1)^2 = 12(3x + 1)^3(4x + 1)^2(7x + 2)$

13. $\dfrac{12(x - 2)^2}{(x + 2)^4}$

14. $y = 3x + 7$

15.

16.

17. 15,000 square feet

18. Price: $170; quantity: 18 per day

19. $\dfrac{dy}{dx} = \dfrac{-x^2 - 3y^2}{6xy + 1}$ (after simplification). Evaluating at (1, 2) gives $\dfrac{dy}{dx} = -1$.

20. $\dfrac{2}{\pi} \approx 0.64$ ft/min

21. $dy = -0.12$, $\Delta y = -0.1208$ so dy is a very good approximation for Δy.

22. ±8 feet, ±8%

Exercises 4.1 page 256

1. a. 7.389   **b.** 0.135   **c.** 1.649

3. a. e^3 **b.** e^2 **c.** e^5

5.   **7.**   **9.** 5.697 (rounded)

11. a. e^x **b.** e^x **c.** e^x **d.** e^x **e.** e^x will exceed any power of x for large enough values of x.

13. a. $2144   **b.** $2204   **c.** $2226   **15. a.** $2196   **b.** $2096

17. The annual yield should be 9.69% (based on the nominal rate of 9.25%).

19. $8629   **21.** $72.65   **23.** $496.97   **25.** 10% compounded quarterly (yielding 10.38%, better than 10.30%)

27. a. $3517   **b.** $15,883   **29.** 8 billion

31. a. 0.53 (the chances are better than 50–50)   **b.** 0.70 (quite likely)

33. a. 0.267 or 26.7%   **b.** 0.012 or 1.2%   **35. a.** 1.3 milligrams   **b.** 0.84 milligram

37. 208   **39. a.** About 153 degrees   **b.** About 123 degrees   **41.** 38   **43.** 6.5%

45. By about 25%   **47. b.** In about 2054 (from $x \approx 42$)   **49.** About 3678% per year   **51.** $59,079

53. The x-axis $(y = 0)$ is a horizontal asymptote.   **55.** e^x **57.** That its growth is proportional to its size.
59. With 5% *monthly* you must wait until the end of the month to receive interest, while for 5% compounded *continuously* you begin receiving interest right away, so the interest begins earning interest without any delay.
61. Depreciation *by a fixed percentage* gives the bigger decrease in the first year, and *straight line* gives the bigger decrease in the last year.

Exercises 4.2 page 271

1. a. 2   **b.** 4   **c.** -1 **d.** -2 **e.** $\frac{1}{2}$ **f.** $-\frac{1}{2}$ **3. a.** 10   **b.** $\frac{1}{2}$ **c.** $\frac{4}{3}$ **d.** 0   **e.** 1   **f.** -3
5. $\ln x$ **7.** $2\ln x$ or $\ln x^2$ **9.** $\ln x$ **11.** $3x$ **13.** $7x$ **15.** Domain: $\{x \mid x > 1 \text{ or } x < -1\}$; Range: $\mathbb{R}$
17. a. 2.9 years (from 35 months)   **b.** 1.7 years (from 20.5 months)   **19. a.** 15.7 years   **b.** 3.2 years
21. 1.9 years   **23.** About 17.1 years   **25.** 77 days   **27.** 0.58 or 58%   **29.** About 4 weeks
31. About 31,400 years   **33.** 4.45 million years   **35.** About 79 days   **37.** About 5 years (from 4.67)
39. a. 35 years   **b.** 55.5 years
41. a. About 11.6 years (from 46.6 quarters)   **b.** About 6.8 years (from 27.2 quarters)
43. a. About 9 days   **b.** About 11 days
45. About 5300 years old. (For reference, this means that Iceman lived 2000 years before King Tutankhamen.)
47. About 6.5 years   **49.** In 7 years and 10 months   **51.** $\ln 1 = 0$ **53.** $\ln \frac{x}{y} = \ln x - \ln y$ **55.** $x = \log y$
57. $\ln(-2) = x$ would mean that $e^x = -2$, which is impossible because e^x is always positive.
59. *Continuous* compounding would give the shortest, and *annual* would give the longest.
61. b. About 39 hours   **63.** Solving $0.94 = (1 - \frac{1}{1000})^x$ gives $x = \frac{\ln 0.94}{\ln 0.999} \approx 61.84$, so about 62 generations.
65. a. 12 years   **b.** 11.9 years   **67.** $\frac{\ln k}{\ln(1 + r)}$

Exercises 4.3 page 285

1. $2x\ln x + x$ **3.** $\frac{2}{x}$ **5.** $\frac{1}{2}x^{-1}$ **7.** $\frac{6x}{x^2 + 1}$ **9.** $\frac{1}{x}$ **11.** $\frac{xe^x - 2e^x}{x^3}$ or $\frac{e^x(x-2)}{x^3}$
13. $(3x^2 + 2)e^{x^3 + 2x}$ **15.** $x^2 e^{x^3/3}$ **17.** $1 + e^{-x}$ **19.** 2   **21.** $e^{1+e^x}e^x$ or e^{1+x+e^x} **23.** exe^{-1} **25.** 0
27. $\frac{4x^3}{x^4 + 1} - 2e^{x/2} - 1$ **29.** $2x\ln x + 2xe^{x^2}$ **31.** $e^x\left(\ln x^2 + \frac{2}{x}\right)$ **33.** $\frac{-2}{x}$ **35.** $6e^{2t}(e^{2t} + 1)^2$
37. $\frac{t + \frac{1}{t}}{\sqrt{t^2 + 2\ln t}}$ **39.** $\frac{e^z(z^2 - 2z - 1)}{(z^2 - 1)^2}$ **41.** $\frac{20e^{-2z}}{(1 + e^{-2z})^2}$ **43.** $\frac{-4}{(e^x - e^{-x})^2}$ **45. a.** $\frac{1 - 5\ln x}{x^6}$ **b.** 1
47. a. $\frac{4x^3}{x^4 + 48}$ **b.** $\frac{1}{2}$ **49. a.** $\frac{e^x - 3}{e^x - 3x}$ **b.** -2 **51.** $y = 2x - 4$ **53.** $y = 2x - 1$
55. a. $5\ln x + 5$ **b.** $5\ln 2 + 5 \approx 8.466$ **57. a.** $\frac{xe^x - e^x}{x^2}$ **b.** $\frac{2e^3}{9} \approx 4.463$
59. $-4x^3 e^{-x^5/5} + x^8 e^{-x^5/5}$ or $x^3 e^{-x^5/5}(x^5 - 4)$ **61.** $f^{(n)}(x) = k^n e^{kx}$
63. $dy = -2e^{-2x}\,dx, -e^{-6} \approx -0.00248$ **65.** $dy = (e^x + xe^x)\,dx, 0.2e \approx 0.544$ **67.** $dy = (2x\ln x + x)\,dx, 0.03e \approx 0.0815$
69.

rel max: $(0, 1)$
IP: $(\frac{1}{2}, 0.61)$
$(-\frac{1}{2}, 0.61)$
on $[-2, 2]$ by $[-1, 2]$

71.

rel min: $(0, 0)$
IP: $(1, 0.69)$
$(-1, 0.69)$
on $[-5, 5]$ by $[-1, 4]$

73.

rel min: $(0, 0)$
rel max: $(2, 0.54)$
IP: $(0.59, 0.19)$
$(3.41, 0.38)$
on $[-1, 8]$ by $[-1, 3]$

75.

rel max: $(-0.37, 0.37)$
rel min: $(0.37, -0.37)$
on $[-2, 2]$ by $[-2, 2]$

77. $\frac{dy}{dx} = \frac{ye^x}{2y - e^x}$ **79.** 9479, so annual salary increases by about $9480 per additional year of calculus (after the first year).

81. a. Increasing by $50 per year   **b.** Increasing by about $82.44 per year

83. Increasing by about 3.5 million people per year

85. a. Decreasing by 0.06 mg/hr   **b.** Decreasing by 0.054 mg/hr

87. a. Increasing by about 81.4 (thousand) sales per week   **b.** Increasing by about 33 (thousand) sales per week

89. $p = \$100$ **91. a.** $R(x) = 400xe^{-0.20x}$ **b.** Quantity: $x = 5$ (thousand); price: $p = \$147.15$

93. a. After 15 minutes the temperature of the beer is 57.5 degrees and increasing at the rate of 43.8 degrees per hour.
  b. After 1 hour the temperature of the beer is 69.1 degrees and increasing at the rate of 3.2 degrees per hour.

95. 1.83. In the 2000s, the percentage of the U.S. population with diabetes was increasing by about 1.83% per year.

97. 2.3 seconds

99. a.   **b.** $f(35) = 132.4, f'(x) = 0.834$. At age 35, the fastest man's time is 2 hours 12.4 minutes and increasing at about 0.834 minute per year.
  c. $f(80) = 234.9, f'(80) = 7.76$. At age 80, the fastest man's time is 3 hours 54.9 minutes and increasing at about 7.76 minutes per year.

101. e^x **103.** 0 (since e^5 is a constant)   **105.** $\dfrac{1}{x}$ **107.** 0 (since ln 5 is a constant)

109. e^x and $\ln x$ are inverse functions, so $e^{\ln x} = x$, and the derivative of x is 1.

111. When $N < K$, $\ln(K/N)$ is positive and the population is increasing, while when $N > K$, $\ln(K/N)$ is negative and the population is decreasing.

113. $r = a/b$ **117.** $e^{0.802x}$ (rounded)

119. a. $(\ln 10)10^x$ **b.** $(\ln 3)(2x)3^{x^2+1}$ **c.** $(\ln 2)3 \cdot 2^{3x}$ **d.** $(\ln 5)6x \cdot 5^{3x^2}$ **e.** $-(\ln 2)2^{4-x}$

121. a. $\dfrac{1}{(\ln 2)x}$ **b.** $\dfrac{2x}{(\ln 10)(x^2 - 1)}$ **c.** $\dfrac{4x^3 - 2}{(\ln 3)(x^4 - 2x)}$

Exercises 4.4 page 299

1. a. $2/t$ **b.** 2 and 0.2   **3. a.** 0.2   **b.** 0.2   **5. a.** $2t$ **b.** 20   **7. a.** $-2t$ **b.** -20

9. a. $\dfrac{1}{2(t-1)}$ **b.** $\frac{1}{10}$ **11.** 0.507, so data traffic is growing by about 51% per year.

13. 0.0071 or 0.71%   **15. a.** 0.012 or 1.2%   **b.** Yes, in about 15.3 years

17. a. $E(p) = \dfrac{5p}{200 - 5p}$ **b.** Inelastic $(E = \frac{1}{3})$ **19. a.** $E(p) = \dfrac{2p^2}{300 - p^2}$ **b.** Unit-elastic $(E = 1)$

21. a. $E(p) = 1$ **b.** Unit-elastic $(E = 1)$ **23. a.** $E(p) = \dfrac{3p}{2(175 - 3p)}$ **b.** Elastic $(E = 3)$

25. a. $E(p) = 2$ **b.** Elastic $(E = 2)$ **27. a.** $E(p) = 0.01p$ **b.** Elastic $(E = 2)$

29. Lower the price $(E = 8)$ **31.** No $(E = 1.25)$ **33.** Yes $(E = \frac{3}{8} = 0.375)$ **35.** Lower its price $(E = 7.2)$

37. $E = 0.112$ **39. a.** $E \approx 0.35$ **b.** Raise the price   **c.** $p \approx \$20,\!400$ **41.** $E \approx 0.52$ (from 8.65%/16.7%)

43. 1.2   **45.** $\dfrac{n}{x}$ **47.** $E(p) = 2p$ **49.** Increase strongly

51. Inelastic for cigarettes (they are habit forming), and elastic for jewelry

53. $E(p) = \dfrac{-pa(-c)e^{-cp}}{ae^{-cp}} = cp$ **55.** $E_s(p) = n$

Chapter 4 Review Exercises and Chapter Test page 303

1. a. $18,845.41   **b.** $18,964.81   **2.** The second bank $(1.015^4 \approx 1.0614 < e^{0.0598} \approx 1.0616)$

3. a. $V(t) = 800,\!000(0.8)^t$ **b.** $327,680   **4.** Drug B   **5.** In about 2060 (from $x \approx 50.6$ years after 2010)

6. $4^{10} = 1,\!048,\!576$ megabits, which is enough to hold more than 200,000 books on one chip.

7. a. 7.1 years (from 14.2 half-years)   **b.** 4.2 years (from 8.3 half-years)

8. a. 9.9 years   **b.** 5.8 years   **9.** 50.7 million years   **10.** 1.85 million years   **11.** 2.3 hours

12. 13.7 years   **13. a.** In about 6.25 years (from $x \approx 25$ quarters)   **b.** In about 6.2 years

14. a. About 11 days   **b.** About 16 days   **15.** $\dfrac{1}{x}$ **16.** $\dfrac{4x}{x^2 - 1}$ **17.** $\dfrac{-1}{1-x}$ or $\dfrac{1}{x-1}$

18. $\dfrac{x}{x^2 + 1}$ **19.** $\dfrac{1}{3x}$ **20.** 1   **21.** $\dfrac{2}{x}$ **22.** $\ln x$ **23.** $-2xe^{-x^2}$ **24.** $-e^{1-x}$ **25.** $2x$

30. $4 - 4xe^{2x} - 4x^2e^{2x}$ **31.** $y = -4x - 3$ **32.** $y = 3x - 5$

33.

rel min: $(0, \ln 4) \approx (0, 1.4)$
IP: $(2, \ln 8) \approx (2, 2.1)$
$(-2, \ln 8) \approx (-2, 2.1)$

34.

rel max: $(0, 16)$
IP: $(2, 16e^{-1/2}) \approx (2, 9.7)$
$(-2, 16e^{-1/2}) \approx (-2, 9.7)$

35. a. Increasing by 136 thousand per week   **b.** Increasing by 55 thousand per week

36. a. Decreasing by 0.12 mg per hour   **b.** Decreasing by 0.08 mg per hour

37. Decreasing by 33.3% per second

38. a. Increasing by 3.5 degrees per hour   **b.** Increasing by 2.1 degrees per hour

39. a. Increasing by 6667 per hour   **b.** Increasing by 816 per hour   **40.** 25 years

41. a. $R(x) = 200xe^{-0.25x}$ **b.** Quantity $x = 4$ (thousand);   price $p = \$73.58$

42. a. $R(x) = 5x - x \ln x$ **b.** Quantity $x = e^4 \approx 54.60$;   price = \$1   **43.** Price = \$50

44.

on $[-2, 10]$ by $[-1, 10]$

rel min: $(0, 0)$
rel max: $(4, 4.69)$
IP: $(2, 2.17)$
$(6, 3.21)$

45.

on $[-2, 2]$ by $[-2, 2]$

rel max: $(-0.72, 0.12)$
rel min: $(0.72, -0.12)$
IP: $(-0.43, 0.07)$
$(0.43, -0.07)$

46. $p \approx \$769$ **47.** $x \approx 130$ (in thousands)   **48.** 0.0033 or 0.33%   **49.** 0.0031 or 0.31%

50. Raise prices $(E = 0.8)$ **51.** Raise prices $(E = 0.8)$ **52.** $E = 0.44$ **53.** 0.104 or 10.4%

54. a. $E \approx 1.29$ **b.** Lower the price   **c.** About \$8700 (from $p \approx 8.7$)

Exercises 5.1 page 317

1. $\frac{1}{5}x^5 + C$ **3.** $\frac{3}{5}x^{5/3} + C$ **5.** $\frac{2}{3}u^{3/2} + C = \frac{2}{3}\sqrt{u^3} + C$ **7.** $-\frac{1}{3}w^{-3} + C = -\frac{1}{3w^3} + C$

9. $2\sqrt{z} + C$ **11.** $x^6 + C$ **13.** $4x^2 - 5x + C$ **15.** $2x^4 - x^3 + 2x + C$ **17.** $4x^{3/2} + \frac{3}{2}x^{2/3} + C = 4\sqrt{x^3} + \frac{3}{2}\sqrt[3]{x^2} + C$

19. $6x^{8/3} + 24x^{-2/3} + C = 6\sqrt[3]{x^8} + \frac{24}{\sqrt[3]{x^2}} + C$ **21.** $6t^{5/3} + 3t^{1/3} + C = 6\sqrt[3]{t^5} + 3\sqrt[3]{t} + C$

23. $-2z^{-2} + 2z^{1/2} + C = -\frac{2}{z^2} + 2\sqrt{z} + C$ **25.** $\frac{1}{3}x^3 - x^2 + x + C$

27. $3x^4 - 4x^3 + C$ **29.** $\frac{2}{3}w^{3/2} + 4w^{5/2} + C = \frac{2}{3}\sqrt{w^3} + 4\sqrt{w^5} + C$ **31.** $2x^3 - 3x^2 + x + C$ **33.** $\frac{1}{3}x^3 + x^2 - 8x + C$

35. $\frac{1}{3}r^3 - r + C$ **37.** $\frac{1}{2}x^2 - x + C$ **39.** $\frac{1}{4}t^4 + t^3 + \frac{3}{2}t^2 + t + C$

41. b.

on $[-3, 3]$ by $[-5, 5]$

43. $C(x) = 8x^{5/2} - 9x^{5/3} + x + 4000$ **45.** $R(x) = 120x - 9x^{4/3}$

47. a. $D(t) = -0.08t^3 + 9t^2$ **b.** 820 feet   **49. a.** $6t^{1/2} = 6\sqrt{t}$ **b.** 30 words

51. a. $P(t) = 16t^{5/2} = 16\sqrt{t^5}$ **b.** 512 tons   **c.** No

53. $P(x) = -\frac{1}{3}x^3 + 25x^2 - 300x + 31,000$; 35,500 thousand (or 35.5 million)

55. $R(x) = 1.6x^2 + 17.4x + 74$; \$274.2 billion

57. $\dfrac{1}{x}$ **59.** C (a constant)   **61.** They differ by a constant (since they are integrals of the same function).

63. a. $\int x \cdot x \, dx = \int x^2 \, dx = \frac{1}{3}x^3 + C$

b. $x \cdot \int x \, dx = x\left(\frac{1}{2}x^2 + C\right) = \frac{1}{2}x^3 + Cx$ [*Note:* The C's in parts (a) and (b) may differ.]

c. No, so you *cannot* move a variable across the integral sign.

65. a. $\int \frac{x}{x} \, dx = \int 1 \, dx = x + C$

b. $\dfrac{\int x \, dx}{\int x \, dx} = \dfrac{\frac{1}{2}x^2 + C}{\frac{1}{2}x^2 + C'}$ [*Note:* The prime is to indicate that constant C' in the denominator may be different from the constant C in the numerator.]

c. No, so the integral of a quotient is *not* the quotient of the integrals.

67. Cost   **69. a.** Differentiate   **b.** Integrate   **c.** Differentiate   **d.** Integrate

Exercises 5.2 page 327

1. $\frac{1}{3}e^{3x} + C$ **3.** $4e^{x/4} + C$ **5.** $20e^{0.05x} + C$ **7.** $-\frac{1}{2}e^{-2y} + C$ **9.** $-2e^{-0.5x} + C$ **11.** $9e^{2x/3} + C$

13. $-5 \ln |x| + C$ **15.** $3 \ln |x| + C$ **17.** $\frac{3}{2} \ln |v| + C$ **19.** $2e^{2x} - 3x^2 + C$ **21.** $\frac{1}{3}e^{3x} - 3 \ln |x| + C$

23. $6e^{0.5t} - 2\ln|t| + C$ **25.** $\frac{1}{3}x^3 + \frac{1}{2}x^2 + x + \ln|x| - x^{-1} + C$ **27.** $250e^{0.02t} - 200e^{0.01t} + C$

29. $\frac{1}{2}e^w - \frac{1}{4}w^2 + C$ **31.** $\frac{1}{2}z^2 + \ln|z| + C$ **33.** $e^x + \ln|x| + C$ **35.** $\frac{1}{2}x^2 + 2x + \ln|x| + C$

37. $t + 2\ln|t| + 3t^{-1} + C$ **39.** $\frac{1}{3}x^3 - 3x^2 + 12x - 8\ln|x| + C$ **41. a.** $360e^{0.05t} - 355$ **b.** About 624 cases

43. a. $50\ln t$ (since $t > 1$, absolute value bars are not needed)   **b.** No (about 170 sold)

45. a. $17,100e^{0.02t} - 17,100$ **b.** In about 2027 (13 years from 2014)   **47. a.** $36,000e^{0.05x} - 36,000$ **b.** About \$10,225

49. a. $60e^{-0.2t} + 10$ **b.** In about 5 hours   **51. a.** $-4000e^{-0.2t} + 4000$ **b.** About $3\frac{1}{2}$ years

53. a. $8000e^{0.05t} - 3000$ **b.** About \$10,190   **55. a.** $100e^{0.05t} - 100$ **b.** In about 2027 (13 years from 2014)

57. a. $25e^{0.086t} - 25$ **b.** About 16.9 million   **59.** $e^x + C$ **61.** $\dfrac{1}{e+1}x^{e+1} + C$ (using the Power Rule for x to a constant power)

63. $e^{-1}x + C$ [e^{-1} is a constant (it has no variable) and the integral of a constant is the constant times x plus C.]

65. $\displaystyle\int e^{ax+b}\,dx = \int e^{ax}e^b\,dx = e^b\int e^{ax}dx = e^b\left(\frac{1}{a}e^{ax} + C\right) = \frac{1}{a}e^{ax}e^b + Ce^b = \frac{1}{a}e^{ax+b} + C'$ (where C' is another constant)

67. No. The box applies only to $\displaystyle\int \frac{1}{x}\,dx$ and not to $\displaystyle\int \frac{1}{x^5}\,dx$.   For the latter, use the Power Rule.

Exercises 5.3 page 340

1. 2.75 square units   **3.** 4.15 square units   **5.** 0.760 square unit   **7. i.** 2.8 square units   **ii.** 3 square units

9. i. 4.411 square units (or 4.412, depending on rounding)   **ii.** $\frac{14}{3} \approx 4.667$ square units

11. i. 0.719 square unit   **ii.** $\ln 2 \approx 0.693$ square unit

13. i. for left rectangles: 2.9, 2.99, 2.999   **15. i.** for left rectangles: 4.515, 4.652, 4.665
     for midpoint rectangles: 3, 3, 3   for midpoint rectangles: 4.668, 4.667, 4.667
     for right rectangles: 3.1, 3.01, 3.001   for right rectangles: 4.815, 4.682, 4.668
   ii. 3 square units   **ii.** $\frac{14}{3} \approx 4.667$ square units

17. i. for left rectangles: 0.719, 0.696, 0.693
     for midpoint rectangles: 0.693, 0.693, 0.693
     for right rectangles: 0.669, 0.691, 0.693
   ii. $\ln 2 \approx 0.693$ square unit

19.

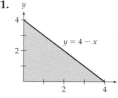

9 square units

21.

8 square units

23.

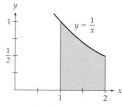

$\ln 2$ square unit

25. 160 square units   **27.** 19 square units   **29.** 2 square units   **31.** 16 square units

33. $\ln 5$ square units   **35.** $\ln 2 + \frac{7}{3}$ square units   **37.** 4 square units (from $2e^{\ln 3} - 2$)

39. $2e - 2$ square units   **41.** 5 square units   **43.** $\frac{15}{32}$ square unit   **45.** $4 + \ln 2$ square units

47. $\frac{111}{100}$ **49.** 13   **51.** $\frac{3}{4}$ **53.** 1 (from $\ln e - \ln 1$)   **55.** $78 + \ln 3$ **57.** $-3\ln 2$ **59.** $4e^3 - 4$

61. $5e - 5e^{-1}$ **63.** 1 (from $e^{\ln 3} - e^{\ln 2}$)   **65.** $\frac{7}{2} + \ln 2$ **67.** 1.107   **69.** 2.925   **71.** 92.744

73. a. 9   **b.** Completing the calculation: $= (\frac{1}{3}3^3 + C) - (\frac{1}{3}0^3 + C) = \frac{1}{3}3^3 + C - \frac{1}{3}0^3 - C = \frac{1}{3}\cdot 27 = 9$
     c. The C always cancels because it is both added and subtracted in the evaluation step.

75. $3\frac{1}{2} + \frac{\pi}{4}$ **77.** 132 units   **79.** 411 checks   **81.** $-300e^{-2} + 300 \approx \$259.40$ **83.** \$2592 billion

85. $300e^{0.5} - 300 \approx 195$, so about \$1.95   **87.** $17,100e^{0.2} - 17,100 \approx 3786$ thousand metric tons

89. $-30e^{-2} + 30 \approx 26$ words   **91. a.** $\dfrac{1}{2}, \dfrac{1}{3}, \dfrac{1}{4}, \dfrac{1}{5}$ **b.** Area $= \dfrac{1}{n+1}$ **93.** 87 milligrams

95. 586 cars   **97.** 4023 people   **99.** Underestimate   **101.** Overestimate

103. The second friend is right: Indefinite integrals are antiderivatives plus constants, and definite integrals are limits of Riemann sums. The Fundamental Theorem of Integral Calculus shows how to evaluate one using the other, but the two ideas are not the same.

105. 0   (since the integral is "from 1 to 1")

107. The Fundamental Theorem of Integral Calculus requires that the function be continuous on the interval, and this function is not even *defined* at $x = 0$, so the theorem cannot be applied. This particular integral cannot be found.

109. $\dfrac{a}{-b+1}B^{-b+1} - \dfrac{a}{-b+1}A^{-b+1} = \dfrac{a}{1-b}(B^{1-b} - A^{1-b})$ **111.** $-4{,}000{,}000e^{-0.6} + 4{,}000{,}000 \approx \$1{,}804{,}753$

Exercises 5.4 page 352

1. 3   **3.** 4   **5.** $\dfrac{1}{5}$ **7.** 5   **9.** $\dfrac{104}{3}$ or $34\dfrac{2}{3}$ **11.** 2   **13.** 3   **15.** $e - 1$ **17.** $10(e^{0.1} - 1) \approx 1.05$

19. $\ln 2$ **21.** $\dfrac{1}{n+1}$ **23.** $a + b$ **25.** About 0.845   **27.** 318   **29.** $70°$ **31.** About 25.6 tons

33. \$3194.53   **35.** \$1.55 billion (from \$1547 million)   **37.** 6 square units   **39.** $\frac{1}{2}e^4 - e^2 + \frac{1}{2}$ square units

41. a.   **43. a.**   **45.** 4 square units   **47.** 32 square units

49. 32 square units   **51.** 32 square units

53. $\frac{1}{12}$ square unit   **55.** 2 square units

57. 1250 square units

59. 5.694 square units (rounded)

    b. 9 square units   **b.** 18 square units

61. About 430 million   **63. a.** 10   **b.** \$6000   **65.** About \$529 billion

67. About \$139 thousand   **69.** About 104,000 lives   **71.** $(2x + 5)e^{x^2 + 5x}$ **73.** $\dfrac{2x+5}{x^2+5x}$

75. The curve rises more steeply in the beginning and so spends "more of the time" higher up, and so its average should be higher than the "middle" y-value.

77. $\dfrac{y_1 + y_2}{2}$ (since the average height of a line is at its midpoint)   **79.** The average value will be greater than 7.

81. $-c$ **83.** The additional sales during the month resulting from the advertising.

Exercises 5.5 page 363

1. \$60,000   **3.** \$10,000   **5.** \$7500   **7.** \$5285 (rounded)   **9.** \$100   **11.** \$160,000

13. a. $x = 500$ **b.** \$50,000   **c.** \$25,000   **15. a.** $x = 50$ **b.** \$2500   **c.** \$7500

17. a. $x \approx 119.48$ **b.** \$10,065   **c.** \$3446 (all rounded)   **19.** 0.52   **21.** 0.35   **23.** 0.2

25. $1 - \dfrac{2}{n+1} = \dfrac{n-1}{n+1}$ **27.** 0.46   **29.** 0.28   **31.** $L(x) = x^{2.13}$, Gini index ≈ 0.36

33. $4(x^5 - 3x^3 + x - 1)^3(5x^4 - 9x^2 + 1)$ **35.** $\dfrac{4x^3}{x^4+1}$ **37.** $3x^2e^{x^3}$

39. When demand is 0, the price is 1000, or equivalently, at a price of 1000 the demand is 0, so 1000 is such a high price that *no one* will buy the product.

41. When demand is 0, the price is 0, or equivalently, at a price of 0 the demand is 0, so whatever they are, you can't even *give* them away.

43. Upward, since at a higher price producers should be willing to make more of them.

45. No. Such a curve would mean, for example, that the lowest paid 10% of the population is making *more* than 10% of the income, but then they can't be the lowest paid 10%.

47. Good thing: Everyone has the same income, so no one is poorer than anyone else. Bad thing: everyone is paid the same, no matter whether they are lazy or hard working, so ambition is not rewarded. (Other answers are possible.)

Exercises 5.6 page 374

1. $\frac{1}{10}(x^2 + 1)^{10} + C$ **3.** $\frac{1}{20}(x^2 + 1)^{10} + C$ **5.** $\frac{1}{5}e^{x^5} + C$ **7.** $\frac{1}{6}\ln(x^6 + 1) + C$

9. $u = x^3 + 1$, $du = 3x^2\,dx$: the powers in the integrand and the du do not match.

11. $u = x^4$, $du = 4x^3\,dx$: the powers in the integrand and the du do not match.   **13.** $\frac{1}{24}(x^4 - 16)^6 + C$

15. $-\frac{1}{2}e^{-x^2} + C$ **17.** $\frac{1}{3}e^{3x} + C$ **19.** Cannot be found by our substitution formulas

21. $\frac{1}{5}\ln|1 + 5x| + C$ **23.** $\frac{1}{4}(x^2 + 1)^{10} + C$ **25.** $\frac{1}{5}(z^4 + 16)^{5/4} + C$

27. Cannot be found by our substitution formulas   **29.** $\frac{1}{24}(2y^2 + 4y)^6 + C$ **31.** $\frac{1}{2}e^{x^2+2x+5} + C$

33. $\frac{1}{12}\ln|3x^4 + 4x^3| + C$ **35.** $-\frac{1}{12}(3x^4 + 4x^3)^{-1} + C$ **37.** $-\frac{1}{2}\ln|1 - x^2| + C$ **39.** $\frac{1}{16}(2x - 3)^8 + C$

41. $\frac{1}{2}\ln(e^{2x} + 1) + C$ **43.** $\frac{1}{2}(\ln x)^2 + C$ **45.** $2e^{x^{1/2}} + C$ **47.** $\frac{1}{4}x^4 + \frac{1}{3}x^3 + C$

49. $\frac{1}{6}x^6 + \frac{2}{5}x^5 + \frac{1}{4}x^4 + C$ **51.** $\frac{1}{2}e^9 - \frac{1}{2}$ **53.** $\frac{1}{2}\ln 2$ **55.** $32\frac{2}{3}$ **57.** $-\ln 2$ **59.** $3e^2 - 3e$

61. a. $u^n u'$ **b.** $\displaystyle\int u^n u'\, dx$ **63. a.** $\dfrac{u'}{u}$ **b.** $\displaystyle\int \dfrac{u'}{u}\, dx$ **65.** $\dfrac{1}{2}\ln(2x+1) + 50$ **67.** $\dfrac{1}{2}$ million

69. $\frac{1}{3}\ln 5 - \frac{1}{3}\ln 2 \approx 0.305$ million **71.** $20\frac{1}{3}$ units **73.** About 346 **75.** $\frac{1}{2}\ln 10 \approx 1.15$ tons

77. 56 pounds **79.** Yes—one is as easy (or as hard) as the other.

81. It simply changes the "name" of the variable, changing, for example, $\displaystyle\int e^{x^2}\, dx$ into $\displaystyle\int e^{u^2}\, du$, which is just as difficult.

83. It ignores the fact that dx and $du = 2x\, dx$ are different, so one cannot be substituted for the other.

85. $\frac{1}{6}(x-5)^6 + \frac{6}{5}(x-5)^5 + C$ **87.** $\frac{1}{8}(x-2)^8 + \frac{2}{7}(x-2)^7 + C$ **89.** $x - 2 + \ln|x-2| + C = x + \ln|x-2| + C'$

91. $\frac{3}{7}(x-4)^{7/3} + 3(x-4)^{4/3} + C$ **93.** $\frac{2}{3}(x+2)^{3/2} - 4(x+2)^{1/2} + C$

Chapter 5 Review Exercises and Chapter Test page 378

1. $8x^3 - 4x^2 + x + C$ **2.** $3x^4 + 3x^2 - 3x + C$ **3.** $4x^{3/2} - 5x + C$ **4.** $6x^{4/3} - 2x + C$

5. $6x^{5/3} - 2x^2 + C$ **6.** $2x^{5/2} - 3x^2 + C$ **7.** $\frac{1}{3}x^3 - 16x + C$ **8.** $x^3 + x^2 + 4x + C$

9. $C(x) = 2x^{1/2} + 4x + 20{,}000$ **10. a.** $P(t) = 200t^{3/2} + 40{,}000$ **b.** 52,800 people

11. $2e^{x/2} + C$ **12.** $-\frac{1}{2}e^{-2x} + C$ **13.** $4\ln|x| + C$ **14.** $2\ln|x| + C$ **15.** $2e^{3x} - 6\ln|x| + C$

16. $\frac{1}{2}x^2 - \ln|x| + C$ **17.** $3x^3 + 2\ln|x| + 2e^{3x} + C$ **18.** $-x^{-1} + \ln|x| - e^{-x} + C$

19. a. $7300e^{0.02t} - 7300$ **b.** In about 2025 (11 years after 2014)

20. a. $2000e^{0.1t} - 2000$ **b.** About 5.6 years **21. a.** $250e^{0.06t} - 250$ **b.** About 2026 (12 years after 2014)

22. a. $200\ln x$ **b.** In about 20 months **23.** 36 **24.** 90 **25.** 4 **26.** $\ln 5$

27. $1 - e^{-2}$ **28.** $2e - 2$ **29.** $20e^5 - 100e + 80$ **30.** $25e^{0.4} - 50e^{0.2} + 25$ **31.** 13 square units

32. 36 square units **33.** $6e^6 - 6$ square units **34.** $2e^2 - 2$ square units **35.** $\ln 100$ square units

36. $\ln 1000$ square units

37. Area ≈ 21.4 square units

on $[-2, 2]$ by $[0, 10]$

38. Area ≈ 2.54 square units

on $[-1, 1]$ by $[0, 3]$

39. 64 pounds **40.** 9 words **41.** About 6.3 degrees **42.** \$1640 **43.** \$653.39

44. About 143 pages **45. a.** 2.28 square units **b.** $\frac{8}{3} \approx 2.667$ square units

46. a. 4.884 square units **b.** $\frac{16}{3} \approx 5.333$ square units

47. a. for left rectangles: 5.899, 7.110, 7.239 **b.** $e^2 - e^{-2} \approx 7.254$ square units
   for midpoint rectangles: 7.206, 7.253, 7.254
   for right rectangles: 8.801, 7.400, 7.268

48. a. for left rectangles: 1.506, 1.398, 1.387 **b.** $\ln 4 \approx 1.386$ square units
   for midpoint rectangles: 1.383, 1.386, 1.386
   for right rectangles: 1.281, 1.375, 1.385

49. $\frac{4}{3}$ square units **50.** 108 square units **51.** $\frac{1}{6}$ square unit **52.** $\frac{3}{10}$ square unit

53. 2 square units **54.** $\frac{1}{2}$ square unit **55.** About 17.13 square units **56.** About 0.496 square unit

57. $\frac{1}{3}\ln 4$ **58.** $\frac{28}{3}$ or $9\frac{1}{3}$ **59.** About 4.72 **60.** About 2.77 **61.** About 457 million

62. About \$5800 **63.** About \$629,000 (from 6.29 hundred thousand dollars) **64.** About \$49.95

65. 27 square meters **66.** About \$934 billion **67.** \$480,000

68. \$160,000 **69.** About \$5623 **70.** About \$611 **71.** About 0.56 **72.** About 0.43

73. About 0.23 **74.** About 0.68 **75.** $\frac{1}{4}(x^3 - 1)^{4/3} + C$ **76.** $\frac{1}{6}(x^4 - 1)^{3/2} + C$

77. Cannot be integrated by our substitution formulas

78. Cannot be integrated by our substitution formulas

79. $-\frac{1}{3}\ln|9 - 3x| + C$ **80.** $-\frac{1}{2}\ln|1 - 2x| + C$ **81.** $\frac{1}{3}(9 - 3x)^{-1} + C$ **82.** $\frac{1}{2}(1 - 2x)^{-1} + C$

83. $\frac{1}{2}(8 + x^3)^{2/3} + C$ **84.** $(9 + x^2)^{1/2} + C$ **85.** $-\frac{1}{2}(w^2 + 6w - 1)^{-1} + C$ **86.** $-\frac{1}{2}(t^2 - 4t + 1)^{-1} + C$

87. $\frac{2}{3}(1 + \sqrt{x})^3 + C$ **88.** $(1 + x^{1/3})^3 + C$ **89.** $\ln|e^x - 1| + C$ **90.** $\ln|\ln x| + C$ **91.** $\frac{61}{3}$ or $20\frac{1}{3}$

92. 2 **93.** 2 **94.** $\frac{5}{12}$ **95.** $\ln 7$ **96.** $-\ln 2$ **97.** $\frac{1}{4}e - \frac{1}{4}$ **98.** $\frac{1}{5}e - \frac{1}{5}$ **99.** 8 square units

100. 5 square units **101.** $\frac{1}{4} - \frac{1}{4}e^{-4} = \frac{1}{4}(1 - e^{-4}) \approx 0.25$ **102.** $\frac{1}{4}\ln 13 \approx 0.64$

103. $C(x) = (2x + 9)^{1/2} + 97$ **104.** $\ln 28 \approx 3.33$ degrees

Exercises 6.1 page 391

1. $\frac{1}{2}e^{2x} + C$ **3.** $\frac{1}{2}x^2 + 2x + C$ **5.** $\frac{2}{3}x^{3/2} + C$ **7.** $\frac{1}{5}(x + 3)^5 + C$ **9.** $\frac{1}{2}xe^{2x} - \frac{1}{4}e^{2x} + C$

11. $\frac{1}{6}x^6 \ln x - \frac{1}{36}x^6 + C$ **13.** $(x + 2)e^x - e^x + C$ **15.** $\frac{2}{3}x^{3/2} \ln x - \frac{4}{9}x^{3/2} + C$

17. $\frac{1}{6}(x - 3)(x + 4)^6 - \frac{1}{42}(x + 4)^7 + C$ **19.** $-2te^{-0.5t} - 4e^{-0.5t} + C$ **21.** $-t^{-1} \ln t - t^{-1} + C$

23. $\frac{1}{10}s(2s + 1)^5 - \frac{1}{120}(2s + 1)^6 + C$ **25.** $-\frac{1}{2}xe^{-2x} - \frac{1}{4}e^{-2x} + C$ **27.** $2x(x + 1)^{1/2} - \frac{4}{3}(x + 1)^{3/2} + C$

29. $\frac{1}{a} xe^{ax} - \frac{1}{a^2} e^{ax} + C$ **31.** $\frac{1}{n + 1} x^{n+1} \ln ax - \frac{1}{(n + 1)^2} x^{n+1} + C$ **33.** $x \ln x - x + C$

35. $\frac{1}{2}x^2e^{x^2} - \frac{1}{2} e^{x^2} + C$ **37. a.** $\frac{1}{2}e^{x^2} + C$ (by substitution)   **b.** $\frac{1}{4}(\ln x)^4 + C$ (by substitution)

c. $\frac{1}{3}x^3 \ln 2x - \frac{1}{9}x^3 + C$ (by parts)   **d.** $\ln(e^x + 4) + C$ (by substitution)   **39.** $e^2 + 1$ **41.** $9 \ln 3 - 3 + \frac{1}{9}$

43. $\frac{2^6}{30} = \frac{32}{15}$ **45.** $4 \ln 4 - 3$ **47. a.** $\frac{1}{6}x(x - 2)^6 - \frac{1}{42}(x - 2)^7 + C$ **b.** $\frac{1}{7}(x - 2)^7 + \frac{1}{3}(x - 2)^6 + C$

49. Using $u = x^n$ and $dv = e^x \, dx$, the result follows immediately.   **51.** $x^2e^x - 2xe^x + 2e^x + C$

53. $R(x) = -4xe^{-x/4} - 16e^{-x/4} + 16$ **55.** \$105.7 million   **57.** $-14e^{-2.5} + 4 \approx 2.85$ milligrams

59. $2 \ln 2 - 1 + \frac{1}{4} \approx 0.64$ square unit   **61.** $-55e^{0.8} + 275 \approx 153$ million

63. About 5.7 liters   **65.** (b) $\int u \, dv = uv - \int v \, du$ **67.** $v = e^x$ **69.** $v = x$ **71.** $u = \ln x$

73. a. The result follows immediately.   **b.** [*Hint:* Think of the C.]   **75.** $-x^2e^{-x} - 2xe^{-x} - 2e^{-x} + C$

77. $(x + 1)^2e^x - 2(x + 1)e^x + 2e^x + C = (x^2 + 1)e^x + C$ (after simplification)   **79.** $\frac{1}{3}x^3(\ln x)^2 - \frac{2}{9}x^3 \ln x + \frac{2}{27}x^3 + C$

81. $2e^2 - 2 \approx 12.78$ **83.** $-x^2e^{-x} - 2xe^{-x} - 2e^{-x} + C = -e^{-x}(x^2 + 2x + 2) + C$

85. $\frac{1}{2}x^3e^{2x} - \frac{3}{4}x^2e^{2x} + \frac{3}{4}xe^{2x} - \frac{3}{8}e^{2x} + C$ **87.** $\frac{1}{3}(x - 1)^3e^{3x} - \frac{1}{3}(x - 1)^2e^{3x} + \frac{2}{9}(x - 1)e^{3x} - \frac{2}{27}e^{3x} + C$

Exercises 6.2 page 401

1. Formula 12, $a = 5$, $b = -1$ **3.** Formula 14, $a = -1$, $b = 7$ **5.** Formula 9, $a = -1$, $b = 1$

7. $\frac{1}{6} \ln \left| \frac{3 + x}{3 - x} \right| + C$ **9.** $-\frac{1}{x} - 2 \ln \left| \frac{x}{2x + 1} \right| + C$ **11.** $-x - \ln|1 - x| + C$ **13.** $\ln \left| \frac{2x + 1}{x + 1} \right| + C$

15. $\frac{x}{2} \sqrt{x^2 - 4} - 2 \ln|x + \sqrt{x^2 - 4}| + C$ **17.** $-\ln \left| \frac{1 + \sqrt{1 - z^2}}{z} \right| + C$

19. $\frac{1}{2} x^3e^{2x} - \frac{3}{4} x^2e^{2x} + \frac{3}{4} xe^{2x} - \frac{3}{8} e^{2x} + C$ **21.** $-\frac{1}{100} x^{-100} \ln x - \frac{1}{10,000} x^{-100} + C$

23. $\frac{1}{3} \ln \left| \frac{x}{x + 3} \right| + C$ **25.** $\frac{1}{8} \ln \left| \frac{z^2 - 2}{z^2 + 2} \right| + C$ **27.** $\frac{x}{2} \sqrt{9x^2 + 16} + \frac{8}{3} \ln \left| 3x + \sqrt{9x^2 + 16} \right| + C$

29. $-\frac{1}{2} \ln \left| \frac{2 + \sqrt{4 - e^{2t}}}{e^t} \right| + C$ **31.** $\frac{1}{2} \ln \left| \frac{e^t - 1}{e^t + 1} \right| + C$ **33.** $\frac{1}{4} \ln \left| x^4 + \sqrt{x^8 - 1} \right| + C$

35. $\frac{1}{3} \ln \left| \frac{\sqrt{x^3 + 1} - 1}{\sqrt{x^3 + 1} + 1} \right| + C$ **37.** $\frac{1}{2} \ln \left| \frac{e^t - 1}{e^t + 1} \right| + C$ **39.** $2xe^{x/2} - 4e^{x/2} + C$

41. $\frac{1}{4} \ln \left| \frac{e^{-x} + 4}{e^{-x}} \right| + C = \frac{1}{4} \ln(1 + 4e^x) + C$ **43.** $\frac{15}{2} - 8 \ln 8 + 8 \ln 4 \approx 1.95$ **45.** $\frac{1}{2} \ln \frac{1}{2} - \frac{1}{2} \ln \frac{1}{3} \approx 0.203$

47. $-4 + 5 \ln 3 \approx 1.49$ **49.** $\frac{1}{2} \ln|2x + 6| + C$ **51.** $\frac{x}{2} - \frac{3}{2} \ln|2x + 6| + C$ **53.** $-\frac{1}{3}(1 - x^2)^{3/2} + C$

55. $\sqrt{1 - x^2} - \ln \left| \frac{1 + \sqrt{1 - x^2}}{x} \right| + C$ **57.** $\frac{1}{2} \left(\ln|x + 1| - \frac{1}{3} \ln|3x + 1| \right) - \frac{1}{2} \ln \left| \frac{3x + 1}{x + 1} \right| + C$

59. $\ln|x + \sqrt{x^2 + 1}| - \ln \left| \frac{1 + \sqrt{x^2 + 1}}{x} \right| + C$ **61.** $x + 2 \ln|x - 1| + C$

63. $-x^2e^{-x} - 2xe^{-x} - 2e^{-x} + 2$ million sales   **65.** 24 generations   **67.** $C(x) = \ln(x + \sqrt{x^2 + 1}) + 2000$

69. Formula 9 with $x = e^{50t}$ **71.** Formula 18 with $x = t^{50}$ **73.** 476 uses

75. By a formula (either 22 or 23) and by integration by parts (with $u = \ln t$ and $dv = dt$)

77. Formula 10 [with denominator $(x + 1)(x - 1) = x^2 - 1$] and formula 15

Exercises 6.3 page 412

1. 0   **3.** 1   **5.** $-\infty$ (does not exist)   **7.** ∞ (does not exist)   **9.** 0   **11.** 0   **13.** ∞ (does not exist)   **15.** 0
17. $\frac{1}{2}$ **19.** $\frac{1}{8}$ **21.** Divergent   **23.** 100   **25.** 20   **27.** $\frac{1}{2}$ **29.** Divergent   **31.** $\frac{1}{3}$ **33.** $\frac{1}{3}$

35. Divergent   **37.** 1   **39.** Divergent   **41.** $\int_0^\infty e^{\sqrt{x}}\,dx$ diverges and $\int_0^\infty e^{-x^2}\,dx$ converges to 0.88623

43. \$200,000   **45. a.** \$10,000   **b.** \$9999.55   **47.** About \$3,963,000 (from 3963 thousand)

49. 1,000,000 barrels (from 1000 thousand)   **51.** 2 square units   **53.** $\frac{1}{a}$ square units   **55.** 0.61 or 61%

57. 0.30 or 30%   **59.** 20,000   **61.** (b) $\lim_{x\to\infty}\frac{1}{x^3}$ exists (and equals zero)

63. (a) $\lim_{x\to-\infty} e^{3x}$ exists (and equals zero)   **65.** False (see Example 5 on page 407)

67. Diverge [*Hint:* Think of the graph.]   **69.** True [*Hint:* Think of the graph.]   **71.** D/r **73.** \$40,992

Exercises 6.4 page 422

Some answers may vary depending on rounding.
1. a. 8.75   **b.** 8.667   **c.** 0.083   **d.** 1%   **3. a.** 0.697   **b.** 0.693   **c.** 0.004   **d.** 0.6%
5. 1.154   **7.** 0.743   **9.** 0.593   **11.** 8.6968   **13.** 2.925   **15.** 0.4772 or about 48%
17. About 42 billion tons   **19.** 8.667   **21.** 0.693   **23.** 1.148   **25.** 0.747

27. 0.593   **29.** 8.69678496   **31.** 2.92530   **33. a.** $-\int_1^0 \frac{t}{1+t^3}\,dt = \int_0^1 \frac{t}{1+t^3}\,dt$ **b.** 0.374

35. 821 feet   **37.** About 845 trillion dollars
39. The second derivative of a linear function is zero, so the error formula gives zero error, making the approximation exact. (Alternatively, the area under the graph of a linear function *is* a trapezoid, so approximating by a trapezoid will give the exact answer.)
41. 9   **43.** $\frac{1}{2}\cdot 0 + 4 + 5 + 3 + \frac{1}{2}\cdot 0 = 12$ square units
45. $n = 2$ (Since the method is exact for *any* n, use the smallest number possible.)
47. The justification follows from carrying out the indicated steps.

Chapter 6 Review Exercises and Chapter Test page 427

1. $\frac{1}{2}xe^{2x} - \frac{1}{4}e^{2x} + C$ **2.** $-xe^{-x} - e^{-x} + C$ **3.** $\frac{1}{9}x^9 \ln x - \frac{1}{81}x^9 + C$ **4.** $\frac{4}{5}x^{5/4}\ln x - \frac{16}{25}x^{5/4} + C$
5. $\frac{1}{6}(x-2)(x+1)^6 - \frac{1}{42}(x+1)^7 + C$ **6.** $\frac{1}{5}(x+3)(x-1)^5 - \frac{1}{30}(x-1)^6 + C$ **7.** $2t^{1/2}\ln t - 4t^{1/2} + C$
8. $\frac{1}{4}x^4 e^{x^4} - \frac{1}{4}e^{x^4} + C$ **9.** $x^2 e^x - 2xe^x + 2e^x + C$ **10.** $x(\ln x)^2 - 2x\ln x + 2x + C$
11. $\frac{1}{n+1}x(x+a)^{n+1} - \frac{1}{(n+1)(n+2)}(x+a)^{n+2} + C$ **12.** $-\frac{1}{n+1}x(1-x)^{n+1} - \frac{1}{(n+1)(n+2)}(1-x)^{n+2} + C$
13. $4e^5 + 1$ **14.** $\frac{1}{4}e^2 + \frac{1}{4}$ **15.** $-\ln|1-x| + C$ **16.** $-\frac{1}{2}e^{-x^2} + C$ **17.** $\frac{1}{4}x^4 \ln 2x - \frac{1}{16}x^4 + C$
18. $(1-x)^{-1} + C$ **19.** $\frac{1}{2}(\ln x)^2 + C$ **20.** $\frac{1}{2}\ln(e^{2x} + 1) + C$ **21.** $2e^{\sqrt{x}} + C$ **22.** $\frac{1}{8}(e^{2x} + 1)^4 + C$
23. $-15,000e^{-0.5} + 10,000 \approx 902$ million dollars   **24.** 6.78 hundred gallons (from $25 - 10e^{0.6}$)
25. $\frac{1}{10}\ln\left|\frac{5+x}{5-x}\right| + C$ **26.** $\frac{1}{4}\ln\left|\frac{x-2}{x+2}\right| + C$ **27.** $2\ln|x-2| - \ln|x-1| + C$
28. $-\ln\left|\frac{x-1}{x-2}\right| + C$ or $\ln\left|\frac{x-2}{x-1}\right| + C$ **29.** $\ln\left|\frac{\sqrt{x+1}-1}{\sqrt{x+1}+1}\right| + C$ **30.** $\frac{2x-4}{3}\sqrt{x+1} + C$
31. $\ln\left|x + \sqrt{x^2+9}\right| + C$ **32.** $\ln\left|x + \sqrt{x^2+16}\right| + C$ **33.** $\frac{z^2-2}{3}\sqrt{z^2+1} + C$ (from formula 13)
34. $e^t - 2\ln(e^t + 2) + C$ **35.** $\frac{1}{2}x^2 e^{2x} - \frac{1}{2}xe^{2x} + \frac{1}{4}e^{2x} + C$
36. $x(\ln x)^4 - 4x(\ln x)^3 + 12x(\ln x)^2 - 24x\ln x + 24x + C$ **37.** $\ln\left|\frac{2x+1}{x+1}\right| + 1000$

38. 1305 (from $750 + 800\ln 80 - 800\ln 40$)   **39.** $\frac{1}{4}$ **40.** $\frac{1}{5}$ **41.** Divergent   **42.** Divergent
43. $\frac{1}{2}$ **44.** $2e^{-2}$ **45.** Divergent   **46.** Divergent   **47.** 5   **48.** $10e^{-10}$ **49.** $\frac{1}{4}$ **50.** $\frac{1}{5}$
51. $\frac{1}{2}$ **52.** $\frac{1}{4}$ **53.** 1   **54.** 1   **55.** $\frac{1}{3}$ **56.** $\frac{1}{2}$ **57.** \$60,000   **58.** 0.35 or 35%   **59.** 240 thousand

60. 13,500 metric tons   **61.** $\int_1^\infty \frac{1}{x^3}\,dx$ converges to 0.5   **62.** $\int_1^\infty \frac{1}{\sqrt[3]{x}}\,dx$ diverges   **63.** 1.102   **64.** 1.09

65. 1.204   **66.** 0.852   **67.** 0.570   **68.** 1.313   **69.** 1.089   **70.** 1.075 **71.** 1.195   **72.** 0.856

73. 0.528   **74.** 1.348   **75.** 1.0894   **76.** 1.0747   **77.** 1.1951   **78.** 0.8556   **79.** 0.5285   **80.** 1.3357

81. 1.089429   **82.** 1.074669   **83.** 1.194958   **84.** 0.855624   **85.** 0.527887   **86.** 1.347855

87. a. $-\displaystyle\int_{1}^{0} \frac{1}{1+t^2}\,dt = \int_{0}^{1}\frac{1}{1+t^2}\,dt$ **b.** 0.783   **88. a.** $-\displaystyle\int_{1}^{0}\frac{1}{1+t^4}\,dt = \int_{0}^{1}\frac{1}{1+t^4}\,dt$ **b.** 0.862

Exercises 7.1 page 441

1. $\{(x, y)\mid x \neq 0, y \neq 0\}$ **3.** $\{(x, y)\mid x \neq y\}$ **5.** $\{(x, y)\mid x > 0, y \neq 0\}$
7. $\{(x, y, z)\mid x \neq 0, y \neq 0, z > 0\}$ **9.** 3   **11.** 4   **13.** -2 **15.** 1   **17.** $e^{-1} + e$ **19.** e^{-1}
21. 0   **23.** 0.0157   **25.** 45 minutes   **27.** 472.7   **29.** Each gives $P = 2700$.
31. $P(2L, 2K) = a(2L)^b(2K)^{1-b} = a2^b L^b 2^{1-b} K^{1-b} = 2aL^b K^{1-b} = 2P(L, K)$ **33.** 20,000 calls

$$\underbrace{\qquad}_{2}$$

35. $C(x, y) = 210x + 180y + 4000$ **37. a.** $V = xyz$ **b.** $M = xy + 2xz + 2yz$
39. a.

  b. A given wind speed will lower the windchill further on a colder day than on a warmer day.
  c. For the lowest curve: $dy/dx \approx -0.63$, meaning that at 20 degrees and 10 mph of wind, windchill drops by about 0.63 degrees for each additional 1 mph of wind. For the highest curve: $dy/dx \approx -0.33$, meaning that at 50 degrees and 10 mph of wind, windchill drops by only about 0.33 degrees for each additional 1 mph of wind.
  d. Yes—the effect of wind on the windchill index is greater on a colder day.
41. 0.64 or 64%   **43.** Coordinate system (b). (The other is a *left*-handed coordinate system.)
45. 5 (3 relative maximum points and 2 relative minimum points, one hidden behind the front most peak).
47. (a) has a relative minimum point and (b) has a relative maximum point.
49. False—the order is backwards. See page 432 for the correct definition.   **51.** Satisfies (a)   **53.** Satisfies (b)

Exercises 7.2 page 454

1. a. $3x^2 + 6xy^2 - 1$ **b.** $6x^2 y - 6y^2 + 1$ **3. a.** $6x^{-1/2}y^{1/3}$ **b.** $4x^{1/2}y^{-2/3}$

5. a. $5x^{-0.95}y^{0.02}$ **b.** $2x^{0.05}y^{-0.98}$ **7. a.** $-(x + y)^{-2}$ **b.** $-(x + y)^{-2}$ **9. a.** $\dfrac{3x^2}{x^3 + y^3}$ **b.** $\dfrac{3y^2}{x^3 + y^3}$

11. a. $6x^2 e^{-5y}$ **b.** $-10x^3 e^{-5y}$ **13. a.** ye^{xy} **b.** xe^{xy}

15. a. $\dfrac{x}{x^2 + y^2}$ or $x(x^2 + y^2)^{-1}$ **b.** $\dfrac{y}{x^2 + y^2}$ or $y(x^2 + y^2)^{-1}$ **17. a.** $3v(uv - 1)^2$ **b.** $3u(uv - 1)^2$

19. a. $ue^{(u^2-v^2)/2}$ **b.** $-ve^{(u^2-v^2)/2}$ **21.** $18, -10$ **23.** $0, 2e$ **25.** $1\frac{1}{2}$
27. a. $30x - 4y^3$ **b. and c.** $-12xy^2$ **d.** $-12x^2 y + 36y^2$
29. a. $-2x^{-5/3}y^{2/3}$ **b. and c.** $2x^{-2/3}y^{-1/3} - 12y^2$ **d.** $-2x^{1/3}y^{-4/3} - 24xy$
31. a. ye^x **b. and c.** $e^x - \dfrac{1}{y}$ **d.** xy^{-2} **33.** All three are $36x^2 y^2$.   **35. a.** $y^2 z^3$ **b.** $2xyz^3$ **c.** $3xy^2 z^2$

37. a. $8x(x^2 + y^2 + z^2)^3$ **b.** $8y(x^2 + y^2 + z^2)^3$ **c.** $8z(x^2 + y^2 + z^2)^3$
39. a. $2xe^{x^2+y^2+z^2}$ **b.** $2ye^{x^2+y^2+z^2}$ **c.** $2ze^{x^2+y^2+z^2}$ **41.** -14 **43.** $4e^6$
45. a. $P_x = 4x - 3y + 150$ **b.** $50 (profit per additional DVD player)   **c.** $P_y = -3x + 6y + 75$
  d. $75 (profit per additional CD player)
47. a. 250 (the marginal productivity of labor is 250, so production increases by about 250 for each additional unit of labor)
  b. 108 (the marginal productivity of capital is 108, so production increases by about 108 for each additional unit of capital)
  c. Labor
49. $S_x = -0.1$ (sales fall by 0.1 for each dollar price increase)
  $S_y = 0.4y$ (sales rise by 0.4y for each additional advertising dollar above the level y)
51. a. 0.52 (status increases by about 0.52 unit for each additional $1000 of income)
  b. 5.25 (status increases by 5.25 units for each additional year of education)
53. a. 97.2 (skid distance increases by about 97 feet for each additional ton)
  b. 12.96 (skid distance increases by about 13 feet for each additional mph)

55. a. 0.001152   **b.** 3.69, so about 4 more games   **57.** 0   **59.** $\dfrac{\partial f}{\partial x} = 3, \dfrac{\partial f}{\partial y} = -2$

61. $\dfrac{\partial f}{\partial x}$ is negative; $\dfrac{\partial f}{\partial y}$ is positive.   **63.** $f(x, y) = -x + y$ **65.** $f(x, y) = \pi x + ey + \ln 2$

67. a. $\dfrac{\partial D}{\partial E} > 0, \dfrac{\partial D}{\partial T} < 0$; interpretation: Your desire to complete the task should increase with the expectation of success and decrease with the time needed to complete the task.

69. a. Rate at which butter sales change as butter prices rise
   b. Negative: As prices rise, sales will fall.
   c. Rate at which butter sales change as margarine prices rise
   d. Positive: As margarine prices rise, people will switch to butter, so butter sales will rise.

71. Harmonic   **73.** Not harmonic

75. From taking the limit as $h \to 0$: $\dfrac{\partial f}{\partial x} = 4x - 3$ and $\dfrac{\partial f}{\partial y} = 6y + 7$

77. From taking the limit as $h \to 0$: $\dfrac{\partial f}{\partial x} = 10x - 3$ and $\dfrac{\partial f}{\partial y} = 7$

Exercises 7.3 page 466

1. Rel min value: $f = 5$ at $x = 0$, $y = -1$ **3.** Rel min value: $f = -12$ at $x = -2$, $y = 2$
5. Rel max value: $f = 23$ at $x = 5$, $y = 2$ **7.** No rel extreme values [saddle point at (2, −4)]
9. No rel extreme values   **11.** Rel min value: $f = 1$ at $x = 0$, $y = 0$
13. Rel min value: $f = 0$ at $x = 0$, $y = 0$
15. Rel max value: $f = 3$ at $x = 1$, $y = -1$ [saddle point at (−1, −1)]
17. Rel max value: $f = 17$ at $x = -1$, $y = -2$ [saddle point at (−1, 2)]
19. No rel extreme values [saddle point at (2, 6)]
21. 10 units of product A, sell for $7000 each; 7 units of product B, sell for $13,000 each.
   Maximum profit: $22,000
23. a. $P = -0.2x^2 + 16x - 0.1y^2 + 12y - 20$
   b. 40 cars in America, sell for $12,000; 60 cars in Europe, sell for $10,000
25. 6 hours of practice and 1 hour of rest
27. a. $x = 1200$, $p = \$6$, $R = \$7200$ **b.** $x = 800$, $y = 800$, $p = \$4$, revenue = $3200 for each
   c. Duopoly (1600 versus 1200)   **d.** Duopoly
29. Sell the sedans for $19,200, selling 12 per day, and sell the SUVs for $23,200, selling 7 per day.
31. a. $P = -0.2x^2 + 16x - 0.1y^2 + 12y - 0.1z^2 + 8z - 22$ **b.** 40 in America, 60 in Europe, 40 in Asia
33. Rel min value: $f = -1$ at $x = 1$, $y = 1$ [saddle point at (0, 0)]
35. Rel max value: $f = 32$ at $x = 4$, $y = 4$ [saddle point at (0, 0)]
37. Rel min value: $f = -162$ at $x = 3$, $y = 18$ and at $x = -3$, $y = -18$ [saddle point at (0, 0)]
39. Saddle point   **41.** True
43. The function must have a relative maximum or minimum at the critical point.
45. Very little. Depending on the sign of D, there may be a relative minimum point or a saddle point.
47. False. No conclusion can be drawn. For example, at (0, 0), $x^4 + y^4$ has a relative minimum, $-x^4 - y^4$ has a relative maximum, and $x^4 - y^4$ has a saddle point, and in each case $D = 0$.

Exercises 7.4 page 476

Note: Your answers may differ slightly depending on the stage at which you do the rounding.
1. $y = 3.5x - 1.67$ **3.** $y = -0.79x + 6.6$ **5.** $y = 2.4x + 6.9$ **7.** $y = -2.1x + 7.6$
9. $y = 2.2x + 5$; prediction: 16 million   **11.** $y = -8x + 125$; prediction: 85 arrests
13. $y = -0.009x + 0.434$; prediction: 0.380   **15.** $y = -1.4x + 29.4$; prediction: about 20%
17. $y = -0.16x + 71.6$ **19.** $y = 1.09e^{0.63x}$ **21.** $y = 17.45e^{-0.47x}$ **23.** $y = 0.98e^{0.78x}$
25. $y = 16.95e^{-0.52x}$ **27.** $y = 1.28e^{0.066x}$ prediction: $2.03 million
29. $y = 41.1e^{1.35x}$ prediction: about $9,100 (rounded)
31. Elderly: $y = 0.232e^{0.548x}$; children: $y = 0.15x + 1.47$; the elderly population is larger at $x = 4$, which corresponds to 2045.
33. False (see, for example, the graph on page 471)   **35.** True
37. $y = ax + b + 5$ (as can be proved from the formulas on page 471)

39. The formulas on page 471 give $a = 0/0$, which is undefined.

41. The third column requires the logarithm of that number, and logs of 0 or negative numbers are *undefined*. Therefore, the y's must be positive.

Exercises 7.5 page 490

1. Max $f = 36$ at $x = 6$, $y = 2$ **3.** Max $f = 144$ at $x = 6$, $y = 4$

5. Max $f = -28$ at $x = 3$, $y = 5$ **7.** Max $f = 6$ at $x = 2$, $y = -1$

9. Max $f = 2$ (from $\ln e^2$) at $x = e$, $y = e$ **11.** Min $f = 45$ at $x = 6$, $y = 3$

13. Min $f = -16$ at $x = -4$, $y = 4$ **15.** Min $f = 52$ at $x = 4$, $y = 6$

17. Min $f = \ln 125$ at $x = 10$, $y = 5$ **19.** Min $f = e^{20}$ at $x = 2$, $y = 4$

21. Max $f = 8$ at $x = 2$, $y = 2$ and at $x = -2$, $y = -2$;
Min $f = -8$ at $x = 2$, $y = -2$ and at $x = -2$, $y = 2$

23. Max $f = 18$ at $x = 2$, $y = 8$; Min $f = -18$ at $x = -2$, $y = -8$

25. a. 1000 feet perpendicular to building, 3000 feet parallel to building
b. $|\lambda| = 1000$; each additional foot of fence adds about 1000 square feet of area

27. $r \approx 3.7$ feet, $h \approx 3.7$ feet

29. End: 14 inches by 14 inches; length $= 28$ inches; volume $= 5488$ cubic inches

31. a. $L = 24$, $K = 15{,}000$, and $P = 33{,}600$
b. $|\lambda| = 2$ *Interpretation:* Each additional dollar of budget increases production by about 2 units.
c. About 600 units of production.

33. a. $L = 28$, $K = 63$, and $P = 7560$
b. $|\lambda| = 3$ *Interpretation:* Each additional dollar of budget increases production by about 3 units.
c. About 150 units of production.

35. Base: 3 inches by 3 inches; height: 5 inches   **37.** Min $f = 24$ at $x = 4$, $y = 2$, $z = -2$

39. Max $f = 6$ at $x = 2$, $y = 2$, $z = 2$ **41.** Base: 10 feet by 10 feet; height: 20 feet

43. False. It should be $F(x, y, \lambda) = f(x, y) + \lambda g(x, y)$.

45. True. A solution can occur only at a critical point.

47. The maximum occurs at one critical point, and the minimum occurs at the other.   **49.** $a \le b$

51. The results follow by direct calculation.

53. The results follow by direct calculation. If increasing C by 1 increases P by λ, then increasing P by 1 must increase C by $1/\lambda$.

55. $\dfrac{w}{r} = \dfrac{35}{140} = \dfrac{1}{4}$ and $MRTS = \dfrac{2/3}{1 - 2/3} \dfrac{6}{48} = \dfrac{1}{4}$

57. $\dfrac{w}{r} = \dfrac{35}{140} = \dfrac{1}{4}$ and $\dfrac{P_L}{P_K} = \dfrac{280}{1120} = \dfrac{1}{4}$

Exercises 7.6 page 502

1. $df = 2xy^3 \cdot dx + 3x^2y^2 \cdot dy$ **3.** $df = 3x^{-1/2}y^{1/3} \cdot dx + 2x^{1/2}y^{-2/3} \cdot dy$ **5.** $dg = \dfrac{1}{y} \cdot dx - \dfrac{x}{y^2} \cdot dy$

7. $dg = -(x - y)^{-2} \cdot dx + (x - y)^{-2} \cdot dy$ **9.** $dz = \dfrac{3x^2}{x^3 - y^2} \cdot dx - \dfrac{2y}{x^3 - y^2} \cdot dy$

11. $dz = e^{2y} \cdot dx + 2xe^{2y} \cdot dy$ **13.** $dw = (6x^2 + y) \cdot dx + (x + 2y) \cdot dy$

15. $df = 4xy^3z^4 \cdot dx + 6x^2y^2z^4 \cdot dy + 8x^2y^3z^3 \cdot dz$ **17.** $df = \dfrac{1}{x} dx + \dfrac{1}{y} dy + \dfrac{1}{z} dz$

19. $df = yze^{xyz} \cdot dx + xze^{xyz} \cdot dy + xye^{xyz} \cdot dz = e^{xyz}(yz \cdot dx + xz \cdot dy + xy \cdot dz)$

21. a. $\Delta f = 0.479$ **b.** $df = 0.4$

23. a. $\Delta f \approx 0.112$ **b.** $df = 0.11$ **25. a.** $\Delta f = 0.1407$ **b.** $df = 0.14$

27. 125 square feet; 250 square feet   **29.** \$4300   **31.** About 113 feet   **33.** 2%

35. 0.5 liter per minute

37. Volume is $1.92\pi \approx 6.03$ cm^3 with a maximum error of $0.496\pi \approx 1.56$ cm^3.   **39.** 0

41. $f(x, y) = x + y + c$ where c is a constant.   **43.** It is independent of y (since $f_y = 0$).

45. a. f is being evaluated at two points along the curve; each of these points gives $f = c$, and $c - c = 0$.
b. Subtracting and adding $F(x)$.
c. Approximating the change $\Delta f = f(x + \Delta x, F + \Delta F) - f(x, F)$ by the total differential $df = f_x \Delta x + f_y \Delta F$.
d. Subtracting $f_y \Delta F$ and dividing by f_y and Δx.
e. Taking the limit as $\Delta x \to 0$ causes $\dfrac{\Delta F}{\Delta x}$ to approach $\dfrac{dF}{dx}$ and the approximation to become exact.

Exercises 7.7 page 514

1. $2x^9 - 2x$ **3.** $6y^4$ **5.** $2x^2$ **7.** 4   **9.** 2   **11.** $\frac{1}{2}$ **13.** 12
15. 14   **17.** $-9e^{-3} + 9e^3$ **19.** 0   **21.** -12 **23.** 0   **25.** 72   **27.** $\frac{1}{2}$

29. a. $\displaystyle\int_1^3\int_0^2 3xy^2\, dx\, dy$ and $\displaystyle\int_0^2\int_1^3 3xy^2\, dy\, dx$ **b.** Both equal 52

31. a. $\displaystyle\int_0^2\int_{-1}^1 ye^x\, dx\, dy$ and $\displaystyle\int_{-1}^1\int_0^2 ye^x\, dy\, dx$ **b.** Both equal $2e - 2e^{-1}$ **33.** 8 cubic units

35. $\frac{4}{3}$ cubic units   **37.** 2 cubic units   **39.** $\frac{1}{2}e^2 - e + \frac{1}{2}$ cubic units   **41.** 45 degrees (from $\frac{540}{12}$)
43. About 180,200 people   **45.** 900,000 cubic feet   **47.** 14   **49.** 10
51. (a) is a double integral and (b) is an iterated integral.

53. 6   **55.** Yes: $\displaystyle\int_c^d e^{f(x)}\, dx \cdot \int_a^b e^{g(y)}\, dy$ **57.** True   **59.** Widgets

Chapter 7 Review Exercises and Chapter Test page 518

1. $\{(x, y) \mid x \geq 0,\, y \neq 0\}$ **2.** $\{(x, y) \mid y > 0\}$ **3.** $\{(x, y) \mid x \neq 0,\, y > 0\}$ **4.** $\{(x, y) \mid x \neq 0,\, y > 0\}$
5. a. $10x^4 - 6xy^3 - 3$ **b.** $-9x^2y^2 + 4y^3 + 2$ **c. and d.** $-18xy^2$
6. a. $12x^3 + 15x^2y^2 - 6$ **b.** $10x^3y - 6y^5 + 1$ **c. and d.** $30x^2y$
7. a. $12x^{-1/3}y^{1/3}$ **b.** $6x^{2/3}y^{-2/3}$ **c. and d.** $4x^{-1/3}y^{-2/3}$
8. a. $\dfrac{2x}{x^2 + y^3}$ **b.** $\dfrac{3y^2}{x^2 + y^3}$ **c. and d.** $\dfrac{-6xy^2}{(x^2 + y^3)^2}$
9. a. $3x^2e^{x^3-2y^3}$ **b.** $-6y^2e^{x^3-2y^3}$ **c. and d.** $-18x^2y^2e^{x^3-2y^3}$
10. a. $6xe^{-5y}$ **b.** $-15x^2e^{-5y}$ **c. and d.** $-30xe^{-5y}$
11. a. $-ye^{-x} - \ln y$ **b.** $e^{-x} - \dfrac{x}{y}$ **c. and d.** $-e^{-x} - \dfrac{1}{y}$
12. a. $2xe^y + yx^{-1}$ **b.** $x^2e^y + \ln x$ **c. and d.** $2xe^y + x^{-1}$ **13. a.** $\frac{1}{2}$ **b.** $\frac{1}{2}$
14. a. 0   **b.** $\frac{1}{2}$ **15. a.** 36   **b.** -24 **16. a.** 216   **b.** -216
17. a. 80: rate at which production increases for each additional unit of labor
     b. 135: rate at which production increases for each additional unit of capital   **c.** Capital
18. $S_x = 222$: rate at which sales increase for each additional $1000 in TV ads.
     $S_y = 528$: rate at which sales increase for each additional $1000 in print ads.
19. Min $f = -13$ at $x = -1,\ y = -4$ **20.** Min $f = -8$ at $x = 4,\ y = 1$
21. Max $f = 8$ at $x = 0,\ y = -1$ **22.** Max $f = 6$ at $x = 1,\ y = 0$
23. No rel extreme values (saddle point at $x = \frac{1}{2},\ y = -3$)
24. No rel extreme values (saddle point at $x = -\frac{1}{2},\ y = 1$)   **25.** Max $f = 1$ at $x = 0,\ y = 0$
26. Min $f = 1$ at $x = 0,\ y = 0$ **27.** Min $f = 0$ at $x = 0,\ y = 0$
28. Min $f = \ln 10$ at $x = 0,\ y = 0$
29. Max $f = 25$ at $x = -2,\ y = -3$ (saddle point at $x = 2,\ y = -3$)
30. Min $f = -20$ at $x = -2,\ y = 2$ (saddle point at $x = 2,\ y = 2$)
31. a. $C(x, y) = 3000x + 5000y + 6000$ **b.** $R(x, y) = 7000x - 20x^2 + 8000y - 30y^2$
     c. $P(x, y) = -20x^2 + 4000x - 30y^2 + 3000y - 6000$
     d. Make 100 18-foot boats, sell for $5000 each, and 50 22-foot boats, sell for $6500 each; max profit: $269,000.
32. a. $P(x, y) = -0.2x^2 + 68x - 0.1y^2 + 52y - 100$
     b. America: sell 170 for $46,000 each; Europe: sell 260 for $38,000 each (since prices are in thousands)
33. $y = 2.6x - 3.2$ **34.** $y = -1.8x + 8.4$ **35.** $y = 4.97x + 15.5$; prediction: 45.3 million
36. $y = 6.4e^{-0.124x}$ (may vary with rounding); prediction: 3   **37.** Max $f = 292$ at $x = 12,\ y = -24$
38. Max $f = 156$ at $x = 10,\ y = 8$ **39.** Min $f = 90$ at $x = 7,\ y = 4$
40. Min $f = -109$ at $x = -3,\ y = 3$ **41.** Min $f = e^{45}$ at $x = 3,\ y = 6$
42. Max $f = e^{-5}$ at $x = 2,\ y = 1$
43. Max $f = 120$ at $x = 2,\ y = -6$; Min $f = -120$ at $x = -2,\ y = 6$
44. Max $f = 64$ at $x = 4,\ y = 4$ and at $x = -4,\ y = -4$;
     Min $f = -64$ at $x = 4,\ y = -4$ and at $x = -4,\ y = 4$
45. a. $40,000 for production, $20,000 for advertising
     b. $|\lambda| \approx 159$: Production increases by about 159 units for each additional dollar

46. a. $\frac{1}{2}$ ounce of the first and 7 ounces of the second

  b. $|\lambda| = 9$: Each additional dollar results in about 9 additional nutritional units

47. a. $L = 18, K = 10$ **b.** $|\lambda| \approx 3.79$;   output increases by about 3.79 units for each additional dollar

48. Base: 12 inches by 12 inches; height: 4 inches   **49.** $df = (6x + 2y) \cdot dx + (2x + 2y) \cdot dy$

50. $df = (2x + y) \cdot dx + (x - 6y) \cdot dy$ **51.** $dg = \frac{1}{x} dx + \frac{1}{y} dy = \frac{dx}{x} + \frac{dy}{y}$

52. $dg = \frac{3x^2}{x^3 + y^3} dx + \frac{3y^2}{x^3 + y^3} dy$ **53.** $dz = e^{x-y} dx - e^{x-y} dy$ **54.** $dz = ye^{xy} dx + xe^{xy} dy$

55. Sales would decrease by about \$153,000 (from $dS = -153$); sales would decrease by about \$76,500.

56. 2%   **57.** $8e^2 - 8e^{-2}$ **58.** 10   **59.** $\frac{4}{3}$ **60.** $\frac{32}{3}$ or $10\frac{2}{3}$ **61.** 40 cubic units

62. 24 cubic units   **63.** $\frac{5}{6}$ cubic unit   **64.** $\frac{8}{9}$ cubic unit   **65.** $12{,}000 \left(\text{from } \frac{192{,}000}{16} \right)$

66. \$640 hundred thousand, or \$64,000,000

Cumulative Review for Chapters 1–7 page 520

1.

2. 4   **3.** $\dfrac{-1}{x^2}$ (but found using the *definition*)   **4.** 4

5. $S'(12) = -2$:   each \$1 price increase (above \$12) decreases sales by 2 per week

6. $3[x^2 + (2x + 1)^4]^2[2x + 8(2x + 1)^3]$

7.

| $f' < 0$ | $f' = 0$ | $f' < 0$ | $f' = 0$ | $f' > 0$ | | $f'' < 0$ | $f'' = 0$ | $f'' > 0$ |

$x = -8$ $x = 2$ $x = -3$

↗   →   ↘   →   ↗   con dn   IP(−3, 50)   con up

rel max   rel min
(−8, 300)   (2, −200)

8.

| $f' > 0$ | f und | $f' > 0$ | $f' = 0$ | $f' < 0$ | f und | $f' < 0$ |

$x = 0$ $x = 2$ $x = 4$

↗   ↗   →   ↘   ↘

rel max
$(2, -\frac{1}{4})$

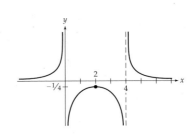

9. 40 feet parallel to wall, 20 feet perpendicular to wall

10. Base: 6 feet by 6 feet; height: 3 feet   **11.** $\frac{2}{\pi} \approx 0.64$ foot per minute   **12. a.** \$1195.62   **b.** \$1197.22

13. In about 14.4 years   **14.** About 6.8 years

15.

| $f' > 0$ | $f' = 0$ | $f' < 0$ | | $f'' > 0$ | $f'' = 0$ | $f'' < 0$ | $f'' = 0$ | $f'' > 0$ |

$x = 0$ $x = -1$ $x = 1$

↗   →   ↘   con up   con dn   con up

rel max   IP $(-1, e^{-1/2})$ IP $(1, e^{-1/2})$
(0, 1)

(See graph on next page.)

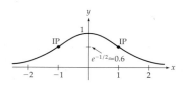

16. $4x^3 - 2x^2 + x + C$ **17.** $900e^{0.02t} - 900$ million gallons

18. $85\frac{1}{3}$ square units   **19.** 16

20. a. $\frac{1}{3}\ln|x^3 + 1| + C$ **b.** $2e^{x^{1/2}} + C$ **21.** $\frac{1}{4}xe^{4x} - \frac{1}{16}e^{4x} + C$

22. $\sqrt{4 - x^2} - 2\ln\left|\dfrac{2 + \sqrt{4 - x^2}}{x}\right| + C$ **23.** $\dfrac{1}{2}$

24. 1.15148 [compared with actual (rounded) value of 1.14779]

25. 1.14778 [compared with actual (rounded) value of 1.14779]

26. a. $\{(x, y) \mid x \geq 0, y > 0, \ x \neq y\}$ **b.** 2   **27.** $f_x = \ln y + 2ye^{2x}, \ f_y = \dfrac{x}{y} + e^{2x}$

28. Min $f = 2$ at $x = 1$, $y = 4$; no relative max   **29.** $y = 1.75x - 4.75$

30. Min $f = 90$ at $x = 4$, $y = 7$ **31.** $df = (4x + y)\,dx + (x - 6y)\,dy$ **32.** 48 cubic units

Exercises 8.1 page 533

1. $90°, \dfrac{\pi}{2}$ **3.** $45°, \dfrac{\pi}{4}$ **5.** $540°, 3\pi$ **7.** $-45°, -\dfrac{\pi}{4}$ **9. a.** $\dfrac{\pi}{6}$ **b.** $\dfrac{5\pi}{4}$ **c.** $-\pi$

11. a. $\dfrac{\pi}{3}$ **b.** $\dfrac{7\pi}{4}$ **c.** $-\dfrac{2\pi}{3}$ **13. a.** 3π **b.** $\dfrac{3\pi}{4}$ **c.** $\dfrac{\pi}{180}$ **15. a.** $45°$ **b.** $120°$ **c.** $-150°$

17. a. $60°$ **b.** $225°$ **c.** $-210°$ **19. a.** $405°$ **b.** $-540°$ **c.** $\dfrac{900°}{\pi}$ **21.** $3\pi \approx 9.4$ inches

23. $\dfrac{3}{2}$ radians   **25.** 3600 miles   **27.** False. The sides are in *proportion* to 1, 2, $\sqrt{3}$.

29. False. The units are reversed.   **31.** 0   **33.** 1   **35.** 0.246 radian

37. i. $(a + b)^2 = a^2 + 2ab + b^2$ **ii.** $c^2 + 4 \cdot \frac{1}{2}ab = c^2 + 2ab$

   iii. $a^2 + \cancel{2ab} + b^2 = c^2 + \cancel{2ab}$ so $a^2 + b^2 = c^2$

Exercises 8.2 page 543

1. $\sin \theta = \frac{24}{25}$ **3.** $\sin \theta = \frac{5}{13}$ **5.** $\sin \theta = \frac{21}{29}$ **7.** $\sin \theta = -\frac{3}{5}$ $\left(\begin{smallmatrix}\text{after}\\\text{simplification}\end{smallmatrix}\right)$

   $\cos \theta = \frac{7}{25}$ $\cos \theta = \frac{12}{13}$ $\cos \theta = -\frac{20}{29}$ $\cos \theta = -\frac{4}{5}$

9. $\sin \theta = -\frac{7}{25}$ $\left(\begin{smallmatrix}\text{after}\\\text{simplification}\end{smallmatrix}\right)$ **11. a.** $\dfrac{\sqrt{3}}{2}$ **b.** $\dfrac{1}{2}$ **c.** $\dfrac{\sqrt{2}}{2}$ **d.** $\dfrac{\sqrt{2}}{2}$

   $\cos \theta = \frac{24}{25}$

13. a. $-\dfrac{\sqrt{2}}{2}$ **b.** $-\dfrac{\sqrt{2}}{2}$ **c.** $\dfrac{1}{2}$ **d.** $-\dfrac{\sqrt{3}}{2}$

15. a. $-\dfrac{\sqrt{3}}{2}$ **b.** $\dfrac{1}{2}$ **c.** $-\dfrac{\sqrt{2}}{2}$ **d.** $\dfrac{\sqrt{2}}{2}$ **17. a.** 0.31   **b.** 0.95   **c.** 0.84   **d.** 0.54

19.   **21.**   **23.**

25.

27. a. $\dfrac{\sqrt{6}}{4} + \dfrac{\sqrt{2}}{4}$ **b.** $\dfrac{\sqrt{6}}{4} - \dfrac{\sqrt{2}}{4}$ $\left(\begin{matrix}\text{after}\\\text{simplification}\end{matrix}\right)$

29. About 28.2 feet   **31.** About 299 feet   **33.** 1219 cones

35. a. 5.3 liters   **b.** 4.5 liters

37. a.   **b.**   **39. a.**   **b.** 14°   **c.** 58°

on $[-2\pi, 2\pi]$ by $[-2, 2]$ on $[-2\pi, 2\pi]$ by $[-2, 2]$ on $[50, 60]$ by $[380, 405]$

41. b. About 402 ppm   **43.** Shifting the cosine curve to the right by $\pi/2$ will turn it into a sine curve.

45. Maximum height is 11, and minimum height is 5.   **47. a.** 0   **b.** 1   **49.** True

51. 1 (since $\sin^2 x + \cos^2 x = 1$)   **53.** The identities follow from the indicated arguments.

Exercises 8.3 page 556

1. $3t^2 \sin t + t^3 \cos t$ **3.** $\dfrac{-t \sin t - \cos t}{t^2}$ **5.** $3t^2 \cos(t^3 + 1)$ **7.** $5 \cos 5t$ **9.** $-6 \sin 3t$

11. $-\dfrac{\pi}{180} \sin\left(\dfrac{\pi}{180} t\right)$ **13. a.** $2t + \pi \cos \pi t$ **b.** π **15. a.** $-2 \sin \dfrac{t}{2}$ **b.** -1

17. a. $-\cos(\pi - t)$ **b.** 0   **19.** $-3(t + \pi)^2 \sin(t + \pi)^3$ **21.** $-3[\cos^2(t + \pi)][\sin(t + \pi)]$

23. $-2(\pi - t) \cos(\pi - t)^2$ **25.** $6t \sin^2 t^2 \cos t^2$ **27.** $-\dfrac{1}{2}(x + 1)^{-1/2} \sin(x + 1)^{1/2}$

29. $4 \cos x + 4 \cos 4x + 4(\sin^3 x)(\cos x)$ **31.** $e^x \cos e^x$ **33.** $(1 - \sin z) e^{z + \cos z}$ **35.** $4(1 - \cos z)^3 \sin z$

37. a. $\dfrac{\cos z}{\sin z}$ **b.** 0   **39.** $\left(1 + \dfrac{1}{z}\right) \cos(z + \ln z)$ **41.** $14 \sin x \cos x \left(\begin{matrix}\text{after}\\\text{simplification}\end{matrix}\right)$

43. $-x^{-2} \cos x^2 - 2 \sin x^2$ **45.** $-e^{-x} \sin e^x + \cos e^x$ **47. a.** $\dfrac{-1}{\sin^2 x} \left(\begin{matrix}\text{after}\\\text{simplification}\end{matrix}\right)$ **b.** $-\dfrac{4}{3}$

49. $4 \sin t \cos t \left(\begin{matrix}\text{after}\\\text{simplification}\end{matrix}\right)$ **51. a.** $\dfrac{1 + \sin t}{\cos^2 t} \left(\begin{matrix}\text{after}\\\text{simplification}\end{matrix}\right)$ **b.** 1   **53.** $-\cos(\cos x) \sin x$

55. $-\left(\cos \dfrac{x}{2} - 1\right) \sin \dfrac{x}{2}$ **57.** $x^2 \cos x$ **59.** $4t \sin(t^2 + 1) \cos(t^2 + 1)$ **61.** $y = x + 1$ **63.** $y = -2x + 2\pi$

65. $2 \cos t - t \sin t \left(\begin{matrix}\text{after}\\\text{simplification}\end{matrix}\right)$ **67.** $-(\sin x) e^{\sin x} + (\cos^2 x) e^{\sin x}$ **69.** $\cos z - z \sin z$

71. Increasing by about 45 sales per week   **73.** $\dfrac{\pi}{2}$ or 90°   **75.** The result follows from the indicated steps.

77. a. $x \approx 0.7854$ (radian)   **b.** 45°

79. a.   **b.** At $t = 0.2$ and $t = 1$ second, blood pressure is 105.

c. At $t = 0.6$ and $t = 1.4$ second, blood pressure is 75.

81. 0 (since you are differentiating a constant)

83. 0 (since $\sin^2 x + \cos^2 x = 1$)   **85.** 4   **87.** $\cos(x + 973)$ **89.** 1

91. The formula follows from the indicated steps.   **93.** Use $f = t$ and $f' = 1$

95. The identity follows by differentiation, as indicated.

Exercises 8.4 page 565

1. $\sin t + \cos t + C$ **3.** $x - \cos x + C$ **5.** $\dfrac{1}{2} \sin(t^2 + 2t) + C$ **7.** $-\dfrac{1}{\pi} \cos \pi t + C$

9. $\dfrac{2}{\pi} \sin \dfrac{\pi t}{2} + C$ **11.** $\cos(\pi - t) + C$ **13.** $\dfrac{365}{2\pi} \sin \dfrac{2\pi(t + 20)}{365} + C$ **15.** $\dfrac{1}{3} \sin^3 t + C$

17. $\dfrac{1}{2} \cos^{-2} t + C$ **19.** $e^{1 + \sin x} + C$ **21.** $-\ln|\cos t + 3| + C$ **23.** $2(1 - \cos w)^{1/2} + C$

25. $(1 + \sin^2 y)^{1/2} + C$ **27.** $\dfrac{1}{3}x^3 - \ln|x| + \dfrac{1}{2}\sin 2x + C$ **29.** $\dfrac{1}{\pi} \approx 0.318$ **31.** $\dfrac{1}{2}\pi^2 \approx 4.935$

33. $\dfrac{1}{2} + \dfrac{2}{\pi} \approx 1.137$ **35.** -1 **37.** 2 square units **39.** 2 square units

41. $\dfrac{d}{dt}\left(-\dfrac{1}{a}\cos at + C\right) = -\dfrac{1}{a}(-\sin at) \cdot a = \sin at$ and similarly for the other formula

43. \$204,000 (from $130 + \dfrac{234}{\pi}$, rounded to the nearest thousand)

45. a. About 101 tons $\left(\text{from } 90 + \dfrac{36}{\pi}\right)$ **b.** About 5.6 tons per month **47.** 3600 birds

49. About \$6668 **51.** About 277 guests per week **53. a.** About 32.3° **b.** About 75.6°
55. 0 (Integrating over a whole period includes an area above the x-axis and an equal (negative) area *below* the x-axis.)
57. -1000 (since $\cos \pi$ is a constant equal to -1) **59.** 0 (since you are integrating over many whole periods)
61. 9 (since, in the long run, $\cos x$ will average out to zero) **63.** 54° Fahrenheit

Exercises 8.5 page 578

1. a. $\dfrac{5}{12}$ **b.** $\dfrac{12}{5}$ **c.** $\dfrac{13}{12}$ **d.** $\dfrac{13}{5}$ **3. a.** $\dfrac{8}{15}$ **b.** $\dfrac{15}{8}$ **c.** $-\dfrac{17}{15}$ **d.** $-\dfrac{17}{8}$

5. a. $\dfrac{1}{\sqrt{3}} = \dfrac{\sqrt{3}}{3}$ **b.** 2 **7. a.** Undefined **b.** -1 **9. a.** 1 **b.** 1 **11. a.** $\dfrac{1}{\sqrt{3}} = \dfrac{\sqrt{3}}{3}$ **b.** -2

13. 1.376 (approx) **15.** 3.864 (approx) **17.** About 56 feet

19. Slope $= \dfrac{\Delta y}{\Delta x} = \dfrac{\text{opp}}{\text{adj}} = \tan \theta$ from diagram **21.** Yes, it will cross the goal post at a height of 14.2 feet.

23. 51.3° (for a height of 20.25 feet)

25.

Period 2π; undefined at $\pm\dfrac{\pi}{2}, \pm\dfrac{3\pi}{2}, \ldots$

on $[-2\pi, 2\pi]$ by $[-4, 4]$

27. Result follows from the hint and the definitions of $\tan t$ and $\sec t$. **29.** $\cot t - t \csc^2 t$ **31.** $2t \sec^2(t^2 - 1)$

33. $\pi \sec(\pi x + 1) \tan(\pi x + 1)$ **35.** $-6z^2 \cot z^3 \csc^2 z^3$ **37.** $y = 4x - 3$ **39.** $y = -2x + 1 + \dfrac{\pi}{2}$

41. $S'(10) \approx -3.5$ (be sure your calculator is set for radians). Shadow shortens by about 3.5 feet for each additional degree of sun elevation (near 10° elevation).

43. $\dfrac{1}{3}\tan(t^3 + 1) + C$ **45.** $-\dfrac{1}{\pi}\csc \pi t + C$ **47.** $\ln|\cos(1 - t)| + C$ **49.** $\dfrac{1}{5}\ln|\csc x^5 - \cot x^5| + C$

51. $-\dfrac{1}{5}\cot^5 x + C$ **53.** $-\ln \dfrac{\sqrt{2}}{2} \approx 0.347$ **55.** $\ln(\sqrt{2} + 1) \approx 0.881$ square unit

57. False. The tangent and cotangent functions have period π. **59.** $\cos \theta$ **61.** There are none.
63. 0. Exercise 28 shows that $\csc^2 t - \cot^2 t = 1$, and differentiating 1 gives 0.
65. The slope is always negative and, in fact, always less than or equal to -1 (agreeing with the graph of $\cot t$).
67. Because of the identity $1 + \tan^2 t = \sec^2 t$ (differentiating each side)
69. Follows from the Generalized Power Rule (and the definitions of $\sec t$ and $\tan t$)

71. a. $\frac{1}{2}\tan^2 t + C$ **b.** $\frac{1}{2}\sec^2 t + C$ **c.** The equality $\tan^2 t + 1 = \sec^2 t$ shows that $\frac{1}{2}\tan^2 t$ and $\frac{1}{2}\sec^2 t$ differ by only a constant, so the two answers are the same (because of the $+C$).

73. The results follow from the hint.   **75.** The results follow from the hint.   **77.** The results follow from the hint.

Chapter 8 Review Exercises and Chapter Test page 583

1. a. $\frac{5\pi}{6}$ **b.** $20°$ **2. a.** $\frac{2\pi}{3}$ **b.** $36°$ **3.** $\frac{104\pi}{3} \approx 109$ feet **4.** $26\pi \approx 82$ feet

5. a. $\frac{\sqrt{3}}{2}$ **b.** $-\frac{1}{2}$ **c.** $-\frac{\sqrt{2}}{2}$ **d.** $-\frac{\sqrt{2}}{2}$ **6. a.** $-\frac{1}{2}$ **b.** $-\frac{\sqrt{3}}{2}$ **c.** $\frac{\sqrt{2}}{2}$ **d.** $-\frac{\sqrt{2}}{2}$

7.

8.

9. 24 feet   **10.** 20 feet

11. a.

on $[0, 365]$ by $[0, 20]$

b. About 14.6 hours   **c.** About 8.7 hours

12. Daylight is maximized on day 172 (June 21) with 16 hours of daylight, and minimized on day 355 (December 21) with 8.5 hours of daylight.

13. $-8\sin(4t - 1)$ **14.** $-3\cos(2 - 3t)$ **15.** $x^3 \cos x$

16. $\frac{\cos t + \sin t + 1}{(1 + \cos t)^2}$ (after simplification)   **17.** $6t \sin^2 t^2 \cdot \cos t^2$

18. $2xe^{x^2}\cos e^{x^2}$ **19.** $\frac{1}{2}(\sin t)^{-1/2}\cos t$ **20.** $\frac{1}{2}t^{-1/2}\cos t^{1/2}$ **21.** $y = -x + 1 + \frac{\pi}{2}$

22. a.

on $[0, 52]$ by $[-100, 600]$

b. Increasing by about 29 sales per week

23. Week 39, sales increasing by about 30 sales per week

24. a.

on $[0, 365]$ by $[0, 90]$

b. Decreasing by about 0.45 degree per day

25. On day 106 (April 16), when temperature is rising by about 0.45 degree per day

26. $\frac{\pi}{4}$ or $45°$

27. Follows from setting $R' = 0$ and solving   **28.** $\frac{5}{\pi}\sin\frac{\pi t}{5} + C$ **29.** $-\frac{1}{2}\cos(t^2 + 1) + C$

30. $\cos(2\pi - t) + C$ **31.** $\cos e^{-x} + C$ **32.** $-e^{\cos x} + C$ **33.** $2(\sin t)^{1/2} + C$ **34.** $\frac{1}{3}\ln(1 + \sin 3t) + C$

35. $\frac{1}{4}x^4 + e^{-x} - \frac{1}{\pi}\sin \pi x + C$ **36.** $\sin(\ln y) + C$ **37.** $\cos z^{-1} + C$ **38.** $\frac{3}{4}$ **39.** $-\frac{1}{24}$ **40.** $\frac{2}{3\pi} \approx 0.212$

41. 200   **42.** $\frac{1}{3}$ square unit   **43.** $\sqrt{2} - 1 \approx 0.414$ square unit

44. a.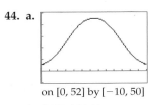

on $[0, 52]$ by $[-10, 50]$

b. $650{,}000$ (from 650 thousand)

45. a.

on $[0, 24]$ by $[10, 70]$

b. About $53.6°$

46. a. $\sqrt{3}$ **b.** $\frac{1}{\sqrt{3}} = \frac{\sqrt{3}}{3}$ **c.** 2 **d.** $\frac{2}{\sqrt{3}} = \frac{2\sqrt{3}}{3}$

47. a. -1 **b.** -1 **c.** $\frac{-2}{\sqrt{2}} = -\sqrt{2}$ **d.** $\frac{2}{\sqrt{2}} = \sqrt{2}$

48. About 503 feet   **49.** About 220 feet

50. $\tan x^2 + 2x^2 \sec^2 x^2$ **51.** $2x \cot x - x^2 \csc^2 x$ **52.** $\csc(\pi - t)\cot(\pi - t)$
53. $\frac{1}{2}t^{-1/2}\sec t^{1/2}\tan t^{1/2}$ **54.** $(\pi \sec^2 \pi t)e^{\tan \pi t}$
55. The identity $1 + \cot^2 t = \csc^2 t$ shows that the functions differ by only a constant, so the derivatives must agree (the derivative of each is $-2\cot t \csc^2 t$).

56. $y = 4x - \pi$ **57.** $\dfrac{2}{\pi} \tan \dfrac{\pi t}{2} + C$ **58.** $-\dfrac{1}{2} \ln|\cos x^2| + C$ **59.** $\dfrac{1}{3} \sec t^3 + C$ **60.** $\csc(\pi - t) + C$

61. $-\dfrac{2}{3}(\cot x)^{3/2} + C$ **62.** $2(\tan x)^{1/2} + C$ **63.** $\ln(\sqrt{2} + 1) \approx 0.881$

64. 1   **65.** $-\ln\dfrac{\sqrt{2}}{2} \approx 0.347$ square unit   **66.** $-\ln(\sqrt{2} - 1) \approx 0.881$ square unit

67.

68.

69. a. $y' = \dfrac{2\cos 2x}{\sin y}$ **b.** -2

70. a. $y' = -\dfrac{4\sin 4x}{\cos y}$ **b.** -4

71. $\dfrac{1}{80}$ radian per second

72. 22.5 radians per hour or about 21 degrees per minute   **73. a.** $2\cos 2x + y\sin x$ **b.** $-\cos x$ **74. a.** $-\dfrac{\sin(x-y)}{\cos(x-y)}$ **b.** $\dfrac{\sin(x-y)}{\cos(x-y)}$

75. a. $(\cos x)e^{\cos y}$ **b.** $-(\sin x)(\sin y)e^{\cos y}$ **76. a.** $3x^2y\cos x^3y$ **b.** $x^3\cos x^3y$

77. $-t\cos t + \sin t + C$ **78.** $t\sin t + \cos t + C$ **79.** $x\tan x + \ln|\cos x| + C$

80. $x\sec x - \ln|\sec x + \tan x| + C$ **81.** $\frac{1}{2}e^t(\sin t - \cos t)$ **82.** $\frac{1}{2}e^t(\sin t + \cos t)$

83. 6.209   **84.** 1.198   **85.** 6.208758036   **86.** 1.198

Exercises 9.1 page 600

1. Check that $(4e^{2x} - 3e^x) - 3(2e^{2x} - 3e^x) + 2(e^{2x} - 3e^x + 2) \overset{?}{=} 4$ **3.** Check that $kae^{ax} \overset{?}{=} a\left(ke^{ax} - \dfrac{b}{a}\right) + b$

5. $y = \sqrt[3]{6x^2 + c}$ **7.** Not separable   **9.** $y = ce^{2x^3}$ Check that $c6x^2e^{2x^3} \overset{?}{=} 6x^2(ce^{2x^3})$

11. $y = cx$ (since $e^{\ln x} = x$) Check that $c \overset{?}{=} \dfrac{cx}{x}$ **13.** $y = \sqrt{4x^2 + c}$ and $y = -\sqrt{4x^2 + c}$

15. Not separable   **17.** $y = 3x^3 + C$ **19.** $y = \dfrac{1}{2}\ln(x^2 + 1) + C$ **21.** $y = ce^{x^4/4}$

23. $y = \left(\dfrac{1-n}{m+1}x^{m+1} + c\right)^{1/(1-n)}$ **25.** $y = (x + c)^2$ **27.** Not separable

29. $y = ce^{x^2/2} - 1$ **31.** $y = ce^{e^x} + 1$ **33.** $y = \dfrac{1}{c - ax}$ **35.** $y = ce^{ax} - \dfrac{b}{a}$

37. $y = \sqrt[3]{3x^2 + 8}$ Check that $(3x^2 + 8)^{\frac{2}{3}} \cdot \dfrac{1}{3}(3x^2 + 8)^{-\frac{2}{3}} \cdot 6x \overset{?}{=} 2x$ and $y(0) = \sqrt[3]{8} = 2$

39. $y = -e^{x^2/2}$ Check that $-xe^{x^2/2} \overset{?}{=} x(-e^{x^2/2})$ and $y(0) = -e^0 = -1$

41. $y = (1 - x^2)^{-1}$ Check that $-(1 - x^2)^{-2}(-2x) \overset{?}{=} 2x[(1 - x^2)^{-1}]^2$ and $y(0) = (1 - 0)^{-1} = 1$

43. $y = 3x$ (using $e^{\ln x} = x$) Check that $3 \overset{?}{=} \dfrac{3x}{x}$ and $y(1) = 3 \cdot 1 = 3$

45. $y = (x + 1)^2$ or $y = (x - 3)^2$ Check that $2(x + 1) \overset{?}{=} 2\sqrt{(x+1)^2}$ and $y(1) = 2^2 = 4$ and that
   $2(x - 3) \overset{?}{=} 2\sqrt{(x-3)^2}$ and $y(1) = (-2)^2 = 4$

47. $y = \dfrac{1}{2 - e^x - x}$ Check that $\dfrac{e^x + 1}{(2 - e^x - x)^2} \overset{?}{=} \left(\dfrac{1}{2 - e^x - x}\right)^2 e^x + \left(\dfrac{1}{2 - e^x - x}\right)^2$ and $y(0) = \dfrac{1}{2 - 1} = 1$

49. $y = 2e^{ax^3/3}$ Check that $2ax^2e^{ax^3/3} \overset{?}{=} ax^2 2e^{ax^3/3}$ and $y(0) = 2e^0 = 2$

51. $D(p) = cp^{-k}$ (for any constant c)   **53.** $y = 20{,}000e^{0.05t} - 20{,}000$

55. a. $y = 28.6e^{-0.32t} + 70$ **b.** About 3.28 hours   **57.** $y = 150 - 150e^{-0.2t}$

59. a. $y' = 3 + 0.10y$ **b.** $y(0) = 6$ **c.** $y(t) = 36e^{0.1t} - 30$
   d. $y(25) = 408.570$ thousand dollars, or \$408,570 (rounded)

61. a. $y' = 4y^{1/2}$ **b.** $y(0) = 10{,}000$ **c.** $y(t) = (2t + 100)^2$ **d.** 15,376

63. a. $y' = 8y^{3/4}$ **b.** $y(0) = 10{,}000$ **c.** $y(t) = (2t + 10)^4$ **d.** 234,256

65. c. $y = \sqrt[5]{10x^3 + 32}$ **67. c.** $y = \sqrt[4]{8x^2 + 16}$ **69. a.**   **b.**

d.   **d.**

73. $y = -2x + C$ **75.** $y = ke^x$ **77.** Not separable   **79.** Separable

81. Quadrant I: positive; II: negative; III: positive; IV: negative

83. a. $p(t) = Ce^{-Kt/R}$ for $t_0 \le t \le T$ **b.** $p(t) = p_0 e^{-K(t-t_0)/R}$ for $t_0 \le t \le T$
c. $p(t) = I_0 R - Ce^{-Kt/R}$ for $0 \le t \le t_0$ **d.** $p(t) = I_0 R - (I_0 R - p_0)e^{K(t_0 - t)/R}$ for $0 \le t \le t_0$

Exercises 9.2 page 614

1. $y' = cae^{at} = a(ce^{at}) = ay$ **3.** Unlimited   **5.** Limited   **7.** None   **9.** Logistic   **11.** Logistic
$y(0) = ce^0 = c$

13. $y = 1.5e^{6t}$ **15.** $y = 100e^{-t}$ **17.** $y = -e^{-0.45t}$ **19.** $y = 100(1 - e^{-2t})$ **21.** $y = 0.25(1 - e^{-0.05t})$

23. $y = 40(1 - e^{-2t})$ **25.** $y = 200(1 - e^{-0.01t})$ **27.** $y = \dfrac{100}{1 + 9e^{-500t}}$ **29.** $y = \dfrac{0.5}{1 + 4e^{-0.125t}}$

31. $y = \dfrac{10}{1 - \frac{1}{2}e^{-30t}} = \dfrac{20}{2 - e^{-30t}}$ **33.** $y = \dfrac{3}{1 + 2e^{-6t}}$

35. $y' = 0.08y$ **37.** $y' = a(100{,}000 - y)$ **39.** $y' = a(5000 - y)$ **41.** $y' = ay(10{,}000 - y)$

$y = 1500e^{0.08t}$ $y = 100{,}000(1 - e^{-0.021t})$ $y = 5000(1 - e^{-0.223t})$ $y = \dfrac{10{,}000}{1 + 99e^{-0.535t}}$

About 22,276   About 7.2 weeks   About 8612 sales

43. $y' = ay(800 - y)$ **45.** $y' = ay(800 - y)$ **47.** $y' = ay(360 - y)$ **49.** $y = 5e^{-0.15t}$

$y = \dfrac{800}{1 + 799e^{-0.558t}}$ $y = \dfrac{800}{1 + 7e^{-0.28t}}$ $y = \dfrac{360}{1 + 0.246e^{-0.286t}}$ About 3.7 mg

About 675 people   About 6.9 years   About 359 sales

51. a. $y' = 0.1(200 - y)$ with $M = 200$ **b.** $y = 200(1 - e^{-0.1t})$ **c.** 30 years   [from solving $200(1 - e^{-0.1t}) = 0.95 \cdot 200$]

53. a. About 17 feet per second   **b.** About 0.6 foot per second   **c.** About 0.006 foot per second
d. About $\dfrac{1}{0.006} \approx 167$ seconds, or about 2.8 minutes

55. a. Limited   **b.** Unlimited   **c.** Logistic   **57. a.** Logistic   **b.** Unlimited   **c.** Limited

59. a. Approaches 0.03   **b.** Approaches 2.5   **c.** Approaches infinity

61. $y = \dfrac{1}{c - at}$ **63.** $y = ce^{a \ln x} = cx^a$ **65.** The solution follows from the indicated steps.

Exercises 9.3 page 627

1. $y = 4 + Ce^{-2x}$ Check: $(-2Ce^{-2x}) + 2(4 + Ce^{-2x}) = 8$ **3.** $y = -\frac{1}{4}e^{-2x} + Ce^{2x}$
Check: $\left(\frac{1}{2}e^{-2x} + 2Ce^{2x}\right) - 2\left(-\frac{1}{4}e^{-2x} + Ce^{2x}\right) = e^{-2x}$ **5.** $y = 3x^3 + Cx^{-5}$ **7.** $y = x^2 + Cx$ **9.** $y = 3 + Ce^{-x^3}$

11. $y = Ce^{x^2}$ **13.** $y = \dfrac{x^2 + C}{x + 1}$ **15.** $y = 3x^4 - 9x^3 + Cx^2$ **17.** $y = -x - 1 + Ce^x$ **19.** $y = 1 + Ce^{-\sin x}$

21. $y = 3e^x + 2e^{-3x}$ **23.** $y = 2x^5 - x^{-2}$ **25.** $y = x^2 \ln x + 3x^2$ **27.** $y = 2 - 2e^{-x^2}$ **29.** $y = Ce^x - 1$

31. a. $y = \dfrac{1}{x^2/2 + C}$ (by separation of variables)   **b.** $y = \dfrac{1}{3}x^3 e^x + Ce^x$ (by using an integrating factor)

33. a. $y = 20te^{0.1t} + 5e^{0.1t}$ **b.** About \$55 million   **35. a.** $y = 125t^3$ **b.** 421,875

37. a. $y(t) = 50e^{0.2t} - 40$　　**b.** 330, 2690, 20,132, and more than 1.1 million investors

39. a. $y = 3e^{t^2/2} - 1$　　**b.** About 21 tons　　**c.** In about 2.3 weeks (2 weeks and 2 days)

41. a. $y' = 2500 - 0.05y$, $y(0) = 8,000,000$　　**b.** $y = 50,000 + 7,950,000e^{-0.05t}$　　**c.** About 40.1 hours

43. a. $y' = 10,000 - 0.05y$, $y(0) = 6,000,000$　　**b.** $y = 200,000 + 5,800,000e^{-0.05t}$　　**c.** About 35 minutes

45. a. $y' = 20 - 0.02y$, $y(0) = 10,000$　　**b.** $y = 1000 + 9000e^{-0.02t}$　　**c.** About 110 hours　　**d.** 1000 grams

47. a. $w(t) = 120 + 50e^{-0.005t}$　　**b.** About 71 days　　**c.** 120 pounds

49. Yes. Just divide through by $u(x)$ to put it in standard form　$y' + \dfrac{v(x)}{u(x)}y = \dfrac{w(x)}{u(x)}$.

51. Unlimited (since it can be written in the form　$y' - ay = 0$)　and limited (since it can be written in the form
$y' + ay = aM$).　　**53.** $I(t) = e^{-at}$　　**55.** 800,000 pCi (the volume times the concentration in the incoming air)

57. Using　$e^{\int p(x)\,dx + C} = e^{\int p(x)\,dx}e^{C}$　multiplies the differential equation by a positive constant e^{C}, which, when divided out,
leaves the same equation as when the C is omitted.

59. a. $y' = 30e^{-3t} - 0.2y$　　**b.** $y = 10.7(e^{-0.2t} - e^{-3t})$　　**c.** 7.1 milligrams　　**d.** $0.97 \approx 1$ hour (8.2 mg)

61. *Hint:* An intermediate step is　$ye^{ct} = \dfrac{ab}{c - b}e^{(c - b)t} + K$

Exercises 9.4 page 636

1. 8　　**3.** $(1, 3)$, slope $= 12$

5. 　　**7.** 　　**9.**

11. a. 2.226
　　b. $y = \sqrt{x^2 + 1}$
　　c. $y(2) = \sqrt{5} \approx 2.236$

13. a. 1.489
　　b. $y = e^{0.2x}$
　　c. $y(2) = e^{0.4} \approx 1.492$

15. a. 7.836　　**b.** $y = e^x + 4e^{-x}$　　**c.** $y(2) \approx 7.930$　　**17.** 1.73　　**19.** 2.24　　**21.** 2.83

23. a. 2.3096　　**b.** 2.3309　　**c.** 2.3327　　**25.** $y(3) \approx 13,490$ (from 13.49 thousand)

27. $y(2) \approx 1981$ tons　　**29.** $q(200) \approx 0.65$

31. Underestimate, since the slopes of the segments will underestimate the true slope on the interval, as in the diagram on page 631.

33. $g(x, y) = \dfrac{f(x)}{g(y)}$　(using $g(x, y)$ and $g(y)$ for two different functions).　　**35.** $g(x, y) = q(x) - p(x) \cdot y$

37. Take the slope to be g evaluated at the *midpoint* of the segment. Alternatively, take the slope to be the *average* of g calculated at the two endpoints of the segment. (Other answers are possible.)

Chapter 9 Review Exercises and Chapter Test page 639

1. $y = \sqrt[3]{x^3 + c}$　　**2.** $y = ce^{x^3/3}$　　**3.** $y = \dfrac{1}{4}\ln(x^4 + 1) + C$　　**4.** $y = -\dfrac{1}{2}e^{-x^2} + C$　　**5.** $y = \dfrac{1}{c - x}$

6. $y = \dfrac{1}{\sqrt{c - 2x}}$ and $y = -\dfrac{1}{\sqrt{c - 2x}}$　　**7.** $y = 1 + ce^{-x}$ or $y = 1 - e^{c-x}$

8. $y = \sqrt{2x + c}$ and $y = -\sqrt{2x + c}$　　**9.** $y = ce^{\frac{1}{2}x^2 - x}$　　**10.** $y = ce^{x^3/3} - 1$　　**11.** $y = \sqrt[3]{3x^3 + 1}$

12. $y = e^{1 - x^{-1}}$ (from $ee^{-x^{-1}}$)　　**13.** $y = e^{\frac{1}{2} - \frac{1}{2}x^{-2}}$ (from $e^{\frac{1}{2}}e^{-\frac{1}{2}x^{-2}}$)　　**14.** $y = \left(\frac{2}{3}x - \frac{2}{3}\right)^{3/2}$ or $y = -\left(\frac{2}{3}x - \frac{2}{3}\right)^{3/2}$

15. a. $y' = 4 + 0.05y$, $y(0) = 10$　　**b.** $y = 90e^{0.05t} - 80$　　**c.** 68.385 thousand or $68,385

16. a. $y' = 4 - 0.25y$, $y(0) = 0$　　**b.** $y = 16 - 16e^{-0.25t}$　　**17.** $y = 106 - 36e^{-2.3t}$

18. $y(1) \approx 102.4$　　**19. c.** $y = \sqrt[3]{x^3 - 8}$　　**20. c.** $y = \sqrt[3]{\frac{3}{2}x^2 - 8}$
$y(2) \approx 105.6$　　**d.**
$y(3) \approx 105.96$

21. $y' = 0.04y$ **22.** $y' = 0.12y$ **23.** $y' = ay(8000 - y)$ **24.** $y' = ay(500 - y)$

$y = 7.85e^{0.04t}$ $y = 132e^{0.12t}$ $y = \dfrac{8000}{1 + 799e^{-2.73t}}$ (t in weeks) $y = \dfrac{500}{1 + 249e^{-3.78t}}$

About \$10.81   About \$438 billion   About 1819 cases   About 443

25. $y' = a(10{,}000 - y)$ **26.** $y' = a(60 - y)$ **27.** $y' = a(500{,}000 - y)$

$y = 10{,}000(1 - e^{-0.051t})$ $y = 60(1 - e^{-0.269t})$ $y = 500{,}000(1 - e^{-0.255t})$ (t in weeks)

About 4577   About 6.7 weeks   About 6.3 weeks

28. $y' = ay(40{,}000 - y)$ **29.** $y = 2x^7 + Cx^5$ **30.** $y = 2x + Cx^{-1/2}$ **31.** $y = 1 + Ce^{-x^2/2}$

$y = \dfrac{40{,}000}{1 + 39e^{-1.47t}}$ **32.** $y = 2 + Ce^{x^2}$ **33.** $y = e^x - x^{-1}e^x + Cx^{-1}$ **34.** $y = \frac{1}{2}x \ln x - \frac{1}{4}x + Cx^{-1}$

About $2\frac{1}{2}$ years   **35.** $y = 3e^{x^2/2}$ **36.** $y = 2e^{-2x^2}$ **37.** $y = 2x - 2x^{-2}$ **38.** $y = -4 + 10x$

39. a. $y = 10te^{0.05t} + 100e^{0.05t}$ **b.** \$192,604 [from $y(5) \approx 192.604$]   **c.** About 7.33 years

40. a. $p = -10e^{0.01t} + 110e^{0.02t}$ **b.** About 123 million people   **c.** About 19.4 years

41. a. $y' = 100 - 0.05y$, $y(0) = 1000$ **b.** $y(t) = 2000 - 1000e^{-0.05t}$

c. About 1528 cubic feet (so the oxygen percentage has risen from 10% to 15.28%)

d. 2000 cubic feet   **e.** About 46 minutes

42. a. $p = 15e^{0.2t} - 10$ **b.** 12 eagles   **c.** About 4 years

43. a.   **b.** $y = 1 + e^{-2x}$ **44. a.**   **b.** $y = \sqrt{4x^2 + 1}$

c. $y(1) = 1 + e^{-2} \approx 1.14$ **c.** $y(1) = \sqrt{5} \approx 2.24$

45. $y(2) \approx 1.70$ **46.** $y(2.5) \approx 6.68$

47. a. $0.1(1 - p) + 0.1p(1 - p) = 0.1(1 - p)(1 + p) = 0.1(1 - p^2)$ **b.** $p(4) \approx 0.61$, or 61%

48. $y(2) \approx 148$

Exercises 10.1 page 652

1. $\dfrac{1}{2} + \dfrac{1}{3} + \dfrac{1}{4} + \dfrac{1}{5} + \dfrac{1}{6}$ **3.** $-\dfrac{1}{3} + \dfrac{1}{9} - \dfrac{1}{27} + \dfrac{1}{81}$ **5.** $1 + 0 + 1 + 0 + 1 + 0$ **7.** $\displaystyle\sum_{i=1}^{\infty} \dfrac{1}{3^i}$ or $\displaystyle\sum_{i=1}^{\infty} \left(\dfrac{1}{3}\right)^i$

9. $\displaystyle\sum_{i=1}^{\infty} (-2)^i$ **11.** $\displaystyle\sum_{i=1}^{\infty} (-1)^i \cdot i$ **13.** $2^{10} - 1 = 1023$ **15.** $4^6 - 1 = 4095 \left(\text{from } 3 \cdot \dfrac{1 - 4^6}{1 - 4}\right)$

17. $1 + 2^7 = 129 \left(\text{from } 3 \cdot \dfrac{1 - (-2)^7}{1 - (-2)}\right)$ **19.** 5   **21.** $\frac{3}{2}$ **23.** Diverges   **25.** 6   **27.** 32   **29.** 25

31. Diverges   **33.** $\frac{2}{3}$ **35.** Diverges   **37.** 100   **39.** 1   **41.** $\frac{4}{11}$ **43.** $2\frac{6}{11}$ or $\frac{28}{11}$ **45.** $\frac{1}{37}$

47. $\frac{1}{81}$ **49. a.** \$26,214,300   **b.** 24 days   **51.** \$597,447   **53.** \$584,541

55. a. \$21,474,836.47   **b.** 27 days   **57.** On the 37th day   **59. a.** 15 units, 3 units   **b.** 4 doses

61. \$500 billion; multiplier = 10   **63.** \$6 billion (from \$6000 million); multiplier = 20

65. 4   **67.** \$20,300   **69. a.** 30 feet   **b.** At the 9th bounce

71. 16,000; 15,200   **73.** 5281   **75.** True   **77.** $1/e^2$; yes

79. Converges ($r = 1/10$)   **81.** $r = 2$, so the sum formula does not apply

83. a. Dose 3   **85. a.** $\dfrac{1}{1 - (-1)} = \dfrac{1}{2}$

b. Dose 5   **b.** No: $r = -1$ violates $|r| < 1$.

c. Partial sums are $1, 0, 1, 0, \ldots$, which in the long run average to $1/2$.

Exercises 10.2 page 664

1. $1 + 2x + 2x^2 + \frac{4}{3}x^3$ **3.** $1 + \frac{1}{2}x - \frac{1}{8}x^2 + \frac{1}{16}x^3$ **5.** $7 + x - 3x^2$ **7.** $1 + x^2$

9. a. $x - \frac{1}{2}x^2 + \frac{1}{3}x^3 - \frac{1}{4}x^4$ **11. a.** $1 - \frac{1}{2}x^2 + \frac{1}{24}x^4$ **13.** $1 + x^2 + \frac{1}{2}x^4$ **15.** $2x - \frac{2^3}{3!}x^3 + \frac{2^5}{5!}x^5$

b.

on $[-2, 2]$ by $[-2, 2]$

b.

on $[-\pi, \pi]$ by $[-2, 2]$

17. a. $1 + x + \frac{1}{2!}x^2 + \frac{1}{3!}x^3 + \frac{1}{4!}x^4$

e.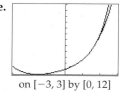

on $[-3, 3]$ by $[0, 12]$

b. 0.60677

c. $\left| R_4\left(-\frac{1}{2}\right) \right| \le \frac{1}{5!} \left| -\frac{1}{2} \right|^5 < 0.0003$

(using $M = 1$, from $f^{(5)}(x) = e^x$ evaluated at the right-hand endpoint 0)

d. $e^{-1/2} \approx 0.6068$ with error less than 0.0003 (actual value: $e^{-1/2} \approx 0.6065$)

19. a. $p_6(x) = 1 - \frac{1}{2!}x^2 + \frac{1}{4!}x^4 - \frac{1}{6!}x^6$ has error $|R_6(x)| \le \frac{1}{7!}|1|^8 < 0.00022$ on $[-1, 1]$.

b. (Actual error for $-1 \le x \le 1$ is less than 0.00003.)

on $[-2\pi, 2\pi]$ by $[-2, 2]$

21. a. $1 + \frac{1}{3}x$ **b.** $[0, 0.3]$ **23. a.** $1 + \frac{1}{2}(x - 1) - \frac{1}{8}(x - 1)^2 + \frac{1}{16}(x - 1)^3$ **25. a.** $-(x - \pi) + \frac{1}{6}(x - \pi)^3$

c.

on $[-2, 2]$ by $[-2, 2]$

b.

on $[-1, 3]$ by $[-1, 2]$

b.

on $[0, 2\pi]$ by $[-2, 2]$

27. a. $1 + \frac{1}{3}(x - 1) - \frac{1}{9}(x - 1)^2$ **b.** 1.09

c. $|R_2(1.3)| \le \frac{10/27}{3!}|1.3 - 1|^3 < 0.002$ $\left(\text{using } M = \frac{10}{27}, \text{ from } f^{(3)}(x) = \frac{10}{27}x^{-8/3} \text{ evaluated at } x = 1\right)$

d. $\sqrt[3]{1.3} \approx 1.09$ with error less than 0.002 (actually, $\sqrt[3]{1.3} \approx 1.091$)

29. a. $(x - 1) - \frac{1}{2}(x - 1)^2 + \frac{1}{3}(x - 1)^3 - \frac{1}{4}(x - 1)^4 + \frac{1}{5}(x - 1)^5 - \frac{1}{6}(x - 1)^6$

b. 0.40469

c. $|R_6(1.5)| \le \frac{720}{7!}|1.5 - 1|^7 < 0.002$

d. $\ln 1.5 \approx 0.405$ with error less than 0.002 (actually, $\ln \frac{3}{2} \approx 0.4055$)

31. 0.78   **33.** $x = \frac{1}{2}$ year   **35.** x

37. $f(x) = \frac{1}{x}$ is not defined at $x = 0$ and so has no linear approximation there.

39. 1 (or any number greater than 1, but the best answer is 1)

41. Left   **43.** $1 + x + x^2 + \cdots + x^n = \frac{1 - x^{n+1}}{1 - x}$

45. The suggested du and v give $-(x - t)f'(t)\Big|_a^x - \int_a^x -(x - t)f''(t)\, dt$ and the formula follows by evaluating at x and a.

47. The calculation proceeds just as in Exercise 45.   **49.** Pull the M outside the integral in Exercise 48 and integrate.

Exercises 10.3 page 676

1. $R = 2$ **3.** $R = 1$ **5.** $R = \infty$ **7.** $R = 0$ **9.** $R = \infty$ **11.** $x - \dfrac{x^2}{2} + \dfrac{x^3}{3} - \dfrac{x^4}{4} + \cdots$

13. $1 - \dfrac{4x^2}{2!} + \dfrac{16x^4}{4!} - \cdots$ **15.** $1 + x - \dfrac{x^2}{2!} + \dfrac{3x^3}{3!} - \dfrac{5 \cdot 3x^4}{4!} + \cdots$ **17.** $1 + \dfrac{x}{5} + \dfrac{x^2}{5^2 \cdot 2!} + \dfrac{x^3}{5^3 \cdot 3!} + \cdots$

19. $1 - \dfrac{1}{2!}\left(x - \dfrac{\pi}{2}\right)^2 + \dfrac{1}{4!}\left(x - \dfrac{\pi}{2}\right)^4 - \cdots$ **21.** $x^2 + x^3 + x^4 + \cdots$ **23.** $x^2 - \dfrac{x^6}{3!} + \dfrac{x^{10}}{5!} - \dfrac{x^{14}}{7!} + \cdots$

25. $1 + \dfrac{x}{2!} + \dfrac{x^2}{3!} + \dfrac{x^3}{4!} + \cdots$ **27.** $\dfrac{1}{3!} - \dfrac{x^2}{5!} + \dfrac{x^4}{7!} - \cdots$ **29.** $x - \dfrac{1}{2}x^2 + \dfrac{1}{3}x^3 - \dfrac{1}{4}x^4 + \cdots$

31. a. $1 - x^2 + x^4 - x^6 + \cdots$ **b.** $R = 1$ **c.** $1 - \dfrac{1}{3} + \dfrac{1}{5} - \dfrac{1}{7} + \cdots$ **d.** The result follows from the indicated steps.

33. $1 + \dfrac{x^2}{2!} + \dfrac{x^4}{4!} + \dfrac{x^6}{6!} + \cdots$ **35. a.** 1.64872   **b.** 1.64583   **c.** 7 terms (the last having exponent 6)

37. $\dfrac{d}{dx}\left(1 - \dfrac{x^2}{2!} + \dfrac{x^4}{4!} - \dfrac{x^6}{6!} + \cdots\right) = -2\dfrac{x}{2!} + 4\dfrac{x^3}{4!} - 6\dfrac{x^5}{6!} + \cdots = -x + \dfrac{x^3}{3!} - \dfrac{x^5}{5!} + \cdots$

$$= -\left(x - \dfrac{x^3}{3!} + \dfrac{x^5}{5!} - \cdots\right) = -\sin x$$

39. a. $1 - \dfrac{1}{2}t^2 + \dfrac{1}{2^2 2!}t^4 - \dfrac{1}{2^3 3!}t^6 + \cdots$ **b.** $0.4x - \dfrac{0.4}{3 \cdot 2}x^3 + \dfrac{0.4}{5 \cdot 2^2 \cdot 2!}x^5 - \dfrac{0.4}{7 \cdot 2^3 \cdot 3!}x^7 + \cdots$

  c. Approximately 0.34, so about 34% have IQ's in the specified range

41. a. $1 - \dfrac{t^4}{2!} + \dfrac{t^8}{4!} - \cdots$ **b.** $x - \dfrac{x^5}{5 \cdot 2!} + \dfrac{x^9}{9 \cdot 4!} - \cdots$ **c.** Approximately 0.905 (thousand), so about 905 sales

43. No. (If the radius of convergence from 0 does not extend beyond 3, it cannot reach 4.)

45. 17   **47.** $e^{-1} = 1/e$ **49.** -1 (from $\cos \pi$)   **51.** 10 (the midpoint of the interval)

53. The absolute ratio of successive terms is $\left|\dfrac{x^2}{(2n + 3)(2n + 2)}\right| \to 0$ as $n \to \infty$.

55. a. $S_n - S_{n-1} = (c_1 + c_2 + c_3 + \cdots + c_{n-1} + c_n) - (c_1 + c_2 + c_3 + \cdots + c_{n-1}) = c_n$
  b. Taking limits: $\lim\limits_{n \to \infty} (S_n - S_{n-1}) = \lim\limits_{n \to \infty} c_n$. Since the series converges, the limit on the left-hand side is
    $S - S = 0$, so the limit on the right-hand side must also be zero: $\lim\limits_{n \to \infty} c_n = 0$.

57. The result follows from the indicated steps.

59. a. If $\left|\dfrac{c_{n+1}}{c_n}\right|$ approaches $r > 1$, then $\left|\dfrac{c_{n+1}}{c_n}\right|$ must eventually remain above 1.

  b. $\left|\dfrac{c_{n+1}}{c_n}\right| \geq 1$ for $n = N$ implies that $|c_{N+1}| \geq |c_N|$. Then, $\left|\dfrac{c_{n+1}}{c_n}\right| \geq 1$ for $n = N + 1$ implies that

    $|c_{N+2}| \geq |c_{N+1}|$, which together with $|c_{N+1}| \geq |c_N|$ implies that $|c_{N+2}| \geq |c_N|$. Similar reasoning leads to the
    other inequalities.
  c. Terms beyond the Nth being (in absolute value) at least of size $|c_N| \neq 0$ means that the nth term cannot approach
    zero, so the series diverges.

61. b. Each parenthesis (and also the next term) is positive, so it is a sum of positive terms.
  c. The series begins with 1, with positive quantities subtracted, and so it must be less than 1.

63. The results of the steps are as follows:
  a. $e^x > \dfrac{x^3}{6}$ **b.** $e^{ax} > \dfrac{a^3 x^3}{6}$ **c.** $e^{-ax} < \dfrac{6}{a^3 x^3}$ **d.** $x^2 e^{-ax} < \dfrac{6}{a^3 x}$ **e.** Letting $x \to \infty$ gives the claimed result.

Exercises 10.4 page 688

1. $x_1 = 1.5, x_2 = 1.39$ **3.** $x_1 = 0.5, x_2 = 0.61$ **5.** 2.236   **7.** 1.513   **9.** -0.814
11. 5.065797019   **13.** 0.486389036   **15.** 3.693441359   **17.** 0.869647810   **19.** 1.106060158
21. 1.029866529   **23.** 0.567143290
25. The iterations give $-1, 1, -1, 1, \ldots$. Reason: $x^2 + 3 = 0$ has no solutions (the left-hand side is always positive).
27. It converges in one step to 2.   **29.** 102 iterations   **31.** \$5.67
33. \$2.61   **35.** 12.6%   **37.** 3.18 weeks   **39.** $-b/m$ **41.** -4
43. It will converge to the zero in one step.   **45.** To the right of the zero

47. The iterations will be exactly the same (because changing the sign of the function will not change the f/f' in the iteration formula).

Chapter 10 Review Exercises and Chapter Test page 691

1. $\frac{3}{7}$ **2.** $\frac{2}{7}$ **3.** Diverges   **4.** $\frac{2}{5}$ **5.** 50   **6.** $\frac{4}{7}$ **7.** 5   **8.** Diverges   **9.** $\frac{6}{11}$ **10.** $\frac{8}{11}$

11. \$736,971   **12.** Follows from summing $D + D(1 + i) + D(1 + i)^2 + \cdots + D(1 + i)^{n-1}$

13. a. 0.4 milligram and 0.14 milligram   **b.** Just after the 4th dose; just after the 6th dose

14. a. Sum $= 10,000$, multiplier $= 10$ **b.** 22 terms; 44 terms

15. a. $3 + \frac{1}{6}x - \frac{1}{216}x^2 + \frac{1}{3888}x^3$ **16. a.** $1 + 3x + \frac{9}{2}x^2 + \frac{9}{2}x^3$ **17. a.** $2x - 2x^2 + \frac{8}{3}x^3$ **18. a.** x^2

b.

on $[-15, 15]$ by $[-6, 6]$ on $[-1, 1]$ by $[-1, 8]$ on $[-1, 1]$ by $[-5, 5]$ on $[-3, 3]$ by $[-2, 2]$

19. a. $1 + \frac{1}{2}x - \frac{1}{8}x^2 + \frac{1}{16}x^3$

   b. 1.4375

   c. $|R_3(1)| \leq \dfrac{15/16}{4!}|1|^4 < 0.04$ (using $|f^{(4)}(x)| = \dfrac{15}{16}(x + 1)^{-7/2} \leq \dfrac{15}{16} = M$ for $0 \leq x \leq 1$)

   d. $\sqrt{2} \approx 1.44$ with error less than 0.04

20. a. $1 - \dfrac{1}{2}x^2 + \dfrac{1}{24}x^4$ **b.** 0.54167   **c.** $|R_4(1)| \leq \dfrac{1}{5!}|1|^5 < 0.009$ (using $|f^{(5)}(x)| = |-\sin x| \leq 1 = M$)

   d. $\cos 1 \approx 0.542$ with error less than 0.009 (actual value: $\cos 1 \approx 0.540$)

21. $1 + \dfrac{1}{n}x$ **22.** $p_7(x) = 1 + x + \dfrac{1}{2!}x^2 + \dfrac{1}{3!}x^3 + \cdots + \dfrac{1}{7!}x^7$ has error $|R_7(x)| \leq \dfrac{3}{8!}|1|^8 < 0.00008$ on $[-1, 1]$

23. a. $1 - \dfrac{1}{2!}\left(x - \dfrac{\pi}{2}\right)^2 + \dfrac{1}{4!}\left(x - \dfrac{\pi}{2}\right)^4$ **24. a.** $8 + 3(x - 4) + \frac{3}{16}(x - 4)^2 - \frac{1}{128}(x - 4)^3$

b.

on $[-2, 5]$ by $[-2, 2]$ on $[-2, 15]$ by $[-5, 60]$

25. a. $1 + \dfrac{2}{3}(x - 1) - \dfrac{1}{9}(x - 1)^2$ **b.** 1.12889   **c.** $f^{(3)}(x) = \dfrac{8}{27}x^{-7/3}$, so $M = \dfrac{8}{27}$

   d. $|R_2(1.2)| \leq \dfrac{8/27}{3!}|0.2|^3 < 0.0004$

   e. $(1.2)^{2/3} \approx 1.1289$ with error less than 0.0004 (actually, $(1.2)^{2/3} \approx 1.1292$)

26. 0.80 gram   **27.** 33 degrees   **28. a.** $200 + 20t^2 + t^4$ **b.** 296 units   **c.** $200e^{0.4} \approx 298$

29. $R = 3$ **30.** $R = 1$ **31.** $R = 0$ **32.** $R = \infty$ **33.** $x^3 + x^5 + x^7 + x^9 + \cdots$

34. $x - \dfrac{2^2 x^3}{2!} + \dfrac{2^4 x^5}{4!} - \cdots$ **35.** $x + x^3 + \dfrac{x^5}{2!} + \dfrac{x^7}{3!} + \cdots$ **36.** $x - x^2 + \dfrac{x^3}{2!} - \dfrac{x^4}{3!} + \cdots$

37. $1 + 2x + 2^2 x^2 + 2^3 x^3 + \cdots$ **38.** $1 - \dfrac{x}{2} + \dfrac{x^2}{2^2 2!} - \dfrac{x^3}{2^3 3!} + \cdots$

39. a. $-\left(x - \dfrac{\pi}{2}\right) + \dfrac{1}{3!}\left(x - \dfrac{\pi}{2}\right)^3 - \dfrac{1}{5!}\left(x - \dfrac{\pi}{2}\right)^5 + \cdots$ **b.**

on $[-3, 6]$ by $[-2, 2]$

40. $\sin(-x) = -x - \dfrac{(-x)^3}{3!} + \dfrac{(-x)^5}{5!} - \dfrac{(-x)^7}{7!} + \cdots = -x + \dfrac{x^3}{3!} - \dfrac{x^5}{5!} + \dfrac{x^7}{7!} - \cdots = -\sin x$,   and similarly for cosine

41. a. $1 - t^2 + t^4 - t^6 + \cdots$ **b.** $x - \frac{1}{3}x^3 + \frac{1}{5}x^5 - \frac{1}{7}x^7 + \cdots$
   c. Approximately 0.465 (thousand), so about 465 cars **d.** 0.464

42. a. $\frac{1}{x^2}(1 + x + x^2 + x^3 + x^4 + \cdots) = \frac{1}{x^2} + \frac{1}{x} + 1 + x + x^2 + \cdots$

   b. $\frac{-1}{x} + \ln x + x + \frac{1}{2}x^2 + \frac{1}{3}x^3 + \cdots$

   c. $\left(\frac{-1}{x} + \ln x + x + \frac{1}{2}x^2 + \frac{1}{3}x^3\right)\Big|_{0.2}^{0.5} \approx 4.36$

   d. $2.5 \cdot 4.36 = 10.9$, so 11 generations are necessary, which agrees with the solution of Example 3 on pages 396–397.

43. 5.099   **44.** 3.107   **45.** 0.818   **46.** 0.269   **47.** 4.641588834   **48.** 3.162277660
49. -0.475711445 **50.** 4.411048937   **51.** 2.435766883   **52.** -1.033503908 **53.** 16.6%
54. \$4.40   **55.** 11.111 (so about 11,111 units)

Exercises 11.1 page 706

1. a. $\frac{2}{3}$ **b.** $\frac{1}{2}$ **3. a.** $\frac{3}{8}$ **b.** $\frac{3}{8}$
5. a. $E(X) = 10$, $\text{Var}(X) = 1$ $\sigma(X) = 1$ **b.** $E(Y) = 10$, $\text{Var}(Y) = 100$, $\sigma(Y) = 10$
7. a. $P(X = 4) = \frac{1}{3}$, $P(X = 7) = \frac{1}{3}$, $P(X = 10) = \frac{1}{3}$ **b.** 7   **c.** 6
9. a. $P(X = 2) = \frac{1}{2}$, $P(X = 4) = \frac{1}{4}$, $P(X = 8) = \frac{1}{4}$ **b.** 4   **c.** 6

11. a. $\frac{1}{12}\left(\text{from } \frac{30°}{360°}\right)$ **b.** $\frac{1}{36}$ **c.** 0

13. $E(X) = e^{-a}\left(0 \cdot 1 + 1 \cdot \frac{a}{1} + 2 \cdot \frac{a^2}{2!} + 3 \cdot \frac{a^3}{3!} + \cdots\right) = e^{-a}\left(a + \frac{a^2}{1!} + \frac{a^3}{2!} + \cdots\right)$

   $= e^{-a}a\left(1 + \frac{a}{1!} + \frac{a^2}{2!} + \cdots\right) = e^{-a}ae^a = a$

15. a. 0.135   **b.** 0.271   **c.** 0.271   **d.** 0.180   **17. a.** 0.406   **b.** 0.677   **c.** 0.857   **d.** 0.143
19. a. {BBBB, BBBG, BBGB, BBGG, BGBB, BGBG, BGGB, BGGG, GBBB, GBBG, GBGB, GBGG, GGBB, GGBG, GGGB, GGGG}

   b. $\frac{1}{16}$
   c. $P(X = 0) = \frac{1}{16}$
   $P(X = 1) = \frac{1}{4}$
   $P(X = 2) = \frac{3}{8}$
   $P(X = 3) = \frac{1}{4}$
   $P(X = 4) = \frac{1}{16}$
   d. $\frac{15}{16}$
   e. $E(X) = 2$, $\text{Var}(X) = 1$

21. \$2800   **23.** 0.238   **25.** 0.70
27. $e^{-1.7} \approx 0.18$, or about 18%
29. 0.053, or less than 6%
31. a. 0.135, or about 14%
   b. About 86% (from 100%–14%)
   c. 0.857, or about 86%

33. a. 0.082, or about 8%   **b.** 0.301, or about 30%   **35. a.** 0.091, or about 9%   **b.** 0.964, or about 96%
37. $\frac{1}{275}$ **39.** $P(X = 5)$ **41.** False. (See Example 9.)   **43.** 105; 10
45. $X = 5$. (In general, if the variance is zero, then the random variable is a *constant*.)
47. 1: Definition of variance   **2:** Expanding $(x_i - \mu)^2$ **3:** Separating into three sums
   4: Taking out constants   **5:** Using $\sum_{i=1}^{n} x_i p_i = \mu$ and $\sum_{i=1}^{n} p_i = 1$
   6: Simplifying and combining the last two terms

Exercises 11.2 page 719

1. $a = 12$ **3.** $a = \frac{1}{9}$ **5.** $a = 1$ **7.** $a = 1$ **9.** $a \approx 0.670$ **11. a.** $\frac{3}{4}$ **b.** $\frac{3}{80} = 0.0375$ **c.** 0.194
13. a. $\frac{3}{5} = 0.6$ **b.** $\frac{1}{25} = 0.04$ **c.** $\frac{1}{5} = 0.2$ **15. a.** 0.441   **b.** 0.079   **c.** 0.280
17. a. $\frac{1}{24}x^2 - \frac{1}{24}$ **b.** $\frac{2}{3}$ **19.** $\frac{1}{2}$ **21.** $\frac{1}{3}$ **23. a.** 0.75   **b.** $f(x) = 2x$ on $[0, 1]$
25. a. $F(A) = P(X \le A) = P(A \le X \le A) = 0$
   b. $F(B) = P(X \le B) = P(A \le X \le B) = 1$ (because A and B are the smallest and largest possible values)
27. 0.590   **29.** 0.296   **31. a.** 3   **b.** $\frac{7}{8} = 0.875$

33. a. $F(x) = \frac{1}{81}x^2$ on $[0, 9]$ **b.** $f(x) = \frac{2}{81}x$ on $[0, 9]$ **c.** 6   **d. and e.** $\frac{1}{3}$

35. a. 1   **b.** $\frac{1}{5} = 0.2$;   0.447   **c.** 0.156, or about 16%

37. a.   **39.** 0.537, or about 54%   **41.** 0.473, or about 47%

43. In *discrete* probability it means that the event *cannot* occur, while in *continuous* probability it means that the event is *overwhelmingly unlikely* (like guessing one number from infinitely many).

45. $F(x + 7)$ (because the graph is shifted 7 units to the left)

b. 0.83 (rounded)

47. There is more area in the *outer* 6 inches of the target than in the *inner* 6 inches, so it is more likely that the dart will land *beyond* 6 inches, making the mean *greater* than 6.

49. True   **51.** 20; 10

Exercises 11.3 page 728

1. a. $f(x) = \frac{1}{10}$ on $[0, 10]$ **b.** 5   **c.** $\frac{25}{3}$ **d.** $\sqrt{25/3} \approx 2.89$ **e.** $\frac{1}{5} = 0.2$

3. a. $F(x) = \frac{x}{B}$ on $[0, B]$ **b.** $\frac{B}{2}$ **5. a.** 15   **b.** 75   **c.** $5\sqrt{3} \approx 8.7$ **d.** $\frac{1}{5} = 0.2$

7. a. $F(x) = \frac{x - A}{B - A}$ on $[A, B]$ **b.** $\frac{A + B}{2}$ **c.** $\frac{(B - A)^2}{12}$ (after simplification)

9. a. $\frac{1}{5}$ **b.** $\frac{1}{25}$ **c.** $\frac{1}{5}$ **11. a.** 2   **b.** 4   **c.** 2   **13.** $f(x) = \frac{1}{3}e^{-x/3}$ on $[0, \infty)$

15. $f(x) = 2e^{-2x}$ on $[0, \infty)$ **17. a.** $F(x) = 1 - e^{-ax}$ on $[0, \infty)$ **b.** $\frac{\ln 2}{a} \approx \frac{0.693}{a}$ $\left(\text{from } \frac{\ln 0.5}{-a}\right)$

19. 0.777   **21.** 0.181   **23. a.** 0.22 or 22%   **b.** 0.78 or 78%   **25.** 0.707   **27.** 1.41   **29.** $\frac{1}{a^2}$

31. 0.35 or 35%   **33.** 0.13 or 13%   **35.** 0.15 or 15%

37. If $X = \begin{pmatrix} \text{Survival} \\ \text{time} \end{pmatrix}$,   then $S(x) = P(X \geq x) = 1 - P(X < x) = 1 - F(x) = 1 - (1 - e^{-ax}) = e^{-ax}$

39. $f(x) = \frac{1}{20}$ on $[0, 20]$ **41.** $f(x) = \frac{1}{6}$ on $[0, 6]$ **43.** $f(x) = 2e^{-2x}$ for $0 \leq x < \infty$

45. A Poisson random variable with parameter 3   **47.** The mean (because there are some very large families)

49. The reporter should have said "mean."

Exercises 11.4 page 737

1. 0.4744   **3.** 0.679   **5.** 0.0782   **7.** 0.001   **9.** 0.9997   **11.** 0.15   **13.** 0.3085

15. 0.909, or about 91%   **17.** 0.2525, or about 25%   **19.** 0.1056, or about 11%

21. 0.851, or about 85%   **23.** 131   **25.** 32.51 oz.

27. Mean 0 and standard deviation 2   **29.** Mean 5 and standard deviation 2   **31.** $P(Y > 4)$

33. Y has the greater mean and the greater standard deviation.   **35.** It has mean 47.8.

37. $x = 0$, $f(0) = \frac{1}{\sqrt{2\pi}}$ **39.** $\left(1, \frac{1}{\sqrt{2\pi}}e^{-1/2}\right)$ and $\left(-1, \frac{1}{\sqrt{2\pi}}e^{-1/2}\right)$ $\left(\text{Note that the } y\text{-coordinates equal } \frac{1}{\sqrt{2\pi e}}.\right)$

41. The steps follow from the given hints.

Chapter 11 Review Exercises and Chapter Test page 741

1. a. 6   **b.** 18   **c.** 4.24   **2. a.** 3   **b.** 5   **c.** 2.24   **3.** $830,000

4. Property B ($150,850 versus $121,800)   **5.** 0.677   **6.** 0.612   **7.** 0.018 or about 2%

8. $e^{-5} \approx 0.0067$ or about seven tenths of 1 percent

9. a. 0.228, or about 23%   **b.** 0.533, or about 53%   **c.** 0.796, or about 80%   **d.** About 20% (using part (c))

10. About 0.05 or 5%   **11.** $a = \frac{3}{4}$ **12.** $a = \frac{1}{18}$ **13.** $a = 2$ **14.** $a = \frac{1}{2}$ **15.** $a \approx 0.637$ **16.** $a \approx 0.578$

17. a. $\frac{7}{27}$ **b.** $\frac{1}{27}x^3$ on $[0, 3]$ **c.** $\frac{7}{27}$ **18. a.** $\frac{1}{4}$ **b.** $\frac{1}{8}x^{-1/2}$ on $[1, 25]$ **c.** $\frac{1}{4}$

19. a. $\frac{3}{8} = 0.375$ **b.** $\frac{19}{320} \approx 0.0594$ **c.** 0.244

20. a. 4   **b.** 12   **c.** 3.46   **21. a.** 5.33   **b.** 7.8 (both in million kilowatt-hours)

22. a. $\frac{11}{16} = 0.6875$ **b.** $\frac{2}{5} = 0.4$ **23. a.** 0   **b.** 0.25   **c.** 0.5   **24. a.** 0.4   **b.** 0.0218   **c.** 0.1477

25. About 0.609   **26.** About 0.530   **27. a.** $\frac{1}{50}$ **b.** 25   **c.** $\frac{625}{3}$ (from $\frac{2500}{12}$)   **d.** 14.4   **e.** $\frac{3}{5} = 0.6$

28. a. 1 **b.** $\frac{1}{3}$ **c.** 0.577 **d.** $\frac{1}{4}$ **29. a.** 100 **b.** 10,000 **30.** $f(x) = 0.25e^{-0.25x}$ on $[0, \infty)$
31. a. 0.0498, or about 5%
    b. 0.00012, or about .012% [*Note:* Although this is a very small proportion, it predicts that of the 7.5 million flights annually, more than 900 will have such runway delays, which agrees with flight records.]
32. 0.393, or about 39% **33.** 1.59 **34.** 0.231 **35.** About 0.22 or 22% **36.** About 0.18 or 18%
37. 0.1655 **38.** 0.7916 **39.** 0.9938 **40.** 0.1587 **41.** 0.7745 **42.** 0.971
43. 0.977 or about 98% **44.** 0.115 or about 12% **45.** 0.938, or about 94% **46.** 556

Cumulative Review for Chapters 1–11 page 743

1. a. $y = 5$ **b.** $x = -2$ **2.** 8 **3.** $f'(x) = 6x - 7$ (but found from the *definition* of the derivative)
4. $T'(4) = 36$, so after 4 hours the temperature is rising at the rate of 36 degrees per hour. $T''(4) = \frac{9}{2} = 4.5$, so after 4 hours the temperature is rising increasingly fast, with the rate of increase rising by 4.5 degrees per hour per hour.
5. $f'(x) = x^2(x^3 + 8)^{-2/3}$ **6.** $f'(x) = 6(2x + 3)^2(3x + 2)^4 + 12(2x + 3)^3(3x + 2)^3 = 6(2x + 3)^2(3x + 2)^3(7x + 8)$

7.

$f' > 0$	$f' = 0$	$f' < 0$	$f' = 0$	$f' > 0$
	$x = -2$		$x = 10$	
↗	→	↘	→	↗

rel max   rel min
$(-2, 464)$ $(10, -400)$

$f'' < 0$	$f'' = 0$	$f'' > 0$
	$x = 4$	
con dn		con up

IP $(4, 32)$

8.

$f' > 0$	f' und	$f' > 0$
	$x = 0$	
↗		↗

neither
$(0, 1)$

$f'' > 0$	f'' und	$f'' < 0$
	$x = 0$	
con up		con dn

IP $(0, 1)$

9.

$f' > 0$	f' und	$f' > 0$	$f' = 0$	$f' < 0$	f' und	$f' < 0$
	$x = -2$		$x = 0$		$x = 2$	
↗		↗	→	↘		↘

rel max
$(0, -1/4)$

10. 7500 square feet **11.** 128 cubic inches **12.** $\frac{dy}{dx} = -5$ **13.** $\frac{3}{4\pi} \approx 0.24$ mile per day
14. a. $6333.85 **b.** $6356.25 **15.** About 6.22 years **16.** $13,803.30
17. a. $\frac{2xe^{2x} - e^{2x}}{x^2}$ **b.** $\frac{3x^2}{x^3 + 1}$ **18.** $p = 25 **19.** 0.039 or 3.9%
20. a. $3x^4 - 2x^2 + 5x + C$ **b.** $-3e^{-2x} + C$ **21.** 2 **22.** 30 thousand dollars or $30,000
23. $800e^{0.15t} - 800$ **24.** 36 square units **25.** 73.8 thousand people
26. a. $(x^2 + 1)^{1/2} + C$ **b.** $-\frac{1}{2}e^{-1} + \frac{1}{2} \approx 0.316$ **27.** $\frac{1}{3}x^3 \ln x - \frac{1}{9}x^3 + C$ **28.** $\frac{2x + 4}{3}\sqrt{x - 1} + C$

29. $\frac{1}{2}$ **30.** 1.98 (rounded) (compare with 2.000 from FnInt)

31. 1.99 (rounded) (compare with 2.000 from FnInt) **32.** $\frac{\partial f}{\partial x} = e^y - y\frac{2x}{x^2 + 1}$; $\frac{\partial f}{\partial y} = xe^y - \ln(x^2 + 1)$

33. Min $f = 3$ at $x = 2$, $y = 2$. No max value.

34. $y = -0.9x + 6$ **35.** Max $f = 10$ at $x = 7,\ y = 4$ **36.** $df = (10x - 2y)dx + (-2x + 6y^2)\,dy$

37. 54 cubic units   **38. a.** $\dfrac{3\pi}{4}$ **b.** $150°$ **39. a.** $\dfrac{\sqrt{2}}{2}$ **b.** $-\dfrac{\sqrt{2}}{2}$ **c.** $\dfrac{1}{2}$ **d.** $-\dfrac{\sqrt{3}}{2}$

40.

41. $\cos^2 x - \sin^2 x$ **42.** $-e^{\cos x} + C$ **43.** 1 square unit

44. a. $-\dfrac{\sqrt{3}}{3}$ **b.** $-\sqrt{3}$ **c.** $-\sqrt{2}$ **d.** $\sqrt{2}$

45. a. $2x \sec^2(x^2 + 1)$ **b.** $\frac{1}{2}\sec(2t + 1) + C$

46. a. $y = Ce^{x^4/4}$ **b.** $y = 2e^{x^4/4}$ **47.** $y = 4e^{-3x^2}$

48.

49. 4   **50.** $1 + 2x + 2x^2 + \dfrac{4}{3}x^3$

51. $x^2 - \dfrac{1}{3!}x^6 + \dfrac{1}{5!}x^{10} - \dfrac{1}{7!}x^{14} + \cdots$ with $R = \infty$ **52.** 10.24695

53. 0.433   **54. a.** $E(X) = 9$ **b.** $\mathrm{Var}(X) = 27$ **55.** 0.777   **56.** 0.383

Index

DEFINITION OF THE DERIVATIVE

$$f'(x) = \lim_{h \to 0} \frac{f(x + h) - f(x)}{h}$$

DIFFERENTIATION FORMULAS

Power Rule: $\dfrac{d}{dx} x^n = n x^{n-1}$

Constant Multiple Rule: $\dfrac{d}{dx} [cf(x)] = cf'(x)$

Sum-Difference Rule: $\dfrac{d}{dx} [f(x) \pm g(x)] = f'(x) \pm g'(x)$

Product Rule: $\dfrac{d}{dx} [f(x)g(x)] = f'(x)g(x) + f(x)g'(x)$

Quotient Rule: $\dfrac{d}{dx} \left[\dfrac{f(x)}{g(x)} \right] = \dfrac{g(x)f'(x) - g'(x)f(x)}{[g(x)]^2}$

Generalized Power Rule: $\dfrac{d}{dx} [f(x)]^n = n[f(x)]^{n-1} f'(x)$

Chain Rule: $\dfrac{d}{dx} [f(g(x))] = f'(g(x))g'(x)$

$\dfrac{dy}{dx} = \dfrac{dy}{du} \dfrac{du}{dx}$ with $y = f(u)$
and $u = g(x)$

Logarithmic Formulas: $\dfrac{d}{dx} \ln x = \dfrac{1}{x}$ $\dfrac{d}{dx} \ln f(x) = \dfrac{f'(x)}{f(x)}$

Exponential Formulas: $\dfrac{d}{dx} e^x = e^x$ $\dfrac{d}{dx} e^{f(x)} = e^{f(x)} f'(x)$

AREA AND VOLUME FORMULAS

Rectangle
Area $= l \cdot w$
Perimeter $= 2l + 2w$

Circle
Area $= \pi r^2$
Circumference $= 2\pi r$

Rectangular solid
Volume $= l \cdot w \cdot h$

Cylinder
Volume $= \pi r^2 l$

Sphere
Volume $= \dfrac{4}{3} \pi r^3$
Surface area $= 4\pi r^2$

PROPERTIES OF NATURAL LOGARITHMS

1. $\ln 1 = 0$
2. $\ln e = 1$
3. $\ln e^x = x$
4. $e^{\ln x} = x$
5. $\ln (MN) = \ln M + \ln N$

6. $\ln \left(\dfrac{1}{N} \right) = -\ln N$

7. $\ln \left(\dfrac{M}{N} \right) = \ln M - \ln N$

8. $\ln (M^P) = P \cdot \ln M$